International best seller. Now read on.....

Airports, like novels, are about people.

And processing people in an airport environment can be a tough, exacting job.

Keeping them happy and getting them off the ground is a complex and sophisticated task.

Ferranti knows about the kinds of problem that face an airport management team in the 1980's.

And they can help.

After all, Ferranti has been involved in airport computer systems since the very beginning.

Using systems that have been 'passenger proved' throughout the world Ferranti can help maximise efficiency and cost savings.

Right from the moment a passenger enters the airport to the moment he heads for the wide blue yonder, Ferranti is at work.

CHECK-IN

Ferranti VDUs speed up the process of checking in by assisting with names, reservations, special seating and meal requirements and security.

LOAD CONTROL

Ferranti systems also keep a watchful eye on all luggage checked in, together with other loads—fuel, cargo, meals and bar stock. They even balance the load and produce all necessary documentation.

FLIGHT INFORMATION

The direct way to keep the public fully up to date with events. Using flapboards, VDU terminals and CCTV monitors that can be updated in real time, they can relay up-to-the-minute information on flight arrivals, departures and delays.

BACK-UP SERVICES

Ferranti not only helps airports help people . . . but also helps with warehouse control; ground traffic; spares, maintenance and servicing control.

Whatever aspect of airport management you're involved in Ferranti can help.

After all, we wrote the book on airport computer systems.

For further information contact:

Ferranti Computer Systems Limited,
Simonsway, Wythenshawe,
Manchester M22 5LA
Telephone: 061-499 3355
Telex: 668084

FERRANTI
Computer Systems

From Ferranti. The automated airport.

CS03/62/052 rb

Complete range of
ATC communications equipment
for stationary
and
mobile applications
Type Series 400
VHF, UHF and VHF/UHF
All units use synthesizers
AM and FM (A3, A9, F3)
High sensitivity and low crossmodulation
Excellent spectral purity of transmitted signal
Modular system, expandable and obsolescence-proof
AC and DC operation with automatic switchover facility
Integrated test functions (display and outputs)
Local, remote and computer control
Antennas
Filters
Multicouplers
Single-channel receivers / transmitters / transceivers
Multichannel receivers / transmitters / transceivers
Amplifiers
Remote-control equipment
For more information please contact our department 4 FV.
Also ask for information on HF communications equipment, ATC direction finders and complete electronic ATC outfit.
Rohde & Schwarz GmbH & Co. KG
Postfach 80 14 69
D-8000 München 80
Federal Republic of Germany
Telex 523 703 (rus d)
Phone internat. +(4989) 4129-1
Independent concern
(established 1933)
represented in 80 countries
ROHDE & SCHWARZ
Electronic measurements · Broadcasting
Radio communications · Radiomonitoring

JANE'S
AIRPORT EQUIPMENT
1982-83

Alphabetical list of advertisers

A

AEG-Telefunken
Industriestrasse 29, D-2000 Wedel (Holst), Federal Republic of Germany .... [5]

B

Belfort Instrument Company
1600 South Clinton Street, Baltimore, Maryland 21224, USA .... [14]

C

CESCOM Electronics Ltd
Harrow Road, Leytonstone, London E11 3QF, England [15]

Claude Société
Protection & Systems Division, 27-29 rue de Sèvres, 92103 Boulogne-Billancourt, France .... [12]

Combined Service Publications Ltd
BAEC Department, PO Box 4, Farnborough, Hampshire GU14 7LR, England .... [2]

D

Defiant Weighing Ltd
Vestry Estate, Otford Road, Sevenoaks, Kent TN14 5EL, England .... [26]

FL Douglas (Equipment) Ltd
Village Road, Arle, Cheltenham, Gloucestershire, England .... *facing inside back cover*

E

Emergency One Inc
PO Box 2710, Ocala, Florida 32678, USA .... [14]

Enertec Schlumberger
1 rue Nieuport, BP 54, 78141 Velizy-Villacoublay Cedex, France .... [15]

F

Ferranti Computer Systems Ltd
Simonsway, Wythenshawe, Manchester M22 5LA, England .... *inside front cover*

H

Heckmann GmbH
Kapellenstrasse 47, D-6200 Wiesbaden, Federal Republic of Germany .... [11]

Houchin Ltd
Chart Road, Ashford, Kent TN23 1DZ, England ... [13]

I

IAL
Aeradio House, Hayes Road, Southall, Middlesex UB2 5NJ, England .... [8] & [9]

International Airport Equipment Group
Edbro Ltd, Lever Street, Bolton, Lancashire BL3 6DJ, England .... [22]

International Airport Expositions
62 Victoria Street, St Albans, Hertfordshire AL1 3XT, England .... [11]

L

Leyland Vehicles Ltd
Bow Lane, Leyland, Preston, Lancashire PR5 1SN, England .... [20]

M

Mather & Platt Ltd
Park Works, Manchester M10 6BA, England ...... [7]

P

Perusyhtymä Oy Ara
PO Box 434, 20101 Turku 10, Finland .... *facing inside back cover*

Philips Electronics Ltd
Telecommunications Division, 601 Milner Avenue, Scarborough, Ontario M1B 1M8, Canada .... [24]

R

Rohde & Schwarz GmbH & Co KG
Postfach 80 14 69, D-8000 Munich 80, Federal Republic of Germany ...... *front endpaper iv*

Rosenbauer
PO Box 176, A-4021 Linz, Austria .... *facing inside front cover*

S

Safety Supply Company Ltd
Airport Division, 9 Gayford Road, London W12 9BY, England .... [12]

Selenia SpA
Civil Radar & Systems Division, Via Tiburtina Km 12.400, 00131 Rome, Italy ...... [18]

T

Teledyne Inet
2750 W Lomita Boulevard, Torrance, California 90509, USA .... [14]

V

Vitroselenia SpA
Via Tiburtina 1020, Rome, Italy .... [3]

Classified list of advertisers

The companies advertising in this publication have informed us that they are involved in the fields of manufacture indicated below:

Aircraft refuellers
International Airport Equipment Group

Aircraft towing tractors
F L Douglas
International Airport Equipment Group

Airfield buses
International Airport Equipment Group
Leyland Bus

Airside buses
International Airport Equipment Group
Leyland Bus

Airport computer systems
Ferranti

Airport departure control systems
IAL

Airport vehicle control systems
Rosenbauer

Air traffic & airport communication systems
IAL

Air traffic control equipment
AEG-Telefunken
Selenia

Ambulances
International Airport Equipment Group
Rosenbauer

Anemometers
Enertec Schlumberger

Approach flash lights
AEG-Telefunken

Approach lighting
AEG-Telefunken

Baggage carts
International Airport Equipment Group

Baggage handling
Defiant Weighing
International Airport Equipment Group

Barometers
Enertec Schlumberger

Cargo handling
Defiant Weighing

Cargo handling & storage systems
International Airport Equipment Group

Catering trucks
International Airport Equipment Group

Cloud/temperature/visibility measuring equipment
Enertec Schlumberger

Commissary trucks
International Airport Equipment Group

Communication & information systems
CESCOM Electronics

Communication, navigation & meteorology
CESCOM Electronics

Communications, including public address systems
CESCOM Electronics

Control towers
International Airport Equipment Group

Conveyor belt loaders
International Airport Equipment Group

Crash/fire/rescue vehicles
Emergency One

Crash tenders
Emergency One
International Airport Equipment Group
Rosenbauer

Data processing equipment
CESCOM Electronics

De-icing & washing vehicles
International Airport Equipment Group

Distance measuring equipment (DME)
Philips Electronics

Dollies
International Airport Equipment Group

Electric energy installations
Houchin
International Airport Equipment Group

Electric surge voltage arrestors
Claude SA

Emergency generating sets
International Airport Equipment Group

Eye protection
Safety Supply Company

Filter/water separators
International Airport Equipment Group

Fire-control equipment
Mather & Platt

Fire-extinguishing systems
International Airport Equipment Group
Rosenbauer
Safety Supply Company

Fresh water carts
International Airport Equipment Group

Fuel/oil/water storage
International Airport Equipment Group

Ground-based air traffic control systems
Rohde & Schwarz

Ground-based navigation aids
Philips Electronics

Ground power units
Houchin

Ground support equipment
International Airport Equipment Group

Ground support systems
Teledyne Inet

Ground-to-air radio communication
IAL

Head protection
Safety Supply Company

Hydrant dispensers
International Airport Equipment Group

Instrument landing systems (ILS)
Philips Electronics

Jet air starter
International Airport Equipment Group

Lifting equipment
Claude SA
International Airport Equipment Group

Lifting platforms
International Airport Equipment Group
Rosenbauer

Lighting equipment
International Airport Equipment Group

Maintenance platforms & bridges
International Airport Equipment Group

Maintenance vehicles
International Airport Equipment Group

Meteorology & flight data equipment
Belfort Instrument Company
Enertec Schlumberger
IAL

Microwave landing systems (MLS)
Philips Electronics

Pallet/container loader
International Airport Equipment Group

Pallet/container transporter
International Airport Equipment Group

Pallet storage systems
International Airport Equipment Group

Passenger & baggage handling
International Airport Equipment Group

Passenger steps
International Airport Equipment Group

Power generators
Houchin
International Airport Equipment Group
Rosenbauer

Radar
Selenia

Refuellers
International Airport Equipment Group

Rescue launches
Rosenbauer

Runway foamers
Emergency One
International Airport Equipment Group
Rosenbauer

Runway foaming systems
Rosenbauer

Runway lighting
AEG-Telefunken

Runway sweepers
Perusyhtymä Oy Ara

Safety equipment
Safety Supply Company

Scales
Defiant Weighing

"DESIGN AN AIRPORT. GET IT BUILT. INSTALL THE EQUIPMENT AND SERVICES. TRAIN THE STAFF. THEN RUN IT."

Green-field site or airport improvement, IAL offers you all the efficiencies of a single source service.

Years of local knowledge, gained in every type of geographical environment, mean we can more readily respond to your requirements. For example . . .

THE COMPLETE GROUNDWORK PACKAGE

IAL can provide a total airport turnkey package or any part of it. We can design your airport, get it built, train personnel and undertake management responsibilities as required.

We install the most advanced systems and computer based communications networks. Several hundred of our ATCOs and engineers are manning international and local airports, heliports and off-shore oil rigs throughout the world. We can also supply meteorology, security, fire and rescue services and the necessary personnel.

AIRPORT DEPARTURE CONTROL

IAL Lopac is a comprehensive airport departure control system. It performs automatic acceptance, inventory control, seat allocation and full weight and balance functions. The system, suitable for all aircraft types and capable of handling in excess of 1,800 passengers an hour, can operate as a 'stand-alone' unit with its own central processor or be integrated into a network operation.

AIRPORT COMMUNICATIONS

IAL Stratus is a complete air traffic and airport communications system. It links all your voice communications into one highly sophisticated control system, providing each operator with a single interface for instant access to multiple radio and telephone channels with interconnecting facilities.

THE WORLD'S MOST ADVANCED SECURITY SCREENING SYSTEM

The IAL Rapiscan baggage screening system uses a new technology that completely eliminates shadow, to give you a perfect picture. With former systems incorporating an x-ray with screen and CCTV, high resolution detail was impossible to achieve due to 'camera noise' and x-ray scatter. The system can scan 1200 bags an hour and its

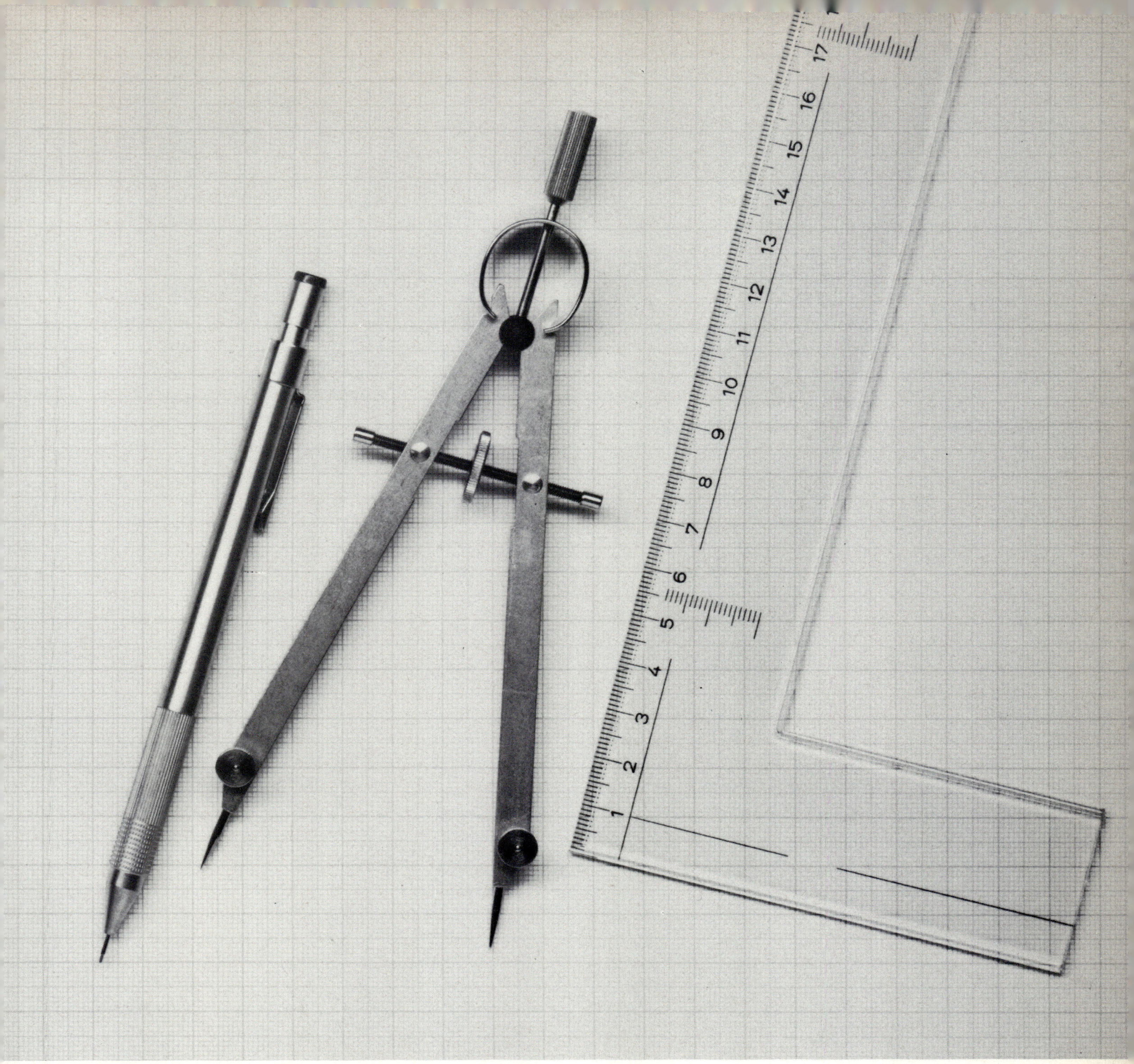

compact design allows it to be moved through standard doorways. Easily interchangeable printed circuit boards reduce maintenance to an absolute minimum.

METEOROLOGY AND FLIGHT DATA

IAL's US subsidiaries, Ocean Data Systems Inc., and Global Weather Dynamics Inc., offer computer based forecast technology utilising sophisticated communications systems including satellites. Their experienced forecasters provide independent international aviation weather data services for both airports and airlines.

GROUND-TO-AIR RADIO COMMUNICATIONS

PAE, a subsidiary company, offers quality VHF solid state transmitters for civil use. Designed to ICAO standards, they can be controlled locally or remotely.

A range of six state-of-the-art fully broad band designs for transmitters for military UHF and VHF requirements is also available, together with a matching receiver and synthesised drive units.

TRAINING

In 21 years over 2,000 ATC students from 75 countries have graduated through IAL. We are in the unique position of having been granted authority by the UK Government not only to operate a sophisticated ATC training unit, but to issue licences in our name.

For the full story phone 01-843 2411 or write to the address below. We can help you in so many ways.

IAL

THE HIGH TECHNOLOGY TASK FORCE

AVIATION · COMMUNICATIONS · COMPUTER · MEDICAL · SYSTEMS AND SERVICES · WORLDWIDE

IAL · Aeradio House · Hayes Road · Southall · Middlesex · England UB2 5NJ · Tel: 01-843 2411 · Telex: 24114

CLASSIFIED LIST OF ADVERTISERS

Screening & alarm systems
International Airport Equipment Group

Security & emergency services
International Airport Equipment Group

Security screening systems
IAL

Special chassis for crash trucks
Emergency One
International Airport Equipment Group
Rosenbauer

Stairs
International Airport Equipment Group

Starting equipment
Houchin
International Airport Equipment Group

Stationary refuelling systems
International Airport Equipment Group

Switch & control board
AEG-Telefunken

Taxiway lighting
AEG-Telefunken

Toilet carts
International Airport Equipment Group

Towbars
International Airport Equipment Group

Tow tractors
F L Douglas
International Airport Equipment Group

Tractors
F L Douglas
International Airport Equipment Group

Trucks
International Airport Equipment Group

Tugs
F L Douglas
International Airport Equipment Group

Vhf omni-range (vor)
Philips Electronics

Vhf omni-range, doppler (dvor)
Philips Electronics

Warning lights
Claude SA

Weighing (aircraft)
Defiant Weighing

Weighing (freight)
Defiant Weighing

inter airport
International fair for airport construction · terminal facilities · ground equipment and cargo facilities
inter airport 1983
1983
TWO TRADE FAIRS GET TOGETHER
inter airport
AIRPORT CONSTRUCTION · GROUND EQUIPMENT
TERMINAL EQUIPMENT · AIR CARGO FACILITIES
26.—29. SEPT. 1983 FRANKFURT/M./AIRPORT
International Airport Expositions
62, Victoria Street, St. Albans,
Herts AL 1 3 XT. England
Telephone: 0727 63213,
Telex 266350
HECKMANN GMBH
Messen + Ausstellungen
Kapellenstr. 47, D-6200 Wiesbaden
Tel.: 06121/524071
Telex 4 186 518

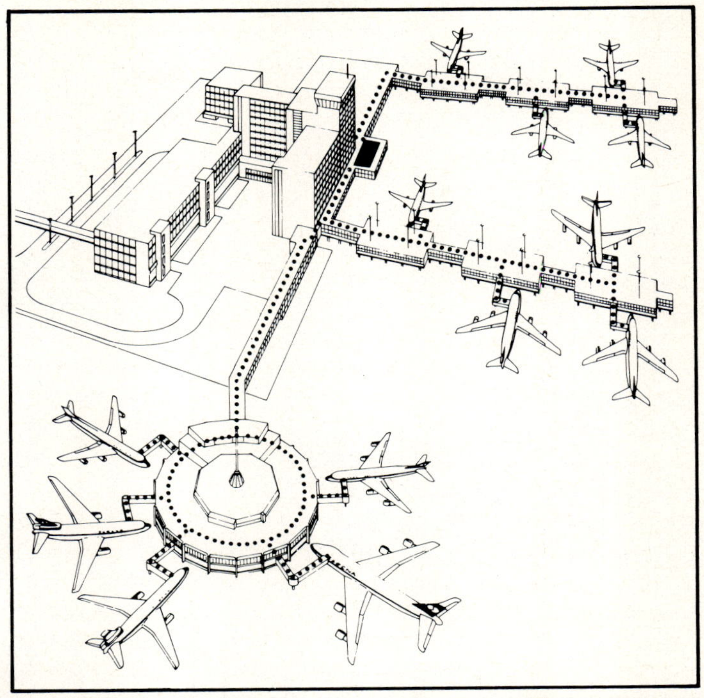

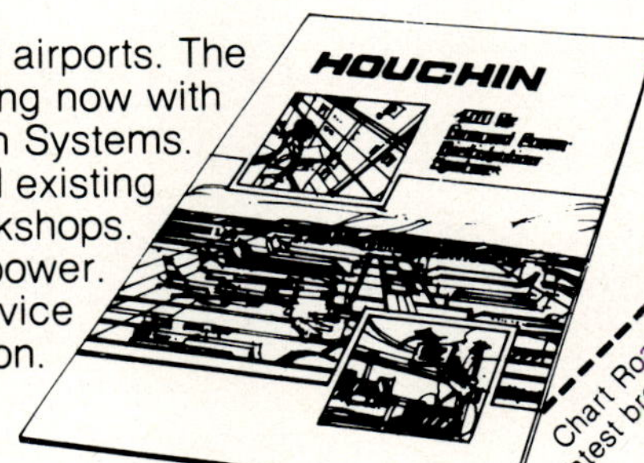

HOUCHIN

Control tower and technical block, Nairobi International Airport, Kenya

JANE'S
AIRPORT EQUIPMENT

FIRST EDITION

EDITED BY
DAVID F RIDER AMRAeS

1982-83

JANE'S YEARBOOKS

Published by Jane's Publishing Company Limited, 238 City Road, London EC1V 2PU, England

ISBN 0 86720-610-1

Distributed in Canada, the Philippines and the USA and its dependencies by
Science Books International Inc, 51 Sleeper Street, Boston, Massachusetts 02210, USA

ATCR-33 S-band TMA radar installation

Internal view of ATCR-33 equipment shelter

CONTENTS

Buses

for airside and rapid passenger transfers

National 100-seater

Artic 160-seater

Tiger 57-seater

Olympian two-axle

Olympian three-axle

Cub

BRE-Leyland Railbus

National
This fully integral rear-engined air-suspended single-decker with fully automatic transmission has doorways on both sides to permit quick loading and unloading. Up to 100 passengers can be carried. Left- or right-hand control is available. Dimensions also allow operation on public roads.

Artic
An articulated airside bus providing high-volume rapid passenger movement, the Artic has an integrally linked turntable rear-steering axle system, fully automatic transmission, full air suspension and doorways on both sides to permit quick loading and unloading. Up to 160 passengers can be carried. Left- or right-hand control is available. Dimensions also allow operation on public roads.

Tiger
Tiger is a powerful, air-suspended luxury coach for fast comfortable transfers between airports, city centre links or hotels, for up to 57 passengers. Left- or right-hand control is available.

Olympian
This is a high-capacity double-decker for fast city centre/airport link services, with full air suspension, automatic transmission and rear-mounted engine, for up to 100 passengers. Left- or right-hand control is available. Olympian can be built to incorporate a large baggage area on the lower saloon.

Olympian three-axle
As an ultra-high-capacity double decker for fast city centre/airport links, the three-axle version can accommodate up to 180 passengers. Technical details are as for the Olympian two-axle; the second axle self-steers.

Cub
This small bus with forward-mounted engine and automatic gearbox is ideal for crew movement, personnel movement and car park services. A rear luggage area can be provided.

Cub is available with right- or left-hand control.

BRE-Leyland Railbus
This low-cost rail vehicle combines advanced road and rail technology in an easily operated, easily maintained self-propelled unit.

The Railbus is available in single-car or twin-car form and is ideal for fast link services between airports, city centres and railheads.

LEYLAND BUS
Leyland Vehicles Ltd
Bow Lane, Leyland, Preston, Great Britain, PR5 1SN
Telephone: Leyland (077 44) 24241 Telex: 67515

FOREWORD

The construction and development of airports continues to show increased activity, and its current level is higher than ever. Forecasts for the level of spending range from $US25,000 million in the next decade to $US85,000 million by the end of the century. This is despite the world recession which has resulted in no significant growth in the world airline industry for the last two years, several major airlines collapsing and others cancelling or slowing down delivery of orders for new aircraft. It is often said that the economic performance of airlines reflect the nature of the economy of the nation which they serve. The airlines' results in 1981 would seem to confirm this point. The economies of Europe and North America are in deep recession and have been since the rapid increase in oil prices following the Iranian crisis in 1979. The airlines serving these continents carry about two thirds of the total world airline traffic, and the growth of airlines in these areas, like the economies of the nations has been stagnant for more than two years.

Air traffic

Preliminary figures from the International Civil Aviation Organisation (ICAO) show that its member airlines carried approximately 728 million passengers in 1981, 2·6 per cent less than in 1980: a decline in passenger traffic for the second successive year. Freight traffic increased by 3 per cent. Financial results of airlines for 1981 also portray a gloomy picture. Preliminary figures for International Air Transport Association members indicate an operating loss of $US440 million with interest payments of $US1,200 million producing a total deficit of $US1,640 million. Although this represents a drop of $US 160 million from 1980 results, there will need to be some steady improvement in 1982 if more airlines are not going to plunge into liquidation.

Despite the sombre picture for the industry as a whole, there are some bright spots. Lack of growth was produced by losses in one area offsetting the gains achieved elsewhere. The growth of air traffic in the Middle East and the Far East in the past decade has been well in excess of that in other areas. In 1980 figures produced by ICAO show that the airlines of the Middle East region performed an estimated 22,000 million passenger kilometres in international scheduled services. This represents 4·7 per cent of world traffic compared with 3·2 per cent in 1970. International passenger traffic carried by the airlines of the region grew at an average annual rate of 15·6 per cent while world traffic rose by 11·1 per cent. Passenger forecasts for airlines registered in the Middle East for the 1980s anticipate an average annual growth in passenger kilometres of about 9·8 per cent per annum. The Far East region of Asia and the Pacific saw tremendous growth in the 1970s with total ton kilometres increasing from 4,500 million in 1970 to 21,000 million in 1980 and it is likely that the growth in this region will continue to be above average during this decade.

But, currently the aviation industry is being strangled by a volume of traffic the existing infrastructure was never intended to handle. Therefore despite the recent decline it is imperative that the many airport authorities throughout the world act now in the planning and implementation of improvements to existing buildings and the construction of new facilities. Most analysts believe that by the mid-1980s air travel will again be expanding and although it is unlikely to be at the 10 to 12 per cent rate experienced in the 1970s, current forecasts indicate an annual growth of between 5 and 7 per cent.

Airport development

Many countries of the Third World have grown to appreciate that air transport is one of the mainsprings of economic growth. This has resulted in plans for new airports to provide for the anticipated growth of air traffic. It is not that the world is currently short of airports. What it is lacking is airports with the appropriate infrastructure of the right calibre equipped to modern standards for scheduled international operations. Several major airports have reached saturation point and air traffic control facilities are similarly congested as evidenced by the rationing of slots in the United States and everyday air traffic control delays in Europe.

ICAO, the aviation technical agency of the United Nations, lists 32,835 civil aerodromes world-wide in its 160-plus member states, but 17,089 of them are listed for private use only, another 490 are heliports and 511 are seaplane bases. This leaves 14,745 land airports of which only 1,025 are listed as suitable for regular international operations at the end of 1980, with another 23 major airports in the advanced stages of planning or under construction. There were about 930 major international airports world-wide in 1972 so that over the last decade approximately 100 new airports have been introduced. In that period scheduled passenger traffic has risen from 450 to 734 million. If world traffic continues to grow as anticipated, it is likely that at least another 100 new airports will be commissioned in the next decade. The new Changi airport in Singapore, the first phase of which was opened in July 1981, is estimated to cost $US935 million and the airports at Jeddah and Riyadh in Saudi Arabia even more money, so on that scale the forecast of $US85,000 million by the end of the century becomes realistic.

Of course, not all airports are going to be built from scratch, and in fact most will be developments of existing facilities by the addition of a further runway, new terminal buildings etc. Very few completely new airports are likely to be constructed.

Aircraft expenditure

Large though the forecast figures for equipment expenditure may seem, they are just a small fraction of the money which is spent and will continue to be spent on new aircraft. Estimates produced by ICAO show that during the coming decade the world's airlines are expected to invest over $US350,000 million on new aircraft, maintenance/parts and fuel. To this should be added $US1,500 million for training programmes. General aviation operations are anticipated to cost $US30,600 million for new aircraft, $US18,300 million for maintenance and parts and $US46,300 million for fuel making a total estimated cost of $US95,200 million. The total figures for airline and general aviation operations for this period are some $US445,000 million. This figure excludes airport construction/maintenance, air traffic control and ground handling systems. The estimates for these are $US12,500 million, $US11,500 million and $US4,800 million respectively, a total of $US28,800 million, or put another way, a mere 6·5 per cent of the total cost. This represents a great disparity between investment in new aircraft and investment in the supporting infrastructure to keep the aircraft airborne. Too often new airlines begin operating wide-body aircraft showing attractively low seat mile costs (if the aircraft can be filled) without similar investment being made on equipment to cope with them. How often in recent years have national airlines been started, to show off the flag, without the governments acknowledging the needs of the supporting infrastructure?

Operational costs

In an attempt to redress the balance in spending and the level of investment, authorities have significantly increased the costs to the operator for airport services eg landing fees, air navigation charges. It is to be hoped that this money is channelled into the appropriate purse for ensuring better airport facilities, and not into a government's general funds whereby only a small percentage of the money is actually spent on the use for which it was intended. Another welcome indication is the general awareness of most airports of the necessity for them to market their services to increase their revenues. At a recent ICAA seminar, Miami International Airport's director, Richard Judy, spoke of the need to get back to basics selling the airport's services to the airlines, travelling public, cargo shipper and the community: 'There is a general rule in the airline marketing world: if a person likes your product he will tell three other people. If he does not like it, he will tell eleven other people. Let us have him tell three people a positive story.' However, M. J. Fischer, Frankfurt's director of air freight planning, sees increased flexibility as being the key. 'Airport authorities have to start to look at airlines as partners rather than customers. Customs too should be seen as a partner not as a superimposed control agency.' Both these views are valid and both give every indication of a healthy recognition of the requirement for an active approach to marketing.

Military equipment

But what of the military? What trends are there likely to be in the next decade? So far this foreword has been addressed to the requirements of the civil airports and airlines. The reason for this is quite simple. The vast majority of ground-based aircraft handling equipment for military operations is commercial equipment

which has been adapted. Obviously there is some equipment which has been developed specifically for military use, for example, weapon carriers, runway repair equipment, and type-orientated maintenance and test equipment. Radar and communications systems for military operations also tend to be specialised, but emergency services, environmental control, fuel services and cargo handling operations use similar systems, whether in the civil or military environment. Recent events in the Middle East and the South Atlantic have once again emphasised the vital importance of air power in modern warfare and thus military budgets in this area are unlikely to grow smaller.

Air traffic control

In the area of air traffic control one of the most significant developments is addressing selectively (Adsel) as it is known in the United Kingdom, or DABS as a similar system is called in the United States. Adsel is a new secondary radar system, which eliminates many of the signal problems experienced with current secondary radar in high density traffic. It does this by a system of selective addressing whereby the interrogator can 'speak' to one aircraft transponder at a time rather than triggering off replies from all transponders within range, as at present. The new system should be adopted in the mid-1980s by universal agreement. A system of monopulse bearing measurement has been developed as part of the Adsel project. It overcomes many of the problems of signal interference but, unlike Adsel, does not require a new transponder to be fitted into the aircraft. Instead of relying on the average of a number of replies from the aircraft transponder to ascertain bearing, monopulse usually requires only one pulse of a single reply: thus the risk of distortion from an interruption to the supply pattern is largely eliminated. The other two major problems associated with secondary surveillance radar display are 'garbling' of aircraft flying close to each other making them hard to distinguish, and false targets resulting from the reflection of radar signals by nearby objects on the ground. Both these problems are expected to be overcome by the system's multi-microprocessor plot extractor.

The rapid improvements of communications may provide a much needed reduction in complexity and overcrowding in air traffic control. As airports have become more congested, aircraft more reliant on information from the ground, their speed and size increased, so air traffic controllers have had to rely more and more on fast and very efficient forms of integrated communications to link them both with the flight crews and with the wide range of essential ground-based services. A major advance in this area has been achieved through modern voice communications switching systems. These provide the operators with flexibility in the selection of radio, telephone and intercom channels. The audio and control channels are separated and microprocessor technology is used to provide distributed intelligence thus removing the need for the complex common circuitry previously required in any computer-controlled communications system. These are modular systems with an inherent expansion capability and include self-diagnostic equipment which isolates faults down to board level permitting simple card replacement. Although these improvements will not solve all of today's problems, they will certainly relieve some of the pressure on air traffic control operations.

Landing systems

The adoption of the microwave landing system (mls) continues slowly. After acceptance by ICAO of the time referenced scanning beam (trsb) system, and the adoption by ICAO of relevant standards and recommended practices, the Federal Aviation Authority (FAA) plans to start implementation of mls in 1984, though procurement activity begins in 1982. Under phase 1 between ten and thirty mls are to be installed over two years to demonstrate capabilities, refine siting and operational requirements and provide guidance for subsequent implementation. About 1250 systems are planned for US airports by 1998. However, mls in the short term, is not designed to replace instrument landing systems (ils) but to be used where ils is unsuitable or non-existent, including localities where the siting of ils is difficult or impossible. Meanwhile ils remains the standard landing aid and there are plenty of signs that this market world-wide is buoyant. Operators and manufacturers are studying ways of overcoming siting problems that could extend the life of the aid and delay its eventual replacement by mls. Indeed Thomson-CSF introduced the ils 381 series in 1981, the fourth generation of such equipment in 15 years, to take advantage of latest technology and Wilcox had the first US-produced category III ils commissioned for use in Kansas City International Airport in September 1981.

Another development in landing aids is the pulse light approach slope indicator (plasi) system for use in vfr weather. The system guides the pilot by a single light signal from an installation parallel to the runway near the intended touchdown point. The light visible to the pilot from approximately four miles on a bright sunny day is designed to provide him with a stabilised approach.

A pilot who is on the glide-path will see a steady white light which is an angular wedge of ⅓ of a degree (this translates to approximately 1¼ dots above or below the glide-path on an ils glide-slope indicator). If the aircraft is above the glide-path, the pilot will see a pulsing white light, if below it a pulsing red light. The further the aircraft deviates from the glide-path, the faster the pulses appear. As the pilot corrects his approach and nears the glide-path, the pulse apparently slows until the pilot intercepts the glide-path and sees one continuous white light.

Plasi was given approval by the FAA in January 1982 for use by fixed wing aircraft as well as helicopters in the National Airspace System.

Meteorology

Experience with integrated meteorological observation systems at airports has demonstrated that they can supply the air traffic service and airlines with comprehensive information updated in real time on meteorological conditions at the airport. Not only can the collection of data be automated but a microprocessor can be used to great advantage in some equipment. For example in a wind monitoring system instantaneous two minute average and a ten minute average of wind speed and direction including maximum and minimum speeds can be displayed. Another major advance in this area is the development of a low level wind shear alert system in the USA to provide advance warning to air traffic control during take off and landings. Since 1974 at least six major airline accidents were caused in part by low level wind shear conditions when the aircraft were near the ground on departure from or on an approach to the airport. The equipment is a real-time microcomputer-controlled, data acquisition, analysis and display system. It collects wind direction and wind speed data from six anemometers; one instrument is placed near the centre of the airport and five boundary anemometers are positioned near the approach and take-off areas of the airport.

Emergency services

A study group within ICAO recently announced its recommendations which included more stringent performance standards of crash/fire/rescue trucks in airports. These recommendations proposed an increase in the required maximum speed of a vehicle from the existing 50 to 62 miles per hour (80 to 100 kilometres per hour) combined with an increase in acceleration performance. The current 45 seconds for 0 to 50 miles per hour (0 to 80 kilometres per hour) would be reduced to 40 seconds. To achieve this, it is estimated that approximately 20 hp (15 kW) per tonne would be needed, though more and better matching transmission stages would compensate for some shortfall in maximum power. A significant number of airport crash tenders currently produced do not conform to this new proposal. Modifications in engine power output and reductions in overall operating weight could permit many to achieve this performance, but diesel engines used in many crash tenders are currently operating at the limits of their design capacity with increased turbo-charging aided by new techniques in intercooling used to provide the highest power to weight ratio. An interesting development from Volvo of Sweden for the F12 chassis is to increase the power output by additional engine fuelling during the emergency period. This is achieved by activating the fuel injection pump when the sirens are operating, to add an extra 50 hp (37 kW) to the 350 hp (258 kW) TD 120 engine.

Following the Air Florida Boeing 737 crash in Washington and the World Airways accident at Boston Logan in January 1982, the US National Transportation Safety Board held a public hearing in Washington 'to consider issues associated with aircraft operations at airports if runways are contaminated with water, snow and ice.' The hearing called upon 60 witnesses organised into 12 panels including representatives of airline pilots, airlines, aircraft manufacturers, equipment manufacturers, airport operators and officials from the FAA, Transport Canada and the British Civil Aviation Authority. As expected, the pilot's representatives argued that the FAA should set standards for snow and ice removal and for the use of friction measuring devices. The result of the hearing has

yet to be published but it is unlikely that nation-wide standards for snow and ice removal will be set, and early signs indicate that the FAA is against the use of friction measurement devices, rather than pilots' braking-condition reports for deciding whether to close a runway.

Passenger handling

The methods used to transport passengers to and from aircraft remain as diverse as ever. The majority of modern airports handling over two million passengers annually (about 100) use loading bridges, provide a covered way to the aircraft, can be used by the handicapped in wheelchairs and provide relatively quick embarkation and disembarkation. Their disadvantages are high capital cost, lack of flexibility, large ground area required and the time taken by passengers on long walkways though this can be reduced by conveyors.

A second method, perhaps the more traditional way, is to take passengers by bus to the aircraft parked at a remote apron location, where they climb steps into the aircraft. This is a relatively cheap method, with flexibility introduced in the number of buses which can be provided. Also, airlines can save a significant amount of fuel by the shorter taxiing distance from the runway. The disadvantages are the time taken to transport the passengers and their exposure to the environment between the aircraft door and the bus.

A third method introduced in recent years is mobile passenger lift lounges. These are installed at Washington Dulles, Montreal Mirabel, Gothenburg Landvetter, and at the new King Abdul Aziz International Airport, Jeddah, Saudi Arabia. As with the bus method, the aircraft is parked at a remote apron location to and from which passengers are transported in large vehicles which elevate themselves to the height of the aircraft door sill enabling passengers to transfer to and from the vehicle without being exposed to the weather, or having to use steps. The main disadvantage of the mobile lounge is cost, the current price being around $US600,000 the large capital cost induces a measure of inflexibility by the very fact that airports are forced to keep the numbers held to a bare minimum.

All these methods of transporting passengers do not affect another major problem of airlines and airport authorities, that of passenger baggage.

Baggage handling

Historically the airlines have looked after the passenger's baggage from the moment he or she checks in at the airport, relieving the passenger of the responsibility but creating an enormous problem for the airlines. It is now seen as part of the accepted service, and in the truly competitive environment created by deregulation it is a difficult, if not impossible time to say to the passenger—'you will be responsible for your baggage until arrival at the gate area. Furthermore, there will be reduction in personal baggage allowance, the remainder to be shipped as cargo.' This would certainly reduce the problem but is unlikely to happen in the near future.

Currently the approach always seems to be same: passenger and bags arrive at airport together, airlines separate passenger from bags, passenger tries to find aircraft, aircraft leaves with or without bags or passenger, theoretically aircraft arrives at destination with passenger and bags, passenger sent to baggage claim area, baggage despatched to carousel, passenger or bags do not arrive and then the problems start.

One new concept being proposed would solve both baggage and passenger transporting problems. This is the passenger and luggage together (palt) airport handling system from Ikarus of Hungary. This new approach to passenger-carrying apron vehicles is basically a bus with an extremely large under-floor space, designed to transport passengers together with their luggage to an aircraft. A height-adjustable, telescopic-covered gangway is provided for aircraft boarding. The first operational tests of a prototype vehicle are currently taking place in Budapest. The vehicle used is the smallest version, the Ikarus 692, with a capacity of 40 seated passengers and between 110 and 140 standing passengers plus a 30 square metre under-floor baggage compartment. The smaller vehicle, as well as the two larger versions, is designed for use between a terminal and the aircraft and to directly transport passenger and baggage between a city-centre terminal or hotel and the aircraft. On the airport speeds of up to 50 kilometres per hour are possible. With a mechanical transmission in the city-to-aircraft version, the maximum speed is extended to 90 kilometres per hour. As with any innovative concept it will take time and a lot of effort by Ikarus if the new proposal is to succeed. How existing airport configurations will accept its unique combined delivery of passenger plus baggage to a single point in the terminal remains to be seen.

Another new concept to reach prototype stage is the MOAB passenger handling system developed by Economical Airport Architecture of West Germany. This system is even more radical in its design, enabling the passenger and baggage to remain together up to the aircraft and for check-in, customs, immigration and security procedures to be carried out on the way to the aircraft. The system comprises a check-in/out handling organisation with a control centre passenger service centre and a technical centre: a pre-check-in assembly point, and the mphu vehicle. Departure from the assembly point is by the mphu vehicle between 40 and 90 minutes before a flight depending on whether it is domestic or international and the distance from the assembly point to the aircraft. A number of pick-up points can be arranged. The assembly point is used for transit transfer from one vehicle to another. The vehicle can accommodate up to 378 passengers, with 8 to 14 check-in clerks and a driver. A separate vehicle is used for arriving passengers with different interior arrangements for baggage claim and customs areas.

Cargo handling

In the cargo handling area a British firm, Listavia, has produced a new concept for tractor/trailer systems for apron operations known as the LA PD 22 roll trailer system which evolved from a lengthy study of airport operations in 1979, following which the company decided to produce a system which would, among other objectives, reduce complexity, lower capital costs and wherever possible use existing equipment. Hydraulic towing converters are fitted to conventional airport tractors. These devices supply a lifting function to the towed load without applying any vertical imposed load on the tractor. This produces only one point of articulation between the towed load and the tractor which allows the vehicle combination to be driven forwards or in reverse. Low bed-roll trailers are designed to connect behind the tractor and towing converter and these will carry any type of unit load devices. The front of the trailer rests on the front frame on the tarmac when disconnected and has no steering gear or brakes. Towable main/lower deck elevators based on standard production roll-trailer chassis complete the system.

Apron equipment

Essential services piped direct to the aircraft on the apron is not a new idea. Fuel hydrants have been used for some time and fixed electric power systems started to become acceptable in the mid-1970s; now fixed air systems are being introduced for engine starting and cabin air conditioning. All fixed services are not practicable unless the parking position of the aircraft can be determined with some certainty. The increased use of the loading bridge, particularly of the fixed type, coupled with nose-in aircraft parking has removed this obstacle. Even with apron-drive bridges, where the exact position of the aircraft is less certain, 'over-the-bridge' delivery has enabled electrical power to be supplied with an outlet at the head of the bridge for connection to the aircraft.

The increasing cost of running an auxiliary power unit (apu) is one of the main factors which is making airlines consider the potential benefits of fixed pneumatic services at airports. However, unlike fixed electrical systems, the decision is not just whether to have the service or not, but also what type of service is required. At present it is uncertain whether the answer lies in supplying hot air for engine start and for running the aircraft cabin conditioning heating and ventilation, or whether it is more practical and economical to supply pre-conditioned air for cabin conditioning. A recent study conducted by Teledyne Inet on fixed air systems for ground operations, showed that:

'Fixed air systems combined with fixed 400 Hz electrical systems save significant fuel compared with apu use to provide these services; engine start using the apu after the aircraft has been 'pushed back' from the bridge is most economical in most cases and appears to be a widely accepted procedure for nose-in parking; cabin conditioning, heating and ventilation is most economically performed by a fixed pre-conditioned (low pressure) air system which costs approximately one tenth of the apu and a quarter of the cost of operating a pneumatic (high pressure) system; capital cost of a fixed pre-conditioned system is

less than a fixed pneumatic system; if a fixed engine start capability is desirable, it can be installed as a separate system for less capital and operating cost than a pneumatic system.'

However, every airport and every airline is different. In some areas only heating air is needed, in others only cooling. Some airlines, for fuel saving and environmental reasons, push aircraft back and turn them before starting engines. At other airports this manoeuvre is not necessary and aircraft just taxi away from the apron. Nevertheless a recent Air Transport Association survey placed potential savings in the order of 350 million gallons annually in the USA by shifting from the use of onboard auxiliary power units to fixed electrical or air systems.

Energy saving will continue to be of prime importance to the airlines as fuel costs are expected to absorb nearly 25 per cent of airline operating revenues in this decade. As well as examining the advantages of fixed systems for ground operations, reduced taxying times is another area of cost saving continually under review by airlines. Meanwhile a number of ground equipment suppliers are making significant strides in producing battery-powered tow tractors for aircraft, freight and baggage trailers, conveyor belt loaders and other types of apron servicing vehicle.

Acknowledgements

To thank all the people who have helped to produce the first edition would almost require a yearbook in itself and so with regret I am only able to mention by name a few individuals who have been outstanding in their efforts. To the remainder I would like to say thank you because without the help and encouragement received from all of you, in government bodies, manufacturing companies and other organisations, this book would certainly not have been possible.

The individuals whom I must mention include Don Parry, Andrew Seacombe, Andy Sweetland, T. F. Taylor (Cossor), C. T. P. Stephenson (GECOS), Gerry Gill and Stewart Wilson (IAL), Alan Diggory (Racal-Decca) and Rob Walker (Thorn Lighting). I am very grateful for advice received from David Woolley (Editor, *Airports International*), Gottfried Hilscher (Editor, *Airport Forum*) and Manfred Momberger (Editor and Publisher, *Airport Forum News*).

The Jane's team of yearbook editors have provided invaluable help, support and encouragement. My thanks to all, especially Chris Foss, Ron Pretty and Bob Raggett. As a mentor, Ron has been most patient and forthcoming in his comments which I have found immeasurably useful.

Within the Jane's organisation there are a host of dedicated people to whom I am indebted. Among these are Tony Hesketh-Gardener and members of his department, in particular Anne McKrill and Gillian Thompson, who have continuously gathered information on my behalf. The book owes a great deal to Sue Walby, who has been a tower of strength and helped me tirelessly. Also my thanks to Ken Harris, David Moyes and Alan Griggs, who along with editorial, accepted the challenge to produce this book in time for it to be launched at Farnborough 1982. My sincere thanks go to Bridget Bowsher who has cheerfully accepted the immense peaks of activity which have gone with typing the initial drafts of the book, and worked extremely hard to produce the copy on time. It would not be fitting to close without mentioning Sidney Jackson, the Managing Director, who has shown tremendous enthusiasm and encouragement during the gestation and birth of his 'baby'.

This is the first international book to be compiled covering airport equipment and, as anyone who has produced a new reference book will know, the first edition is a difficult task. We have tried to build a solid foundation on which subsequent editions will flourish. The subject encompasses a wide variety of equipment and it has been almost more difficult to know what to leave out than what to include. We have arranged, by country of manufacture, brief descriptions, technical specifications and illustrations of civil and military airfield equipment in production and service world-wide. Covering air traffic control, passengers and cargo handling, apron and environmental equipment; emergency, fuel and maintenance services; international aviation organisation airport design and construction agencies.

I would welcome any thoughts or comments on the book's coverage and on subjects which have been omitted that could be covered in subsequent editions.

July 1982 David Rider

GLOSSARY

A0 unmodulated (steady note) continuous wave radio emission

A0/A1 unmodulated (steady note) radio emission identified by morse code in a break period

A0/A2 unmodulated emission identified by morse code heard above unbroken carrier (eg an ndb)

A1 unmodulated but keyed radio emission

A3 am radio transmission with double side-band

A3H am single side-band transmission with full carrier

A3J am single side-band transmission with suppressed carrier

ACC Area Control Centre

ADAP Airport Development Aid Program (US Department of Transportation)

Adsel selective addressing by secondary surveillance radar interrogating each aircraft once only on each aerial rotation

Aeropp aeronautical message switching system

afc automatic frequency control

aff acqueous film forming foam

AFTN Aeronautical Fixed Telecommunications Network

agc automatic gain control, property of radio receiver designed to vary gain inversely with input signal strength to hold approximately constant output

agl above ground level

agnis aircraft guidance nose-in system

air cargo goods; any property carried or to be carried in an aircraft, other than mail or other property carried under the terms of an international postal convention, baggage or property of the carrier, provided that baggage moving is under an air waybill

aircraft container uld which interfaces directly with the aircraft cargo handling and restraint system

aircraft noise aircraft noise monitoring normally involves two types of noise reports: information on each individual noise event (maximum sound pressure level, event exposure and time of occurrence) and long-term assessment of the noise exposure (equivalent sound level, noise statistics etc). For long-term assessment, the actual procedure may depend on local or national regulation; for example, the following ratings are used in different countries: WECPNL (weighted equivalent continuous perceived noise level). NNI (noise and number index), L_{eq}, etc. However, general methods for describing both single events and long-term exposure are described in the International Standard ISO 3891-1978 'Procedure for describing aircraft noise heard on the ground'. The methods which apply to airport noise monitoring are summarised as follows:
the measuring equipment must fulfil the requirements of IEC R 651 for Type 1 sound level meters, and measurements should normally be done using the 'slow' meter characteristics, although 'fast' may be used for low altitude, high speed flyovers. A- or D-weighting may be used (the procedures using D-weighting are also recommended by the Annex 16 of the ICAO Convention). A-weighting should be used when aircraft noise is assessed as one among all other noises affecting a community, while D-weighting should be used when aircraft noise dominates the environment.
Single events are described both by the maximum sound level recorded during the event and by an energy average over the event duration, referred to some reference duration T reference. With A-weighting, the quantities to be measured are therefore L_A max, the max A-weighted sound level, and L_{AX}, the single event noise exposure level (also called SEL, Sound Exposure Level or SENEL), which is the energy average of L_A over the noise event, using T_{ref} = 1 second. With D-weighting, the quantities to be measured are L_{PN} (approximately) max, the maximum approximated perceived noise level, (L_{pn} (approximately) = L_D + 7 dB) and L_{EPN} (approximately), the approximated effective perceived noise level, which is the energy average of L_{PN} (approximately) over the event, using T_{ref} = 10 seconds. L_{AX} and L_{EPN} (approximately) may be derived from the maximum event level using a duration allowance which takes into account the time interval $(t_2\text{-}t_1)$ where the noise level was within 10 dB of the maximum. Long-term noise exposure is assessed using an energy average of the sound level over the period of interest. With A-weighting, the quantity to be measured is L_{eq}, the equivalent continuous sound level. With D-weighting, $L\text{-}_{PNeq}$, the equivalent perceived noise level, should be measured. These quantities may be obtained by continuous integration or by making an average of the values of L_{AX} (for L_{eq}) or L_{EPN} (approximately) (for L_{PNeq}) for all the events recorded

aircraft pallet item of equipment consisting of a flat platform with flat undersurface of standard dimensions on which goods are assembled and secured before being loaded as a unit on to the aircraft. (See *aircraft unit load device*)

aircraft pallet net webbing or rope net for restraining load on to an aircraft pallet; it may also be used in conjunction with an igloo. (See *aircraft unit load device*)

aircraft unit load device assembly of components comprising either an aircraft pallet and pallet net; an aircraft pallet and pallet net over an igloo; or an aircraft container. The purpose of the unit is to enable individual pieces of cargo to be assembled into a standard-sized unit to facilitate rapid loading/unloading on to aircraft having compatible handling and restraint systems which interface directly with the unit

airport category: see *category*

amsc automatic message switching centre

apu auxiliary power unit

ASC II American Standard Code for Information Interchange

atc air traffic control

ATIS Automatic Terminal Information Service

B

BAA British Airports Authority

baggage sizes 'normal passengers' baggage is defined as being not larger than 36 inches (900 mm) long by 15 inches (400 mm) wide by 28 inches (700 mm) high, weight 75 lb (34 kg), nor smaller than 14 inches (350 mm) long by 9 inches (230 mm) wide by 6 inches (150 mm) high, nor articles having loose trailing cord ties or loose bindings, sharp or unusually shaped projections

ball mats horizontal surface consisting of freely rotating balls in housings and protruding above surrounding surfaces. They are used to enable omni-directional movement of units passing over it

bar measurement of pressure, eg foam monitors on air crash tenders. 1 bar = 1·02 kg sq cm

bcf bromochlorodifluoromethane (fire extinguishent)

bit unit of data or information in all digital (binary) systems comprising single character (0 or 1) in binary number

bite built-in test equipment

btm bromotrifluoromethane (fire extinguishent)

C

CAA Civil Aviation Authority

category I decision height 60 m (200 ft) rvr 800 m minute

category II decision height 30 m (100 ft) rvr 400 m minute

category IIIA decision height 0 rvr is 200 metres

category IIIB rvr is 50 metres

category IIIC rvr is 0 metres

ccr constant current regulator used to control intensity of airfield lighting systems

cctv closed circuit television

cfar constant false alarm rate used in radar systems

cfr vehicle crash/fire/rescue vehicle

chap1 papi used for helicopter operations

class of fire: see *fire classifications*

combination aircraft ('Combi') fixed configuration of an aircraft type to carry both passengers and cargo on the main deck

container see *aircraft unit load device*

convertible aircraft aircraft type which can be converted to or from an all-passenger to an all-cargo carrying aircraft

cpu central processing unit

crt cathode-ray tube

D

dB decibel

dBA decibels absolute

df direction finder

dolly platform type towed vehicle equipped with rollers, castors or ball mat, used to accept, discharge and handle ulds on the ground

down-time period during which hardware is inoperative following technical failure

drdf vhf/uhf radio df system (UK)

Dvor Doppler vhf omni-range beacon for aircraft navigation

E

ebrm electronic bearing and range marker used in radar displays

edd electronic data display

elastic limit used to determine the strength of a material (see *yield strength*)

eprom erasable programmable read-only memory

erp effective radiated power

esu emergency supply unit

F

FAA Federal Aviation Administration (USA)

FAR Federal Aviation Regulations: eg FAR Pt 36—noise

fire classifications class A: wood, tissue, paper etc; class B: fuel, petrol, oil; class C: butane, propane, hydrogen

floor locks devices installed in the aircraft floor which overlap on to the aircraft uld base (pallet) for restraint purposes

fm frequency modulated

forklift vehicle used to lift loads vertically and transport them by means of twin tines mounted on the front

forklift capability the unit is equipped with an integral pallet or forklift entries or tineways so that it can be picked up by a forklift

fruiting ssr responses of aircraft to interrogation by other stations etc

fsk frequency-shift keying

G

g expression used to show the magnitude of a force in terms of the standard earth gravitational unit, eg a force of 3 *g* is equal to three times the gravitational force, causing a body to appear to have three times its normal mass

gca ground-controlled approach to a runway

goniometer electrical transformer used with fixed or rotating aerials for determining bearing of radio station

grp glass reinforced plastic

H

handline flexible hose-pipe, stowed on a hose-reel on a cfr vehicle, used for manual fire-fighting operations

I

IATA International Air Transport Association

ICAA International Civil Airports Authority

ICAO International Civil Aviation Organisation

ICM Inter-Console Marker used in atc systems

ICW Independent Carrier Wave

icw interrupted continuous wave

if instrument flying or intermediate frequency

IFR Instrument Flight Rules

igloo, non-structural item of equipment, similar to a structural igloo (see below) but not intended to contribute towards the restraint of the cargo and, therefore, requires the addition of a pallet net to provide restraint capability

igloo, structural item of certified aircraft equipment consisting of a rigid shell attached to an aircraft pallet, shaped to conform to the aircraft envelope, providing weather protection and cargo restraint without the use of a pallet net. (See *aircraft unit load device*)

ils instrument landing system

IMC Instrument Meteorological Conditions

interline intended to mean transfer from one carrier to another

internal tiedowns means of securing cargo inside a container

isb independent side-band

isls interrogation side-lobe suppression

L

L average peak sound pressure level

L_{AX} single event noise exposure level; also known as 'senel'

L_{EPN} effective perceived noise level with tone/duration correction

L_{eq} energy — average sound pressure level

L_N (night) weighted sound pressure level

led light emitting diode

llwsas low level wind shear alert system

loading contour maximum aircraft envelope for purposes of stowage inside the aircraft, having

taken into account the necessary clearances between the aircraft wall and the load
loading gauge rigid framework formed into the shape of an aircraft envelope for the purpose of building up or checking a pallet load, on the ground or in the terminal, to ensure it will fit into a particular position in a specific aircraft type
lower deck container uld shaped to fit the lower deck of the high capacity aircraft. These units come in half sizes and full sizes, related to the width across the aircraft
lpg liquefied petroleum gas

M

maximum gross weight maximum allowable combined weight of the unit and its contents (payload)
maximum net weight maximum allowable weight of the payload, equal to the difference between the maximum gross weight and the tare weight
mcpss microprocessor-controlled page storage system
mcw modulated carrier (or continuous) wave, A2 emission
metar meteorological aeronautical radio code
metrep meteorological report
mfd multi-function display
mfd microfarad
monitor pipe through which the water/foam is projected from a fire tender
mtbf mean time between failures
mtbr mean time between repair
mtd moving target detection on radar displays
mti moving target indication used in radar systems
mttr mean time to repair

N

N noise
narrow-body aircraft commercial transport with fuselage width of approximately 10 feet (3 metres) with single aisle between passenger seat
navaid navigational aid typically an electronic equipment installed at a fixed ground station
ndb non-directional beacon for aircraft navigation
nmt noise monitoring terminal

O

ocr optical character recognition
output power developed by engine or prime mover/total product or yield from a system
overthrow straps straps which secure the load instead of/in addition to the aircraft pallet net on to the pallet/on to the aircraft floor attachments

P

pallet, integral forms the base and is securely attached to a rigid structure unit
pallet, warehouse slaved unit, commonly made of solid, slatted or intersticed material, with or without fixed attachments, and designed to provide forklift capability. Its main function is to facilitate handling by grouping individual items on to a single surface for storage or movement
papi precision approach path indicator
par precision approach radar
pa system public address system
pcm pulse-code modulation
pep peak envelope power
plasi pulse light approach slope
ppi plan-position indicator
ppm parts per million: used in fuel filtration to describe water content
prf pulse-repetition frequency
psr primary surveillance radar
pto power take-off (shaft output)

Q

qdm code: 'will you indicate magnetic bearing for me to steer towards you with no wind?'
qdr code: 'what is my magnetic bearing from you?'
qfe altimeter setting providing aircraft height above ground level at airfield
qnh altimeter setting providing aircraft height above mean sea level

R

ram random-access memory
reils runway-end identification light system (FAA)
real-time absence of delay, except for time required for transmission by electromagnetic energy, between occurrence of event or reception of data, and knowledge of event or reception of data at some other place
resisting bend moment (rbm) used to determine the strength of a vehicle's frame. The rbm of a frame is determined by multiplying the section modulus by the yield strength expressed in cm/gm
restraint system system installed in the floor of an aircraft compartment which secures the aircraft uld on to the floor to prevent its movement during flight
rf radio frequency
riv rapid intervention vehicle for firefighting and rescue, operations
rms root-mean-square/range measurement system (noise)
roller-ball transfer conveyor system either in an aircraft or in terminal facilities consisting of various size balls/rollers over which ulds can be moved
rvr runway visual range. Value representing horizontal distance pilot will see centre-line or edgelights or runway markings down the runway from the approach end

S

scissor lift a lifting mechanism which provides exact vertical movement parallel to the aircraft. Used by high-loaders to ensure food or cargo is not upset or spilled
scr silicon-controlled rectifier, used in the control of electrically powered vehicles
secondary radar radar in which interrogatory pulse is sent to distant transponder which is triggered to send back a different pulse code to originator
section modulus used in determining the strength of a vehicle frame (see *resisting bend moment*). Section modulus is expressed in cubic cm and is the result of multiplying the frame thickness (D) by the flange width (K) and the web depth (B) using the formula

$$SM = \frac{(B \times D^3) - (K^3 \times B - T)}{6D}$$

where T is the thickness of the frame material
senel single event noise exposure level, also known as LAX
shrimpboat small marker of clear plastic on which controller writes flight identity, flight level and other information, subsequently moved by hand to remain adjacent to blip on display
SID Standard Instrument Departure procedure
slug water contamination in fuel systems
snow-clearance machinery apart from the snow-plough, snow-clearing systems fall into four categories: snow-cutter, with cutter drums mounted on lateral drive shafts; snow-blower, with rotors, the shafts of which lie along the vehicle axis; blower-cutter, a combination of these two systems, with the cutter drums in front of the blower rotor; auger-type snow-blower, with transversely-mounted clearing augers in front of the blower rotor; first two clearing systems are referred to as single-stage, the others as two-stage machines
spi short-pulse insertion (ssr)
sre surveillance radar element; portion of gca
ssb single side-band
ssr secondary surveillance radar
stirrup attachments lip-installed in a cut-out in the edge member of an aircraft 9g-pallet to which a 9g overthrow strap is attached
structural igloo see *igloo*
STAR Standard Arrival Route procedures
svr slant visual range
swg standard wire gauge (UK)

T

tacan tactical air navigation, uhf R-theta-type navaid giving bearing/distance of aircraft from an interrogated ground station
TAF Terminal Area Forecast
tare weight weight of empty uld. It includes all liners and/or fittings etc when these are required by the specifications or as registered with IATA
tare weight allowance free weight allowance given to shippers as part of the IATA unitisation incentive programmes for non-member owned registered ulds
tare weight objective recommended but not mandatory target weight for an aircraft uld
tensioning buckles hardware on an aircraft pallet net or overthrow strap used for length adjustment to ensure that all the slack has been taken up and the net/strap fits securely on the load
thermal uld built with insulating walls, doors, floor and roof which retards the rate of heat transmission between the inside and the outside of the uld
tie-down fitting attachment devices designed to transfer forces between a load-bearing device (net/strap/rope/bar) and a tie-down track
tie-down track standardised configuration designed to take tie-down fittings and capable of transferring forces into a uld, or aircraft, structure to which it is attached
tma terminal manoeuvring area ie terminal airspace
tvor terminal vor beacon
twr tower
twy taxiway

U

unit load device (uld) any type of a container, container with integral pallet, aircraft container or aircraft pallet

V

vasi visual approach slope indicator
vdu visual display unit
vfr visual flight rules
vmc visual meteorological conditions
vogad voice-operated gain-adjusting device
volmet routine ground to air broadcast or met information (ICAO)
vor vhf omni-directional radio range; fixed navigational beacon which emits a fixed circular horizontal radiation pattern on which is superimposed a rotating directional pattern at 30 Hz giving an output whose phase modulation is unique for each bearing from the beacon
vor/dme vor steering guidance with dme distance information
vor/ils linkage of vor signals to aircraft ils so that left/right steering guidance is given by ils panel instrument
vortac combination of vor and tacan offering from one fixed station vor azimuth, tacan azimuth and tacan (dme) distance information
vox voice keying (NASA)
vswr voltage standing-wave ratio
vtol vertical take-off and landing

W

wide-body aircraft commercial transport with internal cabin width sufficient for normal passenger seating to be divided by two aisles; in practice this means not less than 4·72 metres, the width of the narrowest wide-body (Boeing 767)

Y

yield strength used to determine the strength of a material (also known as elastic unit). Refers to the maximum stress which can be applied to a material before it takes a permanent set

EMERGENCY SERVICES

Aircraft arresting systems
Aircraft recovery systems: pneumatic elevators, portable roads
Airport crash tenders, crash/fire/rescue vehicles
Ambulances
Fire-fighting equipment, portable and wheeled extinguishers
Fire-protection systems
Hovercraft air crash tenders
Hydraulic fire-fighting and rescue platforms
Rapid intervention vehicles
Rescue/fire-fighting launches

AUSTRIA

Buffalo Series Air-Crash Tenders

The Buffalo series air-crash vehicles vary in size according to water and foam capacity and the chassis on which they are mounted. The Buffalo is available on MAN, Mercedes Benz and Volvo chassis. The MAN 8000 and Volvo 10 000 versions have automatic transmission and all are equipped with six-wheel drive and differential locks on all axles.

The monitor is mounted above and behind the driver's cab and all functions of the vehicle are electro-pneumatically controlled by one operator in the cabin. The control handle and the monitor function in parallel to enable the operator to visualise the position of the monitor. All functions of the fire-fighting installations can be controlled manually. Foam is mixed automatically, the rate of mix being independent of pressure and pump output. The percentage rate of the mix can be adjusted either from the cabin or from the pump compartment. Two hose reels are located, one on either side of the tender, with 30 metres of hose connected to a foam pipe. Vehicle protection is provided by five ground-sweep nozzles; three in front of the front axle, two in front of the rear axle. The cab is designed for a crew of four and has two doors. An alternative version is equipped for six people and has four doors.

Status
Production

Manufacturer
Konrad Rosenbauer KG, Linz, Austria

Characteristics

	7000	8000	8000	10 000
CHASSIS	MAN 26 320	MAN 26 440	Mercedes Benz 2632	Volvo FI2
DRIVE	6 × 6	6 × 6	6 × 6	6 × 6
ENGINE OUTPUT AT 2500 rpm	235 kW	323 kW	235 kW	294 kW
DRIVER'S CAB	1 + 5	1 + 3 or 1 + 5	1 + 3	1 + 3
PUMP	R 480	R 600	R 600	R 600
output	6000 litres/minute at 10 bar 3000 litres/minute at 16 bar	6000 litres/minute at 10 bar	6000 litres/minute at 12 bar	6000 litres/minute at 12 bar
TANK CAPACITY				
water	7000 litres	7000 litres	7000 litres	9000 litres
foam	630 litres	800 litres	1000 litres	1000 litres
FOAM/WATER MONITOR	RM 50J	RM 50	RM 50N	RM 50N
output	3000 litres/minute at 10 bar			
RANGE				
foam	60 m	60 m	60 m	60 m
water	70 m	70 m	70 m	70 m
ROTATION	360°	270°	270°	270°
ELEVATION	80°	−15 to +70°	−15 to +70°	−15 to +70°
HOSES	2 × 30 m	2 × 30 m	2 × 30 m	2 × 30 m
GROUND SWEEP NOZZLES		5 × 130 litres/minute	5 × 160 litres/minute	5 × 130 litres/minute
PERFORMANCE				
ACCELERATION 0–80 km/h	45 s	37 s	45 s	37 s
MAX SPEED	110 km/h	100 km/h		100 km/h
DIMENSIONS				
length	8·65 m	8·8 m	9·4 m	8·45 m
width	2·5 m	2·5 m	2·5 m	2·5 m
height	3·5 m	3·58 m	3·6 m	3·5 m
OPERATIONAL WEIGHT	21 400 kg	22 500 kg	22 000 kg	24 000 kg

Buffalo 8000 on MAN chassis

Buffalo 8000 on Mercedes Benz chassis

Cheetah Rapid-Intervention Vehicle

The Cheetah was developed for airport use as a rapid-intervention vehicle which can reach the scene of disaster well in advance of the heavy tank fire-fighting vehicles. It can be used as the only fire-fighting vehicle at airport categories one and two, or as an additional vehicle at airport categories three to nine.

The Cheetah is based on a Dodge chassis and powered by a 230 hp V8 petrol engine with automatic transmission and four-wheel drive. A foam monitor is mounted on the cab with inside control and two hose reels with high pressure fog nozzle attachments are on either side of the vehicle.

Up to four different extinguishing agents can be carried.

Cheetah rapid-intervention vehicle

Characteristics

CHASSIS: Dodge with 230 hp V8 petrol engine; single tyres 4 × 4 all-wheel drive; automatic differential lock, automatic transmission
ACCELERATION: 0–80 km/h in 17 s (up to 20 s depending on equipment)
MAX SPEED: exceeding 120 km/h
CAB: 2 to 3 people
PUMP: Rosenbauer foam monitor RM 50J; output 1600 litres/minute; range 45 m with foam
TANK: stainless steel containing 1100 litres premixed solution or 940 litres of water plus 160 litres of foam compound, 250 kg powder unit and/or 50 kg BCF-Halon installation
DIMENSIONS: 5·9 × 2·1 × 2·45 m
OPERATIONAL WEIGHT: 4000 kg (approx)

Manufacturer

Konrad Rosenbauer KG, Linz, Austria

Simba Air Crash Rescue Vehicle

The Simba is a development based on an extra-wide Reynolds Boughton chassis with a five-speed Allison automatic transmission and all-wheel drive. It is powered by a rear-mounted Mercedes Benz V12 turbo-charged diesel engine producing 386 kW with acceleration from 0 to 80 in 38 seconds. The vehicle controls and fire-fighting controls are separated but the driving and fire-fighting installation can be controlled by one operator in the cabin. A three-dimensional pilot stick indicates the position of the monitor to the operator. Alternatively a nozzle operator can stand on an elevated platform in the cab with his upper body outside the cabin and behind the monitor. The output which can be 2500 or 5000 litres per minute is selected using an independent pump engine and full output can be maintained while the vehicle is in motion. The throw pattern can be varied between a full stream and a dispersed pattern. Proportioning for the foam mix is automatic, at rates between two and ten per cent and up to 75 000 litres per minute of foam can be produced.
An auxiliary hose reel with 40 metres of hose is on the left-hand side of the vehicle. The tender is fitted with a self-protection system comprising three ground-sweep nozzles and four under-truck nozzles controlled from the cabin. The cabin is designed for a crew of four with storage space. A special seat is provided with integrated breathing apparatus and a heat protection suit. Gull-wing doors are located on both sides behind the front axle combined with escalating entrance steps. The doors are stabilised pneumatically in any position. The vehicle is equipped with a public address system.

Manufacturer

Konrad Rosenbauer KG, Linz, Austria

Characteristics

	10 000	12 000	14 000	16 000
CHASSIS	Reynolds Boughton, Mercedes Benz V12 Diesel – 386 engine			
DRIVE	6 × 6	6 × 6	6 × 6	6 × 6
ENGINE OUTPUT	386 kW	386 kW	386 kW	386 kW
DRIVER'S CAB	1 + 3	1 + 3	1 + 3	1 + 3
PUMP	R 600	R 600	R 600	R 600
output (litres/minute at 10 bar)	6000	6000	6000	6000
TANK CAPACITY				
foam	1000 litres	1000 litres	1500 litres	2000 litres
water	9000 litres	11 000 litres	12 500 litres	14 000 litres
FOAM/WATER MONITOR	RM 80	RM 80	RM 80	RM 80
output	5000 or 2500 litres/minute	5000 or 2500 litres/minute	5000 or 2500 litres/minute	5000 or 2500 litres/minute
RANGE	80 m	80 m	80 m	80 m
ROTATION	270°	270°	270°	270°
ELEVATION	–20 to +70°	–20 to +70°	–20 to +70°	–20 to +70°
AUXILIARY HOSES(S)	1 × 40 m output 400 litres/minute solution			
GROUND SWEEP NOZZLES/UNDER TRUNK NOZZLES	3/4 each producing 130 litres/minute solution			
ACCELERATION				
0–80 km/h	37 s	42 s	46 s	51 s
MAX SPEED	105 km/h	105 km/h	105 km/h	105 km/h
DIMENSIONS				
length	11	11·5	11·5	11·5
width	3·21	3·2	3·2	3·2
height	3·65	3·78	3·78	3·78
OPERATIONAL WEIGHT	19 000 kg	22 000 kg	22 500 kg	23 000 kg

Simba air-crash tender

BELGIUM

Comet Hydraulic Fire-fighting Platform

The Comet hydraulic fire-fighting platform is mounted on a combined telescopic boom/jib system which enables the platform to be manoeuvred from ground level to the maximum height with a rotation of 360 degrees. It is stabilised by means of vertical props which can be extended horizontally from the vehicle on the working side or both sides as required. The design of the telescopic boom/jib system allows the platform to straddle obstructions. When the telescopic arm is lowered as far as possible and the jib is in the vertical position the

Characteristics

	18 T	25 T	30 T
MAX PLATFORM HEIGHT	18 m	25 m	30 m
SIDE OUTREACH	11·7 m	14·6 m	16·3 m
OVERALL LENGTH	±7 m	±9·4 m	±11·3 m
NUMBER OF JACKS	4	4	4
JACK SPREAD	3·1 m	4 m	4·2 m
TOTAL WEIGHT (platform + sub-frame + walkway)	6300 kg	12 000 kg	13 300 kg
TOTAL WORKING LOAD	350 kg	350 kg	350 kg
CAGE DIMENSIONS	1·5 × 0·8 × 1·1 m	1·8 × 1·2 × 1·1 m	1·8 × 1·2 × 1·1 m
GROSS VEHICLE WEIGHT	16 000 kg	18 500 kg	23 000 kg

Comet platform can operate down to 3 metres below ground level.
The platform, which can accommodate up to six people, is equipped with a monitor capable of discharging water at 2000 litres per minute, a control panel, intercom system, spotlights and a cradle mounting for a stretcher.

Status

Comet is in service in Brussels, Paris, Vienna, Algiers, Cairo and Teheran.

Manufacturer

FN Comet, Namur, Belgium

CANADA

Fire Boss and Foam Boss Fire-fighting Systems

These units can be equipped to accommodate one or more of a variety of extinguishing agents.

The Foam Boss skid-mounted units contain a premixed foam system housed in single, double or triple containers which is discharged by means of regulated gas pressure (17 bars) from a dry nitrogen cylinder.

A hose reel is fitted with 30 metres of hose and a CDN variable nozzle capable of producing 455 litres per minute at 13·6 bar.
A combined version of the Foam Boss which is produced in skid form or mounted on a trailer contains 204 kg of Super K dry chemical and 190 litres of afff (premix). The unit has twin hoses and nozzles mounted on a single reel. Pressurised nitrogen produces 17 bars working pressure.

The Fire Boss dry chemical units are designed to fight petrol, oil, paint (class B) and electrical (class C) fires.
The units which can be skid- or trailer-mounted are available with varying capacities up to 2727 kg.

Manufacturer

CDN Research and Development Ltd, Oakville, Ontario, Canada

Foam Boss Airport Crash Tenders

Vehicles range from the heavy-duty tenders such as the ATV 1044-2 to the light-duty Foam Boss 500, which although equipped with four-wheel drive, does not have the cross-country capability of the heavy-duty tender.

The ATV Model 1044-2 designed for travel over rugged terrain is equipped with all-wheel drive, large wheels and wide tyres. The totally enclosed aluminium body is mounted on a six-wheel drive, articulated Foremost Marauder chassis powered by a Detroit Diesel 8 V9 2T engine. The Clark transmission provides eight forward and four reverse speeds. An independent Deutz F10L 4I3F air-cooled diesel engine powers the Hale 60FRQ water pump. A CDN round-the-pump foam proportioning system is fitted, the rate of which can be adjusted between 0 to 10 per cent.
A joystick-type of control column allows all functions of the monitor to be electro-hydraulically operated from the driver's seat including the discharge pattern of the jet stream.
Discharge systems include a roof-mounted monitor on the front of the vehicle, three under truck nozzles, one hose reel and two handlines. Rotation limits can be adjusted to suit the operator's requirements.
The Foam Boss 3000, 1500 and 1000 aluminium bodies are mounted on a Scot-chassis, powered by Detroit diesel engines with all-wheel drive. An independent Detroit diesel engine drives the Waterous pump, and discharge systems include a roof-mounted monitor on the front of the vehicle body, a hose reel with 60 metres of hose attached and two handlines. Three undertruck discharge nozzles provide protection to the wheels, chassis and underside of the vehicle when travelling through ground fires.
The foam proportioning system, monitor and controls are similar to those installed in the ATV tender.

The Foam Boss 500 is mounted on an International Harvester chassis and powered by either the International Harvester MV404 petrol engine or the DT 466 diesel engine.
The vehicle is provided with a foam/water discharge monitor mounted on the cab roof. The monitor is operated remotely by the driver from within the cab. The discharge system is equipped with a positive foam isolation control enabling either foam or water to be discharged from both the monitor and hose lines.
The body of the vehicle is supplied with equipment compartments, steps to the top deck, hand rails, emergency lighting, beacon/siren and other equipment as specified by the operator.

Manufacturer

CDN Research and Development Ltd, Oakville, Ontario, Canada

Characteristics

	ATV MODEL 1044-2	FOAM BOSS 3000	FOAM BOSS 1500	FOAM BOSS 1000	FOAM BOSS 500
CHASSIS	Foremost Marauder	Scot CR2	Scot CR3	Scot	Int Harvester 1824/1854
ENGINE	Detroit diesel 8V 9 2T	Detroit diesel 12V 7 1T	Detroit diesel 8V 9 2T	Detroit diesel 8V 9 2T	Int Harvester MV404 petrol or
	321 kW at 2100 rpm	463 kW at 2500 rpm	321 kW at 2300 rpm	321 kW at 2300 rpm	Int Harvester DT466 diesel
DRIVE	6 × 6	6 × 6	4 × 4	4 × 4	4 × 4
ACCELERATION 0–80 km/h		37 s	35 s	40 s	
MAX SPEED		100 km/h	100 km/h		over 96 km/h
PUMP ASSEMBLY engine	Deutz diesel F10L 413F	Detroit diesel 8V 71	Detroit diesel 6V 71	Detroit diesel 6V 71	Chrysler C318
pump	Hale	Waterous	Waterous		Barton American
OUTPUT	5114 litres/minute at 16 bars	7570 litres/minute at 12 bars	4542 litres/minute at 18·7 bars		2270 litres/minute at 17 bars
TANK CAPACITY foam	730 litres	1514 litres	757 litres	757 litres	295 litres
water	9100 litres	11 350 litres	5678 litres	4542 litres	2340 litres
MONITOR output	4540 litres/minute	5678 litres/minute	3028 litres/minute	2250 litres/minute	1135 litres/minute
range	70 m			60 m	38 m
ROTATION	± 120°	± 120°	± 120°	360°	360°
ELEVATION	−15 to +45°	−15 to +45°	−15 to +45°	−15 to +45°	−15 to +45°
HOSE REEL	1	1 × 60 m	1 × 60 m	1 × 60 m	
OUTPUT	270 litres/minute	454 litres/minute	454 litres/minute		
RANGE	31 m				
HAND-LINES	2	2	2	2 × 30 m	1 × 61 m foam 1 × 46 m water
output	450 litres/minute	454 litres/minute	454 litres/minute	433 litres/minute	
range	31 m			30	
UNDER TRUCK NOZZLES	3	3 × 1146 litres/minute	3 × 114 litres/minute	3 × 1366 litres/minute	
DIMENSIONS length	11·86 m	11·43 m	9·22 m	8·49 m	6·4 m
width	3·35 m	3·05 m	3·05 m	2·59 m	
height	3·81 m	3·66 m	3·58 m	2·65 m	3·07 m
OPERATIONAL WEIGHT	28 576 kg	34 694 kg	22 676 kg		9534 kg

In-Plant Fire-fighting/ Ambulance Vehicle

CDN In-Plant vehicle alongside CDN Foam Boss ATV 1044-2

This vehicle is designed for use in hangars, terminal buildings, underground or multi-deck parking facilities and areas where limited space will not permit access to conventional fire-fighting, ambulance or other emergency vehicles. It may be used to complement existing airport fire-fighting and emergency services.

In order to accommodate the maximum possible volume of water for fire-fighting purposes (1230 litres) the tanks, body and chassis of the vehicle are fabricated together to form one integral aluminium unit. Besides water, containers may be provided for foam concentrate, dry chemical, Halon, carbon-dioxide, etc. Compartments, hose reels, beacon/siren, emergency lights and other fire-associated equipment are mounted on top of the vehicle.

Individual front seats are provided for the driver and one passenger. A folding rear step together with hand rails allows three standing passengers to be carried at the rear. Optional engines are petrol, diesel or propane to power both the vehicle and water pump via hydrostatic drive.

A cylindrical-shaped stowage compartment is incorporated into the body structure. This houses a 8·5 m³ gas cylinder of dry nitrogen, as a pressurising agent for extinguishant (dry chemical) expellant purposes, and oxygen when the basic vehicle is equipped for ambulance duties.

The In-Plant ambulance has the same basic body/chassis construction except that the tank space is compartment space. Two wheel-type stretchers or four basket stretchers can be accommodated on top of the vehicle. The passenger seat is towards the rear of the ambulance so that medical treatment can be administered to patients during transit. Below-deck compartments may house a refrigerator for medical supplies, blanket warmer, escape packs, emergency breathing apparatus etc, according to service requirements.

Fire/ambulance combination units accommodate at least one stretcher and are a compromise between the fire truck and ambulance configurations.

Characteristics

Standard Fire-fighting configuration
CHASSIS: CDN 1059–8
ENGINE: petrol, diesel or propane
OUTSIDE TURNING RADIUS: 3·2 m
STEERING CRAMP ANGLE: 70°
WATER PUMP: Hale CBP or 2 CBP3
output: 227 litres/minute at 55 bars, 1379 litres/minute at 20 bars (2 CBP3); 946 litres/minute at 7 bars (Hale CBP)
capacity: 1230 litres
HOSE PIPES: 2 × 30·5 m
OPTIONAL EQUIPMENT
7·32 m ladder in three sections; 2 × 113·4 kg dry chemical; 1 × 30·5 m dry chemical hose; foam proportioning system; 36·2 kg Halon 1211 extinguishing agent
DIMENSIONS: 4·22 × 1·83 × 1·57 m

Manufacturer

CDN Research and Development Ltd, Oakville, Ontario, Canada

Rapid Rascal Rapid Intervention Vehicle

This small fire-fighting vehicle has been designed for operation on and off the paved surfaces of an airport. The aluminium body is mounted on a four-wheel drive Dodge W350 chassis (or approved equivalent) and is powered by a 360 CID V8 petrol engine with manual transmission.

There are two fire-fighting packages on this vehicle: afff and dry chemical. Each system can be discharged via the twin agent hose reel and/or the dual-nozzle monitor which is mounted on the roof of the cab. The monitor is operated by a remote manual control from within the cab and can be operated by the driver while the vehicle is moving.

The afff container has a capacity of 378·5 litres consisting of three or six per cent afff concentrate and 97 or 94 per cent water. When the system is activated dry nitrogen gas pressurises the afff container causing the premix solution to discharge through the monitor and/or hose where it becomes aerated and converts to foam.

A spherical vessel contains 245 kg of the compound for the dry chemical system. When the system is activated pressurised dry nitrogen gas expels the dry chemical through the monitor and/or hose.

Characteristics

CHASSIS: Dodge W350
ENGINE: 360 CID V8 petrol
DRIVE: 4 × 4
ACCELERATION: 0–80 km/h in 27 s
MAX SPEED: 120 km/h
TANK CAPACITY
afff/water: 378·5 litres
dry chemical: 245 kg
nitrogen: 8·4 m³
HOSE PIPES: 2 × 30 m
TOTAL OUTPUT
afff: 190 litres/minute
dry chemical: 3·6 kg/s
DIMENSIONS:
length: 5·52 m
height: 1·98 m

Manufacturer

CDN Research and Development Ltd, Oakville, Ontario, Canada

FINLAND

Starckjohann Fire-protection Systems

Finnair DC9 hangar at Helsinki Airport safeguarded by Starckjohann fire-protection system

Starckjohann design and manufacture fire-protection systems using a range of extinguishing agents including foam, Halon 1301 and 1201 and carbon-dioxide.

A typical fire protection system produced by Starckjohann was selected by Finnair for the airline's new hangar at Vantaa Airport, Helsinki.

The hangar is capable of housing three DC9 aircraft for maintenance. Two of these can be under heavy overhaul and the hangar is equipped with two separate docks. An afff sprinkler system was chosen for the hangar and the washing and cleaning room. Workshops and offices are protected by water sprinkler systems, apart from the electrical supply rooms which have an automatic Halon 1301 fire-prevention system.

Six automatic afff systems each independently activated to minimise damage in false alarms are capable of discharging a total 1900 litres per minute of afff mixture which is 97 per cent water and 3 per cent foam. These systems are pneumatically controlled by monitored fire-detection units.

HAD fire detectors are installed in the hangar ceiling in six groups and under the dock level in two groups. Three separate HAD detection systems are located in the cable ducts in the hangar floor. These are activated by temperature having a fixed-temperature system and one which measures the rate of change. When one of the HAD groups functions a fire warning is sounded and the appropriate afff sprinkler group is operated. If two HAD groups function a smoke ventilation system is brought into operation. These events are displayed in a control centre which also indicates electrical malfunctions and pneumatic leaks.

Six afff emergency hydrants are located in the hangar. Additional fire-fighting equipment is provided by six-wheeled dry powder units, each filled with 80 kg of Monnex powder for class B, C and E fires.

Status

Starckjohann fixed fire-protection systems are in use throughout the world. Fire-protection systems for aircraft hangars have been supplied to Finnair, Malev and Saudia airlines.

Manufacturer

Starckjohann and Co Ab, Helsinki, Finland

FRANCE

Aerazur Aircraft Arresting System

The Aerazur aircraft-arresting gear is intended to stop an extensive overrun of an aircraft at the end of the runway: military and civil versions vary according to the type of aircraft. The components of the arresting gear are a net constructed of Polyamid 66, a raising system for the net, brakes, a remote-control and monitoring device and a compressed-air generator.
The net consists of a number of separate nets all assembled and connected to the brakes. Each of these nets restrain the aircraft individually so that the strain can be distributed over the front of the fuselage and the breadth of the whole wing.
The net is raised by a system of masts and pneumatic cylinders located on either side of the runway and a number of inflated stanchions in the centre of the barrier.
Three types of remote controls and monitors for the raising system are available; electrical, a manually-operated system controlled from the air traffic control tower, automatic and electronic. The automatic system is based on determining the velocity of an aircraft which, as it is rolling, passes between light beams and two photo-electric cells placed at 180 and 160 metres in front of the barrier installation. If comparison between the true speed of the aircraft and the runway maximum given speed (usually 40 knots) shows the true speed to be high an automatic command raises the nets. Three photo-electric beams are used in the electronic system. Data on the velocity and acceleration of the aircraft is processed by a computer and if both these variables exceed a given level the net is raised automatically. Both automatic systems can be controlled manually.
The time taken to raise the barrier varies between 0·8 and 1·7 seconds according to the type of barrier and weather conditions. The system can be floodlit, the lighting being switched on automatically when the net is raised.

Characteristics

	MILITARY		CIVIL
MULTIPLE NET	Type 30	Type 40	
NO OF INDIVIDUAL NETS	30	40	60
NO OF VERTICAL STRAPS	390	530	
HEIGHT OF NET	3·45 m	3·7 m	8 m (10 m side masts)
WIDTH OF NET	58 m	58 m	160 m (max)
MAX IMPACT SPEED	190 knots	190 knots	
STOPPING DISTANCE	270 m	270 m	300 m
FORCE OF DECELERATION ON AIRCRAFT	2 *g* (max)	2 *g* (max)	1 *g* (150 knots/ 150 ton overall)
AIRCRAFT TYPES	Up to 20 tons	Up to 34 tons	
TIME TO RAISE NET			1·7 s
dry weather	type 4: 0·8 s	type: S 1 s	
wet weather	1 s	1·5 s	
TIME TO LOWER NET	5 minutes	2 s	6 s

Status

Over 200 barriers are installed throughout the world. Military versions are in service with the French and other air forces. The civil version entered service in 1968 when it was installed at Toulouse-Blagnac for Concorde operations.

Manufacturer

Aerazur Constructions Aeronautiques, Issy-les-Moulineaux, France

Biocarbo 5500 Airport Crash Tender

The Biocarbo 5500 heavy-duty airport crash tender is produced in two versions, dry chemical and foam. The vehicles were developed in conjunction with the French Air Force following a series of tests from 1970–73 instituted by them in which it was found that a combination of powder and foam agents was the most effective extinguishing method of dealing with aircraft crashes. Both versions are mounted on a MAN chassis. The tender incorporates the company's B2 system which allows powder or foam to be employed within the same spraying system. To do this a compressed gas system using carbon-dioxide as the propellant has been adopted instead of a mechanical pump system.
To produce powder the Biocarbo equipment comprises a Granito power reservoir and three carbon-dioxide cylinders. In the foam version the equipment consists of a water reservoir, a foam reservoir and two carbon-dioxide cylinders.
Two monitor nozzles, mounted on top of the vehicle to the rear of the driver's cab, are operated manually from behind the monitors. Two flexible hoses are stowed in lockers, each comprising 20 metres of hose with hand nozzles attached.
Protection for the vehicle is provided by four sprayers mounted on the front.

Biocarbo 5500 foam crash tender

Characteristics

CHASSIS
MAN 26 320 DFA with MAN V 10 cylinder engine producing 235 kW; 6 × 6 all-wheel drive, single wheels; 9-speed gearbox
ACCELERATION: 0–100 km/h in 60 s
MAX SPEED: 100 km/h
DRIVER'S CAB: up to 5 people

	POWDER VERSION	FOAM VERSION
TANK CAPACITY:		
water		5500 litres
foam		400 litres
powder	5500 kg	
MONITOR	2 × LC 100G	2 × LC 2000M
unit output	2000 kg/minute	2000 litres/minute
range	30–60 m	30–60 m
rotation/elevation	360°/–15 to +45°	360°/–15 to +45°
AUXILIARY EQUIPMENT	3 × CO_2 cylinders	2 × CO_2 cylinders
HAND NOZZLES	2 × GM 200	2 × GBV 45
unit output	300 kg/minute	300 litres/minute
range	25 m	25 m
SELF-PROTECTION DEVICE	4 × 150 kg/minute	4 × 150 litres/minute

Status

Biocarbo 5500 entered service with the French Air Force in 1976 and 115 vehicles have now been supplied. The vehicle is also in service with other countries.

Manufacturer

Biro SA, Paris, France

Biocarbo 5500 dry chemical crash tender

Biro Fire-fighting Vehicles

Apart from the Biocarbo 5500 which is described separately Biro produce four other vehicles for airport use.

VIC 1: a rapid-intervention vehicle capable of projecting a total 800 kg/mm of foam or dry chemical extinguishing agent through a mixed monitor and two hand-operated nozzles, one for use with foam, the other for dry chemicals.

VLIS 40: an airport crash tender equipped with a single stage pump powered by a Citroen GS 12 flat four engine. Two monitors are installed on the vehicle: one for dry chemical use which is mounted on a pneumatic telescopic jack which extends vertically to a height of 30 metres; the other is a mixed monitor capable of projecting either a premixed solution or a dry chemical agent. A self-protection device is fitted on the front of the vehicle.

MAB 39: an appliance built to the specification of the French Ministry of Defence for use in test-flight centres. A Mk 15 water pump with an output of 2000 litres per minute at 10 bars and a mixed monitor provides a foam/water extinguishing capability.

MAB 67: an airport crash tender mounted on a Mercedes Benz chassis for fast acceleration over rugged terrain. The vehicle is equipped with foam and water.

Characteristics

	VIC 1	VLIS 40	MAB 39	MAB 67
CAPACITIES				
premix	400 litres			
foam		50 litres	400 litres	720 litres
water		650 litres	3500 litres	6000 litres
dry chemical	400 kg	500 kg		
MIXED MONITOR OUTPUT	600 kg/ minute	500 litres/ minute	1200 litres/ minute	3000 litres/ minute
DRY CHEMICAL MONITOR OUTPUT		500 kg/ minute		
HAND NOZZLES/OUTPUT				
dry chemical	1 × 200 kg/minute	1 × 250 kg/minute		
foam	1 × 200 litres/minute		1 × 105 litres/minute	
VEHICLE PROTECTION DEVICE				
nozzles		2	2	5
output		50 litres/ minute	110 litres/ minute	

Status

The vehicles have been supplied to military forces in France, Morocco and Iran and to civil airports in France, Yugoslavia and Zaïre.

Manufacturer

Biro SA, Paris, France

Biro Fire-protection Systems

Biro produce a range of equipment for fixed installations with different extinguishing agents according to the nature of the hazard. Four types of extinguishing agents are available: Granito dry chemical, foam, carbon-dioxide and water.

Granito dry chemical is suitable for fires of class B (petrol, oil and paint) and for class C (butane, propane and hydrogen) while foam is used, generally, for the protection of hangars, hydrocarbon stocks and workshops.

Carbon-dioxide installations are designed for use against fires in electrical equipment and for special fire hazards in the hangars. Sprayed water installations are appropriate for protection against class A fires arising from stocked material in warehouses and other premises not permanently occupied.

The installations incorporate fire-detection devices for automatic or manual operation of the fire-protection systems.

Status

In service with the French Air Force.

Manufacturer

Biro SA, Paris, France

Biro Portable and Wheeled Extinguishers

Biro produce a range of portable and wheeled extinguishers for airport use. Spray water extinguishers are suitable for class A fires (wood, tissue, paper, etc).

Dry chemical extinguishers containing Granito powder can be used on both class B fires (petrol, oil, paint, etc) and class C fires (gas fires such as butane, propane, hydrogen), while dry chemical extinguishers with MG 15 powder are ideal for fires involving light metals such as magnesium, aluminium, plutonium or zirconium.

Carbon-dioxide extinguishers should be used for electrical fires, and can be used for fires in class B where the smothering action provided by carbon-dioxide is required.

Status

In service at civil and military airports in France and other countries.

Characteristics

	CLASS OF FIRE	CAPACITY
GRANITO		
dry chemical	B, C, or ABC	1–200 kg
MG 15		
dry chemical	Metal	9, 25, 50 kg
sprayed water	A	1–200 kg
carbon-dioxide	B/Electrical	2–60 kg

Manufacturer

Biro SA, Paris, France

Sides Light Intervention Vehicles

Sides manufacture three vehicles which are used as first-aid appliances for fighting aircraft fires on the ground or to stand-by during fuelling operations. One vehicle, fitted with the appropriate equipment, has an additional role as a rescue vehicle: the light intervention vehicle is equipped with an additional monitor on a telescopic mounting which extends vertically to 10 metres for use in jet engine fires in DC10- and Tristar-type aircraft.

Status

These vehicles are in service at airports throughout France and in Berlin, Geneva, Moscow and countries of western Africa.

Manufacturer

Sides, Saint Nazaire, France

Characteristics

	LIGHT INTERVENTION VEHICLE		LIGHT FIRE AND RESCUE VEHICLE	LIGHT DUTY INTERVENTION VEHICLE
				Optional: dry powder or mechanical foam
PUMP OUTPUT	800 litres/minute at 10 bar			2000 litres/minute at 11 bar
AUTOMATIC PREMIX SYSTEM	0–8%			6%
TANK CAPACITY				
foam	45 litres			300 litres
water	600 litres			2500 litres
powder	500 kg		500 kg	150 kg
MONITOR	combined:		powder	foam
	premix	powder		
output	500 litres/ minute	450 kg/ minute	250 kg/ minute	1800 litres/ minute
range	30 m	25 m	18 m	50 m
BRANCH-PIPES	portable: foam	powder	powder	foam
output	250 litres/ minute	300 kg/ minute	150 kg/ minute	450 litres/ minute
range	20 m	12 m	10 m	25 m
AUXILIARY EQUIPMENT	powder monitor on telescopic mount		120 kg CO_2 (4 cylinders) Spray gun output: 30 kg/minute CO_2 Rescue kit	dry powder version: 2 × 20 kg CO_2 Vehicle protector device: 2 × 130 litres/ minute nozzles

VMA 54, 67, 70 and 80 Fire-fighting Vehicles

These medium-duty vehicles are designed for use as the prime equipment in airports class five, six and seven or to reinforce heavy-duty vehicles in class eight and nine airports. All are equipped with foam and water and the VMA 54 includes dry chemicals equipment. An automatic premix system controls the foam percentage rate which can be varied between 0 and 10 per cent. A foam spreading ramp can be fitted to all vehicles.

Status

In service at airports in France, Libya, Algeria and Morocco.

Manufacturer

Sides, Saint Nazaire, France

Characteristics

	VMA 54	VMA 67	VMA 70	VMA 80
PUMP OUTPUT	4000 litres/minute at 12 bar	4000 litres/minute at 12 bar	4000 litres/minute at 12 bar	4000 litres/minute at 12 bar
TANK CAPACITY				
water	4800 litres	6000 litres	6250 litres	7000 litres
foam compound	600 litres	720 litres	750 litres	1000 litres
powder	200 kg			
MONITOR – FOAM				
MANUAL CONTROL:				
output	2000 litres/minute	2000 litres/minute	2000 litres/minute	2000 litres/minute
range	30 m	30 m	30 m	30 m
ELECTROHYDRAULIC CONTROL:				
output	3000 litres/minute	3000 litres/minute	3000 litres/minute	3000 litres/minute
range	55 m	55 m	55 m	55 m
PORTABLE FOAM BRANCH-PIPES				
output	460 litres/minute	460 litres/minute	460 litres/minute	460 litres/minute
range	30 m	30 m	30 m	30 m

S 2000 Heavy-duty Foam Crash Tender

The S 2000 is a development based on a Sides' cab and chassis with all-wheel drive and an Allison semi-automatic transmission which has six-forward speeds. A Sides patented hydrostatic transmission provides slow speeds (up to 6 km per hour) forward or reverse during pumping operations. The vehicle is powered by an SSCM Poyaud diesel turbocharged engine. Three versions are produced: VMA 102, 152 and 190. The differences occur in the chassis, engine size, and capacity and output of the monitor.

The driving and fire-fighting controls are located on a single panel in the driver's cabin. A monitor positioning indicator shows the operator the degree of rotation and elevation. The monitor can produce either a straight jet or spray. Hydraulically-assisted and self-balancing it can be controlled manually from a platform which is accessible from the cabin by a stairway. The foam mixing is automatic but it can be adjusted by remote control to two preset positions or by a manual adjustment from 0 to 8 per cent.

Two foam branch-pipes are fitted with revolving couplings and valves and these are in side lockers. A water-spray trigger nozzle attached to 25 metres of semi-rigid hose at the front of the vehicle. The tender is equipped with a self-protecting device comprising eight water/foam generators, four in front and four under the vehicle.

The cab is designed for three people with the driver in the centre seat.

Status

In service at airports in Belgium, France, Spain, Switzerland and the Canary Islands

Manufacturer

Sides, Saint Nazaire, France

Characteristics

	VMA 102	VMA 152	VMA 190
CHASSIS	4 × 4	6 × 6	8 × 8
engine		SSCM Poyaud V12 Diesel	
power at 2400 rpm	500 hp	700 hp	900 hp
transmission		Allison semi-automatic	
ACCELERATION (0–80 km/h)		35 s	
max speed		over 105 km/h	
PUMP		Centrifugal – single stage pto	
output		400 m^3/h at 12 bars	
TANK CAPACITY			
water	9100 litres	14 000 litres	17 300 litres
foam compound	1100 litres	1200 litres	1700 litres
MONITOR			
premix output	5000 litres/minute	adjustable 3000 or 6000 litres/minute	
range	70 m	85–90 m	85–90 m
horizontal	260°	260°	260°
elevation	−12 to +37°	−12 to +37°	−12 to +37°
AUXILIARY EQUIPMENT			
BRANCH-PIPES			
foam: output	2 × 500 litres/minute, range 25 m		
water: output	1 × 25 m 125 litres/minute, range 5–15 m		
SELF-PROTECTING DEVICE	water/foam generators, output: 8 × 130 litres/minute		
DIMENSIONS	10·97 × 3·05 × 12·5 m (empty)		
OPERATIONAL WEIGHT	38 000 kg		

Interior of SIDES S 2000 showing driving and fire-fighting controls

SIDES S 2000 heavy-duty crash tender

TCD 1491/TCD 1360 Medium Fire Crash Tenders

These two airport crash tenders TCD 1491 and TCD 1360 are very similar in chassis type and equipment fit, but the TCD 1491 is a four-wheeled vehicle whereas the TCD 1360 is a six-wheel vehicle. Both have all-wheel drive.

The tenders are equipped with petrol-driven pump units and a monitor is mounted above and behind the driver's cab with internal and external controls. Two hand foam nozzles are provided. An auxiliary hose reel for water contains 40 metres of hose to which a high pressure pistol nozzle is attached.

The TCD 1491 can be fitted with a foam spreading device mounted on the rear of the vehicle.

Manufacturer
Camiva, Saint-Alban-Leysse, France

Characteristics

	TCD 1491	TCD 1360
CHASSIS	Berliet GR 280 4 × 4	Berliet GBH 6 × 6
ENGINE OUTPUT	260 hp SAE at 1800 rpm	250 hp SAE at 2200 rpm
TRANSMISSION	6 forward, 1 reverse	6 forward, 1 reverse
MAX SPEED	87 km/h	80 km/h
PUMP OUTPUT	3000 litres/minute at 11·5 bars	3000 litres/minute at 11·5 bars
TANK CAPACITY		
water	6000 litres	6000 litres
foam compound	900 litres	900 litres
MONITOR FOAM/WATER		
output	2500 litres/minute	2000 litres/minute
range	42 m	40 m
HAND FOAM NOZZLES	2	2
output	450 litres/minute	450 litres/minute
range	20 m	20 m
AUXILIARY HOSE	40 × 22 m	40 × 22 m
VEHICLE PROTECTION	nozzles in front and under vehicle	
DIMENSIONS	8·06 × 2·5 × 3·6 m	8·77 × 2·5 × 3·6 m
OPERATIONAL WEIGHT	19 000 kg	24 000 kg

TCD 1552A Light Crash Tender

The TCD 1552A is a four-wheeled light crash tender mounted on a Saviem chassis with a petrol-driven engine producing 130 hp. The tender is equipped with foam, water and dry chemical extinguishing agents.

A separate petrol engine drives the pump unit which has an output of 1000 litres per minute at 10 bars.

Manufacturer
Camiva, Saint-Alban-Leysse, France

Characteristics

CHASSIS	Saviem SP5 4 × 4
ENGINE	petrol: 130 hp SAE at 4750 rpm
TRANSMISSION	4 forward, 1 reverse
MAX SPEED	80 km/h
PUMP	petrol, output: 1000 litres/minute at 10 bars
TANK CAPACITY	1200 litres water; 150 litres foam; 150 kg dry chemical
FOAM MONITOR	
output	800 litres/minute at 12 bars, range 30 m
HAND NOZZLES	
foam	2 × 450 litres/minute at 12 bars
dry chemical	1 × 150 kg/minute with 20 m hose
range	20 m
AUXILIARY EQUIPMENT	
water hose	1 × 20 m; 1 CO_2 bottle × 6 kg

TCD 1583 Medium Fire Crash Tenders

The TCD 1583 is mounted on a Saviem chassis and powered by a diesel engine developing an output of 265 hp. A four-wheel drive vehicle, with twin-tyres at the rear, the tender has a transmission with six forward gears, giving it a maximum speed of 95 km per hour.

A petrol-driven pump unit supplies a foam/water monitor mounted on top of the vehicle which can be operated hydraulically, from inside or outside the cab. The standard monitor has an output of 2650 litres per minute but an alternative version can be provided to increase the output to 3000 litres per minute. Foam nozzles are fitted to give protection to the front and under the vehicle.

Characteristics

CHASSIS	Saviem MAN 19240 HA 4 × 4
ENGINE	diesel 265 hp SAE at 2300 rpm
TRANSMISSION	6 forward, 1 reverse gear
MAX SPEED	95 km/h
PUMP	petrol, output: 4500 litres/minute at 14 bars
TANK CAPACITY	720 litres foam; 6000 litres water
FOAM/WATER MONITOR	
output	2650/3000 litres/minute
range	50/52 m
MANUAL NOZZLES	2 × 450 litres/minute
AUXILIARY WATER HOSE	40 m × 22 mm
DIMENSIONS	7·64 × 2·5 × 3·22 m
OPERATIONAL WEIGHT	10 260 kg

Manufacturer
Camiva, Saint-Alban-Leysse, France

FEDERAL REPUBLIC OF GERMANY

Binz 2000 G/GK Ambulance

The 2000 G/GK is based on a Mercedes Benz chassis. The 2000 G version is fitted with two stretchers, one mounted on a sliding platform to aid loading and unloading and the other hinged to fold away when not in use. A collapsable seat for the attendant is provided by the partition wall between the cabin and the driver. Medical equipment includes an oxygen unit with flow-meter and moisturising glass, secretion-suction unit and automatic pressure or volume-controlled respiration appliances. The rear door is a single flap-type hinged at the top, using gas spring supports.

The GK version has an additional entrance door on the right-hand side for attendants. The rear doors are double wing-type opening to 270 degrees. Three stretchers are fitted.

Characteristics

CHASSIS:
Mercedes Benz: 240 GD, 300 GD, 230 G or 280 GE

Status

Production

Manufacturer

Binz GmbH, Lorch/Württ, Federal Republic of Germany

Binz RTW/Ambulance

The RTW ambulance produced by Binz is based on a Mercedes Benz chassis with an all-steel body and powered by a petrol or diesel engine. Two large rear doors which open to an angle of 270 degrees give access to the patients' compartment. An additional door for the attendant is fitted on the right-hand side of the vehicle. The driver's cab is separated from the interior of the ambulance by a partition with sliding windows. A hinged roof vent made of tinted insulated glass, one or two side windows and fluorescet lighting provide illumination which can be supplemented by a movable spot lamp with adjustable focus. Two flush-mounted suspensions for transfusion are installed in the roof.

The interior can be fitted as required by the operator. A parallel swivelling platform for Tredelenburg shock treatment is available. Alternatively a stretcher can be placed on the platform and locked in position. The platform can be pushed across from left to right by means of swivel brackets to make more room for the nursing staff to work in. The stretcher with patient can be raised hydraulically or manually to a height of approximately 700 mm. Cupboards for medical equipment, a wash basin with electric water pump and an attendant's seat are fitted at the front of the patients' compartment. Two additional stretchers can be installed on the right hand side of the vehicle which fold away when not in use.

Medical equipment is supplied as required by the operator.

Characteristics

	Mercedes Benz		
CHASSIS	L407D	L409	L508D
ENGINE	4-cylinder diesel	4-cylinder petrol	4-cylinder diesel
transmission	4 gears	4 gears	5 gears
max speed	100 km/h	120 km/h	105 km/h
DIMENSIONS	5·04 × 2·1 × 2·9 m		

Status

Production

Manufacturer

Binz GmbH, Lorch/Württ, Federal Republic of Germany

Binz Unimog Ambulance Type 404/416

The Type 404/416 is a heavy-duty ambulance mounted on a Mercedes Benz chassis with removable all-steel body and powered by a diesel or petrol engine. The patients' compartment, which is accessible through two wing doors in the rear, has capacity for up to four stretchers which can be folded when not in use and four seats. The seating can be varied up to a maximum of 12. A revolving door is fitted on the right-hand side of the vehicle. Medical equipment is supplied as required by the operator.

Status

Production

Manufacturer

Binz GmbH, Lorch/Württ, Federal Republic of Germany

IRELAND

Timoney FCT 2000 Airport Crash Tender

The Timoney FCT 2000 airport crash tender is based on a Timoney double box chassis and powered by a Detroit diesel engine with Allison transmission and all-wheel drive. An independent suspension system is fitted to all wheels, which is reported to give greater stability and more effective cross-country capability than the conventional beam-axle vehicles.

The Godiva pump is driven by a sandwich-type pto and a Feecon foam-making devise is fitted with an adjustable proportioning system. The monitor, branch-pipes and other fire-fighting and rescue equipment are supplied according to the requirements of customers.

Timoney FCT 2000 airport crash tender

Characteristics

MODEL: FCT 2000
CHASSIS: Timoney double box
ENGINE/OUTPUT: Detroit Diesel 8V 92T 492 bhp at 2300 rpm
DRIVE: 8 × 8
ACCELERATION: 0–80 km/h in 43 s
MAX SPEED: 109 km/h
PUMP/OUTPUT: Godiva Mk 20
TANK CAPACITY
foam: 909 litres
water: 9092 litres
MONITOR: to suit individual requirements
DIMENSIONS: 9·13 × 2·5 × 2·9 m
OPERATIONAL WEIGHT: 26 650 kg

Status

This tender is in service with the Irish Airports Authority (Aer Rianta) at Shannon airport. A 4 × 4 rapid intervention vehicle using many of the driveline and suspension components of the FCT is in an advanced state of development.

Manufacturer

Adtec Teoranta, Navan, County Meath, Ireland

ITALY

ABP 400 Rapid Intervention Vehicle

The ABP 400 was developed for airport use and the vehicle complies with the ICAO standard for category one airports. It is mounted on a Land Rover chassis with four-wheel drive and powered by a four-cylinder 2800 cc petrol engine. The water pump is rear-mounted and driven by the engine. It is connected to a 60 metre hose to which a water gun with an adjustable jet is attached.

Characteristics

CHASSIS: Land Rover 109; 4 × 4; 4-cylinder petrol engine 2300 cc developing 70·5 hp DIN at 4000 rpm
DRIVER'S CAB: up to 3 people
PUMP: Chinetti 2T 2220. Output: 2000 litres/minute at 12 bar, 200 litres/minute at 20 bar
TANK: Galvanised steel (stainless steel as option) 400 litres of water
HOSE REEL: 60 m
DIMENSIONS: 4·44 × 1·68 × 2·2 m
OPERATIONAL WEIGHT: 2700 kg

Status

In service at civil and military airports in Italy

Manufacturer

Chinetti SA, Varese, Italy

Chinetti ABP 400 rapid intervention vehicle

AIS 1000/2000 Airport Crash Tender

The AIS 1000/2000 is designed for airport use, to meet ICAO category eight standards. Mounted on a Perlini chassis with four-wheel drive and single tyres, it is powered by two Detroit GM V6 cylinder diesel engines operating in parallel to produce a total 530 hp. The engines are each capable of driving the wheels or the fire-fighting equipment at the operator's choice. In the case of a breakdown of one engine the tender is still operative with the other engine. The foam monitor which is mounted on top of the driver's cab has hydraulic controls in the cab which can be worked either by the operator or the driver. The monitor can be controlled manually. Foam is mixed automatically and the percentage rate of mixing can be adjusted from 0 to 8 per cent. Vehicle protection is provided by four ground-sweep foam nozzles, mounted on the front axle and under the chassis, and ten water-spray nozzles to protect the glass and tyres. Two hose reels are fitted, one on each side, mounted on a hinged frame which pivots outward. Each unit comprises 40 metres of hose with a water/foam nozzle attached.
The cab is designed for a crew of four.

Status

In service at civil and military airports in Italy

Manufacturer

Chinetti SA, Varese, Italy

Characteristics

CHASSIS: Perlini 605 D; 4 × 4; 2 Detroit GM V6 cylinder diesel engines in parallel: total power 530 hp
ACCELERATION: 0–80 km/h in less than 45 s
MAX SPEED: 100 km/h
CAB: up to 4 people
PUMP: Chinetti 2T – 5012 centrifugal – output 6000 litres/minute
FOAM MONITOR: Chinetti CO 72; output 4800 litres/minute; foam range over 70 m; water range over 85 m
ROTATION: 360°
ELEVATION: –20 to +70°
TANK CAPACITY: 1800 litres foam; 10 000 litres water
HOSE: 2 foam hose reels; length 40 m; output 400 litres/minute; range 35 m
VEHICLE PROTECTION: foam: 4 fan nozzles; water: 10 spray nozzles
DIMENSIONS: 9·14 × 3·1 × 3·9 m
OPERATIONAL WEIGHT: 28 490 kg

Chinetti AIS 1000/2000 airport crash tender

Double Power 500 Airport Crash Tender

The Double Power 500 has been developed for rapid intervention and as such is capable of fast acceleration. This is achieved by coupling the pump-unit engine to the vehicle transmission to produce extra power. The tender, designed for a minimum of two crew, is mounted on a Chinetti-type chassis with four of the six wheels being driven. The engine develops a total 475 hp producing an acceleration of 0–80 km per hour in 28 seconds however with 10 000 litres of extinguishing liquids this time is increased to 37 seconds. The pump unit and monitor controls are in the driver's cabin and are activated through pneumatic and oleo-dynamic circuits. A standby system, also oleo-dynamically operated, is similarly located. The monitor, which is situated above the driver's cab, can produce a straight stream or a dispersed jet for short-distance intervention with water or foam. Water can be projected from hand hoses as a straight stream or a dispersed jet.

Manufacturer

Chinetti SA, Varese, Italy

Characteristics

CHASSIS: Chinetti OM 260; 6 × 4; automatic transmission with clutch box for coupling pump engine; engine output 474 hp
ACCELERATION: 0–80 km/h in 28 s
MAX SPEED: 110 km/h
MONITOR
foam: output 32·5 m^3/minute, range 60 m
water: output 2500 litres/minute, range 65 m
ROTATION: +150 to –150°
ELEVATION: +45 to –10°
TANK CAPACITY: 4000 litres water; 600 litres foam, this may be increased to a max of 10 000 litres

NETHERLANDS

Ajax Heavy-duty Airport Crash Tenders

Ajax produce two heavy-duty vehicles for cross-country work, the FLF 48/100-10 and the TRO FLF 28/35-4. Both are mounted on a Faun chassis. The FLF 48 is driven by two diesel engines and the TRO FLF 28 has a single eight-cylinder rear-mounted diesel engine. Both tenders are fitted with automatic transmission and six-wheel drive.

The fire pump is centrally mounted and controllable during drive. Both versions incorporate a central operating station in the driver's cab. In the TRO FLF the monitor, pump, foam and powder units can be controlled from this station. The percentage rate of foam mix can be adjusted from three to ten per cent. At the front a fixed foam fan is fitted for laying a foam carpet. The FLF 48 has an electronically controlled rapid intervention unit.

Status

In service at civil and military airports in the Netherlands

Manufacturer

Ajax Fire Protection, Amsterdam, Netherlands

Characteristics

	FLF 48/100–10	TRO FLF 28/35–4
CHASSIS	Faun LF 910/42 6 × 6	Faun LF 2230 45 6 × 6
ENGINE(S)	2 × 248 kW	8-cylinder diesel 235 kW
ACCELERATION		
0–80 km/h		45 s
MAX SPEED		97 km/h
CREW	4	6
PUMP	FP 48/8–2	FP 28/8
output	6000 litres/minute at 14 bars	2800 litres/minute at 8 bars
TANK CAPACITY		
foam		400 litres
water		3500 litres
powder		750 kg
MONITOR OUTPUT		
water	4000 litres/minute	1000 litres/minute
foam	32 m³/minute	1000 litres/minute
powder		15–25 kg
MONITOR RANGE		
water	77 m	
foam	80 m	40 m
powder		30–45 m
ROTATION		−135 to + 135°
ELEVATION		−15 to + 45°
AUXILIARY HOSES		2 × 30 m hose 2 swing-out powder hose reels fixed foam fan 4000 litres/minute at 10 bars; range 5 m; width 7 m
DIMENSIONS		9·35 × 2·75 × 3·3 m
MAX WEIGHT	31 800 kg	22 000 kg

Ajax Medium-duty Fire-fighting Vehicles

Ajax manufacture give vehicles which fit into this category. Four are mounted on Mercedes Benz chassis and the fifth has a MAN chassis.

Status

In service at civil and military airports in the Netherlands

Manufacturer

Ajax Fire Protection, Amsterdam, Netherlands

Characteristics

	FLF 28/60–6	TRO FLF 28/35–4	SLF 48/00–90	TRO-TLF 28/45–5	TRO LF 4000
CHASSIS	MAN 26 320 DHA 3·85 + 1·35	Mercedes Benz 2626K/3 3·85 + 1·35	Mercedes Benz 2632 K42/135	Mercedes Benz LA 1513/42	Mercedes Benz 1719 AK/38
ENGINE	235 kW	188 kW	235 kW	124 kW	141 kW
CREW	6	3	3	3	
PUMP/OUTPUT	FP 28/8 4800 litres/minute at 10 bars	FP 28/8 rear mounted 4800 litres/minute at 10 bars	FP 28/8 rear mounted 4800 litres/minute at 10 bars	FP 28/8 midship 4800 litres/minute at 10 bars	3000 litres/minute at 8 bars
TANK CAPACITY					
foam	600 litres	1500 litres	1000 litres	450 litres	
water	6000 litres	3500 litres	9000 litres	4500 litres	
powder		2000 kg		750 kg	4000 kg
MONITOR	foam/water	1 × foam/water 1 × powder	foam/water	1 × foam/water 1 × powder	powder
output	1000 litres/minute		2000 litres/minute		
RAPID INTERVENTION UNITS	2	2		2	2
MAX WEIGHT	22 000 kg	26 000 kg	22 000 kg	15 000 kg	16 000 kg

Ajax Portable and Wheeled Extinguishers

Ajax produce a range of portable and wheeled extinguishers for airport use.

Dry chemical extinguishers containing TOTALIT 2000 or SUPER can be used for class B fires (petrol, oil, paint, etc.) and for fires in live electrical installations where the voltage does not exceed 10 000 volts.

Dry chemical extinguishers containing TOTALIT M are designed for use against metal fires including sodium, lithium, calcium and barium. Dry chemical extinguishers containing TOTALIT G are applicable in the presence of smouldering substances and for fires in live electrical installations where the voltage does not exceed 1000 volts at a distance of 1 metre, and 10 000 volts at a distance of 3 metres. TOTALIT G can also be used to combat light alloy (magnesium) fires.

Carbon dioxide extinguishers are suitable for use against fires in live electrical equipment and for class B fires.

Characteristics

	TYPE	CAPACITY
CARBON-DIOXIDE	portable	2·3–9 kg
DRY CHEMICAL TOTALIT 3	M-50 wheeled	50 kg
DRY CHEMICAL TOTALIT G	G-50 wheeled	50 kg
DRY CHEMICAL TOTALIT S/TOTALIT 2000	P 250 Trailer	250 kg
DRY CHEMICAL TOTALIT G	G 250 Trailer	225 kg
DIMENSIONS OF TRAILERS	2·37 × 1·7 × 1·35 m	
OPERATIONAL WEIGHT	650 kg P 250 625 kg G 250	

Status

Civil and military airports in the Netherlands

Manufacturer

Saval-Kronenburg BV, Hedel, Netherlands

Ajax wheeled dry chemical extinguisher

MAC Series Airport Crash Tenders

The MAC series of airport crash tenders are based on three different types of chassis: MAC-06, on a four-wheel drive chassis, is particularly suitable for airports handling ICAO category five traffic; MAC-09, built on a six-wheel drive chassis is designed for category seven and eight airports, while the MAC-12, an eight-wheel drive tender, is capable of meeting the recommendations for category nine airports.
All vehicles are powered by Detroit diesel, rear-mounted engines with an Allison automatic transmission.
The fire-fighting control panel is in a separate section of the dashboard in the cabin: both the driving- and fire-fighting operations can be controlled by one person. The equipment is electro-pneumatically powered. Foam proportioning is automatic or manual and the rate can be varied from 0–10 per cent. A remote-controlled monitor is situated on top of the cabin roof which can produce a variable foam pattern from solid stream to a full-flat fan spray of either foam or water. The water pump is a two-stage high-pressure centrifugal pump driven from the main engine. Capacity is provided by a glass-reinforced-polyester tank which has storage for both water and foam compound. The pump, tank and the main part of the stainless-steel piping are installed on a sub-frame, three-point mounted on to the main chassis, to avoid chassis torsion being transmitted to the superstructure.
Two swing-out type hose reels are installed on either side of the vehicle. These branch-pipes are suitable for jet and spray foam and water fog operation. The vehicles are equipped with a self-protecting device comprising four ground nozzles, one by the front bumper and three under the vehicle.
The cabin has accommodation for a crew of three with an additional three seats available if required.

MAC-09 airport crash tender

Status
Production

Manufacturer
Saval-Kronenburg BV, Hedel, Netherlands

Characteristics

	MAC-06	MAC-09	MAC-12
CHASSIS	Detroit diesel 1 × 8V 92, 4 × 4	Detroit diesel 1 × 8V 92T, 6 × 6	Detroit diesel 2 × 8V 71, 8 × 8
power at 2300 rpm	276 kW	360 kW	2 × 261 kW
TRANSMISSION	Allison 5 speed	Allison 5 speed	Allison 5 speed
ACCELERATION 0–80 km/h	32 s	40 s	40 s
MAX SPEED	112 km/h	112 km/h	112 km/h
CREW	3–6	3–6	3–6
PUMP	2 stage	2 stage	2 stage
output	3000 dm^3 at 12 bar	4500 dm^3 at 15 bar	5400 dm^3 at 15 bar
TANK CAPACITY			
foam	610 dm^3	910 dm^3	1215 dm^3
water	6100 dm^3	9100 dm^3	1215 dm^3
MONITOR			
output	3000 dm^3/minute	4000 dm^3/minute	4000 dm^3/minute
range (foam/water)	40/55 m	55/60 m	55/60 m
AUXILIARY EQUIPMENT			

Handlines 2 × 30 m; output 225 dm^3/minute
Self protection device 1 bumper nozzle × 380 dm^3/minute; 3 under truck nozzles × 40 dm^3 (each)

Saval-Kronenburg Automatic Fire Extinguishing System

Saval-Kronenburg produce an automatic fire-extinguisher system which uses either carbon-dioxide or one of two hologenated hydrocarbons (halons) as the extinguishing agent. The halons are bcf halon 1211 and btm halon 1301.
The systems comprise gas-storage tank or cylinders, opening device for the gas, piping with open sprinklers, detectors and internal or external alarms. The principle of detection is based on one or more of the following phenomena: heat, smoke, combustion gases, infra-red and ultra-violet radiations. The detectors activate the system which starts the alarms and the extinguishing system, shuts down the ventilation and closes doors and/or other openings. Since carbon-dioxide has suffocating properties, sufficient delay after the alarm has been sounded has to be incorporated in any automatic total flooding system in confined spaces so the area can be evacuated. With halon systems, however, the danger is minimal and therefore there is no need for a delay so the extinguishing effect of halons in these situations tends to be greater than carbon-dioxide.
Manually-operated systems based on the same principles are available.

Manufacturer
Saval-Kronenburg BV, Breda, Netherlands

Saval-Kronenburg Rapid Intervention Vehicle

Based on the ICAO requirements, this vehicle is equipped with sufficient amounts of extinguishing agents to deal with a small or incipient fire in large aircraft, or alternatively, to act as the prime fire-fighting vehicle at airports handling general aviation aircraft.
The rapid intervention vehicle is available in three configurations, all based on a four-wheel drive Dodge chassis, with automatic transmission and power-assisted steering. The first variant is equipped with a combined afff/dry chemical installation, 400 cubic decimetre of afff and 200 kg of foam compatible dry powder. Both agents are discharged using nitrogen. Two-wheeled bcf halon units of 50 kg capacity are mounted at the rear of the vehicle. A branch-pipe is fitted with 30 metres of rubber hose and a hand-controlled twin nozzle.
The second version has a foam and dry chemical installation. The foam is discharged by a two-stage centrifugal pump through a foam monitor mounted on the cabin roof. The monitor is operated from within the cab by the driver or commander. Two hand-lines each with 25 metres of hose are supplied, one for foam, the other for dry powder.
The third variant of the rapid intervention vehicle has a foam system with pump and monitor similar to the equipment described above. Two-wheeled bcf halon units are mounted at the rear of the vehicle, each having a capacity of 50 kg. These

Saval Kronenburg rapid intervention vehicle in service at Schiphol airport

units are designed to be unloaded quickly and have rubber wheels of sufficient width and diameter to give good manoeuvrability over rugged terrain.

Status
Production

Manufacturer
Saval Kronenburg BV, Breda, Netherlands

Characteristics

	TYPE 1	TYPE 2	TYPE 3
CHASSIS	Dodge 360–1 V8 4 × 4	Dodge 360–1 V8 4 × 4	Dodge 360–1 V8 4 × 4
ACCELERATION 0–80 km/h	18 s	18 s	18 s
MAX SPEED	120 km/h	120 km/h	120 km/h
CREW	1–3	1–3	1–3
PUMP			
output		1600/10 dm^3/bar	1600/10 dm^3/bar
TANK CAPACITY			
foam	400 dm^3 (pressurised)	300 dm^3	900 dm^3
powder	200 kg	250 kg	
bcf halon	2 × 50 kg		2 × 50 kg
MONITOR			
output (water/foam)		1000 dm^3	1000 dm^3
range (m) foam	35	35	35
water	45	45	45
AUXILIARY EQUIPMENT			
hand-lines	1 × 30 m		
output foam	200 dm^3/minute		
dry powder	2·2 kg/s		
hose reels (m)		2 × 25	1 × 25
output: foam		1 × 200 dm^3/minute	1 × 200 dm^3/minute
dry powder		1 × 2·2 kg/s	
DIMENSIONS	5·55 × 2·1 × 2·3	5·55 × 2·1 × 2·4	5·55 × 2·1 × 2·4
OPERATIONAL WEIGHT	50·199 N	51·61 N	49·257 N

Saval Kronenburg Wheeled Fire-extinguishers

Saval Kronenburg produce a range of wheeled dry powder and halon extinguishers. They can be divided into three categories according to the type of trolley or trailer on which the extinguishing agent is mounted: the M 50, halon 50 and P 250. The M 50 extinguisher consists of a welded-steel tank. The operating pressure is provided by a small carbon-dioxide cylinder. The extinguisher is supplied with ten metres of rubber hose and a heavy-duty brass pistol.
The M 50 can be supplied in three types: M 50-P charged with sodium-bicarbonate powder for fire classes B (petrol, oil, paint, etc.) and C (gas fires as butane, propane) and electrical fires; M 50-G charged with pg powder consisting of ammonium sulphate and ammonium phosphate for fire classes A (solid combustibles such as wood, tissue, paper), B and C; M 50-X charged with Monnex powder for fire classes B and C.

The halon 50 extinguisher is similar in design with a welded-steel tank, ten metres of hose and a nozzle. The latter can be extended by a lance. The extinguishant halon 1211 is pressurised by nitrogen which can be checked periodically on a pressure gauge. Halon 1211 is suitable for fire classes B and C and for the protection of live-electrical equipment. The P 250 consists of a steel tank mounted on a rubber suspension axle of a trailer. The operating pressure is supplied by a nitrogen cylinder connected to the tank by a pressure line. Two 30-metre hoses with pistols attached are stowed in containers on either side of the trailer.

The P 250 can be supplied in three versions: P 250 charged with sodium-bicarbonate powder for fire classes B and C; PG 250 charged with pg powder for fire classes A, B and C; PX 250 charged with Monnex powder for fire classes B and C.

Characteristics

	M 50 P/G	M 50X	HALON 50	P 250/PG 250	PX 250
TANK CAPACITY	50 kg	35 kg	50 kg	250 kg	175 kg
output	1·8 kg/s	1·4 kg/s	0·5*/0·8 kg/s	2 × 4 kg/s	
pressure	18 bar	15 bar	–	14 bar	
range	8 m	8 m	0·5 m*/7 m		
DISCHARGE TIME	30 s	30 s	variable		
HOSE					
length	10 m	10 m	95 m	2 × 30 m	
diameter	19 mm	19 mm	10 mm	25·4 mm	
NITROGEN CAPACITY					
PRESSURE				201 at 150 bar	
contents				3000 litres	
DIMENSIONS					
length (depth)	(0·73 m)	(0·73 m)		2·7 m	2·7 m
width	0·58 m	0·58 m	0·48 m	1·5 m	1·5 m
height	1·07 m	1·07 m	1·21 m	1·37 m	1·37 m
OPERATIONAL WEIGHT	106 kg	91 kg	115 kg	640 kg	

*with lance

Status
Production

Manufacturer
Saval-Kronenburg BV, Hedel, Netherlands

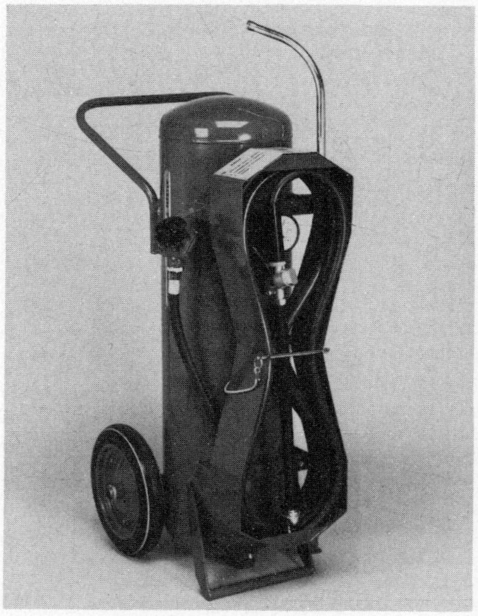

Halon 50 wheeled fire-extinguisher

SWEDEN

Svenska Skumsläcknings Automatic Fire-protection System

Svenska Skumsläcknings design and manufacture foam fire-protection systems for use in aircraft hangars. A recent development is a foam monitor system featuring automatic oscillation controlled by a micro-processor. This system enables the automatic operation of the system to be programmed according to the number, size and type of aircraft housed in the hangar. Trans-Australia Airlines have selected this system for use in a hangar designed for the A300 aircraft introduced into service in Australia in June 1981.
The SAS Hangar II at Stockholm-Arlanda airport is one example where a high-expansion foam system was chosen to protect a volume of 250 000 cubic metres.

Manufacturer
Svenska Skumsläcknings AB, Kungälv, Sweden

Svenska Skumslacknings automatic fire-protection system

UNITED KINGDOM

Acklift 23500 Fire-fighting and Rescue Platform

Acklift 23500 hydraulic lifting platform can be equipped specifically for fire-fighting and rescue duties. It is installed on a chassis with a gross vehicle weight of 16 000 kg.

A water monitor and an electro-mechanical platform control system is mounted on a fabricated tubular steel personnel basket which is fitted with a king post below floor level providing ±45 degrees of vertical slewing. It can be raised to a maximum height of 23·5 metres and has space for two firemen, one at the controls of the platform, the other directing the monitor.

Rescue ladders can be fitted from the personnel basket to the base of the chassis. This allows passengers caught in an aircraft fire on the ground to walk from the aircraft to the personnel basket which is equipped with a special basket extension and then proceed down the ladders to safety.

Status

Manufacture of this system started in 1981

Manufacturer

ACS Engineering Ltd, Southwell, Nottinghamshire, England

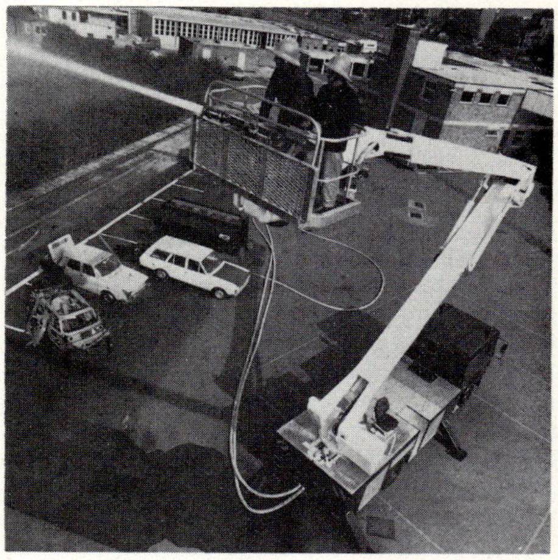

Acklift 23500 fire-fighting hydraulic platform

AFA Fixed Fire Extinguishing Systems

AFA-Minerva produces Amalon fire-fighting systems with conventional methods of detection and modular control equipment.

The System 700 control equipment is designed to monitor fire and/or combustible-gas detection circuits and subsequently to operate associated extinguishing systems. It can be used for the protection of category one hazardous areas provided that the rack-mounted control equipment is situated in a safe area.

The control equipment consists of various types of electronic control modules to meet the requirements of the detection and extinguishing system. The modular design enables a choice of configurations and types of detection and extinguishing equipment for the hazardous areas, including provision for expansion of the overall detection and extinguishing system.

The choice of detection can be made from any combination of combustible gas detectors or smoke, flame and heat detectors for fires.

The control equipment can operate extinguishing systems manually or automatically. It is powered from an external 24 volts dc supply. System 700 control equipment provides the logic functions, user controls, audible and visual alarm and fault indication for a number of detection circuits, the maximum number being governed by the size of the area being protected.

For an alarm or fault condition to be located as well as detected, each detection circuit is linked independently to a detection module. Each detection module has its own indicators and user controls. The control equipment provides for automatic shut-down of equipment on alarm and remote signalling of fire and fault conditions.

A programming facility is also provided which operates selected detection circuits in a predetermined sequence to automatically release the extinguishant with visual and audible indication at the control equipment and at the protected area (or areas) affected.

The control equipment is built up from a selection of standard plug-in modules which are contained in a sub-rack assembly. The sub-racks can be mounted, one above the other, in a standard 19-inch cabinet.

The extinguishant used depends on the equipment to be protected and may be carbon dioxide bcf or halon 1211 or btm or halon 1301.

The fire extinguishing system for each protected area comprises frame-mounted storage cylinders containing gas in liquefied form. Simultaneous extinguishant release from the cylinders is achieved by activating a drop-weight assembly by means of a gas-operated servo-motor. A solenoid valve is activated causing gas to be released from the servo-motor pilot cylinder.

The halon 1211 and 1301 inert gas fire-fighting system is of modular construction.

The Amalon halon container is connected to a predetermined length of pipe having one or two discharge nozzles. The containers forming a system are placed at convenient strategic positions within or adjacent to the compartment being protected requiring only the minimum of piping. All containers are activated simultaneously by means of pressure-operated actuators from the master container which is operated by hand, electrically or pneumatically as required.

Amalon halon containers are available in six different sizes. The size and charge of a given application is selected to suit the type and size of hazard involved. Large hazards are protected with multiple inter-connected modular systems. Container-mounting brackets (vertical, angled or horizontal) are available. All containers are manufactured, tested and marketed in accordance with the appropriate specification. They are fitted with valves which have a high flow rate so systems can discharge in ten seconds or less. Valves may be actuated electrically, pneumatically or manually from a remote or local position. Controls are in unit form, modular- or rack-mounted with outputs for agent release, fan and door control and other functions.

AFA say that rapid and efficient mixing of halon with the atmosphere in the protected space is accomplished by a special nozzle design. Nozzles with 360 or 180 degree discharge patterns are available to suit various hazard configurations. Standard nozzles are made of aluminium. A full line of accessories is available, including remote manual controls (both mechanical and electrical), pressure switches for electrical interlock functions, pressure trips for release of door- or damper-closing hardware, shuttle check valves for interconnection of main and reserve cylinders. Flexible discharge bends to connect cylinder valves to discharge piping may be provided.

Amalon systems are suitable for the fire-protection of engine test areas, machinery spaces, fighting vehicle compartments, computer rooms, communications centres, document stores, cable vaults, transformer rooms, training simulators stand-by power supplies and similar sensitive areas.

Carbon-dioxide systems are available from AFA-Minerva. These are still preferred in many parts of the world because of the low cost of the gas and its wide availability.

Carbon-dioxide acts as an inert diluting agent to reduce the available oxygen concentration in the risk atmosphere below that necessary to maintain combustion. The halons, including 1301 and 1211, have no real diluting effect on the available oxygen concentration but owe their efficiency to a chemical reaction within the combustion process.

Manufacturer

AFA-Minerva (EMI) Ltd, Twickenham, Middlesex, England

Angloco 4300 Air Crash Tender

The 4300 air crash tender is a medium-sized unit designed to meet ICAO requirements for airports. Based on a Unipower chassis with Cummins VT 555 turbo-charged diesel engine and automatic transmission, the tender is capable of acceleration from rest to 80 kph in under 45 seconds. Tanks are provided to carry 4100 litres of water and 275 litres of foam concentrate.

The fire pump is a Godiva UFP Mk 14 single stage light alloy unit, midships-mounted and driven from the chassis power take-off. Foam delivery is via the roof-mounted manual monitor and two sidelines, each of which consists of rubber-lined hose 30 metres long with 50 mm bore terminating in a foam branch-pipe. The monitor solution throughout is up to 2680 litres/minutes at 12 bar, with a jet range of 55 metres.

The complementary agent consists of a 50 kg bcf unit complete with hose reel and 30 metres of hose terminating in an applicator.

Angloco 4300 air crash tender

Accommodation is provided for a crew of four including driver, access via four forward hung doors. A fifth door provides access to the monitor platform.

A total of six equipment stowage lockers are provided, three each side. Accessories include blue rotating beacons, two-tone horns and airfield obstruction lights.

Manufacturer

Angloco Ltd, Batley, West Yorkshire, England

Angloco 4000 Air Crash Tender

The 4000 air crash tender is a medium-duty unit, based on the Shelvoke and Drewry CY chassis, with rear-mounted Cummins VT 555 turbocharged diesel engine and Allison MT 653 automatic gearbox. Tanks are fitted to carry 3600 litres of water and 450 litres of foam compound.
The main pump is a Godiva UFP Mk 10 aluminium alloy unit, midships-mounted and driven from the chassis power take-off which may be engaged while the vehicle is moving. Pump delivery is via the manually operated monitor and two sidelines each of which consists of 30 metres by 45 mm of rubber-lined hose terminating in a foam branch-pipe. The tender is capable of discharging 31 800 litres of expanded foam in approximately one minute through the monitor and sidelines. The complementary agent is 100 kg of bcf.
A large amount of equipment may be carried in the six lockers, three each side, which are enclosed by aluminium alloy roller shutters.
Accessories include rotating beacons, two-tone horns, a searchlight with tripod and cable reel, and a Dale Stemlite extending floodlighting mast.
The 4000 air crash tender can reach 80 kph from rest in under 40 seconds and accelerate to 110 kph.

Angloco 4000 air crash tender

Manufacturer
Angloco Ltd, Batley, West Yorkshire, England

Angloco Jetstream 3500 Medium Foam Tender

Jetstream 3500 is suitable for use at small airfields as well as industrial and municipal sites where petroleum fires occur.

Jetstream 3500 medium foam tender

The appliance illustrated is based on the Bedford M four-wheel drive chassis/cab but the fire-fighting installation may also be fitted to other two-or four-wheel drive chassis of suitable wheelbase and gross vehicle weight.
Accommodation is provided for a crew of two in the vehicle cab with additional crew seating on the monitor platform. Six illuminated lockers, three each side, are provided and enclosed by aluminium alloy roller shutters.
The rear-mounted fire pump is a Godiva UFP Mk 6 unit complete with 100 mm suction inlet and two screw-down delivery valves. Water and foam are drawn from the in-built 3000 litre tanks.
A manually operated monitor is provided complete with blabbermouth to deliver foam solution at a rate of 2000 litres a minute at 10·5 bar, to a maximum jet range of 46 metres at 25 degree elevation.
Two foam sidelines are also included each consisting of 30 metres by 50 mm of rubber-lined hose terminating in a foam branch-pipe. A 50 kg dry powder unit is included as the standard complementary agent complete with 30 metres of hose and an applicator.
Standard accessories include rotating beacons, air horns, a searchlight with tripod and cable reel and nine metre aluminium extension ladder.

Manufacturer
Angloco Ltd, Batley, West Yorkshire, England

Angloco Quester 1000 Rapid Intervention Vehicle

Questor 1000 is designed for airports which need a vehicle to reach accident sites quickly, carry rescue personnel and provide an effective means of fire suppression pending the arrival of major units.
This vehicle is based upon the Reynolds Boughton RB44 four-wheel drive chassis/cab of 3·3 metre wheelbase and 5500 kg gross vehicle weight; accommodation is provided for a crew of three and five illuminated lockers are supplied, enclosed by aluminium alloy roller shutters.
The water pump is a Godiva UFP Mk 6 unit mounted at the rear and driven from the chassis full-torque power take-off. A 100 mm round-thread suction inlet is provided and delivery is through two rear-mounted screw-down valves. Water is supplied from a 1000 litre corrosion-resistant tank.
A manually operated foam monitor is fitted with an output performance of 1500 litres a minute at 7 bar. Foam may also be delivered via two sidelines, each of which consists of 30 metres by 50 mm of rubber-lined hose terminating in a foam branch-pipe with shut-off control.

Quester 1000 rapid intervention vehicle

The complementary agent is a 50 kg bcf unit complete with hose reel, 30 metres of hose and applicator.
Accessories include blue beacons, airfield obstruction light, two-tone horns, and a searchlight with cable reel and tripod.

Manufacturer
Angloco Ltd, Batley, West Yorkshire, England

Angus Fixed Fire-protection Systems

The fixed fire-protection systems produced by Angus use a variety of extinguishing agents according to the type of installation the system is designed to protect, including foam, water, gas and dry powder.
With foam systems low expansion foams are generally adopted for protecting liquid storage tanks such as FP 70, FP 570. Nicerol and Polydol foam compounds are available the last being suitable for water-miscible solvents. Fluoro-chemical foam (Tridol S) can be applied conventionally or as a solution from a water-spray system. The agent is suitable for applications where fast flame extinction is of greater importance than the stability of low expansion foam. High expansion foam, Expandol, is a versatile extinguishing agent for use in enclosed areas or in buildings where it is desirable to flood the fire totally.
To protect particular hazards it is often necessary to provide water simultaneously over a wide area through spray nozzles. These are operated either manually or automatically through a fire-detection system.
The main gases used as fire extinguishing agents are carbon-dioxide and halons. Of the latter, halon 1301 and 1211 are the most widely used. The application of gases varies according to the requirement, but carbon-dioxide is generally used to protect high-voltage switchgear, transformers and other electrical equipment. Halons have a relatively low toxicity and so are suitable for use in applications where personnel evacuation is not practicable.
Dry powder is a specialised extinguishing agent for hazardous materials where cooling of the risk area is undesirable or of secondary importance.
All systems can be operated manually or automatically and distribution varied according to the operational requirement.

Status
Angus fixed fire-protection systems are in use throughout the world

Manufacturer
Angus Fire Armour Ltd, Thame, Oxfordshire, England

Angus Wheeled and Portable Fire-extinguishers

Angus manufacture a wide range of wheeled and portable fire-extinguishers.
Wheeled foam equipment includes the Firefly trailer on which is mounted a foam/water monitor. Two versions of the trailers are available with foam concentrate capacities of 900 and 1800 litres respectively. The AF 20 mobile foam unit is capable of producing 38 000 litres of expanded foam from one charge and is designed to be refilled with foam concentrate while in operation. Wheeled dry chemical extinguishers include a 75 kg mobile dry powder unit.
Portable extinguishers vary in size and include water/carbon-dioxide, foam, dry powder, carbon-dioxide and bcf extinguishing agents.

Manufacturer
Angus Fire Armour Ltd, Thame, Oxfordshire, England

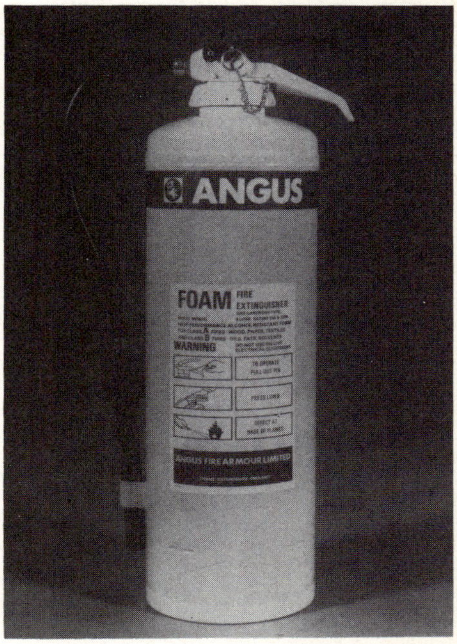

Angus MP 80/9L foam fire-extinguisher

Armfield Comet Hydraulic Fire-fighting Platform

The Armfield Comet hydraulic fire-fighting platform is mounted on a combined telescopic boom/jib system which enables the platform to be manoeuvred from ground level to the maximum height with a rotation of 360 degrees. Stabilisation is achieved by means of vertical props which can be extended horizontally from the vehicle on the working side or both sides as required. The telescopic boom/jib system allows the platform to straddle obstructions. With the telescopic arm lowered and the jib in the vertical position the Comet platform can operate down to three metres below ground level.
The platform, which can accommodate up to six people, is equipped with a monitor capable of discharging water at 2000 litres a minute, control panel, intercom system, spotlights and a cradle mounting for a stretcher.

Status
The Comet system was developed by FN in Belgium and is in service in Brussels, Paris, Vienna, Algiers, Cairo and Teheran. It is built in the United Kingdom by Armfield and marketed as the Armfield Comet

Characteristics

	18 T	25 T	30 T
MAX PLATFORM HEIGHT	18 m	25 m	30 m
SIDE OUTREACH	11·7 m	14·6 m	16·3 m
ROTATION	360°	360°	360°
OVERALL LENGTH	±7 m	±9·4 m	±11·3 m
NUMBER OF JACKS	4	4	4
JACK SPREAD	3·1 m	4 m	4·2 m
TOTAL WEIGHT PLATFORM AND SUB-FRAME AND WALKWAY	6300 kg	12 000 kg	13 300 kg
TOTAL WORKING LOAD	350 kg	350 kg	350 kg
CAGE DIMENSIONS	1·5 × 0·8 × 1·1 m	1·8 × 1·2 × 1·1 m	1·8 × 1·2 × 1·1 m
GROSS VEHICLE WEIGHT	16 000 kg	18 500 kg	23 000 kg

Manufacturer
Armfield Engineering Ltd, Ringwood, Hampshire, England

Cheshire Medium-Duty Air Crash Tender

This medium-duty crash tender is based on a Shelvoke and Drewry CX 4×4 chassis incorporating a Cummins turbo-charged diesel engine. The body is constructed of aluminium alloy with a cab providing seats for a driver and three crew.
The main pump is a Godwin single stage centrifugal fire pump supplying a roof-mounted monitor and hand-lines. Two hand-lines are provided terminating in a foam making branch complete with shut-off valve capable of producing approximately 4545 litres per minute of finished foam. They are operated by valves from the monitor control panel with duplicate controls in the hand-line lockers. A pump control panel is provided on the monitor platform.
Optional equipment includes water hose-reels and a ladder gantry.

Manufacturer
Cheshire Fire Engineering Ltd, Winsford, Cheshire, England

Characteristics
CHASSIS: Shelvoke and Drewry CX
ENGINE: Cummins turbo-charged diesel
DRIVE: 4 × 4
TANK CAPACITY: 3936 litres water or foam
PUMP: Godiva UFP Mk 10
output: 5000 litres/minute at 7 bar
monitor: 1
hand-lines: 2 × 30 m
DIMENSIONS: 7·66 × 2·44 × 3·19 m
OPERATIONAL WEIGHT: 7000 kg

Cheverton Airport Emergency Rescue/Fire-fighting Launch

Cheverton has designed a 17 metre twin-screw emergency launch fitted with rescue apparatus for 250 people and equipment for fire-fighting and oil pollution control.

Propulsion is provided by twin Stewart and Stevenson turbocharged diesel engines with Allison gear boxes. The hull which is divided into five watertight compartments is made of one-piece, heavy-duty grp with an aluminium alloy deck and superstructure.

Two monitors are provided; one on the wheelhouse roof remotely controlled from the wheel house, the other a manually controlled monitor is located on the after deck. A Godiva Mk 21 pump is driven by a separate Stewart and Stevenson diesel engine. Rescue equipment consists of ten 25-man life rafts, 3·15 metre inflatable boat with outboard engine, stern boarding platform, boarding ladder, scrambling net, two line throwers and two diving sets.

Oil pollution control is provided by a dispersant system comprising a pump and two spray booms rigged on each side.

Status

Two units have been supplied to Changi Airport, Singapore

Manufacturer

Cheverton Workboats, Cowes, Isle of Wight, England

Characteristics

ENGINE	2 Stewart and Stevenson GM 8V92 MTI diesel engines
UNITED OUTPUT	484·7 kW at 2300 rpm
STERN GEAR	2 Monel K500 57·15 mm diameter propeller shafts
MAX SPEED	25 knots
FIRE PUMP ASSEMBLY	Godiva Mk 21 pump, Stewart and Stevenson 6V9a diesel engine
output	9000 litres/minute at 10 bars
TANK CAPACITY	foam: 1360 litres
MONITOR(S)	2
DIMENSIONS	length 17 m, beam 4·47 m, draught 1·25 m (loaded)
DISPLACEMENT	26 tonnes

Cheverton rescue and fire-fighting launches

Chubb Medium Airport Foam Tender

The medium airport foam tender produced by Chubb is based on a normal commercial vehicle 4× 4 or 6 × 6 chassis such as Mercedes-Benz, Volvo or Bedford. With a total water tank capacity of 10 000 litres the tender is suitable for use at medium- and high-category airports.

A Chubb round-the-pump foam proportioning system is fitted using a Godiva water pump powered by a shaft drive pto. The roof-mounted monitor is controlled manually from a platform to the rear of the monitor in the case of the FB 70/10 or hydraulically from the cab or platform with the FB 100/10 monitor. Two foam hand-lines are supplied as standard equipment. Optional fittings include manual or power-operated water and/or foam hose-reels, a fixed bumper nozzle capable of producing 643 litres per minute at 14 bar, and five under-truck protection nozzles. Dry chemical units comprising 50 kg bcf units with hose-reel and applicator can also be installed.

Status

Introduced in 1980 six vehicles are in service in Greece and the Cameroons.

Manufacturer

Chubb Fire Vehicles, Sunbury-on-Thames, Middlesex, England

Characteristics

CHASSIS: Mercedes Benz 2632A
or as required by operator
ENGINE: OM 403
POWER: 235 kW at 2500 rpm
DRIVE: 6 × 6
ACCELERATION 0–80 km/h: 45 s
MAX SPEED: 100 km/h
TANK CAPACITY
water: 6000 litres may be increased to 10 000 litres (max)
foam: 720 litres may be increased to 1 000 litres (max)
WATER PUMP: Godiva UFP Mk 14; output 5000 litres/minute
MONITOR: FB 70/10 or FB 100/10
output: 3180 litres/minute at 14 bar or 4540/2270 litres/minute
range: 61 m or 69/53 m
ROTATION: ± 132°
ELEVATION: –5° to +45°
HAND-LINES: 2; output 450 litres/minute at 7 bar
VEHICLE PROTECTION: 5 under-truck nozzles (optional); output 643 litres/minute
DIMENSIONS: 8·84 × 2·5 × 3·68 m
OPERATIONAL WEIGHT: 24 500 kg (basic vehicle)

Chubb P Series Heavy-duty Airport Crash Tenders

The P series consists of the Pathfinder, Patroller and Protector vehicles. All are mounted on a purpose-built chassis powered by a rear-mounted Detroit diesel engine with an Allison automatic transmission and all-wheel drive.

Foam is supplied by the Chubb variable round-the-pump induction system using a Godiva water-pump with shaft drive pto.

The fire-fighting equipment comprises a roof-mounted monitor and branch-pipes. A bumper nozzle capable of producing 643 litres per minute at 14 bar and additional hand-lines for water or foam can be provided. Vehicle protection equipment consists of five optional fixed, under-truck protection nozzles.

Fire-fighting controls are power-operated and allow remote operation of the whole system by the driver if required. The exception to this is the Chubb 70/10 monitor which is manually operated from a platform behind the monitor.

Chubb Pathfinder, Patroller, Protector and Pursuer airport crash tenders

Characteristics

	PATHFINDER	PATROLLER	PROTECTOR
CHASSIS	Chubb	Chubb	Chubb
ENGINE	Detroit Diesel 16V–71N	Detroit Diesel 8V 92T	Detroit Diesel 8V 92T
POWER	454 kW at 2100 rpm	320 kW at 2300 rpm	367 kW at 2300 rpm
DRIVE	6 × 6	4 × 4	6 × 6
ACCELERATION 0–80 km/h	42 s	43 s	30 s
MAX SPEED	100 km/h	100 km/h	
CREW	1 + 3 or 1 + 4	1 + 2 or 1 + 3	1 + 2
TANK CAPACITY water	13 630 litres	6800 or 10 000 litres	6800 or 9100 litres
foam	1363 litres/minute	820 or 1200 litres/minute	408 or 546 litres/minute
WATER PUMP	Godiva UFP Mk 20	Godiva UFP Mk 14	Godiva UFP at 15 bar
output	7273 litres/minute at 16 bar	5000 litres/minute at 15 bar	5000 litres/minute at 15 bar
MONITOR	Chubb FB 140/10	Chubb FB 100/10†	Chubb 70/10*
dual output	6360/3180 litres/ minute at 14 bar	4540/2270 litres/ minute at 14 bar	3180 litres/minute minute at 14 bar
RANGE	76/61 m	89/53 m	61 m
ROTATION	± 132°	± 115°	± 132°
ELEVATION	–5 to +45°	–5 to +45°	–5 to +45°
HAND-LINES	2	2	2
output	454 litres/minute at 7 bar	454 litres/minute at 7 bar	454 litres/minute at 7 bar
UNDER TRUCK NOZZLES	5 × 643 litres/minute	5 × 643 litres/minute	5 × 643 litres/minute
DIMENSIONS	11·37 × 3·05 × 3·73 m	10·03 × 3·04 × 3·81 m	10·35 × 2·5 × 3·38 m
OPERATIONAL WEIGHT	38 000 kg	26 400 kg	23 000/25 200 kg

† Chubb 70/10 can be fitted as an option
* Chubb FB 100/10 can be fitted as an option

Status
The Pathfinder was introduced in 1970 and was the first vehicle to be produced with a centre steering position and a purpose-built chassis: over 80 vehicles have been supplied to 80 countries. The Patroller was developed directly from the Pathfinder in the mid-1970s: 76 tenders have been supplied to eight countries. The Protector Mk 2 was introduced in 1980 and 15 have been built.

Manufacturer
Chubb Fire Vehicles, Sunbury-on-Thames, Middlesex, England

Chubb Pursuer/Pacer Rapid Intervention Vehicles

The Pursuer and Pacer are rapid intervention vehicles which meet the international performance recommendations for this type of vehicle.

The Pursuer rapid intervention vehicle is based on a specially designed, rear-engined chassis with a petrol engine and Allison automatic transmission coupled to a permanent four-wheel drive. The fire-fighting system comprises an independently-powered fire pump set which feeds the foam monitor and the foam liquid inductor enabling the manually-operated monitor to produce foam while the vehicle is stationary or moving. A vehicle-mounted bcf trolley unit is supplied to augment fire-fighting power.

The Pacer rapid intervention vehicle is based on the Dodge Power wagon with a V8 engine. Up to 1000 litres of premix extinguishing agent can be accommodated, with complimentary agents if required.

Status
About 55 Pacer rapid intervention vehicles have been supplied to 14 countries.

Manufacturer
Chubb Fire Vehicles, Sunbury-on-Thames, Middlesex, England

Characteristics
PURSUER

CHASSIS: Chubb
ENGINE: Rolls Royce B81 SV 175 kW at 4000 rpm or Chrysler V8 440–46 168 kW at 4400 rpm
DRIVE: 4 × 4
ACCELERATION: 0–80 km/h: 25 s
MAX SPEED: 105 km/h
TANK CAPACITY: 910 litres water; 91 litres foam
WATER PUMP: Godiva FPD with 4-cylinder petrol engine
MONITOR: Chubb FB 1010; output 4500 litres/minute; range 26 m
AUXILIARY EQUIPMENT: 1 × 1·36 kg bcf extinguisher; 1 × 34 kg bcf trolley
DIMENSIONS: 6·25 × 2·44 × 3·15 m
NET WEIGHT: 6350 kg

Eve Construction Trakway Portable Road System for Aircraft Recovery

Trakway is a portable road system provided by aluminium planks connected by interlocking captive tongue and groove joints. It can be used in aircraft recovery enabling heavy plant machinery and vehicles to cross soft, unstable ground to reach the incident, thereby reducing the possibilities of secondary damage.

The system is available as a light-, medium- or heavy-duty temporary road in roll or mat form. The light-duty Trakway is 3·35 metres wide and assembled in standard lengths of 25 and 50 metre rolls. It is transported on special spooling frames on a lorry- or trailer-mounted turntable, supplied by the manufacturer.

Medium- and heavy-duty Trakway is available in panel form 3·35 metres wide by 2·5 metres long which can be assembled on site.

Manufacturer
Eve Construction Ltd, Sutton-in-Ashfield, Nottinghamshire, England

Fire Fighting Enterprises Portable and Mobile Fire Extinguishers

Fire Fighting Enterprises specialises in the design and manufacture of bcf (halon 1211) equipment for all types of airport use. Equipment available includes the 1·5 and 5·5 kg portable units, 25 and 50 kg BAA model wheeled units and a 100 kg skid-mounted unit for rapid-intervention vehicles. An extending applicator can be supplied for the 50 kg wheeled and skid units.

Manufacturer

Fire Fighting Enterprises (UK) Ltd, Windsor, Berkshire, England

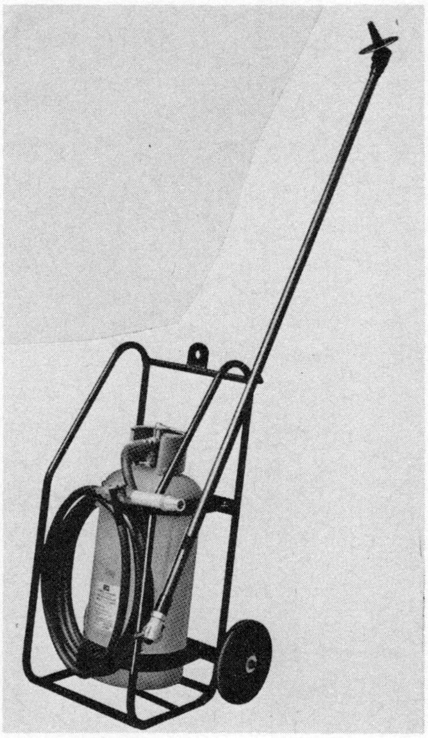

Fire Fighting Enterprises 50 kg mobile halcon 1211 bcf trolley

Gloster Saro Gladiator Medium-duty Foam Tender

The 4 × 4 chassis for this tender can be supplied with either a 300 or 400 6 hp rear-mounted engine. The water tank capacity may also be varied between 2700 and 6000 litres. The tender has been designed to operate in any climate, on or off the airfield and conforms to international road regulations enabling it to operate without any restrictions.
The cab carries a crew of three or four. A remote controlled/manual self-aspirating monitor is fitted together with two foam sidelines. A heavy-duty hose reel is mounted at the rear of the vehicle.

Manufacturer

Gloster Saro Ltd, Gloucester, England

Gloster Saro Heavy-duty Airport Crash Tender

Gloster Saro produce two vehicles which fit into this category, the Javelin and the Gauntlet.

The Javelin is based on a 6 × 6 drive chassis with a rear-mounted engine. The foam producing system comprises separate foam and water tanks and an automatic selecting proportioner fed by a high-performance pump. The maximum water capacity of the tender may be varied between 8000 and 10 000 litres. The fire-fighting equipment consists of a single roof-mounted remote-controlled monitor, two foam lines, two water delivery connections and a hose reel.

The Gauntlet is a development based on the Scammell 6 × 6 rear-engined chassis with Allison automatic transmission. Similar to the Javelin, the design incorporates a four-man cab.
The foam producing equipment comprises a Godiva single-stage centrifugal pump with automatic round-the-pump foam proportioner which can vary in strength between three and eight per cent. The foam and water system provides for 3200 litres per minute fluid flow from the roof-mounted monitor, two foam-lines and two water-lines with a swing-out, rear-mounted hose-reel for domestic use. The monitor can be remotely-controlled from the driver's position or operated using power controls from the monitor platform.

Both vehicles can be fitted with a telescopic aerial cage lift which can provide direct access to fire in high-mounted engines in aircraft such as the DC10 or Lockheed Tristar. For this purpose the cage can either be controlled directly by its occupant or remotely from the monitor deck and used with a fixed-lance applicator. This system of controls also works in reverse and it is possible to direct the cab-mounted foam monitor from an elevated position in the cage.

Gauntlet airport crash tender

Characteristics

CHASSIS: Scammell
ENGINE: Turbo-charged diesel
POWER: 320 kW
DRIVE: 6 × 6
ACCELERATION: 0–80 km/h: 40 s
MAX SPEED: 110 km/h
CREW: 1 + 3
TANK CAPACITY
water 6800 litres
foam 820 litres
PUMP: Godiva Mk 14
MONITOR: 3200 or 3200/1600 or 4500/2250 litres/minute
RANGE: 60 m
ROTATION: ± 270°
elevation −15 to +45°
DIMENSIONS: 9·76 × 2·5 × 3·61 m
OPERATIONAL WEIGHT: 22 350 kgs
AERIAL ACCESS EQUIPMENT
working height agl 10·5 m
horizontal reach 7·92
working load 136 kg
elevation 0–80°
rotation 270°

Status

The Javelin is in service with the British Airports Authority. The Gauntlet entered production in 1981.

Manufacturer

Gloster Saro Ltd, Gloucester, England

Gloster Saro Rapid Intervention Vehicles

Gloster Saro produce two rapid intervention vehicles designed for fire and rescue, one based on the Range Rover 6 × 4 chassis, the other using a Dodge 4 × 4 chassis.

The Range Rover vehicle is capable of 145 km per hour at a gross vehicle weight of 4080 kg with acceleration from 0–80 km per hour in 18 seconds. The cab roof and body structure are constructed in grp incorporating roll-over protection. A crew of four is accommodated in the four-door cab.
A grp sandwich tank contains 910 litres of premix foam which can be discharged at rates of up to 910 litres per minute at 14 bar in jet or spray modes through two hand-lines. Lockers enclosed by roller shutters are provided in the rear bodywork for fire and rescue equipment.

The Dodge Vehicle uses a W 300 chassis with a 3·43 metres wheelbase and left-hand drive. The bodywork is all-metal with lock units on each side enclosed by roller shutter doors. Front and rear towing eyes are fitted.
The vehicle is capable of speeds approaching 145 km per hour with high acceleration and accommodates a crew of three.
The fire-fighting equipment consists of two hand-lines with jet/spray nozzles supplied from a 910 litre grp water tank.
Ladder gantries are provided for the stowage of a 3·89 metre closed two-section ladder.

Manufacturer

Gloster Saro Ltd, Gloucester, England

Perren Wheeled and Portable Fire-extinguishers

Perren range of portable and wheeled fire-extinguishers are suitable for airport duties which vary from air crash cover on helicopter pads and airfields to first-aid protection of fuel dumps, vehicle depots, turbine and generator installations.
The foam trolleys are designed to use premixed solutions of afff although Perren equipment is suitable for both afff and standard foams. They are particularly effective on fires involving flammable liquids where afff gives a rapid flame extinction and spreads a vapour sealing foam over the surface of the burning liquid to prevent flash back and re-ignition.
The expellent can be either carbon-dioxide gas or compressed air. Standard breathing apparatus cylinders are used for compressed air. Perren foam trolleys are available in 90 and 135 litre capacities. Developed to meet the ICAO requirements for fire protection on small airfields, the afff 60 and 150 foam-trailer appliances are used by the Royal Air Force Fire Service at heli-pads. Both appliances are designed for use with premix solutions of afff. The afff 60 (with 270 litre capacity) has a single nozzle supplied through a 20 metre length of 45 mm bore layflat hose giving a discharge rate of 295 litres a minute. The afff 150 (with 685 litre capacity) has twin nozzles each equipped with 30 metres of 45 mm bore layflat hose, giving a total discharge rate of 515 litres a minute. The trailers can be towed by a wide variety of vehicles.

The bcf 25 trolley appliance contains 25 kg of bcf. It is capable of a discharge rate of 0·5 kg a second of bcf and is suitable for major fires involving tyres and high-pressure leakage of flammable liquids. The appliance is equipped with five metres of hose and a lance-type applicator which enables the operator to tackle fires in machinery spaces that might otherwise be inaccessible. The bcf 25 has a performance equivalent of a carbon-dioxide appliance three or four times larger. In addition the wet discharge of bcf liquid helps the bcf 25 appliance to remain effective even under windy and exposed conditions.

The bcf 50, 70 and 100 trolleys have been designed with twice the fire power of the bcf 25. Carrying 50, 70 and 100 kg of bcf respectively these units have a discharge rate of 1·36 kg a second and are capable of tackling major aero-engine fires. The appliances are equipped with 12 metres of hose mounted in a specially designed hose cradle. Swivel connections at the applicator prevent twisting of the hose during fire-fighting. Extending applicators can be supplied for use on fires which are in elevated and otherwise inaccessible positions.

Perren bcf 70, 50 and 25 trolleys

Perren afff 60 foam fire-fighting trailer

Versions of bcf equipment incorporating sun shields, etc have been developed for use in hot and exposed conditions.

Manufacturer

Perren Fire Protection Ltd, Bicester, Oxfordshire, England

Perren Pioneer 700 Rapid Intervention Vehicle

The Pioneer 700 is designed for rough or soft ground carrying a crew of three including the driver. The vehicle fully laden with a complete crew is able to accelerate from 0–80 km an hour in less than 17 seconds. It is based on a Jeep 4 × 4 chassis with a V8 petrol engine and automatic transmission.
The fire-fighting equipment comprises a 720 litre premix-foam (afff) system using compressed air as the expellent and a high-rate discharge bcf system. The foam system uses twin delivery hoses terminating in fully adjustable non-aspirating jet/spray nozzles.

Status

The Pioneer 700 complies with requirements of the United Kingdom CAA for category two airfield protection

Manufacturer

Perren Fire Protection Ltd, Bicester, Oxfordshire, England

Characteristics
CHASSIS: Jeep J 20/46
ENGINE
V8 petrol
175 bhp at 4000 rpm
ACCELERATION: 0–80 km/h under 17 s
TANK CAPACITIES
premix foam: 720 litres
bcf: 225 kg units
HOSES
foam: 230 m × 45 mm diameter; output 545 litres at 7 bars; range 30 m
bcf: 1 × 30 m; output 1·36 kg/s
DIMENSIONS: 5·41 × 2·17 × 1·98 m

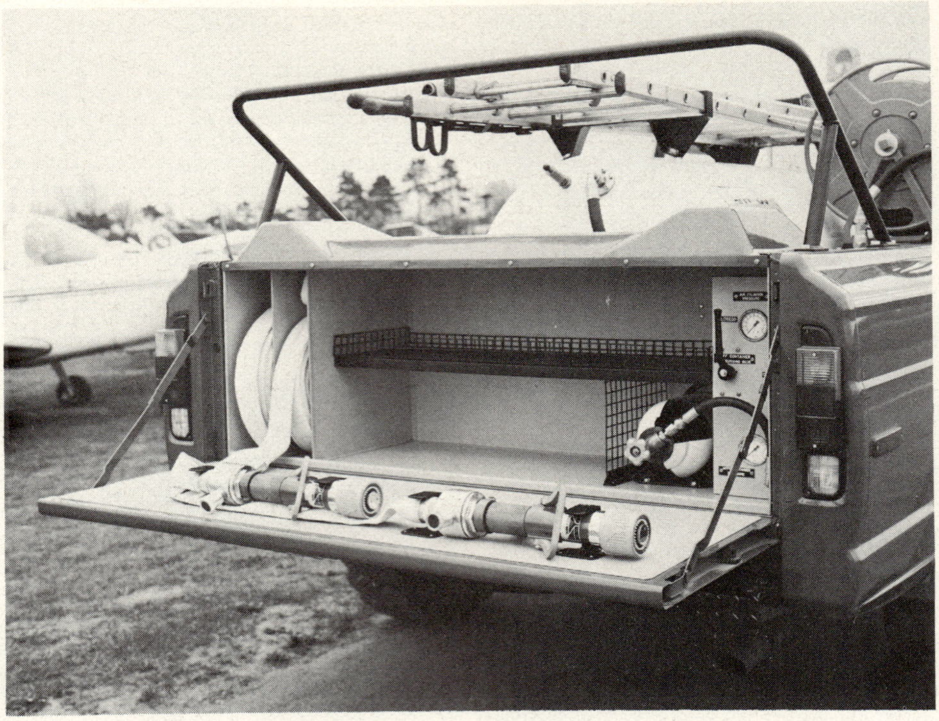

Rear tail-gate of Pioneer 700 in open position showing twin hoses and nozzles, air-pressurisation cylinder, air regulator control panel and central storage area

Pilcher-Greene Ambulances

Pilcher-Greene produce a range of ambulances based on the Land Rover and Range Rover chassis. The 7489 series are mounted on a Land Rover chassis with two body types and three power unit options. Type J body is fully moulded in fire retardant grp; type P body is constructed in aluminium light alloy section with fibre glass panelling. Five interior designs are available to accommodate up to four stretchers or eight sitting patients or various combinations of both.
The Range Rover ambulance has two types of body fits; type SA body is constructed in aluminium alloy sections with 18 swg aluminium sheet exterior; type C5 body uses aluminium alloy sections with fibre glass panelling. Ferno Washington Warwick cot stretchers are used and interiors constructed to suit requirements.

Characteristics

	TYPE 7489	RANGE ROVER
CHASSIS	Land Rover	Range Rover
ENGINE OPTIONS	4-cylinder 2·25 litres petrol 4-cylinder 2·25 litres diesel V8-cylinder 3·5 litres petrol	V8-cylinder 3·5 litres petrol
DIMENSIONS		
length	4·80 m	4·93 m
width	1·80 m (J) 1·70 m (P)	1·78 m
height	2·41 m (J) 2·34 m (P)	2·16 m
WEIGHT	1960 kg	1977 kg

Manufacturer
Pilcher-Greene Ltd, Burgess Hill, West Sussex, England

Pilcher-Greene Branbridge Fire Appliances

The Branbridge range of fire appliances are designed for airport duties in particular for fire-fighting in confined spaces and difficult terrain inaccessible to larger vehicles.
The range is based on the long-wheel base Land Rover chassis with heavy-duty rear suspension. The main feature of the Mk 1 is large capacity lockers for equipment stowage. In addition to the basic specifications the vehicle is equipped with four two-metre lengths of suction hose with aluminium screw couplings. The Mk 2 version has an extended crew cab with extra seating capacity for three people. Two 2·2-metre lengths of suction hose are provided in addition to the basic specification. The Mk 3 is available in various forms offering different capacities of water, dry powder and other extinguishing agents.

Characteristics

CHASSIS	Land Rover long-wheel base
ENGINE OPTIONS	4-cylinder petrol 2·25 litres; 4-cylinder diesel 2·25 litres; V8-cylinder petrol 3·5 litres
DRIVE	4 × 4
PUMP	Godiva UFP Mk 6B/I/T pto
output	2273 litres/minute at 7 bar
TANK CAPACITIES	Mk 1: 360 litres; Mk 2: 315 litres; Mk 3: 450 litres
HOSE REEL	3 × 18 m, 19 mm hose with couplings; 1 × 0·6 m, 19 mm hose with shut-off hose reel branch
SUCTION HOSE	Mk 1: 4 × 2 × 0·1 m; Mk 2: 2 × 2·2 × 0·1 m; Mk 3: 4 × 1·8 × 0·1 m
AUXILIARY EQUIPMENT	two section ladders extending to 7·75 m 1 × 1·3 kg bcf extinguisher
DIMENSIONS	5·18 × 1·67 × 2·08 m
OPERATIONAL WEIGHT	2470 kg

Status
In service at airports in the United Kingdom and other countries

Manufacturer
Pilcher-Greene Ltd, Burgess Hill, West Sussex, England

Pindair Skima Hovercraft Air Crash Tenders

Some airports are either constructed on reclaimed land or are near shallow water, swamp land and or water which freezes in winter. These conditions often make it impossible for rescue teams to reach the scene of the crash quickly using conventional air crash tenders. In these conditions a hovercraft is particularly well suited as a rescue vehicle.
The range of small hovercraft produced by Pindair for use as air crash tenders include 4-, 6-, 12-, 18- and 25-seat designs.
The mid-range Skima 12 for example can be operated in most inshore weather conditions and has been tested over a wide variety of surfaces including obstacles of up to half a metre in height. It can carry six 30-man life rafts and be fitted with a cabin for 12 people, or four stretcher cases and crew of three. Various types of radio and radar can be specified as well as searchlights and other special equipment.

Pindair Skima 12 hovercraft

Manufacturer
Pindair Ltd, Gosport, Hampshire, England

RFD Pneumatic Elevator System

RFD Pneumatic Elevator System for aircraft recovery comprises elevators, inflation equipment, air-control consoles, compressors, air hoses and aircraft tethering equipment.
Three types of pneumatic elevators constructed of woven nylon fabric coated with a neoprene compound are used to lift the aircraft.
The type F with 40-tonnes capacity (at 500 gm/cm^2) consists of 15 superimposed and similarly dimensioned independent elements bonded together to form an assembly. The type G with 25-tonnes capacity has a similar number of elements and the type H, also of 25-tonnes capacity, consists of eleven elements.
An air compressor driven by a diesel engine delivers 2152 litres a minute of air at 7000 g/cm^2 through a six-outlet manifold and semi-rigid hoses. These are connected to air-distribution control consoles. Through a system of valves and a controllable pressure regulator fitted with a gauge, filtered air is distributed at the required pressure to each element of each elevator for any period of time at a constant set pressure. Each console is fitted with a valve for rapid exhaustion of air from the elevator in the event of an emergency. The hose-reel assembly comprises two compartments, one containing ten lengths of 15·2 metre semi-rigid hose and the other ten lengths of 9 metre hose.
The number of individual items vary according to the type of aircraft to be recovered. The equipment required for the recovery of a Boeing 747 for example can be stored in four 200 kg pallets each pallet measuring 320 by 225 by 218 cm.
The system has a wide range of applications for lifting and recovering crashed aircraft with minimum secondary damage from uneven or slanting surfaces or soft ground.

Characteristics
Recommended equipment for the recovery of a Boeing 747 aircraft

DESCRIPTION	QUANTITY	SIZE OF ONE ITEM (cm)	TOTAL VOLUME (m^3)	TOTAL MASS (kg)
ELEVATOR TYPE				
F	2	220 × 104 × 77	3·523	864
G	5	220 × 80 × 66	5·808	1360
H	2	220 × 77 × 66	2·236	472
CONSOLES 10 OUTLETS	16	104 × 71 × 14	2·171	544
HOSES REELS 20 HOSES	8	92 × 92 × 71	4·807	1152
COMPRESSORS	2	236 × 138 × 125	8·142	1218
GROUND PADS	20	338 × 140 × 5	5·442	430
TOP PADS	20	338 × 140 × 1·5	1·42	140
PVC ROLL	2	182 × 30 × 30	0·328	40
SPARES	1	65 × 65 × 65	0·275	68
TIRFOR WINCH	6	75 × 33 × 15	0·573	160
GROUND ANCHORS	12	107 × 25 × 11	0·674	516
CABLES MAXIFLEX + SLING	6	40 × 40 × 15	0·15	252
SHORT HOSES	20	100 × 5 × 5	0·05	50

Status
The equipment is in service with 18 airport authorities, five military forces and seven airlines throughout the world

Manufacturer
RFD Inflatables Ltd, Godalming, Surrey, England

Wormald Fire-protection System

The Wormald International group of companies including Mather and Platt and Ansul Fire Protection provide a range of fire-protection products for the aviation industry.
Fire detection and extinguishing systems for airport buildings and hangars include: halon gas systems for computer rooms and flight simulators; sprinkler systems for general building application and certain special risks; foam systems for bulk fuel storage and aircraft hangar protection. The foam systems comprise roof-mounted foam sprinkler nozzles, fixed foam nozzles for underwing protection and high-level oscillating monitors.
Foam systems use afff. This is available in four grades: three and six per cent concentrate, three per cent freeze-protected concentrate and three or six per cent alcohol resistent concentrate.
Small rapid response vehicles with the twin agent capability, wheeled and portable fire-extinguishing units are manufactured by Wormald International.

Manufacturer
Wormald International, Slough, Berkshire, England
Mather and Platt, Manchester, England
Ansul Fire Protection Ltd, High Wycombe, Buckinghamshire, England

Ansul twin agent applied to fuel in obstacle fire

Carmichael Commando and Jetranger Crash/Fire/Rescue Vehicles

Carmichael produce a range of rapid intervention and crash/fire/rescue vehicles carrying varying quantities to suit all sizes of airports.
The Commando range of rapid intervention vehicles are based on Range Rover, Reynolds Boughton, US Dodge and other suitable chassis. The Commando RR 220 rapid intervention vehicle is mounted on a Range Rover chassis incorporating the Carmichael third-axle conversion. The vehicle is available with either a single tank for pre-mix foam solution or separate water and foam tanks: in the latter case a foam system is also incorporated. The roof-mounted monitor is controlled from within the vehicle.

Carmichael Jetranger S2500 crash tender on Scammell 6 × 6 chassis

The Jetranger range of crash/fire/rescue vehicles are based on 4 × 4 and 6 × 6 chassis produced by special-purpose chassis manufacturers such as Reynolds Boughton and Scammell.
The largest of the Carmichael Scammell crash tenders is based on the 500 bhp 6 × 6 chassis. The vehicle is available with manual- or hydraulically-operated monitors with various outputs: dual output monitors are available. Other optional features include under-truck nozzles, elevating platforms to permit access to rear-mounted aircraft engines and floodlighting systems.

Status
In service at airports throughout the world. Carmichael is supplying the Royal Air Force's latest fire-fighting vehicles, the Truck, Fire Fighting Foam Mk 10

Manufacturer
Carmichael Fire and Bulk Ltd, Worcester, England

Characteristics
Typical specifications of Carmichael vehicles:

MODEL	COMMANDO RR 220	JETRANGER RB 1000	JETRANGER S2500
CHASSIS	Range Rover	Reynolds Boughton Perseus	Scammell Super Major
ENGINE	Rover V8	Cummins VT555C 270	Cummins VTA 903
POWER	97 kW	202 kW	373 kW
DRIVE	6 × 4	4 × 4	6 × 6
ACCELERATION 0–80 km/h	21 s	41 s	38 s
MAX SPEED	112 km/h	100 km/h	108 km/h
CREW	1 + 2	1 + 3	1 + 2
TANK CAPACITY			
water	910 litres	4095 litres	10 000 litres
foam	90 litres	455 litres	1360 litres
WATER PUMP	Single stage centrifugal Output matched to monitor output		
ROOF MONITOR	9100 litres/minute	18 200 or 22 700 litres/minute	31 800 or 45 500 litres/minute
HAND-LINES			
finished foam output at 7 bar	2270 litres/minute	4500 litres/minute	4550 litres/minute
DIMENSIONS			
length	5·79 m	7·67 m	9·95 m
width	1·78 m	2·44 m	2·75 m
height	2·3 m	3·25 m	3·3 m
OPERATIONAL WEIGHT	4000 kg	15 000 kg	28 000 kg

Carmichael Commando RR220 rapid intervention vehicle

UNITED STATES OF AMERICA

Darley Airport Crash Tender

Darley supply a variety of airport crash vehicles which carry water, foam and dry chemical extinguishing agents. Nine tenders have been produced recently for use in Thailand, fitted with roof monitors, branch-pipes and ground sweep foam nozzles.

Manufacturer
W. S. Darley, Melrose Park, Illinois, USA

Darley airport crash tender

El Industries Star-Line Ambulances

El Industries produce a range of ambulances in the Star-Line series. The heavy-duty versions suitable for rescue work are the Super Model EMSV with six wheels and a tandem walking-beam type axle suspension system and the Star Model III.

Both vehicles can be equipped to suit individual operator's requirements. The main features of the EMSV are space for a minimum of two patients and an attendant, incubator facilities, roof-mounted air-conditioner and heater, and a generator producing 110 volts and 220 volts. External compartment space totals 82 cubic feet (2·32 cubic metres). The Star Model III is a four-wheeled vehicle available with an all-aluminium or aluminium/steel body. The patient module measures 12 feet (3·66 metres) long with 5 feet (1·5 metres) headroom. External storage of 35 cubic feet (0·99 cubic metres) is available which can be extended to 57 cubic feet (1·6 cubic metres).

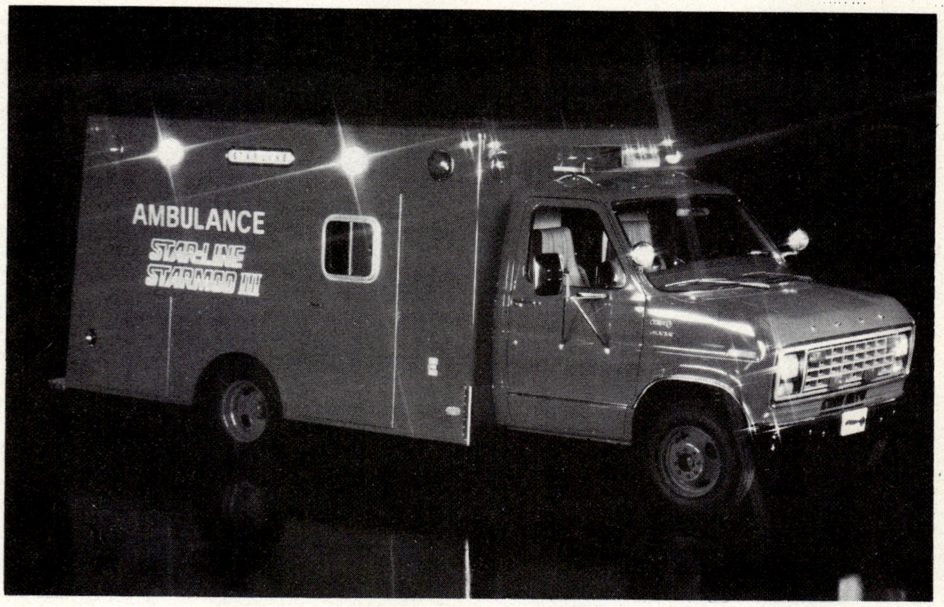

Star-Line Star Model III Ambulance

Characteristics

Super Model EMSV

CHASSIS AND ENGINE
minimum 350 CID V8 engine, as required by operator

DIMENSIONS
7·93 × 2·41 × 2·41 m (body height only)

WEIGHT: 4535 kg

Manufacturer

El Industries Inc, Chicago, Illinois, USA

Garsite GS 25 CFR/GS 11 CFR Airport Crash Tenders

The GS CFR/GS 11 CFR (crash/fire/rescue) vehicles produced by Garsite are mounted on Scammell chassis and powered by a Cummins diesel engine with an Allison semi-automatic transmission. Designed for cross-country operation, the vehicles have a high angle of approach and departure with a 30 degree minimum.

The pump is driven from a power take-off of Scammell design which permits engagement while the vehicle is moving or stationary.

The monitor is mounted above the cab and can be operated remotely from the driver's console or from the top deck. There are manual overrides for all controls. The monitor provides a variable trajectory from a straight stream to a dispersed pattern. Two branch-pipes are fitted each 30·48 metres in length on electric rewind reels. Vehicle protection is provided by two ground sweeps controlled from the cab.

Manufacturer

Garsite Products Inc, New York, USA

Characteristics

DRIVE	6 × 6	6 × 6	4 × 4	4 × 4
CHASSIS	Scammell	Scammell	Scammell	Scammell
ENGINE	Cummins VTA 903 Turbo-charged 373 kW	Cummins VT 903 Turbo-charged 298·4 kW	Cummins VT 903 Turbo-charged 298·4 kW	Cummins V 903 Turbo-charged 225·29 kW
ACCELERATION				
0–80 km/h	35 s	43 s	28 s	38 s
MAX SPEED	106 km/h	106 km/h	106 km/h	106 km/h
PUMP OUTPUT	5676 litres/minute at 17 bar		3785 litres/minute at 15 bar	
TANK CAPACITY				
foam	946 litres	946 litres	570 litres	570 litres
water	9460 litres (dual rate)	9460 litres (dual rate)	4160 litres	4160 litres
MONITOR OUTPUT	2270/4540 litres/minute	2270/4540 litres/minute	2840 litres/minute	2840 litres/minute
range	60 m	60 m	60 m	60 m
rotation	120°	120°	120°	120°
elevation	−20 to +70°	−20 to +70°	−20 to +70°	−20 to +70°
AUXILIARY EQUIPMENT				
hand-lines	2 × 30·48 m	2 × 30·48 m	2 × 30·48 mm	2 × 30·48 mm
unit output	227 litres/minute	227 litres/minute	227 litres/minute	227 litres/minute
VEHICLE PROTECTION EQUIPMENT, GROUND SWEEP	2 × 227 litres/minute	2 × 227 litres/minute	2 × 227 litres/minute	2 × 227 litres/minute
DIMENSIONS	9·8 × 1·48 × 3·2 m	9·8 × 1·48 × 3·2 m	8·41 × 1·48 × 3·2 m	8·41 × 1·48 × 3·2 m
OPERATIONAL WEIGHT	25 647 kg	25 647 kg	17 010 kg	17 010 kg

Gulf & Western 500S/BAK-12 Military Aircraft Arresting System

The 500S rotary friction-brake arresting system produced by Gulf and Western is designated the BAK-12 when conforming to US military specifications. It is suitable for emergency use at forward bases or fixed installations.

The system consists of two rotary friction-energy absorbers joined across the runway by a hook cable or a nylon barricade net. The friction-brake discs on each absorber are driven by a reel which houses the roll of nylon purchase tape that pays out during an arrestment. For tail-hook equipped aircraft the hook cable is held several inches above the runway or over-run by rubber discs or elevated by a tower-operated retractable support system (BAK-14). Aircraft without tail hooks engage the barricade net of the stanchion system (61QS11) see commercial aircraft arresting system which follows. When not in use the net is stretched taut across the over-run surface to allow aircraft roll over and is erected by remote control from the tower. At bases using both types of aircraft the 500S can be equipped with a net/cable interconnect device (62NI) so that either type can be safely arrested. During an arrestment the over-running aircraft catches the 500S cable or net and pulls it along the runway with its attached tapes. As the tapes pay out each rotating reel, mounted on a common shaft with the friction discs of the energy absorbing brakes, drives a velocity-sensitive hydraulic control system which applies a gradual braking force on the aircraft. After coming to a stop the aircraft is disengaged from the cable or net and the purchase tape is rewound by a petrol engine or electric motor rewind system.

The established practice at fixed bases is to bolt each absorber to a concrete pad for surface installation or in concrete pits for below-ground installation.

The 500S is designed to arrest aircraft in either direction which permits hook-equipped aircraft to make precautionary approach-end arrestments as well as the more frequently over-run-end arrestments.

Characteristics

DIMENSIONS OF ABSORBER
length: 120 inch (305 cm) each
width: 52 inch (132 cm) each
height: 68 inch (173 cm) each
WEIGHT OF ABSORBER INCLUDING TAPE: 500S 6500 lb (2940 kg); BAK-12 7000 lb (3178 kg)
MAX ENGAGING VELOCITY
wire rope hook cable: 190 knots (352 kph)
nylon hook cable: 230 knots (426 kph)
AIRCRAFT NOMINAL RUNOUT DISTANCE: 900–1200 ft (274–366 m)
AMBIENT TEMPERATURE OPERATING RANGE: −40 to +125°F (−40 to +52°C)
REWIND SYSTEM POWER: 37 hp (27·5 kW) petrol engine, or 15 hp (11·1 kW) electric motor
NO OF REWIND OPERATORS: 2
REWIND TIME: 3 minutes
CYCLE TIME: 5 minutes

Gulf and Western 500S/61QS11 aircraft arresting system

MIRAGE III aircraft engaging Gulf and Western 61QS11 system at the French Flight Test Centre, Istres, France

Status

Gulf and Western produce a range of arresting equipment for military aircraft. This equipment has been operational since 1965 with aircraft of variable weights and landing speeds.
487 BAK-12 and 256 500S systems have been supplied to military operators. A dual BAK-12 system has been developed for use with F-111 and SR 71 aircraft.

Manufacturer

Gulf and Western, Swarthmore, Pennsylvania, USA

MODEL	500S-4	500S-6 (STANDARD NATO)	BAK-12/500S-8
BREAK ROTORS/ABSORBER	4	6	8
DESIGN ENERGY FOR NOMINAL RUN-OUTS			
900 ft (274 m)	55×10^6 ft lb (75 MJ)	65×10^6 ft lb (88 MJ)	75×10^6 ft lb (102 MJ)
1200 ft (366 m)	65×10^6 ft lb (88 MJ)	75×10^6 ft lb (102 MJ)	100×10^6 ft lb (136 MJ)
NET SYSTEM			
900 ft (274 m)	63×10^6 ft lb (85 MJ)	75×10^6 ft lb (102 MJ)	86×10^6 ft lb (117 MJ)
1200 ft (366 m)	75×10^6 ft lb (102 MJ)	86×10^6 ft lb (117 MJ)	115×10^6 ft lb (156 MJ)
TAPE			
dimensions	7 × 0·2 inch (17·8 × 0·508 cm)	8½ × 0·225 inch (21·8 × 0·572 cm)	8½ × 0·225 inch (21·8 × 0·572 cm)
strength	70 000 lb (31 780 kg)	105 000 lb (47 600 kg)	105 000 lb (47 600 kg)
AIRCRAFT COMPATABILITY	Mirage III, V and F-1, A-4, A-37, F-5, F-16, F-104, G-91, Alpha, Drakken, Jaguar, Lightning and Viggin F-18	All 500S-4 aircraft + A-7, A-10, F-4, F-15	All 500S-4 and 500S-6 aircraft + F-14, F-18, Mirage IV, Tornado

The 500S-4 and 500S-6 can be modified by means of a field-change kit to convert to a higher energy capacity model.

Gulf and Western Commercial Aircraft Arresting System

The commercial aircraft arresting system developed by Gulf and Western uses a multi-element nylon webbing to catch the aircraft. The webbing is manufactured in several sizes for various models of aircraft. It is suspended across a runway over-run on stanchions or masts which are raised and lowered by the tower traffic controller. When not in use the webbing is stretched across the over-run surface to allow aircraft roll-over.
The 61QS11 system consists of two powered mast assemblies on either side of the runway over-run. The masts which support the barricade net are raised and lowered by remote control from the tower with the electric-hydraulic power packages provided with each mast. In the event of system power-loss sufficient hydraulic pressure is stored in the power-package accumulators for manual operation of a complete lifting and lowering cycle. The masts are equipped with winches to maintain the net in a taut condition and prevent tangling in the lower state by jet blast from aircraft ground traffic. The upper net strap is connected to each mast by a shear link which separates under the load of an engaging aircraft. This releases the net and the nylon purchase tape of the arresting equipment attached to the lower net strap. If the system is raised but not used it can be lowered remotely to the battery position for subsequent use. The system can be used with several net configurations or runway spans.
When connected to suitable braking components the webbing stanchions, energy absorbers and controls comprise a self-contained barricade arresting system appropriate to the class of aircraft using the airport. Five systems are available. Two of them, the single and dual 500S-8, use rotary friction mechanisms to provide the retarding force. A description of these devices can be found in the 500S/BAK-12 report above. The dual Model 1300 and the single and dual 700 watts energy absorbers are rotary hydraulic devices.
The rotary hydraulic energy absorber assembly consists of a liquid-filled housing with two sets of eight stator vanes in which a rotor and tape reel have a common shaft so that rotation of the tape reel during an arrestment spins the rotor vanes at high speed. The hydraulic interaction between the rotor and stator vanes causes the retarding force. As the tape pays out from the reel during an arrestment the tape reel diameter decreases and there is a corresponding increase in the rotational velocity of the reel and the attached turbine rotor. This serves to maintain the fluid braking action over the entire run-out distance. After coming to a stop the aircraft is disengaged from the cable or net and the tape is restored on the reel by a petrol engine-powered rewind system.

Status

Gulf and Western has supplied approximately 1250 arresting gear systems to 32 countries.

Manufacturer

Gulf and Western Advanced Development and Engineering Center, Swarthmore, Pennsylvania, USA

Characteristics

ENERGY ABSORBER	SINGLE 500S-8	DUAL 500S-8	DUAL 1300	SINGLE 700W	DUAL 700W
DESIGN ENERGY					
427 m nominal run-out	154 MJ	309 MJ	209 MJ	468 MJ	935 MJ
579 m nominal run-out	177 MJ	355 MJ	241 MJ	540 MJ	1080 MJ
BRAKE TYPE	Friction	Friction	Hydraulic	Hydraulic	Hydraulic
NYLON TAPE DIMENSIONS	8½ × 0·225 inch (21·8 × 0·572 cm)	8½ × 0·225 inch (21·8 × 0·572 cm) (2 per side)	7 × 0·25 inch (17·8 × 0·635 cm) (2 per side)	18 × 0·26 inch (45·7 × 0·927 cm)	18 × 0·26 inch (45·7 × 0·927 cm) (2 per side)
TAPE STRENGTH UTS	47 630 kg	47 630 kg each	45 360 kg each	147 420 kg	147 420 kg each
MAX DECEL LOAD					
427 m nominal run-out	44 230 kg	88 450 kg	64 830 kg	156 490 kg	313 120 kg
579 m nominal run-out	37 650 kg	73 300 kg	55 310 kg	133 360 kg	266 720 kg
WEIGHT PER SIDE					
427 m nominal run-out	3200 kg	6400 kg	7760 kg	6800 kg	13 610 kg
579 m nominal run-out	3350 kg	6900 kg	8260 kg	7530 kg	15 060 kg
AIRCRAFT COMPATABILITY*	B-737, BAe-111, DC-9	B-727, B-720, CV-880	B-737, BAe-111, DC-9	B-707/720, DC-8, C-141, VC-10	B-747, C-5A, DC-10, L-1011

*Each successively higher capacity absorber can accept all aircraft listed for lower capacity absorbers.

DC9 aircraft landing over a Gulf and Western single 500S barricade system at Kuopio, Finland

OMC Lincoln Titan Mini-ambulance

Titan Mini-ambulance is designed for large airports to provide emergency treatment at the airport complex.

Based on the Titan electric vehicle, the ambulance is driven by a 48 volt system with a top speed of 32·3 km per hour providing a range of 45 km between charges.

The vehicle is equipped with a single stretcher, diagnostic and treatment equipment. Seating is available in the front for two people and an attendant's seat is close to the stretcher.

Status

Three Mini-ambulances are in use at San Francisco International Airport

Manufacturer

OMC Lincoln, Lincoln, Nebraska, USA

Oshkosh M series Crash/Fire/Rescue Vehicles

The M series crash/fire/rescue vehicles consist of the M-12, M-15 and M-23 versions. All are based on Oshkosh three-point mounted framed chassis and powered by two Detroit V8 engines with twin automatic transmission. Tandem axles are fitted to the front and rear of the vehicles.

Capacity for the fire system is provided by stainless steel tanks and pipes. The vehicles are equipped with two water and two foam pumps driven by a power take-off arrangement with a disconnect clutch from each engine. An over-centre multiple-disc type clutch is used. A pump, discharge side, automatic balanced pressure proportioning system supplies the foam mixture which can be adjusted from between two and ten per cent.

The fire-fighting equipment consists of two monitors, one mounted on the roof, the other over the front bumper and two 46-metre hand-lines. Discharge patterns of all these units can be varied from straight stream to dispersed jet. Under-truck nozzles can be fitted to provide vehicle protection. Both monitors are operated hydraulically by remote control from the cab. The roof monitor has a dual discharge rate.

All M series vehicles are capable of ascending and descending a 60 per cent gradient, maintaining

Oshkosh M series vehicle in service at Memphis International Airport, USA

traction with a simultaneous diagonally-opposite wheel motion of 508 mm and negotiating a 457 mm high obstacle.
Side-slope stability is in excess of 26 degrees.

Status
In service at airports throughout the USA

Manufacturer
Oshkosh Truck Corporation, Oshkosh, Wisconsin, USA

Characteristics

	M-12	M-15	M-23
CHASSIS	Oshkosh	Oshkosh	Oshkosh
ENGINES	2 × Detroit Diesel 8V 92 T	2 × Detroit Diesel 8V 92 T	2 × Detroit Diesel 8V 92 T
POWER			
at 2300 rpm	2 × 367 kW	2 × 367 kW	2 × 367 kW
DRIVE	8 × 8	8 × 8	8 × 8
ACCELERATION			
0–80 km/h	35 s	40 s	50 s
MAX SPEED	80 km/h	80 km/h	80 km/h
CREW	1 + 2	1 + 2	1 + 2
TANK CAPACITY			
water	12 000 litres	15 000 litres	23 000 litres
foam	1950 litres	1950 litres	1950 litres
WATER PUMPS	2 single stage centrifugal	2 single stage centrifugal	2 single stage centrifugal
DUAL OUTPUT	11 355 litres/minute at 16·5 bar	11 355 litres/minute at 16·5 bar	11 355 litres/minute at 16·5 bar
SINGLE OUTPUT	7192 litres/minute at 14 bar	7192 litres/minute at 14 bar	7192 litres/minute at 14 bar
FOAM PUMPS	2 single stage centrifugal	2 single stage centrifugal	2 single stage centrifugal
dual output	757 litres/minute at 18·6 bar	757 litres/minute at 18·6 bar	757 litres/minute at 18·6 bar
single output	624 litres/minute at 15 bar	624 litres/minute at 15 bar	624 litres/minute at 15 bar
ROOF MONITOR			
output	3407/6813 litres/minute	3407/6813 litres/minute	3407/6813 litres/minute
HANDLINES	2 × 46 m	2 × 46 m	2 × 46 m
output	2 × 303 litres/minute at 6 bar	2 × 303 litres/minute at 6 bar	2 × 303 litres/minute at 6 bar
BUMPER MONITOR	1	1	1
output	1136 litres/minute	1136 litres/minute	1136 litres/minute
UNDERTRUCK NOZZLES	4 × 64 litres/minute	4 × 64 litres/minute	4 × 64 litres/minute
DIMENSIONS	14·22 × 3·05 × 4·19 m	14·22 × 3·05 × 4·19 m	14·22 × 3·05 × 4·19 m
OPERATIONAL WEIGHT	45 903 kg	49 682 kg	59 358 kg

Oshkosh T series Crash/Fire/Rescue Vehicles

The T-6, T-9 and T-12 crash/fire/rescue vehicles are all based on Oshkosh three-point mounted framed chassis and powered by a single Detroit Diesel V8 engine with automatic transmission. The T-9 and T-12 versions are equipped with tandem rear axles.
Stainless steel tanks and piping are fitted and a power take-off with a disconnect clutch arrangement is used to drive both water and foam pumps. A pump, discharge side, automatic balanced pressure proportioning system provides the foam mixture, which can be adjusted from between two and ten per cent.
Fire-fighting equipment includes a roof monitor hydraulically-operated by remote control from the cab, and a 46-metre hand-line with adjustable nozzle. The discharge pattern from both pieces of equipment can be varied between straight stream and dispersed jet. Vehicle protection is provided by two ground sweeps and three under-truck nozzles.
All T series vehicles are capable of ascending or descending a 60 per cent gradient; maintaining

Oshkosh T series vehicle in service at Rock County Airport, USA

Characteristics

	T-6	T-9	T-12
CHASSIS	Oshkosh	Oshkosh	Oshkosh
ENGINE	V8 Detroit Diesel 8V 92 T	V8 Detroit Diesel 8V 92 T	V8 Detroit Diesel 8V 92 T
POWER	367 kW at 2300 rpm	367 kW at 2300 rpm	367 kW at 2300 rpm
DRIVE	4 × 4	6 × 6	6 × 6
ACCELERATION			
0–80 km/h	30 s	40 s	45 s
MAX SPEED	105 km/h	97 km/h	97 km/h
CREW	1 + 3	1 + 3	1 + 3
TANK CAPACITY			
water	6000 litres	9000 litres	12 000 litres
foam	776 litres	1155 litres	1552 litres
MONITOR	Air aspirating	Air aspirating	Air aspirating
dual output	1419/2839 litres/minute	2271/4542 litres/minute	2839/5678 litres/minute
HAND-LINES	1 × 46 m	1 × 46 m	1 × 46 m
output	227 litres/minute	227 litres/minute	227 litres/minute
WATER PUMP	Single stage centrifugal	Single stage centrifugal	Single stage centrifugal
output	4542 litres/minute at 15 bars	6814 litres/minute at 15 bar	6814 litres/minute at 15 bar
FOAM PUMP	Single stage centrifugal	Single stage centrifugal	Single stage centrifugal
output	454 litres/minute at 15 bar	605 litres/minute at 15 bar	605 litres/minute at 15 bar
VEHICLE PROTECTION			
GROUND SWEEPS	2 × 284 litres/minute	2 × 284 litres/minute	2 × 284 litres/minute
UNDER-TRUCK NOZZLES	3 × 64 litres/minute	3 × 64 litres/minute	3 × 64 litres/minute
DIMENSIONS	8·59 × 2·85 × 3·52 m	10·74 × 2·85 × 3·66 m	10·74 × 2·85 × 3·66 m
OPERATIONAL WEIGHT	19 278 kg	26 875 kg	30 277 kg

traction with a simultaneous diagonally opposite wheel motion of 509 mm and negotiating a 457 mm high obstacle.
Side-slope stability is in excess of 26 degrees.

Status
In service at airports throughout the USA

Manufacturer
Oshkosh Truck Corporation, Oshkosh, Wisconsin, USA

Pyrotronics Fixed Fire-protection Systems

Pyrotronics design and manufacture a range of fire-protection systems. These consist of detectors, a universal alarm control and extinguishing systems using halon 1301 extinguishing agent. The detectors operate in a number of ways. Early warning detectors can be of the ionisation type responding to smoke or invisible products of combustion; photo-electric, which are activated by smoke; or a combination of photo-electric and ionisation. Heat-response detectors are operated by a rise in temperature while special application detectors react to infra-red or ultra-violet radiation. The universal alarm control system responds to signals from detectors or to manual alarms. It activates and monitors sprinkler systems and initiates alarm systems.

Manufacturer
Pyrotronics Division, Baker Industries Inc, Cedar Knolls, New Jersey, USA

Walter Kidde Halon 1211 Wheeled Fire-extinguisher

The Halon 1211 wheeled fire-extinguisher produced by Walter Kidde is a portable unit capable of being towed at speeds up to 24 km per hour over level paved surfaces. The extinguisher, which is charged with 68 kg of halon 1211 extinguishing agent and pressurised with nitrogen, is designed to combat flammable liquid, electrical and carbonaceous fires.
The unit consists of a hose cabinet, pistol grip valve, horn/nozzle assembly, the halon 1211 cylinder and the carriage.
The cylinder is equipped with a discharge valve (with pressure gauge), safety outlet, quick-disconnect coupling and a sight glass which indicates the level of the agent in the cylinder.
Two semi-pneumatic tyres and wheels mounted on a steel axle form part of the welded steel carriage assembly. The axle is welded to the carriage and positions the wheels off the ground when the extinguisher is in a vertical position.

Characteristics
HOSE: 15·24 m × 19 mm diameter
DIMENSIONS: 1·21 (depth) × 1·52 × 0·71 m
OPERATIONAL WEIGHT: 181·2 kg

Manufacturer
Walter Kidde, Belleville, New Jersey, USA

FUEL SERVICES

Aircraft refuelling vehicles and trailers
Filter/water separators
Fuelling installations
Fuel monitors
Hydrant dispensers

CANADA

EGW Remtec Fuelling Equipment

Remtec has developed a wide range of fuelling equipment. The refuellers vary from a standard two-axle rigid chassis for smaller airports and private aircraft to train-combination versions with eight-wheel trailers having a total capacity in excess of 77 000 litres with a flow rate of up to 4500 litres per minute.
Heavy-duty articulated refuellers are available for military operations with capacities up to 40 000 litres.
For airports that are equipped with pressurised piping and a tank-farm hydrant servicers can be supplied providing up to 4546 litres of fuel per minute.

Manufacturer

Remtec Inc, Chambly, Quebec, Canada

GERMANY, FEDERAL REPUBLIC

Faudi Filter/Water Separator

Filter/water separators are installed in airport refuellers, dispensers and other mobile aviation refuelling systems and in hydrant refuelling systems at storage tanks, where stationary units are used. Faudi supply horizontal and vertical filter/water separators, using a two-stage separating process. The first stage uses coalescer elements, the second features a permanent separating element.
The first stage is designed to break the fuel/water emulsion and simultaneously remove solids. The cylindrical-coalescer elements contain hydrophilic and hydrophobic materials, for example coated filter papers, fibreglass mats, plastic mats and felt or cotton fabric. The flow through these elements is from the inside to the outside. In the first layer the incoming emulsion passes through the filter paper which is pleated into a star-shape in order to accommodate the maximum possible filter surface in the element. The filter paper retains solids and begins the break-up of the fuel/water emulsion which continues in subsequent layers until the required effect has been achieved. The minute droplets coalesce into larger drops. This process is assisted by the reduction in flow velocity. The drops produced in the outlet layers precipitate into the water sump under gravity.
In contrast to the coalescer elements, the flow through the separator elements in the second stage is from the outside to the inside. These elements, which are covered with stainless steel mesh and coated with Teflon, are designed to stop any water droplets still suspended in the flow but allow the fuel to flow through freely. The water droplets are repelled by the hydrophobic surface of the elements and sink into the water sump. The degree of contamination in the coalescer elements is indicated by a differential pressure gauge. These elements can be replaced individually. Other features incorporated in the separator include an automatic air vent, automatic water-drain valve or automatic control system with shut-off valve and water-drain valve, water-level indicator and an electrical heater for the water sump.

Manufacturer

Faudi Feinbau GmbH, Oberursel, Federal Republic of Germany

Faudi filter/water separators installed in hydrant-refuelling system of an international airport

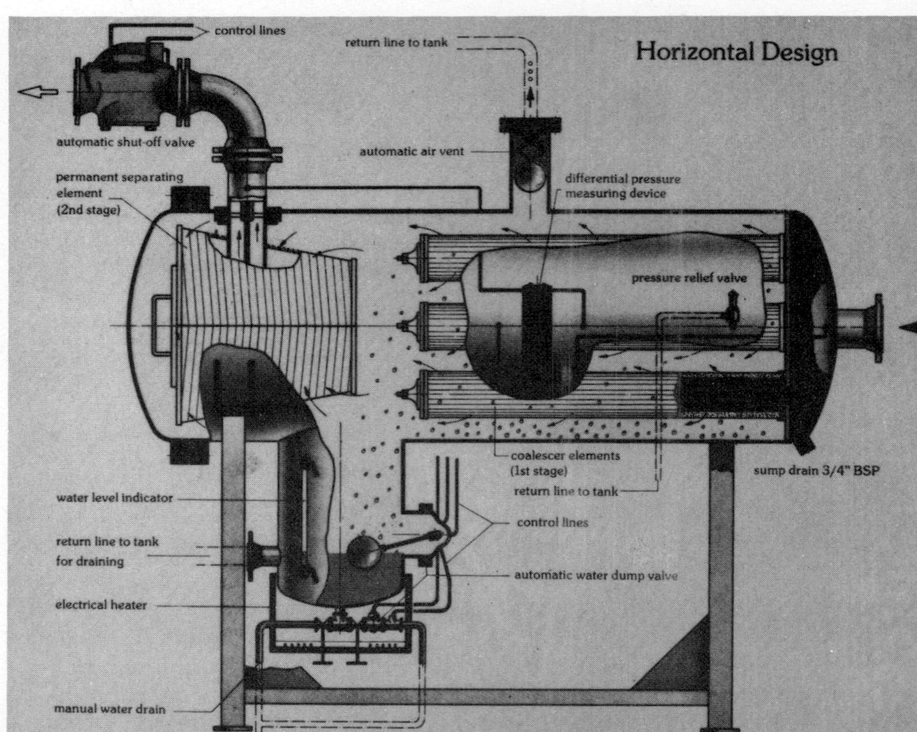

Operation of Faudi filter/water separator

Faudi Refuelling Equipment

In addition to the filter/water separators, for which the company is best known, Faudi produce a range of mobile and stationary units for aircraft refuelling.

The Faudi dispenser is mounted on a two-wheel trolley with steering wheel, parking brake and a hinged tow-bar for pulling by hand. A filter/water separator is fitted with an automatic air vent. Flow and pressure controls are provided with meter and counter, totaliser and flow-meter. Two hoses are supplied; a 10 metre refuelling hose mounted on a self-wind hose reel and an inlet hose 5 metres long with coupling for connection to a hydrant pit. Optional features include, trolley, self-powered or fitted with tow-bar, deadman control and a power-rewind hose reel.
Portable refuelling equipment can be supplied on a towable trailer, skids, trolley or in the case of the small units for private aircraft and helicopters, mounted on a tubular-steel frame for carrying by hand. The trolley has four wheels, towing bar and parking brake. It is fitted with a fuel pump powered by a diesel engine, pump filter and a filter/water separator. The controls include a differential pressure measuring device, counter and automatic shut-off valve which stops the fuel

supply if water is mixed with the fuel. The hose reel contains suction and pressure hoses with couplings.
The small units are capable of transferring fuel at 30 litres per minute using a hand pump. The equipment includes a filter/water separator with differential pressure gauge, counter, suction hose with stand pipe and a refuelling hose with automatic nozzle.

Mobile pump units designed for civil and military use are available for handling aviation fuel at storage points. The military versions incorporate a fuel pump driven by a diesel engine, duplex microfilter for removing solids from fuels that have become heavily contaminated under extreme conditions, and a filter/water separator with automatic air vent and differential pressure measuring device. Other features include a heated-water sump to prevent separated water from freezing, automatic water drain and automatic shut-off valve.

Manufacturer
Faudi Feinbau GmbH, Oberursel, Federal Republic of Germany

Salzkotten Fuelling Installation

Deutsche Gerätebau specialise in the construction of civil and military aviation fuelling installations for general aviation aircraft and helicopters. Standard installations which can be supplied include the Type I – Avgas; Type II, III and IIIa Jet; Type FWA 0-120 Turbo and FWA 300/2 Turbo aviation fuelling plants.
The Type I – Avgas is designed for 20 to 50 cubic metres double-walled tanks with water cup, steel-dome shaft and air venting. Filling the tank is achieved by gravity using a dip stick to indicate the fuel level. Features include a suction device for removing water and impurities at the lowest point, a leakage recording device and a filling safety device.
The dispensing equipment includes a vane pump with motor, a gas separator/microfilter, flow meter, registers, hose reel and a 17-metre hose with nozzle connection and earthing strip. Fuelling rates can be up to 120 litres per minute.
The Types II, III and IIIa Jet fuelling plants incorporate standard tank installations which range from 20 to 50 cubic metres similar to the Type I plant and differ only in the dispensing equipment supplied. The Type II rated at 120 litres per minute is designed for over-wing fuelling as is the Type III, rated at 350 litres per minute. The Type IIIa is equipped for over- and under-wing refuelling at 350 litres per minute.

Status
Salzkotten fuelling installations are in service at civil and military airfields in the Federal Republic of Germany.

Manufacturer
Deutsche Gerätebau GmbH, Salzkotten, Federal Republic of Germany

---

ISRAEL

Hatehof Refuelling Equipment

At the 1981 Paris Air Show Hatehof introduced a heavy-delivery aluminium aircraft refueller Model BR-20. The refueller, which has a 5300 US gallon (20 000 litre) capacity; can be mounted on a Mercedes L-2624/52, 6×4 chassis or any other suitable vehicle equipped with three heavy-duty axles. The rear cabinet housing the equipment and controls enables refuelling and defuelling of aircraft from both sides of the vehicle. The refueller's high capacity centrifugal self-priming pump can handle 2400 litres per minute. The pump is split shaft-mounted and its heavy-duty pressure control system enables efficient refuelling. The refueller is protected by an automatic fire extinguishing system for the rear cabinet and two manual dry powder extinguishers.
Hatehof also manufacture an aircraft refueller Model BR-13, with 13 000-litre capacity, designed for military use. This vehicle is capable of performing all fuelling operations both over- and under-wing. A further model is the BR-68, a small unit with 6800-litre capacity. The company's production list includes: 22 000-litre and 35 000-litre tank semi-trailer fuel transporters Models BT-22 and BT-35; a hydrant fuelling servicer Model BD-6-B and various other tankers and refuellers.

Manufacturer
Hatehof Tankers and Refuellers Builders Ltd, Tel Aviv, Israel

---

ITALY

Romanazzi ART/R 45 000 litres Aircraft Refueller

ART/R aircraft refueller is based on a Fiat six-wheel tractor 300T with a six-wheel semi-trailer designed to carry two 20-foot (6·1-metre) tanks. One tank has a capacity of 27 000 litres the other, which is equipped with the pumping station and an auxiliary engine, contains 18 000 litres. The stainless steel tanks, fitted with baffles to reduce fluid movement during transit, are connected to each other at the top and bottom by flexible hoses. The pumping station which is self-contained can be operated without the cab unit. It contains a centrifugal pump, producing 2500 litres per minute, powered by an auxiliary diesel engine equipped with compressor and alternator. Two 20 metre flexible hoses with couplings mounted on power-rewind hose reels are fitted each capable of delivering over 1000 litres per minute and trigger nozzles are supplied for over-wing fuelling. Controls for the fuelling operation are in the pumping station. They incorporate a dead-man system and a pneumatic safety device which permits transfer operations only when the engine-compartment doors are closed.

Characteristics
CHASSIS: Fiat tractor 300T with semi-trailer
CAPACITY: 45 000 litres
FUELLING RATE: 2 × 1140 litres/minute
HOSES: 2 × 20 m × 76·2 mm
DIMENSIONS
length: tractor chassis: 6·76 m; semi-trailer: 12·5 m
total length: 15·5 m
width: 2·5 m
height: 3·87 m
OPERATIONAL WEIGHT: 59 870 kg

Status
In service with the Italian Air Force

Manufacturer
Officine Romanazzi, Rome, Italy

Viberti Aircraft Refuellers

Officine Viberti produce a wide range of aircraft fuellers mounted on a single chassis, trailer or semi-trailer chassis. The single chassis versions vary in capacity from 12 000 to 20 000 litres. With the latter model, a draw-bar trailer is available to provide an additional 38 500 litres capacity. They are equipped with two 2½-inch (63·5-mm) hoses 20 metres long for refuelling at a rate of 2270 litres per minute and for defuelling operations.

The 18 925-litre capacity refueller has an under-wing, single point refuelling capability through a single 2-inch (50·8-mm) hose 25 metres long providing a rate of flow of 1135 litres per minute. Over-wing refuelling through a 2-inch (50·8-mm) hose is at 700 litres per minute. Defuelling can be carried out using both hoses. The trailer refuellers are independent units equipped with pumping stations. Capacity can vary from 9500 to 18 925 litres. All are equipped for under- and over-wing refuelling and defuelling operations. The semi-trailers coupled to Fiat tractors have capacities ranging from 29 000 to 65 000 litres.

The 45 000-litre capacity refueller is based on a Viberti three-axle semi-trailer drawn by a Fiat tractor Mod 190.26. It is equipped for under-wing refuelling through a 2-inch (50·8-mm) hose 25 metres long at a rate of 1135 litres per minute and over-wing refuelling through a single hose supplying fuel at 700 litres per minute.

The 47 000-litre capacity refueller uses a Fiat 160 PB-HB tractor and has an elevating platform mounted at the rear of the semi-trailer. It is equipped with four hoses; two 2½-inch (63·5-mm) hoses 30 metres long providing 2270 litres per minute mounted on hose reels and two similar bore hoses installed on the elevating platform. Defuelling can be carried out using both systems.

Manufacturer
Officine Viberti, Nichelino, Italy

Viberti Hydrant Dispenser

Two types of hydrant dispenser are manufactured by Officine Viberti.

The 600 US GPM hydrant dispenser is mounted on a Fiat chassis and is equipped with two 2½-inch (63·5-mm) hoses 30 metres long mounted on hose reels. It can deliver fuel at rates up to 2270 litres per minute.

The 1000 US GPM hydrant dispenser is based on a Fiat 80F 13 chassis. The vehicle incorporates an elevating platform which can extend to a maximum height of 3·1 metres above ground. Two refuelling systems are provided. One operates from the elevating platform which can be positioned directly under the wing of the aircraft. The second, a stand-off system uses 2½-inch (63·5-mm) hoses 30 metres long which are mounted on hose reels. The dispenser can refuel at rates of up to 3785 litres per minute. Pressure can be increased as necessary to ensure fuel is supplied at the required rate by means of an auxiliary pump driven by a pto from the drive engine.

Characteristics

	600 US GPM	1000 US GPM
CHASSIS	Fiat 80 F 13	Fiat 80 F 13
HOSES	2 × 30 m	2 short length 2 × 30 m
output	2270 litres/minute	3785 litres/minute
DIMENSIONS		
length	5·37 m	6·5 m
width	2·3 m	2·3 m
lowered platform	2·64 m	2·64 m
elevated platform		3·1 m
OPERATIONAL WEIGHT	8200 kg	8500 kg

Manufacturer

Officine Viberti, Nichelino, Italy

JAPAN

Fuji 800 GPM Hydrant Dispenser

A hydrant dispenser developed jointly by Fuji Heavy Industries and Japan Airport Fueling Service Co Ltd. It has the capability to refuel any type of aircraft at over 800 US gallons (3000 litres) per minute. The vehicle incorporates a lifting platform, operated hydraulically for connecting the fuel nozzle to the under-wing fuelling points of the aircraft.

Characteristics

CAPACITY: 800 US gallons (3028 litres)
DIMENSIONS: 7·23 × 2·6 × 2·09 m

Status

The vehicle entered service at Narita International Airport in November 1980

Manufacturer

Fuji Heavy Industries, Tokyo, Japan

Fuji 800 GPM hydrant dispenser

UNITED KINGDOM

Bowser Petroguard Fuel Monitor

The Petroguard fuel monitor is designed to remove solid contaminants from aviation fuel and provide automatic and positive shut-down of fuel feed if contamination by water exceeds a certain level. The equipment is used in static fuel storage, refuelling, defuelling and fuel transfer. It is available with flow rates from 56 to 4540 litres per minute. The monitor is a cylindrical vessel divided into an inlet and outlet chamber by a transverse deck plate. Check valves are set into the deck plate for mounting filter elements. Up to 40 filter elements may be installed depending on the monitor size and flow rate required. Fuel passes via the inlet chamber from outside to inside the element and from there through the check valve into the outlet chamber. Within the element is a nylon poppet valve spring-loaded against a valve seat but held open by a paper fuse. This fuse is insensitive to aviation fuel but is weakened by water. Occasional traces of water are washed out again by dry fuel but if the water content exceeds a preset level the fuse tears allowing the spring to drive the poppet valve forward, shutting off the fuel flow. The body of the element is made of phenolic resin-impregnated cellulose fibre with an outer wrap of glass fibre. The check valves are automatically opened by the insertion of an element. Failure to insert an element in one or more positions leaves the relevant check valve closed, preventing unfiltered fuel from being supplied.

Characteristics

FILTRATION SELECTIVITY: 1·5 μm
RESPONSE TIME
water contamination:
30 ppm: 20·5 minutes
100 ppm: 5·75 minutes

Status

Petroguard monitors are in use at fuel storage installations at civil and military airfields, heliports and offshore oil rigs. They are used also on aircraft refuelling tankers.

Manufacturer

Bowser Filtration Ltd, Mitcham, Surrey, England

Fluid Transfer Refuelling Equipment

Fluid Transfer design and manufacture aircraft refuelling equipment with transfer rates ranging from 250 to 4500 litres per minute. These include portable refuelling sets, 4500 litres per minute hydrant dispensers mounted on a truck or trailer chassis, and refuellers with capacities up to 2700 litres with draw-bar trailer providing a maximum capacity of 45 000 litres.

The portable refueller is designed for refuelling helicopters and light/medium aircraft by the over-wing or pressure refuelling method, achieving flow rates up to 680 litres per minute. The unit which incorporates a combined filter and water separator, is supplied with hoses fitted with electrical bonding, trigger nozzles or pressure couplings. It operates primarily from air-portable fuel containers or flexible bag tanks and drum stock.

The multi-purpose refueller is suitable for over-wing and pressure refuelling or defuelling military and civil aircraft, operating at pressures from 0 to 4 bars and rates of 0 to 1000 litres per minute. Supplied with combined filter and water separator, hoses, electrical bonding and couplings, the unit has a self-priming, run-dry capability. It can be operated from air-portable fuel containers, flexible storage tanks and underground tanks.

Hydrant dispensers available include the 4500 litre per minute multi-purpose dispenser which can be used for all types of pressure-refuelled aircraft through twin-deck hoses or twin-reel hoses. The deck hoses are installed on a hydraulic platform which has a minimum lowered height of 1·5 metres and a maximum raised height of 3·2 metres. The 30 metre long, 2½-inch (63·5-mm) bore hoses are mounted on twin power-rewind reels, positioned transversely across the rear of the vehicle.

The 3750 litres per minute hydrant dispenser is a simplified version of the multi-purpose model. The refuelling facility is limited to twin-deck hoses operated from a similar type of elevated platform. A single meter is provided to register total flow through the deck hoses.

The 2270 litres per minute hydrant dispenser offers a stand-off refuelling facility through two 30-metre long, 2½-inch (63·5-mm) bore hoses mounted on twin power-rewind hose reels. The flow rate is limited to 1135 litres per minute through each hose. A meter is fitted in each of the reel lines to provide separate records.

In all hydrant dispensers the filtration and pressure control are selected to suit individual operator's requirements and the choice of chassis is dependent on whether a boost pump is required.

Manufacturer
Fluid Transfer Ltd, Stroud, Gloucestershire, England

Fluid Transfer hydrant dispenser with rear-mounted elevating platform

---

Gloster Fuelling Equipment

Apart from the military air transportable refueller which is described separately, Gloster Saro manufacture refuellers ranging in capacity from 4550 to 82 000 litres, hydrant servicers and fueller trailers.

Somerset Mk 3A fueller, adaptable for mounting on a variety of chassis, has a capacity of 18 200 litres with a fuelling rate of 2270 litres per minute. Fuel can be delivered through pressure or open lines. The fueller operates at gross vehicle weight of 27 tons (27 400 kg). Coupled to a trailer, the capacity of the fueller can be increased to 63 560 litres.

Cotswold refueller is a development of the Somerset model. With a pumping rate of 3500 litres per minute the refueller meets the requirements for fuelling wide-body aircraft. The vehicle height has been reduced and an elevating platform installed, equipped with short-length delivery hoses. The refueller has a capacity of 18 200 litres, operating at a gross vehicle weight of 27 tons (27 400 kg).

Gloster hydrant dispenser is available in a range of sizes providing fuelling rates of up to 4550 litres per minute and mounted on a rigid, articulated or drawbar trailer chassis.

Manufacturer
Gloster Saro Ltd, Gloucester, England

Gloster Military Air Transportable Refueller

This vehicle has been developed to a United Kingdom Ministry of Defence specification and carries a NATO classification number. It is capable of fuelling and defuelling all types of aircraft and helicopters and has been designed to be air-lifted by C130 Hercules and C5A Galaxy aircraft. The vehicle is equipped to tow a drawbar trailer of 4500 Imperial gallons (20 500 litres) capacity.
Mounted on a Leyland Bison chassis, the refueller has an aluminium tank stressed to withstand the limited rough terrain capability of the vehicle. Two power-rewind hose-reels are between the cab and front of the tank, one fitted with 120 feet (36 metres) of 2½-inch (63·5-mm) bore hose, the other with 60 feet (18 metres) of 3-inch (76-mm) bore hose.
A filter/water separator, bulk-meters, control valves and two small auxiliary hose reels are in the same cabinet. The pump is a self-priming centrifugal type driven via a toothed-belt drive system from a gearbox-mounted high torque power take-off. Other main items of equipment include a strainer fitted between the tank and pump, suction hoses, an automatic bottom filling system, fuel sampling and water contamination detection units and fire extinguishers. Single- or twin-overhead fuelling booms can be fitted to the vehicle to increase open-line fuelling/defuelling performance. Each boom is fitted with 30 feet (9·1 metres) of 1½-inch (38-mm) bore hose. Fuelling rates up to 90 Imperial gallons (410 litres) are attainable.

Gloster Saro military air transportable refueller

Status
In service with many air forces, operating in the Middle East, Far East, Africa and northern Europe.

Manufacturer
Gloster Saro Ltd, Gloucester, England

Characteristics

CHASSIS: Leyland Bison 3 BN: L 12–24 L2 6 × 4
ENGINE: Leyland L 12 diesel
TRANSMISSION: 6 forward 1 reverse speed

PERFORMANCE	OPERATION	HOSE LENGTH/BORE	OUTPUT/INPUT	PRESSURE
FUELLING	Under-wing	1 × 60 ft × 3 inch (18 m × 76 mm)	2270 litres/minute	3·5 bar (50 psi)
		1 × 120 ft × 2½ inch (36 m × 63 mm)	1135 litres/minute	3·5 bar (50 psi)
		Both hoses simultaneously	2270 litres/minute	3·5 bar (50 psi)
	Over-wing	2 × 25 ft × 1½ inch (7·6 m × 38 mm)	340 litres/minute each	
		Both hoses simultaneously	680 litres/minute	
DEFUELLING	Under-wing	1 × 60 ft × 3 inch (18 m × 76 mm)	4556 litres/minute	
		1 × 120 ft × 2½ inch (36 m × 63 mm)	4556 litres/minute	
		Both hoses simultaneously	680 litres/minute	

DIMENSIONS	FUELLER	TRAILER
length	33 ft 3 in (10·14 m)	29 ft (8·84 m)
width	8 ft 2 in (2·49 m)	8 ft 2 in (2·49 m)
height	8 ft 9 in (2·67 m)	8 ft 2 in (2·59 m)
WEIGHT		
laden	23·50 tons (23 877 kg)	22·02 tons (22 373 kg)
unladen	11·86 tons (12 050 kg)	5·04 tons (5123 kg)

Midland Tankers 1500-Imperial gallon Avtur Helicopter Refueller

A 1500 Imperial gallon (6800-litre) capacity avtur refueller used for helicopter operations, based on a Bedford chassis with a six-cylinder diesel engine, five-speed gear box and single-speed rear axle.
Two Dean electric-rewind hose-reels are fitted, the front one holding 30·48 metres of hose with Emco nozzle, the rear hose reel connected to 18·29 metres of hose and a Carter pressure fitting nozzle. A Gorman Rupp centrifugal pump is installed driven by a gearbox pto with air controls and connected to a Fram filter/water separator with air eliminator. An Avery Hardoll meter is supplied with Veeder Root litre register and rate of flow indicator.

Status

In service at Westland Helicopters Ltd, Yeovil, Somerset, England

Manufacturer

Midland Tankers, Oxford, Oxfordshire, England

Midland Tankers 9000-litre Avgas Aircraft Refueller

A 9000-litre capacity avgas aircraft refueller based on a Ford chassis with a six-cylinder diesel engine, four-speed all synchromesh gearbox with a single-speed rear axle.
A Dean electric-rewind hose-reel is fitted, with 18·29 metres of hose and Emco nozzle. A Drum positive replacement pump is installed, with integral relief valve driven by a gearbox pto with air controls, connected to a Bowser Petroguard filter/water monitor with air eliminator. A Neptune meter with litre register is fitted.

Status

In service with Loganair at Abbotsinch Airport, Glasgow, Scotland

Manufacturer

Midland Tankers, Oxford, Oxfordshire, England

Midland Tankers 13 000-litre Avgas/Jet Al Aircraft Refueller

A 13 000-litre capacity avgas/jet Al aircraft refueller based on a Ford chassis with a six-cylinder, six-litre turbo-charged diesel engine, six-speed all synchromesh gearbox and single-speed rear axles. Alternative chassis are available.
Roberts hose-reels, each with 18·29 metres of hose and Emco nozzles are fitted. A Drum positive displacement pump is installed, with integral relief, driven by a gearbox pto with air controls and connected to a Bowser Petroguard filter/water monitor with air eliminator. Dresser Wayne meters with Veeder Root litre registers are used for fuelling operations.

Status

In service at Norwich Airport, intended for fuelling piston-engined aircraft. However one nozzle was changed for pressure refuelling and the vehicle is used for Jet Al fuelling of turbine-engined aircraft.

Manufacturer

Midland Tankers, Oxford, Oxfordshire, England

Midland Tankers 18 000-litre Jet A-1 Aircraft Refueller

A 18 000-litre capacity aircraft refueller for Jet A-1 fuel, which is available on a Ford chassis with 6 × 4 drive, Perkins V8 diesel engine, six-speed all synchromesh gearbox and single-speed double-drive rear axles. Alternative chassis are available.
Dean hose-reels with electric rewind are fitted, each with 30·48 metres of hose and an Avery Hardoll pressure-fuelling nozzle. A Gorman Rupp centrifugal pump is mounted in and driven by the main vehicle drive line. A Velcon filter and water separator is installed.
The Dresser Ljungman meter with Veeder Root litre register and rate of flow indicator is used for fuelling operations with a Dresser Wayne meter and Veeder Root litre register for defuelling.

Midland Tankers' 18 000-litre Jet A-1 aircraft refueller

Status

In service at Sumburgh Airport, Shetland Islands, Scotland

Manufacturer

Midland Tankers, Oxford, Oxfordshire, England

Midland Tankers Jet A-1 Fuel Dispenser

The Jet A-1 dispenser is mounted on a four-wheel trailer. A Blackmer pump, driven by a Lister ST2 diesel engine, is connected to a Cobham filter/water separator with air release and a Dresser Wayne meter with litre register. Two Dean hose-reels are supplied, one electric rewind holding 15·24 metres of hose the other a manual rewind with 3·05 metres of suction hose each terminating in an Avery Hardoll nozzle.

Status
Used for refuelling and defuelling operations for military aircraft at a British Aerospace airfield in the United Kingdom.

Manufacturer
Midland Tankers, Oxford, Oxfordshire, England

Taymech Fuelling Equipment

Taymech produce refuellers and hydrant dispensers for aircraft refuelling operations. Multi-purpose aircraft refuellers include the 35R/78 version which has 20 000 litre capacity tank and the tactical refueller Mk 5 for military operations with a capacity of 4500 litres. Rigid chassis or articulated units are available for use at civil airports, or with cross-country capability, for military operations.

A range of hydrant dispensers are produced with flow rates varying from 450 to 4500 litres per hour.

Manufacturer
Taymech Aviation Equipment Ltd, Southall, Middlesex, England

Thompson Refuelling Equipment

Thompson produce a wide selection of refuelling equipment suitable for aircraft ranging from wide-body to small private aircraft. The largest vehicle is an articulated high-performance refueller with a 8500 to 9000-Imperial gallon (38 640 to 40 900-litre) capacity tank of double-taper configurations. It has a total fuelling rate of 600 to 700 Imperial gallons (2700 to 3180 litres) per minute through two reeled-hoses and either a rear-mounted fuelling boom or two rear deck-hoses. Pumping, metering and dispensing equipment are all located at the rear of the semi-trailer. The fuelling semi-trailer is mounted on a single- or tandem-axle bogie and is suitable for operation with two-axle motive units of approximately a 10-foot (3·05-metre) wheelbase. It is designed for operation at a gross vehicle weight of 27 tons (27 400 kg).

The Somerset refueller has a 4000 to 4500-Imperial gallon (18 180 to 20 450-litre) capacity tank, with fuelling rate of 500 to 600 Imperial gallons (2270 to 2700 litres) per minute through two-reeled hoses for under-wing refuelling. A third hose-reel is fitted to provide 125 Imperial gallons (570 litres) per minute over-wing refuelling through a 2-inch (50·8-mm) bore hose and nozzle. The Somerset has a gross vehicle weight of 27 tons (27 400 kg).

The Suffolk medium-purpose refueller is fitted with a 1500-Imperial gallon (6820-litre) capacity tank, with a fuelling rate of 250 Imperial gallons (1130 litres) per minute through two reeled hoses. The refueller is mounted on a two-axle rigid chassis with cab, having a gross vehicle weight of 16 tons (16 260 kg).

The Mini refueller has a 700-Imperial gallon (3180-litre) capacity tank with a fuelling rate of approximately 50 to 75 Imperial gallons (220 to 340 litres) per minute under- or over-wing. The vehicle is mounted on a two-axle rigid chassis with cab having a gross vehicle weight of 6 tons (6100 kg).

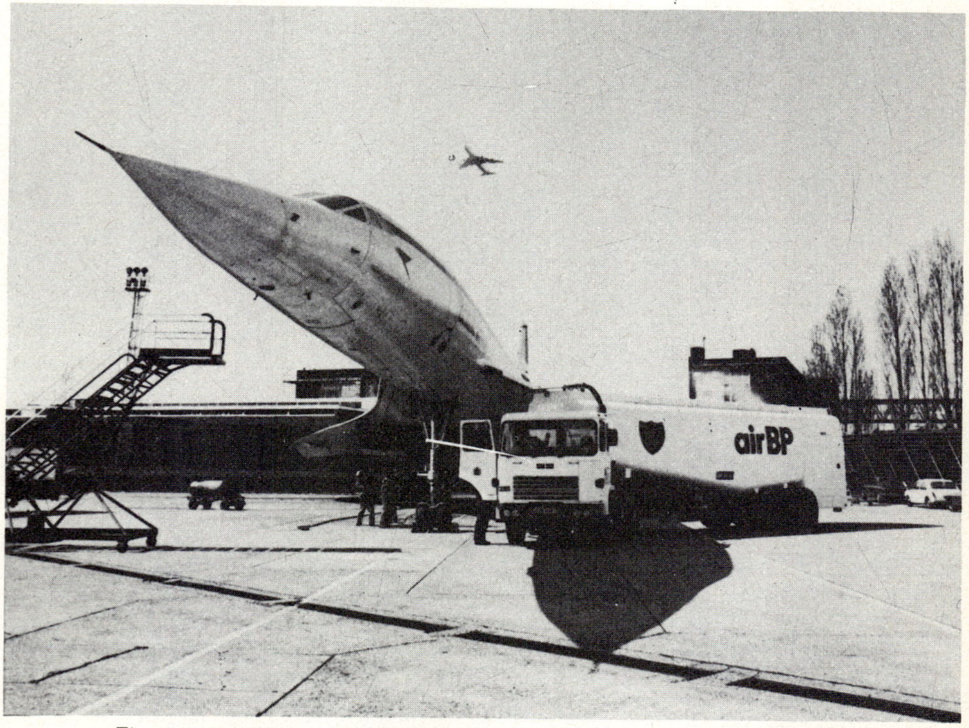

Thompson articulated 9000 Imperial gallon fueller supplying Concorde

A range of mobile hydrant dispensers are available each constructed on a suitable two-axle rigid chassis with cab and having a refuelling capability of between 500 to 1000 Imperial gallons (2270 to 4500 litres) per minute through either two or four hose systems. Normally, twin hose-reels are fitted with either a fixed under-wing access platform or an elevating platform to facilitate coupling of hoses to the aircraft fuelling points. Dispensers can be constructed with or without a boost pump dependent on the hydrant ring main system pressure and on the output required. All the fuelling units can be fitted with deadman control for fuelling operations, hose stowage/brake interlock equipment and a secondary in-line pressure control system.

Status
Thompson tankers have been supplied to all major United Kingdom oil companies.

Manufacturer
NEI Clarke Chapman Ltd, Bilston, West Midlands, England

Western Trailers Refuellers

Western Trailers manufacture refuellers based on a Bedford 4 × 4 or 4 × 2 chassis powered by Bedford diesel engine. Pumps, filtration equipment, registers and hoses are supplied according to the individual operator's requirements.
A wide range of refuelling trailers are available. A typical model, with 1800-litre capacity, is designed for refuelling light aircraft or helicopters. It comprises an epoxy-lined tank, flame-proof lid and a 12/24 volts electric pump with 15·2 metres of hose mounted on a spring-retractable hose-reel. The pump is equipped with an automatic switch to enable operation from either a 12/24 volts supply.

Manufacturer
Western Trailers, Bristol, Avon, England

Parkway refuelling trailer manufactured by Western Trailers

UNITED STATES OF AMERICA

Garsite 10 000-US gallon Aircraft Refueller

Garsite standard aircraft refueller has single-tank capacity of 10 000 US gallons (37 850 litres) and is designed to fuel aircraft through either two reel-hoses mounted on the side of the vehicle or two hoses installed on a lift platform for access to under-wing fuelling points. Simultaneous use of the lift platform hoses permits aircraft refuelling at rates of up to 1000 US gallons (3785 litres) per minute. Refuelling rates of up to 600 US gallons (2270 litres) per minute can be achieved through the two reel-hoses. Pressure control is achieved by an in-line pressure control valve combined with the air deadman system, using a venturi downstream of the valve for sensing the fuelling pressure.

Aircraft defuelling is done by turning the fuel/defuel valve to the defuel position. In this mode fuel from the aircraft flows through the meter, in the reverse direction, through the fuel/defuel valve and then forward through the pump and filter/separator to the tank. The tank is capable of both top- and bottom-loading at flow rates of up to 700 US gallons (2650 litres) per minute.

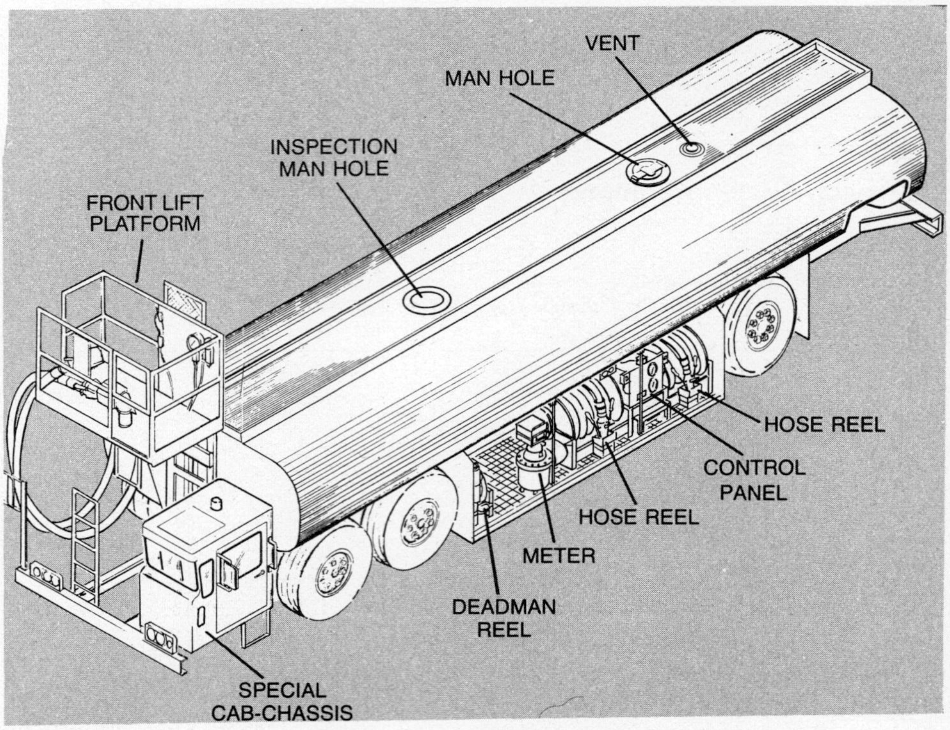

Garsite 10 000-US gallon refueller

Characteristics

CHASSIS: specially designed, with single operator cab
ENGINE: diesel
TRANSMISSION: automatic
HOSE REELS: 2 × 2½ in (63·5 mm) × 75 ft (22·9 m) hose
LIFT PLATFORM HOSES: 2 × 3 in (76·2 mm) × 120 ft (36·6 m) hose
DIMENSIONS: 14·02 × 3·05 × 2·54 m

Manufacturer

Garsite Products Inc, Deer Park, New York, USA

Keene Fuel Filter/Separator

The Model 842F single-stage and Model 852A three-stage filter separator are all-steel units designed for the removal of water and solids from aviation fuels. The three-stage model uses coalescer and separator cartridges. Both models are supplied with a pressure gauge and a liquid-level gauge. A three-way valve is provided for reading inlet and outlet pressure and a pressure relief valve is inserted in the cover. Other features include a manual water drain, complete set of elements, manual air vent and flanged inlet and outlet connections.

The models are designed for full automatic control and function at an operating pressure of 10 bars (150 psi).

The Model 851A is a single-stage filter/separator designed for flows up to 475 litres per minute. A standard unit is supplied with liquid-level gauge, pressure gauges, manual air vent, manual water drain, canisters and a complete set of elements.

Status

These filter/water separators are used in fuel storage installations at civil and military airports. They are manufactured in the USA by the Filtration Division of Keene Corporation and in the United Kingdom by a subsidiary, Bowser Filtration Ltd.

Manufacturer

Keene Corporation, Greeneville, Tennessee, USA

MAINTENANCE BASE EQUIPMENT

Hangars, shelters and engine test cells
Access, lifting and handling
Power supplies, testing and servicing

Hangars, shelters and engine test cells

AUSTRALIA

NEI Operational Aircraft Shelter

Operational aircraft shelters are designed to provide cover for flight-line maintenance of military aircraft, the cantilever design allowing the aircraft to be taxied into the bay. The shelters feature a heavy-duty, angled purlin system and V-shaped columns to provide rigidity, which eliminate the need for mid-span rafters or wind bracing below truss level. The shelter is constructed of steel tubes, sections and sheeting, protected by galvanised or painted zinc. The trusses are fabricated from tubular or rectangular hollow sections, the purlins and girts from steel Z-sections and the roof and end walls from Kliplock profile sheeting. The shelters are modular and available with any even number of bays (ie 2, 4, 6, 8, etc). They can be designed to withstand local conditions of wind, with speeds from 40 metres per second to cyclonic, snow and earthquakes. The equipment is transportable with no site fabrication required.

Manufacturer

New England Industries Pty Ltd, Tamworth, New South Wales, Australia

Characteristics

Six-bay shelters constructed for the RAAF have the following dimensions:
length: 207 m
width: 32 m
height: 8 m

Status

The concept of the shelters arose from the need to provide protection for the Royal Australian Air Force F-111 aircraft at Amberley, Queensland, Australia and two have been constructed there to house six aircraft in each shelter.

CANADA

Richards-Wilcox Aircraft Hangar Doors

The company supplies aircraft hangar doors of various configurations, custom-designed to suit specific requirements.

Three Richards-Wilcox door systems are installed at the new de Havilland Aircraft of Canada production facility in Downsview, Ontario.

The main door is of an eight leaf bi-parting configuration 76 metres wide and 11 metres high and provides two openings of 38 × 11 metres. Two other doors in the same building are six leaf bi-parting with opening dimensions 43 metres wide and 11 metres high. Air Canada has two identical maintenance hangars for wide-body aircraft, one at Dorval International Airport in Montreal and the other at Toronto International Airport. Both four-bay hangars have an opening 175 metres wide and 25 metres high at each end. The opening at each end is closed by two independent electrically-operated leaves that slide one in front of the other. The Wardair hangar at Toronto International Airport features a 80-metre wide and 21-metre high Richards-Wilcox door system at each end of the building. Each is composed of eight electrically-operated door leaves that bi-part in an echelon formation.

For corporate aircraft hangars the company has produced an electrically-operated hangar door. In one example the five-leaves are in formation moving in echelon in one direction to close the 39 metres wide and 12 metres high opening.

Other examples of hangar doors designed and manufactured by Richards-Wilcox are the two CP Air Maintenance hangars in Toronto which are equipped with four-leaf sliding doors, and one in Vancouver fitted with three sliding leaves.

CP air hangar at Toronto International Airport equipped with a four-leaf sliding Richards-Wilcox door system

Manufacturer

Richards-Wilcox of Canada Ltd, London, Ontario, Canada

FRANCE

Bertin Acoustic Systems for Engine Test Cells

The design of these systems developed for engine test cells and aircraft ground tests results from a combination of treatments. The first, the noise-barrier stage, consists of restrictions which may be cooled by air induction. These prevent the generation of low-frequency noise by dissipation of the engine's kinetic energy. This treatment however alters the thrust of the engine and therefore cannot be fitted to a thrust measurement facility.

The second treatment is by induction of cold air. This device is intended to reduce the temperature of the exhaust gases in the silencer and the wall temperature in the entrance zone.

The noise-attenuation zone is acoustically treated by channelling the exhaust gases through an area bounded by sound absorbent panels. These panels

Bertin acoustic system for testing Larzac engine installed at French military base

comprise rockwool blankets combined with glass fibre protected from the exhaust gas by perforated plates and stainless steel mesh. Helmholtz resonators connected in parallel are installed in the exhaust treatment. By shaping the resonator throat the sound attenuation can be made independent of the speed and the attenuation range increased by adjusting the coupling of the resonators.

Characteristics

ACOUSTIC SYSTEM: air induction and exhaust treatment
ENGINE: Snecma/Turbomèca Larzac 04–C5
THRUST SETTING: max
NOISE MEASUREMENT: 80–85 dB
DISTANCE FROM EXHAUST: 15 m

Status

Bertin has installed over 50 acoustic systems for civil and military operators in Europe. The example given above, for the Larzac engine which powers the Alpha Jet, was supplied to the French Air Force in 1981.

Manufacturer

Bertin and Cie, Plaisir, France

Boet Engine Test Cells

These are available for testing jet, turbo-prop and turbo-shaft engines either detached or installed in the aircraft.

The test facility for engines up to 9500 lb st thrust (42·26 kN) designed on the principle of exhaust gas cooling by air induction comprises an air-intake unit, a test cell and an exhaust unit. Sound-absorbent splitters and a filter screen, to prevent the introduction of foreign objects, are installed in the air-intake unit. In order to reduce reverberation and protect the engine against acoustical strain, not caused by the direct sound, the inner concrete walls are partially covered with acoustical lining. In one side wall a folding sound-proof door allows engine access to the test cell area and an inflatable seal reduces the noise levels in the adjacent preparation room. Exhaust gases and the induced air flow move through a bell-mouth to the exhaust unit.

The length of the augmenter tube component of the exhaust unit has been designed to ensure a mixing of the exhaust gas and the induced air. At one end of the augmenter a perforated blast tube allows the dilution of the gas mixture in the exhaust stack. This stack situated above the perforated tube blows the gases high enough to prevent recycling. The unit is sound-proofed through vertical duct lining and absorbing splitters for exhaust noise attenuation.

For jet engines with very high thrusts up to 100 000 lb st (444·8 kN) a U-shaped cell unit, constructed in reinforced concrete, has been developed. Two versions are available. The equipment in the first configuration is similar to the test facility described above but, due to the shape of the air-intake unit, turning vanes are installed in the lower section of the air-intake bend. This cell has been designed to permit exhaust gas temperatures of 250°C with short maximum peaks of 350°C. The second version features turning vanes in the exhaust section only. To ensure a smooth flow of air a deflector is situated in the upper part of the intake bend. The exhaust unit comprises a bell-mouth installed behind the test engine coupled to a mixing basket, a diffuser and acoustic splitters. This version is designed to test Pratt and Whitney JT9D, General Electric CF6 and Rolls Royce RB211 engines.

The test cell for turbo-prop and turbo-shaft engines is L-shaped with a vertical exhaust duct. Splitters and a filter screen are installed in the air-intake section. The engine gases and the propeller-generated air flow are evacuated by means of an exhaust duct. The gases are ejected into the atmosphere via a bend provided with turning vanes and an acoustically-treated stack which consists of a wall lining and a series of absorbing splitters. For re-heated jet engine testing the cell is designed with a secondary air-inlet stack with sound-absorbing splitters above the diffuser to introduce supplementary cold air for increased exhaust temperature cooling. Engines to be tested enter the cell via a mobile air intake duct.

In the combined facility where the engine can be tested and installed independent of the aircraft, the test cell is extended to accommodate the aircraft. The cell is equipped with a front-double sliding door with inflatable seals to provide acoustic security. Sound-proofing ducts are installed to supply the inlet air. A secondary air intake is fitted to reduce the exhaust temperature.

Boet combined F-16 aircraft and F-100 engine test cell supplied to SABCA

Characteristics

ENGINE CATEGORY THRUST	9500 lb st (42·26 kN)	100 000 lb st (444·8 kN)	Turbo-prop/ turbo-shaft	Reheat/ 25 000 lb st (111·2 kN)	Combined 25 000 lb st (111·2 kN)
ENGINE AIRFLOW		1815 kg/s		120 kg/s	120 kg/s
EXHAUST AIR TEMPERATURE	200°C			1800°C	1800°C
ENGINE TYPES		Pratt and Whitney JT3 JT8, JT9; General Electric, CF6; Rolls Royce/ Snecma Olympus	Lycoming 10720 Pratt and Whitney 4ACL, PT6, A27, A28, A45		
NOISE LEVEL AT MAX THRUST (dBA)/DISTANCE IN FRONT OF INTAKE	62/50 m	60/100 m	65/50 m	72/200 m	72/200 m
NOISE LEVEL AT MAX THRUST (dBA)/DISTANCE BEHIND EXHAUST	55/100 m	53/100 m	63/100 m	84/20 m	74/200 m

Status

Boet has installed test cells for Swissair, Air France, HAECO, Sabena, SECA, SNECMA and the Belgian Air Force. Four F-16 aircraft combined test cells have been supplied to the Belgian Army and one to the SABCA.

Manufacturer

A Boet, Villeneuve D'Ascq, France

Boet Military Aircraft Ground-run Exhaust Silencers

This installation comprises an adaptor fitted to a telescopic section, secondary air intake, diffuser, absorbent treatment section with 90 degree bend and a stack.

The light alloy adaptor connects the aircraft to the silencer. The telescopic section of this coupling allows for movement of the aircraft associated with power changes. Secondary air is introduced to the exhaust gases for cooling and mixed in the diffuser. Absorption treatment is used for the attenuation of the middle and high frequencies and gases are then deflected through 90 degrees by a series of turning vanes. A stack completes the sound-proofing and ejects the gases high enough to prevent recirculation. This stack may be supplied with a cap to avoid the corrosion of the internal parts exposed to weather conditions.

Characteristics

AIRCRAFT TYPES: Mirage III, IV, 2000 Super Etendard, F-16, F-5
ENGINE THRUST: 25 000 lb st (111·2 kN)
ENGINE AIR FLOW: 120 kg/s
EXHAUST AIR TEMPERATURE: 1800°C
NOISE LEVEL WITH ONE ENGINE AT MAX POWER 88 dBA
MEASURING POINT: 100 m from silencer

Status

In service with military forces in Belgium, France and Spain

Manufacturer

A Boet, Villeneuve D'Ascq, France

Boet Narrow-body Aircraft Ground-run Exhaust Silencers

The Boeing 727 and Caravelle aircraft installation consists of three exhaust silencers at the outlet of each aircraft engine. These are connected to the engine by an adaptor which may be fixed to the outside of the engine nozzle and move with the aircraft or partially surround the nozzle to allow free displacement of the aircraft. The adaptors are mounted on telescopic sections to simplify connection and removal of the installation.
The body of the silencer is formed by a mixer-diffuser which accelerates the mixing of the two engine flows improving the velocity profile; a resonating unit provided to increase the efficiency of the silencer in the low frequencies and an absorption treatment to reduce the exhaust jet noise level, mainly in the middle and high frequencies. The gases are ejected into the atmosphere through a bend and a stack. The installation for Boeing 707, DC-8 and KC-135 aircraft consists of four exhaust silencers. Adaptors and telescopic sections are fitted. A secondary air inlet follows the adaptor section which consists of a sound-proofed annular duct, provided to enable cool air to mix with the exhaust gases to reduce the temperature before passing through the sound-proofing units. The body of the silencer is similar to the Boeing 727 and Caravelle installation.

Status
In service with Air France

Manufacturer
A Boet, Villeneuve, D'Ascq, France

Air France Concorde using a Boet ground-run exhaust silencer

Characteristics

AIRCRAFT TYPE	Boeing 727/Caravelle	DC-8, B-707, KC-135
ENGINE THRUST	15 000 lb st (66·7 kN)	18 300 lb st (81·4 kN)
CORE AIR FLOW	71 kg/s	120 kg/s
FAN AIR FLOW	73 kg/s	86 kg/s
EXHAUST CORE AIR TEMPERATURE	550°C	510°C
EXHAUST FAN AIR TEMPERATURE	130°C	135°C
NOISE LEVEL FROM ALL ENGINES AT TAKE-OFF POWER	83 dBA	82 dBA
MEASURING POINT	100 m/45°	100 m/45°
UNIT DIMENSIONS		
length	18 m	24 m
diameter	2·5 m	2·8 m
weight	22 tonnes	30 tonnes

Boet Wide-body Aircraft Ground-run Exhaust Silencers

The Boeing 747 ground-run installation consists of four fixed exhaust silencers aligned with the engine centre-lines. The engine exhaust gases are absorbed by each silencer through a bell-mouth, cooled by induced air and silenced according to the principles of absorption and jet deflection through 90 degrees. A series of turning vanes are installed at the bend and the silencer extended by a stack fitted with the same acoustic characteristics as the main cylinder. Acoustic screens can be added to limit the noise radiated by the engine intake and from the gap between the engine and the bell-mouth. The DC-10 and Airbus installation consists of two fixed exhaust silencers aligned with the engine centre-lines. For DC-10 operations the standard installation allows only for testing numbers one and three engines. For the centre engine a third silencer is introduced. This is retractable to enable the equipment to be used for the Airbus aircraft. The engine exhaust gases are absorbed by each silencer through a bell-mouth designed to prevent recirculation. Induced air cools the exhaust gases which are silenced by absorption and are then deflected through 90 degrees by turning vanes. Further absorption treatment occurs before the gases are vented to the atmosphere.
Acoustic screens can be added to limit the noise radiated by the engine intake and from the gap between the engine and the silencer unit.

Status
In service with Air France

Characteristics
ENGINE THRUST: 56 000 lb st (249·1 kN)
CORE AIR FLOW: 120 kg/s
FAN AIR FLOW: 570 kg/s
EXHAUST CORE AIR TEMPERATURE: 850°C
EXHAUST FAN AIR TEMPERATURE: 85°C
NOISE LEVEL FROM TWO ENGINES AT TAKE-OFF POWER: 88 dBA
MEASURING POINT: 100 m and 45° from extended centre-line
UNIT DIMENSIONS
length: 25 m
diameter: 5 m
WEIGHT: 70 tonnes

Manufacturer
A Boet, Villeneuve D'Ascq, France

Pelletier Tent Hangars

Pelletier Industrie have developed a range of air-transportable tent hangars. Constructed of weather-proof fabric on a tubular framework, the hangars are large enough to enable limited maintenance to be undertaken. The side walls incorporate windows for natural light which can be covered at night under operational conditions.
The front is designed with door sections on rails for access to the aircraft. The Type ED 500 is capable of housing two medium-sized helicopters or two Mirage jet aircraft side-by-side, the latter in a nose-to-tail configuration. Other models available include the TD2 tent hangar.

Status
These tent hangars have been supplied to the French Air Force

Manufacturer
Pelletier Industrie, Fresnes, France

Pelletier Industrie ED 500 hangar

GERMANY, FEDERAL REPUBLIC

Mero Aircraft Hangars

Aircraft hangars constructed by Mero are designed to the requirements of individual operators based on three dimensional space-frame supporting structures. The shape of the hangar is adapted to the aircraft contours, reducing the volume of the building and thereby lowering cost for construction and operation. Workshops and maintenance docks are integrated into the design as required.
A hangar constructed on this principle to house a Boeing 747, for example, would measure 72 metres wide and 81 metres deep. The aircraft, parked with its nose facing the front of the hangar, would have maintenance areas behind each wing and with an additional maintenance storage area measuring 72 metres wide and 38·4 metres deep, to the rear of the aircraft. The front elevation of the hangar is tapered on each side with the walls parallel to the contours of the aircraft nose and wings. The hangar roof is extended in the centre to house the tail section of the aircraft.
A multi-purpose hangar of similar design without the maintenance areas on either side of the aircraft would house two Boeing 727–200 and a Boeing 707 aircraft.

Status
Mero have supplied aircraft hangars to Lufthansa/German Airlines. One hangar measuring 83 × 63 metres was built in 1971 at Rhein-Main Airport, Frankfurt

Manufacturer
Mero-Raumstruktur GmbH, Wurzburg, Federal Republic of Germany

JAPAN

Shin Meiwa Aircraft Hangar Doors

Shin Meiwa design and manufacture doors for large, medium and small aircraft hangars. A typical example is a 100 metre wide door produced for a Boeing 747 hangar, built in six leaves, three on either side. Each leaf or panel 25 metres high incorporates windows in the upper half for natural light and personnel doors in the centre panel on each side. The door operates hydraulically through a push-button control system.

Status
Shin Meiwa has supplied aircraft hangar-doors to All Nippon Airways and to other civil and military operators in Japan

Manufacturer
Shin Meiwa Industry Co Ltd, Tokyo, Japan

NETHERLANDS

Hollandia Kloos Wide-body Aircraft Hangars With Universal Docking Systems

Hollandia Kloos International has developed a universal docking system which can be applied to a wide range of aircraft for all types of maintenance. The total system comprises tail, fuselage, wing, nose and stabilo docks. The universal tail-docks are a tower construction enabling maintenance to be carried out simultaneously on the entire tail portion of the aircraft. For the purpose of changing the tail engine on a McDonnell Douglas DC-10 or Lockheed Tristar a special platform, part of the construction, will lift the engine to the required height for installation. When using the complete docking system maintenance of the whole aircraft can be done at the same time including painting or fuselage repairs, engine maintenance and tail inspection and any work to be carried out on the nose section.

Status
Hangars are custom-built and can incorporate the universal docking systems described above. Hollandia Kloos International has developed, fabricated and constructed hangars in many parts of the world

Manufacturer
Hollandia Kloos International, Krimpen a/d Ijssel, Netherlands

SWEDEN

FFV-6068/4666 Engine Test Systems

FFV have developed two engine test systems; the FFV-6068 for medium-sized and large engine test installations, and the FFV 4666 designed for testing small- and medium-sized engines.

The FFV 6068 comprises up to ten sensors to measure engine variables, a micro-computer system with keyboard and a television monitor connected to a printer for hard copy of statistical data, a plotter to produce a graphical presentation of two or more variables, external memory and computer. The micro-processor can govern and control the engine being tested, operating valves, pumps, doors, fire-fighting equipment and throttle actuators, etc. Alternatively the test cycle can be controlled manually. Test results are presented on television monitors and stored in the computer for analysis.

The FFV 4666 consists of nine sensors with digital instruments for each sensor connected to a central computer with printer and x-y recorder. Data is stored on magnetic tape and a hard copy can be produced for statistical analysis. The data is available, also, in graphical form using the x-y plotter. The digital instruments have a hold facility to produce simultaneous instrument readings.
In both systems the sensors have been designed to be connected with the minimum of delay, to facilitate engine changes when several types are to be tested in the same cell.

Status
FFV have installed a FFV-6068 engine test system at the Swedish Army engine test facility at Skövde for automatic testing. The Swedish Air Force performs approximately 2400 engine tests a year at the FFV engine test facility. In addition the company carries out approximately 700 engine tests annually

Characteristics

TYPE	FFV 6068/4666
SENSOR	MEASUREMENT
engine speed:	rpm
flow Q:	m^3/s
temperature:	°K
vibration:	mm/s
differential pressure:	Pa
pressure:	Pa
angle:	rad
position:	μm
thrust:	N
time:	s (only in 6068)

Manufacturer
FFV Maintenance Division, Arboga, Sweden

Pentaport Aircraft Hangars

This is a modular construction system for aircraft hangars and freight terminals. With Pentaport, aircraft are parked nose-in in a circle. For each aircraft a segment extending 67 degrees is necessary to cover the aircraft with a further 2·5 degrees either side of the aircraft to accommodate the pylons of the suspended roof making a total of 72 degrees per aircraft. Thus up to five aircraft hangars can be constructed in each circle.
To provide the most efficient working area the centre-line of each aircraft is used as the centre-line of each triangle. With the aircraft in the nose-in position, an imaginary line may be drawn between reference points in each aircraft's centre-line sufficiently aft of the aircraft to provide clearance for the tips of the horizontal stabiliser. These lines determine the shape of the hangar. The reference point is where the hangar doors meet and is the highest point of the hangar. The lowest point is at the aircraft's wing tips. This is the pylon area and the location of workshops and offices. The

roof slopes down to this line and the pylon-supported cantilever beams are joined together over the aircraft's centre-line. This results in a large open-span area over the aircraft.
It is not necessary to build all five hangars in the circle and only one or two sectors may be constructed initially. In the areas separating the hangars there are workshops and offices on two floors.
To take advantage of the three-way sloping roof special tail docks have been designed. These are suspended from the roof and can be moved in three directions. Personnel access to this maintenance area is provided by a telescopic lift incorporated in the dock.

Status
The Pentaport concept was first shown at the ICAA airport exhibition in Nice in 1979 as a result of preliminary designs for two aircraft hangars for the Norwegian company Fred Olsen Airtransport Limited

Manufacturer
Pentaport Sweden AB, Gothenburg, Sweden

Pentaport aircraft hangar

SWITZERLAND

Sulzer Engine Test Cells

The standard Sulzer engine test cell is a U-shaped reinforced concrete structure comprising the intake, intake sound treatment, engine room, augmenter diffuser, exhaust sound treatment and exhaust sections. A vertical intake stack is used which gives high acoustic efficiency and reduces dirt ingestion. Model tests have shown the need for only one turning vane for the low intake velocity. In locations with high cross winds flow straighteners may be added at the entry section. Intake sound treatment is provided by vertical acoustic panels of perforated mild steel with a mineral or glass fibre infill.
The engine room is the working area of the test cell. The walls are 300 to 500 mm thick depending on the acoustic requirements and incorporate a large, motorised engine door with inflatable seals for maximum acoustic efficiency and an observation window to provide a view of the cell from the control room. The engine is suspended from a ceiling-mounted universal thrust stand. The location of the engine, its centre-line height, the distance between the rear of the intake splitters and the engine bell-mouth and the distance from the exhaust to the augmenter diffuser are all critical in ensuring a uniform flow at the engine intake. Aft of the engine intake, the cell is tapered to accelerate the secondary airflow and prevent recirculation.
Incorporated into the floor of the engine room and under the thrust stand is a work platform operated by scissor-type hydraulic rams.
The augmented diffuser is a mild-steel tube finished with a temperature-resistant coating designed to ensure there is no back pressure on the engine. It incorporates a mixing basket to unite the primary (hot) and secondary (cold) air streams. The mixing basket eliminates the need for variable diffuser-nozzle positions and adds to the acoustic efficiency by changing the low frequency noise into the high frequency range to facilitate treatment. Efficient diffusion of the exhaust gas results in a low average temperature and uniform flow through the vertical acoustic panels similar to those in the intake but with the addition of a temperature-resistant coating.
The exhaust section is at the base of the vertical exhaust stack. It consists of a bank of mild-steel deflector vanes designed to give a uniform exit velocity profile for noise emission control. To install an engine into the thrust stand the pre-rigged engine, complete with adaptor frame, is wheeled into the test cell on a trolley. Two monorail hoists lift the engine off the trolley and transfer it horizontally into the open-ended thrust stand where it is lowered for automatic positioning and locking of the adaptor frame. The upper coupling plate on the thrust stand then mates automatically with the lower coupling plate on the adaptor frame to provide all electrical, temperature and pressure connections for control and measurement. The overhead-mounted thrust stand comprises two U-shaped frames; the upper is fixed to the suspension points, rigidly at the rear and hinged at the front; the lower, a floating frame, is suspended from the fixed frame by four flexure plates. Thrust is measured by two tension load cells mounted on each arm of the U-frame between fixed and floating frames. A third load-cell, mounted in the rear of the thrust stand provides for thrust calibration of the stand. Load cells are selected to give an accuracy of ±0·02 per cent (measurement) or ±0·01 per cent (calibration). The engine adaptor frames which locate the engine in the thrust stand are designed for each particular type of engine and its mounts and include the test harness to link engine connections with those on the lower coupling plate.
The test cell systems and engine control room are in the auxiliary building adjacent to but structurally separate from the test cell itself. The design of the auxiliary building is according to the operator's requirements and can vary from a simple control room to a complete pre-rigging area with offices and control room.

Status
Sulzer a Swiss company with a major subsidiary in the United Kingdom, has been providing engine test cells since the 1950s. Equipment has been installed at Swissair and Garuda Indonesian Airways, Hong Kong Aircraft Engineering Company and engine manufacturer and research sites in India, Italy, Switzerland, the United Kingdom and Federal Germany

Manufacturer
Sulzer Brothers, Winterthur, Switzerland

UNITED KINGDOM

Aberglen Space-frame Structured Aircraft Hangars

Space-frame structures are permanent buildings which are designed for quick assembly and can be dismantled, moved and re-erected if necessary. They consist of extruded aluminium alloy arch frames clad with a pvc-coated polyester fabric which is resistant to fire and corrosion. The Trevira cladding which can be supplied in a range of colours provides a high level of natural lighting. The structures are designed to withstand the appropriate wind and snow loadings at any location and are suitable for extremes of climate. Double-skin cladding can be installed for additional thermal insulation and walls of blockwork, steel or aluminium incorporated for added security. Space-frame structural hangars can be supplied to accommodate wide-body aircraft. The structures may be erected side-by-side under one roof, an area of 2·6 hectares has been covered in this fashion.
The components for a structure covering 865 square metres can be packed into a container measuring 12·1 × 2·4 × 2·4 metres.

Status
Aberglen Orbits have supplied space-frame structures for use as aircraft hangars and workshops to countries in Europe, Africa and the Middle East

Manufacturer
Aberglen Orbits Structures, Glenrothes, Fife, Scotland

Avon Inflatable Engine Intake Blanks

Inflatable blanks are designed to stop the passage of air through jet engine intakes, to protect the engine by preventing windmilling and excluding rain, snow, ice, sand and other foreign objects. Standard blanks comprise hypalon on nylon-coated fabric inflatable tubes. They incorporate a diaphragm constructed in neoprene on a nylon-coated fabric and three separate valves; an inflation/deflation valve for use with a low-pressure air source; an inflation valve to be operated with a high-pressure air source and a preset pressure relief valve. Larger diameter blanks can be provided with the following optional features; a vent flap in the central diaphragm to relieve differential pressures; an extra valve for rapid deflation; D-rings to enable elastic-shock cords to be clipped to the engine nacelle and a nylon case for stowage of the deflated blank.
The blank is partially inflated with a high-volume foot pump or high-pressure air source positioned inside the engine nacelle and inflated fully to become an almost rigid structure.

Status
Blanks have been developed to fit the engines of Boeing 707, 727, 737, 747, Canadair Challenger, Lockheed L-1011, McDonnell Douglas DC-8, DC-9, DC-10, BAe1-11, Trident commercial aircraft and the BAe Nimrod, Harrier and Lockheed C5A Galaxy military aircraft. Civil and military operators who use this equipment include British Airways, Royal Air Force and the United States Air Force

Manufacturer
Avon Industrial Polymers (Melksham) Ltd, Chippenham, Wiltshire, England

---

Conder Aircraft Hangars

Conder aircraft hangars are designed to meet the requirements of individual operators based on a range of standard frame types.

The Mansard frame with high, wide spans, is suitable for wide-body aircraft. Tail docks and elevated sections of the roof to accommodate the tail section of the aircraft can be incorporated with mono-rails and catwalks suspended from the roof structure. Heavy motorised sliding doors can be fitted in the gable end for maximum use of the building.
The Portal frame hangars fall into two categories; the larger versions spanning up to 60 metres suitable for the medium range civil aircraft, namely Boeing 727, McDonnell Douglas DC-9 and Trident aircraft, and military aircraft of similar size eg Lockheed C-130 Hercules. Alternatively it is available with smaller spans of about 20 to 30 metres suitable for light aircraft and helicopters. Like the Mansard this frame will accept cranes in the roof for use with engine removal and repair work. Tail docks can be integrated to reduce the eaves' height which in turn diminishes the volume of the building requiring heat and/or ventilation. Extensions and simple lean-tos on the side of the building can be incorporated in all Conder hangars to create additional space or workshop or maintenance facilities.
The Monitor is a form of roof construction which can be added to the basic frames described above offering natural lighting without excessive heat gain or glare. Each Monitor frame is tailored to suit the geographical conditions of the country, site and orientation.
The standard door has a bolted steel frame clad with 0·56 mm thick, deep profile galvanised steel sheets. Alternatively metal sheeting to match the main hangar can be used. Optional features include internal or external locks and personnel doors which can be set into any leaf.

Status
Conder has supplied hangars to civil and military operators in Europe, Middle East, Far East and the West Indies.

Manufacturer
Conder Export Ltd, Winchester, Hampshire, England

Conder portal-framed aircraft hangar for Dan-Air

Conder 60 metre clear-span portal aircraft hangar in Abu Dhabi: lean-to attached to rear of hangar used for stores

Curran Type J Engine Test Installations

Curran produces a range of facilities for testing any type of pure jet, turbofan or turboprop engine.
The Type J Mk 10 is an aircraft engine thrust-measurement test facility designed to accommodate large pure jet and turbofan engines up to 100 000 lb st (450 kN) thrust capacity. It incorporates engine pre-rigging, hydraulic joist, automatic coupling and remotely-controlled hydraulic clamping.
The Type J Mk 2A aero-engine thrust measuring test facility is a floor-mounted installation for testing pure jet engines up to 20 000 lb st (90 kN) thrust capacity.
The Type J Mk 2F is an overhead-mounted installation, with parameters similar to the Type J Mk 2A.
The Type J Mk 6 test facility is an overhead-mounted installation for testing military jet engines up to 20 000 lb st (90 kN).
Control systems, consoles and interconnecting services can vary according to the operator's requirements and ancillary services such as inhibiting and engine wash can be added.

Status
Curran has supplied engine test cells to civil and military operators. Installations include the test facility up to 80 000 lb st (360 kN) thrust capacity supplied to British Airways at London Heathrow Airport for testing the Rolls Royce RB 211 engines of their Tristar fleet and other turbofan engines

Manufacturer
John Curran Ltd, Cardiff, Wales

Driclad Helicopter Hangars, Shelters and Protective Covers

Driclad produce a system of moisture and vapour-proof sealed accessible protective covers based on the enclosure of valuable equipment in a dry atmosphere contained within a flexible plastic cover. The covers, which are re-usable, are designed to prove an effective barrier to water vapour in extreme conditions of climate, temperature and humidity and to provide protection against damage caused by the presence of chemical gases, salt-laden air, sand, dust, fungi and insects.
Typical uses of the system include protection and preservation of aircraft, aero-engines, palletised loads and miscellaneous equipment. Items stored in Driclad are accessible for systems checks, modification or customs examination.
Semi-permanent storage shelters and helicopter hangars are manufactured by Driclad. These consist of reinforced pvc covering a steel frame with retractable frames which move to each side to provide access.

Status
Driclad systems have been in use for many years in Europe, the Middle East, Africa, Asia and the USA

Manufacturer
Driclad Ltd, Sittingbourne, Kent, England

Driclad Jabarahouse storage hangars

Helicopter protected by Driclad system

Esavian Hangar Doors

Esavian produce two basic types of large doors for maintenance hangars:
Type 126/S is a steel slab door running on multiple tracks and is offered up to 35 metres high and any width. The leaves can be individually motorised or operated by the sheaving method whereby series of leaves are sheaved together and operated by a master leaf containing the motor and control gear. Doors of this type are assembled on site. The top guide members are provided by the steelwork contractor while Esavian are responsible for fabricating and fixing the bottom track assembly.
Type 1200 is an aluminium concertina design which runs on a single track and is available up to 18 metres high and any width. Operation is by power mullions with self-contained motors acting on fixed cables in the head and track. In the case of very wide openings flexibility is offered by dividing the door into sections some of which are unattached ie not connected to the building jambs. In that case power is collected by the mullions using bus bars.
Type 126/S and 1200 can be insulated and contain glazing or double glazing in varying degrees. Unlike Type 126/S which must be assembled on site, Type 1200 is completely finished ex-factory and can be erected without additional on-site labour. The motors are already inserted in the power mullions and all electrical wiring and switches are complete.

Status
Esavian have supplied doors to British Airways, British Caledonian, Laker, British Aerospace and the Royal Air Force in the United Kingdom and completed contracts in Kuwait, Baghdad, Lagos and Budapest.

Manufacturer
Esavian Ltd, Stevenage, Hertfordshire, England

Esavian Type 126/S hangar doors at British Airway's technical block A, London Heathrow Airport, England

Esavian Type 1200 doors installed in British Caledonian's DC10 hangar at London Gatwick Airport, England

---

Henderson Hangar Doors

Henderson design and manufacture doors and door operating gear of all types and sizes. Examples of the company's products include heavy-duty rollers and fittings for large hangar doors and folding shutter doors both fitted at Changi Airport, Singapore. Slide-over sectional doors, which can be constructed in either steel or aluminium were supplied for the fire stations at Penang's international airport. Roller shutters, both standard and fire resistant are produced and these are fitted at Sharjah International Airport.

Manufacturer

P.C. Henderson Ltd, Romford, Essex, England

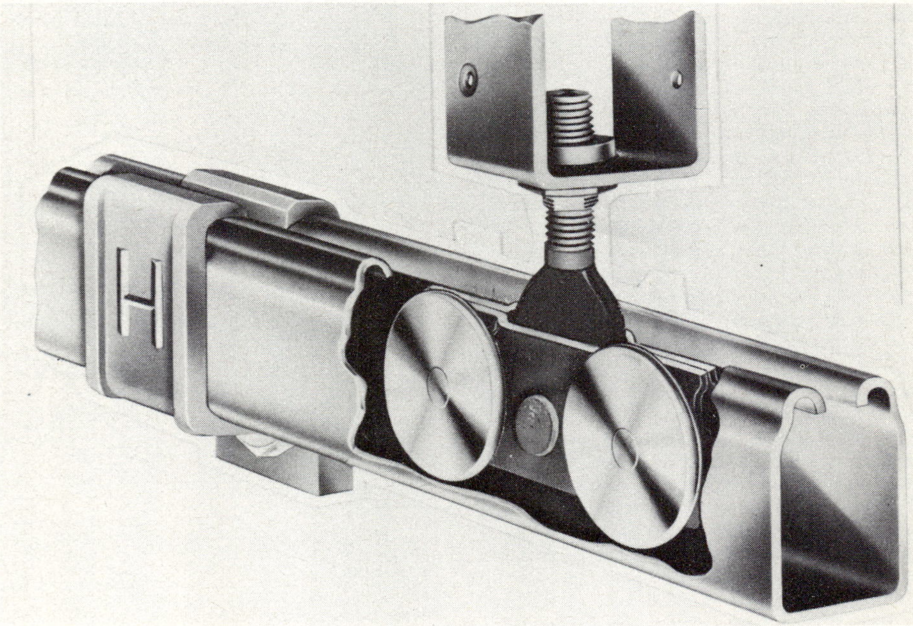

Heavy-duty rollers for hangar doors produced by P.C. Henderson

IAC RB199 Engine Test Facility

RB199 engine test facility at RAF Cottesmore

The Industrial Acoustics Company (IAC) with the Powerplant Systems Division of Marconi Avionics as a major sub-contractor was awarded the contract for the Turbo Union RB199 engine test programme purchased by the Royal Air Force and the Italian Air Force.
The facility, which is demountable to cater for the possibility of a change in location, comprises an acoustic enclosure; intake silencer; augmentor tube; primary diffuser; secondary suppressor; plenum chamber; exhaust splitter elements; exhaust stack; control room; entrance lobby housing bcf extinguishers; and a low-pressure compressor lobby. The engine test bed is of the overhead type and consists of a test stand support-frame assembly which has an integral thrust measuring system and an engine suspension unit. A special purpose trolley has been designed for transporting the pre-rigged engine into the test house. When the trolley is positioned under the bed the engine can be locked into place in a few minutes and, with the aid of battery-plates, all the pneumatic and electronic parameters can be connected simultaneously. These parameters are collected within a central junction area and relayed to the automated powerplant test (apt) equipment housed in the control room. The apt system is based on a Data General Nova Jumbo computer with a five-million byte disc. The operator interface consists of a vdu and keyboard and a hard copy thermal printer. Basically the apt system extracts a frame of data from the engine once a second and displays, on the upper half of the vdu screen, raw data, and data corrected to ISA conditions. The lower half of the screen is used to guide the operator through the test schedule providing instructions to establish various engine conditions while the computer calculates performance data and working lines. Limits can be set up on any of the incoming signals to ensure that if a parameter should exceed the setting then either a caution or shut down alarm will be produced, protecting the integrity of the engine. This is achieved with a micro-processor which controls the input data acquisition, takes ten frames of data a second, performs the alarm and validity checks and passes an averaged data frame to the apt system computer on request once a second.
The control room offers the operators a view of the engine under test and instantaneous display of all measured and calculated engine parameters in engineering units. It houses the engine's electronic fuel control equipment which are slave units incorporated into the control console. General engine and test facility control systems are on or nearby the control console.
The facility has its own low-pressure air start equipment and can carry out five consecutive engine starts. Re-charging the system is carried out automatically.

Manufacturer

Industrial Acoustics Company Ltd, Staines, Middlesex, England

Characteristics

ACOUSTIC SYSTEM: air-cooled, intake and exhaust treatment
ENGINE: Turbo Union RB199
NOISE LEVEL: 65 dBA
DISTANCE FROM EXHAUST: 200 m
ENGINE TURN-ROUND TIME: 30 minutes (from engine-off to ready to test)

Status

The prototype RB199 test facility was commissioned at RAF Cottesmore in 1981. A production contract was awarded in June 1981 to the Industrial Acoustics Company for a further 13 engine test facilities to be based in the United Kingdom and Italy.

Martin Roberts Aircraft Hangar Doors

Martin Roberts specialise in the design, manufacture and installation of aircraft hangar doors to suit specific requirements.
Javelin multi-leaf straight-sliding slab doors are suitable for all aircraft hangars, ranging from those housing light aircraft or helicopters up to the larger wide-body aircraft. The doors are usually electrically-operated but incorporate towing links for use in the event of power failure. Lighter doors may be manually-operated with hand-winding mechanisms. They are constructed from suitable steel sections and bracings to meet specific design considerations such as wind loadings, climatic conditions, frequency of use and budget. In addition, the door construction allows for installation of the required cladding (single- or double-skin insulated types) as well as the incorporation of doors and patent glazing or windows where specified. The doors are self-draining and free from water collection points. Weather-stripping can be provided to all door edges to afford protection against ingress of driving rain, dust or sand, while assisting in maintaining temperature differentials where necessary.
Each door operates by means of a bottom-roller gear system incorporated in the door structure. Top-guide rollers at the head of each door blade locate in guide-channel sections which normally form part of the roof structure and are supplied by the main steelwork contractor. The roller guides are designed to maintain engagement in top channels under specified roof deflections that are a feature of large span buildings. Buffer stops for safety purposes are supplied at high level.

Manually-operated Safeguard sliding doors

Martin Roberts' British Airways helicopter hangar

Power operation of doors is normally required. They may operate as individual door leaves or can be designed to run in cascade, with the leaves moving at different speeds so ensuring all doors reach the open position together. Operating speeds may be fixed or variable up to 60 feet per minute (20 metres per minute) ±15 per cent. The speeds of operation may apply to both directions of door travel. The electrical drive units are housed within the door construction and incorporate suitable brakes, clutch mechanisms and reduction gearings which allow disengagement for tow bar operation. The main electrical control stations are housed adjacent to the drive units. Power supply to the doors is normally 415 volts, 3 phase, which provides the motive power and power for the control and safety system. The power is supplied through overhead conductor rails and collector trolleys. Earthing is by means of the doors themselves with the additional back-up of an earth monitoring system.

Safeguard straight sliding doors are suitable for fitting to helicopter hangars. Openings of any width may be covered by a number of doors sliding to the side of the opening or between opening limits. Doors slide on single or multiple track arrangements by use of top-hung or bottom-roller sliding door gear systems. Single- or double-skin and insulated constructions are available with flat or profiled metal sheet facings. Personnel access doors and vision panels can be incorporated into the door construction where required. Doors may be operated manually but electric-automatic operation can also be supplied.

Each door blade comprises a welded-steel framework designed to suit the size and use of the door, wind loadings and climatic consideration. The framework is braced and reinforced and metal sheeting is applied to one or both faces selected from galvanised flat sheet or profiled, corrugated claddings in steel, aluminium or colour-coated materials. Many applications call for a double-skin construction which may be insulated by using a core of mineral wool, glass fibre, cork or expanded polystyrene. Normally a top-hung gear system is employed but where the lintel construction cannot accept the weight or when doors are very heavy a bottom-roller type may be necessary.

Status

Contracts for aircraft hangar doors include installations in Kuwait, Qatar, Syria, Nigeria and two sets of doors for helicopter hangars in Seeb, Oman and Aberdeen, Scotland

Manufacturer

Martin Roberts Ltd, Sittingbourne, Kent, England

RDL Aircraft Hangars

Wide-body aircraft hangars need exceptionally wide clear spans. A typical hangar requires 100 metres width, with a clear height of at least 25 metres and a depth of about 80 metres. These facilities were provided by Redpath Dorman Long (RDL) for British Airways in a contract to fabricate and erect the structural steelwork for the extensions to the British Airways technical block A at London Heathrow Airport. Each pen required two lattice girder spans 106 metres long, 7 metres high and 3 metres wide to give a ground clearance of 30 metres.

A further example of the type of hangars constructed by RDL is the extension and modification of the wide-body aircraft servicing hangar at Kai Tak International Airport for the Hong Kong Aircraft Engineering Company. The original hangar was built by RDL in the early 1970s to withstand typhoon conditions; the extension has been constructed to the same specification.

Status

RDL have constructed hangars for many civil and military operators including the Royal Air Force and the United States Air Force.

Manufacturer

Redpath Dorman Long International Ltd, Bedford, England

Shelter-Span Modular Buildings

Shelter-Span is a modular building system which combines standard light-metal framing components with one or more layers of thin-skin membrane cladding. The structural frame is of aluminium alloy or composite steel/aluminium dependent on span. Integral grooves on the frame elements allow panels of durable pvc-coated polyester fabric or other material to be fixed and held in position to create a weather-proof enclosure.

Vehicle wash and valet bay

Clear spans of up to 36 metres can be supplied on the standard structures. This can be increased to 60 metres on domes. The covering material provides a high level of natural light. For ventilation, natural airflow occurs through the canopy-covered ridge opening; alternatively the apex opening can be sealed and mechanical ventilation used. A thermal barrier is created by cladding buildings with twin layers of fabric. Alternatively conventional insulation materials can be placed between the skins.

Doors can be at any point around the building perimeter. On side walls, door width is limited to twice the distance between frames; gable ends can be left open or doors of any required width and height installed.

Characteristics

TYPE	MEDIUM SPAN	LONG SPAN
CONSTRUCTION	Aluminium alloy	Composite steel/aluminium
MAX CLEAR SPAN	18 m	36 m
length	Modular in 3·75 m increments	Modular in 5 m increments
width	6–18 m in 3 m increments	18–36 m
MAX EAVES HEIGHT	6 m	6 m
MAX APEX HEIGHT	14 m	14 m

Status

Shelter-Span was established in 1978 and has provided several buildings for airport use, including a vehicle wash and valet bay for British Airways which measures 12 × 22·5 metres with an eaves height of 3 metres. Floor area is 270 square metres. A conventional metal-clad building was not acceptable because of its effect on nearby air traffic control equipment.

Manufacturer

Shelter-Span (UK) Ltd, Maidenhead, Berkshire, England

UNITED STATES OF AMERICA

AeroAcoustic Model ST/B Acoustic Systems for Engine Test Cells

The Model ST tubular jet exhaust treatment consists of an augmentor tube positioned aft of the engine mounting which feeds into a jet energy dissipation and flow straightening area and then is exhausted vertically through an exhaust acoustic treatment. Water-cooled cells are used, except in the case of fan engines, comprising eleven gauge stainless steel tubes covered with stainless steel mesh and glass cloth. For fan engines, $\frac{3}{16}$ inch (4·7 mm) HRS tubes are fitted covered with stainless steel mesh and glass cloth.

The Model B inlet treatment was developed to overcome acoustic control problems encountered with the advent of the fan jet engine. These engines required better high- and low-frequency performance and uniform distribution of air at the engine face. The fat baffle, sinuous passage, inlet treatment used with standard jet engines was deficient in high-frequency noise reduction and the large stall areas at the rear of the fat baffles persist at the engine face causing premature engine stall and low thrust readings. The Model B treatment is a closely-spaced inlet system of moderate thickness designed to overcome these problems and provide a uniform velocity and pressure distribution at the engine face.

Test cells are normally housed in a reinforced concrete construction. They can be supplied also in a heavy-duty, demountable acoustic housing designed for this purpose. Assembled on a concrete slab the cell houses the engine, inlet treatment, augmentor tube and test bed.

Manufacturer

AeroAcoustic Corporation, Jacksonville, Florida, USA

Characteristics

ACOUSTIC SYSTEM: Model ST Exhaust Silencer and Model B inlet treatment
ENGINE: Pratt and Whitney JT 8D
THRUST SETTING: max
NOISE MEASUREMENT: 71 dB
DISTANCE FROM EXHAUST: 2·44 m
HEIGHT ABOVE GROUND LEVEL: 1·52 m

Status

AeroAcoustic have installed Model ST tubular exhaust systems at five sites in the USA including Ozark Airlines, General Electric Company, Pratt and Whitney and one in Japan for Ishikawajima-Harima, Heavy Industries Co Ltd

Aerodome Aircraft Hangar

Aerodome aircraft hangars are developed to house general aviation aircraft and helicopters in minimal space.

It consists of a circular glass fibre or square metal structure housing a rotating floor. To take an aircraft out of the hangar the appropriate aircraft is rotated on the turntable until opposite the hangar door. The bi-fold door operates electrically. When open the aircraft is winched out of the hangar. The process is reversed to house an aircraft. Aircraft maintenance can be carried out while the turntable is operating to move other aircraft. In the event of a power failure the door may be operated manually and the turntable rotated using a tug or other airport vehicle. Optional eyelets can be positioned around the turntable perimeter for cable attachments. Aircraft movement in and out is accomplished in this instance by conventional methods. The turntable is designed to accommodate up to six single-engine, four light twin-engine aircraft or a combination of two twin-engine aircraft and three single engine-aircraft.

Characteristics

TURNTABLE POWER: 3 hp, 220 V electric motor

DIMENSIONS

door: 12·8 × 3·66 m

turntable: 21·34 m diameter

hangar: 24 m diameter (approximately) or 24·38 × 24·38 m

Status

Aerodome hangars have been in use since 1972. Engineering studies are under way to expand the turntable dimensions to house larger executive or corporate aircraft in a 100 × 100 foot (30·48 × 30·48 metres) structure

Manufacturer

Aerodome Industries Inc, Wichita, Kansas, USA

Aero Systems Engine Test Cells

Aero Systems Engineering design and manufacture engine test facilities systems for: engine mounting, engine control and measurement and noise control.

Engine mounting systems available include universal thrust stands turboshaft, turboprop and propeller stands.

For engine control and measurement a comprehensive range of electronic displays, digital and analogue, control equipment, consoles and computer test systems can be supplied including the automatic data acquisition and processing system (adaps), and broad line flow measurement, control and handling equipment.

Force, speed, vibration, positional, flow and temperature data is collected, processed by a 16-bit computer and displayed on a colour crt. Standard day correction is applied where necessary and observed and corrected data, which is updated every second, is available through a hard copy printer and the graphic crt display. A special function keyboard provides test parameter and status information and in the manual mode enables the operator to select data points. The system can provide automatic engine control throughout the test cycles. Other optional features include an expansion of the main computer memory to 128 K words, magnetic tape drives, paper tape equipment, line printer/plotter, video hard copies, x-y plotters, and additional colour or alphanumeric crt displays.

Noise control systems consist of intake and exhaust sound attenuation equipment and run-up silencers. Prefabricated acoustic enclosures can be produced for military aircraft accommodating the aircraft for engine test-runs.

Status

Aero Systems Engineering have supplied complete engine test cells to civil and military operators in countries in North and South America, Europe, Middle East and the Far East

Manufacturer

Aero Systems Engineering, St Paul, Minnesota, USA

Armco Aircraft Hangars

Armco design and manufacture aircraft maintenance hangars. A typical example has been built for Wardair at Toronto International Airport. The building consists of 20 rigid frames and a 76·7 metre truss at each end. Overall it measures 83·5 metres wide, 505 metres long, 24 metres at the eaves and 27·7 metres at the ridge. The structural design of the steelwork was done by Armco. Door clearance is 80·7 metres wide, 21·3 metres high with a design provision to remove material at the ridge to accommodate a 24·4-metre high aircraft tail if necessary.

The hangar has some unusual design features, no cranes or catwalks are included. The 12 616-square metre building has an Armco Steelox CF roof system which uses concealed fasteners to attach roof panels to structural members. The roof has a 150-metre thick blanket insulation and 50 mm thermal break providing an insulation factor of R-20. The hangar is fitted with Armco ARW-IV wall covering. These panels are contoured and rolled to provide 900-metre covering widths. The panels were fabricated from 26-gauge, enamel-coated galvanised steel. Blanket insulation of 125 mm is fitted behind the outside panels for thermal efficiency.

Manufacturer

Armco, Middletown, Ohio, USA

Lynnco/Universal Blast Deflectors

Aircraft are required to run their engines at full take-off power for engine trim or power checks during maintenance engine runs. Blast deflectors provide protection to buildings and personnel positioned behind the aircraft during these engine runs since jet blasts at take off power can reach velocities of 640–800 km per hour at a point 15 metres behind the jet nozzles.

In addition to providing protection, the deflectors can be used for noise control. Experiments have shown that the lowest noise levels are generated along a line directly behind the jet exhaust nozzle and the areas of highest noise level occur along lines approximately 35 degrees on either side of the line of lower noise level. Therefore the tail of the aircraft should be directed towards the building or area where the noise would be most undesirable.

The Lynnco blast deflector is available in heights from 0·9 to 6·1 metres. It consists of curved prefabricated galvanised or cadmium plated structural shapes supported by rear struts which are fixed to the ground by rows of anchor bolts, front and rear. Corrugated, galvanised steel sheets are bolted to the curved structural sections to provide a heat and vibration resistant curved surface.

The Universal blast fence was developed for the wide-body aircraft. It has an extended centre section to cater for the McDonnel-Douglas DC10 and Lockheed Tristar with their high tail-mounted engines.

Status

Lynnco and Universal blast deflectors are in operation at civil and military sites in north America and the Far East.

Manufacturer

Blast Deflectors Inc, Los Gatos, California, USA

Ferguson Aircraft Hangar Doors

Ferguson Door Company provide custom-built hangar doors, to suit different types of operation and environment. Doors can be horizontal rolling; vertical lift; canopy, for the aircraft with high-tail sections; round-the-corner or swing type. Operation of the doors can be pneumatic, hydraulic or electric.

The TWA maintenance base at Los Angeles Airport, California, has an opening 480 feet (146·3 metres) wide, 80 feet (24·4 metres) high. These leaves are on separate rails and operate independently. The outer door leaves can be placed in a closed position at either the right- or left-hand side of the opening. Each door leaf has control stations at both ends. To prevent accidents the right-hand control station moves the door to the left only. A dead-man control feature means that the operator must walk with the door and eliminates the need for safety edges. Limit switches are provided at each terminus of the opening for each door leaf to prevent damage to either the door leaves or the building.

An example of a swing door is provided by the aircraft surface treatment shop for stripping and painting aircraft at North Island Naval Air Station, California. The installation has six openings each 108 feet (32·9 metres) wide and 26 feet (7·9 metres) high. The swing doors which are pneumatically controlled are each 6 feet (1·83 metres) thick and in a closed position operate as a bank of air filters for constant flow filtration. The inside face is a combination of two layers of perforated metal and a skin of air filters.

For the maintenance of aircraft, whose fuselage length exceeds the length of the hangar, aperture closure doors can be fitted into the door configuration, which fit around the extruding tail section.

Status

Ferguson has supplied hangar doors to civil and military operators throughout the world. More recent installations include the Hellenic Aerospace Industry aircraft overhaul service centre at Tanagra, Greece and the British Airways Boeing 747 maintenance facilities at London Heathrow Airport, England

Manufacturer

Ferguson Door Co Inc, Gardena, California, USA

Fleming 500/100 Series Slide and Canopy Hangar Doors

The 500 Series slide or canopy door can be fitted to any size of hangar. Slide doors, either inside- or outside-mounted, stack when open, in pockets, cantilevered guides or at the jambs.
They are designed to move over rails in the floor, steered by top rollers that travel in overhead guides. Three basic types of slide doors are common; independent, uni-directional and bi-parting. The independent versions consist of a number of self-powered leaves which can move freely to a position in the opening. Uni-directional doors comprise a number of leaves all travelling to one side of the opening and bi-parting doors have an even number of leaves, divided at the centre with half opening to the left jamb and half to the right. Uni-directional leaves can be of the anchored group-type in which the jamb leaf is anchored at the building jamb limiting movement to the open and close positions. Bi-parting doors can be of this type or of the floating group-type in which leaves are interconnected but are not tied to the building in any way. They can open in either direction or move as a group abreast of each other. Drive for these doors is provided by unit drive motors mounted on leading leaves, jamb leaves or fixed at the jambs of the opening.
Bottom drives, traction or fixed chain style can be used with uni-directional and bi-parting doors while independent doors have individual traction drives. The traction method drives one or both wheels by chain and sprocket mechanism while the fixed chain style has a positive drive between it and the adjacent leaf. Linear or wrap-around cable systems can be used with the drive systems to connect the idler leaves to the driven leaves and move them to the open or closed positions. Tail doors provide entry for the tails of the largest aircraft. Placed in the centre of the hangar end wall above the slide doors they are produced in three motor-operated types: lift, swing and canopy.
Three types of 500 Series canopy doors are available, telescopic, unbraced and braced.
Telescopic canopy doors are constructed in independent horizontal sections and do not require side guides. A number of sections side-by-side will fill any width opening. The sections, usually 60 to 80 feet (18·29 to 24·38 metres) wide may be operated alone, in combination or in unison. Each consists of a vertically-rising lower and top-hinged upper leaf. The lower leaf moves up behind the top leaf and then both swing out and up to form the canopy. The telescopic canopy is recommended for openings 40 to 80 feet (12·19 to 24·38 metres) high.
Unbraced canopy doors designed for openings up to 130 feet (39·62 metres) wide, 40 feet (12·19 metres) high are single piece doors guided by overhead curved tracks and by vertical jamb rollers. Minimum structural support, no back bracing and small headroom requirements permit the use of these doors on many types of roofs. Power to operate the doors can be located at the jambs, overhead or at floor level, to facilitate servicing.
Braced canopy doors are suitable for any width opening using one or more independent horizontal sections side-by-side. Average sections are between 60 and 100 feet (18·29 and 30·48 metres) wide with a maximum height of 65 feet (19·81 metres). Sections may be operated singly or in combination. The bracing system distributes wind loads into the building, framework and guides door movements. Side guides are not required.
Special canopy steel doors have been constructed for the United States Air Force for their Alert hangars. Each bay has canopy doors front and rear. The door at the front of the hangar incorporates an extension to accommodate the nose section of the interceptor aircraft. The door is suspended on cables and guided at jambs by vertically-rising rollers engaged in straight or curved tracks. Reversible high-torque motors with integral magnetic brakes, either gear head or with separate reducer, are mounted overhead to provide positive drive to the traction sheaves. Cables from the door pass over the traction sheaves and into concrete counterweight slabs which are mounted in guided counterweight racks at the jamb of opening. Manual operation of the doors is also provided.
The 100 Series slide door consists of a matched and balanced system of leaves or panels which can be opened to one side or to both sides as required by a unit-drive mechanism or using the Fleming Mule.
The Mule is a drive mechanism which is bolted on to the inside of any sliding hangar door. Electrically-powered with positive controls the rubber wheel adjusts automatically to uneven floors and is equipped with floor wipers to clear debris in both directions of travel. In the event of an electrical failure the Mule can be converted to manual operation.

Status
Fleming have supplied hangar doors to airlines, NASA and military operators in the USA and to El Al, Lufthansa, PIA and Swissair. Typical examples are an American Airlines hangar at Tulsa, fitted with eight Fleming doors on the west side which opens 520 feet (158·5 metres). Both the north and south sides have openings of 300 feet (91·4 metres) using five doors. All doors are 80 feet (24·4 metres) high.
The Pan Am maintenance hangar at Kennedy Airport, New York has two 265-foot (80·77-metres) wide bays with four coupled slide doors controlled from cabs at each end of the train. Canopy doors 80 feet (24·4 metres) wide and 35 feet (10·67 metres) high accommodate the high tails of the Boeing 747 aircraft. Two vertical lift doors 21 feet (6·4 metres) wide and 16 feet (4·88 metres) high admit service trucks

Manufacturer
The Fleming Steel Company, New Castle, Pennsylvania, USA

Jetway Blast Deflector Fences

Jetway produce a wide range of deflector fences for airports, for deflecting jet blast upwards, reducing exhaust gas temperature and attenuating high frequency noise. Made of solid or perforated galvanised steel panels the product range includes both one- and two-way straight fence sections, angle joints (dog-legs) and end-closure plates together with mesh screens to prevent loose objects from being blown over the fences by jet blast.

Manufacturer
Jetway Division, Abex Corporation, Ogden, Utah, USA

Characteristics

HEIGHT	STANDARD MODEL	INTENSIVE USE MODEL	VERY INTENSIVE MODEL
7 ft (2·133 m)	Single or double face	Single face	
8 ft (2·438 m)	Single or double face	Single or double face	Single face
10 ft (3·048 m)	Single or double face	Single or double face	Single face
12 ft (3·657 m)	Single face	Single face	Single face
14 ft (4·267 m)	Single face	Single face	Single face
20 ft (6·096 m)		Single or double face	Single or double face
33 ft (10·058 m)			Single face

Kuss Para-port Aircraft Hangar Doors

Para-port doors are manufactured from a translucent vinyl/nylon woven fabric, rolled on a rigid steel tube which is at the bottom of the door, in the closed position. The door is raised and lowered by push-button control. An electric motor with gear reduction is connected to the tube, causing it to turn, rolling up the door. The motor and drive mechanism move with the curtain. Vertical guides at the sides of the door opening stabilise the roll as it travels up and down. In the closed position the roll is in contact with a special door seal ensuring a weather-proof connection. The door may be stopped at any point in its traverse and safety locks come into operation to prevent the door from dropping inadvertently.
The translucent nature of the material allows sufficient natural light to penetrate the building even in the fully closed position for normal aircraft maintenance operations.

Status
Para-port hangar doors have been installed at a large number of airports in the United States of America.

Characteristics
MATERIAL: vinyl nylon laminate
MAX WEIGHT: 746 g per m^2
MINIMUM BREAKING STRAIN: 20·7 kg/cm^2
MAX WIND SPEED: 120 km/h (optional: wind protection 160 km/h)
DOOR DIMENSIONS
max width: 145 ft (44·2 m)
max height: 40 ft (12·19 m)
TIME TO RAISE DOOR: 110 × 30 ft (33·5 × 9·14 m) in 2 minutes

Manufacturer
R. L. Kuss and Co Inc, Findlay, Ohio, USA

Access, Lifting and Handling

FRANCE

Air Marrel WAM 3001 Workshop Truck

The WAM 3001 is an elevating workshop mounted on a double-scissors mechanism based on commercial truck chassis. The scissors are operated by a single hydraulic jack. A working platform with hand-rails is fitted on the workshop roof permitting a maximum working height of 9·7 metres. Access is by steps at the rear of the workshop. The latter can be moved forward by 0·65 metres in the elevated position, the platform extended by a further 1·75 metres and lateral extensions to the platform provide an additional width of 0·75 metres on each side. Push-button control panels are on the platform and in the cab to lift and lower the workshop, for forward and backward travel of the workshop (workshop panel only), extension and retraction of the four stabiliser jacks and an emergency stop. Safety features in the design include locking the clutch or the parking brake when stabiliser jacks are extended, non-return safety valves on each hydraulic jack, knotted rope for emergency exit and emergency hand-operated hydraulic pump. Optional features available include an 11-tonne chassis, with 4·5 metre wheelbase, power take-off and power steering.

Characteristics

WORKSHOP CAPACITY: 1400 kg
ELEVATION HEIGHT: 1·5–5·5 m
MAX WORKING HEIGHT: 9·7 m
DIMENSIONS
workshop: 4·6 × 2·3 × 2·1 m
platform: 5·75 × 2·4 m
FORWARD EXTENSIONS: 1·75 m
WIDTH EXTENSION: 0·75 m each side
UNIT DIMENSIONS: 6·4 × 3·9 × 3·6 m
WEIGHT: 8500 kg

Status

In service with Air Afrique

Manufacturer

Air Marrel, Andrezieux Boutheon, France

Reel Aircraft Docks and Maintenance Platforms

Reel specialise in the supply of access equipment for aircraft maintenance. The products range from aircraft nose- and tail-docks to telescopic platforms suspended from the roof as supplied to Air France for maintenance and painting.

Status

Reel access equipment is in service with major airlines in Europe and Africa

Manufacturer

Reel SA, Caluire, France

Reel movable tail-dock around a Boeing 747 aircraft

Telescopic maintenance platforms installed by Reel

Sietam Mobile Elevating Work Platforms

Sietam produces a range of mobile hydraulically-operated elevating work platforms. These may be fitted with single or double bucket-type platforms, which can include dual controls, installed at the platform and turntable, and the provision of air, water and electricity. Equipment has been supplied to major airlines including Aeroflot, Air France, Air Afrique, Gulf Air, Olympic, Saudia and Syrian Arab Airlines.

Manufacturer
Sietam, Viry-Chatillon, France

Siraga Fixed Maintenance Stairs

Low, medium and high series fixed stairs are available. The low series have fixed feet, the medium and high series are supplied either with feet or mounted on swivelling wheels, with a brake in front of the stairway. All are of steel construction with rubber shock absorbers fitted to the platform. Hand-rails are incorporated, the position depending upon the series selected.

Status
Production: supplied to civil and military operators in France and to airlines in Europe, the Middle East and Africa

Manufacturer
Siraga, Buzancais, France

Characteristics

SERIES	LOW	MEDIUM	HIGH
HEIGHT	1, 1·25 m	1·5, 1·75 m	2–3·75 m in 0·25 m steps
NO OF STAIRS	4, 5	6, 7	8–15
WIDTH OF STAIR	0·72 m	0·72 m	0·72 m

Siraga Type 8050 Series Adjustable Maintenance Stairs

The Type 8050 series is designed to fold so that the stairs can be air-lifted if required. The frame, of sectional box construction, is equipped with wheels and solid tyres, two at the rear (four for Model 8090) and two wheels, diabolo-mounted, at the front. Stability is provided by four-screw skids and a draw bar is fitted for towing purposes. The height is adjusted by means of a hydraulic jack actuated by a manual pump with a toothed rack and pawl safety lock controlled by a lever near the pump. Hand-rails are fitted to both sides of the stairs and platform. Optional features include an electric motor with hydraulic-pump unit and installation of a swivelling crane hoist on the platform.

Characteristics

SERIES	8050	8065	8090
DESIGNED WORKING HEIGHT	1·25–6·5 m	1·32–8·05 m	2·92–9·93 m
STEPS			
width	1 m	1 m	1 m
depth	0·3 m	0·3 m	0·3 m
PLATFORM LENGTH	1·2 m	1·2 m	2 m
PLATFORM HEIGHT			
low position	1·25 m	1·32 m	2·92 m
high position	5 m	6·55 m	8·43 m
OVERALL LENGTH			
low position	7·02 m	8·8 m	9·5 m
OVERALL WIDTH	2 m	2·2 m	3 m

Status
Production: supplied to civil and military operators in France and to airlines in Europe, the Middle East and Africa

Manufacturer
Siraga, Buzancais, France

Siraga Types 8115/8130/8140/8155 Adjustable Maintenance Stairs

The Types 8115/8130/8140/8155 adjustable stairs are mounted on a T-shaped steel-frame base equipped with three trains of twin rubber-tyred wheels. Stability is provided by four-screw skids. The height can be adjusted to one of three positions by means of two hydraulic jacks each actuated by a manual pump. Toothed rack and pawls on each jack act as safety locks. A tow-bar is mounted under the first step of the stairway. Hand-rails are fitted to both sides of the stairs and platform. Optional features include the installation of two pumps which enable the stairway to be adjusted from the platform or ground, or simultaneously from both places; electric/hydraulic pump unit and installation of a 150 kg capacity swivelling crane hoist on the platform.

Characteristics

TYPE	8115	8130	8140	8155
DESIGNED WORKING HEIGHT	2·9–13 m	3·5–14·5 m	3·4–15·5 m	3·4–1·17 m
STEPS				
width	1 m	1 m	1·03 m	1·03 m
depth	0·3 m	0·3 m	0·3 m	0·3 m
PLATFORM LENGTH	1·2 m	2 m	1·2 m	1·2 m
PLATFORM HEIGHT				
low	2·92 m	3·5 m	3·4 m	4·9 m
mid-	7·32 m	8·3 m	8·15 m	9·65 m
high	11·5 m	13 m	14 m	15·5 m
OVERALL LENGTH				
low position	12·76 m	14·35 m	14·9 m	14·9 m
OVERALL WIDTH	4·5 m	4·5 m	5·5 m	5·5 m

The four models are similar in design except that the Type 8115 can be configured for air-lifting. The Type 8155 is a Type 8140 equipped with an extension platform 1·5 metres high with a useful width of 0·65 metres.

Status
Production: supplied to civil and military operators in France and to airlines in Europe, the Middle East and Africa

Manufacturer
Siraga, Buzancais, France

SMA Engine Trolley

Soudure et Mecanique Appliquees (SMA) produce three trolleys for aircraft engine changes. These were developed for the Snecma Atar, Rolls Royce/Turbomèca Adour and Pratt and Whitney JTSD engines which power the Mirage, Jaguar and Boeing KC135 aircraft respectively.

Each trolley is fitted with hydraulic arms which provide a parallel lift to position the engine for installation in the aircraft.

Characteristics
ENGINE: Snecma Atar
TROLLEY
capacity: 2·5 tonnes
dimensions: 5 × 2 × 1 m

Status
SMA has supplied equipment to military flight test centres, aircraft and engine manufacturers in France

Manufacturer
Soudure et Mecanique Appliquees, Argenteuil, France

GERMANY, FEDERAL REPUBLIC

Hydro-Geratebau Lifting and Wheel-change Equipment

Hydro-Geratebau manufacture tripod jacks for lifting aircraft with a range of weight capabilities from 3 to 130 tonnes. The company's production programme centres around the requirements of aircraft construction and other aspects of the aviation industry.

The latest device is a tilting dolly which lifts an aircraft by the nose gear and lowers the vertical fin so it can pass through a low hangar door opening or under low beams. The system enables hangars to be used for wide body aircraft which, without the dolly, could not accommodate these aircraft. The dolly is designed for towing and manoeuvring. The vertical angle of the nose gear also is taken into consideration at each lift height. The nose wheels are held by a scoop-shaped unit and secured by webbed straps. The tilting dolly has six solid-rubber tyres, and the bogie at the forward end of the frame is a castoring unit with 180 degrees of freedom in either direction. A supporting hydraulic cylinder is installed near the bogie so that the complete dolly can be raised and the bogie rotated with a crank handle.

The Hydro GSE Combi cart has been on the market since 1980. The unit combines everything needed for an aircraft tyre or brake change: an axle jack, a wheel/brake change dolly, a hoist sling and a box for special tools. The cart also carries up to four nitrogen bottles with three pressure ranges. The tyres are serviced by three self-retracting hoses. Two spare wheels can be carried by the cart which can be towed at speeds up to 40 km per hour. In addition to carts and jacks, Hydro-Geratebau offers more than ten different types of tow-bars. The full complement of tow-bars covers all aircraft types from the DC-9 and B-737 to the B-747.

Characteristics

TILTING DOLLY
capacity: 5–9 tonnes
max height: 4 m

GSE COMBI CART
operating pressure ranges: 0–5 bar, 0–25 bar, 50–150 bar

Status

Air France, Hapag-Lloyd, Lufthansa and Olympic have ordered the tilting dolly as has the MBB/VFW plant in Hamburg for the A300 Airbus.

Alitalia have purchased 19 Combi Carts and other units have been sold to Iberia and Saudia.

The A300 model tow-bar has been purchased by many airlines including Alitalia, Iberia and Lufthansa

Manufacturer

Hydro-Geratebau GmbH and Co, Biberach/Baden, Federal Republic of Germany

---

Mero Maintenance Docks

Mero maintenance docks are constructed in tubular steel with node connectors. They are designed to provide access to all sections of the aircraft for maintenance purposes. Each unit is mounted on wheels or rails to provide mobility and fixed in position with locking devices or brakes adjacent to each wheel. Access is provided by stairs or step ladders with non-skid steps and railings, and hand- and mid-railing is fitted to all sides not adjoining the aircraft. The working platforms are covered with plywood decking with toe boards installed to prevent equipment accidentally being kicked off the platform. Hose-type paddings are attached to the edges of all platforms touching the aircraft. Flaps, sliding platforms and lifting stands are incorporated to enable free movement of the aircraft rudder, flaps and engine cowlings.

For wide-body aircraft the docking can be adjusted in height by 1 metre through an electric drive. With narrow-body aircraft the working height can be altered by jacking the aircraft up to 0·5 metres (0·75 metres for the Boeing 707 aircraft). Tail docks are constructed in two halves which fit together around the tail when the aircraft is in position. With heavy overhaul docks, air cushions are used to move the docking to provide aircraft access. Other platforms are moved, lifted and turned by electric motors and integrated control stations. Installations for power plugs, lighting, compressed air and ventilation systems for painting aircraft can be incorporated in addition to gutters and drip trays for oil and paint waste.

Status

Mero supplies maintenance and overhaul docks for all types of aircraft. The equipment is in service with the West German Army, Air Force and Navy, the Royal Netherlands Navy, the French Navy and the Portuguese Air Force and airlines and aircraft manufacturers in Europe, the Middle East, north and south America and Africa

Manufacturer

Mero-Raumstruktur GmbH, Wurzburg, Federal Republic of Germany

Mero A300 B wing- and tail-dock in service with Lufthansa

Mero tail-maintenance dock for Dominicana Boeing 727 aircraft

Trepel AM 3/62 and AM 5/92 Aircraft Maintenance Platforms

The AM 3/62 and AM 5/92 are mounted on powered chassis for mobility and both platforms can be driven to the desired position at the aircraft. In the AM 3/62, which extends to a maximum height of 61 metres, four vertical stabilisers are fitted for stability while in the larger AM 5/92 developed for wide-body maintenance, four hydraulically-operated outriggers are used, activated by a push button in the driver's cabin. The platform lifts to the desired height by means of a double scissors mechanism operated with two telescopic cylinders, controlled from a panel on the platform. When the working height is reached, the vehicle's engine may be stopped by depressing the engine start/stop control button. In the AM 5/62, the platform includes five lateral sections with a 1 foot 6-inch (0·5-metre) shift and can be fitted with a 7 foot 8-inch (2·4-metre) extension capability, as an optional feature. A power supply of 110/220 volts is installed for power tool operation, and for night operation the platform is equipped with additional light sources. The hand-rails fitted to the platform are removable on all four sides.

For safety reasons, in both models the vehicle can be driven only with lowered platform and fully retracted outriggers/stabilisers.

The maintenance platforms are available also on unpowered chassis fitted with tow-bars.

Characteristics

TYPE	AM 3/62	AM 5/92
PLATFORM		
max height	20 ft (6·1 m)	30 ft (9·15 m)
DIMENSIONS	11 ft 6 in (3·5 m) × 4 ft 3 in (1·3 m)	17 ft 4 in (5·3 m) × 4 ft 3 in (1·3 m)
CAPACITY	300 kg	500 kg
CHASSIS		
GVW	3·5 tons	4·1 tons
type	Mercedes L306D	Mercedes L306 traction head + chassis
ENGINE	DB OM 615	DB OM 615
output	45 kW	45 kW
HYDRAULICS		
pto pump	10 litres/minute	10 litres/minute
MAX OPERATING PRESSURE	13·7 bar	13·7 bar
UNIT DIMENSIONS		
length	18 ft (5·5 m)	24 ft 3 in (7·4 m)
width	7 ft 6 in (2·28 m)	9 ft 6 in (2·28 m)
TRANSIT HEIGHT	8 ft 6 in (2·5 m)	8 ft 6 in (2·5 m)

Status
Trepel aircraft maintenance platforms are in service at Frankfurt, Paris-Orly, Geneva, Rome, Athens and London Heathrow airports

Manufacturer
Trepel AG, Wiesbaden, Federal Republic of Germany

ISRAEL

Mechola Aircraft Hydraulic Jacks

Mechola produces aircraft hydraulic jacks with lifting capabilities from 3 to 20 tons (2·7 to 18·1 tonnes).

Height is adjusted by means of a two-speed hydraulic hand-pump. Spring-loaded, rubber-tyred wheels are fitted adjacent to each leg. A hydraulic-pressure safety valve and double safety-lock nuts are incorporated in the design.

Characteristics
LIFTING CAPABILITY: 9072 kg
MAX EXTENDED HEIGHT: 2·11 m
ADJUSTING BOLT: 0·45 m (max extension)
NOMINAL HEIGHT: 1·1 m
HYDRAULIC FLUID: MIL H-5606 or 83282
WEIGHT: 250 kg

Status
In service with the Israeli military forces

Manufacturer
Mechola Engineering, Bikat Hayardan, Israel

ITALY

OMA Maintenance Equipment

OMA produces a range of custom-built maintenance equipment. These include platforms for the external maintenance of A300 and Boeing 727 aircraft; a trailer designed for the transport and installation of RB 199 engines for the Tornado aircraft and a hydraulic jack developed for lifting the Aeritalia G222 aircraft and providing support for the wings.

Status
The equipment has been manufactured for Alitalia and other airlines of the ATLAS group, Aeritalia and the Italian Air Force

Manufacturer
Officine Meccaniche Aeronautiche (OMA), Foligno, Italy

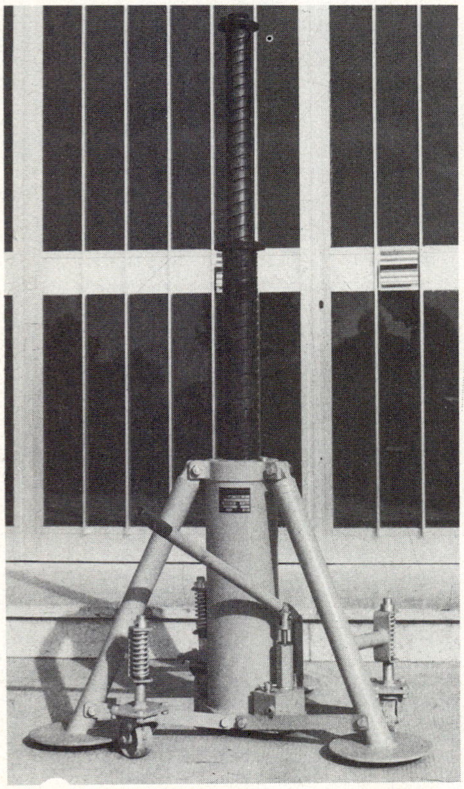

OMA Aeritalia G222 maintenance jack

OMA A300 maintenance platform for Alitalia

Cella Aerial Work Platforms

Cella produce a wide range of aerial work platforms. Hydraulically-operated and fitted with automatic protection devices, the platforms are equipped with dual control panels, one on the aerial cage and one on the turntable.
The 3D/6 series, with two articulated booms, is mounted on medium- and light-weight trucks and tractors, providing 10-, 12-, 13-, 15-, 17- and 18-metres maximum working height. The 3D/3 with three articulated booms, is based on medium and heavy-weight trucks and are available with 18-, 20-, 23-, 25-, 27-, 29- and 31-metres maximum working height. The 3D/4 equipped with three articulated booms and intermediate telescopic extensions, is designed to be mounted on a heavy chassis and is capable of reaching a maximum height of 30-, 40- and 50-metres.
Liftel models are fitted with telescopic boom sections, and can be mounted on standard truck chassis or on a specially-designed, self-propelled base. Maximum working heights range from 10 to 50 metres.
The Liftelmatic series are self-propelled units designed for the single operator with controls in the aerial cage for driving, steering, hoisting, lowering and operation of the outriggers. Electric-propelled versions are available.

Status
In service with Alitalia

Manufacturer
Cella SpA, Mandolossa/Brescia, Italy

NETHERLANDS

Hollandia Kloos Universal Docking System

Hollandia Kloos International have developed a universal docking system which can be applied to a wide range of aircraft for all types of maintenance. The total system comprises tail, fuselage, wing, nose and stabiliser docks. The universal tail-docks are of a tower construction so maintenance can be carried out simultaneously on the entire tail portion of the aircraft. For the purpose of changing the tail engine on a McDonnell Douglas DC10 or Lockheed Tristar a special platform incorporated in the construction lifts the engine to the required height for installation. When using the complete docking system, maintenance of the whole aircraft can be performed simultaneously including painting or fuselage repairs, engine maintenance and tail inspection plus any work to be carried out on the nose section.

Status
Hollandia Kloos International has supplied docking systems to several major airlines including KLM and Philippine Airlines

Manufacturer
Hollandia Kloos International, Krimpen a/d Ijssel, Netherlands

SPAIN

Equisa Maintenance Platforms

Equisa design and manufacture maintenance platforms for a range of aircraft. An example of the type of equipment available is the platform produced for the Spanish Air Force to provide access to the tail-section of the Boeing Vertol CH47C helicopter. The top level enables maintenance personnel to work on the rotor head and the intermediate level permits access to the two Avco Lycoming engines mounted on either side of the tail. The levels are connected by stairs equipped with hand-rails and the platform is mounted on four wheels. A hinged tow-bar is fitted which can be restrained in the vertical position when the platform is in place.

Status
Service with the Spanish Air Force

Manufacturer
Equisa, Madrid, Spain

Equisa maintenance platform providing access to Boeing CH47C helicopter

UNITED KINGDOM

Access AAP Series Adjustable Maintenance Platform

The AAP range of adjustable maintenance platforms is based on a scissor mechanism with variable platform heights of up to 4 metres and a safe working load of up to 750 kg. All are hydraulically operated either by hand- or power-pump. The platforms are mounted on castors with the exception of the AAP 4·0 which is supplied on a towable trailer fitted with road wheels. Outriggers are installed on two models to provide additional stability. All platforms are supplied with hand-rails and toe-boards.

Characteristics

MODEL	AAP 2·4	AAP 3·9	AAP 4·0	AAP 3·9 Mk 3
PLATFORM DIMENSIONS (inside guard-rails)	2 × 0·79 m	2 × 0·79 m	2·13 × 0·92 m	2 × 0·79 m
minimum height	1·9 m	1·43 m	1·55 m	1·43 m
max height	2·4 m	3·9 m	4 m	3·9 m
CAPACITY	200 kg	200 kg	750 kg	750 kg
OVERALL DIMENSIONS				
length	2·4 m	2·53 m	3·54 m	2·53 m
width	1·01 m	1·12 m	1·91 m	1·02 m
OUTRIGGERS		2·14 m		2·38 m

Manufacturer
Access Equipment Ltd, Hemel Hempstead, Hertfordshire, England

Access AAP maintenance platform

Access Giraffe Hydraulic Maintenance Platform

The Giraffe range of maintenance platforms provide mobile access to under-wing mechanisms, engine installations and the fuselage with a capacity for a safe working load of 270 kg. A Junior version designed for helicopters and small aircraft is rated at 204 kg. The platforms are mounted on castors and fitted with screw-down jacks to provide additional stability while the equipment is in use. Stepfast jacks are fitted to Models AA and D4. These are lowered to the ground by foot pressure on the foot wheel and the platform is stabilised by turning the wheel. To move the equipment the wheel is turned in the opposite direction and the jack returns to its original position. The platform is raised by the operator working a pump lever on the chassis until the platform is at the required height. To lower, the operator pulls the hydraulic-pressure release control near the pump. A mechanical device prevents descent unless the lock/release control is operated. A combined hand and power operation is available as an optional feature on all models except the Junior Giraffe which is provided with a hand or power operation.
A jig can be mounted on the platform of all models except the Junior version with a capability of lifting 113·4 kg.

Access Giraffe hydraulic maintenance platform

Characteristics

MODEL	JUNIOR	AA	B2	C2	D4
MAX WORKING HEIGHT	5·63 m	7·16 m	9·14 m	10·97 m	12·8 m
MAX PLATFORM LEVEL	3·81 m	5·33 m	7·31 m	9·14 m	10·97 m
MINIMUM PLATFORM LEVEL	1·82 m	2·28 m	4·26 m	4·87 m	6·7 m
OVERALL HEIGHT EXTENDED	4·72 m	6·24 m	8·22 m	8·83 m	10·66 m
PLATFORM DIMENSIONS (m)	1·22 × 0·68	1·22 × 0·91	1·22 × 0·91	1·22 × 0·91	1·22 × 0·91
CHASSIS DIMENSIONS (m)	2·36 × 1·82	3·65 × 1·82	4·14 × 2·13	4·14 × 2·13	4·9 × 2·43
WEIGHT	457·2 kg	914·4 kg	1065·9 kg	1225·3 kg	1814·4 kg

Manufacturer

Access Equipment Ltd, Hemel Hempstead, Hertfordshire, England

Access Mountie Fixed Maintenance Steps

Mountie steps are of tubular cross-section steel A-frame, all-welded construction with hand-rails and flat steel treads to which a slip-resistant surface is fitted. In sizes up to five steps, spring-mounted castors of 0·05 metres diameter are supplied. Rubber feet take over automatically from the castors as soon as the user's weight is placed on the first step. With the larger sizes a pedal is fitted which is depressed when the steps are in position for use, lowering them on to the rubber feet for firm anchorage. When the steps are ready to be moved further pressure on the pedal brings the castors into action. A variety of heights are available with two widths and a choice of platform depths in all but the two largest sizes.

Characteristics

RANGE OF STEPS: 3–12
HEIGHT OF TOP STEP: 0·77–2·89 m
OVERALL HEIGHT: 1·41–3·75 m
OVERALL LENGTH: 0·69–2·11 m
OVERALL WIDTH: 0·51–0·78 m
TREAD DEPTH OF TOP STEP: 0·25 or 0·76 m
ALL OTHER TREAD DEPTH: 0·18 m

Manufacturer

Access Equipment Ltd, Hemel Hempstead, Hertfordshire, England

Access Mountie steps

Access Zipper Elevated Maintenance Access Platform

The Zipper range is a trailer-mounted, battery-operated system designed for lifting two men in a glass fibre basket by means of articulated arms. The FM 30 is fitted with an additional internal manual telescoping member to provide an extra 1·52 metres reach. The trailer is an all-steel welded fabrication with hinged outrigger arms to form a square base. Screw-down jacks are provided for levelling and increased stability. The steel superstructure with extending arms is mounted on an enclosed ball bearing slewing ring which provides continuous slewing through 360 degrees. Power is supplied by four heavy-duty batteries and a mains-operated battery charger is incorporated. The motor and hydraulic pump are housed in a single unit embodying a hydraulic tank and integral relief valves. Normal maximum working pressure is 173 bar (2500 psi). The arms are operated by hydraulic cylinders with standard seals, chromed piston rods and integral lock valves for safety in the event of a line failure. Push-button control is provided for the operator from the working position. Additional safety features include: manually-operated valves, operated from the ground level to enable the upper and lower arms to be lowered in the event of a power failure or accident to the operator; safety interlocks to prevent the operator inadvertently reaching an unsafe condition and an automatic brake fitted to the slew pinion to stop side drift. Optional features include hydraulic levelling jacks, aluminium basket and dual control from ground level and from the basket.

Characteristics

MODEL	FM 30	FM 38	FM 45
MAX WORKING HEIGHT	8·08/9·44 m*	11·58 m	13·56 m
MAX OUTREACH	4·04/5·57 m	5·18 m	6·4 m
BASKET DIMENSIONS (m)	0·76 × 1·07 × 0·99	0·76 × 1·07 × 0·99	0·76 × 1·07 × 0·99
CAPACITY	300 kg/165 kg (with extension)	175 kg	250 kg
MAX SLEW SPEED	1 rpm	1 rpm	1 rpm
OVERALL DIMENSIONS IN TRANSIT			
length	4·95 m		6·55 m
width	1·45 m		
height	1·83 m		2·1 m
WEIGHT IN TRANSIT	1150 kg		2115 kg

Access Zipper 45 elevated maintenance access platform

Manufacturer

Access Equipment Ltd, Hemel Hempstead, Hertfordshire, England

Access Zip-Up Aluminium Staging System

Zip-Up is a welded construction aluminium staging system mounted on locking castors for mobility. Two types of staging are available, Span and Stairway.

Span consists of base frames which provide a starting height of 2·05 metres onto which other frames of varying heights are slotted together to build a tower to the required height. Suitable outriggers must be used for heights of up to 15·24 metres, after which towers should be tied onto buildings according to local regulations. All frames (base frames included) are available in two widths, 0·73 and 1·34 metres and there is a choice of three lengths, 1·91, 2·5 and 3·1 metres. Clip-on stairs provide access. Platform boards are placed at every level with trap-doors in the platforms and two boards at every working level. Safe-working load per platform is 225 kg distributed and 675 kg per structure. Two towers can be joined to span a greater working area by means of bridging beams. Outriggers of varying types are attached according to tower height and terrain. Both staging and outriggers have adjustable legs to compensate for uneven ground or height differences on stairs. This leg adjustment is not for use as additional working height. Where the site makes the use of outriggers impractical ballast weights may be substituted. Guard-rails, diagonals, stairs, platforms and outriggers are attached with locking hooks which ensure against accidental release.

Stairway consists of folding sections in which frames, braces and an integral stairway are incorporated. There are lower, upper and guard-rail sections and also half sections. The sections are hinged together and platforms drop into the sections. The safe working load per platform is 225 kg distributed and 675 kg per structure. This type of tower is particularly rigid and mostly used for repeatable maintenance work where heights can be predetermined and are fixed.

Manufacturer

Access Equipment Ltd, Hemel Hempstead, Hertfordshire, England

Access Zip-Up staging system

ACS Acklift 9000 Hydraulic Access Platform

The Acklift 9000 is a trailer-mounted hydraulic-access platform which can be operated with a variety of power packs. Petrol, diesel or electric drives can be supplied or dual-power packs, for example, petrol engine drive and ac mains supply, so the equipment can be used in an enclosed area where noxious fumes would be undesirable and at sites where mains supply is not available. A heavy-duty dc battery power pack is a further option.

Characteristics

MAX WORKING HEIGHT: 9 m
MAX OUTREACH: 4·75 m
CAPACITY: 160 kg
OVERALL DIMENSIONS: 5·3 × 1·6 × 2·1 m
WEIGHT: 1350 kg

Manufacturer

ACS Engineering Ltd, Southwell, Nottinghamshire, England

ACS Acklift 9000 hydraulic access platform

ACS Acklift 2100 Elevating Maintenance/access Platform

The Acklift 2100 is a hydraulic-access platform with a capability for inspecting aircraft up to and including the Boeing 747. The unit may be mounted on any chassis with a gross vehicle weight of 12 000 kg. Full hydraulic controls are provided at the base of the king post and in the personnel basket. The standard fibreglass personnel basket is 2·08 metres long and is capable of accommodating three people and their tools.

Characteristics

MAX WORKING HEIGHT: 21 m
MAX OUTREACH: 10·7 m
CAPACITY: 510 kg
OVERALL DIMENSIONS: 11·2 × 2·27 × 3·35 m
WEIGHT: 1350 kg

Status

ACS produces a range of vehicle-mounted hydraulic access platforms which vary in maximum working height from 9·5 to 31·5 metres

Manufacturer

ACS Engineering Ltd, Southwell, Nottinghamshire, England

ACS Acklift hydraulic access platform

Aero-Docks Maintenance Equipment

Aero-Docks Ltd design and manufacture facilities for the maintenance of aircraft. Facilities designed range from basic access steps to tail-docks and complete aircraft maintenance hangars for all major passenger aircraft. Most structures are designed by Aero-Docks to meet specific requirements and to integrate with existing or new hangars. Equipment fitted can include ancillary services, power transmissions and motivation for both main equipment and operating levels as required by various aircraft service checks.

The company produces both internal and external tail-docking facilities. In the latter case the dock forms an environmental seal to the hangar and can incorporate all services which are normally available within the main building. In the case of aircraft fully docked within the hangar tail-docks can be mounted and motivated either on the ground or from the hangar roof depending on the proposed docking method and other factors. Access platforms for nose, wings, fuselage and engines can be either fixed or mobile.

Status

Aero-Docks' equipment is in service with British Airways, British Caledonian Airways and other major airlines. A tail-dock built for the McDonnell Douglas DC-10 is 78 feet (23·77 metres) high. Motor-driven, it runs on rails set into the apron across two bays of the British Caledonian aircraft hangar at Gatwick Airport, London. It has five working levels set at different heights to provide access to various parts of the tail section and each level has extending floor sections which fit close to the aircraft contours

Manufacturer

Aero-Docks Ltd, Southampton, Hampshire, England

Armfield 105L/TR Aerial Work Platform

This elevating, articulated-boom maintenance platform is based on a two-wheel trailer which can be towed by any vehicle with a tow bar capacity of 15 tonnes. The general specification is similar to the L-series produced by Armfield except for the stabilisers and power source. The power to drive the hydraulic pump is supplied by a 5 hp petrol engine. Alternative power packs available are an electric motor which can be 12 or 24 volts dc, single or 3-phase ac; or a 5 hp diesel engine. Optional features include a remote-control to operate the engine from the cage, an intercom system between the cage, turntable and the cab, and the provision of air, water and electric services.

Characteristics

MODEL: 105L/TR
MAX WORKING HEIGHT: 10·5 m
MAX OUTREACH: 5·85 m
MAX WORKING LOAD: 200 kg
TOTAL BOOM SLEWING: 360° continuous
NETT WEIGHT OF PLATFORM: 1·45 tonnes (subframe and decking)
OPERATOR'S CAGE
length: 1·1 m
width: 0·6 m
CAGE INSULATION: 600 V
NUMBER OF STABILISERS: 4

Status

Armfield aerial access equipment is in service with several major airlines including British Airways and Gulf Air

Manufacturer

Armfield Engineering Ltd, Ringwood, Hampshire, England

Armfield Comet Aerial Work Platform

The Armfield Comet hydraulic platform is mounted on a combined telescopic boom/jib system which enables the platform to be manoeuvred from ground level to the maximum height with a rotation of 360 degrees. Stabilisation is achieved by means of vertical props which can be extended horizontally from the vehicle on the working side or both sides as required. The telescopic boom/jib system allows the platform to straddle obstructions. With the telescopic arm lowered and the jib in the vertical position the Comet platform can operate down to three metres below ground level.

Characteristics

MODEL	18 T	25 T	30 T
MAX PLATFORM HEIGHT	18 m	25 m	30 m
SIDE OUTREACH	11·7 m	14·6 m	16·3 m
ROTATION	360°	360°	360°
OVERALL LENGTH	±7 m	±9·4 m	±11·3 m
NUMBER OF JACKS	4	4	4
JACK SPREAD	3·1 m	4 m	4·2 m
TOTAL WEIGHT PLATFORM AND SUB-FRAME AND WALKWAY	6300 kg	12 000 kg	13 300 kg
TOTAL WORKING LOAD	350 kg	350 kg	350 kg
CAGE DIMENSIONS	1·5 × 0·8 × 1·1 m	1·8 × 1·2 × 1·1 m	1·8 × 1·2 × 1·1 m
GROSS VEHICLE WEIGHT	16 000 kg	18 500 kg	23 000 kg

Armfield Comet 25T aerial work platform

Status

The Comet system was developed by FN in Belgium. It is built in the United Kingdom by Armfield and marketed as the Armfield Comet

Manufacturer

Armfield Engineering Ltd, Ringwood, Hampshire, England

Armfield H Series Mobile Aerial Work Platform

An elevating, articulated-boom platform mounted on a standard commercial chassis which is available in six models with maximum working height from 45 to 70 feet (13·7 to 21·34 metres). The structure comprises two booms fabricated in box sections from Corten high-strength steel. The turntable and the sub-frame are manufactured from mild steel. The latter houses the oil reservoir and provides direct attachment points for the turntable, boom rest and stabilisers so that operational loads are transmitted independently of the chassis.

Rotation is provided by a large externally-geared bearing of the crossed roller type sealed with external lubrication points. A hydraulic motor provides the power through a worm-wheel reduction gear box. From the gear box a vertical shaft with pinion meshes with the external teeth of the bearing providing controlled, continuous rotation in either direction with inherent self-locking characteristics. Four telescopic hydraulically-operated stabilisers are incorporated in the sub-frame. They are operated independently from a control valve mounted on the sub-frame and possess over-travel for levelling on uneven ground. Lock valves are incorporated to prevent creep during operation and to maintain the ram position in the event of hose failure.

Dual-levelling rods are provided. These give smooth levelling while holding the cage rigidly and require no adjustment. Automatic safety stops limit the movements of the booms to prevent the levelling system exceeding its working limits.

Hydraulic power is provided from a pneumatically-operated power take-off from the carrying vehicle's gear box. Holding valves are fitted to both elevating rams so that boom positions are maintained in the event of power failure. The elevating rams incorporate internal restrictors which limit the speed of descent should the holding valves be accidentally damaged. A hand-pump is supplied enabling all boom movements to continue in the event of the primary source of power failing.

Dual controls are fitted with one valve bank on the turntable and one in the cage. The cage-valve bank has priority and is recessed to prevent accidental operation.

Optional features include workshop cab; boom/stabiliser interlock; handbrake/pto interlock; remote cage/turntable/cage spotlight, safety netting for the operator's cage, auxiliary engine, air, water, electrical services and an hour meter.

Armfield 152H and 183H are both available with a telescopic upper boom, designated 152L and 183L respectively.

Characteristics

MODEL	137H	152H	168H	183H	200H	213H
MAX WORKING HEIGHT	13·7 m	15·2 m	16·8 m	18·3 m	20 m	21·34 m
MAX WORKING OUTREACH	7·25 m	8 m	8 m	8 m	8 m	8 m
MAX WORKING LOAD	360 kg	360 kg	300 kg	250 kg	250 kg	225 kg
TOTAL BOOM SLEWING	360° continuous	360° continuous	360° continuous	360° continuous	360° continuous	360° continuous
NETT WEIGHT OF PLATFORM, SUB-FRAME AND DECKING	4·15 tonnes	4·25 tonnes	4·375 tonnes	4·5 tonnes	7·28 tonnes	7·38 tonnes
OPERATOR'S CAGE LENGTH	1·5 m	1·5 m	1·5 m	1·5 m	1·5 m	1·5 m
OPERATOR'S CAGE WIDTH	0·9 m	0·9 m	0·9 m	0·9 m	0·9 m	0·9 m
CAGE INSULATION	600 V	600 V	600 V	600 V	600 V	600 V
NUMBER OF STABILISERS	4	4	4	4	4	4
RECOMMENDED CHASSIS	8128 kg	9134 kg	10 160 kg	10 160 kg	12 550 kg	11 330 kg
wheelbase	3·33 m	3·83 m	3·96 m	4·24 m	4·902 m	4·902 m

Status

Armfield aerial access equipment is in service with several major airlines including British Airways and Gulf Air

Manufacturer

Armfield Engineering Ltd, Ringwood, Hampshire, England

Armfield L Series Mobile Aerial Work Platform

An elevating, articulated-boom maintenance platform mounted on a standard commercial chassis available in three models with maximum working heights ranging from 8·5 to 13 metres.

The equipment comprises two articulated booms, turntable and sub-frame, stabilisers, dual-levelling rods, hydraulic system, operator's cage and controls. The booms are manufactured from standard steel sections and the turntable and sub-frame are constructed in mild steel. The sub-frame houses the oil reservoir and provides direct attachment points for the turntable, boom rest and stabilisers so that operational loads are transmitted independent of the chassis. The telescopic, hydraulically-controlled stabilisers are incorporated in the sub-frame. They are operated independently from a control valve mounted on the sub-frame and possess over-travel for levelling on uneven ground. Lock valves are incorporated to prevent creep during operation and maintain the ram position in the event of hose failure. External dual-levelling rods are provided which give smooth levelling while holding the cage rigidly and require no adjustment. Automatic safety stops limit the movement of the booms to prevent the levelling system exceeding its working limits.

Rotation is provided by a large bearing of the crossed-roller type. Power is supplied by a hydraulic motor driving a worm which meshes with a worm wheel.

The hydraulic pump is driven from the carrying vehicles gear-box power take-off. The simple hydraulic circuit includes rams fitted with adjustable safety restrictors and flexible hose. An emergency descent valve is fitted to allow the top boom to be lowered at a controlled rate of descent in the event of an emergency. A hand-pump is installed to enable all movements of the booms to be made if the primary power source fails.

A manual valve bank mounted in the cage allows the operator to control upward, downward and slewing movements. These movements are duplicated on the turntable though this is an optional feature in the 85L. Safety stops are provided to prevent the booms being moved to unsafe positions.

The cage is manufactured from steel-reinforced glass fibre and can hold two workers.

Optional features include interlocks for the boom/ stabilisers, handbrake/stabiliser and handbrake/pto; remote-engine controls fitted to the cage; intercom system between cage, turntable and cab; auxiliary engine and the provision of air, water and electrical services.

Characteristics

MODEL	85L	110L	130L
MAX WORKING HEIGHT	8·5 m	11 m	13 m
MAX OUTREACH	4·25 m	5·85 m	6·8 m
MAX WORKING LOAD	175 kg	200 kg	200 kg
TOTAL BOOM SLEWING	360° continuous	360° continuous	360° continuous
NET WEIGHT OF PLATFORM, SUB-FRAME AND DECKING	2·4 tonnes	2·8 tonnes	2·9 tonnes
OPERATOR'S CAGE			
length:	1·1 m	1·1 m	1·1 m
width:	0·6 m	0·6 m	0·6 m
CAGE INSULATION	600 V	600 V	600 V
NUMBER OF STABILISERS	2	2	2
RECOMMENDED CHASSIS	3455 kg	5791 kg	7498 kg
wheelbase:	2·8 m	3·35 m	3·35 m

Status

Armfield aerial access equipment is in service with several major airlines including British Airways and Gulf Air

Manufacturer

Armfield Engineering Ltd, Ringwood, Hampshire, England

Didsbury Minilift Aircraft Hoist

The Minilift is a light and portable system comprising a winding mechanism to which a wide range of ancillary attachments can be fitted to meet specific lifting requirements. The mechanical power for lifting is supplied by the manually-operated winding mechanism with a 9-inch (12·8-cm) handle. Two lifting speeds are provided; slow for relatively heavy loads and fast for rapid winding and unwinding of the unloaded cable or for carrying light loads. Safety in the operation is ensured by a deadman-type brake which is normally in the on position. The mechanism consists of a light-alloy casing enclosing three main assemblies interconnected by roller chains and chain-wheels. These are the cable drum, brake- and counter-shaft assemblies, each mounted on ball bearings within the casing. The lifting action is obtained when the winding handle engages the dogs fitted to the protruding ends of the counter- and brake-shafts. A simple device prevents rotation of the winding handle when lowering. In its basic form the winding mechanism with cable is attached to a hook through extension tubes with a tube-coupling unit between each section and a top sheath which is attached to a strong point on the aircraft itself or to a convenient outside structure. The system provides for the forced feed of cable from the mechanism, guided through tubing integral with the aircraft structure to emerge at points remote from the winding mechanism. For this reason extension tubes include couplings which can change the direction of the hoist.

A range of sheaves are available. All types are similar in construction comprising a light alloy casting housing a pulley which is ball-bearing mounted on a fixed spindle.

The method of suspension from the strong-point differs between models but each has a standard socket for attaching to an extension tube and provision is made for the cable to be fed through without dismantling the sheath.

With all capacities other than the 25 cwt (1270 kg) model sheaths are designed to act in conjunction with a bottom sheath as a double-purchase block handling up to twice the normal capacity of the hoist. Where this facility is provided the design safety factors allow for double loading. Another attachment is the Minilift Servicer which provides a strong-point to which the top sheath can be attached. The standard 5 cwt (254 kg) Minilift, which can be removed for other uses, provides the mechanical lifting and controlling mechanism. The chassis is equipped with fully swivelling wheels. To each wheel is fitted a clamping brake device to prevent movement and a directional lock. A handle is supplied at the rear of the Servicer for manual manoeuvring. The pillar is designed to fold for transit and a wheel unit is available, fitted with two road wheels, to provide towing facilities.

Characteristics

MODEL	CAPACITY SINGLE FALL	MAX CABLE EXTENSION	WEIGHT
41/1	2½ cwt (127 kg)	24 ft 9 in (7·54 m)	22 lb (10 kg)
60/1	2½ cwt (127 kg)	44 ft 3 in (1·27 m)	40 lb (18 kg)
42/1	5 cwt (254 kg)	24 ft 8 in (7·52 m)	28 lb (12·7 kg)
43/1	10 cwt (508 kg)	21 ft 9 in (6·63 m)	41 lb (18·6 kg)
54/1	18 cwt (914 kg)	21 ft 6 in (6·55 m)	47 lb (21 kg)
44/80	25 cwt (1270 kg)	15 ft 9 in (4·8 m)	58 lb (26·3 kg)

Status

In extensive use with the Royal Air Force and the Royal Navy the Minilift system is in service also with commercial airlines.

Manufacturer

Didsbury Engineering Co Ltd, Manchester, England

EAI Aircraft Jacking Systems

These jacking units are purpose-built for application to third generation and wide-body aircraft and are designed to sustain their rated loads for extended periods under diverse conditions.

The jacking systems involve primarily the use of a number of general-purpose tripod and quadruped hydraulically-actuated jacks with modified units available for special requirements.

The jacks are divided into three categories: the primary jacks are the units that lift the aircraft and remain in position throughout, giving main support and are usually at points that correspond to the aircraft's longitudinal and lateral axes and the load-bearing keel and centre-wing structural boxes; the secondary jacks are those that are deployed to augment the primary jacking system during engineering and landing-gear retraction tests which may induce aircraft movement and centre of gravity imbalance. The auxiliary jacks are normally used as stabilisers and are only used for jacking when the aircraft is at maximum all-up weight, thereby exceeding the maximum single-point jacking loads laid down by the facility planning manual pertaining to the specific aircraft.

Primary elevation of the jacks is by means of single- and double-action hydraulic rams which when raised to the required height can be locked by a device that enables the weight of the aircraft to be carried for a long period of time without risk of failure. This consists of a locking collar which is screwed down the ram of the jack and locks on to the head of the hydraulic cylinder.

The jacks are manually operated as standard but have attachments to accept either pneumatic-, electric- or hydraulic-pressure rig facilities and can be supplied to suit varying combinations of jacking sets.

Standard operation is achieved by two manually-operated pumps: a low-pressure one for raising the head without load to the required height to enable engagement with the aircraft jacking point and a high-pressure one for jacking the aircraft. Lowering is effected by the manual operation of the jack-lowering valve the speed of which is variable. Apart from the main hydraulic extension further variation can be attained by use of screw extensions and by alternate lug positions on the main-jack body.

Characteristics

JACK MODEL	RATED LOAD (tons)	CLOSED HEIGHT A (cm)	CLOSED HEIGHT B* (cm)	HYDRAULIC LIFT (cm)	SCREW EXTENSION (cm)	MAX EXTENSION A (cm)	MAX EXTENSION B* (cm)
T-100-66L/33/S	254	335·3		167·6	8·38	586·7	
T-100-66L/33/S	190·5	335·3		167·6	8·38	586·7	
Q-60-66L/33/S	152·4	170·2	193	99·1	33·02	269·2	292·1
Q-50-39L/13/S	127	170·2	193	99·1	33·02	269·2	292·1
Q-40-39L/13/S	101·6	170·2	193	99·1	33·02	269·2	292·1
T-30-36L/9/S	76·2	139·7		91·4	33·02	231·1	
Q-25-41L/9/S	63·5	139·7	162·6	104·1	33·02	243·8	266·7
T-15-41L/9/S	38·1	139·7	162·6	104·1	33·02	243·8	266·7
Q-5-18L/10/S	12·7	61	91·4	45·7	25·4	106·7	137·2

*Heights A and B refer to the two alternative positions of the trestle lugs

Status

EAI aircraft jacking systems are in service with seven airlines including Air Malta, Britannia, British Caledonian, BMA and Cathay Pacific (Hong Kong Engineering Company)

Manufacturer

Eastern Airmotive Industries (USA) Ltd, St Albans, Hertfordshire, England

Edgehill Maintenance Access Equipment

Edgehill produce a range of access equipment designed for aircraft maintenance and servicing. These comprise full wing, fuselage and tail maintenance staging and scissor lift workstands, towable and self-propelled with platform heights ranging from 0·75 to 7·5 metres. Giraffe access steps are available with platform heights ranging from 1·6 to 7·3 metres. Other maintenance equipment supplied includes aircraft tow bars and undercarriage wheel-change rigs, for narrow- and wide-body aircraft.

Status

Edgehill maintenance equipment is in service with airlines in Europe, the Middle East and the United States of America

Manufacturer

H W Edgehill Equipment Ltd, Hook, Hampshire, England

Edgehill EDG,391-1 Giraffe access step supplied to Gulf Air

Kwikform Access Towers and Personnel Lifts

The aluminium Kwiktower provides access to all types of maintenance and construction work. Lightweight, it is fitted with guard-rails for security at platform level.

Frame modules of 4 feet 8 inches (1·45 metres), 7 feet (2·15 metres), and 9 feet 3 inches (2·85 metres) sizes are available. Kwiktowers may be mobile or static. Ladder and trap door units can be supplied.

The Kwiktower personnel lift with telescopic mast simplifies access. Three models provide working heights up to 19 feet (5·8 metres), 25 feet (7·6 metres) or 31 feet (9·5 metres). Each is powered by a hydraulic unit running from a small electric motor. Safety locks ensure the security of personnel. The aluminium construction provides a lightweight unit which can be handled by one person.

Manufacturer

Kwikform Ltd, Birmingham, England

Kwiktower personnel lift

Aluminium-framed Kwiktower with vertical ladder access

Laser 456-1 and 237-1 Undercarriage Ground Handling Units

These two units have been developed for ground handling of the Boeing 747 main undercarriage and nose-wheel systems.
The 456-1 is a main undercarriage four wheel-bogie system dolly providing two degrees of angular motion through an air-powered jacking system. The whole assembly carries air-cushion hover pallets which permit lateral motion in any direction.
The 237-1 for nose-wheel equipment provides vertical movement through a manually-operated hydraulic jacking system and lateral movement using two groups of three hover pallets in a similar manner to the main undercarriage unit. Each of the groups are lowered or raised by two hydraulic jacks and either pair of jacks can be operated independently to the other, thus providing a small degree of angular movement to locate the undercarriage when fitting. The normal hangar-servicing air-line pressure of approximately 55 lkN/m^2 is adequate to operate the hover systems and also to operate the vertical jacking system on the main undercarriage equipment.

456-1 Main undercarriage hover pallet servicing unit

Status
This equipment is in service with British Airways and Kuwait Airways

Manufacturer
Laser Engineering (Development) Ltd, London, England

237-1 Nose-wheel hover pallet servicing unit

Laser L391 Aircraft Wheel Handling Trolley

The L391 Aircraft Wheel Handling Trolley has been designed for the removal and replacement of aircraft wheels and brakes.
The rotating wheel-support forks have adjustable spread and are mounted on an articulated frame, the height of which can be positioned using a manually-operated hydraulic jack. The main structure of the trolley has exceptionally low clearance height to give the maximum possible access to the aircraft being serviced.
Four-castored wheels are provided, of which the two rear ones may be locked in any of four 90-degree quadrants.
The trolley may be towed by a standard tractor on normal airfield surfaces in either the unloaded or loaded condition. A chain is provided to retain the wheel load in position.
A special attachment can be supplied to permit the trolley to remove, replace or carry a brake unit as an alternative to the wheel, according to the customer's brake stock handling methods. A choice of two standard brake stack adaptors is available.
The equipment is designed to be suitable for all aircraft wheels up to 4 feet 6 inches (1·37 metres) diameter. These include Boeing 747, McDonnell Douglas DC-10, Lockheed L011, Airbus A300/310, Concorde.

Characteristics
LENGTH
(excluding tow-bar): 63 in (160 cm)
WIDTH: 35 in (89 cm)
MINIMUM ROLLER HEIGHT: 2½ in (6 cm)
MAX ROLLER SPACING: 28 in (71 cm)
LIFT: 11 in (28 cm)
MINIMUM FRAME HEIGHT
(without extensions): 22½ in (57 cm)
EXTENSION LENGTH: 9 in (23 cm)
MAX TYRE SIZE: 54 in (13 cm) (nominal)
ROLLER LENGTH: 18 in (46 cm)
LIFTING CAPACITY: 500 lb (227 kg)
ADAPTORS FOR
brake units: up to 500 lb (227 kg)
max diameter: 24 in (61 cm)
WEIGHT: 375 lb (170 kg)

Status
In service with British Caledonian Airways, European and the Middle East Airlines

Manufacturer
Laser Engineering (Development) Ltd, London, England

Mingay's Aircraft Jacks

Two types of aircraft jacks are produced.
The Tripod jack with capacities up to 5 tons (5080 kg) is suitable for use with the B Ae 125 aircraft, S61 helicopter and other similar sized aircraft.
Axle jacks are available with capacities up to 30 tons (30 482 kg). The largest version is designed for use with Boeing 757 and 737 aircraft. A 10-ton (10 160 kg) capacity axle jack can also be used for Boeing 737 maintenance. The 11-ton (11 176 kg) rated jack is suitable for BAe 1-11, 748 and other similar aircraft, while the 6-ton (6096 kg) capacity jack is designed for lifting smaller aircraft such as the BAe 125.
In addition to those aircraft listed above, Mingay jacks are available for maintenance on BAe 146, Jetstream, Nimrod, Trident, Boeing 707, Dassault-Breguet Falcon 50, Fokker F28, Lockheed Tristar, McDonnell Douglas F4 Phantom, Pilatus B-N Islander, Trislander and the Short Skyvan.

Status
Mingay's have supplied aircraft jacks to major commercial airlines and military forces throughout the world

Manufacturer
Mingay's (Chigwell) Ltd, Stansted, Essex, England

Power Lifts Air Worker Mobile Servicing Platform

Air Worker mobile platforms are powered by a pneumatic motor supplied with air from the hangar compressed air system through a hose mounted on a self-reeling drum. The motor, installed in the centre of the driving wheel, powers a hydraulic pump which supplies the lifting cylinders and wheel motors. The driving wheel includes a tensioning device to ensure positive contact with the ground. it is mounted centrally between two swivel wheels. The operator controls speed and direction from the pivoted control arm, which may be used also to tow the vehicle. The two rear wheels are fixed.
The platform is based on a scissor mechanism and uses four jacks one mounted on each corner for stability. The platform can be adjusted to any desired height within its working range. For underwing fuselage work, models can include an angled platform, to follow the contours of the aircraft, while for rear stabilisers, the lift section is mounted on suitable extended framework.
Lift controls are installed on the platform. Access is provided by an extended vertical ladder and/or integral steps. Each platform, which is designed to carry four men and equipment, up to a total of 1000 kg, has an open steel mesh surface with removable drip trays. Steel hand-rails and a spring-closing gate are provided for security. Other safety and aircraft protection features include: a buffer hose fitted to all edges of the top platform and to the outer edge of the staircase; control mechanism to prevent the equipment being extended without an operator in control on the platform; emergency lowering valve fitted at chassis level in case of an air supply failure and an emergency hand-pump which can be used to raise the platform if required. Pressure sensitive guards, fitted beneath the platform, will stop descent in the event of an obstruction and valves incorporated in the lifting cylinders, prevent collapse due to hydraulic failure.

Characteristics

MODEL	UNDER-WING ACCESS	WING TIP AND LOWER FUSELAGE	HORIZONTAL STABILISER
PLATFORM			
dimensions	3·05 × 2·44 m	3·05 × 2·44 m	3·05 × 2·44 m
height	Variable 1·5 to 4 m	Variable 1·5 to 4 m	Variable 5 to 8 m
ADJUSTABLE SLOPE	Up to 4°		Up to 7°
CAPACITY	1000 kg	1000 kg	1000 kg
AIR HOSE LENGTH	22 m	22 m	22 m

Status
The basic models were designed for Boeing 747 and Concorde aircraft, but the range of adjustment enables the equipment to be used with other aircraft. In service with British Airways

Manufacturer
Power Lifts Ltd, Watford, Hertfordshire, England

Power Lifts Skyworker Mobile Elevating Maintenance Platform

The Skyworker is a self-powered mobile elevating maintenance platform. Power is supplied by a 48 volt battery-driven hydraulic power unit which provides forward and reverse traction with a capacity for 45 lifts in an eight-hour shift. A battery charger is incorporated to connect to a mains supply for recharging purposes. The platform constructed in steel with plywood decking is based on triple-acting scissors. Steering and elevation control is from the platform using a traversing and steering joystick and deadman-type push buttons. Additional safety features include: toe-boards, hand-rails and safety gate, trip-bar safety guards on all four sides which stop the platform automatically in the event of an obstruction and valves incorporated in the lifting cylinders to provide a controlled descent, should a hydraulic failure occur. An overriding manual valve is fitted to ensure controlled descent in an emergency.
Optional features include an extending powered cantilever platform; self-levelling hydraulic stabilisers; fixed height; demountable, extension platform; and platform-mounted jib crane.

Characteristics
PLATFORM DIMENSIONS: 3·81 × 1·98 m
minimum height: 1·68 m
max height: 9·14 m
stroke: 7·47 m
SPEED
vertical: 89 mm/s
forward: 3·2 km/h with platform up to 4·88 m high
1·6 km/h with platform above 4·88 m
reverse: 1·6 km/h
CAPACITY: 508 kg
OPERATIONAL WEIGHT: 3175 kg

Status
In service with British Airways

Manufacturer
Power Lifts Ltd, Watford, Hertfordshire, England

Skyworker mobile maintenance platform providing access to a Boeing 747 horizontal stabiliser

Power Lift Versalift Elevating Maintenance/Access Platform

The Versalift is an elevating maintenance/access bucket-shaped platform designed for the single operator. It comprises a square fibreglass bucket attached to a telescopic arm mounted on a commercial van chassis. The upper boom of the arm is an aluminium alloy tube, the lower boom and pedestal are constructed of steel tubing and weldment. Rotation is by hydraulic motor through a speed reducer and the boom is raised and extended by hydraulic cylinders. Hydraulic power operating at 138 bar (2000 psi) is produced by either an electric-powered pump or from a belt drive pto from the vehicle. Control is provided by a 12 volt dc control system, operating a solenoid valve, installed in the personnel bucket. Manual override actuators are fitted at the pedestal. Lateral stabiliser jacks are mounted, one on each side of the vehicle. Safety features include counter balance holding valves to prevent creep and lock the cylinder in the event of a hydraulic line failure, crossover-valve to limit rotation torque, self-locking worm drive to stop rotation drift and bucket dashpot and limit stops to control rate and extent of bucket travel. An emergency hydraulic power system can be supplied as an optional feature. A double bucket version is available mounted on a Dodge 546 commercial van.

Characteristics

CHASSIS	Commercial van eg Ford Transit	
BOOM TRAVEL	−6 to +84°	
ROTATION	360°	
SINGLE BUCKET MODEL	Van-Tel 24	Van-Tel 28
HEIGHT TO BOTTOM OF BUCKET	7·62 m	8·84 m
WORKING HEIGHT	9·14 m	10·36 m
OUTREACH	5·18 m	6·4 m
STOWED TRAVEL HEIGHT	2·74 m	2·74 m
BUCKET CAPACITY	136 kg	136 kg

Status
Production

Manufacturer
Power Lifts Ltd, Watford, Hertfordshire, England

Versalift Van-Tel 24B elevating platform

Flying Carpet Elevating Maintenance Platforms

The Flying Carpet range of self-propelled work platforms is based on a hydraulically-operated scissor mechanism powered by batteries. All functions of the equipment (raise, lower, drive, reverse and steer) are operated from the platform with deadman controls on the joystick. Cantilever cages can be fitted to the main platform allowing access over ground obstructions.
The Flying Carpet is available in three categories; 5000 series slab-type and 8000 series, slab-type and rough terrain. Key features of the 5000 series are four 6-volt batteries rated at 250 ampere hours which can be recharged through an integral charger from 110/220 volt supply, and platform guard-rails which fold inboard for transit. The 8000 series (slab-type) is available with propane or diesel engine or battery power options. It is fitted with solid tyres and twin-drive wheels designed for use on hard level floors. The rough terrain version, which can be powered by either a propane or a diesel engine, is supplied with large flotation tyres. Powered outriggers with interlock controls, prevent travel with the platform raised unless outriggers are within 2 inches (50·8 mm) of the ground.

Characteristics

SERIES	5000	8000
CAPACITY	320–500 kg	900–1100 kg
MAX PLATFORM HEIGHT	5·5–7·28 m	7·5–11 m
MINIMUM PLATFORM HEIGHT	1·57–1·65 m	1·35–1·8 m
PLATFORM		
width	1–1·2 m	2·25 m
length	1·75–2·44 m	4 m
CAGE		
fitted	2·44–3·13 m	
extended	3·55–3·94 m	
BATTERY CAPACITY	40–60 elevations	40–60 elevations (slab-type only)
WEIGHT	1690–1950 kg	battery-powered: 4100–4600 kg propane engine: 3600–4800 kg diesel engine: 3700–4850 kg
with cage fitted	1750–2010 kg	

Status
In service with airlines in Australia, Federal Republic of Germany, Kuwait and the United States of America

Manufacturer
John Rusling Ltd, Newport, Shropshire, England

Flying Carpet mobile work platform with extending cantilever cage providing access to a DC-10 aircraft at Hanover Airport

Instant Rib-Grip Maintenance Scaffold

Rib-Grip light alloy maintenance scaffold comprises a rationalised three frame access system using interchangeable seven, four and three rung end frames for use in two widths and three standard length modules of 1·9, 2·5 or 3 metres. These are linked vertically or arranged in parallel with interconnecting bridges.
The 1·9 module offers the option of inclined stairs with banister guard-rails as an alternative to the integral vertical ladder frame, enabling multi-level maintenance platforms to be constructed with stairs and guard-rails in every extension section. Other optional features include high clearance base sections permitting spanning of machines, personnel doors etc.

Manufacturer
John Rusling Ltd, Newport, Shropshire, England

Instant Rib-Grip maintenance scaffold

Rusling Maintenance Steps

A new development from John Rusling is a range of portable aluminium steps for use by maintenance personnel providing access to aircraft doors and for servicing and inspection purposes.

Status
In service with British Airways

Manufacturer
John Rusling Ltd, Newport, Shropshire, England

Rusling maintenance steps

Simon CA Series Hydraulic Maintenance Platforms

The CA series is a development from Simon Engineering of four models with maximum working heights from 13·5 to 21·5 metres.
The elevating platform is mounted on a twin-boom system installed on a welded main frame with four integral jack housings containing four telescopic jacks with integral lock valves. The turntable rotates on a ball bearing slewing ring with hydraulic motor drive. Rotation is continuous in either direction. Identical large double-acting cylinders power each boom. The cylinders have built-in restrictor valves preventing sudden descent even if a hydraulic line should fail. Cushioning is provided at each end of the stroke. The booms are of rectangular hollow sections with the first boom closed for maximum rigidity and the second boom open to provide access and protection to the hydraulic hoses and levelling cables. A 'compensated' operating linkage of the second boom is designed to reduce operating time. The platform is insulated against 3000 volts and all sides are fully enclosed with a self locking door for access. The platform levelling system consists of dual steel cables 16 mm diameter with a safety factor of 32–1. A fully hydraulic control system is fitted so that at each control position the operator directly controls the flow of oil, with controls on the platform and the vehicle. In both positions valve handle movements correspond to the platform movements.
The lower controls are grouped in a panel at the rear of the decking immediately adjacent to the platform access door. In its lowered position, the operator has direct access to the platform from ground level. The hydraulic power requirement is very low, usually requiring only tick-over speed of the vehicle engine to drive the power take-off unit on the gearbox.
Safety devices fitted include: mechanism to ensure that booms are stowed before platform is lowered to dismounting position; stop to prevent operator leaving the cage when booms are elevated; jack/boom hydraulic interlock; boom cylinder lock valves; upper boom bleed-down system, operable from the cage and dual insulation at the turntable.
Optional features comprise: hydraulic, pneumatic and water services at the platform, intercom, spotlight, vehicle stop/start controls from the platform, and electrical slip rings on the centre post to permit continuous rotation movement.

Characteristics

MODEL	CA 120	CA 160	CA 190	CA 200
MAX WORKING HEIGHT	13·5 m	17·5 m	20·5 m	21·5 m
MAX PLATFORM HEIGHT	12 m	16 m	19 m	20 m
MAX OUTREACH		9 m	10 m	10·65 m
at height		9·5 m	13 m	11·77 m
MAX WORKING LOAD		465 kg	265 kg	500 kg
PLATFORM DIMENSIONS		9·15 × 2·5 × 3·5 m	10·65 × 2·5 × 3·5 m	11 × 2·5 × 3·5 m
RECOMMENDED VEHICLE				
minimum gvw		11·8 tonnes	12 tonnes	13 tonnes
wheelbase		4·2 m	4·5 m	5·2 m

Manufacturer
Simon Engineering Dudley Ltd, Dudley, West Midlands, England

Simon DH 167 Hydraulic Maintenance Platform

A twin-hydraulic boom system the DH 167 is very similar in operation to the S series produced by Simon Engineering. A glass-fibre cage insulated against 600 volts available in two sizes replaces the light alloy maintenance platform used in the S series. Other differences comprise the control system, which uses open-centre, four way, double-acting hydraulic spool valves with control levers directly connected to the valve spools; the hydraulic pipes to the cage, which are incorporated within the booms; and the stabilising jacks. The latter act directly and unlike outriggers when fully extended are little wider than the vehicle.
Optional features include jack/boom interlock, upper boom lock-valves, lighting and compressed air services at the cage.

Simon DH 167 hydraulic maintenance platform

Characteristics
MAX WORKING HEIGHT: 18·2 m
MAX PLATFORM HEIGHT: 16·7 m
MAX OUTREACH: 10·3 m
MAX WORKING LOAD
standard cage: 585 kg
large cage: 510 kg
ROTATION: 360° in either direction
CAGE DIMENSIONS
standard: 1·07 × 0·61 m
large: 1·52 × 0·91 m
NET WEIGHT OF PLATFORM BOOM AND MOUNTING: 4978 kg (excluding vehicle)
RECOMMENDED CHASSIS: 8140 kg (minimum gvw)

Status
In service with British Airways

Manufacturer
Simon Engineering Dudley Ltd, Dudley, West Midlands, England

Simon S Series Hydraulic Maintenance Platforms

The S series are three-boom construction hydraulic platforms, which can extend to a maximum working height of 31·5 metres, mounted on either a custom-built or a standard commercial chassis. The equipment comprises three articulated booms, turntable, and sub-frame, stabilising jacks, dual-levelling rods, hydraulic system, operator's platform and controls. The steel booms are connected by pivot pins in bored holes with pre-lubricated bearings, and installed on a steel turntable with a double-row ball bearing slewing ring. Rotation is by hydraulic-motor drive through an oil-immersed reduction unit worm gear. The main frame is a rigid welded structure, with integral stabilising jacks at each end. Construction of the jacks is telescopic with rectangular box-sections enclosing the operating cylinders. Large swivelling end-plates are fitted to spread the load evenly. Each jack is locked in position by a hydraulic-lock valve and by wedge action under load.
The platform is a light-alloy structure with a lifting eye located underneath. Services which can be provided at the platform include: intercom, electric power, compressed air and water. Hydraulic power take-off points are supplied. Electric lighting comprising a 60 watt spotlight and illumination of the controls is operated from the vehicle's batteries.
A variable delivery hydraulic pump is driven from the vehicle's power take-off. The hydraulic circuit is fitted with full flow suction and pressure filters. A stand-by engine driving an auxiliary hydraulic pump allows low-speed operation of all movements in the event of failure of the main engine or pump.
Other safety features include automatic safety stops at the limits of boom movement; restrictor valves in the boom cylinders to control the rate of descent; hydraulic-lock valves to maintain the ram position in the event of a pipe failure; a second set of controls at the turntable which can override the platform controls in an emergency; and an interlock to prevent operation or rotation of the booms until all four stabilising jacks are in position. The jacks cannot be retracted until the booms stowed for transit.

Characteristics

MODEL	S 220	S 263	S 300
MAX WORKING HEIGHT	23·5 m	27·8 m	31·5 m
MAX PLATFORM HEIGHT	22 m	26·3 m	30 m
MAX OUTREACH	12·59 m	14·66 m	15·75 m
at height	11·73 m	14 m	16·1 m
MAX WORKING LOAD	510 kg	365 kg	365 kg
ROTATION	360° continuous	360° continuous	360° continuous
PLATFORM DIMENSIONS			
length	2·29 m	2·29 m	2·29 m
width	1·23 m	1·23 m	1·23 m
NET WEIGHT OF PLATFORM BOOMS AND MOUNTING (excluding vehicle)	8·13 tonnes	8·53 tonnes	9·55 tonnes
RECOMMENDED CHASSIS (minimum gvw)	14·73 tonnes	15·3 tonnes	16·26 tonnes

Status
The S 70 in this series is being replaced by the CA 200. Simon hydraulic platforms are in service with airlines in Europe, the Middle East, Africa and the United States of America, aircraft manufacturers and European airport authorities

Manufacturer
Simon Engineering Dudley Ltd, Dudley, West Midlands, England

Simon S series hydraulic platform in service with Air France

Simon S 70 hydraulic platform providing access to the fin of an Air Portugal-TAP Boeing 707 aircraft

---

Transequip Engine Change Equipment

Transequip designs and manufactures special-purpose mobile engineering equipment. The engine change equipment is designed to complement the fifth POD system and enable an engine change to be effected at an off-line station. All the necessary equipment, including fan blades which are not flown on the fifth POD engine, tools and accessories area arranged in air-portable palletised containers which can be transported on the special dollies supplied.

Status
Transequip has delivered eleven sets of equipment for the Boeing 747 to British Airways, Qantas, and Saudia. Designs have been completed for engine change equipment for most major aircraft types and engines

Manufacturer
Transequip (London) Ltd, Feltham, Middlesex, England

Skyhi Aircraft Hydraulic Jacks

Skyhi design and manufacture a range of hydraulic jacks for all types and sizes of civil and military aircraft. The jacking systems are available for routine maintenance, servicing and recovery of aircraft and helicopters. All jacks can be fitted with a pneumatically-operated power pack in addition to the standard manual operating pumps.

Status

Skyhi has supplied aircraft jacks to the Ministry of Defence in the United Kingdom

Manufacturer

Skyhi Division Zwicky Engineering Ltd, Wokingham, Berkshire, England

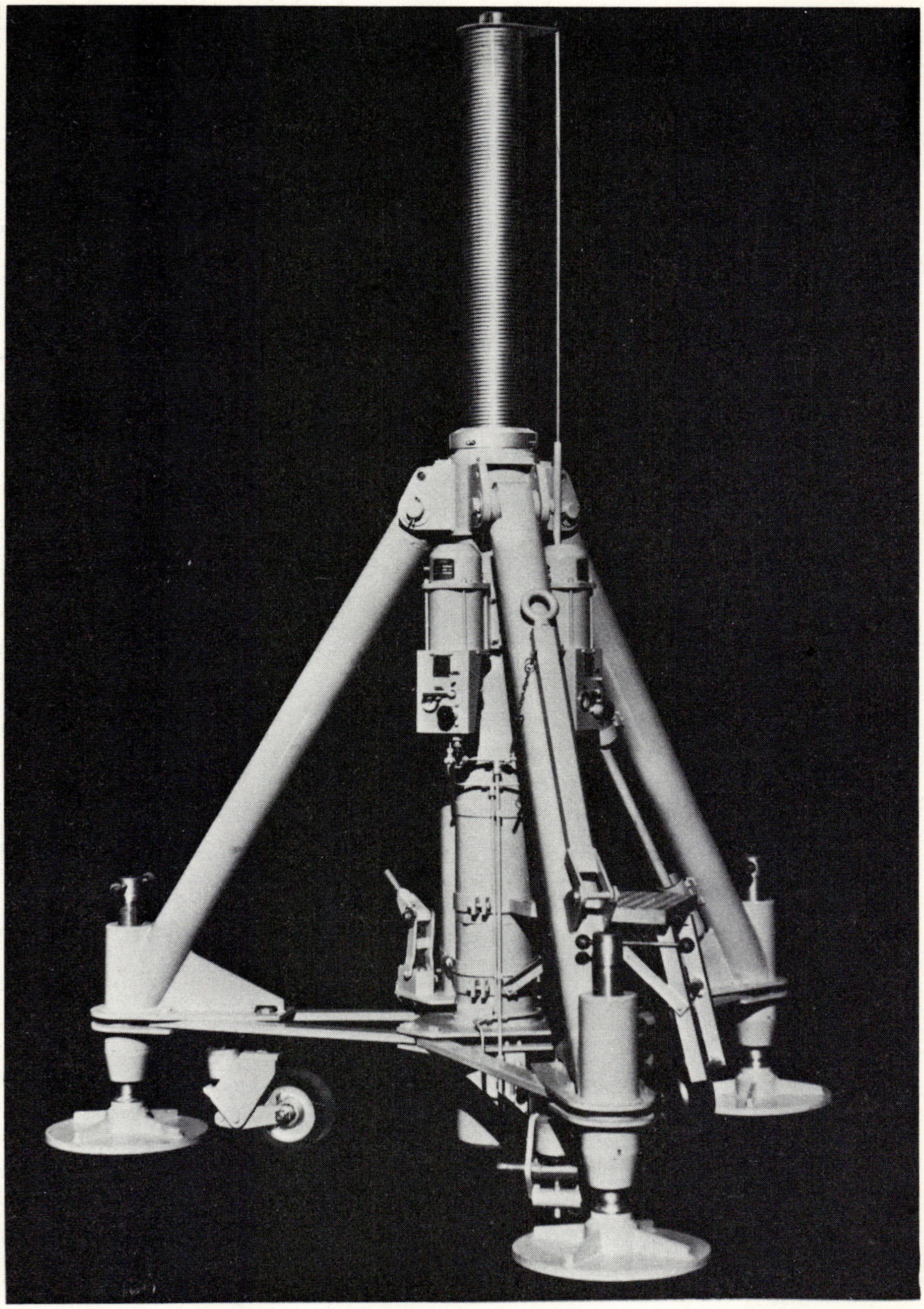

Skyhi 100-ton capacity aircraft jack with preset height indicator

UNITED STATES OF AMERICA

AM 1022 Adjustable Work Stand

A mobile general purpose unit designed to provide access to various parts of the aircraft, the AM 1022 is adjustable in height and is available with manual or power options. Construction is of structural steel tubing with 1 foot (30·48 cm) diameter polyurethane castors. A towbar is provided. Dual actuators, floor locks and bumpers, to protect the aircraft, are standard. Optional features include pneumatic jackscrews, and manual hydraulic cylinders or pneumatic/hydraulic cylinders.

Characteristics

PLATFORM DIMENSIONS: 3 × 4 ft (0·91 × 1·22 m)
CAPACITY: 1000 lb (454 kg)
OVERALL LENGTH/WIDTH: 20 ft (6·1 m)/8 ft 6 in (2·59 m)
HEIGHT: 25 ft (7·62 m) max; 11 ft (3·35 m) minimum

Status

In service with airlines in the USA

Manufacturer

Advanced Ground Systems Engineering Corporation, Long Beach, California, USA

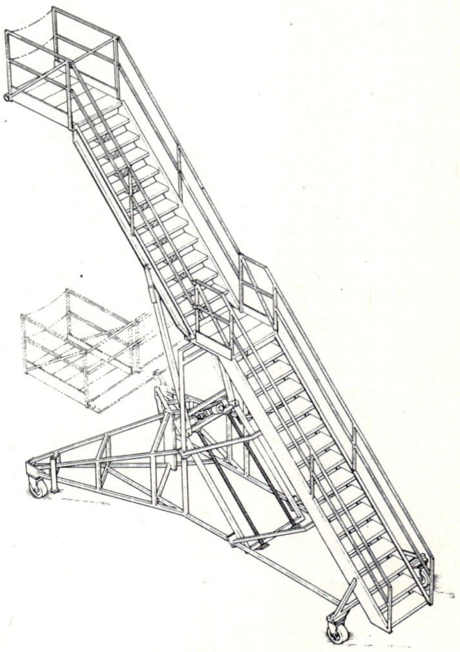

AM 1022 adjustable work stand

AM 1241 Engine Overhaul Gantry

The AM 1241 is a T-type gantry design fabricated steel structure providing two independent stations for engine work. Each station is equipped with a three-rail overhead frame attached to a powered inverted scissor lift. The lift using two Duff-Norton jack screws, each rated at 10 tons (9070 kg) is capable of raising or lowering the engine a total of 4 feet (1·22 metres) to provide convenient working heights for mechanics. A separate hydraulic safety system is installed to control the load in the event of a mechanical system failure. Each station is equipped with independent controls, air and electrical outlets, module trollies and hangars, lighting and power hoists.

Characteristics

NO OF STATIONS: 2
UNIT RATED CAPACITY: 12 000 lb (5443 kg)
VERTICAL SCISSOR TRAVEL: 4 ft (1·22 m)

Status

The equipment was designed as a handling system for the General Electric CF6 engine during overhaul. It is capable of simple conversion to handle RB 211 and JT 9 engines

Manufacturer

Advanced Ground Systems Engineering Corporation, Long Beach, California, USA

AGSE Engine Cradle/stand and Trailer

Advanced Ground Systems Engineering (AGSE) supply a range of cradles and stands to fit specific engine types, for example the AM 1423A engine cradle for the CFM-56 engine. These are mounted on appropriate rail trailers for transport. Alternatively the engine stand, as in the case of the AM 1454, for the JT 9D-59/70 engine, may be mounted on castors, with a tow-bar attached.

The AM 1072 is a dual rail 7 foot 6-inch (2·29-metre) trailer constructed of structural steel tubing and equipped with shock mountings and pneumatic tyres. Mechanical brakes are fitted to all four wheels. The trailer, designed to be towed and steered from either end, is supplied with extendable tow bar for use with and without the nose cowl attached. The rails are set at a height of 3 feet 2 inches (96·5 cm) to interface with various other types of engine handling equipment. Alignment guides and positive stops are provided at each end of the rails.

Characteristics

MODEL: AM 1072
CAPACITY: 20 000 lb (566·3 kg)
LENGTH: 10 ft 6 in (3·2 m)
MAX WIDTH: 8 ft (2·44 m)
HEIGHT: 3 ft 2 in (0·96 m)
MAX TOWABLE SPEED: 5 mph (8 km/h)

Status

In service with airlines in the USA

Manufacturer

Advanced Ground Systems Engineering Corporation, Long Beach, California, USA

AGSE Maintenance Docks and Access Equipment

Advanced Ground Systems Engineering (AGSE) produce maintenance docks and access equipment for wide- and narrow-body commercial aircraft. The equipment is specific to type. Standard equipment available includes nose docks for the Boeing 747, A300B, wing stands for the Boeing 747, DC10 and L1011; side engine stands for the DC9-80 and L1011; vertical stabiliser platform for the Boeing 747 and tail docks for Boeing 707, 727, DC9-80, DC10 and L1011 aircraft. Typical aircraft maintenance/access arrangement for the A300B aircraft consists of a nose dock, forward fuselage stands on both sides. Leading- and trailing-edge wing stands, aft cargo door access stand on the starboard side incorporating a rolling platform, aft fuselage stand, port side and a horizontal stabiliser stand. The horizontal stabiliser stand incorporates an elevating bucket-type platform to give access to the upper parts of the fin. The Boeing 767 arrangement is similar except that inboard leading and trailing edge stands replace the forward fuselage stands required for the A300B, and engine stands are provided. For the Boeing 767, a nose stand is fitted with a jib crane.

A typical tail dock such as developed for the DC9-80 is a towable unit providing three levels of access. A section is cut out to accommodate the vertical fin, and fitted with sliding units supplied with bumpers to give complete access to the fin when the aircraft has been docked. The top level of the tail/dock covers the entire underside of the horizontal stabiliser with a cut out section for rudder swing. The middle level is used primarily for access to the lower half of the stabiliser and rudder. Access to the aft bulkhead through the tail cone is provided at the lower level. Removal of the central stair is permitted without repositioning the stand. The dock is constructed of bolted and welded steel sections with Aero Deck stair treads and flooring. Hand-rails and toe-plates are provided on all open sides and air and electrical service outlets are installed at each level. Mobility is achieved by large diameter castors with brakes and swivel locks. A tow bar is supplied. Optional features include lighting at each level and galvanised flooring.

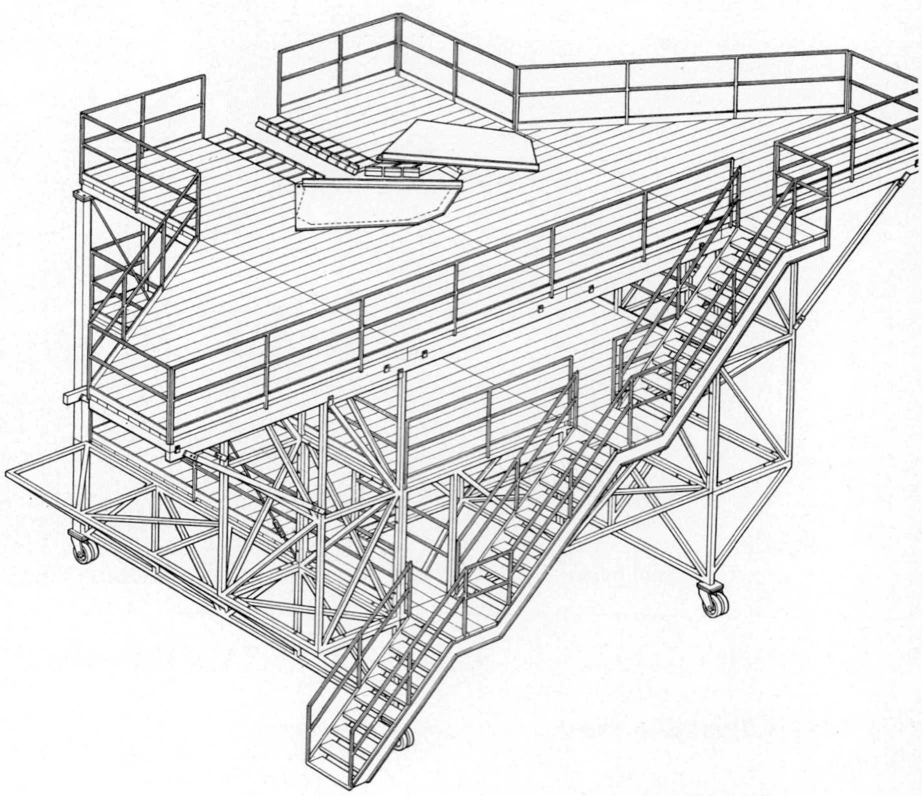

DC9-80 tail maintenance dock supplied by AGSE

Characteristics

DC9-80 TAIL DOCK
MODEL: AM 1351
DIMENSIONS
overall width: 48 ft (14·63 m)
overall length: 24 ft (7·32 m)
PLATFORM HEIGHTS
upper: 22 ft (6·71 m)
middle: 14 ft (4·27 m)
lower: 7 ft (2·13 m)
WEIGHT: approx 40 000 lb (18 144 kg)

Status

Maintenance docks and access stands have been supplied to the United States Air Force and to airlines in north and south America, Europe and the Middle East

Manufacturer

Advanced Ground Systems Engineering Corporation, Long Beach, California, USA

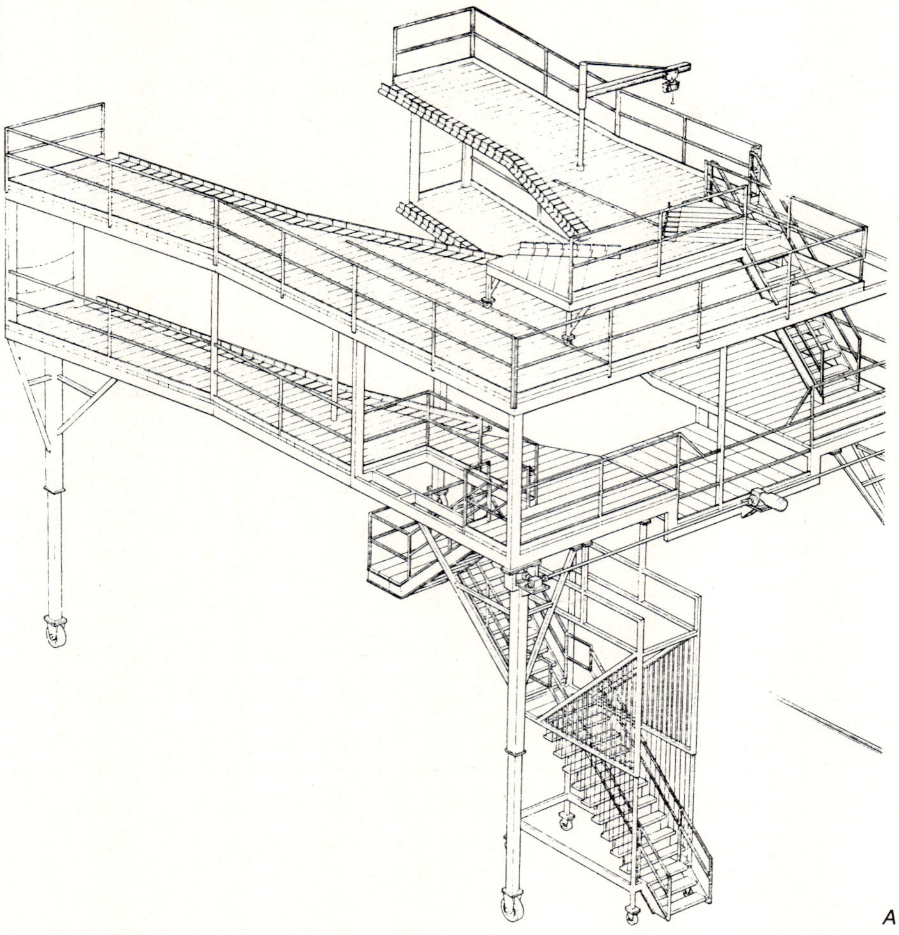

AGSE nose dock for Boeing 747 maintenance

Bauer/Electro Universal Wheel Build-up Stand

Two control levers are used to operate this stand. One lever controls the position (angle) and the height of the adaptor housing, the other lever controls clamping of the wheel halves and tyre, and release of the assembled wheel. The ability to change the height of the adaptor housing is one feature that makes the stand adaptable to many different wheel sizes.
A basic wheel build-up unit consists of the stand, appropriate main wheel or nose wheel adaptor plate and corresponding bearing adaptor, a wheel clamp and clamp plate assembly, and a trolley.
The stand comprises hydraulic and electrical components which control the operation of the base, tilt and clamping cylinders. Almost any aircraft tyre can be changed on the stand by changing the adaptor plate and bearing adaptor. The wheel adaptor plate holds the floating sockets which match the bolt configuration of the aircraft wheel. The floating feature enables the tie bolt heads to seat themselves automatically in the sockets while the tie bolt nuts are being loosened. The bearing adaptor fits into the inboard wheel bearing cup and supports the weight of the wheel assembly during several steps in the assembly procedure. The wheel clamp and clamp plate is inserted through the wheel hub and is locked in place in the adaptor housing so that the wheel halves will be drawn together during the clamping process.
The trolley enables an assembled wheel or tyre to be rotated so that it can be aligned with and pressed on to the adaptor plate sockets or inboard wheel half on the adaptor housing.

Characteristics

AVAILABLE ADAPTORS: Commercial Aircraft – Boeing 707, 727, 747, DC8, DC9, DC10, F27, F28, L1011 A300
Military Aircraft – CP140, 137, 130E/H, and 109
POWER REQUIREMENTS: 220/440 V 3 phase 50/60 Hz

Status

This equipment has been supplied to 16 airlines including Air Canada, Alitalia, Canadian Pacific Air, Garuda and Varig

Manufacturer

Bauer/Electro Inc, Farmington, Connecticut, USA

Brackett PJ-1 Nose Jack

The PJ-1 is a nose jack for general aviation aircraft which fits under the propeller blade and lifts the nose while the aircraft pivots on the main under carriage. The jack fits both two- and three-blade propellers and is equipped with soft rubber cups to prevent damage.

Characteristics

TYPE: PJ-1
MINIMUM HEIGHT: 3 ft 10 in (1·17 m)
MAX HEIGHT: 5 ft 9 in (1·75 m)
WEIGHT: 31 lb (14 kg)

Manufacturer

Brackett Aircraft Co, Mesa, Arizona, USA

Clyde Machines Maintenance Equipment

A range of maintenance equipment is produced by this company including maintenance stands, engine stands and tow bars.
The C 519 maintenance stand, which is mounted on castors, can be adjusted by a manually-operated hydraulic pump to provide platform heights from 5 to 19 feet (1·52 to 5·79 metres). Constructed in steel, the stand is fitted with hand-rails on both sides.
The 15F267 maintenance platform, mounted on rubber-tyred wheels, can be adjusted in height hydraulically by a hand-pump operated from the platform. Constructed in steel, the platform which is 10 feet 3 inches (3·18 metres) long and 5 feet 1 inch (1·55 metres) wide is fitted with hand-rails.

Status

Clyde Machines maintenance equipment is in service with airlines in the USA

Manufacturer

Clyde Machines, Glenwood, Minnesota, USA

Columbus Jack Lifting Equipment

Columbus Jack manufacture a wide range of aircraft maintenance equipment including aircraft jacks, tow bars, hydraulic pumps and cylinders. A typical example of the aircraft jacks produced is the Stanchion Jack developed by the company in conjunction with Boeing for the Boeing 747. The jack raises hydraulically to stabilise the tail section of the aircraft during loading and unloading manoeuvres. It is portable and is designed to be carried in the aircraft and to be unloaded and set up by one or two members of the ground personnel team.

Status

Columbus Jack is the largest producer of aircraft jacks in the United States. The company supplies equipment to the Department of Defense and commercial airlines in the USA

Manufacturer

Columbus Jack Corporation, Columbus, Ohio, USA

Duff Norton Coffing Aircraft Hoists

A wide range of Coffing hoists is available for aircraft maintenance. Over 1200 different models of the Worm drive hoist are produced with capacities up to 15 tons (13 600 kg). The hoist operating from a trolley mounted on an overhead beam is powered by a 230 or 460 volts electric motor. Duff Norton say the key features are fewer moving parts than the conventional spur-gear mechanism, double gearing and tapered roller bearings operating in an enclosed oil bath.

Electric wire rope hoists provide capacities from 0·125 tons (113·4 kg) to 2 tons (1814 kg). Most units can be supplied with single or two speed operation using 115/230 volts, single-phase, 60 Hz or 230/460 volts, 3-phase, 60 Hz motors. The 2 ton (1814 kg), 1 ton (90 kg) and 0·5 ton (454 kg) capacity hoists are equipped with dual-braking systems, provided by a motor brake and a self-adjusting load brake. Limit switches are fitted to prevent over-runs and over-wrapping of the rope on the drum. Trolley versions are available either powered, or manually-operated by gearing mounted on an overhead beam.

Electric chain hoists with capacities ranging from 0·25 tons (227 kg) to 5 tons (4536 kg) incorporates a three-stage transmission with alloy gearing lubricated in a sealed oil bath. Helical gearing is used in the first stage in all models and in the third stage in certain 2 ton (1814 kg) models and all 3 ton (2722 kg), 4 ton (3629 kg) and 5 ton (4536 kg) versions. An overload limiting attachment is fitted, a wet cone clutch-type device in the hoist transmission activated when the preset capacity is exceeded. A mechanical load brake is installed which operates even when the power is off. Forty-seven models of the single-speed electric chain hoist are produced using a 230/460, 208, or 575 volt, 3-phase power supply. Two-speed versions are also available with similar capacities. The second speed, one-third of the normal speed, is achieved by the use of an additional separate winding in the hoist motor. When the push-button control is pushed to the first contact, the hoist operates at the slow speed. Depressing the button fully, automatically increases the speed to the normal setting. The two-speed version is available in 18 models using 208, 230, 460 or 575 volts, 3-phase, 60 Hz motors.

Hand chain hoists are produced with capacities ranging from 0·125 (113·4 kg) to 25 tons (22 680 kg) using a variety of mechanisms. Lever chain hoists similarly employ a range of mechanisms and are available with capacities from 0·25 tons (226·8 kg) to 15 tons (13 608 kg).

Manufacturer

Duff Norton Company, Charlotte, North Carolina, USA

Lift-a-Loft Elevating Maintenance Platforms

Lift-a-Loft maintenance platforms are self-propelled, battery operated units available with working heights from 7 feet (2·13 metres) to 70 feet (21·34 metres). Outriggers are not required in any of the models.

The VAL series is designed for maintenance work in crowded facilities, with a base measuring 5 × 3 feet (1·55 × 0·91 metres) and a turning radius of 6 feet 6 inches (1·98 metres). All movements are controlled from the platform which rotates 360 degrees around a central mast to provide a 9-foot (2·74-metre) working diameter.

The MCL series has similar features to the VAL series but is capable of extending to a greater height and reach. Models in this series are equipped with a platform capable of supporting two men and their tools.

Requirements for maximum height, and reach, are met by the LAL series. Up to three men can be lifted to a height of 65 feet (19·8 metres) on a platform which traverses on a tilted mast and rotates through 360 degrees to provide a lateral reach of 21 feet (6·4 metres) and a working diameter of 45 feet (13·7 metres).

In the RAL series an articulating mast swings the operator's platform out and up or out and down, for use in situations where access is blocked by machinery or obstacles exist overhead. The 5-foot (1·52-metre) long articulating mast arm extension moves the platform through a 180 degree arc up or down providing a maximum lateral reach of 8 feet (2·44 metres) over the chassis. All movements are controlled from the platform.

The MRT series provides mobility between maintenance tasks and models are equipped with a driver's position for transit and industrial pneumatic tyres. The mast and platform design is similar to the MCL models.

The NLU series is designed to meet requirements for a mobile work platform providing the capability to drive and steer the vehicle while in the elevated position. The platform extends to 20 feet (6·1 metres). A hydraulically controlled traversing platform which offers 2-foot (0·61-metre) additional reach as an optional feature.

Characteristics

SERIES	VAL	MCL	LAL	RAL	MRT	NLU
NO OF MODELS	2	6	13	9	6	2
PLATFORM HEIGHT						
max:	7·26 m	9·17 m	19·81 m	10·97 m	9·17 m	6·1 m
minimum:	0·61 m	1·4 m	1·42 m	0·61 m	1·4 m	0·64 m
CAPACITY	226·8 kg	226·8 kg	340·2 kg	226·8 kg	226·8 kg	226·8 kg
PLATFORM DIMENSIONS						
length:	0·91 m	2·29 m	3·04 m	0·91 m	2·29 m	0·91 m
width:	0·91 m	0·76 m	0·84 m	0·91 m	0·76 m	0·86 m

Manufacturer

Lift-a-Loft Corporation, Muncie, Indiana, USA

Stinar Adjustable Maintenance Platforms

Stinar produce self-propelled and towable maintenance stands. Hydraulically-operated they can be adjusted to the correct height as required. The WS 10 model is a self-propelled high lift version with integral stairs. The platform has an open steel mesh surface with removable drop trays. Steel hand-rails are provided for security.

Status

Stinar has supplied maintenance platforms to North West Airlines and other USA airlines in addition to engine stands and docks

Manufacturer

Stinar Corporation, St Paul, Minnesota, USA

Stinar WS 10 adjustable maintenance platform

Wollard L-549F Mobile Access Work Platform

The Model L-549F provides personnel with a mobile work platform for all aircraft having access points requiring a platform from 3 feet (0·91 metres) to 14 feet (4·27 metres) high. The platform size and fold-down extension, coupled with the overall height of 6 feet 2 inches (1·88 metres), allow the platform to be raised inside the B 747 wheel well.

The platform is designed to accommodate two people or a total load of 500 lb (226·8 kg). It is equipped with fixed guard-rails, and toe-board. By pulling up on safety pins through the sides of the platform, the rear side folds down, forming an auxiliary platform. This auxiliary platform contains an additional open deck platform that can be extended, giving a total working area of 9 feet 6 inches (2·92 metres) × 4 feet 2 inches (1·27 metres). The elevating mechanisms incorporate safety features to prevent sudden lowering of the platform. Built-in hydraulic jacks are used to maintain stability under varying conditions.

Characteristics

CHASSIS
Ford F-250 vehicle
Standard transmission
6-cylinder engine

PLATFORM HEIGHT
low position: 3 ft (0·91 m)
high position: 14 ft (4·27 m)

WORK PLATFORM DIMENSIONS
length: 3 ft 10 in (1·17 m)
width: 3 ft 10 in (1·17 m)

FOLD-DOWN PLATFORM
length: 2 ft 11 in (0·89 m)
width: 3 ft 6 in (1·07 m)

SLIDE-OUT PLATFORM
length: 2 ft 10 in (0·86 m)
width: 3 ft 6 in (1·07 m)

OVERALL DIMENSIONS
length: 17 ft 11 in (5·46 m)
width: 6 ft 8 in (2·03 m)
height: 6 ft 3 in (1·96 m)

OPERATIONAL WEIGHT: 7700 lb (3494 kg)

Manufacturer

Wollard Airport Equipment Inc, Miami, Florida, USA

Power supplies, testing and servicing

FRANCE

AMS Fuel Systems Test Stands

Various models are available from AMS including the B1 641B fuel nozzle test stand. This test equipment is designed to function in a room subject to explosion-proof standards. The electrical power is not explosion-proof and is suitable for installation in a non-hazardous area. A piston pump feeds the nozzle in tests at different pressures regulated through two self-balanced valves. These valves are controlled by a set of five pilot regulators and five distributors preset at five pressures between 7 and 550 psi (0·5 and 28 bar).
The fluid temperature is maintained at 27°C ±1° by a pneumatic regulator which adjusts the water flow in the exchanger.
The nozzle fuel flow is measured by two turbines covering the range from 0·09 to 9 litres per minute.

Characteristics

MAX DYNAMIC PRESSURE: 600 psi (40 bar)
MAX FLOW: 2·4 US gallons/minute (9 litres/minute)
MAX STATIC PRESSURE: 1500 psi (100 bar)
POWER REQUIREMENTS: 380 V, 3 phase, 50 Hz
WATER REQUIREMENTS
capacity: 800 US gallons/h (3000 litres/h)
pressure: 30–60 psi (2–4 bar)
PNEUMATIC SUPPLIES
nitrogen or dry air pressure: 500 psi (35 bar)
dry and clean air pressure: 80–100 psi (5–7 bar)

DIMENSIONS	TEST STAND	ELECTRICAL CABINET
length	8 ft 2 in (2·48 m)	1 ft 9 in (0·53 m)
width	5 ft 7 in (1·7 m)	10 in (0·25 m)
height	6 ft 2 in (1·88 m)	2 ft (0·6 m)

Status

The B1 641B is designed to test fuel nozzles in the General Electric CF6-50 engines. Other fuel test systems available include BR 1282 for the fuel control system in the Snecma/Turbomèca Larzac engines which power the Alpha Jet aircraft; RSP 951 for testing the fuel-injection regulators in the Snecma Atar 9C and 9K50 engines and the PFC 596B test stand for the fuel pumps and controls installed in the GE CF6 -50 engines

Manufacturer

Appareilles et Materials de Servitudes (AMS), Fresnes, France

---

AMS Hydraulic Component Test Stand

AMS produce several models for servicing hydraulic components. A typical example is the BAH 1662 designed for tests on wide-body aircraft hydraulic accessories. The stand provides for checking pumps, hydraulic engines, jacks, valves and static testing of hydraulic components and comprises three sub-assemblies: test panel, hydraulic system and electrical cabinet. The test panel is a welded structure containing the engine, valves, hydro-pump, etc. The display instruments are on three vertical panels behind a work bench; the latter consisting of two trays with appropriate hydraulic couplings. The electrical and hydraulic controls are positioned in front of the work bench at the same height. The hydraulic system comprising vacuum pump, high-pressure unit, tank-boost unit and electrical-junction cabinet are installed in a separate room with the electrical cabinet. The stand is equipped with a variable-speed unit, adjustable hydraulic flow and pressure; static-pressure circuit and an electrical-test circuit of 0 to 30 volts dc. Pneumatic, electrical and hydraulic safety devices are incorporated.

Characteristics

MODEL: BAH 1662
OPERATING PRESSURE: adjustable 14·5–4000 psi (1–275 bar)
FLOW RATE: adjustable 0·3–39 US gallons/minute (1–148 litres/minute)
VARIABLE SPEED UNIT
setting 1 power: 70 kW at 4000 rpm
speed: adjustable 0–4000 rpm
setting 2 power: 44 kW at 9000 rpm
speed: adjustable 0–9000 rpm
MAX STATIC PRESSURE: 10 000 psi (700 bar)
HYDRAULIC TANK CAPACITY: 105 US gallons (400 litres)
HYDRAULIC FLUID: Hyjet
OIL DECONTAMINATION: Class 5
POWER REQUIREMENTS: 380V, 3 phase, 50 Hz

DIMENSIONS	PANEL	HYDRAULIC SYSTEM	ELECTRICAL CABINET
length	13 ft 4 in (4 m)	8 ft 4 in (2·5 m)	6 ft 8 in (2m)
width	10 ft 8 in (3·2 m)	8 ft 4 in (2·5 m)	5 ft 4 in (1·6 m)
height	6 ft (1·8 m)	2 ft (0·6 m)	6 ft 8 in (2 m)

Status

BAH 1662 is designed to meet the requirements for servicing A300B aircraft. Other similar design stands available include BAH 1403 for Breguet 1150, Mirage F1 and Alpha Jet-type aircraft; and the GEE 746

Manufacturer

Appareilles et Materials de Servitudes (AMS), Fresnes, France

---

AMS Mobile Hydraulic Test Rigs

AMS design and manufacture a wide range of mobile hydraulic test rigs for aircraft maintenance. A typical product is the EMH 1569, a mobile rig developed for servicing high-volume hydraulic systems associated with wide-body aircraft. This is a self-contained test rig requiring external electrical supplies of 380 volt, 3 phase, 50 Hz to operate the hydraulic-pump motor. The rig is assembled on a four-wheel chassis fitted with a tow-bar, and drum brakes equipped to the rear axle. The body is constructed of steel sheeting with removable panels and provision is made on the sides for stowage of the hydraulic hoses and electrical cable.
Pressure and flow are adjustable and electrical and hydraulic safety devices are incorporated in the system.

Optional features available include static circuit for tests up to 8700 psi (600 bar); bleeding circuit for hydraulic fluid de-aeration of the aircraft system; function for checking the locking and unlocking pressures of the landing gear by remote control; flushing and replenishment of aircraft hydraulic system; and measuring circuit for identifying hydraulic leaks in the aircraft system.

Characteristics

MODEL: EMH 1569
OPERATING PRESSURE: adjustable 600–3000 psi (40–210 bar)
FLOW RATE: adjustable 3–40 US gallons/minute (10–150 litres/minute)
HYDRAULIC FLUID: MIL-H-5606B, Oronite, or Skydrol
OIL DECONTAMINATION: Class 6
POWER REQUIREMENTS: 380 V, 3 phase, 50 Hz
TOWING SPEED: 15 mph (25 km/h)
DIMENSIONS
length: tow-bar retracted 10 ft 6 in (3·19 m)
tow-bar extended 14 ft (4·26 m)
width: 5 ft 10 in (1·78 m)
height: 4 ft 10 in (1·47 m)
OPERATIONAL WEIGHT: 4180 lb (1900 kg)

Status

Designed for servicing the A300B aircraft

Manufacturer

Appareilles et Materials de Servitudes (AMS), Fresnes, France

ISRAEL

Mechola Mobile Hydraulic Test Rig

Four standard models are available from Mechola, with single or triple systems capability; two are diesel-engine driven, two are powered by an electric motor. The rig designed for MIL-H-83262 fluid, though other options are available, is assembled on a four-wheel chassis fitted to a tow-bar. The rear wheels are equipped with mechanically operated parking brakes. Adjustable hydraulic shock absorbers are installed in all instrument lines.

The equipment features automatic and manual bleed-valves to eliminate air bubbles trapped in the hydraulic fluid and a manual relief valve to permit a rapid reduction of supply pressure. A separate fluid replenishment system is fitted, supplying hydraulic fluid through a three micron-rated filter, for filling and topping-off operations and fluid sampling ports are provided for fluid contamination analysis. Automatic shut-down occurs when the fluid temperature is too high or when there is insufficient boost pressure. The diesel-powered models have additional protection devices activated when the engine temperature is too high or the engine-oil pressure is too low.

Characteristics

MODEL	MHD-1	MHD-3	MHEL-1	MHEL-3
TYPE	Diesel engine Single system	Diesel engine Triple system	Electric motor Single system	Electric motor Triple system
RATED CAPACITY (each system) at 3000 psi (21 bar)	15 US gallons/ minute (56·8 litres/ minute)	30 US gallons/ minute (113·6 litres/ minute)	15 US gallons/ minute (56·8 litres/ minute)	30 US gallons/ minute (113·6 litres/ minute)
ENGINE/MOTOR OUTPUT	60 hp (44·7 kW)	60 hp (44·7 kW)	40 hp (29·8 kW)	60 hp (44·7 kW)
INPUT REQUIREMENT	Diesel fuel	Diesel fuel	380 V, 50 Hz	380 V, 50 Hz
DIMENSIONS	2·1 × 1·57 × 1·4 m	2·1 × 1·57 × 1·4 m	2·1 × 1·57 × 1·4 m	2·1 × 1·57 × 1·4 m
DRY WEIGHT	2000 kg	2100 kg	1800 kg	2000 kg

Status

In service with the Israeli military forces

Manufacturer

Mechola Engineering, Bikat Hayardan, Israel

ITALY

Aeronautica Macchi Mobile Hydraulic Test Rig

Aeronautica Macchi produces five models of mobile and hydraulic test rigs.

The Model S4-1A supplies hydraulic power for filling, flushing and testing aircraft systems. Two systems can be handled simultaneously and the aircraft can be tested using the test-stand reservoir or the aircraft reservoir. A vacuum system is incorporated evacuating the test-stand reservoir to 1 foot (0·3 metres) Hg to assist in de-aerating the hydraulic fluid. Sampling valves are provided for checking fluid contamination. The rig is powered by an electric motor but a diesel-powered version is also available.

The Model 26-1E, electric motor drive, and the Model 12-1EF, petrol engine driven are similar. Both are designed for testing military aircraft hydraulic systems and components.

The Model 16-1E was developed for military aircraft testing. Powered by an electric motor it is equipped with a pneumatic system with two outlets, one at high the other at low pressure.

Characteristics

MODEL	54-1A	26-1E/12-1EF	16-1E
ENGINE	Electric 93 kW 1450 rpm	Electric: 41 kW, 2850 rpm Petrol: 66 kW, 2850 rpm	Electric 15 kW 1450 rpm
OPERATING PRESSURE	Continuous 150 litres/minute at 275 bar or 75 litres/minute at 345 bar	Single system 94 litres/minute at 207 bar double system 47 litres/minute at 207 bar	32 litres/minute at 207 bar or 191 litres/minute at 345 bar
FLUID		MIL-H-5606, Skydrol (option)	MIL-H-5606, Skydrol (option)
HYDRAULIC RESERVOIR	225 litres	140 litres	
FILTERS	5 micron		
AIR TANK CAPACITY			20 litres
AIR PRESSURE HEIGHT			high: 193 bar low: 11 bar
POWER REQUIREMENTS	380 V, 3 phase, 50 Hz	380 V, 3 phase, 50 Hz	220/380 V, 3 phase, 50 Hz
DIMENSIONS	3·43 × 1·8 × 1·63 m	2·05 × 1·46 × 1·02 m	
EMPTY WEIGHT	2745 kg	1300 kg	

Status

In service with the Italian Air Force. Designed in conjunction with Teledyne Sprague. The 54-1A was developed for servicing the Tornado aircraft

Manufacturer

Aeronautica Macchi, Varese, Italy

Aeronautica Macchi Models 49-1/29-1 Fixed Installation Hydraulic Component Test Stands

The models 49-1 and 29-1 are test stands developed for testing non-rotating hydraulic components.

The 49-1 is dual system stand with a two station configuration designed for use with Skydrol 50 A hydraulic fluid. Each hydraulic system is independent; but the two systems may be connected resulting in an increased flow and pressure.

All pumps are in a sound-proofed cabinet separated from the main test-bench structure. Each system comprises a boost pump to supercharge the main pump; low pressure filter; main pump; pulsation dampener; relief valve; high pressure filters; and a separate oil circulating pump and heat exchanger for maintaining oil temperature. All motor starters are in the pump cabinet with push-button stations and pilot lights at each test station on the test stand.

An additional boost pump system provides supercharged pressure for external hydraulic pump and motor testing.

The 29-1 is designed with multiple circuits to enable many types of aircraft hydraulic system components to be tested. Six pressure gauges are included in the instrumentation.

A reduced pressure circuit allows simultaneous testing at two pressures.

Characteristics

MODEL	49-1	29-1
OPERATING PRESSURE		
system A	95 litres/minute at 345 bar	57 litres/minute at 207 bar
system B	65 litres/minute at 345 bar	34 litres/minute at 345 bar
A + B	165 litres/minute at 345 bar	
system C		95 litres/minute at 103 bar
STATIC PRESSURE	1035 bar	690 bar
TEST FLUID	Skydrol 500A	Skydrol 500A

Status

This equipment was developed in conjunction with Teledyne Sprague

Manufacturer

Aeronautica Macchi, Varese, Italy

NETHERLANDS

Verolme Elektra Aircraft Engine Test Equipment

Verolme Elektra design and manufacture a wide range of specialist and multi-purpose equipment for testing electric, pneumatic, fuel, lubricating oil, and hydraulic systems and their components. The multi-purpose accessories test stands are applicable to all types of aircraft, but all stands are specially designed to meet the specific conditions and requirements of individual users. The company is engaged on a development programme for computerised test stands equipped with data processing.

Status

Operators of Verolme Elektra equipment include the Royal Netherlands Air Force with test stands for J79/J85 engines, Turkish Air Force, with J79/J85 and Tyne engine test stands, Braathens SAFE, Fokker, FFV Sweden, KHD Germany and MTU. Equipment currently under manufacture includes a complete set of test equipment for Air Tanzania, a fuel and Skydrol test stand for Lufthansa and a components test stand for the United States Air Force.

Manufacturer

Verolme Elektra BV, Maassluis, Netherlands

UNITED KINGDOM

Auto Diesels Mogens M272 Frequency Converters

Mogens M272 frequency converters have been designed to meet the requirement for 200 volt, 3 phase, 400 Hz electric supply for systems for civil and military aircraft and related electronic equipment. The range covers outputs from 15 to 140 kVA rated to British Standards. The basic models are designed for use in hangars and workshops. They can be supplied with weatherproof housing for outdoor installation or mounted on a trailer similar to the Auto Diesels 700/900 series diesel-driven ground power units. For supplying aircraft with 28 volt dc, a transformer rectifier pack can be supplied with suitably rated motor generators. Alternatively if a 28 volt dc is the sole output required, a special version of the alternator, designated M271, is available that can be directly coupled to a rectifier without a transformer.

The 400-cycle alternator is driven by a 50 Hz motor, chosen from standard commercial ranges, with speed and voltage and protection matched to the operator's intended use. The four basic types of unit are: open set, canopy set, indoor mobile sets and weather-proof towable units.

The open set is for use in hangars or workshops. The alternator and electric drive motor are open coupled on a steel skid. They may be fixed, free standing on a level floor or positioned on resilient mounts. The control panel may be mounted on the top of the motor generator or wall mounted.

In the canopy set, the basic open set is fully enclosed in a sheet metal canopy. This makes it suitable for indoor use or, if it is weatherproof, for installation out of doors.

The indoor mobile sets are essentially open and canopy sets, mounted on castors for mobility within a workshop area.

In the weatherproof towable units, the motor generator is mounted on a chassis developed from the 700/900 series diesel driven ground power unit, allowing use outdoors in all climates and towable at up to 30 km per hour. The units comprise a fan-cooled cage induction motor, brushless alternator, input section and control panel. The input section consists of a fused isolator switch, Star Delta starter with overload provision and motor start and stop push-buttons.

The control panel incorporates output contactor with on-off/reset buttons, voltmeter, ammeter and frequency meters, inbuilt protection units covering high and low voltage, frequency and overload. The protection modules are individually removable and have adjustable settings. The annunciator unit on the control panel incorporates a lamp test pushbutton. An aircraft-type interlock circuit can be supplied to control the output contactor.

Input and output cables, the latter with or without aircraft type connectors, can be supplied to customer requirement.

Mogens M272 frequency converter

A 28 volt dc output is obtained by feeding the 200 volt, 400 Hz supply to a transformer and silicon-diode bridge rectifier mounted on an extruded aluminium alloy heat sink. The output voltage is regulated by the ac automatic voltage regulator and will maintain 28 volts ±1 volt up to 1500 amperes falling to better than 24 volts at 2000 amperes. The maximum rating on this unit is 700 amperes continuous, 2000 amperes peak. A minimum 75 kVA, 400 Hz unit is required to generate this full power rating, the standard control system is such that the motor generator must be operated in 28 volts dc or 400-cycle mode. For starting smaller turbine engines a 'soft start' current limiting facility can be built into the 28 volt dc system.

Manufacturer

Auto Diesels Braby Ltd, Uxbridge, Middlesex, England

Characteristics

MODEL	M272
RATED OUTPUT	15–140 kVA 200/115 V, 3 phase, 400 Hz
POWER FACTOR	0·8 nominal continuous kVA 0·4 2 × nominal kVA
ACCURACY	
voltage	±2% (±1% available on request)
frequency	+0, −1·7%
POWER REQUIREMENTS	380–440 V, 3 phase, 50 Hz
OPERATIONAL TEMPERATURE RANGE	−10° to +50°C

MODEL	M272-15	M272-30	M272-50	M272-75	M272-90	M272-110	M272-140
RATED OUTPUT	15 kVA	30 kVA	50 kVA	75 kVA	90 kVA	110 kVA	140 kVA
DIMENSIONS							
length	1·67 m	1·75 m	1·8 m	1·8 m	1·85 m	1·85 m	2 m
width	0·58 m	0·58 m	0·71 m	0·83 m	0·83 m	0·83 m	0·93 m
height	1·14 m	1·15 m	1·19 m	1·59 m	1·59 m	1·59 m	1·7 m
WEIGHT	510 kg	600 kg	700 kg	800 kg	1000 kg	1000 kg	1100 kg

Auto Diesels Mogens M271 Transformer Rectifier Unit

This unit has been designed for use when 28 volt dc is the only output required. A special motor generator consisting of a low voltage alternator with direct rectification is available. The basic rating is 750 amperes continuous with 2000 amperes peak. All other details are as for the transformer rectifier unit described in the M272 report above.

Characteristics

MODEL	M271-500	M271-600	M271-750
RATED OUTPUT	28 V dc	28 V dc	29 V dc
continuous	500 A	600 A	750 A
peak	1300 A	1500 A	2000 A
DIMENSIONS	1·8 × 0·71 × 1·19 m	1·8 × 0·71 × 1·19 m	1·8 × 0·83 × 1·59 m
WEIGHT	700 kg	700 kg	800 kg

Manufacturer

Auto Diesels Braby Ltd, Uxbridge, Middlesex, England

Defiant Aircraft Weighing System

The prime system in this series is designed for the weighing of medium to heavy aircraft on a wheel-by-wheel basis.

It uses strain gauge load-cell weight platforms in conjunction with an electronic package capable of monitoring the load applied to each wheel, total aircraft weight and providing a centre of gravity determination.

The whole system has been designed specifically for the aviation industry and takes into consideration the environment into which it will probably be placed. Each platform is protected from hydraulic fluids and aviation fuel splashes and, while having a working capacity of 25 000 kg with inbuilt overload, only weighs approximately 60 kg. Connection of the platforms to the electronics is by means of aviation-type connectors and the platforms are compatible with existing ramps and spacers.

The electronic package can be powered from a conventional mains supply or, by means of an invertor option, by batteries thereby providing a mobile system if required.

Further flexibility of the system is provided by each platform having its own channel within the electronic package. These channels are fully interchangeable each having its own zero control. Other systems with this range are available for light aircraft and helicopter weighing applications.

Aircraft weighing system produced by Defiant Weighing

Characteristics

UNIT CAPACITY: 25 000 kg
DIMENSIONS: 36 × 32½ in (915×826 mm)
adjustable height 4½–5 in (125 mm ±5)
WEIGHT: 60 kg

Status

In service at airports and aircraft manufacturing plants

Manufacturer

Defiant Weighing Ltd, Sevenoaks, Kent, England

Electro Dynamic Hangar Power Supply - Type M400 Frequency Converters

The Type M400 frequency converter produces a 400 Hz supply for hangar power distribution with a continuous output ranging from 8 to 50 kVA. The brushless motor-alternator, mounted on a single shaft, is manufactured and rated in accordance with British Standards and MIL Standard 704. Controls and switchgear include Star Delta starter for the induction motor output and excitation circuit contactors, voltmeter, ammeter, frequency meter, elapsed hours meter and output voltage regulator to control the voltage to within ±1 per cent. Safety and protection features comprise motor and generator overload relays, over/under voltage and under frequency protection units. An aircraft interlock 28 volt dc supply is provided. Special solid state sensing and switching equipment may be included in the control gear for automatic paralleling and load sharing of the 400 Hz outputs of two or more motor generator sets, thereby permitting the use of an automatically- or manually-switched redundant supply system. The control gear panel may be mounted on top of the motor generator or fixed to a hangar wall, as required. Vertically-mounted frequency converters are available for use where there are severe space limitations.

Characteristics

MODEL	M400/8	M400/16	M400/30	M400/50
RATED OUTPUT	8 kVA 200/115 V*, 3 phase, 400 Hz	16 kVA 200/115 V*, 3 phase, 400 Hz	30 kVA 200/115 V*, 3 phase, 400 Hz	50 kVA 200/115 V*, 3 phase, 400 Hz
POWER FACTOR (lagging)	0·8	0·8	0·8	0·8
ACCURACY				
voltage	±1%	±1%	±1%	±1%
frequency	+0, −1·25%	+0, −1·25%	+0, −1·25%	+0, −1·25%
POWER REQUIREMENTS	415 V*, 3 phase, 50 Hz	415 V*, 3 phase, 50 Hz	415 V*, 3 phase, 50 Hz	415 V*, 3 phase, 50 Hz
HORIZONTAL MOUNTING	0·68×0·42×0·98 m	0·84×0·42×0·98 m	1·03×0·6×1·28 m	1·05×0·6×1·32 m
weight	280 kg	380 kg	830 kg	990 kg
VERTICAL MOUNTING	0·42×0·49×0·92 m	0·42×0·49×0·92 m	0·6 × 0·6 × 1·1 m	0·6 × 0·6 × 1·1 m
weight	200 kg	296 kg	700 kg	840 kg
OPERATIONAL TEMPERATURE RANGE	−10 to +55°C	−10 to +55°C	−10 to +55°C	−10 to +55°C

*Alternative voltages can be provided

Status
Field trials were carried out successfully with the Royal Air Force on both aircraft and helicopters, and as a result 15 kVA units have been supplied to RAF Cottesmore for servicing the Tornado aircraft.
The 30 kVA machine is in service with British Airways Helicopters for servicing the Chinook helicopter at London Gatwick Airport, England

Manufacturer
Electro Dynamic Construction, Orpington, Kent, England

Type M400 15 kVA frequency converter supplied to Royal Air Force for servicing Tornado aircraft

Ferranti Vdu Terminals for Maintenance Data Systems

Ferranti have provided vdu terminal systems which are connected to British Airways drive information system. The documentation review into video entry system (drive) deals with aircraft maintenance, spares, management and stock control, workshop scheduling and component tracking. All information relating to aircraft servicing is stored in a central bank of information to which staff have immediate access via vdus. A complete history of each component in every aircraft is kept, along with a record of the component's lifespan. The central computer mainframe uses this information to predict those parts which need to be replaced. The maintenance technician can see what is required and can request the appropriate spare from the store. Levels of stock are monitored automatically, shortages are reported and orders for new parts are placed by the central computer system.

A total of over 70 PT7 controllers with more than 500 connected vdus, keyboards and printers, are linked to the central IBM 370 mainframe by a video-data cable communications system. The system is principally at London Heathrow Airport but extends to other British Airways sites in the United Kingdom. The PT7 terminal systems use standard software written in Coral 66, a high-level language designed primarily for real-time applications. The terminal controller, PT7-2S, is based on the Ferranti Argus 700F series processor using 32K words of main memory. Each is capable of supporting up to 12 vdu/keyboards terminals and six printer units. Controllers and terminals may be separated by up to 1·2 km. All keyboards are modelled on the IBM layout with an N-key rollover. A light sensitive pen and/or bar code reader are standard optional units which can be added to supplement data entry from the keyboard.

At Heathrow the communication links between the central mainframe and terminal controllers are provided by the video-data cable system. Videodata is designed to carry parallel and simultaneously active audio, video and data channels over a single wide-band coaxial cable network. This particular system provides for a 9·6 K baud synchronous communications channel between each terminal controller and the central system.
Other British Airways sites, remote from Heathrow, where there are PT7-2S terminals are connected into the central system via conventional Post Office communication circuits.

Manufacturer
Ferranti Computer Systems Ltd, Wythenshawe, Manchester, England

HML Hydraulic-ring Mains System

Hydraulic-ring mains systems have been developed to provide a fixed maintenance hangar method for servicing hydraulic systems. An external power unit supplies a hydraulic-ring main connected to system test points in the hangar and component test stands in the workshops. The system is controlled by start/stop buttons and protected by automatic warning and shut-down safety devices. The supply and return lines are fed under floor level to man-hole terminations in the hangar floor. Mobile test consoles provide the hydraulic interface between the ring main and the aircraft. The return lines feed into a sound-proofed return tank and pump unit installed, with the control units and panels, at the side of the hangar.

Status
HML has installed several of these systems at air bases in the United Kingdom

Manufacturer
HML (Engineering) Ltd, Isleworth, Middlesex, England

HML Hydraulic Test Equipment

HML manufacture hydraulic test equipment for the commercial and military operator.
For commercial operations electrically powered, sound-proofed trolleys are provided for hydraulic system tests. Individual hydraulic components can be tested on hydraulic test benches, each powered by its electric motor. Alternatively all the test requirements can be met by a single comprehensive hydraulic test system comprising component test stands and system test consoles, all of which are supplied by a single external power unit.
For en route servicing, smaller hydraulic trolleys are available. These can be petrol-, diesel- or electrically-powered. Electric versions can be provided for a variety of supply voltages.
To meet the needs of the smaller operator who requires equipment suitable for routine tests on a variety of small aircraft, there are general-purpose trolleys, petrol-, diesel- or electrically-powered, and electrically-powered universal hydraulic component test benches. The benches can incorporate a number of optional facilities such as an air/hydraulic intensifier for extra high pressure tests up to 20 000 psi (1379 bar), a separate air-blast fluid cooler and facilities for testing solenoid valves.
For military operations, HML will assist an air force in planning the hydraulic test facilities required for carrying out tests on a range of aircraft and components. In the case of advanced defence projects this may involve the design, development and manufacture of correspondingly advanced hydraulic test equipment. The hydraulic test trolleys developed for defence work may have to meet additional requirements such as weight limitations and being able to withstand a drop test, which normally do not apply to commercial equipment. A typical example is the Mk 8 lightweight general hydraulic servicing trolley with a maximum output of 8 Imperial gallons per minute (36 litres per minute). Weighing 1150 lb (520 kg), the trolley is transportable by air. Hydraulic component test benches for military use do not differ greatly from similar commercial benches. Various sizes of standard benches are provided with outputs from 5 Imperial gallons per minute (23 litres per minute) to 32 Imperial gallons per minute (145 litres per minute). In most cases the maxium operating pressure of the main dynamic supply is 5000 psi (345 bar).

Status
HML has supplied commercial aircraft hydraulic test equipment for BAe 1-11, Britannia, Concorde, VC10, Boeing 707, 727, 737, McDonnell Douglas DC-8, DC-9 and wide-body and executive jet aircraft. In the military field, the company is involved in the provision of Harrier support equipment for the US Marine Corps.

Manufacturer
HML (Engineering) Ltd, Isleworth, Middlesex, England

Houchin Hangar Power Distribution Systems

Houchin Model 674 16 kVA frequency converter

A range of frequency converters from 16 to 100 kVA capacity are manufactured by Houchin, which can be installed in the service hangar area to form the nucleus of a 400 Hz distribution system. The 400 Hz output from the frequency converter can be relayed to a number of strategically placed supply cubicles by underground, wall or overhead cables. Alternatively, the smaller frequency converters can be mounted on a trailer with a push-bar, for mobility in the hangar area. Three standard models are produced: Model 648, 663, and 647 rated at 100, 40, and 16 kVA respectively. The 648 and 663 are brushless synchronous motor-alternators, whereas the 674 is fitted with a brushless alternator and squirrel cage motor. All have a split frame and single horizontal shaft and are manufactured and rated in accordance with British Standards and MIL Standard 704.

Controls and switchgear, housed in a fabricated sheet steel cabinet, include a push-button operated output contactor, voltmeter and voltage trimmer. Output voltage is maintained to ±1 per cent by an automatic voltage regulator. Devices are installed for over/under voltage, under frequency and over-current protection. In the event of a malfunction the output contactor is opened automatically. The control cabinet may be removed for remote wall mounting. If hangar mobility is required the 648 and 663 models can be mounted on a heavy-duty chassis with tow-bar. The smaller 674 is available fitted to a four-wheeled trailer with a push bar.

Status

Houchin have supplied hangar power distributor systems to commercial and military operators throughout the world

Manufacturer

Houchin Ltd, Ashford, Kent, England

Characteristics

MODEL	648	663	674
RATED OUTPUT	100 kVA 115/200 V 3 phase, 400 Hz	40 kVA 115/200 V 3 phase, 400 Hz	16 kVA 115/200 V 3 phase, 400 Hz
POWER FACTOR	0·8	0·8	0·8 lagging
POWER REQUIREMENTS	380–440 V, 3 phase, 50 Hz	380–440 V, 3 phase, 50 Hz	380–440 V, 3 phase, 50 Hz
INPUT CABLE LENGTH	15/30 m	15/30 m	
OUTPUT CABLE LENGTH	15 m	10 m	
BASE-MOUNTED	1·9 × 1·3 × 1·8 m	1·3 × 0·8 × 1·6 m	0·92 × 0·48 × 0·93 m
weight	1951 kg	1125 kg	425 kg
TRAILER-MOUNTED	3·3 × 1·6 × 1·7 m	3 × 1·3 × 1·5 m	
weight	3155 kg	1703 kg	

Houchin Model 659 Transformer Rectifier Unit

The Model 659 transformer rectifier unit is designed to supply electrical power for servicing 28 volt dc aircraft systems from the standard mains supply voltage. The unit comprises two transformers, a fan-cooled double bridge rectifier, controls and instruments.

A voltmeter, ammeter and output switches are incorporated in a steel control panel protected by a perspex shield. A mains input isolation switch is mounted at the opposite end of the control unit. The transformer rectifier unit installed in a weather-proof canopy, fabricated from zinc coated sheet and incorporating a storage tray for input and output cables, can be supplied either trailer- or base-mounted for hangar installation. The trailer is manufactured from steel angle and channel section and complete with waist-high push bar, four rubber-rimmed castors including two swivel type and a handbrake.

Manufacturer

Houchin Ltd, Ashford, Kent, England

Characteristics

RATED OUTPUT
28 V dc, 8000 A continuous
24 V dc, 2000 A intermittent
POWER REQUIREMENT: 380–440 V, 3 phase, 50 Hz
INPUT CABLE LENGTH: 15/30 m
OUTPUT CABLE LENGTH: 5 m
DIMENSIONS: 1·7 × 0·9 × 1·2 m trailer-mounted
WEIGHT: 600 kg

Sperry Vickers Hydraulic Test Stands

Sperry Vickers design and manufacture specialist hydraulic test equipment to suit individual operator's requirements.

The power supply for aircraft component test stands is provided by Sperry Vickers variable displacement engine-driven pump as used in Boeing B-747, McDonnell Douglas DC-10, Lockheed L-1011 and Airbus Industrie A-300 B aircraft; while the hydraulic pump and motor test stands have a thyristor-controlled dc-motor for variable direct drive of the pump to be tested up to 4000 rpm. This dc-motor is used also as a brake-generator for hydraulic motor testing.

The hydraulic components such as control valves, relief valves, directional valves are manufactured by Sperry Vickers. High flow/high pressure systems are fitted with cartridge valve blocks to reduce noise level and the plumbing is equipped with O-ring sealed welded cone fittings. Housing of the control stand is constructed from stainless steel with tubes and supporting frames galvanised for corrosion protection.

A typical example of the Sperry Vickers product is the Universal Component Test Stand. This is designed to meet the special test requirements of engine-driven hydraulic pumps, flap, slat and stabiliser-trim motors, power transfer units and electric motor pumps installed in Boeing B-707, B-727, B-737, B-747, McDonnell Douglas DC-8, DC-9, DC-10, Lockheed L-1011 and the European Airbus A-300 B aircraft. Valves and actuators within the pressure and flow range can also be tested. The main drive is thyristor-controlled with digital and analogue speed and torque indication. Flow measurement is provided by four turbine flow-meters, each with digital indication.

Status

The universal component test stand described above was supplied to THY, Turkish Airlines. Other airline operators of Sperry Vickers equipment include KLM, Lufthansa, Sabena in Europe, Iran Air and Indian Airlines

Manufacturer

Sperry Vickers, Portsmouth, Hampshire, England

Characteristics

MAIN DRIVE: 134 hp (100 kW)
SPEED (DIRECT TO PUMP/MOTOR UNDER TEST)
4000 rpm
MAX SPEED (WITH GEARBOX ADDED): 9000 rpm
HYDRAULIC POWER SYSTEM 1 FLOW: 0–19 US gpm (0–72 litres/minute)
MAX PRESSURE: 3500 psi (250 bar)
HYDRAULIC POWER SYSTEM 2 FLOW: 30 US gpm (115 litres/minute)
MAX PRESSURE: 3500 psi (250 bar)
DIMENSIONS
width: 12 ft (3·65 m)
height: 6 ft 6 in (2 m)
depth: 5 ft 11 in (1·8 m)
WEIGHT: 8816 lb (4000 kg)

Sperry Vickers hydraulic test stand

Rees Instruments Multi-purpose Closed-circuit Television Inspection Systems

Rees Instruments manufacture a range of closed-circuit television inspection systems which are used for the inspection of airframes and aircraft engines. A high resolution system, the Tees 90 series includes the facility for interfacing both rigid endoscopes and flexible fibre-optics.

Systems are available with cameras having an outer dimension of 17·3, 34 or 40·5 mm. Various vidicon tubes can be selected, depending on application. For multiple applications where space is not the main criterion, the R93 system has complete remote-control over focus, iris and mirror rotation when side viewing. All systems are portable.

Manufacturer
Rees Instruments Ltd, Old Woking, Surrey, England

TCC Fixed Installation Hydraulic Test Rig

This hydraulic test rig is installed as a fixed servicing bay installation to provide facilities to service aircraft. Each rig is a facility providing an alternative power source for each aircraft hydraulic system. The rigs are connected to the respective systems by flexible hoses, fixed pipework on the servicing staging and flexible hoses on reels within the staging. In addition to the hydraulic servicing facilities, a pneumatic unit is incorporated in each rig to supply compressed air to the pneumatic system in the aircraft associated with the hydraulic system. This facility requires an air supply to the rig taken from an external source capable of delivering 100 psi (7 bar).

Fluid is drawn from the aircraft system reservoir into the test rig via a suction filter unit, pressure compensated pump, control valves and pressure filter unit to a delivery point from which cleansed fluid is returned to the aircraft. Flow of fluid from the aircraft depends on a preload pressure of 45 psi (3·1 bar) in the aircraft system reservoir and suction pressure from the test rig axial piston-type pump. An air-blast cooling system is installed in the rig to control fluid temperature below 70°C. A low hp motor drives a fan which forces air over a radiator formed by a series of gilled tubes to accelerate dissipation of heat generated in the fluid.

Provision is made for establishing contamination and cleansed levels of the fluid through sampling points in the suction and delivery lines to which fluid sampling set is adapted. The filtration rating of both suction and delivery filter units is three micron. Each of these filter units is installed with pressure switches across the filter unit whereby limiting differential pressures cause the switches to operate and automatically shut-down the rig through an Aish protection unit. This protection unit functions to shut-down the pump motor in the event of loss of pneumatic air pressure; low-suction line pressure, low limit 5 psi (0·35 bar); high fluid temperature; forced air fan motor overheat; pump motor overheat; suction filter unit clogged; limited to 15 psi (1 bar); and pressure filter unit clogged; limited to 35 psi (2·5 bar).

The electrical supply required for each rig is 415 volts, 3 phase, 50 Hz. Control circuits provided have a supply of 110 volts through a transformer in the starter cabinet. The pump loading valve solenoid, hour meter and the Aish protection unit are supplied with 24 volt dc through rectifiers.

The 100 hp pump motor is started by an automatic Star Delta starter in the rig starter cabinet which carries the related push button switches. The forced air-cooling motor has a direct-on-line starter, the operation with which is linked to the main pump motor starter but may be run separately. This means that both motors start instantaneously, but when the main pump motor is shut-down a time delay operates to ensure the cooling fan runs for a further pre-determined period.

Four identical hydraulic test rigs are installed to provide alternative power sources when servicing a Boeing 747 aircraft hydraulic system and to cleanse the system fluid by re-circulation.

Characteristics
FLUID: Skydrol
OPERATING PRESSURE: 1000–3000 psi (69–207 bar)
DELIVERY: 0–38 US gallons/minute (0–145 litres/minute)
FLUID TEMPERATURE: Up to 70°C
MODE OF OPERATION: Electrical
PNEUMATIC SYSTEM: 100 psi (7 bar)
ELECTRICAL SYSTEM
415 V, 3 phase, 50 Hz,
110 V via transformer for control circuits,
24 V dc rectified
DIMENSIONS
width: 7 ft 10 in (2·39 m)
height: 8 ft 1 in (2·46 m)
depth: 6 ft 7 in (2·01 m)

Status
Supplied to British Airways to service Boeing 747 aircraft at London Heathrow Airport, England

Manufacturer
Thermal Control Company Ltd, Hove, East Sussex, England

TCC Hydraulic Test Stand

The unit is designed for the testing of aircraft components, and enables both static and dynamic testing to be carried out. The facility comprises hydraulic supply modules and a combined test bench and control console.

The dynamic system provides cooled and filtered hydraulic fluid with variable flow. The flow direction can be reversed. The static system has a maximum pressure of 6000 psi (414 bar).

Flow, pressure and temperature can be monitored during testing and a variable dc supply is fitted for electrically-operated components, incorporating a timer.

Manufacturer
Thermal Control Company Ltd, Hove, East Sussex, England

Characteristics
FLUID: Skydrol
OPERATING PRESSURE
dynamic: 5–4500 psi (0·35–310 bar)
static: 5–6000 psi (0·35–414 bar)
flow: 0–30 US gallons/minute (0–136 litres/minute)
MAX OPERATING TEMPERATURE: 70°C
RESERVOIR CAPACITY: 60 Imperial gallons (273 litres)
FILTRATION CAPACITY: 5 micron
PUMPS: Fixed displacement piston type
POWER REQUIREMENT: 415 V, 3 phase, 50 Hz

Thermal control hydraulic test facility

TCC Mobile Hydraulic Test Rig

The mobile hydraulic test rig provides for servicing high volume hydraulic systems associated with wide-body aircraft.

This is a self-contained test rig requiring external electrical supplies of 415 volt, 3 phase, 50 Hz to operate the hydraulic pump motor and 110 volt, single phase, 50 Hz for the rig control circuit. An electrical connector 36 metres long is reeled on a mobile cable drum which can be towed in train with the test rig. The complete rig is assembled on a four-wheel chassis fitted with a tow-bar. The rear wheels are equipped with brakes manually operated by a ratchet type lever. An inner chassis, on shock-mounts, carries the high-pressure hydraulic test system. Acoustic panels enclose the test system except for inlet and outlet apertures for a through flow of air induced by an air-blast cooler to control the hydraulic-fluid temperature.

Fluid is drawn from the aircraft system by the rig hydraulic pump through a flexible hose, the air blast cooler and a boost filter fitted with a differential switch controlling a fault light. The hydraulic pump is an axial piston-type fitted with a pressure compensator set to the operating pressure.

On the discharge side of the hydraulic pump is a relief valve which relieves back to the air blast cooler. The delivery rate of flow is controlled by a flow valve.

Downstream of the flow control valves is a delivery filter embodying a disposable element. A differential pressure switch fitted across the filter operates a fault light. A non-return valve downstream of the delivery filter safeguards against flow reversal should the rig hoses be incorrectly coupled to the aircraft hydraulic system. Both suction and delivery hoses have self-sealing couplings and are carried on reels at the forward end of the test rig.

Pressures are monitored on gauges on the control panel to indicate return pressure, pump discharge pressure and the delivery pressure. Fluid sampling points enable a fluid sampling set to be adapted to monitor the fluid contamination level at the pre-filtering stage, and cleanliness downstream of the delivery filter.

Characteristics

OPERATING PRESSURE: 3200 psi (220 bar)
RELIEF VALVE SETTING: 3600 psi (248 bar)
FLOW RATE: variable 51 Imperial gallons/minute (0–232 litres/minute)
FILTERS
boost filter: Differential pressure limit 16 psi (1·1 bar)
filtration rating: 3 microns
delivery filter: Differential pressure limit 105 psi (7·2 bar)
filtration rating: 5 microns
HYDRAULIC FLUID: Skydrol 500B and Phosphate ester-based fluids
FLUID TEMPERATURE: Up to 70°C
HOSES: 2 × 45 ft (13·7 m)
POWER REQUIREMENTS: 415 V, 3 phase, 50 Hz
TOWING SPEED: 10 mph (16 kmh)
TURNING CIRCLE: 25 ft (7·62 m)
DIMENSIONS
length: 16 ft 8 in (5·08 m) tow-bar lowered
13 ft 1 in (3·99 m) tow-bar raised
height: 6 ft 2 in (1·88 m)
width: 6 ft 8½ in (2·04 m)

Status

Supplied to British Caledonian Airways for McDonnell Douglas DC-10 servicing and to British Airways for maintenance of Lockheed Tristar aircraft

Manufacturer

Thermal Control Company Ltd, Hove, East Sussex, England

Thermal control mobile hydraulic test rig servicing a British Caledonian Airways DC-10 aircraft

Valair Scamp Aircraft Skin Cleaner

The Scamp is a mobile cleaning device for aircraft skins. The equipment consists of a manipulator installed on an aerial platform which can be elevated according to the aircraft to be cleaned. The vehicle on which the platform is mounted incorporates a diesel-driven generator to provide the hydraulic power. The manipulator uses an anthropomorphous mechanical arm with forced feedback so that force from the boom is felt in the operator's handle. The master-arm system used by the operator, provides simultaneous or independent motions in hoist, extension and azimuth. Movement from the master arm is transmitted through a boom to a positioner which holds the tooling. Brush and spray equipment can be handled in addition to the air-knife drying system.

There are two ways to clean an aircraft skin, one is a rubbing/scrubbing action of the surface, the other is a liquid jet process. Tools are provided with rotating brushing heads, through the centre of which passes either the detergent liquids, or polishes used for this type of cleaning. The other alternative is to use a high-pressure jet system, with a water-based, emulsion combination 'grease-eater'. This system enables the emulsion to be discharged at high pressure on to the surface and allowed to stand while its action combines with the contamination prior to being rinsed clean with clear water.

Whichever cleaning system is used the final drying is by an air-knife system; air pressure is converted to kinetic energy by passing the air through an elongated air chamber via an accurately machined slot; air leaves this slot with a velocity of between 10 and 45 000 feet per minute (3·05 and 13 700 metres per minute), removing the liquid from the aircraft's surface.

To enable the Scamp to circumnavigate the chosen aircraft's footprint in safety, a variety of guidance systems are available. The driver himself is able to control the vehicle manually around the aircraft; alternatively, the unit can be guided automatically using electro-magnetic sensing. This is achieved by burying a cable in the hard standing. Signals passing through this wire are sensed by the antennas beneath the vehicle, steering the unit along the buried cable's path; thus allowing the vehicle to circumnavigate the aircraft. An electric route-control system inside the scanning head measures and compares voltages. The magnetic field is between the two antennas: when the voltages in each antenna are the same the unit maintains its course. Approaching bends the different distances of the antennas from the magnetic field of the guide-wire creates voltages of different force and polarity. The electronic course-control system adjusts the travel direction through an amplifier and the steering motors of the unit. If permanent stops are to be incorporated around the aircraft's footprint, permanent magnets are installed in the floor. When the unit reaches these points it is brought automatically to a standstill; allowing the telemanipulator to carry out its cleaning operation.

To prevent aircraft damage and collision with other airport vehicles three detection systems are installed. At low level, tactile sensing is used; at mid- and upper-levels, ultrasonic detection, and at higher levels magnetic sensing. All the information received by these sensing systems is fed back to the on-board computer for comparison and collision avoidance correction.

Status

Under development

Manufacturer

Valair Division, Vaughan Associate Group, Shrewsbury, Shropshire, England

Vickers Hydraulic Ring-mains System

Hydraulic ring-mains systems have been developed to provide a fixed maintenance hangar method for servicing hydraulic systems.

The system feeds pressure and return lines under floor level from the pump house to man-hole terminations in the hangar floor. The control consoles can be plugged into any one of the man-hole positions.

In the pump house the ring-main is fed by high pressure, variable-volume pumps specifically designed for continuous operation on aircraft fluids, DTD 585B, Skydrol and other aircraft approved specifications. The pumps are sized to provide maximum flow and pressure requirements for specific aircraft applications and can be adjusted to suit changes in aircraft types. The pump house contains associated electrical controls, logic units, oil reservoir, boost pumps, filtration and cooling units. On-line de-aeration is available on selection from each console.

The ring-main pipework is installed under the hangar floor either in ducts or tunnels. Welded connections are used, and pipework can be designed to ensure there are no cross connections between separate aircraft systems.

Each control console provides a hydraulic interface between the main ring and the aircraft and includes electrical controls for the remote operation of appropriate high-pressure pumps and associated equipment. Visual indication of hydraulic fluid conditions is provided together with a digital flow indicator, temperature and pressure gauges, a flow-control valve and other necessary controls. A dump valve is incorporated to rapidly reduce pressure in the aircraft system in the event of an emergency operation.

The noise level inherent in a ring-main installation is well below currently accepted limits for aircraft maintenance operations.

Status

The first of three systems was installed at the British Aerospace facility at Warton for servicing the Tornado aircraft. A number of installations have been completed since, for military operators including the Royal Air Force

Manufacturer

Vickers Fluid Power, Swindon, Wiltshire, England

UNITED STATES OF AMERICA

Avitech Aircraft Test Equipment

Avitech design and produce specialist test equipment for fuel, hydraulic, electrical and pneumatic system and components. Fuel test equipment comprises fuel control, pump, nozzle, accessory and booster stands, temperature simulators and various test consoles. For hydraulic testing, stands have been developed for Skydrol hydraulic systems components, including pump, accessory and motor test stands; brake testers and other miscellaneous equipment. Electrical test equipment consists of generator and constant speed drive stands while pneumatic test facilities have been designed for various types of aircraft accessories. These include pressure ratio bleed-control test stands and consoles, bench-mounted test consoles, fixtures for sub-assembly testing and high-pressure stands for proof testing.

Status

Operators of Avitech test equipment include the US military, Pan Am and Western Airlines, Alitalia, Japan Air Lines, Vasp and Varig

Manufacturer

Avitech, Port Chester, New York, USA

Bauer Aircraft Engine Component Test Stands

Bauer design and fabricate engine component test stands to meet the specific requirements of individual operators. Temperature simulators and turbine blade, Vane and Duct test stands have been developed for the General Electric CF6, J79, J85 engines and Pratt and Whitney JT8D, JT9D and TF30 engines and supplied to engine manufacturer and airlines including General Electric, Pratt and Whitney, Air Portugal-TAP, American Airlines, Ansett and Scandinavian Airlines System. Other specialist engine stands, produced for military aircraft engines include equipment supplied to Pratt and Whitney to provide a means of post-overhaul testing of components employed on the F100-PW-100 engines in the F-14 and F-15 aircraft.

Manufacturer

Bauer/Electro Inc, Farmington, Connecticut, USA

Bauer Fuel Systems and Accessories Test Stands

Bauer specialise in the design and manufacture of test equipment for fuel nozzle control and pump tests. An example of their equipment is the BEI 1579 universal fuel nozzle test stand. A new development, the equipment measures and displays the performance of individual fuel nozzles under controlled simulated operating conditions. It provides a means of observation and measurement of the spray cone for angle conformity, spray deviation from true nozzle tip, mist diffusion and uniformity.

Characteristics

FACILITY REQUIREMENTS
power: 220/440 V, 3 phase, 60 Hz
water: 25 psi, 70°F max, 20 US gallons/minute (1·7 bar, 21°C max, 76 litres/minute)
air: 90–100 psi (6–7 bar)
DIMENSIONS
length: 7 ft (2·13 m)
width: 5 ft 4 in (1·63 m)
height: 6 ft 6 in (1·98 m)
WEIGHT: 3500 lb (1588 kg)

Status

Military operators include the US military and the Royal Australian Air Force. Airline operators of Bauer equipment include British Airways, KLM and Air Portugal-TAP in Europe, PIA, JAL, Varig, Air New Zealand, and in north America, Air Canada, Eastern and United Airlines. The United installation in San Francisco is a computerised fuel test unit for CF-6, JT9-D, JT8-D and JT 3D engines. The unit has the capability of handling, simultaneously, two different fuel control programmes. The computer module is designed to be able to accept control from a third test stand

Manufacturer

Bauer/Electro Inc, Farmington, Connecticut, USA

Bauer BE1 1579 universal fuel nozzle test stand supplied to Airmotive, Ireland

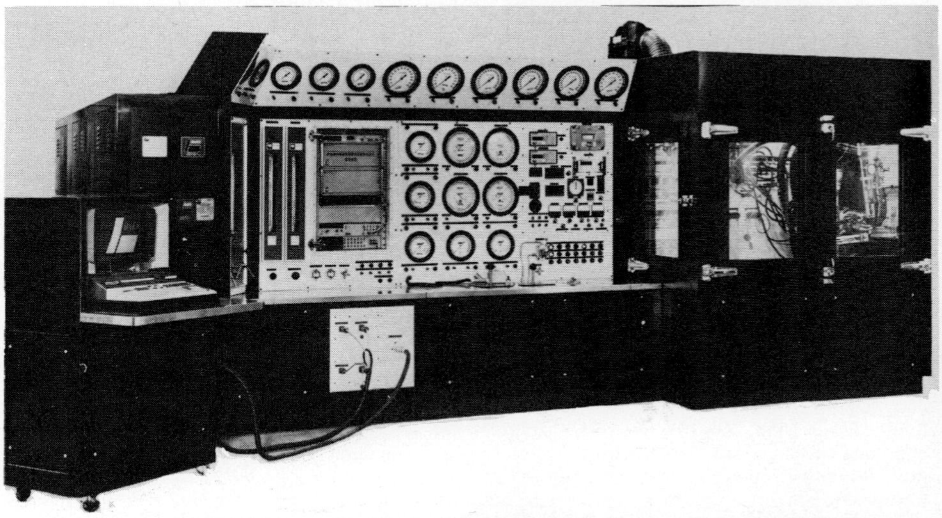

Bauer BEI automated main fuel accessories test stand built for Eastern Airlines for testing CF6-50, RB211 and JT8D main fuel accessories

Bauer Mobile Hydraulic Ground Power Unit

This unit is designed to deliver the necessary hydraulic power required to operate an aircraft's hydraulic system during periodic maintenance and overhaul checks or tests. Using Skydrol LD/Aerosafe hydraulic fluids the equipment delivers: 5 to 50 US gallons per minute at 3000 psi (19 to 198 litres per minute at 210 bar) through a variable-displacement pump with remote pressure compensators. The unit can be operated from any 3-phase power supply, the voltage and frequency being specified by the user. Alternatively a diesel-powered version is available. Constructed in steel with a roof walkway accessible by an integral ladder the unit is mounted on a four-wheeled trailer fitted with tow-bar, swivelling front wheels and a manual brake operating on the two rear wheels.

Other features include: 5 micron filtration, air cooler which returns the fluid at 120°F (48·9°C) and dual discharge and return circuits.

Capable of either open- or closed-loop operation, the system is protected by temperature and pressure cut-out switches and relief valves.

Characteristics

HYDRAULIC FLUIDS: Skydrol LD/Aerosafe
OPERATING PRESSURE: adjustable 5–50 US gallons/minute (19 to 198 litres/minute) at 3000 psi (210 bar)
RETURN TEMPERATURE: 120°F (48·9°C)
HOSES LENGTH: 50 ft (15·2 m) with connectors
UNIT DIMENSIONS
length: 10 ft 10 in (3·3 m)
width: 5 ft 8 in (1·72 m)
height: 5 ft 4 m (1·63 m)
WEIGHT: 6500 lb (2948 kg)

Status

In service with major airlines for servicing Boeing 707, 727, 737 and 747, McDonnell Douglas DC-8, DC-9, DC-10 and Lockheed L-1011 aircraft

Manufacturer

Bauer Electro Inc, Farmington, Connecticut, USA

Cryocart Model TS-40-2D Liquid Nitrogen Pneumatic Unit

The Model TS-40-2D is a liquid nitrogen pneumatic unit designed for servicing all sizes and pressures of aircraft tyres. The unit can be used, also, as a source for large volumes of nitrogen gas to operate pneumatic tools, paint spray units, purge and inert fuel systems. The equipment comprises two Dewar liquid storage vessels, and a control cabinet with regulator, valve gauges, and two 25-foot (7·62-metre) hoses mounted on retractable hose reels.

The unit is installed on a chassis fitted with four 6-inch (15·2-centimetre) metal castors. The unit uses ambient temperature as a heat source in conjunction with a high capacity vaporiser system, for the conversion of liquid nitrogen to nitrogen gas.

Characteristics

MODEL	TS 40-2D	TS 50-2D
OPERATING PRESSURE	Up to 250 psi (17 bars)	Up to 250 psi (17 bars)
CAPACITY	7500 ft³ (212 m³)	9310 ft³ (264 m³)
DIMENSIONS		
length	7 ft 8 in (2·34 m)	7 ft 8 in (2·34 m)
width	2 ft 6 in (0·76 m)	2 ft 6 in (0·76 m)
height	5 ft 4 in (1·63 m)	4 ft 7 in (1·4 m)
DRY WEIGHT	950 lb (430 kg)	950 lb (430 kg)

Manufacturer

Cryogenic Applications, San Gabriel, California, USA

Hobart CL Model 28·5 Volt Dc Transformer Rectifier

This unit has been designed for use when 28·5 volts dc is the only output required. Two models are available, one for 50 Hz supply, the other for 60 Hz, both rated at 400 amperes continuous, 1500 amperes peak. The units, mounted on a four-wheeled chassis with parking brake, feature ferro-resonant circuitry, convection cooling, separate input and output contactors with indicating lights and an output voltage independent of ±10 per cent input variations.

Characteristics

MODEL	5FR 28-400 CL	6FR 28-400 CL
RATED OUTPUT	28·5 V dc	28·5 V dc
CONTINUOUS	400 A	400 A
PEAK	750–1500 A	750–1500 A
POWER SUPPLY	230/400 V, 3 phase, 50 Hz	208/240/480 V, 3 phase, 60 Hz
UNIT DIMENSIONS	50 × 36 × 48 in (1·27 × 0·91 × 1·22 m)	50 × 36 × 48 in (1·27×0·91×1·22 m)
WEIGHT	1700 lb (771 kg)	1575 lb (715 kg)

Manufacturer

Hobart Brothers Company, Power Systems Division, Troy, Ohio, USA

Metric Systems Models 8300/8697 Corrosion Control Units

Two models are available; the model 8697 which can be powered by either a petrol or diesel engine and the 8300, driven by a petrol engine. Both models are mounted on a four-wheel trailer fitted with a tow-bar. The engines drive a compressor to produce separately-regulated water and oil sprays, for corrosion control of aircraft jet engines and aircraft components. In-line filters are incorporated to prevent particle contamination of the sprays.

Model 8697 corrosion control unit

Characteristics

MODEL	8300	8697
ENGINE	Petrol 4·7 hp at 2200 rpm	Petrol: 47 hp at 2200 rpm Diesel: 6·5 hp at 2200 rpm
COMPRESSOR	110 psi (7·6 bar) at 700 rpm	110 psi (7·6 bar) at 700 rpm
OIL FILTER	10 microns	10 microns
WATER FILTER	10 microns	10 microns
CAPACITIES		
air	30 gallons (114 litres)	30 gallons (114 litres)
oil	10 gallons (38 litres)	10 gallons (38 litres)
water	200 gallons (760 litres)	200 gallons (760 litres)
MAX TOWING SPEED	20 mph (32 km/h)	20 mph (32 km/h)
DIMENSIONS		
length	8 ft (2·44 m)	8 ft (2·44 m)
width	5 ft (1·52 m)	5 ft (1·52 m)
height	5 ft 6 in (1·68 m)	5 ft 6 in (1·68 m)
DRY WEIGHT	1500 lb (680 kg)	1700 lb (680 kg)

Status

In service with the US Air Force, US Navy, Canadian military forces and 15 other military forces

Manufacturer

Metric Systems Corporation, Fort Walton Beach, Florida, USA

Hobart Super Jet Frequency Converters

Hobart produce frequency converters for converting 50 or 60 Hz supply to 400 Hz for aircraft servicing. The larger units have a capacity from 90 to 200 kVA; the smaller units are rated at 37·5 and 60 kVA. All models are brushless, with two sealed bearings and the motor and generator on a common shaft.

The 90F, 125F, 160F and 200F models are similar in design with all control components, metering and protective devices housed in pull-out trays equipped with military standard plugs and receptacles. Each tray contains a small test panel with insulated test jacks. The top tray contains a frequency meter, voltmeter, ammeter, and automatic/manual voltage-control switch and manual-voltage control for the generator. The second tray contains the protective monitor system, plug interlock relays and test-bank switch. The protective monitor has SCR-controlled, fault-indicating lights functioning with protective relays to open the contactor when there is an abnormal condition. The lights for each function will remain on until the reset switch is pushed and the abnormality corrected. The overload system will disconnect the electrical system of the aircraft from the ground-power generator when the output in any phase exceeds 125 per cent rated current for approximately 5 minutes. The system is operated by a solid-state device with inverse time characteristics and a hermetically sealed reed relay output. The third tray comprises a 3-phase averaging magnetic amplifier and adjustable regulator which provides automatic voltage line drop compensation. Controls permit adjustments for output cable size and length.

The power module contains: load contactor(s), load contactor push-button(s), pilot light, panel lights, and four 400 Hz, 115-volt receptacles with circuit breakers.

The motor controls consist of an automatic wye-delta motor starting switch, motor start-stop push-button switch, pilot light and running time meter. Optional features available include motor start counter and automatic shutdown timer, to shut down the motor generator automatically if left running unattended without a load for more than three hours.

The 37F and 60F models are fitted with similar equipment to the larger models but without a frequency meter in the generator monitor tray. The automatic voltage line drop compensation is an optional feature.

Characteristics

50 Hz motor, 1500 rpm

MODEL	RATING	kW	GENERATOR	MOTOR	WEIGHT
37FH51	37·5	30	108 A	50 hp (37 kW)	2825 lb (1281 kg)
60FH51	60	48	173 A	75 hp (56 kw)	3125 lb (1418 kg)
90F-H51	90	72	260 A	125 hp (93 kW)	6620 lb (3003 kg)
90F-H52	90	72	260 A	125 hp (93 kW)	6670 lb (3026 kg)
125F-H51	125	100	361 A	150 hp (112 kW)	7500 lb (3402 kg)
125F-H52	125	100	361 A	150 hp (112 kW)	7550 lb (3425 kg)
160F-H51	160	128	462 A	200 hp (149 kW)	9000 lb (4082 kg)
160F-H52	160	128	462 A	200 hp (149 kW)	9050 lb (4105 kg)
200F-H51	200	160	577 A	250 hp (187 kW)	9600 lb (4355 kg)
200F-H52	200	160	577 A	250 hp (187 kW)	9650 lb (4377 kg)

60 Hz motor, 1200 rpm

MODEL	RATING	kW	GENERATOR	MOTOR	WEIGHT
37FH61	37·5	30	108 A	50 hp (37 kW)	2825 lb (1281 kg)
60FH61	60	48	173 A	75 hp (56 kW)	3125 lb (1418 kg)
90F-H61	90	72	260 A	125 hp (93 kW)	6620 lb (3003 kg)
125F-H61	125	100	361 A	150 hp (112 kW)	7500 lb (3402 kg)
125F-H62	125	100	361 A	150 hp (112 kW)	7550 lb (3425 kg)
160F-H61	160	128	462 A	200 hp (149 kW)	9000 lb (4082 kg)
160F-H62	160	128	462 A	200 hp (149 kW)	9050 lb (4105 kg)
200F-H61	200	160	577 A	250 hp (187 kW)	9600 lb (4355 kg)
200F-H62	200	160	577 A	250 hp (187 kW)	9650 lb (4377 kg)

DIMENSIONS

Models 37F, 60F	60 × 40 × 46 in (1·52 × 1·02 × 1·17 m)
Models 90F, 125F, 160F, 200F	126 × 40 × 68 in (3·2 × 1·02 × 1·73 m)

Status

In service with major airlines

Manufacturer

Hobart Brothers Company, Power Systems Division, Troy, Ohio, USA

Teledyne Sprague 83138 Mobile Hydraulic Test Rig

The 83138 is a single-system mobile hydraulic test rig designed for servicing high-volume hydraulic systems associated with wide-body aircraft. It is self-contained after connection of a 460 volt, 3 phase, 60-cycle power supply, by means of a 50 foot (15·2 metre) electrical cable. The complete rig, housed in a weather-proof steel cabinet is mounted on a four-wheel chassis fitted with a tow-bar. The rear wheels are equipped with mechanical brakes operated by foot lever. Fluid is drawn from the aircraft hydraulic system by the rig pump through a flexible hose. Two pressures are available. A three micron-rated filter is used in the high pressure system and a ten micron-rated filter when operating at the low-pressure setting. Controls include a flow-meter, switches for start/stop, system pressure release bypass to flow-meter and light test.

Characteristics

OPERATING PRESSURE
low: 3000 psi (207 bar)
high: 5000 psi (345 bar)

FLOW RATE
30 US gallons/minute (113·6 litres/minute) at 3000 psi (207 bar)
50 US gallons/minute (189·3 litres/minute) at 5000 psi (345 bar)

FILTERS
low pressure: 10 micron
high pressure: 3 micron

HYDRAULIC FLUID: Skydrol 500, Hyjet or Stauffer

HOSES: 50 ft (15·2 m) high and low pressure

POWER REQUIREMENTS
460 V, 3 phase, 60-cycle
380 V, 3 phase, 50-cycle (option)

TOWING SPEED: 10 mph (16 km/h)

DIMENSIONS
length: 10 ft 10 in (3·3 m)
width: 5 ft 6 in (1·68 m)
height: 6 ft 2 in (1·88 m)

Status

The Model 83138 is one of a range of mobile hydraulic test rigs, electrical or diesel powered, produced by Teledyne Sprague to meet the requirements of commercial and military operators

Manufacturer

Teledyne Sprague Engineering, Gardena, California, USA

Teledyne Sprague Fuel Systems Test Equipment

The model 1500–100 is a fixed installation test facility which supplies the drive, fuel supply and instrumentation for functional tests on engine-driven fuel pumps. The equipment housed in a steel cabinet, requires electric, pneumatic and water power supplies.

Characteristics

FUEL VISCOSITY: 0·730 to 0·745 specific gravity at 60°F (15·6°C)

DRIVE SPEED: 500–3600 rpm (pump test pad)

MAX FUEL PRESSURE: 30 psi (2·06 bar)

MAX AIR PRESSURE: 30 psi (2·06 bar)

RESERVOIR CAPACITY: 17·5 US gallons (66·2 litres)

FILTRATION: 10 micron

PLANT FACILITY REQUIREMENTS
electrical power: 440 V, 3 phase, 60 Hz
air: 100 psi (6·87 bar)
coolant: Cold water, supply and drain lines

DIMENSIONS
length: 3 ft (0·91 m)
width: 3 ft (0·91 m)
height: 5 ft 4 in (1·63 m)

Status

This unit is one example of a variety of fuel systems and component test equipment designed and manufactured by Teledyne Sprague to meet individual operator's requirements

Manufacturer

Teledyne Sprague Engineering, Gardena, California, USA

Teledyne Sprague Pneumatic Test Equipment

Fixed installation pneumatic test stands provide facilities for testing performance and operating characteristics of aircraft pneumatic devices and components. Static and leakage tests in addition to vacuum tests can be conducted. Other pneumatic test equipment available includes: test stands for portable oxygen breathing systems, aircraft oxygen regulators, anti-icing air valves and regulators, engine air flow, and aircraft cabin pressure systems. The equipment may be a fixed installation or portable, as required by the operator.

Status
A variety of pneumatic test equipment is available, designed and manufactured to meet individual customers' requirements

Manufacturer
Teledyne Sprague Engineering, Gardena, California, USA

Teledyne Sprague S 169 Fixed Installation Universal Hydraulic Test Stand

The S 169 universal test stand is designed for the simultaneous testing of several aircraft hydraulic components. A reservoir supplies the hydraulic test fluid for the equipment while micronic fluid filtration is provided by a low pressure filter in the boost system. An air-bleed valve or plug in the top cover of the low-pressure filter permits purging of air from the test stand system. A filter drain is provided to allow draining the filter for maintenance purposes.

An electric motor-driven, pressure compensating pump supplies operating pressures up to 6000 psi (345 bar). A volume control on the control panel permits manual adjustment of pump delivery from zero to the maximum-rated capacity of the pump. The main pump is super-charged by a belt-driven boost pump which supplies fluid at low pressure from the reservoir to the suction port of the main pump. Boost pressure to the main pump is controlled by an adjustable, low-pressure relief valve. An optional accumulator-type pulsation dampener can be provided in the high-pressure system at the main pump which absorbs pump pulsations in the high pressure system. A similar type dampener performs the same function in the gauge system. An adjustable high-pressure relief valve controls pressure in the high-pressure system. An additional high-pressure relief valve is provided in the pump test system to regulate maximum test pump pressures. The flow system consists of a four-way valve with inlet and outlet shut-off valves plus a single pressure outlet controlled by a shut-off valve. All pressure outlets are along the front edge of the work tray. Four return inlets with shut-off valves are on the instrument panel, and three flow-meters are incorporated in the flow system. The latter provide sufficient range for flow tests up to the delivery capacity of the main pump. Three flow-meter outlets are provided to permit direct connection from test component to the required flow-meter. An accumulator is installed in the high-pressure system to facilitate cycling tests on unloading valves. A shut-off valve can isolate the accumulator from the system when not required for test purposes.

Hydraulic pumps can be tested to manufacturers' specifications. The pump test system is comprised of a variable speed pump drive pad with a standard capacity of approximately 20 hp at 4000 rpm in either direction of rotation plus all necessary controls.

Optional features available include: air-driven booster pump, timer, variable speed drive, dc power supply, pulsation dampener, high pressure 10 micron filter and remote power supply.

Characteristics
OPERATING FEATURES
DYNAMIC: up to 5000 psi (345 bar)
STATIC: up to 10 000 psi (690 bar) with hand-pump
up to 30 000 psi (2070 bar) with pneumatic hydraulic pump
FLOW RATES: optional ranging from 10–60 US gpm (37·9–227·1 litres/minute)
PRESSURE GAUGES
0–30 psi (0–2·1 bar)
0–160 psi (0–11 bar)
0–3000 psi (0–207 bar)
0–5000 psi (0–345 bar)
HIGH-PRESSURE RELIEF VALVE: 200–5000 psi (13·7–345 bar)
RESERVOIR CAPACITY: 110 US gallons (416·4 litres)
POWER REQUIREMENT: 440 V, 3 phase, 60 Hz

Status
Teledyne Sprague produce a wide range of hydraulic test stands. The S 169 is a universal model which is capable of testing the majority of aircraft hydraulic components in current use

Manufacturer
Teledyne Sprague Engineering, Gardena, California, USA

Unitron GFC Series Frequency Converter

The GFC series frequency converter is a solid-state switching-type ac unit producing a 400 Hz supply for hangar power distribution at 7·5, 15 and 30 kVA power ratings. Up to 20 GFC series units, mounted in standard 19-inch cabinets, may be operated in parallel to produce the required amount of power. Controls include: power on/off, selections for variable frequency, input/output voltage, output current, parallel operation, and adjustments for current limit, and output voltage. Input/output voltmeter, output ammeter and frequency meters are incorporated. An output voltage regulator maintains the voltage, under balanced-load conditions, to ±0·5 per cent. Protection devices include over/under voltage, loss of input phase, over current and short circuit.

Status
Production, the equipment is designed to meet US military standards

Characteristics

MODEL	GFC-07	GFC-15	GFC-30
RATED OUTPUT	7·5 kVA	15 kVA	30 kVA
adjustable	0–132/288 V, 3 phase,	120/208 V, 3 phase,	120/208 V, 3 phase,
selectable	360, 380, 400, 420, 440 Hz	400 Hz	400 Hz
POWER FACTOR	0·0 lagging–0·0 leading	0·0 lagging–0·0 leading	0·0 lagging–0·0 leading
ACCURACY			
voltage	±0·5%	±0·5%	±0·5%
frequency	±0·1%	±0·1%	±0·1%
POWER REQUIREMENTS	120/208 V, 45–66 Hz	120/208 V, 45–66 Hz	120/208 V, 45–66 Hz
DIMENSIONS			
length	43·2 cm	43·2 cm	43·2 cm
depth	61 cm	68·6 cm	68·6 cm
height	26·7 cm	53·3 cm	66 cm
WEIGHT	63 kg	104 kg	171 kg
OPERATIONAL TEMPERATURE RANGE	–20° to +55°C	–46° to +55°C	–46° to +55°C

Manufacturer
Unitron Inc, Garland, Texas, USA

ENVIRONMENTAL CONTROL

Airport and runway sweepers
Automatic spreading equipment
Bird dispersal units
Grass mowers
Ice-warning systems
Noise monitoring equipment
Runway rubber removal units
Runway surface-friction measuring devices
Snow and ice clearance machinery

CANADA

Frink Snow-ploughs

Three models are produced by Frink suitable for snow-clearance operations of runways, taxiways and apron areas. These are the Reversible Polymer Plough, Trailblazer and the FTD-20 Ramp-Hog.
The Reversible Polymer plough is offered in two sizes, 19 feet (5·8 metres) in length and 27 feet (8·2 metres), both ⅜ inch (9·5 mm) thick. Constructed of polymer, the mold board is secured to a heavy-duty steel frame consisting of an angle reinforced with a U-shaped channel and 12 ribs with a box section on top. The plough, which is supported by two pneumatic-tyred swivel castors, is reversed using two hydraulically-operated cylinders.
The Trailblazer is supplied with a mounting kit to suit most 4 × 4 chassis. The moldboard, constructed in polymer with a ½ × 8 inch (12·7 × 203 mm) carbon-steel cutting edge, is coupled to a tubular drive frame with thrust from reversing rams or braces applied directly to plough connecting lugs. A safety trip mechanism is incorporated in the design and the moldboard remains upright with the blade in the tripping position. An electric-hydraulic pump is installed under the bonnet (hood) of the vehicle, operated by levers and control rods from the cab. Optional features include angling rams, with a cross-over relief valve to protect the cylinders from shock loads, and a head lamp kit comprising twin headlamps and direction indicators installed on top of the mounting bracket.
The FTD-20 Ramp-Hog is designed for ramp or apron snow clearance with a moldboard length of 20 feet (6·1 metres). The steel moldboard with a ½ × 8 inch (12·7 × 203 mm) carbon steel cutting edge, incorporates side panels supported by runners providing a capacity of 270 cubic feet (7·64 cubic metres). It is reinforced by 11 vertical ribs and horizontal mouldings. Two parking legs are fitted to support the snow plough when not in use.

Frink FTD-20 Ramp-Hog snow-plough

Characteristics

TYPE	Reversible Polymer	Trailblazer	FTD-20
LENGTH	19/27 ft (5·8/8·2 m)	8 ft (2·44 m)	20 ft (6·1 m)
HEIGHT	3 ft 6 in (1·07 m)	2 ft 6 in (0·76 m)	4 ft 8 in (1·42 m)
CUTTING PATH AT 32°	12 ft 8 in (4·88 m)		
SIDEPLATES DEPTH			2 ft 8 in (0·81 m)
CAPACITY			270 ft³ (7·64 m³)
WEIGHT			3300 lb (1500 kg)

Status

In service at airports in Europe and North America

Manufacturer

Frink Canada Ltd, Cambridge, Ontario, Canada

SMI Detachable Snow-blowers

Two models of detachable snow-blowers are produced by SMI for mounting on the front of a loader or other suitable vehicle. They are the Series 3000 and the RHS 1500 L.
The Series 3000 is a self-contained unit which uses twin augers, a separate impeller, a rotating turbine to permit regulated long distance casting and a telescopic rotating chute for control of the snow. It displaces up to 20 tons (20·3 tonnes) of snow per minute and casts up to 150 feet (45·72 metres) to the right. The unit can be mounted on a large front end loader, grader or any other wheeled vehicle designed for slow speed operation with a capacity of 10 000 to 18 000 lbs (4535 to 8165 kg) on the front axle.
The RHS 1500 L is designed for mounting on an all-wheel drive front end loader. The engine including the fuel tank, is mounted on the rear of the loader. The unit is fitted with hydrostatic drive in which power is transferred to the rotary head through high pressure hoses and a planetary gearbox. Flexible hoses are used to accommodate the articulated steering of the loader. The blower engine operates at a constant rpm.
The speed of the helical cutters and the impeller is variable within design limits, and determines both the tonnage of snow the rotary head will handle, and the casting distance. The operator controls the cast with a single lever. The unit operates with a two stage rotary system; the first stage cuts and gathers snow from in front of the loader. The second stage accelerates and discharges the snow through an impeller. Bucket-shaped impeller blades, rotating at twice the speed of the helical cutters, throw snow at a rate of up to 1750 tons per hour (1778 tonnes per hour) over distances from 15 to 100 feet (4·6 to 30·5 metres). The impeller casing rotates to permit snow casting to either side of the blower head, or vertically through an optional spot casting chute.

Characteristics

TYPE	Series 3000	RHS-1500 L
ENGINE	Diesel	Detroit Diesel/ Cummins/ Caterpillar
	225 hp (168 kW)	6 V 71 N or 6 V 71 T/ VT 903 306
CLEARING width	9 ft 1 in (2·77 m)	8 ft 6 in (2·59 m)
performance	1200 tons/h (1088 tonnes/h)	1750 tons/h (1600 tonnes/h)
speed		20 mph (32 km/h)
HELICAL CUTTER		3 ft 8 in (1·12 m)
AUGER/DIAMETER	2 × 1 ft 8 in (0·508 m)	
IMPELLER DIAMETER	3 ft (0·91 m)	3 ft 8 in (1·12 m)
HEIGHT to top of	5 ft 10 in (1·78 m)	6 ft (1·83 m)
cutting knives		optional 10 ft (3·05 m)
CASTING DISTANCE from chute	12 ft 6 in (45·72 m)	5–50 ft (1·5–15·5 m)
from impeller		15–100 ft (4·6–30·5 m)
DIMENSIONS length	6 ft 6⅔ in (2 m)	
width	9 ft 1 in (2·77 m)	
height	11 ft 5 in (3·48 m)	
WEIGHT	9600 lb (4355 kg)	13 000 lb (5900 kg)

Manufacturer

SMI Industries, Montreal, Quebec, Canada

SMI Self-propelled Snow-blowers

Two of the models of self-propelled snow-blowers produced by SMI, are the Series 5000 and the larger M6. Both are four-wheeled vehicles powered by Detroit diesel engines, through an automatic transmission and a torque converter in the series 5000; in the M6 hydrostatic transmission is coupled with a mechanical four-speed drop transmission equipped with a synchroniser. Air brakes are fitted in the M6, and a dual-hydraulic braking system in the Series 5000. Both models feature power-assisted steering, and similar snow-clearance machinery with two horizontal augers, a six-bladed impeller and a spot casting chute. With the latter, the snow stream can be controlled in any direction over a 190 degree (215 degrees in the M6) horizontal arc. The M6 is equipped with two steering vanes for directing the snow. These are controlled hydraulically from inside the cab.

Characteristics

MODEL	Series 5000	M6
ENGINE	Detroit diesel Allison 4-53	Detroit diesel 6 V 92 TA
OUTPUT	102 kW at 2800 rpm	250 kW at 2100 rpm
CLEARING		
width	2·29 m	2·54 m
performance	1705 tonnes/h	2150 tonnes/h
SPEED		
max	69 km/h	48·3 km/h
operational	1·6 km/h	
IMPELLER DIAMETER	0·91 m	1·07 m
CONVEYOR DIAMETER	0·51 m	0·66 m
CHUTE HEIGHT	3·76 m	3·76 m
diameter	0·56 m	0·56 m
casting distance	1·2–15 m	1·2–15 m
TURNING RADIUS	8·6 m	7·15 m
WHEELBASE	2·92 m	2·11 m
DIMENSIONS	6·93 × 2·29 × 3·68 m	7·21 × 2·54 × 3·61 m
WEIGHT	12 680 kg	11 700 kg

Status
Both snow-blowers have been supplied to major airports. The Series 5000 is in service with the British Airports Authority

Manufacturer
SMI Industries, Montreal, Quebec, Canada

SMI Series M6 snow-blower

SMI Series 5000 snow-blower

SMI Series 300 Airport Runway Sweeper

The Series 300 is a towable unit which provides a swept path 12-feet (3·66-metres) wide at speeds up to 25 metres per hour (40 km per hour). It is available powered by a petrol or diesel engine. The power plant is housed in a steel sheet enclosure equipped with side-mounted louvre doors. A pto with clutch mounted at the engine flywheel housing, transmits power to the air impeller, the hydraulic pump and the broom drive arrangements. The broom is 13 foot 10 inches (4·22 metres) long with a diameter of 3 feet (0·91 metres) made of tufts fastened to rings mounted on a tubular arc. It is mounted on a tubular frame with detachable chain cases supported by two pneumatic castors which are connected to the main frame by means of the top torque rod and the bottom push frame. The broom is free to move up and down to follow irregularities of the runway and can be swung 30 degrees to either side into the working position. A broom cover consisting of a fixed-cover plate bolted to the broom frame and the hinged snow deflector, keeps forward swept snow and dust down and seals off the gap between the broom and the cover plate. The broom is controlled in height by a crank-operated adjustable stopper built into the broom lifting cylinder permitting the broom to be adjusted to the required working height through push buttons on the control panel. When fixed in this position, by the stoppers, the broom can be lifted 4 inches (101·6 mm) to the travelling position. The adjustment compensates for broom wear from 3 feet (0·91 metres) to 1 foot 6 inches (0·46 metres) diameter or a total of 9 inches (0·23 metres) brush wear. In the air system, a centrifugal blower rated at 6000 cubic feet per minute (170 cubic metres per minute) is driven by the engine through a belt arrangement. The air-ducts can be raised 6 inches (152 mm) to provide 8 inch (203 mm) ground clearance in the travelling position. During the sweeping operation one nozzle is always in line with the leading edge of the broom; the other one directs air towards the trailing edge.

Hydraulic actuators turn both nozzles through 180 degrees when the broom is swung to the opposite working side, and hinged flaps at the nozzle opening can shut off the air flow if required. Control of the sweeper is via a cable equipped with a pull-out safety connector at the tow-bar end which connects the machine and the remote control box in the cab of the towing vehicle. The box contains a tachometer, emergency engine stop button, engine speed switch and four-way toggle switch controlling the position of broom and air nozzles. The engine speed control switch operates a four-way solenoid valve attached to the engine which directs engine oil pressure to a hydraulic cylinder working on the carburettor throttle lever.

For cleaning runway lights a selector valve on top of the engine cowling above the main control panel reverses the rotation of the nozzles relative to the broom. With the broom out of the way, the machine can be driven with one air nozzle close to the line of lights.

Characteristics
ENGINE: petrol (model 314G)/Diesel (model 314D)
TOWING SPEED: 25 mph (40 km/h)
BROOM DIMENSIONS
length: 13 ft 10 in (4·22 m)
diameter: 3 ft (0·91 m)
SPEED: 520 rpm/optional: 400 rpm
SWEPT WIDTH: 12 ft (3·66 m)
SWEEPING PERFORMANCES: 1 200 000 ft²/h (111, 480m²/h)
CENTRIFUGAL BLOWER CAPACITY: 6000 ft³/minute (170 m³/minute)
NOZZLE AIR VELOCITY: 225 mph (362 km/h)
OVERALL UNIT DIMENSIONS
length: 30 ft (9·14 m)
width: 15 ft 1 in (4·6 m)
height: 7 ft 10 in (2·39 m)
WEIGHT
314D: 8975 lb (4071 kg)
314G: 8310 lb (3769 kg)

Manufacturer
SMI Industries, Montreal, Quebec, Canada

DENMARK

Brüel and Kjaer Airport Noise Monitoring Systems

Brüel and Kjaer supply a wide range of equipment designed for aircraft noise measurement. These can be divided into two categories: portable systems for short-term airport noise monitoring and computer-controlled systems for simultaneous multi-channel monitoring required at large international airports.

The portable systems are battery-operated instruments suitable for aircraft noise measurements at selected places around an airport for periods of from a few hours up to one day. They may be equipped with rechargeable Ni-Cd cells and many of them have provision for external power supply.

The impulse precision sound level meter Type 2209 or 2210 is used when both A and D weighted measurements are required. The slow meter characteristics are selected except for low altitude, high-speed flyovers where the fast response may be necessary to obtain the true level.

Type 2210 has a maximum hold feature which can be used with either time constant. It is obtained using the manual clock of the digital display. Each time the manual push button is pressed the maximum sound level measured since the previous read-out is displayed. The maximum values can be recorded manually but an alphanumeric print-out of the time of occurrence and the maximum level can be given if the 2210 is connected to the alphanumeric printer Type 2312. The system can also be adjusted to give one print-out per second.

The precision integrating sound level meter Type 2218 combines a precision sound-level meter with analog read-out and an L_{eq}-meter with digital display. If an L_{eq} measurement is started at the beginning of a flyover the L_{AX} value may be obtained at the end of the event by setting the instrument to stand-by and pressing the L_{AX} push-button. Alternatively the overall L_{eq} value may be measured over a pre-selected time interval. For later signal processing the sound signals may be recorded using the Types 7003/7004 tape recorders.

The noise level analyser Type 4426 can be used for a wide range of measurements for aircraft noise assessment. Input can be from a microphone preamplifier, a sound level meter or a tape recorder. The signal is A-weighted (preamplified input only) and fed to a log root mean square detector which has the standardised fast, slow and impulse time constants. The output of the detector is sampled (selectable sampling rate) and classified according to the level. The sample may represent either the instantaneous level or the maximum level recorded over the sampling period. Without affecting the measurements, the digital display can indicate at any time the sound level, the amplitude distribution in two dB-wide classes over a 64 dB dynamic range, the cumulative distribution, any statistical level L_N (level exceeded for N per cent of the measurement time, $1 \leqslant N \leqslant 99$) and L_{eq}.

Analogue outputs are provided for recording L_N or the sound level as a function of time as well as cumulative or distribution histograms. The digital output allows direct connection to a printer or a digital cassette.

Computer-controlled systems are designed for multi-channel monitoring. The system comprises the noise monitoring terminals (nmt), which collect noise data at selected locations; the transmission system and the central station which controls the system operation, processes the data received, produces a noise-event report and stores the data necessary for statistical analysis. The transmission systems uses modems for digital transmission on conventional telephone lines.

The noise monitoring Terminal 5667 which measures 40–130 dB in two ranges, consists of an out-door microphone unit and an electronic section containing a signal-conditioning unit, modem and air-conditioning unit. The complexity and size of the central station depends on the maximum number of nmts the system can handle and on user requirements for data processing and report presentation.

Typically the central station includes one or several modems for communication with the nmts, a computer and various peripherals for data storage and print-out and for communication with the computer (disk systems, printers, video terminals, etc). The NOVA computer from Data General has a 32 K-word MOS memory (expansible in eight K modules) with automatic program load, battery back-up, power monitor and auto-restart. A controller is included for communication with the modem.

The status of each nmt is individually defined and can be checked and/or modified at will without interrupting monitoring. The status defines the operation mode of the nmt (fast or slow time constant, maximum or effective let-time constant, maximum or effective level) and the limits defining a noise event. One level L_1 is the event limit while the L_2 can for example be the absolute noise limit which must not be exceeded.

Data requests are sent once per second. The

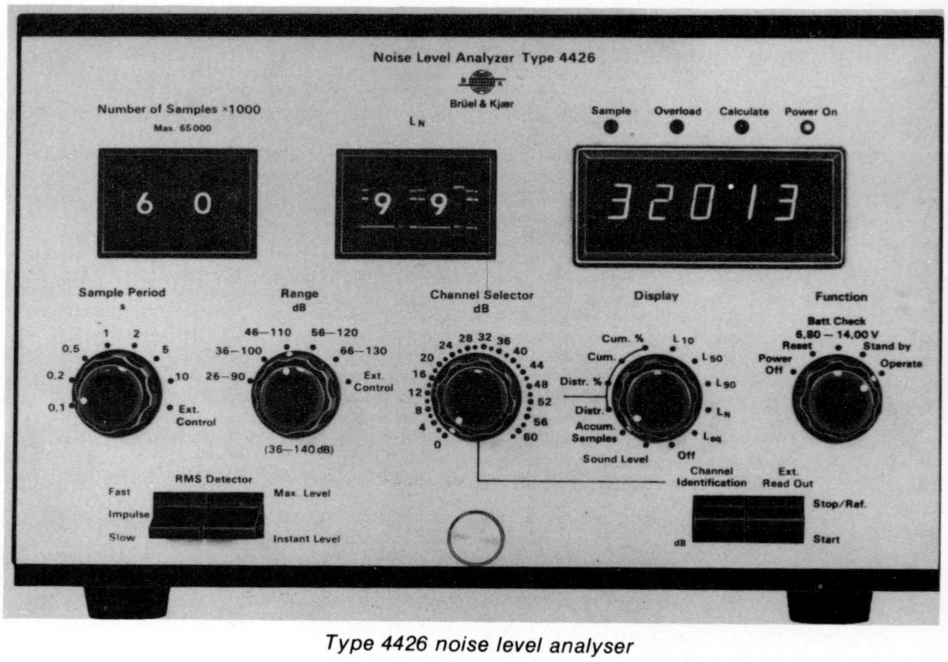

Type 4426 noise level analyser

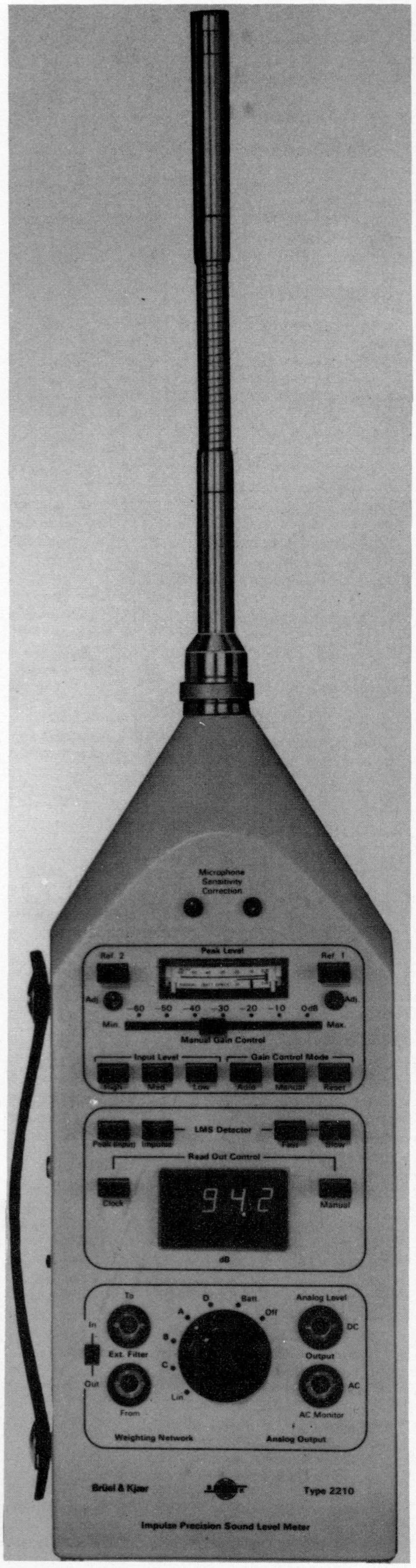

Type 2210 impulse precision sound level meter

responses are checked and if valid are used for a running L_{eq} calculation. For each nmt, the measured level is compared to the corresponding event threshold. If L_1 is exceeded the program initiates event-processing procedures. When the level goes below L_1, an event report is printed giving:

- nmt identification
- time of day of maximum
- event exposure level
- maximum noise level
- L_1 and L_2
- duration of noise above L_1 and L_2
- excess of noise above L_2

If the event lasts longer than a user-defined duration it is considered as a prolonged event (eg an engine test run) and will be reported as follows: nmt identification; time of event start; equivalent level of event; maximum level and time of maximum L_1 and L_2; number of times level went above L_2; time of event end.

Three different time periods may be defined for periodic reports eg hourly, 8-hourly and 24-hourly periods. At the end of these intervals a report is printed out which includes for each nmt L_{eq} for the interval and a selected number of L_N values.

At midnight a daily report will be printed. Automatic calibration checks are performed twice a day.

Status

Brüel and Kjaer have supplied noise monitoring systems to many European airports including Basle-Mulhouse Airport in Switzerland

Manufacturer

Brüel and Kjaer, Naerum, Denmark

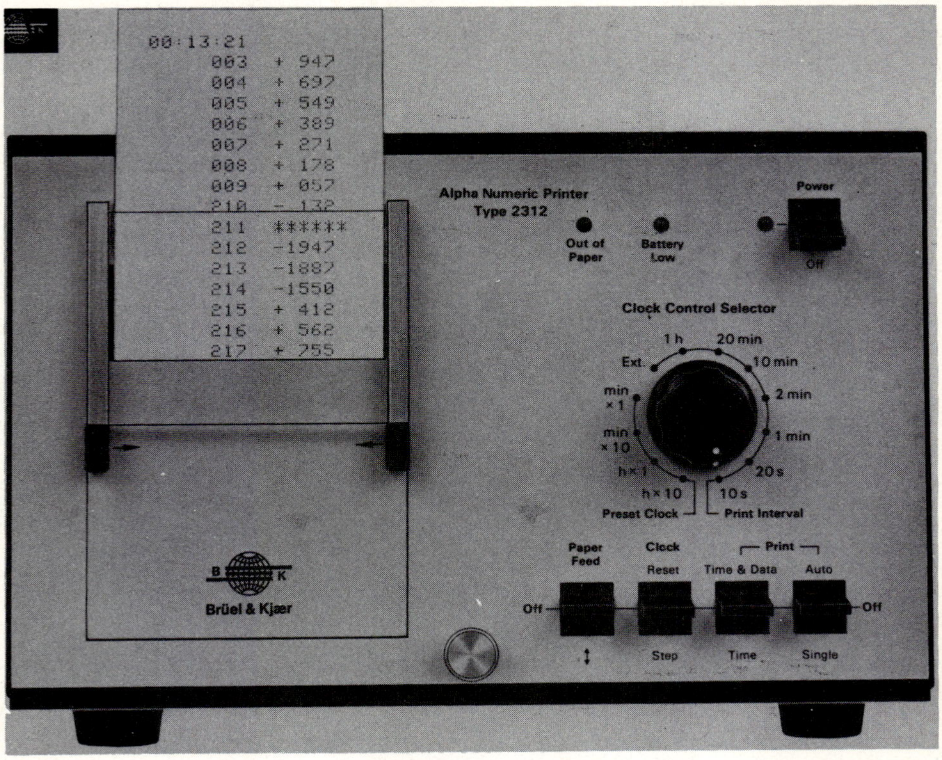

Type 2312 alphanumeric printer

Thomsen Epoke Runway Maintenance Equipment

Epoke sand and urea spreading equipment can be mounted on the back of most commercial trucks. The capacity of the hopper can range from three to eight cubic metres and the equipment provides a spread width of between 6 and 27 metres. The spreading quantity varies from a minimum of 5 g/m² for urea, up to a maximum of 150 g/m² for sand. The system allows for remote control of the spreading symmetry.

Type PWD 58 spreader, supplied to Düsseldorf airport, contains a small road wheel installed behind the vehicle on which the rear of the spreading unit is mounted. This additional wheel provides ground speed data for synchronisation of spreading rates. A moistening system can be fitted to the spreaders to spray a calcium chloride solution over the dry salt. This helps to bind the salt to accelerate the thawing process.

Epoke Triplex is a hydraulically-powered flail grass mower designed for high-speed driving producing a cut width of 5 metres. The side mowers can be adjusted individually for cutting on slopes.

Manufacturer

A/S Alfred Thomsen, Vejen, Denmark

FINLAND

Perusyhtyma ARA HP2 Runway Snow Removal Equipment

The ARA HP2 is a towed machine for clearing snow for runways. Developed in 1962 to meet the requirements of the Finnish and Swedish civil and military aviation authorities the current model, the HP2F, features the same basic design. The triangular chassis is of welded-profile construction mounted on three twin wheels which carry the machine's total weight. The rear wheels have a steering capability, turning the machine ± 30 degrees, while the front wheels behind the broom are turned to the towing direction by the tow-bar. The broom, which is a rotating multi-section brush on curled zinc wire is driven by a V-belt/triplex chain drive from the main shaft connected to the engine clutch. The blower, a high pressure fan designed for high rpm, is driven by the same main shaft using fixed air tubing and a turning blow nozzle to eject air at 120 metres per second. Power is by an industrial type water-cooled super-charged diesel engine designed for continuous use and governed by a remote-controlled constant speed regulator. Electrically-operated hydraulic controls are provided for lifting and locking the brush, steering the rear wheel, changing the sweeping angle and adjusting the blow nozzle. The sweeper is electrically-powered and independent of the towing vehicle. Amber beacon, side indication lamps and working lights are fitted.

Runway sweepers in operation at Helsinki-Vantaa International Airport

Characteristics

ENGINE: Scania DS11 diesel engine developing 240 hp (179 kW) at 1900 rpm

BRUSH
- diameter: 3 ft (0·91 m)
- length: 13 ft 11 in (4·24 m)

ROTATING SPEED: 600 rpm

BLOWER
- velocity: 120 m/s
- capacity: 27 000 m³/h

DIMENSIONS
- length: 30 ft 2 in (9·2 m) including towing bar
- width: 15 ft 3 in (4·63 m)
- height: 6 ft 11 in (2·1 m)

SWEEPING WIDTH AT 30°: 11 ft 4 in (3·45 m)

WEIGHT: 8700 kg

Status

ARA HP2 runway sweepers have been supplied to airports in Canada, Denmark, Finland, Poland, Sweden, United Kingdom and the USSR.

Helsinki-Vantaa airport uses 11 plough-truck-sweeper combination units. In normal winters the machines are on a 24-hour alert for a period of five months. In the summer only two sweepers are used to keep the runway clear of sand and deposits

Manufacturer

Perusyhtyma Oy ARA, Turku, Finland

FRANCE

Bertin and Cie TS 2 Runway Snow and Ice Clearance Machine

The TS 2 machine is designed to clear runways of snow, ice, water and debris using the thermo-kinetic energy of a turbo-jet engine whose gases are ejected by four nozzles in a direction at right angles to the machine and parallel to the runway surface. Two semi-circular nozzles are at the front in the centre. The other two are located at the end of arms which extend laterally on either side of the machine. Each nozzle is fitted with vector controls to direct the blast. The exhaust gases create a slight depression at ground level which causes the snow and ice to separate from the surface. The kinetic energy blows all snow, ice, water and debris from the runway while thermal energy dries the tarmac. The machine is designed to clear an operational width in one sweep. With fresh snow the TS 2 will clear a track 40 metres wide at 15 to 20 km per hour. With ice the speed is reduced to a maximum of 5 km per hour and the machine's swept width to 25 metres.

The TS 2 consists of a chassis powered by a 100 hp diesel engine, cab, a turbo-jet engine and hot pipe units. An Elastimatic torque converter enables the diesel engine to start the turbo-jet. Fuel for the turbo-jet comes from a kerosene tank which is connected to a tank trailer.

TS 2 runway clearance machine

Characteristics

ENGINES: Turbo-jet developing 3500 kg of static thrust at cruise (2/3 power); 100 hp diesel engine

MAX SPEED: 45 km/h

CREW: 2

OPERATIONAL WEIGHT: 10 000 kg

Status

The TS 2 machine is in service at 23 airports

Manufacturer

Bertin and Cie, Plaisir, France

Bertin and Cie Turboclair Fog Clearance System

Turboclair is a thermo-kinetic fog clearance system which has been developed jointly by Bertin and Aeroport de Paris, employing out-of-commission jet engines as generators of hot gases. The engines supply both thermal and kinetic energy. The thermal energy vaporises the water found in fog in the form of fine droplets, and warms the air to a sufficient temperature in order for it to become unsaturated and clear air. The kinetic energy enables the flow of hot air to remain at ground level over a distance of approximately 150 metres and ensures a good mixture of the hot gases with the ambient fog-laden air while avoiding heat losses due to uncontrolled thermal lift.

The jet engines are installed in concrete pits built alongside the runway with the equipment required to operate the ejection of the gases. The number and siting of the pits depends on the characteristics of the fogs encountered, the prevailing wind conditions in foggy weather, the type of jet engines used and the performances required. A control station enables surveillance and monitoring of the installation. This station is connected to a control desk normally situated in the control tower. The operation and monitoring of each gas generator is controlled remotely. Other systems include a medium- and low-voltage electrical network to supply energy to the pits and the control station, a fuel system including storage and pumping station, and a drainage network to prevent rain water accumulating in the pits.

Runway at Charles de Gaulle Airport, Paris, showing Turboclair installations on right-hand side

Status

The first underground device was installed at Orly Airport, Paris, France and trials took place between 1965/69. As a result of this trial the number of pits at Orly was extended to eight for further trials during the winters 1970/71 and 1971/72. Tests were continued until in 1974 official approval was obtained for the system from the French authorities.

The initial aim during these trials was to develop an installation to work in conjunction with radio communication and category two runway lighting. The trials showed that even with the runway visual range reducing to 150 metres with unmeasurable cloud ceiling a visual runway reference was always acquired at an altitude of 30 metres or more. Furthermore the turbulence induced by the hot air emitted by the installation did not cause any problems in the handling of the aircraft.

The Orly Turboclair was increased to 14 gas generators in 1974 which permits a runway length of more than 1300 metres to benefit from fog clearance. An installation comprising 13 generators was installed alongside 09 runway at Charles de Gaulle Airport, Paris, France in 1973/74.

Manufacturer

Bertin and Cie, Plaisir, France

Turboclair installation for fog clearance

GERMANY, FEDERAL REPUBLIC

Pietsch Ice-X Automatic Spreaders

Pietsch produces truck-mounted automatic spreaders for applying grit, sand, urea and salt to runway taxiway and apron surfaces. Spreading widths of 75 feet (22·9 metres) can be achieved from hoppers with capacities ranging from 4 to 8 cubic metres. Smaller hoppers are produced with capacities from 0·5 to 2·5 cubic metres.

The hoppers are available equipped with a separate engine or powered hydraulically by a speed-related drive from the pto shaft of the truck. A V-shaped girder attached to chains inside the top of the hopper, swings on top of the load as it descends in the hopper to prevent arching of wet material. A rubber conveyor belt is at the bottom which transfers the spreading material through a gate to regulate the required spreading density. The gate is equipped with a crushing roll and can be controlled remotely from the driver's cab. It can be split for spreading two different materials simultaneously. The spread is achieved by one or two discs or spreading units suspended from the rear of the truck. Synchronisation of the driving speed and the conveyor belt is controlled hydraulically and the spreading width and quantity can be adjusted by means of servo-motors. A pre-wetting system is available as an optional feature.

Status

In service at major airports in Europe

Manufacturer

Max Pietsch KG, Hanover, Federal Republic of Germany

Schmidt TS Series Turbostream Snow-blower

The TS series Turbostream is an auger-type snow-blower with vertical clearing augers mounted either side of the blower rotor. The augers function to feed snow into the rotor preventing bridging in wet snow and enabling heavy, hard-packed snow to be cleared. The snow is driven out through a chute which can be adjusted to function on either side of the snow-blower. The rotor consists of four protruding and overlapping blades the ends of which are strengthened with a protective ring. This prevents large objects from entering the rotor which is also protected by shear pins to prevent damage to the drive. Counter-rotating flights are fitted to the augers at the upper end. When there is little snow the augers can be covered by hydraulically-operated cylindrical casings.

A turbo-charged auxiliary engine for the blower drive is installed in the carrier vehicle and the required accessory units, such as the hydraulic system for the raising and lowering of the snow-blower, are mounted on the engine. For the auger drive, variable hydraulic pumps have been used which enable auger speeds to be adjusted individually according to the snow conditions. The pumps are driven by a pto fitted in the drive train after the clutch.

Characteristics

MODEL	TS 3	TS 5	TS 7
CHASSIS	Unimog U 1100	Unimog U 1700	Kaelble
MAIN ENGINE/POWER	81 kW	Mercedes Benz/ OM 352A/124 kW	117 kW
AUXILIARY ENGINE/POWER	191 kW	MB OM 404A 386 kW or MB OM 424 LA 441 kW	590 kW
CLEARING WIDTH	2·45 m	2·7 m	3 m
PERFORMANCE	3000 tonnes/h	5000 tonnes/h	7000 tonnes/h
SPEED	15 km/h	25 km/h	35 km/h
ROTOR DIAMETER	1·4 m	1·65 m	2·25 m
BLOWER HEAD HEIGHT	1·6 m	1·75 m	2·5 m
DIMENSIONS		7·64 × 2·7 × 2·75 m	
OPERATIONAL WEIGHT		12 000 kg	

Status

The TS 5 is in service at Munich Airport, Federal Republic of Germany

Manufacturer

Alfred Schmidt GmbH, St. Blasien/Schwarzwald, Federal Republic of Germany

Schmidt Type MF Snow-plough

The Type MF is a spring-loaded, multi-section snow-plough. The main feature of the plough is its ability to operate close to the ground at high clearing speeds. When it encounters an obstacle on the ground a flexible shock-absorber rail between the section plate and the cutting edge absorbs the impact and springs back. At the same time the section turns upwards and slides over the obstacle. A pivoting skid-mounted frame behind the plough, one on either side, supports the plough when a section is deflected and stabilises the operating height until the obstacle is passed, when the cutting edge returns to the clearing position and a return spring pushes the section back against the ground. With obstacles up to a height of 17 cm only the appropriate section will lift the remainder of the plough remaining close to the ground. Other features of the plough are section guiding elements to maintain the rigidity of the system, a stabilising spring which enables the plough to adapt to the surface profile when clearing on a slope and a protection shoe to prevent damage to the plough when driven sideways into an obstruction.

The cutting edge is constructed of steel but rubber or Vulkollan cutting edges are available. Other optional features include a hydraulic-angling device to change the lateral direction of the plough, a snow deflector fitted to the top of the plough to prevent snow accumulations on the windscreen of the vehicle and castor wheels to reduce noise and enable the distance between the cutting edge and the ground to be finely adjusted.

Characteristics

MODEL	MF 1	MF 3	MF 5	MF 7
LENGTH OF PLOUGH				
bottom	2·5 m	2·97 m	3·29 m	3·63 m
top	2·68 m	3·18 m	3·52 m	3·91 m
HEIGHT	0·93 m	0·93 m	1·1 m	1·3 m
CLEARING WIDTH	2·2 m	2·6 m	2·9 m	3·2 m
WEIGHT	200 kg	277 kg	299 kg	340 kg
SUITABLE CARRIER VEHICLE	Unimog 421	Unimog 403/406/416	Truck 140–200 hp	Truck over 200 hp

Manufacturer

Alfred Schmidt GmbH, St. Blasien/Schwarzwald, Federal Republic of Germany

SWEDEN

Saab Runway Surface Friction Test Unit

The Saab friction tester for producing data on runway surface friction uses a measuring wheel acting with a slip device mounted behind the rear axle of a Saab 900 turbo car, allowing measurements to be made at speeds up to 165 km per hour. The forces measured by the equipment are processed in real-time by a computer system which displays and records the friction values on paper. Runway programmes are made for the two opposite directions of the runway. The time, runway number and average friction readings are recorded on the same printer tape. Taxiways and curves can be measured as well as the aquaplaneing speed. A built-in self wetting system is incorporated which provides friction data on wet runways.

Status

The Saab friction tester has given satisfactory service in winter conditions on both civil and military airports in several countries over a number of years. Over 80 friction testers have been sold including ten with the built-in wetting system. A recent order was made by Aer Rianta for use at Dublin Airport, Ireland

Manufacturer

Saab Scania AB, Saab Car Division, Nykoping, Sweden

Skiddometer BV11 Runway-surface Friction Test Unit

The Skiddometer BV11 is a runway-surface friction measurement device designed to be towed behind a car or any other convenient towing vehicle. The unit is mounted on a single-axle trailer fitted with road wheels. The data is computed automatically in the towing vehicle producing a separate recording of average values for each third of the runway, in accordance with IACO recommendations.

Status

The Skiddometer was developed originally by the Swedish Board of Civil Aviation and the National Road Research Institute and has been supplied for use in Scandinavia, Federal Republic of Germany, Switzerland, Austria, the USA and Canada

Manufacturer

Airport Equipment Company, AB, Vallingby, Sweden

SWITZERLAND

Boschung Ice-warning System

The ice early-warning system is designed to provide data for airport authorities of the likelihood of icing on runways and taxiways to enable spreading or spraying operations to commence in time to prevent icing. The effectiveness of these thawing agents can be monitored by the system. A ground-probe assembly is installed in the runway or taxiway flush with the surface. This is connected to electronic gear for measurement and processing, housed in a splash-proof casing on a mounting frame adjacent to the tarmac. Sensors for air temperature and precipitation measurements are attached to the casing.

Data is transmitted by wire or radio to a microprocessor and displayed on a vdu in the control room. The measurements relating to the air and runway-/taxiway-surface conditions are shown and an alarm system is incorporated to alert the operator. The data is recorded automatically over two-hour periods.

Status

Over 250 systems are in operation in Europe including installations at Zurich and Basle airports in Switzerland and airports in Federal Republic of Germany

Manufacturer

Boschung Mecatronic, Schmitten, Switzerland

Boschung ice-warning system screen terminal

UNITED KINGDOM

Aviasweep Airfield Sweeper/Snow-plough

The Aviasweep is a compact suction sweeper designed for cleaning pavements, roadsides, inside buildings and small tarmac areas of airfields to prevent foreign object damage. A main feature of the vehicle is its narrow width of 1·1 metres for passing between obstructions with a sweeping width of 1·2 to 1·7 metres. Vehicle drive is by hydrostatic transmission to all four wheels. Two brushes are fitted at the sides and two extra brushes can be installed at the front to increase the swept width to 1·7 metres. Alternatively the vehicle can be equipped with a snow-plough. A conversion for the vehicle to be driven by liquid petroleum gas is available.

Aviasweep sweeper/snow-plough

Characteristics

DIMENSIONS
sweeper: 0·5 m diameter
vehicle: 3 × 1·1 × 2 m
WEIGHT (UNLADEN): 1000 kg
PAYLOAD: 1500 kg

Status

Production

Manufacturer

Avialift Products Ltd, Southend-on-Sea, Essex, England

Bomford and Evershed Bandit/Brigand Flail Mowers

The Bandit 2500 is suitable for airport grass control. It fits any tractor with three-point linkage and can cut long, short, wet or dry grass covering up to 4 acres (1·6 hectares) per hour. Maintenance time is limited to five minutes twice a day.

The Brigand is a long-reach mower which can extend to 7·2 metres for cutting the sides of storage bunkers and clearing unwanted vegetation from ditches with a downward reach of 5·6 metres and other areas inaccessible to conventional mowers. Forward speeds of up to 6·5 km per hour can be achieved. It can be fitted to most 52 kW tractors. A high capacity hydraulic-drive system ensures that up to 38·8 kW of power is transmitted from the power take-off.

Characteristics

MODEL	Bandit 2500	Brigand
CUTTING WIDTH	2·5 m	1·15 m
CUTTING HEIGHT	19·05 mm–15·2 m	
ROTOR SHAFT SPEED	2000–2200 rpm	2500 rpm
HYDRAULIC POWER REQUIREMENT	21–22·5 kW	38·8 kW
WEIGHT	440 kg	13·5 kg

Status

In service at airports in the United Kingdom. The Bandit 2500 has been supplied to the Kuala Lumpur Airport Authority

Manufacturer

Bomford and Evershed Ltd, Evesham, Worcestershire, England

Bunce R-type Runway Snow-plough

The R-type Snow-plough manufactured by Bunce, is a 5-metre blade mounted on castor wheels which is installed on the front of a suitable vehicle. The main features are a rubber bottom edge which gives clearance for runway lights and an articulated action which accommodates surface undulations. The blade can be raised, lowered and angled hydraulically by controls operated from the cab. A mechanical stop prevents undue strain on the hydraulic system during snow-clearance operations. The snow-plough is powered by its own hydraulic pump.

Bunce R-Type runway snow-plough

Characteristics

LENGTH: 5 m
DEPTH: 0·95 m

Status

Developed in co-operation with the British Airports Authority the equipment has been supplied to major airports in the United Kingdom

Manufacturer

Bunce (Ashbury) Ltd, Swindon, Wiltshire, England

Bunce Runway Sweepers

Bunce produce two sweepers which are suitable for airport use. Both sweepers can be towed behind any appropriate airport vehicle, Land Rover, tractor, dumper or truck. The standard runway sweeper has a brush made of steel wire or polypropylene material which is raised or lowered by adjusting a screw. It can be regulated for road camber and the driving gear can be disengaged when not sweeping. The machine may be fitted with water spray equipment comprising a 470 litre water tank and sprinkler. The water is sprinkled in front of the brush and the supply regulated by a valve on the tank.

The two-way towed sweeper uses a different type of sweep system which enables the brush to be angled from one side to the other in seconds by releasing a locating pin on one side and locking the pin on the other. The brush has a floating action which adjusts automatically for undulation, camber and wire. Drive is taken from both rear wheels and is engaged by a pawl into a ratchet on each wheel. Because ratchet are used a differential action is obtained on the drive to the generator. Two oil-bath gear boxes transmit drive through two double-universal joints.

Characteristics

	STANDARD	TWO-WAY SWEEPER
BRUSH UNIT		
length	2·6 m	2·5 m
diameter	0·5 m	0·51 m
SWEEPING WIDTH	2·28 m	2·2 m
OVERALL DIMENSIONS		
length	3·04 m	3·89 m
width	2·82 m	2·64 m
HEIGHT		
basic	1·24 m	0·94 m
with water tank	1·5 m	1·42 m
OPERATIONAL WEIGHT		
basic	750 kg	780 kg
with water tank	810 kg	840 kg

Status

Supplied to civil engineering contractors involved in airport projects and to airports in the Middle East for sweeping sand from runways

Manufacturer

Bunce (Ashbury) Ltd, Swindon, Wiltshire, England

CEL 200 PET Noise Monitoring System

The CEL 200 PET system has a number of remote noise-monitoring terminals at various noise-sensitive sites around the airport. Each of these uses a CEL 194 outdoor microphone system which feeds a CEL 195 A/1 remote-noise monitoring terminal. These terminals have a frequency encoded log root mean square (rms) output which is transmitted over conventional telephone lines to a CEL 199 decoder in the central station. This is a standard, single 19-inch rack of modular construction that can handle up to 16 data lines. The signal received on each line is decoded back into log rms form and then digitised and multiplexed into a standard PET 8032 computer which is supported by a PET 8050 disc unit and PET 3032 printer. A vdu enables the system to be monitored. Each sample received is compared with the measurement range of the respective out-station and if out of range the sample is rejected. Should a succession of erroneous data be received the appropriate out-station is automatically disconnected and an alarm sounded. All valid samples are copied into registers for each out-station and this data is used to calculate the hourly environmental-noise rating parameters that have been selected. Typically the system provides L_{eq}, L_{max}, L5, L10, L50, L90, L95 and L_{min} and these results are stored for subsequent print-out. Simultaneously each valid sample is compared against the preset thresholds for the appropriate out-station and if in excess of the threshold a noise event routine is initiated. Noise events are defined as signals that exceed the first threshold for a predetermined time-window, which is normally between 5 to 50 seconds, but may be varied by the operator. Hence events of less than five seconds will be discarded, but events within the time-window are logged on the disc and printer with the date and time of occurrence, threshold levels, duration above thresholds plus the measured L_{max} and L_{AX} values of the event. Events in excess of the time-window will be logged on the printer with the location and time that the long event started and stopped and an alarm will sound to alert the operator.

Central station in CEL 200 PET noise monitoring system

A hard copy of all data is produced on the printer as it occurs. In addition a complete print-out of the environmental noise rating parameters is produced for the previous day at 0300 hours immediately prior to an automatic system calibration and logic check.

Manufacturer

Computer Engineering Ltd, Hitchin, Hertfordshire, England

Danline Sweepers

Three types of sweeper attachments are produced by Danline suitable for airport use. These are the Towsweep, Brushmate and the Scavenger.

The Towsweep can be towed by any vehicle with a standard draw-bar. The brush height and pressure can be adjusted and the unit is available with a choice of mechanical or hydraulic drive. A water sprinkler is an optional feature.

The Brushmate is designed for mounting on a tractor universal three-point linkage system, driven from the tractor power take-off unit. Control of brush height and sweeping pressure is by adjustable castor wheels. The equipment is offered with a wire, polypropylene or mixed brush assembly.

The Scavenger is a sweeper collector, for mounting on a tractor capable of sweeping 9·289 square metres at a speed of 8 km per hour. Refuse is collected in a 1·56 cubic metre hopper which can be emptied automatically. A water sprinkler system is provided for dusty conditions.

Manufacturer

Danline Brushes Ltd, Llanrwst, Gwynedd, Wales

Everall Runway De-icing Unit

The Everall runway de-icing unit consists of chemical spraying equipment mounted on a two-wheeled towable trailer. The spraying equipment comprises a 2045-litre capacity tank, a Briggs and Stratton petrol engine and a Firth Cleveland pump connected to a 15·2 metres spray boom. Spray nozzles are fitted on the boom at 305 mm intervals, and the boom can be adjusted in height from the ground between 254 and 635 mm. The wing sections are hinged and can be folded forward to the stowage position. The boom wings are designed to hinge forwards, backwards or upwards on striking the ground or an obstruction. Spray controls are mounted on the front of the tank, and isolating valves are fitted for the centre section and each wing section of the spray boom. A separate on/off control is provided for installation in the towing vehicle.

The de-icing chemicals used are Konsin and Imsol A. Konsin is a glycol-based, water soluble, de-icing fluid. Imsol is a distillate of isopropanol and among other uses may be employed as a dilutent for Konsin. Normal handling of Konsin does not present a hazard with a flashpoint above 100°C but Imsol is a highly flammable liquid. The equipment is capable of clearing ice up to 6·5 mm thick from a runway of 67 000 square metres in one hour.

Characteristics

ENGINE: Briggs and Stratton air-cooled, 4-stroke petrol, 5·15 kW
PUMP: Firth Cleveland single stage self-priming centrifugal
output: 365 litres/minute
TANK CAPACITY: 2045 litres
SPRAY NOZZLES: Capable of operating down to 0·5 bar
DIMENSIONS
length: 4·9 m
width spray boom unfolded: 15·2 m
spray boom folded: 2·5 m
height: 2·4 m
WEIGHT: 1207 kg

Status

The Everall sprayer is in use with the Royal Air Force

Manufacturer

Evers and Wall Ltd, Newbury, Berkshire, England

Findlay, Irvine Icelert Mk 4

The Icelert Mk 4 provides information on runway conditions. Out-stations monitor runway surface temperature, air temperature and surface moisture and calculate in advance the freezing point of the moisture, or chemical solution, before freezing occurs. Surface wind and direction can also be monitored. All data collected at the out-station is transmitted to the base station by either a two-wire telemetry link or a radio transmitter/receiver.

The base or master station, which is usually in the operations control tower, is in three parts; control unit, information display equipment and alarm unit. The control unit accepts coded signals from up to 25 out-stations. After decoding the data is transferred to a memory bank contained in the vdu.

The vdu, besides giving the ground temperature and air temperature, is programmed to show whether the runway is wet or dry and indicate the status of the runway 'normal alert, critical or ice'. The audio warning sounds automatically when the runway is wet and about to freeze or when it becomes icy. A graph of the temperature over the past hour at any out-station is displayed on the vdu enabling the operator to observe the temperature gradient. A printer is used for recording data at regular intervals or on demand by the operator.

Characteristics

SENSORS OPERATING RANGES
SURFACE TEMPERATURE: −20 to +20°C
AIR TEMPERATURE: −20 to +20°C
WIND SPEED: 0–100 knots
WIND DIRECTION: 0–360°
POWER REQUIREMENTS: 110, 220 or 240 V, 50/60 Hz, 150 W

Status

Systems in use at Edinburgh (a prototype), and Aberdeen Airports, Scotland and Toronto airports, Canada, and at RAF Leuchars

Manufacturer

Findlay, Irvine Ltd, Penicuik, Midlothian, Scotland

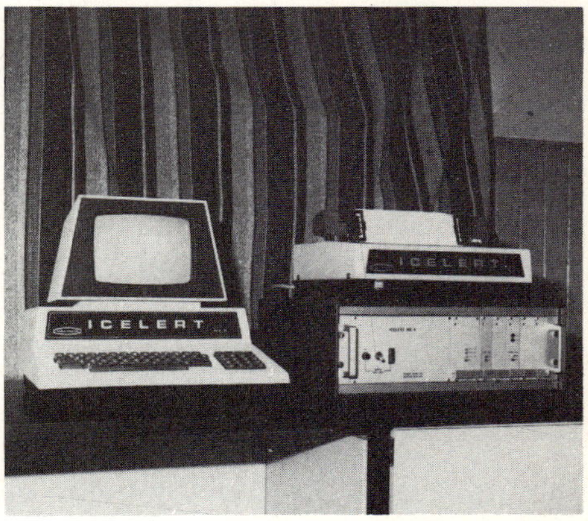

Icelert Mk 4

Lacre Airport Compact Sweeper

The Airport Compact sweeper manufactured by Lacre is designed to complement the TRS Runway Sweeper (see following entry) for, unlike the TRS, the Compact can cope with cambers, curves, uneven ground and obstacles.

The Airport Compact is available with a choice of Bedford, Ford, British Leyland or Dodge chassis and either a three- or four-cylinder auxiliary engine to drive the exhauster.

The sweeping equipment consists of a fully floating vertical scarifying brush and collector nozzle permitting a 1·07-metre swept path. Additional equipment which can be supplied includes an extension brush, which increases the sweeping width to 2 metres and full width suction nozzles. With normal sweeping procedure the brushes direct the debris into the path of the suction nozzle. However in certain instances it is more efficient to remove it by suction only and a full width, 2·29 metres, set of nozzles is available.

An automatic body prop is brought into operation when the body is tipped for refuse disposal.

Visibility from the cab may be augmented by overhead rear mirrors and, as in the TRS, sweeping operations can be controlled from the cab.

Status

In service at civil and military airports throughout the world

Manufacturer

Lacre Ltd, St. Albans, Hertfordshire, England

Lacre TRS Suction Runway Sweeper

The TRS is designed specifically for runway sweeping operations. Mounted on a Bedford truck chassis, the vehicle is powered by a Bedford six-cylinder diesel engine. The sweeping equipment is housed centrally under the vehicle, suction being provided by an exhauster driven by a Perkins six-cylinder diesel auxiliary engine.

The brush and suction canopy injects all loose matter in a 2·28-metre swathe, and the refuse is airborne through a system of separate intakes for large and small particles. Solid matter falls by gravity into a collection compartment. In dry conditions, fine grit and dust continues upwards through a complex of 99 fine-mesh nylon filters. A by-pass compartment, behind and above the filtration compartment, is provided for wet conditions. Selection of the air path is by a hydraulic ram-controlled flap.

The body can be tipped to a maximum angle of 40 degrees for dumping refuse, by an Edbro four-stage telescopic tipping-ram mounted on the body frame, which is operated from the side of the chassis. Power is provided by a gearbox mounted hydraulic pump.

Visibility from the cab may be augmented by overhead rear mirrors. In addition to the normal controls, the driver has an auxiliary engine control panel to control sweeping operations.

Characteristics

CHASSIS: Bedford SJ R3BCO
ENGINE: Bedford 6-cylinder diesel
AUXILIARY ENGINE: Perkins 6·354 diesel
BRUSH
diameter: 0·4 m
effective swept width: 2·28 m
REFUSE COMPARTMENT CAPACITY: 4·22 kg
SWEEPING SPEED: Up to 40 km/h
DIMENSIONS: 7·15 × 2·5 × 3·2 m
WEIGHT: 12 548 kg

Status

The TRS runway sweeper is in service at military and civil airports throughout the world

Manufacturer

Lacre Ltd, St. Albans, Hertfordshire, England

Lacre TRS suction runway sweeper

Metallite Bird Dispersal Equipment

The system, developed by Metallite, involves the siting of a number of propane gas guns in areas where the birds are a danger to aircraft. Each gun can be programmed to fire a single shot as required. Trials have shown that two individual shots are sufficient to disperse birds, one to get the birds into the air and the second shot to scatter the flock.

Control can be by cable or vhf radio. On application of a pulse from the main- or remote-control panel, a timing sequence is initiated, culminating in the transmission of a sequence of pulses via the control cables or radio transmitter. After each transmission the system automatically resets itself and is ready for the next operation. On receiving the firing signal a solenoid valve operates allowing a regulated amount of gas to enter the combustion chamber. After a preset time the timer module operates cutting off power to the solenoid valve which switches off the gas supply. A contact switches on the spark generator causing a continuous stream of sparks within the combustion chamber until combustion occurs, the cumulative effect being a controlled explosion. After a further programmed delay the inset pulse is removed automatically switching off the timer module and re-setting the system for the next operation.

Characteristics

COMBUSTION AGENT: Propane gas
pressure: 0·56 kg/cm²
capacity: Standard 9 kg cylinder will service one gun for 9–12 months
GUN WEIGHT: 9·5 kg

Metallite bird dispersal gun

SOUND PATTERN		
in 6-knot wind	65 dB	75 dB
downwind	300 m	135 m
crosswind	125 m	60 m
upwind	75 m	45 m

Status

A pilot system using six guns has been operating successfully at Coventry Airport, England for two years

Manufacturer

Metallite Aviation Lighting, Birmingham, England

Schmidt SK 150 Airport Suction Sweeper

The SK 150 uses a suction pick-up system. Side brushes are mounted in front of the machine which are fully floating and adjust to follow the surface contours. Suction created by a slow-running radial fan lifts material swept to the centre of the machine by the side brushes through a suction nozzle and conveys the sweepings up an inclined flexible suction hose to a hopper. The nozzle adjusts automatically to allow large objects to be lifted.

A water-recirculation system is installed for dust control. As soon as the radial fan starts, the water pump operates automatically, feeding water to the suction nozzle. The water is atomised by the high-speed air mixing the water droplets with dust particles. These drop into the hopper and the water is filtered before returning to the re-circulation water tank.

The standard high-tip hopper is designed for dumping the refuse in containers.

Characteristics

ENGINE: VW diesel
power: 29 kW at 2900 rpm
speed: up to 25 km/h
BRUSH DIAMETER: 0·75 m
maximum swept width: 1·95 m
sweeping performance: 20 000 m²/h
HOPPER CAPACITY: 1·5 m³
DIMENSIONS: 3·14 × 1·3 × 1·96 m
OPERATION WEIGHT: 4000 kg

Manufacturer

Schmidt Manufacturing and Equipment (UK) Ltd, Ely, Cambridgeshire, England

ML Mu Meter Mk 3 Runway Surface Friction Meter

The ML Mu Meter Mk 3 is a compact and comparatively lightweight trailer unit which measures, records and presents information on runway surface friction conditions. This data can be relayed without further processing to pilots prior to take-off and landing. The Mu Meter, which is designed for one-person, all-weather operation can be towed by most vehicles fitted with a suitable hook attachment. It presents friction data in simple digital form on a remote-display unit inside the towing vehicle. This data consists of average friction values based on samples taken every 3 metres of distance travelled. During friction measuring runs a recorder on the trailer produces a paper trace and any particular hazards observed by the driver, such as oil patches, can be highlighted on the trace by pressing an event button on the display unit. This unit supplies on demand an instant print-out of the data, including the date, time and runway number.

The Mu Meter can be supplied with a self-watering system for the classification (initial calibration) of runways.

Characteristics

DIMENSIONS: 1533 × 810 × 862 mm
OPERTIONAL WEIGHT: 254 kg

Status

The ML Aviation Mu Meter has been developed with the collaboration of the ICAO, and is fully approved by the British Civil Aviation Authority and the US Federal Aviation Administration. In service at airports throughout the world

Manufacturer

ML Aviation Co Ltd, Maidenhead, Berkshire, England

Mu Meter Mk 3

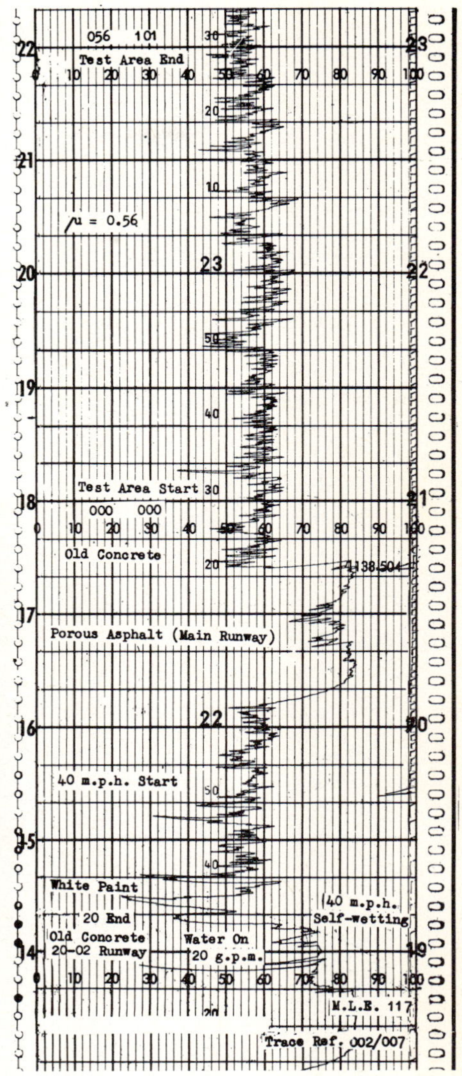

Typical trace produced by Mu Meter Mk 3 of smooth concrete runway with porous asphalt crossing under wet conditions

UNITED STATES OF AMERICA

Alco K-1000 Runway-rubber Removal Unit

A runway-rubber removal unit has been developed by Alco Engineering.

The unit consists of a trailer, on which is mounted an 11 hp Briggs and Stratton four-cycle petrol engine, a 14 hp Kawasaki with an electric start and a generator with 120/240 volts, 15/30 ampere outlets. Two tanks are fitted, a 12 686 litre fibreglass tank and a polythethylene sand tank. A Hypo Triplex piston pump capable of discharging 18·9 litres per minute at 137 bars is installed with two guns, a chemical downstream injector and three 240 volts, 2·5 kW electric heaters. A set of lights is provided for use at night.

Characteristics

TRAILER DIMENSIONS: 3·05 × 1·83 × 2·65 m

Status

Production

Manufacturer

Alco Engineering International, Minneapolis, Minnesota, USA

K-1000 runway-rubber removal unit

DA 607P Environmental Noise Monitor

The DA 607P environmental noise monitor is a portable unit comprising microprocessor with keyboard and light emitting diode led display, alphanumeric printer, clock with its own auxiliary rechargeable battery, and two independent rechargeable power packs with built-in charger. In addition to the 16-key data entry panel and the led display, three selection switches are contained in the control unit. These are power on/off, weighting flat/A and response fast/slow. A microphone can be supplied as an optional feature.

When power is first switched on, the battery voltage is shown on the led display, all read and print memories are cleared and certain parameters are set to pre-programmed values the most significant of which are: noise values of L_N to 99, 90, 50, 33·33, 10, 5, 1, 0·1 dB; senel (L_{AX}) threshold to 65 dB; interval duration 1 hour.

Each of these parameters may be altered for a specific measurement, but the instrument will not permit any setting to be made until an acoustic calibration has been performed. Once this is achieved by applying any sound level calibrator whose reference level is pre-programmed into the DA 607P or by entering the calibration factor of the microphone in use, other data required for the print-out may be programmed, such as: real-time clock: month, day, year; interval starting time; interval duration; L_{AX} threshold; N values in L_N and the print mode required.

The standard print-out lists the initial conditions which have been preset together with the print-out modes selected for each interval. The independent manual L_{eq}, L_{max}, L_{min}+ can be printed at any time by command.

The senel (L_{AX}) of any noise event over the threshold can be printed automatically together with its maximum dB value, duration and the time of the maximum. At the end of each measurement period, print-out may be made of L_{eq}, L_{max}, L_{NP}, the eight L_N selected, L_{min} plus a bar histogram and cumulative distribution over a 50 dB dynamic scale. If required one of six chart speeds can be selected to provide a graphical record of the dBA value with time. If this record is found too erratic to indicate the general trend, a smoothed chart mode can be selected.

Characteristics

MEASUREMENT RANGE: 26–140 dBA
SAMPLE RATE: 8/s
POWER SUPPLY: 2 × 6 V rechargeable packs
BATTERY LIFE: approximately 8 days

Status

The DA 607P was chosen by the United States Air Force for use in the Noisecheck Programme

Manufacturer

Digital Acoustics Inc, Santa Ana, California, USA

Fair Snocrete Blowers

Snocrete blowers are designed for snow-clearance operations and can be mounted on the front or rear of tractors. The sizes range from 3 feet (0·91 metres) to 10 feet (3·05 metres) long.

An alternative version developed for installing on trucks and end loaders is powered by an auxiliary engine. This version is available in sizes ranging from 6 feet (1·83 metres) to 10 feet (3·05 metres) long. In the larger versions, twin 2-foot (1·22-metres) blowers are fitted, capable of removing 30 tons (27 200 kg) of snow per minute.

Manufacturer

Fair Manufacturing Co, Menno, South Dakota, USA

Fair Snocrete blower

FMC Runway Sweepers

Three models of runway sweepers are available from FMC, the 707, 74 and 12.

The Model 707 runway vacuum-sweeper can clean up to one million square feet (9290 square metres) per hour. The sweeper has a vacuum pickup system with a 7 foot 4 inch (2·24 metres) intake nozzle and an agitator broom. A heavy-duty engine drives a high-speed vacuum fan, moving filtered air at speeds of 100 to 200 miles per hour (160 to 320 km per hour). The unit sweeps wet or dry surfaces and is designed for cleaning airfield runways, taxiways, aprons and jet aircraft operation areas. Diesel power is available as an optional feature.

The Model 74 airport maintenance sweeper is designed for sweeping in all areas of airport maintenance: access roads, streets, maintenance areas and runways. The gutter broom is spring suspended for contact with the surface. An auxiliary engine powers the filtering system and brooms and the sweeper features a high velocity air system. This system is cleaned automatically every 72 seconds during operation, returning only clean air to the atmosphere. A diesel engine is offered as optional equipment.

The Model 12 four-wheel sweeper is suitable where sweeping and fast travel speeds are a requirement. The unit is equipped with an industrial V-8 petrol engine with a diesel engine available as an option. The sweeper has a high speed broom system with dual gutter brooms. High- or low-dump features are offered. A high-dump unit allows dumping in high, low or intermediate positions to heights of 10 feet (3·05 metres).

Status

FMC sweepers are in service with the United States Air Force and with airport authorities in USA

Manufacturer

FMC Corporation, San Jose, California, USA

FMC Model 707 runway sweeper

FMC Model 74 airport maintenance sweeper

FMC Model 12 airport sweeper

Klauer Snogo Rotary Snow-clearance Equipment

Klauer produce four types of rotary snow-clearance equipment.
The model TU 3 is designed for mounting on heavy-duty four-wheel drive trucks. Powered by a turbo-charged diesel engine developing 380 hp, the three rows of 18 inch (457·2 mm) augers are capable of clearing 1800 tons (1633 tonnes) of snow an hour.
The model KSR-A is designed specifically for airport use. It is powered by a diesel engine developing 450 hp. Equipped with helical ribbon cutters, the equipment is capable of moving 2400 tons (2177 tonnes) of snow an hour.
The model MTR is a medium size snow-remover with a diesel engine producing 235 hp. It has a capacity in excess of 1400 tons (1270 tonnes) per hour.
The model MP-3D is a turbo-charged diesel-powered, two-stage unit designed for attaching to tractors or trucks. The equipment is capable of clearing up to 1400 tons (1270 tonnes) per hour. Controls for engine output, raising or lowering the cutters and direction of snow discharge are in the cab.
A loading chute can be fitted to any of these models capable of being rotated, extended or retracted by hydraulic controls also operated from the cab.

Manufacturer

Klauer Manufacturing Company, Dubuque, Iowa, USA

Law Model 690D Runway Surface Profilometer

The model 690D surface friction meter or profilometer measures and records the runway-surface profiles of two wheel paths. The profiles are processed by a digital computer housed in the operator's section of the air-conditioned measuring vehicle. The profile signals can be displayed on a strip chart and recorded on digital magnetic tape for data processing. The system operates at speeds of up to 88 km per hour with automatic speed-adjusted filtering. Via the simulator programs it computes and prints May's Meter, PCA Meter, and BPR Roughmeter data from the true profiles obtained. Pavement management computer programs include: May's meter calibration, bituminous fill, Rotomill or Grind level and smooth straight edge simulation.

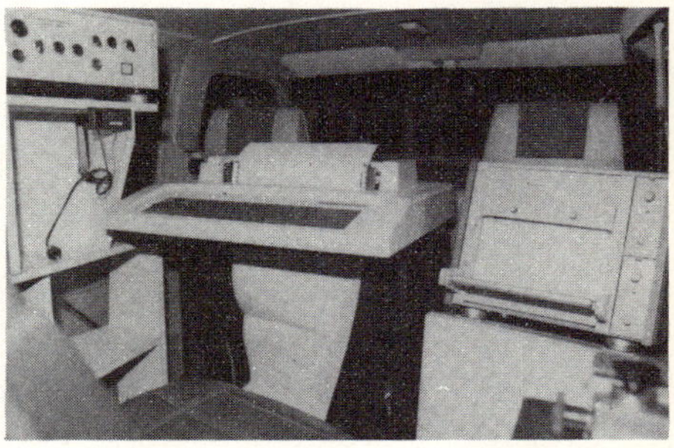

Model 690D runway surface profilometer

Manufacturer

K.J. Law International, Farmington Hills, Michigan, USA

Meyer M9/M10 Snow-ploughs and Spreading Equipment

The M9 and M10 snow-ploughs are designed for mounting on medium-duty vehicles with a gross vehicle weight of between 6260 and 8400 kg. The blade, constructed of high tensile steel with a 6 inch (152·4 mm) cutting blade at the base, is mounted on two heavy-duty runner shoes. Coned trip-springs attached to the top of the blade and the base frame enable the blade to trip forward and ride over immovable objects. The snow-plough is positioned by means of hydraulic controls operated from the cab which lower or raise the blade and move it laterally to change the angle. An additional feature is the Sno-flo control, an aerofoil-shaped projection bolted to the top of the blade which reduces the snow flying over the top of the plough and on to the windscreen by trapping it and driving it downwards behind the plough.

Meyer produce a wide range of snow-ploughs suitable for mounting on forklift trucks, tractors and four-wheel drive vehicles, Land Rover, Jeep, etc. Spreaders which can be mounted under the tailgates of trucks are offered by the company. Three widths are available; the EZ-8 with either a 96- or 88-inch (2·44- or 2·24-metres) width spread and the dual auger UTG-700 with a 94-inch (2·39-metre) spread.

Meyer EZ-8 tail-gate spreading equipment

Characteristics

MODEL	M9	M10
BLADE		
length	2·74 m	3·05 m
height	0·81 m	0·81 m
PLOUGHED WIDTH		
AT MAXIMUM ANGLE	2·35 m	2·62 m
WEIGHT	500 kg	544 kg

Status
Meyer snow-ploughs have been supplied to many airports in the USA

Manufacturer
Meyer Products Inc, Cleveland, Ohio, USA

Oshkosh Model H-2218 Rotary Snow-plough

The Model H-2218 is equipped with a rotary head with cutter bars and snow-blower. This snow-clearance machinery is mounted on the front of a formed channel carbon manganese steel chassis with all-wheel drive. The axles are both full floating with alloy steel shafts. The front, manufactured by Oshkosh, is equipped with cage ring-type steering ends. The rear axle is produced by Rockwell. Hotchkiss semi-elliptic spring-type suspension and a dual air operated brake system are used. The vehicle is powered by a Detroit Diesel with Allison automatic transmission providing five-forward and one-reverse speed. A separate Detroit Diesel drives the rotary cutters and the snow-blower. The cutters are raised and lowered hydraulically the pump being driven by the chassis engine.

The heavy-duty steel cab which is three point rubber-mounted on the chassis is equipped with a reverse slope windscreen to prevent snow accumulation.

Oshkosh H-2218 rotary snow-plough

Characteristics
CHASSIS ENGINE: Detroit diesel 8·2T
 output: 153 kW at 3000 rpm
AXLE LOADS AT HUB:
 front: 9979 kg
 rear: 8351 kg
 total: 18 370 kg
SNOW-CLEARANCE EQUIPMENT
 truck loading height: 3·05 m
 working height of snow-blower: 1·53 m
 working width of snow-blower: 2·6 m
ROTARY ENGINE: Detroit diesel 8V–92T
 output: 369 kW at 2300 rpm
ROTARY HEAD: 2-stage Auger ribbon, 1·32 m diameter
 capacity: 36·4 tonnes/minute
 max speed: 32 km/h
DIMENSIONS:
 length: 8·75 m
 width: 2·59 m
 height with plough raised: 4·26 m
 height with plough lowered: 3·96 m
OPERATIONAL WEIGHT: 14 528 kg

Manufacturer
Oshkosh Truck Corporation, Oshkosh, Wisconsin, USA

Oshkosh P Series Snow-removal Vehicles

The P series vehicles are all-wheel drive trucks designed for primary use in snow-removal conditions. The series consists of four configurations with various models within each classification to meet the requirements of specialised operations. These trucks may be equipped with one-way-, V-, roll-over-, or levelling wings-type snow-ploughs. Rotary packages are also available.
The vehicles are mounted on a formed channel frame constructed of carbon manganese steel.
Three models have two axle 4 × 4 drive. A non-driving tag axle can be fitted to the rear of these vehicles if required, to supplement the standard Rockwell International full floating single rear axle. The fourth model, P 2344, is a six-wheel drive vehicle with a Rockwell International SSHD tandem rear axle. All are equipped with Hotchkiss front and rear suspension.
The Oshkosh front axle is a full floating single reduction-type with cage ring steering ends providing a 30 degrees maximum turning angle.
A dual air-brake system is fitted, one for the front and one for the rear. A modulated split-type secondary emergency system operates in the event of a failure of a main system.
Various transmissions are available to suit the type of operation and snow-plough. Manual transmissions are installed with a two plate pull-type clutch.
The heavy-duty steel cab is three point rubber-mounted on the chassis and equipped with a reverse slope windscreen to prevent snow accumulation.

Status
Over 1000 Oshkosh displacement and rotary snow-ploughs are operated by the US Military forces

Manufacturer
Oshkosh Truck Corporation, Oshkosh, Wisconsin, USA

Oshkosh P Series snow-plough

Characteristics

MODEL	P 2323	P 2327	P 2330	P 2344
ENGINE	Cat 3208 V8/ Cat 3306T-IL6/ Cummins PT-270-IL6	Cummins NTC 300-IL6	Cummins NTC 350-IL6	Cummins NTC 300-IL6
VEHICLE RATING				
front axle	10 433 kg	10 433 kg	10 433 kg	10 433 kg
rear axle	10 433 kg	12 247 kg	13 608 kg	19 958 kg
total	20 866 kg	22 680 kg	24 041 kg	30 391 kg
DIMENSIONS				
length	6·19 to 6·5 m*	6·29 m	6·29 m	7·28 m
height	2·87 m	8 m	3·09 m	2·88 m

*with tag-axle arrangement

Root ASR-16/ASR-22 Airport Runway Snow-ploughs

The ASR-16 and ASR-22 are reversible ploughs designed specifically for high-speed operation to move the snow from the runways and taxiways to the side for removal by a snow-blower. The ploughs are not designed to plough high banks at the runway edge. Both versions are mounted on castor wheels and are fitted with a simple hitch arrangement to couple the plough to the driving vehicle. The plough can be moved laterally to reverse the angle by remote-controls operated from the driver's cab. Rubber shock-protection is provided for safety against hidden obstructions under the snow. Both ploughs can be equipped with either steel or rubber blades. The ASR-22 is available also with Tungsten blades.

Root manufacture a range of snow-ploughs including V-shaped, one-way trip or rigid versions and snow-blowers.

Characteristics

MODEL	ASR-16	ASR-22
DIMENSIONS		
length	4·88 m	6·71 m
height	0·97 m	1·4 m
PLOUGHED WIDTH		
at 38°	4·32 m	5·28 m
at 45°	4·01 m	4·93 m
PLOUGH AND HITCH WEIGHT	1470 kg	2994 kg
RECOMMENDED VEHICLE		
capacity	2722 kg minimum	
weight		20 412 kg minimum

Status
In service at airports in the USA

Manufacturer
Root Spring Scraper Co, Kalamazoo, Michigan, USA

Root RB-8/RB-9 Snow-blowers

The RB-8 and RB-9 snow-blowers are designed for mounting on front-end loaders and motor graders. The cutters are powered by Detroit or Cat diesel engines through a chain drive protected by a Rockwell sheer-pin installation. The loading chute is controlled hydraulically from the cab and can be positioned through a 360-degree arc.

Characteristics

MODEL	RB-8	RB-9
ENGINE	Detroit 4-53	Cat 3208/ Detroit 6V53
DIMENSIONS		
height	1·4 m	1·42 m
height with chute	3·05 m	3·45 m
width	2·46 m	2·77 m
length without adaptor brackets	2·03 m	2·13 m
cutting width	2·45 m	2·74 m
cutting height	1·04 m	1·22 m
WEIGHT	1941 kg (without adaptor)	2722 kg

Status
In service at airports in the USA

Manufacturer
Root Spring Scraper Co, Kalamazoo, Michigan, USA

Snowblast RAHS-3000A Snow Clearance Vehicle

The RAHS-3000A has a two-stage rotary head with hydrostatic drive. The first stage cuts away and gathers the snow. In the second stage a high-volume impeller system discharges the snow from behind the operator's cab. The rotary head features a helical cutter with two separate cutter halves, driven by a bevel gearbox mounted between the two cutter assemblies. The casting distance is variable through an arc of 140 degrees.

A single 600 hp diesel engine is used for both vehicle propulsion and blower operation. The unit features hydrostatic drive, with two hydrostatic transmission systems; one to drive the rotary head and the other to drive the vehicle. Double reduction axles are used for airport applications, with four-wheel drive. The cab is mounted above the rotary head, forward of the casting chute. It features a driver's heated windscreen and for particularly harsh operating conditions, an optional passenger's heated windscreen, heated rear window with air wiper and heated mirrors are available. Speed is controlled by a single foot pedal, equipped with a dead-man safety feature which brings the vehicle to a halt if the operator's foot leaves the pedal. The use of a hydrostatic steering system provides power steering. The vehicle is designed so that the casting angle can be changed hydraulically to any angle from minimum to maximum without stopping the machine. The Snowblast can operate at speeds up to 30 miles an hour (48·3 km per hour) and is capable of removing 3000 tons (2720 tonnes) of snow an hour.

Status
The RAHS-3000A has been in operation at airports in the USA since 1971

Manufacturer
Snowblast Division, Caelter Industries Inc, Denver, Colorado, USA

Valk Snow-ploughs

Valk manufacture a wide range of snow ploughs which include reversible, one-way, fixed angle and V-shaped versions.

Reversible ploughs can be adjusted for discharging snow from either side of the plough. A dual trip-spring assembly permits the blade to trip forward when the plough strikes an obstruction and returns the blade to the normal position when the obstruction has passed. Adjustable castor assemblies equipped with skids stabilise the plough, which can be set for different cutting-widths and ploughing angles. The pitch of the blade can also be adjusted. Power-reverse versions of this plough are available using hydraulic cylinders to change the angle of the blade.

One-way snow-ploughs, designed for high speed snow removal, are constructed with a smooth rolled, tapered blade. They are equipped with similar features to the reversible plough.

Fixed angle snow-ploughs operate at an angle of 35 degrees. They are suitable for high speed ploughing operations.

The V-shaped ploughs feature a smooth rolled, tapered blade which discharges snow on both sides simultaneously. The plough is equipped with a nose guard, cutting edge and three skids.

Steel or rubber cutting edges are available to fit all plough models.

Manufacturer
Valk Manufacturing Company, Carlisle, Pennsylvania, USA

W-W Grinder MRS1-72 Magnetic Runway Sweeper

The MRS1-72 magnetic runway sweeper employs two heavy-duty permanent magnets each capable of attracting metal foreign objects weighing up to 40 kg from the runway surface. The magnets are mounted to the rear of a two-wheeled towable trailer and can be adjusted for clearance between 50 and 150 mm above the runway to obtain maximum attraction under varying conditions. The sweeper is equipped with collector plates and self-contained carrying pans for collection and disposal.

Characteristics
SWEPT WIDTH: 1·83 m
RUNWAY CLEARANCE: 50 to 150 mm
MAX SWEEPING SPEED: 16 km/h
DIMENSIONS
length: 1·35 m
width: 2·13 m

Status
In service at airports in the USA

Manufacturer
W-W Grinder Inc, Wichita, Kansas, USA

PASSENGER HANDLING EQUIPMENT

AUSTRALIA

AWA Flight Information Display Systems

Flight information display systems produced by AWA are designed to meet the needs of the respective user. A base system will service one airport facility only and additional features may be included as required. The base system comprises the display system which controls television displays throughout the airport, flap-board displays at arrivals, departures, gates and carousels, and recording or retrieval of information. Optional features include automatic display of flight details, automatic allocation of gates, carousels, airport co-ordination, linking to the flight reservations computer and information display to caterers, refuellers and maintenance personnel. The display system is co-ordinated by the display system controller which is a mini- or micro-computer depending on the application, programmed to carry out the display requirements of the system. Programmes are modular enabling additional features to be incorporated. Television display monitors are mounted throughout the airport complex in positions which are best suited for public viewing. Additional flight information detail is made available to staff on separate television channels. Displays can be colour, monochrome or a mixture of both. High resolution character generators can produce a variety of founts. Information is sent to the television displays from the display system controller via a hf cable television distribution system. Up to six channels of vision and six of sound can be provided in each cable. Channel selection is made at each television display by a rotary switch. Features of the television displays include variable size characters, graphics, logos, blink feature on characters, words or fields and message scroll. The colour monitors can reproduce up to seven colours with a colour-on-colour capability.

Flap-board displays are positioned throughout the airport complex to present arrivals and departures information, flight numbers at relevant departure gates, flight numbers for arrivals at baggage carousels and indicate flights 'now boarding'. They can be updated locally or remotely, and rolled-up or down to accommodate new information or delete old information.

Public address announcements or music can be distributed to various outlets with loudspeakers housed in television monitors or in separate enclosures.

Status

The company has developed and installed several airport display systems for airline companies in Australia and New Zealand. The initial system produced for Trans-Australia Airlines carried out the basic airport display requirements with additional capability for airline staff to use the system for local-operations control and for co-ordination between major airports. Currently a network of three computers, one in each of the three larger airports, is installed. The computers communicate with each other and the central reservations computer. Ansett Airlines of Australia's system is based on all displayed information originating from a central operations and reservations area. The system acts as a display system only.

Manufacturer

Amalgamated Wireless (Australia) Ltd, North Ryde, New South Wales, Australia

Hastings Deering Passenger Boarding Steps

Hastings Deering has developed steps to service wide- and narrow-body aircraft ranging from the Boeing 747 with a maximum sill height of 18 feet (5·5 metres) to the Boeing 737 with a minimum sill height of 7 feet 8 inches (2·4 metres) above the tarmac. These steps, which are mounted on a standard chassis with a petrol or diesel engine, use a hydraulically-operated special platform which can be raised or lowered to cater for smaller jets. For larger aircraft the steps are raised to a normal position and appear to be similar to other mobile aircraft steps. The base frame carries the step assemblies and is bolted to the truck chassis. During normal operating procedures, the unit is stabilised by jacks secured to the main frame of the step assembly. Two vertical-stabiliser jacks are mounted at the rear of the unit and two on the front-jack frame. The jacks are controlled by one self-centring manual level mounted in the driver's cab. Pilot-operated check valves are fitted to each stabiliser jack.

The open top and bottom step assemblies, including the platform vestibule area, are fabricated from various fold-form steel sections, square tube and angle iron sections welded and braced to form a rigid framework covered with light gauge aluminium sheeting riveted and weather-sealed. The bottom step assembly including the hinged section houses 17 steps. The top assembly and vestibule area accommodate ten. Lighting is provided along each side wall with 17 flush-mounted incandescent lights to illuminate all steps and landing areas.

The special low-level vestibule platform is operated by two mechanical self-centring levers. By activating this device the unit will service all aircraft from B727 to B737 aircraft. This lower level device consists of an integral step assembly with a sliding platform and folding gates. A micro-slope adjustment jack-frame assembly is fitted to the unit to provide a final height adjustment to the aircraft sill changing the step assembly slope from 35 to 41 degrees. This is controlled by two self-centring toggle switches positioned in the driver's cab and the top vestibule control panel. A pilot-operated check valve is fitted to the hydraulic ram in the slope-adjustment mechanism.

In the event of failure of an engine, gearbox or main pump a hand-operated hydraulic pump fitted to the unit feeds the complete hydraulic circuit. Two hitches with one pin are installed to the front and rear of the unit. A tow-bar is available as an option.

Characteristics

MODEL	Mk IV	Mk V
CHASSIS	Bedford J4/J5 petrol engine	Standard commercial petrol or diesel
AIRCRAFT FLOOR HEIGHT		
max	18 ft (5·5 m)	17 ft 6 in (5·33 m)
minimum	9 ft 7 in (2·94 m)	8 ft (2·44 m)
with low level option	7 ft 8 in (2·33 m)	–

STEP ASSEMBLY DIMENSIONS	
width	
bottom	3 ft 8 in (1·12 m)
top	4 ft 4 in (1·32 m)
TOP LANDING	4 ft 8 in (1·42 m) × 3 ft (0·92 m)
PLATFORM	5 ft 10 in (1·78 m) × 6 ft (1·83 m)
CAPACITY	
per step	2 people
top landing	4 people
platform	12 people
UNIT LENGTH	26 ft 2 in (7·98 m)

Status

In service with Qantas and other airlines in Australasia and the Far East

Manufacturer

Hastings Deering Ltd, Lidcombe, New South Wales, Australia

AUSTRIA

OAF Airfield Bus

OAF, a subsidiary of MAN-VW, has developed an airfield bus for transporting passengers between the airport terminal building and the aircraft parked on a remote area of the apron. The bus features a driver's station at both ends with power steering operating dual-air suspended steering axles which give a 39 foot 4 inch (12-metre) radius turning circle. Power is supplied by a MAN 6-cylinder diesel engine with an automatic gearbox. The body is of monocoque construction with 1·25 mm sheet-steel side walls. Access is provided by two pneumatic folding double doors 4 feet 11 inches (1·5 metres) wide on both sides with a further double door at the rear of the vehicle. Standard accessories include a Webasto heating system, four electric blowers for ventilation and a public address system or two-way radio.

Characteristics

MODEL: G 5 FH 240 M 14
ENGINE: MAN D2566 6-cylinder diesel
OUTPUT: 177 kW at 2200 rpm
CAPACITY: 12 seats and 120 standing passengers (4·5 persons per m^2)
DIMENSIONS
length: 46 ft 11 in (14·3 m)
width: 11 ft 6 in (3·5 m)
height: 10 ft 5 in (3·17 m)

Status

The bus which made its United Kingdom début at the Airport 81 exhibition in Manchester late 1981 is in service at Vienna and Los Angeles airports

Manufacturer

OAF-Graf and Stift AG, Vienna, Austria

Rear of OAF airport bus showing double doors and alternative driving position

Interior of OAF airfield bus

BELGIUM

Balteau Controlix 200 Baggage Screening Unit

Controlix 200 is a baggage screening system which uses a 200 kilovolt X-ray unit controlled by a micro-processor to view the inspected item as it passes through the unit on a belt conveyor. A roller-type exit conveyor is connected to the end of the belt conveyor to facilitate collection of the baggage. Large bags are inspected automatically by the unit in several sections. With very small bags the unit operates an automatic pan and tilt system and the optical zoom divides the surface of the radioscopic screen into nine zones each pre-selectable via push buttons.
Each zone selected is observed on the television monitor. The Controlix system can adapt the X-ray energy to the nature of the objects being examined.
Four levels of pulsed X-rays are provided. All normal security checks are done at one level and it is only necessary to change to another level if there is doubt, when it may be useful to increase or decrease the X-ray energy. The pulsed emission of the X-rays is regulated to $\frac{1}{200}$th of a second and the radioscopic image produced is captured by a camera and stored in a digital memory which transfers the pictures on to a television monitor for examination.
In addition to the pulsed modes continuous emission is available for inspection of the more difficult objects. The function of the Controlix is automatic with three modes of operation each selected by push-button control.
Mode 1 permits the familiarisation of the facilities and the functioning of the unit. It is the appropriate mode to use for maximum security. In this mode the conveyor stops when the baggage is in the control zone, the pulsed X-ray is taken automatically and the picture appears immediately on the television monitor. If there is nothing suspicious the operator presses a push button and the baggage moves out of the control zone. If there is doubt the operator may make use of the following features: a choice of three other levels of X-ray penetration, continuous X-ray emission, enlargement of the suspicious area by using the automatic zoom feature or press the bag reject button.
In Mode 2 the stopping of the conveyor and the X-ray pulse operate automatically as in Mode 1. However, after a pre-set time delay the conveyor operates moving to the next bag and the X-ray picture changes automatically. Once again if the operator is in doubt he may hold the bag by operating a push button for further inspection as in Mode 1.
In Mode 3 the baggage passes through at the rate of 12 metres a minute. The pulsed radioscopic images appear on the screen simultaneously with the passing of the baggage. This mode permits the examination of baggage at the rate of 1500 bags per hour. At this speed the average time per bag is 2 to 4 seconds. If the operator is in doubt he can stop the conveyor and inspect the baggage more thoroughly as in Modes 1 and 2. If the operator sees a baggage he considers suspicious he may press the reject bag push button. When the reject bag appears on the conveyor exit a visual and/or audible signal operates. By operating a change-over switch it is possible to leave the conveyor running or for it to stop when the alarm sounds.

Characteristics

X-RAY GENERATOR: 200 kV capable of penetrating 10 mm steel
PULSE: 0·005 s
DIGITAL MEMORY: 512 × 512 points, 5 bits, 64 levels of grey
MAX BAGGAGE SIZE: 0·7 m wide × 0·5 m high, indeterminate length
CONVEYOR
speed: 12 m/minute
height: 0·65 m
INSPECTION RATE: 1000–1500 bags/h
BELT CONVEYOR: 3 × 0·65 m
EXIT CONVEYOR: 1·5 × 0·65 m
UNIT DIMENSIONS: 4·5 × 1 × 2·24 m
WEIGHT: 1165 kg

Manufacturer

Balteau SA, Beyne-Heusay, Belgium

Van Hool AP Series Airport Bus

The airport bus AP series comprises three vehicles of standard length and varying widths which are designed for apron use transporting passengers between the aircraft and the terminal building. The engine is a front-mounted water-cooled MAN diesel with several ratings available.
Mercedes, Fiat and Leyland are alternative power sources which can be specified. Standard features are: automatic transmission, steering-driving axle and fixed-rear axle, integral air suspension and dual independent-circuit air brakes. Four electro-pneumatic double doors are fitted, two each side, which are independently controlled with rotating doors on either side of the vehicle to provide access to the separate crew compartment. The interior layout can be arranged to suit individual operator's requirements with luggage units installed above the front and rear axles. The vehicles heating system is designed for a total capacity of 25 200 kcal per hour with a pre-heater producing 20 000 kcal per hour.

Status
In service with Sabena, Belgian World Airlines

Manufacturer
Van Hool NV, Lier, Belgium

Characteristics

MODEL	AP 1130	AP 1132	AP 1137
CAPACITY varies according to seat layout	130–150 passengers	140–160 passengers	175–200 passengers
DIMENSIONS	13·95 × 3 × 2·75 m	13·95 × 3·2 × 2·75 m	13·95 × 3·75 × 2·75 m

Van Hool AP series airport bus in service with Sabena

CANADA

Ferranti Packard Baggage Display Systems

The BDS-270/410 display system is a simple pre-packaged group of display hardware capable of servicing up to six baggage dispensers and six baggage loading points. The system includes single- or double-faced master signs at the baggage dispensers, single faced slave signs at the baggage loading points, and a small desk-top control unit. The operator selects the desired baggage dispenser, inputs the flight data by thumb-wheel switches, and presses the display push-button. Automatically the data is displayed at the desired dispenser and the corresponding load point. When all baggage is cleared, the signs can be blanked. The control unit maintains a readout of the contents of each display. One operator can control all baggage dispensing and load points from a central location. The simplest display format for a dedicated baggage dispenser is four numeric digits displaying in-coming flight number. If a dispenser is shared among several airlines, a six character display showing the IATA code followed by a four-digit flight number can be used. Multiple line displays in a single enclosure can be provided. Display character sizes can be either 2·7 inch (70 mm) or 4·1 inches (104 mm) suitable for viewing distances of 125 feet (38 metres) or 200 feet (60 metres) respectively. The control input unit is capable of entering up to six different characters of data to a maximum of six sign systems. Each display character consists of a 5 by 7 matrix of small discs which are rotated to form a coloured character on the matt black background.

Ferranti Packard baggage claim area displays at Toronto Airport

Characteristics
CHARACTER SIZE:
Series BDS-270: 2·7 in (70 mm) high
Series BDS-410: 4·1 in (104 mm) high
POWER REQUIREMENTS:
master sign: 115 V ac 50/60 Hz 1A or
230 V ac 50/60 Hz 0·5A
SIGN DIMENSIONS:
single line:
length 3 ft (914 mm)
width 11 in (280 mm)
height 1 ft 4 in (406 mm)
additional lines: + 6 in (150 mm) of height per line
WEIGHT
single line: 50 lb (23 kg)
additional lines: 10 lb (4.5 kg) per line
CONTROL UNIT DIMENSIONS
9 in (225 mm) ×
8 in (200 mm) ×
6 in (150 mm)

Status
In service at airports in Canada

Manufacturer
Ferranti Packard Electronics Ltd, Mississauga, Ontario, Canada

Ferranti Packard Flight Information Display Systems

The company offers the PDS 100 pre-packaged flight information display or custom-built systems operated by the TDS 100 control unit and the FP 500 series display controllers. Electromagnetic character display panels are available with either 2·7 inch (70 mm) or 4·1 inch (104 mm) high characters. The display systems are modularised and can be expanded as required using the same basic components, only the number of components and their intercabling differ from system to system. The TDS 100 information display system is a prepackaged group of hardware and software comprising an input crt terminal with keyboard, a controller driver unit and a range of display boards to suit the particular terminal plan. These include arrival/departure boards, gate/baggage signs, information boards for those meeting passengers, and baggage handling signs. The basic system will accommodate approximately 16 lines of arrivals and 16 lines of departures plus the gate and baggage sign requirements consistent with this number of operations. The system is controlled from a desk-mounted crt display terminal, which can be installed in any office environment. The microprocessor controlled terminal stores the information in a protected format identical to the sign configuration. Data can be recalled, displayed, edited and transmitted to the signs on keyboard command. The operator enters schedule information via the terminal keyboard. The protected format permits the operator to enter data only in fields corresponding to the actual signs. Schedule data are entered, visually checked on the terminal, transmitted to the display and committed to crt memory. Up to 120 flights can be stored at one time. Changes can be made or comments added to schedule data by calling the appropriate page of data, inserting the new message and retransmitting to the signs. Comments are not limited by fixed-word selection. The operator calls the appropriate page, deletes the top line and transmits the modified information to the signs. This automatically moves each succeeding line up one space. Emergency messages can be displayed by overriding the format protection feature, and typing directly on the signs for instant communication to the travelling public. The system is capable of operating multiple display boards displaying identical information and the bottom lines of the boards can be formatted for advertising messages. To accommodate system expansion, options can be added to the basic system comprising clock displays, automatic scheduling and an additional crt display system. The time is displayed from a crystal-controlled internal clock. For automatic schedule loading, information for up to 800 flights can be stored on a tape cassette and input to the system on a single command. Schedule tapes can be prepared off-line via the crt terminal. To service small areas such as lounges, dining rooms and operations rooms, a two-channel television monitor system can be added to the basic TDS 100 system to display arrivals/departures data. Operating information channels can also be added.

Ferranti Packard flight information display system, Mexico City

Ferranti Packard International flight arrivals display, Toronto airport

The basic display module consists of a 5 by 7 matrix of small discs; electromagnetically positioned to form the desired character. The display modules are supplied either as individual units for small signs or as pre-wired 40-character panels to make up large signs. Panels are mounted on a pre-fabricated aluminium framework suitable for wall or ceiling mounting. The modules are driven by the controller drive. This wall- or cabinet-mounted unit is capable of addressing and writing up to 6400 character modules at a rate of 30 characters per second. Several signs can share a common driver. The display boards weigh approx. 3 lb per square foot (15 kg per square metre) and can be wall-mounted or suspended from most ceilings without additional structural support. The controller driver is housed in an industrial class cabinet for wall mounting in any electrical equipment room. The controller weighs approximately 100 lb (45 kg). An air-conditioned environment is usually not required. Connection from the input crt to the controller driver is by a two wire circuit that can be installed in any convenient cable tray. Connection from the controller to the signs is via three common-data/address cables. Standard 110/220 volts, 50/60 Hz power is required to the input and driver units. All power to each sign is derived from the controller driver and is distributed through the data cable.

Status

Ferranti Packard flight information systems are installed at Toronto and Vancouver airports in Canada, the National Airlines Terminal at the JF Kennedy airport, New York, USA and at Mexico City airport

Manufacturer

Ferranti Packard Electronics Ltd, Mississauga, Ontario, Canada

Mathews' Cres-Flight Baggage Distribution Loop

Mathews' Cres-Flight baggage distribution loop is a single, endless, non-reversible conveyor which can be loaded manually from apron service vehicles external to the baggage claim area and can conform to any appropriate configuration within the baggage hall. The Cres-Flight may be installed as an island within the hall to be fed automatically by subterranean or overhead conveying systems. The frame is a formed channel mild-steel construction with the exposed surfaces, in public areas, of stainless steel designed to fit over the top of the pallets to act as a baggage guard and finger protector. The drive is floor-mounted under the conveying tread and consists of two strands of precision roller chain with hardened pushers designed to engage a hardened abutment on the underside of each pallet. This drive mechanism is coupled to an enclosed speed reduction unit which in turn is connected to an electric motor via a multiple v-belt/centrifugal clutch drive combination.

The conveying tread consists of a series of crescent-shaped, urethane-covered pallets fastened together by link connectors. This tread is designed to form a flat endless conveying surface supported by continuous rows of plastic self-lubricating rollers.

The supports are floor-type, with threaded vertical adjustment for accurate levelling.

Cres-Flight conveyors in the Eastern Airlines terminal at JFK Airport, New York, USA

Characteristics

CONVEYOR WIDTH: 2 ft 10 in (0·86 m)
CAPACITY: 30 lb (13·6 kg) per linear foot
UNIT WIDTH: 3 ft 4 in (1·02 m)
MINIMUM HEIGHT: 1 ft (0·3 m)
SPEED
standard: 80 ft/minute (24·4 m/minute)
optional: 100 ft/minute (30·5 m/minute)

Status

Over 300 Cres-Flight units have been installed at 122 airports in 30 countries

Manufacturer

Mathews Conveyor Company, Port Hope, Ontario, Canada

Mathews' Rotary Carrousel Luggage Dispensing Unit

The Carrousel luggage dispensing unit is a low profile, inclined circular conveyor installed in the baggage claim area and fed subterraneously or overhead, or from a segment of the periphery accessible to apron service vehicles external to the baggage hall. The frame of the unit is constructed of structural steel, rigidly braced and welded into segmental sections bolted to each other and to the circular steel structure at both the inside and outside rim to form a continuous support for slide surface panels and the outside bumper. The luggage slide is constructed of no fewer than 30 segments of particle board surfaced with stainless steel on the conveying surface. A continuous, tubular-type rubber, secured by concealed fasteners to the frame, is designed to cushion the impact of luggage on the slide.

The drive for the conveyor consists of an electric motor and an enclosed, speed reduction unit connected to a spring-loaded drive wheel. The supports consist of no fewer than eight wheel assemblies for carrying and four for guiding the rotary portion of the unit. All wheels have hard rubber or urethane tyres and run on anti-friction bearings. A non-rotating guard is provided around the periphery of the rotating unit, constructed of stainless steel.

The circular central panel is constructed of particle board supported by a structural steel frame fitted to the Carrousel receive conveyor with a hinged trap door for access to the drive unit.

Mathews' rotary Carrousels at Winnipeg International Airport

Characteristics

LUGGAGE SLIDE PANELS
length: 6 ft 4 in (1·93 m)
thickness
0·625 in (15·88 mm) particle board
0·625 in (15·88 mm) stainless steel
UNIT DIAMETER
standard: 25 ft 4 in (7·72 m)
options
20 ft 4 in (6·2 m)
22 ft 4 in (6·81 m)
30 ft 4 in (9·25 m)
UNIT CAPACITY: 4000 lb (1814 kg)
UNIT SPEED
standard: 80 ft/minute (24·4 m/minute)
optional: 100 ft/minute (30·48 m/minute)

Status

Over 300 Carrousel units have been installed at 74 airports in 16 countries

Manufacturer

Mathews Conveyor Co, Rexnord Canada Ltd, Port Hope, Ontario, Canada

FINLAND

Kone CD 555 Flight Information Display System

This system uses monochrome or colour video monitors to present flight schedule information to passengers in the airport building. The pictures are produced by CD 555A image generators and the system is controlled by a micro-computer which has the airport's weekly and daily flights programmed in the memory. Changes to schedules as well as other instructions can be effected manually by means of an ASC II keyboard and display terminal. The displays can be single monitors for use at gates and check in counters or multiple monitors for main displays in the terminal building.

Kone information display at Turku airport

Characteristics

DISPLAY: Intel 2732, 8 channels, alphanumeric, semigraphic and trend, 40 lines, 72 columns per line, 384 characters and symbols
POWER REQUIREMENTS: 220 V ac ±10%; 50 Hz ±3 Hz; 200 W
OPERATING TEMPERATURE RANGE: +10 to +40°C

Status

The system is installed and operational at Turku Airport, Finland

Manufacturer

Kone Corporation, Espoo, Finland

Metor Model 114 Metal Detector

The Metor Model 114 is a walk-through metal detector suitable for use in airport terminals. The unit comprises the archway through which people pass and the electronics cabinet, linked by an inter-connecting cable with snap-action connectors. The archway consists of four parts. Two wooden side-sections each incorporate a copper wire winding (transmitter and receiver). These sections are covered in leatherette with the lower parts protected by hard panels. Two beech boards hold the two side-sections at a fixed distance and make the assembly rigid. One of the boards contains a cable and connectors for the electrical connection of the two side-sections. The connector at the lower part of one of the side-sections is recessed to prevent damage from the outside. The electronics are enclosed in a metal unit. The front panel holds two lamps, red for alarm and green for stand-by, and the rear panel includes all connectors, controls, switches and fault indicators.
The Metor Model 114 operates on pulsed-field technology. Transmitter pulses create slowly decaying eddy currents in a metal object near the transmitter winding. The decay rate of the eddy currents after each successive pulse is measured in the receiver section of the electronics unit by sampling and processing the signal induced in the receiver winding producing an estimate of the size of the metal object. Each metal object larger than the threshold size, carried through the coil will cause an alarm, which is shown by a red light on the front panel of the electronics unit, and by an audio signal. In addition the output relay is energised. The duration of the alarm indicates the relative size of the metal object.

Characteristics

PERFORMANCE: capable of detecting 10 × 80 mm iron rod
POWER REQUIREMENTS: 110/220 V ac ±20% 50–60 Hz, 50 VA
OPERATING TEMPERATURE RANGE: −10° to +50°C
ARCHWAY DIMENSIONS
height 2·175 m
width 0·873 m
depth 0·519 m
WEIGHT: 35 kg
ELECTRONIC UNIT DIMENSIONS: 0·32 × 0·223 × 0·11 m
WEIGHT: 6 kg

Status

In use at Changi Airport, Singapore and other major airports

Manufacturer

Outokumpu 07, Espoo, Finland

FRANCE

Ateliers Reunis Caddie Baggage Trolleys

Ateliers Reunis produce six models suitable for use as airport baggage trolleys.
The Caddie Horizon is a stackable, baggage trolley for individual use which can be supplied with or without rear-wheel brakes. It is constructed in steel rod and tube and finished in chrome. The baggage platform which has capacity for three to four suitcases slopes down towards the rear. Two swivelling 125 mm diameter rubber wheels are fitted at the front of the trolley and two fixed 200 mm diameter rubber wheels at the back. Accessories include a basket for hand luggage, bag hook and a frame for an advertising sign.
The Caddie Standard is available with or without rear-wheel brakes. The trolley, which can be stacked, is made from steel rod and tube with a chrome finish. The trolley is mounted on four rubber wheels, two swivelling 120 mm diameter wheels at the front and two fixed 160 mm at the rear. Front and side bumper guards are fitted. The luggage platform and the other accessories are similar to those provided with the Horizon model.
The Caddie Compact trolley is supplied in two sizes both can be stacked and constructed in steel rod and tube and finished in chrome. The larger size trolley is mounted on three rubber wheels, a single swivelling 125 mm diameter wheel at the front and two fixed 200 mm wheels at the rear. The luggage platform has capacity for three or four airline cases. The smaller compact model also has three wheels with two fixed 200 mm at the front and a swivelling 125 mm wheel at the rear. The luggage platform which is underneath the extended steering handle can hold three to four airline cases. Both models include a basket for hand baggage and a frame for an advertising sign. The larger model has additional features comprising front-swivel corner bumper guards and a bag hook.

Characteristics

MODEL	Horizon	Standard	Compact	Compact
NUMBER	17.01/96.01	07.01/91.01	39.01	37.01
CAPACITY	200 kg	200 kg	150 kg	150 kg
PLATFORM DIMENSIONS				
length	0·62 m	0·7 m	0·61 m	0·45 m
front width	0·32 m	0·31 m	0·35 m	0·37 m
rear width	0·4 m	0·45 m	0·48 m	0·25 m
TROLLEY DIMENSIONS				
length	0·95 m	0·86 m	0·94 m	0·82 m
width	0·64 m	0·55 m	0·69 m	0·56 m
height	1 m	0·93 m	1 m	0·9 m
WEIGHT	20 kg	25 kg	19 kg	15 kg
STACKED UNITS				
number	10	10	10	10
length	3·75 m	3·2 m	3·37 m	3 m
FLOOR SPACE	2·3 m²	1·63 m²	2·33 m²	1·7 m²

Status
In service at over 30 airports in Europe, Middle East, Africa, and Central America

Manufacturer
Ateliers Reunis SA, Schiltigheim, France

Caddie Standard luggage trolley

IER Airline Ticket, Boarding Pass and Baggage-tag Printers

IER has developed printers to produce airline tickets, boarding passes and baggage tags.
The IER 321 is designed to produce self-adhesive, ready-cut tags printed with date, final destination, stop-overs, reservation number and the passenger's name.
The printer operates in conjunction with the reservation computer which simultaneously prints out the passenger's boarding pass through an IER 221 printer.

Status
The IER 312 has been selected by Air France. IER claims to supply 80 per cent of the market outside the USA for ticket and boarding pass printers

Manufacturer
Impression Enregistrement des Resultats, Courbevoie, France

Sietam Baggage Handling and Sorting Systems

Sietam design, manufacture and install baggage handling and sorting systems. The equipment for baggage handling includes: check-in desks, electronic and electro-optical weighing scales, and weighing and collecting belt conveyors. For baggage sorting systems, conveyors for elevating, descending and sorting are used, the latter equipped with deviating arms. Coding and monitoring of the baggage is carried out with system control performed by relay cabinets, programmable controllers or computers according to the complexity of the system.

Status
Sietam baggage handling systems have been installed at the following airports: Charles de Gaulle and Orly, Paris; Lille, Abu Dhabi, Athens, Dacca, Damascus, Erevan, Riyadh, Sirte, Tallin and Tashkent

Manufacturer
Sietam, Viry-Chatillon, Cedex, France

Sietam check-in facility at Charles de Gaulle Airport, Paris

Siraga Passenger Stairs

Siraga produces a range of towable and self-propelled steps.
Four types of towable steps are available. The Fokker is a simple fixed-height stairway with a steel framework and tubular hand-rails mounted on front castors and fixed rear wheels. The steps and platform are covered with a non-skid material. Lighting can be supplied as an optional feature.
The 816 is designed for aircraft with door-sill heights between 2·1 and 3·7 metres. The frame is constructed of rectangular steel-tube side-members with two stabilising skids incorporated, a steel-sheet platform covered in aluminium sheeting, and aluminium sheet steps. Plastic hand-rails are fitted and a rubber bumper is mounted on the railing and platform edges in contact with the aircraft. Changes in elevation are effected by a screw-nut system acting on the platform which is manually controlled by a fly-wheel. Lighting can be supplied as an optional item.
The 817 is similar in design and construction to the 816, but is fitted with two hydraulic jacks which are adjusted by a hand-pump.
The 822 is a towable unit designed to service aircraft with door-sill heights between 2·9 and 5·15 metres. Similar to the 817 this model incorporates two mechanical stabilisers in lieu of the skids.
Two models of self-propelled stairs are available, 946 and 922. The 946 for aircraft door-sill heights between 2·2 and 3·7 metres is mounted on a steel-plated chassis powered by a Peugeot J9 petrol engine with a turning radius of 8·1 metres. The platform is constructed in steel sheeting covered by chequered aluminium sheeting with steps in chequered aluminium sheeting. Plastic hand-rails are fitted and a rubber bumper is mounted on the front of the platform. Height adjustment of the platform is operated by a screw-nut system controlled manually by means of a fly-wheel. A diesel-powered version of the 946 is available as an option.
The 922 is designed for aircraft-sill heights between 2·9 and 5·15 metres. The chassis which consists of rectangular tube side-members is powered

by either a Berliet 90B 6XL or a Saviem JK 60 AXL four-cylinder diesel engine with fuel injection. Four hydraulic stabilisers are fitted. Construction is similar to the 946 though height adjustment is effected through powered-hydraulic jacks. A manual control is fitted in case of a hydraulic system malfunction.

Status
In service with Air Afrique, Air Guinee, Air Inter and Sabena

Manufacturer
Siraga, Buzancais, France

Characteristics

TYPE	Fokker	816	817	822	946	922
PLATFORM						
height	1·4 m	2·1–3·7 m	2·05–3·96 m	2·9–5·15 m	2·25–3·6 m	2·6–5·15 m
width	0·8 m	2·378 m	1·5 m	2·38 m	1·6–1·54 m	2·3 m
STEP						
width	0·8 m	1·235 m	1 m	1·2 m	1·25 m	1·25 m
height	0·202 m	0·135–0·22 m	0·11–0·21 m			
UNIT						
length (tow-bar up)	2·28 m	6·8 m	7 m	9·1 m	7·33 m	8·9 m
width	1·25 m	2·5 m	2·05 m	2·48 m	2·33 m	2·5 m
minimum height	2·2 m	3·07 m	3·05 m	3·88 m		
WEIGHT	100 kg	3000 kg	1900 kg	3500 kg	3000 kg	6100 kg

Sovam Passenger Transfer and Baggage Handling Equipment

Passenger transfer equipment manufactured by Sovam comprises passenger-loading bridges, the Shuttle Lift and the Shuttle Lounge.
The Safeway passenger-loading bridges are manufactured under licence from Steelcons of Sweden. They can be supplied by Sovam with single- or double-bridge heads. A detailed description of this equipment can be found in this section under Sweden.
The Shuttle Lift is a mobile passenger lounge mounted on a scissor-lift chassis designed to carry up to 170 passengers between the terminal building and the aircraft parked on a remote part of the apron. At the aircraft the passenger compartment, with the driver, elevates to the height of the aircraft-door sill. The front of the platform is equipped with an extending section and a canopy which can swivel laterally ±35 degrees to ensure a close fit with the aircraft.
The lounge body is constructed of a double skin of grp, the internal skin incorporating seat sections which are designed to be interchangeable. The driver's compartment is at the front and centre of the vehicle with passenger doors to the lounge area on either side of the compartment. A closed-circuit television camera is mounted on the rear of the vehicle with a control monitor fitted above the dashboard to enable the driver to check on traffic to the rear of the vehicle before reversing.
The Shuttle Lounge is a low-slung, high-capacity coach designed to carry up to 250 passengers between the terminal building and the aircraft.

Characteristics
SHUTTLE LIFT
CAPACITY: up to 170 passengers
SEATING CAPACITY: 60–100 passengers
MAX SPEED: 45 km/h
TURNING RADIUS: 14·5/15·6 m
FLOOR HEIGHT OF BODY
  when lowered: 2 m
  when elevated: 5·6 m
FLOOR HEIGHT (telescopic step)
  lowered position: 1·94 m
  elevated position: 5·54 m
LIFT STROKE: 3·6 m
OVERALL DIMENSIONS
  length: 15·15 m
  width (pneumatics included): 4·7 m
  height in lowered position: 4·5 m
WEIGHT
  empty: 55 000 kg
  max permissible with 170 passengers: 68 600 kg

Status
Shuttle Lift and Shuttle Lounge vehicles are in operation at Charles de Gaulle airport, Paris

Manufacturer
Sovam Division Aeroportuaire, Parthenay, France

Teleflex Baggage Handling and Sorting Systems

Teleflex systems cover the whole range of baggage handling from the check-in point to aircraft departure and from aircraft arrival to the baggage claim area.
For check-in, manual and automatic check in facilities are available with one, two or three belt counters incorporating electronic weighing machines. In-line X-ray facilities for baggage screening can be provided. Transfer can be by either the belt-type or the flexible slat Diplodocus conveyor. Maximum floor space can be obtained by suspending the conveyors from the ceiling. Transfer systems use mechanised diverters or pushers. For elevating and descending sections inclined belt conveyors can be used, capable of gradients up to 17 degrees, or the Hippocampe which is a continuous vertical system with a pressing belt. Manual or automatic baggage sorting systems can be installed. For manual systems belt conveyors can feed accumulating free roller tables or sorting loops can be supplied with flat or inclined Diplodocus or Dinosaure carousels. The latter provide greater capacity by re-circulating the baggage.
Automatic sorting systems available range from the slat-conveying loop with moving diverters to the gull-wing sorter. The latter is a high-speed sorting machine equipped with tilting trays and specialised loading stations. In the baggage claim area, the Diplodocus inclined carousel or the Dinosaure flat carousel can be installed in various configurations to suit the traffic flow and available space for passengers and storage of baggage.

Status
Teleflex baggage systems have been installed at 73 airports in 43 countries

Manufacturer
Tissmetal Lionel-Dupont, Ivry-sur-Seine, Cedex, France

Velec Sefat Flight Information Display System

Velac Sefat systems for flight information display are controlled by a display centre consisting of a central processing unit, floppy disc memory and interface. A console provides the interface between the cpu and the operator. It comprises a QW keyboard, an operations keyboard and a vdu and enables the operator to create pages of data which can be either displayed or stored. The operator can also make changes in storage contents, distribute the data and check the operation of the central unit.
Data can be either composed manually on the console keyboard or called out of the memory in which it has been stored. Transmission of data to the display units is effected in two ways: video distribution by coaxial cable is used when the distance does not exceed 500 metres. The data processed by the central processing unit is edited in a character generator which delivers a video signal to the 625-line standard. This signal is transmitted by a coaxial cable permitting diffusion of pictures from camera or video-tape recorder.
Multi-point distribution is used in long-range communication and when it is desirable to route several messages on a single cable. The data processed by the central-processing unit is coded and transmitted sequentially on a two-wire line through an appropriate interface. Each message is addressed and each diffusion unit has its address decoder and its RAM store. Thus one display unit (monitor or panel) is receptive only to the message intended for it. Transmission speed varies in terms of the quality of the line (110 to 9600 bauds) allowing for long-distance diffusion.
Two types of display may be used, monochrome/colour monitors and alphanumeric panels. The television monitors can be supplied in a variety of formats. In the alphanumeric panels each character is formed by a 5 by 7 point matrix. Each point on pivoting presents two stable states: one reflecting state which is white, yellow, green or red, the other non-reflecting which is black. The pivoting of the elementary points is triggered by a brief current impulse. The displaying panel is regarded as a matrix of characters and the writing is achieved by successive scanning of the lines. A microprocessor built into the panel decodes the data from the display centre and pilots the character control circuits. Each panel is addressable and connected to the central unit by the multi-point technique which requires two two-wire lines, one for the data the other to control the transmission. These panels are used when the dimensions of the monitors prove to be insufficient. They are organised in lines of text formed by the juxtaposition of all-purpose characters which make it possible to compose all the standard alphanumerical symbols and certain graphics.

Characteristics
TELEVISION MONITORS: 17, 24, 31, 44, 51, 61 cm black and white, red, yellow or green phosphorus crts
OPERATING TEMPERATURE RANGE: −20°C to +70°C
ALPHANUMERIC
  character height: 7, 10, 22, 37 or 45 cm
  display speed: up to 100 characters/s

Status
Velec Sefat flight information display system is installed at Lille Lesquin airport

Manufacturer
Vandeputte Fils and Cie, Tourcoing, France

GERMANY, FEDERAL REPUBLIC

Bizerba Check-in 3000 Baggage Weighing Machine

The Check-in 3000 is an automatic baggage weighing machine incorporating a digital display and controlled by a micro-processor. The equipment comprises the display unit, electronic unit, control panel and weighing machine. The display unit provides a digital weight presentation for the operator and the passenger with a 13 mm high led seven-segment display the weight of the luggage being shown in increments of 0·1 kg. Signals which indicate correct functioning of the equipment and fault identification are included in the operator's display. Control of the weighing machine is by a micro-processor in the electronic unit, with a capability of distributing baggage weight values to a printer, booking machine or central computer. A memory store for recording cumulative weight can be supplied if required. The weighing machine consists of bending force weighing cells with a stainless steel platform, welded frame and built-in load lever action. Provision is made for the fitting of a moving belt. Weighing and checking keys with a pilot lamp are installed in the control panel.

Characteristics

WEIGHING RANGE: 5–150 kg
PLATFORM DIMENSIONS
sizes: 600 × 600 mm; 800 × 600 mm; 1000 × 600 mm; 1200 × 600 mm
height: 150 mm
DISPLAY UNIT: 70 × 155 mm
DESK-TOP SECTION: 60 × 145 × 35 mm
ELECTRONIC UNIT: 400 × 234 × 120 mm
CONTROL PANEL: 60 × 145 × 55 mm
POWER REQUIREMENTS: 220 V ac, 50 Hz

Status

In use at airports in Federal Republic of Germany

Manufacturer

Bizerba Weighing Machine Works, Balingen, Federal Republic of Germany

Bizerba Check-in 3000 baggage weighing machine

Conrac Elektron Flight Information Display Systems

Conrac Elektron, a subsidiary of Conrac USA, design and install flight information display systems for airports in Europe and the Middle East. The company produce systems using video and flap board displays or a combination of both presentations.

For desk-top monitors and computer terminal displays normal character height is 5 mm and up to 2000 characters may be displayed on a 12-inch (31 cm) crt. For arrival and departure displays for small audiences, 17-inch (43 cm) and 23-inch (59 cm) monitors are used. Normal character height is $\frac{1}{2}$ inch (1·25 cm). The Macrofont video-display is designed for larger audiences. Macrofont is a method of electronically-generating large characters for presentation on video-monitors. Instead of using the standard dot matrix to form the characters the display features solid, undotted block characters. This method of character generation provides increased brightness with reduced phosphor intensity. Three basic character sizes are available with attendant line, page capacities and up to 15 Macrofont channels can be driven by a single controller and its associated power supply. The controller chassis contains an interface card, a sync card, a data buffer card and up to 15-character generator cards. Video output for each channel is raster scan, non-interlaced, 1·4 volts peak-to-peak, at 75 ohms. To allow interface with existing systems, in remote applications, input to the controller chassis is RS 232 transmission using modified ASCII Code. By grouping several Macrofont video-monitors together behind a special tinted-plastic face-plate more information can be displayed. Mosaic displays are typically in clusters of three or six television monitors to provide extended line capacity and between six or twelve lines of information. The special face-plate covers the front of each monitor and gives the appearance of a continuous line display.

Split-flap display boards are available with characters $1\frac{3}{4}$ inches (4·5 cm), $2\frac{1}{2}$ inches (6 cm) or 4 inches (10 cm) high. Character style, choice of characters and national alphabets are available as optional features in addition to several widths up to 2 foot 6 inches (76 cm). Automatic and manual electronic control of split-flap boards is available and in systems using a combination of flap boards and video simultaneous control of both display medium is accomplished automatically.

Status

Conrac Elektron has produced flight information systems for Schiphol Airport, Amsterdam, Netherlands; Arlanda Airport, Stockholm and Landvetter Airport, Gothenburg, Sweden and eight airports in Saudi Arabia

Manufacturer

Conrac Elektron, Weikersheim, Federal Republic of Germany

Dornier Info 80 Airport Information System

Info 80 is an airport information system to be produced by Dornier for Frankfurt Rhein-Main Airport replacing the existing system installed in 1970. The new system is to be introduced in several phases. The first phase will cover all data relating to flight information including gates and baggage details. The data provided by the system will be used by various control centres in the airport area for day-to-day airport operation and management including personnel planning and assignment and the supply of resources (fuel, catering, airport vehicles etc).

It is anticipated that the first phase will take approximately two and a half years to complete.

Status

Dornier System completed a contract for an automatic seat reservation system for the West German Railway in 1980 but this is the first system to be designed by the company for airport use. Dornier are supplying the computer and routine software with Tandem Computers GmbH as a sub-contractor providing the hardware elements and the operational software

Manufacturer

Dornier System GmbH, Friedrichshafen, Federal Republic of Germany

Dahms Mobile Passenger Stairs

Dahms produce four types of passenger stairs: Universal, Luna, Hamburg and Tele.
The Universal is a self-propelled unit available in two versions 2·8 to 5·6 metres and 3 to 5·4 metres. Height adjustment is by telescopic extension of the upper section of the stairs, set at 34 degrees with an intermediate platform between the two sections. Accessories include four stabiliser jacks, two rear head-lamps, chains and a light barrier to prevent unauthorised access to the stairs.
The Luna is another self-propelled unit. Two models are produced, the Luna MT and the newer Luna FT. The FT, which was developed in conjunction with Boyd Airport Equipment Ltd, is mounted on a Ford Transit chassis. Height is adjusted telescopically by extension of the upper section from an intermediate platform. Capable of serving the low sill BAe 1-11 and Boeing 737 the stairs are designed for field modification to enable them to be used for larger aircraft. Additional features are: four stabilising jacks, steps 4 foot (1·2 metres) wide and simplified hydraulics fitted with an emergency pump and an auto-selection capability to prevent misalignment of the step height.
The Hamburg and the Tele are both towable stairs, the former available in two sizes. The height of the Hamburg steps is adjusted by means of a hydraulic jack and hinged outriggers are fitted for stability. In the Tele stairs the upper section extends telescopically from an intermediate platform.

Status
Dahms mobile passenger stairs are in service with military forces in 4 countries, 34 airlines and 29 airport authorities

Manufacturer
Dahms and Co, Luneburg, Federal Republic of Germany

Dahms Luna FT mobile passenger stairs

Characteristics

SERIES	Universal	Luna	Hamburg	Tele
NO OF MODELS	2	2	2	1
AIRCRAFT SILL HEIGHT				
max	5·4/5·6 m	3·5 m (MT) 2·7 m (FT)	5·7/5·2 m	3·4 m
minimum	3/2·8 m	2·3 m (MT) 1·78 m (FT)	4·6/3 m	2·1 m
STEP WIDTH	1·25 m	1·22 m	1·25 m	1·25 m
UNIT DIMENSIONS				
length	9 m	6·2 m	12 m	7·35 m
width	2·6 m	2·17 m	1·65 m	2·56 m

EAA MOAB Passenger Handling System

The MOAB passenger handling system is designed to enable the passenger and baggage to remain together up to the aircraft and for check-in, customs, immigration and security procedures to be carried out on the way to the aircraft. The system comprises Rollerair check-in/-out handling organisation, Expressway assembly point and the MPHU 378 vehicle.
The Rollerair check-in/-out handling organisation consists of a control centre, passenger service centre and a technical centre. The control centre co-ordinates all movements including transit and establishes radio communication with the MPHU drivers and other personnel. The centre controls customs and passport authorities, allocation of personnel and passenger reservations in conjunction with a central computer. The passenger service centre supplies information, controls people meeting passengers, taxis and security measures. Equipment maintenance, technical monitoring and baggage handling control is co-ordinated by the technical centre.
The Expressway is a pre-check-in assembly point for departing passengers and disembarkation point for arriving passengers. Departure from this point is by an MPHU vehicle, between 40 and 90 minutes before a flight depending on the type of flight, domestic or international, and the distance from the Expressway to the aircraft. A number of pick-up points can be arranged. The Expressway area is used for transit transfer from one vehicle to another achieved by connecting the two vehicles with a 2-metre satellite bridge operated at the second level.
The MPHU 378 is used for transferring passengers with their baggage from the Expressway to the aircraft. The vehicle mounted on eight steerable axles, with a capacity of 115 000 kg is driven by hydrostatic power supplied by hydraulic motors coupled to two diesel engines. The vehicle constructed in steel with an aluminium superstructure will accommodate 378 passengers with 8 to 14 check-in clerks and a driver. Use of space and operation differs between departure and arrival units.
For departures passengers, on boarding the vehicle through doors 3 metres wide, face two to four check-in desks. There is space for up to 56 passengers in this area. Baggage is taken by conveyor to a storage area where it is transferred to containers.
Stairs 0·75 metres wide lead to the second level which is a standard area. The stairway forms customs and passport control with a special system to segregate passengers who have checked in previously. The second level is fitted with seats and bar facilities for 32 people, two toilets, a transit room, waiting lounge with 90 seats and a VIP room. Transfer to the aircraft is by telescopic loading bridges extending to a maximum height of 5·4 metres from the second level via security control. Baggage is transported in containers by conveyor to be loaded into the aircraft.
For arriving passengers the interior arrangement is different. The lounges are similar but the vehicle is fitted with a 51·3 square metre baggage claim area, with baggage conveyors on both sides 54·6 metres long. A separate belt is provided for small baggage. The unit is furnished with customs desks and passengers can leave the vehicle through two exits both 3 metres wide.

Characteristics
MODEL: MPHU 378
ENGINES: 2 × KHD diesel F8L413F
OUTPUT: 165 kW each at 2500 rpm
MAX DRIVING SPEED: 40 km/h
HYDRAULIC PUMPS
drive: 2 × 420 litres/minute at 320 bar
lifting/steering: 100 litres/minute at 250 bar
lifting speed: 2 m/minute
TURNING RADIUS (OUTSIDE): 10·8/11·6 m
MAX PASSENGERS: 378
seats: 212
standing (3 m^2): 166
UNIT DIMENSIONS: 19·6 × 8·9 × 5·5 m
GROSS VEHICLE WEIGHT: 115 000 kg

Status
The MOAB passenger handling system was designed by EAA and the idea launched at Frankfurt in October 1980. A prototype vehicle is under construction

Manufacturer
Economical Airport Architecture, Wunstorf, Federal Republic of Germany

Expresso Type 894871 Baggage Trolley

The Expresso 894871 is a three-wheeled baggage trolley with a single castor at the front constructed in galvanised steel with rubber wheels. Deflectors are fitted on each side above the fixed-rear wheels to prevent damage to doors and injury to people and a nylon-covered rubber bumper is mounted on the front to facilitate movement through swing doors. The loading area slopes down to the rear to prevent suitcases from tipping forward and to reduce the weight on the front castor. A deposit basket and automatic brakes are incorporated in the design; the latter are released by pressing down on the trolley handle or by pulling upwards. The plastic tapered brake-shoes which operate on the V-belt principle are adjustable. Optional features include a 1·9 metre flagpole with a plastic flag fitted to the trolley to aid identification, collapsible front guard, scooter wheel fitted to the rear of the trolley and a 1·7 by 0·75 metre framework for hanging clothes.

Characteristics

TYPE: 894871
CAPACITY: 200 kg
DIMENSIONS: 1·1 × 0·5 × 1·1 m
WEIGHT: 25 kg
LENGTH OF 10 TROLLEYS STACKED: 3·7 m

Status

Expresso baggage trolleys are in use at the following airports: Graz and Vienna, Austria; Copenhagen, Denmark; Dusseldorf, Hamburg, and Munich, Federal Republic of Germany; Athens, Greece; Jakarta, Indonesia; Amsterdam, Netherlands; Lagos, Nigeria and Geneva and Zurich, Switzerland

Manufacturer

Expresso Material Handling Equipment, Kassel, Federal Republic of Germany

Henschel Passenger Loading Bridges

Two basic types of passenger loading bridges are produced by Henschel, the telescopic models with apron drive and the nose-loading models with fixed bridges. Both are capable of handling all types of narrow and wide-body commercial aircraft from BAe 1-11 to Boeing 747.

The telescopic loading bridge has a tunnel with extensible sections which pivots around the fixed main tunnel to the terminal building. The telescopic tunnel can comprise up to three sections with the largest section at the aircraft end. It is installed on a support column which provides the vertical adjustment and mounted on a single axle pivoting chassis fitted with large volume pneumatic wheels. The telescopic tunnel is manoeuvred to the aircraft by means of a variable dc main drive providing simultaneous swivelling and telescoping movement. Integrated steps enable access from the ground to the tunnel. Up to two aircraft doors can be accommodated by this type of bridge.

The Nose-loader bridges are designed to provide access to aircraft parked in the nose-in position. Aircraft manoeuvring is far more critical with this type of bridge since there is little room for lateral misalignment. For this reason a guidance system is available to help the pilot position the aircraft correctly and the bridge is fitted with a wide cabin at the bridge head and an adjustable split cabin sill to compensate for any slight misalignment. An electro-mechanical drive system powers the elevating mechanism.

The NLS is a version of the Nose-loader bridge. With this model the differences in height between the terminal building and the aircraft door are compensated by vertical movement of the main tunnel which is arranged parallel to the aircraft. This enables the telescopic tunnel with the bridge-head cabin to remain in a horizontal position for access. The NLS-T variant provides for two-door handling of wide-body aircraft.

Henschel telescopic loading bridges in operation at Düsseldorf Airport

Henschel Nose-loader bridges at Tripoli Airport

Status

Henschel telescopic loading bridges are installed at the following airports: Brussels (2), Belgium; London Gatwick (12), England; Düsseldorf (20), Frankfurt (67), Hamburg (2) and Hanover (12), Federal Republic of Germany; Baghdad (18), Iraq; Nairobi (4), Kenya; Edinburgh (2), Scotland; Moscow (19), USSR Sharjah Eriwan (3), United Arab Emirates and Belgrade (2), Yugoslavia.

Nose-loader bridges are fitted at Brussels (12), Belgium; Tripoli (7), Libya; Edinburgh (1), Scotland; Madrid (12), Spain; Stockholm (14), Sweden; Damascus (3), Syria and Tallinn (3), USSR

Manufacturer

Thyssen Henschel, Kassel, Federal Republic of Germany

Henschel Nose-loader special loading bridge at Stockholm-Arlanda Airport, pilot guidance system on right

MBB Airport Vehicle Control System

Airport vehicle control systems monitor and control the operation of ground vehicles and equipment. By analysing route logistics the system dispatches and controls ground traffic between the place of operation and the aircraft for example, ground traffic between the aircraft parking position, the fuel supply bay, freight area, ramp equipment station, apron and passenger terminals. The vehicles keep the control headquarters informed of their position between their starting point, the depot and their destination, the aircraft parking position. As a result, the headquarters may reroute vehicles between their starting point and their destination which may be up to 2½ miles (4 km) apart or may commission them to take over additional functions on the way. Due to long distances at most airports some synchronising points may be provided on the apron to allow vehicle locations to be established more accurately.

The system comprises operation control centre and sub-centres: stationary and vehicle terminals (INFO-terminals), portable terminals (INFO-datamobiles) and system software. The programming system supervises all functions, compares actual and target figures on the basis of data files and then gives the necessary instructions. This programming system comprises two sub-systems: one for control and organisation, one for applications and data files.

The system handles dialogue-mode operation between the control officer and the control computer via data displays, system operation via keyboard and display of data and automatic distribution of incoming reports to individual control positions.

Absolute priority is given to emergency calls. Other processes handled by the system include print-out of reports in the event of errors; data storage in a mass memory for statistical evaluation; and management of audio channels and data transmission channels and an individual paging system. The operation control centre provides the link to all ground traffic vehicles. This is where decisions are made regarding vehicle operation, special tasks and individual requirements. Within the entire system, data is transferred via the control centre only. The operation control centre consists of a main processor which controls vehicle dispatching and data transmission to and from stationary terminals, vehicle and portable terminals. Data computed by the main processor is then conveyed to sub-processors, formulated and transferred to the individual terminals.

Video terminals are provided for vehicle dispatching input, presenting vehicle operating conditions (such as defects, distance covered, position); signifying the number of vehicles currently in operation, the number of vehicles available (together with their location), the planned route, place of operation and purpose (such as delivering mail, water supply), the ground staff currently occupied, the ground staff available, etc, showing errors and defects (such as no report for five minutes) and indicating the transport instructions automatically determined by the system and the transport requests by the control officer. A typewriter terminal is supplied for computer operation. A high-speed printer compiles the accounting report, documenting each order given the system to a vehicle, a ramp agent or any other recipient. In this process it indicates the date, time, flight, gate, parking position, destination numbers, airline and the driver's code when a job is commissioned and when it is completed.

Following the close down of operations in the evening the high-speed printer can also print out the daily statistics showing all the procedures and missions conducted. A tape recorder provides an external memory for operating programmes, statistics and documentation. The system uses a disc as working memory for planning and control.

The main processor is connected to two sub-processors. The first sub-processor is able to handle all data transmissions with the vehicles and the portable INFO-datamobile. Data is modulated in the semi- or full-duplex mode at either 1200 or 1400 bits per second transmission rate, for a radio communication interface. The sub-processor also handles routine and emergency inquiries. The control officer's display terminal shows which vehicle or ramp agent wants to talk to the control centre.

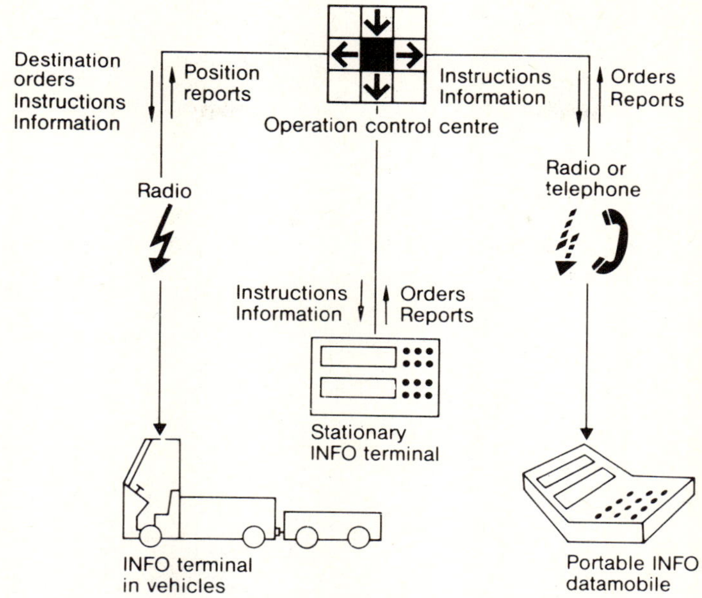

Operation of the system

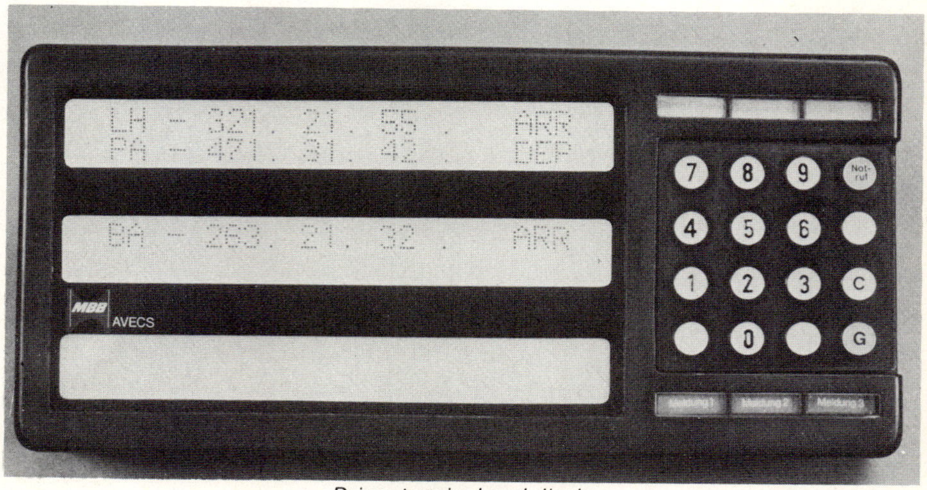

Driver terminal and display

The function of the second sub-processor is to handle all the communication with permanently linked INFO-terminals or portable INFO-datamobiles. Using airport lines or PTT, the system can cover distances up to 12 miles (19·3 kms). There are two kinds of INFO-terminals: stationary INFO-terminals and driver's terminals, that is those installed in the vehicles. In addition the INFO-terminals are available in two different versions: either the waterproof version for outdoor installation such as baggage vehicles or the standard version for installation in offices and closed buses. Each INFO-terminal consists of two components; the display unit and the electronic unit. Both components are separate and may therefore be installed together or as individual units. Terminal-/control-centre communication is either via PTT line or radio.

Special broadcasting frequencies are available for digital data transmission and audio channels. All vehicles broadcast and receive either on the same or on different frequencies, depending on official regulations.

The INFO-datamobile is a portable, battery-operated input and output unit for the radio transmission of data to the operating control centre. It receives its intelligence from a microprocessor and the specially developed software.

Characteristics

FREQUENCY RANGE: 80, 160, 460 MHz
CHANNEL PATTERN: 20 or 25 kHz
MODULATION: fm, differential phase shift keying
MODE OF OPERATION: Simplex, semi-, half- and full-duplex
TRANSMITTER POWER: up to 6 W
TRANSMITTER KEYING: electronic keying filter
ANTENNA CHANGE-OVER: electronic
TRANSMISSION RATE: 1200/2400 bits/s switchable
TERMINAL ADDRESS: 10 bits (< 1024)
MESSAGE LENGTH
operations control centre → vehicle 69 bits
vehicle → operations control centre 60 bits
POWER REQUIREMENTS: 12 V/30 V/approx 2 A
TEMPERATURE RANGE: 0 to +45°C
DRIVER'S INPUT/OUTPUT TERMINAL
height: 140 mm
width: 300 mm
depth: 60 mm
weight: 5 kg
VEHICLE ELECTRONICS
height: 140 mm
width: 300 mm
depth: 300 mm
weight: 10 kg

Status

Airport vehicle control system entered a trial operation at Frankfurt-Main airport at the end of 1981 where the system is applied to the control and co-ordination of baggage transport vehicles by means of a computer program and digital-data communication

Manufacturer

Advanced Transport Systems, Messerschmitt-Bolkow-Blohm GmbH, Munich, Federal Republic of Germany

Neoplan N980 Galaxy Aero Lounge Mobile

Neoplan N980 Galaxy Aero Lounge

A new development for Neoplan, the Galaxy, is designed to accommodate up to 342 passengers on two decks with additional space on the upper deck for a VIP lounge providing seating for ten passengers.

The gangway is hydraulically operated and controlled from the driver's position within the tunnel section. It has adjustable entrance heights between 2 and 5·45 metres reaching up to the level of the Boeing 747 door sill.

Passenger entry to the Galaxy is either through the elevating tunnel section set at the terminal gate height or via pneumatically-operated folding doors on each side of the lower deck, depending on the layout of the terminal. Passenger movement between decks is by stairs positioned at each end so if the elevated tunnel section to the upper deck is used passengers use both sets of stairs. With 150 passengers or less the upper deck only may be used.

The Galaxy is a self-supporting, space frame construction welded from square section steel tubes and angle sections with exterior panelling of 1 mm thick sheet steel. The vehicle is powered by a Daimler Benz ten-cylinder diesel engine and features Voith automatic transmission, ZF power steering, air brakes and air suspension. The driver is assisted by all-round vision from an infra-red camera system designed for day and night operation. An air-conditioning system is fitted or alternatively six electric fans can be installed on each deck and six mechanically-operated lifting roofs provide ventilation.

Characteristics
ENGINE: Daimler-Benz Diesel engine OM403
OUTPUT: 239 kW
AIRCRAFT DOOR-SILL HEIGHT
max: 5·45 m
minimum: 2 m
CAPACITY
lower deck: 60 seats, 96 standing places (6 per m²)
upper deck: 90 seats, 96 standing places (6 per m²)
VIP lounge: 10 seats
DIMENSIONS: 15 × 4·5 × 4·5 m

Status
The vehicle was introduced at the 1981 Frankfurt Motor Show. In recent trials conducted by Neoplan 200 passengers were handled in seven minutes and with its capacity the makers claim the Galaxy is likely to be large enough to deal with the majority of the passengers of a wide-body aircraft arriving 70 to 80 per cent full

Manufacturer
Neoplan Gottlob Auwarter GmbH, Stuttgart, Federal Republic of Germany

Neoplan Single Decker Airfield Buses

The N912, N922 and N940 are single decker buses suitable for transferring passengers between the airport terminal building and the aircraft parked on the apron. Each model varies in capacity but is similar in basic design. The bus is a self-supporting, space-frame construction, electrically welded from seamless drawn square-section steel tubes and angle sections. The exterior panelling is made of senzimir-galvanised sheet steel 1 mm thick. Power steering and a Daimler Benz five-gear synchron transmission are fitted. Air suspension provides a minimum ground clearance of 20 cms. Three pneumatic double-winged passenger doors are installed (two in the N922 model) each 1·1 metres wide which open outwards with the moving elements under the floor. Access for the driver and the driver's assistant is provided by revolving doors fitted on either side of the vehicle. The tinted side windows incorporate flapping elements in the upper part to give additional ventilation. Accessories include five electric ventilator fans and passenger heating equipment can be supplied as an optional feature.

The Neoplan VIP bus is a small vehicle which can provide transport for VIPs, the crew or handicapped people. For VIPs and the crew the vehicle offers up to 16 seats. When used for handicapped people the bus can be lowered to the ground for entry in wheel-chairs.

Neoplan single decker airfield bus

Status
Neoplan buses are in service at 42 airports in 27 countries

Manufacturer
Neoplan Gottlob Auwarter GmbH, Stuttgart, Federal Republic of Germany

Characteristics

MODEL	N912	N922	N940
ENGINE	Benz OM401	Benz OM401	Benz OM401
OUTPUT	141 kW	141 kW	141 kW
VEHICLE TURNING CIRCLE RADIUS	13·5 m	13 m	14 m
CAPACITY			
seats	5	7	12 or 27 or 33
standing			
at 4 persons per m²	65	84	100 80 70
at 6 persons per m²	102	126	156 120 105
DIMENSIONS	12 × 2·5 × 2·69 m	12·26 × 3·15 × 2·81 m	12·66 × 3·75 × 2·69 m

MODEL	VIP bus	
ENGINE	Benz 300 D	Peugeot J7
OUTPUT	59 kW	50 kW
CAPACITY	16 seats or 6 wheel-chairs and 4 attendants	or 5 wheel-chairs and 10 attendants
DIMENSIONS	6·69 × 2·3 × 2·42 m	

Scharfenstein Aircraft Passenger Stairs

Five models of aircraft passenger stairs are produced by Scharfenstein. Three are towable and the other two models are self-propelled. The towable stairs under the designations T1, T2 and T3 are available to service aircraft with door-sill heights from 1·8 to 5·7 metres. The self-propelled stairs, S1 and S2, both based on a Mercedes Benz chassis will service aircraft with door-sill heights between 2·6 and 5·7 metres.

Characteristics

MODEL	T1	T2	T3	S1	S2
TYPE	towable	towable	towable	self-propelled	self-propelled
ELEVATING HEIGHT RANGE	1·8–3·4 m	2·8–4 m	4·6–5·7 m	2·6–4 m	3·1–5·7 m
CHASSIS				Mercedes Benz L407/D	Mercedes Benz 1113/B

Status
In service at airports in Federal Republic of Germany

Manufacturer
Fritz Scharfenstein, Nister/Hochenburg, Federal Republic of Germany

Siemag BCS Baggage Handling Systems

Siemag plan and build complete systems for baggage handling and control. The equipment comprises mechanical or electronic scales; labelling and weighting belts; conveyor lines with curved sections for horizontal and inclined flow; and straight and circular conveyors for distribution. The circular conveyor or carousel, fitted with crescent-shaped rigid mounting pallets, is a combined transport and storage unit. Loading is effected manually or using a conveyor belt. The conveyor consists of four sub-assemblies: straight and curved sections, driving and tension stations.

Characteristics
CONVEYOR WIDTH: 0·8 m
FRAME WIDTH: 0·95 m
AVERAGE CURVE RADIUS: 1 m
OVERALL HEIGHT
baggage claim: 0·4 m
baggage distribution: 0·5–0·6 m
OPERATING SPEED
baggage claim: 0·4 m/s
baggage distribution: 0·5–0·6 m/s
SYSTEM CAPACITY: 50 or 100 kg/minute
ELECTRICAL POWER: 3–4 kW

Status
Siemag has obtained contracts in 25 countries for over 150 carousels from such airports as Berlin/Tegel, Cologne-Bonn, Düsseldorf, Frankfurt, Hamburg, Hanover and Stuttgart, Federal Republic of Germany; Amman, Jordan; Findel Airport, Luxembourg; airports in Mozambique; Oslo, Norway; Abu Dhabi, United Arab Emirates; Moscow, USSR; and Zagreb, Yugoslavia

Manufacturer
Siemag Rosenkaimer GmbH, Leichlingen, Federal Republic of Germany

Siemag baggage check-in point at Düsseldorf Airport showing weighing equipment, labelling and collecting conveyor belts

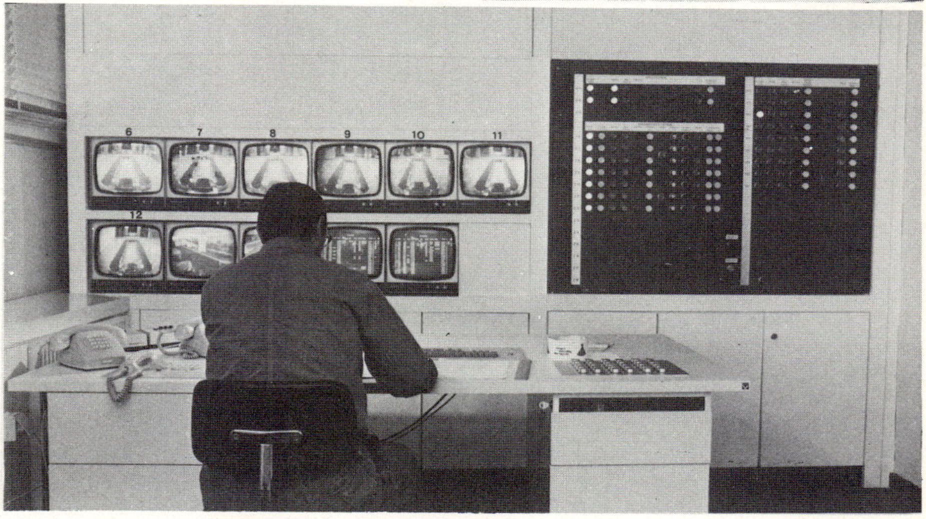

Control desk for Siemag baggage claim and check-in equipment at Düsseldorf Airport

TN Annax 100 Flight Information Display

The TN Annax 100 flight information display system uses alphanumeric display modules with bi-stable indicator elements in a 5 by 7 matrix. These permit the display of optional texts formed by numerical or alphanumeric characters and symbols. The display system can be controlled by tele-typewriters, vdu terminals and large capacity computers.

A display module comprises 35 bi-stable indicator elements mounted on a circuit board, and the electronic circuit. The character height is approximately 10 cm. The special features of the bi-stable indicator elements are a long service life (several 100 million reversals), low power consumption, legibility and quick information change; the display can also be in colour. The serial pulse signal from the computer or the terminal is picked up by optocouplers in the receiver section. The transmission speed ranges from 300 to 9 600 bauds. When the board contents have been completely stored the presentation of information on the display board begins; an optional number of lines with 20 positions each requires only 1 second.

The power supply is accommodated in the aluminium profile housing of the display board. The mains unit converts the ac voltage supplied on the primary side to 24 volts dc.

TN Annax 100 flight information display

Characteristics
OPERATING VOLTAGE: 24 V
IMPULSE DURATION FOR INFORMATION CHANGE: ⩾ 10 ms
DIMENSIONS: 14 × 14 × 36 cm
WEIGHT: 9 g
POWER REQUIREMENTS: 220 V, 50/60 Hz
POWER CONSUMPTION IN QUIESCENT STATE: 3 VA
POWER CONSUMPTION DURING INFORMATION CHANGE: 5 VA
CHARACTER HEIGHT: approx 10 cm
VISIBILITY RANGE: 40–50 m
OPERATING TEMPERATURE RANGE: −20 to +80°C

Status
Installed at airports and railway stations in the Federal Republic of Germany

Manufacturer
Telefonbau und Normalzeit, Frankfurt, Federal Republic of Germany

Trepel Lift Lounge

The Trepel Lift Lounge is a special-purpose vehicle designed to transport up to 150 passengers between the terminal building and the aircraft. The scissor-lift mechanism is capable of raising and lowering the lounge body to service aircraft from the McDonnell Douglas DC9 to the Boeing 747.

The unit comprises a lounge body with an extensible connecting platform and canopy incorporating a driver's compartment, coupled to a large passenger cabin. An air-conditioning unit is installed at the rear of the vehicle. The body, mounted on a scissor-lift mechanism, is fitted on a four-axle chassis equipped with a main and auxiliary engine. The main engine provides power for the vehicle drive and lifting mechanisms while the auxiliary engine drives a generator for the air-conditioning unit and illumination. The extending and lifting mechanism is push-button controlled by the driver at the centre front of the lounge body with doors to the passenger cabin mounted on either side of the driver's compartment. The lounge body can be raised during the final approach to the aircraft or terminal. The suspension is stabilised automatically when both motions are activated simultaneously. In an emergency the body can be lowered by a manually-operated by-pass system. Accessories include fire extinguishing systems, incorporated in the cabin and the chassis, and a television monitoring system providing the driver with a view behind the vehicle for reversing manoeuvres.

Characteristics
ENGINES: 2 × 160 ps
AIRCRAFT DOOR-SILL HEIGHT
max: 5·6 m
minimum: 2 m
CAPACITY: 150 passengers, 90 seats
LIFTING TIME TO MAX ELEVATION: 30 s
AIR-CONDITIONING: 30 000 kcal/h
DIMENSIONS
length: 14·4 m
width: 4·3 m

Status
In service at Düsseldorf Airport, Federal Republic of Germany; Jeddah Airport, Saudi Arabia and Gothenburg-Landvetter Airport, Sweden. Jeddah Airport ordered an additional ten units in October 1981 to augment the existing fleet of ten vehicles

Trepel Lift Lounge at Düsseldorf Airport

Manufacturer
Trepel AG, Wiesbaden, Federal Republic of Germany

Vetter Airport Buses

Three basic types, A, B and C, are produced by Vetter suitable for transporting passengers from the terminal building to the aircraft. Additionally a shortened version of type A is available for use as a crew bus or, furnished with a suitable interior, for transporting VIPs.

The types A, B and C have the same total length and differ only in respect to width, number of doors and seating arrangements. Capacity ranges between 70 and 150 people. The doors and interior fit are adapted according to individual operator's requirements.

Characteristics

TYPE:	A	B	C
NO OF PASSENGER DOORS	2	2	3
DIMENSIONS			
length:	12·7 m	12·7 m	12·7 m
width:	2·5 m	3 m	3·75 m
height:	2·64 m	2·64 m	2·64 m

Status
In service at major airports including Stuttgart, Federal Republic of Germany and Hong Kong

Manufacturer
Walter Vetter GmbH, Fellbach, Federal Republic of Germany

ITALY

Alba 2 Passenger Handling Data System

Alba 2 is an improved version of an aircraft loading and balancing automation system for co-ordinating the turn-round operations in small and medium sized airports. The system configuration is tailored to individual airport's requirements. The main functional units of the system are the co-ordination of the passenger handling, aircraft loading and balance, monitoring of flight information and the telex network intercommunication. All data concerned with passenger check-in, including SITA telexes, is handled by the system which provides a display of flight information on television monitors throughout the terminal. For cargo handling the system monitors the freight from acceptance in the warehouse to on-board cargo distribution, providing loading instructions and a load sheet print-out with trim indications for different aircraft types.

Characteristics
COMPUTER: DEC PDP 11 with memory from 128 kW–1 MW
DISC CAPABILITY: 56 MBy–512 MBy
TERMINAL: DEC VT 100
NO OF TERMINALS: 8–64
PRINTER: DEC LA 120 (180 cps)
NO OF PRINTERS: 1–16

Status
The first Alba system was installed in the Caselle Airport, Turin in 1975. Alba 2 replaced the original system in October 1980. A new version of the Alba 2 is under development with full message switching capability

Manufacturer
Aeritalia, Turin, Italy

Aviogei Compact Series Passenger Bus

Three models of the compact series passenger bus are available: the CBP 90S, 135S and 175S, having capacities for 90, 135 and 175 passengers respectively.

The CBP 90S is fitted with double folding doors either side and at the rear of the vehicle operated pneumatically. Power is provided by a Deutz air-cooled diesel engine and the bus is equipped with Allison automatic transmission, Danfoss power steering and air brakes.

Standard features include: electric fans in the driver's cabin and passenger area and a public address system. Air-conditioning can be installed as an option. The other models incorporate similar features and provide additional capacity.

Characteristics
TYPE: CBP 90S
PASSENGER CAPACITY
standing (6 per m^2): 82
sitting: 8
ENGINE: Deutz F6L 912 developing 112 hp at 2500 rpm
MAX SPEED: 40 km/h
TURNING RADIUS: 15 m
GROUND CLEARANCE: 0·18 m
DIMENSIONS: 11·7 × 2·5 × 2·75 m
MAX WEIGHT: 10 100 kg

Status
In service with Air Zaire

Manufacturer
Airport Ground Equipment Italia, Campoleone, Italy

Aviogei CBP 90S passenger bus

Aviogei Passenger Steps

Two types of passenger steps are produced by Aviogei operating on the telescopic principle with the upper section extending from an intermediate platform providing steps with a 35-degree angle of inclination. The 707 series is designed for narrow-body aircraft with a maximum height extension of 4 metres and the 747 series for wide-body aircraft operating up to a maximum height of 5·6 metres. The 707 series comprises two models. The ST 707 is a towable unit equipped with a hand-operated hydraulic pump which controls the height of steps between 2·1 and 4 metres. Four stabiliser jacks provide stability during operation. Safety locks ensure that these cannot be retracted while the steps are in use and that the steps cannot be moved with the jacks extended. A mechanical lock prevents the steps from descending in the event of hydraulic failure. The SI 707 is a self-powered version of the ST 707 with a 40 hp diesel engine, automatic transmission and power steering. Operation of the steps is via a motor-driven hydraulic pump.

The 747 series consists of two models both self-powered and equipped with 1·6-metre wide treads providing a double gangway with a hand-rail in the centre. The SI 747 is a diesel-powered unit with automatic transmission and power steering. The steps are operated through a variable flow pump and a hydraulic motor with axial pistons. Four stabiliser jacks are fitted to provide stability during operation with safety locks similar to those described above. A mechanical lock prevents inadvertent operation of the steps due to hydraulic failure. The upper platform is articulated with the ability to swivel ±15 degrees to fit the contours of the aircraft. Eight lamps are fitted to provide suitable illumination of the staircase and the aircraft door during night operation. The SA 747 is similar to the SI 747 but mounted on a Fiat chassis with a 122 hp diesel engine. All steps can be covered as an optional feature.

Characteristics

TYPE	ST 707	SI 707	SI 747	SA 747
AIRCRAFT DOOR-SILL HEIGHT				
max	4 m	4 m	5·5 m	5·5 m
minimum	2·1 m	2·1 m	3·1 m	3·1 m
TREAD WIDTH	1·4 m	1·4 m	1·6 m	1·6 m
POWER UNIT		VM 1052 40 hp	VM 1052 40 hp	Fiat 90 NC 122 hp
TURNING RADIUS	6·5 m	6·5 m	8 m	5·6 m
MAX SPEED	towable up to 50 km/h	25 km/h	25 km/h	80 km/h
UNIT DIMENSIONS				
length	5·65 m	5·65 m	6·25 m	6·9 m
width	2·25 m	2·25 m	2·5 m	3·3 m
WEIGHT	1800 kg	3000 kg	5500 kg	6000 kg

Status

In service with Alitalia and other major airlines and Rome Airport Authority

Manufacturer

Airport Ground Equipment Italia, Campoleone, Italy

Aviogei PB 6000 S Mobile Lounge

Aviogei has designed a mobile lounge to service all types of commercial aircraft with a capacity for 135 passengers. A double-lift scissor-mechanism raises the lounge body to a maximum height of 5·5 metres. The lounge body, with driver's compartment in the centre at the front, is mounted on a six-wheel chassis with a single driving axle, forward and twin axles at the rear. Access to the lounge from the terminal is by double doors fitted to the rear of the body providing an open width of 2 metres. The vehicle is driven by a 120 hp diesel engine and equipped with power steering.

Characteristics

AIRCRAFT DOOR-SILL HEIGHT
max: 5·5 m
minimum: 1·4 m
ENGINE: VM 1056 120 hp air-cooled diesel
max speed: 28 km/h
LIFT SPEED: 8 m/minute
TURNING RADIUS: 12·8 m
DIMENSIONS
length: 12·2–13 m
width: 3·5 m
TRANSIT HEIGHT
front: 3·8 m
rear: 3 m
WEIGHT: 16 000 kg

Manufacturer

Airport Ground Equipment Italia, Campoleone, Italy

Reggiane Passenger Loading Bridges

Reggiane passenger loading bridges are available in various types but generally two basic types are used. These are the apron drive and pedestal versions.

The apron drive loading bridge consists of a tunnel with extensible sections which pivots around a fixed main tunnel connected to the terminal building. The front end of the telescopic tunnel is mounted on a wheel system actuated by a Ward Leonard device which provides freedom of movement with ±95 degrees of rotation, elevation and depression, extension and retraction. Access stairs are supplied for service personnel from the ground to the telescopic tunnel. Two sets of telescopic tunnels are used for wide-body aircraft.

The fixed pedestal loading bridge is designed to service aircraft parked in the nose-in position. It comprises a telescopic tunnel with two sections, which can be extended, retracted, elevated and depressed. The bridge head which is fitted with a canopy can be rotated ±15 degrees to compensate for aircraft lateral displacement. Service personnel access steps are incorporated in the design.

Characteristics

	MINIMUM LENGTH	EXTENSION	MAX LENGTH
APRON DRIVE	9·9 m	20·1 m	30 m
	11·1 m	22·5 m	33·6 m
	14·2 m	29·9 m	44·1 m
	15·1 m	32·6 m	47·7 m
	17·4 m	39·2 m	56·6 m
PEDESTAL	11 m	8·5 m	19·5 m
	13·7 m	10 m	23·7 m
	15·5 m	17 m	32·5 m
	16 m	12·75 m	28·75 m

Status

Reggiane loading bridges are installed at airports in Italy. The bridges are manufactured under licence from Abex Corporation of the USA

Manufacturer

Officine Meccaniche Italiane Spa, Reggio Emilia, Italy

Solari Flight Information Display Systems

Solari supply, install and commission systems involving the dissemination and display of information to the public and airport staff under computer control.

Information is presented on flap boards and television displays. Split flap indicator boards are available with script 35, 60 and 100 mm high. The number of lines of information and the composition of each line is virtually unlimited with a choice of scripts and special graphics such as airline logos.

The video-monitor displays provide page formats ranging from eight rows of eight characters to 24 rows of 80 characters.

Character generators are available for foreign languages and alphabets such as arabic and cyrillic.

Direct reading digital clocks which can be driven from Solari's quartz-controlled master clock or synchronised with other master-clock systems can be incorporated in the display.

The systems are controlled by central processing units of the 90 series which provide a range of sophistication from the simple 90/20 to the powerful 90/60 unit. A range of peripherals is available to support each system including vdu terminals, high-speed printers, fixed and moving-head hard discs, floppy discs and magnetic tape memories. A library of proven, standard software has been developed specifically for handling the display of information to staff and public in airport environments. Software routines allow direct input of schedule data and output of statistical data from and to other system computers. They can be programmed to drive automatic announcement systems provided by Solari or by others.

A typical flight information display system was produced by Solari for Leonardo da Vinci Airport, Rome. This system is controlled by a computer with a high-capacity magnetic disc memory. A twin computer is always on stand-by and intervenes immediately if the main one goes out of service. Operators can accede to the memory and interact with the computer by means of seven video terminals. They can control simultaneously the information displayed on the public network by means of television monitors installed in all offices and centres concerned.

Summary tables containing all information relating to the daily traffic can be obtained from two high-speed printers.

The public announcer has a monitor divided into four sectors, each one of which displays automatically the first flights indicating the status of each flight such as 'last call', 'now boarding', 'landed', and those for which the expected time of departure has been fixed. All the incoming and outgoing domestic and international flights relating to a whole season are contained in the mass memory. Their handling is automatic and controlled by means of an operational program stored on a magnetic disc. The computer automatically controls the correct distribution of the information on all networks, displaying them in chronological order following their scheduled timetable.

By means of the video terminals the operators can, at any moment, introduce new flights and cancel or modify any information already displayed. They can intervene on the baggage and gate networks and ask for print-outs and working conditions of the computer and summaries of all incoming, outgoing and transit flights which will immediately appear on the video terminals.

All the internal personnel, the offices, services, coordination centres etc are constantly updated on the domestic and international traffic by means of the operational network, composed of over 200 television monitors, that also serve all airline company offices in the airport.

Characteristics

Leonardo da Vinci Airport information network

INTERNATIONAL AREA

- general departures board: 2 with 30 lines, type 60
- transit board: 2 with 30 lines, type 35; 1 with 20 lines, type 35
- general arrivals board: 2 with 30 lines, type 35
- gate indicator: 48, type 60
- repeater baggage board: 2 for public, type 60; 1 for staff, type 100
- baggage indicator: 7 for public, type 60; 7 for staff, type 60

Solari flight information display system at Charles de Gaulle Airport

DOMESTIC AREA

- general departures board: 2 with 20 lines, type 60
- general arrivals board: 1 with 20 lines, type 60
- gate indicator: 14, type 60
- baggage indicator: 5 for public, type 60; 5 for staff, type 60

Status

Solari flight information display systems have been supplied to over 130 airports in 53 countries. Airports provided with the equipment in 1980–81 were: Barbados; Sao Paolo, Brazil; Ajaccio, Corsica; Charles de Gaulle, Roissy II, France; Rimini, Italy; Kimpo, Republic of Korea; Manila, Philippines; South, North and Haj Terminals, Jeddah, Saudi Arabia; Changi, Singapore; Cape Town, South Africa; Ankara, Turkey; Braniff terminal, Dallas Fort Worth and J F Kennedy, New York, USA

Manufacturer

Solari and c/udine SpA, Udine, Italy

Trastecnica Rapistan Baggage Handling Equipment

Trastecnica Rapistan systems cover baggage handling requirements from check-in facilities to the baggage claim area. For check-in, conveyors are provided for weighing and transfer and closed-loop conveyors are used in the sorting area. Recirculating loops, capable of handling peak periods of activity enable baggage to be loaded into carts for transfer to the aircraft. For arrival baggage, the carts are unloaded and the baggage fed on to a single closed-loop conveyor and moved into the baggage-claim area.

Two types of conveyor are produced, the Crescent and the Triplanar.

The Crescent conveyor is a continuous loop unit, which provides maximum claiming frontage. It is particularly adaptable in the areas with limited space and can be fed manually or automatically. The conveyor consists of a series of plates, which are crescent-shaped and connected through a chain made of rigid lengths hinged at the plate centre. The plates are of steel construction 6 mm thick with a 4 mm Adiprene or similar chemical covering to resist abrasion and wear. They are guided by one central wheel covered with a Vulkollan surface and supported by two lateral, rubber-surfaced, pivoting wheels.

The shrouding of the conveyors can be of stainless steel or painted depending on the application and desired appearance. Caterpillar drives, with 2 or 3 hp gear motors are used to provide conveying speeds of up to 35 metres per minute.

The Triplanar is a flexible slat conveyor which can traverse all three planes in one continuous integrated line operating from a single-power source. It combines the conveying characteristics of a belt conveyor with the operating characteristics of a chain conveyor.

The Triplanar is particularly suited for use in applications where one drive and a moving conveyor surface is preferred to a number of short-power conveyors and powered curves.

Flexible, overlapping slats connected to a universal chain provide a conveying surface capable of handling materials and containers of various sizes and weights simultaneously. The slats are designed to maintain product orientation over long distances. The Triplanar can be adapted to a number of flow configurations. It provides a conveying surface capable of vertical and horizontal bends allowing flow in any direction. The continuous chain makes it useful in applications requiring controlled conveyor movement and sorting or where perfectly synchronous control is essential. In common with most chain conveyors applications can, within wide limits, be as long as needed by simply adding more drive units to distribute the chain transport load. The Triplanar has a special caterpillar-type drive unit which is available in a variety of speeds.

Characteristics

	CRESCENT	TRIPLANAR
PLATE PITCH	0·5 m	—
CURVE RADIUS		
horizontal	1·015 m	1·15 m
vertical		1·15 m
WIDTH	1·015 m	1·15 m
CONVEYOR SPEED	35 m/minute	As required
NORMAL CAPACITY	50 kg/minute	60 kg/minute

Status

Trastecnica Rapistan conveyors are manufactured under licence from Rapistan. They have been installed at 15 airports in Italy, and airports in Jeddah, Saudi Arabia; Moscow, USSR; and Sanaa, Yemen Arab Republic

Manufacturer

Trastecnica Rapistan, Cologno Monzese, Milan, Italy

JAPAN

Shin Meiwa Airporter Passenger Baggage Claim Conveyor

The Airporter Type BC2F is a friction roller-type conveyor for baggage claim which can be arranged in a T, L or oval shape. The Type RC1A is similar to the BC2F but comprises sloped panels in the baggage-claim area to facilitate collection of the baggage. The system can be arranged in either an oval or a triangular shape.

Characteristics

CONVEYOR
width: 1 m
height: 0·35 m
CAPACITY: 60 kg/minute
MINIMUM RADIUS: 1·5 m
POWER REQUIREMENTS
400/440 V, 3-phase, 1·5 kW;
200/220 V, 3-phase, 1·5 kW

Status

In operation at airports in Japan

Manufacturer

Shin Meiwa Industry, Tokyo, Italy

Shin Meiwa Paxway Loading Bridges

Four types of Paxway passenger loading bridges are produced by Shin Meiwa: apron drive, elevating, pedestal and radial. The basic design of each type is similar with steel sheet monocoque structure and an electro-mechanical control system. The control panel is set in the cab at the bridge head, which features a flexible rubber bumper at floor level and a nylon canopy to ensure a weather-proof seal between aircraft and bridge. Access is provided for service personnel by stairs between the bridge head and the ground. The bridges are fitted with an automatic leveller using friction rollers to ensure the bridge floor remains level with the aircraft door sill during embarkation and disembarkation.

The apron-drive loading bridge has a telescopic tunnel which can comprise up to three sections extending from a rotunda which enables the telescopic tunnel to move laterally. In this model the bridge head swivels ±65 degrees to match the contours of the aircraft. The tunnel is installed on a support column which provides the vertical adjustment mounted on a pivoting chassis fitted with two aircraft wheels, each driven by an independent dc motor. This is the most versatile of the four Paxway loading bridges and can be used for all types of commercial aircraft permitting nose-in, parallel or angular aircraft parking positions.

The elevating loading bridge is for aircraft parked in the nose-in position. The tunnel with the bridge head which connects to the aircraft is at right angles to the passage way in the terminal building and is capable of extending up to 3 metres towards the aircraft. The bridge head swivels ±15 degrees to ensure a close fit. The main feature of the bridge is the sliding tunnel to the aircraft which is always kept horizontal since adjustment in height for the aircraft type is compensated by vertical movement of the passage way to the terminal building.

The pedestal-type of loading bridge is used for aircraft parked in the nose-in position. A telescopic tunnel mounted on a fixed pedestal enables the bridge to be adjusted vertically, extended and retracted. Since the vertical movement is limited the aircraft type is not flexible and it can only be used for either wide- or narrow-body aircraft.

The radial loading bridge is equipped with a driving unit and a rotunda in similar fashion to the apron-drive version. However the tunnel does not extend or retract and is limited to rotation. It can be used for either nose-in or angular-parked aircraft. Again vertical adjustment is limited and the bridge may be used only for aircraft with door sills between 1·9 and 3·5 metres high.

Characteristics

TYPE	Apron drive		Elevating	Pedestal	Radial
MODEL	A2-262PS	A3-401PW	E-200PW	P-180PS	R-170PW
LENGTH					
extended	26·2 m	40·1 m	main tunnel 20 m	18 m	
retracted	15·4 m	20·9 m	extension tunnel 9·5 m	14·6 m	
BRIDGE ROTATION	180°	180°			180°
BRIDGE-HEAD ROTATION	±65°	±65°	±15°	±15°	±15°
RATE OF TRAVEL					
horizontal	0–0·5 m/s	0–0·5 m/s	0·06 m/s	0·06 m/s	0–0·3 m/s
vertical	0·8 m/s	0·8 m/s	0·8 m/s	0·8 m/s	0·8 m/s
BRIDGE-HEAD FLOOR HEIGHT					
wide-body aircraft only	3–5·3 m	3–5·3 m		4–5·3 m	
narrow-/wide-body aircraft	1·9–5·3 m	1·9–5·3 m	1·9–5·3 m		1·9–3·5 m
narrow-body aircraft only				1·9–2·7 m or 2·5–3·5 m	
WIND VELOCITY					
max	60 m/s	60 m/s	60 m/s	60 m/s	60 m/s
in operation	26 m/s	26 m/s	26 m/s	26 m/s	26 m/s

Status

Paxway loading bridges are in operation at the following airports: Addis Ababa, Ethiopia (3); Naha (8), Kumamoto (3), Matsuyama (3), Osaka (8), Haneda (7), Nanta (48) and Sendai (2) in Japan; Manila, Philippines (1); Changi, Singapore (34); Don Muang, Thailand (4)

Manufacturer

Shin Meiwa Industry Co Ltd, Tokyo, Japan

NETHERLANDS

Berkel Baggage Weighing Units

Berkel produce baggage weighing units which are incorporated in the check-in counter. Optical and electronic scales are available installed under the weighing belt. Analogue displays with a luminous scale or digital displays are built into the counter indicating the baggage weight to both the operator and the passenger through two displays. Digital displays with characters 16 mm high have data outputs for a calculator, printer or data-processing units.
Berkel, in conjunction with Nederlandse Kumststof Industrie and Rapistan van der Lande, have produced a standard check-in desk as illustrated.

Manufacturer
Berkel International BV, Rotterdam, Netherlands

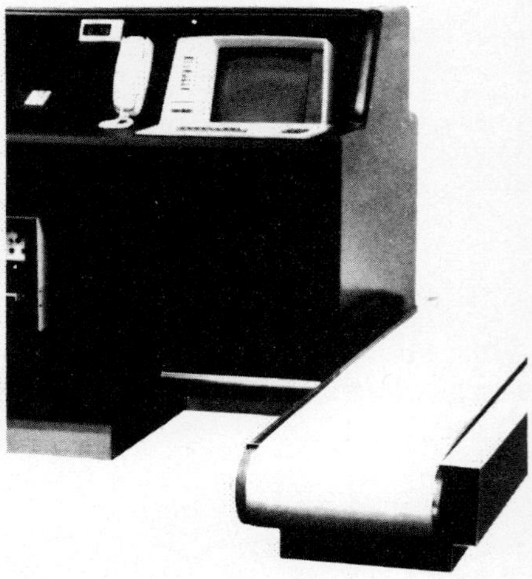

Standard check-in facility

Luxalon Building Products for Terminal Facilities

The Luxalon range of aluminium building products for use at terminal buildings includes ceilings, flush- and open-faced facades, vertical and horizontal sun louvres and sandwich walls. Five types of aluminium ceilings are available. These are acoustic, ventilated, radiant, screen and grid ceilings. The sandwich walls are supplied in sections ready for installation. All products are available in a range of stove-enamelled colours.

Status
Luxalon aluminium products have been supplied to the following airports: Brunei, South East Asia; Bratislavia, Czechoslovakia; Rio de Janeiro, Sao Paulo, Brazil; Copenhagen, Denmark; Düsseldorf, Frankfurt, Hamburg, Hanover, Munich, Federal Republic of Germany; Bayonne, Bordeaux, Hyeres, Lyons, Marseilles, Nice, Nimes, Paris/Orly, Paris/Roissy, and Strasbourg, France; Budapest, Hungary; Jakarta, Indonesia; Osaka, Japan; Kuala Lumpur, Malaysia; Amsterdam, Maastricht, Netherlands; Singapore; Alicante, Barcelona, Gerona, Las Palmas, Madrid, Malaga, Palma de Mallorca, Spain; Boston, Detroit, Grand Rapids, Indianapolis, Jacksonville, Lincoln, Philadelphia, USA; and Dubrovnik, Ljubljana, Sarajevo, Zadar and Zagreb, Yugoslavia

Manufacturer
Hunter Douglas, Rotterdam, Netherlands

NKI Airport Interior Design

Nederlandse Kumststof Industrie (NKI) design, manufacture and produce airport interior installations. The company designs the layout of terminal buildings and produces glass fibre reinforced polyester counters for: check-in, ticket sales, information, banking/cashier, concession/showcase, duty-free shop, immigration, security, gate position, transfer baggage service and customs.
With a flexible shape and modular design counters can be grouped in various combinations for these specific purposes. Electronic equipment, weighing scales and conveyors can be incorporated in the individual counter or counter groups.
The company has co-operated on various airport projects with Nolte Airport Signs, Rapistan van der Lande and Berkel International and in conjunction with the latter two companies, produces a standard check-in facility.

Status
NKI has supplied interior installations to major airports in Europe, Africa, Middle East and Far East, including Schiphol, Amsterdam, Murtala Muhammed, Lagos, Riyadh International Airport Saudi Arabia and Changi Airport, Singapore

Manufacturer
Nederlandse Kumststof Industrie, Dongen, Netherlands

NKI check-in facilities at Saudi terminal, Riyadh International Airport

Nolte Eurorouting Sign and Routeing Devices

Eurorouting is a versatile, modular signposting system suitable for buildings where changes have to be made quickly and efficiently.
The parts have been designed so that they can be combined, and an unlimited number of additions or changes can be made. The letters are engraved on the signs in lower case Helvetica lettering and the standard colour combinations are: dark brown characters on beige, beige characters on dark brown and dark brown characters on yellow.
Directional signs have wording on one or both sides. The size is dependent on character height. The signs are made of silver- or brown-anodised aluminium holders, plastic plates with engraved wording, arrow plates with arrow pointing vertically, horizontally or diagonally and blank plates. The separate arrow plates enable any direction to be indicated. The blank plate permits words to begin in line regardless of which end of the sign the arrow is placed. A pictogram or a colour code can take the place of a blank plate.

Status
Nolte have co-operated with Nederlandse Kumststof Industrie and Rapistan van der Lande to produce airport interior facilities for many major airports

Manufacturer
Nolte EMI, BV, Eindhoven, Netherlands

Rapistan Lande Baggage Handling Equipment

Rapistan van der Lande design, manufacture and install baggage handling systems in accordance with terminal operations and passenger movements.

The equipment supplied varies depending on the operation, whether domestic or international, short- or long-haul, single airline or common-user airport terminal. Check-in systems for baggage at departure range from simple hand loading behind counters to mechanised herringbone check-in islands to automated baggage read-out stations and check-in units. Application depends on initial data such as plotting timegraphs, volume handled in peak hours and versatility of operation.

Mechanised belt curves are used primarily to negotiate horizontal curves and belt inclination limited to 18 degrees for vertical displacement enabling baggâge to remain upright during handling. Special side-guards and belt-curves to prevent delay are required during movement in each system. The company have designed, manufactured and installed linear conveyors, turntables, and recirculating devices in horizontal and incline positions for sorting operations in accordance with the terminal requirements. Linear conveyors are used for the quick release of baggage at airports where a small number of simultaneous destinations must be served so that sorting operations are limited. Oval, triangular or L-shaped recirculating devices accommodate baggage for various destinations, while giving accumulation capacity for early arrivals. Special feeding stations are supplied for transit baggage. Belly-containers require special treatment and can be mobilised by accumulating units.

Besides turntable and crescent-type racetracks, Rapistan van der Lande has developed the Triclaimar. Flexible both horizontally and vertically it is designed to meet most architectural demands for new and existing terminals.

Status

In service at 128 airports in 31 countries. In conjunction with NKI and Berkel, the company offer a standard check-in unit

Manufacturer

Rapistan van der Lande BV, Veghel, Netherlands

SPAIN

Equisa EA Series Passenger Stairs

The EA series comprises three models of self-propelled passenger stairs. The EAA and EAB are identical in construction and design. Both are based on a telescopic upper section extending hydraulically from a fixed lower section with an intermediate platform. The upper platform is fitted with protective side panels which slide on runners latched in the appropriate position. The steps and platforms are constructed of aluminium alloy with lights installed for night operation. Four hydraulic jacks are fitted for stabilisation. Mechanical inter-locks prevent the collapse of stairs as the result of a hydraulic failure. The eight-wheel chassis is powered by a Perkins diesel engine coupled to a hydraulic pump with an emergency hand-pump fitted to operate the stairs in the event of a failure. Optional features include an all-weather drivers' cab and a canopy for the stairs and platform. The EAA is suitable for providing access to narrow-body aircraft, the EAB for wide-body aircraft.

The EAB-C mounted on a commercial truck chassis is similar except for a forward support system consisting of two outriggers and sliding hand-rails on the upper platform which are latched in position. The stairs will service all aircraft with door-sill heights ranging between 2·6 and 5·5 metres. Optional features include rear jacks, automatic retractable hand-rails, mobile control box on the upper platform, preselecting device for height adjustment, rotating upper platform for automatic alignment to the aircraft fuselage, and a minimum operational height of 2·2 metres.

Equisa EAB self-propelled passenger stairs

Equisa EAB-C truck-mounted passenger stairs

Characteristics

MODEL	EAA	EAB	EAB-C
AIRCRAFT DOOR-SILL HEIGHT			
max	3·8 m	5·3 m	5·3 m
minimum	2·4 m	3·1 m	2·6 m
PLATFORM DIMENSIONS	1·7 × 1·8 m	2 × 3 m	2·2 × 1·6 m
STEP DIMENSIONS			
width	1·2 m	1·2 m	1 m
depth	0·34 m	0·34 m	0·3 m
height	0·18 m	0·19 m	0·2 m
ENGINE	Perkins 4108 diesel	Perkins 4108 diesel	optional type
output	36 kW at 3600 rpm	36 kW at 3600 rpm	optional type
MAX SPEED	25 km/h	25 km/h	60 km/h
TURNING RADIUS	8·16 m	10·4 m	7 m
UNIT DIMENSIONS			
length	7 m	8·2 m	
width	2·9 m	3·4 m	
height max	4·8 m	6·3 m	
minimum	3·4 m	4·1 m	
WEIGHT	4560 kg	7150 kg	5700 kg

Status

In service with major airlines and airport authorities

Manufacturer

Equisa, Madrid, Spain

SWEDEN

Satt Automatic Baggage Handling and Sorting Control System

Satt Electronics design and produce control systems for automated baggage handling and sorting. The system installed at Stockholm-Arlanda Airport consists of two sub-systems, one for charter traffic and the other for the scheduled flights. Each of these sub-systems comprises: a main conveyor, along which there is a series of baggage pockets (16 in one case and 32 in the other), two primary conveyors and a row of check-in points (22 and 23 respectively). There are three short conveyors at each of the check-in desks used for weighing and marking cases and for holding before feeding on to the primary conveyor. The primary conveyor passes down through the floor into the baggage-handling terminal where the items of baggage are transferred to the main conveyor. From this conveyor they are carried to the pockets where they are sorted by folding guides. Although there are two separate sub-systems to be controlled this is done by a single common mini-computer, a PDP 11/05. Two additional micro-computers are used as reserves and these can take over control if the main computer should fail, although at a lower level of automation.

A computer terminal at each check-in desk allows the operator to feed in the flight number and destination of the item of baggage to which a tag is attached. Provided that the code has been programmed into the computer's time-table, a clearance signal is sent to the check-in desk terminal and the item is moved to a waiting position at the primary conveyor. As a vacant zone passes the computer starts the 'waiting' conveyor and the baggage is transferred to the primary conveyor.

Both the primary and main conveyors are divided up into zones; the computer keeps a check on the baggage present on each of these zones by the use of pulse generators mounted on the drive shafts of the conveyors. In order to space the cases one of the primary conveyors passes cases on to alternate zones of the main conveyor, while the other feeds on to the remaining zones.

The baggage is sorted into the pockets according to the flight numbers, destinations and departure times that have been programmed into the time-table in the computer. A pocket is opened to receive baggage for a flight as soon as the preceding flight has departed. Baggage which has been checked in early is collected in a special pocket and is fed back into the system later at another position which also handles transit baggage.

Photocells are installed above the conveyors at crucial points, at transfer stations, before the baggage pockets and in the pockets, so that the computer can check the baggage situation and compare it with the anticipated situation. If there is a difference the computer tries to correct its information. Baggage for which the destination is uncertain is re-routed to a special pocket for unsorted items, the last one in the row.

The photocells in the baggage pockets are used to pass on cases to the last pocket if the correct pocket is already full and to check that the baggage has been carried down into the pocket. Alarm signals and fault messages occur if there is any disturbance in the flow and this draws the attention of the chief operator to any problems in the mechanical or electrical systems.

The speeds of the main and primary conveyors can be altered to suit the capacity required. This is done by speed adjustment of the thyristor-controlled dc motors. Speed can be varied between 20 and 100 per cent of the full speed, 1 metre per second. Feeding on to the primary conveyors and removal of baggage into the pockets is carried out by variable-speed belt conveyors. These are driven by motors which are supplied with current from thyristor-controlled converters at a frequency and voltage proportional to the speed of the main conveyor. It is essential for the primary conveyors always to be synchronised with the main conveyor, since the baggage must be transferred from a particular zone on the primary conveyor to a particular zone on the main conveyor. This is carried out by an electronic synchronising system which regulates the speed of the primary conveyors accurately to give a maximum deviation of ±5 cm from the speed of the main conveyor. CMOS circuits are used throughout to allow for reserve power supplies in case of power failure and so avoid loss of counter memories.

A modified RSX 11A is used in this system as a real-time monitor and a cassette driver is added. The functions of the monitor are to start an operation at a given time; to deal with priority setting; to allocate storage to the core memories available and to administer inputs and outputs. The monitor also controls a number of programme modules which carry out communication with and control of check-in desks; control and monitoring of baggage transport; maintenance of time-tables and associated operator communications; and test and servicing programmes.

The computer stores a daily time-table in its core memories comprising flight numbers, destinations, departure times and allocated baggage pockets. All daily programmes are stored on cassettes. The time-tables are updated by the chief operator through a display terminal. The baggage handling computer has been designed so that it can communicate with the information computer, to allow automatic updating of time-tables.

Manufacturer

Satt Electronics AB, Malmo, Sweden

Steelcons Safeway Type 2 Passenger Loading Bridge

The Safeway Type 2 is a fixed pedestal T-bridge designed for single door service of nose-in parked aircraft using a fixed centre-line. It provides a convenient, weather-protected and fire resistant passageway between the terminal building and the aircraft for embarking airline passengers. The bridge can accommodate all existing and known future jet aircraft from F-28 and DC-9 to wide body DC-10, L-1011, Airbus and B747 as well as Concorde. The bridge comprises three tunnel-like structures of welded-steel framework design, a fixed support column, a fixed portal (pedestal) and a hydraulic-lift cylinder. The fixed tunnel connects the terminal building with the hinged tunnel and in most cases serves as a bridge over the ramp service road. At its inner end it is mounted on the terminal building wall structure or on a separate support column close to the wall. The outer end is mounted on a fixed support column. An access door, a platform and a fixed staircase can be installed at the outer end of the fixed tunnel. This installation can be used as an emergency exit and for passengers transported by bus between aircraft and the terminal building. The hinged tunnel is connected to the fixed tunnel permitting the necessary vertical movement of the telescopic tunnel to interface with various aircraft-sill heights. The outer end of the hinged tunnel is connected to the telescopic tunnel by a hinge system, similar to the connection to the fixed tunnel. A crew door with platform and service staircase is installed at the outer end on the right side of the hinged tunnel.

The telescopic tunnel rides on rollers on a frame structure mounted on top of the hydraulic-lift cylinder. It is normally positioned at right angles to the hinged tunnel thus forming a T-shape. The outer end has a widened section, the bridge head, which is equipped with an electrically-operated roll-up door, a hood forming a weather-seal between aircraft and bridge head and a swivelling floor segment for optimum fitting with the aircraft fuselage contour and to compensate for aircraft misalignment in relation to the fixed centre-line. The weather-hood provides sealing between the bridge head and the aircraft by its own weight without springs or steel pressure rods. A soft

Steelcons Safeway loading bridge at Gothenberg-Landvetter International Airport

bumper runs along the edge of the bridge-head floor and is equipped with a micro-type sensor which stops the forward travel of the telescopic tunnel when the bumper touches the aircraft and ensures that the contact pressure does not exceed 0·1 kg per square metre. The telescopic tunnel can be equipped with proximity sensors as an optional feature. By means of an auto-levelling device installed on the right side of the bridge head, the bridge automatically follows the vertical movement of the aircraft caused by loading and unloading. The bridge-control panel is in the bridge head. A hydraulic cylinder installed under the floor provides the telescopic movement. The fixed support column, which has a rectangular cross-section of approximately 2 square metres, provides space at ground level for an equipment room containing the hydraulic package and main electrical system components.

The portal horizontally supports the hinged telescopic tunnel assembly by rails providing guidance to rollers mounted on a rectangular frame enclosing and linked to the hinged tunnel. Limit switches keep the vertical movement of the bridge within a specified height range. A height indicator scale, mounted on the portal, guides the bridge operator when pre-setting the bridge for the appropriate aircraft type. The elevating and telescopic movements of the bridge are powered by a hydraulic system. This is electrically operated and self-contained. System functions are controlled by solenoid-operated control valves which direct hydraulic flow to the elevating and the telescopic cylinders. The control valves are operated by a joy stick installed on the bridge-control panel.

Characteristics

MAX LOADINGS
personnel: 200 kg/m^2
snow: 100 kg/m^2
wind: 100 kg/m^2

TUNNEL DIMENSIONS
fixed: 12 × 1·8 m
hinged: 19·5 × 1·8 m
telescopic extension: 3·6 m

BRIDGE-HEAD FLOOR HEIGHT
max: 5·2 m
minimum: 1·6 m

AIRCRAFT MISALIGNMENT: ±5°

BRIDGE OPERATING SPEEDS
vertical movement: 1 m/minute
telescopic movement: 2 and 4 m/minute

POWER REQUIREMENTS
380 V ac, 3-phase, 50 Hz, 25 A
220 V ac, single phase, 50 Hz, 10 A

Status

Installed at Gothenberg-Landvetter International Airport, Sweden

Manufacturer

Steelcons AB, Trelleborg, Sweden

SWITZERLAND

Autophon Flight Information Display Systems

The Autophon flight information display system provides information for the general public and service personnel as required. The system comprises the following equipment: input device; display boards and/or monitors and a control unit.

The system proposed for airports provides for simultaneous information display both on flap boards and television monitors by means of a control unit accessible by an operator. The system is designed to offer good conditions of visibility from any position with information legible under extreme light conditions, either dim or intense.

Manufacturer

Autophon AG, Zurich, Switzerland

Beumer Baggage Handling and Sorting Systems

Beumer design, manufacture and install baggage handling and sorting systems to suit specific requirements. The baggage-sorting plant installed at Zurich Airport, Switzerland, has a sorting capacity of 7000 pieces an hour. Baggage check-in takes place in the departure hall two floors above the sorting system where 60 counters are grouped around four conveyor belts leading down below to four coding stations. A second set of four stations is connected to another two conveyor belts which deliver the transfer baggage from the respective unloading points for encoding. Here the operators read the flight number on the baggage tag and enter it into a keyboard. Then a computer takes over the control of baggage flow on to the sorting machines. The system is comprised of two machines, one for check-in and one for transfer baggage. The sorting is accomplished on the tilting-slat principle into 25 collection boxes lined up on either side of the machines. If necessary two flights can be assigned to each box simultaneously thus increasing the number of destinations to 50. Out of these 25 boxes, 16 are specially designed for mechanised container handling. A conventional race-track sorting device which is also part of the back-up system can accommodate another 12 sorting destinations.

On a track running between the two sorting machines a semi-automatic vehicle driven by four linear motors distributes empty containers from the container store to the container loading stations. The overall system is centrally controlled by a 16 K computer.

A baggage conveying system with a sorting capacity of more than 7200 pieces per hour was put into service in the extended apron hall V 3 at the Frankfurt Rhein/Main Airport, Federal Republic of Germany. This system consists of two sorting machines working independently of each other each with a quadruple coding section at the feed-in position. The baggage is fed to the coding sections by a belt-conveying system. At the coding sections the coder reads the flight number on the baggage tag and enters it into an alphanumeric keyboard, then a computer takes over the control of the baggage flow on to one of the two sorting machines. Sorting is made according to the tilting-slab principle into 70 collection boxes with each box connected with the sorting machines by a brake and timing belt.

Each box is equipped with a monitor on which the flight number, the apron position and the final sorting time are indicated. The control of the feed belts is performed by six programmable controllers working independently of each other. The sorting machines with the coding heads are operated by two independent machinery controls. The total system is centrally controlled by two 80 K PDP 11 computers, with automatic switch-over.

Status

Beumer has produced three airport baggage handling and sorting systems. Two for Swissair, at Zurich terminal B and Geneva were installed in 1975. The third, for Frankfurt Airport, Federal Republic of Germany, was supplied in 1980

Manufacturer

Beumer Maschinenfabrik KG, Beckum, Switzerland

Patek Philippe MIC Electronic Time Measurement Systems

The MIC is a modular integrated electronic time measurement system employing three master-time bases. Incorrect information transmitted from any of the three bases is rejected by the system. Input voltage is monitored and an alarm signal activated in the event of a power failure. An electronic memory repeats the polarity of the last pulse if a line is cut which prevents any variation developing between the secondary clocks. Other features include semi-automatic correction to summer or winter time, digital display of voltage and current levels and a multi-voltage capability. The T33 equipment developed for airport use operates with coded-time generators. Three BS 10 time bases are used each incorporating a monitoring mechanism. These are synchronised and simultaneously forward their information to LS-LM or LE modules which, when necessary, make a choice between the different information provided.

Characteristics

ACCURACY
10^{-6} = ±0·1 s/day, standard version
10^{-7} = ±0·01 s/day, with thermo-compensated oscillator
10^{-8} = ±0·001 s/day, with thermostatically-controlled oscillator
10^{-11} = ±0·0003 s/year, with atomic-frequency standard
absolute in the long term, with radio-synchronisation on one of the following transmitters:
HBG, Prangins (CH) 75 kHz
DCF 77, Mainflingen (D) 77·5 kHz
MSF, Rugby (GB) 60 kHz
OMA, Prague (CS) 50 kHz

Status

The MIC system is a development of the MI system introduced in 1969. It is in service at airports in Switzerland

Manufacturer

Patek Philippe, Geneva, Switzerland

Riwosa MD-15 Metal Detector

The MD-15 is a lightweight hand-held detector which indicates acoustically the presence of all metallic objects, ferrous or non-ferrous, through a loudspeaker or headset. The signal is variable in pitch and volume to enable the operator to estimate the size of the object. Power is provided by alkali batteries and the on/off switch includes an illuminated diode to indicate proper functioning of the battery. The unit is specially designed to generate a low magnetic field and so prevent any effect on the operation of heart pacemakers.

Characteristics

BATTERIES: 9V; up to 40 hours' operation
UNIT WEIGHT: 260 g

Manufacturer

Riwosa SA, Zurich, Switzerland

SSIH Omega Flight Information Display Systems

The Omega systems supply the public with visual and audible information on the state of arriving and departing flights.

In a typical situation information for large crowds is displayed by means of flap boards whereas television monitors supply the same information at a closer range. Announcements initiated by the system are transmitted through the terminal's public address system. The master clock generates the exact time which is distributed to the slave clocks throughout the terminal and the communication/meteorological building.

All controlling equipment is in the equipment and control room. Flap boards are in the main public areas such as departure lounges, at each of the gates and at each baggage conveyor on the public and the service side. Television monitors are installed in the VIP pavilion, the domestic departure passage, both international and domestic holding lounges and in the restaurant.

In the normal operating mode the computer will interrogate the operator at intervals of five minutes if certain operations are to be executed. For example the computer will ask 30 minutes before the scheduled departure time of a flight if it is boarding time. If the operator answers in the negative no further action will take place immediately but the question will be repeated five minutes later.

If the operator answers affirmatively the computer will perform all necessary operations to make this information available at the flap boards, the monitors and the automatic announcement device.

Interfaces between the computer and the peripherals modify the information into a format acceptable to the devices. These interfaces are the Omega video generator (Ovg 2), the video switching unit, the automatic announcement device interface, and the flap board control logic.

The television monitors are connected to the video switching unit via coaxial cables. The transmission standard is the normal composite video signal, 1 volt peak-to-peak. Each flap board is connected to the flap board control logic via two coaxial cables, each carrying information in one direction. The information currently displayed on any of the flap boards is available to the operator. This feature is called 'read back' and can be initiated manually or under independent computer control.

The software programs can be divided into two categories: the programs for flight and file management and the programs for the driving of the different display units. In the 'flight and file management' program 'dialogue' programs drive the communication between the system operator and the computer. Through these programs the system operator is able to inquire on file contents, updates flight status etc.

'Command stored flight' programs insert the scheduled flights. The parameters are: flight number; flight-type (departure, transit, domestic etc); scheduled time; destination/origin; via, flight running day; and date validity of the scheduled time.

Other programs for flight and file management are: 'stored flights' where all the scheduled flights are stored; 'command today flights' programs which insert the unscheduled flights (charter, special flights); a 'today flight' file which contains all the flights operating today; and transfer programs. The 'transfer' programs extract, according to the current date, flights from the 'stored flight' file and introduce them to the 'today flight' file. The 'command today flight' allows the updating

Omega flight information display system at Moscow Airport

Omega television monitor of flight information

of any 'stored flight' information and written remarks on the display units.
The other main category for driving the displays contains programs for update, transfer, flap display, video display and automatic announcement. The update programs are used to extract the ten first flights to be displayed from the 'today flight' file. This function can be executed automatically at regular intervals or at operator intervention. Transfer programs dispatch the data to the different display units and flap display programs transfer the data from the computer to the flap board control logic.
A software dictionary translates the names (destination, origins etc) to be displayed into their representation in ASCII code and data from English into Arabic, or any other language required. The video display programs transfer the data from the computer to the video generator (Ovg 2) and on to the monitors. Automatic announcement programs transmit the relevant data from the computer to the automatic announcement device.

Status
In operation at many international airports. Recent installations include: Algiers, Algeria; London Gatwick, England; Teheran, Iran; Tripoli, Libya; Kuching, Malaysia; Damascus, Syria and Moscow, USSR

Manufacturer
SSIH Equipment SA, Bienne, Switzerland

UNITED KINGDOM

AI Entry Scan Model 75 Explosives Detection Unit

The Entry Scan Model 75 is an automatic system for the screening of personnel as they are channelled individually, controlled by lights, past the unit's air sampling grille. The air which surrounds the body is extracted, from which an analysis is made for the presence of nitro-based explosives. In the event of an emergency a visual and audible alarm alerts the security guard.
A stand-by unit reduces the level of contamination, such as dust entering the detector by switching off the laminar air-flow curtain and vacuum-extraction system automatically after a preset period of time. The process, which is switched on automatically by passing personnel, takes 45 seconds before giving a ready indication. Manual control of the unit is by a remote hand-held unit which comprises a push-button to cancel alarm signals, a toggle switch which switches off the audible alarm and indicators for ready, not ready and system fault.

Entry Scan Model 75 explosives detection unit

Characteristics

RESPONSE TIME	5 s (single unit), 3 s (double unit)	
SENSITIVITY	greater than 1 × 10 in (254 mm) for EDGN-based explosives	
ARGON SUPPLY	internal replaceable argon-cylinder	
consumption	0·44 ft³/h (0·125 m³/h)	
DETECTOR	10 millicurie Ni 63 heated detector	
MAIN CONSOLE DIMENSIONS	Single unit	Double unit
height	1·73 m	1·73 m
width	0·7 m	1·6 m
depth	0·445 m	0·7 m
REQUIRED FLOOR AREA	1·2 × 0·7 m	1·6 × 0·7 m
STANDBY UNIT	0·69 × 0·34 × 0·12 m	
CONTROL UNIT	0·17 × 0·07 × 0·032 m	
POWER REQUIREMENTS	108–128 V, or 200–250 V, 50–60 Hz, 1·5 kW	

Manufacturer
AI Security, Pampisford, Cambridge, England

AI Triscan 200 and 125 Baggage Screening Equipment

The Triscan 200 is a high-security conveyor-type screening system. It incorporates a specially designed low dose X-ray with selected high and low operational voltages. The television system has a low light level camera and monitor. For close examinations electronic aim and zoom is incorporated. Baggage is moved via a reversible conveyor feed, which is adjustable in height, through the examination tunnel. Operator controls have been kept to a minimum to prevent confusion and misinterpretation.
The Triscan 125 provides the same screening qualities as the 200 but its operation is completely automatic and incorporates a single X-ray operational voltage with a fixed-size image.
The control unit for both systems comprises a television monitor and the relevant camera and X-ray controls which can be positioned on each side of the X-ray console. It can be revolved through 180 degrees to suit airport conditions and the height of the unit is adjustable to suit the operator. A support shelf is incorporated and when the X-ray equipment is not in permanent use the control-unit support shelf converts into a lockable cover for the television monitor and controls. When the equipment is not in constant use power to the camera and X-ray tube is reduced. This facility can be selected automatically after a period of five minutes if a bag has not been X-rayed. In the event of a suspect article being viewed the operator has the option to instigate a bag reject alarm. A signal is given by the reject bag cancel/alarm unit, audible and visual, sited near the manual search operator. The alarm will only occur when the suspect bag is leaving the X-ray tunnel. The alarm should be cancelled by the manual search operator, but if not the audible alarm will occur once more in the event of another suspect item being viewed and alarm instigated by the operator. If the alarm is not cancelled each subsequent bag will cause an alarm until action is taken.

Characteristics
MODEL: Triscan 200 C
X-RAY GENERATOR: 125 kV, 200 kV
RESOLUTION: 28 gauge (0·373 mm) insulated wire
ZOOM MAGNIFICATION: 5:1
MAX BAGGAGE SIZE: 0·65 × 0·5 m, unlimited length
CONVEYOR
speed: 9 m/minute
height: variable 0·64–0·81 m
INSPECTION RATE: zup to 1200 items per hour
DIMENSIONS
display unit: 0·78 × 0·63 × 1·15 m
CONVEYOR AND X-RAY UNIT: 2·72 × 0·92 × 2 m
weight: 600 kg
POWER REQUIREMENTS: 240 V ± 10%, 50 Hz ± 2%, single phase
MODEL: Triscan 125 C
As for Triscan 200 C with following variations:
X-RAY GENERATOR: 125 kV only
VIDEO DISPLAY: zoom and camera arm facility not included
CONVEYOR FEED: reversible feed not available

Status
The Triscan 125 was designed in co-operation with the British Airports Authority and is in use at airports in the United Kingdom

Manufacturer
AI Security, Pampisford, Cambridge, England

Triscan 125 baggage screening unit

Audix Flight Information and Passenger Announcement Systems

A comprehensive range of mixing/switching and amplification equipment is produced by Audix for public address systems, supplemented by microphones, horn-loudspeakers etc which are purchased from specialist companies. A typical flight information and passenger announcement system designed and installed by Audix is operating at the intercontinental terminal 3 at London Heathrow airport. This is an acoustically-matched 800 watt audio system which includes automatic noise sensing, announcement level adjustment devices, multi-access and routeing facilities, emergency back-up power supply and a range of loudspeakers including the Super Cardiod Golf Ball Speaker. The system is designed for expansion up to eight inputs and ten outputs with flexibility of access and priority.

The public address system for the British Airports Authority's northern satellite terminal building at London Gatwick Airport, features automatic noise sensing, limiting, solid state switching for zone selection and priority, and a bank of 120 watt power amplifiers. A standby battery with automatic charging is fitted and the system is provided with indications of mains power failure and low battery voltage.

Status

Audix has supplied flight information and passenger announcement systems to airports in Malaysia and South Africa; Abu Dhabi, Ras Al Khaimah and Sharjah airports in the United Arab Emirates and London Heathrow and Gatwick, Aberdeen, Edinburgh and Glasgow airports in the United Kingdom

Manufacturer

Audix Ltd, Stansted, Essex, England

Audix flight information and passenger announcement equipment at terminal 3 London Heathrow Airport

Avery Passenger Baggage Handling Units

Avery supplies baggage weighers for all types of terminal and baggage-handling arrangements. The standard range includes models which can be built into the counters and/or equipped with motorised conveyors to link up with the main baggage handling system. Free-standing weighers are also available.

All are designed to give a weight reading which can be seen by passenger, clerk and porter from their normal positions. Mechanical weighers indicate by optical projection, revolving chart or dial. The electronic machines register weights on a digital display which can be mounted in the desk or on the scale itself.

Suitable for integration into counters, the Avery 32N35 is a 136 kg capacity weigher giving optically projected readings through two windows, one for the passenger, the other for the clerk. It has a fully dead-beat action resistant to temperature changes.

The Avery 32S is a 128 kg capacity digital-electronic weigher which can be free standing or built into the counter with or without a motorised weight conveyor. It can be engineered to provide a dual digital display and to totalise individual baggage loads when required.

Other models include the 136 kg capacity Avery 32N88 and 100 kg capacity 33N21 (both indicating by revolving chart) and the bench-mounted 3303COB (100 kg) and floor-mounted 3205COE (250 kg) both dial-indicating scales with a graduated tare feature.

Status

Avery baggage weighing units are in use at London Heathrow Airport

Manufacturer

W and T Avery Ltd, Warley, West Midlands, England

Buko Baggage Trolleys

Buko manufacture two models suitable for conveying passenger baggage at airport terminals. The larger, heavy-duty version is an all-welded construction of rectangular section steel, mounted on four 125 mm diameter castors, two swivelling castors at the front, two fixed castors at the rear. Accessories include bumper protection on the front uprights, rear wheels and handle ends, fitted basket and a frame for an advertising/property panel.

The second model is constructed in tubular steel. Two fixed 200 mm diameter wheels are mounted at the rear with two swivel 125 mm castors at the front. Accessories comprise a fitted basket and space for an advertising/property panel.

Characteristics

MODEL	A	B
CAPACITY	300 kg	150 kg
PLATFORM LENGTH	0·78 m	0·65 m
STACKED UNITS		
number	10	10
length	3 m	3 m

Status

Buko trolleys are in service at Jeddah International Airport, Saudi Arabia

Manufacturer

Buko Wire Goods Ltd, Glenrothes, Fife, Scotland

Defiant Luggage Weighing Equipment

Initially developed from industrial weigh platforms this range of equipment is capable of being integrated into single or multi-channel baggage systems. The prime weight measurement is derived from a four-strain gauge load cell configuration which is an entirely static system having no moving parts and a minimal deflection together with a high shock and overload capability. Installation of the system does not require side guards. The instrumentation can vary slightly from one application to the next as the equipment is designed to be incorporated in the airport counter. The weight indicator is approved to the OIML standards and includes micro-electronic integrated circuitry. A dual display provides the weight reading for the passenger and the hostess. Outputs are available from the equipment to enable the system to be integrated into a computerised and price computing system.

Manufacturer
Defiant Weighing Ltd, Sevenoaks, Kent, England

Luggage weighing unit produced by Defiant Weighing

Digivision Flight Information Display Monitors

Video monitors produced by Digivision have been designed for use in flight information display systems.
In the system produced for Newcastle Airport, the monitors have a capability of displaying 80 characters per line over 24 lines. They are controlled by a System 800 disc-based micro-processor which stores all departure and arrival flight data for the airport. The data is accessed by operators at vdu screens which are linked to a character generator to provide relevant information on the monitors. The system offers wipe on/wipe off and scrolling facilities.

Manufacturer
Digivision Ltd, Leicester, England

Digivision video monitors in service at Newcastle Airport

Dunlop Starglide Passenger Conveyors

Starglide moving walkways are a continuously available means of transport for passengers and their luggage around many airports, for instance from car parks to check-in, along aircraft loading piers and as links from main terminal buildings to satellite terminals. Inclined walkways overcome the problem of movement between floor levels, as passengers may take child push-chairs, invalid wheel-chairs or baggage trolleys on to the walkway. The transition from inclined to horizontal movement is made on a smoothly curved, continuous surface. By using ribbed trolley wheels fixed automatic brakes can be incorporated. As the trolley-wheel ribs drop into the belt grooves this brake rests on the belt surface and prevents relative movement between belt and trolley.
The main feature of the walkways is the Starglide belt which is supported on rollers along each edge under the balustrade. The belt comprises longitudinal textile reinforcement sandwiched between two transverse steel cords. The top steel cord, on which is mounted a ribbed, rubber treadway, is under compression while the bottom cord is under tension.
Conveyors are available in 0·8- and 1-metre wide treadway units or with the extra width of 1·4 metres. The latter enables passengers in a hurry to pass standing passengers even when both have baggage trolleys.

The conveyors require an installation depth of 0·55 metres under the walkway, with the machine pits at either end 1·25 metres deep. Balustrades can be of toughened glass, panelled in stainless steel or with coloured Melamine finish.
Access for inspection of belt-support rollers and hand-rail tension adjustment is available from the treadway surface. The balustrade is designed so that cladding panels can be removed, individually. Access to machinery pits is via hinged and lifting floor plates. Pit lighting is provided built into the trusses which carry the machinery. The control cubicles, which are also built into the end trusses, incorporate a power supply for hand tools and a vulcanising press.

Dunlop Starglide passenger conveyors at London Gatwick Airport

Characteristics

TYPE	0·8 m wide	1 m wide	1·4 m wide
SPEED		PASSENGER CAPACITY PER HOUR	
0·5 m/s	6750	9000	11 250
0·6 m/s	8100	10 800	13 500
0·7 m/s	10 125	13 500	16 875

HORIZONTAL CONVEYORS
tread widths: 0·8, 1, 1·4 m
INCLINED CONVEYORS
tread widths: 0·8, 1, 1·4 m
max angle of inclination: 12°

Status
16 Starglide walkways are installed in United Kingdom airports and approximately 40 walkways in overseas airports

Manufacturer
Dunlop Transportation Systems Ltd, Hayes, Middlesex, England

Edghill Passenger Boarding Steps

Edghill produce a series of passenger boarding steps suitable for all types of commercial aircraft. The range includes six self-propelled stairs and seven towable versions.
The Multistair passenger boarding step is a self-propelled steel-box girder construction unit designed to service wide-body aircraft and most narrow-body aircraft with sill heights down to 3·2 metres. These include Boeing 707, BAe VC 10 and McDonnell Douglas DC 8 aircraft.
The Universal is a towable or self-propelled passenger step designed for use with all wide- and narrow-body aircraft down to a sill height of 2·44 metres. These include the Boeing 737.

Status
In service with many major airlines including British Airways, British Caledonian Airways and Singapore Airlines

Manufacturer
H.W. Edghill Equipment Ltd, Hook, Hampshire, England

Multistair boarding step in service at London Heathrow Airport

Fenamec Baggage Reclaim Conveyors

Fenamec has developed a stainless-steel clad baggage reclaim conveyor which accepts passenger baggage from a series of belt conveyors manually loaded from baggage trains. Spacing of baggage on the conveyors feeding the claim conveyor is controlled by a series of photo-electric cells.
Baggage handling systems using conveyors, baggage claim devices, chutes and weigh scales, designed, supplied and installed by Fenamec are operated at London Heathrow and other airports in the United Kingdom and in other countries including Denmark, Sweden, Cameroon, Taiwan, Kenya and the Seychelles.

Manufacturer
Fenamec Ltd, Hull, England

Fenamec baggage reclaim conveyor in operation at terminal 3, London Heathrow Airport

Ferranti Flight Information Display System

The system is designed to provide flight information to airport staff, operators and passengers. Typically the configuration uses three component sub-systems. These are: the Argus central processing system, the data distribution cable network, and the display unit.

The Argus central processing system is based on a standard multi-processor configuration which enables the processor load on the system to be distributed and balanced between the various tasks. The configuration can be duplicated and can either be fitted with automatic or manual changeover facilities depending upon the operator's requirements. Data to the central processor is inserted using vdus from a host cpu. All data inputs are recorded on a local logging printer.

The design of the system allows the data to be stored for two days' flight schedules, with up to 500 departures and arrivals catered for each day. The details of each flight are stored on a flight card which is held on disc and copied to the standby system for security purposes. Each flight card can hold a maximum of 512 bytes of data. The system is capable in its maximum configuration of storing 1000 information pages, the pages being generated from data held on the flight cards. Any change to data held on a particular flight card may affect up to 50 of these information pages and at its peak load the system will cater for a change to a flight card every four seconds. The information pages are updated automatically when any change occurs and other automatic features include roll on, roll off, flashing of changed data and alarms.

As current pages of information are generated, they are passed to one or more data output cards. These are configured to form part of the central processor unit. The maximum number that can be fitted into a central processor unit is dependent on the overall configuration of the system but three is a typical quantity. For security purposes a similar quantity is fitted to the standby system. Each data output card can be configured to hold up to 200 pages of information. The pages can be distributed via a single output port or alternatively can be switched to distribute up to 100 pages of data on each of two ports. Each page is formed using a basic 16 lines × 58 character format. Codes contained within the format can be decoded by a receiver unit to enable a single character size or combinations of the character sizes to be displayed. Pages of information are number 0–999 and pages can be held on the data output card in that range, with a total of up to 200 pages on any one data output card. Normally the first few pages on the output port will include an index defining the pages available on that network. One of the pages is designated as a broadcast page and up to 16 messages may be set up on this page. With each message the number of the pages which are to be broadcast is included with the facility to send to all pages on any sub-set. The broadcast messages are up to 58-characters long and appear on the second line on the selected page. There exists a facility to rotate certain selected pages in a fixed sequence. Each port of the data output card in use is connected to a separate cable network. For example, a configuration with three data output cards could connect to three cable networks each network being capable of supporting 200 pages of information. Alternatively the three data output cards could connect to six cable networks each network in this case being capable of supporting 100 pages of information. It is not mandatory that all data output cards are configured in the same way.

A single cable network usually consists of a main trunk cable. This uses heavy-duty, ½ inch (1·27 cm) co-axial cable and may be up to 2000 metres long. Spurs can connect to this main trunk using tap units. A maximum of ten tap units can be supported by the main trunk cable. Each tap can support three spur units each of which may be up to 500 metres long. A single spur may contain a further ten devices; these may again be tap units (a maximum of eight) or receiver units. Each tap unit inserted in a spur cable can support three secondary spur cables up to 500 metres long capable of supporting ten receiver units.

The display unit consists of a commercially-available monitor unit and a receiver unit. The receiver unit is designed to interface the monitor unit to the cable network.

A receiver unit is fitted with a page-selection device which allows the user to select a page number circulating on the associated cable network. All the pages held on a data output card are continually circulating around the associated network, taking approximately four seconds for the 200 pages to complete one cycle.

The receiver unit uses a micro-processor and incorporates a number of features. For example if a page selected is not available on that network, the micro-processor is sent the message page not available on the monitor display. A keylock is fitted to each receiver unit which permits certain page numbers to be non-selectable from the keyboard. This enables privileged information to be accessed only by authorised personnel. This facility can be enhanced if used in conjunction with the network/page allocation. In the event of a broadcast message being put out to the network all receiver units will display the information contained in that message on the second line of the monitor. Only action from the central system can clear this message. Once a page number has been selected the receiver unit will continue to monitor the page information on the cable network. Should changes occur in the content of a page, these will be reflected in the displayed information during the next cycle of that page on the network. For staff purposes the receiver unit can be mounted directly onto the side of the monitor unit. A monitor unit with a 9-inch (22·9-cm) diagonal tube is usually suitable for this purpose. Where the system is used to provide data for public display the receiver unit can be supplied as a free standing unit, driving a monitor at a distance of up to 25 metres. The video output of the first monitor is usually capable of supporting other monitor units, all of which will display the same information. A 23 inch (58·4 cm) monitor is used in public information areas. Both staff and passenger information displays may exist on the same cable network. If the appropriate pages for public information are selected initially, these will be updated automatically and controlled by the computer.

The central-processor system can support remote stations using a small Argus processor system at the site connected to the central-processor system using conventional connections. Data generated at the central processor is copied on to a dummy data output card and transmitted across the communication link to the remote processor. The data is transferred to a second data output card in the remote processor which is used to transmit pages of information around the remote site by the same method employed at the main site. Only data updates are passed over the communication link as they occur. This enables a relatively slow link, usually 2400 bps, to maintain a relatively high rate of remote system updates.

It is possible to operate a number of satellites from a central processor, the number being dependent on the overall system configuration.

Status

Ferranti has installed a flight information system for British Airways at London Heathrow Airport

Manufacturer

Ferranti Computer Systems Ltd, Wythenshawe, Manchester, England

Ferranti Passenger Handling Data Systems

Passenger handling data systems developed by Ferranti using the PT7 terminal equipment include the departure control and check-in system and an airline seat reservation system.

The departure control and check-in terminal system forms part of an overall system and operates in conjunction with the appropriate application's software on a host cpu. A suitable host cpu can support a number of departure control systems locally or at other airports. The connection between the host cpu and each departure control system uses conventional data circuits operating at up to 2400 bps.

Passengers are checked in on arrival at the airport desk by staff using vdus with keyboard units operating in a variety of modes adapted to the different requirements of different airlines on long- and short-haul flights. A security check can be carried out if required at the boarding gate using a hard copy printer unit. This enables boarding passengers to be checked and helps identify any who are missing. As a user option the departure control terminal computers can also support boarding-pass printers.

The host computer keeps a running tally of all passengers and baggage checked in and of all other loads, fuel, cargo and meals, entered at other input terminals as it is accepted. It assists also in planning the distribution of load in the aircraft to ensure suitable trim settings. When the load has been finally accepted the system produces the necessary load and balance documents and despatches any load and passenger signals required.

Both check-in and load control are undertaken by the host computer using twin Argus 700F controllers. Linked over a dual-communications circuit to the main-frame, the twin controllers are in continuous and simultaneous operation with visual displays, printers and boarding-pass printers, shared between them. In the event of a failure of one controller all its terminals are immediately and automatically switched to the working controller.

A failure message is sent automatically to the host computer to allow maintenance organisations to be alerted to take necessary repair action. The return of the failed controller on-line is announced, enabling a log of all malfunctions to be kept. By addressing the relevant device the host computer can communicate with specific displays or printer terminals independent of the status of their principal controller.

PT7 terminal systems are used on the British Airways reservation system, BABS. PT7 systems are operated where large clusters of vdus and printers are necessary and each controller may have up to 60 devices. These clusters connect to the British Airways host main frame computers (dual Amdahl V7s) at London Heathrow Airport using conventional synchronous circuits operating at 2400 bps, using as a basis Sabre 1006 line procedures.

The IBM International Programmed Airlines Reservations System software suite, a British Airways project, modified and upgraded by them to its present level provides a comprehensive range of facilities which include seat reservations, car rental and hotel booking.

Status

The host software of the departure control system has been developed by British Airways for use on IBM compatible host main-frames and has been installed at Manchester and Hong Kong International airports. Depending upon the final network required it may be possible to connect a number of Argus 700F departure control terminal systems to a central host main-frame.

In the BABS system British Airways are operating 57 Argus PT7 cluster controllers and more than 2000 Ferranti vdus and printers forming a significant part of the overall terminal reservation network. Each PT7 cluster controller connected to the network has been configured to control approximately 44 vdus and four printers

Manufacturer

Ferranti Computer Systems Ltd, Wythenshawe, Manchester, England

Ferranti Public Address Systems

Ferranti Offshore Systems Limited are specialists in the production of public address systems.
A typical system supplied can include the following features: public address, message annunciation, passenger information incorporating pre-recorded message stores and access points as required; visual displays ranging from simple lamp arrays to comprehensive vdu formats; supply of the full range of loudspeakers distributed as determined by an acoustic survey.

Manufacturer

Ferranti Offshore Systems Ltd, Aberdeen, Scotland

GEC Baggage Conveyors

The unit handling division of GEC Mechanical Handling Ltd specialises in the design, supply and installation of conveying systems for mechanical handling and sorting purposes.
The division provides complete baggage handling systems which, although custom-built for specific airports, incorporate standardised components wherever possible for ease of maintenance.
Equipment includes weigh, waiting, collector belt and baggage reclaim conveyors of various types including the Safeway Racetrack, tray- or belt sorting-conveyors, accumulation and gravity roller conveyors, check-in desks, spiral or straight chutes, vertical elevators and supporting structures and complete electrical control systems.

Status

Baggage handling systems have been installed by GEC Mechanical Handling Ltd in several major airports including Mehrabad International Airport, Teheran, Iran

Manufacturer

GEC Mechanical Handling Ltd, Melksham, Wiltshire, England

GEC baggage conveyor at Mehrabad Airport, Teheran

Glasdon Olympic Modular Building System

The Olympic modular building system, constructed in grp, is designed to meet standard and specialised building requirements within the transport and travel industry. The buildings are composed from a range of 59 panels supported by a variety of additional features. Five unit widths of the buildings are available from 1 to 5·4 metres. Examples of current airport applications include security posts, car park kiosks, ticket and reservation points, information booths, souvenir stalls, control posts and telephone kiosks at airports in the United Kingdom including Aberdeen, Edinburgh and London Heathrow airports.

Manufacturer

Glasdon, Modular Buildings Division, Blackpool, Lancashire England

Glasdon Olympic Design Centre Shop at London Heathrow Airport

IAL Rapidex Passenger and Baggage Screening Systems

IAL market a range of passenger and baggage screening systems.
The IAL Rapiscan baggage-screening system uses a linear X-ray sensor to view the inspected item in slices as it moves through an extremely narrow X-ray beam. These slices are converted directly from X-ray to video signal through solid-state circuitry stored in a digital memory unit and displayed on a television monitor. This system results in a high-resolution X-ray image with high contrast. It is claimed wire as fine as 34 gauge can be discerned.
The Rapidex 100 is a pulsed X-ray screening unit incorporating a conveyor system, capable of processing up to 1100 bags per hour. The main features of the system are sequential examination of long items of luggage; sensors to ensure that corners of tall objects are screened; variable pictures display time and variable intensity X-rays. The system is equipped with a fail-safe device which shuts down the machine automatically in the event of radiation being emitted for more than the preset period.
The Rapidex 3000 metal-detection gateway is the latest edition to the Rapidex range of gateways. Incorporating a proximity metal overload arm, digital-display counter and tamper-proof sensitivity level control, the model 3000 gateway will detect the smallest quantity of both ferrous and non-ferrous metals. When the field is broken by any type of metal object the red alarm light is immediately activated on the control panel. Units may be installed side-by-side without interference to control systems. Performance levels are unaffected by temperature changes and there is no adverse influence on heart pacemakers. Operation is fully automatic.
For personal searches IAL markets the lightweight MD 160 and MD 199 metal detectors. Battery-powered, each unit incorporates variable sensitivity controls and encapsulated maintenance-free electronics and will detect both ferrous and non-ferrous metals.

Status

Rapidex passenger and baggage screening systems are currently in use at 16 airports in the United Kingdom and in over 100 airports world-wide

Manufacturer

IAL Security Systems, Southall, Middlesex, England

IAL Rapidex 100 X-ray screening unit

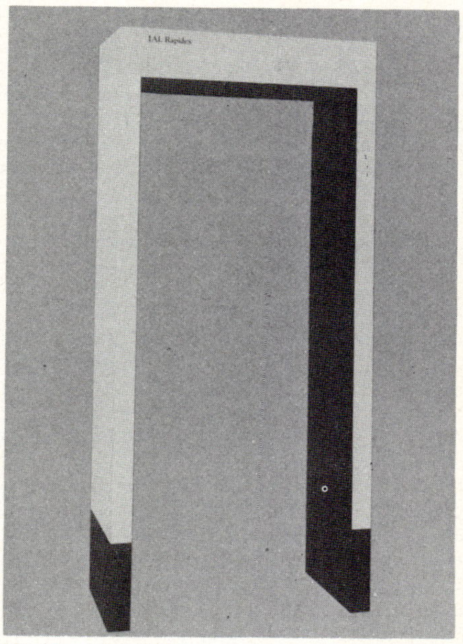
IAL Rapidex 3000 metal-detection gateway

IBM Passenger Handling Data Systems

IBM 3650 point-of-sale terminal in operation at Brussels Airport

IBM equipment is used in many major airports to control various functions of passenger handling. These include flight information display systems, reservation equipment, duty-free shopping and airport planning.

At Rio International Airport, two IBM System/370 Model 145 central processors, plus a range of IBM peripheral equipment, form the data-processing nucleus for the airport's operations control centre. One major function of the computers is to distribute and update passenger information for display on the numerous flap boards as well as various internal and public video-display screens. Basic information displayed on the boards combined with other relevant details such as load information, aircraft type and provisioning data is used by most of the organisations and groups working within the airport.

An example of IBM reservation equipment is the computerised passenger reservation system for Aerolineas Argentinas which is handled by one of the three IBM 3031 processors installed at the airline's Buenos Aires base. Aerolineas branch offices in some 20 cities are on-line to the airline's computer offering immediate booking and confirmation to passengers. Aerolineas agents in seven Argentine cities are linked to the system to provide them with the same capability, their terminals being linked to the system via IBM 8100 Information Systems which act as satellites to the main computer. Reservations software used is the IBM International Programmed Airlines Reservations System.

Brussels and Rome airports use the IBM 3650 point-of-sale system to record transactions in their extensive duty-free shopping areas.

These systems, capturing sales data at the point of sale, can automatically update stock levels, generate receipts and convert from local currency to any other that the passenger may request. The assistant may enter an item's price and details without keying, operating a wand, reading a preprinted bar code on the item which allows immediate data entry.

An IBM System/370 computer installation was used in planning Jeddah's King Abdulaziz International Airport which is expected to be handling over four million scheduled passenger boardings by 1985.

One function of the computer has been to provide information on the likely impact of over a million pilgrims on airport services during the annual Hadj period.

Manufacturer

International Airlines Support Centre, IBM (UK) Ltd, Feltham, Middlesex, England

Jasmin Flight Information Display System

The Jasmin large-format message display system has been developed to display airport flight information.

Special character founts suitable for most of the world's languages are available to special order.

A special wipe feature allows the display of more information than could otherwise be put on the screen at any one time. When the wipe action is in operation the displayed message, or part of it, changes every few seconds to display some additional information and then reverts to the original message. In this way information concerning, for example, the delay of an aircraft or its expected time of arrival can be superimposed on the basic flight information without need to reduce the number of flights being displayed on the screen.

The display system includes a facility for generating grey lines or areas on the screen. This feature is used to group message data together within a border and so avoid any confusion between single- and multi-line messages.

The character generator is designed to accept data from a special purpose micro-processor controlled visual display terminal. Alternatively the unit can be coupled via a serial link to a data processing system and the data may then be input automatically as a computer generated message.

Status

Jasmin displays are installed at London Heathrow Airport, England

Manufacturer

Jasmin Electronics Ltd, Leicester, England

Leyland Airport Buses

Two types of buses are available from Leyland with right- or left-hand drive, suitable for transporting passengers from the terminal building to aircraft on the apron. A third, the Cub, is ideal for VIP or crew bus applications.

The Artic is a 59-foot (18-metre) long single-deck articulated bus with a Leyland DAB chassis with a body comprising a modified Leyland National front end married to the modular sections used in 38-foot (11·6-metres) and 34-foot 9-inches (10·6-metres) long National buses. The vehicle can be built with front entrance and centre and rear exit/entrances on both sides, and supplied with Sundym windows and a public address system. The seating, whether fully sitting, all standing or a combination, can be built to suit individual operator's requirements. Driven by a Leyland turbocharged engine, the Artic bus is fitted with a ZF 500 torque converter transmission with foot-operated hydrodynamic retarder. Power steering, dual-circuit air brakes and air suspension are included in the standard specification.

The National 2 is an integral construction, rear-engined single-deck bus with semi- or fully-automatic transmission and air suspension. Two sizes are available. Front doorways and centre exit/entrances can be fitted on both sides. The internal trim can be for fully seated, all standing or a combination. Accessories include luggage pen, twin roof-mounted heaters and a public address system.

The Cub has a small psv chassis suitable for use as a VIP transport or a crew bus. Equipped with a forward mounted Leyland six-cylinder diesel engine and manual or automatic transmission, the Cub is available with an 11 foot 2 inch (3·4 metre), 12 foot 10 inch (3·9 metre) or 14 foot 5 inch (4·4 metre) wheelbase. The internal trim can be fitted to suit operator's requirements and the vehicle can be supplied with right- or left-hand drive.

Leyland Artic bus

Characteristics

MODEL	Artic	National 2
ENGINE	Leyland 690	Leyland 680
OUTPUT	157 kW at 2000 rpm	127 kW at 2000 rpm
TYPICAL CAPACITY	151 people (27 seated, 124 standing)	98 people (33 seated, 65 standing)
OVERALL LENGTH	59 ft (18 m)	38 ft (11·6 m) or 34 ft 9 in (10·6 m)

Status

Seven Artic buses are in service with British Airways at London Heathrow Airport, while Manchester International Airport selected the National 2 bus for their airfield requirements. Eight of these buses have been supplied

Manufacturer

Passenger Vehicle Division, Leyland Vehicles Ltd, Preston, Lancashire, England

Leyland National 2 airfield bus in service at Manchester International Airport

Leyland Cub with Vanguard body by Wadham Stringer

Middlehurst Mk 3 Baggage Trolley

The Mk 3 baggage trolley is an all-welded unit constructed in square-section steel tube with a nylon finish and mounted on four Flexello 125 mm diameter swivelling wheels. A simple hand-operated brake, fitted basket and advertising/property panel are incorporated in the design. Pvc buffers on polyurethane mouldings can be supplied and fitted to the front uprights and rear baggage retainer loop.

Characteristics

CONSTRUCTION: 35 × 16 mm and 25 × 25 mm square section steel tube
CAPACITY: 300 kg
LUGGAGE PLATFORM
length: 0·78 m
front width: 0·33 m
rear width: 0·52 m
TROLLEY DIMENSIONS: 0·895 × 0·52 × 1 m
STACKED DIMENSIONS
no of units: 10
length: 2·7 m

Status

The Mk 3 version was designed by the British Airports Authority for use at airports in the United Kingdom

Manufacturer

Middlehurst Ltd, Maidenhead, Berkshire, England

Rediffusion Flight Information Display Systems

Rediffusion produce a standard flight information display package controlled by a Data General MP200 computer with a dual 8-inch (20·32-cm) disc and dual-line driver. The system uses a Megadata four-channel character generator and the package is complete with interfaces to connect all peripheral and terminal equipment, a mains distribution panel, blower unit and ancillary fittings. The equipment is housed in two 6 foot (1·83 metre) equipment racks. The distribution package is contained in a single equipment rack of similar dimensions. It comprises a video/data switcher; four hf modulators (MO 105); four hf amplifiers (AR 121 J); mounting frame (MF 104); distribution panel (CP 104); mains distribution unit; video/hf monitor and ancillary fittings. Peripheral equipment, included in the standard package, which provides operator and maintenance facilities, consists of two Dasher vdu/keyboard units and two ksr printer terminals.

Manufacturer

Rediffusion Industrial Services Ltd, Surbiton, Surrey, England

Rediffusion Velward Clock Systems

Velward clock systems are suitable for public displays at airport terminals. A wide range of standard equipment is available. Basic to all Velward systems is an accurate quartz time source generating real-time data in coded form. The encoded data is refreshed continuously throughout the system and decoded at each digital clock to show the correct time updated simultaneously at each location. A feature of the distribution is that loss of local power or a break in cabling will not affect the accuracy of the displayed time once power is restored or the line fault is corrected. When the line fault is corrected each clock in the system will immediately display the correct time as set by the central-master source without any manual intervention or re-setting process. Systems can be supplied ranging from simple master/slave configurations to complex arrangements using high accuracy master units with automatic control of line monitoring, changeover, standard-time change and other facilities, driving a variety of time displays, recording systems and local submaster clocks.

Status

A Velward clock system is installed at Sanaa International Airport, Yemen Arab Republic

Manufacturer

Rediffusion Industrial Services Ltd, Surbiton, Surrey, England

Reditronics Public Address System

Public address systems at airports require automatic volume control, whose instantaneous level may rise and fall, depending on the ambient noise level at any given time during the announcement. The CU 107 is a variable attenuator unit, the attenuation being determined by the ambient noise level in the area covered. The unit differentiates between the signal and the ambient noise, and reacts only to the background noise. The system gain increases directly in accordance with the level of ambient noise.

The system gain can be made directly proportional to the ambient noise level subject to the setting of the preset switches for maximum and minimum. The unit's design allows adjustment of response time to optimise performance to suit a particular environment (reverberation time, noise weighting etc). It incorporates also a preset-level switch for emergency announcements, which over-rides the sensing circuit, so that the announcement may be as loud as desired.

Reditronics' range of sound communications equipment includes a choice of microphones, amplifiers, loudspeakers, volume controls and other specialised items. Any of these can be specified for an integrated Reditronics' sound communications system tailored to a specific requirement, including the provision of public address and paging facilities, airport information and background music services. Included in this range are two public-recorded commentary systems: the Listening Post and the Talking Post. Both provide pre-recorded information in public places such as airport public-viewing galleries. The Listening Post has plug-in ear-pieces which can be hired or sold. The Talking Post has loud-speaking commentaries. Both can be operated by push-button or coin-operated by an add-on coin box.

Characteristics

MODEL: CU 107
APPLICATION: automatic noise sending audio control
DIMENSIONS
width: 0·483 m
depth: 0·225 m
height: 0·088 m
WEIGHT: 4·5 kg

Manufacturer

Rediffusion Reditronics Ltd, Jersey, Channel Islands

Sovex Marshall Baggage Handling and Sorting Systems

Sovex Marshall's product range includes 50 automatic systems capable of handling baggage from check-in through to the airside container, and back again to baggage reclaim.

The tilted belt conveyor provides a facility for the automatic routeing of baggage and packages from check-in to individual assembly points for outgoing flights. The system comprises a belt tilted at an angle of up to 45 degrees with belt spurs leading away from the lower edge and connecting up with other conveyors. Items travel along the tilted belt's lower edge and are discharged automatically at the appropriate spur. Capacity can vary according to the number of discharge points though 1800 items per hour would be typical.

The Carousel 70s baggage reclaim unit is designed to cope with the baggage content of high passenger capacity aircraft. It features a rotating carrying surface sloping down towards the outer perimeter with the angle of slope varying from 20 degrees on the straight sections to 45 degrees on the bends. The units can be virtually any shape but are normally elongated to save floor space and to allow size increases at minimum cost and inconvenience. Baggage slides down to the outer edge where it can be easily reached.

Race-track conveyors incorporate a coated metal, continuous carrying surface and can be assembled to form baggage reclaim units offering flexibility in layout. The baggage handling installations can maintain luggage in the attitude in which it is placed on the unit. Height, width and speed can all be suited to the needs of a particular situation.

Baggage reclaim turntables designed specifically for the smaller passenger terminal buildings, provide an economical and efficient means of displaying baggage for passenger reclaim. Their overall diameter and peripheral speed can be adjusted to meet specific requirements, as can their height and width.

The baggage sorting system designed by Sovex Marshall for terminal 3 at London Heathrow Airport consists of two seven-way sorting systems each capable of handling 2000 units on a continuous basis with a maximum of 2400 units for ten minute periods.

The sequence for each of the two systems which replace airside conveying systems is as follows:

baggage is routed from the departure hall on two parallel high-level conveyors (preceded by a low-level section which subsequently ascends to the required height) to queueing conveyors which lead directly to a coding and control cubicle equipped with two coding positions, one for each line. A bag held at the coding station initiates sequential operation of the queueing conveyors to control baggage flow. Should the whole queueing section become full, a progressive shut-down of all upstream conveyors takes place. A cross-over conveyor section positioned at the coding station provides inter-connection between the two parallel conveyors. This accepts baggage from an early-arrivals store and feeds it into the system for coding, the starting and direction of this section being activated by a coding station operator. Interlocks are provided to ensure that the conveyor feed from the store cannot be started unless the cross-over section is also running. The T junctions of the section with the two main lines are controlled by photo-electric cell units on an arrival-time basis with an over-ride facility provided for use by the operator as appropriate. These junctions are supplied with powered diverters by which the operators can direct all baggage to either of the coding positions.

Photo-electric cell units are arranged at the coding positions to hold the flow of baggage when a destination code has not been entered for the bag, the preceding bag has insufficient headway to permit accurate sorting or if the code entered is for a loading spur which is full or not in use. The local control panel provides for the direction of baggage to any one of seven loading spurs or back to the early-arrivals store.

Two high-level sorting conveyors lead away from the coding area. These are supplied with a series of powered diverters which deflect baggage along selected descending spurs leading to the baggage accumulators in between the main conveyors lines but at ground level. Should a spur and accumulator become full this is indicated at the control console via a photo-cell and it is not possible to code further items. A solid state, programmable controller is used for sorting and all other control functions.

Sovex Marshall baggage claim units

Characteristics

AVERAGE BAG: 15 kg
MAX WEIGHT: 35 kg
MAX DIMENSIONS: 91 × 74 × 35 cm
HEADROOM OF SYSTEM AIRSIDE CONVEYORS: 2·3 m except baggage accumulators

Status

Sovex Marshall has installed baggage handling systems at London Heathrow Airport, England

Manufacturer

Sovex Marshall, Carlton, Nottingham, England

Shelter-Span Transfer Baggage Hall

A transfer baggage hall developed by Shelter-Span has been constructed at terminal 3, London Heathrow Airport. It comprises a standard Shelter-Span structural frame of aluminium beam extrusion. A single skin of white pvc-coated polyester forms the roof cladding. The lower exterior wall is covered with plastic-coated profiled steel sheeting with an inner hard wall of plastic laminate chipboard. The building is unheated, natural ventilation being provided by the roof apex opening. It is secured by chemical resin anchor bolts direct to the existing concrete apron. A guttering system has been fitted at the eaves and a crash barrier and guard-rail system is incorporated both inside and outside the building. Ten pairs of flexible rubber doors provide access for baggage vehicles.

Characteristics

DIMENSIONS: 525 × 18 m
VERTICAL WALL HEIGHT: 3 m
APEX HEIGHT: 6·6 m
FLOR AREA: 945 m^2
ROOF SLOPE: 1:10

Shelter-Span transfer baggage hall, terminal 3, London Heathrow Airport

Status

Installed in May 1980 the building has an intended life of five years. It was procured by British Airways on behalf of the Heathrow Airport Airline Operators' Committee baggage consortium comprising Pan Am, TWA, Air Canada, Aer Lingus and other operators

Manufacturer

Shelter-Span (UK) Ltd, Maidenhead, Berkshire, England

Universal Scholtz Baggage Conveyor

The Universal-Scholtz powered-belt bend conveyor, available in any angle up to 180 degrees, is suited to the handling of luggage at airport terminals. Its principal feature is the heavy-duty construction of the belt which is built up from segments designed to ensure that the tensile direction remains in the direction of the warp of the fabric so that the belt will not distort or wrinkle. Spring steel members are embedded within the belt to stabilise it in the radial direction. The carrying strand slides on a flat bed and the return strand is carried on ball-bearing rollers. Belt tracking is by spring-loaded guide rollers bearing on the solid rubber edge of the belt with drums of taper design mounted on bright steel shafts running in ball bearing plummer blocks. The conveyor is powered by geared electric motor, which can be reversible if required, either direct coupled or through roller chain dependent on layout. Side guides are provided which cover the guide rollers, and guards are fitted to the drums and exposed guides. A range of belt widths are produced and the belt is available to operate in either or reverse direction.

Status

Manufactured in the United Kingdom under licence from Scholtz-EFS GmbH

Manufacturer

Universal Conveyor Co Ltd, Leicester, England

Characteristics

CURVE BELT DIMENSIONS

inner radius (mm)	belt width (mm)
875	480
875	580
875	680
875	880
1000	750
1000	900
1000	1050
1050	700
1150	1100
1150	1300
1200	800
1200	1475

INDIVIDUAL LOAD CAPACITY: 56 kg

Vaughan Flight Information Display System Software

Vaughan is a major supplier of flight information display system software to the British Airports Authority.

The British Airports Information Retrieval (BAIR) system using over 20 4M micro-processor sub-systems provides flight information for all existing London Heathrow Airport terminals and extensive facilities at central control to aid the allocation of stands and to record all flight details. The system runs from a stored flight schedule that is controlled centrally.

The central system holds the flight schedule for up to 12 months ahead, permitting the forecast schedule for any day to be printed. A day's flights are updated with current information on delays, arrivals, departures, etc. All this information is distributed to the terminal public display systems and two staff systems. The latter use the teletext system. One system permits 400 pages, which include a continuously updated schedule, to be distributed to just under 200 subscribers throughout the airport. The average time to select a new page is less than one second. The other system provides more detailed information to the operators including a stand chart in colour.

The central system is duplicated with two machines on different sites. The terminal systems provided by Vaughan use a combination of flap boards and television screens from a variety of suppliers to provide a full public display service showing arrivals concourse, baggage reclaim and summary, departures concourse and lounge, and pier gate and routeing displays.

The terminal software is a single programme package which is set up for a particular terminal's arrangement of displays by the use of data base management techniques.

Status

The company installed the Heathrow terminal 3 system in 1971, the Gatwick arrivals system in 1973, the Gatwick departures system in 1976 and the Heathrow BAIR system in 1980

Manufacturer

Vaughan Systems and Programming Ltd, Ware, Hertfordshire, England

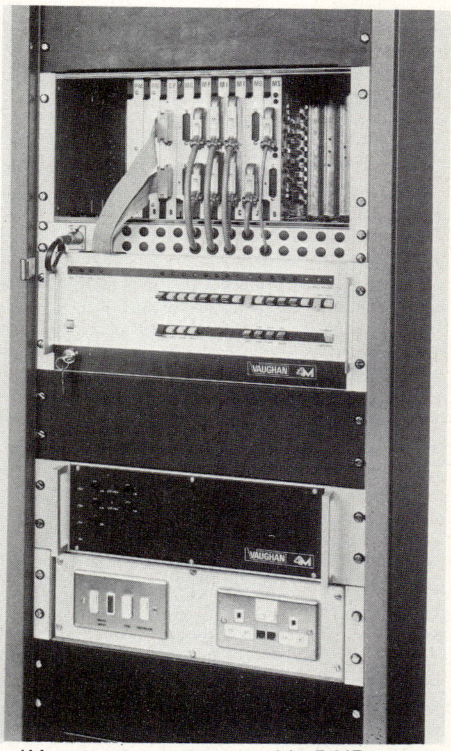

4M computer system used in BAIR system

Television displays used in Vaughan flight information system at London Gatwick Airport

UNITED STATES OF AMERICA

Abex Jetway Passenger Loading Bridges

A range of five types of passenger loading bridges are offered by Jetway. Two are fixed-position bridges, the others are able to move towards and away from the parked aircraft on the apron. In the former, the aircraft can use its own power to arrive at the docking position, but a common centre-line is necessary for all types of aircraft for each gate so apron space may not be fully used. A tug is required for push-out operations. In the Apron Drive, Full Service Radial (FSR) and the GL-1 ground level bridge, aircraft may use their own power to enter and leave the docking area.

The Pedestal is a fixed-position loading bridge which comprises three tunnels, lift column, service door and stairs, cab and aircraft closure. Tunnel A is adjacent the terminal and provides a level transition from the terminal to the bridge. It is supported by a large column which is designed to prevent load or vibration transfer from bridge to terminal. The telescoping tunnels, B and C, are rectangular in cross-section, with the tunnel having the smallest cross-section closest to the aircraft. Pedestal-loading bridges are available in either short or long versions, and with single- or twin-lift columns of varying ranges of vertical travel. This allows accommodation of all types of commercial jet aircraft.

The aircraft cab is able to rotate laterally through 30 degrees, 15 degrees either side of the centre-line so that it can be aligned for contact with aircraft at various aircraft parking arrangements. It is designed to allow completion of the docking manoeuver from a fully-enclosed operator station. The cab is rotated by a linear activator through a ball-screw assembly. An override clutch and physical stops control the extremes of rotation. It is equipped with a forward-facing control console behind a wire-glass window. Additional visibility is obtained through the windows to the left and right sides of the control console. A manually-operated roll-up weather door is at the cab front adjacent to the control console to seal the bridge when not in use. A full-width bumper is positioned at the aircraft end of the cab floor.

The outer end of the cab is fitted with a folding accordian bellows-type aircraft closure. Each side of the aircraft closure is independently adjustable to conform to the aircraft contours.

Pressure-sensitive limit switches are incorporated into each side of the closure mechanism to prevent excessive pressure on the aircraft. A service door, landing and stair provide apron access to and from the bridge for authorised personnel.

The passenger loading bridge is equipped with an automatic levelling system which allows it to automatically follow small changes in aircraft door-sill elevation. The auto-level circuit includes a safety timer which permits the vertical travel motor to operate continuously for a maximum of six seconds (adjustable), at which time it cuts all power to this system and energises an audible alarm and warning light at the control station. The control station incorporates a key switch, switches for extending and retracting the telescopic section of the bridge, raising and lowering, cab rotation, left or right, adjustment of the bellows aircraft cover, and operation of the floodlights and the auto-level wheel arm. In addition to the controls there are indicators and lights to show correct operation of the bridge.

The Elevating bridge consists of the following components in order progressing from the terminal: tunnel A, lift column, tunnel C, service door and stairs, cab and aircraft closure. Tunnel A is a variable-slope tunnel adjacent to the terminal and is the variable transition in slope and length, between the terminal and aircraft elevations. The length of tunnel A is determined by terminal floor height and aircraft to be serviced. Tunnel C is a sliding tunnel, rectangular in cross section, always horizontal, and constructed to support design loads in both fully-extended and retracted positions. The connection between the horizontal tunnel C and variable tunnel A is accomplished by an assembly of rolled curtain slats. Other features are similar to the Pedestal version.

The Apron Drive comprises the following components in order progressing from the terminal: rotunda and corridor, tunnel A, tunnel B, tunnel C, drive column, cab bubble, service door and stairs, cab and aircraft closure. The rotunda and rotunda corridor are adjacent to the terminal and are supported by a large column. As the main pivot for the entire bridge, the rotunda allows the bridge to swing a total of 175 degrees, 87½ degrees either side of the centre-line. Dual limit switches are provided to prevent overtravel.

Jetway Elevating passenger loading bridge

Jetway Radial FSR passenger loading bridge

The telescoping tunnels, A, B and C are rectangular in cross-section with the tunnel having the largest cross-section closest to the aircraft. Apron drive loading bridges are available in either two or three telescoping tunnel models and in a variety of lengths to allow accommodation of all commercial jet aircraft.

The aircraft cab rotates 125 degrees (+92½ degrees, −32½ degrees) from the centre to align with the aircraft. It is designed to allow completion of the docking manoeuver from an enclosed operator station. The cab is rotated at a speed of 19 feet (5·8 metres) per minute by a gear motor and chain drive operating on the circumference of a fixed circular floor section or cab bubble at the end of the outboard tunnel. Limit switches and physical stops control the extremes of rotation.

Other features, except for the drive column and its control, are similar to the bridges previously described. The motors and mechanisms for vertical and horizontal motion including extend, retract and swing of the passenger bridge are an integral part of the drive column. The design permits simultaneous vertical, horizontal and swing (steering) travel which allows the bridge to move on a varying curved path, a straight line or at a constant radius.

The GL-1 is designed specifically for ground level terminals operating Boeing 727, 737 and McDonnell Douglas DC-9 aircraft. The rotunda corridor can be interfaced with the terminal at any angle between 0 and 90 degrees, and to a building aperture with an elevation of up to 1 foot (0·3 metres) above the apron. The corridor can be extended outward from the building to any desired length. The cab is fitted with front and side windows. Variable apron speed operation and the horizontally-adjustable cab can correct aircraft parking deviations. Rotunda rotation and horizontal drive make it possible to store the bridge against the terminal when not in use. Design features provide a 20-foot (6-metre) minimum clearance during aircraft manoeuvring.

Jetway Pedestal passenger loading bridge

The Jetway Radial model FSR passenger boarding bridge is their latest development.

The FSR bridge is capable of handling door heights of all current and planned commercial jet aircraft. It permits aircraft power-in and power-out or push-out operations. It consists of five basic components: rotunda and support column, tunnel, drive column, cab and aircraft closure, service door and stairs. The rotunda is adjacent to the terminal and is supported by a single foundation and column. As the main pivot for the bridge, the rotunda allows free swing of the unit up to 175 degrees. A sloping tunnel is hinged to the rotunda and is rectangular in cross section. It may be ordered in a variety of lengths. The end of the tunnel is mounted on the drive column which is capable of allowing both horizontal and vertical motion simultaneously. The cab connected to the sloping tunnel is designed to maintain a near level or horizontal attitude at all times as the unit raises or lowers. It is capable of rotating 92·5 degrees to the left and 32·5 degrees to the right of the bridge centre-line for alignment with the aircraft. The operator stations and controls, which are fully enclosed, provide full visibility for the docking interface. The console is forward-facing with digital horizontal and vertical position indicators and the bridge is equipped with an automatic levelling system which allows it to follow changes in aircraft door-sill elevation. A manually-operated roll-up door can seal the bridge from weather when not in use. A full-width bumper is positioned at the aircraft end of the cab floor. The aircraft closure is a folding accordian bellows-type. Either side may be individually actuated, left and right, to conform and seal against the aircraft contours. A service door, landing and stairs are available at the cab end to provide apron access to and from the bridge. The stairs are equipped with self-adapting risers and are level with the cab floor so that they can be used at all heights and positions of the boarding bridge.

Optional equipment includes automatic bridge prepositioning, heating and ventilation systems, ice melting equipment and the Jetpower 400 Hz ground-power system.

Jetway Apron Drive passenger loading bridge

Jetway GL-1 passenger loading bridge

Status
Jetway Bridges are installed at airports throughout the world

Manufacturer
Jetway Division, Abex Corporation, Ogden, Utah, USA

Characteristics

MODEL	Pedestal	Elevating	Apron Drive	GL-1	Radial FSR
CONNECTING TUNNEL(S) LENGTH	As required	As required	13 two tunnel models 6 three tunnel models	63 ft 1½ in (19·24 m)	29 ft (8·84 m)–131 ft (39·93 m) in 2 ft (0·61 m) increments
INTERNAL MINIMUM DIMENSIONS					
width	5 ft 10 in (1·78 m)	5 ft 10 in (1·78 m)	4 ft 10 in (1·47 m)	5 ft ½in (1·52 m)	5 ft 10 in (1·78 m)
height	7 ft (2·13 m)	7 ft (2·13 m)	7 ft (2·13 m)	7 ft ½in (2·15 m)	7 ft (2·13 m)
ROTUNDA ROTATION			±87½°	±90°	175°
CAB					
extension/retraction				2 ft 6 in (0·76 m)	
rotation	±15°	±15°	+92½°, −32½°	±15°	+92½, −32½°
speed			19 ft/minute (5·8 m/minute)	+2°/s	19 ft/minute (5·8 m/minute)
CAB FLOOR HEIGHT	variable	variable	variable	6 ft 3 in–9 ft 9 in (1·9 to 2·97 m)	variable
VERTICAL SPEED	2·5 ft/minute (0·76 m/minute)	2·5 ft/minute (0·76 m/minute)	2·5 ft/minute (0·76 m/minute)	5 ft/minute (1·52 m/minute)	2·5 ft/minute (0·76 m/minute)
HORIZONTAL DRIVE	17 ft/minute (5·2 m/minute)	17 ft/minute (5·2 m/minute)	90 ft/minute (27 m/minute)	variable 0–120 ft/minute (36·58 m/minute)	90 ft/minute (27 m/minute)
HORIZONTAL RANGE	10 ft (3·05 m)	10 ft (3·05 m)			
CAPACITIES					
max live load	40 lb/ft^2 (195 kg/m^2)	40 lb/ft^2 (195 kg/m^2)	40 lb/ft^2 (195 kg/m^2)	40 lb/ft^2 (195 kg/m^2)	40 lb/ft^2 (195 kg/m^2)
max wind load	25 lb/ft^2 (122 kg/m^2)	25 lb/ft^2 (122 kg/m^2)	25 lb/ft^2 (122 kg/m^2)	25 lb/ft^2 (122 kg/m^2)	25 lb/ft^2 (122 kg/m^2)
operational	12·5 lb/ft^2 (61 kg/m^2)	12·5 lb/ft^2 (61 kg/m^2)	12·5 lb/ft^2 (61 kg/m^2)	12·5 lb/ft^2 (61 kg/m^2)	12·5 lb/ft^2 (61 kg/m^2)
max roof load	25 lb/ft^2 (122 kg/m^2)	25 lb/ft^2 (122 kg/m^2)	25 lb/ft^2 (122 kg/m^2)	25 lb/ft^2 (122 kg/m^2)	25 lb/ft^2 (122 kg/m^2)
POWER REQUIREMENTS	480 V, 3-phase, 60 Hz	480 V, 3-phase, 60 Hz	480 V, 3-phase, 60 Hz	440 V, 3-phase, 60 Hz	480 V, 3-phase, 60 Hz
OPERATING TEMPERATURE RANGE	−25° to +127°F (−35° to +53°C)	−40° to +125°F (−40° to +52°C)	−25° to +125°F (−35° to +52°C)	−25° to +125°F (−32° to +52°C)	−31° to +127°F (−35° to +53°C)

ASE Models 400 and 50 Micro-Dose Baggage Screening System

The Micro-Dose Model 100 X-ray inspection system which is designed for the rapid screening of baggage and parcels in high-volume security areas, is available with the operator's console and image display mounted either as an integral part of the system or situated as far as 50 feet (15·2 metres) from the main unit. Its standard 19-inch (48·3-cm) wide conveyor can be replaced by an extra wide 30-inch (76·2-cm) conveyor for heavy-duty operation. Radiation generated by an X-ray tube is collimated by a long stationary slit and a rotating wheel with four slits, the combination of which causes a scanning pencil-size X-ray beam to move rapidly along a lineal-shaped radiation detector. The entire suitcase is inspected as it moves on a conveyor past the scanning beam. The detector signal is sampled, digitised and stored in a digital display for review on a high-resolution video monitor.

Among the system's standard features are the 512 × 512 element digital display, the ability to scroll the image across the screen which accelerates the inspection process, and a long source-to-detector distance to minimise image distortion. The unit has a full digital image zoom capability and steady high kilovoltage through the inspection process to facilitate thick object penetration.

The Micro-Dose Model 50 X-ray inspection

system is a hand-operated X-ray system designed for use in low-volume security areas such as small airports. It offers the same 512× 512 picture element image as the Model 100 and is based on the same principle of operation but uses an electrically-operated drawer, with X-rays orientated vertically rather than horizontally. The complete system can be moved from one area to another through any standard doorway. Parcels or bags are loaded into the drawer for scanning. Both the Model 100 and Model 50 offer the option of an automatic threat alert component which alerts the system operator audibly and visually when a suspicious object passes through the X-ray system.

Characteristics

MODEL: 100/100E
X-RAY GENERATION: 90 kV capable of penetration through 6 mm of steel
RESOLUTION: 36 awg gauge, single, copper wire 0·005 in (0·13 mm) diameter
X-RAY EMISSION: less than 0·1 mr per inspection
CONTRAST SENSITIVITY: 16 levels of grey
MAX BAGGAGE SIZE
standard conveyor: 19× 24 in (48·3× 61 cm) length unlimited
optional conveyor: 30× 30 in (76·2× 76·2 cm) length unlimited
CONVEYOR SPEED
reversing capability
Model 100: 30 ft/minute (9·14 m/minute);
Model 100E: 25 ft/minute (7·2 m/minute)
INSPECTION RATE: up to 1500 bags per hour
CONVEYOR DIMENSIONS
length: 10 ft (3·05 m)
width: 19 in (48·3 cm) standard 30 in (76·2 cm) optional
height: 1 ft 1 in (0·33 m)
UNIT DIMENSIONS WITHOUT CONVEYOR
length: 4 ft 5 in (1·35 m)
width: 5 ft 10 in (1·78 m)
height: 4 ft 6 in (1·37 m)
WEIGHT: 2200 lb (998 kg)
POWER REQUIREMENTS
Model 100: 120 V ac ± 10%, 60 Hz 1·9 kVA, single phase
Model 100E: 220/240 ac ± 10%, 50 Hz 1·9 kVA, single phase

MODEL: 50/50E as for 100/100E except:
X-RAY GENERATION: 70 kV
MAX BAGGAGE SIZE
height: 1 ft (0·31 m)
width: 2 ft 1½ in (0·65 m)
depth: 2 ft 3 in (0·69 m)
INSPECTION TIME: 5 s Model 50; 6 s Model 50E
UNIT DIMENSIONS
height: 4 ft 3 in (1·3 m)
width: 3 ft 10 in (1·17 m)
depth: 2 ft 6 in (0·76 m)
WEIGHT: 800 lbs (363 kg)
POWER REQUIREMENTS
Model 50: 120 V ac ± 10%, 60 Hz 0·5 kVA, single phase
Model 50E: 200/240 V ac ± 10%, 50 Hz 0·5 kVA, single phase

Baggage being screened by Micro-Dose Model 100

Micro-Dose Model 50 baggage screening unit

Status

This is a product developed by American Science and Engineering. The company has supplied X-ray screening equipment to American Airlines, Chilean Air Force and airport authorities in: Columbia, Dubai, Finland, France, China, Portugal, Saudi Arabia and Switzerland

Manufacturer

American Science and Engineering Inc, Cambridge, Massachusetts, USA

ASE Unex Mk 411 Flight Information Display System

The Unex Mk 411 is a full matrix display system which can be used for flight information and internal and external communications. The displays are capable of presenting different languages on the same board and basic information in any foreign language can be programmed into the system where the information is stored and available for recall and display at any time. With the addition of a video-image reader and an interface unit, the system has a graphics capability in which airline logos and other graphics can be shown with appropriate flight information. Information can be presented in a variety of letter sizes and styles and for special emphasis, letters, words or whole frames can be shown in reverse. Back lighting enables the data to be displayed in a variety of colours by inserting colour sleeves over the fluorescent tubes. A further development permits copy and graphics to be presented in colour combinations that produce a four-colour effect. To transfer information to each panel Unex uses a seven-wire data cable. The system is controlled by the 411 keyboard controller which uses a simplified command system. The operator can create new messages and direct them to any board in the system, and schedules that are continually repeated can be stored. Messages can be copied, amended, erased or made to roll up and at any time the operator can inspect and edit the file.

Unex systems can include crt and a range of monitors are available; a typical installation comprises main board departure and arrival display, gate, baggage and video displays.

Status

Unex displays have been supplied to major airports including Copenhagen, Denmark; Manchester, United Kingdom; Kuala Lumpur, Malaysia; Chiang Kai-Shek, Taiwan and Miami International, USA

Manufacturer

American Sign and Indicator Corporation, Spokane, Washington, USA

Unex flight information display system at Miami International Airport, Florida

---

Conrac Flight Information Display Systems

The flight information display systems produced by Conrac use video- and split-flap displays to present the information. For desk-top monitors and computer terminal displays normal character height is 5 mm and up to 2000 characters may be presented on a 31-cm crt. For arrival and departure displays for small audiences, 43 and 59 cm monitors are used. Normal character height is 1·25 cm. The Macrofont video display is designed for larger audiences. Macrofont is a method of electronically generating large characters for presentation on video-monitors. Instead of using the standard dot matrix to form the characters, the display features solid, undotted block characters. This method of character generation provides increased brightness with reduced phosphor intensity. Three basic character sizes are available with attendant line and page capacities and up to 15 Macrofont channels can be driven by a single controller and its associated power supply. The controller chassis contains an interface card, a sync card, a data buffer card and up to 15 character generator cards. Video output for each channel is raster scan, non-interlaced, 1·4 volts peak-to-peak at 75 ohms. To allow interface with existing systems, in remote applications, input to the controller chassis is RS 232 transmission using modified ASCII Code. By grouping several Macrofont video-monitors together behind a special tinted plastic face-plate, more information can be displayed. Mosaic displays typically are in clusters of three or six television monitors to provide extended line capacity and up to six or twelve lines of information. The special face-plate covers the front of each monitor and gives the appearance of a continuous line display.

Split-flap display boards are available with characters in 4·5, 6 and 10 cm heights. Character-style, choice of characters and national alphabets are available as optional features in addition to several widths up to 76 cm. Automatic- and manual-electronic control of split flap boards is available and in systems using a combination of flap boards and video, simultaneous control of both display mediums is accomplished automatically.

The flight information display system installed by Conrac in terminal 1 of the Changi International Airport, Singapore, in 1981, was developed for the processing of the passengers of more than 30 airlines, ticketing, catering, refuelling, baggage handling, flight operations, air freight, public address, maintenance and apron control. The total visual communication system consists of output and display devices for both public and internal viewing, input terminals for manual-data entry, controls and processing equipment and uninterruptible power supplies. The system includes more than 8000 split-flap alphanumeric display modules mounted in 67 signs installed at various locations throughout the terminal building. A total of 208 television monitors display arrival/departure information to the public and 284 television monitors present operational information for internal viewing. The system is designed for the operational information system to be expanded to 640 positions, each able to select up to 128 separate display channels.

The central computer room contains input terminals with full access to the data base. Twenty-six other data-input terminals, assigned to the individual airlines, the apron tower controller, the baggage hall supervisor, aircraft ground handlers and other internal functions, provide restricted access to specific portions of the data base. These restrictions are an off-line adjustable feature of the system so that access authority at any given input terminal may be altered as an operator's duties change. The data input system, which is expansible to 64 positions, initially includes two portable terminals which can be connected to the system through any telephone.

The computer-control system is of a distributed processing configuration. Twenty-one computers are used on-line or in a standby mode, in addition to numerous micro- and pico-processors. Computer sizes range from the dual 64K machines that manage the data base and their respective rigid discs through 32K machines controlling display terminals and printers to the micros and picos used for discrete unit control. A dual-redundant system, it can be operated also with manual inputs. The data base cpu are connected through an automatic switch to minimise down-time. The computer software is divided into many small tasks and modules, with the result that the growth of Changi International Airport into terminals 2 and 3 will require only minor software changes.

Characteristics

VIDEO-MONITOR DISPLAY			
size of audience	single	small	medium
monitor size	12 in (31 cm)	17 in (43 cm), 23 in (59 cm)	Macrofont 23 in (59 cm)
MAX CHARACTERS			
total	2000		320, 108, 52
per line			32, 18, 13
height	0·2 in (0·5 cm)	½ in (1·25 cm)	1 in (2·5 cm), 2 in (5·1 cm), 3 in (7·6 cm)
SPLIT FLAP BOARD DISPLAY			
height of characters	1¾ in (4·5 cm), 2½ in (6 cm), 4 in (10 cm)		
flap widths	up to 30 in (76 cm)		

Status

Conrac has installed flight information display systems at 32 airports in 12 countries. Conrac Elecktron, the Federal German subsidiary, handles all European and Middle East contracts

Manufacturer

Conrac Corporation, Duarte, California, USA

CTS Blue Max Airline Ticket Printer

The Blue Max airline ticket printer was introduced by Computer Terminal Systems in 1981. Operated through a micro-processor the unit uses a tractor feed mechanism and ballistic-type printing to produce standard IATA multi-page tickets at 20 columns per second. The printer, which is capable of storing up to 1000 tickets, can be interfaced with any display/controller system.

Optional features include optical character recognition, bar code and magnetic strip encoders and readers.

Manufacturer

Computer Terminal Systems, Plainview, New York, USA

Custom Conveyor Baggage Handling Systems

Custom Conveyor specialise in the design, manufacture and installation of baggage handling systems which are built to individual operator's requirements. Systems offered range from manually-operated conveyors to computer-controlled baggage sorting systems. To produce these systems the equipment available includes slider bed conveyors, mergers and diverters (powered or non-powered), turns, radius breakovers, baggage claim units and loading devices, vertical conveyors, container conveyors and baggage racks.

Status
The company, formed in the early 1970s has installed systems for: Air Canada, Air Florida, American, Braniff, Continental, Delta, Eastern, Frontier, North West, Ozark, Piedmont, Rio, South West, Texas International, TWA, United, US Air and Western Airlines

Manufacturer
Custom Conveyor Company, Dallas, Texas, USA

Datom NXT Passenger Direction System

The NXT Service Directer is an electronic system designed to organise queues, such as at airline ticket counters and to provide those facilities with a streamlined customer flow. The system consists of the Service Directer, an electronic master display column of station numbers and directional points. When a station is free to serve the next customer a floor switch or desk-top button is activated. Simultaneously the station number is illuminated, a soft chime sounds and the directional pointer indicates the open station. The system is used in conjunction with appropriate posts, ropes and signs to direct the passengers to an orderly queue.

Status
This system is used by American, Air California, Continental, Pan Am, TWA, United and Western Airlines in the USA

Manufacturer
Datom Directers Inc, Redundo Beach, California, USA

NXT passenger direction system in use at Los Angeles International Airport

---

Dayton Daycom Flight Information Display Systems

Dayton Communications specialise in the design, production and installation of flight information display systems. Primarily, Daycom supplies video flight information display systems, but integrated systems using large dynamic signs such as flap boards and Unex boards have been provided. Audio as well as visual information can be provided, the audio information being primarily public address and flight announcement systems. The Daycom systems are based on the microprocessor controlled page storage system (mcpss) that the company has developed over the last three years. These are 8- and 16-bit, micro-processor systems specifically designed for flight information display application and range from a small system configuration for an airline with very few flights per day to a configuration for a complete airport system with many hundreds of flights each day.
A typical installation by Daycom is the British Airways terminal at the J F Kennedy airport, New York. The mcpss, which operates the automated passenger information system comprises four subsystems: the main system, the public/domestic display system, check-in display system and gate/apron display system.
The primary mcpss control unit has a floppy-disc data storage system, control electronics and the power supplies. A back-up system is provided in the event of main system and floppy-disc failure. The daily data base is copied automatically on to a stand-by floppy-disc during normal operation. If a failure should occur, the total system is back-online within two minutes by depressing one switch.
The present system has 32 channels with six outputs per channel. These channels are broken down into 16 check-in, one public arrivals, one public departure, six gates, two apron and six domestic. The main file can store data for 500 flights as well as 24 hours of flight information. Features include an automatic roll-down and roll-off capability based upon an adjusted time sequence and a keyboard for manual roll-back. A crawl-message capability is provided at the check-in and apron channels, and is displayed across the bottom line of the video monitors. Separate messages may be shown on each channel or all channels may display the same message, up to 150 messages of 900 characters may be used. The rate of change is adjustable and the time period for repeat messages can be mixed.
The mcpss provides video output signals to 32 channels with a 540-line interlaced scan with composite sync and video signals. There is a digital output capability for the transmission of stored data to distant places via telephone lines at a selectable baud rate of 300–9600 baud per second.
The displays consist of 68 23-inch and 24 10-inch high-resolution monitors. To compensate for changing light conditions, a photo-electric cell on the front of the 23-inch television operates in conjunction with the monitor's brightness circuitry to provide optimum readout.
Three 95-character keyboards are provided for changes to the program using shifted and unshifted functions, generating an eight-bit parallel ASCII code. Extensive editing functions are available. Print-out is provided by an electric printer which supplies information on schedules, flight history and general information.

Daycom flight information monitor

Future developments will include a graphics input for foreign language symbols, logos and any special character fount requirements.

Status
Since the company was founded in 1968, Dayton Communications has provided over 150 flight information systems to airports in Brazil, Greece, Republic of Korea, Lebanon, Saudi Arabia, the United Kingdom and the USA

Manufacturer
Dayton Communications Corporation, Dayton, Ohio, USA

DI-AN Airline Ticket and Boarding-pass Printers

Three models are available, the 8100 and 8170 for airline tickets and the 8500 for boarding-pass printing. The information to be printed on each document can be originated from a local source such as a keyboard, tape, or computer or from a remote site via a communication network. The information may arrive at the printer complete and full-formatted. Alternatively only the raw data may be sent and the printer formats the document, adds sequence numbers or generates identification data.
Print time for a document varies with data format and character sizes. Character founts may be mixed from line-to-line by size and symbols. Typical printing rates average 20 columns per second.
The 8170 is a teleprinter fitted with an airline ticket fount which uses nylon or fabric ribbon.
The Model 8500 boarding-pass unit is a drum printer using an ink roll similar to the 8100.
In addition to the normal fount all three models can print composed characters with heights of ½ inch (12·7 mm) to 2 inches (50·8 mm).

Characteristics

MODEL	8100	8170	8500
TYPE OF PRINTER	drum	tele	drum
APPLICATION	airline tickets	airline tickets	boarding passes
ELEMENT LIFE	100 000 h	2000 h	100 000 h
TYPE FACE	airline ticket fount	airline ticket fount	$\frac{1}{8}$ in (3·17 mm), $\frac{1}{6}$ in (3·81 mm), $\frac{1}{5}$ in (5·08 mm) or mixed 0·085 (2·15 mm) and $\frac{1}{6}$ in (3·81 mm)
COPY SIZE	IBM card (18 lines × 72 columns)	—	—
AVAILABLE INTERFACES	RS-232C; 20 ma current loop or misa	RS-232C; 20 ma current loop, bit serial	RS-232C 20 ma serial

Status

Di-An airline ticket and boarding pass printers are in use with airlines and travel agencies in the USA

Manufacturer

Di-An Controls Inc, Boston, Massachusetts, USA

Flomaster Belt Power Turn and Spiral Lift Baggage Conveyors

The Flomaster Belt Power Turn can be used to convey baggage around curves from 30 to 180 degrees. The turns can be combined to form s-shaped curves or curves greater than 180 degrees. All turns are reversible and can be supplied in steel, stainless steel or fibreglass construction.
The standard belt supplied with the Flomaster Belt Power Turn is of nylon/nitrile construction with clipper brand lacing available in widths from 4 to 48 inches (10·2 to 121·9 cm). The power source is a shaft-mounted gear reducer connected to a flange-mounted motor which may be coupled direct to the motor or driven by a V-belt. Control is from the outside edge of the belt which is installed in frames of slider bed construction. The Flomaster Spiral Lift enables baggage travelling on a roller or belt conveyor to be conveyed vertically. The modular design of this conveying system allows vertical lifts from 2 to 35 feet (0·61 to 10·7 metres) with a single unit. Standard belting is available in widths of 6 to 48 inches (15·24 to 121·9 cm), 2- or 3-ply, black rough-top with steel clipper lacing. Depending on size, the slider bed frame is deep box steel construction. End rolls are tapered and sprocket-driven by a single strand sidebow roller chain with special attachments to the belt. Motor-reducer or customised drive arrangements are variable to meet the operator's requirements.

Status

Flomaster Belt Power Turns and Spiral Lift baggage conveyors are in operation at many major airports

Manufacturer

Flomaster Division, Portec Inc, Canon City, Colorado, USA

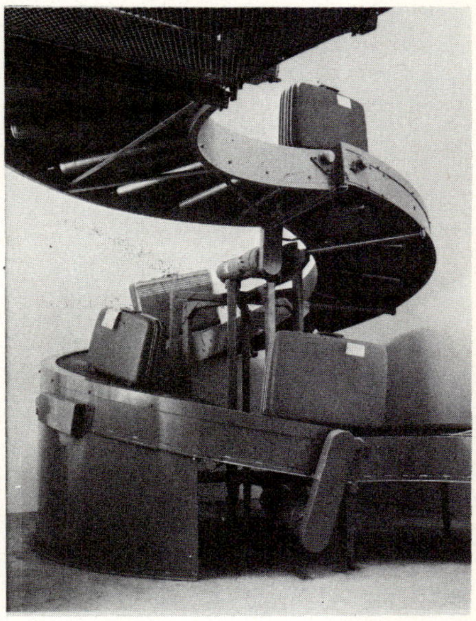

Flomaster Spiral Lift baggage conveyor

Friskem AF 500 Metal Detector

The Design AF 500 is a walk-through metal detector. Overhead an illuminated sign is mounted which instructs the passenger to wait or walk. Entry into the archway, under a walk indication, breaks an infra-red beam to notify the next person to wait and activates the detection system. During the search phase, any distortions of the transmitted field are registered by a multi-channel receiver and analysed. The system incorporates logic programming to reduce false alarms. Screening is above ankle height. When the passenger leaves the archway a second infra-red beam is broken and the result of the search phase shown on the display as clear or alarm. In the latter case a horn sounds. This second beam recycles the system for the next search. Between searches the unit is on standby. Key features of the detector are a remote-operator display, on-board electronics installed on top of the archway; self-supporting outriggers, open-side field coils, built-in shielding and a dead side to prevent those not being screened from activating the system. Optional features include: shoe-sole screening capability, electronic calibration, synchronising capability to permit detectors to operate side-by-side in close proximity, digital meter read-outs and a printer for signature checks fitted with an optional memory input to prevent collusion.

Characteristics

FLOOR SPACE: 4 × 4 ft (1·2 × 1·2 m)
MINIMUM CEILING: 7 ft (2·13 m)
STANDARD AISLE: 2 ft 6 in (0·76 m)
OVERALL DIMENSIONS
  length 2 ft 8 in (0·82 m)
  width 3 ft 5 in (1·04 m)
  height 7 ft (2·13 m)
WEIGHT: 150 lb (68 kg)
POWER REQUIREMENTS: 100–130 V; 50–60 Hz 1·5 A
  export model 90–250 V

Status

In service at major airports in the USA and in Moscow, USSR and Peking, China

Manufacturer

Friskem/Infinetics Inc, Wilmington, Delaware, USA

Friskem Design AF500 metal detector

Harper Executive I Baggage Trolley

The Executive I trolley is designed for baggage handling.
Constructed in aluminium with two solid rubber tyres, the Executive has nylon carpeting fitted to pad the back-plate and the folding nose extension. A retractable garment hook is incorporated in the design.

Characteristics

CAPACITY: 600 lb (272 kg)
DIMENSIONS OF FOLDING
  height: 1 ft 1 in (0·33 m)
  width: 2 ft 6 in (0·76 m)
NOSE EXTENSION
  height: 4 ft 4 in (1·32 m)
  width: 1 ft 6 in (0·47 m)
WEIGHT: 33½ lb (15·2 kg)

Manufacturer

Harper Trucks Inc, Wichita, Kansas, USA

Honeywell Computer Terminals

Honeywell Information Systems introduced three products to their computer line in 1981. Two micro-processor based terminals have been added to their series 1500 family, the Model 1510 stand-alone video display system and the Model 1530 large cluster terminal processing unit. Both terminals are compatible with the industry's airline reservations systems (PARS) for use in the airline reservation centres, airport and city ticket offices, and travel agencies.
The 1510 is programmable by the user and offers up to 16 K bytes of erasable programmable read-only memory (eprom) or random-access memory (ram). It has a 12-inch screen and keyboard and can display up to 2000 characters in four formats. The 1530 terminal processing unit connects up to 16 VDU 1501 video display terminals to a host computer. An optional multi-plexer controller enables one modem to support two terminal processing units for connecting a maximum of 32 work stations. The unit is programmable by the user and can have up to 64 K bytes of eprom and ram, depending upon the specific configuration. The self-service terminal is a modular system that can be programmed to meet individual airline's requirements. Software programming options enable passengers to do their own ticketing, to offer flight reservations, seat selection, issue boarding passes and provide tickets with ocr characters. Passengers can make use of the machine by inserting a valid credit card into the system. Transactions take between one and a half to three minutes.

Manufacturer
Honeywell Information Systems Inc, Minneapolis, Minnesota, USA

Intex Lymps 76 Metal Detector

Lymps 76 is a lightweight, hand-held metal detector which indicates acoustically the presence of all metallic objects, ferrous and non-ferrous through an adjustable loudspeaker incorporated in the handle. Alternatively, an earpiece may be used to replace the speaker. The signal is variable in pitch to enable the operator to locate the object. Power is provided by a rechargeable nickel-cadmium battery. The unit sounds continuously when the battery needs recharging.
The low magnetic field generated by the unit does not affect heart pacemakers.

Characteristics
BATTERY: 7·2 V, 500 ma/h rechargeable
POWER CONSUMPTION
minimum: 3 ma/h
max: 75 ma/h
AUDIO FREQUENCY: 1–4500 Hz
DIMENSIONS
length: 0·467 m
diameter: 0·022–0·44 m
UNIT WEIGHT: 625 g
OPERATIONAL TEMPERATURE RANGE: –30 to +70°C

Status
Approved for airport use by the FAA

Manufacturer
Intex Inc, Bethesda, Maryland, USA

Intex Sentrie Metal Detection System

The Sentrie, a walk-through detection system, is designed to detect any metal by using pulsed magnetic fields to excite transient eddy currents in target metal objects and to sense and process secondary signals during the on- and/or off-time of the excitation pulses. Alarm levels are adjustable over a broad range. The processor is available with various pattern-recognition programmes to meet specific discrimination requirements for weapons or other types of objects.
The unit comprises sensor coils encapsulated in strengthened panels with a formica finish. The console is constructed of heavy-gauge aluminium with walnut end-panels with a locking cabinet providing access to all circuitry. Circuit controls and assembly screws are accessible only from inside the locked cabinet. One red alarm lamp, one green standby/normal lamp and an audio volume control for the internal audible alarm unit are on the front panel of the console. Remote-alarm devices and shielded cables may be connected for operation up to 200 feet (60 metres) from the console. Remote-alarm contacts are provided on an alarm relay accessible through the rear of console for connection as required.

Characteristics
REQUIRED FLOOR AREA: 3·5 ft^2 (0·325 m^2)
PASSAGEWAY CLEARANCE
width: 2 ft 6 in (0·76 m)
height: 6 ft 6 in (1·98 m)
SENTRIE DIMENSIONS
height: 6 ft 8½ in (2·04 m)
width: 3 ft (0·91 m)
depth: 1 ft 1½ in (0·34 m)
WEIGHT: 70 lb (31·8 kg)
ELECTRONIC CONSOLE
depth: 6¼ in (0·16 m)
width: 1 ft 5½ in (0·44 m)
height: 1 ft ½ in (0·32 m)
WEIGHT: 20 lb (9 kg)
POWER REQUIREMENTS: 115/230 V, 50/60 Hz, less than 100 VA
OPERATING TEMPERATURE RANGE: 0 to +55°C

Status
Sentrie walk-through metal detectors are in use in over 20 countries

Manufacturer
Intex Inc, Bethesda, Maryland, USA

ITI Model 75 Entry-Scan Explosives Detector

The ITI model 75 is an explosives detector containing its own replaceable argon supply which continuously flows through the detection system. When the unit is not operational, this flow is reduced to conserve argon and to maintain a purge through the detector in stand-by status. During use the argon flow is adjusted automatically to the optimum operational flow-rate. The system is ready for use approximately 15 minutes after the electrical supply is switched on. The ready indicator light on the local/remote operational monitoring unit is illuminated as soon as the unit is operational.
In use, the system automatically signals people through, one at a time, by a system of lights. If the signals are not followed the operator is alerted and the person can be directed to pass through the entry-scan again. The system may be operated remotely from a central guard station using an extension to the local/remote operational monitoring unit.

An alarm output can be provided for incorporating into an existing access control system or for data logging purposes.

Characteristics
SENSITIVITY: greater than 1×10^{11} for EDGN-based explosives
ARGON SUPPLY: internal replaceable argon cylinder
consumption: 0·44 ft^3/h (0·125 m^3/h)
DETECTOR RESPONSE: 1·5 s
DETECTOR: 10 mCi Ni63 heated detector
DIMENSIONS
height: 5 ft 8 in (1·73 m)
width: 2 ft 3½ in (0·7 m)
depth: 1 ft 5½ in (0·45 m)
WEIGHT: 250 lb (114 kg) approx
POWER REQUIREMENTS: 110 V, 60 Hz or 240 V, 50 Hz

Manufacturer
Ion Track Instruments Inc, Burlington, Massachusetts, USA

ITI Model 75 Entry-Scan explosives detector

Magline Airport Luggage Trolley

The Magline luggage trolley has a hinged, 2 foot 6 inch (0·76 metre) folding nose with a full-perimeter, grey bumper. The folding nose and backplate are covered by carpet. A baggage-check holder with compartments is fitted at the rear of the trolley with a hinged cover providing a flat surface for writing or marking claim checks. A wide aluminium back-strap surrounds the baggage check folder which serves as a floor support for level loading with the trolley in a horizontal position.

Other features of the model include two 8-inch (204-mm) pneumatic-tyred wheels, retractable garment hanger hook and hand grips.

Status

This product was introduced in 1981

Manufacturer

Magline Inc, Pinconning, Michigan, USA

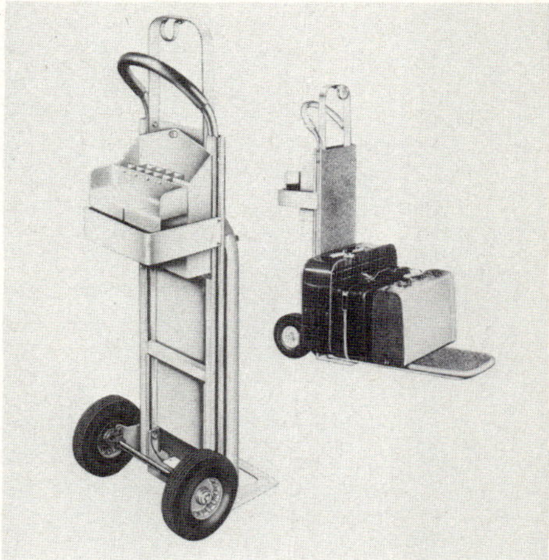

Magline airport baggage trolley

Raytheon PTS 100 Passenger Handling Data System

The PTS 100 data terminal is used by airlines to make and confirm reservations, issue tickets and boarding passes, handle all passenger check-in functions, and calculate the weight and balance of the aircraft.

A PTS 100 system includes its own intelligent controller capable of communicating with the displays or with an IBM, Univac or Honeywell host computer. Seven different line and column formats can be generated on the 15-inch (38·1-cm) screen, using either a 64- or 96-character display set. Eight standard keyboard arrangements are available, all featuring an n-key roll-over to ensure that data is not lost if two or more keys are depressed at the same time. The status indicators can be programmed to suit individual applications.

The PTS 100 system includes software emulators for IBM, Univac and Honeywell protocols.

Application software modules are provided in the PTS 100 to perform functions involved in the printing of a boarding pass and a ticket based on data received from the host computer. These functions include communicating to the host, line-to-column conversion and character checking.

Model 3430 boarding-pass printer is intended for airlines that use computer based passenger reservation systems. The printer is designed for single part forms equal in size to an IBM card. The printing format is one to four lines of nine characters to the inch.

Model 3440 ticket printer is intended to complement Model 3430 boarding-pass printer. The ticket printer will print simultaneously up to ten part standard flight coupons in the IATA transitional ticket format. The print cycle for a typical airline ticket (which are removed by hand) takes approximately six seconds. The ticket printer is designed to handle single carrier multi-part forms. With both printers, in a typical operating environment, all passenger (and other) information is stored in the host computer. The appropriate information is presented to the agent for verification on request. The agent checks the data and if correct a message is sent to the host computer requesting that the ticket be printed. All the information is still retained by the host computer for subsequent processing and retrieval.

Characteristics

MODEL	3430	3440
PRINTING APPLICATION	boarding pass	airline ticket
TYPEFACE	news gothic bold	airline ticket fount
PASS/TICKET SIZE	80 column $3\frac{1}{4} \times 7\frac{3}{8}$ in (8·25 × 18·73 cm)	$8\frac{1}{2} \times 4\frac{1}{4}$ in (2·59 × 10·79 cm)
INTERFACE	PTS 100	PTS 100

Status

PTS 100 Data terminals have been supplied to over 140 airlines

Manufacturer

Raytheon Data Systems Co, Norwood, Massachusetts, USA

Rens Minisquealer Model 25 Metal Detector

The Rens Electronic Squealer Model 25 is a hand-held, battery-operated scanner which is capable of detecting both magnetic and non-magnetic metal objects. The control arrangement consists of an on/off momentary switch and the sensitivity control knob. Metal detection is indicated by a change in the audible signal. The highest sensitivity (for special cases involving paper clips, razor blades etc) is found when the calibration control is set to the left of the threshold point (where chirping sound stops). Setting the calibration control a one hour segment to the left of the threshold point will yield a sensitivity such that a set of keys or a wristwatch are detected at about 1 inch (2·54 cm) distance and a cigarette pack at about 4 inches (10·16 cm) distance. The sensitivity range is adequate for most hand-held metal detector applications.

Characteristics

BATTERY: 9 V
CONSUMPTION
minimum: 6·5 ma
max: 11 ma
FREQUENCY: 130 kHz
DIMENSIONS
length: $17\frac{3}{8}$ in (44·1 cm)
width: 6 in (15·2 cm)
WEIGHT: 22 oz (620 g)

Status

Minisquealer is in use at airports in the USA

Manufacturer

Rens Manufacturing Co, Creswell, Oregon, USA

Scanray Linescan System One Baggage Screening Unit

Linescan System One is a baggage screening system with a vertical linear X-ray sensor to view the inspected item in sections as it moves through a narrow vertical fan-shaped X-ray beam on a conveyor. The linear X-ray sensor consists of an array of 480 silicon photo-diodes. The signal from each is read sequentially and the X-ray picture of the item passing through is assembled in the memory boards of a digital storage unit. An output board converts the video information to standard television data and the picture is presented on a television monitor. The conveyor can be stopped or reversed if necessary. The image remains on the television screen until the next item to be inspected is sensed by the photo-cell, or the machine is switched off. Manufacturers say the image quality using this technique is higher though the X-ray emission level is lower than that required in a wide-screen fluoroscopic system.

Characteristics

RESOLUTION: elements $\frac{5}{8} \times \frac{5}{8}$ in (15·88 × 15·88 mm) able to detect 34 gauge (awg) bare solid copper wire
X-RAY EMISSION: 0·15 mr per inspection
CONTRAST SENSITIVITY: 30:1 or 14 levels of grey with aluminium
TUNNEL OPENING: 28 in (71·1 cm) wide × $19\frac{1}{2}$ in (49·5 cm) high step wedge
BAGGAGE SIZE: 26 × 10 in (66 × 25·4 cm) without corner cut-offs
CONVEYOR SPEED
48 ft/minute: 60 Hz (14·6 m/minute)
40 ft/minute: 50 Hz (12·2 m/minute)
INSPECTION RATE: up to 2200 items per hour
POWER REQUIREMENTS
105–125 V ac, 60 Hz, 10 A
210–250 V ac, 50 Hz, 5 A

Status

Production in 1980, over 150 systems in operation

Manufacturer

Scanray Corporation, Harbor City, California, USA

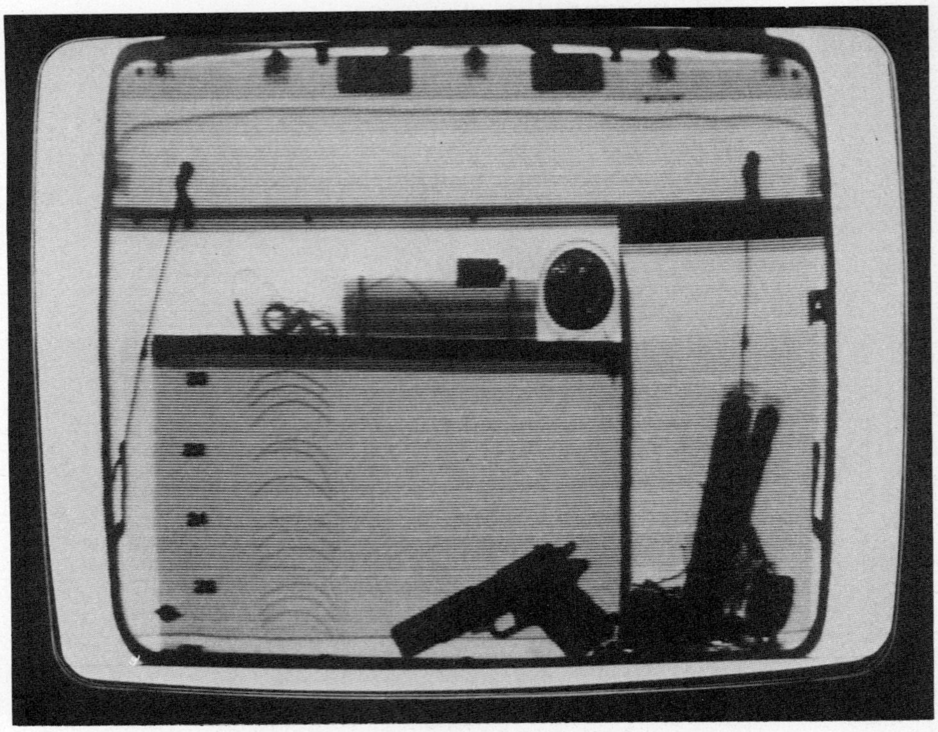

Linescan System One baggage screening unit

Scanray Linescan System Two Baggage Screening Unit

The Scanray Linescan System Two is a larger capacity version of the Linescan System One. Items up to 38 inches (96·5 cm) wide, 26 inches (66 cm) high and of any length can be handled by the unit at a conveyor speed of 60 feet (18·29 metres) per minute. As an optional feature the System Two can be equipped with a 160 KVCp X-ray generator replacing the standard 80 KVCp generator for inspecting very large, dense packing cases and boxes. The height of the conveyor is adjustable from 14 to 29 inches (35·6 to 73·7 cm) for loading or to mate with existing conveyor systems.

Status
Production

Manufacturer
Scanray Corporation, Harbor City, California, USA

Linescan System Two baggage screening unit

Characteristics
RESOLUTION: elements $^5/_8 \times ^5/_8$ in (15·88 × 15·88 mm) able to detect 34 gauge (awg) bare solid copper wire
X-RAY EMISSION: 0·15 mr per inspection
CONTRAST SENSITIVITY: 30:1 or 14 levels of grey with aluminium
TUNNEL OPENING: 26 in (66 cm) high × 38 in (96·5 cm) wide
BAGGAGE SIZE: $28^1/_2 \times 10$ in (72·4 × 25·4 cm) without corner cut-offs
CONVEYOR SPEED: 60 ft/minute: (18·29 m/minute)
POWER REQUIREMENTS
115 V, 60 Hz, 10 A
220 V, 50 Hz (10 A max at 115 V)

Scanray Mk 100 Metal Detector

The Scanray 100 is a portable, walk-through metal detector. By adjustment of the sensitivity level in the control console, the unit is designed to detect metal objects of weapon size 0·22 mm, hand gun, knives etc. Coins, keys, pens and other small metal objects should be ignored by the equipment.
The unit comprises an archway with sensor coils encapsulated in panels with a formica finish, connected to an aluminium construction control console.
The console is fitted with alarm and standby indicator lights, an alarm volume control with provision for a remote alarm facility which can be sited up to 200 feet (51 metres) away.
Optional features include remote alarm panels; led display giving unit increments of 1 to 10; shielding panels for instances when personnel not being screened must pass close to the coil structure; and slaving/blanking functions for synchronising operation in multi-unit installations and providing blanking of alarms which may be generated by pulsed X-ray or other electrical devices in the vicinity of the archway.

Characteristics
REQUIRED FLOORSPACE: 3·5 ft² (0·32 m²)
PASSAGEWAY CLEARANCE: $13^1/_2 \times 30 \times 78$ in (34 × 76 × 198 cm)
OVERALL ARCH ASSEMBLY: $13^1/_2 \times 36 \times 80^1/_2$ in (34 × 91 × 204 cm)
ELECTRONIC CONSOLE: $12^1/_2 \times 17^1/_2 \times 6^1/_4$ in (32 × 44 × 16 cm)
POWER REQUIREMENT
115 V, 60 Hz, or 230 V, 50 Hz
OPERATING TEMPERATURE RANGE: −10°C to +55°C

Status
Production, the equipment is in use at airports in the USA

Manufacturer
Scanray Corporation, Harbor City, California, USA

Smarte Carte Baggage Trolley

The Smarte Carte trolley is designed as a luggage and child carrier and can be converted to an emergency wheel-chair. The trolley comprises a chrome tubular frame mounted on a single axle with two 8-inch (20·3-cm) diameter wheels and semi-pneumatic tyres at the front with a single 5-inch (12·7-cm) diameter swivel castor at the rear.
The trolleys are designed to stack in a control unit equipped with automatic vending which includes a cash deposit, so that a trolley can be rented from the unit as required. When the trolley is returned to the control unit, the cash deposit is refunded to the user.

Characteristics

TROLLEY DIMENSIONS
length: 3 ft (0·91 m)
width: 2 ft 1 in (0·64 m)
height: 3 ft (0·91 m)
WEIGHT: 48 lb (21·77 kg)
CAPACITY: 500 lb (226·8 kg)
TROLLEY CONTROL UNIT
STANDARD UNIT DIMENSIONS
length: 15 ft (4·57 m)
width: 2 ft 9 in (0·84 m)
height: 3 ft $7\frac{1}{2}$in (1·1 m)
WEIGHT: 750 lb (340·2 kg)
NO OF TROLLEYS: 17
ADDITIONAL STORAGE SECTIONS
length: 5 ft (1·52 m)
weight: 150 lb (68 kg)
TROLLEY CAPACITY: 7

Status

In service at airports in the USA and Canada

Manufacturer

Smarte Carte Inc, White Bear Lake, Minnesota, USA

Stearns Maxiclaim Baggage Carousel

The Maxiclaim heavy-duty baggage carousel has a large canting moving surface capable of negotiating inside and outside turns which permits installation to any configuration. Remote loading can be accommodated from below the floor or from overhead.
The system employs a series of elongated hexagon-shaped pallets linked together with pivot pins and enclosed with hinged closure panels. The linkage pivot pins, pallets and closure panels permit the pallet chain to conform to a conical shape while articulating through outside or inside turns. Wheels and centre-line guide rollers support and guide the pallet chain throughout a closed-loop configuration.
The conveying surface is made of molded Vykan plastic material and the hinged closure panels are constructed of steel. The lower edge of the canting pallet is equipped with a rubber bumper that serves as a baggage stop and guard-rail. This bumper travels with the pallet at the same velocity and has an accordion-type cross-section to contract and expand as required through the turns.
The unit is driven normally by one motor with gear reduction mounted on the inside main frame. Driving power is transmitted to the conveying surface only in the drive section by means of a pair of parallel roller chains equipped with a series of driving lugs secured between the chains. These lugs sequentially engage and disengage the centre guide rollers resulting in a direct, positive mechanical link to the drive motor. The driving chains are supported vertically at each drive lug by four guide rollers the axle shafts of which also secure the drive lugs to the chains. A screw-type chain takeup is provided with the idler shaft bearing mounts to compensate for normal wear.
The complete unit is mounted above the floor and rests on levelling screws and pads fitted with a rubber floor contact surface.
Normal operating speed is 90 feet per minute (27·4 metres per minute) with a counter-clockwise direction of travel. Travel can be clockwise if required.
An electric control panel and two push button stations are normally furnished with each unit. This panel consists of a fused disconnect switch, reduced voltage magnetic reversing starter, control transformer, relays and terminals. Each push-button station is equipped with a start and stop push-button. A forward-reverse selector switch is included at the main panel for servicing equipment.

Characteristics

CONVEYING SURFACE: textured Vykan
width: 4 ft 9 in (1·45 m)
slope: $22\frac{3}{4}$°
speed: 20 ft/minute (27·4 m/minute)

Status

In service at 25 airports including Miami, Kansas City and Dallas Fort Worth airports in the USA

Manufacturer

Stearns Airport Equipment Division, Stearns Manufacturing Co, Crowley, Texas, USA

Stearns TB-500 TransiBridge Loading Bridge

The TB-500 TransiBridge loading bridge has been developed to service all types of commercial aircraft currently in service. A single structure supports both the tunnel to the aircraft and the corridor to the terminal building. The aircraft tunnel at right angles to the corridor is connected to the aircraft through a swivelling vestibule or bridge head, fitted with an air-seal canopy which can be rotated through 40 degrees to allow for any misalignment of the aircraft.
All aircraft are serviced from a single centre-line. The loading bridge is controlled hydraulically through a solid-state control panel operated within full view of the aircraft. Service access is provided as an optional feature by stairs from the ground to the loading bridge is from 7 to 17 feet (2·13 to 5·18 metres). Vertical travel of the bridge and an automatic levelling device adjusts the height as the aircraft load changes.

Status

Installed at airports in the USA

Manufacturer

Stearns Airport Equipment Division, Stearns Manufacturing Co, Crowley, Texas, USA

The TB-500 TransiBridge Loading Bridge

Stearns TransiTread Baggage Conveyor

The TransiTread closed-loop conveyor system consists of a series of crescent-shaped steel pallets, lined with a polyurethane tread, which are linked to form a continuous moving surface capable of turning corners without a break. Pallets are guided along the system centre-line by cam rollers and the conveyor is driven by a caterpillar-type drive. Both the conveyor and the drive are mounted on the floor surface.

Characteristics

TREAD
thickness: 0·25 in (6·35 mm)
width: 2 ft 8 in (0·81 m)
usable width: 2 ft 5 in (0·3 m)
UNIT HEIGHT: 1 ft (0·74 m)
STANDARD SPEED: 87 ft/minute (26·5 m/minute)
STANDARD TURN ANGLES: 90°, 180°
ROTATION: reversible
POWER REQUIREMENTS: to suit operator

Status

TransiTread baggage conveyors are in use in many airports in the USA including Detroit-Metro, Miami International and San Diego Airports

Manufacturer

Stearns Airport Equipment Division, Stearns Manufacturing Co, Crowley, Texas, USA

Stearns TransiVeyor Baggage Conveyor

The TransiVeyor is a heavy-duty belt conveyor system comprising slider design beds, strengthened cross bearings, heavy steel mounting brackets, crowned face pulleys and anti-friction pulley bearings. The beds are of U-shaped construction, incorporating stationary side guards, except in the load and unload areas. Power turns are of steel or fibreglass construction. Drives are sized in accordance with the belt pull and speed required, three sizes are available; normal, intermediate and heavy-duty.
The drives are shaft-mounted reduction gearing powered from the motor through a series of V belts and sheaves. For inclines and declines, an automatic braking device is incorporated to prevent over-run or roll-back after the motors have been stopped.

Characteristics

DIMENSIONS: as required by operator
DRIVES: MAX BELT PULL
normal: 500 lb (226·8 kg)
intermediate: 1000 lb (453·6 kg)
heavy-duty: 1500 lb (680·4 kg)
POWER REQUIREMENTS: to suit operator

Status

In service at airports in the USA

Manufacturer

Stearns Airport Equipment Division, Stearn Manufacturing Company, Crowley, Texas, US/

Stearns TransiWalk Passenger Conveyor

TransiWalk moving walkways can be used for moving passengers and luggage around airport terminals from car parks to check-in along aircraft loading piers and from the main terminal buildings to satellite terminals. The TransiWalk surface comprises crescent-shaped steel treads with a grooved rubber-tread surface.
Conveyors are available in four widths and can be installed on top of existing floor surfaces or at floor level. The continuous conveyor moves people in one direction down one side of the corridor and in the reverse direction on the other side. Alternatively the walkways can be arranged alongside each other.
Multiple drives are available for extended runs and the conveyor is capable of changing direction. The conveyor requires an installation depth of approximately 12 feet (0·3 metres) under the walkway with machine pits on each side 2 feet (0·61 metres) deep.

Characteristics

TREAD
thickness: ¾ in (19·05 mm)
steel: ⅜ in (9·53 mm)
grooved rubber: ⅜ in (9·53 mm)
TREAD WIDTH: 2 ft 4 in (0·71 m), 3 ft (0·91 m), 3 ft 4 in (1·02 m), 3 ft 8 in (1·12 m)
OVERALL WIDTH OF SINGLE CONVEYOR: 4 ft 4 in (1·32 m), 5 ft (1·52 m), 5 ft 4 in (1·63 m), 5 ft 8 in (1·73 m)
OVERALL WIDTH IN SIDE-BY-SIDE CONFIGURATION: 8 ft 8 in (2·64 m), 10 ft (3·05 m), 10 ft 8 in (8·25 m), 11 ft 4 in (3·45 m)
WALKWAY FLOOR RECESS DEPTH: 1 ft (0·3 m)
MACHINE PIT
depth: 2 ft (0·61 m)
length: 12 ft (3·66 m)
SPEED: as required by the operator

Status

TransiWalk passenger conveyors are installed a Miami International Airport, USA

Manufacturer

Stearns Airport Equipment Division, Stearns Manufacturing Co, Crowley, Texas, USA

Stinar Passenger Stairs

Three types are produced by Stinar, PS 813, PS 1018 and SPS 2513.
The PS 813 is a towable model available in four sizes to service narrow-body aircraft from the BAe 1-11 to the Boeing 757, and the wide-body Boeing 767. The main frame is constructed of aluminium-covered tubular steel with steps formed from aluminium diamond tread-plate and hand-rails from aluminium tubing. The platform is steel-ribbed covered with aluminium overlay 5 feet (1·52 metres) wide, 5 feet (1·52 metres) deep with two sliding panels. The height of the platform is adjusted manually. A heavy-duty castor and dual wheels provide mobility and a detachable tow-bar is supplied. Screw jack-type stabilisers on both sides provide stability during use.
The PS 1018 is a self-propelled passenger boarding stair designed for wide-body aircraft. This stair is engineered to service all aircraft with door-sill heights between 8 feet (2·44 metres) and 17 feet 7 inches (5·36 metres). An 18-hp petrol engine drives a hydraulic pump which operates two hydraulic orbit motors. These motors are connected directly to each drive wheel. The pump also supplies power to the main stair cylinder, stabilisers and the A-frame. A foot-operated valve controls forward and reverse movements and provides a positive stop in neutral position. A hand-operated valve controls the main cylinder, A-frame and stabilisers. The main frame is constructed of aluminium-covered tubular steel with steps formed from aluminium diamond tread-plate, and hand-rails from aluminium tubing. The platform is steel ribbed covered with an aluminium overlay, 5 feet (1·52 metres) wide, 6 feet (1·83 metres) deep. Incorporated are two sliding panels with locks and rubber bumpers mounted on the platform's forward edge. The steps and upper platform are illuminated and additional spotlights on the upper section are provided for operating personnel. The unit is equipped with a mechanical locking mechanism for the main stair section. Optional features comprise a 6-foot (1·83-metre) wide upper platform, adjustable front platform, rain canopy over the stairs and platform, 4-foot (1·22-metre) and 5-foot (1·52-metre) wide steps, diesel engine, battery-powered or towable unit without engine.
An alternate unit is available for Boeing 737-767 aircraft.
The SPS 2513 passenger boarding stair, designed to service aircraft with door-sill heights between 7 feet (2·13 metres) and 12 feet (3·66 metres) is hydraulically-operated and controlled from the driver's seat. The stairs are mounted on a 3/4-ton truck equipped with a six-cylinder petrol engine, manual transmission, heavy-duty alternator and battery and heavy-duty springs with rear auxiliaries. A number of vehicle options are available. An engine-driven hydraulic pump with electric clutch is the standard power source for the hydraulic system. A manual directional control and control panel with warning lights is in the cab. Stabilisers, front and rear, and an auxiliary hand-pump are fitted. The steps and upper platform are illuminated and additional spotlights on the upper section are provided for operating personnel. The main frame is constructed of aluminium-covered tubular steel with steps formed from an aluminium diamond tread-plate and hand-rails from aluminium tubing. The platform is steel ribbed covered with an aluminium overlay, 5 feet (1·52 metres) wide, 6 feet (1·83 metres). Incorporated are two sliding panels with locks and rubber bumpers and rubber bumpers mounted on the platform's forward edge. Safety features comprise a mechanical locking mechanism for the main stair section, lock valves which are mounted on each hydraulic cylinder; each stabiliser is equipped with a warning light to indicate when they are down. Optional features include a mechanical locking mechanism for the A-frame; 6-foot (1·83-metre) wide upper platform, adjustable front platform; rain canopy over the steps and platform and automatic transmission. An alternate model SPS-2513A is available to service aircraft with door-sill heights ranging from 6 feet 4 inches to 11 feet 3 inches.

Characteristics

MODEL	PS 813A	PS 813B	PS 813C	PS 813D
AIRCRAFT-TYPE	BAe 1-11 to DC-9	B 737 to B 727	B 707 to DC-8	B 757 to B 767
AIRCRAFT DOOR-SILL HEIGHT				
minimum	7 ft (2·13 m)	8 ft (2·44 m)	9 ft 11 in (3·02 m)	12 ft 5 in (3·78 m)
max	7 ft 10 in (2·39 m)	9 ft 9 in (2·97 m)	11 ft 3 in (3·43 m)	14 ft 8 in (4·47 m)

MODEL	PS 813	PS 1018	SPS 2513
AIRCRAFT DOOR-SILL HEIGHT			
minimum	7 ft (2·13 m)	8 ft (2·44 m)	7 ft (2·13 m)
max	14 ft 8 in (4·47 m)	17 ft 7 in (5·36 m)	12 ft (3·66 m)
STEP			
width	3 ft 6 in (1·06 m)	3 ft 6 in (1·06 m)	3 ft 6 in (1·06 m)
depth	10 in (25·4 cm)	10 in (25·4 cm)	10 in (25·4 cm)
RISER HEIGHT	7¾ in (19·69 cm)	7¾ in (19·69 cm)	7¾ in (19·69 cm)
UNIT DIMENSIONS			
transit length		20 ft 6 in (6·25 m)	20 ft 2 in (6·15 m)
width		8 ft (2·44 m)	6 ft 4 in (1·93 m)
WEIGHT			6800 lb (3084 kg)

Status

Stinar passenger steps are in service with airlines in the USA

Manufacturer

Stinar Corporation, St. Paul, Minnesota, USA

Wollard TLPH-252 SMT Passenger Stairs

The Wollard Model TLPH-252SMT truck-mounted passenger stair is designed to load and unload passengers from aircraft having door-sill heights of between 8 feet 4 inches (2·54 metres) and 16 feet 8 inches (5·08 metres). The stair consists of a main section, stationary unit, an adjustable section, telescoping unit, and an A-frame. The telescoping section and A-frame contain an adjustment capability to allow precise positioning of the top platform at the aircraft door-sill. The unit has illuminated stairs and aluminium hand-rails. All controls are fitted inside the cab of a Ford truck which is equipped with a six-cylinder petrol engine, automatic transmission and power steering. A diesel engine is available as an optional feature.

Characteristics

AIRCRAFT DOOR-SILL HEIGHT: 8 ft 4 in (2·54 m)–16 ft 8 in (5·08 m)
TOP PLATFORM DIMENSIONS: 6 ft 3 in (1·91 m) × 4 ft 11 in (1·5 m)
INTERNATIONAL LANDING DEPTH: 1 ft $7\frac{3}{4}$ in (0·5 m)
STAIR
width: 3 ft 6 in (1·07 m) or 3 ft 11 in (1·19 m)
depth: $10\frac{5}{8}$ in (0·27 m)
RISER HEIGHT: $7\frac{4}{5}$in (0·2 m)
CHASSIS: Ford F 350, 6-cylinder petrol engine
UNIT DIMENSIONS
length: 23 ft 1 in (7·04 m)
width: 7 ft 1 in (2·16 m)
height: 12 ft (3·66 m)
GROSS VEHICLE WEIGHT: 10 000 lb (4536 kg)

Status

In service with airlines in the USA

Manufacturer

Wollard Airport Equipment, Miami, Florida, USA

Wollard WS Series Passenger Loading Bridge

Wollard produces three types of loading bridge in the WS series; a telescopic model with apron-drive and two fixed-pedestal loading bridges suitable for the full range of commercial aircraft.

The WS 900 is an apron-drive loading bridge which has an extensible tunnel comprising two sections which pivot around the fixed main corridor to the terminal building. The telescopic tunnel is installed on a support structure which provides the vertical adjustment mounted on a single-axle pivoting chassis fitted with large volume pneumatic wheels. The mobile tunnel is manoeuvred to the aircraft by means of a variable drive providing swivelling and telescopic movement. Service steps are installed from the tunnel to the ground.

The WS 750T is a fixed-pedestal loading bridge for aircraft parked in the nose-in position. Differences in height between the terminal building and the aircraft door are compensated by vertical movement of the main tunnel which is arranged parallel to the aircraft. A telescopic tunnel positioned at right angles to the main tunnel is mounted on rollers on a frame structure. The outer end has a widened section, the bridge head, which is equipped with a canopy to form a weather-proof seal with the aircraft.

The WS 500 L-type loading bridge is another fixed-pedestal bridge for nose-in operations.

Status

Wollard loading bridges have been installed at airports in the USA

Manufacturer

Wollard Airport Equipment, Miami, Florida, USA

Zem-Air Counters and Magnetic Signs

Zem-Air design and manufacture ticket and gate counters and backwalls to meet individual operator's requirements. These are finished in the appropriate colours for the airline and decorated with the airline logo. Magnetic arrival and departure boards, gate signs and information boards are supplied in the airline colours and portraying the company's logo.

Status

Zem-Air has supplied passenger handling furniture to airlines and airport authorities in the USA

Zem-Air ticket counter and magnetic sign board for Midway Airlines, St. Louis Airport, Missouri, USA

CARGO HANDLING EQUIPMENT

AUSTRALIA

Hastings Deering Conveyor Belt Loader

This unit is designed to be used for loading, unloading and transferring of loads to and from aircraft holds. The height of both ends of the belt is variable. The chassis, fabricated by Hastings Deering is equipped with a Perkins diesel engine and manual or automatic transmission. Orbitrol power steering is fitted as standard equipment. Lifting and lowering the boom and the conveyor belt drive is hydraulically operated. The front lifting arrangement is via two single-acting two-stage rams. The rear height adjustment is through two single-acting, single-stage rams incorporating a stabiliser gate. The speed of the belt is variable. Controls are at the front and rear of the conveyor boom for the operation of the belt only. The top control is installed with a cable and hand control. A 6-inch (152-cm) diameter rubber buffer is fitted to the front of the boom.

Characteristics

CHASSIS: Hastings Deering with Perkins 4·108 diesel engine
CONVEYOR BOOM
front height: 2 ft (0·61 m)–15 ft 10 in (4·82 m)
rear height: 1 ft 6 in (0·71 m)–7 ft (2·13 m)
length: 26 ft 8 in (8·13 m)
capacity: 2205 lb (1000 kg)
speed: variable 10–26 m/minute
Since the boom is rigid, the front and rear heights are interacting. Therefore the minimum front height of 2 feet (0·61 metres) can only be achieved with a rear height of 7 feet (2·13 metres).

Hastings Deering conveyor belt loader

Status

In service with Qantas and other airlines in Australasia and the Far East

Manufacturer

Hastings Deering Ltd, Lidcombe, New South Wales, Australia

Hastings Deering Tow Tug Mk V

The tow tug has been designed to meet general haulage requirements, and has a maximum rated draw-bar pull of 2000 kg. The unit can be fitted with either a petrol or a diesel motor and with automatic or manual transmission. It is equipped with power-assisted four wheel hydraulic drum brakes and a mechanically-actuated hand brake and tow hitches, both front and rear are provided. The rear axle is of the double-reduction type with various ratios to suit the specific requirement for speed and draw-bar pull. Power steering is available as an optional feature.

Characteristics

ENGINE: Holden 6-cylinder
OUTPUT: 101 kW at 4400 rpm
MAX DRAW-BAR PULL: 2000 kg
TOWING HOOK HEIGHT
front: 0·36, 0·46 m
rear: 0·22, 0·3, 0·38, 0·46 m
WHEELBASE: 1·58 m
TRACK: 1·06 m
UNIT DIMENSIONS
length (with tow hitches): 3·33 m
length (without tow hitches): 2·79 m
width: 1·32 m
height: 1·54 m
WEIGHT: 2870 kg

Status

In service with Qantas Airways

Manufacturer

Hastings Deering Ltd, Lidcombe, New South Wales, Australia

CANADA

Moody Pallet/Container Trailers

The four models of trailers available for transporting pallets and containers, are the ODPC, DSL, SLET, and the SLT.
The ODPC is designed to carry two LD-1 or LD-3 containers, one LD-7 or LD-11 container, one igloo, or one pallet measuring up to 96 by 125 inches (2·44 by 3·18 metres). It has a load capacity of 15 000 lbs (6800 kg) at speeds up to 25 miles per hour (40 km per hour).
The carrier frame is of structural steel and formed steel checker plate walkways with 3-inch (76·2-mm) diameter swivel castors mounted between walkways to facilitate handling in any direction. A cam activated combination guide and stop provide both vertical and lateral restraint of LD-1 or LD-3 containers, while allowing for positive side loading. Ten spring-loaded pin stops permit appropriate spacing for pallets and LD-3 containers, while four dual-action stops provide container or pallet hold down and restraint. The unit is equipped with two swivel dual-front wheels with compression springs and solid rubber-tyres mounted on demountable wheels, rear rigid units supplied with spring suspension, E-shaped rear hitch with spring-loaded pin and a 4-inch (102-mm) tow-bar ring. Parking brakes are activated by raising the tow-bar.
Designed to carry two LD-1 or LD-3 containers or one LD-11, the DSL has a load capacity of 8000 lbs (3630 kg) at speeds of up to 25 miles per hour (40 km per hour). The heavy-duty reinforced frame extends beyond the container for protection and the flared opening of the roller deck and distance between fore and aft restraints will accept a reasonable amount of container misalignment. Containers are supported on four rows of 2½-inch (63·5-mm) diameter rollers, with the heavy-duty feed-in rollers set ¼ inch (6·35 mm) lower for container loading. Hand-operated container locks are fitted on either side of the unit with a spring-loaded central guide for LD-1/LD-3 containers. The trailer features two-wheel automotive steering, mechanical parking brakes engaged by raising the tow-bar, an E-shaped rear hitch spring-loaded pin and a 4-inch (102-mm) tow-bar ring.
The SLET model trailer provides for side loading one LD-1 or LD-3 container from either side of a central elevating turntable. The trailer has a capacity of 3500 lbs (1588 kg) at speeds of up to 25 miles per hour (40 km per hour). The cam activated central elevating turntable and rollers are mounted in a deep formed steel frame which extends beyond the container and the wheels to provide protection to both the load and running gear. The trailer is equipped with four wheel steering and all other features are similar to the DSL model.
The SLT trailer has a side loading capability from either side of a 360 degree rotatable roller deck. Designed to carry one LD-1 or LD-3 container, the SLT has a load capacity of 3500 lbs (1588 kg) at speeds of up to 25 miles per hour (40 km per hour). The turntable with 360 degrees of rotation is mounted on a large support ring and flanged bearing pivot point which is integral to the steel frame. Turntable locking is provided by spring-loaded locks that engage the dollies' frame and are operated from either side of the unit. All other features are identical to the SLET trailer.

Characteristics

MODEL	ODPC	DSL	SLET	SLT
CAPACITY	15 000 lb (6840 kg)	8000 lb (3629 kg)	3500 lb (1588 kg)	3500 lb (1588 kg)
TOWING SPEED	25 mph (40 km/h)	25 mph (40 km/h)	25 mph (40 km/h)	25 mph (40 km/h)
DECK HEIGHT	1 ft 8 in (0·508 m)	1 ft 8 in (0·508 m)	1 ft 8 in (0·508 m)	1 ft 8 in (0·508 m)
WHEELBASE	8 ft 5 in (2·57 m)	7 ft 6 in (2·29 m)	5 ft 2 in (1·58 m)	5 ft 9 in (1·75 m)
TRACK	5 ft (1·52 m)	4 ft 10 in (1·48 m)	5 ft (1·52 m)	4 ft 10 in (1·48 m)
TURNING RADIUS/ CRAMP ANGLE	65°	14 ft 6 in (6·42 m)	10 ft 4 in (3·15 m)	10 ft 3 in (3·12 m)
UNIT DIMENSIONS				
length (tow-bar raised)	12 ft 11 in (3·54 m)	13 ft 9 in (4·2 m)	10 ft 6 in (3·2 m)	10 ft (3·05 m)
width	8 ft 6 in (2·59 m)	6 ft 4 in (1·93 m)	6 ft 1 in (1·86 m)	6 ft 1 in (1·85 m)
WIDTH	2800 lb (1270 kg)	2300 lb (1043 kg)	1360 lb (617 kg)	1060 lb (481 kg)

Status

The trailers are in service with Air Canada, CP Air, Pacific Western and Wardair airlines, handling agents and freight forwarders at airports in Canada and the USA

Manufacturer

Moody SI Ltd, Terrebonne, Quebec, Canada

FRANCE

Air Marrel Pallet/Container Loaders

Three models of self-propelled pallet/container loaders are available: LAM 7000, LAM 9000 and LAM 12000. All are powered by a Deutz F4 air-cooled diesel engine with an axial-hydraulic pump supplying 250 bar pressure.

The LAM 7000 is fitted with two platforms each elevated by a hydraulically-operated scissors mechanism to a maximum height of 3·72 metres. The unit has a capacity of 7000 kg and is suitable for loading 88 by 108 inch (2·24 by 2·64 metre), 88 by 125 inch (2·24 by 3·18 metre) and 96 by 125 inch (2·44 by 3·18 metre) pallets and LD-1, LD-2 and LD-3 containers. It can service all types of conventional cargo aircraft and the front and rear cargo doors of wide-body aircraft.

An additional scissors mechanism rated at 6500 kg mounted on top of the forward platform enables the loader to reach the main decks of Boeing 747, DC-10 and A300B aircraft. The operator's controls are fitted to the forward platform which is equipped for pallet transfer and rotation. The main platform with automatic transfer facilities can be loaded from either side or from the rear. Five stabilisers are installed which function hydraulically.

The LAM 9000 with a capacity of 9000 kg can handle two pallets 88 by 125 inches (2·24 by 3·18 metres) mounted on a single 9-tonne pallet or two pallets 96 by 125 inches (2·44 by 3·18 metres) also on a single pallet or two LD-1, LD-2 or LD-3 containers. It features a single platform mounted on a scissors mechanism which can be elevated 3·5 metres for lower-deck loading, and combined with a detachable fixed height extension fitted to the platform, up to the main decks of wide-body aircraft. Both the platform and extension are equipped with automatic pallet stops and transfer facilities and pallets/containers can be rotated on the extension. Four hydraulic stabilisers are fitted.

The LAM 12000 has a capacity of 12 000 kg. The unit can handle a similar type of load to the LAM 9000, though the pallets need not be mounted on a single loading pallet. The main platform and operator's bridge function independently, the platform using a scissors mechanism, the bridge with the power station, fitted on a hydraulic elevating arm. A detachable fixed height extension is available for main deck loading. The platform features three-position guide-rails, front and rear automatic block stops and two independent pallet drives, the front with castors, the rear using rollers. A double row of non-driven rollers is installed on each side of the roller driver. The loader which is driven hydrostatically by two hydraulic motors acting on the rear wheels, is fitted with power steering and rear disc brakes.

Characteristics

MODEL	LAM 7000	LAM 9000	LAM 12000
CAPACITY	7000 kg	9000 kg	12 000 kg
ELEVATING HEIGHTS			
main platform	0·48–3·72 m	0·497–3·5 m	0·508–3·5 m
with extensions	5·4 m	5·95 m	5·95 m
DIMENSIONS			
main platform			
length:			
width	5 m	6·6 m	6·59 m
front	2·24/2·44 m	3·92 m	3·4 m
rear	2·615 m		
forward platform	3·2 × 2·615 m		0·8 × 1 m
(operator's bridge)			
extensions	4·655 × 3·75 m	7·365 × 4·408 m	7·365 × 4·408 m
WIDTH BETWEEN GUIDES			
pallet	2·28/2·5 m	3·33 m	
container	1·56 m		
LENGTH BETWEEN STOPS			
pallet		6·422 m	6·422 m
container			5·608 m
ENGINE	Deutz F4L 912/ Perkins 4236	Deutz F4L 912	Deutz F4L 912
OUTPUT	48·5/52 kW	48·5 kW	48·5 kW
MAIN/FORWARD PLATFORM TIME MINIMUM TO MAX HEIGHT			
raising	15/35 s	15 s	20 s
lowering	20/35 s	20 s	25 s
SPEEDS			
conveying	18 m/minute	18 m/minute	18 m/minute
driving	12 km/h	10 km/h	10 km/h
TURNING RADIUS	6·5 m	13·7 m	15 m
UNIT DIMENSIONS			
length	8·17 m	8·02 m	6·72 m
width	3·64 m	3·972 m	4·33 m
height			1·785 m
WEIGHT	9800 kg	10 000 kg	12 000 kg

Status

In service with airlines and airport authorities in Europe, Africa and the Middle East

Manufacturer

Air Marrel, Andrezieux Boutheon, France

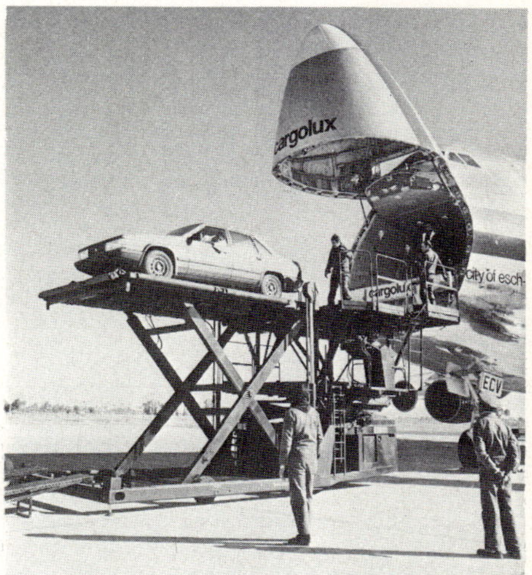

Air Marrel LAM 7000 pallet/container loader

Sietam Cargo Handling Equipment

Sietam design, manufacture and install freight handling equipment and systems. The equipment available includes electronic and electro-optical weighing scales and loading/unloading pallet systems with weigh scales and elevating platforms. Storage facilities are provided by benches equipped with rollers and retractable stops and static shelving with free running rollers. For transferring loads within the terminal, Sietam produces conveyors for pallets/containers, weighing and elevating transporters.

Status

Sietam handling systems are installed at major airports in Europe, the Middle East, Africa and the USSR

Manufacturer

Sietam, Viry Chatillon, Cedex, France

Sietam accumulation tables

Siraga Baggage/Freight Trailer

The Siraga baggage/freight trailer is an open-sided unit with boards at the front and back of a steel platform, which is mounted on four solid rubber-tyred wheels. The tow-bar, which activates the parking brake when lifted or lowered, can be restrained in the raised position by means of a hook. Optional features available include front- and rear-axle suspension, floor board, tarpaulin and flanged wheels.

Characteristics

PLATFORM DIMENSIONS: 1·5 × 3·02 m
HEIGHT OF HEAD AND BACK BOARDS: 0·485 m
LOADING HEIGHT: 0·585 m
TRAILER DIMENSIONS: 3·34 × 1·5 × 1·07 m
WEIGHT: 350 kg

Manufacturer

Siraga, Buzancais, France

Siraga baggage/freight trailer

Siraga Freight Terminals

Siraga specialise in the design, manufacture and installation of freight terminals from simple storage facilities to automated terminals. Examples in this area are the UTA terminal at Lyons with storage capacity for 250 pallets, the Air France freight terminal at Charles de Gaulle Airport which handles 300 000 tons of freight annually, and the Air Afrique terminal at Abidjan designed to process 100 000 tons of freight a year. Within the terminals storage racks available comprise the Aerolift and the 1000/1100 series for LD-1/LD-3 containers and pallets; and the CO 410/CO 400 for container storage. All are similar in design and construction with galvanised rollers mounted on ball-bearings fitted to a steel frame. The racks feature side guides for end loading, walkways and a retractable thrust block for manual control.

Three Aerolift models are available, the 1086 and 1186 with different sized rollers, and the 1286 which features lateral or longitudinal loading by swivelling of removable frames. The legs of the rack are adjustable in height between 330 and 520 mm. The 1000/1100 series consists of four models: 1044 and 1146 rated at 4500 and 7000 daN (4588/7137 kg) for lateral loading, and the 1144 and 1146 at similar ratings equipped for longitudinal loading. The CO 410/CO 400 are designed for the storage of full containers (CO 410) and empty containers (CO 400).

Manufacturer

Siraga, Buzancais, France

Siraga UTA freight terminal at Lyons

Characteristics

MODEL	Aerolift	1000/1100 series
UTILISATION	LD-1/LD-3 container	LD-1/LD-3 container
pallets	88 × 125 in (2·24 × 3·18 m) 96 × 125 in (2·44 × 3·18 m)	88 × 125 in (2·24 × 3·18 m) 96 × 125 in (2·44 × 3·18 m)
ROLLER HEIGHT	0·508 m 0·33–0·52 m (1286 model)	0·508 m
CAPACITY	7000 daN (7137 kg)	4500/7000 daN (4588/7137 kg)

MODEL	CO 400	CO 410
UTILISATION	Empty container	Full container
ROLLER HEIGHT	0·675 m	0·51 m
width between guides	1·61 m	1·58 m
length between thrust block	1·61 m	1·55–1·575 m
FRAME DIMENSIONS		
length	1·733 m	1·778 m
width	1·61 m	1·69 m
WEIGHT		148 kg

Siraga Pallet/Container Trailers

The FT 2001, FT 2011 and FT 2500 are designed for LD-1/LD-3 containers. The units comprise a steel frame equipped with castors at the front and fixed wheels at the rear. Galvanised rollers mounted on ball-bearings and walkways form the deck. Other features include a standard hook tow-bar, rear coupling hook, and parking brakes which are activated by lowering the tow-bar. The FT 2001 and 2011 are designed for lateral loading; the FT 2500 is equipped with a turntable.

The series 1000/1100 are suitable for transporting 88 by 125 inch (2·24 by 3·18 metre) and 96 by 125 inch (2·44 by 3·18 metre) pallets. The steel frame, incorporating galvanised rollers on ball-bearings and walkways, is mounted on swivelling wheels at the front and fixed wheels at the rear. Retractable thrust blocks which are manually controlled are incorporated in the deck. The trailer is fitted with a tow-bar with a standard ring, which activates the brakes when lowered in most models, but when raised in models 1008 and 1108, and a towing hook at the rear.

The FT 4012, 4016 and 4312 are similar in design to the 1000/1100 units, but rated at 14 000 daN (14 274 kg), and are capable of transporting 96 by 125 inch (2·44 by 3·18 metre), 88 by 125 inch (2·24 by 3·18 metre) pallets and 20-foot (6·1 metre) containers. The FT 4012 and 4016 are designed for lateral loading, the 4016 comprising two coupled trollies. The FT 4212 is equipped for longitudinal loading and the FT 4312 is capable of both lateral and longitudinal loading by orientating removable frames.

Siraga pallet/container trailer

Characteristics

MODEL	2001/2011/2500	1000/2000	4012/4016/4212/4312
UTILISATION	containers LD-1/LD-3	pallets 88 × 125 in (2·24 × 3·18 m) 96 × 125 in (2·44 × 3·18 m)	pallets 88 × 125 in (2·24 × 3·18 m) 96 × 125 in (2·24 × 3·18 m) 20 ft (6·1 m) container
ROLLER HEIGHT			
loaded	0·508 m	0·508 m	0·508 m
unloaded	0·515 m	0·515 m	0·515 m
CAPACITY	3510 lb (1600 kg)	4500, 7000, 8500 daN (4588, 7137, 8666 kg)	14 000 daN (14 274 kg)

Manufacturer

Siraga, Buzancais, France

Siraga Towable Baggage/Freight Conveyor

The towable conveyor suitable for loading baggage and similar sized freight has a capacity of 500 kg with an incline of 20 degrees, and 350 kg at the steeper gradient of 30 degrees. The height of the conveyor can be adjusted at either end. It is powered by a four-stroke petrol engine with Bernard electric starter and Leduc hydraulic pump. Other features include a parking brake activated by lowering the tow-bar, 64-ampere hours battery and push-button controls at the lower end of the boom.

Characteristics

ENGINE: W112 petrol engine
CONVEYOR BOOM:
FRONT HEIGHT: 1·2–4·5 m
REAR HEIGHT: 0·7–0·51 m
BELT WIDTH: 0·6 m
BELT SPEED: 20 m/minute
CAPACITY: 350/500 kg
CONVEYOR DIMENSIONS: 8·2 × 1·51 × 1·3 m
WEIGHT: 2200 kg

Manufacturer

Siraga, Buzancais, France

Sovam Ground Handling Equipment

A wide range of ground handling equipment for cargo operations is offered by Sovam. For carrying ulds between the aircraft and terminal building, Sovam produces a self-propelled short-haul transporter with 7-tonne capacity which is available configured with a through deck for side and front loading, or as a front loader with the operator's position at the side and the power unit at the back. For long-hauls a semi-trailer is available fitted with an automatic mechanical transfer deck which can carry up to four pallets.

For aircraft loading five types of self-propelled elevating loaders are produced. The PE 50 rated at 5 tonnes, the PE 70 at 7 tonnes and the PE 100 with a capacity of 10 tonnes are suitable as lower-deck loaders extending to a maximum height of 3·5 metres. The PE 140 with a loading capacity of 14/16 tonnes and capable of a maximum height of 5·5 metres can be used for main-deck loading. Designated PE 140 the unit can be equipped with a bridge. The largest loader is the PEB 180 which is fitted with a mini-bridge elevated by a separate scissors mechanism. Capable of extending to 5·5 metres the loader has a capacity of 18/20 tonnes.

Sovam PE70 loader

Sovam PE140 loader

Characteristics

MODEL	PE 50	PE 70	PE 100	PE/PEB 140	PEB 180
UTILISATION	LD Loader	LD Loader	LD Loader	MD Loader	MD Loader
CAPACITY	5 tonnes	7 tonnes	10 tonnes	14/16 tonnes	18/20 tonnes
MAXIMUM ELEVATION	3·5 m	3·5 m	3·5 m	5·5 m	5·5 m

Status

In service with major airlines including Air France, and UTA and with the French airport authorities

Manufacturer

Sovam, Parthenay, France

Tracma Tow Tractors

Tracma produce a wide range of tractors suitable for airport duties, available with a choice of petrol or diesel engines or electric motors. The petrol and diesel tractors have draw-bar pull capacities ranging from 3000 to 60 000 lb (1360 to 27 215 kg). The electric-powered vehicles produce a draw-bar pull of 3000 lb (1360 kg).

Status

In service at 133 airports world-wide

Manufacturer

Saxby, Department Tracma, Paris, France

GERMANY, FEDERAL REPUBLIC

Bizerba MCA Freight Weighing Machine

The MCA is an automatic air cargo weighing machine incorporating a digital display and controlled by a micro-processor. The equipment comprises the display unit, electronic unit, control panel and weighing machine. The display unit provides a digital weight presentation for the operator with a 13 mm high led seven-segment display, showing the weight and tare value of the freight in increments of 0·1 kg. Signals which indicate correct functioning of the equipment and fault location are included in the operator's display. Control of the weighing machine is by a micro-processor in the electronic unit, with a capability of distributing cargo weight values to a printer or central computer.

A memory store for recording cumulative weight can be supplied if required. The weighing machine consists of bending force weighing cells with a stainless steel platform, welded frame and built-in load lever action. Provision is made for the fitting of a moving belt. Weighing and checking keys with a pilot lamp are installed in the control panel.

MCA freight weighing machine

Characteristics

WEIGHING RANGE: 5–150 kg
DIMENSIONS
platform sizes: 600 × 600 mm, 800 × 600 mm, 1000 × 600 mm, 1200 × 600 mm
height: 150 mm
DISPLAY UNIT: 70–155 mm
ELECTRONIC UNIT: 400 × 234 × 120 mm
CONTROL PANEL: 60 × 145 × 55 mm
POWER REQUIREMENTS: 220 V ac, 50 Hz

Status

In use at airports in Federal Republic of Germany

Manufacturer

Bizerba Weighing Machine Works, Balingen, Federal Republic of Germany

BMA Pallet/Container Transporters

Three models of transporters are produced by BMA for carriage of pallets and containers between the aircraft loader and dollies. These are the universal TPC-070, the Star Carrier TPC-070D with a through deck and the heavy-duty TPC-180D also equipped with a through-deck loading system.

The TPC-070, with a 7-ton capacity is suitable for moving one pallet measuring 88 by 125 inches (2·24 by 3·18 metres), 88 by 108 inches (2·24 by 2·74 metres), or 96 by 125 inches (2·44 by 3·18 metres). Alternatively, one 10-foot (3·05-metre) cargo container can be carried or two LD-1/LD-3 luggage containers. The unit is powered by an air-cooled Deutz diesel engine, with hydrostatic drive to the rear wheels and power steering. The engine and hydraulic unit are arranged behind the loading area across the width of the transporter. Driving, lifting and loading operations are controlled from the operator's seat on an elevated stand attached to the left hand side of the vehicle. The front end is provided with loading 'fingers' which can be adjusted in height hydraulically between 0·47 and 6 metres, for alignment with the dolly or trailer. Cargo transfer is by three transport wheels in the loading fingers driven hydraulically by chains, with cylindrical supporting rollers and a fixed cargo drive on the bed. Two retractable pallet stops are at the front end. Spring-loaded and hydraulically-operated these stops are retracted automatically when the pallet drive system is actuated and are returned to their normal position by the action of the springs. The lateral divided guide-rails can be changed from 96 to 88 or 60½ inches (2·44 to 2·24 or 1·84 metres) for LD-1 and LD-3 containers.

Optional equipment available includes roofed operator's position glazed on three sides, hydraulically extensible operator's stand and extension of cargo drive from two to four rows.

The Star Carrier TPC-070D transporter is equipped for through-deck loading with vertical adjustment at both ends. Two stops which retract automatically are at either end of the unit. All other features are similar to the model TPC-070.

The TPC-180D is a heavy-duty transporter rated at 8-tons capacity, suitable for carrying two pallets measuring 88 by 125 inches (2·24 by 3·18 metres), 88 by 108 inches (2·24 by 2·74 metres) or 96 by 125 inches (2·44 by 3·18 metres). Alternatively, two 10-foot (3·05-metre) cargo containers or a single 20-foot (6·1-metre) cargo container can be carried. The unit powered by a 70 kW diesel engine has a through-loading system and features similar equipment to the TPC-070D, though vertical adjustment is limited to the front end only.

BMA Star Carrier container/pallet transporter

Status

In service with freight forwarders, major airlines and airport authorities in Europe, the Middle East, Africa, and north America

Manufacturer

BMA, Braunschweig, Federal Republic of Germany

Characteristics

MODEL	TPC-070	TPC-070D	TPC-180D
CAPACITY	7 tons	7 tons	18 tons
ROLLER DECK	5·5 × 2·5 × 0·508 m	5·4 × 2·5 × 0·508 m	6·4 × 2·5 × 0·508 m
VERTICAL ADJUSTMENT			
front	0·47–0·6 m	0·46–0·6 m	0·46–0·6 m
rear	–	0·46–0·51 m	–
ENGINE	Deutz diesel F4L 912	Deutz diesel F4L 912	Deutz diesel F6L 912
output	40 kW	40 kW	70 kW
travelling speed	10–24 km/h	10–24 km/h	10–24 km/h
turning radius	9 m	8·6 m	11·5 m
UNIT DIMENSIONS	7·25 × 3·28 × 2·5 m	5·54 × 3·65 × 2·2 m	8·5 × 3·85 × 2·2 m
WEIGHT	5 tons	4·6 tons	8 tons

BMA Pallet/Container Lower Deck Loaders

BMA 7-ton Star Loader servicing A 300B aircraft

The HBPC-070B Star Loader is designed for handling lower deck pallets and containers of all wide-body aircraft, in addition to DC-8, DC-9, Boeing 707, Ilyushin 18 and 62 aircraft, for threshold heights of up to 3·7 metres and maximum pallet weights of 7 tons. The loader is powered by a Deutz diesel engine, with hydrostatic drive and power steering. It features two working platforms. The operator's stand is on the right-hand side of the bridge platform accessible by a ladder. Elevation of this platform is achieved by a scissors mechanism hydraulically-operated and cargo transfer by means of ball castors and adjustable guide rails. For lateral adjustment of the loads it is possible to lower the cargo drive hydraulically. Folding adaptors are fitted either side of the 1·54-metre wide central unit to enable the loader to accommodate the loading hatches of different aircraft. Stops at the rear end in an extended position secure the load. When the bridge level has been reached lifting is terminated automatically and these stops are retracted by the central guide of the main platform, the front stops of the main platform being extended at the same time. Lifting of the main platform is performed by a scissors mechanism with stabilisation by telescopic cylinders. Cylindrical supporting rollers and fixed-cargo drive wheels provide longitudinal transport on the main platform. In the ground transfer (at 0·508 metres) the rear mechanical stops are lowered, and extend again when the main platform is raised. Optional features available are roll adjustment to the bridge platform, lowerable longitudinal uld transfer or transverse uld transfer provisions on the main platform.

The HBP-070 with one working platform is a simplified version of the Star Loader. Apart from the folding adaptors fitted in this loader either side of the 1·54-metre wide central unit for accommodating narrow aircraft doors, the features are similar to the TPC-070 main-deck loader which is described separately.

The HBC-020 is suitable for loading heavy mixed cargo into the belly compartments of Boeing 727-200, Boeing 707-320 and -420, BAe 1-11 and the DC-8. To accommodate these aircraft, the operator's stand on the single platform loader is arranged on the left-hand side and the unit is equipped with a 1·2 metre adaptor width and a folding adaptor measuring 0·4 by 0·35 metres on the left-hand front corner. The adaptors are fitted with a Delta rubber section in the front. Height adjustment of the platform between 508 and 2000 mm is controlled either from the operator's stand or from a portable remote-control box via a penduline telescopic cylinder acting on a bearing scissor. The platform, provided with ball castors, is lifted at a speed of approximately 0·1 metres per second. The rear platform edge is equipped with reinforced transfer rollers and, additionally, with a mechanical cargo stop which in a lowered position is automatically retracted. In the central adaptor area a manually-operated stop is fitted.

Characteristics

MODEL	HBPC-070B		HBP-070	HBC-020
PLATFORM	Main	Bridge	Main	Main
CAPACITY	7 tons	7 tons	7 tons	2 tons
DIMENSIONS	3·3 × 2·5 m	2·7 × 2·5 m	4·7 × 3·3 m	3·55 × 1·4 m
ELEVATING RANGE	0·508/3·7 m	1·875/3·7 m	0·508–3·7 m	0·508–2 m
ELEVATING TIME	28 s		45 s	20 s
LOWERING TIME	20–30 s		30 s	20 s
ENGINE	Deutz diesel F4L 912		Deutz diesel F4L 912	Deutz diesel F3L 912
output	40 kW		40 kW	31 kW
travelling speed	0–15 km/h		0–10 km/h	20 km/h
turning radius	8·3 m		7·5 m	6·5 m
UNIT DIMENSIONS	7·3 × 3·45 × 3·5 m		6·7 × 3·5 × 2 m	4·1 × 2·4 × 1·3 m
WEIGHT	9·5 tons		8·5 tons	3·6 tons

Status

BMA equipment is in service with major airlines and airport authorities world-wide

Manufacturer

BMA, Braunschweig, Federal Republic of Germany

BMA HBP-180/HBPC-180B Pallet/Container Main Deck Loader

BMA HBPC-180B main-deck loader servicing a Boeing 747

Two types of loaders are produced by BMA for main-deck loading of wide-body aircraft, the HBP-180 and the larger HBPC-180B unit. The HBPC-180B can handle the main- and lower-deck and lateral servicing of all wide-body aircraft and the freight compartments of narrow-body aircraft. The loader is equipped with two working platforms. Lifting of the bridge platform is performed by a solid single-scissor lifting mechanism assisted by four hydraulic cylinders acting on the outside of the bridge frame. Load transfer is effected by ball castors and twin transport wheels driven by chain and mounted in pairs on a wing which can be lowered hydraulically. Adjustable guide rails are fitted. For lower deck container handling, loading 'fingers' to the left and right of the rigid central member, which is 5 feet 1 inch (1·54 metres) wide, are folded down. For lower-deck pallet handling, the loading fingers are raised again subject to loading conditions at the aircraft doors. The cargo is secured by stops in an extended position at the rear end. These are retracted automatically when the main platform is aligned with the bridge platform, the main platform stops adjacent to the bridge being simultaneously retracted. Lateral inlets at the bridge for the main platform guide rolls facilitate load transfer. A portable remote-control box is positioned on a hinged bracket at the bridge platform railing where the transfer system in the aircraft is within the operator's reach. The main platform is equipped with cylindrical supporting rollers. Twin wheels arranged in pairs, driven hydraulically by chains, carry out longitudinal cargo movements. Folding guide rails are provided at both sides for 20-foot (6·1-metres) cargo containers.

At a transfer height of 1 foot 8 inches (0·508 metres) the rear mechanical stops are retracted. They extend when the main platform is elevated. The operator's cab is below the bridge platform at its right-hand corner. The frame narrowing towards its rear end carries the external height-adjustable wheels. The undercarriage of the bridge platform, welded to the frame, holds the driving and steering axle with differential gear and the diesel engine. Optional equipment available comprises roll adjustment to the bridge platform; main platform with lowerable longitudinal pallet/container transfer and main platform with transverse pallet/container transfer.

The HBP-180 a heavy-duty loader with one working platform is a simplified version of the HBPC-180B. The hydraulically-adjustable operator's seat is at the right-hand side of the working platform. For loading/unloading operations the platform is raised to a level approximately 100 mm below the loading/unloading height. This height estimated by the operator, having been reached, the loader proceeds automatically to the correct height, subsequent adjustments being made at very slow speed.

Logical interlocking allows pallets and containers to be handled only when the loader stabilisers are extended. Elevation of the platform is by a scissor-type lift ram system, supported by telescopic cylinders, and transfer is effected by cylindrical supporting rollers, fixed-drive wheels, and adjustable guide rails. A lowerable conveying system and ball castors, available as an option, are required behind the operator's stand to be able to load the platform laterally.

At a transfer height of 1 foot 8 inches (0·508 metres) the rear mechanical stops are lowered. They extend when the platform raises from the frame. The front end stops do not extend before the aircraft door has been reached and the conveying system been actuated.

Characteristics

MODEL	HBPC-180B		HBP-180
PLATFORM	Main	Bridge	Main
CAPACITY	18 tons	18 tons	18 tons
DIMENSIONS	6·4 × 3·3 m	4·3 × 3·3 m	6·4 × 3·3 m
ELEVATING RANGE	0·508–5·6 m	2·2–5·6 m	0·508–5·6 m
ELEVATING TIME	50 s	40–60 s adjustable	85 s
LOWERING TIME	50–60 s	40–60 s adjustable	50 s
ENGINE	Deutz diesel F6L 912		Deutz diesel F4L 912
output	70 kW		44 kW
travelling speed	0–15 km/h		0–10 km/h
turning radius	13·5 m		9 m
UNIT DIMENSIONS	11·5 × 3·96 × 3·2 m		8·3 × 3·5 × 2 m
WEIGHT	22·5 tons		10 tons

Status

In service with major airlines and airport authorities world-wide

Manufacturer

BMA, Braunschweig, Federal Republic of Germany

Linde Lift Trucks

Linde produce a wide range of lift trucks powered by diesel, electric motor and liquid propane gas. The diesel powered trucks vary from 1·2 to 8·0 tonne lifting capacities with Deutz air-cooled engines which have a two-stage direct fuel injection system. The trucks feature a hydrostatic transmission system which transmits power by the positive displacement of oil. It comprises a variable hydraulic drive unit with a speed control which selects engine speed and power output automatically.

A range of six electric-powered lift trucks are available, with capacities ranging from one to three tonnes. The units are equipped with hydraulic-regenerative braking systems and incorporate a design which enables each truck to be able to turn within its own length. The smaller trucks, those with up to 1·5 tonne capacities feature an electric drive motor on each front wheel, providing a total of 4 kW drive power. A similar arrangement on the larger models results in a total of 11·2 kW drive power. Hydraulic motors fitted to electric drive trucks are rated at 4·5 kW capacity for the smaller models, 10·5 kW for the 2 and 2·5-tonne units and 11·8 kW for the 3 tonne version. An SCR control system is installed which prevents roll-back on gradients and operates the truck's

Linde 2·5-tonne electric lift truck

travel direction switch by solid-state photo-electric cells. Optional six-stage contactor systems are available on the three smaller models in the range. The 24-volt battery systems have a capacity of 700 ampere hours in the 1 and 1·2-tonne models and 800 ampere hours in the 1·5 tonne model. Larger models are equipped with 72-volt battery systems with a choice of 430 to 600 ampere hours capacities for the 2-, 2·5- and 3-tonne units. Two pedal drive and single lever lift/lower controls coupled with forward, reverse creeping and inching control are standard features. Five liquid propane gas powered lift trucks are available with 1·2-, 1·5-, 2-, 2·5- and 3-tonne capacities. The units share identical drive and gear pump systems with the company's diesel range, the only major difference being in the engine assembly. Three liquid propane gas trucks feature the same VW engine producing 35 kW at 2650 rpm.

Manufacturer

Linde AG, Aschaffenburg, Federal Republic of Germany

Linde 1·2-tonne electric lift truck with triple height mast

Lodige Freight Terminals

Lodige specialise in the design, manufacture and installation of freight terminal equipment. A typical example is the one built for Changi International Airport Services in Singapore. The system is fully automated and can handle up to 150 000 tonnes of freight annually with an expansion capability of up to 260 000 tonnes. The heart of the terminal is a pallet/container handling system (pchs). Outbound freight transported to the terminal is fed into the pchs. Storage and retrieval of cargo is performed by an elevating transfer vehicle with a capacity of 14 000 kg, capable of travelling 25 metres per minute vertically and 100 metres per minute horizontally with the ability to handle 20-foot (6·1-metres) ulds. The three-level storage area features two work station areas for 10-foot (3·05-metres) uld build up/breakdown, ten automatic accumulating input/output lanes, 104 storage positions holding one 20-foot container or two 10-foot containers, one 20-foot work station and a truck interface point. The work station areas for 10-foot ulds are equipped with 12 working positions interconnected by automatic conveyor systems each containing five right-angle decks and a turntable.

Status

The Changi terminal was completed in 1981. Other areas where Lodige equipment is in operation include the Sabena freight terminal at Brussels International Airport, Belgium and the KLM freight terminal at London Heathrow Airport, England

Manufacturer

A Lodige GmbH, Paderborn, Federal Republic of Germany

MBB Airport Vehicle Control System

MBB's airport vehicle control system monitors and controls the operation of all ground vehicles and equipment. By analysing all route logistics the system dispatches and controls ground traffic between the place of operation and the aircraft. For example, ground traffic between the aircraft parking position, the fuel supply bay, freight area, ramp equipment station, apron and passenger terminals. The vehicles keep the control headquarters informed of their position between their starting point (eg the depot) and their destination (eg aircraft parking position). As a result, the headquarters may re-route vehicles between their starting point and their destination which may be up to 4 km apart or may commission them to take over additional functions on the way. Due to the large distances at most airports, some synchronising points may be provided on the apron, as this permits vehicle locations to be established more accurately.

Driver terminal installed in airport vehicle

Characteristics

FREQUENCY RANGE: 80, 160, 460 MHz
CHANNEL PATTERN: 20 or 25 kHz
MODULATION: FM, differential phase shift keying
MODE OF OPERATION: Simplex, semi-duplex, half-duplex, full-duplex
TRANSMITTER POWER: up to 6 W
TRANSMITTER KEYING: electronic keying filter
ANTENNA CHANGE-OVER: electronic
TRANSMISSION RATE: 1200/2400 bits/s switchable
TERMINAL ADDRESS: 10 bits (1024)
MESSAGE LENGTH
operations control centre to vehicle: 69 bits
vehicle to operations control centre: 60 bits
TEMPERATURE RANGE: 0 to +45°C
UNIT DIMENSIONS
driver's input/output terminal
140 × 300 × 60 mm
vehicle electronic unit
140 × 300 × 300 mm
WEIGHT: 5 and 10 kg

Status

This system entered a trial operation at Frankfurt-Main Airport, Federal Republic of Germany at the end of 1981 where it is applied to the control and co-ordination of baggage transport vehicles by means of a computer program and digital data communication. A detailed description of the equipment is contained in the Passenger Handling section under Federal Republic of Germany

Manufacturer

Advanced Transport Systems, Messeschmitt-Bolkow-Blohm GmbH, Munich, Federal Republic of Germany

Roder Prazision RP 23 B Mobile Conveyor

Roder Prazision RP 23 B mobile conveyor

The RP 23 B mobile conveyor can service aircraft with sill heights from 1·25 to 4·85 metres. Power is provided by a Perkins diesel engine with Borg Warner 65 automatic transmission and the vehicle is equipped with power steering and a servo-assisted dual braking system. The 1-metre wide conveyor will accommodate loads to a maximum load density of 200 kg per square metre.

Characteristics

ELEVATING HEIGHTS
front: 1·25–4·85 m
rear: 1·63–0·6 m
CONVEYOR BOOM: 7·95 × 1 m
BELT WIDTH: 0·65 m
CAPACITY AT 15°/33°: 1500/100 kg
MAX LOAD DENSITY: 200 kg/m^2
ENGINE: Perkins 4108 (petrol V4/V6 optional)
DRIVING SPEED: 70 km/h
WHEELBASE: 2·8 m
TRACK
front: 1·67 m
rear: 1·59 m
TURNING RADIUS: 14·2 m
UNIT DIMENSIONS: 7·95 × 2 × 1·48 m
WEIGHT: 2600 kg

Status

Introduced at the Frankfurt airport equipment exhibition in 1980, the RP 23 B is the result of a joint development between Roder Prazision and Boyd Airport Equipment Ltd of England

Manufacturer

Roder Prazision, Kreis Offenback, Engelsbach, Federal Republic of Germany

Rubner Electric Tractor

Two models are produced by Rubner with towed-load capacities of 6000 and 10 000 kg. Mounted on a front-steering axle and a rear-driving axle powered by a series-type motor the open driver's compartment contains two seats, the driver's being hydraulically adjustable. Steering is via a standard linkage with ball and socket joints, and variable speed control in forward and reverse is attainable through a foot pedal. Hydraulic internal expanding brakes are fitted to all four wheels operated by a foot pedal with a hand lever parking brake acting on the driving axle. The tractor is equipped with a multiple stage insert coupling for different tow-bar heights in the rear and a trailer-handling coupling in the front. Standard accessories include head and tail lights and reflectors, electric horn, ignition key and battery charging plug.

Manufacturer

Gotz Rubner, Mellrichstadt/Bayem, Federal Republic of Germany

Characteristics

MODEL	EFZ 6000	EFZ 10000
TOWING CAPACITY	6000 kg	10 000 kg
ENGINE	electric	electric
output/h	3 kW	5 kW
BATTERY	80 V	80 V
5-h rating	280 A/h	280 A/h
DRIVING SPEED	7–12 km/h	6–10 km/h
DRAW-BAR PULL		
max 5-minute rating	250 kg	400 kg
continuous	150 kg	302 kg
TURNING RADIUS	2·06 m	2·06 m
UNIT DIMENSIONS	2·12 × 0·87 × 1·25 m	2·12 × 0·87 × 1·3 m
WEIGHT	1470 kg	1487 kg

Rubner Electric Truck

Rubner produces three electric trucks with capacities 1000, 2000 and 3000 kg. Battery-powered with electronic-pulse control using a series wound motor, the trucks are of steel construction. The open driver's position equipped with two seats is protected from the loading area by a steel panel. Hydraulic internal expanding brakes are fitted, to two wheels in EPW 1000 and to all four wheels in the other two models. These are pedal-operated with a hand-lever acting as a parking brake, which disconnects the current when engaged. Standard accessories include towing point, battery charging plug, lighting and signalling system including a push-button electric horn and an ignition key. As an optional feature the driver's compartment can be enclosed.

Characteristics

MODEL	EFW 1000	EFW 2000	EFW 3000
CAPACITY	1000 kg	2000 kg	3000 kg
ENGINE	Electric	Electric	Electric
output/h	2 kW	5 kW	5 kW
BATTERY	24 V	80 V	80 V
5-hour rating	200	280	280
DRIVING SPEED	9–12 km/h	13–16 km/h	13–16 km/h
DRAW-BAR PULL			
max 5-minute rating	260 kg	350 kg	360 kg
continuous with nominal load	62 kg	90 kg	65 kg
without nominal load	86 kg	130 kg	125 kg
TURNING RADIUS	2·7 m	3·35 m	3·21 m
PLATFORM DIMENSIONS	1·65 × 1·1 m	2·25 × 1·25 m	2·25 × 1·25 m
LOADING HEIGHT	0·5 m	0·75 m	0·77 m
UNIT DIMENSIONS	2·75 × 1·1 × 1·22 m	3·45 × 1·25 × 1·3 m	3·6 × 1·25 × 1·4 m
WEIGHT	720 kg	1780 kg	1860 kg

Manufacturer

Gotz Rubner, Mellrichstadt/Bayem, Federal Republic of Germany

Rubner Pallet Truck

The twin forks of the Rubner pallet truck are raised hydraulically by operation of a foot lever which also controls the descent. A load safety valve prevents an overload of more than 10 per cent. The forks are mounted on twin rubber wheels at the rear, to which a steering arm is attached, with wheels underneath each fork at the front. Maximum capacity of the truck is 2500 kg. It is available in two models. Alternative fork lengths and widths can be supplied.

Manufacturer

Gotz Rubner, Mellrichstadt/Bayem, Federal Republic of Germany

Rubner Pedestrian-controlled Electric Truck

Two types of pedestrian-controlled electric truck are available from Rubner. The type EGWA with a capacity of 1500 kg is equipped with an asymmetrically-positioned drive unit. The battery-driven motor which features electronic pulse control is operated by a toggle switch on the control column. The platform can be supplied with or without sides.
The EGW type consists of two models with differing capacities. The platform is mounted on two axles, the steering axle at the front with an encapsulated ball-bearing slewing ring and a jointed cross-shaft axle of box-type design at the rear to which the transmission is connected. Power is from two parallel 24-volt batteries which supply a series electric motor providing a driving speed from 0 to 5 km per hour by means of a variable electronic-pulse control. Controls are fitted at the end of a driving shaft connected to the steering axle, and the elevation of the shaft operates a drum-type motor brake producing a driving or braking action according to the shaft's position.

Manufacturer

Gotz Rubner, Mellrichstadt/Bayem, Federal Republic of Germany

Characteristics

MODEL	EGWA 1500	EGW 1500	EGW 2000
CAPACITY	1500 kg	1500 kg	2000 kg
ENGINE	Electric	Electric	Electric
output/h	1 kW	1 kW	2 kW
BATTERIES	2 × 24 V	2 × 24 V	2 × 24 V
5-h rating	140/160 A/h	160 A/h	180 A/h
DRIVING SPEED	4–5 km/h	4–5 km/h	4–5 km/h
TURNING RADIUS	2·1 m	2·45 m	2·45 m
PLATFORM DIMENSIONS	2 × 1 m	2·5 × 1·25 m	2·5 × 1·25 m
LOADING HEIGHT	0·5 m	0·65 m	0·65 m
UNIT DIMENSIONS	2·6 × 1 × 1·47 m	2·75 × 1·25 × 1·5 m	2·75 × 1·25 × 1·5 m
WEIGHT	465 kg	500 kg	580 kg

Scharfenstein Baggage/Cargo Trailer

Scharfenstein produce two types of trailer suitable for transporting baggage and freight between the aircraft and the terminal building. Both feature similar undercarriage arrangements with tow-bar actuated steering and brakes, rear towing point, front and rear axles and solid rubber wheels. The baggage trailer with a 1250 kg capacity has open sides, while the cargo trailer capable of carrying 2000 kg with an option for 3000 kg, is equipped with a roof frame and a weather-proof cover.

Characteristics

TYPE	Baggage	Cargo
PLATFORM DIMENSIONS	2 × 1 m	3·4 × 1·5 m
LOADING HEIGHT	0·57 m	
CAPACITY	1250 kg	2000/3000 kg
WHEELBASE		2·9 m
TRACK	1·25 m	1·25 m
TOTAL LENGTH		3·9 m including tow-bar

Status

Both types of trailer are in service at airports in the Federal Republic of Germany

Manufacturer

Fritz Scharfenstein, Nister/Hachenburg, Federal Republic of Germany

Scharfenstein Baggage/Freight Mobile Conveyors

Three models of mobile conveyors are available from Scharfenstein designed to load and unload baggage and freight from the aircraft.
The conveyor belt vehicle 9/1 is based on a Mercedes Benz chassis with enclosed driver's cab. A Mercedes Benz hydraulic torque-converter is fitted with power for the hydraulic pump being taken off the crankshaft via a pulley and fan belt. The conveyor boom, equipped with side guards elevates to a height of 4·5 metres at the front only, the rear end being at a fixed height of 0·7 metres. The model 5/2 is similar to the 9/1 but both ends of the boom can be adjusted in height.
The conveyor belt baggage loader is based on an enclosed van body mounted on a Mercedes Benz chassis, with a five-speed gear box and a pto driving a hydraulic pump. A Mercedes Benz hydraulic torque converter can be supplied as an optional feature. The elevating boom extends from the van body, parallel to the driver's cab.

Status

In service at airports in the Federal Republic of Germany

Manufacturer

Fritz Scharfenstein, Nister/Hachenburg, Federal Republic of Germany

Characteristics

MODEL	9/1	9/2	Baggage Loader
ELEVATING HEIGHT RANGE			
front	1·35–4·5 m	1·35–4·5 m	
rear	0·7 m	0·7–1·5 m	
CONVEYOR WIDTH	0·9 m	0·9 m	
CAPACITY	1000 kg	1000 kg	2500/3000 kg
CHASSIS	Mercedes Benz 407/D	Mercedes Benz 407/D	Mercedes Benz L608/D
INSIDE DIMENSIONS			
length			4·5/5 m
width			2·35 m
height			1·95 m

Scharfenstein Pallet/Container Dollies and Rack

Three types of dollies are produced by Scharfenstein for transporting pallets and containers and a rack for the storage of pallets.
The container dolly is designed for LD-1 and LD-3 containers and features an automatic loading capability with a rotating roller bed and A-type steering. Two models of the pallet dolly are available with capacities of 4500 or 6800 kg. They are mounted on castors, with tow-bar steering and are suitable for transporting pallets measuring from 88 by 108 inches (2·24 by 2·74 metres) to 96 by 125 inches (2·44 by 3·18 metres), 10-foot (3·05 metres) pallets and several types of containers.
The combined pallet and container dolly has similar capacities to the pallet dolly but is equipped with a rotating platform and A-type steering. The dolly can be configured for automatic loading using a friction roller drive as an optional feature. The pallet rack is a fixed base version of the pallet dolly with a capacity of 4500 or 6800 kg.

Characteristics

MODEL	Container dolly	Pallet dolly	Combined dolly	Pallet rack
CAPACITY	1600 kg	4500/6800 kg	4500/6800 kg	4500/6800 kg
LOAD				
pallets		From 88 × 108 in (2·24 × 2·74 m) to 96 × 125 in (2·44 × 3·18 m) 10 ft (3·05 m)	From 88 × 108 in (2·24 × 2·74 m) to 96 × 125 in (2·44 × 3·18 m) 10 ft (3·05 m)	From 88 × 108 in (2·24 × 2·74 m) to 96 × 125 in (2·44 × 3·18 m) 10 ft (3·05 m)
CONTAINERS	LD-1/LD-3	LD-1/LD-3/LD-5 LD-6/LD-7/LD-9	LD-1/LD-3/LD-5 LD-6/LD-7/LD-9 LD-11/LD-21 20 ft (6·1 m)	LD-1/LD-3/LD-5 LD-6/LD-7/LD-9

Status

Scharfenstein equipment is in service at airports in the Federal Republic of Germany

Manufacturer

Fritz Scharfenstein, Nister/Hachenburg, Federal Republic of Germany

Schenck Freight Handling Systems

Schenk have developed several types of modular freight handling systems which can be integrated as necessary to suit individual operator's requirements.

The carrier system designed for automatic or manual operation uses slave pallets, or carriers within the system to overcome the problems of the different sizes of ulds in operation with airlines. A unit load from the aircraft arriving into the system is automatically connected with a slave pallet carrier (3·3 by 2·6 metres), capable of accommodating a 10-foot (3·05 metre) unit load weighing up to 6·8 tons. For a 20-, 30- or 40-foot (6·1-, 9·1-, 12·2-metre) unit load, two, three or four carriers may be used coupled together for conveyance through the system.

The unit load and carrier remain together within the installation, so that, after initial identification and correlation between the unit load and the carrier, the latter can be coded for automatic identification of the load. All stationary system elements of this system have a standard unit length of 3·6 metres, matched to the size of a 10-foot (3·05 metre) unit load.

The important elements within the carrier system comprise: long- or short-side flow decks; right-angle transfer decks; unit load to carrier mating units; unit-load-from-carrier separating units; weighing decks and vehicles for horizontal and/or vertical travel (transfer vehicles and elevating transfer vehicles).

The roller conveyor system has been designed for small- to medium-sized installations which may be operated using manual or simple automatic control techniques. The basic element of this system is a roller deck for 10- and 20-foot (3·05- and 6·1-metre) aircraft pallets and containers. These are available as an active element within the conveyor section equipped with its own drive, or as a passive element within the storage and conveyor sections. The deck may be chain driven from a neighbouring deck or powered via a friction roller drive activated from a transfer vehicle or elevating transfer vehicle. Weighing stations, lift tables, right-angle transfer decks, transfer vehicles and elevating transfer vehicles can be incorporated in the system as required.

The roll box system is designed to handle and store, automatically, small- to medium-sized shipments with a maximum total weight of 600 kg in a screen-type roll box measuring 1·1 by 1·4 metres, 1·2 metres high. Used around the build-up and break-down areas, the roll box can be moved around on its own wheels, while within the system it is moved on specially designed driven flanged rubber wheels. These are incorporated in the guides which operate within the track for guiding and handling the cargo. As in other systems, the roll box system is designed in modular blocks. The most important elements of this system are: flow decks, right-angle decks, turntables, weighing decks, elevators and exit and inlet stations, including stacker cranes.

The sorter system is used to sort small shipments in large numbers by destination. This applies, for instance, to transit shipments, which must be sorted as quickly as possible for the departing flights. The heart of the system is a sorting conveyor with tilting slats, which guides the individual shipments, coded on the inlet feeder belt, to the respective target belt. The shipments entered into this system are immediately sorted. From the target belts the shipments are then transferred to shelves, or immediately placed in roll boxes or baggage carriers for further transport.

Status

Schenck sorter systems are in operation in the Lufthansa air cargo terminal at Frankfurt. In its initial phase the facility is able to handle more than 500 000 tonnes of cargo per year. Specially designed carriers are used for pallets and containers and in the first phase more than 750 are in operation in the system. Approximately 3800 roll boxes are used mostly along the 1800 metre continuous handling section and there are 3516 roll box slots in the vertical racking system which is fed by ten stackers.

A vertical container racking system measures 103 metres long by 23 metres high which can hold 496 of the 10-foot (3·05-metre) containers and 122 20-foot (6·1-metre) containers. There are two stackers for the 10-foot (3·05-metre) units (7·5-tonne payload) and one stacker for 20-foot (6·1-metre) containers (15-tonne payload). The sorter system for mini-shipments has approximately 9000 stow locations, 62 feed-out positions and a tilt-tray conveyor 268 metres long

Manufacturer

Carl Schenck AG, Dormstadt, Federal Republic of Germany

Still Electric Pallet Truck/Stacker

Still produce two units for pallet transfer and stacking, the type EGU pallet truck and the type EGV stacker.

The type EGU pallet truck is equipped with tiller steering by means of a spring-balanced column to the drive unit, which is mounted on a four-point bearing. Travel speeds for forward and reverse are controlled by a selective cam-operated switch in the head of the column. Travel direction is changed from forwards to reverse by pressure on the safety column head and when the safety switch is released the truck comes to a stop. The standard truck has three speeds forward and reverse operating through a resistance control. As an option, an electronic-pulse control can be fitted which allows variable increase and decrease of the travelling speed. Power is provided by a 1 kW traction motor, with a 2 kW hydraulic-pump assembly supplying two lifting cylinders. Braking is automatic when the column is in the horizontal or vertical position.

The type EGV pedestrian controlled stacker is similar in design to the pallet truck except for the hydraulic equipment. In the stacker, an enclosed motor drives a high-pressure geared pump. Lifting and lowering movements are operated by control valves and the system is protected from overloading by a pressure relief valve.

Characteristics

MODEL	EGU 16	EGU 20	EGU 30	EGV 125	EGV 160
CAPACITY	1600 kg	2000 kg	3000 kg	1250 kg	1600 kg
MAX LIFT HEIGHT	0·115 m	0·115 m	0·115 m	2·68/2·98/ 3·58 m	2·58/2·88/ 3·48 m
ELECTRIC MOTORS					
drive	1 kW	1 kW	1 kW	1 kW	1 kW
pump	2 kW	2 kW	2 kW	2 kW	2·5 kW
battery	24 V	24 V	24 V	24 V	24 V
capacity	210 A/h	210 A/h	210 A/h	210/180 A/h	210/180 A/h
TURNING RADIUS	1·52/1·68 m	1·52/1·68 m	1·52/1·68 m	1·55 m	1·55 m
FORK DIMENSIONS					
width	0·18 m	0·18 m	0·2 m	0·18 m	0·18 m
overall width	0·56 m	0·56 m	0·58 m	0·56 m	0·56 m
length	0·9/1·12 m	0·9/1·12 m	0·9/1·12 m	0·9/1·12 m	0·9/1·12 m
UNIT DIMENSIONS					
length	1·69/1·85 m	1·69/1·85 m	1·69/1·85 m	1·7/1·92 m	1·7/1·92 m
width	0·71 m	0·71 m	0·71 m	0·86 m	0·86 m
height (column mast lowered)	0·77 m	0·77 m	0·77 m	1·81/1·96/ 2·26 m	1·81/1·96/ 2·26 m
WEIGHT	625 kg	635 kg	650 kg	1200/1140 kg	1200/1140 kg

Manufacturer

Still GmbH, Hamburg, Federal Republic of Germany

Still R6/R7 Tow Tractors

The two types of tractors suitable for towing baggage and freight trailers are the series R6 electric and the series R7 diesel and electric tractors.

The Series R6 rated at six tonnes towing capacity, is based on a three wheeled configuration with traction drive to the rear wheels via a 2·8 kW 24-volt motor and Stilltronic SCR control. Hydraulically-operated brakes and independent suspension are fitted on all wheels and a mechanical parking brake operates on the rear drive wheels. The vehicle is equipped with a multi-position, rear-mounted towing hitch with a captive hitch pin.

The Series R7, with a 20-tonne towing capacity, is available powered either by an electric motor or a dc generator coupled to a diesel engine. A four-wheel vehicle, the unit features a regenerative braking system with hydraulically-operated brakes and independent suspension on all four wheels. The parking brake is operated by cable from a hand lever. Type 244D and K2D Rockinger automatic towing couplings, are fitted at the rear of the tractor, with push-button inching control to position the truck accurately for coupling, from the towing position. The differential rear wheel drive is via a 12 kW, 80-volt traction motor in both versions. An enclosed driver's cab is fitted as standard equipment in the diesel engined unit.

Characteristics

MODEL: R6
MOTOR: electric
OUTPUT: 2·8 kW
BATTERY: 24 V
capacity: 500 A/h (5-h rate)
DRAW-BAR PULL
max: 330 kg (5 minutes)
continuous: 120 kg
WHEELBASE: 1·15 m
TRACK: 0·79 m
OUTSIDE TURNING RADIUS: 1·35 m
UNIT DIMENSIONS: 1·7 × 0·91 × 1·3 m
WEIGHT: 880 kg

Status

In service at Hamburg and Frankfurt airports, Federal Republic of Germany

Manufacturer

Still GmbH, Hamburg, Federal Republic of Germany

Still R6 tow tractor

Still R50/R60 Electric Fork Lift Trucks

Two types of electric fork lift trucks are available from Still, the R50 series rated at 1, 1·2 and 1·5 tonnes and the R60 series comprising six models with capacities ranging from 1·5 to 4 tonnes.
The R50 models have a four-wheel configuration, with a single motor rear-wheel drive and optional electronic pulse-controlled power steering. With this form of steering, control is by a solid-state logic unit via an electrical transmitter on the steering column. The speed of the motor powering the steering pump is proportional to the steering effort required and the motor will not operate unless the steering wheel is moved. Dual servo-assisted brakes are fitted to the front wheels.
The vehicle is available with a choice of three masts, the telescopic for warehouses or freight terminals of orthodox construction, the Hi-Lo full free mast for use in areas with a low headroom and a three stage mast suitable for high stacking operations but permitting passage through low doorways. A towing hitch is mounted to the rear of the lift truck.
The R60 models incorporate similar features to the R50 though the electronic steering is standard equipment and the dual-hydraulic brakes are coupled to a regenerative braking system. With this system, during the first third of the brake pedal stroke, the traction motor acts as a generator feeding the energy back into the battery. When the brake pedal is depressed further hydraulic braking commences.

Status

In service at airports in the Federal Republic of Germany

Manufacturer

Still GmbH, Hamburg, Federal Republic of Germany

Still Series 8 Electric Platform Truck

The Series 8 electric platform truck rated at 1·2 and 2 tons carrying capacity is suitable for the carriage of baggage and lightweight freight and for towing trailers/dollies. A four-wheeled vehicle, the unit is equipped with a differential rear-wheel drive via a 4 kW, 80-volt traction motor and Stilltronic SCR control. Independent suspension and a regenerative braking system with hydraulically-operated brakes are fitted to all wheels, and a mechanical parking brake operates on the drive wheels. A type 244D Rockinger automatic coupling is mounted on the rear of the vehicle.

Status

In service at Stockholm Airport, Sweden and Hamburg and Frankfurt Airports, Federal Republic of Germany

Manufacturer

Still GmbH, Hamburg, Federal Republic of Germany

Characteristics

MODEL: R8
MOTOR: electric
OUTPUT: 4 kW
BATTERY: 80 V
capacity (5-h rate): 165/220 A/h
DIMENSIONS: 2·21 × 1·25 m
DRAW-BAR PULL
max — laden: 322/304 kg
(5-minute rating) — unladen: 346/344 kg
continuous — laden: 64/46 kg
unladen: 88/86 kg
WHEELBASE: 1·9 m
TRACK: 0·99 m
OUTSIDE TURNING RADIUS: 3·12 m
UNIT DIMENSIONS: 3·54 × 1·25 × 1·35 m
WEIGHT: 1430/1555 kg

Trepel PCL 21540 Pallet/Container Loader

This Trepel loader is standardised for handling baggage/cargo containers and pallets up to a width of 8 feet (2·44 metres) on two independent operating platforms. All conventional cargo aircraft, as well as the lower-deck holds of wide-body aircraft can be serviced. In operation the loader approaches the aircraft and the front platform is aligned to the cargo door sill. Adjustment of the leading edge of the front platform, to suit lower deck holds, is by means of hinged sections. A remote-control panel, located on the front platform, contains all the push-buttons for lifting and conveying required for one-man operation. Adjustable side guides restrain ulds during through loading and the main and front platforms align automatically during each lift cycle. Load transfer is by rubber-drive wheels with a ball mat surface for load adjustment.
The 21540 can transfer ulds of all standard sizes up to 96 by 125 inches (2·44 by 3·18 metres) to and from dollies and/or transporters.

Trepel PCL 21540 7-tonne loader at London Gatwick Airport

Status
In service at London Gatwick Airport, England and with Bremen, Frankfurt and Hamburg airport authorities, Federal Republic of Germany and Alia, Olympic, THY-Turkish Airlines and Nigerian Airways

Manufacturer
Trepel AG, Wiesbaden, Federal Republic of Germany

Characteristics
CAPACITY: 7000 kg
LIFTING HEIGHTS
main platform: 0·508–3·7 m
front platform: 1·9–3·7 m
MAX LOAD DIMENSIONS: 96 × 125 in (2·44 × 3·18 m)
ENGINE: 4-cylinder diesel air-cooled
OUTPUT: 46 kW
DRIVING SPEED: 0–15 km/h
LIFTING SPEED: 200 mm/s
CONVEYING SPEED: 300 mm/s
UNIT DIMENSIONS: 8·15 × 3·245 × 1·9 m
WEIGHT: 12 tons

Trepel PCL 21560 Pallet/Container Loader

The PCL 21560 Trepel self-propelled loader is designed to handle cargo containers up to 20 feet (6·1 metres) long and weighing 18/20 tons, pallets up to 125 inches (3·18 metres) transverse loading and all lower-deck containers on two operating platforms. It will service the main deck and lower deck compartments of Boeing 747, DC-10, L-1011 and A 300 aircraft, as well as cargo aircraft with low door-sill heights down to the DC-9 aircraft.
In operation the loader approaches the aircraft and aligns the front platform to the cargo door sill. Hinged sections on both sides of the bridge can be raised or lowered to dock with the cargo doors of various types of aircraft. Loading and unloading is a one-man operation by means of remote-controls from the front platform. Stabilisers extend/retract automatically and the ulds on both platforms are transferred by means of drive wheels. Automatic- and manually-operated side rails and stops restrain cargo units of various dimensions. Ball mats ensure alignment of pallets/containers on the front platform and omni-directional transfer on the main platform for side loading of 10- and 20-foot (3·05- and 6·1-metre) containers.

Characteristics
CAPACITY: 20 tons
LIFTING HEIGHT:
main platform: 0·508–5·6 m
front platform: 1·9–5·6 m
MAX LOAD WIDTH: 125 in (3·18 m)
ENGINE: 6-cylinder diesel air-cooled
OUTPUT: 70 kW
DRIVING SPEED: 0–15 km/h
LIFTING SPEED: 400 mm/s
CONVEYING SPEED: 300 mm/s
UNIT DIMENSIONS: 12·31 × 3·8 × 1·9 m
WEIGHT: 26 tons

Manufacturer
Trepel AG, Wiesbaden, Federal Republic of Germany

Trepel PCL 21560 main deck loader

Status
In service with airport authorities at Cologne/Bonn, Frankfurt and Hamburg, Federal Republic of Germany, Linz, Austria and with Alia, Sabena, Finnair, South African, Nigerian and Olympic Airways

---

Trepel PCT 20570 Pallet/Container Transporter

The PCT 20570 is a diesel-engine powered transporter fitted with hydrostatic drive and power steering for transporting ulds up to 10 feet (3·05 metres) long between the aircraft loader and the uld dollies. Fixed-side guides are set for 96-inch (2·44-metre) wide loads while lower-deck containers are restrained by retractable side guides. Other features include hydraulic-powered end stops and remote height adjustment at both ends. All operations are controlled from the cab by the driver. Manual controls select drive and uld transfer, foot-operated pedals control speed and actuate the dual hydraulic brake system. In operation, the transporter approaches the loader or the dolly and aligns its bed height for uld transfer. The driver preselects the stops which are actuated automatically in accordance with the selection transfer direction. The load is moved by seven sets of powered-drive wheels and idler rollers.

Characteristics
CAPACITY: 7000 kg
PLATFORM HEIGHT ADJUSTMENT
front: 465–625 mm
rear: 475–520 mm
MAX LOAD DIMENSIONS
width: 2·44 m
length: 5·36 m
ENGINE: 4-cylinder diesel air-cooled
OUTPUT: 46 kW
DRIVING SPEED: 0–25 km/h
CONVEYING SPEED: variable 0–300 mm/s
TURNING RADIUS: 7·5 m
UNIT DIMENSIONS: 5·58 × 3·7 × 2·35 m
WEIGHT: 6000 kg

Status
In service with the Frankfurt Airport Authority, Federal Republic of Germany

Trepel 20570 uld transporter

Manufacturer
Trepel AG, Wiesbaden, Federal Republic of Germany

Wiedeman Tow Tractors

Five models of tractors are produced by Wiedeman suitable for towing baggage and cargo trailers. All are available with diesel-, petrol- or liquid propane gas-fuelled engines and automatic transmission. The DW 13 and T 40 models are equipped with a loading platform behind the enclosed driver's cab.

Characteristics

MODEL	DS 40	DS 60	DS 45/75/100	DW 13	T 40
max draw-bar pull	2550 kg	2960 kg	4445 kg	1585 kg	1000 kg
TOWING CAPACITY	55 tons	85 tons	110 tons	50 tons	24 tons
PAYLOAD				2000 kg	4000 kg

Manufacturer
Wiedeman, Wedemark, Hanover, Federal Republic of Germany

ITALY

Aeritalia ALBA 2 Cargo Handling Data System

ALBA 2 is an improved version of an Aircraft Loading and Balancing Automation System (ALBA) for co-ordinating the turnround operations in small- and medium-sized airports. The system configuration is tailored to individual airport's requirements. The main functional units of the system are co-ordination of the passenger handling, aircraft loading and balance, cargo handling, monitoring of flight information and the telex network intercommunication. For cargo handling, the system co-ordinates acceptance of export goods and outgoing mail in the cargo terminal and produces a final print-out or display of the warehouse situation and a cargo manifest for the various flights. Import goods are accepted for storage and the system provides a print-out of the invoices for goods collected from the warehouse, with a daily print-out or display of the incoming cargo and the customs register. For aircraft balancing, the system monitors on-board cargo distribution and provides a check of the balancing index. Loading instructions and a load sheet print-out are produced together with trim indications for the different types of aircraft.

Manufacturer
Aeritalia, Caselle Torinese, Italy

Characteristics
COMPUTER: DEC PDP 11 with memory ranging from 128 kW–1 MW
DISC CAPABILITY: 56–512 MBy
TERMINAL: DEC VT 100
NO OF TERMINALS: 8–64
PRINTER: DEC LA 120 (180 cps)
NO OF PRINTERS: 1–16

Status
The first ALBA system was installed in the Caselle Airport, Turin in 1975. ALBA 2 replaced the original system in October 1980. A new version of the ALBA 2 is under development, with full message switching capability

---

Aviogei Cargo Loaders

Four models are available, ranging from 11 000 to 33 000 lb (5000 to 14 970 kg) capacity capable of loading cargo into all types of commercial aircraft. All four models are powered by diesel engines and employ a hydraulically-operated scissor lift mechanism to raise the loading platform and the driver's platform to the required height. Automatic loading can be carried out from the side or the rear of the vehicle, though the side loading is an optional feature in the 2P 525 model. The conveyor is a continuous belt drive, or in the case of the 2P 525 model a roller drive can be fitted. All the models are equipped with dual controls on the operator platform and the rear of the vehicle. Other standard features include dual-main electrical circuits, with immediate change over in the case of a failure in a single circuit; a hydraulic system comprising several independent sub-systems to reduce the possibility of a failure of the entire hydraulic system; safety cocks incorporated to isolate a faulty sub-system, and a hand-pump installed for operation of the hydraulic system in the event of an engine failure. Hand-rails are fitted to both sides of the loader.

Aviogei cargo loader

Characteristics

HANDLING RANGE				
main deck	B 707, B 727, B 737, DC-8, DC-9, F 28, Mercure aircraft		B 747F, DC-10F, L-1011, A 300B aircraft	
lower deck	B 747, DC-10, L-1011, A 300B aircraft			
LOAD				
containers	2 × LD-1/LD-3		4 × LD-1/LD-3	
pallets				
	96 × 125 in (2·44 × 3·18 m)		2 × 96 × 125 in (2·44 × 3·18 m)	
	88 × 125 in (2·24 × 3·18 m)		2 × 88 × 125 in (2·24 × 3·18 m)	
	88 × 108 in (2·24 × 2·74 m)		2 × 88 × 108 in (2·24 × 2·74 m)	
LOADER TYPE	2P 525	2P 725	IP 1200	2P 1500
capacity	11 000 lb (5000 kg)	16 580 lb (7500 kg)	26 400 lb (12 000 kg)	30 800 lb (14 000 kg)
bridge platform				15 400 lb (7000 kg)
LOADING HEIGHT				
max front and rear	11 ft 8 in (3·57 m)	11 ft 8 in (3·57 m)	18 ft 1 in (5·5 m)	18 ft 1 in (5·5 m)
minimum front	5 ft 7 in (1·7 m)	5 ft 7 in (1·7 m)	1 ft 8 in (0·508 m)	6 ft 5 in (1·9 m)
rear	1 ft 6 in (0·46 m)	1 ft 6 in (0·46 m)	1 ft 8 in (0·508 m)	1 ft 8 in (0·508 m)
ENGINE	VM 1053, 60 hp	VM 1054, 60 hp	VM 1054, 80 hp	VM 1056, 120 hp
TRAVEL SPEED	9 mph (15 km/h)	9 mph (15 km/h)	9 mph (15 km/h)	9 mph (15 km/h
TURNING RADIUS	21 ft 4 in (6·5 m)	32 ft 10 in (10 m)	42 ft (12·8 m)	46 ft (14 m)
CONVEYING SPEED	59 ft/minute (18 m/minute)	59 ft/minute (18 m/minute)	59 ft/minute (18 m/minute)	59 ft/minute (18 m/minute)
LIFT SPEED	45 ft/minute (13·7 m/minute)	45 ft/minute (13·7 m/minute)	33 ft/minute (10 m/minute)	33 ft/minute (10 m/minute)
DIMENSIONS				
driver's bridge platform	8 ft 5 in × 8 ft 2 in (2·57 × 2·5 m)	10 ft 6 in × 8 ft 5 in (3·2 × 2·56 m)		12 ft 7 in × 11 ft 2 in (3·84 × 3·4 m)
main platform	16 ft 5 in × 4 ft 10 in (5 × 2·5 m)	21 ft × 8 ft 5 in (6·4 × 2·56 m)	25 ft 2 in × 11 ft 2 in (7·67 × 3·4 m)	25 ft 2 in × 11 ft 2 in (7·67 × 3·4 m)
OVERALL DIMENSIONS				
length	24 ft 10 in (7·57 m)	32 ft 2 in (9·8 m)		38 ft ½ in (11·6 m)
width	10 ft 10 in (3·3 m)	10 ft 2 in (3·1 m)		11 ft 2 in (3·4 m)
WEIGHT	19 800 lb (9000 kg)	28 142 lb (10 000 kg)	24 250 lb (11 000 kg)	31 966 lb (14 500 kg)

Status
Production. Aviogei cargo loaders have been supplied to Alitalia

Manufacturer
Airport Ground Equipment Italia, Campoleone, Italy

Aviogei Cargo Transporters

Aviogei produce two models of cargo transporters, the TV 500 and TV 1000 which can handle up to 22 046 lb (10 000 kg) of 96 by 125 inch (2·44 by 3·18 metre) pallets. Based on a twin-axle low-level chassis with driver controls at the front, left-hand side, both models are powered by diesel engines, with power steering and hydraulic transmission. The TV 500 can be front and side loaded; the TV 1000 from the front only. As an optional feature an enclosed driver's cab can be supplied.

Status
Production. Aviogei cargo transporters have been supplied to Alitalia

Manufacturer
Airport Ground Equipment Italia, Campoleone, Italy

Characteristics

TYPE	TV 500	TV 1000
LOAD		
containers	2 × LD-1/LD-3/LD-7	4 × LD-1/LD-3/LD-7
pallets	96 × 125 in (2·44 × 3·18 m) 88 × 125 in (2·24 × 3·18 m) 88 × 108 in (2·24 × 2·74 m)	96 × 125 in (2·44 × 3·18 m) 88 × 125 in (2·24 × 3·18 m) 88 × 108 in (2·24 × 2·74 m)
CAPACITY	11 023 lb (5000 kg)	22 046 lb (10 000 kg)
LOADING HEIGHT	1 ft 8 in (0·508 m)	1 ft 8 in (0·508 m)
ENGINE	VM 1053 60 hp	VM 1054 80 hp
TRAVEL SPEED	19 mph (30 km/h)	22 mph (35 km/h)
TURNING RADIUS	18 ft 1 in (5·5 m)	20 ft 8 in (6·3 m)
DIMENSIONS		
platform	10 ft 9 in × 8 ft 6 in (3·28 × 2·6 m)	10 ft 8 in × 10 ft 8 in (3·25 × 3·25 m)
overall length	19 ft 8 in (6 m)	18 ft 1 in (5·5 m)
overall width	8 ft 6 in (2·6 m)	13 ft (3·95 m)
WEIGHT	7275 lb (3300 kg)	12 787 lb (5800 kg)

Aviogei Dollies

Seven models of dollies are produced by Aviogei for transporting baggage, pallets, igloos and containers. Six are mounted on four-wheeled towable trailers, the seventh, the PS500 is for pallet storage capable of supporting pallets and LD-1/LD-3 containers.

The RC2 and RC3 models are baggage trailers, both fitted with tow-bar actuated parking brakes. The RC2 is open, with drop sides while the RC3 is a covered trailer which being larger can be used also for cargo.

The RC150 is designed for transferring LD-1 and LD-3 containers and features a central turntable mounted on ball bearings, which enables the containers to be rotated through 360 degrees with stops every 90 degrees. The dolly is equipped with tow-bar actuated parking brakes and steering is effected through co-ordination of the four wheels linked to the tow-bar.

The PD500 and PP500 are pallet/container trailers, mounted on four 12-inch (304-mm) castors, swivelling at the front and rigid at the rear. the PD500 features automatic loading with a friction drive roller.

The SP500 is a pallet/container trailer, fitted with tow-bar actuated parking brakes. The front wheels are steered through a linkage to the tow-bar. Castors are incorporated in the floor of the trailer for loading.

Characteristics

MODEL	RC2	RC3	RC150
TYPE	Baggage trailer	Baggage/cargo trailer	Container trailer LD-1/LD-3 (2) half-size
CAPACITY	1764 lb (800 kg)	4409 lb (2000 kg)	3307 lb (1500 kg)
TURNING RADIUS	9 ft 10 in (3 m)	11 ft 6 in (3·5 m)	7 ft 10 in (2·4 m)
DIMENSIONS			
length	6 ft 7 in (2 m)	9 ft 10 in (3 m)	10 ft 2 in (3·09 m)
width	3 ft 7 in (1·1 m)	5 ft 11 in (1·8 m)	6 ft (1·83 m)
LOADING HEIGHT	2 ft (0·6 m)	1 ft 8 in (0·5 m)	1 ft 8 in (0·508 m)
WEIGHT	794 lb (360 kg)	1146 lb (520 kg)	882 lb (400 kg)
MODELS	PD500, PP500, SP500, PS500		
TYPE	Pallet/container trailer/storage		
LOAD			
pallets	96 × 125 in (2·44 × 3·18 m) 88 × 125 in (2·24 × 3·18 m) 88 × 108 in (2·24 × 2·74 m)		
CONTAINERS	LD-1/LD-3 (2)		
CAPACITY	13 228 lb (6000 kg)		
DIMENSIONS	PD500	PP500	SP500
length	10 ft 9 in (3·28 m)	11 ft 3 in (3·42 m)	11 ft 1 in (3·37 m)
width	8 ft (2·44 m)	8 ft (2·43 m)	8 ft 7 in (2·61 m)
LOADING HEIGHT	1 ft 8 in (0·508 m)	1 ft 8 in (0·508 m)	1 ft 8 in (0·508 m)
WEIGHT	2425 lb (1100 kg)	2160 lb (980 kg)	2425 lb (1100 kg)

Status
Production. The dollies are in service with several major airlines including Alitalia and Varig

Manufacturer
Airport Ground Equipment Italia, Campoleone, Italy

Aviogei Self-powered Conveyor Belts

Four types of self-powered conveyor belts are available covering operating heights from 2 feet (0·60 metres) up to a maximum of 18 feet (5·5 metres) and load capacities from 1102 to 3306 lbs (500 to 1500 kg). All are mounted on four-wheeled chassis powered by a diesel engine and use a single or double hydraulic cylinder to adjust the height of the belt. In the models NS 450 and NS 550 telescopic cylinders are used to provide the extra height required for servicing wide-body aircraft. A driver's cab is fitted to the models NS 350 and NA 350. In the other two models the driver's area can be covered as an optional feature.

Status
In service with several major airlines including Alitalia and Aero Trasporti Italiani

Manufacturer
Airport Ground Equipment Italia, Campoleone, Italy

Characteristics

TYPE	NS 350	NA 350	NS 450	NS 550
LIFT RANGE				
front	8 ft 3 in–11 ft 6 in (1–3·5 m)	2 ft–11 ft 6 in (0·6–3·5 m)	3 ft 3 in–14 ft 9 in (1–4·5 m)	3 ft 3 in–18 ft 1–5·5 m)
rear	2 ft 2 in–1 ft 8 in (0·65–0·5 m)	2 ft 9 in (0·84 m)	2 ft 11 in–5 ft 10 in (0·9–0·45 m)	2 ft 11 in–4 ft 11 in (0·9–1·5 m)
CONVEYOR				
capacity	1543 lb (700 kg)	1102 lb (500 kg)	2866 lb (1300 kg)	3307 lb (1500 kg)
speed	13–59 ft/minute (4–18 m/minute)	26–82 ft/minute (8–25 m/minute)	26–82 ft/minute (8–25 m/minute)	26–82 ft/minute (8–25 m/minute)
width	1 ft 8 in (0·5 m)	2 ft (0·6 m)	2 ft (0·6 m)	2 ft (0·6 m)
CHASSIS	Aviogei	Fiat		
ENGINE	Fiat AIFO, 30 hp	OM 55	VM 1052, 40 hp	VM 1053, 40 hp
TRAVEL SPEED	31 mph (50 km/h)	50 mph (80 km/h)	16 mph (25 km/h)	16 mph (25 km/h)
TURNING RADIUS	13 ft 2 in (4 m)	18 ft 9 in (5·7 m)	19 ft 9 in (6 m)	19 ft 9 in (6 m)
VEHICLE				
DIMENSIONS				
length	12 ft 2 in (3·72 m)	16 ft 7 in (5·06 m)	18 ft 4 in (5·6 m)	18 ft (5·5 m)
width		7 ft 6 in (2·3 m)	7 ft 6 in (2·3 m)	7 ft 6 in (2·3 m)
height	5 ft $2\frac{1}{2}$ in (1·59 m)	11 ft 6 in (4·78 m)		
WEIGHT	2866 lb (1300 kg)	7451 lb (3380 kg)	5511 lb (2500 kg)	6349 lb (2880 kg)

Caproni Vizzola PB 2000/PB 500 Baggage/Cargo Trailers

Two models of trailers for baggage and cargo handling are available. The PB 2000 with a capacity of 2000 kg and the smaller PB 500 with a 500 kg capacity. Both trailers are constructed of welded steel plates, C-profiles and rectangular tubes with cadmium plated bolts and screws. To protect the luggage a covering material, made of green hemp and cotton, is provided on a tubular frame roof. This is retractable in the PB 500. The platform is mounted on a front-steering axle connected to a tow-bar, and a rear-fixed axle. Rubber wheels are fitted on hubs with tapered roller bearings and the trailer is equipped with parking brakes activated by the tow-bar in the vertical position. A towing hook is fitted to the rear of the trailer.

Status
In service at Milan Airport, Italy

Manufacturer
Caproni Vizzola, Vizzola Ticino, Italy

Characteristics

MODEL	PB 2000	PB 500
CAPACITY	2000 kg	500 kg
PLATFORM DIMENSIONS	2·9 × 1·5 m	1·8 × 0·8 m
LOADING HEIGHT	0·525 m	0·5 m
UNIT DIMENSIONS		
length (with tow-bar raised)	3·6 m	2·15 m
width	1·5 m	0·8 m
height	1·85 m	1·5 m

Caproni Vizzola PP 7000 Pallet/Container Trailer

The PP 7000 trailer is suitable for handling pallets measuring 88 by 108 inches (2·24 by 2·74 metres), 88 by 125 inches (2·24 by 3·18 metres) and 96 by 125 inches (2·44 by 3·18 metres) with a capacity of 7000 kg.

Construction is of welded steel plates, C-profiles and rectangular tubes with cadmium plated bolts and screws. A castored loading floor is mounted on a front steering axle connected to a tow-bar and a rear fixed axle. The trailer is equipped with rubber wheels fitted on hubs with tapered roller bearings and parking brakes which are activated by the tow-bar in a vertical position. A towing hook is fitted at the rear of the trailers.

Status
In service at Milan Airport, Italy

Manufacturer
Caproni Vizzola, Vizzola Ticino, Italy

Characteristics
CAPACITY: 7000 kg
PLATFORM DIMENSIONS: 3·68 × 2·96 m
LOADING HEIGHT: 0·508 m
TURNING RADIUS: 4 m
DIMENSIONS
length (with tow-bar up): 4·05 m
width: 2·96 m

Cella Pallet/Container Main Deck Loaders

Cella aircraft loaders can service all wide- and narrow-body aircraft. Self-propelled units equipped with two working platforms, the units can handle pallets and containers up to a maximum weight of 60 000 lb (27 200 kg). The platforms, extended hydraulically by two individual scissors mechanisms, can reach a maximum height of 5·5 metres and can be adapted for containers up to 40 feet (12·2 metres) long. Control of the unit is from an operator's position mounted on the right-hand side of the bridge platform, accessible by an extensible ladder. Loads are transferred longitudinally by drive wheels and rollers, the sides of the platforms being equipped with hand-rails. The central unit of the bridge platform incorporating drive wheels and end rollers can be extended to couple with the cargo loading equipment in the aircraft.

Status
Cella main deck loaders are in service with Malev Hungarian airline and at Venice Airport, Italy

Manufacturer
Cella SpA, Mendolossa/Brescia, Italy

Cella main deck pallet/container loader

CESAB Fork Lift Truck

CESAB produce four-wheel fork lift trucks powered by battery or diesel engine. The battery-driven vehicles are available with capacities up to 8 tons and diesel-engined trucks can be supplied with capacities of 4 tons. Different lift heights are available in duplex and triplex versions with or without free lift, from the standard lift height of 3·2 metres to a maximum height of 8 metres.

The ECO series are powered by battery-driven motors available in various forms with different outputs and speed ranges. They are of the series wound type, providing a high torque, having optimum switching and a large commutation area. The electrical installation includes an electronic modulated frequency speed control, switch panel, micro-switch housing, instrument panel, fuses, safety micro-switch circuit. The SCR logic box, the power circuit and the contactors are housed in the right-hand side compartment, protected by an inspection cover.

The chassis, constructed of heavy-gauge steel, is mounted on front and rear axles. In the front axle the drive wheels run on taper roller bearings inside a spheroidal single piece axle casing secured to the chassis. The reduction gearing unit is of the double reduction type, operating on a crownwheel and pinion. Steering is through the rear axle, achieved through a recirculatory ball and nut steering box. Power steering is standard on some models and available as an option on others. It is operated from a separate electric motor and pump unit to ensure that the truck may still be steered in the event of a failure to the vehicle drive unit.

The dash panel is secured to the chassis and forms an anchorage for the accelerator housing, braking equipment, hydraulic control valve bank, and the hydraulic and electrical connections.

The mast section is of alloy steel and configurations of duplex or triplex with or without free lift are available to provide a choice of lift heights to within 50 mm. A series-wound hydraulic pump motor operates an hydraulic geared pump with a

CESAB ECO series fork lift truck

lift ram of the single action type using double action tilt rams. The hydraulic circuit incorporates a maximum pressure safety valve, a lift ram check valve and a forward tilt speed restriction. Driver protection is provided by heavy-duty overhead guard. It can be removed for battery changing using an overhead hoist.

The SID series includes many similar features to the ECO series. Power is provided by VM HR 488H or Perkins 4.203.2 diesel engines.

Status

All vehicles comply with the prescribed stability tests of UNI table 4520-66 and the standards prescribed by the European Mechanical Handling Federation

Manufacturer

CESAB Carrelli Elevatori, Bologna, Italy

FIT Pony Tow Tractors

Pony tractors are three-wheeled vehicles with an open-body produced in five models with draw-bar pull capacities ranging from 550 to 1000 kg. All are powered by a four-stroke, single cylinder diesel engine and are equipped with manual transmission, pedal operated hydraulic brakes and a mechanical hand-brake operating on the two rear wheel shoes. The T20L and T30L feature Bosch starter motors, the other models use starting handles with electric starters available as optional equipment.

Characteristics

MODEL	T20L	T20	T30L	T30	T40
MAX DRAW-BAR PULL	550 kg	600 kg	700 kg	800 kg	1000 kg
ENGINE	Lombardini LDA 510	Farymann L 25	Lombardini 720	Farymann LD	Farymann LG
OUTPUT	7·5 kW	6 kW	10 kW	8 kW	10·5 kW
MAX SPEED	18 km/h	18 km/h	16 km/h	10·3 km/h	10·3 km/h
TURNING RADIUS	1·2 m	1·2 m	1·4 m	1·4 m	1·7 m
UNIT DIMENSIONS					
length	1·64 m	1·64 m	1·64 m	1·64 m	1·95 m
width	0·97 m	0·97 m	1·12 m	1·12 m	1·2 m
WEIGHT	600 kg	630 kg	740 kg	740 kg	980 kg

Status

In service at Marco Polo Airport, Venice, Italy

Manufacturer

Fabbrica Italiana Trainatori, Milan, Italy

Papa FP/276 Mobile Conveyor

The FP/276 is a self-propelled conveyor for loading baggage and freight into the aircraft hold which can be used to service all types of narrow- and wide-body aircraft. The unit is powered by a 1904 cc diesel engine developing 40 hp and features power steering, independent suspension, dual braking system, parking brake and a fully enclosed driver's cab. The conveyor is adjustable in height hydraulically with a roller platform mounted on the front which folds down by means of a hydraulic cylinder to a rest position for transit.
The primary hydraulic system is supplemented by a low pressure secondary system and a hand-pump is fitted to adjust the conveyor height in the event of a breakdown of both systems.

Characteristics

CONVEYOR		
HEIGHT	front	rear
max	3·5 m	0·78 m
minimum	1·05 m	0·6 m
ENGINE	1904 cc diesel – 40 hp at 28 800 rpm	
TURNING RADIUS:	5 m	
UNIT DIMENSIONS		
length:	7 m	
width:	2·16 m	

Status

In service at Rome Airport, Italy

Manufacturer

Fratelli Papa, Cormano, Milan, Italy

Papa FP/276 mobile conveyor

Papa WV 12-16-20 Pallet Transporters

Three models of the WV pallet transporters are available from Fratelli Papa. Twin forks which are inserted underneath the pallet are raised hydraulically by the combined jacking arm and push/pull bar. An adjustable pedal lowers the pallet when in the required position. The unit is mounted on twin rubber wheels at the rear which can swivel through 220 degrees and rubber front wheels with pallet guide rollers.

Characteristics

MODEL	WV-12	WV-16	WV-20
CAPACITY	1200 kg	1600 kg	2000 kg

Manufacturer

Fratelli Papa, Cormano, Milan, Italy

Papa Withforce Fork Lift Trucks

Fratelli Papa produce a range of hydraulically-operated fork lift trucks for cargo handling in the freight terminal either power-driven or manually-operated.
The WL series comprises four models, fitted with a manually-operated drive handle and a vertical return spring. The units are mounted on swivelling rubber wheels at the rear, and fixed rollers at the front. The hydraulic lift is operated by a 1500-watt electric motor powered by a rechargeable 12 volt battery. Other features are: a removable battery charger module, safety guard and a parking brake.
The AW-AWP series consists of seven models with capacities ranging from 800 to 1800 kg. The units are equipped with electric traction with three forward and three reverse gears, operated by a person walking behind the fork lift. Apart from the drive-wheel mounted between the two rear wheels, and an electro-magnetic brake with automatic clutch, this equipment is similar to the WL series.

Manufacturer

Fratelli Papa, Cormano, Milan, Italy

Characteristics

SERIES	WL	AW-AWP
MAX CAPACITY	600–100 kg	800–1500 kg
MAX HEIGHT RANGE	1·6–3·2 m	1·6–3·2 m
TRACTION	manual	750 W, 12 V motor
LIFT POWER	1500 W–12 V	1500 W–12 V
LIFT SPEED		
unloaded	7·4 m/minute	7·4 m/minute
loaded	5 m/minute	5 m/minute
BATTERY	92 A/h, 12 V	158 A/h, 12 V
WORKING PERIOD	4–5 h	4–5 h
RECHARGEABLE UNIT	220 V, 50 Hz	220 V, 50 Hz

Papa WL 16/10 lift truck

PGS Lift Trucks

PGS produce a wide range of lift trucks and pallet trucks suitable for cargo handling within the freight terminal or warehouse. Pallet trucks are available in four sizes with capacities from 1200 to 3000 kg. The twin forks of the truck are raised hydraulically by operation of a foot lever which also controls the descent. The forks are mounted on a single wheel at the rear to which a steering arm is attached and wheels underneath each fork at the front. Self-propelled electric powered pallet trucks are produced with capacities from 1200 to 2000 kg and elevating versions, both manual and self-propelled, can be offered with capacities from 1000 to 1500 kg.
Electric powered lift trucks produced by PGS range from 1000 kg capacity to 6000 kg, and diesel engine powered models from 1200 kg up to the model M450, rated at 45 000 kg.

Characteristics

TYPE	NO OF MODELS	POWERED DRIVE	POWERED LIFT	CAPACITIES
PALLET TRUCK	4	Manual	Hydraulic	1200–3000 kg
	3	Electric	Hydraulic	1200–2000 kg
PALLET ELEVATOR	2	Manual	Hydraulic	1000–1500 kg
	3	Electric	Electric	1000–1500 kg
FORK LIFTS				
battery	13	Electric	Electric	1000–6000 kg
diesel	6	Diesel	Hydraulic	1200–3500 kg
diesel automatic	18	Diesel	Hydraulic	4000–45 000 kg

Manufacturer

PGS, Cadeo Piacenza, Italy

Pirelli Storage Equipment for Automated Freight Terminals

Pirelli specialise in the design, manufacture and installation of storage and retrieval systems for automated warehouses and freight terminals and a wide range of lift equipment is available.
The ET 12 turret truck is suitable for storing pallets in high rise warehouses or freight terminals with narrow aisles. Based on a frame constructed from sheet steel, the telescopic mast is in three sections. The fork is lifted by two steel wire ropes and a winch consisting of two drums keyed directly on to two outlet shafts of a special reduction gear equipped with a Hyndley's screw. The 12·5 kW motor is electronically-controlled and has two lifting speed ranges, one fast and the other for positioning. The rear wheel is driven by a 5-watt motor producing speeds up to 9·5 km per hour. The steering wheel controls the steering action with a 28:1 reduction ratio and a maximum leading wheel steering angle of 120 degrees. Between the aisles, the truck is driven by lateral contrast wheels which run along special rails fixed to the floor on both sides of the gangway.
The unit can be programmed for automatic selection of the floors through a push-button board and proximity switches. The fork stops, after a slowing down action in the correct position for collecting or storing the load. Alternatively a parallel-ray collimeter may be fitted to provide the automation. Other optional features include a device with telescopic forks which permits storage of pallets on both sides of the aisle; and cantilevered or hanging rotating forks which rotate and move the load cross-wise for loading and unloading the shelves, placing the load on the floor.

Characteristics

MODEL: ET 12
CAPACITY: 1200 kg
MAX LIFTING HEIGHT: 8·92 m
LIFTING MOTOR: 80 V 12·5 kW 2300 rpm
LIFTING SPEED: 3 m/minute adjustable / 16 m/minute
TRAVEL MOTOR: 80 V, 5 kW, 1800 rpm
TRAVEL SPEED: 9 km/h
TURNING RADIUS: 2·37 m
BATTERY: 80 V – 500 A/h/80 V – 400·A/h
OPERATING RANGE: 10/6 h
UNIT DIMENSIONS
length: 3·37–3·77 m depending on equipment fitted
width: 1·25 m
minimum height: 4 m

SERIES	CHARACTERISTICS	MODEL	CAPACITY
KA	Single mast, fixed forks	KA 600	300 kg
	Sliding operator cab	KA 1000	600 kg
KB	Single mast	KB 1000	400 kg
	Loading cab	KB 1500	800 kg
		KB 2000	1200 kg
KD	Single mast, telescopic forks	KD 1000	350 kg
	Cantilevered sliding operator cab	KD 1500	750 kg
		KD 2000	1150 kg
KE	Single mast SRM, telescopic forks	KE 300	100 kg
	Fixed operator cab	KE 600	300 kg
		KE 1000	650 kg
		KE 1500	1100 kg
		KE 2000	1500 kg
HD	Double mast, telescopic forks	HD 2000	1150 kg
	Sliding operator cab	HD 3000	1750 kg
		HD 4000	2550 kg
HE	Double mast, telescopic forks	HE 2000	1500 kg
	Fixed operator cab	HE 3000	2250 kg
		HE 4000	3000 kg

Manufacturer

Dima SpA, Pirelli, Balsamo, Italy

Trastecnica Cargo Handling Systems

Trastecnica/Rapistan cargo handling equipment is available for accumulating, sorting and transporting air freight within the airport warehouse. The systems, which can be automated by computer, are operated from a central console, and can be designed to handle both individual packages and pallets.

Status

Trastecnica/Rapistan equipment is installed at airports in Italy, North Yemen, Saudi Arabia and the USSR

Manufacturer

Trastecnica/Rapistan, Cologno, Monzese, Italy

NETHERLANDS

Alreso Sophia Cargo Handling Data System

The Sophia system has been designed to enable transfer and subsequent memory storage of information from a central supervisory station to a number of receivers. These can be stationary but in most cases are mobile within the operational area. The information transmitted from the central station, at a rate of 2400 baud, is received by the mobile units and stored in ram memory. A keyboard enables the operator to gain instant access to the memory store on punching the appropriate code.
The information is transmitted from the central supervisory station by any recognised carrier: since, in most cases, the receptors are mobile radio transmission in conventional frequency bands is used.
At the central station, the equipment comprises: an encoder of data transmitted on the radio channel, processor, prom floppy disc/cassette recorder, modulator and transmitter.
At each receiving station the equipment consists of: a receiver and demodulator, processor, keyboard and alphanumerical display.
Typical applications include cargo handling at airports with loading and unloading of ulds and any stores management where, in rapidly changing situations, the goods have to be transported to the appropriate storage area. In cargo handling and despatch storage functions, the supervisory centre has a continuous input of information to be transferred to the receptors.
For example, flight numbers, destination and cargo areas. The information is processed and sent out in a digital form, to any number of units where it is stored in a memory and available for retrieval. In total, the memory can hold 16 000 bytes, the equivalent of 16 000 letters or figures, or six pages of written information. The frequency of updating information is variable, depending upon operational requirements and the number of receiving units. For example, the operator of a fork lift truck may wish to know where a load should be taken for a certain flight. Upon entry of the cargo destination code via the keyboard the appropriate information is displayed informing the operator where the load should be delivered.

Status
In service with KLM at Amsterdam Schiphol Airport for cargo handling and marshalling

Manufacturer
Alreso Elektronica BV, Zaandam, Netherlands

Berkel Cargo Weighing Systems

Two types of cargo scales are available from Berkel. The digital display version carries out integrated calculation using a micro-processor which can be linked to a printer for hard copy and to a central processing unit for co-ordinating functions. The analogue display cargo scales are fitted with a dial head and mechanical printer for data registration.

Manufacturer
Berkel International, Rotterdam, Netherlands

SPAIN

Equisa Baggage/Freight Conveyors

Three models of conveyors suitable for baggage and cargo loading and unloading operations are produced.
The CAA is a self-propelled model, powered by a Perkins diesel engine with manual gearbox and power steering. Belt drive is controlled by a hydraulic motor, enabling the speed to be varied and height adjustment is provided at both ends by independent hydraulic cylinders. Railings are installed on either side to prevent smaller loads from falling off the belt and these can be folded down for loads that are wider than the boom. A rubber bumper is fitted to the front of the conveyor boom.
The CAA-2 is an improved version of the CAA with automatic transmission and a torque converter replacing the manual gearbox. Optional features available are: collapsible side-rails; manual emergency pump; power steering; all weather driver's cab and air-conditioning unit.
The FCA is a truck-mounted conveyor based on Edbro 60/70 D chassis with self-containing bodywork, forming a solid unit with the driver's cabin. Power is provided by a Perkins diesel engine, using a Hardy Spicer transmission for vehicle drive, and via an oleohydraulic gear pump and a Cardan transmission unit for operation of the conveyor belt and height adjustment. The boom which extends from the van body can be raised to a height of 3·29 metres. Baggage and non containerised packages can be transported in the unit. Optional features available include Ford, Mercedes or similar chassis and an air-conditioning unit.

Manufacturer
Equisa, Madrid, Spain

Equisa FCA baggage/freight conveyor

Characteristics

MODEL	CAA	CAA-2	FCA
BOOM DIMENSIONS	7·1 × 0·87 m	7·7 × 0·87 m	
BELT			
width	0·7 m	0·7 m	
capacity	100–200 kg/m	100–150 kg/m	2500 kg
ELEVATING HEIGHTS			
front	0·9–4 m	0·75–4·3 m	0·38–3·29 m
rear	0·28–1·45 m	0·28–2·1 m	
ENGINE	Perkins 4108 diesel	Perkins 4108 diesel	Perkins G 305
OUTPUT	36 kW at 3600 rpm	36 kW at 3600 rpm	
MAX SPEED	20 km/h	35 km/h	
BELT SPEED	23–30 m/minute	10–30 m/minute	
UNIT DIMENSIONS	7·5 × 2·25 × 1·3 m	8 × 1·95 × 1·3 m	8·49 × 2·5 × 3·25 m
WEIGHT	4500 kg	2950 kg	4900 kg

Equisa Baggage/Freight Utility Trailers

Equisa produces a range of utility trailers for transporting cargo and lightweight packages. There are no major differences between the various models: each unit features aluminium grid-platform mounted on front and rear axles, with a ratchet brake fitted to the front wheels, tow-bar and towing hook. The TEA II is the standard model. The TEA III is equipped with collapsible side-rails and the TEA IV and the TEC are similar to the TEA III but fitted with a canvas cover to protect the loads. The TEA V incorporates a roller bed for palletised cargo. It can be adapted to the TEA II configuration for carrying baggage and similar-sized cargo. The CBA is designed for the collection and movement of waste and has six compartments. It has a metal body-work with upper doors at both sides and double lower doors. Optional features available to all models are: solid wheels, galvanised steel platform and drum brakes fitted to the front wheels.

Equisa CBA utility trailer

Manufacturer
Equisa, Madrid, Spain

Characteristics

MODEL	TEA II	TEA III	TEA IV	TEA V	TEA VI	TEC	CBA
PLATFORM DIMENSIONS							
length	2·5 m	2·5 m	3 m	3 m	2·6 m	2·5 m	2·5 m
useful width	1·3 m	1·22 m	1·5 m	1·3 m	1·3 m	1·22 m	1·22 m
height	0·5 m	0·5 m	0·5 m	0·5 m	0·55 m	0·5 m	0·5 m
CAPACITY	2000 kg	2000 kg	4000 kg	3000 kg	1000 kg	2000 kg	2000 kg
WHEELBASE	1·68 m	1·68 m	2·04 m	2 m	1·5 m	1·61 m	1·68 m
TRACK							
front	0·59 m	0·59 m	0·59 m	0·59 m	0·88 m	0·59 m	0·59 m
rear	1·09 m	1·09 m	1·09 m	1·09 m	1 m	1·09 m	1·09 m
UNIT DIMENSIONS							
length (with tow-bar raised)	3·31 m	3·31 m	3·82 m	3·8 m	3·16 m	3·31 m	3·31 m
width	1·3 m	1·4 m	1·58 m	1·3 m	1·3 m	1·3 m	1·3 m
height	1·15 m	1·15 m	1·3 m	1·15 m	1·19 m	1·23 m	1·5 m
WEIGHT	380 kg	540 kg	550 kg	600 kg	275 kg	500 kg	600 kg

Equisa MPC/MCC Pallet/Container Racks

The MPC pallet rack has been designed for the storage of pallets of up to 88 by 125 inches (2·24 by 3·18 metres). The units can be joined together laterally by means of bolts to form stands. The platform height of the racks is standardised to facilitate their use with other ground support equipment, and the racks are mounted on adjustable supports to compensate for differences in the ground surface. They are equipped with two load stops, one at either end of the platform which are activated manually.

The MCC container rack has been designed for the storage of semi-containers and incorporates similar features to the MPC model.

Characteristics

MODEL	MPC	MCC
CAPACITY	4500 kg	2000 kg
ROLLER HEIGHT	0·453–0·553 m	0·453–0·553 m
UNIT DIMENSIONS		
length	3·467 m	2·172 m
width	2·38 m	1·704 m
WEIGHT	530 kg	270 kg

Manufacturer
Equisa, Madrid, Spain

Equisa MCC container rack

Equisa Pallet/Container Dollies

The TMA is designed to transport semi-containers. It features a rotating platform with braking system, equipped with eight rollers; six of galvanised steel mounted on ball bearings and two covered with synthetic rubber. The rollers are linked by chains to automate the transfer system and the load stops can be operated automatically or manually. The dolly is equipped with tow-bar steering and braking with ratchet brakes on the front wheels activated when the tow-bar is in the raised position.

The TMA-G rotating dolly is suitable for transporting LD-1 or LD-3 containers with a load capacity of 3500 lbs (1588 kg). Transfer is by means of a system of 16 rollers arranged in two rows.

The TMB series for pallets are fitted with undercarriages similar to the TMA series. The TMB has a load capacity of 1500 lbs (6800 kg) and is equipped with a platform comprising 45 rollers arranged in three rows. Two load stops are fitted either end, which are manually operated. Optional features available include additional stops and guides for LD-1 and LD-3 containers, load height increased to 1 foot 9 inches (0·533 metres), articulated steering on the front wheels and a bulldog-type safety towing hook. TMA-G is similar but is equipped with a 360 degree rotating load platform. An automatic transfer system can be incorporated as an optional item.

Status
In service with Iberia

Manufacturer
Equisa, Madrid, Spain

Equisa TMB-G pallet/container dolly

Characteristics

MODELS	TMA	TMA-G	TMB	TMB-G
CAPACITY	2000 kg	1600 kg	4500/7200 kg	4500/7200 kg
PLATFORM USEFUL DIMENSIONS	1·6 × 1·6 × 0·5 m	1·6 × 1·55 × 0·508 m	3·26 × 2·5 × 0·508 m	3·26 × 2·5 × 0·508 m
UNIT WEIGHT	676 kg	470 kg	800 kg	1250 kg

Equisa Pallet/Container Loaders

The self-propelled loader PEA-16E has been designed to load and unload containers of the types LD-1, LD-2, LD-7 and pallets up to 96 × 125 inches (2·44× 3·18 metres), with a maximum load of 16 000 lbs (7200 kg). It is suitable for the main decks of all conventional freighters and the lower decks of wide-body aircraft. The unit handles pallets and containers both longitudinally and transversally. Four sets of rollers on the main platform and one set on the bridge platform independently operated, permit the automatic transfer of loads. These systems are activated by hydraulic motors with the operation of the loader through a portable control unit.

To ensure stability during loading or unloading operations, the unit is fitted with six stabilising jacks, the front four jacks being used to adjust the level of the loader to the aircraft hold. The vehicle is powered by a Perkins diesel engine and is equipped with automatic transmission and power steering. As an optional feature the loader can be adapted for pallets of 96× 125 inches (2·44 × 3·18 metres) to be transferred longitudinally.

The PEA-16MA is a military self-propelled loader designed to load and unload pallets up to 88 by 108 inches (2·24 by 2·74 metres) with a maximum load of 16 000 lbs (7528 kg). This loader can be used as a transporter and is suitable for use with military transport aircraft such as the Hercules C-130, Caribou and Aviocar. The dimensions of the loader allow the PEA-16MA to enter under its own power the hold of the C-130 Hercules and be transported by air. The loader is equipped with levelling cylinders which can achieve a variety of inclinations, longitudinally and transversally with adjustment up to 80 mm, to adapt the platform to the floor of the aircraft hold. The load transfer to the hold of the aircraft is automatic and can be carried out either longitudinally or transversally, by means of omni-directional rollers activated by hydraulic motors. Operation of the loader is through a portable control box. The vehicle is propelled by automatic transmission of the hydrostatic-type formed by a variable flow pump and a slow-speed high torque engine coupled to the differential of the drive shaft providing wheel-band torques of 500 kilogram per metre. Optional features comprise automatic load transfer capability from pallet dolly or rack and a driver's cabin.

Equisa PEA-16E container/pallet loader

Characteristics

MODEL	PEA-16E		PEA-16MA
CAPACITY	7200 kg		7200 kg
PLATFORM	Main	Bridge	Main
ELEVATING HEIGHT			
max	3·5 m	3·5 m	1·6 m
minimum	0·5 ± 0·6 m	119 m	0·5 ± 0·6 m
USEFUL DIMENSIONS	3·6 × 2·49 m	2·95 × 2·49 m	2·9 × 2·3 m
LENGTH BETWEEN STOPS	3·6 m		
TILTING CAPACITY			
longitudinal			± 80 mm
lateral	± 50 mm		± 80 mm
ENGINE	Perkins 4236 diesel		Perkins 4108 diesel
OUTPUT	58 kW at 2600 rpm		36 kW at 3600 rpm
DRIVING SPEED	14 km/h		20 km/h
WHEEL BASE	6·07 m		3·5 m
UNIT DIMENSIONS	8·1 × 3·5 × 2·3 m		5·1 × 2·5 × 1·5 m
WEIGHT	12 000 kg		4800 kg

Status

The PEA-16E has been supplied to Iberia and the military loader is in service with the Spanish Air Force

Manufacturer

Equisa, Madrid, Spain

Equisa TAA-2 Pallet/Container Transporter

The TAA-2 self-propelled transport has been designed to transfer and transport palletised loads. The dimensions of its load platform make it operative with pallets of up to 96 by 125 inches (2·44 by 3·18 metres) and containers, semi-containers and igloos with a maximum payload of 16 000 lb (7258 kg). The transporter is powered by a Perkins diesel engine with hydrostatic transmission and power steering. The main servo-assisted brakes act on the front wheels. A mechanical parking and emergency brake operated-manually is connected to the rear wheels. The unit is equipped with a height levelling and adjusting system acting on the front part of the vehicle carried out by two hydraulic cylinders installed between the front axle and the chassis, thus permitting load transfers with vehicles of non-standardised height. The load is carried along the platform by means of two rows of independently controlled drive rollers and load transfer operations can be carried out from the front end of the platform as well as from the rear end. The transfer of loads from the transporter to other load vehicles and vice versa is effected automatically by friction rollers which turn and make the rest of the rollers of the load platform move in the same direction. To limit the movement of the load during transfer and transport operations the transporter is provided with load stops in the platform which are operated from the driver's position. The stops are fitted with a mechanical device activated by a roller which lowers the stop automatically when the pallet passes over the roller.

Equisa TAA-2 pallet/container transporter

Characteristics

ENGINE: Perkins 4203 water-cooled diesel
OUTPUT: 44 kW at 2600 rpm
MAX SPEED: 22 km/h
PLATFORM DIMENSIONS
length between stops: 3·28 m
width: 2·48 m
height: 0·58 ± 0·1 m
CAPACITY: 7270 kg
WHEELBASE: 2·97 m
TRACK
front: 3·63 m
rear: 3·5 m
UNIT DIMENSIONS: 5·1 × 3·94 × 1·52 m
WEIGHT: 3800 kg

Manufacturer

Equisa, Madrid, Spain

Equisa Tractors

Three tractor models are produced by Equisa, the TE-3, R3-E and R5-2.
The TE-3 is an electric tow tractor suitable for baggage and freight trailers and dollies. The chassis constructed of steel plate and laminated profiles incorporates front-wheel drive, powered by a 9 kW series motor and rechargeable batteries, using thyristor control. A double reinforced braking circuit enables electrical braking to all four wheels, the parking brake acting on the drive axle only. A rear towing hook is fitted to the vehicle. Optional features available comprise driver's cab, front towing hook and spring/shock absorber suspension replacing the standard rubber suspension system.
The R3-E is a diesel engine powered tractor rated at 3000 lbs (1360 kg) draw-bar pull, equipped with automatic transmission torque converter, front wheel drive and servo-assisted drum brakes on all four wheels. A rear towing hook is fitted. Optional features include: all weather driver's cab, air-conditioning, passenger seat and swivel light for night operations.
The R5-2 is the largest vehicle in the range with a draw-bar pull of 4000 lb (1814 kg) sufficient to tow smaller aircraft such as the Fokker 27, in addition to auxiliary units. Powered by a Perkins diesel, the tractor is equipped with automatic transmission driving the rear axle, power steering and three independent braking systems. The main hydrostatic brake acts on the two traction wheels, with a drum-type parking brake, and hydro-mechanical drum brakes are fitted to the front wheels. The towing hook at the rear of the vehicle is of variable height to accommodate a variety of towable units.

Manufacturer
Equisa, Madrid, Spain

Equisa R5-2 tractor

Characteristics

TYPE	TE-3	R3-E	R5-2
MOTOR/ENGINE	Electric 72 V	Perkins 4103 diesel	Perkins 4203 diesel
OUTPUT	9 kW		44 kW at 2600 rpm
MAX SPEED		25 km/h	25 km/h
DRAW-BAR PULL		1360 kg	1814 kg
TOW-BAR HEIGHT	0·42–0·51 m	0·42–0·51 m	0·37–0·55 m
OUTER TURNING RADIUS	2·3 m	2·3 m	3·2 m
WHEELBASE	1·4 m	1·4 m	1·52 m
TRACK			
front	1·1 m	1·1 m	1·36 m
rear	1·1 m	1·1 m	1·15 m
UNIT DIMENSIONS	2·5 × 1·3 × 1·45 m	2·5 × 1·3 × 1·45 m	2·78 × 1·63 × 1·6 m
WEIGHT	2700 kg	2370 kg	2800 kg

SWEDEN

Kalmar TA 900 Tow Tractor

The TA 900 is a four-wheel tractor suitable for towing baggage and cargo trailers. Available with a choice of diesel or lpg/petrol engines, the tractor is equipped with manual transmission, hydraulic brakes on all four wheels, mechanical parking brake, all-weather steel cab and a rear towing point. A passenger seat is offered as an optional feature. Rated at 900 kg draw-bar pull, a higher value can be obtained with additional ballast on the rear axle.

Characteristics
MAX DRAW-BAR PULL: 900–1500 kg
ENGINE: Ford 227E lpg/petrol, Volvo 1320 lpg/petrol or Perkins 4108 V diesel
OUTPUT: 43/60/35 kW
MAX TRAVELLING SPEED: 26 km/h
WHEELBASE: 1·4 m
TURNING RADIUS: 2·7 m
UNIT DIMENSIONS: 2·54 × 1·48 × 2·1 m
WEIGHT: 1550 kg

Manufacturer
Kalmar Motor AB, Kalmar, Sweden

Kalmar TA 900 tow tractor

Pentaport Cargo Terminal

This is a modular construction system for aircraft hangars and freight terminals. With this system aircraft are parked nose-in in a circle. For each aircraft, a segment extending 67 degrees is necessary to cover the aircraft with a further 2·5 degrees either side of the aircraft to accommodate the pylons of the suspended roof making a total of 72 degrees per aircraft. Thus up to four aircraft can be accommodated with freight handling, offices and workshops in the fifth segment. A full description of the concept for hangar construction is contained in the Maintenance section. The cargo terminal is equipped with lifts and powered conveyors. A large throughput of cargo may require the installation to be computer controlled. It is anticipated however that the majority of installations will not require such sophisticated equipment. Workshops are on the ground floor of the office building. The container terminal is situated so that containerised goods are inspected by customs and passed directly into the cargo handling area for sorting and loading. An area for transit goods is provided by-passing the customs area and air-tight doors are provided where required. To cater for side-loading aircraft only it is not necessary to provide the central tower. A system can be devised where mixed aircraft types are used which will incorporate both side- and nose-loading facilities. The central tower is of concrete construction with a centrally-positioned telescopic cylinder on which a platform can be raised and lowered and rotate about a vertical axis. An adjustable height platform is provided at each bay to accommodate the varying heights of different aircraft. Aircraft can be towed into position and the towing vehicle driven away through the central tower. Both nose- and side-loading aircraft are covered. The diameter of the central tower is between 20 and 25 metres in an average installation.

Status
A Pentaport was first shown at the ICAA airport exhibition in Nice in 1979 as a result of preliminary designs for two aircraft hangars for the Norwegian company Fred Olsen Airtransport Ltd. The cargo terminal is phase two of the design

Manufacturer
Pentaport Sweden AB, Gothenburg, Sweden

Tellus BL 257 Belly Compartment Loader

The Type BL 257 is designed to travel to and from aircraft and for the loading and unloading through the aircraft's cargo hatch goods that are not palletised or containerised. The load surface consists of a roller deck-mounted on a scissor-lift table which can be raised by means of a hydraulic hand-pump on one side of the loader from 550 mm to 1650 mm. The roller deck comprises 15 steel rollers with permanently lubricated precision ball bearings at both ends of the rollers. The rollers have a diameter of 76 mm are 900 mm long and have steel tubing 2·9 mm thick. The roller axles have 16 mm diameters and are equipped with lock pins at both ends for dismantling and assembly. Each roller permits a maximum load of 200 kg. To prevent loads from sliding off the surface the loader is equipped at both ends with spring-loaded collapsible security barriers that are controlled from the side of the load surface. At its front end the loader has a towing bar connected to the front wheels with an eye ring for coupling to a tractor. Both rear wheels are equipped with brakes which are operated by a lever on one side of the loader.

Characteristics

CAPACITY: 1500 kg
PLATFORM HEIGHTS
max: 1·65 m
minimum: 0·55 m
UNIT DIMENSIONS: 2·175 × 0·90 × 0·55 m

Manufacturer

Tellus Maskin AB, Vallentuna, Sweden

Tellus D206 Container Dolly

The Tellus D206 Container Dolly is designed for transporting LD-1/LD-3 containers between the aircraft and the cargo terminal, with base dimensions of 1·54 by 1·52 metres. To permit loading and unloading from the side, the dolly's roller deck can be rotated 360 degrees, with locking positions every 90 degrees. The load platform consists of a roller deck with 26 precision ball-bearing steel rollers in two rows, separated by a catwalk with three large steel rollers at each end. The rollers in the load platform have a diameter of 51 mm, are 600 mm long, constructed of steel tubing with a thickness of 1·5 mm. The roller axles have diameters of 10 mm and are equipped with lock pins at both ends for dismantling and assembly. The distance between the centre lines of the rollers is 135 mm and each roller permits a maximum load of 95 kg. The shock-absorbing rollers have diameters of 102 mm with 25 mm axles.

To prevent loads from sliding off the roller deck during transportation, the dolly is equipped with fixed side guides with wheels and spring-loaded tip-over lockings, as well as with spring-loaded security barriers at each end. During loading, the first barrier is depressed by the container itself and held in this position until the container reaches the barrier at the other end; the first barrier is then raised automatically, fixing the load in a longitudinal position. To unload, one of the barriers is depressed by operating a pedal on the outside of the dolly.

During loading and unloading, spring-loaded locking arms are moved away from the cargo, but in the fixed travelling position between the barriers the arms automatically lock into grooves in the container bottom, securing the container to the dolly.

The dolly has parallel-steered solid rubber wheels in both front and back, so that all wheels on the same side of coupled dollies move in the same track. Four-wheel steering and the relatively small distance between axles provides a small turning radius. At the front end the dolly has a towing bar, with eye ring, that is connected with the wheels. At the rear a hook with a lock bolt is fitted for coupling to the towing bar of another dolly. A parking brake attached to the left-rear wheel is activated when the towing bar is placed in an upright position.

Optional features available include loading height of 530 mm and zinc-plated rollers.

Characteristics

CAPACITY: 1600 kg
LOADING HEIGHT: 0·508 or 0·530 m
MAX LOAD: 1·56 × 1·53 m
DIMENSIONS
TOWING SPEED: 15 km/h
OUTSIDE TURNING RADIUS: 4·3 m
UNIT DIMENSIONS: 3·92 × 1·72 × 0·508/0·530 m (including towing bar)
WEIGHT: 650 kg

Status

In service with Scandinavian Airlines System

Manufacturer

Tellus Maskin AB, Vallentuna, Sweden

Tellus MT 270 Mobile Belt Loader

Tellus mobile belt loader Type MT 270 is designed for wide- and narrow-body aircraft for the loading and unloading of passenger baggage and other similar sized cargo.

The front and rear ends of the belt conveyor can be raised and lowered separately. The conveyor is fitted with an 800 mm wide rubber belt with three-cord reinforcement and a grip-top surface. The belt is driven, via a chain gear, by an hydraulic motor. Between the drive and turning drums, the belt runs on roller supports fitted with ball bearings.

The frame of shaped sheet steel is equipped with strong stiffeners and the conveyor has rails running the length of the belt. The outer rail can be lowered for wide loads and the inner rail can be lowered in front of the loading space to facilitate cargo handling.

The front of the loader is fitted with a soft-rubber fender to prevent damage to aircraft. The conveyor, which is mounted in a linkage, is raised and lowered hydraulically. The belt is started and stopped, and its forward and reverse movement controlled, by push buttons at the operator's position and at each end of the conveyor.

Travelling between terminal and aircraft, goods are placed on a low platform behind the operator's cab and on the conveyor, which is locked in a horizontal position. Eight square metres of load space is available. In addition, the vehicle can be used to tow a train of trailers. Personnel can be carried to and from aircraft on the footboard on the front right-hand side which can accommodate four or five persons. The vehicle is powered by a four-cylinder diesel engine.

Characteristics

FRONT LOADING HEIGHT: 1·25–3·8 m
REAR LOADING HEIGHT: 0·508–1·5 m
BELT
length: 700 kg at 15° incline
capacity: 500 kg at 20° incline
speed: 12 m/minute
TOTAL CAPACITY
1200 kg/minute at 15° incline
855 kg/minute at 20° incline
VEHICLE DIMENSIONS
length: 7·25 m
width: 2·15 m

Status

In service with Scandinavian Airlines System

Manufacturer

Tellus Maskin AB, Vallentuna, Sweden

Tellus MT 274 Pallet/Container Transporter

The MT 274 is designed primarily for the movement of a single 125-inch (3·18-metre) pallet or two LD-1/LD-3 containers between the aircraft and container dollies, pallet trailers or cargo terminals. It can be used with loads up to 7000 kg. The height of the front end of the loading surface can be adjusted from the operator's position, from 468 to 568 mm above the ground for use with dollies, trailers and loading docks of various heights or to compensate for uneven terrain. The load platform consists of a roller deck with powered rollers and a safety walk between the rollers. There are 24 steel rollers, each with a diameter of 127 mm on 35 mm axles. The direction of the rollers' rotation is controlled from the operator's position, which operates at speeds up to 27 metres per minute. To prevent the load from sliding off the platform, the transporter is equipped with side guides and front and back end security barriers that can be raised and lowered from the operator's position. The outer-side guide is collapsible. A diesel engine drives a large and small hydraulic pump. The large pump powers the rear wheels. The smaller one operates in an open loop to power the servo-controls, the rollers, the raising and lowering of the front end of the loading surface, and the cargo safety devices. The speed of the diesel engine is set by a manual control and the transporter can be driven both forwards and backwards at variable speeds up to 20 km per hour. All controls are operated from the driver's cab which with the engine compartment and fuel tank are along the left side of the transporter.

Characteristics

CAPACITY: 7000 kg
LOADING HEIGHT: adjustable 0·468–0·568 m
PLATFORM DIMENSIONS: 5·328 × 2·63 m
DRIVING SPEED: 20 km/h
CONVEYING SPEED: 27 m/minute
TURNING RADIUS: 11 m
UNIT DIMENSIONS
length: 5·5 m
width: 3·4 m

Manufacturer

Tellus Maskin AB, Vallentuna, Sweden

Tellus 77 Series Pallet Trailer

The Tellus 77 series Pallet Trailer is designed for the transportation of pallets up to 96 inches (2·44 metres) by 125 inches (3·18 metres). In its standard configuration, Type 77A, the trailer roller deck is mounted crosswise, for lateral loading and unloading. The trailer can be supplied as Type 77B, with the roller deck rotated 90 degrees, to permit end loading and unloading. The load platform consists of a roller deck with 39 precision ball bearing steel rollers in three rows, separated by catwalks. The rollers have a diameter of 76 mm, are 500 mm long and have a steel tubing with a thickness of 2·9 mm. The roller axles have 16 mm diameters and are equipped with lock pins at both ends for dismantling and assembly. Each roller permits a maximum load of 330 kg. To prevent loads from sliding off the roller deck while moving, the trailer is equipped with fixed 100 mm side guides and a removable security barrier at each short end. One of the removable barriers can be placed in two alternative positions, of which the innermost is suitable for cargo 108 inches (2·74 metres) long and the outer for cargo 125 inches (3·18 metres) long.

When the barriers are not in use, they are hung on wires along the outside of the frame. The wheels, fitted with solid soft-rubber tyres, are made of cast iron, with double ball bearings in the hub. At the front end the trailer has a towing bar, with eye ring, connected to the front wheels. At the rear a hook with a lock bolt is fitted for coupling to another trailer. A parking brake in one of the front wheels is activated when the bar is placed in an upright position.

Optional features available include four-wheel steering, loading height of 530 mm, spring-loaded barriers, galvanised rollers and safety walk between rollers in the same row.

Characteristics

TYPE: 77A/B
CAPACITY: 6000 kg
LOADING HEIGHT: 0·508 m or 0·53 m
MAX LOAD DIMENSIONS: 3·18 × 2·44 m
TOWING SPEED: 25 km/h
UNIT DIMENSIONS: 4·65 × 3·4 × 0·508/0·63 m (including tow-bar)

Status

In service with Scandinavian Airlines System

Manufacturer

Tellus Maskin AB, Vallentuna, Sweden

UNITED KINGDOM

Avery Cargo Scales and Weighbridges

Avery supplies a wide variety of standard electronic and mechanical weighers suitable for cargo handling. They range from small bench-mounted dial scales for weighing packages up to 60-tonne capacity weighbridges for weighing vehicle loads. In addition to these standard models, the company designs special-purpose systems for duties such as weighing scissor-lifts, motorised cargo handlers and lift trucks.

The company has introduced a series of free-standing and fixed platform scales suitable for transit sheds. Available in capacities from 200 to 5000 kg, these digital electronic machines can be integrated into control or data-processing circuits, the weight signals being transmitted as binary-coded outputs through an integral socket. The digital display which register to 1/2500th part of capacity can be mounted remotely from the weight platform on wall or desk.

Both mechanical and load-cell weighbridges are supplied for airport duties. The electronic types are manufactured in standard capacities up to 60 000 kg and are equipped with digital electronic instrumentation giving a maximum of 6000 weight increments. Optional features include interfaces for digital output, active filters for suppressing surging loads, dual graduations, and electronic printers for recording gross, tare and net weights on tickets. Other products for cargo weighing include dial and optically indicating scales for bend- and floor-mounting (some with facilities for printing weights and related data), a selection of small-capacity dial and fan scales and a series of digital electronic counting machines for use in store rooms.

Manufacturer

W and T Avery Ltd, Smethwick, Warley, West Midlands, England

Avialift APL 750/900 Baggage/Freight Conveyors

APL 750 and 900 conveyors for baggage and freight loading and unloading operations are available driven by a battery-powered electric motor or a diesel engine. Adjustable hydraulic rams at the front and rear of the conveyor boom allow a range of conveyor angles and heights. Maximum working height of the APL 750 conveyor is 4·32 metres and the APL 900 can reach 52 metres. Conveyor-belt direction and speed can be controlled at three remote stations; two at the rear end of the vehicle and one from within the aircraft using the remote control box on its wander cable. Adjustable side-rails are fitted to keep normal loads on the conveyor but fold down to accommodate non-standard bulk loads on the wide belt.

Both models are equipped with road lights, four-wheel braking system, power steering, emergency hydraulic hand-pump for conveyor elevation and in the diesel-powered version a Borg Warner automatic transmission. A battery charging facility is provided in the battery-powered version. Optional features include enclosed driver's cabin, safety hand-rails on the conveyor boom and wing mirrors.

Avialift APL 750E electric baggage/freight conveyor

Characteristics

ENGINE: Perkins 4108 diesel/battery-driven electric motor
OUTPUT: 38·8 kW at 4000 rpm
MAX DRIVING SPEED: 64 km/h
CONVEYING SPEED: 12–36 m/minute
BELT WIDTH: 0·61 m
CAPACITY: 907 kg at 15°; 450 kg at 30°
ELEVATING TIME front/rear: 35/6 s
LOWERING TIME front/rear: 29/6 s

TYPE:	APL 750	APL 900
CONVEYING HEIGHTS		
max		
front:	4·32 m	5·2 m
rear:	1·625 m	1·6 m
minimum		
front:	0·965 m	0·81 m
rear:	0·229 m	0·122 m
OUTSIDE TURNING RADIUS	0·701 m	1·219 m
DIMENSIONS	7·465 × 2·032 × 1·26 m	9·09 × 2·032 × 1·26 m
WEIGHT	2360 kg	2750 kg

Status

In service with major airlines at airport authorities

Manufacturer

Avialift Products Ltd, Southend-on-Sea, Essex, England

Avialift Super Hylo Container/Pallet Loaders

Three models are available for container and pallet loading and unloading operations, the 15, 18 and 102·3 metre.

The Super Hylo 15 metre features rear wheel steering and front-wheel drive. Powered by a Perkins diesel engine the loader can handle pallets sized 96 by 125 inches (2·44 by 3·18 metres), a full-size belly container, two LD-1 (A2) or two LD-3 (A1) containers to aircraft door-sills on a wide range of aircraft including main decks of B707, B727, B737, DC-8, DC-9, F28 aircraft and the lower decks of B747, L-1011 Tristar, DC-10, A 300B aircraft.

Positioning the loader at the aircraft door is done by the driver in the forward drive cab. Stability during loading operations is provided by use of six hydraulic stabiliser jacks. Hinged drop-down sections on the front-bridge platform permit loader positioning to any width of aircraft door. The rear main platform can be raised and lowered with a load of 7000 kg, while the front bridge platform holds 550 kg. An emergency-power system allows operation of all functions such as pallet transfer, lowering of main platform and raising of stabiliser jacks for removal of the loader from an aircraft in the event of engine failure. Safety interlocking devices prevent drive engagement when stabiliser jacks are lowered or when the main rear platform is raised.

Optional equipment includes powered-side loading/transfer on the rear main platform and latch-on adaptor mechanism to the front bridge with automatic levelling to adjust to varying aircraft door-sill heights.

The Super Hylo 18 metre was developed for servicing wide-body upper and lower decks of wide-body aircraft and the main decks of narrow-body aircraft. The loader is similar to the 15 metre but incorporates programmed logic control, a microprocessor based system which provides continuous control of the machine by means of a pre-selected programme and includes fault diagnostics. The loader can handle two pallets sized 88 by 108 inches (2·24 by 2·74 metres), 88 by 125 inches (2·24 by 3·18 metres) or 96 by 125 inches (2·44 by 3·18 metres) or two full-size belly containers, four LD-1/LD-3 containers or one 20-foot (6·1-metre) container.

The Super Hylo 102·3, designed and built to meet NATO equipment standards, is capable of elevating a load of up to 20 000 lb (9000 kg) to a height of 11 foot 4 inches (3·45 metres). It can transport its load while towing up to four trailers carrying a further 20 000 lb (9000 kg), the total load approaching the maximum load of a Hercules C-130.

The platform will take two pallets with a 9 foot (2·74 metre) dimension along or across the platform. In addition to standard fore and aft through loading, a side-loading capacity is included; both facilities having powered-drive rollers. The machine is air transportable and can be prepared for flight in four hours and reassembled in the same time.

It can be carried in most military transport aircraft. An emergency power system is fitted to the machine for use in the event of an engine failure. This system will operate for 15 minutes to enable all functions to be carried out and clear the load.

Status

In service with major airlines, freight forwarders and airport authorities

Manufacturer

Avialift Products Ltd, Southend-on-Sea, Essex, England

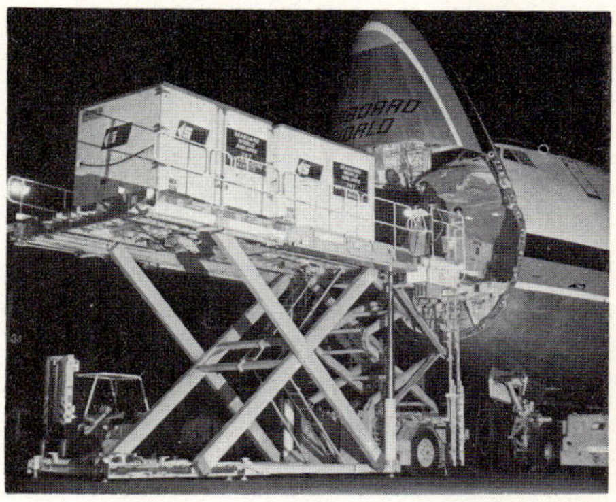

Avialift Super Hylo 18-metre container/pallet loader

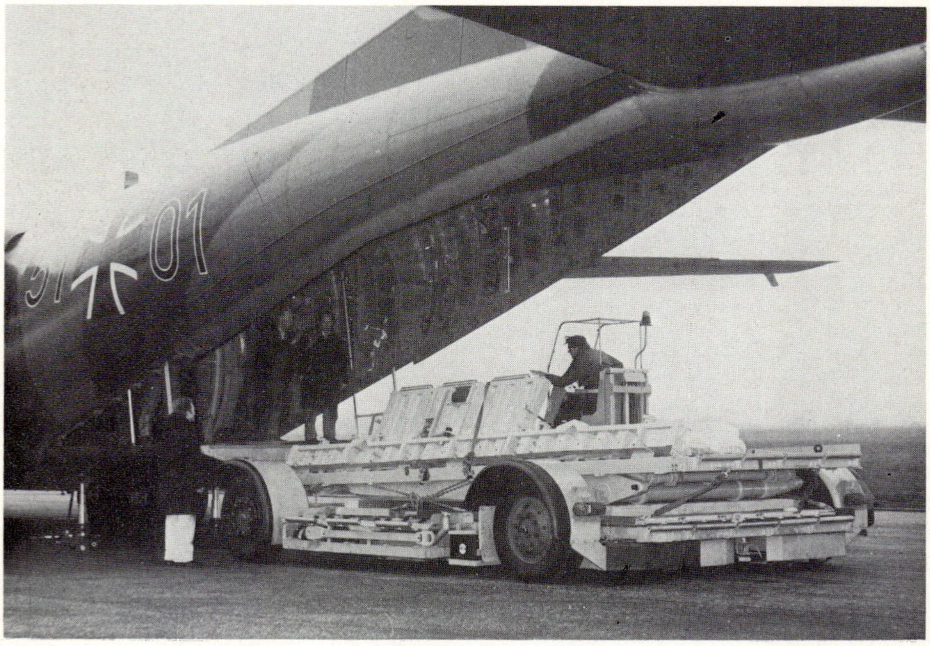

Avialift Super Hylo 102·3-metre loader with a Transall C160 aircraft

Characteristics

TYPE	15 metre		18 metre		102·3 metre
PLATFORM	Main (rear)	Bridge (front)	Main (rear)	Bridge (front)	
CAPACITY	7000 kg	5500 kg	13 610 kg	6806 kg	9000 kg
HEIGHT					
max	3·53 m	3·53 m	5·518 m	5·518 m	3·45 m
minimum	0·508 m	1·768 m	0·508 m	2·083 m	0·508
WIDTH	3·06 m	3·13 m	4·32 m	3·328 m	3·46 m
between pallet guides	2·286 m	2·286 m	2·47 m	2·47 m	2·24, 2·44, 2·74 m
between container guides	2·489 m	2·489 m	3·23 m	3·23 m	
ELEVATING TIME	15 s	40 s	30 s	30 s	25 s
LOWERING TIME	22 s	45 s	35 s	90 s	34 s
ENGINE	Perkins 4236 diesel		8-cylinder diesel		Perkins 4236 diesel
OUTPUT	53 kW at 2100 rpm		127 kW at 2500 rpm		60 kW at 2800 rpm
DRAW-BAR PULL					1150 kg
SPEEDS					
main drive	19·5 km/h		14 km/h		28 km/h
transfer	12 m/minute		18 m/minute		18 m/minute
OUTSIDE TURNING RADIUS	6·93 m		10·5 m		12·2 m
UNIT DIMENSIONS					
length	7·09 m		10·65 m		6·12 m
width	3·131 m		4·76 m		4·36 m
WEIGHT	8373 kg		23 500 kg		9096 kg

Becker Scissor-lifts and Dock Levellers

Becker has developed a range of scissor lifts suitable for cargo handling with lifting capacities up to 10 000 kg. Standard platform sizes extend up to 3 by 2·4 metres and lifting heights range from 0·81 to 3·8 metres.

The scissor tables are fabricated from heavy-rolled steel sections and decked with plain mild steel plate or Durbar-pattern plate with a minimum thickness of 6 mm. All platforms in the raised position can accept full safe working load evenly distributed or half that load over half the platform length or one third the load over half the platform width. Options are available for point and axle loads. The scissor arms are profiled from heavy-duty steel sections with a minimum safety factor of 5. The hydraulic rams are solid rod manufactured from chromed ground steel running in honed cylinders, fitted with oil seals and scraper rings and fitted with a pipe-rupture valve.

Power units incorporate a 3-phase motor, pump, pressure relief valve, non-return valve, solenoid down valve, constant down speed valve, mounted on a common base with a hydraulic oil tank.

The controls are deadman-type raise/lower push-buttons.

Becker offer Serco manual, semi- and fully-automatic dock levellers to act as a bridge between the loading/unloading dock and road vehicle. Their standard single axle loading capacity is 9000 kg. A variety of sizes caters for different widths and lengths. Manual- and semi-automatic levellers are operated by helical springs. The semi-automatic model has a manually-operated release ring to raise the platform and extend the lip automatically. Walking on to the raised platform positions it on the vehicle. When the lorry pulls away, the lip drops vertically. Automatic levellers are hydraulic, operated by push-button control. As the ramp reaches its maximum height the hinged lip extends and locks into position. The ramp lowers until the lip engages on the vehicle and will return automatically to its flush dock position when the vehicle pulls away. The lip can be withdrawn and the ramp returned to the flush position without moving the vehicle.

Characteristics

SCISSOR LIFT
CAPACITY: up to 10 000 kg
MINIMUM PLATFORM SIZE: 3 × 2·4 m
MAX HEIGHT RANGE: 0·81–3·8 m
POWER UNIT: 1 or 2·5 gallon/minute (4·5 or 11·4 litres/minute)
POWER REQUIREMENTS: 400/330 V, 3 phase, 50 Hz

Status

Alitalia uses the 6000 kg twin scissor lift at London Heathrow and Rome airports to lift air-cargo igloos from the warehouse floor to trailer level

Manufacturer

Becker Lifts Ltd, Wembley, Middlesex, England

Crompton Battery-electric Towing Tractors/Platform Trucks

The B 2000 range produced by Crompton Electricars, with a three-wheel configuration and front-wheel drive are available as towing tractors or platform trucks. The tractors can tow up to 2000 kg at speeds from 5 to 8 miles per hour (8 to 13 km per hour) on level surfaces and can be fitted with an insulated cab to provide protection for the driver. The platform version carries loads up to 2000 kg. It can be applied to general duties or fitted with an elevated platform for the transport of pallets.

The Brook Victor (BV) range uses thyristor control systems and trucks are fitted with hydraulic brakes, all-round suspension and floating double reduction rear axles. The BV tow tractors pull trailed-loads of between 4 and 15 tons. Other tractors available include the C range, a four-wheeled vehicle with a towing capacity up to 10 tonnes; the low profile K tractor which has an overall height of 6 feet (1·85 metres) for operation in areas with limited headroom, and the D-10 where the driver is seated on top of the main frame enabling the wheels to be placed near the four corners to produce a turning radius of 7 feet (2·14 metres). Loads of over 20 tons can be trailed by this tractor.

Status

B 2000 tractors are used for towing baggage trailers at Cardiff Airport, Wales. BV K8 tractors are employed by Aer Lingus for towing baggage and freight between the aircraft and terminals at Dublin Airport, Ireland

Crompton B 2000 tow tractors at Cardiff Airport

Manufacturer

Crompton Electricars Ltd, Tredegar, Gwent, Wales

Crompton BV K8 tow tractors at Dublin Airport

Defiant Freight Weighing Equipment

Defiant produces low-profile weigh platforms with capacities ranging from 1000 to 20 000 kg. The platforms, which incorporate strain-gauge load cells to provide resistance to shock and overload conditions, are designed for the electronic weighing of containerised and palletised freight and are compatible with centralised computer facilities.

Status

In service with UTA

Manufacturer

Defiant Weighing Ltd, Sevenoaks, Kent, England

Defiant pallet/container weighing platform

Dexion Speedlock Container Handling System

The container handling system developed and manufactured by Dexion uses speedlock structures of upright frames and boxed beams to support 18-level roller conveyor lanes each 105 feet (32 metres) long. The conveyors are 20 inches (508 mm) clear of the ground and each lane can accommodate 14 cargo containers. Complete segregation of cargo according to flight is possible, since each lane is designated for a particular aircraft. Containers are brought from the cargo terminal and fed into the system from a hydraulic-scissor lift at the loading end. Each one is then moved along a transfer table to the appropriate lane down which it is directed. When required for loading, the containers are fed directly on to low-loading transporters for transfer to the aircraft.

Dexion Speedlock container handling system installed at Pan Am cargo terminal, New York-Kennedy Airport

Status

This system was developed for Pan American World Airways for use with Boeing 747 aircraft and is installed at New York, Kennedy Airport

Manufacturer

Dexion-Comino International Ltd, King's Langley, Hertfordshire, England

Douglas Tugmaster DC4 Tow Tractor

The DC4 is designed for towing baggage and freight trailers. The unit is powered by a Perkins 4236 diesel engine coupled to a Borg Warner automatic transmission incorporating a torque converter. The enclosed full width cab is mounted on a four-wheel chassis with front and rear leaf suspension, a dead beam steering front axle and a double reduction drive axle. The tractor is equipped with orbitrol hydraulic steering, air/hydraulic braking to all four wheels and a mechanical brake operating on the rear wheels. Standard towing pintles are mounted front and rear.
Optional features available include left- or right-hand drive, alternative towing attachments and a flashing warning beacon.

Characteristics

ENGINE: Perkins diesel 4·236
OUTPUT: 60 kW at 2500 rpm
DRIVING SPEED: 20 mph (32 km/h)
DRAW-BAR PULL: 6000 lb (26·7 kN)
WHEELBASE: 5 ft (1·52 m)
TURNING RADIUS: 12 ft (3·65 m)
UNIT DIMENSIONS
length: 8 ft 4 in (2·54 m)
width: 5 ft 3 in (1·6 m)
height: 6 ft 3 in (1·91 m)
WEIGHT: 8800 lb (4000 kg)

Status

Introduced in 1980

Manufacturer

F.L. Douglas (Equipment) Ltd, Cheltenham, Gloucestershire, England

Douglas Tugmaster DC4 tow tractor

Edghill (Cargomatic) Cargo Handling Equipment

Cargo handling equipment developed by Cargomatic comprises pallet/container transporters, dollies, roller-bed trucks, baggage/freight trailers, lower-deck loaders and baggage/freight conveyors. It is manufactured by Edghill after Cargomatic was liquidated early 1982.
The CM 2420 transporter is the latest model produced. It is capable of handling two 96 by 125 inch (2·44 by 3·18 metre) pallets or one 96 by 240 inch (2·44 by 6·1 metres) container with a total load capacity of 12 500 kg.

Status

Cargomatic supplied equipment to Alitalia, Syrian Arab Airlines and the British Airports Authority

CM 3800 Electric-powered baggage conveyor

CM 2804 Electric-powered transfer transporter

Manufacturer

H. W. Edghill Equipment, Hook, Hampshire, England

Edghill EDG 419-1 Roller Bed Vehicle

Edghill Equipment produces a range of self-propelled and towable container handling equipment. The self-propelled system is based on the EDG 419-1 roller-bed vehicle, consisting of a standard commercial vehicle chassis fitted with self-contained powered conveyor modules giving full mechanical handling of aircraft LD containers and pallets. A full range of towable container/pallet dollies are supplied covering all airport handling requirements. The company also builds and installs full warehouse systems to customers' specifications.

Status

The EDG 419-1 roller-bed vehicle is in service with Air Canada at London Heathrow Airport, England

Manufacturer

HW Edghill Equipment, Hook, Hampshire, England

Edghill EDG 419-1 roller-bed vehicle

Hamech Electric Tow-tractors

Three models of electric tractors are produced by Hamech with components supplied by Ransomes suitable for towing baggage and cargo trailers. These are the T12M and a low profile version of the T12M developed for wide-body aircraft operations, both rated at 1200 kg draw-bar pull, and the T25M with a maximum draw-bar pull of 2500 kg.
The tractors are equipped with a Ransomes' electric motor and power steering (manual steering on the T12 model) and an automatic-transmission multiplier is fitted to the two T12M models. Mounted on all-round leaf suspension with a dual-braking system and front- and rear-towing hitches, the units incorporate an opportunity charging facility which enables the battery to be recharged at convenient moments throughout the day. Optional features offered include a high-speed version of the T25M; an external inching button fitted to the rear of the vehicle enabling the driver to adjust the position of the tractor while standing behind it to align towing points. Also available are remote towing hitches, automatic battery cut-out at 80 per cent discharge, key-operated high/low speed variations and provision for trailer air-braking systems.

Characteristics

MODEL	T12M	T12M low profile	T25M
MAX DRAW-BAR PULL (5-minute rating)	1200 kg	1200 kg	2500/1940 kg (high-speed version)
CONTINUOUS DRAW-BAR PULL (1-h rating)	160 kg	160 kg	680 kg/220 kg
BATTERY	36 cell, 72 V	36 cell, 72 V	2 × 24 cell, 96 V
CAPACITY	329/378 Ah	329/378 Ah	561/680 Ah
UNLADEN TRAVEL SPEED	24 km/h	24 km/h	18·5/28·8 km/h
TURNING RADIUS	2·23 m	2·9 m	2·96/4·35 m
UNIT DIMENSIONS	2·19 × 1·31 × 2·04 m	2·49 × 1·31 × 1·76 m	2·88 × 1·5 × 2 m
WEIGHT	2537 kg	2740 kg	5080 kg

Status
Hamech tractors are in service with Aer Lingus, Air Canada, British Airways, Iberia, KLM, Qantas, Swissair and TWA

Manufacturer
Hamech Ltd, Staines, Middlesex, England

Hamech T12M low-profile tractor

Hamech T25M tractor

Hydraroll Pneumatic Roller Track

The Hydraroll pneumatic roller track is operated by low-pressure compressed air, through flexible reinforced air bags under the tracks. In the raised position, the track is live and ready for loading and unloading. Deflated, the rollers sink beneath the surface of the tracks, or into the vehicle floor and the load is supported along its entire length during transit. Tracks may be mounted on the surface or fitted flush with the vehicle floor. The compressed air is supplied either from a reservoir fed by the vehicle servo-supply system, or by a diaphragm compressor driven by a 12- or 24-volt electric motor. Vertical movement of the rollers is 0·43 inch (11 mm). In the event of the vehicle air supply not being available and the reservoir being empty, it is possible to pressurise the reservoir through an integral external valve.
Control is through a switch which operates the roller lift. A pressure regulating valve allows working pressure to be varied according to the load and the system is equipped with a pressure relief valve to ensure the integrity of the vehicle braking system.

Characteristics
MAX RECOMMENDED STATIC LOAD: 600 lb/ft (892 kg/m) single-track run
TRACK CONFIGURATION: supplied in standard modular lengths 8 ft 1 in (2·46 m)
SAFETY LATCHES: at 48 in (1200 mm) intervals, can be varied
DIMENSIONS
Track height (with rollers down): 2·79 in (71 mm)
Max width: 4·1 in (104 mm)
WEIGHT: 6·1 lb/ft (9·1 kg/m)
WORKING PRESSURE: 20–40 psi (1·4–2·8 kg/cm²)
ROLLER PITCH: 3–6 in (76·2–152·4 mm) as required
ROLLER LIFT: 0·43 in (11 mm) deck clearance
ROLLER WIDTH: 2 in (50·8 mm)
ROLLER DIAMETER: 2 in (50·8 mm)

Status
The Hydraroll system is in service with many airlines and airline freight carriers. Airlines who operate the system include Air France, British Airways, British Caledonian Airways, Finnair, KLM, Lufthansa, Pan Am and Sabena

Hydraroll pneumatic track

Manufacturer
Hydraroll Ltd, Anglesey, Wales

Hyster Lift Trucks

Hyster produces three ranges of lift truck suitable for airport operations; the Challenger, Electric and the Spacesaver.
The Challenger range of 33 models extends from small three-wheel models with a 1000 kg capacity to trucks capable of handling loaded 40-foot (12·2-metre) containers rated at 37 000 kg capacity. For general air cargo, trucks from the H150-275H range are most suitable with capacities up to 12·5 tonnes at 0·6 metre load centre. A Monitrol directional-control system is fitted to this range which offers also the choice of diesel, petrol or lpg engines with manual or automatic transmissions.
The Electric range consists of 22 models with capacities varying from 1000 to 5000 kg at 0·5 metre load centre. Battery-powered in a three- or four-wheel configuration, these units can be supplied with cushion tyres at all capacities or pneumatic tyres up to 3000 kg.

Hyster Electric lift truck

The Spacesaver is a cushion-tyred unit designed for warehouses and freight terminals. Twelve models are available with capacities from 1000 to 6750 kg at 0·5 metre load centre. Options include petrol, lpg or diesel engines, automatic or manual transmissions, and a choice of upright types, sizes and attachments.

Hyster Challenger 130F lift truck

Status
In service with major airlines including Air Canada, American Airlines, and British Airways

Manufacturer
Hyster Europe Ltd, Basingstoke, Hampshire, England

IBM Cargo Handling and Storage Data Systems

IBM equipment is used by many airlines for cargo handling and storage. A typical example is the system used by Alitalia where visual display terminals and printers, installed in the airline's cargo handling areas at various stations in the airline's route network are all linked to the Alitalia Rome data-processing centre which is equipped with IBM system 370 Model 168 central processors. Using Alitalia's cargo handling application software, operators at remote stations for example, London Heathrow, can respond to freight-agent demands and answer queries using their terminals to reserve cargo space, track consignments and produce the documents required.

Manufacturer
International Airlines Support Centre, IBM (UK) Ltd, Feltham, Middlesex, England

Operators at console of IBM System/370 Model 168 computer

Iso-Speedic Battery-electric Towing Tractors/Platform Trucks

Iso-Speedic manufactures a wide range of electrically-powered vehicles for carrying and towing loads.
The PPT 1000P three-wheeled platform vehicle can carry a load of 100 lb (455 kg) on level ground or a load of up to 500 lb (227 kg) on gradients not exceeding 1:5. The vehicle has built-in battery charger, two speeds forward and reverse with controlled acceleration from rest and tiller arm steering with finger tip control of speed and direction. If required, wire-mesh sides can be fitted.
The PPT 2500 is a heavy-duty version of the PPT 1000P, with a load capacity of 2500 lb (1136 kg) on the platform or towing effort. Various traction batteries and custom-built platforms can be supplied.
The RPT 800P is a four-wheeled rider-controlled vehicle which can carry a platform load of 800 lb (360 kg) or tow a load of 1250 lb (546 kg). It has a built-in battery charger, two speeds forward and reverse, with manual controls and independent foot and hand brakes. It is of all-steel construction with platforms of various sizes and can attain speeds of up to 8 miles per hour (13 km per hour) on level ground with suitable gearing. Trailer units are also available.
Model RT 7000P is a four-wheeled towing tug which can haul loads of up to 6600 lb (3000 kg) on level ground at speeds up to 5·2 miles per hour (8·5 km per hour). The tug has stepless electronic speed control. It is of all-steel construction fitted with normal automotive batteries. Options include special batteries, an automatic battery charger, hazard warning lights, a fully-enclosed metal or canvas cab and trailer units. A specially extended model with a carrying platform is also available.

Manufacturer
The Iso-Speedic Company Ltd, Warwick, England

JCB 520/525 Telescopic Handler

JCB telescopic handler unloading drill equipment from a CL 44 Yukon freighter

Two models of telescopic handlers are produced by JCB which can be used for freight loading operations.
The JCB 520 is equipped with a two-piece hydraulically-operated telescopic boom providing a lift capability of two tonnes to a height of 5·74 metres and a maximum forward reach of 3·04 metres with a 0·8 tonne load. Available with two- or four-wheel drive powered by a 53 kW engine and automatic transmission, the standard model is fitted with a protective canopy for the operator. A weather-proof cab is an optional feature.
The JCB 525 which is produced in two-wheel and four-wheel drive models, can lift a 2·5 tonne load to a height of 6·4 metres and reaches 3·62 metres forward with a load of 0·82 tonnes. A choice of carriage builds is available. The standard carriage is 1·2 metres wide. Alternatively it can be fitted with a 2·3 metre masted carriage or a hydraulically-operated ± 12 per cent side-tilt facility. Weather-proof cabs are fitted to the 525 series as standard equipment and instrumentation in all models includes an electronic safe working load indicator with progressive visual signals and an audible warning at 95 per cent.

Status
In service at Stornoway Airport, Scotland

Manufacturer
J.C. Bamford Excavators Ltd, Rocester, Staffordshire, England

King and Taylor LAPD 12 Pallet/Container Transporter

The LAPD 12 carrier available on a choice of chassis is designed to transport pallets and containers from the aircraft loader to destinations off the airport using public highways. The rear end of the vehicle incorporates a turntable which can rotate containers through 90 degrees to stow them transversely on the vehicle loading bed and at the same time move them sideways to place them on the pallet guides of the bed. When loaded the vehicle floor is raised above the rollers, ensuring the load is secured.

Characteristics
CHASSIS: Dodge Commando or as required by operator
WHEELBASE: 19 ft (5·8 m)
LOAD: 4 AVK/AVE containers or 2 UK2 and 2 UAK containers 88 × 125 in (2·24 × 3·18 m) pallets
CAPACITY: 8·3 tonnes
LOADING HEIGHT: 4 ft 9 in (1·45 m)
UNIT DIMENSIONS
length: 33 ft 7 in (10·24 m)
width: 8 ft 1 in (2·46 m)

Status
Developed in conjunction with British Airways and in service with the airline, the transporter is marketed by Listavia Ltd

King and Taylor LAPD 12 ULD carrier

Manufacturer
King and Taylor (Commercial) Ltd, Godalming, Surrey, England

King and Taylor PD11 Conveyor Loader

The PD11 conveyor loader is designed primarily for baggage loading. Based on a Dodge RG11 chassis powered by a 6354 cc diesel engine, the conveyor boom projects from an enclosed alloy body and frame. A horizontal conveyor is incorporated within the lockable van body which feeds the aircraft loading conveyor. Aircraft protection buffers are fitted to the outer end of the elevating conveyor.
Optional features available include an extended boom height, tropical roof replacing the standard translucent fibreglass roof, ac power in the cab, and urethane filled tyres.

Characteristics
CHASSIS: Dodge RG11, 6354 cc diesel
MAX LOADING HEIGHT: 3·354 m
FLOOR HEIGHT: 1·067 m
CONVEYOR
speed: 12 m/minute, forward/reverse
capacity: 204 kg
width: 0·61 m
HORIZONTAL CONVEYOR: 4·726 × 0·61 m
UNIT DIMENSIONS: 8·46 × 2·464 × 3·303 m
GROSS VEHICLE WEIGHT: 11·382 kg

Status
In service with British Airways, the loader is marketed by Listavia Ltd

Manufacturer
King and Taylor (Commercial) Ltd, Godalming, Surrey, England

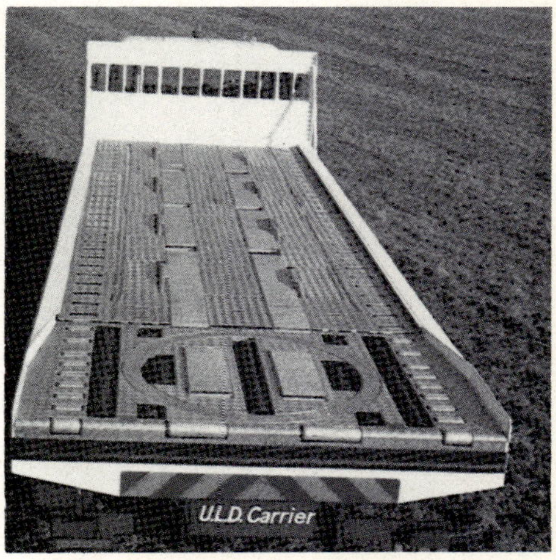

King and Taylor PD11 conveyor loader

Lansing Electric Tow Tractors

Lansing produces two basic models of electric tow tractors, the TOER 8·1 and TOER 4·2.

The TOER 8·1 chassis is steel fabrication of open box construction with a steel-plate scuttle bolted to the front and a towing bracket and battery securing device welded to the rear.

Power is provided by a series wound motor which drives the road wheels via a universal jointed transmission shaft, double reduction gears and differential unit. A hytronic-pulse control system provides variable speed control. Alternative axle ratios are available giving different towing capacities and speeds.

Steering is through a recirculating ball-type steering box connected by a sealed ball-jointed linkage to a centrally-pivoted front axle. The unit is equipped with a double-reduction rear axle, helical primary gears and spiral bevel secondary gears, mounted on rubber-bushed trailing arms and located by a Panhard rod. Suspension of the vehicle is provided by Aeon rubber springs. Hydraulically-operated expanding shoe-type brakes are fitted on all four wheels operated by a foot-pedal. The parking brake operates on the rear wheels only. The battery on standard mountings and quick-release latches can be supplied shaped to allow the operator to see the towing attachment from his seat. In order to meet the requirements of travelling on public roads, two headlights, side lights and flashing indicators are fitted and a standard power take-off provides for trailer lighting and other requirements.

The smaller single seat TOER 4·2 has many similar features to the larger model. It is mounted on three wheels with a single steering wheel at the front and two rear-drive wheels. A canopy with windscreen and wipers can be fitted to the tractor.

Low-profile version of the Lansing TOER 8·1 tractor in service with Continental Airlines

Characteristics

TYPE	42/4500	4·2/3000	8·1/11000	8·1/8000
MOTOR	Electric	Electric	Electric	Electric
OUTPUT	2·5 kW	3·3 kW	8·9 kW	8·9 kW
BATTERY	48 V	48 V	72 V	72 V
AMP/H (5-h rate)	187·5	187·5	329	329
DRIVING SPEED	12·5 km/h	17·7 km/h	20 km/h	23·5 km/h
DRAW-BAR PULL				
max (5-minute rating)	4500 N	3000 N	11 000 N	8000 N
continuous	900 N	700 N	2500 N	1700 N
TURNING RADIUS	1·475 m	1·475 m	2·4 m	2·4 m
UNIT DIMENSIONS	1·62 × 0·915 × 1·17 m	1·62 × 0·915 × 1·17 m	2·495 × 1·27 × 1·45 m	2·495 × 1·27 × 1·45 m
WEIGHT	800 kg	800 kg	2040 kg	2040 kg

Lansing TOER 8·1 and 4·2 tractors

Status

In service with major airlines world-wide

Manufacturer

Lansing Ltd, Basingstoke, Hampshire, England

Lansing Pallet Transporters and Lift Trucks

Lansing offers a wide range of lift trucks, including simple pallet transporters, counterbalanced fork lift trucks (electric or diesel engine-powered), narrow-aisle reach trucks and turret trucks.

Pallet transporters can be either pedestrian- or rider-controlled and provide capacities of up to 3 tonnes. The pedestrian models can be fitted with contactor or electronic controls. They are used for the horizontal movement of palletised goods over moderate distances such as within a freight marshalling area.

Lansing Bagnall electric counter-balanced lift trucks used for general materials handling duties are available with lifting capacities up to 5 tonnes. The narrow-aisle reach trucks and turret trucks are designed to provide access to high density storage systems in freight warehouses or transit sheds. Reach trucks are available with contactor or electronic control and double or triple motors providing capacities up to two tonnes. The turret trucks developed for high level storage with narrow aisles are equipped with a turret mechanism which turns the load through 90 degrees for stacking enabling the truck to remain in line with the aisle.

Diesel-, petrol- or lpg-powered lift trucks are manufactured by Lansing Henley in the Lansing group. Capacities available in a wide range of models vary from 2000 to 26 000 kg.

The Hercules which is produced with a long-wheel base version, is fitted with a six-cylinder engine developing 90 kW, torque converter transmission, servo-assisted brakes and power steering. A choice of mast lift heights is offered. This model is in service with the Royal Air Force. The long wheelbase meets the stringent floor loadings of the C130 aircraft for transporting by air.

Lansing Hercules lift truck

Status

In service with Royal Air Force, England

Manufacturer

Lansing Ltd, Basingstoke, Hampshire, England

Listavia LA PD22 Tractor/Trailer System

Listavia PD22 tractor/trailer system

The PD22 system comprises prime movers, hydraulic towing converters, low bed-roll trailers and main-/lower-deck elevators. The prime movers are conventional 3000/8000 lb (1360/3620 kg) rated draw-bar hard ground airport tractors, either diesel, electric- or petrol-/lpg-powered, usually incorporating an automatic or torque converter transmission.

Hydraulic towing converters are fitted to the rear of the tractors and operate from a hydraulic pto from the tractor-power unit or a separate power unit mounted on the tractor. This device supplies a lifting function to the towed load, without applying any vertical imposed load to the tractor. It produces only one point of articulation between the towed load and the tractor which allows the vehicle combination to be driven either forward or in reverse. The absence of any vertical load on the tractor is of great importance, as the majority of draw-bar tractors are already ballasted correctly to apply their tractive effort to the ground surface and would be adversely affected by any additional ballasting.

Low bed-roll trailers are designed to connect behind the tractor and towing converter and will carry any uld. They are supplied in three basic sizes to accommodate two, three or four pallets or equivalent, or eight/ten or sixteen LD-3 containers or equivalent, or 10-, 20-, 30- or 40-foot (3·05-, 6·1-, 9·1-, 12·2-metre) containers or equivalent. Most trailers are decked with roller, castor or ball mat beds, with the vast majority as castor beds. They are fitted with retractable stops and locks for load retention and location. Frames are rigid beamed or box-girder structures fitted with single or twin fixed or rocker axles mounted at the rear only. The wheels are heavy fabricated steel units with solid-rubber tyres mounted on roller bearings. The front of the trailer rests on the front frame directly on the tarmac when disconnected from the hydraulic towing converter unit. This trailer has no steering gear, or brakes and can be butt or herringbone parked to produce a continuous static stillage platform. The trailers have lead on rollers for transfer of pallets and loads and the loads can be aligned in any desired direction by rotating on the castor beds.

The towable main-/lower-deck elevators are based on a standard production roll trailer chassis, and are positioned on the tarmac by the tractor/hydraulic towing converters combination. The elevator is equipped with a power drive for the elevating and roller bed and will handle both lower and upper deck operations with elevating heights ranging from 20 to 218 inches (0·508 to 5·54 m). A three direction auto aligning electronic device allows the operator to place the leading edge of the elevator within one metre of the sill line on the tarmac. When the elevator is raised, the auto aligning device adjusts the reach and level of the elevator platform automatically. The elevator adjusts automatically each time it is elevated to allow for the variance of aircraft altitude during the loading/unloading process.

Two similar elevators placed in tandem fore and aft can be operated as a single shelf compensating lift to raise the 30 or 40 foot (9·14 or 12·2 metre) load which may be necessary for nose loading the B747 F.

Characteristics

HYDRAULIC TOWING CONVERTER

MODEL	HTC	HTC A	HTC HD	HTC M
CAPACITY	6000 kg	6000 kg	20 000 kg	12 000 kg
TYPE OF TRAILER	PD22·2/3	PD22·2/3/4	PD22·4	Military
DECK HEIGHT	0·225 m	0·225 m	0·4 m	0·4 m
LENGTH	1·7 m	1·7 m	2 m	2 m
WIDTH	1·025 m	1·025 m	1·025 m	1·025 m

ROLL TRAILERS

MODEL	PD22·2/3 CB	PD22·2/3 RB	PD22·4 CB	PD22·4 RB
CAPACITY	15 000 kg	15 000 kg	28 000 kg	28 000 kg
TYPE	Castor bed	Roller bed	Castor bed	Roller bed
DECK HEIGHT	0·52 m	0·52 m	0·52 m	0·52 m
LENGTH	8·305 m	8·305 m	13·05 m	13·05 m
WIDTH	3·47 m	3·47 m	3·47 m	3·47 m

ELEVATORS

MODEL	LDE	MDE	MDE 2	MDE 2A	MDEM
CAPACITY	3628 kg	3628 kg	13 620 kg	13 620 kg	1800 kg
ELEVATING HEIGHT					
minimum	0·51 m	0·51 m	0·51 m	0·51 m	0·51 m
max	3·7 m	5·54 m	5·54 m	5·54 m	5·54 m
LOAD	1 × 88/96 × 108/125 in (1 × 2·24/2·44 × 2·74/3·18 m) pallet	1 × 88/96 × 108/125 in (1 × 2·24/2·44 × 2·74/3·18 m) pallet	1 × 20 ft (1 × 6·1 m) container	1 × 20 ft (1 × 6·1 m) container	3 × 88/96 × 108/125 in (3 × 2·24/2·44 × 2·74/3·18 m) pallet
TRAILER				Non-powered	Military air portable

Status

The PD22 tractor/trailer system is in service with Gatwick Handling Services, at London Gatwick Airport, England

Manufacturer

Listavia Ltd, Godalming, Surrey, England

Listavia PD4/PD9 Pallet Dollies

The PD4 heavy-duty pallet dolly is a rigid frame with chamfered corners mounted on 300 by 150 mm nylon or rubber-solid castors capable of transporting a 108 by 125 inches (2·74 by 3·18 metres) pallet. Reverse ackerman steering is fitted with adustable track rods and the unit is equipped with kick-on safety brakes, walkways and stops. Optional features include pneumatic tyres, highway lighting, linked front-/rear-steering, extensible rear tow-bar and over-run cable operated brakes.

The PD9 is similar to the PD4 but fitted with an elevating deck, which is hydraulically-operated, either manually- or remotely-powered. As an optional feature (PD9A) hand-rails can be fitted to the deck.

Listavia PD4 pallet dolly

Characteristics

TYPE	PD4	PD9/9A
NO OF MODELS	15	
CAPACITY	11 200/17 636 lb (5080/8000 kg)	11 200/17 636 lb (5080/8000 kg)
LOAD HEIGHT	1 ft 8 in (0·508 m)	1 ft 8 in (0·508 m)–6 ft (1·83 m)
UNIT LOAD	108 × 125 in (2·74 × 3·18 m) pallet	108 × 125 in (2·74 × 3·18 m) pallet
BED	Roller/castor/ball mat/turntable	
LOADING	Side/end	

Status
In service with major airlines

Manufacturer
Listavia Ltd, Godalming, Surrey, England

Listavia PD5 Baggage/Freight Trailer

The PD5 baggage/freight trailer is mounted on cushion-solid tyres with turntable or reverse ackerman steering, and a foot-operated parking brake. The V-shaped deck with fork lift pockets incorporated underneath, has open sides, and a hinged tail-gate. Other features include: safety locks, self-draining canopy box, side protection strakes and captive pin rear hitch. Optional extras comprise castor steering, cable-operated over-run brakes with tow-bar parking brake, pneumatic tyres and a weather canopy.

Listavia PD5 baggage/freight trailer

Characteristics

PD5 with turntable steering

MODEL	15 TT	20 TT	30 TT
CAPACITY	1500 kg	2000 kg	3000 kg
DECK DIMENSIONS	1·83 × 0·92 × 0·6 m	2·44 × 1·22 × 0·6 m	2·44 × 1·22 × 0·6 m

PD5 with ackerman steering

MODEL	15 AS	20 AS	30 AS	50 AS	100 AS
CAPACITY	1500 kg	2000 kg	3000 kg	5000 kg	10 000 kg
DECK DIMENSIONS					
length	1·83 m	2·44 m	2·44 m	3·5 m	4·5 m
width	0·92 m	1·22 m	1·22 m	2 m	2·5 m
bed height	0·6 m	0·6 m	0·6 m	0·63 m	0·7 m

Status
In service with major airlines

Manufacturer
Listavia Ltd, Godalming, Surrey, England

Listavia PD7 Container Dolly

The PD7 container dolly is a rigid frame mounted on cushion-solid tyres equipped with reverse ackerman steering, a kick-on safety brake, walkways, rear hitch and captive pin, deadlock safety stops and automatic rotation stops. The dolly has a 50 per cent overload capacity at 6 km per hour. Optional features available include: turntable steering with a trailer link for continuous conveying in train, increased deck height, over-run/flip-up tow-bar cable brakes, highway lighting, pneumatic tyres, castor steering and rear lead rollers.

Listavia PD7 container dolly

Characteristics

TYPE	PD7/35	PD7/70
NO OF MODELS	7	7
CAPACITY	3500 lb (1588 kg)	7000 lb (3175 kg)
LOAD HEIGHT	1 ft 8 in (0·508 m)	1 ft 8 in (0·508 m)
UNIT LOAD	1 × LD-3 container	2 × LD-3 container
BED	Roller/castor/ball mat	Roller/castor/ball mat
LOADING	End/side	End/side

Status
In service with major airlines

Manufacturer
Listavia Ltd, Godalming, Surrey, England

Listavia Tractors

Listavia offers a range of three tractors suitable for airport duties.

The PD6 draw-bar tractor powered by a diesel engine with automatic transmission, is rated at 7500 lb (3400 kg) draw-bar capacity. The body and cab are an integral steel fabrication containing three seats. Optional features available include cab doors, remote hitch coupling and power take-off for ancillary equipment. This tractor is used as the prime mover in the Listavia PD22 aircraft loading system (see separate entry).

The La Ranger 2500 can be driven by either a 1100 cc petrol engine or a lpg engine. It is equipped with four-wheel drive, independent-suspension, and four wheel steering. Pintle tow hitches are installed at the front and rear end. Rated at 2425 lb (1100 kg) draw-bar capacity, the vehicle is fitted with a two seat cab.

The PD3 electric tractor has a flat platform for cargo behind the driver's seat, capable of carrying up to 2860 lb (1300 kg). Twin chloride batteries providing an eight-hour capacity, supply a 2·5 hp motor with variable thyristor control. A battery charger is incorporated. A rear pintle towing hitch is installed, enabling the tractor to tow loads of up to 13 200 lb (6000 kg).

Listavia PD8 electric tractor

Characteristics

MODEL	PD6	2500	PD3
DRAW-BAR PULL	7500 lb (3400 kg)	2425 lb (1100 kg)	
towing capacity		7·6 tons on 1:13 gradient 12 tons on 1:20 gradient	13 200 lb (6000 kg)
carrying capacity			2860 lb (1300 kg)
ENGINE	4236 cc diesel	1100 cc petrol/lpg	2·5 hp 24 V electric
driving speed	25 km/h		10 km/h
turning radius (outside)	12 ft (3·66 m)	7 ft 4 in (2·25 m)	6 ft 6 in (2 m)
DIMENSIONS			
length	8 ft 4 in (2·54 m)	8 ft 5 in (2·57 m)	5 ft 7 in (1·7 m)
width	5 ft (1·52 m)	3 ft 11 in (1·19 m)	3 ft 2 in (0·97 m)
height	6 ft (1·83 m)	3 ft 8 in (1·12 m)	3 ft 4 in (1·02 m)
WEIGHT	10 140 lb (4600 kg)	2865 lb (1300 kg)	1058/1322 lb (480/600 kg)

Status

In service with major airlines. The La Ranger 2500 is manufactured by Turner Engineering and marketed by Listavia

Manufacturer

Listavia Ltd, Godalming, Surrey, England

Massey Ferguson MF40/MF50 Tow Tractors

Two tractors are produced by Massey Ferguson, the MF40 and MF50, rated at 6500 lb (2950 kg) and 9000 lb (4080 kg) draw-bar pull capacity respectively. They can be used for a variety of applications including towing cargo and baggage trailers. The two models give power outputs of 47 bhp (33 kW) and 63 bhp (47 kW) and a larger model of 76 hp (57 kW) is available to order. Power is provided by three- and four-cylinder Perkins diesel engines driving through a torque converter transmission system. The units feature a steel driver's cab with seating for ground crew members, power-assisted steering, disc brakes, and five level towing hitches mounted at the front and rear of the vehicle. Optional equipment available includes a three line air brake system and for operation in hot climates the units are offered without a cab and with optional sun protection.

Characteristics

TYPE	MF40	MF50
ENGINE	Perkins AD 3.152	Perkins A 4.236
OUTPUT	33·5 kW at 2250 rpm	47 kW at 2000 rpm
MAX DRIVING SPEED	17 mph (28 km/h)	17 mph (28 km/h)
MAX DRAW-BAR PULL	6500 lb (2950 kg)	9000 lb (4080 kg)
OUTSIDE TURNING RADIUS	12 ft 6 in (3·81 m)	14 ft (4·27 m)
WHEELBASE	6 ft 3 in (1·91 m)	6 ft 10 in (2·08 m)
TRACK		
front	4 ft 4 in (1·32 m)	4 ft 6 in (1·37 m)
rear	5 ft 4 in (1·63 m)	5 ft 4¾ in (1·64 m)
UNIT DIMENSIONS		
length	12 ft 4 in (3·76 m)	12 ft 11⅛ in (3·94 m)
width	8 ft (2·44 m)	8 ft (2·44 m)
height	8 ft (2·44 m)	8 ft 4 in (2·54 m)
WEIGHT	7900 lb (3583 kg)	10 800 lb (4900 kg)

Status

Production. The MF50 was developed for the UK Ministry of Defence and is in service with the Royal Air Force

Manufacturer

Massey Ferguson (UK) Ltd, Coventry, England

Montgomerie Reid 3000/6000 Tow Tractors

Available in two models, the 30-ton capacity 3000 and the 60-ton capacity 6000, can be supplied with diesel, lpg or petrol-fuelled engines. All are available with an engine preheater for starting in adverse conditions and an exhaust gas purifier system for operating in environments where fumes could cause problems. Manual transmission is standard and the diesel-powered 3000 and all 6000 models can be fitted with an optional torque converter. Other standard equipment includes hydraulic braking on all four wheels and multi-position towing hitch. A sound-proofed steel cab, heater/demister unit and headlights, sidelights and indicators for night and public road use are optional features.

Characteristics

MODEL	3000	6000
MAXIMUM DRAW-BAR PULL	66 150 lb (30 000 kg)	132 250 lb (60 000 kg)
ENGINE	As required by operator	As required by operator
TURNING RADIUS	9 ft 9 in (2·97 m)	3·25 m (128 in)
TOWING POINT HEIGHT	1 ft (0·305 m)/1 ft 4 in (0·406 m)	14/18/22 in (0·356/0·457/0·559 m)
WHEELBASE	3 ft 7½ in (1·11 m)	4 ft 2 in (1·27 m)
TRACK		
front	3 ft 5 in (1·04 m)	4 ft 4 in (1·32 m)
rear	3 ft 5 in (1·04 m)/3 ft 9½ in (1·16 m)	3 ft 7 in (1·09 m)/3 ft 11 in (1·19 m)
UNIT DIMENSIONS		
length (with towing point)	7 ft 4 in (2·24 m)	9 ft (2·74 m)
width	4 ft 1 in (1·25 m)	5 ft 2 in (1·58 m)
height (with roof exhaust)	7 ft (2·13 m)	7 ft 1 in (2·15 m)
WEIGHT (WITH/WITHOUT CAB)	3700 lb (1680 kg) 3400 lb (1540 kg)	6770 lb (3070 kg) 6470 lb (2940 kg)

Status
The 3000 model is in service with British Airways

Manufacturer
Montgomerie Reid Ltd, Basingstoke, Hampshire, England

Montgomerie Reid 6000 DT tow tractor

Montgomerie Reid 3000 LT tow tractor

Power Lift Baggage/Freight Conveyor Cars

Power Lift conveyor cars for loading/unloading baggage and freight are available in two sizes; 7·5 and 9 metres long. The conveyors are based on a purpose-built car chassis powered by a Perkins or Ford diesel engine with Borg Warner automatic transmission and a hydraulic pump pto. Two independent single stage rams, one at the front and one at the rear, raise or lower the conveyor to the correct inclination and levels. Both are fitted with safety valves in the event of a break in a hydraulic line. The conveyor, equipped with side guards to prevent spillage, rubber buffer on the front end and amber side reflectors, is controlled from the driver's position. This features a Bostrom driver's seat, independently suspended with a hydraulic damper, deadman conveyor inclination control, hand-throttle to increase engine speed for the conveyor and a gear selector lever fitted with mechanical stop bar to make the change from forward to reverse a two-handed operation. A safety interlock prevents the vehicle from being driven with the conveyor in operation. Optional features available include; auxiliary hand-pump for emergency use, foot brake locking pedal, manual transmission and weatherproof cab.

Characteristics

TYPE	7·5 m unit	9 m unit
CONVEYOR HEIGHTS		
max: front	14 ft 8 in (4·47 m)	17 ft (5·2 m)
rear	5 ft 5 in (1·65 m)	4 ft 7 in (1·4 m)
minimum: front	3 ft 7¾ in (1·11 m)	3 ft 8 in (1·12 m)
rear	10 in (0·254 m)	10 in (0·254 m)
BELT		
width	2 ft (0·61 m)	2 ft (0·61 m)
speed	39–95 ft/minute (12–29 m/minute)	39–95 ft/minute (12–29 m/minute)
CAPACITY		
at 15°	2000 lb (907 kg)	2000 lb (907 kg)
at 30°	1000 lb (454 kg)	1000 lb (454 kg)
ELEVATING TIME		
front	15 s	26 s
rear	5 s	4 s
LOWERING TIME		
front	35 s	55 s
rear	12 s	10 s

ENGINE: Perkins 4/108 diesel / Ford York diesel
OUTPUT: 32·06 kW at 300 rpm / 456 kW at 3600 rpm
DRIVING SPEED: 35 mph (56·3 km/h)
OUTSIDE TURNING RADIUS: 47 ft (14·33 m)
UNIT DIMENSIONS
length (including boom): 7·5 m/9 m
width: 6 ft 8 in (2·03 m)

Power Lifts 9-metre model baggage conveyor car

Status
This equipment is in service with airlines and airport authorities in Europe, Middle East and Africa

Manufacturer
Power Lifts Ltd, Watford, Hertfordshire, England

Power Lifts Elevating Platforms

Two models of elevating platforms are available. The 26-ton capacity machine will accommodate one 20-foot (6·1 metre), or two 10-foot (3·05 metre) containers on the 22-foot (6·71-metre) by 8-foot 4-inch (2·54 metre) platform which is fitted with a conveyor in two sections, each of which is hydraulically powered by an individual drive unit. The machine comprises dual sets of heavy-duty scissors mounted one behind the other, with the cylinders fitted in the scissors providing a rigid unit capable of accepting offset loads. A side-shift mechanism allows the conveyor deck to move across the platform 9 inches (0·229 metres) either side of the centre line to provide alignment with transfer vehicles and dollies. The 6·7 ton capacity machine will accommodate one full size or two half size aircraft pallets on its 10 foot (3·05 metre) by 8 foot 4 inch (2·54 metre) platform which is fitted with a hydraulically powered conveyor. The design of this equipment is similar to the larger capacity platform with cylinders incorporated in the scissors mechanism and a side-shift mechanism for alignment with transfer vehicles.

Status
In service with the Royal Air Force, airlines, freight forwarders and airport authorities in Europe, Middle East and Africa

Manufacturer
Power Lifts Ltd, Watford, Hertfordshire, England

Power Lifts 26-ton capacity elevating platform

Reliance Mercury Tow Tractors

Three models of tractors suitable for baggage- and cargo-towing operations are produced by Reliance Mercury. These are the RM 301, RM 601 and the RM 801.

The RM 301 is available in left-hand drive with a choice of diesel or lpg engine equipped with manual or automatic transmission. The integral body and chassis is mounted on a front-steering axle and a rear-driving axle with foot-operated hydraulic service brakes on all four wheels. A four-position towing jaw is fitted at the rear of the vehicle. The driver is seated in an open-sided steel cab. Optional equipment available includes cab doors, passenger seat and front towing position.

The RM 601 and 801 models can be supplied with either diesel or lpg engines and automatic transmission. Mounted on a front steering axle and rear driving axle the vehicle is equipped with foot operated servo-assisted hydraulic service brakes on all four wheels. Other features include a three position towing joint fitted at the rear of the tractor, and a sound-proofed driver's cab. As in the RM 301, a passenger seat and a front towing position can be added as optional equipment.

Status

The RM 301 and the 601/801 tractors are in service with major airlines including Aer Lingus, British Airways and British Caledonian Airways

Manufacturer

Reliance Mercury Ltd, Halifax, West Yorkshire, England

Characteristics

MODEL	RM 301	RM 601/RM 801
MAX DRAW-BAR PULL	3000 lb (1360 kg)	6000 lb (2722 kg)/8000 lb (3630 kg)
ENGINE OUTPUT/MAX SPEED	Perkins 4108 diesel: 33 kW Ford 227E lpg: 33·5 kW	Perkins 4236 diesel: 81·2 kW S.I.4 lpg: 57·4 kW
OUTSIDE TURNING RADIUS	11 ft 8 in (3·56 m)	11 ft 4 in (3·45 m)
WHEELBASE	5 ft 3½ in (1·62 m)	5 ft 6 in (1·68 m)
TRACK		
front	3 ft 4¼ in (1·02 m)	4 ft 3 in (1·3 m)
rear	3 ft 1⅔ in (0·96 m)	3 ft 8½ in (1·13 m)
UNIT DIMENSIONS		
length	8 ft 6¾ in (2·61 m)	10 ft 2 in (3·12 m)
width	4 ft 1¼ in (1·25 m)	5 ft 1 in (1·57 m)
height	5 ft 9¼ in (1·76 m)	6 ft 6¾ in (2 m)
WEIGHT	4356 lb (1980 kg)	9020 lb (4093 kg)/9820 lb (4456 kg)

Reliance Mercury 301 tow tractor

Reliance Mercury 601/801 tow tractor

Weldwork Cargo Handling Systems

Weldwork designs and manufactures cargo handling equipment and installs handling systems to meet individual operator's requirements. In the warehouse powered or friction driven storage/retrieval bays, and manual or powered build-up/tear-down bays can be installed with weigh scales and indexing accumulator lines.

These can be supported by powered roller beds with powered roller bed rotation units, omni-directional transfer tables, rope or chain elevators, manual/powered ball mats, and scissor lifts. Slave pallets are available for cargo handling.

Also for freight transfer within the warehouse, Weldwork provides elevating/fixed travelling uld pallet vehicles, either mounted on rails, or solid-rubber tyres.

For cargo transfer between the aircraft and the freight terminal the company produces dollies for LD-1/LD-3 containers, and uld trailers.

A typical cargo handling system designed by Weldwork is installed at the Trans Mediterranean Airways facility at Frankfurt Airport, Federal Republic of Germany.

The system comprises a two-tier elevating travelling vehicle feeding spur-mounted, friction-driven storage and retrieval bays. Consolidated freight is transferred from 'landside' via a truck dock scissor lift to an accumulator line of powered rollers, which index to an omni-directional powered transfer mat and weigh scale. This mat will then transfer palletised loads/ulds to the elevating travelling vehicle, which is mounted on rails and runs parallel to the storage/retrieval bays. Each storage bay is friction driven from the elevating travelling vehicle, as are the build-up/tear-down bays, which are spur-mounted to the track. Loads are transferred at the 'airside' door via a powered roller transition mat to a powered buffer loading dock for onward transmission to the aircraft transporter vehicle. Additional dock levellers are mounted at the 'landside' doors to receive random freight prior to loading on the build-up bays.

Weldwork powered weigh scale, omni-directional transfer and split pallet/uld beds

Weldwork 7-ton trailers for multi-unit loading

Status

Weldwork has supplied equipment to many freight forwarders, Liverpool and Manchester Airport authorities and to the following airlines: British Airways, British Caledonian Airways, Air Canada, Aer Lingus, Air India, Cargolux Airlines, Japan Air Lines, KLM, Kuwait Airways, Lufthansa, Malev, Middle East Airlines, Nigeria Airways, Pakistan International Airlines, Seaboard World Airlines, Trans Mediterranean Airways and Trans World Airlines

Manufacturer

Weldwork Cargo Systems, Hounslow, Middlesex, England

UNITED STATES OF AMERICA

Arvico Loadfast Mechanical Cargo Handling System

In the Loadfast system pre-packed pallets are received by a 20 000 lb (9090 kg) capacity dock lift which is adjustable in height. The roller deck is mechanised to assist movement in and out of the warehouse. The lift is hydraulically-powered and all movements are controlled from a push-button pendant control.

Weighing of the pallets is achieved by four 'weigh bars' under each corner of a gravity roller deck. The weight can be read in pounds or kilograms and a tare/gross weight selector switch is provided. A printer records the weight time and date on the airway bill. A right-angled transfer assembly deck placed directly after the weigh scales enables the pallets to be either discharged sideways onto a pallet dolly, or moved forward into the warehouse. This assembly can be manual or powered in either direction. The build-up/break-down stations are roller decks mounted on lifts positioned in pits in the warehouse floor. The pallet height can be adjusted to different heights above or below the warehouse floor to ease the build-up or break-down. Movement in the system is performed by standard gravity roller sections to the shuttle car interface. At this point, the pallet can be discharged across the shuttle car deck, or received by the shuttle car for transfer to the other storage positions. The shuttle car can receive pallets in one direction, rotate them through 90 degrees and move them into storage positions. These other storage positions are used to transfer pallets to and from dollies or receive pallets that have been built up in the warehouse.

Arvico Loadfast cargo handling system

Manufacturer

Arvico International Inc, Miami, Florida, USA

Blue Giant Cargo Handling Equipment

The equipment produced by Blue Giant is for freight handling operations within the warehouse or terminal area. The range comprises pallet transporters, tow-tractors, fork-lift trucks, hydraulic-lift tables, dock levellers and elevating docks.

The pallet transporters can be either manual or powered. The manual version consists of eight models. Hydraulically-operated, control is through a foot pedal. The unit is equipped with a spring-loaded handle connected to twin rear wheels capable of swivelling through 260 degrees. Guidance rollers are fitted under the forks at the front to assist pallet entry and exit. Optional features available include an adaptor to enable the transporter to lift both pallets and skids, hand-operated brake, and a back stop to prevent loose packages from falling over the rear of the unit. The powered units are pedestrian-controlled but can be supplied as an optional feature with a rider platform. The units are equipped with a spring loaded handle on the drive unit which provides a 186 degree steering arc, automatic deadman braking and power cut-off on releasing or lowering the handle. A two speed forward/reverse control and lift and lower push-buttons are fitted on the handle.

The tow tractors can be either pedestrian controlled or fitted with a rider platform. The unit's features are similar to the powered pallet transporter with the addition of a pin-type coupling. Automatic couplings are available as an optional feature.

The fork lift trucks are pedestrian controlled. Fourteen standard models are produced with single-, double- or triple-stage masts. The basic features are similar to those fitted in the powered pallet transporter described above.

Hydraulic lift tables are incorporated in freight handling systems. Blue Giant produces three types, the Series FS, Series TS and Series DS. The Series FS consists of 48 models which are open on all four sides and usually stationary or installed in a pit, though some models can be made portable with a dolly. The Series TS is a portable type, open on three sides, with a 12 volts dc power pack and a floor lock. Nine standard models are produced. The Series DS comprises 14 models equipped with

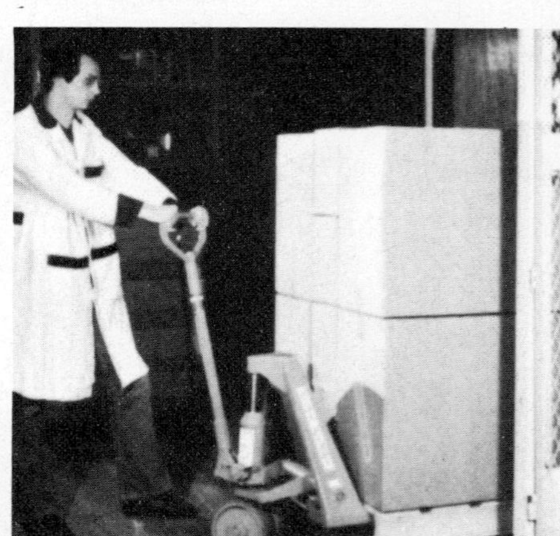

Blue Giant cargo handling equipment

double scissors. The equipment is similar to the Series FS but capable of higher lifts. This series is designed for situations where space limitations or other factors limit the deck size, but where vertical travel is an important consideration.

Dock levellers provide a ramp between loading/unloading vehicles and the deck of the terminal or warehouse. Ten mechanical and eight hydraulic models are produced for installation in a pit with a further model which can be either mechanical or hydraulic for adding to an existing freight handling installation. A more recent model has a 90 000 lb (40 824 kg) handling capacity with widths up to 9 feet 8 inches (2·95 metres). Basic features include a positive hold-down, a swing-away lip and safety stops.

Elevating docks can be used when the height differential is too large for a dock leveller. Stationary style elevators of which 112 models are available, feature a flush floor pit installation. They are suitable for cargo handling with all types of equipment from manual pallet transporters to heavy duty lift trucks. The semi-portable style is offered in 28 models. They can be moved using a dolly, or fitted with an optional power-pack mounted on the deck.

Manufacturer

Blue Giant Corporation, Buffalo, New York, USA

Characteristics

EQUIPMENT	Pallet transporter		Tow tractor	Fork lift truck
TYPE	Manual	Powered		
CAPACITY	5000 lb (2268 kg)	4000 lb (1814 kg) 6000 lb (2722 kg) 8000 lb (3629 kg)		2000 lb (907 kg) 3000 lb (1361 kg) 4000 lb (1814 kg)
FORK OR PLATFORM length	32, 36, 42, 48 in (0·81, 0·91, 1·07, 1·22 m)	36, 42, 48 in (0·91, 1·07, 1·22 m)		42 in (1·07 m)
width	20·5, 27 in (0·52, 0·69 m)	20·5, 27 in (0·52, 0·69 m)		14–30 in (0·36–0·76 m)
TILT FORWARD/BACK				2°/10°
LIFT HEIGHT				Up to 12 ft 6 in (3·81 m)
DRAW-BAR PULL continuous			225 lb (102 kg)	
max			500 lb (227 kg)	
TURNING RADIUS			2 ft 10 in (0·86 m) pedestrian	4 ft 5 in (1·35 m)
			3 ft 5 in (1·04 m) rider	4 ft 10 in (1·47 m)

EQUIPMENT	Hydraulic	Lift table		Elevating dock	
TYPE	Series FS	Series TS	Series DS	Stationary	Semi-portable
CAPACITY	1000 lb (454 kg) 20 000 lb (9072 kg)	2000 lb (907 kg) 4000 lb (1814 kg)	1000 lb (454 kg) 4000 lb (1814 kg)	3000 lb (1316 kg) 20 000 lb (9072 kg)	3000 lb (1361 kg) 6000 lb (2722 kg)
DECK SIZE	18 × 42–84 × 144 in (0·45 × 1·07–2·13 × 3·66 m)	24 × 48–36 × 72 in (0·61 × 1·22–0·91 × 1·83 m)	24 × 48–48 × 72 in (0·6 × 1·22–1·22 × 1·83 m)	48 × 84–84 × 144 in (1·22 × 2·13–2·13 × 3·66 m)	48 × 96–84 × 120 in (1·22 × 2·44–2·13 × 3·05 m)
LIFT vertical travel	2 ft (0·61 m)–4 ft 11 in (1·5 m)	2 ft 7 in (0·79 m)	As required	3 ft 10 in (1·17 m)–4 ft 11 in (1·5 m)	4 ft 5½ in (1·36 m)–4 ft 7½ in (1·41 m)

EQUIPMENT: dock leveller
ROLL-OVER/CROSS-OVER CAPACITY: 12 000/20 000 lb
(option up to 40 000 lb and 90 000 lb)
((5443/9072 kg (option up to 18 144 kg and 40 824 kg))
DECK DIMENSIONS
width: 5 ft (1·52 m)–9 ft 8 in (2·95 m)
length (including lip): 5 ft 2 in (1·57 m)– 10 ft 3 in (3·12 m)
TRAVEL
up: 2 ft 4 in (0·71 m)/2 ft 6 in (0·76 m)
down: 8 in (0·2 m)/10 in (0·25 m)

CAS Series 300 Main Deck Loaders

The Cochran Airport Systems (CAS) Series 300 loaders are four post, cable lift units suitable for loading the main-deck cargo compartments of both wide-body and conventional freighter aircraft. Models 314A, 315A and 316 are designed for transporting by air, after partial disassembly.

All units will lift one 20-foot (6·1-metre) container, or two 10-foot (3·05-metre) containers or two pallets measuring 96 by 125 inches (2·44 by 3·18 metres) between 20 inches (0·508 metres) and 216 inches (5·49 metres). Load transfer is achieved by hydraulically-powered Bearcat tyres and galvanised rollers at a speed of 60 foot per minute (18·3 metres per minute). The platform can be adjusted in both pitch and roll by ± 2 inches (50·8 mm) at the forward end of the bed. All units are self-propelled by a petrol engine with hydrostatic drive and tow-bar steering. A diesel engine is available as an optional feature.

The Model 320 was developed for regular aircraft operations and can load containers up to 20 feet (6·1 metres) long. Like the other models the loader has a height range between 20 (0·508 metres) and 216 inches (5·49 metres), and can lift container or pallet loads measuring up to 8 by 8 by 20 feet (2·44 by 2·44 by 6 metres). Total capacity is 60 000 lbs (27 216 kg). The bed of the loader can be adjusted in pitch and roll in a similar manner to the other models. It is equipped with inverted castors and retractable drive wheels and can be loaded from the end or right side using standard 20-inch (0·508 metres) high or truck height transporter units. Power is provided by a petrol or optional diesel engine and the unit is fitted with hydrostatic drive, power steering and air brakes.

Characteristics

MODEL	314A	315A	316	320
CAPACITY	30 000 lb (13 608 kg)	30 000 lb (13 608 kg)	40 000 lb (18 181 kg)	60 000 lb (27 216 kg)
ELEVATING HEIGHT RANGE	20–216 in (0·508–5·49 m)	20–216 in (0·508–5·49 m)	20–216 in (0·508–5·49 m)	20–216 in (0·508–5·49 m)
PLATFORM dimensions	256 × 128 in (6·5 × 3·25 m)	256 × 128 in (6·5 × 3·25 m)	256 × 128 in (6·5 × 3·25 m)	276 × 131 in (7·01 × 3·33 m)
PITCH/ROLL ADJUSTMENT	±2 in (±50·8 mm)	±2 in (±50·8 mm)	±2 in (±50·8 mm)	±2 in (±50·8 mm)
ENGINE	25 hp petrol/diesel 2-cylinder	62·5 hp petrol/diesel 4-cylinder	62·5 hp petrol/diesel 4-cylinder	Ford 6-cylinder petrol/GM 6-53 diesel
ENGINE SPEEDS lift: fast	17 ft/minute (5·1 m/minute)	32 ft/minute (9·7 m/minute)	30 ft/minute (9·1 m/minute)	30 ft/minute (9·1 m/minute)
slow	8 ft/minute (2·4 m/minute)	18 ft/minute (5·4 m/minute)	15 ft/minute (4·57 m/minute)	15 ft/minute (4·57 m/minute)
transfer	60 ft/minute (18·3 m/minute)	60 ft/minute (18·3 m/minute)	60 ft/minute (18·3 m/minute)	60 ft/minute (18·3 m/minute)
driving	2 mph (3·4 km/h)	2 mph (3·4 km/h)	2 mph (3·4 km/h)	7 mph (12 km/h)
UNIT DIMENSIONS length	23 ft 3 in (7·09 m)	23 ft 3 in (7·09 m)	23 ft 3 in (7·09 m)	30 ft (9·14 m)
width	16 ft 5 in (5 m)	16 ft 5 in (5 m)	16 ft 10 in (5·13 m)	17 ft 8 in (5·38 m)
height	20 ft 8 in (6·3 m)	20 ft 8 in (6·3 m)	20 ft 8 in (6·3 m)	23 ft (7·01 m)
WEIGHT	16 000 lb (7260 kg)	17 500 lb (7940 kg)	17 500 lb (7940 kg)	38 000 lb (17 240 kg)

Manufacturer

Cochran Airport Systems, Castroville, California, USA

Status

CAS Series 300 main loaders are in service with the following operators:

314A Air France (17 units)
315A UTA (3), Air Afrique (6), Avianca (1) and the French Air Force (1)
316 Pan Am (2), Flying Tigers (1), Seaboard (2), Cargolux (1), PIA (1), Lufthansa (2), JAL (1), Garuda (1), Sabena (1) and Genoa Airport Authorities (1)
316 (A) capacity 25 000 lb (11 340 kg) US Air Force (13)
320 Pan American, Lufthansa

CAS Model 320 main-deck loader

CAS Series 300 air transportable main deck loader

CAS Model 360 Container/Pallet Lower Deck Loader

CAS Model 360 pallet/container loader

The Model 360 is a container/pallet loader with a 15 000 lb (6800 kg) capacity designed to service all narrow body aircraft and the lower decks of wide-body aircraft. An adaptor is available to enable the unit to reach the Boeing 747, DC-10 and L-1011 main decks. The standard model features a rear platform on a scissors mechanism with the capability of fore/aft and angular adjustment for aircraft interface. Transfer is by a flat top chain system on the elevator and the bridge. Powered rollers on either side of the rear platform permit full length side loading and containers can be rotated on the elevator.

The loader is powered by a Ford six-cylinder petrol engine with hydrostatic drive to the front wheels and Charlyn power steering providing ±90 degree steering. The hydraulic system is supplied by a fixed displacement pto pump operating at 2000 psi (140 bar). Optional equipment available includes, a hook-on bridge with hydraulic adjustment for coupling to aircraft door spools, diesel engine, powered bridge guides, rear control station, powered elevator stops and personnel hand-rails.

Characteristics

PLATFORMS	Rear	Forward
CAPACITY	15 000 lb (6800 kg)	15 000 lb (6800 kg)
ELEVATING RANGE	1 ft 6 in (0·457 m)–11 ft 4 in (3·45 m)	4 ft 4 in (1·32 m)–11 ft 8 in (3·56 m)
speed	50 ft/minute (15·2 m/minute)	15 ft/minute (4·6 m/minute)
DISTANCE BETWEEN STOPS	13 ft 4 in (4·06 m)	10 ft 8 in (3·25 m)

ENGINE: Ford 300 CID petrol / GM3-53 diesel
DRIVING SPEED: 7 mph (12 km/h)
WHEELBASE: 20 ft 8 in (6·3 m)
UNIT DIMENSIONS
length: 29 ft 2 in (8·89 m)
width: 11 ft 4 in (3·45 m)
height: 8 ft 2 in (2·49 m)
WEIGHT: 26 000 lb (11 794 kg)

Status

Introduced in 1979, in service with major airlines

Manufacturer

Cochran Airport Systems, Castroville, California, USA

CAS Model 370 Lower Deck Container Loader

The Model 370 is designed for container only loading and unloading of current wide-body aircraft and the new generation Boeing 757 and 767 aircraft. When pallet handling is not required, Model 370 offers dimensional advantages while retaining all the technical and operational characteristics of the pallet container loader Model 360. Featuring ±90 degree steering, the basic design consists of two elevating platforms with a dual track flat top chain container transfer system. Rated at 10 000 lb (4536 kg) capacity (15 000 lb/6800 kg optional) the Model 370 can be supplied with petrol or diesel engine and hook-on or non-hook-on bridge/aircraft interface. The loader is equipped with powered side loading along the length of both right and left sides of the elevator platform. A fixed operator cab and platform are mounted on the right side of the bridge. This area contains the operators' station and pendant control box that controls a hydraulically-actuated system for the aircraft interface.

CAS Model 370 container loader

Characteristics

PLATFORM	Rear	Forward
CAPACITY	10 000 lb (4536 kg)	10 000 lb (4536 kg)
ELEVATING RANGE	1 ft 6 in (0·457 m)–11 ft (3·35 m)	5 ft 7½ in (1·71 m)–11 ft (3·35 m)
lift speed (adjustable)	50 ft/minute (15·2 m/minute)	15 ft/minute (4·6 m/minute)
transfer speed (adjustable)	60 ft/minute (18·3 m/minute)	60 ft/minute (18·3 m/minute)
DISTANCE BETWEEN STOPS		
length	13 ft 4 in (4·06 m)	8 ft 4 in (2·54 m)
width	5 ft 5 in (1·65 m)	5 ft 5 in (1·65 m)

ENGINE: Detroit Diesel Series 53
DRIVING SPEED: 7 mph (11·3 km/h)
OUTSIDE TURNING RADIUS: 25 ft (7·6 m)
WHEELBASE: 17 ft 11 in (5·46 m)
UNIT DIMENSIONS
length: 26 ft 7 in (8·1 m)
width: 8 ft 7 in (2·62 m)
height: 9 ft 2 in (2·79 m)
WEIGHT: 22 000 lb (9980 kg)

Manufacturer
Cochran Airport Systems, Castroville, California, USA

CAS Mobile Belt Conveyors

Cochran Airport Systems produces a range of three self-propelled belt conveyors, the 660, 661 and 666 and one towable unit, Model 670.

The Model 660 features a central conveyor mounted on a chassis based on the Ford E-350 truck. The conveyor is elevated at both ends by lift arms actuated by single stage, double-acting cylinders. The position of the driver and driving controls are similar to other mobile belt conveyors, with a two lever Morse gear shift and a throttle interlock being fitted as standard equipment. The area directly behind the driver is open for stowing last minute baggage. Alternatively, a seat can be provided for carrying additional personnel. The unit is equipped with belt controls operable from either end of the conveyor and an interlock prevents the engine throttle from being advanced for adjustment of belt speed unless the automatic transmission is in neutral. A retractable guide is provided on each side of the conveyor assembly which may be retracted manually if a clear conveyor surface is required. As an option, a collapsible 2 foot 8 inch (0·8 metre) high safety rail is available for either side of the conveyor. In an intermediate position these safety rails function as guides, and when fully collapsed a clear conveyor is provided. The design of the guides enables safety rails to be retrofitted at a later stage. Other optional features available include: electrical belt controls, emergency hand-pump, Perkins diesel engine, spotlights and a parking brake.

Both the 661 and the 666 are of similar design to the 660, the 661 being longer and able to elevate to 15 feet (4·59 m) and the 666 is battery powered.

The Model 670 is a towable vehicle designed to load and unload baggage, light freight and mail to/from the lower holds and bulk compartments of conventional and wide-body aircraft. Rated at 2000 lb (907 kg) capacity at maximum angle, the vehicle is built on a heavy duty, structural channel type chassis. The conveyor assembly, identical to the Model 660, is attached to the chassis through a front lifting frame and rear pivot. Conveyor belt controls are the manual type. Ackerman type steering is provided. The power pack mounted on the chassis comprises on Onan two-cylinder petrol engine with controls and instrumentation, 12-volt alternator, hydraulic pump and reservoir.

A rear operator station contains the controls required for engine and conveyor operation. The front station has controls for conveyor operation only.

Characteristics

MODEL	660/666	661	670
LIFT RANGE			
front	2 ft 8 in (0·81 m)–14 ft 2 in (4·32 m)	2 ft 4 in (0·71 m)–15 ft 1 in (4·6 m)	2 ft 8 in (0·81 m)–14 ft 2 in (4·32 m)
rear	5 ft 1 in (1·55 m)–1 ft 1 in (0·33 m)	5 ft 10 in (1·78 m)–1 ft (0·3 m)	5 ft 1 in (1·55 m)–2 ft 6 in (0·79 m)
CONVEYOR			
capacity	2000 lb (907 kg)	2000 lb (907 kg)	2000 lb (907 kg)
speed	40–120 ft/minute (12·2–36·6 m/minute)	40–120 ft/minute (12·2–36·6 m/minute)	
length	24 ft 6 in (7·47 m)	28 ft 6 in (8·69 m)	24 ft 6 in (7·47 m)
BELT WIDTH	2 ft (0·61 m)	2 ft (0·61 m)	2 ft (0·61 m)
CHASSIS	Ford E-350	Ford E-350	Towable
ENGINE	660: Ford 300 petrol, 6-cylinder; 666: 8 hp dc electric motor 72V	Ford 300 petrol, 6-cylinder	Onan 2-cylinder petrol
BATTERIES	666: 2 × 36 V, 340 Ah		
DRIVING SPEED	666: 10 mph		
WHEELBASE	11 ft 6 in (3·51 m)	11 ft 6 in (3·51 m)	
UNIT DIMENSIONS			
length	25 ft (7·62 m)	29 ft (8·84 m)	
width	6 ft 8 in (2·03 m)	6 ft 8 in (2·03 m)	
height	4 ft 10 in (1·47 m)	4 ft 10 in (1·47 m)	
WEIGHT	660: 6480 lb (2940 kg); 666: 8800 lb (3992 kg)	6480 lb (2940 kg)	

CAS Model 660 mobile belt conveyor

Status
CAS mobile belt loaders are in service with major airlines

Manufacturer
Cochran Airport Systems, Castroville, California, USA

Clyde Machines MFB Trailers

The MFB series designed for mail, freight and baggage, are four-wheeled towable trailers with capacities from 2500 to 12 000 lb (1134 to 5443 kg). All are equipped with either fold-down sides or chain gates with curtains. The baggage trailers are fitted with a shelf.

Characteristics

TYPE	120	117B	144	151	535	580
USE	Baggage	Cargo	Cargo	Cargo	Interline freight	Cargo
DIMENSIONS						
length	10 ft (3·09 m)	12 ft (3·66 m)	12 ft (3·66 m)	10 ft (3·09 m)	12 ft (3·66 m)	6 ft 6 in (1·98 m)
width	5 ft (1·52 m)	5 ft (1·52 m)	5 ft (1·52 m)	4 ft 8 in (1·42 m)	5 ft (1·52 m)	4 ft (1·33 m)
CAPACITY	4000 lb (1814 kg)	8000 lb (3629 kg)	4000 lb (1314 kg)	8000 lb (3629 kg)	6000 lb (2722 kg)	2500 lb (1134 kg)

TYPE	711	1189	1402
USE	Cargo	Cargo	Cargo
DIMENSIONS			
length	12 ft (3·66 m)	10 ft (3·09 m)	10 ft (3·09 m)
width	6 ft (1·83 m)	5 ft (1·52 m)	5 ft (1·52 m)
CAPACITY	12 000 lb (5443 kg)	4000 lb (1814 kg)	4000 lb (1814 kg)

Status
In service with airlines in the USA

Manufacturer
Clyde Machines, Glenwood, Minnesota, USA

Clyde MFB-151 trailer

Clyde Machines Uld Dollies and Trailers

Clyde Machines produce a wide range of dollies and trailers for transporting ulds, incorporating dual-wheel swivel castors, double ackerman steering and tow-bar actuated brakes. Latest in the range is the CT 1800, 5400 lb (2450 kg) capacity side-loading trailer designed to meet the requirements of the Boeing 767 aircraft. The unit which will handle LD-1, LD-2, LD-3, LD-4 containers is equipped with four-wheel ackerman steering, automatic stops with integrated vertical restraints, and an optional turntable feature providing 300 degree rotation for the desired orientation of LD-1, LD-2 and LD-3 containers to either an aircraft loader or storage rack.

Characteristics

EQUIPMENT	Castor bed dolly	
MODELS	CBD 1457	CBD 1468
DIMENSIONS		
length	12 ft 2 in (3·71 m)	10 ft 6 in (3·2 m)
width	8 ft (2·44 m)	7 ft 5 in (2·26 m)
CAPACITY	15 000 lb (6804 kg)	15 000 lb (6804 kg)
EQUIPMENT	Cargo/Pallet trailer	
MODELS	CPT-396	CPT-E-15000
DIMENSIONS		
length	11 ft 1 in (3·38 m)	12 ft 2 in (3·71 m)
width	7 ft 8 in (2·34 m)	8 ft (2·44 m)
CAPACITY	15 000 lb (6804 kg)	15 000 lb (6804 kg)

EQUIPMENT	Container trailer		
MODELS	CT 1245	CT 1571	CTT 1563
TYPE	End/side loading	Side loading	With turntable
DIMENSIONS			
length	13 ft 8 in (4·17 m)	9 ft 9 in (2·97 m)	9 ft 9 in (2·97 m)
width	5 ft 6 in (1·68 m)	5 ft 7 in (1·7 m)	5 ft 7 in (1·7 m)
CAPACITY	10 000 lb (4536 kg)	3500 lb (1588 kg)	3500 lb (1588 kg)

Status
In service with airlines in the USA

Manufacturer
Clyde Machines, Glenwood, Minnesota, USA

Clyde CT-1245 double container trailer showing side to end loading

Cochran Western Model 600 U Mobile Conveyor

The Model 600 U mobile conveyor is specifically designed to meet the minimum and maximum lower-lobe cargo bin door-sill requirements of all modern aircraft from the Boeing 707 to the Boeing 767, DC-8, DC-9 and DC-10; the L-1011; A 300 and A 310. Load rated at 2000 lb (907 kg), the conveyor has a belt width of 2 feet (0·61 metres) and belt speeds from 40 to 120 feet per minute (12·2 to 36·6 metres per minute). Adjustable in height at both ends, the conveyor features belt controls front and rear, automatic belt alignment, personnel walkways on both sides and folding conveyor side rails. The standard model is powered by a Ford six-cylinder petrol engine and automatic transmission with a hand throttle interlock to restrict conveyor operation unless the transmission is in the neutral position. Optional power plants available are the Perkins 4236 diesel engine and the GE Electric 8·2 hp dc/evi motor.

Characteristics
LIFT RANGE
- front: 3 ft 3 in (0·99 m)–14 ft (4·27 m)
- rear: 5 ft 2 in (1·57 m)–10 in (0·25 m)

CONVEYOR
- capacity: 2000 lb (907 kg)
- speed: 40–120 ft/minute (12·2–36·6 m/minute)
- length/width: 24 ft (7·32 m)/2 ft 10 in (0·86 m)

BELT WIDTH: 2 ft (0·61 m)
ENGINE: Ford 6-cylinder 3276 cc
TURNING RADIUS: 23 ft (7·01 m)
WHEELBASE: 9 ft 2 in (2·79 m)
UNIT DIMENSIONS
- length: 24 ft 9 in (7·54 m)
- width: 6 ft $8^{1}/_{4}$ in (2·04 m)
- height: 4 ft $3^{1}/_{2}$ in (1·31 m)

WEIGHT: 5200 lb (2360 kg)

Status
In service with airlines in the USA

Manufacturer
Cochran Western Corporation, Salinas, California, USA

Cochran Western Model 600 U mobile conveyor

Cochran Western Model 818 Lower Deck Container/Pallet Loader

Capable of lifting 15 000 lb (6804 kg) the Model 818 lower deck loader can handle LD-3 containers and pallets up to 125 inches (3·18 metres) wide. The unit features powered side and end loading, powered retractable guides and a hydraulically-actuated rear pallet stop on the elevator, and manual container rotation. The elevator height for load transfer to/from the transporter or dolly can be adjusted hydraulically between 1 foot 6 inches (0·46 metres) and 1 foot 10 inches (0·56 metres). Transfer is by dual flat top chains on both the elevator and bridge platforms. Height adjustment is by scissors mechanisms fitted to both platforms which are operated by dual lift cylinders. Power is provided by a Ford petrol engine with a hydrostatic drive system and a dual section hydraulic pump supplying the lift cylinders. The loader is designed to service the lower lobes of the Boeing 747, 767, L-1011, DC-10, A-300 and A-310 wide-body aircraft, and the main decks of Boeing 737 and DC-8 aircraft.

Optional features available include GM 3-53 Detroit or Perkins 4238 diesel engine, 125 by 125 inch (3·18 by 3·18 metres) pallet capability, universal aircraft attaching bridge and powered LD-8 container rotation.

Cochran Western Model 818 container/pallet loader

Characteristics

PLATFORMS	Elevator (rear)	Bridge (front)
CAPACITY	15 000 lb (6804 kg)	15 000 lb (6804 kg)
ELEVATING RANGE	1 ft 8 in (0·508 m)–12 ft (3·66 m)	5 ft 10 in (1·78 m)–12 ft (3·66 m)
DIMENSIONS	17 ft 6 in (5·33 m) × 8 ft 6 in (2·59 m)	12 ft ½ in (3·67 m) × 8 ft 6 in (2·59 m)
WIDTH BETWEEN GUIDES	5 ft 2 in (1·57 m)–8 ft 2 in (2·49 m)	5 ft 2 in (1·57 m)–8 ft 2 in (2·49 m)
LIFT SPEED	40 ft/minute (12·2 m/minute)	24 ft/minute (7·3 m/minute)
TRANSFER SPEED	60 ft/minute (18·3 m/minute)	40 ft/minute (12·2 m/minute)
ROLL ADJUSTMENT		±0·5°

ENGINE: Ford 300 CID petrol
DRIVING SPEED: 0–7 mph (11·3 km/h)
TURNING RADIUS: 25 ft (7·6 m)
UNIT DIMENSIONS
length: 29 ft 10 in (9·09 m)
width: 11 ft 8 in (3·56 m)
height: 7 ft 10 in (2·39 m)

Status
In service with airlines in the USA

Manufacturer
Cochran Western Corporation, Salinas, California, USA

Cochran Western Model CT 5501 Container/Pallet Transporter

The model CT 5501 is designed for transporting two LD-1/LD-3 containers or one LD-1/LD-11 container or pallet between the aircraft loader and the dolly, or freight terminal. Powered by a six-cylinder petrol engine with hydrostatic drive and power steering, the deck incorporates end rollers and pallet drive wheels, the latter being pneumatic tyres which are driven through a vane-type hydraulic pump. To facilitate loading, the front end of the roller deck can be adjusted in height between 1 foot 6 inches (0·46 metres) and 1 foot 10 inches (0·56 metres). Optional equipment available comprises Perkins 4236 diesel engine, 96 inch (2·44 metre) pallet capability, and three additional conveyor drive wheels across the aft end of the transporter for loading from the rear end.

Characteristics

CAPACITY: 15 000 lb (6804 kg)
ROLLER DECK DIMENSIONS
width between guides: 7 ft 6 in (2·29 m)
length between guides: 10 ft 5 in (3·28 m)
ROLLER HEIGHT: 1 ft 8 in (0·508 m)
HEIGHT ADJUSTMENT: 1 ft 6 in (0·46 m)–1 ft 10 in (0·56 m)
ENGINE: 6-cylinder petrol
TURNING RADIUS: 21 ft (6·4 m)
WHEELBASE: 8 ft 1 in (2·46 m)
UNIT DIMENSIONS
length: 13 ft 5 in (4·09 m)
width: 10 ft (3·05 m)
height: 5 ft 1 in (1·55 m)
WEIGHT: 7250 lb (3290 kg)

Status

In service with airlines in the USA

Manufacturer

Cochran Western Corporation, Salinas, California, USA

Crest ECT-2500 Electric Tractor

This is a new development by Crest Manufacturing. Rated at 3000 lb (1363 kg) draw-bar pull, the ECT-2500 is mounted on an all-welded chassis of heavy plate and powered by a 36-cell acid industrial battery. The tractor is of low profile design with an optional all-weather insulated cab equipped with heater-defroster, wiper, safety glass windscreen, Lexan windows, roof window and removable doors.

Manufacturer

Crest Manufacturing, Overland Park, Kansas, USA

Crest ECT-2500 electric tractor

Cushman Pick-up Trucks and Tow Tractors

The Cushman range of small petrol engined and electric-powered vehicles suitable for cargo handling include the Haulster, the Full-Ton Flatbed and the Titan class. All vehicles can be fitted with towing points.

The Haulster is a four-wheeled vehicle which carries a driver and a passenger and 1000 lb (454 kg) of cargo. It is powered by an 18 hp petrol engine and has a flat platform behind the crew area which can be equipped as an optional feature with sides and a tailgate. An enclosed cab is available also as an optional item. Two differentials can be supplied, one with 10·25 to 1 axle ratio, the other with 6·5 to 1 ratio.

The Full-Ton Flatbed, a three wheeled vehicle, with 2000 lb (907 kg) capacity is powered by an 18 hp petrol engine, which provides speeds up to 18 mph (29 km/h). Similar optional features are available as in the Haulster model.

The Titan 4000 and 6000 are battery powered electric vehicles which can be adapted for a variety of tasks, with capacities of 4000 and 6000 lb (1814 and 2722 kg) respectively. The larger model is powered by two 48 volt dc series motors each rated at 8 hp, the smaller version by two 36 volt dc series motors producing 5·4 hp each. Both models are based on a heavy-duty steel frame, mounted on multi leaf spring suspension at the rear and coil spring/shock absorber suspension in the front. The power train features a gear drive with a worm gear differential and the vehicle is equipped with reciprocating ball steering. Eight batteries are fitted, and a battery balance system switches the power demand created by each motor from battery to battery in a pre-established sequence enabling all batteries to remain in equal balance during a discharge cycle. An automatic battery charger is incorporated in the system.

Status

The Haulster vehicle is in service with Federal Express, the freight airline, at Memphis, Tennessee, where 52 units are in operation, each capable of towing seven loaded baggage trailers. Full-Ton Flatbed units are operated by Piedmont at Winston-Salem for towing baggage and cargo trailers and Titan vehicles are used by TWA at Phoenix airport for carrying maintenance crew

Manufacturer

OMC Lincoln, Lincoln, Nebraska, USA

OMC Lincoln Titan unit operated by TWA at Phoenix Airport

Characteristics

MODEL	Haulster	Full-Ton Flatbed	Titan 6000	Titan 4000
CAPACITY	1000 lb (454 kg)	2000 lb (907 kg)	6000 lb (2722 kg)	4000 lb (1814 kg)
ENGINE/MOTOR	Petrol	Petrol	2 × 48 V dc	2 × 36 V dc
OUTPUT	18 hp (13·4 kW)	18 hp (13·4 kW)	16 hp	10·8 hp
BATTERIES			8 × 6 V	8 × 6 V
SPEED				
10·5:1 axle	18 mph (29 km/h)	18 mph (29 km/h)	9 mph (14 km/h)	9 mph (14 km/h)
6·5:1 axle	30 mph (48 km/h)			
RANGE			23–40 miles (37–64 km)	23–40 miles (37–64 km)
max (10·5:1 axle)	615 lb (279 kg)	800 lb (363 kg)	2028 lb (920 kg)	1188 lb (599 kg)
MAX DRAWBAR PULL				
10·5:1 axle	615 lb (279 kg)	800 lb (363 kg)	2028 lb (920 kg)	1188 lb (599 kg)
6·5:1 axle	800 lb (363 kg)			
LOAD PLATFORM SIZE	4 ft 4 in (1·32 m) × 3 ft 10½ in (1·18 m)	6 ft (1·83 m) × 4 ft 3 in (1·3 m)	6 ft 4 in (1·93 m) × 3 ft 9 in (1·14 m)	6 ft 4 in (1·93 m) × 3 ft 9 in (1·14 m)
WHEELBASE	7 ft (2·13 m)	7 ft (2·13 m)	5 ft 10 in (1·78 m)	5 ft 10 in (1·78 m)
TRACK				
front	3 ft ½ in (0·93 m)		3 ft 2 in (0·97 m)	3 ft 2 in (0·97 m)
rear	3 ft 11½ in (1·21 m)	3 ft 6 in (1·07 m)	3 ft 2½ in (0·98 m)	3 ft 2½ in (0·98 m)
TURNING RADIUS	8 ft (2·44 m)	8 ft 6 in (2·59 m)	11 ft 6 in (3·51 m)	11 ft 6 in (3·51 m)
UNIT DIMENSIONS				
length	9 ft 7½ in (2·93 m)	10 ft 11 in (3·33 m)	9 ft 10 in (3 m)	9 ft 10 in (3 m)
width	4 ft (1·22 m)	4 ft 3 in (1·3 m)	3 ft 10 in (1·17 m)	3 ft 10 in (1·17 m)
height	3 ft 11½ in (1·21 m)	3 ft 8¾ in (1·14 m)	6 ft 7½ in (2·02 m)	6 ft 7½ in (2·02 m)
WEIGHT	1025 lb (465 kg)	1080 lb (490 kg)	2850 lb (1293 kg)	2850 lb (1293 kg)

Drexel SwingMast Lift Trucks

The SwingMast is an electrically-powered lift truck which rotates an integral mast and fork assembly to pick up loads from the front, side or any intermediate position. As a side-loader, the SwingMast lifts the load from the side and carries it in either a side or front position. With long loads, it can lift the load, retract it in line with the truck and rest it on a load-bearing surface to the driver's right. The unit incorporates a side-shift feature which enables it to perform reach truck operations. In very narrow aisle operations, the lift truck can side-load a pallet; rotate the pallet in the aisle, and transport the load within the warehouse or transit shed.
Models are available with capacities up to 8000 lb (3630 kg) and elevating heights up to 30 feet (9·1 metres). Standard features include electronic (SCR) control, full mast tilt and rear axle articulation. Minimum aisle measurements range from 4 foot 8 inches (1·4 metres) to 5 foot 6 inches (1·68 metres) with cross-aisle measurements from 10 feet (3·05 metres) to 12 feet (3·66 metres).

Characteristics

MODEL	SL 33-ESS	SL 33-ESS HP	SL 44-ESS	SL 44-ESS HP
CAPACITY	3000 lb (1360 kg)	3000 lb (1360 kg)	4000 lb (1815 kg)	4000 lb (1815 kg)
MAX ELEVATION	18 ft 2 in (5·54 m)	22 ft 9 in (6·93 m)	19 ft 7 in (5·97 m)	25 ft (7·62 m)
DRIVING SPEED	6 mph (9·7 km/h)	6 mph (9·7 km/h)	4·5 mph (7·2 km/h)	7 mph (11·3 km/h)
UNIT DIMENSIONS				
length	8 ft (2·44 m)	8 ft 3 in (2·5 m)	7 ft 10½ in (2·4 m)	7 ft 10½ in (2·4 m)
width	3 ft 6 in (1·07 m)	3 ft 6 in (1·07 m)	4 ft 2 in (1·27 m)	4 ft 2 in (1·27 m)
height	6 ft 11 in (2·11 m)	6 ft 11 in (2·11 m)	7 ft 4 in (2·24 m)	7 ft 4 in (2·24 m)

MODEL	SL 55-ESS	SL 66-ESS	SL 77-ESS	SL 88-ESS
CAPACITY	5000 lb (2270 kg)	6000 lb (2720 kg)	7000 lb (3175 kg)	8000 lb (3630 kg)
MAX LIFT	26 ft 8 in (8·13 m)	30 ft (9·14 m)	30 ft (9·14 m)	30 ft (9·14 m)
DRIVING SPEED	7 mph (11·3 km/h)	6·9 mph (11·1 km/h)	6·8 mph (10·9 km/h)	6·7 mph (10·8 km/h)
UNIT DIMENSIONS				
length	8 ft (2·44 m)	8 ft 3 in (2·5 m)	8 ft 3 in (2·5 m)	8 ft 4 in (2·53 m)
width	4 ft 2 in (1·27 m)	4 ft 2 in (1·27 m)	4ft 2 in (1·27 m)	4 ft 8 in (1·42 m)
height	7 ft 4 in (2·24 m)	7 ft 4 in (2·24 m)	7 ft 4 in (2·24 m)	7 ft 4 in (2·24 m)

Status

In service with Eastern Airlines, and major aircraft manufacturers, the SwingMast is available in commercial or military configurations

Manufacturer

Drexel Industries Inc, Horsham, Pennsylvania, USA

Eagle Bob Tail F5000 Series Tractors

The F5000 series of Bob Tail tractors are produced at 4000 lb (1815 kg) and 5000 lb (2270 kg) rated draw-bar pull, suitable for towing baggage and freight trailers. Equipped with single or twin rear wheels the tractors are available with petrol, diesel or lpg fuelled engines. The tractors feature automatic transmission, four-wheel brakes and a weather-proof cab accommodating three people. Speeds of up to 30 mph (48 km/h) are attainable. These tractors can be used also for towing small aircraft and helicopters.

Status

In service with the US Air Force, Air France, SAS, Air Canada, CP Air, Qantas and 20 airlines in the USA. The larger F6000 series are to be found in the Apron Equipment section

Manufacturer

Eagle Bob Tail Tractors Inc, Grosse Ile, Michigan, USA

Eagle Bob Tail F5000 tractor

FMC Air Freight Handling Systems

FMC specialises in the design, provision and installation of air freight handling systems for terminal operations to meet individual operator requirements.
A typical example is an air freight terminal which receives about 22 million lb (10 million kg) of palletised and loose freight each month. Pallets are transferred on to scissors lifts with powered roller conveyor decks and are directed immediately into the system. A right-angle transfer unit, controlled by a console operator, transfers pallets and containers from the receiving conveyors through 90 degrees on to an electric-powered conveyor racetrack. Included in the conveyor racetrack is a load-cell scale conveyor that provides weight tickets for the container/pallet load before it is transferred on to the travelling loader. Loose freight is assembled on makeup/breakdown scissors-lift stations. These stations can be lowered 2 foot 2 inches (660 mm) below the floor surface for pallet makeup, then raised 1 foot 8 inches (508 mm) above the floor surface for transfer on to the travelling loader. The transfer of pallets and containers from the receiving conveyor to the storage racks or staging racks is accomplished by an electrohydraulic travelling loader. These loaders move on tracks interconnected by turntables throughout the terminal and out to the airfreighter loading area. Each loader has a lift capacity of 13 500 lb (6124 kg) and can be raised from the normal carrying position to the storage rack third level.
They are capable of moving 300 feet per minute (91·4 metres per minute) throughout the terminal system. The actual transfer of pallets and containers is accomplished by friction drive wheels on the loader, which extend to engage rubber-tyred wheels on the storage rack and in turn activate chain-driven live rollers to move the pallets.
An aircraft loading bridge is used to transfer pallets and containers between the airfreighter and the travelling loader at each aircraft loading station. The unit bridges the gap between the aircraft door and the travelling loader through a series of three live conveyors and can hold two pallets in a line. To initiate pallet/container transfer, the travelling loader is stopped in line with the aircraft loading bridge and raised to the required level. The pallet is moved on to a live roller holding conveyor on the bridge. Then, the operator transfers the load to a shorter holding conveyor, or if ready, activates automatic transfer into aircraft.

FMC air freight handling systems

Status

FMC has designed and installed air freight handling systems in major airline terminals

Manufacturer

FMC Corporation, San Jose, California, USA

FMC Baggage/Cargo Trailers

FMC produces a range of eight models of trailers to transport cargo and baggage between the aircraft and the terminal building.

The EBC-2TSA and EBC-4TSA convoy baggage/cargo trailers are capable of transporting 4000 lb (1814 kg) of baggage or cargo at speeds up to 15 mph (24 km/h). Both EBC models can be supplied with an optional heavy-duty, sliding chain-link gate and nylon curtain installed on one or both sides. An optional solid wall can be provided opposite a single chain or curtain. The 4TSA can be equipped with a fixed baggage shelf. The steel body incorporates a metal top with a 4 inch (10 cm) drop, an eaves trough and reinforced end panels, mounted on a welded steel undercarriage with a shock-mounted suspension system. The front and rear axles are connected by a reach pole. Axle members consist of hardened spindles on which the hub and wheel assembly are mounted. Pneumatic tyres are standard, with optional solid tyres available. Positive parking brakes are applied to the rear wheels by raising the tow-bar to the vertical latch-lock position. A jaw-type towing point is mounted at the rear. A pintle-type hook is available as an optional feature.

The EBC-2SA Economy baggage/cargo trailer is equipped with a heavy-duty reach pole undercarriage and fifth wheel steering. Capable of carrying 4000 lb (1814 kg) of baggage and cargo at speeds up to 15 mph (24 km/h) the welded steel unit is fitted with fold-down side gates on free swinging hinges. All other features are similar to the EBC-2TSA described above.

The BT 5 × 12 and CT 5 × 12 extra capacity baggage/cargo trailers are equipped with sliding chain gate curtains constructed of heavy-duty chain links and nylon curtain material which permit baggage loading from both sides on Model BT 5 × 12 and single-side cargo loading on the Model CT 5 × 12. A centre steel baggage shelf is installed in the BT 5 × 12. A facing solid wall on the CT 5 × 12 is available on either side, embossed with longitudinal beads. This model is fitted also with an interior plywood backstop and drain holes in the bottom shelf. Apart from the standard pin-type towing point at the rear of the trailer all other features are identical with the EBC-2TSA.

The CBC is a convertible baggage/conveyor trailer similar to other FMC trailers but offered in various design configurations, fitted with a V-type top shelf when used as a baggage trailer and without the shelf for transporting cargo. Both shelf and bottom bed are constructed of heavy-gauge steel with drain holes in the bottom bed. Side door openings have sliding chain gates made of heavy-duty chain links and nylon curtains.

The ECC and ECC-T are heavy-duty freight trailers capable of transporting up to 8000 lb (3629 kg) of freight or baggage at speeds of 15 mph (24 km/h). The ECC-T has a metal top welded as an integral part of the body and a sliding chain-and-nylon curtain door opening. Both ECC models have the extra heavy-duty undercarriage constructed of all-welded steel, with rectangular-tubing axles and channel stringers. Folded-down steel side gates on both models are secured by latches during transporting operations. They drop down on free-swinging hinges for loading and unloading operations. Stops are provided on the steering to limit tow-bar cramp and prevent damage to the steering linkage. Positive parking brakes are applied to the rear wheels by raising the tow-bar to the vertical latch-lock position. The beds of both trailers are constructed of 10 gauge steel fitted with drain holes to prevent water collection, with a welded steel body and reinforced end panels. A pintle-type coupling is mounted on the rear of the trailer.

Status

More than 10 000 trailers for airlines world-wide

Manufacturer

FMC Corporation, Airline Equipment Division, San Jose, California, USA

FMC model CRC trailer

FMC model ECC-T freight trailer

Characteristics

MODEL	EBC-2TSA	EBC-4TSA	EBC-2SA	CBC
CAPACITY	4000 lb (1814 kg)	4000 lb (1814 kg)	4000 lb (1814 kg)	4000 lb (1814 kg)
TOWING SPEED	15 mph (24 km/h)	15 mph (24 km/h)	15 mph (24 km/h)	15 mph (24 km/h)
OUTSIDE TURNING RADIUS	9 ft 3 in (2·82 m)	9 ft 3 in (2·82 m)	9 ft 3 in (2·82 m)	9 ft 3 in (2·82 m)
INSIDE MEASUREMENTS				
length	9 ft (2·74 m)	9 ft (2·74 m)	9 ft (2·74 m)	8 ft 11½ in (2·73 m)
width	4 ft 4½ in (1·33 m)	4 ft 4½ in (1·33 m)	4 ft 4½ in (1·33 m)	4 ft 6 in (1·37 m)
height	3 ft 8 in (1·12 m)	4 ft 3 in (1·3 m)		4 ft 5½ in (1·36 m)
TRAILER DIMENSIONS				
length including tow-bar	13 ft 8 in (4·17 m)	13 ft 8 in (4·17 m)	13 ft 8 in (4·17 m)	14 ft 2 in (4·31 m)
with tow-bar raised	10 ft 3 in (3·12 m)	10 ft 3 in (3·12 m)	10 ft 3 in (3·12 m)	10 ft 11½ in (3·34 m)
width	4 ft 8 in (1·42 m)	4 ft 8 in (1·42 m)	4 ft 8 in (1·42 m)	4 ft 11½ in (1·51 m)
height	6 ft (1·83 m)	6 ft 7 in (2·01 m)	3 ft 4 in (1·02 m)	6 ft 8 in (2·03 m)
WEIGHT	1200 lb (544 kg)	1325 lb (601 kg)	830 lb (376 kg)	1200 lb (544 kg)

MODEL	BT 5 × 12/CT 5 × 12	ECC	ECC-T
CAPACITY	4000 lb (1814 kg)	8000 lb (3629 kg)	8000 lb (3629 kg)
TOWING SPEED	10 mph (16 km/h)	15 mph (24 km/h)	15 mph (24 km/h)
OUTSIDE TURNING RADIUS	13 ft (3·96 m)	15 ft 6 in (4·72 m)	15 ft 6 in (4·72 m)
INSIDE MEASUREMENTS			
length	11 ft 2 in (3·4 m)	11 ft 4 in (3·45 m)	10 ft 3 in (3·12 m)
width	4 ft 6 in (1·37 m)	5 ft 7 in (1·7 m)	5 ft 7 in (1·7 m)
height	4 ft 5 in (1·35 m)		4 ft 11 in (1·5 m)
TRAILER DIMENSIONS			
LENGTH			
including tow-bar	16 ft 3½ in (4·97 m)	16 ft (4·88 m)	16 ft (4·88 m)
with tow-bar raised	12 ft 5 in (3·78 m)	13 ft 2 in (4·01 m)	13 ft 2 in (4·01 m)
width	4 ft 10 in (1·47 m)	6 ft 1 in (1·85 m)	6 ft 1 in (1·85 m)
height	6 ft 8 in (2·03 m)	3 ft 9 in (1·14 m)	7 ft 11 in (2·41 m)
WEIGHT	1800 lb (816 kg)	1545 lb (701 kg)	2260 lb (1025 kg)

FMC CPT-3 Container/Pallet Transporter

The CPT-3 container/pallet transporter incorporates a flow-through design which enables containers and pallets to be loaded or unloaded from either end, and long loads to be accommodated. The conveyor deck features full-width, hydraulically-powered rollers with ramp-type pallet stops at both ends of the deck to secure the load during transport. For unloading, these stops are retracted by individual hydraulic cylinders. The front height of the transporter is adjustable to permit alignment with trailers, storage racks, and aircraft loaders. The unit is powered by a six-cylinder petrol engine or an optional diesel engine coupled with a semi-automatic transmission and heavy-duty rear-axle assembly.

An optional container side-shift system permits the operator to position and retain a container against the operator's side bulkhead of the cab in preparation for unloading.

FMC CPT-3 container/pallet transporter

Characteristics

CAPACITY: 15 000 lb (6804 kg)

LOAD
2 × LD-1/LD-3 containers or
1 × LD-7/LD-11 container or pallet

ROLLER DECK
length: 10 ft 7 in (3·23 m)
width: 8 ft 2 in (2·49 m)
height: 1 ft 8 in (0·508 m)

FRONT END ADJUSTMENT: 1 ft 6 in (0·457 m)–1 ft 11 in (0·584 m)

ENGINE: 4916 cc 6-cylinder petrol/4162 cc petrol/diesel

MAX SPEED: 25 mph (40·3 km/h)

TURNING RADIUS: 25 ft (7·62 m)

UNIT DIMENSIONS
length: 17 ft $1\frac{1}{2}$ in (5·22 m)
width: 11 ft $6\frac{1}{2}$ in (3·52 m)
height: 4 ft $7\frac{1}{2}$ in (1·41 m)

WEIGHT: 11 000 lb (4990 kg)

Status

FMC has supplied 151 CPT-3 transporters to 30 operators at 37 airports

Manufacturer

FMC Corporation, San Jose, California, USA

FMC CST Container/Pallet Semi-trailers

The CST semi-trailers transport conventional airborne containers and pallets, between aircraft and terminal or storage facilities. Two models are available: the CST-23 for loads up to 20 feet 10 inches (6·35 metres) long and the CST-44 for loads up to 41 feet 8 inches (12·7 metres) long.

Standard semi-trailer chassis are used with a powered conveyor system and associated container/pallet latches, guides and stops. The system can transport 96 inch (2·44 metre) pallets by road without guides projecting beyond the overall pallet width and carry vertical LD-1 and LD-3 containers.

Unit loads are transferred across three rows of heavy-duty rollers by a bi-directional shuttle-drive mechanism. The shuttle is powered by a reversible electric gearmotor. Powered rollers at the rear of the semi-trailer provide an interface and bring the load into position for the shuttle operation.

Any standard, commercial fifth-wheel tractor with the required power range can be used to tow CSTs. Electrical power is needed to operate the semi-trailers at the aircraft loader or terminal. The power source can be provided on FMC Model MDL and JC/PL loaders or at terminal facilities. Alternately, CSTs can be supplied with a non-powered conveyor system.

CST semi-trailers can handle mixed loads of LD-1/ LD-3 containers and standard aircraft pallets as well as partial loads.

FMC CST-44 semi-trailer

Characteristics

	CST-23	CST-44
CAPACITY	30 000 lb (13 608 kg)	60 000 lb (27 216 kg)
TOWING SPEED	As required by law	As required by law

	CST-23	CST-44
UNIT LOAD CAPACITY		
8 × 8 × 40 ft (2·44 × 2·44 × 12·2 m) containers	0	1
8 × 8 × 20 ft (2·44 × 2·44 × 6·1 m) containers	1	2
8 × 8 × 10 ft (2·44 × 2·44 × 3·05 m) containers	2	4
88 × 125 in pallet/igloos (2·24 × 3·18 m)	2	4
LD-1 or LD-3 containers	4	8

	CST-23	CST-44
CONVEYING SYSTEM SPEED	60 ft/minute (18·3 m/minute)	60 ft/minute (18·3 m/minute)
POWER REQUIREMENTS	230 V, 3-phase, 60 Hz	230 V, 3-phase, 60 Hz
DIMENSIONS		
length	24 ft (7·32 m)	45 ft 6 in (13·87 m)
width	8 ft (2·44 m)	8 ft (2·44 m)
height	5 ft 1 in (1·54 m)	5 ft 1 in (2·44 m)
WEIGHT	10 060 lb (4563 kg)	16 800 lb (7620 kg)

Status

In service with airlines in the USA

Manufacturer

FMC Corporation, Airline Equipment Division, San Jose, California, USA

FMC JCL-2 Container Loader

The JCL-2 container loader has been designed for servicing the lower lobes on B 747, B 767, DC-10, L-1011, L-1011-500, and A 300 aircraft. Similar in appearance to the FMC JC/PL-2 loader, the JCL-2 is smaller and lifts and transfers containers only. The loader operates with a sill-supported bridge or an optional unattached bridge. Using the sill-supported bridge, the JCL-2 is connected to the aircraft by a bridge that adjusts to pitch movements of the aircraft. The rear platform can be loaded from either side or the rear. Automatic side guides provide simultaneous or individual transfer of two LD-1, LD-2 or LD-3 containers, or one LD-4 or LD-11 container. The load is received on the rear platform, elevated until the two platforms are aligned, and transferred across the bridge into the aircraft. The unattached-bridge method of transfer can be used when the loader is not to be coupled to the aircraft.

The unit comprises a heavy-duty steel framework chassis which is supported by a propulsion and steering system at the front and by two non-driven, low-profile wheels at the rear. The loader is powered by a four-cylinder diesel engine. During loading/unloading operations, the unit is supported by six stabilisers.

The bridge is raised and lowered by a hydraulically-powered, cylinder actuated scissor mechanism. The conveying bridge is attached to the scissor mechanism through a system of horizontal tie-rods which allows the conveying bridge to move vertically and independent of the scissor mechanism when interfacing with the aircraft and/or the rear platform. Engagement with the rear platform raises the rear edge of the forward platform until it is level with the rear platform, at which point the rear platform is stopped automatically. The operator monitors the forward scissors elevation indicator and makes elevation adjustments caused by vertical excursions of the aircraft. Inverted castors support the load, and a powered-belt module provides load transfer. Lead-in rollers on the front and rear of the bridge prevent castor damage.

An adaptor bar, supported by tow extension arms, is mounted on the front of the bridge. Adaptor fittings permit attachment to the various aircraft, and extendable guides, when connected to the adaptor, align with the guides in the aircraft.

An extensible driver/operator platform on the right side elevates with the platform. All controls necessary to drive the loader with the forward platform at any elevation are provided at the driver's position. The operator's position is forward of the driver's position with a control panel which incorporates all controls necessary for loading operations. Hand-rails are provided in the operator's and driver's areas for personnel safety. Access to the platform at any elevation is provided by a ladder on the right side of the loader. The platform is extended for servicing wide doors and retracted for servicing narrow doors, the extension and retraction being hydraulically-powered and controlled by a push-button. The rear platform is raised and lowered by a scissor mechanism powered by two hydraulic cylinders. Safety interlocks, one-way stops, and automatic container guides are incorporated. Minimum elevation is powered and adjustable by a switch between 1 foot 6½ inches (0·47 m) and 1 foot 9½ inches (0·54 m). Powered-edge rollers move the load on to the platform from the end or either side. Castors in the platform deck support the load, and powered-belt modules and lagged rollers provide longitudinal and transverse movement. The operator controls the hydraulically-operated guides.

In the hydraulic system the fixed-displacement, gear-type pump is connected directly to the engine through an adaptor mounted on the bell housing. All hydraulic functions are controlled by solenoid valves and where possible, the valves are equipped with manual actuators to permit operation in the event of electrical failure. Pilot-operated check valves are mounted on the ports of the platform-lift cylinders. In event of a hydraulic failure, these valves prevent uncontrolled cylinder movement. Using an optional emergency pump, the operator can lower the rear platform, raise the bridge to disconnect it from the aircraft, and retract the stabilisers if hydraulic power fails.

Characteristics

CAPACITY: 10 000 lb (4536 kg)
ELEVATION RANGE
bridge: 5 ft 10 in (1·78 m)–11 ft (3·35 m)
rear platform: 1 ft 6½ in (0·47 m)–11 ft (3·35 m)
DIMENSIONS
width:
bridge – max: 12 ft 2 in (3·71 m) operator's platform extended
minimum: 9 ft 4 in (2·85 m) operator's platform retracted
rear platform: 7 ft (2·16 m)
width between guides:
bridge: 5 ft 8 in (1·73 m)–5 ft 1 in (1·55 m) forward
rear platform: 5 ft 8 in (1·73 m)
ENGINE: Perkins 4-cylinder 236-CID diesel
SPEEDS
lift, rear platform: 45 ft/minute (13·7 m/minute)
conveying: 60 ft/minute (18·3 m/minute)
drive: 0–8 mph (13 km/h)
TURNING RADIUS
RT – 20 ft 5 in (6·22 m)
LT – 27 ft 4 in (8·33 m)
UNIT DIMENSIONS
length: 28 ft (8·53 m)
width: 9 ft 4 in (2·86 m)
height: 8 ft 7 in (2·74 m)
GROSS VEHICLE WEIGHT: 19 800 lb (8981 kg)

Status

The JCL-2 was announced in October 1981. The new loader has been developed primarily to service the Airbus A 310 and the Boeing 767 aircraft

Manufacturer

FMC Corporation, San Jose, California, USA

FMC JC/PL-2 Container/Pallet Loader

FMC JC/PL-2 container/pallet loader

The JC/PL-2 is a development of the JC/PL loader designed to service the lower-lobe compartments of B 747, B 767 (container-width door), DC-10, L-1011 and A 300-B aircraft, as well as the main deck of the DC-8F, DC-9F, B 707C, B 727C, B 727QC, B 737C and the B 737 QC aircraft.

The main lift platform is raised by a scissors lift with vertically mounted hydraulic cylinders. A separate scissors lift with conventionally-mounted hydraulic cylinders raises the forward platform to aircraft interface level. The combination driver's/operator's platform is mounted on the forward scissors lift and is raised to working height with the forward platform. Driving controls are at the front of the platform. The forward platform is supported by extension arms which are attached to the adaptor face plate. Aircraft attachment fittings are mounted on the face plate and when in place on the aircraft provide a positive match of the loader roll plane with aircraft roll plane on every transfer cycle. An automatic servo-mechanism provides for accommodation of aircraft height changes during the loading cycle and aircraft pitch changes are accommodated by the torsional flexibility of the forward platform. The rear platform mechanically engages the forward platform on each cycle. The forward platform then becomes a floating bridge between the main platform and the roller system in the aircraft. Powered belts and rollers move containers and pallets across heavy-duty inverted castors. Longitudinal transfer is accomplished by powered-belt modules and powered rollers and transverse movement is effected by powered rollers on both sides and in the centre of the rear platform. Belt modules and rollers are driven hydraulically and provide pressure to the container/pallet bases. Automatic stops and guides keep the containers under control.

The welded tubular steel frame forms the chassis and, at the rear and side transfer area, serves as a guide-rail. Driving and steering is through a heavy-duty drive axle with standard pneumatic tyres. Propulsion is by a hydraulic motor coupled to a two-speed gearbox. Solid rear tyres are mounted directly to the frame. Power is provided by either a six-cylinder petrol or four-cylinder diesel engine.

Characteristics

CAPACITY: 15 000 lb (6804 kg)
ELEVATION RANGE
bridge: 6 ft 3 in (1·91 m)–11 ft 8 in (3·56 m)
rear platform: 1 ft 7 in (0·48 m)–11 ft 8 in (3·56 m)
ENGINE: 300 CID Industrial petrol
SPEEDS
fast lift: 45 ft/minute (13·7 m/minute)
slow lift: 15 ft/minute (4·6 m/minute)
conveying: 60 ft/minute (18·3 m/minute)
CONVEYOR
driving: 8·5 mph (13·7 km/h)
OUTSIDE TURNING RADIUS: 40 ft (12·2 m)
UNIT DIMENSIONS
length: 27 ft 3 in (8·31 m)
width: 11 ft 8 in (3·56 m)
height: 9 ft 7 in (2·92 m)
GROSS VEHICLE WEIGHT: 25 800 lb (11 703 kg)

Status

In production since May 1977 supplied 748 units to 147 places in 86 countries

Manufacturer

FMC Corporation, San Jose, California, USA

FMC JC/PL-W Wide Container/Pallet Loader

The Model JC/PL-W wide container/pallet loader is a Model JC/PL-2 container pallet loader with extensions provided on either side of the unit to allow the transfer of 88 by 125 inch (2·24 by 3·18 metre) or 96 by 125 inch (2·44 by 3·18 metre) pallets with the long dimension cross-wise on the deck.
The Model JC/PL-W provisionally has the capability to handle the lower lobe cargo compartments of the B 747, DC-10, L-1011 and A 300 aircraft, as well as the B 767, and the main decks of DC-9F, B 707C, B 727C, B 727QC, B 737C and B 737QC aircraft. With the B 767 lower lobe, the JC/PL-W is designed primarily to allow servicing of the wide forward cargo door available with this aircraft. The loader is also applicable for transfer of the last container/pallet on the DC-9F aircraft. All controls and functions of the JC/PL-2 Loader are the same on the JC/PL-W with additions to the interface, guides and stops, and powered side transfer edge rollers on the rear platform.
Manually deployed filler sections are provided on either side of the forward platform for interface with either the B 767 lower lobe wide cargo door or the main deck of the DC-9F.
An additional control is provided to lower both powered side guides of the rear platform simultaneously. This is required to transfer pallets laterally over the standard powered guides of the JC/PL-2. Manually actuated cam-over type side guides are provided at either edge of the extended rear platform.
Powered side transfer edge rollers on the rear platform are at the edges of the JC/PL-2 loader, but are mounted inboard on the JC/PL-W rear platform. This is due to the critical dimensioning of the rollers for LD-3 container side transfer. Additionally these rollers retract when the belt modules are actuated for longitudinal container transfer.

Status

Production when the prototype has been completed. Orders for 35 units

Manufacturer

FMC Corporation, San Jose, California, USA

Characteristics

CAPACITY: 15 000 lb (6805 kg)
ELEVATION RANGE
bridge: 6 ft 3 in (1·91 m)–11 ft 8 in (3·56 m)
rear platform: 1 ft 7 in (0·48 m)–11 ft 8 in (3·56 m)
WIDTH BETWEEN GUIDES: 10 ft 9 in (3·28 m)
ENGINE: 300 CID Industrial petrol
SPEEDS
FAST LIFT: 45 ft/minute (13·7 m/minute)
SLOW LIFT: 15 ft/minute (4·6 m/minute)
CONVEYING: 60 ft/minute (18·3 m/minute)
DRIVING: 8·5 mph (13·7 km/h)
UNIT DIMENSIONS
length: 27 ft 3 in (8·31 m)
width: 14 ft 8 in (4·47 m)
height: 9 ft 7 in (2·92 m)
GROSS VEHICLE WEIGHT: 30 300 lb (13 744 kg)

FMC MDL 40 and MDL 60 Main Deck Loaders

The MDL 40 and the MDL 60 were developed to service the nose and side doors of the B 747F main deck and the main-deck side doors of DC-10C, DC-8F, B 707F, and B 727F aircraft. They can be used also to service the lower lobe cargo doors of the B 747 and the front cargo door of the DC-10 and L-1011 aircraft.
The MDL is a dual-platform self-propelled loader with two front drive wheels and two pairs of tandem rear wheels. The two platforms of the loader each include an inverted-castor deck which is lifted hydraulically by a scissors assembly. Both platforms are controlled electro-hydraulically from the operator's station located at the right side of the forward platform, directly behind the driver's cab. The driver's cab and operator's station elevate with the forward platform. All loader driving and loading functions can be performed by a single operator. The rear platform is raised and lowered by two hydraulic cylinders (four in the MDL 60) that control lift chains located at each front corner of the platform. The platform is supported and maintained in a level position by the rear scissors assembly. Powered-edge rollers on the rear platform provide for load transfer at either side or at the rear. Two rows of belt modules convey loads longitudinally, and three sets of rollers convey loads transversely, at a rate of 60 feet per minute (18·3 metres per minute). The unit is powered by an eight-cylinder petrol or an optional diesel engine.
Driving the loader, approaching the aircraft, and aligning the loader with the aircraft are performed at the driver's station. Vehicle steering is accomplished using a hydraulic Orbitrol-type system. Forward or reverse direction is controlled by a toggle switch and vehicle speed by a foot pedal. Driving speeds are variable from 0 to 8 mph (13 km/h) with an inching control for final positioning when adjacent to the aircraft. Attachment fittings are positioned on the loader and, after proper aircraft alignment, are manually engaged on aircraft door hooks or spools.
The rear platform accommodates containers and pallets with dimensions up to 8 by 8 by 20 feet (2·44 by 2·44 by 6·1 metres). Pallet stops and side guides ensure that cargo is transferred in the proper sequence. Load conveying rollers and belts are raised automatically when the operator selects load-movement direction. Power-retracted guides along each side and across the rear of the platform are also controlled by the operator. When the rear platform is lifted and interfaced with the forward platform, a sensing circuit isolates the rear platform lift circuit and maintains the correct level for load transfer to the forward platform. As cargo is transferred into the aircraft, the servo-mechanism

Characteristics

	MDL 40	MDL 60
CAPACITY		
ELEVATION RANGE		
forward platform	8 ft 4 in (2·54 m)–18 ft 4 in (5·59 m)	8 ft 9 in (2·57 m)–18 ft 4 in (5·59 m)
rear platform	1 ft 7 in (0·48 m)–18 ft 4 in (5·59 m)	1 ft 8 in (0·508 m)–18 ft 4 in (5·59 m)
DIMENSIONS		
forward platform	14 ft 2 in (4·32 m) × 10 ft 8 in (3·25 m)	14 ft 2 in (4·32 m) × 10 ft 8 in (3·25 m)
rear platform	23 ft 3 in (7·09 m) × 10 ft 8 in (3·25 m)	21 ft 3 in (6·48 m) × 10 ft 8 in (3·25 m)
ENGINE	CID V8 petrol/diesel	8-cylinder diesel
SPEEDS		
LIFT		
loads to 30 000 lb (13 600 kg)	45 ft/minute (13·7 m/minute)	45 ft/minute (13·7 m/minute)
loads above 30 000 lb	15 ft/minute (4·57 m/minute)	22 ft/minute (6·7 m/minute)
CONVEYING	60 ft/minute (18·3 m/minute)	60 ft/minute (18·3 m/minute)
DRIVING	0–8 mph (15 km/h)	0–8 mph (15 km/h)
UNIT DIMENSIONS		
length	39 ft 5 in (12·01 m)	40 ft (12·19 m)
width	14 ft 8 in (4·47 m)	14 ft 8 in (4·47 m)
height	12 ft 6½ in (3·82 m)	12 ft 6 in (3·81 m)
GROSS VEHICLE WEIGHT	65 740 lb (29 820 kg)	85 000 lb (38 556 kg)

FMC MDL 40 main-deck loader

monitors aircraft movement and automatically adjusts the forward platform position.

Limit-switch circuitry is used to ensure correct operational sequence of loader functions. Pilot-operated check valves on the platform lift cylinders maintain platform elevation in the event of electrical-, hydraulic- or prime-power failure. Relief valves prevent pressure surges in the hydraulic system. Emergency-stop buttons on the operator's panel, driver's dashboard and main control panel are used to shut down prime power in an emergency. Circuit breakers prevent an electrical overload in automotive, light and control circuits.

Status

Supplied 156 MDL 40 loaders to 68 operators in 48 countries, and six MDL 60 units to four operators in four countries

Manufacturer

FMC Corporation, San Jose, California, USA

FMC MDL 60 main-deck loader

FMC Pallet/Container Trailers

FMC offers a wide range of trailers for transporting pallets and containers.

The PD-R3 is designed for powered or non-powered transfer of pallets and containers up to 88 by 125 inches (2·24 by 3·18 metres) weighing up to 10 000 lb (4536 kg). These trailers can be towed in trains of four units at a maximum speed of 15 mph (24 km/h). PD-R3 trailers can be loaded at the freight terminal, staged near the aircraft and individually interfaced with the aircraft loader by a tow tractor with a side-hitch capability or a transporter with friction drive. The power-transfer lead-in roller is friction driven from the loader or transporter. All interior rollers are chain-driven from the connecting transfer roller. Pallets and containers can be transferred manually. Side guides provide pallet/container restraint and permit removal of structural igloo doors. The trailer is mounted on two swivel castors at the front and two non-swivel castors at the rear. Both front and rear castor units are fitted with rubber tyres and the trailer is equipped with a parking brake. Tineways are provided for moving trailers with a fork-lift truck. The tow-bar is hinged for vertical stowage and a captive-pin rear hitch permits in-train towing. To use the power-transfer capability of the PD-R3 trailer, friction-drive options are required for the JC/PL-2 loader and the CPT-2B and CPT-3 transporters.

The STT side-transfer turntable trailer has a capacity of 3500 lb (1588 kg) and is suitable for carrying a LD-1 or LD-3 container, at speeds up to 15 mph (24 km/h). The turntable is mounted on large diameter wheels and pivots on a ball-bearing mount fixed to the centre of the welded steel frame. The sides and ends of the turntable are reinforced and extend beyond the container to protect against damage. Two-position turntable locking is provided by spring-loaded locks that engage striker plates on the running gear to hold the turntable in either of the two positions. The lock latch can be released by either of two recessed handles at opposite corners of the deck. Containers are carried on two rows of 2½ inch (68·5 mm) diameter rollers, with the edge rollers set ¼ inch (6·4 mm) lower. The trailer features four wheel co-ordinated steering, mechanical parking brakes which are actuated by raising the tow-bar and a cone-shaped rear towing point. Optional types of towing points available include pintle-hook and pin-hitches.

The OTT omni-directional trailer has the capability of transferring LD-1 or LD-3 containers across its ends or sides. All functions can be controlled by a single operator. The transfer deck is mounted on a turntable, which permits 360-degree rotation, in 90 degree increments. Heavy-duty inverted castors form the primary transfer surface and edge rollers at the four transfer interfaces accommodate reasonable container misalignment. During transport, the container is restrained from movement by container stops at the sides and ends. Hooks engage the lip of the container for vertical retention. These retention devices accept containers automatically without being preset. They are retracted manually. The turntable rotates on four steel roller-bearing wheels and pivots about a ball-bearing fixed to the running gear frame. Latch striker plates are positioned at 90 degree intervals for rotation control and a special rocker carriage permits container transfer from either side. For end transfer, the turntable deck is rotated 90 degrees, the appropriate container stop is retracted, and the container is manually pushed off the trailer. The trailer, mounted on a welded steel frame with four heavy-duty kingpins installed with bronze bushings, is equipped with co-ordinated four-wheel steering.

FMC ETT container trailer

FMC PCD-R1 pallet/container trailer

The ETT end transfer turntable trailer has the capability to load/unload LD-1 or LD-3 containers at either end. The deck is mounted on a turntable with rotation striker plates positioned at 90 degree intervals to secure the turntable in the correct loading/unloading or towing position. The roller-bed frame is constructed of formed steel sections welded to a rolled angle ring. A centre gripstrut cover plate between the rows of rollers serves as a walkway. Four metal wheels on each side of the frame support the turntable to pivot on a fixed bearing secured to the centre of the frame. The turntable is rotated by raising either of the two actuating handles at either end of the trailer.

Spring-loaded container stops remain in the raised position until depressed by raising the handle on each side of the turntable deck. Releasing the handle allows the end stops to raise. Vertical restraint is provided by retainer plates that are installed in the sides of the trailer. These retainer plates are spring-loaded to provide side-to-side motion, pivoting, and flexing to accept skewed containers. Four optional deflection compensation ramps can be provided, two at each end, to provide a ramping effect when the container is being off-loaded. The welded steel trailer frame is carried on rubber tyres on demountable wheels with four-wheel co-ordinated steering. Parking brakes are provided on the front wheels of the trailer and are applied by raising the tow-bar. A cone shaped rear towing point is fitted. Optional towing points such as pintle-hook and pin-hitches are available.

The DST double side transfer container trailer provides a side-transfer capability between trailer and aircraft loader, carrying two LD-1/LD-3 containers or one LD-series full-width container. The container deck is constructed of heavy-duty steel with two expanded metal deck walkways. A container guide, which is operated manually, is provided for container separation when two LD-3 containers are being transported. Four container stops, positioned at each corner of the end-roller bays, automatically retract as containers are positioned on the trailer and return to the raised position to provide horizontal and vertical restraint once the containers are in place. Containers are carried on four rows of 2½ inch (63·5 mm) diameter rollers with edge rollers set ¼ inch (6·4 mm) lower. For unloading, the stops are manually retracted by recessed handles located in the deck frame on each side. Other features are as for the STT model.

The PCD-R1 Pallet/Container trailer is designed to carry two LD-1 or LD-3 containers, one LD-7 or LD-11 container, one igloo, or one pallet measuring up to 96 by 125 inches (2·44 by 3·18 metres). The trailer has a load capacity of 15 000 lb (6800 kg) at speeds up to 25 mph (40 km/h). The trailer deck frame is constructed of formed steel channel and reinforced plates with 3 inch (76·2 mm) diameter swivel castors mounted on the deck. Four removable stops, with two positions, provide both vertical and lateral restraint of LD-1 or LD-3 containers, permitting either side or end loading. Two cushioned pin stops provide appropriate spacing for 108-inch (2·74-metre) pallets and LD containers. Retractable side guides are provided for directional control of containers and pallets and serve to secure containers on the trailer deck. A retractable end guide permits end loading and unloading. The PCD-R1 is equipped with heavy-duty castors with solid rubber tyres fitted to demountable wheels. The front units are swivel castors with spring suspension and a spring-loaded mechanism to prevent high-speed shimmy. The rear units mounted in a fixed position are fitted also with spring suspension.

The PCD-R2 pallet/container trailer is suitable for transporting two LD-3 containers, one LD-7 container, one 88 by 125 inch (2·24 by 3·18 metre) pallet, or one 88 by 125 inch (2·24 by 3·18 metre) A container. The trailer has a load capacity of 13 600 lb (6169 kg) and can travel at speeds up to 15 mph (24 km/h). The trailer-deck frame is constructed of formed steel channel and reinforced plates with 3 inch (76·2 mm) diameter swivel castors mounted on the deck. A covering on the castor deck serves as an operator walkway and protects the castors from damage. The castor deck has no fixed stops or guides, which permits 360 degree rotation of pallets or containers. Manually removable cushioned pins provide front and rear restraint for pallets and containers. Side and vertical restraint is supplied by pin stops, which can be lowered below the castor roll plane for side transfers. The pin stop is positioned by raising the pin and rotating it 90 degrees, so that the vertical restraint plate hooks over the lip of the pallet. Heavy-duty 2½ inch (63·5 mm) diameter entry rollers are positioned at the sides and ends of the trailer for container/pallet movement. LD-3 container retention is provided by four retractable stops on the interior of the transfer deck. These stops can be positioned by using handles on either side of the trailer. All pins and stops are designed to withstand the pressure from a 13 600 lb (6169 kg) container/pallet moving at 60 feet per minute (18·3 metres per minute) from any direction. The trailers are fitted with two-wheel steering and positive parking brakes are applied to the rear wheels by raising the tow-bar to the latch-lock position. PCD-R2 trailers may be towed in trains of four units at speeds up to 15 mph (24 km/h).

Status

Over 10 000 trailers for airlines world-wide

Manufacturer

FMC Corporation, Airline Equipment Division, San Jose, California, USA

Characteristics

MODEL	PD-R3	STT	OTT	ETT	DST	PCD-R1	PCD-R2
CAPACITY	10 000 lb (4536 kg)	3500 lb (1588 kg)	3500 lb (1588 kg)	3500 lb (1588 kg)	7000 lb (3175 kg)	15 000 lb (6804 kg)	13 600 lb (6169 kg)
TOWING SPEED	15 mph (24 km/h)	15 mph (24 km/h)	15 mph (24 km/h)	15 mph (24 km/h)	20–45 mph (32–72 km/h)	25 mph (40 km/h)	15 mph (24 km/h)
TURNTABLE DIMENSIONS							
length		8 ft 2 in (2·49 m)	8 ft 2¼ in (2·5 m)	8 ft 4 in (2·54 m)			
width		5 ft 7 in (1·7 m)	5 ft 8 in (1·73 m)	6 ft (1·84 m)			
ROLLER/CASTOR DECK DIMENSIONS							
length	10 ft 8 in (3·25 m)				10 ft 10 in (3·3 m)	10 ft 7 in (3·23 m)	10 ft 7 in (3·23 m)
width	7 ft 6½ in (2·3 m)	5 ft 6 in (1·68 m)			6 ft (1·83 m)	7 ft 6½ in (2·3 m)	7 ft 4½ in (2·25 m)
height – empty	1 ft 8¼ in (0·514 m)	1 ft 8 in (0·508 m)	1 ft 8 in (0·508 m)	1 ft 8 in (0·508 m)	1 ft 8 in (0·508 m)	1 ft 8 in (0·508 m)	1 ft 8⅔ in (0·524 m)
loaded	1 ft 7¾ in (0·502 m)						
TRAILER DIMENSIONS							
length							
tow-bar up	13 ft 8¾ in (4·19 m)				16 ft 5 in (5 m)	12 ft 1 in (3·68 m)	12 ft 10 in (3·91 m)
tow-bar down	16 ft 9 in (5·11 m)				19 ft 8 in (5·99 m)	15 ft 6½ in (4·74 m)	16 ft 9 in (5·11 m)
width	8 ft 3 in (2·52 m)	5 ft 7 in (1·70 m)			6 ft (1·83 m)	8 ft 6 in (2·59 m)	7 ft (2·34 m)
height	2 ft (0·61 m)	1 ft 10 in (0·56 m)	1 ft 10½ in (0·57 m)	1 ft 11 in (0·58 m)	1 ft 10 in (0·56 m)	1 ft 11½ in (0·6 m)	1ft 8⅔ in (0·525 m)
WEIGHT	2600 lb (1179 kg)	1100 lb (499 kg)	1530 lb (694 kg)	1700 lb (771 kg)	1940 lb (880 kg)	2400 lb (1089 kg)	2300 lb (1066 kg)

Harlan HTA/HTA-H Towing Tractors

Two versions are available from Harlan, the HTA, a petrol- or diesel-powered tractor and the HTA-H, a hybrid version which is fitted with both petrol/diesel engine and a battery-driven motor. The HTA, which can be powered by a Ford Industrial petrol engine, a Deutz diesel engine or a Perkins diesel engine, is capable of producing between 2000 lb (907 kg) and 6000 lb (2722 kg) draw-bar pull. The tractor is equipped with automatic transmission and torque converter, dual air-braking system, full-floating drive axle and semi-elliptical spring suspension on both the steering and the driving axle. A Harlan E-type towing hitch is fitted. Optional features include an all-weather cab, passenger seat, dual-drive wheels, power steering with Saginaw hydraulic gear, and an lpg fuel system. The hybrid version HTA-H features two independent power systems which can be selected by the operation of a switch. It is fitted with the additional electric motor with SCR control supplied by nine heavy duty 6-volt industrial batteries, rated at 220 ampere hours. A 24-cell industrial battery rated at 350 ampere hours is an optional feature. The tractor includes an engine driven 72 volt, 100 ampere Leece Neville alternator with a solid state regulator to recharge the batteries while the vehicle is under engine power. Since the same braking system is used in engine drive or electric-drive mode, a 1300 cubic inch (2130 cc) vacuum reservoir is provided evacuated to 20-inch Hg by the engine when running and by a 12 volt motor driven auxiliary vacuum pump when the engine is stopped. This pump is controlled by a pre-set vacuum switch to turn on and off as required. Safety interlocks are incorporated to prevent operation in electric mode when the engine is running, and starting the vehicle in gear. The operator is provided with an emergency switch to disconnect the electric power supply, if required.

Characteristics

MODEL	HTA	HTA-H
ENGINE	Ford Industrial 300 CID 6-cylinder petrol/Deutz Diesel F4L 912 4-cylinder/ Perkins 4236 4-cylinder diesel	Petrol or diesel engine as HPA and Prestolite series wound motor
OUTPUT	103 hp (77 kW) at 2800 rpm	20 hp (15kW) at 2000 rpm

DRAW-BAR CAPACITY	2000–6000 lb (907–2722 kg)	2000–6000 lb (907–2722 kg)
TURNING RADIUS	9 ft 7 in (2·92 m)	5 ft 7 in (2·92 m)
UNIT DIMENSIONS		
length	8 ft 6 in (2·59 m)	8 ft 6 in (2·59 m)
width:		
single-drive types	4 ft 7½ in (1·41 m)	4 ft 7½ in (1·41 m)
dual-drive types	5 ft 5½ in (1·66 m)	5 ft 5½ in (1·66 m)
height	5 ft 3 in (1·6 m)	5 ft 3 in (1·6 m)

Manufacturer
Harlan International Corporation, Kansas City, Kansas, USA

Harlan HTA-H towing tractor

Kornylak Low Lift Karry-All Transporter/Loader

The Karry-All embodies a 96 by 260 inch (2·44 by 6·6 metre) adjustable-height cargo deck which has two independently-powered rows of conveyor rollers. Driver-controlled from the cab, the Karry-All can be loaded with up to ten 42 by 48 inch (1·07 by 1·22 metre) pallets and used as a cargo carrier on the public roads.
At the destination, the Karry-All driver swings the cab to the side, adjusts the height of the cargo deck to match the receiving dock level, actuates the powered rollers, and discharges the cargo automatically.
The cargo deck is hydraulically-adjustable from 3 feet 9 inches (1·14 metres) to 6 feet (1·83 metres) above ground level, depending upon design requirements. The deck may also be angled forward or backward, or rolled to left or right.
The vehicle is powered by a Chrysler petrol engine with Allison automatic transmission, power steering and air brakes. A van body is available as an optional feature.

Characteristics
CAPACITY: 15 000 lb (6804 kg)
LIFTING HEIGHTS: 3 ft 9 in (1·14 m)–6 ft (1·83 m)
LIFTING SPEED: variable 0–10 ft/minute (0–3·05 m/minute)
PLATFORM DIMENSIONS: 96 × 260 in (2·44 × 6·6 m)
MAX TILT: 2 ft 9 in (0·84 m) pitch
MAX ROLL: 3 in (76 mm) either side
ENGINE: Industrial 5916 cc Chrysler V8 petrol
MAX DRIVING SPEED: 50 mph (80 km/h)
TURNING RADIUS: 25 ft 3 in (7·7 m)
UNIT DIMENSIONS
cab in front
length: 26 ft 3 in (8 m)
width: 8 ft (2·44 m)
cab at side
length: 21 ft 8 in (6·6 m)
width: 11 ft (3·35 m)
GROSS VEHICLE WEIGHT: 30 000 lb (13 608 kg)

Status
In service with airlines in the USA

Manufacturer
Kornylak Corporation, Hamilton, Ohio, USA

Kornylak KH-14 High Lift Transporter/Loader

The High Lift, Karry-All loader embodies a 96 by 260 inch (2·44 by 6·6 metres) deck with two independently-powered roller conveyors. Controlled by the driver from the cab or by remote control the unit can be fully loaded with two igloo pallets or air cargo pallets and may be used as a cargo carrier on the public road.
To load an aircraft, the High-Lift is driven up to the aircraft. The driver swings the cab to one side, inches up to the fuselage, elevates the deck to the necessary height, then actuates the power rollers. The cargo is loaded automatically by the two rows of 3 foot 6 inch (1·07 metre) rollers from the cab or by remote control. In some situations it may be desirable to operate the unit solely as a cargo elevator. The deck may be elevated 17 feet 7 inches (5·36 metres) above ground level and shifted 4 inches (102 mm) from side to side to allow alignment with pallet locking mechanisms in the aircraft floor. When the deck is elevated pilot-operated check valves within the lifting cylinders prevent sudden descent of the deck through hydraulic malfunction or operator error. In addition, built-in stabilisers at each rear wheel provide protection from vehicle movement during loading or unloading. Safety side-rails are fitted and end-rail stops prevent accidental discharge of cargo. The loader is powered by a Chrysler petrol engine, with Allison four-speed automatic transmission, power steering and air brakes. A van body can be supplied as an optional feature.

Kornylak KH-14 transporter/loader

Characteristics

CAPACITY: 14 000 lb (6350 kg)
LIFTING HEIGHTS: 3 ft 9½ in (1·16 m)–17 ft 7 in (5·36 m)
LIFTING SPEED: 45 s full travel
PLATFORM DIMENSIONS: 96 × 260 in (2·44 × 6·6 m)
ENGINE: Industrial 5916 cc Chrysler V8 petrol
MAX DRIVING SPEED: 50 mph (80 km)
TURNING RADIUS: 25 ft 3 in (77 m)
UNIT DIMENSIONS
cab in front:
length: 26 ft 7 in (8·1 m)
width: 8 ft (2·44 m)
height: 7 ft 4 in (2·24 m)
cab at side:
length: 22 ft (6·71 m)
width: 11 ft 10½ in (3·62 m)
height: 7 ft 4 in (2·24 m)
GROSS VEHICLE WEIGHT: 30 000 lb (13 608 kg)

Status

In service with airlines in the USA

Manufacturer

Kornylak Corporation, Hamilton, Ohio, USA

Kornylak KH-14 transporter/loader

Metric Model 8880 Cargo Handling Assembly Line System

The Metric cargo handling assembly line and system comprises parallel assembly lines 150 feet (45·7 metres) long and 9 feet (2·7 metres) wide. Each contains a gravity roller system which feeds a combined weigh table and scissor-lift elevator for transferring the equipment from the assembly line to the aircraft loading vehicle. The weigh table and elevator is equipped with catwalks on either side of the platform and guardrails are fitted to the outside edge. Two types of line are available, one with a capacity of 12 000 lb (5445 kg) the other with a capacity of 35 000 lb (15 900 kg).

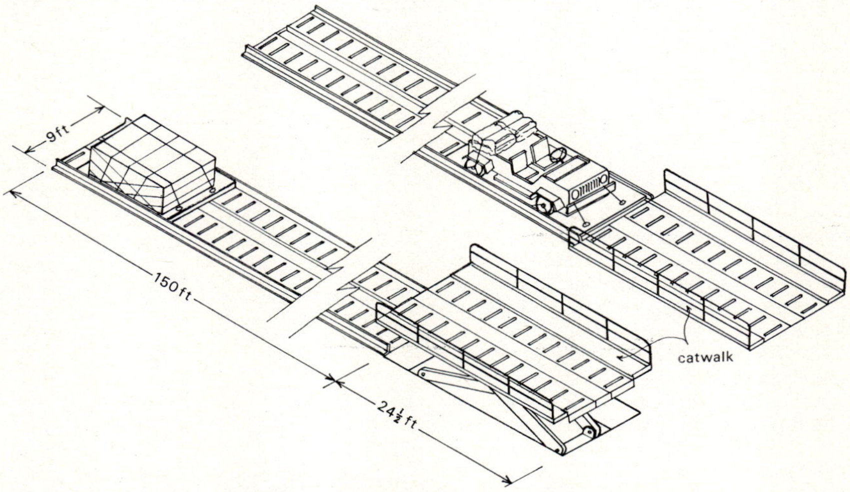

Metric cargo handling assembly line system

Characteristics

TYPICAL PAYLOAD
16 × 108 in (2·74 m) × 88 in (2·24 m) pallets
12 × 108 in (2·74 m) × 144 in (3·66 m) metric platforms
2 × 108 in (2·74 m) × 92 in (2·34 m) metric platforms
2 × 108 in (2·74 m) × 240 in (6·1 m) metric platforms
UNIT DIMENSIONS
assembly line: 150 ft (45·72 m) × 9 ft (2·74 m)
weigh table and elevator: 24 ft 6 in (7·47 m) × 9 ft (2·74 m)
capacity: 12 000 lb (5445 kg)/35 000 lb (15 900 kg)

Status

Prototype production

Manufacturer

Metric Systems Corporation, Fort Walton Beach, Florida, USA

NVF Cargo Handling Systems

NVF designs, manufactures and installs cargo handling equipment to meet individual operator's requirements.

Within the terminal, the heart of the system is the OMNI deck which comprises powered- or manual-roller bed conveyors with powered- or manual-roller ball mats at the junctions for distribution of freight from one conveyor system to another. Storage bays use heavy-duty conveyors to hold LD-1/LD-3 containers. For transfer to transporters or trucks, NVF manufactures a powered travelling dock lifter which enables loading or unloading from any type of truck. Standard-Terminal Units mounted on dollies are used to transport cargo between the terminal and the aircraft.

NVF has produced an LD-3 container customer service dolly with push-bar which is available with or without castors. Tineways are provided to permit the dolly to be handled by a fork-lift truck.

Manufacturer

NVF Company, Kennett Square, Delaware, USA

NVF cargo handling system

NVF cargo handling system

NVF cargo handling system

Oshkosh 40K Military Cargo Loader

The 40K military cargo loader is a self-propelled diesel engine vehicle equipped with a lift platform, deck roller system and a winch to assist in cargo loading. The unit can handle cargo up to 40 000 lb (18 144 kg) and can be transported by air in a C5A or C141 aircraft. It is mounted on tandem steering axles with dual wheels at the front, and tandem axles and dual wheels at the rear, one axle being driven, the other acting as a trailing axle. All driving and positioning adjustments are controlled from the one man cab. Power is provided by a Caterpillar diesel engine with a single-speed transmission and a hydraulically-operated lift system. Other features include air brakes, power steering, two headlights, six floodlights, two rear lights and four clearance lights, two on each side.

Characteristics

CAPACITY: 40 000 lb (18 144 kg)
ELEVATING HEIGHTS: 3 ft 5 in (1·04 m)–13 ft (3·96 m)
WINCH CAPACITY: 2300 lb (1043 kg)
ENGINE: Caterpillar diesel Model 3208
OUTPUT: 210 hp (157 kW) at 2800 rpm
MAX DRIVING SPEED: 15 mph (24 km/h)
WHEELBASE: 25 ft (7·62 m)
INSIDE TURNING RADIUS: 40 ft (12·2 m)
UNIT DIMENSIONS
length: 41 ft 5 in (12·62 m)
width: 12 ft 11 in (3·94 m)
WEIGHT: 43 210 lb (19 600 kg)

Oshkosh 40K military cargo loader

Status

Production for the US Air Force

Manufacturer

Oshkosh Truck Corporation, Oshkosh, Wisconsin, USA

Pettibone Airliner EPL-33 Battery-Powered Tractor

The Airliner PL-33 is a battery-powered tractor designed for towing baggage and cargo trailers. Powered by a 72 volt motor, the vehicle has a rated draw-bar pull of 600 lb (273 kg), with a front steering and rear driving axle. Service drum brakes are fitted to all four wheels, foot operated via a dual braking system. The unit features SCR control which provides variable speeds within its speed range. Battery power consumption is proportional to truck speed. When the tractor reaches approximately 80 per cent of full motor speed a contactor bypasses the SCR to provide full motor torque. When the current subsides, another contactor inserts a parallel resistance with the motor field to achieve top speed. Should the torque demand increase, for example when ascending a gradient, this contactor is automatically de-energised to enable full motor-field to be provided thereby reducing the motor current. When the operator leaves the spring-actuated-seat the travel circuit is interrupted. A passenger seat is provided and a low-profile cab is available as an optional feature.

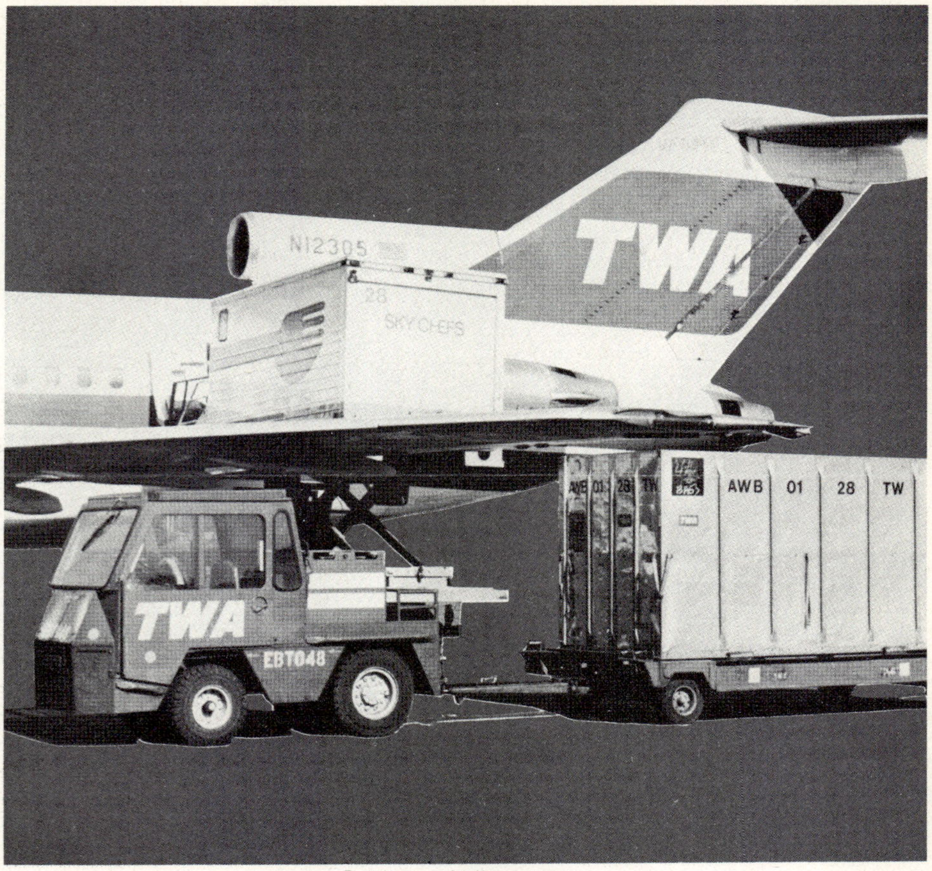

Pettibone Airliner tractor

Characteristics

MOTOR: 72 V, dc
BATTERY: 36–55 cell, 17 plate
CAPACITY: 440 Ah
MAX DRIVING SPEED: 15 mph (24 km/h) (unladen)
MAX DRAW-BAR PULL: 3000 lb (1361 kg)
WHEELBASE: 4 ft $1\frac{1}{2}$ in (1·26 m)
TRACK
front 3 ft $4\frac{1}{2}$ in (1·03 m)
rear 3 ft 3 in (0·99 m)
TURNING RADIUS: 9 ft 2 in (2·79 m)
UNIT DIMENSIONS
with cab:
length 8 ft $10\frac{1}{4}$ in (2·7 m)
width 4 ft 4 in (1·32 m)
height 5 ft 6 in (1·68 m)
without cab:
length 8 ft $10\frac{1}{4}$ in (2·7 m)
width 4 ft 4 in (1·32 m)
height 3 ft 11 in (1·19 m)
WEIGHT
with cab: 6632 lb (3008 kg)
without cab: 5327 lb (2870 kg)

Status

Production. In service with TWA

Manufacturer

Pettibone Mercury Corporation, Chicago, Illinois, USA

Raytheon Raycargo Data Systems

Raycargo is the collective name for several systems Raytheon has developed for automated air cargo processing.

Raycargo 1 has been designed to provide inventory control of shipments and uld for use by one or more airlines at a single airport. Its main functions are flight schedule maintenance, reservations, air waybill data input, waybill preparation, allocation of shipments to flights, tally and manifest preparation, warehouse splits, house shipments, notification of arrival, release, tracing, uld control and reporting and local accounting.

Raycargo 2 has been developed to provide faster information handling — including accounting functions. For use by one or more cargo agents, the system also has communications facilities with related systems. The main functions are air waybill and house air waybill preparation, carrier sales reports, carrier remittances, customer invoicing, inventory control, tracing, uld control and reporting, consolidations, customer record maintenance, Customs declarations and operational accounting.

Raycargo 3 is designed to provide facilities for air cargo import/export control for customs through communications with airline and agent systems. Its main functions are: rate-table maintenance, network support, declarations, agent accounting, documentation printing, inventory, audit support for airline inventory systems, management of random and other checks and statistics.

Raycargo 4 provides facilities for cargo revenue accounting, including the collection of data from a network of both mechanised and non-mechanised stations to be used by an airline. Its main functions are historical schedule maintenance, rate-table maintenance, manifest input, waybill input, load sheet input, reconciliation and pro-rating assistance, transfer manifest reference, reporting and interline billing.

Raycargo 5 is designed to speed the information handling associated with export cargo for one or more airlines at a single airport. Its functions include consignment data input, uld data input, manifest printing, manifest message transmission, consignment status enquiries and concurrent conventional low-speed message operation.

Raycargo systems support data communications including the ability to send and receive automatically all messages defined in the IATA 'Cargo Automation Standard Manual' which are appropriate for the application using either X.25 or 5-level teletype protocols.

Status

In service with major airlines

Manufacturer

Raytheon Data Systems Company, Boston-Providence Turnpike, Norwood, USA

SAGE Cargo Handling Equipment

Stevens Aircraft Ground Equipment (SAGE) Corporation produces a range of cargo handling equipment which includes two Bob Tail tow tractors, a 30 000 lb (13 600 kg) pallet transporter, freight and baggage carts, and container/pallet trailers and dollies. The F350 Bob Tail tow tractor is rated at 5000 lb (2268 kg) draw-bar pull capacity and the larger F600 at a draw-bar pull capacity of 10 000 lb (4536 kg).

A new development is the cargo loader dolly which permits cargo loaders to be transported over long distances by placing the rear tyres in each dolly tray. This enables the loader to be pulled by a tow tractor. The dolly can carry a 25 000 lb (11 340 kg) load at speeds up to 25 mph (40 km/h).

Status

In service with major airlines including American Airlines, Braniff, Eastern, Lufthansa, Pan American, TWA, United and Varig

Manufacturer

Stevens Aircraft Ground Equipment Corporation, Lynbrook, New York, USA

SAGE pallet transporter

Sperry USAS Cargo Handling Data System

Sperry Univac in conjunction with the airlines has developed the Univac Standard Airline System (USAS) Cargo System. Information contained in the waybill is entered into the data base and supplemented by other data entries to assemble an electronically-controlled record for each shipment. The availability of the data permits other computer programs to build load lists and manifests, send advisories to down-line stations and expediate invoices and interline billings.

The system comprises 25 modules:

Data collection: provides the freight agent with transactions to enter, edit and validate the information contained on a freight bill

Rating Assist: produces automated rate and cost calculations of air freight transport charges

Terminal/warehouse control: permits on-line control of import and export functions within freight terminals

Shipment status: supplies real-time information about specific shipments, identifying the controller, the place and the route progress

Document issue: prints and displays cargo information for conventional cargo documents such as waybills and manifests

Ground facility scheduling: supplies the information needed to schedule special handling for freight — including restricted articles and exceptional loads

Storage and miscellaneous charges calculation: provides automated calculations of non-weight freight charges

Station cash control: permits a real-time record of station and agent accounts for cash freight bills

Audit: validates basic air waybill data fields for such data as weight, volume and total charges

Management information: supplies management with data for statistical, financial and operational reports on a periodic or demand basis

Marketing information: provides data on cargo rates, aircraft flight scheduling and sales strategy

Training: for staff training, using both computer aided instruction and practice training

System integration: provides the cargo system with support and service functions for the cargo application programmes, data base maintenance, recovery and contingency control and transaction control

Routeing: provides a flight schedule display of the segments required to move a shipment between cities

Space availability: permits inventory control of cargo space by weight and volume at the flight segment level

Uld control: supplies inventory, status and condition information of ulds at the carrier, station and interline level

Load planning: provides the information required to preplan the loading of aircraft, generates a load distribution plan and computes aircraft weight and balance

Credit control: creates invoices for charges associated with the movement of freight

Interline accounting: controls the settlement of accounts for interline freight activity

Pro-rating: supplies the joint-rate information for interline accounting

Accounts receivable: applies payment, controls accounts and provides balance-due and past-due notices

Miscellaneous charge accounting: controls and settles accounts receivable and payable associated with shipment charges

Messages processing: accepts, processes, assembles and distributes cargo messages in IATA standard formats

Customs interface: accepts, processes, formats and transmits basic customs clearance information.

Sperry Univac has designed the system using database technology in a high-level language and in accordance with the recommendations set forth by IATA in the 'Cargo Automation Research Report'.

USAS Cargo can be implemented as a stand-alone application or run on the same computer system with other USAS applications. It has multiple functions and an airline may select those required.

Status

In service with many major airlines

Manufacturer

Sperry Univac Division, Blue Bell, Pennsylvania, USA

Transact TA40 Main Deck Loader

The TA40 loader which can be transported by air in a Boeing 747 freighter, is designed to service all types of freighter aircraft. Rated at 40 000 lb (18 144 kg) capacity, the main platform can be elevated to a height of 18 feet (5·49 metres). The lift mechanism comprises a leaf chain over an hydraulic cylinder at two central points of the platform, the cylinder being synchronised with a scissors linkage. Load transfer is by stainless steel balls forming a castor mat 5 feet (1·52 metres) long with an independent retractable pallet drive at the front, and by 5 inch (127 mm) diameter heavy-duty rollers at the rear of the platform. With a powered lead-in roller on the right hand side at the front this combination permits front and side loading operations. The 22-foot (8·66-metre) platform can accommodate two 10-foot (3·05-metre) or one 20-foot (6·1-metre) container. Optional equipment available includes, choice of petrol or diesel engine, a front bridge using a single cylinder and leaf chain, fire extinguishing system for the engine compartment, and a tow-bar for front or rear towing.

Characteristics

CAPACITY: 40 000 lb (18 144 kg)
PLATFORM LENGTH: 22 ft 8 in (6·91 m)
WIDTH BETWEEN GUIDES: 10 ft 9 in (3·28 m)
LIFT RANGE: 1 ft 8 in (0·508 m)–18 ft (5·49 m)
SPEED
lifting: 26 ft/minute (7·9 m/minute)
lowering: 32 ft/minute (9·8 m/minute)
ENGINE: Ford 6-cylinder 300 CID
SPEED: 0–4 mph (0–6·4 km/h)
WHEELBASE: 20 ft (6·1 m)
TURNING RADIUS: 25 ft (7·6 m)
UNIT DIMENSIONS
length: 27 ft (8·23 m)
width: 12 ft 7 in (3·84 m)
height: 11 ft 11 in (3·63 m)

Transact TA40 main deck loader

Status

In service with Flying Tigers

Manufacturer

Transact International, Stamford, Connecticut, USA

United Tow Tractors

United Tractors produces a wide range of tow tractors, the larger models being featured under Apron equipment, for aircraft towing operations. The smaller models SM-340, 40, 50 and 60, rated with draw-bar pulls from 3000 to 6000 lb (1361 to 2722 kg) are suitable for towing baggage and cargo trailers.

Available with petrol, diesel or lpg engines, the tractors are equipped with automatic transmission, two-wheel brakes fitted to the rear driving axle and a separate mechanical parking brake. The open driving position is mounted on a single unit construction frame and steering is via the front axle with a worm and roller steering gear. A pin and clevis type towing point is fitted to the rear of the tractor.

Optional features available include: cab, heater/defroster, passenger seat, and four-wheel split power-assisted brakes.

United SM-340 tow tractor

Characteristics

MODEL	SM-340	SM 40/50/60
MAX DRAW-BAR PULL	3000 lb (1361 kg) 4000 lb (1814 kg)	4000 lb (1814 kg) 5000 lb (2268 kg) 6000 lb (2722 kg)
ENGINE	Chrysler H225 petrol	Chrysler H225 petrol
OUTPUT	105 hp (78 kW) at 2800 rpm	105 hp (78 kW) at 2840 rpm
TRAVEL SPEEDS	13 mph (21 km/h)	13 mph (21 km/h)
TURNING RADIUS	9 ft 5 in (2·87 m)	9 ft 5 in (2·87 m)
WHEELBASE	4 ft 10 in (1·47 m)	4 ft 10 in (1·47 m)
TRACK		
front	3 ft 11 in (1·19 m)	3 ft 11 in (1·19 m)
rear	3 ft $11\frac{1}{2}$ in (1·21 m)	4 ft 10 in (1·47 m)
UNIT DIMENSIONS		
length	8 ft 3 in (2·51 m)	8 ft 3 in (2·51 m)
width	4 ft 11 in (1·5 m)	5 ft $5\frac{1}{2}$ in (1·66 m)
height	4 ft 11 in (1·5 m)	4 ft 11 in (1·5 m)
WEIGHT	4800 lb (2177 kg) 5250 lb (2381 kg)	5100 lb (2313 kg) 5800 lb (2631 kg) 7050 lb (3198 kg)

Manufacturer

United Tractor Company, Chesterton, Indiana, USA

UNIT LOAD DEVICES AND RESTRAINT GEAR

Manufacturers

Australia

Aluminium Alloy Pty Ltd
286 Woodpark Road, Smithfield, Sydney, New South Wales 2164, Australia
TELEPHONE: (02) 604 5233

Belgium

Sabena World Airlines
Aeroport National, B-1530 Zaventem, Belgium
TELEPHONE: (02) 751 81 80
Containers, pallets, igloos

Westland EMG SA
Kruineikestraat 99, B-2991 Tildonk, Belgium
TELEPHONE: (016) 60 26 96
TELEX: 23759
Containers

France

L'Aiglon
BP 926, 49009 Angers Cedex, France
TELEPHONE: (41) 43 98 53
Pallet nets, special nets, retention slugs, tensioning buckles, strap hooks, fittings

Carel Fouché
55 rue d'Amsterdam, 75366 Paris Cedex 08, France
TELEPHONE: 874 01 71
TELEX: 290985
Containers, pallets, igloos

Compagnie Europeenne d'Emballages Spéciaux (CEES)
85 bis, rue Faubourg-St-Honoré, F-75008, Paris, France
TELEPHONE: (1) 266 62 51
TELEX: 760947F
Containers

Comptoir Linier
79 rue Tocqueville, Paris 17e, France
Specialised containers

Societe Charentaise d'Equipements Aeronautiques (SOCEA)
Ancien Arsenal BP 109, F-17303 Rochefort Cedex, France
TELEPHONE: (46) 99 44 44
TELEX: 790784F
Containers

Star Pack SA
21 du Coudrier, F-95650, Boissy-L'Aillerie, France
TELEPHONE: (3) 442 18 18
TELEX: 695006F
Containers

Stratifies SA
10 boulevard Malesherbes, F-75008, Paris, France
TELEPHONE: (1) 266 37 55
TELEX: 250827F
Pallets, igloos, containers, nets

Societe Nouvelle d'Albret
374 Chemin de Moularès, BP 1084, 34007 Montpellier Cedex, France
TELEPHONE: (67) 92 92 26
TELEX: 480629F
Containers

Societe de Construction des Aviones Huvel-Dubois
39/43 rue Grange-Dame-Rose, BP 6, F-92360 Meudon-la-Forêt, France
TELEPHONE: (1) 630 21 15
TELEX: 250955F
Containers

Pelletier Industrie
22–24 Allee des Jachères, Sofilic 431, 94263 Fresnes Cedex, France
TELEPHONE: (01) 668 76 76
TELEX: 201094F
Pallets, nets, tie-down equipment, on-board cargo handling equipment

Canada

Abercom Aero (1974) Ltd
70F Brunswick Boulevard, Dollard Des Ormeaux, Quebec H9B2C5, Canada
TELEPHONE: (514) 683 8424
TELEX: 05 822669
Containers, nets

Barber Industries
PO Box 5280, Station A, 9625 Shepard Road, South East Calgary, Alberta, Canada
TELEPHONE: (403) 279 7511
TELEX: 038 25721
Pallets, air cargo rollers, livestock transport units

Aeroquip (Canada) Ltd
287 Bridgeland Avenue, Toronto, Ontario, Canada
TELEPHONE: (416) 781 5261
Cargo nets

False Creek Industries Ltd
860 Beech, Vancouver, British Columbia V6Z IC9, Canada
TELEPHONE: (604) 683 8371
Nets

Gourock Industries Ltd
1311 Ampere Street, Boucherville, Quebec J4B 5J6, Canada
TELEPHONE: (514) 655 3200
Pallet nets, helicopter cargo nets

Mactainair Systems Ltd
708 River Road, Richmond, British Columbia V6X 1X5, Canada
TELEPHONE: (604) 273 8301
Containers

MF Containers Ltd
PO Box 23289, Vancouver International Airport, Richmond, British Columbia V7B IV7, Canada
TELEPHONE: (604) 278 7242
Container

Plastal Inc
476 Edouard Street, Granby, Quebec J2G 3J3, Canada
TELEPHONE: (514) 378 8439
TELEX: (514) 866 3224
Containers

Federal Republic of Germany

Aluminium Walzwerke Singen GmbH
Alumsingen-Platz 1, D-7700, Singen/Hohentwiel, Federal Republic of Germany
TELEPHONE: (07731) 80695
TELEX: 07 93811
Containers, pallets, igloos

BMA Braunschweigische Maschinen-bauanstalt AG
PO Box 3225, Am Alten Bahnhof 5, D-3300 Braunschweig, Federal Republic of Germany
TELEPHONE: (0531) 804 0
TELEX: 952456
Pallets, containers

Bruggemann and Brand KG
5802 Wetter/Ruhr 4, Oberwengenstrasse 209, Federal Republic of Germany
TELEPHONE: (02335) 7487
TELEX: 823254
Containers, pallets, nets and restraint gear

Deutsche Lufthansa AG
Von-Grablenz-Strasse 2/6, D-5000 Cologne, Federal Republic of Germany
TELEPHONE: (0221) 8268
TELEX: 8873485
Containers

Dornier GmbH
Werke Munchen, Trimburgstrasse, PO Box 2160, 8000 Munich 60, Federal Republic of Germany
TELEPHONE: (089) 871 5480 1
TELEX: 0523543
Containers

Felton and Guilleaume Carlswerke AG
D-5000 Cologne 80, Federal Republic of Germany
TELEPHONE: (0221) 6761
TELEX: 8873261
Containers

Messerschmitt-Bolkow-Blohm GmbH
PO Box 801109, D-8000 Munich 80, Federal Republic of Germany
TELEPHONE: (089) 6000 1
TELEX: 052870
Containers

R F Schulz
2 Hamburg 70, Kedenburgstrasse 53-59, Federal Republic of Germany
Containers, pallets, igloos

Seibelwerk GmbH
Inustriestrasse 4, 885 Donauwoerth, Federal Republic of Germany
Containers

VFW GmbH
Hunefeldstrasse 1-5, D-2800 Bremen, Federal Republic of Germany
TELEPHONE: (0421) 5381
TELEX: 245821

Israel

Fedco Israel Ltd
Maskit Street, Industrial Zone, Herzlia B, PO Box 416, Israel
TELEPHONE: 052 70150
TELEX: 35770
Unit load devices, restraint gear

Italy

Irvin Manifatture Industriali SpA
Via Nettunense Km 22300, 0411 Aprilia (Latina), Italy
TELEPHONE: (06) 922881
TELEX: 68019
Container, pallets, slings, cargo restraint gear

Japan

Fuji Heavy Industries Ltd
Fuji Jukogyo KK, 2–7 Nishi-Shinjuku-Ku 1, Subaru Building, Shinju-ku, Tokyo 160, Japan
TELEPHONE: (03) 347 2500/2529
TELEX: 2322268
Containers

Kawasaki Heavy Industries Ltd
Aircraft Manufacturing Division, 1 Kawasaki-Cho, Kamigahara, Gifu 504, Japan
TELEPHONE: (0583) 82 5111
TELEX: 4722368
Unit load devices

Mitsubishi Heavy Industries Ltd
Nagoya Aircraft Works, 10 Oyecho, Minato-Ku, Nagoya, Japan
TELEPHONE: (052) 611 2111
TELEX 22381, 22443
Containers

Shin Meiwa Industry Co Ltd
Aircraft Division, 1-1 Ohgi-Higashinada-Ku 1, Kobe 658, Japan
TELEPHONE: (078) 431 4151
TELEX: 2222431
Unit load devices

Nippon Sharyou Seizo Kaisha Ltd
2-33 Nihonbashi Kabuto Choo Chuuoo-ku, Tokyo, Japan
Containers, igloos

Showa Hikoki Kogyo KK
600 Tanaka-Machi, Aki-Shimi-Shi, Tokyo 196, Japan
TELEPHONE: (0425) 41 2111
TELEX: 2223766
Containers

Suzuei Co Ltd
Fukushima Building, 15/10 Uchikanda 1 chome, Chiyoda-ku, Tokyo, Japan
Igloos

Netherlands

Belgraver BV
Amsterdamsestraatweg 9, PO Box 5012, NL-1410, Naarden, Netherlands
TELEPHONE: (02159) 45514
TELEX: 43248
Nets, tie-down straps

Fokker BV
PO Box 1065, NL-1000 BB Amsterdam, Netherlands
TELEPHONE: (020) 5449111
TELEX: 12227
Containers

Paul and Van Weelde NV
Nieuwerkerk AD Ijssel, Netherlands

Norway

AS Norcem
N-3470 Siemmestad, Norway
Containers, igloos, pallets

Nordisk Aluminium A/S
Wiedemannsgt 8, PO Box 222, N-3081, Holmestrand, Norway
TELEPHONE: (033) 51480
TELEX: 21373
Containers

Sweden

International Cooltainer AB
Sandhamnsgaten 38, 115 28 Stockholm, Sweden
Containers

Wedaverken
Sodertalje, Sweden
Containers

Switzerland

Schweizensche Aluminium AG
Feldeggstrasse 4, CH-1034, Zurich
TELEPHONE: (01) 258 22 22
TELEX: 52487
Containers

United Kingdom

Aeroquip (UK) Ltd
Yates Barco Division, Marple, Cheshire SK6 6AR, England
TELEPHONE: 061-427 3613/6
TELEX: 668053

Bridport Aviation Products
Bridport Gundry, Bridport, Dorset DT6 3QU, England
TELEPHONE: (0308) 56666
TELEX: 41132
Nets, tie-down straps and fittings

Bristol Air Containers Ltd
Unit 7, Industrial Estate, Winterstoke Road, Weston-Super-Mare, Avon BS24 9AH, England
Containers, pallets

Brownline Ltd
Tamian Way, Green Lane, Hounslow, Middlesex TW4 6BL, England
TELEPHONE: 01-572 0321
TELEX: 262184
Containers, igloos, pallets, nets, straps, tie-down fittings and associated cargo restraint hardware

Cargo Aids Ltd
Brett Drive, Bexhill-on-Sea, Sussex TN40 2JR, England
TELEPHONE: (0424) 216611/6
TELEX: 95285
Helicopter cargo release hooks, cargo tie-down equipment, quick-release couplings, freight lashings

DES (Rollason) Ltd
2 Commerce Way, Croydon CR9 4PA, England
Cargo fittings

Concargo Ltd
Winterstoke Road, Weston-Super-Mare, Avon BS24 9AH, England
TELEPHONE: (0934) 28221
TELEX: 449173
IATA profile insulated air freight containers

Labcourt Air Freight Development Products
1 Dunraven Street, London W14 3FG, England
TELEPHONE: 01-408 1677
TELEX: 8953652
Pallets, containers

Lambourn Coachbuilders
Division of Lambourn Engineering Ltd, Lambourn, Newbury, Berkshire, England
TELEPHONE: (0488) 71111
TELEX: 848045
Palletised horse stalls

W H Sheridan & Son Ltd
Eagle Works, Invicta Industrial Estate, Hunt's Lane, London E15 2QE, England
TELEPHONE: 01-555 9454
TELEX: 886684

Metair Aircraft Equipment Ltd
West Street, Erith, Kent DA8 1BE, England
TELEPHONE: (03224) 42321/3
TELEX: 896261
Baggage containers, on-board cargo handling equipment, tie-down gear

Fedco unit load device

Transequip Ltd
Space Waye, North Feltham Trading Estate, Feltham, Middlesex TW14 0TH, England
TELEPHONE: 01-890 0788
TELEX: 264993
Containers, pallets, nets and restraint gear

Triwall Containers Ltd
Tri-Wall House, High Beech Road, Loughton, Essex IG10 4DA, England
TELEPHONE: 01-508 5656
TELEX: 895 2018

United States of America

Acco — Integrated Handling Systems Division
Bailes Road, Frederick, Maryland 21701, USA
TELEPHONE: (301) 662 4181
Containers

Aerial Machine & Tool Corporation
38-27 30th Street, Long Island, New York 11101, USA
Tie-down gear

Air Cargo Equipment Corporation
1550 North 105th East Avenue, PO Box 15630, Tulsa, Oklahoma 74112, USA
TELEPHONE: (918) 836 8846
TELEX: 497541
Air cargo baggage containers, on-board conveyor systems

Air Carrier Interiors Inc
Building C, 8167 Perimeter Road South, Seattle, Washington 98108, USA
TELEPHONE: (206) 767 3255
Containers, nets, restraint gear

AGM Cargo-Ties Inc
3526 East Ft Lowell Road, PO Box 27346, Tucson, Arizona 85726, USA
TELEPHONE: (602) 881 2130
TELEX: 165531
Cargo restraint gear

Air Logistics Corporation
3600 East Foothill Boulevard, Pasadena, California 91109, USA

Ancra Corportion
2233 East Grand Agenue, El Segundo, California 90245, USA
TELEPHONE: (213) 322 7365
Nets

Atti Engineering Corporation
108-110 36th Street, Union City, New Jersey 07087, USA

Avco Systems Division, Avco Corporation
201 Lowell Street, Wilmington, Massachusetts 01887, USA
TELEPHONE: (617) 657 5111
TELEX: 947448

William Bal Corporation
947 Newark Avenue, Elizabeth, New Jersey 07208, USA
TELEPHONE: (201) 354 9625
ATA 300 Container

Breeze Corporations Inc
700 Liberty Avenue, Union City, New Jersey 07083, USA
TELEPHONE: (212) 686 4000
Aircraft cargo moving equipment

Brooks & Perkins Inc
Advanced Structures Division, 12633 Inkster Road, Livonia, Michigan 48151, USA
TELEPHONE: (313) 522 2000

Brownline Division
2950 Lomita Boulevard, Torrance, California 90505, USA
TELEPHONE: (213) 539 0320
Air cargo containers, pallets, nets, restraint gear, on-board cargo handling systems

Clyde Machines Inc
Soo Hill, Glenwood, Minnesota 56334, USA
TELEPHONE: (612) 634 4504
Pallet, containers

Coordinated Equipment Co (CEC)
1707 East Anaheim Street, Wilmington, California 90744, USA
TELEPHONE: (213) 775 6431
Cargo slings

Container Research Corporation
Hollow Hill Road, Glen Riddle, Pennsylvania 19037, USA
TELEPHONE: (215) 459 2160
Containers, pallets, igloos

Container Systems Corporation
145–80 228th Street, Springfield Gardens, New York 11413, USA
TELEPHONE: (212) 276 6500
TELEX: 125061
Air freight containers

Craig Systems Corporation
Walter Kidde & Co Inc, 360 Merrimack Street, Lawrence, Massachusetts 01842, USA
TELEPHONE: (617) 688 6961
Containers, pallets, igloos

Criton Corporation
10800 North East 8th Street, Bellevue, Washington 98004, USA
TELEPHONE: (206) 453 9400
Containers, pallets, nets

CTL-Dixie Inc
1240 Glendale, Milford Road, Cincinnati, Ohio 45215, USA
TELEPHONE: (513) 771 3200

Davis Aircraft Products Inc
Scudder & Woodbine Avenues, Northport, New York 11768, USA
TELEPHONE: (516) 261 1000
Cargo restraint gear

Doskocil Manufacturing Co Inc
PO Box 1246, Arlington, Texas 76010, USA
TELEPHONE: (817) 467 5116
Animal containers and other specialised containers

Eastern Rotocraft
PO Box 110, Route 313, Doylestown, Pennsylvania 18901, USA
TELEPHONE: (215) 345 0300
Cargo tie-down gear, helicopter cargo hooks

Eon Corp
2425 San Fernando Road, Los Angeles, California 90065, USA
TELEPHONE: (213) 223 1241
Cargo restraint, straps, nets

Fleet Equipment Corporation
13600 South Prairie Avenue, World Way Center, Box 90452, Hawthorne, California 90009, USA
TELEPHONE: (213) 679 8136
Containers, pallets, nets

Flite Line Equipment Corporation
PO Box 59311, AMF, Miami, Florida 33159, USA
TELEPHONE: (305) 871 1736
TELEX: 515609

Fruehauf Corporation
15 Hackensack Avenue, South Kearny, New Jersey 07032, USA
TELEPHONE: (201) 344 5551

General Air Services Inc
3200 North West 59th Avenue, Building 1033, Miami International Airport, Miami, Florida 33122, USA
TELEPHONE: (305) 526 4415
Aircraft roller system, nets and restraint systems

General Veneer Manufacturing Co
8652 Otis Street, South Gate, California 90280, USA
TELEPHONE: (213) 564 2661

M.C. Gill Corporation
4056 Easy Street, El Monte, California 91731, USA
TELEPHONE: (213) 283 4786
TELEX: 677467
Containers

Hitco
1600 West 135th Street, Gardena, California 90249, USA
TELEPHONE: (213) 321 8080
TELEX: 677163

HTL Electrokinetics
402 East Gutierrez Street, PO Box 1500, Santa Barbara, California 93102, USA
TELEPHONE: (805) 963 2055

Irvin Industries Inc
900 Bedford Street, Stamford, Connecticut 06901, USA
TELEPHONE: (203) 357 9700
TELEX: 965817
On-board cargo systems, containers, nets, restraint gear

Marino Systems Inc
17 Battery Place, New York 10004, USA
TELEPHONE: (212) 425 1460
Containers, pallets, igloos

McElfish Parachute Service
2615 Love Field Drive, Dallas, Texas 75235, USA
TELEPHONE: (214) 351 5343
Cargo nets, tie-down devices

Myton Industries Inc
CSM Systems, Building 12, Brooklyn, New York 11205, USA
Containers

Neushul Corporation
1911 West 139th Street, Gardena, California 90249, USA
TELEPHONE: (213) 323 3153
Containers

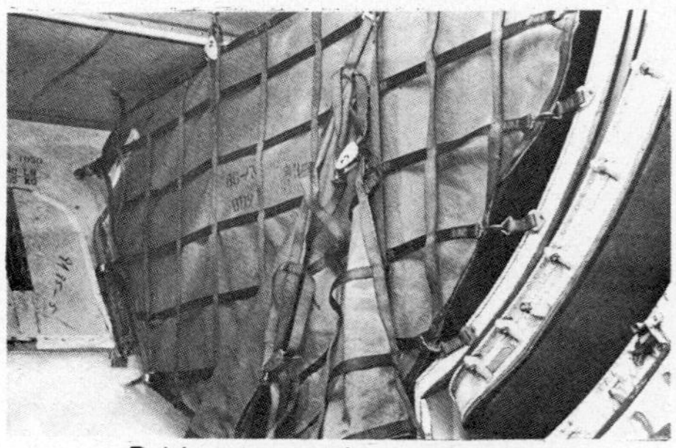

Bridport restraint equipment

NVF Company
Container Division, Kennett Square, Pennsylvania 19348, USA
TELEPHONE: (215) 444 2800
Specialist containers

Palletower Inc
70 West Oakland Avenue, Doylestown, Pennsylvania 18901, USA
TELEPHONE: (215) 348 1250
Containers, pallets

Pedley-Knowles & Co
538 Second Street, San Francisco, California 94107, USA
TELEPHONE: (415) 421 9500
Helicopter cargo nets

Pemco Engineers Inc
PO Box 5, 10762 Chestnut Street, Stanton, California 90680, USA
TELEPHONE: (714) 826 4910
(213) 860 1678
TELEX: 692388
On-board cargo handling and restraint systems

Royal Industries
2040 East Dyer Road, Santa Ana, California 92705, USA
TELEPHONE: (714) 540 3210
Containers

Satco Inc
1509 East El Segundo Boulevard, El Segundo, California 90245, USA
TELEPHONE: (213) 322 4719
Cargo restraint systems, container, pallets, nets

Sicma Aero Seats Inc
6929 North West 69th Street, Miami, Florida 33166, USA
TELEPHONE: (305) 592 7428
TELEX: 807067
Containers

Summit Products Co
146 South Beeson Avenue, Uniontown, Pennsylvania 15401, USA
TELEPHONE: (412) 437 7251
Containers, pallets

Texas Trunk Co Inc
1024 South Laredo Street, PO Box 7268, San Antonio, Texas 78207, USA
TELEPHONE: (512) 227 7353
Containers, transit cases

UOP Inc
Aerospace Division, Route 202, Bantam, Connecticut 06750, USA
TELEPHONE: (203) 567 9441
Pallets

Versa Ferguson Conveyor Corporation
28 North Clark, PO Box 152, Mount Sterling, Ohio 43143, USA
TELEPHONE: (614) 869 2738
TELEX: 246687
Containers, pallets

Western Gear Corporation
Flight Structures Division, PO Box 629, Jamestown, Maryland 58401, USA
TELEPHONE: (701) 252 6250
On-board cargo handling systems

APRON EQUIPMENT

Aircraft docking guidance systems
Aircraft tow bars
Aircraft towing tractors
Apron security and surveillance systems
Crew bus
Fixed services for air-conditioning, electrical power supply and engine starting
Ground power units
Ground service units
Lavatory service vehicles
Mobile air-conditioning, air-starting and heating systems
Portable hydraulic starter unit
Potable water vehicles
Powered service units
Scissors-lift vehicles for aircraft interior cleaning, ambulance duties, catering, de-icing and washing operations

AUSTRALIA

Fox Aircraft Towing Tractors

The chassis of the tractors are fabricated from steel plate and all the components are accessible through removable panels, and can be lifted out of the tractor through the top. The general configuration of all Fox tractors is a General Motors Detroit diesel engine or engines driving all four wheels through six-speed Allison automatic transmissions and Hardie Spicer driving axles. Steering is to all four or two wheels depending on the size of the vehicle. The largest model, the F2-300, and the single-engined F1-350 will tow a fully-laden Boeing 747. Successively smaller tractors are the F-250, F-200, F-180 and the F-100TC.

Status
Fox tractors are used by Qantas Airways at its stations worldwide, by Trans Australia Airlines and by the Royal Australian Navy at sea and on shore

Manufacturer
Fox Manufacturing, Smithfield, New South Wales, Australia

Hastings Deering Invalid Wheel-chair Hi-lift Vehicle

This specialised scissors lift vehicle is designed to meet the special requirements for the transfer of incapacitated passengers whether in wheelchairs or on stretchers. The vehicle can handle up to four handicapped people in wheel-chairs or two on stretchers, plus two attendants. The unit is capable of servicing both narrow- and wide-body aircraft and can be mounted on any commercial truck chassis, with a suitable wheelbase and chassis rails configuration.
The vehicle features a hydraulic system incorporating electro-hydraulic solenoid valves and manual control valves for the various operations. Interlocks are fitted to prevent the vehicle being driven when pto is engaged, stabiliser jacks are not retracted, the body is not in the lowered position; neither of the two rear doors are closed and locked nor the rear tailgate lifting platform housed correctly. There is a provision also that the engine will stop if the power take off is selected without moving the gearshift lever into the neutral position.
Two main lifting rams are fitted to the scissors mechanism with check valves for raising and lowering of the van body and the vehicle is stabilised by four hydraulic jacks. The rear door step combination is opened and closed via one double acting hydraulic ram controlled from inside and outside the van body. This mechanism is interlocked with the van body to ensure the door can only be operated with the van body in the fully down position and that it can be raised only with the door in a closed and locked position.
The van body is constructed of steel frame at both ends and the floor is lined with aluminium plate. An aluminium roller shutter is fitted to the front opening of the van body. The rear doors are hinged in one part, the other part being the step door assembly and lighting is provided inside the van body to illuminate the step assembly. Windows and vents are fitted and the unit is equipped with stretcher brackets, attendant seats and wheel-chair lock-down points.

Characteristics
ELEVATING HEIGHTS: 2·44–5·64 m
DIMENSIONS: 4·47 × 2·3 m
EXTENSION: 0·61 × 1·17 m
REAR LOADING HEIGHT: 1·45 m
WHEELBASE: 3·72 m
UNIT DIMENSIONS
length: 6·62 m
width: 2·3 m

Status
In service with Qantas Airways

Manufacturer
Hastings Deering Ltd, Lidcombe, New South Wales, Australia

Hastings Deering Mobile Lavatory Servicing Unit

The Hastings Deering mobile unit is a self-contained aircraft lavatory servicing unit with the capacity to service a Boeing B747 aircraft or several smaller aircraft. The unit comprises three stainless steel tanks with an operator's station at the rear of the truck on an elevating platform. This can be operated at heights of 1 foot 9 inches (0·53 metres) to 11 feet (3·35 metres) from the vehicle. The unit is built on a standard chassis and the three stainless steel tanks provide waste, water and chemical stowage capacity. Level gauges are supplied on all tanks. A spray ring is installed inside the waste tank for cleaning with flush water from the unit or an external source and large hatch covers are installed on all tanks. The waste tank is fitted with a waste filling port, drain plug and a dump valve underneath. A relief valve is fitted into the flushing circuit to protect the flushing pump from damage if flow blockage should occur. The elevating platform provides for servicing at points on the aircraft up to 16 feet 6 inches (5·03 metres) from the ground. Safety rails and a gate are provided. A 4 inch (102 mm) dump hose and a 1 inch (25·4 mm) flush hose are installed on the platform, and all elevating, dumping and flushing operations are performed from the platform. At the completion of dumping and flushing operations all fluids left in the hoses automatically drain down to the tanks. A reset flowmeter graduated from 0 to 100 Imperial gallons (0 to 455 litres) on the platform indicates the volume of water or chemical pumped.
Safety features offered include an interlock system fitted to the vehicle gearbox to prevent inadvertent movement of the vehicle while the pto is engaged or while the platform is in the elevated position; controls fitted at the base of the platform to lower the platform in an emergency; and pressure relief valves and auxiliary hand pumps incorporated in both oil and water systems for use in the event of engine failure.

Hastings Deering lavatory servicing unit

Characteristics
TANK CAPACITIES
waste: 600 Imperial gallons (2728 litres)
water: 175 Imperial gallons (796 litres)
chemicals: 125 Imperial gallons (568 litres)
UNIT LENGTH: 12 ft 4 in (3·76 m)

Status
In service with Qantas Airways and other airlines in the Far East and Australasia

Manufacturer
Hastings Deering Ltd, Lidcombe, New South Wales, Australia

Hastings Deering Mobile Potable Water Unit

Hastings Deering potable water unit

The Hastings Deering mobile unit is a self-contained potable water servicing unit with the capacity to service two Boeing 747 aircraft or several smaller aircraft. It can be mounted as a towable unit or based on any self-propelled chassis of suitable loading capacity and general measurements to conform to aircraft ground heights.

It is fitted with a 600-Imperial gallon (2728-litre) tank of single compartment design with two internal transverse baffles. Three bolted hatch covers are provided with one containing a vacuum and pressure vent. An opening flap lid is fitted to the rear compartment for the instrument and control panel and hand-pump, together with 15 feet (4·57 metres) of delivery hose and hose reel. A ladder is provided at the rear of the vehicle to gain access to the top of water tank from ground height to 5 foot 2 inches (1·57 metres) to connect the Roylyn coupling into the aircraft water inlet. The unit is equipped with a water pump to give 30 gallon per minute (136 litres per minute) maximum output, combined with an emergency hand pump.

A reset water meter graduated 0 to 100 Imperial gallons (0 to 455 litres) and a vertical-sight liquid level gauge is installed at the operator's station at rear of the vehicle. The tank is filled through a 3-inch (76·2-mm) screw cap fitting mounted in the centre at the rear of the unit. Other models are available with a 350-Imperial gallon (1590-litre) capacity. A rear lift platform to provide access to McDonnell Douglas DC-10 aircraft can be fitted to the rear of the vehicle as an optional feature.

Status

In service with Qantas Airways and other airlines in the Far East and Australasia

Manufacturer

Hastings Deering Ltd, Lidcombe, New South Wales, Australia

Hastings Deering Series III Super Hi-Lift Catering Vehicle

The Hastings Deering Hi-Lift loader can be used for catering or baggage operations. It is designed to service aircraft with a floor height from 2·61 to 5·64 metres and has a payload capacity of 4500 kg. The unit is based on a standard chassis, petrol or diesel engine, and 4·68 metre standard or 5·33 metre extended wheelbase. The Hi-Lift mechanism is designed as a separate unit, bolted to the truck chassis and the van body is attached to the Hi-Lift unit. During lifting and lowering operations the unit is stabilised through hydraulic jacks which are welded to the Hi-Lift frame and become part of the Hi-Lift unit. The lift mechanism is a welded construction with its own chassis to carry the scissor arms. Two lifting rams are mounted between these arms. The platform is capable of handling a 900 kg load within 450 mm of the forward end with the extension in the retracted position. The extension operates from the front of the main platform section and takes place by means of a hydraulic system controlled through a manually-operated hand lever. The extension slides out 0·91 metres from the front edge of the platform which is 1·2 metres wide. A fixed vertical ladder provides access to the platform at a height of 2·61 to 3·07 metres. The van body, manufactured from welded steel tube frame, is covered with 1 mm plain aluminium. Hand-operated roll-up doors are fitted at both ends, with inside and outside door handles. The unit operates with manual control valves for jacks and platform extension and self-centring toggle switches for the Hi-Lift van body. The controls for jacks and van body are mounted beside the driver's cab. One dial-control switch for the Hi-Lift and one control lever for platform extension are installed inside the van body and the Hi-Lift van body can be operated from both stations between 1·57 and 5·64 metres floor height. A dual engine start and stop system is also mounted inside the van body.

A mechanically-operated over-riding device is incorporated in the pilot-operated check valves. By engaging this device and opening a needle valve the jacks can be raised manually with lever bar and chain and the unit towed clear from the aircraft. Hi-Lift and stabiliser rams are fitted with hydraulically-operated check valves which require fluid pressure to open the return line and so give controlled operation to lower the Hi-Lift and lift the stabilisers from the ground.

Characteristics

ELEVATING HEIGHTS: 2·61–5·64 m
CAPACITY: 4500 kg
DIMENSIONS
van body: 7·31 × 2·13 × 2·09 m
front platform: 1·89 × 2·33 m
extensions: 0·91 × 1·16 m
wheelbase: 5·33 m
track: 2·18 m
UNIT DIMENSIONS: 9·37 × 2·33 × 3·66 m

Status

The Series III Super Hi-Lift has been supplied to Qantas and other airlines in the Far East and Australasia

Manufacturer

Hastings Deering Ltd, Lidcombe, New South Wales, Australia

AUSTRIA

Zellinger Potable Water and Lavatory Servicing Units

Zellinger has introduced two units for supplying potable water and servicing aircraft lavatories. Based on a Magirus Deutz chassis, the units are equipped with an elevating cage which can be raised to a height of 3·6 metres to service wide-body aircraft, mounted at the rear of the vehicle. The water hoses are stored on reels, fitted with a return spring, to prevent the supply line from getting distorted. On the lavatory servicing unit, a telescopic pipe is provided to transfer waste from the lifting cage. The disinfectant solution is prepared in the vehicle itself. A small disinfectant tank of 100 litres contains the concentrated solution. If necessary, it is added to the rinse water via an injector with an adjustment ratio of 1 to 5 per cent. The injector is operated from the control panel in the lifting cage. On the left and right side of the cage, there are two compartments, in which the hydraulic installation, the cable and hose-reel and the water pump and injector are stored. These compartments feature a thermostatically-controlled warm air heating blower which protects the fittings and the lower part of the tank against freezing in winter. On the lavatory service vehicle, a water meter with electronic digital indication is fitted, displaying the quantity of water that has been discharged. The meter can be installed on the potable water vehicle as an optional feature.

Zellinger service vehicle

Status

The units were developed for Swissair in 1981

Manufacturer

Zellinger Fahrzeugfabrik, Traun, Austria

BELGIUM

Atlas Copco Air Partner Mobile Air Starting Systems

Atlas Copco produces a range of mobile Air Partner compressors which can be skid-, trailer- or truck-mounted. The units incorporate rotary dry-screw compressors with two asymmetrical helical rotors. Air is compressed between the rotors and these spaces become progressively smaller as the rotors rotate. Both inlet and outlet ports are automatically covered and uncovered by the ends of the rotors.

A balancing piston arrangement and a constant spring pressure combine to equalise loads on the ball bearings. Ball and roller bearings are used in the compression stages and gearbox which permit starting and stopping of the unit without pre-lubrication.

Three Air Partner compressors all offering the same features provide output air capacities from 130 to 300 lb per minute (59 to 136 kg per minute). The MA3, supplies an airflow of 130 lb per minute (59 kg per minute) and is ideal for starting many short- and medium-haul aircraft such as the Boeing 727 and 737, McDonnell Douglas DC-9, BAe Trident and the 1-11. The middle capacity model is the MAS6, which provides an airflow of 180 lb per minute (82 kg per minute). It has been designed for service with larger aircraft such as the McDonnell Douglas DC-10 and Airbus A300 B.

The largest model is the MA8 Super Air Partner. It has a capacity of 300 lb per minute (136 kg per minute), sufficient for servicing the compressed air needs of the Boeing 747, and the new range of Boeings, the B 757 and B 767.

All models are available in silenced versions which have an average noise level of 85 dBA ± 3 dBA at a distance of 15 feet (4·5 metres) and 5 feet (1·5 metres) off the ground.

Atlas Copco Air Partner MAS6

Atlas Copco Air Partner MA8

Characteristics

MODEL	MA3Gd	MAS3Gd	MA3Dd	MAS6Gd	MA8Cud
ENGINE	GM 8V 71 N70 V8	GM 8V 71 N70 V8	Deutz F12 L413F V12	GM 12V 71 N70 V12	Cummins KTA 1150 6-cylinder
OUTPUT	236 kW	236 kW	247 kW	353 kW	448 kW
COMPRESSOR					
max air flow	130 lb/minute (59 kg/minute)	130 lb/minute (59 kg/minute)	128 lb/minute (58 kg/minute)	180 lb /minute (82 kg/minute)	300 lb/minute (136 kg/ minute)
WORKING PRESSURE	43·5 psi (3 bar)	43·5 psi (3 bar)	43·5 psi (3 bar)	43·5 psi (3 bar)	40 psi (2·8 bar)
DUCT TEMPERATURE	390°F (200°C)	390°F (200°C)	390°F (200°C)	390°F (200°C)	355°F (180°C)
DIMENSIONS					
trailer units					
length — tow-bar raised	12 ft 1 in (3·69 m)	12 ft 1 in (3·69 m)	12 ft 1 in (3·69 m)	13 ft 9 in (4·19 m)	
width	6 ft ½ in (1·84 m)	6 ft ½ in (1·84 m)	6 ft ½ in (1·84 m)	6 ft ½ in (1·84 m)	
height	6 ft 6 in (1·98 m)	7 ft 7 in (2·31 m)	6 ft 6 in (1·98 m)	7 ft 7 in (2·31 m)	

Status

In service with Sabena and Cargolux airlines

Manufacturer

Atlas Copco Air Power NV, Wilrijk, Belgium

Atlas Copco Z Series Centralised Compressed Air Systems

The centralised compressed air systems are based on the Z series compressors. These dry-rotary screw compressors supply air to each aircraft loading station either through a network of underground pipes or over the passenger loading bridges. In the latter method, a hose-cart coupled to the bridge can be used to service aircraft with tail pneumatic connections. Pipes can be connected to underground service units on the apron which contain air supply controls and flexible hoses for connecting to the aircraft for engine starting or cabin air conditioning. Within the electrically-powered Z series a number of product variants are offered. At the lower pressure end of the range (3·5 bar or 51 psi), some 20 compressor variants, designated ZA, offer free air delivery ranging from 165 litres per second (350 cubic feet per minute) to 5600 litres per second (11 865 cubic feet per minute). By operating the ZA compressor elements in parallel these coupled units can deliver increased compressed air supplies — 2682 litres per second (5683 cubic feet per minute) to 3950 litres per second (8370 cubic feet per minute). Higher in range are two stage ZR pack models. These operate at pressures up to 10·5 bar (152 psi) at similar air delivery rates to the single stage ZA range.

Principle of centralised air supply system

Status

Atlas Copco centralised systems are installed at Salt Lake City and Atlanta airports. Eleven aircraft gates at the new Salt Lake City terminal are each equipped with hose storage and control manholes served from a central installation of ZA 6 low pressure compressors

Manufacturer

Atlas Copco Airpower NV, Wilrijk, Belgium

Atlas Copco compressed air supply installation at Salt Lake City Airport

Hovey AC 460 Aircraft Ground Heater

The Model AC 460 is a diesel-driven electric generator providing heat through electric resistance elements and coolant and exhaust gas heat exchangers. Powered by a Perkins six-cylinder engine, heat control is operated by the aircraft or through a unit thermostat as selected by the operator. Air is pumped through a 1-foot (0·305 metres) diameter collapsible air duct equipped with a standard coupling. Automatic shut-down occurs in conditions of low fuel, low coolant and overheating. The unit is designed to preheat narrow-body aircraft such as the DC-9 at ambient temperatures down to −40°C. This unit can be used for wide-body aircraft with ambient temperatures above −15°C. Under colder conditions two units would be used.

Characteristics

HEAT OUTPUT: 460 000 Btu/h (485 Mj/h)
AIR OUTPUT: 200 lb/h at 20 in pressure (90·7 kg/h at 50·8 cm)
ENGINE: Perkins 6-354 diesel
FUEL CAPACITY: 40 gallons (182 litres)
FUEL CONSUMPTION: 4·7 gallons/h (21 litres/h)
UNIT DIMENSIONS
length: 12 ft (3·05 m)
width: 6 ft 5 in (1·96 m)
height: 6 ft 3 in (1·91 m)

Status

In service with Air Canada

Manufacturer

Hovey and Associates (1979) Ltd, Ottawa, Ontario, Canada

Hovey AF 800 Mobile Aircraft De-icer Unit

The AF 800 is a mobile aircraft de-icer unit with a hydraulic platform which can be elevated to a maximum height of 20 feet (6·1 metres). It is mounted on a chassis powered by a Wisconsin petrol engine fitted with hydrostatic transmission. The personnel basket has a 500 lb (227 kg) capacity and is supplied with a de-icer unit driven by a pto from the vehicle engine. The boom can be rotated through 95 degrees either side of the centre line. Deadman controls are in the basket and driver's cab and lock valves are fitted to maintain the boom's position in the event of a pipe failure.

Characteristics

BASKET
max height: 20 ft (6·1 m)
max reach: 11 ft (3·35 m)
capacity: 500 lb (227 kg)
ENGINE: Wisconsin 4-cylinder petrol
OUTPUT: 28 hp (21 kW)
MAX DRIVING SPEED: 15 mph (24 km/h)
LOW RANGE SPEED: 5 mph (8 km/h)
DE-ICER
pump: 15 gallons/minute (68 litres/minute)
tank capacity: 166 gallons (755 litres)
WHEELBASE: 10 ft (3·05 m)
TRACK: 6 ft 8 in (2·03 m)
UNIT DIMENSIONS
length: 18 ft 8 in (5·69 m)
width: 8 ft 4 in (2·54 m)
height: 9 ft (2·74 m)

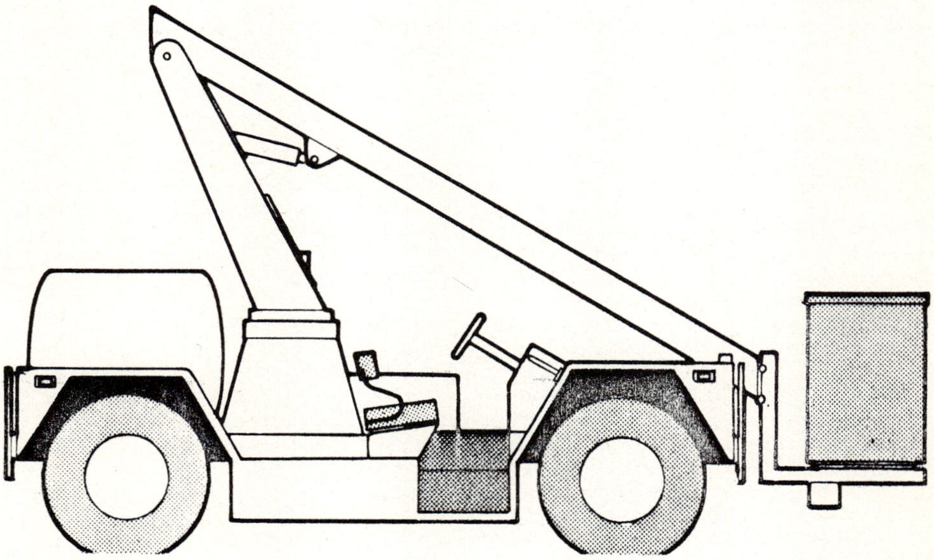

Hovey AF 800 aircraft de-icer unit

Status

In service with the Royal Canadian Air Force

Manufacturer

Hovey and Associates (1979) Ltd, Ottawa, Ontario, Canada

FRANCE

Aerzen VM Series Aircraft Air Starters

The VM series consists of three models of jet air starters truck or trailer-mounted, which incorporate rotary dry-screw compressors with two asymmetric helical rotors. At the beginning of the cycle air enters the chamber through an inlet port. As the compressor rotates it forces the air to decrease in volume as it is pushed towards the outlet port. The steel rotors which mesh without lubrication or contact are positioned by hardened and polished steel timing gear. The male rotor is connected through a torsion bar to the gearbox. Rollers and ball bearings are used throughout and gears and bearings are cooled and lubricated by oil injection. Double seals and ventilated intermediate chambers prevent oil from contaminating the rotors. The compressor is air cooled and driven by a General Motors two-stroke diesel engine. A Deutz engine is offered as optional equipment. Aircraft protection devices include a pressure switch set at 45 psi (3·1 bar) which opens an air bypass valve and a mechanical relief valve which operates at 48 psi (3·3 bar). A two-step cut-off system protects the air starter components, and is automatically by-passed as soon as the jet engine start has been initiated, preventing an interrupted engine start. Normal unit protection is immediately resumed upon completion of the aircraft engine start. An optional silencing canopy is offered which induces noise levels to 75–80 dBA at 15 feet (4·57 metres).

The model VM 310-3S and VM 310-3S Duo were specially designed for wide-body jet application but can be used with conventional jet aircraft. Model VM 310-3S Duo consists of 2 VM 310-3S compressors mounted side by side and driven through a dual output gearbox. These units are equipped with two hoses which can be used simultaneously or separately depending on the aircraft to be handled.

Status
In service with Air Inter and Iberia, Aerzen starters are marketed through Babb Co SA, Paris, France

Aerzen VM series air starter

Characteristics

MODEL	VM 310-3S	VM 410-3S	VM 310-3S-Duo
ENGINE	GMC 8V 71 T	GMC 12V 71 T	GMC 12V 92 T
OUTPUT	360 hp	510 hp	655 hp
MAX AIR FLOW	150 lb/minute	220 lb/minute	300 lb/minute
PRESSURE	42·7 psi (3·2 bar)	42·7 psi (3·2 bar)	42·7 psi (3·2 bar)
OUTLET TEMPERATURE	340°F (170°C)	340°F (170°C)	340°F (170°C)
DIMENSIONS			
without trailer			
length	13 ft 6 in (4·1 m)	15 ft 5 in (4·7 m)	16 ft 4 in (5 m)
width	5 ft 7 in (1·7 m)	6 ft 3 in (1·9 m)	7 ft 3 in (2·2 m)
height	5 ft 3 in (1·6 m)	6 ft 1 in (1·8 m)	5 ft 11 in (1·8 m)
height on trailer	6 ft 9 in (2·05 m)	7 ft 5 in (2·25 m)	
WEIGHT			
without trailer	11 700 lb (5300 kg)	15 500 lb (7000 kg)	17 600 lb (8000 kg)

Manufacturer
Aerzen, Paris, France

Air Marrel CAM SUD 3001 Catering Vehicles

The CAM SUD 3001 catering vehicle mounted on a scissors-lift mechanism is designed to meet the requirements of those airlines operating the Boeing 747 stretched upper deck (SUD) aircraft or using the lounge area of the Boeing 747 as an additional passenger cabin. To meet this requirement the vehicle is capable of extending to a height of 8·3 metres. In its standard form, the CAM 3001 extends to 5·5 metres for servicing all wide-body aircraft. The van body is constructed of horizontal corrugated aluminium sheeting with a translucent polyester fibreglass roof. In the elevated position with the stabilisers extended the van body, with front platform and hand-rails, can be moved forward a distance of 0·6 metres towards the aircraft. Controls inside the van comprise lifting and lowering, forward and backward travel of the van, emergency stop and starting/stopping the engine. Safety features include an emergency hand-pump and locking of the clutch or hand-brake when stabiliser jacks are extended. Optional equipment available comprises: truck of 24 300 lb (11 000 kg) gross weight, and a 15-foot (4·57-metre) wheelbase, pto, power steering, rear tail-gate loader and lateral, removable gangway either manually- or hydraulically-powered.

Characteristics
ELEVATION
SUD version: 18 ft/27 ft 3 in (5·5/8·3 m)
CAPACITY: 6600 lb (3000 kg)
INSIDE DIMENSIONS
length: 19 ft 2 in (5·84 m)
width: 7 ft 8 in (2·34 m)
height: 6 ft 11 in (2·1 m)
FRONT PLATFORM
length: 4 ft 11 in (1·5 m)
width: 7 ft 9 in (2·36 m)
WHEELBASE: 148 ft (45 m)
UNIT DIMENSIONS
length: 25 ft 3 in (7·7 m)
width: 8 ft (2·45 m)
height: 11 ft 6 in (3·51 m)
WEIGHT: 18 800 lb (8500 kg)

Status
Supplied to catering services and major airlines in Europe, the Middle East and Africa

Air Marrel CAM 3001 catering vehicle

Manufacturer
Air Marrel, Saint Etienne, Boutheon, France

Augereau Aircraft Catering Vehicles

Two models are produced by Augereau, the B 747 standard which services aircraft with door sills between 2·5 and 5·6 metres, and the Boeing 747 half cab which operates between 1·4 and 5·6 metres to cater for aircraft with low door sills such as the BAe 1-11.

The B 747 standard model is based on a standard commercial chassis with a gross vehicle weight of 11 000 to 13 000 kg. The scissors-lift mechanism and sliding front platform are operated by a monobloc hydraulic plant with controls in the cab and the van body. Four stabiliser jacks provide stability with the body elevated. Stainless steel telescopic rails are fitted to the platform.

The B 747 half cab is mounted on a modified commercial chassis. The front platform of the unit in the lower positions is adjacent to the half width driver's cab, which contains two seats fitted in tandem. Other features include: scissors-lift mechanism with monobloc hydraulic power plant, controls in cab and van body, stainless steel telescopic rails and four stabiliser jacks.

Characteristics

MODEL	standard	half cab
CHASSIS	commercial	commercial
CAPACITY	3500–4500 kg	3000–4500 kg
ELEVATING HEIGHTS		
front platform	2·5–5·6 m	1·4–5·6 m
van body	1·5–5·6 m	1·4–5·6 m
DIMENSIONS		
interior van body	5·8 × 2·35 × 2 m	6·2 × 2·35 × 2 m
overall	8·3 × 2·5 × 3·7 m	9·1 × 2·5 × 3·6 m
GROSS VEHICLE WEIGHT	1100–13 000 kg	1100–13 000 kg

Status

Both units are in service with Air France

Manufacturer

SA Augereau (Heuliez), Brou, France

Microturbo GCAT-100 Ground Services Trolley

The GCAT-100 trolley has been developed specifically to meet the needs of advanced combat aircraft, particularly when operating from remote airstrips at short notice. It provides three services simultaneously: dc and ac electrical power, air for the cockpit and electronics cooling and heating.

The four-wheeled trolley comprises a small gas turbine power unit driving dc and ac power systems and supplying compressed bleed air for the refrigeration system. A turbo-compressor air cycle refrigeration system is employed, with its associated heat exchangers, water separator, temperature control system and flexible air supply ducting. The air-cycle cooling system is capable of supplying cold or hot air on demand. Although specifically intended to supply cold air, the temperature control system can be adjusted to supply air at any controlled temperature up to 80°C for equipment pre-heating, cockpit heating or de-icing the aircraft. Lightweight hoses are provided with the specified quick release couplings. The controls comprise: master switch, start and stop buttons, ac power switch, air bleed valve, on/off switch and air temperature control knob. The starting sequence and engine speed control are fully automatic and appropriate interlocks and safety devices are fitted for safety purposes in the event of malfunction. The temperature of the air entering the aircraft, the engine speed and the fuel level are indicated on the control panel which is fitted with an extension lead for remote operation. The trolley carries its own batteries and charging system for instant starting and sufficient kerosene fuel for up to four hours running. Intake and exhaust silencing systems are incorporated. The trolley is equipped with a tow-bar with front wheel steering, parking brake which is applied automatically when the tow-bar is on the ground or in an upright position and a rear towing hook.

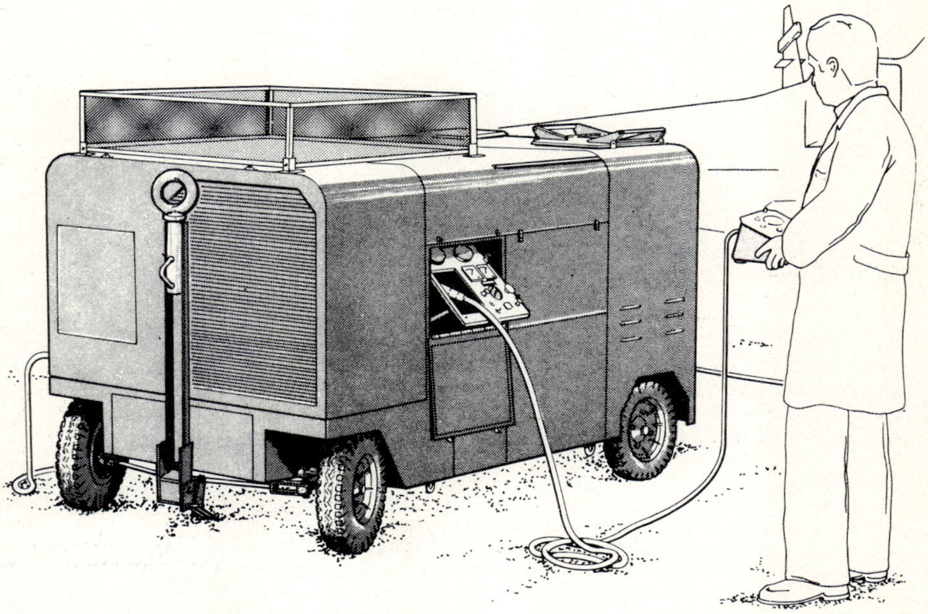

Microturbo GCAT-100 ground services trolley

Characteristics

ENGINE: Microturbo Dragon Model 058-2
COOLING PACK: British Aerospace Turbo Compressor; air-cycle-type
POWER SUPPLIES
28 V dc 5 kW
115/200 V 400 Hz ac 20 kVA
AIR SUPPLY: 21 kg/minute at 5°C
AIR PRESSURE: 0·5 bar
NOISE LEVEL: 80 dBA at 2 m
TOWING SPEED: 25 km/h
TURNING RADIUS: 3 m
DIMENSIONS: 2·3 × 1·3 × 1·3 m
DRY WEIGHT: 850 kg

Status

The GCAT-100 is the result of a joint development by Microturbo and British Aerospace Dynamics Group who produce the cooling pack

Manufacturer

Microturbo, Toulouse, France

Siraga RMS 9500 Lavatory Servicing/RMS 10 000, Water Supply Vehicle

This vehicle is available configured either for lavatory servicing as Model RMS 9500 or as a water supply vehicle, Model RMS 10 000. Based on a Saviem chassis and cab, the flat stainless steel tank unit provides a working platform, which can be extended by an optional retractable ladder to a maximum height of 2·8 metres. An access ladder to the tank platform is fitted at the rear of the vehicle. A selector valve pressure gauge and tapped valve are incorporated in the delivery circuit, with an emergency hand-pump in case of engine or main pump failure. Optional equipment available includes low profile cab 5 foot 3 inches (1·6 metres) high, one or two draining hoses of a 4-inch (102-mm) diameter with straight or 'Y' aircraft coupling (RMS 9500), water softener (RMS 10 000) and diesel engine.

Siraga RMS 9500 lavatory servicing vehicle

Characteristics

MODEL	RMS 9500	RMS 10 000
TANK CAPACITIES		
waste	400 US gallons (1500 litres)	
fresh water	240 US gallons (900 litres)	660 US gallons (2500 litres)
disinfectant	80 US gallons (300 litres)	
PUMP OUTPUT	925 US gallons/h (3·5 m³/h at 3 bar)	1190 US gallons/h (4·5 m³/h at 3 bar)
WORK LEVEL HEIGHTS	1 ft 2½ in (0·37 m); 3 ft 9 in (1·15 m); 9 ft 2 in (2·8 m)	1ft 2½ in (0·37 m); 3 ft 9 in (1·15 m); 9 ft 2 in (2·8 m)
DIMENSIONS		
length	20 ft 10 in (6·35 m)	19 ft 6 in (5·95 m)
width	7 ft 6½ in (2·3 m)	7 ft 6½ in (2·3 m)
height	7 ft 5 in (2·27 m)	7 ft 5 in (2·27 m)

Status
The RMS 9500 lavatory servicing vehicle is operated by Air Inter and Sabena

Manufacturer
Siraga, Buzancais, France

Siraga Towed Lavatory Servicing Vehicles

The Models 8000, 8045 and 8150 are trailer units which contain stainless steel tanks for waste, fresh water and chemical products. A rinsing circuit is equipped with a Bernard petrol fuelled pump producing a flow rate of 3 cubic metres an hour. The flat top profile of the trailer unit provides a suitable working platform, 1 metre high, to facilitate coupling to the aircraft using a 6-metre long flexible hose. An access ladder is mounted at the rear of the unit. Other features include manometer, selective tap for water or chemical product, and a valve calibrated at 1·4 bar with return into selected tanks. The parking brake is actuated by lifting or lowering the tow-bar. Optional equipment available comprises emergency hand-pump, one or two draining hoses of 4-inch (102 mm) diameter with straight or 'Y' aircraft coupling, provision for rinsing waste tank with water and chemical products, and folding ladder mounted on top extending to 2·65 metres above ground level in the VT 8150 model only.

Siraga 8150 lavatory servicing trailer

Characteristics

MODEL	8000	8045	8150
TANK CAPACITIES			
waste	1000 litres	450 litres	1500 litres
fresh water	360 litres	360 litres	360 litres
chemical product	180 litres	180 litres	180 litres
POWER UNIT/FLOW RATE: Bernard petrol engine and pump unit: 3 m³/h			
DIMENSIONS			
length (tow-bar raised)	3·37 m	3·37 m	4·46 m
width	1·44 m	1·44 m	1·44 m
height	1 m	1 m	1 m

Status
In service with major airlines

Manufacturer
Siraga, Buzancais, France

Siraga Towed Water Supply Vehicles

Three models of trailers designed for the supply of drinking water are produced by Siraga. These are the 7000, 7020 and he 7200. The 7000 and 7020 are three-wheeled units, the latter being of simplified manufacture with a maximum capacity of 200 litres. The 7200 is fitted with four pneumatic wheels. All are equipped with a stainless steel tank, Bernard petrol-engined pump unit, manometer, pressure valve calibrated at 1·4 bar and a 6-metre long flexible hose for coupling to the aircraft. The tank unit forms a working platform, access being provided by a ladder at the rear of the vehicle. On the 7200 model, a folding ladder is available with a maximum height of 2·6 metres above ground level. The tank unit can be fitted under pressure with the appropriate coupling or by gravity. Optional features offered include: emergency hand-pump, water softener and electric starter motor.

Characteristics

MODEL	7000	7020	7200
TANK CAPACITY	1000 litres	200 litres	2000 litres
UNIT DIMENSIONS			
length	2·68 m	2·68 m	4·38 m
width	1·44 m	1·44 m	1·44 m
height	1 m	1 m	1 m

Status
In service with major airlines

Manufacturer
Siraga, Buzancais, France

Siraga 7200 potable water trailer

Tracma Tow Tractors

Tracma produces a wide range of tractors suitable for airport duties, available with a choice of petrol or diesel engines or electric motors. The petrol and diesel tractors have draw-bar pull capacities ranging from 3000 to 60 000 lb (1360 to 27 215 kg). The electric-powered vehicles produce a draw-bar pull of 3000 lb (1360 kg).

Status
In service at 133 airports world-wide

Manufacturer
Saxby, Department Tracma, Paris, France

GERMANY, FEDERAL REPUBLIC

Dahms Potable Water Supply/Lavatory Servicing Vehicles

Dahms produces self-propelled and trailer-mounted vehicles for providing potable water and for lavatory servicing.

For wide-body aircraft, the self-propelled models DATRI 3000 M (potable water) and the DAFA 3200 M (lavatory servicing) are mounted on a Mercedes Benz chassis equipped with tank platform and integrated steps at the rear of the vehicle. A collapsible step ladder on the tank platform provides a platform 2·78 metres above ground level to couple the hose to the aircraft servicing point. The DATRI 1500 M (potable water) and the DAFA 500 M (lavatory servicing) both based on a Mercedes Benz chassis with tank platform and integrated rear steps, are suitable for servicing narrow-body aircraft. An additional 0·45 metres in height is available using a folding ladder mounted on the tank platform.

The trailer units are designated DATRI 1500 T for potable water supply and DAFA 1500 T for lavatory servicing.

Status
Dahms potable water and lavatory servicing units have been supplied to the Federal German Air Force, 16 airlines and 12 airports

Manufacturer
Dahms and Co, Luneburg, Federal Republic of Germany

Characteristics

TYPE	DATRI 3000 M	DAFA 3200 M	DATRI 1500 M	DAFA 1500 M
APPLICATION	Potable water	Lavatory servicing	Potable water	Lavatory servicing
TANK PLATFORM HEIGHT	1·57 m	1·57 m	1·65 m	1·35 m
STEP LADDER PLATFORM HEIGHT	1·21 m	1·21 m	0·45 m	0·45 m
WHEELBASE	4·2 m	4·2 m	3·35 m	3·35 m
TRACK	1·75 m	1·75 m	1·6 m	1·6 m
UNIT DIMENSIONS				
length	6·75 m	7·45 m	5·35 m	5·65 m
width	2·1 m	2·1 m	1·8 m	1·8 m
height	2·48 m	2·48 m		

Dahms DAFA 1500M lavatory servicing vehicle

Dahms DAFA 3200M lavatory servicing vehicle

Dornier Radobs Aircraft Security Unit

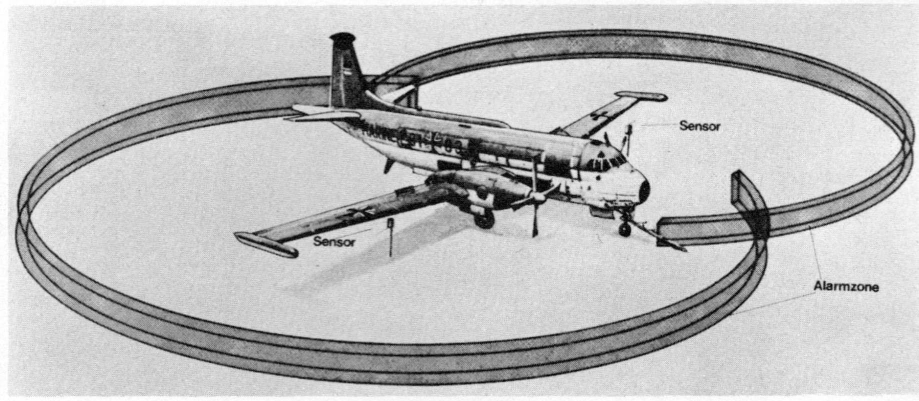

Dornier Radobs aircraft security unit

The Dornier Radobs radar intruder detection system is a portable unit which can be used to provide security to aircraft parked on the apron or hardstanding. The system comprises a sensor with omni-directional or sectorial antenna, display and control unit and a power supply. In the case of an attempted intrusion, the hf signal which is emitted via an antenna is reflected within a given alarm zone and evaluated. Outside the alarm zone, also within the protected area, it is possible to move freely. The alarm zone is strictly defined, its form being determined by the layout of the antenna and the depth and the radius of the zone can be varied as required. The signals are processed in the display and control unit which indicates alarm and fault signals. Up to four sensor units can be controlled by the display and control unit.

Radobs can be used independently or within a security system when corresponding signal outputs are provided. In operation the functions of the unit are monitored by an automatic test facility.

Manufacturer

STS Zweigniederlassung der Dornier System GmbH, Friedrichshafen, Federal Republic of Germany

Characteristics

ALARM ZONE
radius: 25–70 m
width: 5 m

POWER REQUIREMENTS: 24 V dc 40 W (for 2 antenna system)

OPERATING TEMPERATURE RANGE: −30 to +70°C

DIMENSIONS
sensor unit: 180 × 330 × 480 mm
antenna: 190 × 55 mm
display and control unit: 350 × 530 × 220 mm

WEIGHT
sensor unit: 8 kg
control unit: 10 kg

Garrett Jet Air Starter Unit

Garrett RSS-184 jet air starter, skid-mounted on pick-up van

Garrett produce two basic models of jet air starters which are available in a variety of mountings. In addition to compressed air, units are offered with electrical power supplies of 75 kVA 120/208 volt 400 Hz. Power is provided by a Garrett Airesearch gas turbine/compressor which features an integrated low- and high-pressure filtered fuel system and an air-cooled closed lubricating oil system. Starting of the unit is by two 12 volt 66 ampere an hour batteries which are recharged automatically. Connection to the aircraft is via a 20-foot (6·61-metre) flexible hose and coupling.

Safety features include automatic overload, high temperature and overspeed controls, prevention of start with low oil and minimum fuel conditions, and an emergency shut-down switch. The units fitted with electrical power supplies incorporate over and under-voltage and over-current protection.

The models offered comprise skid, trailer and VW van mountings, van-kit installation and air transportable versions mounted on a cart, skid or trailer.

Characteristics

MODEL	RSS–90	RSS–184
AIR FLOW	124 lb/minute (56 kg/minute)	155 lb/minute (70 kg/minute)
pressure	51 psi (3·5 bar)	51 psi (3·5 bar)
OUTLET TEMPERATURE	360°F (185°C)	400°F (205°C)
ELECTRICAL POWER	75 kVA 120/208 V 400 Hz	75 kVA 120/208 V 400 Hz

Status

Garrett air starter units have been supplied to the US Navy and Air Force, American. Eastern, Delta, Flying Tigers, Pan Am, and Western Airlines in USA, and Air New Zealand, Korea, Philippine Airlines, Thai International and Varig airlines

Manufacturer

Garrett GmbH, Raunheim am Main, Federal Republic of Germany

Garrett RSS-184/75 air starter/electrical power unit

Scharfenstein Potable Water and Lavatory Servicing Vehicles

Four models are produced by Scharfenstein, two for lavatory servicing and two for the supply of potable water, all based on a Daimler Benz chassis, and capable of servicing both wide- and narrow-body aircraft.

Type 1700-2·4 is a lavatory servicing vehicle which is offered with a choice of Benz L407 D diesel or L 409 petrol engine. Two stainless steel tanks are fitted for water and effluent and a hydraulically-operated lift platform extends from the centre of the vehicle to a height of 3·3 metres. The platform is housed between the two tanks when not in use. Access to the tank platform is by a ladder mounted on the side of the vehicle behind the driver's cab.

Type 2800-4·2 is a larger version of the 1700-2·4, powered by a Daimler Benz LP 809 engine and equipped with two stainless steel tanks providing a 1400 litres capacity for water and 2800 litres for effluent.

The potable water supply unit with a 1500 litre capacity stainless steel tank is available with a Daimler Benz L 407 D diesel or an L 409 petrol engine. Access to the tank platform is by a vertical ladder installed on the side of the vehicle behind the driver's cab. A folding ladder mounted on the tank platform provides an additional 1·5 metres. A larger model with a Daimler Benz L 608 diesel engine and a 2800 litre water tank is also produced.

Characteristics

TYPE	1700-2·4	2800-4·2	1500 litre	2800 litre
APPLICATION	Lavatory servicing	Lavatory servicing	Potable water	Potable water
ENGINE	Daimler Benz L407 D (diesel)/ L 409 (petrol)	Daimler Benz LP 809 petrol	Daimler Benz L407 D (diesel)/ L 409 (petrol)	Daimler Benz L 608 D (diesel)
TANK CAPACITIES				
water	700 litres	1400 litres	1500 litres	2800 litres
effluent	1700 litres	2800 litres		
TANK PLATFORM HEIGHT	1·55 m	1·6 m	1·55 m	1·55 m
ELEVATING PLATFORM/ FOLDING LADDER				
height above ground	0·4–3·3 m	0·4–3·45 m	3·05 m	3·05 m
CAPACITY	100 kg	100 kg		
WHEELBASE	3·5 m	4·2 m	3·5 m	
UNIT DIMENSIONS				
length	6·1 m	7·6 m	6·25 m	7·4 m
width	2·26 m	2·3 m	2·15 m	2·15 m
height	2·45 m	2·45 m	2·45 m	2·45 m

Scharfenstein 1700-2·4 lavatory servicing unit

Status
In service at airports in Federal Republic of Germany

Manufacturer
Fritz Scharfenstein, Nister/Hachenburg, Federal Republic of Germany

Scharfenstein 1500 litre potable water unit

Scharfenstein 2800-2·4 lavatory servicing unit

Schroder High Lift Ambulance

Schroder has designed an ambulance which can be elevated to a maximum height of 5·8 metres for transferring stretchers and handicapped people in wheel-chairs. The manufacturer has taken the standard chassis used for catering vehicles with a 3·25 metre wheelbase and a maximum height of 4·5 metres and added a chain suspension system to provide the extra height. The van body is raised above the upper frame of the scissors mechanism on telescopic tubes positioned at each corner, operated by the chain suspension equipment. Four outrigger stabilisers fitted to the chassis are hydraulically-operated. The hydraulic system is supplied by a hydraulic pump driven by an engine pto with electrically and manually-operated valves; the hand-pump is incorporated for operation in the event of a power failure. The van body is of sandwich construction with polyester and 50 mm of insulating material for the walls and roof, and an aluminium floor. It comprises a passenger cabin with a sliding door at the front and a two-leaf swing door at the rear. A sheltered front platform provides lateral access from ground level. The doors and front wall are fitted with windows. Wheel-chairs and stretchers are loaded by means of a swing lift fitted to the rear entrance. Transfer to and from the aircraft proceeds along a special front lift which can be moved laterally and incorporates an extensible section. Collapsible bridge sections fitted to each side of the front lift enable transfer to passenger steps or passenger loading bridges. As an option, a single driver's cab can be fitted enabling the front lift to be lowered to a transfer height of 1·3 metres. Optional medical equipment available includes: two stretchers with supports, wheel-chairs with locks, unit for transferring patients to aircraft with narrow aisles, oxygen equipment, medical instrument locker and seats with safety belts. A heater can be installed for winter operation.

Characteristics

CHOICE OF CHASSIS	CAPACITY	WHEELBASE
Magirus 90 M.7 Fl	7490 kg	3·2 m
MAN 8·90 F	7490 kg	3·6 m
Daimler Benz LP 809	7490 kg	3·6 m
Ford 0808	7490 kg	3·38 m
Volvo F 408	7500 kg	3·25 m

ELEVATING RANGE	2·35–5·8 m
optional	1·3–5·8 m
DIMENSIONS	
passenger cabin	3·4 × 2·35 × 2 m
front platform	1 × 2·35 m
front lift	1·5 × 1 m
EXTENSION LENGTH	0·5 m
LATERAL MOVEMENT	0·45 m either side
UNIT DIMENSIONS	6·3 × 2·49 × 3·5 m

Schroder High Lift ambulance configured as general-purpose vehicle for LSG

Status
Launched in July 1980, the prototype vehicle is in service with LSG, the Lufthansa catering subsidiary, configured as a general-purpose vehicle with refrigerated van body

Manufacturer
Eisenwerk Schroder, Munich, Federal Republic of Germany

Schroder Hilocab Catering Vehicle

The Hilocab universal catering vehicle is the result of a development by Schroder in conjunction with Lufthansa, its LSG catering subsidiary and Swissair. Available with capacities of 2⅔, 3½ and 4 tons, vans can be constructed in aluminium or plastic material using a sandwich construction. Special tropicalised-versions are offered. The van has a lift height ranging from 1·35 to 5·6 metres which makes the vehicle suitable for all types of civil aircraft. It can be equipped with a traversible and extensible front platform and a tail-gate loader, and can be mounted on any suitable commercial chassis chosen by the operator from manufacturers such as Bedford, Daimler Benz, Fiat, Ford, General Motors, Magirus-Deutz, MAN and Volvo. A special feature of the vehicle is the provision of a low transfer height to cater for DC-9 and other aircraft with low door sills. To service these aircraft the van body is raised to the operating position and after the driver and any passenger have left the cab, the upper section of the cab is lowered by the front platform to the correct position for servicing the aircraft. On completion the cab section is returned to its normal position by the operation of four pneumatic springs.

The scissors lift is a welded frame construction operated by two hydraulic telescopic cylinders with safety valves. The van body is mounted on the upper frame of the lift. The inside walls are equipped with ramming and lashing strips, and two transparent top sections in the van body ceiling are available. Four outrigger stabilisers fitted to the chassis are hydraulically-operated. The front platform is fitted to transfer containers from the van body to the aircraft door. It is guided by rollers in two vertical rails attached to the van body and joins its vertical movement when raised above the driver's cabin. The front of the platform is equipped with a soft rubber fender to protect the aircraft. A hand-pump is installed to enable the hydraulic system to operate in the event of a power failure.

As an optional feature the front platform can be made of a lightweight aluminium/steel construction provided with a traversing front bridge furnished with an extensible top section. Lateral movement is manually-operated and the platform is mechanically locked in the operating position. The extension is hydraulically-operated by a hand-pump mounted on the hand-rail.

Other optional equipment includes rear tail gate loader, aluminium ladder, magnetically-locked swing gate, knotted rope for leaving the elevated front platform in an emergency and telescopic hand-rails with plastic bumpers on the front platform.

Characteristics

CHASSIS: Magirus Deutz 130M 13 FL
(or any suitable commercial chassis)
ELEVATING RANGE
van body: 1·35–5·6 m
front platform: 2·1–5·6 m
LIFT SPEED FAST/SLOW: 0·15 m/s/0·075 m/s
CAPACITY
van body: 3350 kg (can vary with chassis)
front platform:
rigid section: 600 kg
sliding section: 600 kg
extension: 200 kg
tail loading ramp: 600 kg
tail-gate loader: 750 kg
DIMENSIONS
van body internal: 5·8 × 2·35 × 1·95 m
FRONT PLATFORM
standard: 2 × 2·33 m
optional
rigid section: 0·9 × 2·33 m
sliding section: 1·1 × 1·2 m
extension length: 0·5 m
tail-loading ramp: 1·4 × 2·28 m
WHEELBASE: 4·6 m
UNIT DIMENSIONS: 8·2 × 2·5 × 3·6 m

Status

Developed in 1979 the Hilocab with its low-loading capacity is in service with Swissair. The standard version with elevating range from 2·5 to 5·6 metres is used by Lufthansa. Other airlines using Schroder loaders are: Sabena, Air France, Austrian, Pan Am and Syrian Arab Airlines

Schroder Hilocab catering vehicle

Manufacturer

Eisenwerk Schroder, Munich, Federal Republic of Germany

Trepel Catering Vehicles

Three types of catering vehicles are produced by Trepel to service all types of commercial aircraft. The TAE 36/54 747 commissary truck suitable for servicing wide-body aircraft, is based on a Mercedes or Ford chassis with a van body, front bridge and tail-gate loader. The main platform, on which the body is mounted, has a 3600 kg capacity. The front bridge can be extended and traversed hydraulically by controls in the van body. The unit features an integrated design with the chassis incorporated in the lift mechanism, and wide base outriggers for stability. After approaching the aircraft, the outriggers are extended by controls from the driver's cab. The body is then raised to the pick-up point of the front bridge before the operator uses the ladder to reach the van body, from where the approach manoeuvre is completed. Final adjustment to align the van body with the aircraft door sill can be accomplished by moving the front bridge. Adjustable tie-down straps hooked to the side walls, hold the load in place. Optional features available include engine start/stop controls from the body, a 1000 kg capacity tail-gate loader and additional remote controls for the tail-gate loader. Two other models are offered in this range, the TAE 40/54 with a 4000 kg capacity and 12 ton gross vehicle weight, and the TAE 45/54 rated at 4500 kg and 13 ton gross vehicle weight.
The ProfiCom incorporates a design which enables the vehicle to service both BAe 1-11 aircraft with low door sills and Boeing 747 aircraft with high door sills. This is achieved through a swept-over front axle chassis with the cab positioned ahead of the front axle providing the vehicle with a front platform height from 2·08 to 5·6 metres. Standard features include automatic transmission and power-assisted steering. The diesel engine is installed under the van body and the scissors mechanism between the two axles. The van body lift system which can handle a 4-ton payload, comprises a single-lift cylinder, mounted on widely-spaced scissor arms, to produce an integrated assembly.
The Mini Commissary Truck, type TAE 10/35 is designed for servicing narrow-body aircraft with loads up to 1 ton. The feature of the unit is the low driver's cab which allows servicing of the DC-9, BAe 1-11, F28 and Boeing 737 aircraft with low door-sill heights. The unit is loaded via the loading ramp on the rear of the body. Tie-down straps, adjustable and hooked to the bars along the body side-walls, keep the load in place. Stabilisers and the lift mechanism level the van body floor with the building ramp height. The truck approaches the aircraft and stops in the correct position. The operator extends the stabilisers, lifts the body to the height of the front bridge and, using the ladder, reaches the bridge. There the body can be adjusted to its final loading position.

Manufacturer

Trepel AG, Wiesbaden, Federal Republic of Germany

Status

In service with major airlines and catering services. The TAE 10/35 is an improved version of the original Mini unit developed jointly by Trepel and Boyd Aircraft Equipment Co, their United Kingdom distributors. The new model has been supplied to SAS in Frankfurt, ACS, Trusthouse Forte in Paris, Lufthansa Services in Berlin and Nigerian Airways at Lagos

Trepel TAE 10/35 Mini catering vehicle

Characteristics

TYPE	36/54, 40/54, 45/54	ProfiCom	TAE 10/35
CHASSIS	Mercedes Benz LP 1113/42 / Ford D 1210	MAN 100 kW diesel	Ford A 0609/0610 or Mitsubishi Canter 55
ELEVATING RANGE			
van body	1·33–5·4 m	1·275–5·6 m	1·25–4 m
front platform	2·87–5·4 m	2·08–5·6 m	2·11–4 m
CAPACITY			
van body	3600/4000/4500 kg	4000 kg	1000 kg
front platform	500 kg	1000 kg	500 kg
EXTENSIONS			
tail-gate loader	500 kg (10 000 kg option)	400 kg	
lifting/lowering speed		0·15 m/s	
DIMENSIONS			
van body	5·8 × 2·38 × 2·05 m	5·87 × 2·38 × 2·035 m (internal)	4·1 × 2·32 × 2·03 m (internal)
front platform length	1·88 m	1·85 m	1·3 × 1·9 m
extension length	0·5 m	0·6 m	0·5 m
tail-gate loader length	1·2 m		
WHEELBASE	4·2 m	3·9 m	3·28 m
TURNING RADIUS		16·3 m	
UNIT DIMENSIONS	7·9 m (length)	7·86 × 2·5 × 3·57 m	6 × 2·39 × 3·2 m
GROSS VEHICLE WEIGHT	11 000/12 000/ 13 000 kg	13 000 kg	6000 kg

---

Trepel Lift-O-Mobil Vehicle for Handicapped Persons

The Lift-O-Mobil has been specially designed for people who must rely on wheel-chairs or other devices and is applicable to service narrow- and wide-body aircraft by lifting its body up to the aircraft door sill. Capable of handling up to four handicapped passengers in wheel-chairs or two on stretchers and two attendants, the vehicle is equipped for travelling on public roads. Furnished with a first-aid kit and oxygen supply, it can be fitted for emergency cases with stretchers as an optional feature. The unit can be entered through a folding staircase at the rear of the vehicle and an additional wheel-chair lift transports people from ground level to van body. Locking devices prevent the wheel-chairs from moving during travelling and each cabin wall is furnished with foldable stretcher supports. The S-type scissor-lift mechanism with two telescopic cylinders, is controlled from either inside the van body or from the driver's cabin. Transition from vehicle to aircraft is via a revolving shutter-type door on the front of the vehicle and an extensible front platform.

Characteristics
CHASSIS: Mercedes Benz L 809 B or equivalent
ENGINE: Mercedes Benz OM 314
OUTPUT: 85 hp (63 kW)
ELEVATING RANGE
 van body: 1·39–5·6 m
 front platform: 2·5–5·6 m
CAPACITY
 front platform: 400 kg
 van body: 6 persons (700 kg)
 rear lift: 500 kg
WHEELBASE: 3·6 m
UNIT DIMENSIONS: 5·97 × 2·5 × 3·45 m
WEIGHT: 6500 kg

Manufacturer
Trepel AG, Wiesbaden, Federal Republic of Germany

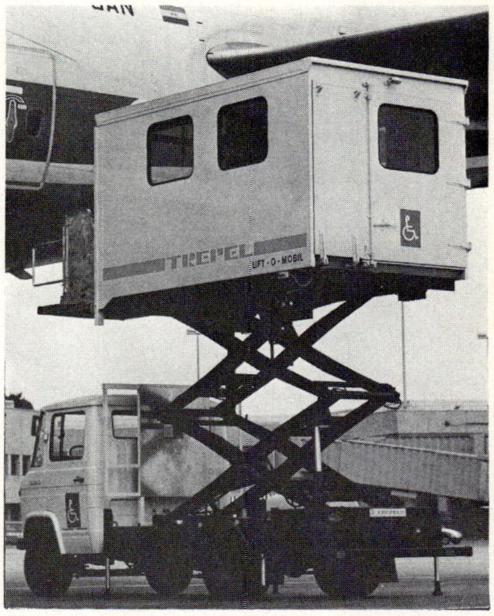

Trepel Lift-O-Mobil vehicle

Trepel Type CCL/55 Aircraft Cabin Cleaning Vehicle

The Type CCL/55 mounted on a scissors-lift mechanism is designed to provide cleaning facilities for narrow- and wide-body aircraft. The lifting mechanism is controlled from inside the body and reaches all aircraft door-sills from 2·5 to 5·5 metres high. The body is equipped with a sink unit, water and waste tank, hot and cold water supply, vacuum cleaners with accessories, cable drums, waste trolleys and racks, trunks and benches to seat people and store equipment. A 220 volt/50-cycle generator supplies power for a water boiler and power sockets at the front platform for vacuum cleaning independent of the aircraft's auxiliary power unit.

In the lowered position the cleaning staff can enter the vehicle on steps under the rear-body door. Four tilting outriggers are fitted to provide stability with the body in the raised position.

Optional features available include: choice of chassis and engine start/stop controls from inside the van body.

Characteristics
CHASSIS: Mercedes LP 808/42, Ford D 0807 or equivalent
GENERATOR: diesel 220 V, 5 kVA
ELEVATING RANGE: 2·5–5·5 m
CAPACITY
 van body: 2000 kg
 front platform: 450 kg extending 0·4 m
TANK CAPACITIES
 fresh water: 110 litres
 waste tank: 110 litres
CABIN CLEANING EQUIPMENT
 2 vacuum cleaners 402/670 W;
 2 cable drums;
 2 trolleys with waste bags
UNIT DIMENSIONS: 7·8 × 2·4 × 3·3 m
WEIGHT: 6000 kg

Status
In service with major airlines

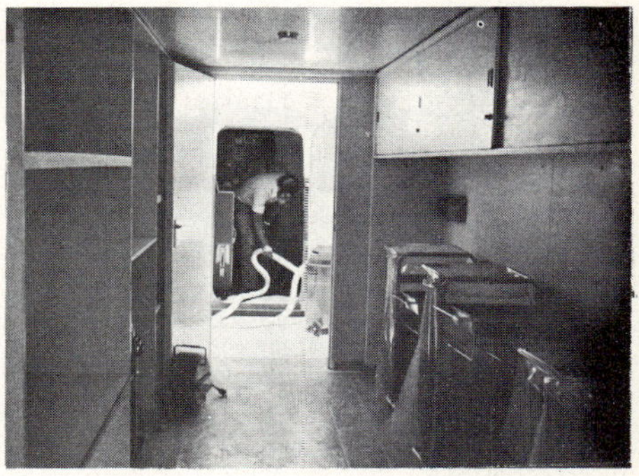

Trepel Type CCL/55 aircraft cleaning vehicle

Manufacturer
Trepel AG, Wiesbaden, Federal Republic of Germany

ISRAEL

Mechola MHS-18 Portable Hydraulic Helicopter Starter

The MHS-18 was designed as an air-transportable emergency starter for Sikorsky helicopters equipped with hydraulic starters: S-62A (HH-52A), S-64 (CH-54A-B), and the S-65 (CH-53A-D, HH-53B-D). Powered by an air-cooled petrol engine the unit features a stainless-steel oil-tank, and is mounted on a two-wheeled trolley with a tow-bar.

Status
In service with the Israeli military forces

Manufacturer
Mechola Engineering, Bikat Hayarden, Israel

Characteristics
ENGINE: Petrol
OUTPUT: 65 hp (48·5 kW)
PUMP
 capacity: 18 US gallons/minute (68 litres/minute)
 pressure: 4000 psi (275 bar)
HYDRAULIC FLUID: MIL-H-5606 or equivalent
FILTRATION: 3 microns
TANK CAPACITY: 15 litres
HOSE LENGTH: 8 m
UNIT DIMENSIONS: 1 × 1·4 × 1·35 m
WEIGHT: 500 kg

Mechola MHS-18 portable hydraulic helicopter starter

Mechola Portable Hydraulic Fluid Replenishment

Mechola produces a portable unit for topping-off hydraulic fluid tanks and replenishing brake system reservoirs. Constructed in stainless steel, it is suitable for flight line operations and features a hand-pump with telescopic lever, double filtration of the liquid and a sight glass.

Characteristics

CAPACITY: 5 litres
FILTRATION: 3 microns
HOSE LENGTH: 2·5 m
DIMENSIONS: 0·15 × 0·15 × 0·25 m
WEIGHT: 10 kg

Status

In service with the Israeli military forces

Manufacturer

Mechola Engineering, Bikat Hayarden, Israel

Mechola portable hydraulic fluid replenishment

ITALY

Aeronautica Macchi Ground Power Unit

Aeronautica Macchi produces two diesel-powered mobile units for ac and dc power supply. Both are mounted on four-wheeled trailers equipped with tow-bars.

The model 26-6D generates a 30 kVA, 400 Hz, 3-phase output at 115/200 volts which can be adjusted up to 120/208 volts. A 28 volt, dc 30 amperes supply is also available.

The 12-6B is designed for aircraft starting supplying 28 volt dc with a peak output of 1500 amperes. The unit incorporates an electronic regulator to allow preset current control. The starting current may be adjusted between 500 and 1500 amperes. It is powered by a 6-cylinder V-type diesel engine which is air cooled and operated using diesel or JP4 fuel.

Characteristics

TYPE	26-6D	12-6B
ENGINE	diesel	6-cylinder diesel
POWER SUPPLY		
ac	400 Hz, 3-phase 115/200 V	
dc	28 V, 30 A	28 V, 1500 A
VOLTAGE ADJUSTMENT	Up to 120/208 V	26–30 V
HARMONICS	0·8% max	
MAX OVERLOAD	90 kVA for 15 s	

Status

In service with the Italian Navy

Manufacturer

Aeronautic Macchi, Varese, Italy

Aeronautica Macchi 12-B dc generator with Agusta Bell 205 helicopter

Aeronautica Macchi 26-6D ac generator with Sikorsky SH 3 D helicopter

Aviogei Catering Vehicle

Two models are produced by Aviogei for catering services. The EC 707, with an elevation range from 2·1 to 4 metres, is suitable for narrow-body aircraft while the EC 747 is capable of servicing aircraft with door-sill heights between 2·5 and 5·6 metres. Both are based on a Fiat chassis.
An optional feature available for the EC 747 is a hydraulically-operated tail-gate loader.

Characteristics

MODEL	E 707	E 747
CHASSIS	Fiat 50 NC	Fiat 110 NC
AIRCRAFT HANDLING RANGE	DC-8, DC-9, B707, Mercure, BAC1-11	DC-8, DC-10, B707, B747, L-1011, A300B, IL62, Concorde
ELEVATING RANGE	2·1–4 m	2·5–5·6 m
CAPACITY	1200 kg	2000/3000/ 5000 kg

Manufacturer
Aviogei, Campoleone-Aprilia, Italy

Aviogei Lavatory Service Vehicles

Three models of vehicles for lavatory servicing are produced by Aviogei, the VST 707, VST 747 and the VST 1000. All models are equipped with two Yabsco electric pumps, Royling connections and tank level indicators and are capable of rapid discharge by remote control.
The VST 707 is based on an Aviogei chassis powered by a Fiat engine with manual transmission. Access to the aircraft servicing point is provided by the tank platform which is 1·4 metres high.
The VST 747 features a Fiat chassis with a diesel engine, manual transmission and dual-rear wheels. The vehicle is equipped with an elevating tank unit which can be raised to 3 metres above the ground.
The VST 1000 is mounted on an Aviogei chassis with a VM 1052 diesel engine and automatic transmission. A hydraulic lift platform measuring 1·5 by 1·5 metres is fitted at the rear of the vehicle capable of extending to a height of 3·5 metres above ground level.

Aviogei potable water and lavatory service vehicles

Characteristics

MODEL	VST 707	VST 747	VST 1000
CHASSIS	Aviogei	Fiat 50 NC	Aviogei
ENGINE	Fiat Aifo petrol	Fiat diesel	VM 1052 diesel
OUTPUT	30 hp (22 kW)	81 hp (60 kW)	40 hp (30 kW)
TANK CAPACITIES			
waste	1000 litres	2000 litres	2000 litres
water	400 litres	600 litres	800 litres
disinfectant	200 litres	400 litres	400 litres
WHEELBASE	2·16 m	2·6 m	3·5 m
TRACK	1·32 m		2 m
UNIT DIMENSIONS	3·72 × 1·6 × 1·59 m	5·08 × 2·23 m (height)	5 × 2·3 × 1·6 m

Status
In service with Aero Trasporti Italiani and Air Malta

Manufacturer
Aviogei, Campoleone-Aprilia, Italy

Aviogei Potable Water Vehicle

Aviogei manufactures two models of potable water vehicles, the VSA 707 for servicing narrow-body aircraft, and the VSA 747 for wide-body aircraft. Both models are equipped with a stainless-steel tank (two tanks in the VSA 747), Yabsco water pump, 1-inch (25·4-mm) diameter supply hose and a water level indicator.
The VSA 707 is based on an Aviogei chassis with Fiat petrol engine and manual transmission. The flat top of the tank unit provides the working platform with access by a stair unit adjacent to the driver's cab.
The VSA 747 is mounted on a Fiat chassis and powered by a diesel engine with manual transmission. Dual wheels are fitted to the rear axle. Foldable steps with hand-rails are installed on top of the tank unit, providing a working height of 3 metres.

Characteristics

MODEL	VSA 707	VSA 747
CHASSIS	Aviogei	Fiat 50 NC
ENGINE	Fiat Aifo petrol	Fiat diesel
OUTPUT	30 hp (22 kW)	81 hp (60 kW)
WATER TANK CAPACITY	1000 litres	2000 litres
UNIT DIMENSIONS	3·72 × 1·6 × 1·4 m	5·08 × 2·23 m (height)

Status
The VSA 747 is in service with Air Malta

Manufacturer
Aviogei, Campoleone-Aprilia, Italy

Cella Aircraft Washing and De-icing Systems

Cella produces a range of aerial work platforms, all of which can be equipped with washing and de-icing systems, comprising inox thermally insulated reservoirs, pump units, heating elements, and spraying nozzles. Hydraulically-operated and fitted with automatic protection devices the platforms are equipped with dual-control panels; one on the aerial cage, the other on the turntable.
The 3D/6 series with two articulated booms are mounted on medium- and light-weight trucks and tractors, providing 10, 12, 13, 15, 17 and 18 metres maximum working height. The 3D/3 with three articulated booms are based on medium and heavy weight trucks, and are available with 18, 20, 23, 25, 27, 29 and 31 metres maximum working height. The 3D/4 equipped with three articulated booms and intermediate telescopic extensions, are designed to be mounted on a heavy chassis and are capable of reaching a maximum height of 30, 40 and 50 metres.
Liftel models are fitted with telescopic boom sections, and can be mounted on standard truck chassis or on a specially-designed self-propelled base. Maximum working heights range from 10 to 50 metres.
The Liftelmatic series are self-propelled units designed for a single operator, with controls for driving, steering, hoisting, lowering and operation of the outriggers situated in the aerial cage. Electric-propelled versions are available.

Status
In service with Alitalia

Manufacturer
Cella SpA, Mandolossa/Brescia, Italy

Cella aircraft washing and de-icing systems

Cella Apron Equipment

Potable water supply equipment, lavatory servicing units, and vehicles for catering and transporting handicapped persons are included in the range of apron equipment offered by Cella.
The potable water and lavatory servicing units are available mounted on standard commercial truck chassis or based on the Cella low profile chassis, equipped with hydrostatic transmission. Both types of vehicle are equipped with work platforms on top of the tanks and an additional hydraulically-operated elevating platform at the rear of the vehicle. Various tank capacities are available.
Catering vehicles produced by Cella are mounted on a scissors lift mechanism based on a standard commercial chassis. They can service wide- and narrow-body aircraft and are capable of extending to 5·5 metres. The hydraulic systems for lifting, operating the four outriggers, and movement of the front platform are powered by a pto from the vehicle's engine.
Optional equipment offered includes tail-gate loader and hydraulically-operated traversing and extending bridge platform to facilitate alignment with the aircraft. The van body can be furnished to cater for the transport of handicapped persons.

Status
The potable water and lavatory servicing vehicles have been supplied for use at airports in Italy. The Cella catering vehicle is in service with Alitalia

Cella potable water supply vehicle

Manufacturer
Cella SpA, Mandolossa/Brescia, Italy

Fresia SP Series Aircraft Towing Tractors

The SP series comprises four low-profile aircraft tractors, SP 700, SP 500, SP 300 and SP 200 and a tractor designed for push-out operations, the SP 300 PO.
The SP 700 with a maximum draw-bar pull of 56 800 kg is suitable for all types of wide-body aircraft. Powered by two supercharged Fiat diesel engines, two semi-automatic Allison transmission systems and diesel torque converters, the tractor features four-wheel power steering, hydro-pneumatic independent suspension and drum brakes on all four wheels. The steering can be selected for two, four-wheel co-ordinated or four-wheel crab-type operations. A single seat driver's cab is fitted at either end, which when in use can be elevated to provide a rear view. A dual passenger compartment is installed at one end and the tractor is provided with a housing for a 90 kVA ground power unit.
The SP 500 is a derivative of the SP 700 with a single Fiat diesel engine, Allison semi-automatic transmission and torque converter. Rated at 40 000 kg draw-bar pull when ballasted up to 50 tons, the unit can handle Boeing 747, McDonnell Douglas DC-10 and Lockheed Tristar aircraft. The standard version has a single seat driver's cab and twin seat passenger compartment at the front end only. This can be elevated in a similar manner to the SP 700. A rear cab and stowage for a ground power unit are offered as optional equipment.
The SP 300 is equipped with a full-width, driver's cab at the front, which can accommodate three people. This can be elevated to provide the driver with good rear visibility. It is driven by a Fiat diesel engine through an Allison automatic transmission and torque converter. Other features include standard tow couplings front and rear, half-elliptic spring suspension and drum brakes on all four wheels operated through two independent circuits. Optional equipment comprises additional ballast and compartment to house a ground power unit.
The SP 200 is designed for handling narrow-body aircraft. Similar to the SP 300, this tractor has a draw-bar pull capacity of 16 000 kg when ballasted to 20 tons or 20 000 kg without the additional ground power unit which is offered as an option.
The SP 300 PO is a derivative of the SP 300 but designed for push-out operations, with a draw-bar pull of 40 000 kg when ballasted up to 50 tons or 33 600 kg without ballast.

Fresia SP 300 towing tractor

Status
All units are in service with Alitalia

Manufacturer
Officine Fresia, Millesimo (Savona), Italy

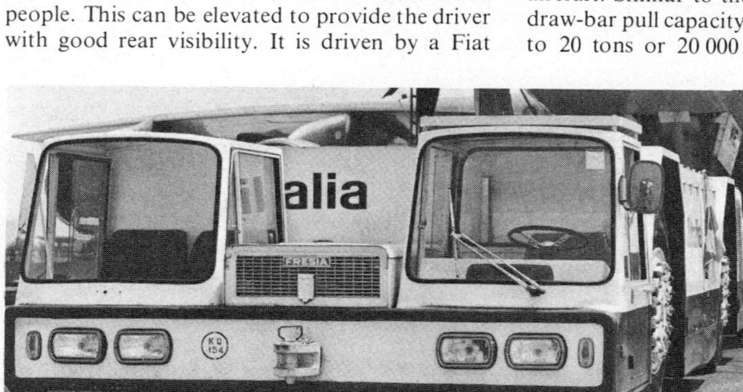

Fresia SP 500 towing tractor

Fresia SP 200 towing tractor

Characteristics

TYPE	SP 700	SP 500	SP 300	SP 200	SP 300 PO
ENGINE	2 × Fiat 828 diesel	Fiat 828 diesel	Fiat 828 diesel	Fiat 8200AM diesel	Fiat 828 diesel
output	380 hp (284 kW) at 2400 rpm	380 hp (284 kW) at 2400 rpm	380 hp (284 kW) at 2400 rpm	218 hp (163 kW) at 2500 rpm	380 hp (284 kW) at 2400 rpm
max driving speed	35 km/h	26 km/h	32 km/h	35 km/h	30 km/h
max draw-bar pull					
with ballast	56 800 kg	40 000 kg	32 000 kg	20 000 kg	40 000 kg
without ballast	40 000 kg	29 000 kg	24 000 kg	16 000 kg	33 600 kg
adhesion factor	0·8 mμ	0·8 mμ	0·8 mμ	0·8 mμ	0·8 mμ
minimum turning radius	7·61 m	7 m	6 m	5·18 m	
wheelbase	4·55 m	3·8 m	2·99 m	2·5 m	4·2 m
UNIT DIMENSIONS					
length	9 m	7·2 m	6·57 m	5·74 m	7·6 m
width	3 m	3 m	2·7 m	2·48 m	2·7 m
height					
cab lowered	1·57 m	1·58 m	1·58 m	1·48 m	1·58 m
cab elevated	1·87 m	1·88 m	2·08 m	1·85 m	2·08 m
WEIGHT					
basic	50 000 kg	36 000 kg	30 000 kg	20 000 kg	42 000 kg
ballasted	70 000 kg	50 000 kg	40 000 kg	25 000 kg (with gpu)	50 000 kg

SPAIN

Equisa CAP Potable Water Supply Vehicle

The CAP model has been designed to supply potable water to all types of commercial aircraft. It consists of a tank with a capacity of 2000 to 2500 litres mounted on a commercial chassis with a rear work platform where the control panel for the hydraulic system is installed and steps are incorporated for access to the tank platform. The stainless steel tank equipped with a baffle to prevent surge and insulation to prevent freezing in extreme temperatures is protected by an aluminium plate casing. Two manholes are in the upper part of the tank with two non-slip working platforms on each side of the tank platform. A service ladder is incorporated in one of these platforms which provides an additional 0·9 metres in height to facilitate coupling the hose to the aircraft servicing point. A sight glass fitted to the left-hand side of the tank gives an indication of the water level.

The control panel comprises a switch for the supply of water and emptying the hose, manometer, water outlet pressure regulating switch, water meter and an emergency manual pump.

The supply of water to the aircraft is carried out by the main hydraulic pump operated from an engine pto. This supply is activated by means of a lever above the parking brake. As a safety measure this operation can only be carried out when the vehicle is stationary.

Equisa CAP potable water vehicle

Status

In service with Iberia

Manufacturer

Equisa, Madrid, Spain

Characteristics

WATER CAPACITY: 2000–2500 litres

ELEVATING PLATFORM
- height: 0·9 m (2·1 m above ground level)
- dimensions: 0·45 × 0·35 m

TANK PLATFORM
- dimensions: 3·75 × 2 m

Equisa FEM Catering Vehicle

The universal catering truck type FEM is designed to service all commercial aircraft from F-28 to Boeing 747. Based on an Edbro chassis the van body is mounted on a scissors-lift mechanism, with a front platform adjacent to the driver's cab which enables the unit to service aircraft with door-sill heights from 1·7 to 5·6 metres. The front platform can be extended to connect with the aircraft. Controls are in the cab and the van body. Four hydraulic jacks are fitted to provide stability with the van body in the raised position and safety locks to prevent the vehicle being moved with the jacks lowered. Alternatively the van body cannot be elevated without the jacks being down. The engine can be stopped from inside the van body. Optional features available include: cabin air-conditioning, side safety-rails on the front platform, removable roller floor and engine starter controls in the body, rear lift platform and a hydraulic alignment and automatic coupling facility for the front platform.

Status

In service with Iberia

Manufacturer

Equisa, Madrid, Spain

Characteristics

CHASSIS: Edbro 110-8

ENGINE: Perkins diesel

ELEVATING RANGES
- van body: 1·5–5·6 m
- front platform: 1·7–5·6 m

CAPACITY
- van body: 3500 kg
- front platform: 1500 kg
- extensions: 500 kg

LIFTING/LOWERING SPEED: 0·15 m/s

VAN BODY DIMENSIONS: 6 × 2·2 × 2·45 m

SWEDEN

Kalmar Aircraft Towing Tractors

Kalmar BM 355 aircraft towing tractor

Kalmar BM 220 aircraft towing tractor

Two models of tractors suitable for aircraft towing duties are produced by Kalmar, the BM-355, BM-220 and in addition the company has developed a helicopter tug in a joint venture with Boyd Airport Equipment Ltd of England.

The BM 355 is built on a Volvo-BM bucket loader chassis complete with diesel engine, powershift gearbox, two drive axles and centre-pivot power steering. Disc brakes are fitted to all wheels. Under the unit, two ballast frames are attached adding 5000 to 11 000 kg to the weight; the welded steel frames filled with the ballasting material are anchorages for the front and rear hooks. The driver's cabin comprises a frame of heavy steel tubing protected by a bumper projecting from the front ballasting frame. It is mounted on twin hydraulic arms capable of lifting the cabin by 2 metres to increase rear visibility when towing from the rear hook.

The BM 220 is built on a Volvo-BM chassis complete with diesel engine, converter, powershift and gearbox and with two drive axles. Around this unit a ballast frame is built weighing 1000 to 9000 kg to which a driver's cabin is attached on rubber mountings. The frame encloses the drive unit and is anchorage for the front and rear hooks. The tractor is equipped with selectable two- and four-wheel drive.

The helicopter tug is a four-wheeled vehicle primarily designed for military helicopters but suitable as a multi-purpose vehicle. The drive wheels are fitted with an anti-spin lock for operation on snow-covered runways and the draw-bar pull of 2400 kg is produced from a Ford petrol/lpg engine and C3 automatic transmission. A small gpu is fitted on the flat bed behind the driving cab, with a 12/24 volt 63 amperes output from the NATO standard socket.

Other features include: insulated batteries and four cab heaters, fitted for Arctic operations, and an option of Volkswagen diesel and Borg-Warner automatic transmission in lieu of the Ford equipment.

Kalmar helicopter tug

Status

These tractors entered production in 1981 and the BM 355 has been supplied to the Swedish Board of Civil Aviation. The BM 220 is in service at four airports in Sweden. The first helicopter tug was delivered to the Royal Swedish Air Force in October 1981

Manufacturer

Kalmar Motor AB, Kalmar, Sweden

Characteristics

TYPE	BM 355	BM 220	Helicopter tug
ENGINE	Volvo BM TD 70H 6-cylinder turbo-charged diesel	Volvo BM D 42 4-cylinder diesel	Ford petrol/lpg
OUTPUT	137 kW at 2200 rpm	62 kW at 2300 rpm	
MAX DRIVING SPEED	36 km/h	23·6 km/h	
MAX DRAW-BAR PULL			
2-wheel drive		6000 kg	2400 kg
4-wheel drive	16 000 kg	10 000 kg	
OUTER TURNING RADIUS			
2-wheel drive		4·6 m	
4-wheel drive	5·6 m	5·25 m	
WHEELBASE	3 m	2·24 m	
UNIT DIMENSIONS	6·7 × 2·7 × 2·3 m	3·7 × 2·5 × 2·4 m	
WEIGHT	16 000–25 000 kg	6000–14 000 kg	3000 kg

Safegate Docking Guidance System

The Safegate docking system is designed to help pilots to position the aircraft correctly when using loading bridges. The system comprises sensor loops, scanning unit, operator's panel, visual display board with central unit, and serial interfaces connecting the units. The sensor coils are laid in slots in the apron concrete along the centre parking line. The slots are 10 mm wide and 20 mm deep. Each loop is 1000 by 300 mm and the distance between each loop is 500 mm. The sensors are protected in the slots by a flexible gasket to avoid the wires breaking due to movement in the concrete and each slot is filled with epoxy concrete. The scanning unit is a micro-processor connected to the sensor loops and to the central unit by a serial channel. Power is received from the central unit. The purpose of the scanner is to examine the sensor loops in the apron for faults and to activate the right number and positions of loops. All sensors are checked when the docking system is not in use and the system cannot be used until the fault has been repaired. When an aircraft type is selected on the operator's panel, the three first loops to be used are activated. As the aircraft proceeds, a new loop is activated in front of the nose-wheel so that three loops are always in use up to the stop position. If an error should appear (eg cable interruption) this is indicated with the loop number on the visual display board. Should a fault appear during a docking procedure the system will allow the docking to continue to the stop position and thereafter indicate the fault and block the system. Up to 48 sensor loops can be connected to the scanning unit.

The operator's panel houses a 24-position switch for selecting aircraft type to dock. One of the positions is reserved for testing all the lamps in the display. An emergency push button is fitted. When activated, red lamps will be lit and the red stop signal will start to flash on the display seen by the cockpit crew. A key switch is installed to lock the system so that only authorised people can operate the system. The display board is constructed of fibreglass. The top part of the display indicates the aircraft type using five alphanumerical displays. Below a fixed red stop sign is placed between two red stop lights. At the bottom of the display are two green lights which indicate to the pilot that the gate is ready. Between the bottom green and top red lamps 14 pairs of yellow lamps are fixed to provide closing rate information to the pilot. The centre-line guidance consists of a 750 mm high illuminated yellow bar which is fitted 2 metres in front of an illuminated aircraft configuration on the display board. The central unit is micro-computer based and inside the visual display board. The unit co-ordinates and controls the communications with the other parts of the system. The information is checked with a parity bit, which prevents incorrect information from being accepted. Transmission distance up to 500 metres is possible. An inter-locking and sequence program for evaluation of sensor loops and operator signals and control of correct lamp and text combination is stored in E PROM memories. This module can be reprogramed if additional aircraft types, new stop positions or other changes are required. The central unit adjusts the intensity of the lamps automatically in the visual display board depending on the surrounding light sensed by a photocell. Up to five different levels can be selected.

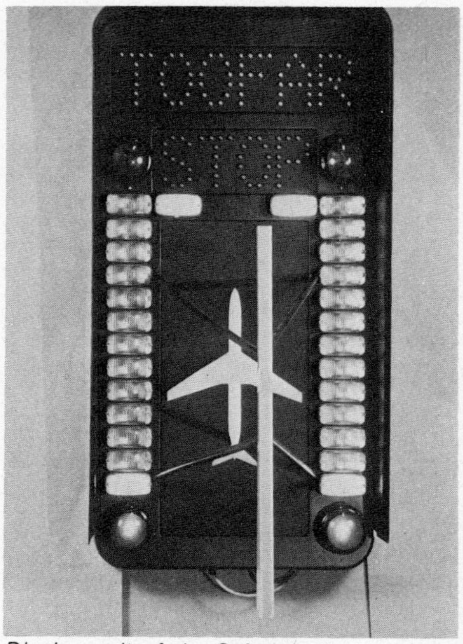

Display unit of the Safegate docking guidance system

When an aircraft approaches the terminal gate the pilot is guided by the azimuth centre beacon on the display board which is fixed to the terminal building and facing the cockpit of the aircraft. The display confirms to the flight deck crew that the correct aircraft type has been selected for the guidance system and that the gate is clear for docking. During the last 5·5 metres of the approach to the stop position, the aircraft's nose-wheel or main landing gear activates the inductive sensors embedded in the apron concrete along the parking centre-line. As the aircraft wheel passes over the sensors, each inductive loop transmits a signal to the central micro-computer unit installed in the display board. These signals activate pairs of moving lights and thus give the pilot closing rate information up to the stop position. When the correct position has been reached two red lamps and a stop signal light up on the display board.

Characteristics

AIRCRAFT TYPES: up to 99 different types
POWER SUPPLY: 240 V ac, 50–60 Hz
OPERATING TEMPERATURE RANGE: −40° to +80°C

Status

The Safegate guidance system, which has been in operation since 1976, complies fully with ICAO recommendations outlined in Annex 14. The system is installed at Amsterdam-Schiphol airport in the Netherlands, Gothenburg-Landvetter, Malmo-Sturup, and Stockholm-Arlanda airports in Sweden and Seattle-Seatac airport, USA

Manufacturer

Steelcons AB, Trelleborg, Sweden

SWITZERLAND

Limess Dynalif Aircraft Tow Tractors

Dynalif low-profile tractors are designed for towing aircraft weighing up to 30 tons. Powered by a Volkswagen diesel engine with hydrostatic transmission and four-wheel drive, the units are equipped with power assisted four-wheel steering. The latter can be selected to produce two-, four-wheel co-ordinated or crab steering. The brakes are operated through the hydrostatic system with an independent parking brake. Tow-bar coupling points are mounted at both ends of the tractor, for push and pull operations. The flat profile of the tractor enables it to be used as a platform for servicing the aircraft. The steering wheel is designed to be lowered below the upper edge of the tractor to reduce the height of the unit if required.

Limess Dynalif aircraft tow tractor

Characteristics

TYPE: Dynalif 15B
ENGINE: Volkswagen diesel 068·3
OUTPUT: 37 kW at 4000 rpm
MAX SPEED: 25 km/h
TOW CAPACITY: 28 000 lb (12 700 kg)
TURNING RADIUS
front-wheel steering: 11·9 m
4-wheel steering: 7·6 m
WHEELBASE: 1·92 m
TRACK: 1·35 m
UNIT DIMENSIONS
length: 2·9 m
width: 1·68 m
height: 1·02 m (0·8 m with steering wheel lowered)
WEIGHT: 1070 kg

Manufacturer

Limess Maschmenfabrik AG, Kloten, Switzerland

UNITED KINGDOM

Armfield Mobile De-icer Units

The Armfield range of aerial access equipment is available for anti-icing and de-icing operations and for exterior valeting of the aircraft. Each unit can be equipped with tanks for water, de-icer fluid, and detergent with mixing pumps and other equipment according to individual operator requirements.

Status

In service with British Airways and Gulf Air for anti and de-icing duties and exterior valeting of the aircraft. A full description of these units can be found in the maintenance equipment section

Characteristics

MODEL	MAX WORKING HEIGHT	MAX OUTREACH	CAPACITY
85 L	8·5 m	4·25 m	175 kg
105 TR	10·5 m	5·85 m	200 kg
110 L	11 m	5·85 m	200 kg
130 L	13 m	6·8 m	200 kg
137 H	13·7 m	7·25 m	360 kg
152 H	15·2 m	8 m	360 kg
168 H	16·8 m	8 m	300 kg
183 H	18·3 m	8 m	250 kg
200 H	20 m	8 m	250 kg
213 H	21·34 m	8 m	225 kg

Manufacturer

Armfield Engineering Ltd, Ringwood, Hampshire, England

Auto Diesels Ground Power Units

The range of ground-power units (gpu) offered by Auto Diesels Braby comprises 700 and 900 series available vehicle-, trailer-mounted or as a skid unit, the 825 Mk 2 and the Tugpac model T 900/72.

In the 700/900 series, the standard trailer-mounted gpu comprises a brushless, rotating-field alternator driven by a Deutz air-cooled, four stroke, direct injection diesel engine. The complete power unit is installed on a steel monocoque chassis. Front and rear axle beams bolted to the chassis carry the independently sprung, pneumatic-tyred wheels, brakes being fitted on the rear wheels. The front wheels are steered through the towing arm which has a standard tractor-ring eye with an overrun braking device.

The 200 volt ac output is taken directly from the alternator, a transistorised automatic voltage regulator (avr) holding the voltage to within ±0·5 volt. The 28 volt dc output is obtained by feeding the 200 volt, 400 Hz, supply to a 3-phase transformer and a silicon-diode bridge mounted on an extruded aluminium alloy heat sink. The output voltage is regulated by the ac avr and will maintain 28 volts ±1 volt up to 1500 amperes falling to better than 24 volts at 2000 amperes. The 112 volts dc output is determined by the regulation of the ac supply and will maintain 118 volts under no-load conditions, falling to better than 100 volts at 1000 amperes. With running gear, wheels, suspension and steering removed the gpu can be installed on its own chassis in any suitable vehicle.

The Model 825 Mk 2 is offered with a single 28 volt dc output, 750 amperes continuous and a peak of 2000 amperes.

A 90 kVA Tugpac gpu is produced for mounting on suitable towing tugs. Designated the model T 9090/72, it comprises an air-cooled Deutz six-cylinder diesel engine and a Type 269 brushless 200/115 volt, 400 Hz alternator complete with controls.

Auto Diesels Tugpac ground-power unit

Characteristics

SERIES	700	900	825	Tugpac
MODELS	700, 740, 745	900, 940, 945, 990, 994	Mk 2	T900/72
POWER SUPPLY	200 V 400 Hz	200 V 400 Hz		200/115 V 400 Hz
RATING	60/75/100 kVA	75/90/140 kVA		90 kVA/180 kVA
POWER FACTOR	0·8	0·8		0·8 / 0·4
DC POWER SUPPLY	28 V dc	28 V dc	28 V dc	28 V dc
peak	2000 A	2000 A	2000 A	
continuous	750 V	750 V	750 V	
	112 V dc	112 V dc		
peak	1000 A	1000 A		
continuous	400 A	400 A		
TRAILER				
WHEELBASE	6 ft 8 in (2·04 m)	6 ft 8 in (2·04 m)		
TRACK	5 ft 5 in (1·64 m)	5 ft 5 in (1·64 m)		
DIMENSIONS				
length	9 ft 11 in (3·03 m)	9 ft 11 in (3·03 m)		6 ft 5 in (1·96 m)
width	5 ft 11 in (1·8 m)	5 ft 11 in (1·8 m)		2 ft 8 in (0·81 m)
height	4 ft 4 in (1·32 m)	4 ft 4 in (1·32 m)		3 ft 6 in (1·06 m)
WEIGHT	5040 lb (2280 kg)	5040 lb (2280 kg)		3014 lb (1370 kg)

Status

Auto Diesels gpus are in service with major airlines. The Tugpac model is designed for installation on Douglas Tugmaster 11/12/14, Reliance Mercury 450 and similar aircraft towing tractors

Manufacturer

Auto Diesels Braby Ltd, Oxbridge, Middlesex, England

Auto Diesels Model 990/140 ground-power unit

Auto Diesels LPA5/LPA6 Air Starter Units

The low-pressure air starters have been designed to meet the requirements of both civil and military aircraft having an air-start system requiring compressed air at 40 psi (2·75 bar) gauge. Two models are offered, the LPA5 providing 120 lb per minute (54 kg per minute) and LPA6 providing 280 lb per minute (127 kg per minute).

The trailer comprises a fabricated steel bedframe and sheet-steel canopy. A rotary air compressor driven by an in-line six-cylinder diesel engine is mounted on the trailer. Suspension is fully independent using flexitor (rubber in torsion) suspension units with wheels incorporating solid rubber or pneumatic tyres. At the front the flexitors are mounted on a turntable steering unit which is steered by a tow-bar incorporating an over-run brake bar and hand-brake. These operate the front-wheel drum brakes. At the rear the flexitors are connected to the bedframe. The filtered air from the compressor is passed through a pressure relief valve and a pressure reducing valve either to the aircraft or to atmosphere depending on the mode of operation. The unit has three modes of operation: idle; jet start and air conditioning. In the first mode the air is dumped directly to atmosphere. In modes two and three the air is supplied to the aircraft at 40 and 30 psi (2·75 and 2 bar) respectively via the pressure reducing valve. This valve is automatically-controlled by a dual mode pilot valve and a single, hand operated, three position valve on the control panel. The compressor is an Atlas Copco oil-free, screw-type unit using twin rotors. It is driven by a Cummins, turbo-charged, diesel engine. The unit, with the removal of the suspension, wheels and steering, may be mounted on any suitable vehicle to the operator's requirements.

Auto Diesels LPA5 air starter unit

Characteristics

MODEL	LPA 5	LPA 6
ENGINE	Cummins NT 855/C280 diesel	Cummins KTA 1150/C600 diesel
OUTPUT	280 hp (209 kW) at 2100 rpm	600 hp (448 kW) at 2100 rpm
COMPRESSOR	Atlas Copco ZA 4B	Atlas Copco ZA 6C
AIR FLOW	120 lb/minute (54·4 kg/minute)	280 lb/minute (127 kg/minute)
PRESSURE	40/30 psi (2·75/2 bar)	40/30 psi (2·75/2 bar)
OUTLET TEMPERATURE	338°F (170°C)	320°F (160°C)
DIMENSIONS		
TRAILER UNIT		
length	12 ft 2 in (3·7 m)	12 ft 2 in (3·7 m)
width	6 ft 7 in (2 m)	6 ft 7 in (2 m)
height	6 ft 11 in (2·1 m)	6 ft 11 in (2·1 m)
VEHICLE UNIT		
length	20 ft 5 in (6·22 m)	23 ft (7 m)
width	7 ft 3 in (2·21 m)	8 ft 3 in (2·5 m)
height	8 ft 1 in (2·45 m)	9 ft 11 in (3 m)

Status
In service with Gatwick Handling at London Gatwick Airport

Manufacturer
Auto Diesels Braby, Uxbridge, Middlesex, England

Douglas Super Taskmaster Mk II Aircraft Towing Tractor

The Super Taskmaster Mk II is powered by a Perkins diesel engine with Clark powershift transmission incorporating a torque converter. Equipped with power steering and air-assisted hydraulic brakes to all wheels, the unit is mounted on an L-section beam-steering front axle and a double reduction drive rear axle with front and rear leaf suspension. A pintle-type single towing attachment is fitted at the front of the tractor with a five position pintle towing point at the rear. Optional equipment available includes: cab complete with screen and no doors; flashing beacon, swivelling floodlights, shunting plates and alternative towing attachments.

Characteristics
ENGINE: Perkins diesel 4·236
OUTPUT: 60 kW at 2800 rpm
DRIVING SPEED: 18 mph (29 km/h)
DRAW-BAR PULL: 10 000 lbf (44·5 kN)
WHEELBASE: 6 ft 9 in (2·06 m)
TURNING RADIUS: 18 ft 3 in (5·56 m)
UNIT DIMENSIONS
length: 11 ft 1 in (3·37 m) without pintles
width: 6 ft (1·83 m)
height with cab: 8 ft 3 in (2·51 m)
without cab: 6 ft 2 in (1·88 m)
WEIGHT: 12 768 lb (5792 kg)

Status
The Super Taskmaster has been supplied to the United Kingdom Ministry of Defence and other military forces

Manufacturer
FL Douglas (Equipment) Ltd, Cheltenham, Gloucestershire, England

Douglas Tugmaster Aircraft Towing Tractor

Seven models of aircraft towing tractors are available in the Tugmaster series with draw-bar pulls ranging from 6·8 to 28 tonnes and all up weights from 11·7 to 70 tonnes. The DC-12, 12–44 and the DC-14 are all capable of handling Boeing 747 aircraft. All the models are diesel-engine powered and incorporate a torque-converter transmission. All are equipped with Orbitrol hydraulic steering, the smaller model having conventional two wheel steering, the larger models featuring four-wheel steering. The DC-11 and the PIII push-out can be fitted with four-wheel steering as optional equipment. Other optional features offered with the DC-11 are a ground power unit which can be mounted at the rear of the tractor increasing the vehicle's length to 21 feet 6 inches (6·55 metres), an integral-powered hydraulic jacking system, a fixed low-profile cab 5 feet 3 inches (1·6 metres) high or an elevating low-profile cab. The latter can be raised to a height of 7 feet 3 inches (2·21 metres). In this configuration, the tractor is suitable for under-belly operation with the Airbus A300/A310 aircraft. The DC-12 is fitted with two-wheel steering and features an enclosed steel cab with seating for driver and two crew. As optional equipment the tractor can be fitted with an additional full width cab at the rear with similar accommodation and limited driving controls. Other options comprise hydraulic elevation of the forward cab by 18 inches (46 cm) to increase visibility; integral hydraulic jacking and a mounting platform at the side for a gpu with fuel and 24 volt dc supply.

Douglas Tugmaster DC-12 tractor with front and rear low-profile cabs

Douglas Tugmaster DC-11 aircraft towing tractor

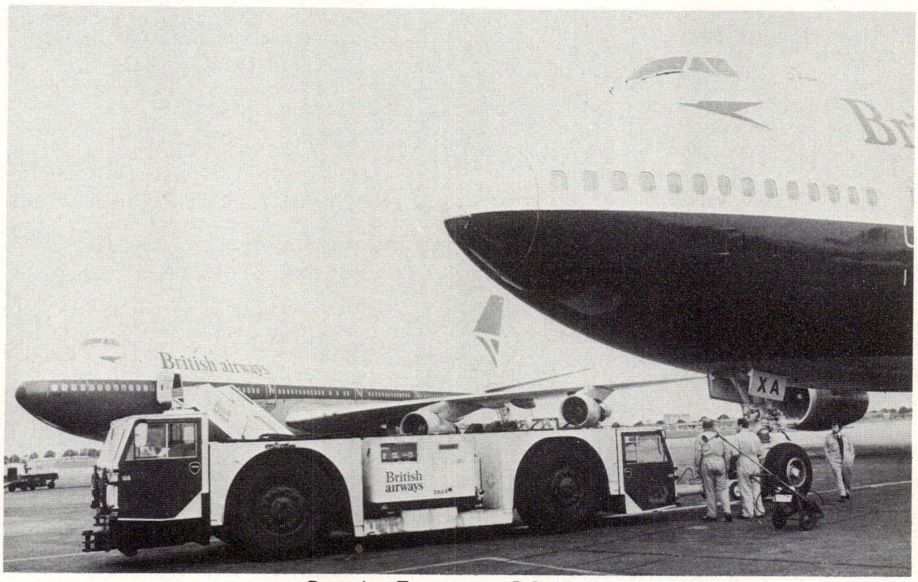

Douglas Tugmaster DC-14 tractor

The DC-12-44 is based on the DC-12 but has an increased all up weight of 120 000 lb (54 431 kg) and four-wheel steering. Low-profile cab, elevation of cab and integral hydraulic jacking options are offered on the DC-12-44. An optional gpu is fitted at the rear of the vehicle and a choice of power plants is available.
The DC-14 incorporates similar features to the model 12-44, but the elevated cab can be raised from a height of 5 feet 5 inches (1·67 metres) to 7 feet 6 inches (2·29 metres).

Status
Douglas Tugmaster range are in service with Air Malta, Alitalia, British Airways, British Caledonian, KLM and Malaysian Airlines System

Manufacturer
FL Douglas (Equipment) Ltd, Cheltenham, Gloucestershire, England

Characteristics

MODEL	DC-6	DC-8	P111 push out	DC-11	DC-12	DC-12-44	DC-14
ENGINE	Perkins diesel 6·3544	Perkins diesel V8·540	Perkins diesel V8·540	Perkins diesel V8·540	690 6-cylinder turbo-charged diesel	Deutz diesel F8L 413F or Cummins diesel V903C	Deutz diesel F12L 413F
OUTPUT	92 kW at 2800 rpm	132 kW at 2600 rpm	125 kW at 2500 rpm	132 kW at 2600 rpm	192 kW at 2200 rpm	188 kW at 2650 rpm or 198 kW at 2600 rpm	277 kW at 2500 rpm
DRIVING SPEED	23 mph (37 km/h)	22 mph (35 km/h)	21 mph (34 km/h)	25 mph (40 km/h)	19/16 mph (30/25 km/h)	20 mph (32 km/h)	20 mph (32 km/h)
DRAW-BAR PULL	15 000 lbf (66·7 kN)	25 000 lbf (111·25 kN)	35 000 lbf (155·75 kN)	35 000 lbf (155·75 kN)	47 200/57 000 lbf (210/254 kN)	60 000 lbf (267 kN)	62 000 lbf (276 kN)
WHEELBASE	6 ft (1·83 m)	8 ft 9 in (2·67 m)	7 ft 6 in (2·29 m)	8 ft 5 in (2·59 m)	12 ft 3 in (3·73 m)	12 ft 6 in (3·8 m)	14 ft 9 in (4·5 m)
TURNING RADIUS							
2-wheel steering	18 ft 6 in (5·64 m)	22 ft 3 in (6·78 m)	26 ft 3 in (7·99 m)	29 ft 5 in (8·96 m)	36 ft (10·97 m)		
4-wheel steering			18 ft (5·49 m)	19 ft 7½ in (5·98 m)		23 ft (7 m)	25 ft (7·62 m)
UNIT DIMENSIONS							
length	12 ft 6 in (3·81 m)	16 ft 11 in (5·16 m)	15 ft 6 in (4·72 m *)	18 ft 4 in (5·59 m)	22 ft 6 in (6·86 m)	24 ft 7 in (7·5 m)	27 ft 4 in (8·33 m)
dual cab					26 ft 7 in (8·1 m)	28 ft 10 in (8·8 m)	31 ft 6 in (9·6 m)
width	7 ft 6 in (2·29 m)	8 ft (2·44 m)	8 ft 6 in (2·59 m)	8 ft 2½ in (2·5 m)	9 ft 7 in (2·92 m)	10 ft (3·05 m)	10 ft (3·05 m)
height	7 ft (2·13 m)	8 ft (2·44 m)	8 ft 6 in (2·59 m)	6 ft 11 in (2·11 m)	5 ft 6 in (1·68 m)	5 ft 3 in (1·6 m)	5 ft 6 in (1·68 m)
WEIGHT	25 760 lb (11 685 kg)	40 320 lb (18 289 kg)	56 000 lb (25 402 kg)	55 000/65 000 lb (24 948/29 484 kg)	100 000/108 000 lb (45 360/48 989 kg)	120 000 lb (54 432 kg)	155 000 lb (70 308 kg)

*over towing pintles

EAI Orbitair Gas Turbine Air-start Unit

The EAI 120 Orbitair gas turbine air-start is designed as a medium light-weight trolley-mounted unit suitable for fixed apron operations. Its dimensions permit it to be transported by air in the belly hold of a narrow-body transport aircraft. It is fitted with roller shutters for weather proofing while not in use and protection during transit. The unit has a Artouste gas turbine power plant producing a continuous flow of compressed air which is bled from the front compressor and is suitable for starting the majority of narrow-body aircraft producing 120 lb of air per minute at 35 psig. The complete unit is mounted on a custom-built trolley with removable cover. The Artouste gas turbine power plant when adapted for use in an airstart role is designated the Artouste Mk 120 or Mk 122, the latter being a modified version of the Mk 120 incorporating an air intake with a steel containment lining, solid state speed sensing switches instead of the centrifugal switch, and an exhaust duct with six thermo-couples for overheat sensing instead of a Firewire sensing element. The basic engine comprises a single-sided centrifugal compressor driven by a two stage axial flow turbine through a one-piece rotor shaft. This rotating assembly is supported by two bearings and is enclosed in flanged casings bolted together to form the outer shell of the engine.
Two accessory gearboxes are mounted on the front of the engine one above and other below the centre line. These gearboxes are driven from the front of the rotating assembly and provide the drive for the accessories.
Excess air compressed by the rotating assembly is by-passed via an air delivery valve, into the aircraft ducting for aircraft starting and ancillary services. When not required, this air is expelled to atmosphere. The gas turbine is started by an electric starter in conjunction with a high energy torch lighter and is automatically-controlled throughout its operating range by a fuel control unit and a mechanically driven centrifugal switch (Mk 120) or speed switches (Mk 122). A dry sump lubrication system incorporates pressure and scavenge pumps.

EAI Orbitair Airstart Unit

Characteristics
POWER PLANT: Artouste Mk 120/122
OUTPUT: 473 shp
AIR FLOW: 120 lb/minute (54·5 kg/minute)
PRESSURE: 35 psi (2·4 bar)
TEMPERATURE: 320°F (160°C)
TIMES
start to governed speed: 25–27 s
stop from governed speed: 90 s
TOWING SPEED: 12 mph (20 km/h)
DIMENSIONS
length tow-bar raised: 8 ft 8 in (2·64 m)
tow-bar lowered: 11 ft 3 in (3·43 m)
width: 4 ft (1·22 m)
height on wheels: 4 ft 8 in (1·42 m)
on skids: 4 ft 3 in (1·3 m)
DRY WEIGHT: 1250 lb (567 kg)
OPERATING WEIGHT: 1470 lb (667 kg)

Status
Production. The Orbitair was launched at the Airport 81 exhibition in Manchester late 1981. A 'Magnum Air' jet start unit for servicing wide-body aircraft is under development

Manufacturer
Eastern Airmotive Industries (USA) Ltd, St. Albans, Hertfordshire, England

Edbro Scissor Lift Aircraft Servicing Equipment

Edbro produces a range of aircraft servicing equipment using the scissor-lift principle. The Hi-Lift system can be fitted to any standard truck chassis of suitable specification. The Edbro Hi-Lift originally designed for commissary purposes can be adapted for various tasks, for example, aircraft washing, cleaning and servicing. The Hi-Lift is a scissor-lift loader, the main feature of the equipment being the hydraulically-powered scissor mechanism which provides a vertical and parallel lift up to the body of the aircraft. Variants have been designed for such specific purposes as mobile control towers and loading baggage beneath the fuselage, for example on Concorde.

The Edbro Jumbo Hi-Lift is designed specifically to serve wide-body aircraft.

The Edbro Universal Hi-Lift, which can load aircraft ranging from DC-9 to Boeing 747, is based on a variety of specially modified standard truck chassis (eg Bedford and Ford) to facilitate spares availability.

Smaller Hi-Lifts, again based on a variety of standard truck chassis, are offered where neither the Edbro Jumbo nor Universal versions are required. When the Hi-Lift is positioned at the aircraft, hydraulic stabilising jacks are lowered. A folding stabiliser can be fitted to the rear of the lifting mechanism to give additional lateral stability. The lifting mechanism is operated hydraulically by telescopic cylinders powered by a combination power take-off and hydraulic pump and controlled via solenoid control valves from push button panels located in both the chassis cab and the van body. The body consists of either aluminium or glass fibre reinforced plywood panels with an aluminium planked floor. A tail lift can be fitted to the rear of the vehicle. At the front end, over the driver's cab, a loading bridge constructed on steel or aluminium frames covered with an aluminium sheet is installed.

The bridge assembly is cantilevered on arms with enclosed roller guides secured to the front pillars of the body. On the front of the main bridge is an extended section which can be moved forward and/or sideways by hydraulic rams operated by an independent power pack controlled from the van. The main lifting rams are held in position to ensure that a sudden loss of pressure will not cause an uncontrolled drop of the body. A standby emergency system is provided to lower the scissor mechanism and to enable the hydraulic jacks to be raised. This allows the vehicle to be moved in the event of failure of the main pump, transmission or engine.

Edbro Hi-Lift for aircraft washing and cleaning

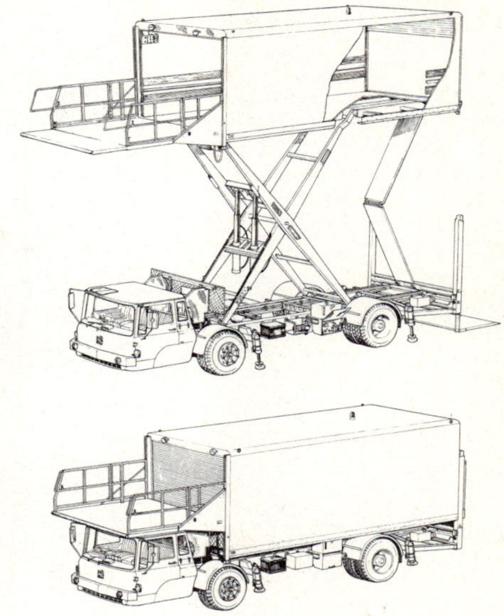

Operation of Edbro Universal Hi-Lift

Edbro Hi-Lift vehicle

Characteristics

MODEL	Standard Hi-Lift	Jumbo Hi-Lift	Universal Hi-Lift
CHASSIS	variable	variable	variable
WHEELBASE	variable	variable	variable
GROSS VEHICLE WEIGHT	9000 kg	14 740 kg	14 740 kg
OUTSIDE BODY LENGTH	4300 mm	6045 mm	6045 mm
HEIGHTS			
ground to floor in lowered position	1300 mm	1450 mm	1450 mm
ground to floor in full raised position	3950 mm	5500 mm	5500 mm
ground to bridge floor in lowered position	2390 mm	2566 mm	2100 mm
PAYLOADS			
fixed bridge	1000 kg	1000 kg	1000 kg
extended bridge	400 kg	400 kg	400 kg
vehicle	3000 kg	4500 kg	4500 kg

Status

The company has been producing High-Lift vehicles since the 1920s. The Universal Hi-Lift was introduced in September 1980. Edbro Hi-Lift vehicles are in service with many airlines and airports including British Airways, British Airports Authority, Pan Am, Sabena, KLM, Manchester International Airport Authority, SAS, Garuda Indonesian Airways, Air Canada and British Caledonian Airways

Manufacturer

Edbro Ltd, Bolton, Lancashire, England

Edghill Lavatory Servicing and Potable Water Units

Edghill produces self-propelled and towable potable water and lavatory servicing units ranging in size from 800 to 4100 litres. Lavatory servicing units are available to both gravity and vacuum specifications. In addition to the lavatory servicing and potable water units Edghill offers a range of equipment for use on apron areas, including crew and service personnel access steps and aircraft tow-bars for narrow- and wide-body aircraft.

Status

The lavatory servicing and potable water units are in service with major airlines

Manufacturer

H.W. Edghill Equipment, Hook, Hampshire, England

EDG411-1 self-propelled lavatory servicing vehicle

Forest Apron Parking Guidance System

The system comprises three units mounted together on a common frame. The top unit is the stand number indicator; the middle unit is the stopping beacon while the bottom unit is an Agnis unit.
The stand number indicator is a fibre-optic illuminated numeral to indicate the aircraft stand. It has an input of 250 volts 50 hertz to a transformer which energises the 12 volt, 50 watt TH lamps. The indicator is illuminated when the mains supply is switched on.
The stopping beacon houses the electronics, stand-by battery pack and the red and green fibre-optic beacons. The rate of flashing can be manually adjusted inside the unit. The signal to stop is made by releasing the two push-buttons. If at any time the deadman push-button is released then the beacon will signal red.
Should there be a lamp or mains failure the complete unit will shut down and the battery pack will illuminate a red failure lamp. A dimmer unit reduces the brilliancy for night time viewing.
The Agnis unit is a fibre-optic illuminated aircraft docking centre-line display. This sign is illuminated by switching on the mains supply, and in the event of lamp failure the complete system will shut down.

Manufacturer

Forest City Signs Ltd, Altrincham, Cheshire, England

Hallam Aircraft Towing Tractors

Four models are produced by Hallam suitable for aircraft towing duties, the Bulldog series being developed in conjunction with County Tractors. The models are the Bulldog 764/1164, which can be supplied as a standard tractor or as a push back unit for narrow-body commercial aircraft, and the HE 50T/HE 50 SP tractors designed for aircraft up to 5500 kg all-up weight. All models are fitted with five position towing points at the front and rear.
The Bulldog tractor is powered by a County four-wheel drive with a Brockhouse 11F torque converter transmission in the 764, and a Brockhouse 33F torque converter with reverse powershift transmission in the 1164.
A choice of engines is offered with a Ford four-cylinder 78 hp diesel in the 764 model and a six-cylinder 120 hp diesel in the 1164. The units are equipped with hydrostatic full power steering through the rear wheels and air brakes. A Boughton model 2H Quickfit hydraulic recovery winch rated 12-ton line pull, and an all-weather cab are available as optional equipment.
The HE 50T/HE 50 SP are driven by a Perkins diesel engine through a torque converter with an instant reverse capability. Dual-disc brakes are fitted on each half shaft. The HE 50 T model features Orbitrol full hydraulic power steering, driver and passenger adjustable seats and space for an additional crew of four behind the driving position. A six-seater cab and an air braking system are offered as options.
The HE 50 SP has power-assisted steering with a manual over ride when the engine is stopped. The driver's position is in the centre with space for equipment or additional crew to the rear of the tractor.

Status

The Hallam/County Bulldog tractors were introduced at the Airport 81 exhibition in Manchester late 1981. Both the HE 50 T and the HE 50 SP models were used at the 1978 Farnborough International Show for towing the display aircraft. The HE 50 T was tested by the Military Vehicles and Engineering Establishment at Christchurch in Dorset in October 1978 where, with a total weight of 11 033 lb (5015 kg), the unit produced a maximum draw-bar pull of 8714 lbf (38·76 kN) on level asphalt

Hallam/County Bulldog tractor

Hallam HE 50 T tractor

Characteristics

TYPE	Bulldog 764	Bulldog 1164	HE 50 T/HE 50 SP
ENGINE	Ford 4-cylinder diesel	6-cylinder diesel	Perkins A 4.212
OUTPUT	78 hp (58 kW) at 2100 rpm	120 hp (89 kW)	60 hp at 2000 rpm (45 kW)
MAX DRAW-BAR PULL	18 000 lb (8100 kg)*		95 000 lb (4318 kg)*
WHEELBASE	5 ft 11½ in (1·82 m)	5 ft 11½ in (1·82 m)	6 ft 9½ in (2·07 m)
TURNING RADIUS			13 ft 3 in (4·04 m)
UNIT DIMENSIONS			
length	14 ft 3 in (4·34 m)	16 ft 1 in (4·9 m)	11 ft 11¾ in (3·65 m)
width	7 ft 3 in (2·26 m)	7 ft 3 in (2·26 m)	6 ft 6 in (1·98 m)
height with cab	8 ft (2·44 m)		8 ft (2·44 m)
height without cab	6 ft 3 in (1·91 m)	6 ft 5½ in (1·97 m)	5 ft 6 in (1·68 m)
WEIGHT (WITHOUT CAB)	18 000 lb (8182 kg)	22 150 lb (10 050 kg)	11 069 lb (5020 kg)

*manufacturer's calculated draw-bar pull

Manufacturer
EC Hallam Engineering (Leicester) Ltd, Smeeton Westerby, Leicester, England

Hants and Sussex Powered Service Trolley

The company offers a powered trolley unit for a variety of aircraft servicing duties. The unit is powered by the Hants and Sussex Aviation Type A250 gas turbine engine. The engine comprises a direct-coupled centrifugal compressor and turbine assembly operating from a servo-controlled fuel system. Management and operation of the engine are automatic following start initiation by means of a single button. In the lower pressure air-start from the engine has a rated standard temperature and pressure air output of 55 kg per minute delivered at 3·8 bar and 220°C. The rated continuous power is 186 kW at 650°C and stp. The fuel can be kerosene or diesel oil. Mobility can be provided by any suitable prime mover vehicle and the unit may be mounted on any suitable trolley or trailer. The unit can be supplied also with various electric-power generators or pumps to special order, to meet a range of requirements.

Status
The unit is in use for more than 26 civil and military aircraft types and variants. Civil aircraft include DC-8, DC-9, Boeing 707, 727 and 737, Trident, BAe 1-11, CL44, VC10, Caravelle 12, Electra and Concorde. Military types include Phantom F5, F104G, Hercules, Buccaneer, Nimrod, Gnat, Vulcan, Victor, VC10, Harrier and Lightning

Manufacturer
Hants and Sussex Aviation Ltd, Portsmouth, Hampshire, England

Harford Engineering Ground Servicing Units

The Harford AS-80 range of de-icer units is suitable for all types of aircraft, including F28, BAe1-11 to B707/VC10 sizes, DC-10, B747 and other wide-body aircraft. These towable units comprise 900 litre capacity lagged main tanks of mild-steel fabrication. These are water jacketed with six heavy-duty immersion heaters, three thermostatically-controlled, three in line, which are arranged two per phase of a three phase supply. The heating system maintains the de-icing solution at the required temperature for correct lance discharge characteristics. A self-contained propane gas bottle provides heating in tow. The main flooded suction pump is diecast aluminium, driven by a four-stroke air-cooled petrol engine rated at 2·8 kW at 3000 rpm. A bronze-cased loading/night circulation pump has temperature compensated clearances and a vane impeller. The 20 watt single phase supply is controlled by a thermostatic switch which cuts out as the de-icer solution reaches 50°C. The unit can unload undiluted compound from ground level containers direct into the main tank.

The AS-80 has a lightweight lance with vented handles and modulating nozzle. A safety valve within the main body permits recirculation to main tank on lance shut-off.

The Harford WS-81 water servicing unit is a towable machine with capacity to permit servicing of aircraft from BAe1-11 to B707/VC10 size.

The power source for the water pump is either an internal combustion engine or 24 volt motor and batteries. Both power sources can be fitted if required. Suitable switching permits batteries to be charged through a charging point; the charger is not included. All pump controls, main tank contents gauges and a stand-by hand pump are at the front end of the unit. A supply hose is fitted to a reel within one of the rear side doors. The drinking water tank is of 1360 litres capacity and is an all-welded stainless steel construction equipped with mainhoses for interior cleaning and drain valves on the underside. Delivery pressures are variable up to 1·7 bar with delivery capacity from 45 to 90 litres per minute. A small stainless chlorinator tank regulates the chlorination of water during the filling procedure.

The unit's detachable superstructure is fitted with light alloy tread plates which enable operators to reach aircraft water connections. Access to all controls is through sliding side door panels. Flame-proof alternatives are available. A patay self-priming hand pump is available as an option. The shipping weight of the engine model is 680 kg.

The Harford WE-81 lavatory servicing unit is mechanically identical to the WS-81 except that an aromatic extraction device replaces the chlorinator unit.

Manufacturer
Harford Engineering Co Ltd, Norwich, England

Houchin Fixed 400 Hz Ground Power Distribution System

The Houchin fixed 400 Hz ground power distribution system has been designed to provide electrical power at airports for operating avionics, lighting, galley equipment etc on aircraft during turnround at multiple service points from a single centralised source of power. The system may be installed as a complete airport facility or limited to a specific airline terminal building. Power is transmitted from a centralised 400 Hz generating plant room through cables to each aircraft gate at a voltage, dependent on application, between 600 and 1000 volts. At each gate the voltage is transformed to 115/200 volts, three phase, 400 Hz to within a voltage tolerance of ±2 per cent.

A range of high capacity, 50/400 Hz frequency converters are available. The vertical motor-alternator sets can be paired with additional machines if required. A special load sensing device is incorporated to allow only the minimum number of frequency converters to be run continuously. During off peak periods machines automatically cut out to save power; conversely where a high number of aircraft require power, additional machines will start up to share the load. Each motor-alternator is provided with an instru-

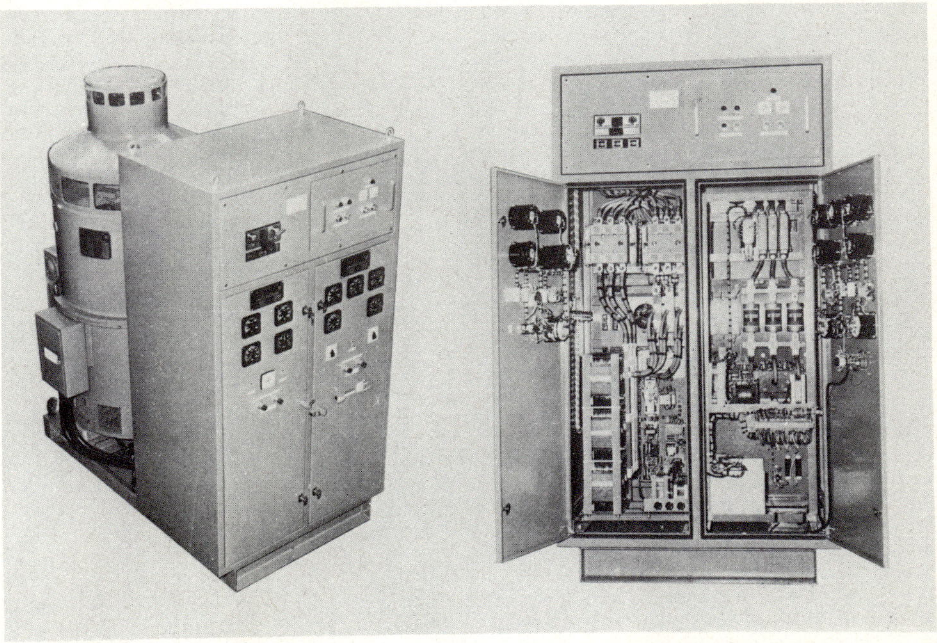

Houchin 50/400 Hz frequency converter for fixed ground power distribution

mented controls cabinet mounted together on a common base. The control system for the generator function consists of solid-state modules, governing power output with protection for over and under voltage, under frequency and overload conditions. The output from each frequency converter is fed into a 400 Hz supply distribution panel which contains circuit breakers and input/output cable connectors. From this panel the power is transmitted to any number of aircraft docking points via high voltage distribution cables; the quantity and size being governed by the kVA required and length of cable to each outlet. Each aircraft gate is provided with a cubicle containing a step-down transformer, switchgear and instrumentation. Supply from the gate unit to the aircraft is effected by a surface cable carrier or by a telescopic cable support system integrated with the pier structure. In the event of a requirement for 28 volts and/or 112 volts dc, the appropriate transformer rectifier units can be incorporated in the gate unit or as mobile trolleys with trailing input cables. Excessive voltage drop along supply cables due to high line impedance at 400 Hz is prevented by a line drop compensator incorporated in the control system. This device enables the voltage at each aircraft to be controlled independent of any loads drawn from other gates.

Status

Installed by Houchin for the British Airport Authority at London Heathrow and Gatwick Airports, England and are also in service at the following airports: Sydney, Australia; Brussels, Belgium; Montreal and Vancouver, Canada; Budapest-Ferihegy Airport, Hungary; Oslo-Fornebur Airport, Norway; Jeddah, Saudia Arabia; Abu Dhabi, United Arab Emirates and Moscow/Sheremetyevo Airport, USSR.
Fixed base 400 Hz installations in hangars for maintenance purposes are covered in the maintenance section.

Manufacturer

Houchin Ltd, Ashford, Kent, England

Houchin Model 615 Aircraft Pneumatic Starter

The Model 615 is a jet engine starting unit which supplies a controlled flow of low pressure air to the aircraft for ground checks, engine running-up, de-icing and air conditioning. The unit featuring a turbo-charged diesel engine coupled to a Howden rotary screw compressor is designed to operate over a range of climatic conditions and has a standard rated output of 120 lb per minute (54·4 kg per minute) continuous at a pressure of 42 psig (2·9 bar). It is equipped with a 30-foot (9·1-metre) supply hose and is available in trailer form or can be mounted on a commercial vehicle chassis selected by the operator.

The trailer comprises a heavy-duty steel chassis fitted with a tow-bar, spring suspension and either pneumatic or solid-tyred wheels. An over-run and parking brake mechanism operates on the rear wheels.

Characteristics

ENGINE: Cummins 6-cylinder diesel
COMPRESSOR: Howden positive displacement screw-type
AIR FLOW: 120 lb/minute (54·4 kg/minute)
PRESSURE: 42 psi (2·9 bar)
TRAILER-MOUNTED DIMENSIONS
length: 14 ft 2 in (4·32 m)
width: 6 ft 9 in (2·06 m)
height: 6 ft 5 in (1·96 m)
WEIGHT: 11 200 lb (5080 kg)

Houchin Model 615 aircraft pneumatic starter

Manufacturer

Houchin Ltd, Ashford, Kent, England

Houchin Mobile Ground Power Units

Houchin design and manufacture a range of mobile ground power units for civil and military applications.
The models 604 and 620 are suitable for supersonic and wide-body aircraft. The 604 is a self-propelled unit with traction and generating power produced by a Cummins eight-cylinder diesel engine. A brushless alternator is coupled to the engine providing an output of 100 kVA continuous and 180 kVA peak power. The steel chassis is fitted with four radial or cross-ply tyred wheels, spring suspension, cam and peg steering, manual gearbox, hydraulic brakes to all four wheels and a parking brake. An optional glass fibre driver's cab can be fitted. The rear drive axle is rated to provide a drawbar pull of 1000 lb (454 kg) suitable for towing other ground support vehicles.
The output cable is 25 feet (7·6 metres) of four core 400 Hz cable complete with standard aircraft socket. Optional 800/2500 amperes 28 volts dc and 300/1200 amperes 112 volts dc units with cables and sockets can be incorporated on an extended chassis.
The 620 is mounted on a four wheeled trailer with spring suspension, towbar and turntable steering. An overrun and parking brake mechanism operates on the front wheels. Other features are similar to the 604 model.
For operation of helicopters, business jets and piston-engined aircraft, Houchin produces a 28 volts dc only gpu, model 641. Mounted on a trailer, the unit is powered by a Ford six-cylinder diesel engine capable of 2000 amperes for starting engines and 800 amperes for electrical servicing.
The model 624 is suitable for mounting on aircraft towing tractors or tugs, and has a rated output of 40 kVA, 115/200 volts, 3 phase, 400 Hz at 0·75 power factor continuous or twice full-load current at 0·4 power factor for five seconds. The unit features a Houchin alternator coupled to an industrial diesel engine, which derives its fuel and electrical starting supply from the tractor fuel tank and battery.
The model 680 is designed for use in a military environment. Mounted on a trailer, the unit is equipped with a Cummins V8 engine and brushless alternator, the latter being suppressed to meet the electromagnetic requirements of military aircraft.

Houchin Model 620 ground power unit

Houchin Model 604 ground power unit

The chassis is constructed in heavy-duty steel and incorporates lugs for a four-point sling lift and channels for a fork lift. Noise is suppressed to 75 dBA at 4 metres and optional features include self-propulsion system for single-handed manoeuvring, internal automatic fire detection and extinguishing system and remote control for ac output. All units incorporate devices to protect against over/under voltage, over/under frequency and over current.

Characteristics

MODEL	604	620	624	641	680
ENGINE	Cummins 8-cylinder diesel	Cummins V555 diesel	Industrial diesel	275 IE 6-cylinder diesel	Cummins V8
output	182 hp (156 kW) at 2400 rpm	166 hp (124 kW) at 2400 rpm	63 hp (47 kW) at 2000 rpm	120 hp (90 kW)	105 hp (78 kW) at 1500 rpm
ac power supply 400 Hz 3-phase	115/200 V	115/200 V	115/200 V		115/200 V
continuous	100 kVA at 0·8 pf lagging	100 kVA at 0·8 pf lagging	40 kVA at 0·75 pf lagging		60 kVA at 0·8 pf lagging
peak	180 kVA at 0·4 pf lagging	180 kVA at 0·4 pf lagging	80 kVA at 0·4 pf lagging		90 kVA at 0·6 pf lagging
power supply 28 V dc	28 V dc	28 V dc		28 V dc	28 V dc
continuous	800 A	800 A		800 A	350 A
peak	2500 A	2500 A		2000 A	500 A
power supply 112 V dc	112 V dc	112 V dc			
continuous	300 A	300 A			
peak	1200 A	1200 A			
UNIT DIMENSIONS					
length	14 ft (4·27 m)	10 ft 8 in (3·25 m)	6 ft (1·83 m)	9 ft 2 in (2·79 m)	15 ft 9 in (4·8 m)
width	6 ft 6 in (1·98 m)	5 ft 8 in (1·73 m)	2 ft 3 in (0·69 m)	3 ft 11 in (1·19 m)	6 ft (1·83 m)
height	4 ft 5 in (1·35 m)	5 ft 4 in (1·63 m)	3 ft 11 in (1·19 m)	5 ft (1·52 m)	4 ft 11 in (1·5 m)
WEIGHT	6700 lb (3039 kg)	6160 lb (2794 kg)	2742 lb (1244 kg)	4256 lb (1931 kg)	7252 lb (3290 kg)

Status
In service with civil and military operators worldwide

Manufacturer
Houchin Ltd, Ashford, Kent, England

Houchin Model 680 ground power unit

Hunting Transportable Power Units

The transportable power units produced by Hunting are the result of a joint development between Hunting Hivolt and Microturbo of Toulouse, France. The self-contained units provide ac and dc current and/or a compressed-air supply for starting aircraft and helicopter main engines. Starting the units, which have a multi-fuel capability, is by battery which is recharged automatically. Connection to the aircraft is via a 7 metre power cable. Installation of a silencer reduces the noise levels to 75 dBA at 10 metres. Units are offered mounted on a trailer or in a lightweight form on skids with two separate packs mounted one above the other for transport by air and capable of being carried by one or two people.

Characteristics

MODELS	M-171	2 generators 9 kW
	M.171-2	2 generators 9 kW
	M.172	1 generator 13·5 kW
	M.173	1 generator 9 kW + 1 alternator 15 kVA
	M.175	1 generator 10·5 kW + compressed air bleed

TYPE	Main Portable Unit	Trailer Unit
FUEL TANK CAPACITY	5·5 gallons (25 litres)	60 gallons (268 litres)
DIMENSIONS		
length	4 ft (1·22 m)	5 ft 2 in (1·58 m)
length including tow-bar	7 ft 10 in (2·39 m)	
width	1 ft 4 in (0·41 m)	4 ft (1·22 m)
OVERALL HEIGHT		4 ft (1·22 m)
POWER PACK HEIGHT	2 ft 4 in (0·71 m)	
SERVICES PACK HEIGHT	1 ft 4 in (0·41 m)	
WEIGHT		1400 lb (622 kg)
POWER PACK	198·4 lb (90 kg)	
SERVICES PACK	150 lb (57 kg)	
OPERATING CONDITIONS		
altitude	sea level to 10 000 ft (0–3000 m)	
temperature	-50°C to +50°C	

Cold weather trials — Hunting gpu — man portable section

Manufacturer
Hunting Hivolt Ltd, Shoreham-by-Sea, Sussex, England

Hydro-Hygiene Ground Servicing Trailers

Hydro-Hygiene Services offer a range of standard and purpose-built apron service truck and trailers for servicing lavatory and drinking water systems, exterior cleaning of aircraft surfaces and anti-icing protection of aero surfaces. The ground equipment is designed for commercial, military and private users. The smaller units are suitable for medium and small aircraft operators and airports.

The anti-icing unit is a trailer-mounted self-contained 400 litre pressure system for spraying an anti-icing barrier compound to give the aircraft protection against snow and ice until after the aircraft has taken off. Pressure is maintained at working level by a compensating diesel-driven compressor with automatic pressure control. The trailer is a four-wheel chassis with turntable steering. Fitted with collapsible handrails, it provides a working platform for two persons 1·3 metres above ground level.

The chemical spray trailer is designed for cleaning exterior surfaces of the aircraft and is based on a pressure system for spraying a liquid jet, mist or foam of an approved alkaline water-soluble cleaner. Depending on the type of chemical used, the unit can be used also as a general cleaner or as a cleaner for stubborn carbonated areas. This cleaning equipment is in current use on aircraft up to B 747, Concorde and by air defence forces.

The lavatory servicing unit is designed for medium and small airline and airport operators. It can also be used as a back-up unit by international airlines during emergency and peak periods. The trailer-mounted tank has two compartments for 730 litres sewage and 270 litres of disinfected flush water. The sewage is loaded and unloaded by gravity and the flush water is pumped into the aircraft by a battery-driven pump with flow-meter control.

Manufacturer
Hydro-Hygiene Services Ltd, Maidenhead, Berkshire, England

IAL Rapidex PRC1 Portable Radar Cone

The IAL Rapidex PRC1 portable radar cone has been developed to meet the aviation industry's requirement for a means of providing protection of parked aircraft against the approach of unauthorised personnel. The PRC1 is a portable battery-powered Doppler radar sensor designed to detect intruders within a prescribed radius of the unit. The sensor operates by transmitting a microwave signal uniformly in all directions. Any moving object within a 20 metre radius causes a signal to be reflected back with a Doppler shift in frequency. This shift is detected by the sensor, processed and used to generate an alarm. This alarm may be generated in a number of ways depending upon the user's requirements. The two most common options are: opening or closing relay contacts to control floodlights, switch on sirens or provide remote indication by landline or alternatively using a radio telemetry link to provide an alarm indication to a remote point which must be provided with a monitor unit. This monitor can handle up to 12 sensors on the same radio link frequency. As varying numbers of sensors may be used to protect different aircraft types, each sensor is assigned an identity code which is displayed on the monitor unit whenever that sensor is activated.

Manufacturer
IAL, Southall, Middlesex, England

IAL Rapidex PRC1 portable radar cone

ISO Service Units

ISO units provide a means of supply of electricity, hydraulic power, compressed air and water to aircraft on the apron or in the hangar. Provision can be made for a telephone connection and for earthing facilities. The units manufactured from steel with nylon bearings are in modular form so a single installation can be used to provide a number of different combinations of services. Access to the unit is achieved by lifting the cable aperture. The main cover is assisted in opening, and controlled when closing, by a gas-spring strut. A foot-operated pressure pad is provided to maintain the main cover in the open position leaving both hands free for other operations. Removing the foot from the pressure pad allows the main cover to close, with the cable aperture cover remaining in a partly open position resting on the withdrawn cable.

Characteristics
UNIT DIMENSIONS
length: 1·27 m
width: 0·33 m
depth: 0·19 m
WEIGHT: 179 kg
CAPACITY: up to 20 m of cable
WHEEL LOADING: 13 kg/cm^2 over an area of 2030 cm^2

Manufacturer
Industrial Service Outlets Ltd, Horsham, Sussex, England

ISO service unit

Lansing TOER 8·1 Electric Tractor

The TOER 8·1 is an electric-powered tractor rated at a draw-bar pull of 11 000 newtons which enables it to be used for towing helicopters, and small commuter/business jet aircraft, such as the Islander and the Learjet. A four-wheeled unit with a 1·53 metre wheelbase the tractor has a 1·5 metre turning radius. It features rubber spring suspension, electronic pulse control and an all-weather cab, and is available with a special shaped battery case which permits visibility of the towing pin from the driver's seat.

Characteristics

MODEL	TOER 8·1/ 11 000	8·1/8000
MOTOR	Electric	Electric
OUTPUT	8·9 kW	8·9 kW
BATTERY	72 V	72 V
CAPACITY	329 Ah	329 Ah
DRIVING SPEED	20 km/h	23·5 km/h
DRAW-BAR PULL		
MAX (5 h rating)	11 000 N	8000 N
CONTINUOUS	2500 N	1700 N
WHEELBASE	1·53 m	1·53 m
TURNING RADIUS	2·4 m	2·4 m
UNIT DIMENSIONS	2·5 × 1·27 × 1·45 m	2·5 × 1·27 × 1·45 m
WEIGHT	2040 kg	2040 kg

Lansing TOER 8·1 towing a Royal Navy Sea King helicopter

Status
In service with major airlines, the TOER 8·1 and the smaller TOER 4·2 are used generally for towing baggage and cargo trailers. A detailed report of both types can be found in the Cargo Handling section

Manufacturer
Lansing Ltd, Basingstoke, Hampshire, England

Laser L 544 Hydraulic Pump

Designed and built by Laser Engineering to effect a pressure delivery of Phosphate Ester hydraulic fluids from standard 200-litre drums direct to the aircraft reservoirs. The unit is particularly suitable for use with the McDonnell-Douglas DC-10 and similar aircraft where the reservoir is constantly at a pressure of at least 4 bar. The complete unit is mounted on a standard drum trolley for operation on the apron, or in the hangar. A dip tube with a coarse particle strainer fitted is inserted through the drum filler neck using an airtight threaded plug, which is fitted with a 15 μm nominal air-breather to prevent entry of dust into the drum. A flexible hose connects the dip tube to the pump unit. The pump is a twin rotor gear-type with integral pressure relief valve. All fluid passes through a 2 μm nominal filter before reaching the outlet coupling to which the delivery hose and a 7 bar pressure gauge is fitted. A by-pass valve is installed so that after each filling operation the delivery hose pressure can be vented to the tank.

Status
In service with British Caledonian Airways

Manufacturer
Laser Engineering (Development) Ltd, London, England

Laser Engineering L 544 Skydrol pump

Leyland Cub Crew Bus

The Leyland Cub based on a small psv chassis is suitable for crew bus applications. The vehicle is fitted with a forward-mounted Leyland 6.98 NV six-cylinder diesel engine and a Turner five speed synchromesh gearbox with an overdrive top gear. Designed for operation at 8·5 tons gross weight, the vehicle has air-hydraulic split circuit brakes, leaf spring suspension and power steering. The bus, with internal trim furnished to operators' requirements, is available with 3·4, 3·9 and 4·4 metre wheelbases.
Other optional features include automatic transmission and left- or right-hand drive.

Status
In service with British Airways

Manufacturer
Leyland Vehicles Ltd, Passenger Vehicle Division, Preston, Lancashire, England

(JAEQ AE64A (1 col.))

Leyland Cub crew bus

Massey Ferguson MF40/MF50 Aircraft Towing Tractors

The MF40 and MF50 tractors are designed for airport applications and can be used for towing aircraft or ground equipment. The two models give power outputs of 47 bhp (33 kW) and 63 bhp (47 kW) and a larger model of 76 hp (57 kW) is available to order. Power is provided by three and four-cylinder Perkins diesel engines driving through a torque converter transmission system. The units feature a steel driver's cab with seating for ground crew members, power-assisted steering, disc brakes, and five level towing hitches mounted at the front and rear of the vehicle. Optional equipment available includes a three-line air brake system and for operation in hot climates the units are offered without a cab and with optional sun protection.

Status
Production. The MF50 was developed for the British Ministry of Defence and is in service with the Royal Air Force

Massey Ferguson MF40 aircraft tug

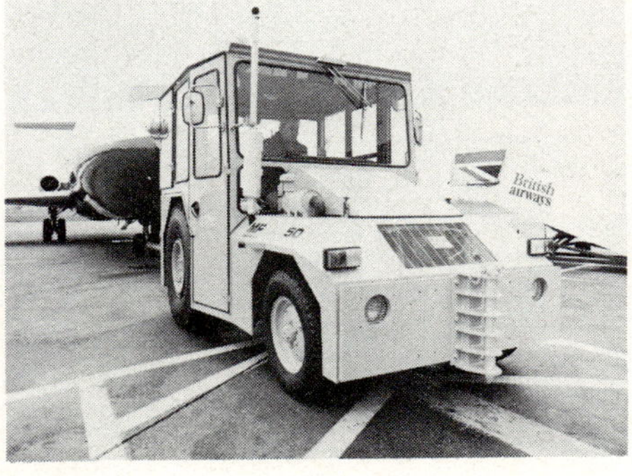

Massey Ferguson MF50 aircraft tug

Characteristics

TYPE	MF40	MF50
ENGINE	Perkins AD3.152	Perkins A 4.236
OUTPUT	33·5 kW at 2250 rpm	47 kW at 2000 rpm
MAX DRIVING SPEED	17 mph (28 km/h)	17 mph (28 km/h)
MAX DRAW-BAR PULL	6500 lb (2950 kg)	9000 lb (4080 kg)
OUTSIDE TURNING RADIUS	12 ft 6 in (3·81 m)	14 ft (4·27 m)
WHEELBASE	6 ft 3 in (1·91 m)	6 ft 10 in (2·08 m)
TRACK		
front	4 ft 4 in (1·32 m)	4 ft 6 in (1·37 m)
rear	5 ft 4 in (1·63 m)	5 ft 4¾ in (1·64 m)
UNIT DIMENSIONS		
length	12 ft 4 in (3·76 m)	12 ft 11 in (3·94 m)
width	8 ft (2·44 m)	8 ft (2·44 m)
height	8 ft (2·44 m)	8 ft 4 in (2·54 m)
WEIGHT	7900 lb (3583 kg)	10 800 lb (4900 kg)

Manufacturer

Massey Ferguson (UK) Ltd, Coventry, England

Negretti Cryogenic Liquid Replenishment Trolley

Two models are available from Negretti and Zambra designed for replenishment of cryogenic liquid to aircraft on the line. Both have similar features; with a maximum height of one metre to enable unrestricted access to the aircraft. The differences are in the overall capacities of liquid oxygen. Equipped with single-acting valves the dispenser features an automatic pressure controller which maintains the selected charging pressure throughout the dispensing cycle. The main wheels are mounted on an independent rubber-torsion suspension system which enables the dispenser to be towed up to 50 km per hour on secondary roads and 8 km per hour on rough terrain. Dynamic braking during towing is applied progressively by an over-run mechanism mounted within the towing eye coupling. The dispenser is stressed for air transportation by fixed wing aircraft or as an externally-slung load by helicopter.

Negretti 200 litre cryogenic liquid replenishment trolley

Characteristics

CAPACITY	100 litres	200 litres
EVAPORATION RATE	Better than 0·25%/h	Better than 3·75% in 24 h
DELIVERY RATES		
at 2 bar pressure	3·5 litres/minute	3·5 litres/minute
at 3·5 bar pressure	5 litres/minute	5 litres/minute
at 6 bar pressure	9 litres/minute	9 litres/minute
SAFETY RELIEF VALVE	7 bar	7 bar
blow-out discs	9 bar	9 bar
TURNING RADIUS	1·5 m	2 m
TRACK	1·15 m	1·27 m
DIMENSIONS	2·65 × 1·15 × 1 m	3·34 × 1·47 × 1 m
UNLADEN WEIGHT	333 kg	464 kg

Status

The 100 litre model was designed to meet the requirements of the British Ministry of Defence

Manufacturer

Negretti and Zambra (Aviation) Ltd, Croydon, Surrey, England

Norcolift Catering Vehicles

Norcolift small series catering vehicle

Norcolift large series catering vehicle

In addition to the Universal Hi-Loader, Norwich Coachworks produces two types of catering vehicles mounted on a standard commercial chassis. The large series extends to 18 feet (5·47 metres) based on a 5¼-ton (5333 kg) chassis and the small series is mounted on a 2-ton (2032 kg) chassis.

The main body underframe is constructed with outrigger and side rave members fixed to the scissor mechanism main members to form a rigid assembly. The side and roof assemblies are constructed with extruded aluminium section riveted at the intersections forming a framework panelled on the exterior with aluminium alloy sheeting. Both front and rear assemblies are fabricated with steel box section pillars and cross members welded to form a frame to which the side and roof assemblies are fixed. The lower frame cross members on both the rear and front assemblies are fixed to the Hi-Lift mechanism to form additional cross members. The front and rear apertures are fitted with upward opening doors. The Hi-Lift mechanism is mounted on the vehicle chassis and comprises a scissor-action lift controlled by hydraulic rams and the necessary tanks and controls to enable both electrical and manual emergency operation. Hydraulic outriggers permit the weight of the chassis cab and body to be supported, providing stability. The platform floor assembly is constructed of aluminium channel bearers and aluminium alloy tread plate, supported by steel posts, incorporating large flitch plates. The loading platform is mounted in steel guides each side of the front frame assembly and movement is by means of steel spindles and bearings. Safety rails are fitted, the front section of which is hinged to provide variable access from the front of the loading platform. Controls are installed in both the cab and the body interior which enable the body to be lifted and lowered from either position.

The platform operating heights are subject to some variation depending on the chassis make on which the vehicle is mounted. Chassis cab modifications are offered which enables a half width platform to be fitted achieving operating heights from 5 feet 10 inches (1·78 metres). Adjustable angle and forward moving platforms are offered as alternatives to the standard fixed bridge.

Characteristics

TYPE	Large series	Small series
OPERATING HEIGHTS		
minimum	8 ft 4 in (2·54 m)	8 ft (2·44 m)
max	18 ft (5·49 m)	13 ft (3·96 m)
CAPACITY		
loading platform	1550 lb (703 kg)	1550 lb (703 kg)
vehicles	5¼ tons (5333 kg)	2 tons (2032 kg)
LOADING PLATFORM DIMENSIONS		
length	6 ft 6 in (1·98 m)	6 ft 6 in (1·98 m)
width	7 ft 4 in (2·24 m)	7 ft 1 in (2·16 m)
INTERIOR BODY DIMENSIONS		
length	24 ft (7·32 m)	13 ft 8 in (4·17 m)
width	7 ft 8 in (2·34 m)	7 ft 5 in (2·26 m)
height	7 ft 6 in (2·29 m)	6 ft 10 in (2·08 m)

Status

In service with major catering service companies and airlines, the vehicles can be furnished for use as aircraft cleaning units or for the transport of handicapped persons

Manufacturer

Norwich Coachworks Ltd, Norwich, Norfolk, England

Norcolift Universal Hi-Loader

The Norcolift Universal vehicle has been developed to meet the needs of the aircraft catering services permitting the servicing of narrow and wide-body aircraft with one type of vehicle. Additionally, the unit can be furnished for use as an aircraft cleaning vehicle or for the transport of handicapped persons.

The chassis, developed in association with Shelvoke and Drewry, incorporates a low-profile cab. This low cab allows the height of the front loading platform to be much lower than on conventional vehicles ranging from 6 feet 5½ inches (1·87 metres) with a thin fixed steel bridge to 18 feet 3 inches (5·56 metres) fully raised. A composite plywood/grp panelled body with an aluminium planked floor is fixed to a frame of steel channel. This frame and body can be raised and lowered by a scissor mechanism fixed at the front of the chassis and rolling on upper and lower tracks at the rear of the vehicle. Angled ground rams fitted to outrigger arms extending from the chassis are used to stabilise the vehicle on the ground while the body is being raised or lowered. A rear stabiliser which automatically extends upwards as the body is raised, is fitted between the upper frame and the vehicle chassis, holding the rear of the body in position under adverse wind conditions. The thin steel bridge, supported just above the cab roof acts as the loading platform for the aircraft. This bridge fitted with rollers running in a track attached to the body, is picked up by the body when both floors are level, and the bridge and body are raised together to the required loading height. Rear end loading can be achieved either by using the body raising facility to elevate it to a loading bay height, or a hydraulically-operated tail lift can be fitted. Controls fitted in the cab in the body and valve compartment at the side of the vehicle are used to operate the functions of the lifting mechanism.

The unit is equipped with Load Lok rail and straps on each side of the body to secure the trolley loads and high-level shelves which can be lowered are fitted on each side of the body to store lighter loads. Lockable up and over shutters are fitted at both ends. Body internal lighting, bridge floodlighting, roof marker lights and a flashing beacon are installed as standard equipment.

Optional equipment available includes extending or traversing/extending bridge, hydraulically-operated tail lift, engine start control in the van body and additional indicating/warning lights.

Norcolift Universal Hi-Loader

Characteristics

CHASSIS: Norwich Coachworks/Shelvoke & Drewry

ENGINE: Ford 2714 E

ELEVATING HEIGHTS

van body: 4 ft 6 in (1·37 m)–18 ft 3 in (5·56 m)

front platform: 6 ft 5½ in (1·97 m)–18 ft 3 in (5·56 m)

traversing/extending bridge: 6 ft 8 in (2·03 m)–18 ft 3 in (5·56 m)

DIMENSIONS

interior van body

length: 20 ft 8 in (6·3 m)–23 ft 10 in (7·26 m)

width: 7 ft 8 in (2·34 m)

height: 7 ft 6 in (2·29 m)

CAPACITY

standard

4500 kg (no tail lift)

4000 kg (with traversing/extending bridge)

3750 kg (with tail lift and traversing/extending bridge)

Status

Introduced in 1980. The Norcolift Universal Hi-loader has been supplied to SAS Catering, Gatwick Flight Catering and Trusthouse Forte

Manufacturer

Norwich Coachworks Ltd, Norwich, England

Photoscan Cctv Apron Surveillance System

The Photoscan closed-circuit television (cctv) system can be used to monitor aircraft apron parking, refuelling and baggage handling areas. The system comprises three low-light level cctv cameras incorporating auto iris and pan, tilt and zoom facilities. Each camera is fitted in a weatherproof housing with windscreen wiper and thermostatically-controlled screen heater. Cameras are remotely controlled and monitored from the apron control tower, and the baggage control office.

The system is designed to serve two main purposes. It allows the apron controller to monitor the correct parking and positioning of aircraft in the apron area and it permits the baggage controller to synchronise the despatch of handlers and vehicles for unloading. Similarly baggage control can observe when refuelling has been completed and the fuel tanker has cleared the aircraft so that outgoing baggage can be loaded without delay.

Status

Installed at Aberdeen Dyce Airport, Scotland in 1980

Manufacturer

Photoscan Ltd, Sunbury on Thames, Middlesex, England

Photoscan cctv apron surveillance system at Aberdeen Dyce Airport

Piller Fixed 400 Hz Ground Power Supply System

Piller designs, manufactures and installs fixed 400 Hz ground power distribution systems. Each system consists of a central power room, from which cables are fed to major service areas, terminals or parkways. Within the power room is at least one rotary frequency converter, producing 400 Hz power from a standard 3-phase mains input. For distribution, higher voltages (typically 900 volts) are used to compensate for cable losses. The 400 Hz cables are run in an underground formation to the various terminal points, where a transformer steps down the voltage to that required by aircraft systems. Special cables are provided here for connecting the output of the small transformers into each aircraft via the standard ground supply socket. The standard output voltage is 115/200 volts, 3 phase. The Piller system is passive, and does not incorporate line-drop compensators or boosters, but operates with low impedance cables and transformers.

The proposed Manchester International Airport Authority's system will consist of 12 aircraft stand positions supplying power to passenger or freight aircraft and ranging in size from the Boeing 737 to the Boeing 747. The equipment consists of three 312 kVA horizontal brushless frequency/voltage converters, complete with control cubicles. These contain a load-dependent paralleling system which is designed to conserve power. The main control cubicle incorporates a master switch which selects the master frequency converter this being the first machine to start when 400 Hz power is required at the aircraft. The operator connects the aircraft to the system via the cable supply. As more aircraft arrive at stand positions and the power demand increases, then the second frequency converter will automatically start. The electronic control circuits synchronise the two outputs, and the load will be shared equally between frequency converters. If the power demand increases to its maximum the third frequency converter will start in the same way. As the demand decreases the sensing circuits will then shutdown the frequency converters in a similar fashion until one remains on line. This will continue until a state of no-load has been timed, for instance a quarter of an hour, or any pre-selected setting. The voltage is distributed from the frequency converter at 950 volts ac and reduced to the required 200/115 volts at each stand position.

Piller three 250 kVA horizontal motor alternator at Frankfurt Airport

Characteristics

INPUT VOLTAGE: all standard voltages available
FREQUENCY: 50/60 Hz
POWER SUPPLY: 115/200 V 400 Hz 3-phase at aircraft plug
RATINGS: 63, 125, 200, 315 kVA
OPERATING TEMPERATURE RANGE: −10 to +40°C (optional +50°C)
NOISE LEVEL: 73 dBA at 1 m
DISTRIBUTION RANGE: up to 1000 m

Status

A Piller centralised power system using three 250 kVA horizontal single shift motor alternators is installed at Frankfurt Airport, Federal Republic of Germany, for Lufthansa. A similar system was ordered by Manchester International Airport Authority, England in autumn 1981.

Manufacturer

Anton Piller (UK) Ltd, Cirencester, Gloucestershire, England

Power Lifts Ambulift

This unit provides wheel-chair-bound passengers with access to all aircraft with sill heights ranging from 4 feet 3 inches to 18 feet 3 inches (1·3 to 5·56 metres). The ambulance body, manufactured by Pilcher Greene, is capable of taking four wheelchairs and two attendants or two stretcher cases with two attendants (or a combination of the two). Access from ground level to the ambulance body is via a hydraulically-operated tail lift which folds into the ambulance body when not in use.

Elevation is provided by a hydraulically-activated, double scissor mechanism, running off the vehicle's pto. Access from the ambulance to the aircraft is by means of an extending bridge. The custom built chassis is based upon the Power Lifts' baggage conveyor car chassis driven by a Perkins 4/108 diesel, four-cylinder, water-cooled engine with a Borg Warner 65 automatic transmission. The chassis incorporates four hydraulic stabilising jacks.

Power Lifts Ambulift

Power Lifts Ambulift

Characteristics

CHASSIS: Power Lifts
ENGINE: Perkins 4/108 diesel
ELEVATING RANGE: 4 ft 3 in (1·3 m)–18 ft 3 in (5·56 m)
VAN BODY INSIDE DIMENSIONS
length: 11 ft 2 in (3·4 m)
width: 6 ft 3 in (1·9 m)
height: 5 ft 11½ in (1·8 m)
UNIT DIMENSIONS
length: 16 ft 6 in (5 m)
width: 6 ft 6 in (2 m)
height: 10 ft 9 in (3·28 m)

Status

The Ambulift was launched at the Airport 81 exhibition in Manchester in October 1981

Manufacturer

Power Lifts Ltd, Watford, Hertfordshire, England

Racal Checkpoint CR-20 Aircraft Security System

Checkpoint is a portable battery-powered Doppler radar sensor designed to detect intruders within a prescribed radius of the sensor. It is primarily intended for the safeguarding of aircraft parked unattended on tarmac or concrete surfaces. The sensor operates by transmitting a microwave signal uniformly in all directions. Any moving object near to the sensor will cause a signal to be reflected back to the sensor with a Doppler shift in frequency. This Doppler shift is detected, processed and used to generate an alarm. The alarm may be implemented in a number of ways depending on the user's requirement. The two most common options are: opening or closing relay contacts to control floodlights, switch on sirens or provide remote indication by landline; or alternatively a radio telemetry link to provide an alarm indication at a remote point. The remote point must be provided with a monitor unit which can handle several sensors on the same radio link frequency. Each sensor is assigned an identity code which is displayed on the monitor unit whenever that sensor is activated.

Sensors are provided with removable batteries to permit operation to be maintained on a continuous basis. An automatic charger is provided as standard equipment so that batteries can be recharged at the correct rate and to the correct level.

Racal Checkpoint CR-20 aircraft security system

Characteristics

RADAR FREQUENCY: 1·4–1·7 G Hz
BANDWIDTH: 12 M Hz
POWER OUTPUT: 10 mW
POWER SOURCE: 16 cell Ni Cd battery capacity: 7 Ah
PERSONNEL DETECTION CAPABILITY: 20 m
DIMENSIONS: 1·07 m high, 3·25 wide at base
WEIGHT: 15 kg
OPERATING TEMPERATURE RANGE: −20° to +50°C

Status

In service with British Airways

Manufacturer

Racal-MESL Radar Ltd, Linlithgow, Lothian Region, Scotland

RM Aircraft Towing Tractors

Four models of aircraft towing tractors are produced by Reliance Mercury with draw-bar pull capacities ranging from 16 500 lb (7486 kg) to 57 000 lb (25 850 kg). They are the RM 450, RM 350 and RM 150 and a further addition, the RM 2501. Two other models, the RM 601/801 and the RM 301 can be used for towing light aircraft and helicopters though their primary duties are concerned with freight and baggage handling.

The RM 450, a low profile tractor designed for push/pull and under-belly operation, is suitable for handling all types of wide-body aircraft. Powered by a Cummins V8 diesel engine, the unit incorporates four-wheel drive, power steering, power brakes and power shift transmission. A ground-power unit compartment is fitted and a rear driving station is offered as optional equipment.

The RM 350 is suitable for push/pull aircraft handling operations. Fitted with a Perkins V8 diesel engine, the unit has a draw-bar pull capacity of 35 000 lb (15 876 kg) and is capable of manoeuvring aircraft up to McDonnell-Douglas DC-10 size. The tractor features power shift transmission, four-wheel drive, power steering, power brakes, with an optional ground power unit fitted at the rear of the vehicle.

The RM 150, designed to handle medium range aircraft, is rated at 16 500 lb (7486 kg) draw-bar pull capacity for pull-out and 11 500 lb (5217 kg) for push-out operations. The unit is powered by a six-cylinder diesel engine with power steering and power shift transmission.

Reliance-Mercury RM 350 aircraft towing tractor

The RM 2501 rated at 33 500 lb (15 225 kg) is designed for handling aircraft such as the Boeing 767 and Airbus A310. Powered by a Deutz turbocharged diesel engine, the unit is equipped with power shift transmission, four wheel drive, disc brakes on both axles and front wheel Orbitrol power steering. The full width cab can be elevated to provide good rear visibility. A two position pintle towing bracket is fitted to both front and rear of the tractor. Optional features offered are: four-wheel steering in crab and/or conventional modes, integral hydraulic jacking system, spotlights to illuminate tow pintles, tractor-mounted 30 kVA gpu and additional ballasting to provide a maximum draw-bar pull of 39 000 lb (17 695 kg).

Status

The RM 450, 350 and 150 are in production and operational with major airlines. The RM 2501 is under development. Reports on the RM 601/801 and the RM 301 can be found in the Cargo Handling section

Manufacturer

Reliance-Mercury Ltd, Halifax, West Yorkshire, England

Reliance-Mercury RM 150 aircraft towing tractor

Characteristics

TYPE	RM 450	RM 350	RM 150	RM 2501
ENGINE	Cummins V8-V903 diesel	Perkins V8-540 diesel	Perkins 6.354 diesel	Deutz BF 6L913
OUTPUT	190 kW at 2600 rpm	134 kW at 2600 rpm	90 kW at 2800 rpm	118 kW at 2500 rpm
MAX SPEED UNLADEN	18 mph (29 km/h)	19 mph (30 km/h)	25 mph (40 km/h)	23 mph (37 km/h)
MAX DRAW-BAR PULL	57 000 lb (25 850 kg)	35 000 lb (15 876 kg)	16 500 lb (7486 kg)	33 500 lb (15 225 kg) option: 39 000 lb (17 695 kg)
ADHESION FACTOR	0·735 μ	0·7 μ	0·75 μ	0·75 μ
TOWING POINT HEIGHT				
front	1 ft 8¼ in (0·52 m), 2 ft ¼ in (0·62 m)	1 ft 6½ in (0·47 m), 1 ft 10½ in (0·57 m)	1 ft ¼ in (0·44 m), 1 ft 5¼ in (0·54 m)	1 ft 6 in (0·46 m), 1 ft 10 in (0·56 m)
rear	1 ft 8¼ in (0·52 m), 2 ft ¼ in (0·62 m)	1 ft 7 in (0·48 m), 1 ft 11 in (0·58 m)	1 ft 7½ in (0·5 m), 1 ft 11½ in (0·6 m)	1 ft 6 in (0·46 m) 1 ft 10 in (0·56 m)
OUTER TURNING RADIUS				
2-wheel steering	34 ft 6 in (10·52 m)	26 ft 3 in (8·01 m)	15 ft 6 in (4·72 m)	60 ft 7 in (182·46 m)
4-wheel steering				38 ft 6 in (112·74 m)
WHEELBASE	13 ft 7½ in (4·15 m)	9 ft (2·74 m)	7 ft 2 in (2·19 m)	10 ft 3 in (3·13 m)
TRACK				
front	6 ft 4 in (1·93 m)	6 ft 5 in (1·96 m)	6 ft 2½ in (1·89 m)	6 ft 4 in (1·93 m)
rear	6 ft 10 in (2·08 m)	6 ft 4 in (1·93 m)	6 ft (1·83 m)	6 ft 4 in (1·93 m)
UNIT DIMENSIONS				
length	30 ft 5 in (9·27 m)	20 ft 2 in (6·15 m)	15 ft 11¾ in (4·87 m)	21 ft 2 in (6·45 m)
width	8 ft 9 in (2·67 m)	8 ft (2·44 m)	7 ft 8 in (2·34 m)	8 ft 2 in (2·5 m)
height	5 ft 2 in (1·58 m)	7 ft 2½ in (2·20 m)	6 ft 10 in (2·07 m)	5 ft 2 in (1·58 m) cab lowered 7 ft 9½ in (2·38 m) cab raised
WEIGHT	77 500 lb (35 140 kg)	49 500 lb (22 455 kg)	29 900 lb (13 564 kg)	44 754 lb (20 300 kg) 52 910 lb (24 000 kg)

Shorrock Microwave Fence Aircraft Security Unit

The Shorrock microwave fence unit is a portable system which can be used to provide security to an aircraft parked unattended on tarmac or concrete surfaces.

It is an external alarm system comprising four sets of microwave transmitters and receivers which produce a microwave beam between transmitter and receiver units. The receiver is sensitive to changes in incident microwave power which occur if the beam is interrupted or obstructed. This change will cause an alarm condition to be generated if it exceeds the threshold defined by the receiver's sensitivity control. In addition the microwave beam between the transmitter and receiver can be used to communicate information between the units. The state of each microwave fence unit is encoded and the coded information relayed along the microwave beam between each transmitter and receiver.

The microwave alarm system is capable of producing a sensing zone covering an area of up to 120 square metres.

Shorrock microwave fence aircraft security unit

Characteristics

EQUIPMENT

4 microwave fence units
10 encoder units and 12 V rechargeable battery
1 control unit
1 battery charger, mains operated
8 lightweight collapsible tripods
1 cable reel
9 connecting cables (fences and control)
5 connecting cables (battery charger)

Manufacturer

Shorrock Security Systems, Blackburn, Lancashire, England

Shorrock microwave fence aircraft security unit in situ

VJ T272V Lavatory Servicing Vehicle

The VJ T272V operates a vacuum system to service aircraft lavatories with service points up to 16 feet 6 inches (5 metres) above ground level. The unit, offered with a choice of commercial chassis, incorporates a fixed-sided lift platform at the vehicle which houses all the service controls and hoses. These controls cover vacuum, metered water/disinfectant delivery, waste evacuation and platform movement. An emergency lowering valve which enables the platform to be lowered is operated from ground level. The water filling point, full depth gauges and the pressure selector for rapid tank discharge are built into the bodywork.

The tanks are rigid steel assemblies with the waste tank galvanised, and the water tank epoxy-lined. Water and vacuum pumps are housed in one compartment of the body together with all major circuit valves, situated beneath large hatches in the canopy roof. Steps at the rear of the vehicle provide access to the tank platform.

Characteristics

AIRCRAFT SERVICING CAPABILITY: service points up to 16 ft 6 in (5 m)

TANK CAPACITY

waste: 435 Imperial gallons (1978 litres)
water mix: 20 Imperial gallons/minute (90 litres/minute)

WATER LINE PRESSURE: 30 psi (2 bar) adjustable
WASTE TANK PRESSURE: 10 psi (0·7 bar)
AIR DISPLACEMENT: 2825 ft^3/h (80 m^3/h)
TANK PLATFORM HEIGHT: 5 ft 6 in (1·67 m)
ELEVATING PLATFORM MAX HEIGHT: 10 ft 6 in (3·2 m)
WHEELBASE: 13 ft (3·96 m)

UNIT DIMENSIONS

length: 24 ft 10 in (7·57 m)
width: 7 ft 7 in (2·3 m)
height: 8 ft 5 in (2·56 m)

Status

Production. In service at London Heathrow and Gatwick Airports, England

Manufacturer

VJ Engineering, Leamington Spa, Warwickshire, England

VJ W320 Potable Water Service Vehicle

The unit is designed to provide a large volume of potable water to service all aircraft with service points up to 12 feet (3·66 metres) above ground level. It is offered with a choice of commercially available chassis. The low-profile stainless steel tank, fitted with baffles to prevent surge during travel, provides a flat working platform with steps at the rear of the vehicle for access. A protected console which houses the controls required to replenish the aircraft drinking water tanks, is incorporated into the rear of the body above a full width, stepped, work platform covering a range of reach heights. The 25 feet (7·62 metres) of 1-inch (25 mm) bore delivery hose can be used from the cab roof, work platform or from the ground. The body-work houses the water fill point, full depth contents gauge, water sample tap and gravity drain valve.

VJ W320 potable water service vehicle

Characteristics

WATER TANK

capacity: 650 Imperial gallons (2955 litres)
delivery: 20 Imperial gallons (90 litres/minute)
pressure: 30 psi (2 bar) adjustable
hose: 300 ft × 1 in (7·62 m × 25 mm) bore

TANK PLATFORM HEIGHT: 5 ft 6 in (1·66 m)
CAB PLATFORM HEIGHT: 7 ft $10\frac{1}{2}$ in (2·4 m)
WHEELBASE: 13 ft 11 in (4·24 m)
TRACK: 7 ft 4 in (2·24 m)
WEIGHT: 11 240 lb (5100 kg)

Status

Production. In service at London Heathrow and Gatwick Airports, England

Manufacturer

VJ Engineering, Leamington Spa, Warwickshire, England

UNITED STATES OF AMERICA

Abex Jetpower 400 Hz Ground Power System

The Jetpower ground power system is designed to provide 400 Hz auxiliary power for aircraft during ground operation. It is intended to function as an integral part of Jetway passenger boarding bridges. Using solid-state high frequency inverters the unit is mounted on the passenger boarding bridge along which line and power service this system operates. Aircraft connection is a plug-in operation that is interlocked to the loading bridge to prevent damage. Metered operation can be provided to allow charges to be made on a consumption basis for gates used by more than one airline.

Jetpower 400 Hz ground power supply

Characteristics

OUTPUT: 116/200 V, 3-phase, 400 Hz
CAPACITY: 40 kVA
POWER REQUIREMENTS: 208–600 V, 3-phase, 50/60 Hz
UNIT DIMENSIONS
length: 5 ft 4 in (1·63 m)
width: 4 ft 6 in (1·37 m)
height: 1 ft 4 in (0·41 m)

Status

Production. Current production models of Jetway passenger loading bridges offer the Jetpower system as optional equipment

Manufacturer

Abex Corporation, Jetway Division, Ohio, Utah, USA

Jetpower 400 Hz ground power supply

ACE 497 Series Aircraft Lavatory Servicing Vehicle

The ACE 497 series vehicles provide servicing of the aircraft lavatory system for all types of aircraft. With fibreglass waste, flush and fill tanks they are designed to service several aircraft before dumping and refilling. Available in a variety of sizes and configurations, the units can be mounted on standard cab-type chassis or trailer depending on operator preference and requirements. The units have a single servicing location at the rear and can be equipped with an optional telescopic lift platform elevating to 10 feet (3·05 metres) above ground level. Tank manholes are provided with vents, level indicators, and hinged portholes and a spray jet with shut-off valve and port is supplied for flushing of the waste tank with flush water from the unit or from an external source. Dual 4-inch (102-mm) diameter waste hoses are installed 7 feet (2·13 metres) long and 14 feet (4·27 metres) on the telescopic lift platform, with fittings for waste servicing, and a 7 foot (2·13 metre) length, 1-inch (25·4-mm) diameter hose at the rear the platform with fittings for flush and fill tank servicing. Other features include heat in the rear compartment and under the tanks supplied by an automotive-type engine coolant heater to prevent freezing of fluids in winter; drain valves for waste, flush and fill tanks; stainless steel compartments and flow-meter installed near the control panel for all levels of operation to indicate the amounts of flush and fill fluids dispensed through the hose to the aircraft.

Characteristics

MODEL	203	303	781	803	903
CHASSIS	Ford/GMC F600 C5 D0 42	Ford/GMC F600 C5 D0 42	Trailer ACE 550-106	Trailer ACE 550-108	Trailer ACE 550-111
TANK CAPACITIES					
waste	700 US gallons (2650 litres)	450 US gallons (1703 litres)	200 US gallons (757 litres)	700 US gallons (2650 litres)	450 US gallons (1703 litres)
deodorant	250 US gallons (946 litres)	80 US gallons (303 litres)	30 US gallons (114 litres)	250 US gallons (946 litres)	80 US gallons (303 litres)
water	250 US gallons (946 litres)	170 US gallons (644 litres)	100 US gallons (376 litres)	250 US gallons (946 litres)	170 US gallons (644 litres)
PUMP CAPACITY	30 US gallons/ minute at 50 psi (114 litres/minute at 3·4 bar)	30 US gallons/ minute at 50 psi (114 litres/minute at 3·4 bar)	30 US gallons/ minute at 50 psi (114 litres/minute at 3·4 bar)	30 US gallons/ minute at 50 psi (114 litres/minute at 3·4 bar)	30 US gallons/ minute at 50 psi (114 litres/minute at 3·4 bar)
WHEELBASE	12 ft 9 in (3·89 m)/ 12 ft 5 in (3·78 m)	10 ft 9 in (3·28 m)/ 10 ft 5 in (3·18 m)			
TRAILER CAPACITY			6000 lb (2727 kg)	18 000 lb (8182 kg)	12 000 lb (5454 kg)
DIMENSIONS					
length	23 ft 9 in (7·24 m)	20 ft 2 in (6·15 m)	8 ft 6 in (2·59 m)	15 ft 4 in (4·67 m)	11 ft 6 in (3·51 m)
width	7 ft 6 in (2·29 m)	7 ft 6 in (2·29 m)	4 ft (1·22 m)	7 ft (2·13 m)	7 ft (2·13 m)
height with lift	6 ft 11 in (2·11 m)	6 ft 11 in (2·11 m)	5 ft 4 in* (1·70 m)	5 ft 3 in (1·6 m)	5 ft 3 in (1·6 m)
gross vehicle weight	11 180 lb (5082 kg)	8320 lb (3782 kg)	2600 lb (1182 kg)	5775 lb (2625 kg)	4000 lb (1818 kg)

*without lift platform which is not available on this model

Manufacturer

Accessory Controls and Equipment Corporation, Windsor, Connecticut, USA

ACE 499 Series Potable Water Servicing Vehicle

The ACE 499 series provide servicing of potable water systems for all types of aircraft. Using a totally enclosed, self-contained module incorporating a fibreglass tank the equipment can service several aircraft before refilling. Available in a variety of sizes, the module can be mounted on a standard cab chassis or trailer, depending on operator preference and requirements. With a single servicing location in the rear of the vehicle and equipped with an optional telescopic lift platform elevating to 10 feet (3·05 metres) above ground level, this system is capable of servicing aircraft ports at all levels.

ACE 303 potable water servicing vehicle

Characteristics

MODEL	303	781	903
CHASSIS	GMC/Ford C5D042	Trailer ACE 550-106	Trailer ACE 550-111
WATER TANK CAPACITY	700 US gallons (2650 litres)	330 US gallons (1250 litres)	700 US gallons (2650 litres)
PUMP CAPACITY	30 US gallons/minute at 50 psi (114 litres/minute at 3·4 bar)	30 US gallons/minute at 50 psi (114 litres/minute at 3·4 bar)	30 US gallons/minute at 50 psi (114 litres/minute at 3·4 bar)
WHEELBASE	10 ft 5 in (3·18 m)/ 10 ft 9 in (3·28 m)		
TRAILER CAPACITY		6000 lb (2727 kg)	12 000 lb (5454 kg)
DIMENSIONS			
length	20 ft 2 in (6·15 m)	8 ft 6 in (2·59 m)	11 ft 6 in (3·51 m)
width	7 ft 6 in (2·29 m)	4 ft (1·22 m)	7 ft (2·13 m)
height	6 ft 11 in (2·11 m)	5 ft 4 in (1·7 m)	5 ft 3 in (1·6 m)
WEIGHT	8640 lb (3945 kg)	2600 lb (1182 kg)	4000 lb (1818 kg)

Manufacturer

Accessory Controls and Equipment Corporation, Windsor, Connecticut, USA

ACE 500 Series Pneumatic Jet Engine Starter Units

The ACE 500 series provide medium pressure air sources for jet aircraft engine starting. Air pressure from 15 to 45 psig (1–3 bar) at 0–200 lb per minute (90·7 kg per minute) air flow are available to start all types of jet aircraft engines. These units are offered in a variety of sizes and configurations to meet individual operator requirements based on aircraft to be serviced and frequency of use. The units are mobile, mounted on a truck or trailer or stationary. A suitably rated piston-type compressor, ASME-cylinder, petrol, diesel or electric motor comprise the basic components for this equipment which uses an ACE regulator control valve for starting operations. This is normally in the closed position. Signal pressure controls the valve to produce a slow opening and fast closing operation.

Manufacturer

Accessory Controls and Equipment Corporation, Windsor, Connecticut, USA

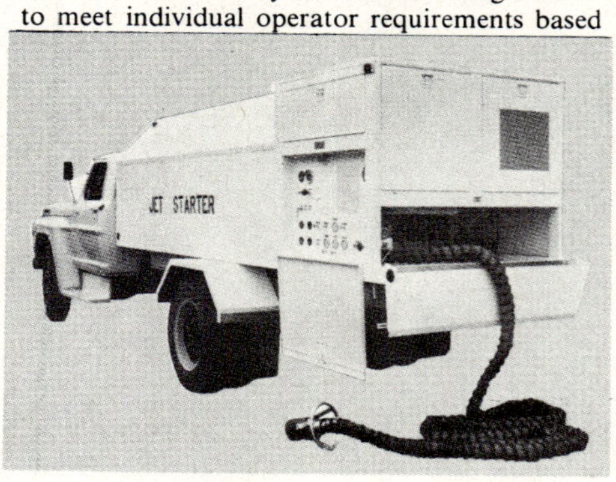

Characteristics

NO OF MODELS	3	3	3
CHASSIS	Trailer	Trailer	Trailer
POWER PLANT	Petrol/diesel/electric	Petrol/diesel/electric	Petrol/diesel/electric
STORED AIR CAPACITY	150 lb (68 kg)	200 lb (91 kg)	315 lb (143 kg)
AIR PRESSURE	385 psi (26 bar)	385 psi (26 bar)	385 psi (26 bar)
NO OF MODELS	9	9	9
CHASSIS	Trailer/truck/fixed base	Trailer/truck/fixed base	Trailer/truck/fixed base
POWER PLANT	Petrol/diesel/electric	Petrol/diesel/electric	Petrol/diesel/electric
AIR CAPACITY	280 lb (127 kg)	400 lb (182 kg)	540 lb (245 kg)
AIR PRESSURE	500 psi (34·5 bar)	500 psi (34·5 bar)	500 psi (34·5 bar)

ACE 500 series pneumatic starter unit

ACE Air-Conditioning and Heater Units

The ACE air-conditioning and air-conditioning/heating units are available in a variety of capacities and configurations. The units employ an air-to-air type refrigeration system and for operators requiring heat this equipment can be supplied with a reverse-cycle heat-pump system. Provisions are made for engine coolant heat to augment the reverse-cycle heat mode when required. These diesel- or petrol-powered units are fitted with an air-operated, automatic clutch between the engine and compressor. This system allows the blower to operate alone for the ventilating mode. The automatic clutch, coupled with a reversing valve operates the cooling, heating and ventilation modes. The change from one mode to another can be performed by the operator at the aircraft. The refrigeration equipment uses a suction control system which regulates the suction pressure to the compressor and isolates the suction side of the compressor upon shut-down. A discharge check valve isolates the discharge side of the compressor upon shut-down. All units are self-contained and

ACE air-conditioning/heating unit

Characteristics

SERIES	401	405	404	400	406
AIR-CONDITIONING CAPACITY	30 tons (27·2 tonnes)	40 tons (36·3 tonnes)	60 tons (54·4 tonnes)	85 tons (77·1 tonnes)	110 tons 99·8 tonnes)
MAX AIR FLOW	200 lb/minute (91 kg/minute)	250 lb/minute (113 kg/minute)	350–500 lb/minute (159–227 kg/minute)	400–600 lb/minute (181–272 kg/minute)	500–700 lb/minute (227–318 kg/minute)
AIR TEMPERATURE	35–50°F (2–10°C)	35–50°F (2–10°C)	35–50°F (2–10°C)	35–50°F (2–10°C)	35–50°F (2–10°C)
HEATING CAPACITY	360 K Btu/h (380 Mj/h)	400 K Btu/h (422 Mj/h)	500 K Btu/h (528 Mj/h)	660 K Btu/h (696 Mj/h)	750 K Btu/h (791 Mj/h)
AIR TEMPERATURE	100–135°F (38–57°C)	100–135°F (38–57°C)	100–135°F (38–57°C)	100–135°F (38–57°C)	100–135°F (38–57°F)

can be supplied mounted on a chassis or trailer. A model was introduced at the Airport 81 exhibition at Manchester in October 1981. This is the 806 series which uses a rotary screw compressor and variable capacity control. The operation of the unit is in two steps: starting the unit and setting the desired delivered air temperature. The pump-down cycle is eliminated.

Manufacturer

Accessory Controls and Equipment Corporation, Windsor, Connecticut, USA

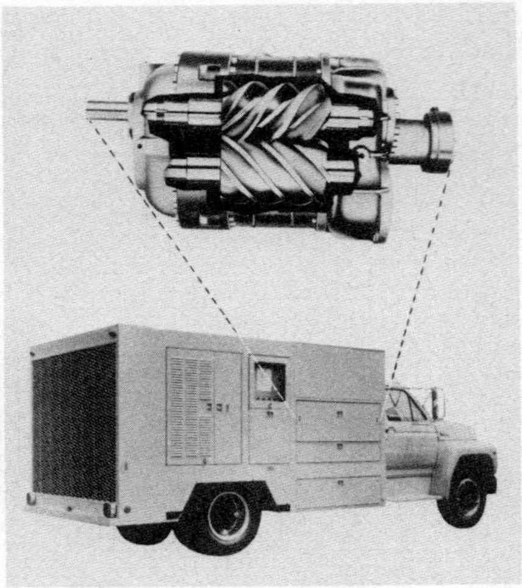

ACE 806 air-conditioning/heating vehicle with rotary screw compressor

---

Aeroenergy AirStart 125/200 Starter Systems

Two models of pneumatic starting systems are produced by Aeroenergy, the AirStart 125 and AirStart 200.

The AirStart 125 system entails the accumulation of high pressure (500 psi) air in receivers in sufficient quantity to provide the instantaneous high mass flow required for jet engine starting. The receiver air is provided by small displacement compressors. Ambient air is compressed to 500 psi, dried and delivered to the accumulator. A pressure reduction system provides 150 psi air to the distribution system which is on the surface or underground. Final system pressure is further reduced within the control console to the specific aircraft requirement. The start sequence is controlled by ground personnel. Control console configuration may be electric or all-pneumatic depending on operator preference and can be a fixed base unit, portable or installed in a service pit. The compressor skid comprises a two stage reciprocating compressor, closed loop air cooling system, mechanical refrigeration dryer and coalescant filtration. The receiver skid where the air is accumulated consists of three receiver tanks each holding 113 cubic feet (3·2 cubic metres) of air.

The AirStart 200 features a diesel-powered compressor skid with dual regenerative dehydration towers, but otherwise is similar to the 125 model.

Manufacturer

Aeroenergy Systems Inc, Chatsworth, California, USA

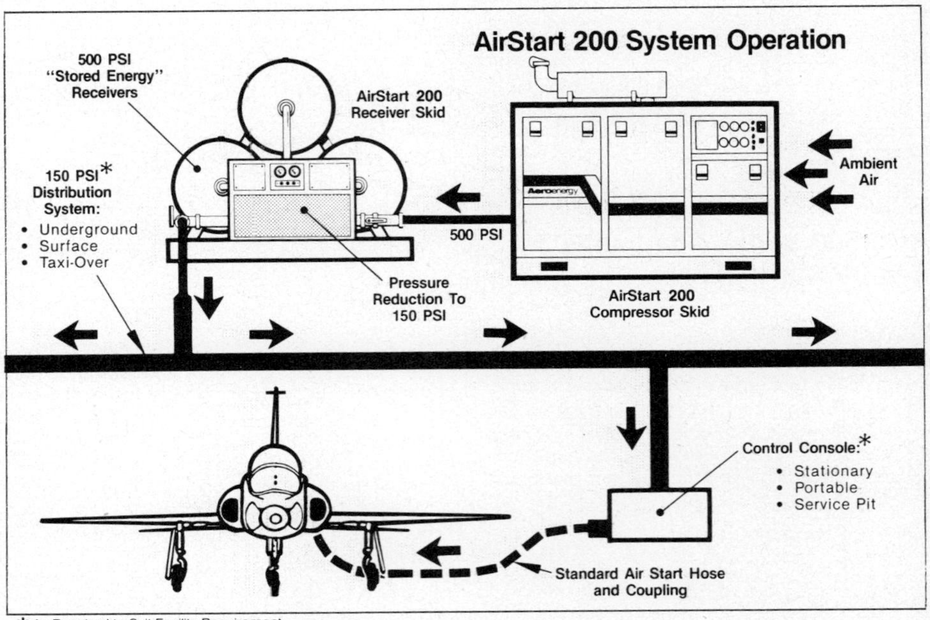

* As Required to Suit Facility Requirement

Aeroenergy AirStart 200 system operation

Characteristics

MODEL	AirStart 125	AirStart 200
COMPRESSOR SKID POWER REQUIREMENTS	230/460 V, ac 3-phase 50/60 Hz	Diesel-powered
DIMENSIONS		
length	6 ft (1·83 m)	8 ft (2·44 m)
width	5 ft (1·52 m)	6 ft (1·83 m)
height	4 ft 6 in (1·37 m)	6 ft (1·83 m)
WEIGHT	5000 lb (2268 kg)	11 000 lb (4990 kg)
RECEIVER SKID DIMENSIONS		
length	20 ft 6 in (6·25 m)	20 ft 6 in (6·25 m)
width	6 ft 9 in (2·06 m)	6 ft 9 in (2·06 m)
height	7 ft 2 in (2·18 m)	7 ft 2 in (2·18 m)
WEIGHT	16 500 lb (7484 kg)	16 500 lb (7484 kg)

Air-A-Plane Air-Conditioner/ Heater Units

Air-A-Plane produces a range of fixed base and mobile units for supplying air-conditioning to aircraft cabins on the ground units.

The 625-D is an air-conditioning/heater unit powered by a Detroit diesel 6 V 71N engine offered mounted on a truck chassis on skids. It is equipped with an air-conditioner rated at 110 tons, eight-cylinder compressor, a heater and 20 feet (6·1 metres) of 12-inch (0·305-metre) diameter hose. An air-dump system is fitted to protect duct systems and to facilitate connection and release. A ventilation mode can be selected to circulate air when neither heating or cooling is required.

The 952-D mounted on a four wheeled trailer with towbar steering, is fitted with similar equipment to the 652-D model but is furnished with 30 feet (9·14 metres) of 14-inch (0·356-metre) diameter hose for connection to the aircraft.

The 6600-D heater is offered with a choice of F350 truck chassis, skid or trailer mounting. Powered by a Detroit diesel 6V53, the unit is equipped with thermostat heating control, air-dump system and 20 feet (6·1 metres) of 12-inch (0·305-metre) diameter hose.

Two fixed base units available are the GHIE-2 with a 75 ton air-conditioning system and the GH2E, rated at 110 tons. Both are electrically-operated and offer heating equipment as an optional feature. All units are equipped with a motor-driven iris control valve to adjust air flow, a ventilation mode to circulate air when neither heating nor cooling is required and 30 feet (9·14 metres) of 14-inch (0·356-metre) diameter hose.

Characteristics

TYPE	652D	952-D	6600-D	GHIE-2	GH2E
MODELS	652-D – C600 truck chassis 652D-3 – Skid mount 652D-4 – F600 truck chassis		6600D-2 – trailer 6600D-3 – skid 6600D-4 – F350 truck chassis		
POWER PLANT	Detroit Diesel 6 V 71N	Detroit Diesel 6 V 71	Detroit Diesel 6 V 53	1 × electric blower/ 1 × electric condenser	1 × electric blower/ 1 × electric condenser
OUTPUT	220 hp (164 kW) at 1800 rpm	220 hp (164 kW) at 1800 rpm	180 hp (134 kW) at 2200 rpm	60 hp (45 kW) at 3600 rpm 15 hp (11 kW) at 1170 rpm	100 hp (75 kW) at 3600 rpm 25 hp (18·7 kW) at 1300 rpm
HEATER					
air flow	400 lb/minute (181 kg/minute)	425 lb/minute (193 kg/minute)	300 lb/minute (136 kg/minute)	320–500 lb/minute (145–227 kg/minute)	500–640 lb/minute (227–290 kg/minute)
capacity	635 K Btu/h (670 Mj/h)	635 K Btu/h (670 Mj/h)	750 K Btu/h (791 Mj/h)	550 K Btu/h (580 Mj/h)	800 K Btu/h (844 Mj/h)
duct temperature	130°F (54°C)	130°F (54°C)	140°F (60°C)	140°F (60°C)	140°F (60°C)
AIR-CONDITIONER					
air flow	400 lb/minute (181 kg/minute) at 1900 rpm	425 lb/minute (193 kg/minute) at 1900 rpm		500 lb/minute (227 kg/minute) (max)	640 lb/minute (290 kg/minute) (max)
cooling capacity	110 tons (100 tonnes)	110 tons (100 tonnes)		75 tons (68 tonnes)	110 tons (100 tonnes)
COMPRESSOR	L 7322-80	L 7322-80		L 7322-60 Electric motor 100 hp (75 kW) at 1750 rpm	L 7322-80 Electric motor 125 hp (93 kW) at 1750 rpm
POWER REQUIREMENTS				440 V, 3-phase, 60 Hz, 150 kW	440 V, 3-phase, 60 Hz, 210 kW
DIMENSIONS					
	652D		6600D-2		
length	24 ft 1 in (7·34 m)	17 ft (5·18 m)	8 ft 8 in (2·64 m)	12 ft (3·66 m)	12 ft 6 in (3·81 m)
width	8 ft 2 in (2·45 m)	7 ft 5 in (2·26 m)	6 ft 2 in (1·88 m)	7 ft (2·13 m)	8 ft (2·44 m)
height	10 ft 9 in (3·28 m)	6 ft 3 in (1·91 m)	6 ft 9 in (2·06 m)	6 ft 4 in (1·93 m)	7 ft 6 in (2·29 m)
	652D-3		6600D-3		
length	17 ft 4 in (5·28 m)		8 ft 8 in (2·64 m)		
width	8 ft 2 in (2·45 m)		6 ft 2 in (1·88 m)		
height	9 ft 7 in (2·92 m)		6 ft 4 in (1·93 m)		
	652D-4		6600D-4		
length	25 ft 7 in (7·8 m)		16 ft (4·88 m)		
width	8 ft (2·44 m)		6 ft 2 in (1·88 m)		
height	10 ft 7 in (3·23 m)		7 ft 2 in (2·18 m)		
WEIGHT					
	652D-2		6600D-2		
	21 660 lb (9825 kg)	12 000 lb (5443 kg)	5000 lb (2268 kg)	10 000 lb (4536 kg)	10 000 lb (4536 kg)
	652D-3		6600D-3		
	14 600 lb (6623 kg)		4800 lb (2177 kg)		
	652D-4		6600D-4		
	10 350 lb (9231 kg)		10 000 lb (4536 kg)		

Status

Air-A-Plane air-conditioning and heating units have been supplied to 73 airlines, 11 airport service companies and the Portuguese and United States Air Forces

Manufacturer

Air-A-Plane Corporation, Norfolk, Virginia, USA

Air-A-Plane 952-D air-conditioner/heater

Alco Aircraft De-icing and Power Cleaning Equipment

Alco Equipment International produces a range of de-icing and power cleaning equipment designated D-ICR and POWR-KLEENR, used to service aircraft from single-engine general aviation types to the Boeing 727 aircraft.

The model 50 is a trailer-mounted unit equipped with a 335 US gallon (1268 litres) tank of water mixed with a chemical solution. It features a 7 hp (5·2 kW) Briggs and Stratton Motor which has an output of 10 US gallons per minute at 400 lb per square inch (38 litres per minute at 27·5 bar) and a down-stream injector. Three 2·5 kilowatt, 240 volt heaters are fitted which can reach a temperature of 140°F (60°C) in 2½ hours. The unit is suitable for commuter and corporate aircraft.

The model 100 is a trailer unit, powered by a 7 hp (5·2 kW) Briggs and Stratton engine that produces 10 US gallons per minute (38 litres per minute) at 400 lb per square inch (27·5 bar) and is fitted with a 400 US gallon (1514 litre) fibrelass tank with three 2·5 kW, 240 volt heaters. The top of the platform is 9 feet (2·74 metres) above ground level, or 11 feet (3·35 metres) when guard-rails are installed. This model is designed for fixed base operators, commuter and executive operators.

The model 250 is mounted on a one ton (907 kg) GMC truck and incorporates a 550 US gallon (2082 litre) stainless steel tank and a hydraulic pump capable of 25 US gallons per minute (95 litres per minute) at 400 lb per square inch (27·5 bar). A hydraulic-powered personnel lift platform provides a working height of 39 feet (11·9 metres). Other features include a heavy-duty hose reel, and a gas heating system with a recovery time of 1 hour up to 180°F (82·2°C). The model 250 is designed also for use as a maintenance truck or as a power cleaning unit during the summer months.

Status
D-ICR and POWR-KLEENR units introduced in 1980 are in service with operators in the USA

Manufacturer
Alco Engineering International Inc, Golden Valley, Minnesota, USA

Alco model 50 de-icer unit

Alco model 250 de-icing and power cleaning unit

Arvico Stordair Engine Starting Systems

Stordair systems are fixed-base installations for starting jet engine aircraft. A typical system consists of four major sections, the compressor, receiver/storage, air distribution and aircraft starting stations. In the compressor section one or more compressor skids are provided to meet the system requirements. The compressor is driven by a diesel engine or electric motor. Each skid is fitted with the appropriate coolers, filters and a control console with instrumentation and malfunction warning systems. The compressor section provides air compressed to 500 psig (35 bar) for the receiver/storage section. In this section one or more receiver skids are supplied to store the compressed air so that it will be available for immediate use. The primary pressure regulator is part of this section and permits use of all stored air down to 60 psig (4 bar) system pressure. The low pressure distribution system is designed to meet the starting requirements at any particular airfield. The piping is normally buried beneath the ground with protection against corrosion.
A 400 Hz ac power supply system may be installed to parallel the air starting system. Diesel-driven generators are installed in the same place as the compressors and a power distribution cable follows the air pipe network. Control switches, instruments, and the aircraft cable and connector are at each starting station. A 24 volt dc power supply is available furnished by two 12 volt dc batteries at the starting station. An alternate portable control console is offered which houses a 400 Hz generator driven by an air motor. An air coupling is provided at each starting station so that the portable unit may be plugged into the stored air supply. The unit is self-contained and includes all controls and instruments necessary for electrical system testing. Batteries for 24 volts dc power are included in the console.

Arvico Stordair receiver unit

Status
Stordair can be used in apron or ramp areas, alert pads, maintenance hangars and engine test facilities. It is applicable to the A4 Sky Hawk, C130 Hercules, F4 Phantom, F5 Tiger, F14 Tomcat. T38 Talon aircraft and other aircraft using external air for starting. Over 300 units have been installed

Manufacturer
Arvico International Inc, Miami, Florida, USA

Brackett TR-3 Aircraft Tow-bar

The TR-3 is a universal tow-bar for general aviation aircraft fitted with nose-wheels. Using a telescopic hitch, aircraft from the Cessna 150 to King Air and Learjet aircraft may be handled. The heavy-duty TR-3 is $8\frac{1}{2}$ feet (2·6 metres) long and weighs 27 lb (12 kg).

Manufacturer
Brackett Aircraft Co, Mesa, Arizona, USA

Christie Rectodyne Dc Power Trolleys

Rectodyne trolleys are designed to supply regulated dc ground power for aircraft avionic, radio and air-conditioning and for aircraft starting systems. The unit is fitted in a two-wheel mobile cabinet and is solid-state using hermetically sealed silicon diodes. A six-position tap switch on the front panel compensates for high or low input voltage, with a separate switch on the RB models for operator selection of current limits. Start and stop push buttons and dc meters are fitted on the front panel and a pilot lamp indicates when the unit is energised. Safety devices provide thermal and dual fan failure protection.

Manufacturer
Christie Electric Corporation, Torrance, California, USA

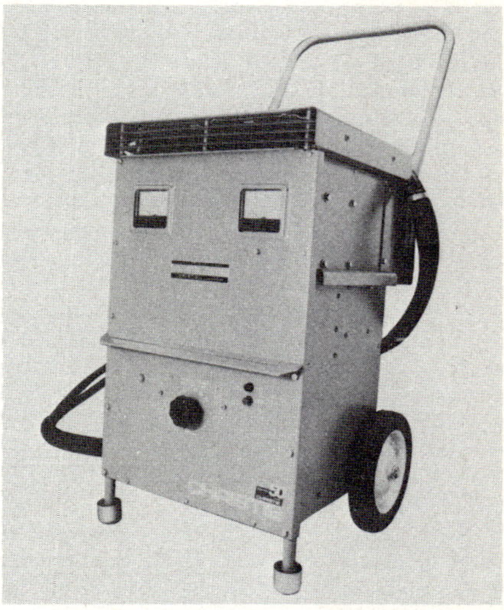

Christie Rectodyne dc power trolley

Characteristics

SERIES	MH 28	MH 28	MH 28	H 28	H 28	A 28
TYPE	100RA2M	200RA24M	300RB24M	400RB24M	600RA4M	100TJA4M
OUTPUT	28 V dc ±2%	28 V dc ±2%	28 V dc ±2%	28 V dc ±2%	28 V dc ±2%	28 V dc ±2%
MAX CONTINUOUS	100 A	200 A	300 A	400 A	600 A	1000 A
ENGINE STARTING	600 A	1000 A	1500 A	2000 A	3000 A	6000 A
POWER REQUIREMENTS	208/230 V 10 A	208/230 V 18 A 460 V 9 A	208/230 V 26 A 460 V 13 A	208/230 V 35 A 460 V 18 A	460 V 25 A	460 V 43 A
DIMENSIONS						
length	1 ft 10 in (0·56 m)	1 ft 10 in (0·56 m)	1 ft 10 in (0·56 m)	2 ft (0·61 m)	2 ft (0·61 m)	2 ft 7½ in (0·71 m)
width	1 ft 8 in (0·51 m)	1 ft 8 in (0·51 m)	1 ft 8 in (0·51 m)	1 ft 10¼ in (0·57 m)	1 ft 10¼ in (0·57 m)	2 ft 4 in (0·71 m)
height	2 ft 7½ in (0·8 m)	2 ft 7½ in (0·8 m)	2 ft 7½ in (0·8 m)	3 ft 1 in (0·94 m)	3 ft 1 in (0·94 m)	3 ft 7⅝ in (1·11 m)
WEIGHT	140 lb (63·5 kg)	235 lb (106·5 kg)	260 lb (118 kg)	315 lb (143 kg)	465 lb (211 kg)	575 lb (261 kg)

Dole Aero Tractor

The Aero Tractor is suitable for moving general aviation/business jet aircraft and helicopters. Powered by a petrol or electric engine, the unit is pedestrian-controlled by a single operator. Four models are produced, each equipped with hydrostatic drive, deadman controls on the steering arm and manual or electric starters. The tractor uses a clamping mechanism to couple the unit to the nose-wheel of the aircraft, or tail wheel of the helicopter. The mechanism is designed to pivot to accommodate the rake angle of the nose or tail strut. Adaptors are offered where necessary. The 302 series are suitable for single engine and medium size twin executive aircraft. The 402 with a larger engine can handle aircraft weighing up to 24 000 lb (10 886 kg). For the military, units are available for servicing aircraft up to 70 000 lb (31 750 kg).

Characteristics

TYPE	302 series	402 HE
ENGINE	Petrol/electric	Petrol
OUTPUT	5 hp (302 H) 5 hp (302 HAC) 6 hp (302 HE)	8 hp
CAPACITIES		
max lift:	1400 lb (635 kg)	2000 lb (907 kg)
tow	12 000 lb (5443 kg)	24 000 lb (10 886 kg)
lift height	25 in (0·64 m)	11 in (0·28 m)
UNIT DIMENSIONS		
length	6 ft 3 in (1·91 m)	8 ft 8 in (2·64 m)
width	2 ft 9 in (0·84 m)	3 ft 11 in (1·19 m)
height	3 ft 1½ in (0·95 m)	3 ft 4½ in (1·03 m)

Status
In service at airports in the USA

Dole Aero Tractor 302 series

Manufacturer
Dole Division, Newport Corporation, Redwood City, California, USA

Eagle Bob Tail F6000 Series Tractors

The F6000 series of Bob Tail tractors are designed for aircraft towing duties and are available at draw-bar pull ratings of 6000, 8000 and 12 000 lb (2722, 3629 and 5443 kg). Equipped with dual rear wheels and towing points front and rear, the vehicles are offered with a choice of petrol; diesel- or lpg-fuelled engines. The tractors feature automatic transmission, four wheel-brakes and a weather-proof cab accommodating three people. Speeds of up to 30 mph (48 km/h) are attainable.

Status
Eagle Bob Tail tractors are in service with the US Air Force, Air Canada, Air France, CP Air, Qantas, SAS and 20 airlines in the USA

Manufacturer
Eagle Bob Tail Tractors Inc, Grosse Ile, Michigan, USA

F6000 Eagle Bob Tail tractor

FMC Aircraft De-icing/Washing Units

FMC produce four models for aircraft de-icing and washing operations. Three are truck-mounted, the TM-1800, TM-1800E and the LA-1000. The fourth SA-600 is mounted on a trailer. The TM-1800 is suitable for washing and de-icing wide-body aircraft. It is equipped with a basket mounted on a telescopic rotating boom and articulated yoke, providing a working height of 48 feet (14·6 metres). The basket fitted with latching gate, 42-inch (1·07-metre) sides and metal floor, can be entered from the ground in front of the truck. It contains self centring controls, swivel-mounted fluid gun, lighting for night operations and an intercom system for communications with the cab. A second fluid gun is fitted at ground level with a 50 foot (15·2 metre) reel-mounted hose. Stainless-steel tanks provide a capacity of 1800 US gallons (6800 litres) of fluid. Optional equipment available includes proportional mixing system for de-icing/anti-icing operations, tank selection system for independent selection of fluids required for de-icing/anti-icing, soap unit for washing aircraft, boom controls in cab, tank insulation and choice of fluid guns and chassis.

The TM-1800E based on Ford C800 truck chassis is designed for European operations. Similar in design to the TM-1800 this unit is equipped with a Euromix proportional mixing system. For de-icing operations the proportioning system allows the operator in the aerial basket to select the mixture of de-icing fluid and hot water, based on existing conditions. The mixture can be varied from 100 per cent water with no de-icing fluid to 40 per cent water and 60 per cent de-icing fluid. The mixture is applied using a turret-mounted spray nozzle. A purge system is provided to protect the system from freezing. Anti-icing consists of applying a protective coating either on a clean aircraft

before it undergoes icing conditions or just after de-icing to extend the period of protection against icing. The problems associated with switching from hot-fluid de-icing to applying unheated anti-icing liquid have been solved by isolating the hot water and anti-icing fluid tanks. In addition, the anti-icing fluids are protected from overcirculation by an electrically-operated clutch. The pump operates on demand. Optional equipment offered includes a 9-kW recirculating heater in the water tank, soap system and a fire protection system.
The LA-1000 is equipped with a telescopic boom and gravity levelling basket, providing a working height of 42 feet (12·8 metres). The basket contains self centring and boom controls, turret-mounted fluid gun and lighting for night operations. Stainless-steel tanks are fitted with a capacity of 1000 US gallons (3785 litres) and the truck is fitted with individual braking to the dual-rear wheels. With the addition of a diesel power plant with heater, the optional equipment offered is similar to the 1800 model options.
The SA-600 can be used to de-ice and wash aircraft with either hot or cold fluid. The unit is mounted on a two-wheeled trailer. A 600 US gallon (2271 litre) capacity fluid tank stores water or with a partition inserted, glycol and water. Fluid is pumped from the tank by a self-priming centrifugal pump, through a petrol-operated heater if desired to the discharge gun(s). A standard four-cylinder industrial petrol engine powers the pumping system. A fluid-discharge gun is mounted on the platform which when erected is 22 feet (6·7 metres) above ground level. A second gun is connected to a 50 foot (15·2 metre) reel at ground level. Optional equipment includes a proportional mixing system that allows the operator to vary the water/glycol mixture, a soap system with a 50 US gallon (189 litre) detergent tank for washing operations, a re-filler assembly for filling the de-icer fluid tank from a drum or bulk storage and an engine-driven air compressor to power pneumatic tools for aircraft maintenance.

Status

The TM-1800 has been supplied to the US and Iranian Air Forces and 13 airlines, and the 1800E is in service with Lufthansa. The LA-1000 and the SA-600 are used by airlines in the USA. Production was suspended in 1981 because of company overload but recommenced in 1982 at a new manufacturing facility at Orlando, Florida, USA

Manufacturer

FMC Corporation, San Jose, California, USA

FMC SA-600 trailer-mounted de-icer/washer

FMC LA-1000 maintenance master de-icer/washer

FMC TM-1800 de-icer/washer

Characteristics

TYPE	TM-1800/TM-1800E	LA-1000	SA-600
CHASSIS	Ford C800/General Motors C7 DO 42	industrial 6-cylinder petrol engine hydrostatic propulsion	
BASKET			3 × 3 ft (0·91 × 0·91 m) platform
CAPACITY	450 lb (204 kg)	500 lb (227 kg)	500 lb (227 kg)
MAX ELEVATION	42 ft (1·28 m)	36 ft (10·97 m)	22 ft (6·71 m)
MAX SIDE REACH	26 ft (7·93 m)		
BOOM ROTATION	± 185°		
AUXILIARY ENGINE	Industrial 6-cylinder petrol/diesel		4-cylinder petrol
FLUID PUMP	2-stage centrifugal	2-stage centrifugal	centrifugal self-priming
CAPACITY	40 US gallons/minute at 200 psi (151 litres/minute at 14 bar) 110 US gallons/minute at 90 psi (416 litres/minute at 6 bar)	40 US gallons/minute at 200 psi (151 litres/minute at 14 bar) 110 US gallons/minute at 90 psi (416 litres/minute at 6 bar)	70 US gallons/minute at 125/130 psi (265 litres/minute at 8·6 bar)
HEATER(S)	2 in parallel	1	1
OUTPUT	2000 K Btu/h - petrol (2110 Mj/h - petrol) 1700 K Btu/h - diesel (1794 Mj/h - diesel)	1000 K Btu/h (1055 Mj/h)	500 K Btu/h (528 Mj/h)
TANK CAPACITIES			
fluid	1800 US gallons (6800 litres)	1000 US gallons (3780 litres)	600 US gallons (2271 litres)
detergent	160 US gallons (605 litres)	50 US gallons (190 litres)	50 US gallons (190 litres)
TURNING RADIUS	26 ft 8 in (8·13 m)	16 ft (4·88 m)	
WHEELBASE	11 ft 3 in (3·43 m)	12 ft 10½ in (3·32 m)	
DIMENSIONS			
length	27 ft 10 in (8·48 m)	27 ft (8·23 m)	27 ft 9⅜ in (8·47 m)
width	8 ft (2·44 m)	8 ft (2·44 m)	7 ft 11½ in (2·43 m)
height	12 ft 9 in (3·89 m)	9 ft (2·74 m)	7 ft 2 in (2·18 m)
WEIGHT WHEN TANKS EMPTY	24 500 lb (11 113 kg)	17 160 lb (7784 kg)	6932 lb (3134 kg)

Garsite Aircraft Ground Service Units

Garsite units provide a means of supply of electricity, fuel, heating and air-conditioning to aircraft on the apron, hardstanding or in the hangar. Standard pits are supplied in round or square styles and available in steel or fibreglass construction. Both styles can be supplied as cover and frame only; cover, frame and shell; cover, frame, shell and bottom; or as complete units with fittings as specified by the operator. The covers are aluminium, designed to withstand up to 245 psi (16·8 bar) tyre pressure on a 200 square inch (1290 square centimetres) tyre print area. Four types of covers are available. The cover for round pits has a long skirt which fits in a steel ring fitted with neoprene gasket. A retaining chain can be supplied as optional equipment. A three-lug twist-lock cover for round pits locks in a steel ring with a neoprene gasket which seals against surface water. The cover is handled with a sparkproof wrench. A retaining chain is optional. Round hinged covers are furnished in bolted steel frames. Double action hinges are supplied to permit the cover to lie flat when open. Rectangular hinged covers can be supplied with mechanical retaining bars or with adjustable hydraulic snubbers.

Other features which are available include a retaining bar to hold the cover in the open position fitted as standard on some models, or alternatively a hydraulic snubber, counterbalance spring, or an air counterbalance reservoir and snubber. Standard units offer a 2·5:1 safety factor, the heavy-duty units a 4:1 ratio. Covers are plain but can include operator's insignia or pit service identification.

Garsite Aircraft ground service pit

Status

In service at airports in the USA

Manufacturer

Garsite Products Inc, Deer Park, New York, USA

Hobart 400 Hz Central Station System

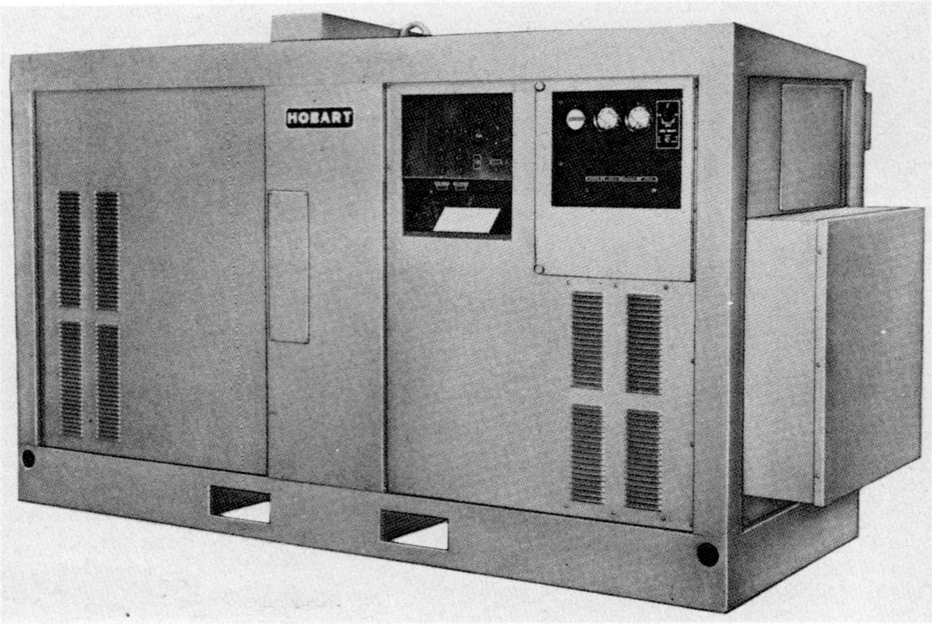

Hobart 400 Hz frequency converter

The Central Station System comprises the required number of 400 Hz 575 volts automatic paralleling alternators to service the proposed aircraft load with two paralleling frequency converters at a central stationary place. Power is supplied to the aircraft via line drop compensators installed in each circuit to prevent inductive line losses and a transformer/gate box. The gate box contains a step-down transformer and safety controls which reduce the 575-volt transmission to 115/200 volts which is passed to the aircraft via the aircraft cable. Distribution can be from passenger loading bridges or from an underground service point on the apron. Where apron-drive passenger bridges are used, a telescopic, flexible cable system and cable storage unit are available.

Status

The first installation was at St. Louis Airport, USA in 1976. Hobart central stations are in service at the following airports for apron (gates) and hangar (positions) distribution:

- Finland: Helsinki Vaanta (7 gates)
- France: Charles de Gaulle (70 gates, Orly (10 gates)
- Morocco: Casablanca (Royal Air Maroc 5 positions)
- Saudi Arabia: Jeddah International (Saudia 6 positions)
- Singapore: Changi (SIA 7 positions)
- Republic of Korea: Seoul (8 gates)
- Taiwan: Tao Yuan International (22 gates and China Airlines 10 positions)
- USA: New York, La Guardia Eastern (16 gates), New York JFK (Seaboard 4 positions), Bradley (Canadair 6 positions)

Manufacturer

Hobart Brothers Company, Power Systems Division, Troy, Ohio, USA

Hobart Jet-Ex Dc Ground Power Units

The Jet-Ex II provides ground electrical power characteristics tailored to the requirements of turbo-jet, turbo-prop and piston engine executive aircraft with 28·5 volt dc electrical systems. Current may be preset to any limited value in two ranges of 700 to 1300 and 900 to 2000 amperes as recommended by the aircraft manufacturers. A special 400 ampere current limiting system is provided for helicopter and small turbine starting. The current limiting system provides constant voltage up to a pre-selected current value at which time the voltage drops to maintain the selected current desired for each aircraft or helicopter. The unit features an all-welded steel frame with a canopy and hinged panels for access to the engine and generator. The generator and engine controls are protected by a hinged Plexiglass door. It is powered by a Perkins 354CID, six-cylinder industrial diesel engine, developing 108 hp, governed at 2400 rpm. A lifting eye, cable hangars, four clearance lights, triplex reflectors and fire-safety fuel cap are furnished as standard equipment. The self-contained unit can be mounted on an optional trailer with tow-bar.

The Jet-Ex 3 has similar features with a current limiting facility which can be adjusted between 300 and 1500 amperes. The unit is powered by a Ford industrial petrol engine producing 92 hp at 2400 rpm and is available as a fixed-base self-contained unit or mounted on a trailer equipped with a tow-bar.

Manufacturer

Hobart Brothers Company, Power Systems Division, Troy, Ohio, USA

Characteristics

MODEL	Jet-Ex D II	Jet-Ex 3
ENGINE	Perkins 354 CID diesel	Ford 300 CID petrol
POWER SUPPLY		
VOLTAGE	28·5 V dc	28·5 V dc
AMPERES		
peak	1900/2000 A	1500 A
range	adjustable 700–1300 A/ 900–2000 A	adjustable 300–1500 A
continuous	530 A	530 A
optional voltage	14·25 V dc	14·25 V dc
TRAILER MOUNTING		
WHEELBASE	5 ft (1·52 m)	4 ft 2½ in (1·28 m)
TRUCK	4 ft 8 in (1·42 m)	4 ft 10 in (1·47 m)
DIMENSIONS		
length	7 ft 9 in (2·36 m)	7 ft 7 in (2·31 m)
width	5 ft 6 in (1·68 m)	5 ft 6 in (1·68 m)
height	5 ft 1 in (1·55 m)	4 ft 7 in (1·4 m)
FIXED BASE DIMENSIONS		
length	6 ft 8 in (2·03 m)	6 ft 11 in (2·11 m)
width	3 ft (0·92 m)	3 ft 7 in (1·09 m)
height	3 ft 11 in (1·2 m)	4 ft (1·22 m)
WEIGHT	2980 lb (1352 kg)	2100 lb (953 kg)

Hobart Jet-Ex dc ground power unit self-contained

Self-contained Jet-Ex II

Truck mounted on customer's vehicle

Hobart Superjet Mobile 400 Hz Ground Power Units

Hobart produces a variety of 400 Hz ground power units in the Superjet series which are available with a choice of Cummins or General Motors diesel power plants. They can be mounted on a specialist self-propelled chassis, truck, trailer or as a fixed-base self-contained unit. Additionally 60 and 90 kVA units are offered suitable for installation on International Harvester Hough T-500 and T-800 tractors. Superjet ground power units are engineered to support all 400 Hz jet aircraft including the Boeing 747, McDonnell-Douglas DC-10 and the Lockheed L-1011. Each model features a brushless alternator; diesel engine; isochronous governor; solid-state magnetic amplifier regulator, metering and protective devices in removable trays with modular controls. Alternator controls and meters are contained in the top tray. In the second tray the protective monitor system has scr controlled indicating lights and protective relays. The protective system will disconnect the electrical system of the aircraft from the ground power unit when overload, over-, under-voltage, over- or under-frequency conditions occur. The third tray contains the magnetic amplifier, and the adjustable regulator which provides automatic voltage line drop compensation with adjustable settings to match the size and length of cable being used.

Characteristics

MODEL RATING	60 kVA	90 kVA	140 kVA	160 kVA
ENGINE	Cummins V6-378/ GM 3-71 N diesels	Cummins V8-504/ GM 4-71 N diesels	Cummins NT-280/ GM V8-71 N diesels	Cummins NT-310/ GM V8-71 N diesels
POWER SUPPLY	48 kW 400 Hz	72 kW 400 Hz	112 kW 400 Hz	128 kW 400 Hz
RATED AMPERES	173 A	260 A	404 A	462 A
AVAILABLE MOUNTING	Self-propelled truck/trailer tractor, or fixed base	Self-propelled truck, trailer tractor, or fixed base	Truck/trailer	Truck/trailer

Manufacturer

Hobart Brothers Company, Power Systems Division, Troy, Ohio, USA

International Harvester PAY Mover Aircraft Tow Tractors

Four models of the PAY mover tractor are produced for towing aircraft; the T-225 SL, T-300 SL, T-5003 and the T-8005. All models are offered with a choice of diesel or petrol engines.

A torque converter delivers multiplied engine power to either an industrial type automatic or construction type power-shift transmission. Both drive axles are equipped with a power-transfer differential, so that increased torque is shifted automatically to the wheels with best traction, and construction-type planetary final drives. The tractor is fitted with three power-assisted steering modes: conventional two-wheel steering, four-wheel co-ordinated steering reducing the turning radius to a minimum and four-wheel crab steering that allows oblique travel to position the tractor for coupling to the aircraft. Other features common to all PAY mover models are standard pin-type couplers which incorporate pin-retainer cams, power-assisted four-wheel service brakes and drive line parking brakes. Optional equipment includes a platform to accept a gpu and a choice of internal power cabling for supplying gpu output to a front receptacle and a full-width steel cab.

The T-225SL rated at 22 500 lb (10 206 kg) draw-bar pull, and the T-300SL rated at 30 000 lb (13 608 kg) are designed to handle conventional-body aircraft. A choice of power plants is available for both models either a 153 kW petrol or a 147 kW diesel engine.

The four-speed automatic transmission provides a top speed of 17 mph (27 km/h) for petrol engine machines and 15 mph (24 km/h) for diesel powered units. The integral multi-phase torque converter provides both torque multiplication and a direct-drive phase. Heavy-duty planetary axles with power transfer differentials and standard tyres complete the drive train. Both units can be equipped with additional ballast for dealing with heavier aircraft or adverse apron conditions. For the T-300SL, the standard gross vehicle weight of 40 000 lb (18 144 kg) can be increased to 53 000 lb (24 040 kg). For the T-225SL, gross vehicle weight can be boosted from the standard 32 000 lb (14 516 kg) to a total of 40 000 lb (18 144 kg).

The low profile T-500S with a draw-bar pull capacity of 50 000 lb (22 680 kg) is designed for under-aircraft operation and, with few exceptions, can handle all commercially available aircraft in all types of weather. The unit is fitted with a 253 kW diesel engine; a four-speed power-shift transmission providing a top speed of 21 mph (34 km/h) and two front and rear axles, with a torque multiplication ratio of 25·5:1. Other features include power transfer differentials and planetary final drives; three-mode steering and a standard operating weight of 65 000 lb (29 484 kg).

The T-500S is equipped with both front and rear operator's station and a front passenger compartment. The front station is self-elevating and telescopes 1 foot 6 inches (457 mm) to facilitate positioning the tractor. A vertical wand sensory warning system automatically lowers the operator's station for operation under the aircraft. The unit can be counterweighted to a total gross vehicle weight of 110 000 lb (49 896 kg). Optional equipment available includes cushioned hitches to protect the nose gear, a fire extinguisher system activated from either operator station which automatically shuts down the tractor and gpu engines; and an auxiliary steering system that gives the operator power steering control in the event of engine failure.

The T-800S rated at 80 000 lb (36 288 kg) drawbar pull, can handle all types of aircraft. The unit has a standard weight of 108 000 lb (48 988 kg) and a total operating weight of 150 000 lb (68 039 kg) with maximum counterweight. Two independent, synchronised drive trains power the T-800S: two six-cylinder diesel engines developing a total of

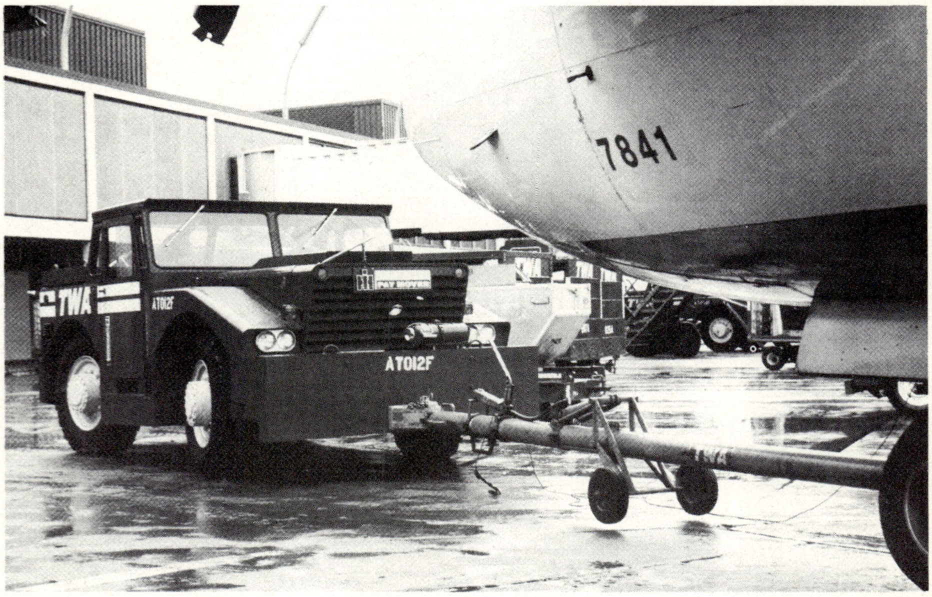

International Harvester T-300 SL PAY tractor

International Harvester T-500 S PAY tractor

348 kW; two torque converters; two full power-shift transmissions with three forward speeds; and planetary axles with power-transfer differentials. Other standard features include three-mode steering, three operator's stations, and a front-elevating cab. Optional equipment includes jack adaptor brackets, cushioned hitches and an integral fire extinguishing system.

International Harvester T-800 S PAY tractor

Status
In service with major airlines including Air France, Air India, China Airlines, Continental, Delta, Northwest, Pan Am, Singapore Airlines, TWA and United Airlines

Manufacturer
International Harvester, Schaumburg, Illinois, USA

Characteristics

MODEL	T-225 SL	T-300 SL	T-500 S	T-800 S
ENGINE	IH VS-549 Petrol/Detroit Diesel	6 V-53 N	Detroit Diesel 8 V 71 T	Detroit Diesel 6 V 53 T
OUTPUT	153 kW at 3200 rpm/147 kW at 2800 rpm		253 kW at 2200 rpm	174 kW at 2500 rpm
MAX SPEED	17 mph (27 km/h)	15 mph (24 km/h)	21 mph (34 km/h)	17 mph (28 km/h)
DRAW-BAR PULL	22 500 lb (10 206 kg)	32 000 lb (13 608 kg)	50 000 lb (22 680 kg)	80 000 lb (36 228 kg)
OUTSIDE TURNING RADIUS				
TWO-WHEEL STEERING	27 ft 11 in (8·51 m)	27 ft 11 in (8·51 m)	34 ft 6 in (10·52 m)	40 ft (12·2 m)
FOUR-WHEEL STEERING	17 ft 11 in (5·46 m)	17 ft 11 in (5·46 m)	22 ft (6·71 m)	25 ft (7·6 m)
COUPLER HEIGHT				
front	1 ft 6 in (0·457 m)	1 ft 6½ in (0·472 m)	1 ft 4 in (0·406 m)	1 ft 6 in (0·457 m)
rear	1 ft 4 in (0·406 m)	1 ft 4½ in (0·422 m)	1 ft 4 in (0·406 m)	1 ft 4 in (0·406 m)
WHEELBASE	8 ft 5½ in (2·58 m)	8 ft 5½ in (2·58 m)	11 ft 10 in (3·6 m)	13 ft 7 in (4·14 m)
RACK	6 ft 4½ in (1·94 m)	6 ft 4½ in (1·94 m)	7 ft 2 in (2·13 m)	7 ft 2 in (2·13 m)
DIMENSIONS				
length	15 ft (4·57 m)	15 ft (4·57 m)	26 ft (7·92 m)	30 ft (9·15 m)
width	8 ft (2·44 m)	8 ft (2·44 m)	9 ft (2·74 m)	9 ft (2·74 m)
height	7 ft 4 in (2·24 m)	7 ft 5 in (2·26 m)	6 ft 8 in (2·03 m)	6 ft 8 in (2·03 m)
WEIGHT				
standard	32 000 lb (14 516 kg)	40 000 lb (18 144 kg)	65 000 lb (29 484 kg)	108 000 lb (48 988 kg)
ballasted	40 000 lb (18 144 kg)	53 000 lb (24 040 kg)	110 000 lb (49 896 kg)	150 000 lb (68 039 kg)

Kersey Battery-powered Aircraft Tractors

Three models of battery-powered aircraft tractors are produced by Kersey, the K-600, K-350 and the K-300. All feature a low-profile design with flat top and vertical sides, hydrostatic steering and stepless scr control.

The K-600 is powered by two 40 hp (30 kW) motors, with an additional 10 hp (7·5 kW) motor operating the hydraulic systems. Both axles are driving and steering axles with heavy-duty leaf suspension and disc brakes. The overall height of 4 feet 3 inches (1·3 metres) with an open driver's position, permits operation under the fuselage of L-1011 and A300 aircraft. A two passenger bench seat is mounted over the 12-volt battery compartment.

The K-350 has many similar features to the larger K-600 model, with two 20 hp (15 kW) motors driving front and rear axles, four-wheel steering and a 10 hp (7·5 kW) motor powering the hydraulic systems. Drum brakes are fitted which are power-assisted in operation. The driver's position is open and a single passenger seat is installed in the passenger compartment. Additional ballast can be supplied to increase the unit weight to 40 000 lb (18 144 kg).

The K-300 is a more recent development from Kersey. Rated at 13 125 lb (5925 kg) draw-bar pull capacity, additional counterweights are available to increase the tractive effort to a maximum of 16 000 lb (7258 kg). The tractor is driven by one 40 hp (30 kW) motor with a 10 hp (7·5 kW) motor operating the hydraulic system, and is equipped with four-wheel drive and steering and two wheel-drum brakes which are power-assisted.

A driver's cab is offered as optional equipment on all models.

Kersey K-600 aircraft tractor

Kersey K-350 aircraft tractor

Characteristics

TYPE	K-600	K-350	K-300
MOTOR	Electric	Electric	Electric
DRIVE	2 × 40 hp (30 kW)	2 × 20 hp (15 kW)	1 × 40 hp (30 kW)
HYDRAULIC	1 × 10 hp (7·5 kW)	1 × 10 hp (7·5 kW)	1 × 10 hp (7·5 kW)
BATTERY	2 × 600 Ah	2 × 350 Ah	
CAPACITY	1200 Ah	700 Ah	340/480 Ah
CHARGING TIME	8 h	8 h	
MAX DRAW-BAR PULL	36 000 lb (16 330 kg)	21 000 lb (9526 kg)	13 125 lb (5956 kg)
option A ballast			14 700 lb (6668 kg)
option B ballast			16 000 lb (7258 kg)
ADHESION FACTOR	0·6 μ	0·6 μ	0·6 μ
OUTSIDE TURNING RADIUS	22 ft (6·7 m)	20 ft (6·1 m)	20 ft (6·1 m)
WHEELBASE	13 ft 4 in (4·06 m)	10 ft 6 in (3·2 m)	6 ft 11 in (2·11 m)
TRACK			
UNIT DIMENSIONS			
length without hitches	23 ft 6 in (7·16 m)	19 ft 10 in (6·05 m)	14 ft (4·27 m)
length with hitches	27 ft 4 in (8·3 m)	22 ft 5 in (6·83 m)	15 ft 10 in (4·83 m)
width	8 ft 8 in (2·64 m)	8 ft (2·44 m)	8 ft (2·44 m)
height without cab	4 ft 3 in (1·3 m)	3 ft 9 in (1·14 m)	4 ft 2 in (1·27 m)
height with cab	5 ft 2 in (1·57 m)	5 ft (1·52 m)	5 ft 8 in (1·73 m)
WEIGHT	60 000 lb (27 216 kg)	35 000 lb (15 876 kg)	25 000 lb (11 340 kg)
with ballast		40 000 lb (18 144 kg)	
option A			28 000 lb (12 700 kg)
option B			32 000 lb (14 515 kg)

Status
Production: the K-600 and the K-350 units are in service with Eastern Airlines

Manufacturer
Kersey Manufacturing Co, Bluefield, Virginia, USA

McCormick-Morgan Fixed 400 Hz power and Pneumatic Distribution Systems

McCormick-Morgan specialise in the design, manufacture and installation of fixed 400 Hz ground power distribution and fixed pneumatic systems. The company provided the 400 Hz power distribution for the Midfield Project at Hartsfield International Airport, Atlanta, Georgia, USA. This consisted of several generating rooms with power distributed over apron drive bridges for 123 gates. Fixed-pneumatic systems supply air for aircraft air-conditioning and starting at 28 gates.

Status
Apart from Atlanta where in addition to those mentioned above McCormick-Morgan are supplying pneumatic systems for Piedmont, Braniff and Republic Airlines, the company has completed 400 Hz installations at 24 airports in USA and two for Air Canada at Calgary and Toronto. Fixed pneumatic systems by McCormick-Morgan are also in service at Houston and Pittsburgh airports

Manufacturer
McCormick-Morgan, San Francisco, California, USA

RLG Aircraft Docking System

The RLG automated guide-in system has been developed to guide and stop aircraft in a precise position at airport terminals. Current airport design generally requires nose-in docking with the correct positioning of aircraft to loading bridges and fuel pits. The system can accommodate all aircraft types on a single docking device. It consists of a metal enclosure attached to the terminal building or other support in a position perpendicular to and 21 inches (0·53 metres) left of the aircraft J-line or centre-line, this dimension aligns to the centre of the pilot's left hand seat. The bottom of the enclosure contains three neon tubes, a green tube flanked by two yellow tubes (red tubes are optional). The green tube is encased in 18-inch (0·46 metre) deep baffles, the yellow tube is mounted behind flanges. When the pilot taxis in and can see only the green tube, that is a correct centre position to ±3 inches (76·2 mm). If a yellow line is shown on either side, the aircraft is off line in that direction. The upper half of the enclosure housing contains digital aircraft indicators — 747, 10, 707, etc, up to 17 different types as required for a particular position for that aircraft mix. This part of the housing may as the mounting requires be situated to the side of the light housing. Below the aircraft indicator are three sets of lamps; top, green — centre, amber — bottom, red. The green lamp indicates that the apron is clear ready for aircraft and system programmed for aircraft indicated; amber, 15 feet (4·6 metres) to stop; red, stop. As each colour is activated the other colours are automatically turned off.

The control unit is switched on by operating the start switch on a control box at base of ramp. The system is automatic and the red and amber lamps are activated by the aircraft as its nose wheel passes over induction loops in the ramp. These loops are turns of wire in 0·25-inch (6·4-mm) wide by 1·5-inch (38·1-mm) deep saw cuts in a rectangular configuration 1 foot (0·305 metres) by 6 feet (1·83 metres). Since aircraft are of different dimensions from nose wheel to passenger door, separate sets of loops are required for different types. However where apron-drive loading bridges are available and fuelling pits are flexible it is possible to accommodate various types of aircraft with common loops.

A selector switch is fitted on the control box with a position for each type of aircraft. In order for the ground crew to operate the start switch to indicate gate readiness to receive a particular aircraft the proper switch position must be selected. After docking; recycling is not necessary since controlling circuitry automatically switches the system off approximately 10 seconds after the red lamps have been activated. The system can accommodate two separate centre lines with a single control box if a loading bridge is not versatile enough to service wide- and narrow-body aircraft on a common centre line or due to the position of the fuel pits.

An emergency stop switch is mounted on the control box which activates the red stop lights. This may be required if auxiliary vehicles obstruct the path of parking aircraft or loading bridges are improperly stowed. Semi-automatic and manual switches are on the controller for remote or temporary parking positions.

A deadman switch is optional. Other safety features include a search system which de-activates the entire system should both lamps fail, and printed circuit check system which automatically checks the system controls and shows the condition by indicator lights on the control panel.

Status
The RLG docking system is installed at the following airports: Helsinki, Finland; Seoul, Republic of Korea; Penang, Malaysia; Manila, Philippines; Jeddah, Saudi Arabia; Singapore airports; Guam and 20 airports in the USA

Manufacturer
RLG and Co, Studio City, California, USA

S and S Model SS30 AV Tow Trac Tracker

S and S Model SS30 AV Tow Trac is a battery-powered tractor designed for aircraft towing duties. Powered by an integral gearhead dc motor supplied from 128 volt battery source, the unit is equipped with a stepless solid-state control system with an inching control between 0 and 4 mph (0 and 6·4 km/h). A dual hydraulic disc-brake system acts on all four wheels and the tractor is fitted with planetary steerable axles and hydraulic power steering. Other features include an enclosed driver's cab with passenger seat, front and rear towing hitches and a built-in battery charger. The latter is designed for 3 phase, 440 volt supply and is rated at 30 kVA. Optional equipment offered includes power-phase monitor for battery charger supply, removable driver's cab, alternative towing attachments, fire extinguishing systems and an emergency steering capability.

Characteristics
TYPE: SS30 AV
MOTOR: Electric
OUTPUT: 45 hp (34 kW)
BATTERY: 128 V
CAPACITY: 675 A/h
MAX DISCHARGE RATE: 6 h
RECHARGE CYCLE: 8 h
MAX PUSH-OUT SPEED: 4·2 mph (6·8 km/h)
MAX DRAW-BAR PULL: 30 000 lb (13 608 kg)
TURNING RADIUS: 8 ft 1½ in (2·48 m) minimum
WHEELBASE: 10 ft 4 in (3·15 m)
UNIT DIMENSIONS
length: 18 ft 6 in (5·64 m)
width: 8 ft 5 in (2·57 m)
height: 5 ft 6 in (1·68 m)
UNBALLASTED WEIGHT: 42 730 lb (19 382 kg)
BALLASTED WEIGHT: 60 000 lb (27 215 kg)

Status

Field trials were conducted with the SS30 AV at Chicago O'Hare Airport, USA in 1981 by TWA, Delta and Continental, and production is scheduled for early 1982. In addition to this model S and S are developing two other variants, the SS 16 AV rated at 16 000 lb (7256 kg) draw-bar pull and a larger version the SS 50 AV to be capable of 50 000 lb (22 676 kg) draw-bar pull. The SS 16 A prototype was completed in early 1982 and full production is scheduled to follow. The SS 50 AV is expected to enter production in 1983.

Manufacturer

S and S Corporation, Cedar Bluff, Virginia, USA

S and S Model SS30 AV aircraft tug

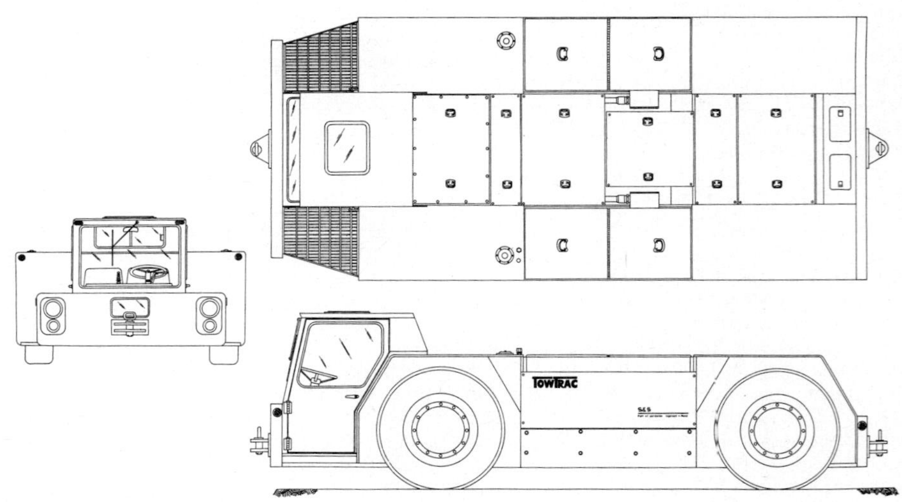

Stewart and Stevenson Aircraft Pneumatic Starters

These are available in truck or trailer-mounted models with delivery rates from 120 to 270 lb per minute (54 to 122 kg per minute). The prime mover is a GM Detroit Diesel 8V-71, 8V-92T, 8V-92 (two), or 16V-71 depending on the model. All models can be used for aircraft air-conditioning or jet-start purposes. A further development is a truck-mounted unit with an 8V-92T engine with pilot command air control for air-conditioning or air-starting wide-body aircraft.

Status

Production. In service with major airlines world-wide

Manufacturer

Stewart and Stevenson Services Inc, Houston, Texas, USA

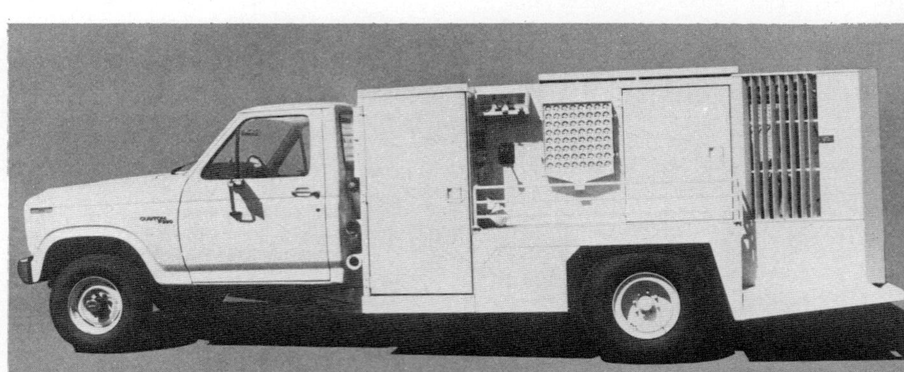

Stewart and Stevenson aircraft pneumatic starter

Stewart and Stevenson Aircraft Towing Tractor

The tractor produced by Stewart and Stevenson features an all-steel welded body with a tractive effort from 14 000 to 40 000 lb (6350 to 18 150 kg). The prime mover is a GM Detroit Diesel 3-53, 4-53 or 4-53T depending on the model, driving through an Allison automatic transmission Model AT-540 or TT 3420. A battery-powered model is available as a further option.

Status
Production. In service with major airlines world-wide

Manufacturer
Stewart and Stevenson Services Inc, Houston, Texas, USA

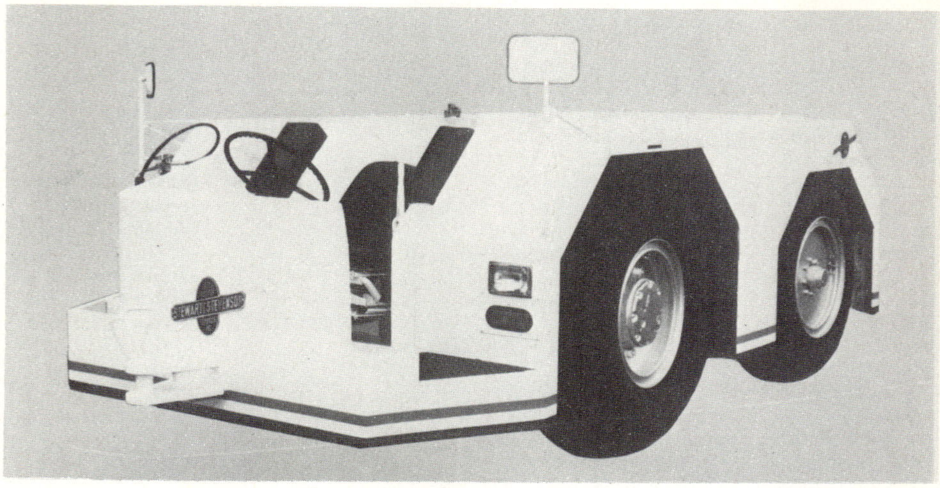

Stewart and Stevenson aircraft tug

Stewart and Stevenson Electrical Ground Power Unit

Eight models are produced which range from 30 kW to 160 kW; 400 Hz with 0·8 power factor, and 115/200 volt ac or optional 28·5 to −112 volt dc. All models are available in truck, trailer, skid or tow tractor configurations. The prime power is a GM Detroit Diesel engine model 2-71, 3-71, 4-71, 6V-71, or 8V-71.

Status
Production: in service with major airlines world-wide

Manufacturer
Stewart and Stevenson Inc, Houston, Texas, USA

Stewart and Stevenson electrical ground power unit

Stinar Lavatory Service Vehicles

Three models of vehicles for servicing aircraft lavatories are offered by Stinar. They are the SLS-250, SLS-350 and the SLS-500.
The SLS-250 is designed and built for servicing narrow-body and commuter aircraft. This model can be equipped for warm or cold weather operations. Although the primary function of the unit is servicing the smaller aircraft, a lift system can be installed to provide servicing capabilities to wide-body aircraft.
The model is based on a ¾-ton pick-up or cab and chassis equipped with six-cylinder petrol engine and manual transmission. A V-belt-driven hydraulic pump with electric clutch powers the hydraulic motor which operates the water pump. An engine-driven water pump with electric clutch is available as a warm weather configuration. All pumps and fittings are either brass or stainless steel. A meter indicating US gallons which can be reset to zero is incorporated.
Stainless-steel tanks are mounted on a plywood-covered framework attached to the vehicle frame. The water/charge tank is equipped with an access panel, overflow lines, sight gauge, baffles, and vents. A proportional water/charge tank is available for optional dual tank configuration. The waste tank is equipped with sight gauge, baffles, man-hole and 4-inch (102-mm) drain. Access plates are provided for maintenance and inspection on both tanks.
Pumps, valves, hoses with couplings and associated plumbing are in the rear of the vehicle in the pump compartment which is illuminated for night operation. A stainless-steel hose storage pan at the rear of the shroud is attached directly to the waste tank. This allows any spillage to drain back into the waste tank. A waste hose with coupling is stowed in this area. A storage nipple for the waste-hose coupling is attached to a bulk-head which enables any spillage from the hose to drain into the waste tank. The top deck is constructed in heavy-gauge steel plate and an access plate is provided on top for entrance into the waste tank. Optional equipment available includes: scissor-type hydraulic lift for servicing wide-body aircraft,

Stinar SLS-350 lavatory service vehicle

Stinar SLS-500 PL lavatory service vehicle

larger tank capacity, hose reel and a closed dump system.
The SLS-350 based on a 1-ton chassis is equipped with a 200 US gallon (757 litre) water/combined charge tank and a waste tank with 450-US gallon (1703-litre) capacity. All other features are similar to model SLS-250.
The SLS-500 is designed for wide- and narrow-body aircraft operations where large capacities are needed for peak operating periods. Lift systems are available to provide access to wide-body servicing points. The chassis on this model of lavatory truck is a 2-ton truck with a short wheel base to support the large capacities. The 2-ton cab and chassis is equipped with a V8 engine and manual transmission.
Optional equipment available comprises two types of hydraulic lift for servicing wide-body aircraft, larger tank capacities, hose reel, closed dump system, diesel engine, automatic transmission, power steering and a suction pump. All other features are similar to model SLS-250.

Characteristics

MODEL	SLS-250	SLS-350	SLS-500
ENGINE	6-cylinder petrol	6-cylinder petrol	V8 petrol
TANK CAPACITIES			
water/charge	160 US gallons (606 litres)	200 US gallons (757 litres)	300 US gallons (1136 litres)
waste	300 US gallons (1136 litres)	450 US gallons (1703 litres)	550 US gallons (2082 litres)
WATER PUMP	16–20 US gallons/minute at 40 psi (61–76 litres/minute at 2·75 bar)	16–20 US gallons/minute at 40 psi (61–76 litres/minute at 2·75 bar)	16–20 US gallons/minute at 40 psi (61–76 litres/minute at 2·75 bar)
WHEELBASE	11 ft 1 in (3·38 m)	11 ft 5 in (3·48 m)	12 ft 9 in (3·89 m)
UNIT DIMENSIONS			
length	18 ft (5·49 m)	19 ft 6 in (5·94 m)	21 ft (6·4 m)
width	6 ft 1 in (1·85 m)	7 ft (2·13 m)	7 ft 6 in (2·29 m)
height	6 ft (1·83 m)	6 ft (1·83 m)	7 ft 3 in (2·21 m)
WEIGHT	6300 lb (2857 kg)	7500 lb (3402 kg)	9000 lb (4082 kg)

Status
In service with airlines in the USA

Manufacturer
Stinar Corporation, St. Paul, Minnesota, USA

Stinar Potable Water Vehicles

Three models of potable water vehicles are available from Stinar. They are the SPW-250, SPW-350 and SPW-500.
The SPW-250 is designed and built for servicing narrow-body and commuter aircraft. This model can be equipped for warm or cold weather operation and although the primary function is for servicing the smaller aircraft, a lift system can be installed to provide servicing capabilities to wide-body aircraft. The unit is based on a ¾-ton pick-up or cab and chassis equipped with six-cylinder petrol engine, and manual transmission. A V-belt driven hydraulic pump with electric clutch powers the hydraulic motor which drives the water pump. An engine driven water pump with electric clutch is available as a warm weather configuration. All pumps and fittings are either brass or stainless steel. Stainless-steel tanks are mounted on a plywood covered framework which is attached to the vehicle frame. They are equipped with overflow lines, level gauge, baffles and vents. Access plates are provided for maintenance and inspection. Pumps, valves and hose reel with 25 feet (7·62 metres) of hose and coupling are located at the rear of the vehicle. This compartment is illuminated for night operation. The top deck is constructed of heavy-gauge steel plate incorporating a hinged inspection cover. Access steps allow the deck to be used as a work platform. Optional equipment available includes scissor type hydraulic lift for servicing wide-body aircraft, larger tank capacity, V8 engine, automatic transmission and power steering.
The SPW-350 based on a 1-ton chassis, is equipped with 500-US gallon (1893-litre) capacity water tank. All other features are similar to the model SPW-350.
The SPW-500 is designed and built primarily for both wide- and narrow-body aircraft where large capacities are needed for peak operation periods. Lift systems are available to provide access to wide-body servicing points. A 2-ton short-wheel base truck chassis has been chosen to support the required capacities, equipped with a V8 engine, and manual transmission.
Optional equipment available includes two types of hydraulic lifts for servicing wide-body aircraft, larger tank capacity, diesel engine, automatic transmission and power steering. All other features are similar to the SPW-250.

Stinar SPW-500 potable water truck

Status
In service with airlines in the USA

Manufacturer
Stinar Corporation, St Paul, Minnesota, USA

Characteristics

MODEL	SPW-250	SPW-350	SPW-500
ENGINE	6-cylinder petrol	6-cylinder petrol	V8 petrol
TANK CAPACITY	350 US gallons (1325 litres)	500 US gallons (1893 litres)	700 US gallons (2650 litres)
WATER PUMP	20–22 US gallons/minute at 40 psi (76–83 litres/minute at 2·75 bar)	20–22 US gallons/minute at 40 psi (76–83 litres/minute at 2·75 bar)	20–22 US gallons/minute at 40 psi (76–83 litres/minute at 2·75 bar)
WHEELBASE	11 ft 1 in (3·38 m)	11 ft 4 in (3·48 m)	12 ft 9 in (3·89 m)
UNIT DIMENSIONS			
length	18 ft (5·49 m)	19 ft 6 in (5·94 m)	20 ft 8 in (6·3 m)
width	6 ft 1 in (1·85 m)	7 ft (2·13 m)	7 ft 6 in (2·29 m)
height	6 ft (1·83 m)	6 ft (1·83 m)	7 ft 3 in (2·21 m)
WEIGHT	6000 lb (2722 kg)	7000 lb (3175 kg)	8000 lb (3629 kg)

Teledyne Centralised 400 Hz Ground Power Distribution Systems

A central 400 Hz power system consists of three main parts: the central 400 Hz power source, the distribution network, and the service cable storage and handling devices. The power source in the Teledyne system is a 400 Hz vertical motor generator. This is a synchronous brushless motor-generator available up to 313 kVA. Multiple units can be used to support the varying load demands and they can be installed in parallel to increase system reliability. A voltage regulator simultaneously controls both the motor and generator fields keeping power factor on input ac power which can be either 50 or 60 Hz near unity. As the load demand on the generator decreases, the input ac power to the motor decreases until at zero generator load the motor current is 10 per cent of the full load amperage. The voltage regulator is fully solid-state with protection against operating at overvoltage, undervoltage and underfrequency. To achieve automatic paralleling, a circuit is provided which will start a second generator when the first generator has reached a pre-determined load-share point. When the load reaches the second pre-determined load-share point, a signal is given and the second generator will automatically parallel to come on line with the master generator resulting in a sharing of the load demand. This unit will remain on-line as indicated by the load demand or for a specified time in order to preclude cycling due to load surges of short durations. This concept may be extended to three or more generators.
Line-drop compensators automatically operate for reactive voltage drops of up to 20 per cent in lines connecting generator to load. These passive devices may be installed in a distribution panel on the same base as the motor generator, in a separate free-standing cabinet, or at the load point in each service cabinet. Teledyne systems distribute 400 Hz power at 575, 1000 or 4160 volts. Distribution voltage drops are reduced at higher voltages so that power transfer for greater distances from the generator to the aircraft are practical. A step-down transformer reduces the distributed voltage to 115/200 volts for use by the aircraft. These are designed for mounting either on the side of a passenger loading bridge or on the apron of a remote aircraft parking area. Multiple outputs and various transformer ratings are available.
Cable distances are measured from the central source to the aircraft receptacle. Many distribu-

tion networks are a combination of line, ring and spoke systems. The nominal aircraft voltage 115/200 volts individual spoke distribution is the simplest and most economical form of distribution for cable distances up to 490 feet (150 metres). Reactive losses are equalised by the line drop compensators. In the medium voltage spoke distribution systems cable sizes are reduced considerably with step-down transformers near the aircraft. With 575 volts, distances may be extended up to 2000 feet (609 metres). The same distribution may be used employing 960 volts with the use of line drop compensators placed on the opposite side of the step-down transformer and extending transmission distances up to 5000 feet (1524 metres). Line and ring distribution of standard cable utilising 960 volts, which is higher than the nominal aircraft voltage, may be extended up to 2460 ft (750 metres). 4160 volts distribution may be employed economically for distances measured in miles or kilometres. Depending on the airport layout, a ring bus distribution may be preferred or combinations of spoke, line and ring configurations.

Storage and handling of the 400 Hz service cable and aircraft duct may be performed in many standard fashions dependent on the individual requirements of the airport. The 400 Hz power may be brought to an aircraft from a fibre-glass pit with a lightweight aluminium alloy hinged-cover strong enough to support a Boeing 747 wheel. Power is supplied to the pit by underground cables. A 30 foot (10 metre) service cable with aircraft plug is stowed in the pit and pulled out through rollers when aircraft power is required. Control of 400 Hz power is contained just inside and below the lightweight cover.

Another method is to use a padded counter-balanced boom to support the service cable. The boom is stored alongside the passenger bridge tunnel. When in operation it is swung under the aircraft fuselage to permit the short service cable and plug to reach the aircraft receptacle. Power controls are at the end of the boom. With a cable retriever reel an electric motor with gear reduction turns two reels to lift the service cable for storage alongside of a passenger bridge tunnel. The central control switch for the cable reel is near the aircraft plug. When this is operated to release sufficient service cable, the operator walks to the aircraft receptacle and plugs in 400 Hz power. Control of the power contractor is at the control centre. The cable is reeled in for storage when use of power is no longer needed.

Teledyne frequency counter

Teledyne time drop compensator

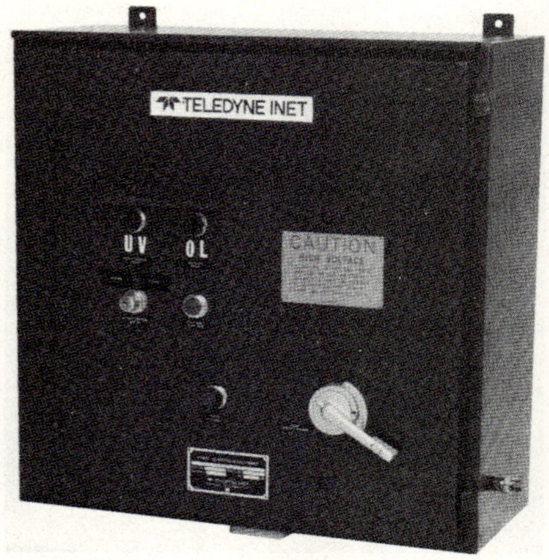

Teledyne gate service cabinet

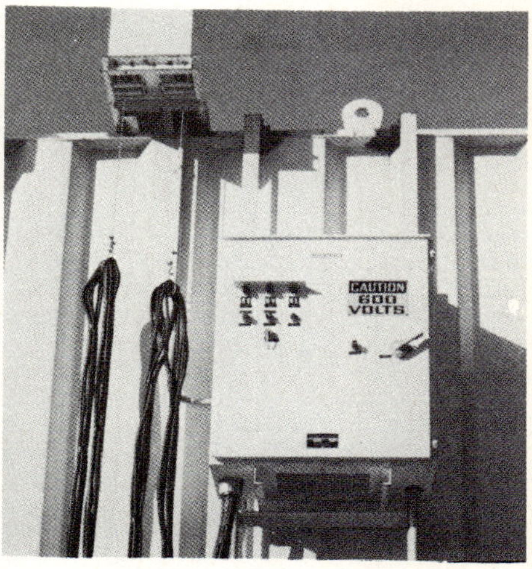

Teledyne cable carrier

Over-the-bridge power delivery is a further method for either fixed- or apron-drive type telescoping passenger bridges. This allows the cable to be extended or retracted with the movement of the bridge. The devices are mounted on the sides of the bridges.

Status

Teledyne Inet 400 Hz distribution systems have been installed at 44 airports in 18 countries

Manufacturer

Teledyne Inet, Torrance, California, USA

Trilectron Auxiliary Power Units

Five models are produced by Trilectron for use as auxiliary power units. They are the dc Lectropak, the REV-IT III, IV-H, V-FH, and VI-S.

The dc Lectropak is a solid-state dc power supply which can be used as a stationary or mobile unit for most fixed- and rotary-wing aircraft requiring up to 300 amperes continuous for dc servicing and up to 1200 amperes starting current. The unit features a voltage adjustment facility between 22 and 30 volts, and a current regulating device which enables the unit to be used as a battery charger for on-board batteries.

REV-IT III is a 28-volt dc ground power source for aircraft starting and servicing powered by a 21 hp diesel engine. It can be used for most fixed and rotary wing aircraft requiring up to 1600 amperes peak starting current and up to 300 amperes continuous for operation and servicing of dc systems. The unit is equipped with an automatic battery charging facility and can be operated in a hangar from an ac power source. It is mounted on a leaf-spring suspension system with a single axle and pneumatic tyres.

REV-IT IV-H is a diesel-engined dc power unit designed for helicopter starting and servicing. It can provide up to 500 amperes peak starting current and 300 amperes continuous for operating and servicing of on-board dc systems. The engine can be operated on Jet A or JP4 fuels. It incorporates

Trilectron REV-IT V-FH auxiliary power unit

an adjustable device for limiting output current and/or a soft-start feature to prevent over-torque. The unit is mounted on a swivelling front wheel and two fixed wheels at the rear, and is equipped with a tow-bar.
REV-IT V FH is a petrol-engined dc power source. It can be used for helicopters or fixed-wing aircraft requiring up to 1000 amperes peak starting current and up to 120 amperes continuous current. The unit can be run on 220 volts single phase for use in the hangar as a power source. An automatic recharge facility provides repeated starts with a dual 14/28 volts dc option available. It is mounted on pneumatic tyres in a tricycle arrangement.
REV-IT VI-S is similar to the V-FH but can be powered by either a diesel or petrol engine, providing a 28 volts dc supply, up to 1600 amperes peak and 600 ampere continuous current. An optional-powered drive system is available.

Manufacturer
Trilectron Industries Inc, Hawthorne, New Jersey, USA

Characteristics

MODEL	Dc Electropak	REV-IT III	REV-IT IV-H	REV-IT V-FH	REV-IT VI-S
ENGINE		Lombardini 1-cylinder diesel	2-cylinder diesel	1-cylinder petrol	6-cylinder petrol/4-cylinder diesel
OUTPUT		21 hp (16 kW)	24 hp (18 kW)	10 hp (7·5 kW)	60 hp (45 kW)
DC POWER SUPPLY	28 V ±5% adjustable 22–30 V	28·5 V ±2%	28·5 V ±2%	28·5 V ±2%	28·5 V dc ±2%
peak	1200 A	1600 A	500 A	1000 A	1600 A
continuous	300 A	300 A	300 A	120 A	600 A
POWER REQUIREMENTS	220 V, 3-phase, 60 Hz or 440/480/380 V 3-phase, 50 Hz				
DIMENSIONS					
length	1 ft 10 in (0·56 m)	4 ft 8 in (1·42 m)	4 ft (1·22 m)	3 ft 6 in (1·07 m)	5 ft (1·52 m) (1·52 m)
width	1 ft 8 in (0·51 m)	4 ft 4 in (1·32 m)	2 ft 6 in (0·76 m)	2 ft 10 in (0·86 m)	3 ft 4 in (1·02 m)
height	3 ft (0·91 m)	4 ft (1·22 m)	4 ft (1·22 m)	2 ft 11 in (0·89 m)	4 ft 9 in (1·45 m)
WEIGHT	300 lb (136 kg)	750 lb (340 kg)	650 lb (295 kg)	500 lb (227 kg)	1900 lb (862 kg)

Trilectron Power-Tug Combined Tractor and Ground Power Unit

The Power-Tug is a combined aircraft towing tractor and ground power unit. Five models are available: 16/20, D-38, D-6000 and D-8000.
The 16/20 models are similar with a 10 000 lb (4536 kg) tow capacity from the 16 variant and 30 000 lb (13 608 kg) towing capacity from the larger Power-Tug 20. Both feature hydrostatic drive, and a front towing hitch. The electrical power pack mounted behind the driver's seat comprises 28 volt dc generator/battery system which supplies up to 1300 amperes starting current and 120 amperes dc on a continuous basis. An automatic charging facility is incorporated.
The DC-38 is a 38 hp diesel tug fitted with four-wheel drive, manual transmission, front towing point and wide heavy-duty tyres. The electrical power pack with a 28-volt dc generator/battery system is installed at the rear of the vehicle. Models are available with 1000, 1300 and 1600 amperes peak starting current, a choice of 120 or 300 amperes continuous supply and an automatic charging facility. The tractor which is suitable for use with aircraft weighing up to 68 000 lb (30 844 kg) can be operated on JP4 or JET A fuel.
The D-6000 is a 90 hp diesel tug rated at 6000 lb (2722 kg) draw-bar pull. The tug can be operated on JP4 or JET A fuel and is equipped with hydrostatic four-wheel drive and four-wheel steering which provides a choice of front/rear or four-wheel crab operation. Dual seats are fitted for driver and passenger and towing points can be mounted front and rear. A brushless dc generator system is installed with a solid state regulator and automatic over voltage protection. For use with general aviation aircraft up to 75 000 lb (34 020 kg) weight, the unit is offered with an optional 40 kVA, 400 Hz, 3-phase, supply with 28 volts dc and an optional dual 14/28 volt dc output.
The D-8000 powered by a 60-hp diesel engine, and rated at 8000 lb (3629 kg) draw-bar pull features similar equipment to the D-6000 but is equipped with hydraulic brakes on all four wheels compared to two-wheel hydraulic braking on the smaller model.

Manufacturer
Trilectron Industries Inc, Hawthorne, New Jersey, USA

Trilectron D-38 Power-Tug

Characteristics

MODEL	16	20	D-38	D-6000	D-8000
ENGINE	1-cylinder petrol	2-cylinder petrol	2-cylinder Lombardini diesel	219 CID-4 diesel	diesel
OUTPUT	16 hp (12 kW)	20 hp (15 kW)	38 hp (28·4 kW)	90 hp (67 kW)	60 hp (45 kW)
MAX DRAW-BAR PULL				6000 lb (2722 kg)	8000 lb (3629 kg)
tow capacity	10 000 lb (4536 kg)	30 000 lb (13 608 kg)	68 000 lb (30 844 kg)	75 000 lb (34 020 kg)	100 000 lb (45 360 kg)
DC POWER SUPPLY	28 V dc ±1%	28 V dc ±1%	28 V dc	28 V dc ±0·5 V	28 V dc ±0·5 V
peak	1300 A	1300 A	1000/1300/1600 A	1600 A	1600 A
continuous	120 A	120 A	120/300 A	600 A	600 A
AC POWER SUPPLY (optional)			400 Hz 115/200 V 750 kVA 3-phase	400 Hz 115/200 V 40 kVA 3-phase	400 Hz 115/200 V 40 kVA 3-phase
DIMENSIONS					
length	8 ft 10 in (2·69 m)	10 ft 3 in (3·12 m)	10 ft 7 in (3·25 m)		10 ft 10 in (3·3 m)
width	3 ft 1 in (0·94 m)	4 ft (1·22 m)	3 ft 11 in (1·19 m)		5 ft (1·52 m)
height	3 ft 4 in (1·02 m)	4 ft 9 in (1·45 m)	4 ft (1·22 m)	4 ft (1·22 m)	4 ft (1·22 m)
WEIGHT	1200 lb (544·3 kg)	1800 lb (816·4 kg)	2600 lb (1179·3 kg)		8000 lb (3628·8 kg)

Tronair Aircraft Tow Tractors

Three types of tow tractors designed for general aircraft are available from Tonair providing draw-bar capacities from 3000 to 12 000 lb (1360 to 5440 kg). In addition to the towing capability several models can provide 28 volt dc power for starting and servicing.
The Series 5 and 15 are four-wheel tractors with draw-bar pull capacities from 3000 to 8000 lb (1360 to 3630 kg). Both series are offered with Chrysler petrol or diesel engines and automatic transmission. The Series 15 models are equipped with a dual-hydraulic braking system fitted to all four wheels while power-assisted braking is fitted to the Series 5 vehicles.
The Series 4 are larger four-wheel tractors with draw-bar pull capacities ranging from 8000 to 12 000 lb (3630 to 5440 kg). Chrysler petrol or diesel engines are offered with automatic transmission and a powered-braking system operating on two-drive wheels.
Optional features available include a dual petrol/lpg power-pack, lpg-powered engine, all weather cab, passenger seat and front towing point.
The Starting versions are 50 PT, 80-5 or 100-5 models powered by either petrol or diesel engine fitted with a 28 volt power-pack. Power is supplied by both a 300 ampere generator and a nickel cadmium battery providing a peak of 1600 amperes. The generator charges the battery whenever the vehicle is running.

Manufacturer
Tronair, Holland, Ohio, USA

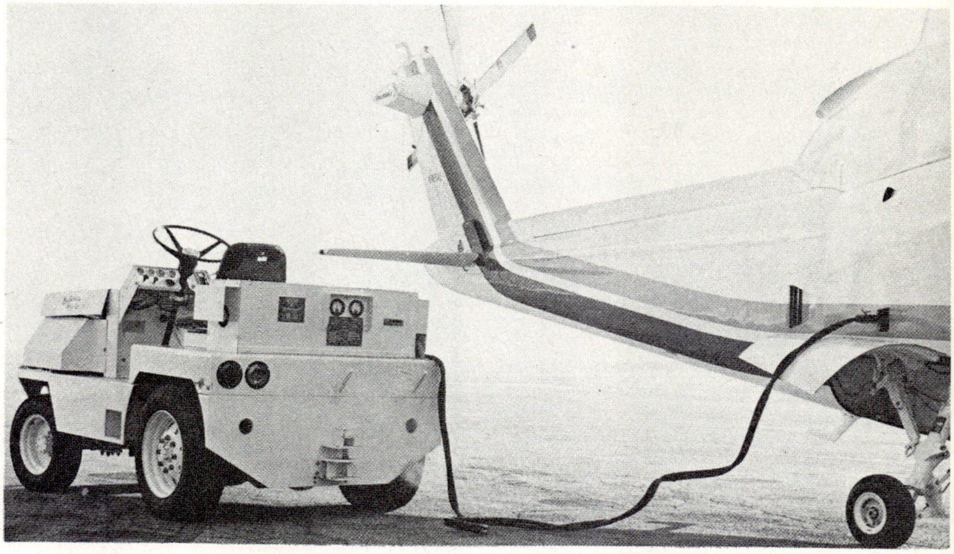

Tronair Series 15 aircraft tow tractor

Characteristics

SERIES	15	5	4
MODELS	30 PT/40 PT/50 PT	80-5/100-5	80-4/100-5/120-5
DRAW-BAR PULL	3000/4000/5000 lb (1360/1814/2268 kg)	8000/10 000 lb (3630/4536 kg)	8000/10 000/12 000 lb (3630/4536/5440 kg)
ENGINE	Chrysler 225-6 petrol	Chrysler C 441-4 diesel	Chrysler H 318-V8 petrol/Chrysler C1-655-6 diesel
UNIT DIMENSIONS			
length	8 ft 5 in (2·58 m)	9 ft 4 in (2·84 m)	10 ft 4 in (3·15 m)
width	4 ft 7 in (1·4 m)	5 ft 6 in (1·68 m)	5 ft 6 in (1·68 m)
height	5 ft (1·52 m)	5 ft 2 in (1·57 m)	
WEIGHT	4300/5300/6300 lb (1950/2404/2858 kg)	10 000/12 800 lb (4536/5806 kg)	10 000/12 800/15 000 lb (4536/5806/6804 kg)
POWER PACK	28 V dc		
OUTPUT	300 A continuous; 1600 A peak		
VOLTAGE REGULATOR	±2%		
SUPPLY CABLES LENGTH	16 ft (4·57 m)		

Tronair Tow-bars

Tronair produce tow-bars to suit all types of general aviation, corporate and commuter aircraft ranging from the Canadair Challenger to the Piper Navajo and for several types of helicopter. The tow-bars available fall into four basic designs: custom aluminium, custom steel, multi-head and snap-bak.
The Custom aluminium tow-bar constructed of aluminium with plated steel attachment fittings is equipped with a double-spring shock absorber with heavy-duty wheels and castors, positive locking mechanism to the aircraft and a free swivelling 2⅜-inch (60 mm) diameter forged steel eye. The Custom steel design is an all-steel tow-bar with individual head attachments designed for each type of aircraft. The tow-bar is fitted with handles at both ends, a positive locking mechanism to the aircraft and a free swivelling 2⅜-inch (60 mm) diameter forged steel eye. A shock absorber is available on all models as optional equipment.
The Multi-head tow-bar features interchangeable attachment heads, fitting business jets, turboprops and most helicopters. The heads are attached to the tow-bar by a single stainless steel pin. Other features include a shock absorber, positive locking mechanism to aircraft and a free swivelling 2⅜-inch (60 mm) diameter forged steel eye.
The Snap-Bak design is a universal tow-bar which fits most light aircraft. Measuring 80 inches (203 mm) long and weighing 20 lb (9 kg), the tow-bar is equipped with a 3-inch (76·2 mm) eye. The Snap-bak feature permits the tow-bar to bend at the head and this bending occurs at any time there is an attempt to exceed the aircrafts towing limits.

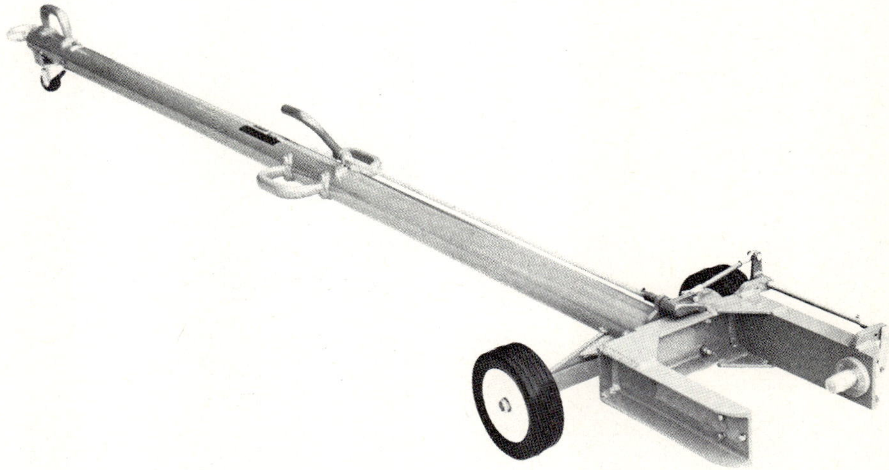

Tronair custom aluminium tow-bar

Manufacturer
Tronair, Holland, Ohio, USA

Trump Aircraft De-icer Vehicles

Two models of aircraft de-icer vehicles are produced by Trump, the D-30 and the D40-D. Both utilise the Trump method that separates the de-icing and anti-icing stages into two separate operations. In the de-icing operation, a water solution free of glycol is heated to a temperature between 175°F (85°C) and 195°F (95°C). The water is heated both in the tank and before reaching the nozzle and is then sprayed on the skin of the aircraft. The heat of the water warms the wings slightly and causes any ice to break loose and to be washed away by the stream of water. With atmospheric conditions suitable for evaporation the thin film of water left will evaporate into the air chilling the skin and in so doing leaves a cold, dry aircraft skin not susceptible to accumulating cold precipitation.
After the aircraft has been de-iced, anti-icing may or may not be necessary. If wet precipitation is falling, a thin film of glycol may be sprayed on the aircraft to provide an anti-icing film on the aircraft's upper surface. The proportioning system offers 20 possible glycol-water ratios. However, if weather conditions are such that dry snow is falling, or there is no snow at all, there is no need for anti-icing solution to be sprayed on the aircraft's surface.
The D-30 de-icer has been designed specifically for the Boeing 757 and 767 aircraft but has the capability to service current generation narrow- and wide-body aircraft. The unit is mounted on a single rear axle vehicle, the chassis and power

plant being chosen by the operator. A three-point articulating boom with basket provides the aerial platform. The lower boom is a parallelogram connected at its base to the turntable and at the top to the elbow frame. The elbow frame is kept in a constant relationship to the ground as the angle of the lower boom changes. Since the cylinder operating the upper boom is in the elbow frame, the angle of the upper boom relative to the horizon remains constant as the lower boom and elbow frame are elevated or lowered. Rigid steel levelling rods provide automatic levelling of the work basket, the tubular rods acting as conduits for carrying fluids to the basket. Hydraulic control valves are located at the work basket for operator control. Hydraulic spring lockouts eliminate the need for outriggers and the boom cannot be operated unless the lockouts are engaged. Nozzle controls which enable the operator to adjust the volume and pattern of the jet and glycol controls are installed in the basket. The boom is mounted on a turntable driven by a hydraulic motor and incorporates a priority selector valve to enable the lower controls to override the basket controls. Water is heated by the M1 closed-flame heater which has a vertical double-pass system in which the flame and/or hot gases pass down around and up two concentric coils. Heat flows from the burner nozzles to the opposite end through the centre coil. It is then turned against a water-cooled flame deflector to pass back up around the outer coil. The temperature of the heater's exterior wall remains cool from a combination of cold fluid entering the outer coil first and a cooling air passage which surrounds the heated elements. Controls monitor the operational cycle and lights indicate a malfunction. The system is programmed to shut down automatically until the cause of malfunction is corrected.

An auxiliary six-cylinder industrial engine provides power for the water and glycol delivery pumps and the hydraulic pump. An emergency hydraulic system is fitted driven by a 12-volt dc battery. This lowers the booms, operates the turntable and spring lockouts in case of auxiliary engine failure. For a short period of storage lines can be purged upon completion of de-icing. With the engine at idle the purge switch is activated as the nozzle is opened allowing the glycol to enter the system. This procedure replaces the water in pump, heater and lines with a 50/50 glycol/water mixture. For long periods of storage, the water tank can be drained or the heater can be fired for a short time during each shift. An optional standby heater, which prevents freezing of water in the tank down to −30°F (−34°C), is also available. Other features of the D-30 include an intercom system for communication between the operator's basket and the driver's cab, a ground hose, 50 feet (15 metres) long fitted with nose gun and a hand operated hose-reel. A remote clutch can be provided as optional equipment. This provides clutching or de-clutching of the water and glycol pumps in response to controls at basket and ground control station and enables purging to be carried out by a single operator. Activation of any automatic safety or emergency shut-down switch (including low tank level switches) will trigger de-clutching of the pump drive.

The D40-D is similar to the D-30, but designed primarily for wide-body operations. The M2 heater is fitted as standard equipment (available as an option with the D-30).

Status

The D-30 was introduced to the Trump range in 1980; 288 D40-D units are in service with 24 airlines, the US Air Force and the US Navy

Manufacturer

Trump Inc, Plattsmouth, Nebraska, USA

Trump D-30 de-icer unit

Characteristics

TYPE	D-30	D40-D
MAX WORKING HEIGHT	38 ft 6 in (11·7 m)	46 ft (12·5 m)
SIDE REACH	22 ft 6 in at 21 ft 6 in (6·9 m at 6·6 m)	28 ft at 28 ft (8·5 m at 8·5 m)
ROTATION	370° (non continuous)	370°
OPERATOR'S BASKET		
dimensions:		
length	2 ft (0·61 m)	4 ft (1·22 m)
width	2 ft (0·61 m)	4 ft (1·22 m)
height	3 ft 6 in (1·07 m)	3 ft 6 in (1·07 m)
capacity	300 lb (136 kg)	450 lb (200 kg)
TANK CAPACITIES		
de-icer fluid	200 US gallons (757 litres)	300 US gallons (1100 litres)
water	1100 US gallons (4164 litres)	1500 US gallons (5700 litres)
FLUID PUMP	100 US gallons/minute at 350 psi (380 litres/minute at 24 bar)	100 US gallons/minute at 350 psi (380 litres/minute at 24 bar)
GLYCOL PUMP	25 US gallons/minute at 500 psi (95 litres/minute at 34·5 bar)	25 US gallons/minute at 500 psi (95 litres/minute at 34·5 bar)
NOZZLE CAPACITY	20–100 US gallons/minute (76–380 litres/minute)	20–100 US gallons/minute (76–380 litres/minute)
HEATER CAPACITY		
M1 petrol	1 080 000 Btu/h (272 156 kcal/h)	1 080 000 Btu/h (272 156 kcal/h)
jet fuel	1 280 000 Btu/h (347 754 kcal/h)	1 280 000 Btu/h (347 754 kcal/h)
M2 petrol	1 800 000 Btu/h (453 592 kcal/h)	1 800 000 Btu/h (453 592 kcal/h)
jet fuel	2 300 000 Btu/h (579 590 kcal/h)	2 300 000 Btu/h (579 590 kcal/h)
WHEELBASE	9 ft 3 in (2·82 m)	11 ft 3 in (3·43 m)
UNIT DIMENSIONS		
length	24 ft 3 in (7·39 m)	27 ft 6 in (8·38 m)
width	8 ft (2·44 m)	8 ft (2·44 m)
height	13 ft 2 in (4·01 m)	13 ft 5 in (4·09 m)

Trump D40-D de-icer vehicle

Tug-a-Lug Helicopter and Aircraft Handling Dollies

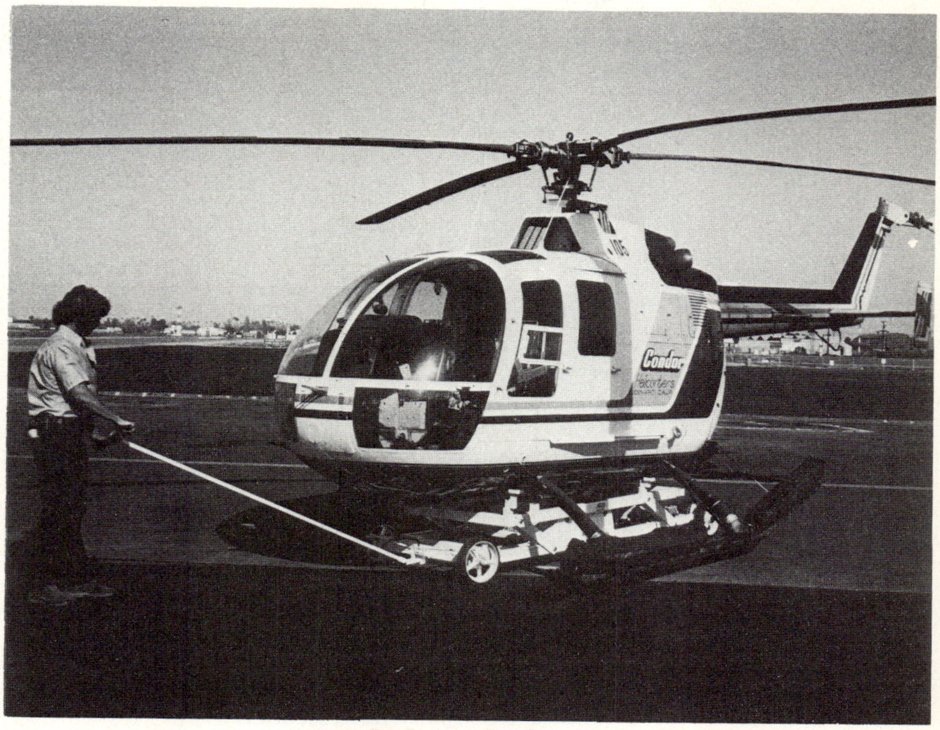

Tug-a-Lug D1 dolly

Tug-a-Lug has developed three models of powered dollies for moving skid-mounted helicopters on the ground. The dolly is manoeuvred under the helicopter between the skids and lifts the helicopter off the ground before transporting it to the required position.

The D1 and D2 models offer similar features, the D1 being able to lift a Bell Long Ranger weighing 4050 lb (1830 kg), the D2 capable of moving the Bell 214 ST helicopter, weighing 11 200 lb (5080 kg). The dollies are equipped with electro-hydraulic lifting and lowering with two speeds, the first to cradle the craft and the second to raise it. For transporting the helicopter, electric drive supplies two speeds forward and reverse. All controls operate through momentary speed and function switches on the handle and at the rear of the dolly. Two switches must be kept in the on position to activate the dolly. Other features include: adaptors offered for different types of helicopters, built-in battery charger, and in the D2 Dolly power steering to assist the operator with the heavier craft.

The D3, designed to handle a Bell 214 ST is fitted with hydrostatic front wheel drive providing speeds up to 5 mph (8 km/h) powered by an 18 hp Briggs and Stratton engine. The handle design allows the operator to control the dolly from any position in front of the unit. Additional operating stations are fitted at the middle and rear of the unit to help the operator position the dolly under the helicopter and emergency stop switches are located at all the operating stations.

The 'T' models are designed to enable the front wheel of the aircraft or rear wheel of the helicopter to ride on the unit, held in position by a towing strap or gate. The weight of the aircraft or helicopter is carried over the drive axle of the tug. The T2 model is pedestrian-controlled, with a two speed, 12 volt electric drive and self-contained charger. The T6 features an open driver's position with seat, steering wheel and foot controls. It is powered by an 18 hp Briggs and Stratton engine with hydrostatic drive providing speeds up to 5 mph (8 km/h).

Status

Tug-a-Lug units are in service at airports in the USA. The models are suitable for use with Aerospatiale, Agusta, Bell, Hiller, Hughes and Sikorsky helicopters

Manufacturer

Tug-a-Lug Inc, Lancaster, California, USA

Turbomach MEP-414A Dc Ground Power Supply Unit

The 10 kW, 28 volt dc ground power supply unit developed by Turbomach permits direct coupling of a turbine and an alternator on a common shaft operating at 93 500 rpm. The alternator is a high-frequency, solid-rotor, brushless type. Use of a voltage transformer and solid-state rectifier provides 23 to 35 volt dc output at full rated load of 10 kW. The Turbomach turbine set will operate on a variety of fuels including combat and compression ignition/turbine engine fuel. Space requirement for the unit is 9·2 cubic feet (0·26 cubic metres). The power unit produces its 10 kW rated output at temperatures up to 125°F (52°C) at sea level and 95°F (35°C) at 8000 feet (2438 metres) altitude, and can incorporate an electric starting system.

Characteristics

ENGINE: turbine
POWER SUPPLY: 28 V dc, 10 kW
ADJUSTABLE RANGE: 23–35 V
DIMENSIONS
length: 2 ft 9 in (0·84 m)
width: 2 ft (0·61 m)
height 1 ft 10 in (0·56 m)
WEIGHT: 300 lb (136 kg)
OPERATING TEMPERATURE RANGE: −65 to +125°F (−54 to +50°C)

Status

Developed for United States Army Mobility Equipment Research and Development Command, Fort Belvoir, Virginia, and in production for the US Army Firefinder programme. A 10 kW, 400 Hz version has also been developed

Manufacturer

Turbomach Division, Solar Turbines International, San Diego, California, USA

Turbomach MEP-414A dc ground power unit

United Aircraft Towing Tractors

United SML-120 aircraft towing tractor

United Tractors produce 12 models of tow tractors, the smaller versions (Models SM-340, 40, 50 and 60) being suitable for general towing duties, are featured in the Cargo Handling section. All except the SM-300D, which has an enclosed cab and four-wheel drive, feature an open driver's position, dual-driving wheels and offer optional diesel engines or lpg fuel systems.

The SM-80 and SM-100 are powered by Chrysler engines with Allison automatic transmission and torque converter driving the rear axle. Power steering operates on the front axle and hydraulic power-brakes act on all four wheels. A pin and clevis towing attachment is fitted as standard equipment.

The SML-80, SML-100 and SML-120 are driven by either a Ford or Chrysler petrol engine through

a Ford C6 or Chrysler A727 automatic transmission and torque converter. Two wheel shoe brakes are fitted to the rear axle and power steering and pin and clevis towing point are standard equipment.
The SML-160 and SML-200 models feature a Chrysler petrol engine equipped with electronic ignition and Allison transmission with a torque converter. Other equipment fitted is similar to the SM-80 model.
The SM-300D is powered by a John Deere diesel engine via a Funk 1000 transmission system and incorporates hydrostatic four-wheel steering. Two wheel, four-wheel co-ordinated or four-wheel crab modes can be selected. Other features include enclosed cab, passenger seat and front and rear towing points.

Status
In service with major airlines

Manufacturer
United Tractor Company, Chesterton, Indiana, USA

Characteristics

TYPE	SM-80/SM-100	SML-80/SML-100/SML-120	SM-160/SM-200	SM-300D
ENGINE	Chrysler LH-318	Ford 302/Chrysler LH-318	Chrysler LH-318	John Deere 6359D
OUTPUT	170 hp (127 kW) at 2400 rpm	153 hp (114 kW) at 3400 rpm/ 170 hp (127 kW) at 2400 rpm	170 hp (127 kW) at 3000 rpm	diesel
MAX DRIVING SPEED	15 mph (24 km/h)	15 mph (24 km/h)	20 mph (32 km/h)	
MAX DRAW-BAR PULL	SM-80: 8000 lb (3629 kg) SM-100: 10 000 lb (4536 kg)	SML-80: 8000 lb (3629 kg) SML-100: 10 000 lb (4536 kg) SML-120: 12 000 lb (5443 kg)	SM-160: 16 000 lb (7258 kg) SM-200: 20 000 lb (9072 kg)	30 000 lb (13 608 kg)
WHEELBASE	5 ft 3 in (1·6 m)	5 ft 6 in (1·68 m)	6 ft (1·83 m)	8 ft 4 in (2·54 m)
TRACK				
front	4 ft 7¾ in (1·42 m)	4 ft 8½ in (1·44 m)	5 ft 7·1 in (1·7 m)	
rear	3 ft 2½ in (0·98 m)	3 ft 2½ in (0·98 m)	4 ft 10⅔ in (1·49 m)	
OUTSIDE TURNING RADIUS	14 ft 7½ in (4·46 m)	11 ft 8 in (3·56 m)	13 ft 4 in (4·06 m)	15 ft 6 in (4·72 m)
UNIT DIMENSIONS				
length	9 ft 8½ in (2·96 m)	10 ft 4 in (3·15 m)	11 ft 3 in (3·43 m)	15 ft (4·57 m)
width	5 ft 5½ in (1·66 m)	5 ft 6 in (1·68 m)	7 ft 10 in (2·39 m)	8 ft (2·44 m)
height	5 ft 6 in (1·68 m)	3 ft 7 in (1·09 m)	4 ft 9 in (1·45 m)	8 ft 4 in (2·54 m)
WEIGHT	SM-80: 9300 lb (4218 kg) SM-100: 11 300 lb (5126 kg)	SML-80: 9100 lb (4128 kg) SML-100: 11 500 lb (5216 kg) SML-120: 12 600 lb (5715 kg)	SM-160: 21 000 (9526 kg) SM-200: 24 100 lb (10 932 kg)	37 000 lb (16 783 kg)

Wollard Aircraft Tow-bars

A range of tow-bars are available from Wollard to suit most types of commercial aircraft.
The TB-357A designed for Boeing 707 aircraft, is constructed of aluminium tubing with shear pins for protection of the nose-gear. A steel head is fitted with a Boeing quick head available as an option. The 357S is similar, constructed in steel.
TB-373 is suitable for DC-6, DC-7 and DC-8 aircraft. Fabricated in steel with a pivot arm operated by a hand-wheel, machined towing pins and retractable undercarriage, the tow-bar is fitted with power and intercom cable brackets. The 373A is the same except that the nose-wheel security pins are free to rotate. These pins turning on needle bearings permit the tow-bar to be fitted also to Convair 880 and 990 aircraft.
TB-449 is a tubular steel tow-bar for Boeing 727 aircraft, equipped with shear bolts for protection of the aircraft nose-gear. Optional features available are the Boeing quick head (TB-449A) and a retractable undercarriage (TB-449AR).
TB-473 designed for the DC-9, is engineered in square steel tubing, with resting support and towing eye. Nose-wheel security pins are fitted to the tow-bar with shear bolts to protect the aircraft nose-gear.
TB-531 is a heavy-duty tubular construction suitable for towing the Boeing 737-100, fitted with a one-piece steel head and shear bolts. The TB-631 is similar but has a quick disconnect head for the Boeing 737-200 or a 737 aircraft that has been modified with the cross pin towing kit.
TB-561 for the Boeing 747 consists of an aluminium tube, lunette towing eye, tow-bar head and retractable undercarriage. The retractable undercarriage is designed to support the tow-bar while being towed and to serve as a vertical adjustment while attaching the end connections. The tow-bar head is designed for quick connect and disconnect from the B747 nose-landing gear. Overload of the tow-bar will shear bolts in the head of the tow-bar.
The TB-571 is similar in design to the TB-561, but suitable for towing DC-10/L-1011 aircraft.

Characteristics

TYPE	TB-357	TB-373	TB-449	TB-473
AIRCRAFT	B707	DC-6, -7, -8 Convair 880, 990	B727	DC-9
CONNECTING LENGTHS	21 ft 3 in (6·48 m)	13 ft 6 in (4·11 m)	15 ft 4 in (4·67 m)	10 ft 2 in (3·1 m)
CONNECTING HEIGHTS				
fixed undercarriage	1 ft 11 in (0·58 m)		1 ft (0·3 m)	9 in (0·23 m)
retractable undercarriage	1 ft 4 in–2 ft 10 in (0·41–0·86 m)	1 ft 3 in–1 ft 10 in (0·38–0·56 m)		
UNIT DIMENSIONS				
length	21 ft 8 in (6·6 m)	14 ft (4·27 m)	15 ft 8 in (4·78 m)	10 ft 7 in (3·23 m)
width	2 ft 6 in (0·76 m)	3 ft 6 in (1·07 m)	2 ft (0·61 m)	3 ft (0·91 m)
height	2 ft 3 in (0·69 m)	2 ft (0·61 m)	1 ft 2 in (0·36 m)	1 ft 3 in (0·38 m)
WEIGHT	180/465 lb (82/211 kg)	465 lb (211 kg)	215 lb (98 kg)	165 lb (75 kg)

TYPE	TB-531/631	TB-561	TB-571
AIRCRAFT	B737	B747	DC-10/L-1011
CONNECTING LENGTHS	15 ft 5 in (4·7 m)	13 ft 4 in/10 ft (4·06/3·05 m)	10 ft (3·05 m)
CONNECTING HEIGHT			
fixed undercarriage	11¾ in (0·3 m)		
retractable undercarriage		11½ in (0·29 m)–2 ft 1 in (0·64 m)	11½ in (0·29 m)–2 ft 1 in (0·64 m)
UNIT DIMENSIONS			
length	15 ft 5 in (4·7 m)	13 ft 10 in/10 ft 6 in (4·22/3·2 m)	10 ft 4 in (3·15 m)
width	2 ft (0·61 m)	2 ft 8 in (0·81 m)	2 ft 8 in (0·81 m)
height	16½ in (0·42 m)	2 ft 8 in (0·81 m)	2 ft 8 in (0·81 m)
WEIGHT	230 lb (104 kg)	600/650 lb (272/295 kg)	700 lb (318 kg)

Status
In service with airlines in the USA

Manufacturer
Wollard Airport Equipment, Miami, Florida, USA

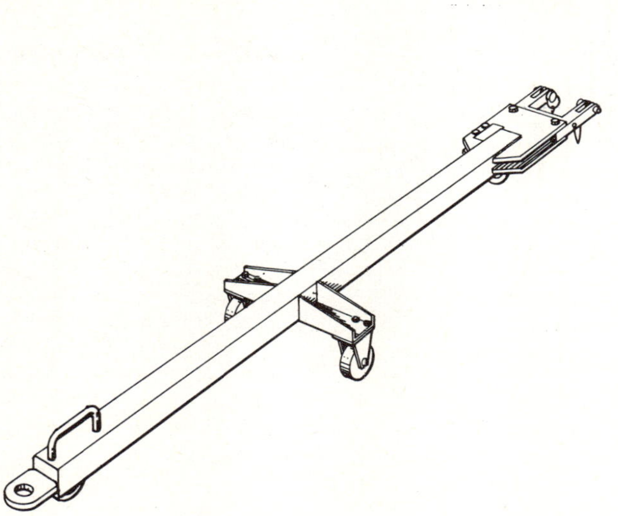

Wollard TB-5311 631 tow-bar

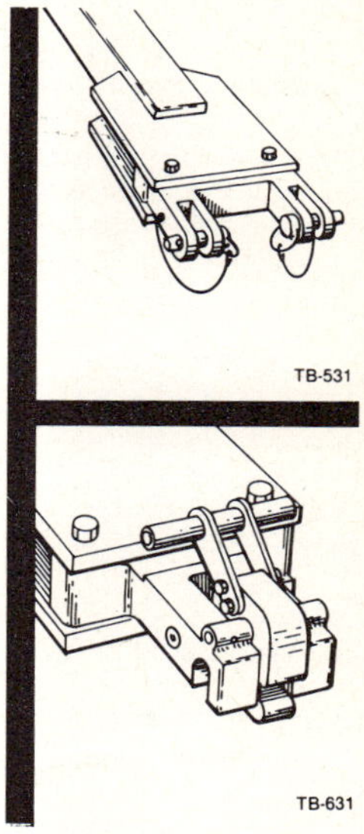

Wollard LS-259F Lavatory Service Cart

The model LS-259F lavatory service cart is a towable unit equipped with a stainless-steel tank and a 4 inch (102 mm) bronze-fitted Everlast waste-drain valve. The unit is operated by a petrol engine-driven pump with a bronze hand-pump provided in case of power failure.

Characteristics
TANK CAPACITIES
waste: 150 US gallons (568 litres)
flush water: 75 US gallons (284 litres)
UNIT DIMENSIONS
length: 8 ft 4 in (2·54 m)
width: 4 ft 1 in (1·24 m)
height: 4 ft 9 in (1·45 m)
WEIGHT: 1385 lb (628 kg)

Status
In service with airlines in the USA

Manufacturer
Wollard Airport Equipment Inc, Miami, Florida, USA

Wollard LS-259F lavatory service cart

Wollard TLS-770A Lavatory Service Truck

The model TLS-770A lavatory service truck is designed to service commercial aircraft. It is equipped with stainless-steel waste and flush water tanks, pump and waste-hose compartments and hydraulic lift platform for access to the aircraft servicing points. A hydraulic-pumping system is used for pumping flush water into the aircraft. A foot-operated water pump control switch is on the top deck between the aft compartments, as is a foot-operated flush hose drain valve actuator. A ladder is provided on the driver's side for access to the top deck. A disc-type meter with a zero-setting feature measures the amount of flush water dispensed into the aircraft. Standard equipment includes a relief valve and suction-line filter, and a compartment and exterior floodlight for night operation. Optional tank immersion heaters and compartment space heaters are available for winter operation.

Characteristics
CHASSIS: Ford F 350, 6-cylinder engine automatic transmission, power steering
TANK CAPACITIES
flush water: 260 US gallons (984 litres)
waste: 400 US gallons (1514 litres)
LIFT PLATFORM DIMENSIONS
length: 4 ft (1·22 m)
width: 2 ft (0·61 m)
MAX HEIGHT: 11 ft 8 in (3·57 m)
UNIT DIMENSIONS
length: 19 ft 10 in (6·04 m)
width: 7 ft (2·13 m)
height: 6 ft 4 in (1·93 m)
WEIGHT: 7950 lb (3606 kg)

Status
In service with airlines in the USA

Manufacturer
Wollard Airport Equipment Inc, Miami, Florida, USA

Wollard TWS-402 Potable Water Vehicle

The model TWS-402 potable water tank is constructed of stainless steel, welded and tested at 3 psi pressure. Tank capacity is 450 US gallons (1703 litres) of potable water. The unit is designed to accommodate all types of commercial aircraft, including the Boeing B-747 and the Lockheed L-1011. Equipped with an optional personnel-lift platform (model TWS 402-HL), the unit is compatible with the McDonnell-Douglas DC-10 aircraft. Other features include baffles to prevent water surging, a man-hole on top of the tank for cleaning access, bronze water-pump and an enclosed pump compartment.

Characteristics

CHASSIS: Ford F 350, 6-cylinder petrol engine, automatic transmission
TANK CAPACITY: 450 US gallons (1703 litres)
WHEELBASE: 18 ft 9 in (5·72 m)
TRACK: 6 ft 6 in (1·98 m)
LIFT PLATFORM DIMENSIONS
length: 4 ft (1·22 m)
width: 2 ft (0·61 m)
MAX HEIGHT: 11 ft 5¼ in (3·49 m)
Model TWS 402-HL
UNIT DIMENSIONS
length: 18 ft 9 in (5·72 m)
width: 6 ft 6 in (1·98 m)
height: 5 ft 11 in (1·8 m)
WEIGHT: 6820 lb (3094 kg)

Status

In service with airlines in the USA

Manufacturer

Wollard Airport Equipment Inc, Miami, Florida, USA

Wollard TWS-402 potable water vehicle

Wollard WC-86 Potable Water Service Cart

The model WC-86 water service cart comprises a stainless-steel tank with a capacity of 120 US gallons (454 litres) mounted on a towable unit. It is designed to dispense and fill from a single hose. Water is transferred by means of a Tokheim Hi-Vacuum hand pump. The cart has a flat bed with dimensions of 2 foot 1 inch (0·64 metres) by 5 foot (1·52 metres) and incorporates fifth-wheel steering, rubber tyres and automatic safety brakes.

Characteristics

TANK DIMENSIONS
length: 4 ft (1·22 m)
width: 2 ft (0·61 m)
height: 2 ft 2 in (0·66 m)
TANK CAPACITY: 120 US gallons (454 litres)
UNIT DIMENSIONS
length: 6 ft (1·83 m)
width: 2 ft 2 in (0·66 m)
height: 4 ft 2 in (1·27 m)
WEIGHT: 690 lb (313 kg)

Status

In service with airlines in the USA

Manufacturer

Wollard Airport Equipment Inc, Miami, Florida, USA

Wollard WC-86 potable water service cart

AIR TRAFFIC CONTROL

Radar
Navaids
Airfield lighting
Communications
Meteorology

RADAR

BRAZIL

ASR-BRI Air Traffic Control Surveillance Radar

This S-band airfield surveillance radar is the result of PEA's Meta 2 project and the equipment is being procured in quantity for the Brazilian atc system.

The ASR-BRI has a conventional transmitter with a magnetron modulator and delay-line pulse generation. Apart from the magnetron and a thyratron switching tube, the transmitter is an all solid-state system providing adjustment of the prf within ±10 per cent of its normal value. All transmitter components have automatic protection circuits against malfunctions. Receiver characteristics include: low-noise figure due to a parametric input amplifier; self-radiation protection; synthesiser local oscillator with automatic frequency control (afc) and if pre-amplifier with time-variable gain. The last feature equalises the returns from identical targets at different ranges, as well as protecting the if amplifier against saturation caused by excessively strong re-run echoes from short range targets.

Signal processing is based on the use of a digital mti which rejects fixed echoes, and the prf can be varied to eliminate blind speeds. A conventional ppi crt display is provided, and there is a video-map facility which generates synthetic video representing the local geography of the airfield including such features as towers, tall buildings, mountains etc.

Characteristics

FREQUENCY: 2700–2900 MHz
PEAK POWER: 450 kW (typical)
PULSE WIDTH: 0·75 μs
PRF: 1000/s (variable ±10%)
ANTENNA BEAM-WIDTH: 1·5° azimuth, $\cos^2$ 5·5–40° elevation

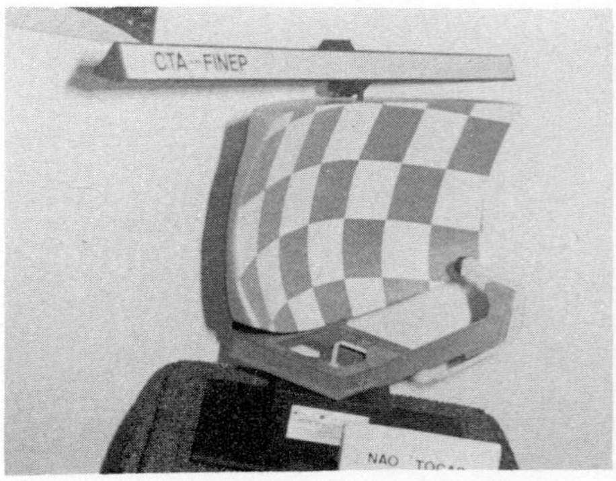

Model of Brazilian ASR-BRI airfield surveillance radar (Ronaldo S Olive)

ANTENNA GAIN: 32 dB (max)
SIDE-LOBE ATTENUATION: 25 dB
ROTATION RATE: 15 rpm
RECEIVER NOISE: 2 dB (at parametric amplifier input)
IF: 30 MHz
MAX RANGE: 110 km (2 m^2 target)
DETECTION PROBABILITY: 0·9
FALSE ALARM PROBABILITY: 10^{-5}
SUB-CLUTTER VISIBILITY: 25 dB
FIRST BLIND SPEED: 3500 km/h

Status

A prototype has been completed and has undergone certification testing preparatory to production by private industry. Twenty-four sets have been ordered for the Brazilian atc network, and the first two installations will be at the Porto Alagre (Rio Grande do Sul State) Airport and the Sao Pedro da Amdeia (Rio de Janeiro State) Naval Air Force Base

ASR-BRI transmitter receiver cabinets (Ronaldo S Olive)

ASR-BRI display mounted in atc console (Ronaldo S Olive)

Manufacturer

IPD/PEA — Departamento de Electronica, São Paulo, Brazil

CANADA

Goodwood Air Traffic Control Systems

Goodwood Data Systems develops and manufactures airport-processing systems for atc. The operational information display system consolidates operational information on a display to provide the controllers with data on a need-to-know basis. The following information is available: time; automatic terminal information service; approach aids and landing runway; approach and runway light settings; wind direction and speed; altimeter settings; weather sequences; flow control restrictions; navigational facility unserviceabilities; notams; runway and field conditions and gate assignments. The flight plan processing and display system separates the flight plan processing, printing, display and transmission as a stand-alone feature. Goodwood claim it allows a modern, low-cost flight strip print system and is a step towards flight plans on crt displays.

Status

Goodwood systems are in use at international airports at Montreal, Toronto, Winnipeg, Edmonton and Vancouver, Canada

Goodwood airport information distribution system

Manufacturer

Goodwood Data Systems Ltd, Carleton Place, Ontario, Canada

Goodwood flight plan processing and display system

Raytheon Canada ASR 8000 Series Primary Surveillance Radars

The ASR 8000 series comprises three types of D- and E/F-band radars developed primarily for terminal and en route applications in civil aviation. The system employs a moving target detector signal processor which continuously tailors the parameters of the radar at each point in its coverage to reject clutter while retaining wanted targets. The approach is to separate the received signals into individual velocity channels, process each individually and add the resultant outputs. The signal processor also automatically stores the long-term average clutter level by means of a 65-cell adaptive clutter map and uses this to adjust target detection thresholds. This combination of doppler filtering and adaptive thresholding is known as moving target detection (mtd). The output of the mtd processor is similar to that of a conventional mti in that it is compatible with conventional ppi displays, scan converted displays, or existing primary radar digitisers. Using quadrature mtd, the mtd signal processor provides improved detection of tangential and slow moving targets and eliminates blind speeds. Adaptive clutter filters provide residue suppression and an adaptive threshold integrator is incorporated to minimise weather integration. A block stagger is used to eliminate transmitter fm and an interference and saturation detector blanks interference and suppresses scanning modulation residue. Other features include: a cfar processor to provide inter-clutter visibility, prf mini-stagger to suppress second-time-around returns, and a sweep-to-sweep, quadrature and diversity integration to improve signal-to-noise ratio and to provide single pulse interference suppression.

Characteristics

TYPE	ASR 8300	ASR 8400	ASR 8800
FREQUENCY BAND	D	D	E/F
PEAK POWER	500 kW	500 kW	500 kW
PULSE LENGTH	0·7–1·5 μs	1·5–3 μs	0·7–1·5 μs
PRF	600–1000 pps	350–600 pps	800–1000 pps
RANGE	60–120 nm (111–222 km)	120–200 nm (222–371 km)	60–80 nm (111–148 km)

Status
Production. A forerunner of the ASR 8000 series, the ASR 803-2 is installed at Subang International Airport, Kuala Lumpur, Malaysia

Manufacturer
Raytheon Canada Ltd, Waterloo, Ontario, Canada

Raytheon Canada D-band ASR installation in Kuala Lumpur

CZECHOSLOVAKIA

Tesla OPRL-4 Surveillance Precision Approach Radar

The OPRL-4 is a combined surveillance and precision approach radar (par) providing for atc of aircraft within 50 km of an airfield and recovery and landing guidance to individual aircraft by means of gca-type procedures. The equipment is housed in a cabin with two co-mounted azimuth and elevation antennae on top. The system operates in the J-band and features of the radar include mti with prf change-over facility, antenna feeds with continuously adjustable polarisation (linear, elliptical, circular), solid-state design, and remote-display and control facilities.

Characteristics
FREQUENCY: 9250–9450 MHz
PULSE WIDTH: 0·45 μs
PRF: 2000/2400 pps, pulse-to-pulse switched
RECEIVER
NOISE FIGURE: 7 dB
IF: 30 MHz
LOGARITHMIC RESPONSE: 60 dB
ANTI-CLUTTER GAIN CONTROL: 30 dB
FIXED ECHO REJECTION: 20 dB
AZIMUTH ANTENNA: 3·2 × 1·2 m
BEAM-WIDTH: 0·7° azimuth; 2 × 4·3° cosec^2 elevation (3 dB)
SIDE-LOBE LEVEL: 24 dB
GAIN
39·5 dB lower beam
37·5 dB upper beam
ROTATION RATE: 15 or 30 rpm
POLARISATION: continuously adjustable
ELEVATION ANTENNA: 0·37 × 3 m
BEAM-WIDTH: 6° azimuth; 0·7° elevation (3 dB)
SIDE-LOBE LEVEL: 24 dB
GAIN: 37·5 dB
POLARISATION: circular
SCAN RATE: 30/minute

Status
Operational. The Tesla OPRL-4 was developed in co-operation with various Soviet organisations

Manufacturer
Tesla, Pardubice, Czechoslovakia

OPRL-4 surveillance par

Tesla RL-41 Surveillance Radar

The Tesla RL-41 is a general-purpose atc air search radar operating in the E/F-band and has a range of about 150 km. Good high cover is provided by the horn-fed 5 by 3·6 metre antenna, reaching over 15 000 metres at 30 to 40 degrees elevation. Beam-width is about 1·5 degrees and the vertical cover extends from 0·3 to 45 degrees. Two rotation rates, 7·5 or 15 rpm, are available. The receiver incorporates mti facilities and dual 40 cm display consoles can be remotely sited up to 200 metres from the radar head. Resolution is 500 metres in range and 1 degree in azimuth. Dual 800 kW transmitters permit diversity operation.

Status
In service with Warsaw Pact countries

Manufacturer
Tesla, Pardubice, Czechoslovakia

RL-41 search radar

Tesla RL-42 Surveillance Radar

Derived from the OR-1, OR-2 and RL-41 radars, this latest development is a 10 cm radar with a range of 120 km. Its technical specifications are claimed by the manufacturers to meet CMEA standards for Group V 2 airport radars and as a sre radar is capable of meeting the requirements of ICAO Annex 10. The radar operates in the diversity mode on two frequencies with an auxiliary lobe, digital moving target indication, digital target detector and target extractor. Displays available include the UZJ-4 modular unit which presents data from primary and secondary surveillance radars alphanumerically and has an auxiliary graphic capability.

Status
Announced in 1981 this radar was believed to be in combined civil/military use in Warsaw Pact countries

Manufacturer
Tesla, Pardubice, Czechoslovakia

Tesla RL-2 Surveillance Radar

The RL-2 is an airfield surveillance radar operating in the J-band producing a range of approximately 40 km. The radar features an mti facility and uses a UPA-4 display with a 40 cm ppi. Antenna rotation can be selected for either 7·5 or 15 rpm.

Status
In service in Warsaw Pact countries

Manufacturer
Tesla, Pardubice, Czechoslovakia

Tesla RL-2 surveillance radar

DENMARK

Terma ASTI Airport Surface Control Radar

Airport Surface Traffic Indicator (ASTI) is an I-band radar developed to provide air traffic controllers with information about the position of aircraft and vehicles on airport runways, taxiways and apron areas. The unit uses a narrow beamwidth antenna featuring high gain and a slotted waveguide. Two rotational speeds are available, 48 and 96 rpm. The transmitter/receiver is fully solid-state using a high pulse repetition frequency of 8000 Hz with a short pulse width of 50 nano seconds. The display unit is a 16-inch (406 mm) fixed-coil crt and all timing and control functions use digital integrated circuits. A joy-stick operates an off-centring facility for intensive study and magnification of critical sections of the manoeuvring area. Eight different range scales can be selected by means of push-buttons. The standard system includes one master and one slave display, and more displays are offered as optional equipment. The ASTI display unit can be used as a separate control position and when free-standing it can be tilted from the horizontal to the vertical position. Alternatively the ppi can be built into the controller's console with the display controls placed according to the controller's requirements. A programmed blanking unit is incorporated in the system, based on the geographical layout of the airport. When activated, this unit suppresses all irrelevant echoes from buildings and areas between and surrounding the runways and taxiways. Only the runways and taxiways are visible. The unit contains a 65 K bit digital memory which is programmed with data obtained from a map of the airport area.

Terma ASTI airport surface control radar

ASTI antenna

Characteristics
TRANSMITTER/RECEIVER
- frequency: 9375 ±30 MHz
- peak power: 20 kW
- prf: 8000 Hz
- pulse width: 50 μs
- noise factor: 9 dB

DISPLAY UNIT
- range scales: 0·5, 0·8, 1·1, 1·4, 1·7, 2, 2·5, 3 nm
- rings: 0·1, 0·25, 0·5 nm

PROGRAMMED BLANKING UNIT
- BLANKING SECTOR NUMBERS
  - range: 512
  - azimuth: 1024
- BLANKING SECTOR RESOLUTION
  - range: 8–30 m
  - azimuth: 0·35°
- BLANKING: 15 dB/40 dB

ANTENNA
- type: slotted waveguide
- horizontal beam-width: 1°
- vertical beam-width: 20°
- polarisation: horizontal
- attenuation of side-lobes: >25 dB
- gain: 32 dB
- rotation speed: 48/96 rpm
- weight: 54 kg

POWER REQUIREMENTS: 220/115V, 50/60 Hz, 1200 VA

Status
Production

Manufacturer
Terma Electronic AS, Lystrup, Denmark

FRANCE

Sintra STRAPP Air Traffic Control System

STRAPP is an air traffic control system developed from the STRIDA air defence and traffic control data processing and display system operated by the French Air Force. It is designed for the automation of air traffic control in the approach area and is suitable for civil or military operations. The system provides automatic extraction and displaying of data from primary and secondary radars. An electronic map is generated and up to 200 aircraft are marked and tracked automatically. Track labelling supplies: information on speed, secondary codes and the trajectory followed, and these data are displayed on the ppi screen according to the selections made by the operator. In addition to the data, the operator can limit automatic tracking facilities to certain sectors of this display, create tracks manually, select displayed elements either individually or by characters and label each track in a different way. The data can be presented on a colour graphic display, colour data display and transmission terminal or a large coloured screen. Additionally STRAPP can transmit the data to a central system through a simple digital link. The equipment can receive information through a digital link and correlate local data to present a complete picture of the situation and via the same digital link can receive and display flight plans. For aircraft transfers from one area of responsibility to another, STRAPP equipment also provides inter-marking between controllers at the centre or by centres interconnected with digital links.

STRAPP can be connected to any conventional primary or secondary radar and to any display system compatible with radar signals and having a synthetic input. The air pattern can be presented in a synthetic form either at 50 Hz to be compatible with a daylight environment or at 16 Hz to provide for the superimposition of the real radar video.

Sintra has developed a suitable colour graphic display system, which has a collective display-capability and presents the same image in several different places, on a flat-face 23-inch crt in four colours. The operator uses a keyboard, rolling ball or electronic wand to provide inputs. An outline generator capable of 10 000 vectors provides the map display.

Characteristics

COLOUR GRAPHIC DISPLAY SYSTEM
display: 23 in crt
resolution: 2048 × 2048 dots
colours: red, orange, yellow, green
symbols: character, dot, dash, circle, 2 dash and 4 character/flash symbol sizes
OUTLINE GENERATOR: 10 000 vectors

Status

STRAPP is in production and has been connected to the following radars in France: SRE, Tiger, Centaure, Aladin, Marconi 5650 and Cerbere

Manufacturer

Sintra-Alcatel, Asnieres, France

Thomson-CSF 10 Series E/F-Band Approach Radars

The 10 series of E/F-band approach radars comprises the TA 10M and TA 10K, the former is magnetron-powered, the latter uses a piloted klystron transmitter. Both feature fixed echo and weather clutter suppression, and an mti system based on signal phase analysis and linear doppler filtering. An optional system offered analyses the amplitude of the echoes and provides correlation over several antenna revolutions to remove any spurious echoes. The TA 10M has a peak output of 650 kW though it is operated at 500 kW peak power and 0·5 kW mean power to increase service life. The TA 10K produces a peak power of 1500 kW but operates at 1250 kW with a mean power of 1·9 kW. Both radars employ the ac 316 antenna and can be supplied pre-assembled in transportable cabins.

Thomson-CSF 10 series radar in Bali, Indonesia

Characteristics

MODEL	TA 10M	TA 10K
FREQUENCY BAND	E/F	E/F
POWER TUBE	Magnetron	Klystron
TRANSMITTER TYPE	ER 712S	ER 820S
PEAK POWER	500 kW	1·25 MW
PULSE WIDTH	0·8 μs	0·8/1·7 μ or 0·75 + 0·75 μs
PRF	1000 Hz	750/1000 Hz
MINIMUM RANGE	60 nm (111 km)	80 nm (148 km)
ANTENNA	ac 316	ac 316
SPEED	15 rpm	12/15 rpm
GAIN	34 dB	34 dB
HORIZONTAL BEAM-WIDTH	1·5°	1·5°
DIMENSIONS	2·3 × 4·7 m	2·3 × 4·7 m

Status

Production. The latest version of the TA 10 M was introduced in 1979. Both radars are in service in France and foreign countries

Manufacturer

Thomson CSF, Division DRS-TVT, Bagneux, France

Thomson-CSF 23 Series Surveillance Radars

The 23 series comprises five models of surveillance radars: LP 23K, LP 23M, TR 23K, TR 23MR and TR 23MA.

The LP 23K is a coherent, klystron-powered, long-range surveillance radar operating in the D-band of the radar spectrum. It has been evolved for air traffic surveillance roles and similar applications. It embodies the latest concepts in steerable pattern null, coherent double-pulse operation and auto-regulating signal processing. This dual-channel 23 cm radar is claimed to be able to detect small aircraft at ranges of up to 220 nautical miles (408 km), and superior clutter see-through capabilities permit detection down to low levels. An integral plot extractor enables the radar to be connected to a data processing system.

The dual beam antenna uses off-set feeds for reduced masking of the energy radiated by the reflector, thereby reducing side-lobes, and there is an elongated lower edge to the reflector to intercept radiation from the primary sources so that back-lobes are reduced. The feeds are designed to ensure that the reflector edges are only illuminated by low-energy radiation, thus reducing spillover to a level comparable to the background level. A cosecant-squared radiation pattern, boosted at high elevation angles is employed, providing a good contrast between short range targets and ground clutter. There is a sharp roll-off at low elevation angles, which also improves the signal-to-clutter contrast, and this is obtained by employing a reflector with a larger-than-usual vertical dimension. Compact stacking of the main and auxiliary feeds, gives good focusing for the high beam and reduced boresight separation between the two beams, while still maintaining a target-to-clutter enhancement of 16 to 18 dB. Programmable range/azimuth gated beam mixing (steerable pattern null) further improves clutter rejection.

The transmitter uses a ER 850L klystron which is stated to have a saturation gain of over 50 dB. It is a pulse-driven tube with five cavities giving a peak output of 5 MW, although in the present application it is deliberately under-driven to increase its service life. It is operated at 3·5 MW peak power and 4·7 kW mean power. The transmitter is crystal-controlled. At each pulse repetition interval, the transmitter generates a coherent pulse train consisting of a long duration pulse followed immediately by a short pulse. This provides a means of obtaining simultaneously in a single radar the benefits of both long- and short-pulse operation. The latter gives a precise radar cell and so improves both target resolution and signal-to-clutter contrast, these characteristics being especially valuable at short and medium ranges. At greater ranges, the longer pulse results in an increased useful detection range.

Signal processing and remoting is provided by a TPL 800 processor with enhanced capability thanks to the coherent signals received. Video processing includes log/ftc/anti-log conversion to suppress slow-moving clutter patches. In the uhf spectrum, stc is controlled through digital programming with a matrix of 64 × 64 cells. The receiver input stage is a low-noise solid-state, wide-band amplifier, and the linear amplifier which drives the mti processor prevents clutter spectrum spreading. If gain is automatically controlled by a dynamic clutter memory having a capacity of 128 × 128 azimuth/range cells (3° × 2 or 1 nautical mile), and 10-bit phase-coding of in-phase and quadrature channels eliminates mti blind speeds. The mti filter can be either a three-pulse filter or a three-stage recursive filter with programmable negative feedback. Video diversity summing is performed digitally.

Other members of the family include the TR 23K which uses the same transmitter/receiver as the LP 23K but operates with the THD 286 antenna, and the TR 23MR which is a magnetron-powered alternative to the TR 23K with a similar antenna. The LP 23M is a magnetron-powered version of the LP 23K. The fifth model, TR 23 MA is lower-powered variant employing the ER 711L magnetron-transmitter producing a peak power of 500 kW.

Status

In production and exported. The LP 23K was introduced in 1978, the remainder in 1979

Manufacturer

Thomson CSF, Division DRS-TVT, Bagneux, France

TR 23 radar at Berlin Tegel Airport

LP 23 long-range surveillance radar

Characteristics

MODEL	LP 23K	LP 23M	TR 23K	TR 23MR	TR 23MA
FREQUENCY BAND	D	D	D	D	D
POWER TUBE	Krystron	Magnetron	Krystron	Magnetron	Magnetron
TRANSMITTER TYPE	ER 850L	ER 722L	ER 850L	ER 722L	ER 711L
PEAK POWER	3·5 MW	2 MW	3·5 MW	2 MW	500 kW
PULSE WIDTH	1 μs + 3 μs	3 μs	2 μs	1·5 μs	1 μs
PRF	340 Hz	340 Hz	450 Hz	500 Hz	750/1000 Hz
MINIMUM RANGE	220 nm (408 km)	200 nm (371 km)	140 nm (259 km)	120 nm (222 km)	80 nm (148 km)
ANTENNA	AT 432	AT 432	THD 286	THD 286	THD 286
SPEED	5/6 rpm	5/6 rpm	6/12 or 7·5/15 rpm	6/12 or 7·5/15 rpm	6/12 or 7·5/15 rpm
GAIN	35 dB	35 dB	33 dB	33 dB	33 dB
HORIZONTAL BEAM-WIDTH	1·1°	1·1°	1·7°	1·7°	1·7°
DIMENSIONS	7 × 13 m	7 × 13 m	5 × 9 m	5 × 9 m	5 × 9 m

Thomson-CSF Aircat Air Traffic Control System

Aircat are modular air traffic control systems for the detection, transmission, processing and operation of radar and flight plan data. Each system's architecture is defined according to the desired degree of automation and the needs to be met: approach or en route control, medium or dense traffic, centralised or non-centralised network.

The design involves the separation of the chains processing data of different types: radar data, flight plans, general data and weather. This is combined with a distributed processing architecture within the processing chain.

For terminal control, it is still a frequent practice to use raw video data from at least one nearby radar, sometimes associating this data with digital information from other centres or from remote radars.

However, en route control mostly uses processed data, 'tracks' sent in digital form over narrow-band channels and operated as synthetic images.

Data processing can be allocated to various levels in the detection and processing chain. On the radar site, the primary, secondary (and if necessary weather receiver) signals are extracted combined, cleaned, smoothed and sent to the control centres, in the form of tracks. In the control centre, the tracks from various stations are correlated and compared with flight data. Each console has a processor to manage the displayed picture and commands from the controller.

The primary radars extraction is integrated in the TPL 800 digital signal processing unit. This unit calculates a true position for each target, from the multiple echoes received, to produce a digital message adapted to the subsequent transmission and processing systems.

The extractor EV 760 ensures correlation, on three interlaced modes, of secondary surveillance radar responses. It reconstructs split codes, eliminates asynchronous, overlapped or incomplete responses and supplies digital messages which define the position and validated code for each aircraft.

The weather channel data is quantised and stored by azimuth/range cells. The map obtained is analysed by the EVM 800 extractor which produces the contours and intensity of the atmospheric phenomena encountered.

The processor PR 800 included in modern radar stations associates corresponding primary and secondary plots, comparing positions from one antenna revolution to another, in order to smooth out tracks which are freed from spurious or unwanted echo residue, thus defining the flight path of each aircraft. The PR 800 controls the remoting of these tracks and the weather map on a telephone line, to the control centres. Stations without PR 800 equipment supply primary and secondary radar plots, which are correlated by a computer in the control centre. The tracks and plots are digital data which can be routed from the radar stations to the various user centres by telephone. Some centres wish to use the raw video of a nearby radar (whether this radar belongs to a larger network or not) and the signals from the radar are thus multiplexed and sent to the centre before extraction, by microwave link.

In the control centre tracks from several radar stations are compared after correction of geographical coordinates in order to: validate one track per aircraft if detection by several radars takes place (multi-radar tracking) and calculate certain data to complete track labels (ground speed, direction of level change). After correlation with flight data, all the control data is submitted to the appropriate controllers presented on their radar screens and on strips (or on tubular displays).

A computer records and analyses the flight plan data and position reports and combines them with the radar data in order to: replace the secondary code by the flight number in the labels and add the authorised level, estimated time over, etc, and update the flight data and strips, detect and signal conflicts and prepare handovers. The computer can predict sector loads, programme flight course, indicate secondary codes and available flight levels to controllers, and aid them in choosing a course. The information useful to the controller is distributed selectively to the interested sectors in the two customary formats: labels supplying the radar picture and electronic or printed strips.

The displays available comprise the IP 570R, IP 670S and the bright display. The IP 570R displays are panoramic plan-position indicators (ppis) using persistent phosphor for raw radar video display in attenuated lighting conditions. The 12 inch version was designed particularly for radar station monitoring. The 16 and 21 inch versions have a synthetic display capability.

The IP 670S is an autonomous graphic display controlled by a built-in processor for bright display of fully synthetic images or dim light display of raw radar video (compressed), overlaid on the synthetic picture. This unit which offers many management functions can be used for writing in four colours on the penetration tube and permits the use of dual persistence tubes. It has a capacity of 300 tracks, 100 vectors, and 1200 characters.

The bright displays are intended for use in control towers. They offer daylight presentation of the radar picture with or without synthetic markings. The image is transformed into television signals by a scan converter with a storage tube (IPT 14) or a digital memory (TDI 800) and is presented on 12 or 16 inch bright displays. The association of a synthetic generator will endow the system with the same facilities as the IP 670S.

Aircat atc systems at Brest

Aircat atc system at Karlsruhe

Characteristics

AIRCAT		100			200	
RADAR STATION		1		1		1
TYPE	primary radar	SSR	primary radar	SSR	primary radar	SSR
EXTRACTOR				EV 760	TPL 800	EV 760
DATA TRANSMISSION	broad-band		broad-band	telephone	telephone	
CONTROL CENTRE OPERATION	Raw video	Passive/ active decoding	Raw video	Automatic decoding	Automatic; primary and secondary tracking	
PROCESSING				Special computer	Special computer or screen processor	
SCREENS	IP 570R/bright display		IP 570R/bright display		IP 570R/ bright display	IP 670S/ IP 670S/

AIRCAT		300			400		500	
RADAR STATION		2–4			2–4		more than 4	
TYPE	primary radar (near-by)	primary radar	SSR	primary radar	SSR	primary radar	SSR	Met radar
EXTRACTOR		TPL 800	EV 760		EV 760	TPL 800	EV 760	EVM 800
PROCESSOR		PR 800				PR 800	PR 800	PR 800
DATA TRANSMISSION	broad-band	telephone		broad-band	telephone		telephone	
CONTROL CENTRE OPERATION	Raw video	Automatic; primary and secondary tracking		Raw video	Multi-radar SSR tracking	Multi-radar primary and secondary tracking and weather map		
PROCESSING		Special computer			General purpose computers	General purpose computers		
SCREENS	IP 570R/bright display IP 670S			IP 670S/bright display			IP 670S	

Status

Production. Installed at atc centres in Brazil, Denmark, France and Federal Republic of Germany

Manufacturer

Thomson-CSF, Division DRS-TVT, Bagneux, France

Thomson-CSF ASTRE Airport Surface Control Radar

ASTRE radar antennae

ASTRE is a high resolution radar capable of detecting 1 square metre radar cross-section targets in a radius of 10 km in clear weather. In heavy rainfall, 16 mm an hour, the range is a minimum of 4 km, sufficient to cover a large airport.

The radar has a resolution cell, measuring 6 by 8 metres at 1500 metres, which provides separation of close-moving aircraft or service vehicles and its high antenna rotation speed (60 rpm) permits continuity of the presentation.

The inverse cosecant squared antenna pattern is designed to reduce cloud illumination and the radar uses a self-adaptive signal processing system to limit ground clutter, highlight target contrast and draw the map of the airport.

The display is presented on high-brightness, high definition monitors which can be viewed in the bright light environment of a control tower, after radar to television image transformation by a scan converter with 1000-line by 1000-point resolution.

Characteristics

FREQUENCY: 15·7/16·7 GHz
PEAK POWER: 30 kW
MEAN POWER: 0·012 kW
PULSE: 0·04 μs
PRF: 10 000 Hz
ANTENNA
FOCAL LENGTH: 1·5 m
HORIZONTAL BEAM-WIDTH: 0·33°
GAIN: 41 dB
ROTATION SPEED: 60 rpm
DIMENSIONS: 4 × 0·55 m

Status

Production. Introduced in 1977, the radar is installed at the Charles de Gaulle Airport, Paris, France

Manufacturer

Thomson-CSF, Division DRS-TVT, Bagneux, France

Thomson-CSF RS 770/870 Secondary Surveillance Radars

The secondary surveillance radars RS 770 and RS 870 comply with ICAO specifications. They are derived from a long line of interrogators and iff equipment. The combination of three types of interrogators and three antennae means that ranges from 60 to 200 nautical miles can be obtained, covering en route or approach needs.

The RS 870 is designed for adaptation to selective mode operation (ADSEL or DABS) and to monopulse reception. All the secondary surveillance radars can be equipped with systems inhibiting transponders in clutter areas and suppressing side-lobe responses.

The antennae are designed to permit co-mounting on the primary radar antenna. The antennae types AS 304 and AS 309 comprise a reflector with a dipole array at the front and an omni-directional radiating element at the rear.

The AS 809 is adjustable in elevation and comprises an array of 36 columns of 10 dipoles. For use under a radome, secondary radar interrogation signals can be radiated by the primary radar antenna.

Characteristics

INTERROGATOR MODEL	RS 770	RS 870-100	RS 870-200
TRANSMITTER/RECEIVER	Solid-state + triode	Solid-state	Solid-state + triode
POWER	2·5 kW	0·5 kW	2·5 kW
MODES	1, 2, 3/A, B, C, D	1, 2, 3/A, B, C, D	1, 2, 3/A, B, C, D
DIMENSIONS	24 in cabinet or 19 in unit	19 in unit	19 in unit
ANTENNA	AS 304	AS 309	AS 809
SPAN	4·2 m	8·4 m	8·7 m
HEIGHT	0·6 m	0·6 m	1·9 m
GAIN	19·5 dB	22·5 dB	24 dB
HORIZONTAL BEAM-WIDTH AT 3 dB	4·8°	2·2°	2·3°
SIDE-LOBES	–24 dB	–24 dB	–26 to –28 dB

Status

Production

Manufacturer

Thomson-CSF, Division DRS-TVT, Bagneux, France

Thomson-CSF TRS 2060 Approach Radar

The TRS 2060 radar station is an improved version of the THD 1098 which can be supplied with an integrated iff antenna and is now available as either a mobile or a static installation. It has been designed for easy transport and setting up, and is capable of being used in a variety of operational conditions. The manufacturers claim that its mobility and performance make it well suited for use as an autonomous surveillance station.

Characteristics

ANTENNA
TYPE: AC 316
POLARISATION: linear/circular
REFLECTOR DIMENSIONS: 2·3 × 4·7 m
HORIZONTAL BEAM-WIDTH: (3 dB): 1·4° ±10%
COSECANTED PATTERN: in excess of 40°
GAIN: 34 dB
ROTATION SPEED: 7·5/15 rpm
TRANSMITTER/RECEIVER TYPE: THD 047

TRS 2060 approach radar

FREQUENCY RANGE: (covered with 2 tunable magnetrons); 2900–3200 MHz
PEAK POWER: 1 MW
AVERAGE POWER: 2 kW
PULSE DURATION: 4 or 2 μs
PRF: 500 Hz; 1000 Hz
RECEIVER: parametric amplifier incorporated
NOISE FIGURE: less than 5 dB
RECEPTION CHAINS: log with differentiation/expansion and cfar chain
DIGITAL MTI (phase coding): double canceller loop 3 prfs; sub-clutter visibility: 25 dB
POWER SUPPLY: 220/380 V, 50 Hz, 20 kVA
DETECTION RANGE: 130 km on modern fighter, for a detection probability of 80% with single T/R unit

Status
Production

Manufacturer
Thomson-CSF, Division DRS-TVT, Bagneux, France

Thomson-CSF TRS 2310 Precision Approach Radar

This J-band equipment replaces the THD 1012. Specific improvements are: the equipment's mobility; performance in the presence of heavy rainfall improved by the use of velocity filters in the processing chain; and the possibility of controlling several approaches simultaneously by optimisation of antenna radiation patterns.
The equipment can be delivered for static installation with the same constituent elements. Approach control room equipment also is available in a static and in mobile versions, with a remoting capability of up to 3000 metres.
The par TRS 2310 meets ICAO recommendations. Fully solid-state design for reliability and ease of maintenance. The operation detection range is 20 nautical miles (37 km) for Mirage aircraft with 90 per cent detection probability.

TRS 2310 precision approach radar

Characteristics
FREQUENCY BAND: 9000–9200 MHz
ANTENNAS: mounted on orientable turret allowing angular displacement of 200 or 270° (at least 2 QFUs)
3-dB BEAM-WIDTHS: 0–6° (elevation antenna); 1·1° (azimuth antenna)
GAIN (HIGH-RESOLUTION PLANE): 39·5 dB (elevation antenna); 38·5 dB (azimuth antenna)
ATTENUATION OF FIRST SIDE-LOBES: < –25 dB
POLARISATION: circular
TRANSMITTER: high-gain TWT stage fed by high-stability pilot
AVERAGE POWER: 100 W (bad-weather mode); 50 W (clear weather mode)
PRF: 3 prf values in mode 2; 2 prf values in mode 1
RECEIVER: digital doppler processing
NOISE FIGURE: 3·5 dB
PERFORMANCE
OPERATING MODES: bad weather and clear weather
RANGES: 10 and 20 nm
SCANNING RANGES: elevation from –1 to +9°; azimuth 20°
DATA RATE: 1 azimuth and elevation picture every half second
IMPROVEMENT FACTOR
40 dB (fixed echoes)
30 dB (heavy rain echoes)
PRECISION: according to ICAO standards

Status
Production for several countries

Manufacturer
Thomson-CSF, Division DRS-TVT, Bagneux, France

Thomson-CSF Volex IV-3D Surveillance Radar

Volex is a fixed, E/F-band, three-dimensional surveillance radar providing continuously and simultaneously the azimuth, range, and altitude of any target within its cover volume.
The equipment consists, primarily, of an antenna system and dual transmitter-receivers. The antenna system comprises two antennas mounted side-by-side on a common turntable. Both are equipped with a Robinson elevation rapid-scan primary feed. The electronics are cabin-mounted and include two transmitter/receivers operating at a single-knob-controlled variable frequency, and equipped with constant-false-alarm-rate anti-jamming receivers. These solid-state equipments are fully interchangeable. Other features are side-lobe blanking and new-generation technology processing and data extraction facilities which result in high precision and resistance to ecm. The Volex IV radar is intended to feed an automatic exploitation system housed in the same cabin as the transmitter/receiver. Height-finding accuracy is ±460 metres at a range of 240 km.

Volex IV-3D surveillance radar

Characteristics

FREQUENCY	tunable through 300 MHz in E/F-band (with 2 magnetrons)
PEAK POWER	1 MW
AVERAGE POWER	2 kW
PULSE DURATION	4 μs
PRF	500 Hz
RECEIVER	low noise, wide band, parametric amplifier
NOISE FIGURE	less than 4 dP
RANGE	270 km on modern fighter

ANTENNA	VT 359 (lower elevations)	VT 150 (higher elevations)
SPAN OF REFLECTOR	3·3 m	3·8 m
HEIGHT OF REFLECTOR	6·8 m	3·4 m
WEIGHT	900 kg	350 kg
BEAM-WIDTH	azimuth: 2° elevation: 1°	azimuth: 2° elevation: 1°
GAIN	40 dB	37 dB
POLARISATION	linear +circular (optional)	linear +circular (optional)
DIMENSIONS	3·3 × 6·8 m	3·8 × 3·4 m
WEIGHT	900 kg	350 kg

Status
Production. Twelve systems are in service in France and other countries

Manufacturer
Thomson-CSF, Division DRS-TVT, Bagneux, France

GERMANY, FEDERAL REPUBLIC

AEG-Telefunken Surveillance Radar Equipment

Four types of radar are manufactured by AEG-Telefunken for airport surveillance operations, SRE-A5, -A7, -LL1 and -M5. The SRE-A5 is an E/F-band surveillance radar with a maximum range of 60 nautical miles (111 km). SRE-A7 is under development. It is a D-band short range radar designed to operate up to 100 nautical miles (185 km). SRE-LL1 is a D-band medium range radar with a maximum range of 150 nautical miles (280 km). The fourth radar SRE-M5, a medium-/long-range L-band radar using a klystron transmitter is equipped with circular polarisation, constant false alarm rate, and clutter-locked mti. It has a maximum range of 250 nautical miles (463 km).

All radars operate using antennae with a variable coverage diagram. A modular radar plot extractor is available which consists of a pr plot extractor, an ssr plot extractor and a correlator. All these components centre around a micro-programmable data processor. Depending on the type of application, each of the extractor components can be used autonomously. The mts display family is a variable modular system with a large number of different data displays. The display head is available with 16- and 23-inch screen, using a one-colour phosphor. A 23-inch penetron for multi-colour presentation is being developed. Data displays available for coupling with computers within a radar system are the SIG 50, SIG 51 and the microprocessor-controlled Terminal 52. On the data processing side, the AEG 80-20, AEG 80-40 and the AEG 80-60 computers are offered.

Characteristics

RADAR TYPE: SRE-M5
FREQUENCY BAND: L
PEAK POWER: 2500 kW
MEAN POWER: 2·5 kW
PRF: 400 pps
MAX RANGE: 250 nm (463 km)
MAX ELEVATION: 40°
ANTENNA TYPE: 2 beams variable scan, lower cut-off
ROTATION SPEED: 5 rpm
REFLECTOR DIMENSIONS
wide: 14·5 m
high: 9 m
HORIZONTAL BEAM-WIDTH: 1·05°

Status

Production

Manufacturer

AEG-Telefunken, Radio and Radar Systems Operating Group, Ulm, Federal Republic of Germany

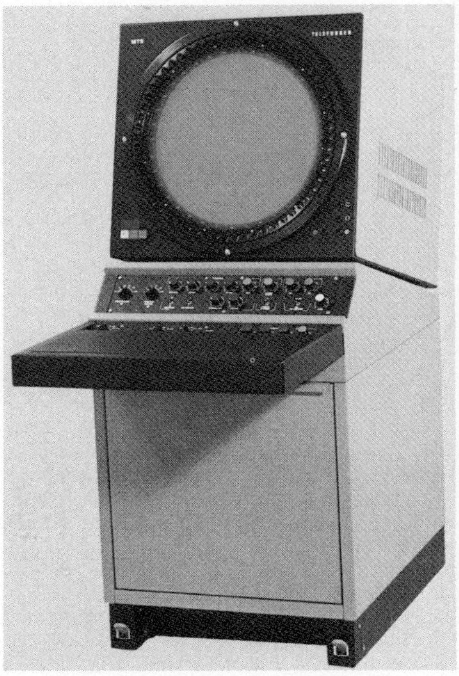

AEG-Telefunken Terminal 52 data display

AEG-Telefunken Air Track Air Traffic Control Systems

AEG-Telefunken produces a family of three modular systems for air traffic control, Air Track I, II and III. In Air Track I, the primary radar signals are transmitted to the processing centre via a cable or radio link using broad-band transmission and are presented on displays. The ssr signals, on the other hand, are decoded and extracted in the ssr extractor at the radar site, with the extractor also serving as defruiter. Target messages are transmitted to the processing centre via the telephone network. From these ssr target messages, ssr tracks are generated by an automatic tracking procedure. In the case of small systems, the tracks can be generated in the various control computers of the display equipment or in an AEG 80-20 central computer. The ssr target messages are also used to complement synthetically the raw video.

In Air Track II, the configuration is designed for en route control and comprises, in addition to the ssr target extraction, a primary radar extractor for automatic pr target identification, and a correlator for combining extracted ssr and pr information into joint target messages. This provides the basis for the computer to process all radar data and to prepare fully synthetic air situation displays which are constantly updated. The analogue radar video can serve as background or redundant information whenever the digital channel is defective. On the evaluation side, the target messages transmitted on narrow band are processed further in a double computer system.

The Air Track III system is used for evaluating information from several radar stations with overlapping coverage. By contrast to systems I and II, the flight tracks are generated at the radar sites (local tracking) and then transmitted to the joint processing centre. Local radar data processing contributes substantially toward relieving the central computer and reduces the volume of data to be transmitted since false targets are eliminated on the radar side during track generation. The co-ordinates of tracks generated locally are converted to the co-ordinates of the joint system before transmission to the central computer. The central computer therefore receives only tracks that relate to a uniform co-ordinate system, and calculates from the incoming data the most likely target positions, taking into account radar errors.

The flight plan processing and co-ordination system (flipco) can be implemented as an independent system or as a component of Air Track II or III. In either case, AEG 80-60 computers are used. The flipco system processes flight plan data and enables the atc controller at any time, in conjunction with the actual position reports of aircraft, to identify actual and specified aircraft positions. It comprises input, storage, up-dating and co-ordination of flight plans on data displays, flight control strip printing and co-ordination between the individual operators' positions. A notices to airmen (Notam) Information System is available which provides storage capacity for Notams which include instructions from the atc authority and any important information and messages affecting regular operations in aviation.

The DÜV data transmission and distribution system allows several atc centres to be linked for the purpose of data exchange. The distribution of data is effected by message switching. Normal telephone lines connecting the aftn can be used for the exchange of flight plan, Notam and flight control strip information.

For the purpose of reconstructing situations of conflict or, in extreme cases, accidents, all tracks and inputs into the Air Track system are stored on tape. Extractor messages, track data, operators' inputs as well as errors and status reports of the computer are all recorded. The system can be expanded so that the content of any memory area can be recorded. If special problems are to be investigated, a data reduction procedure can be used to select and output important data from the overall data volume recorded. This selection is effected with punched cards, the user being able to generate his own sub-routines for special tasks. This form of data reduction can be carried out during normal operation.

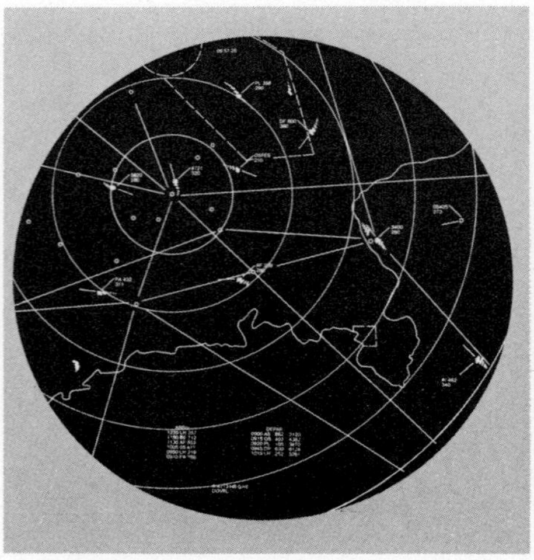

AEG-Telefunken Air Track I atc system

Status

Production

Manufacturer

AEG-Telefunken, Radio and Radar Systems Operating Group, Ulm, Federal Republic of Germany

ISRAEL

Elta EL/M-2215/2220 Surveillance Radars

The Elta EL/M-2215/2220 is a general-purpose surveillance radar, evolved from the successful EL/M-2205 search radar, and many of the engineering features of the latter radar have been retained in its successor. The radar is designated -2215D in the atc configuration and -2220 when operated as a mobile air surveillance radar. Improvements made possible by advances in the state-of-the-art have been incorporated, such as more elaborate signal processing facilities and increased transmitter output power. The transmitter/receiver is a redesigned and repackaged version of the former EL/M-2205. Modular design allows the second-generation EL/M-2215 to be configured in either single or dual transmitter/receiver versions, the latter offering a further choice between a dual-channel arrangement for redundancy or dual-channel with diversity operation. Other features include high resolution, a dual high/low beam antenna; high clutter suppression; quadrature (I/Q) digital mti and fast frequency tuning. Sixteen inch ppi displays with five range scales, electronic cursor and digital read-outs, or crt displays with digital scan converters are offered with both versions. An automatic tracking facility using EL/S-8650 radar extractor and the EL/S-8610 mini-computer iff and communication equipment are available as further optional features. Built-in test equipment (bite) is incorporated as standard equipment. The EL/M-2215 employs the AT 104 dual (high/low) beam antenna. The EL/M-2220 uses the AT 103 single beam antenna for fixed and mobile applications of air surveillance (-2220S, -2220D) and lightweight AT 101 when configured as the EL/M-2220F fast deployment radar system.

Elta AT-103 atc antenna

Characteristics

TRANSMITTER			
frequency	3·1–3·3 GHz (2·7–2·9 GHz optional)		
output power	425 kW peak		
pulse width	0·95/1·4 μs		
prf	1000/750 Hz, 4 staggered or random variable		
RECEIVER			
noise figure	2 dB (at transistorised amplifier input)		
sensitivity	107 dBM		
PROCESSING			
digital mti	quadrature (I/Q) 3-pulse canceller with or without feedback		
improvement factor	33 dB		
blind speeds	none below Mach 3		
antenna	AT 104	AT 103	AT 101
gain	34 dB	34·5 dB	28 dB
azimuth beam-width	1·4°	1·4°	3°
elevation beam-width	5·5° low beam 8° high beam	5·5°	10°
pattern	cosec² to 45°	cosec² to 30°	cosec² to 30°
polarisations	horizontal/circular	horizontal/circular	horizontal/circular

ELTA AIR SURVEILLANCE RADARS

MODEL	EL/M-2215D	EL/M-2220D	EL/M-2220S	EL/M-2220F
APPLICATION	fixed air traffic control	mobile long range air surveillance	mobile air surveillance	mobile fast deployment
AIRCRAFT DETECTION RANGE	80 nm	80 nm	60 nm	30 nm
AZIMUTH RESOLUTION	1·4°	1·4°	1·4°	1·4°
RANGE RESOLUTION	150 m	220 m	150 m	150 m
HEIGHT COVERAGE	30 000 ft	30 000 ft	25 000 ft	18 000 ft
PRF	1000 Hz staggered	750 Hz staggered	1000 Hz staggered	1000 Hz staggered
ANTENNA ROTATION SPEED	12·5 rpm	7·5 rpm	7·5/15 rpm	12/24 rpm

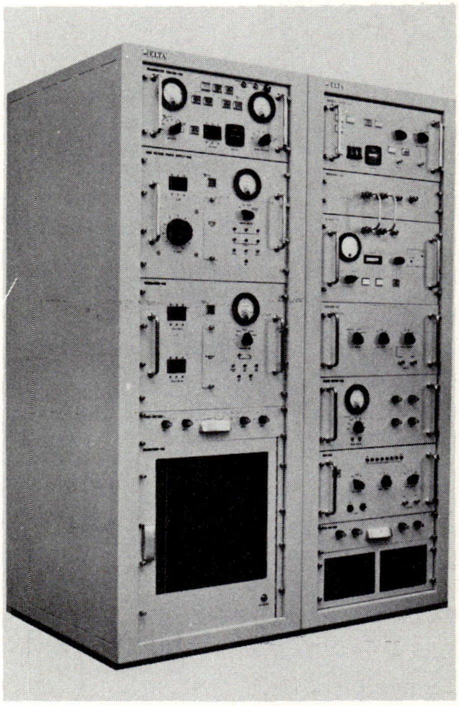

Elta EL/M-2215 transceiver

Manufacturer

Elta Electronics Industries Ltd, Ashdod, Israel

ITALY

Selenia ATCR-22 Primary En-Route Radar

Selenia ATCR-22 radar installation at Fiumicino Airport, Rome

The ATCR-22 is a high-power 23 cm wavelength, fully solid-state (except for the magnetron and thyratron) search radar. In an atc configuration the radar consists normally of an antenna group, two transmitter units, two receiver units and the common wave-guide system. In the dual channel version, a dual frequency diplexer allows the radar to work in frequency diversity. A third single channel can be used as a hot stand-by to any of the two main channels. Two antenna systems are available, the G-14 high gain, extra long range antenna and the smaller G-21 antenna, equipped with a dual feed and multiple feed respectively.

A tunable, highly stable magnetron, thermo-stabilised by a water-vapour cooling system is used; six different repetition periods of prf staggering give the first actual blind speed well outside the speed capabilities of modern aircraft, while ensuring a flat velocity response curve.

In the receiver the signals from the main and auxiliary feeds are amplified in separate low noise, high gain, broad band, solid-state rf amplifiers. The resulting very low noise figure is continuously checked by a test equipment. A clutter fix system optimises the signal/clutter ratio at the receiver input and a programmable adaptive clutter attenuator ensures linear dynamic characteristic of the receiver, within its full range.

The digital processor includes a fully digital, I and Q, double canceller mti with 10-bit a/d converter and a moving window video integrator resulting in a high improvement factor, and hence an extremely high sub-clutter visibility against a very high clutter/noise ratio.

A clutter mapper is also provided: this device automatically defines areas in which mti video is desired because of clutter, but in other areas normal raw video is displayed. A video correlator eliminates second time round echoes and rejects non-synchronous returns and noise. It can be automatically introduced, checking on the level of residues. Log-ftc is introduced when needed in both mti and normal operation. A special output displays rain contours. A fan circuit guarantees a constant false alarm rate of 10^{-6}. Provision for on-mounting ssr is standard and the antenna has a triple channel rotating point.

The ATCR-22 can be controlled and checked out remotely. A performance monitor continuously monitors noise figure, transmitted and received power. Alarms are initiated (remotely, if required) when these pass a preset threshold.

Built-in test equipment performs on-line test of the whole radar. A two digit led display indicates failed pcbs or sub-assemblies.

Characteristics

TRANSMITTER	Magnetron M 5051/M 5052	
FREQUENCY		
M5051	1250–1310 MHz	
M5052	1350–1355 MHz	
PEAK POWER	1·8 MW	
AVERAGE POWER	2·25 kW	
PRF	300–800 pps staggered in 6 different repetition periods	
RECEIVER DIGITAL PROCESSOR		
RF STAGES		
NOISE FIGURE	3 dB (typical)	
DYNAMIC RANGE	95 dB	
PACA	programmed adaptive clutter attenuator	
CLUTTER FIX	adaptive, in accordance with 5 different combinations	
CLUTTER FIX MEMORY	automatic data store in 3 antenna revolutions in 1024 range-azimuth sectors	
PHASE DETECTORS	2-phase detectors with 0°/90° phase reference	
DIGITAL STAGES		
ANALOGUE-TO-DIGITAL CONVERTER	10 bits	
CANCELLER	dual 3-phase canceller	
NORMAL-MTI SELECTOR	automatically controlled by adaptive clutter mapper	
CLUTTER MAPPER:	automatic data store in 1 antenna revolution in 2048 sectors (64 azimuth and 32 range divisions)	
LOG-FTC	10-bit dynamic range logarithmic characteristic and dc suppression for rain signal normalisation	
RAIN SENSOR	rain data storer, working in real time	
MOVING WINDOW INTEGRATOR	dual threshold and moving window matched to the number of hits per scan	
FAN	scan-to-scan adaptive adjustment of integrator thresholds and clutter fix setting	
CORRELATOR	second-time around echo of targets and anti-anomalous propagation device against second and third time around echoes	
RAIN CONTOUR	extractor for trailing and leading edges of the rain for ppi presentation	
MTI PERFORMANCE		
IMPROVEMENT FACTOR	29 dB	
CFAR DYNAMIC RANGE	37 dB	
SUB-CLUTTER VISIBILITY	30 dB	
ANTENNA	G-14	G-21
operating frequency	1250–1350 MHz	1250–1350 MHz
polarisation	horizontal and circular (switchable)	horizontal through elliptical to circular
turning rate	5, 6, or 7·5 rpm (other rates available upon request)	5–10, 6–12 or 7·5–15 rpm (other dual rates available upon request)
feed type	dual horn	integrated 8 horns

	G-14	G-21
MAIN BEAM		
3 dB horizontal beam-width	1·2° ± 0·15°	1·25°
vertical pattern	cosecant squared with added high angle coverage	modified cosecant squared with added high angle coverage
antenna gain	36·5 dB typical	32·5 dB
side-lobe level	−26 dB typical	−25 dB
measured integrated cancellation ratio	23 dB typical	20 dB
AUXILIARY BEAM		
3 dB horizontal beam-width	1·25° ± 0·15°	1·25°
antenna gain	36 dB typical	31 dB
side-lobe level	−28 dB typical	−25 dB
measured integrated cancellation ratio	20 dB typical	
OPERATING TEMPERATURE RANGE		
indoor units	0–50°C	
outdoor units	−40 to +70°C	

Status
Production. In service in Italy, Sweden and other countries

Manufacturer
Selenia Industrie Elettroniche Associate SpA, Rome, Italy

Selenia ATCR-33 Primary Terminal Area Radar

This equipment is an E-band 10 cm medium power radar suitable for approach or GCA surveillance. It employs the same receiver as the ATCR-22 radar and is equipped with the G-33 antenna. This turns at 15 rpm, and with the transmitter operating at 1000 pps the coverage achieved on a small jet aircraft with the single channel ATCR-33 is about 45 nautical miles (83·4 km) in range and over 6000 metres high. Short range low cover properties make the radar extremely suitable for aircraft recovery.

Portable version of ATCR-33 radar

ATCR-33 radar installed at Palermo Airport, Sicily

Characteristics
TRANSMITTER
- magnetron: 5586
- cooling: air
- peak power: 0·5 MW minimum
- average power: 500 W (minimum output of the magnetron)
- frequency: 2700–2900 MHz
- prf: staggered in 6 different repetition periods

RECEIVER DIGITAL PROCESSOR
- noise figure: 2·1 dB measured at receiver input

IF STAGES
- paca: 3 different attenuation laws selected according to the clutter level by pre-programmed memory setting
- phase detectors: 2-phase detectors

DIGITAL STAGES
- analogue-to-digital converter: 10 bits
- canceller: dual 3-pulse canceller
- normal-mti selector: automatically controlled by adaptive clutter mapper
- clutter mapper: automatic data store in 1 antenna revolution in 2048 sectors (64 azimuth and 32 range divisions)
- log-ftc: 10-bit dynamic range logarithmic characteristic and dc suppression for rain signal normalisation
- rain sensor: rain data storer, working in real time
- moving window integrator: dual threshold and moving window matched to the number of hits per scan
- fan: scan-to-scan adaptive adjustment of integrator thresholds and beam combiner setting
- correlator: anti-anomalous propagation device against second-time around echo of targets and second and third time around echoes
- rain contour: extractor of trailing and leading edges of the rain for ppi presentation

MTI PERFORMANCES
- improvement factor: 39 dB
- cfar dynamic range: 37 dB
- sub-clutter visibility: 30 dB

G-33 ANTENNA GROUP
- operating frequency: 2700–2900 MHz
- polarisation: variable from linear to circular through all degrees of ellipticity
- type of coverage: modified cosec^2 with improved high elevation angle coverage
- turning speed: turning rate is 15 rpm (other speeds are available upon request)

MAIN BEAM
- 3 dB horizontal beam-width: 1·45° ± 0·05°
- gain: 33·5 dB
- main side-lobes: better than −25 dB

AUXILIARY BEAM
- 3 dB horizontal beam-width: 1·45° ± 0·05°
- gain: 32 dB
- main side-lobes: better than −25 dB

OPERATING TEMPERATURE RANGE
- indoor units: 0–50°C
- outdoor units: −40 to +70°C

Status
Production. In service in Italy, Thailand, Turkey and other countries. As a result of an agreement signed in October 1981 the Scottish Group of Ferranti Ltd is to manufacture this radar under licence

Manufacturer
Selenia Industrie Elettroniche Associate SpA, Rome, Italy

Selenia ATCR-44 Primary Radar

This radar is a medium powered version of the ATCR-22 employing a peak power of 500 kW and is especially suitable as an approach radar or for GCA surveillance. Either the G-21 or the G-14 antenna can be used according to operation requirements. With the antenna rotating at 15 rpm, the range coverage of the ATCR-44 on a small jet aircraft is over 60 nautical miles (111 km) in single channel, with a height coverage of over 9000 metres. In dual diversity the range coverage on a small jet aircraft is over 75 nautical miles (139 km), with a height coverage of over 12 000 metres. In its normal version, the radar has a prf of 1000 pps with a pulse width of 1 microsec giving an average power output of 500 watts. The pulsed magnetron is tunable over the frequency band 1280–1350 MHz. The receiver/mti chain is identical to that of the ATCR-22 and features the same digital processor resulting in a system which is flexible, reliable, and easier to maintain. As with the ATCR-22, a clutter mapper is provided to delineate segments in which mti video is desired: in all other segments the video is normal. Video correlator and video integrator are also identical to those of the higher-powered radar.

Status
Production. In service in many countries including Italy, Hong Kong, USSR and Venezuela. A contract to supply the radar to Spain was announced in February 1982. It will be employed for the atc of the Madrid terminal area and will be installed at Paracuellos

Characteristics
TRANSMITTER
- magnetron: 5J26 (air-cooled)
- tuning range: 1280–1350 MHz
- peak power: 0·5 MW
- average power: 500 W
- prf: up to 1000 Hz

RECEIVER-DMTI: as for ATCR-22

ANTENNAE: as for ATCR-22

OPERATING TEMPERATURE RANGE
- indoor units: 0–50°C
- outdoor units: −40 to +70°C

Manufacturer
Selenia Industrie Elettroniche Associate SpA, Rome, Italy

Selenia SATCAS-80 Air Traffic Control System and Displays

SATCAS is a fully digital system, including primary and secondary video extractors, digital central units for generation of radar sweep, lines, vectors, and symbols, digital indicators, keyboards and designation devices. Arranged around a general purpose digital computer, the system can provide aircraft labels, automatic handover, sectoring, digital mapping, microtabulars, and other features required by a modern atc centre.

The system employs distributed processing techniques using the Selenia NDC-160 minicomputer in the radar head for extraction, tracking and narrow band transmission; in the display, for man/ machine interface; as a radar data processor for data distribution, configuration control, and mosaic display; and as flight data processor with interface to aftn and other centres, for strip printing, data distribution and flow control.

The design philosophy permits a flexible approach, the system being composed of independent, but interacting, digital modules. These modules can be assembled in different configurations in order to meet requirements from simple to complex atc centres.

The software is adaptable to site requirements, and emphasis has been placed on achieving effective presentation of essential data to the controller, requiring minimum manual intervention.

Three types of display in current production are the IDM-7, DDS-80 and the CDS-80 bright display. The first two are digital displays suitable for mixed raw radar and synthetic/alphanumeric presentation. The CDS-80 is a television monitor which is a high-intensity daylight display for use in the atc tower, technical radar room and other daylight environments where the presentation of bright raw radar video is required. The IDM-7 is designed to operate either as stand-alone monitor or as indicator in a centralised display system. It is available with 16 or 22 inch crt and various types of phosphors according to the type of presentation required: raw radar; synthetics, or a mixture of raw radar and synthetics.

The DDS-80 is an advanced, dedicated 22 inch display unit designed to meet the requirements of a wide range of automated atc centres up to multi-radar, complex systems having a high traffic load. Basic characteristics of this display are a dedicated data processor, and a high refresh rate.

Selenia SATCAS-80 atc system

Status

SATCAS-80 is in service in Bangkok, Thailand

Manufacturer

Selenia Industrie Elettroniche Associate SpA, Rome, Italy

Selenia SIR Secondary Surveillance Radar

The SIR secondary surveillance radar is the latest development from Selenia. A fully solid-state D-band radar, SIR can be operated on a single or dual channel configuration.

The radar is equipped with power programming which reduces the incidence of confusing reflections by ensuring that a smaller amount of power is radiated in the critical directions. Therefore the reflected energy is much lower and does not trigger the transponder, thereby eliminating the erroneous reply. In this way the effects of site-related phenomena such as reflections and interference and fruit between adjacent stations can be minimised.

SIR includes 10-mode interlace selector thumbwheel switches which allow a wide range of interlace patterns, including different triple interlace in three azimuths, varying interlace on successive antenna revolutions, and also external interlace.

The transmitter consists of four to eight power modules (each module has a 500 Watt drive capability) all in parallel. The power transistors in each module are also in parallel, so that in the event of failure only a small power reduction occurs.

Characteristics

TRANSMITTER

frequency: 1030 ± 0·1 MHz
peak power: 1·8 kW, 3·4 kW (option)
power regulation: 3, 6 and 12 dB steps
pulse duration: 0·8 μs
modes: 1, 2, 3/A, B, C, D, test, external
interlace types: circular, azimuth, scan/scan, test
interlace modes: 3 groups of 3 thumb-wheels select any modes
azimuth gate: 4 sectors covering 360° with 0·08° boundary definition

RECEIVER

frequency: 1090 MHz
band-width: 8 MHz
tangential sensitivity: not less than −87 dBM
gain time control: 8 range gates with ± 10 dB attenuation

ANTENNA

Length	Beam-width	Gain
8·2 m	7°	21 dB
4·1 m	4·5°	19 dB
2·8 m	2·35°	19 dB

POWER REQUIREMENTS: 220 V ± 10%, 45–400 Hz, 350 W single channel

OPERATING TEMPERATURE RANGE
indoor units: 0–50°C
outdoor units: −40 to +50°C

Status

Production. The unit is compliant with ICAO and STANAG recommendations

Manufacturer

Selenia Industrie Elettroniche Associate SpA, Rome, Italy

JAPAN

NEC Air Traffic Control Radars

A range of radars covering most aspects of civil aircraft movement is manufactured by the Nippon Electric Co Ltd (NEC). These radars are made primarily to comply with ICAO and other civil aviation specifications: however, some of them have been supplied to the Japanese Defence Agency.

The air route surveillance radar (arsr) is a D-band long-range surveillance radar intended for en route atc.

NEC provides two types of arsrs, NPG-434 (magnetron transmitter version) and the NPG-630 (klystron transmitter version) arsr, which are combined with NPG-542 secondary surveillance radar (ssr), a radar data transmission system such as NPG-359 radar microwave link (rml) or NZG-594 radar video band-width compression system(vbc) and associated equipment.

The NPG-434/NPG-630 arsr are characterised by the application of modern computer-aided design and beam antenna, solid-state circuits and advanced digital signal processing technique providing extended coverage of more than 200 nautical miles (370 km) and 70 000 feet (21 336 metres). The NPG-359 all solid-state rml is operated in the 7 GHz band and has a standard transmission range of 50 km.

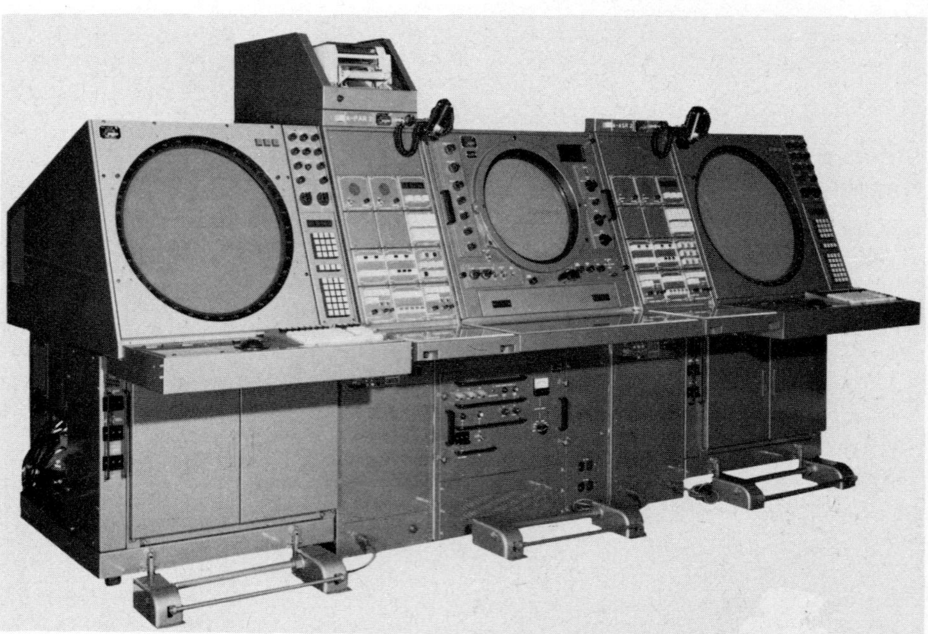

Surveillance radar and par display of NEC gca NPG-864

NEC has developed the NZG-594 vbc system which makes it possible to transmit arsr raw radar video and ssr digitised radar data on narrow-band 12-voice telephone channels (48 kHz) and one channel of the telecommunication network as a transmission means. Six of these systems have been supplied to the Japanese Civil Aviation Bureau.

NEC manufactures an E/F-band surveillance radar for use in terminal area atc. Two versions exist: the NPG-360 which has a range of 64 nautical miles (118 km) and 40 000 feet (12 200 metres) on small commercial aircraft and the NPG-460 which has a range of 80 nautical miles and 50 000 feet (15 240 metres) on a jet fighter. About 20 dual installations of one or other or both of these radars have been supplied to the Defence Agency. The NPG-360 is installed at Hiroshima Airport, Japan. Both radars have beam-widths of 1·2 degrees, data renewal rates of 15 rpm and digital mti.

NPG-360 has a 500 kW transmitter and a transistor rf amplifier giving an overall noise figure of 3·5 dB; NPG-460 has a 3·5 MW transmitter with a transistor rf amplifier, as in the NPG-360.

NPG-435 is a precision approach radar which has been deliberately built to a standard above the ICAO requirements so that it can be used on fighter airfields. The system provides detection coverage of 7 degrees in elevation and 20 degrees in azimuth with a detection range of more than 10 nautical miles (18·5 km). A dozen dual installations have been supplied to the Japanese Defence Agency.

NPG-542 is a complete ICAO-specification ssr system with interrogator/responsor decoder and defruiter and with sif capability. About 100 of these systems have been supplied to the Japanese military and civil aviation authorities.

NPM-554 is the system number given to an integrated installation comprising asr, par, and ssr/sif elements together with communications subsystems. The component elements are essentially those described above but slightly reconfigured to suit the integrated installation. The secondary radar antenna is in a linear array mounted on the primary radar antenna. Six of these systems have been supplied to the Defence Agency.

NPM-554 mobile gca system

NPG-864 is a fixed type gca system featuring a high performance alpha-numeric radar display that automatically detects and tracks aircraft equipped with a beacon transponder, and then correlates them with flight-plans previously registered at a terminal atc. It is provided not only with a data processing system to conduct the above function, but also with the conventional gca component elements such as NPG-460 asr, NPG-435 par, NPG-542 sif, ccu and associated equipment. These equipments are installed in shelters at the asr, par and ACC operational areas. The radar has been supplied to the Defence Agency.

Two other radars developed by Nippon Electric are the high-power NPG-434 arsr magnetron version, and NPG-630 (arsr-klystron version), which have been bought by the Civil Aviation Bureau.

Manufacturer
Nippon Electric Co Ltd, Tokyo, Japan

NEC Automated Air Traffic Control Systems

Automated atc systems produced by NEC comprise flight data processing, radar data processing for en route control and terminal radar data processing systems.

The flight data processing system (fdp) controls the all IFR flight-plans in the atc area. The system features a flight handling capacity of more than 6400 flights per day with three computers, an automatic file recovery facility and automatic fault detection. In this system all flight-plan messages filed prior to the departure of the aircraft are sent to flight data processing via input data processing. Flight data is updated by the input data on actual time of departure, transfer, estimated time of arrival, aircraft altitude or flight route. Flight progress strips are produced by a strip printer, and advisory messages are displayed on a crt monitor for controllers. Flight data is progressively exchanged with other related atc facilities. The system permits statistical processing of journal data for each acc and provides conflict search on oceanic flights.

The radar data processing system (rdp) processes the radar data fed by air route surveillance radars and provides air traffic controllers with current air traffic information. The system features aircraft tracking by means of radar augmented beacon tracking, fail-safe operation based upon full dual system configuration, and a fail-soft operation based upon a distributed processor configuration. Three display modes are provided for the controllers' selection: full digital, television time-shared and television mode. A radar raw video centralisation facility using a narrow bandwidth radar raw video transmission system is employed and data is presented on a flat faced 23-inch plan view display in monochrome or colour.

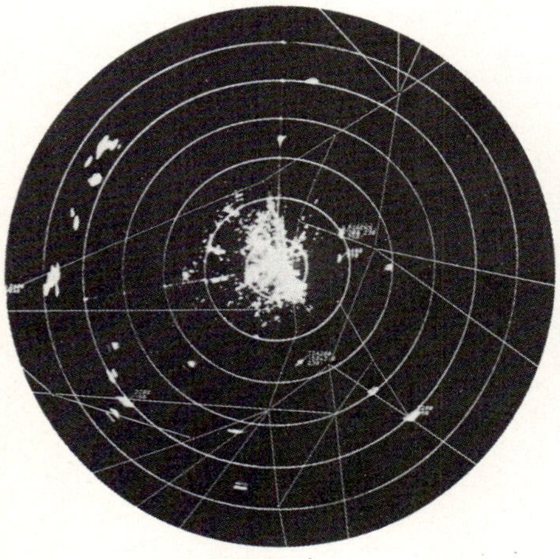

NEC pvd time-share mode

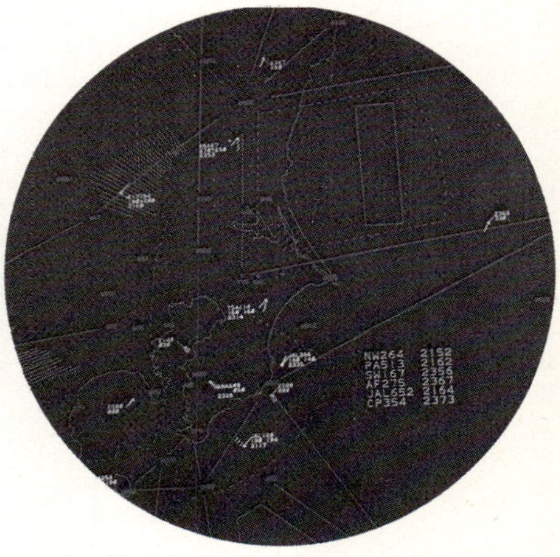

NEC pvd full digital mode

Basic functions of the equipment are ssr/arsr data detection, ssr data tracking; correlation of rdp data and fdp flight-plan and alphanumeric display of aircraft position, altitude, call sign and forecasting lists.

Optional features available include automatic tracking of arsr data, flight-plan updating derived from rdp data, weather contour display and conflict alert prediction.

Since 1964, NEC has gained experience in long distance radar video transmission for the radar processing systems of the Japan Civil Aviation Bureau. The standard system for the transmission is the broad-band radar microwave link (rml). Also, NEC uses a narrow band-width radar raw video transmission system adopting video bandwidth compression technology. The system enables long distance transmission up to approximately 2500 km via only 12 telephone channels (one group 48 kHz channel).

The radar microwave link provides radar video multiplexing for high fidelity transmission and telephone channel multiplexing, with normal mti, ssr video, a supervisory and remote control signal and telephone channels. Standard transmission distance per span at 7 GHz is 50 to 80 km.

In the narrow band-width radar video transmission system arsr video band-width of 400 kHz is compressed into about 40 kHz through digital signal processing technology. Only one group (12 telephone channels) is needed for transmission.

Radar displays offered by NEC comprise a plan view display and a bright display. The plan view display presents mass data on a 23 inch flat-faced beam penetration crt screen in monochrome or colour, with three display modes. A ppi time share-mode can also be provided at the customer's request.

Features of the display are electrical pin-cushion correction and dynamic focusing; uniform brightness of displayed characters and lines by means of digitally-controlled writing speed with a random access deflection; and a 30 Hz data refresh rate. The colours produced are red, orange, yellow and green.

The bright display is designed to present bright and legible symbols, graphics and raw video for viewing in a daylight environment and is therefore suitable to be used for airfield control in control towers. The display consists of a digital scan converter with built-in map generator and a television display using a 16-inch crt.

NEC has developed a terminal radar data processing system which has a target tracking capacity up to 256 targets. The system provides asr augmented ssr tracking for decreasing target-coasting probability and real-time flight-information exchange with an fdp system. Two modes can be selected. In the duplex mode one processor is used for on-line operation and the other one for stand-by. When the on-line processor fails, the stand-by processor replaces it within 5 seconds by manual operation. In the dual modes, two processors treat input data in parallel and check each other. When the master processor fails, the slave processor automatically replaces it within one or two seconds. The equipment features built-in test equipment, a fail-soft operation with a distributed data processing configuration which permits a degraded mode operation when the central mini-computer fails.

Other features of the system are a time-shared ppi display with a 23 inch flat-face crt, mass digital presentation, automatic call sign conversion and flight-plan list presentation control, automatic flight progress strip printing, semi-automatic hand-over facilities, digital map and cursor presentation by a display control unit and an on-line flight plan data.

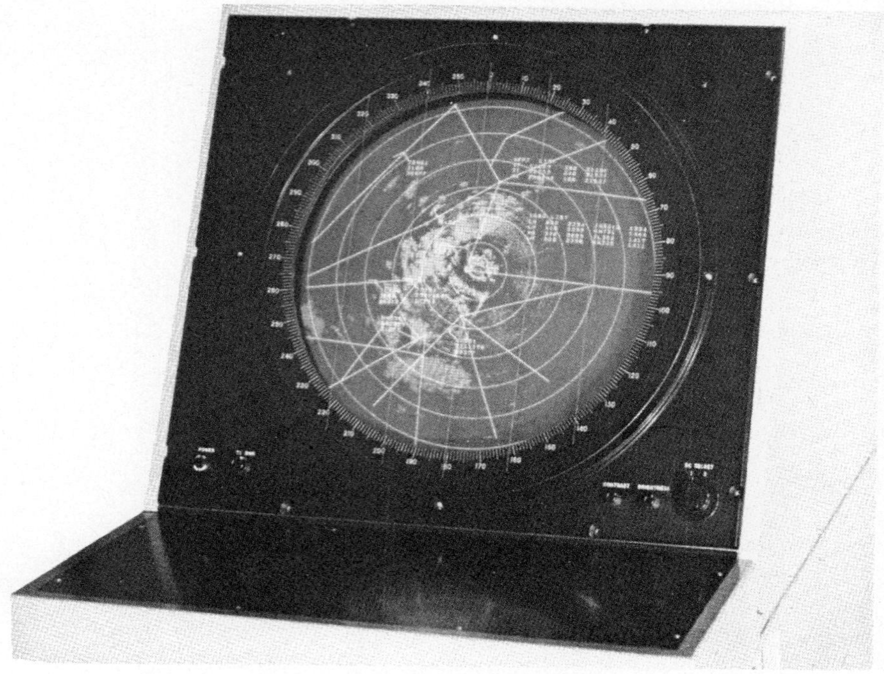

NEC bright display

Status

The equipment was developed for the Japan Civil Aviation Bureau. Phase I of these systems was initiated in 1965 and phase II including the radar data processing systems in 1972 and the equipment is now in service

Manufacturer

Nippon Electric Co, Tokyo, Japan

NETHERLANDS

Signaal ASDE Airport Surface Detection Equipment

ASDE is designed to provide accurate information on moving aircraft and vehicles, as well as stationary objects, at and around the runways, taxiways and aprons of an airport, and to present this information under high ambient lighting conditions to the tower and ground controllers. In its basic configuration ASDE is composed of a high-definition J-band radar, using an antenna with a very narrow horizontal beam-width, a solid-state digital scan converter and a bright display with a 20 inch crt.

The main features of the equipment are the digital scan converter with modular multiport random access memory, video presentation with high resolution, built-in digital video map and extensive built-in test equipment. As a duplicated system, it is extensible to six displays per digital scan converter and is suitable for handling TMA-radar information. A limited amount of computer-generated data can be displayed and the unit has an output to 625/896/ 1295-line television-displays.

The radar is designed to operate under all-weather conditions, with a range of 10 km on a 3 square metres fluctuating target under normal conditions, and 5 km during 16 mm per hour rainfall. Range discrimination is 7·5 metres with an angular discrimination of 0·25 degrees.

Characteristics

ANTENNA-TYPE: parabolic reflector
FREQUENCY BAND: J
BEAM-WIDTH (–3 dB)
  horizontal: 0·25°
  vertical: 3°
GAIN: 40 dB
ROTATIONAL SPEED: 60 rpm
TRANSMITTER
  frequency: 16–16·3 GHz
  peak pulse power: 60 kW
  pulse length: 50 μs
  prf: 4096 Hz
RECEIVER:
  noise figure: 10 dB
  if frequency: 60 MHz
  digital scan converter memory capacity: 1024 × 1024 elements

Signaal ASDE radar

Status

Production. Introduced in 1977 several systems have been sold

Manufacturer

Hollandse Signaalapparaten BV, Hengelo, Netherlands

Signaal Alpha Air Traffic Control Flight-plan Processing Aid Strip Printing System

Alpha is an autonomous, computer controlled flight-plan processing and flight progress strip printing system. It is designed for those atc centres where the growth of traffic makes automation a necessity and is an ideal first step towards automation. Alpha is designed to be expansible, capable of integration with existing or future radar data-processing systems and to be linked with Philips' AEROPP aeronautical message switching system for automatic flight data input.

The system comprises a Philips P800-type processor with two keyboards providing inputs and six printers. A dual processor configuration with automatic or manual change-over facilities, and electronic data displays with keyboard are available as optional equipment. Flight data are fed into the system in the standard ICAO flight-plan format. The system calculates flight paths using stored aircraft-type characteristics and airspace structure, the current meteorological situation, runway in use, and stored standard instrument departure and standard arrival route procedures. The flight strip is printed in red or black on standard strips using a layout defined by the customer. Optional features available include repetitive flight-plan storage, aftn coupling, message switch coupling (AEROPP) and conflict detection.

Characteristics

PROCESSOR: Philips P800 series
INPUTS: 2 × Diablo Hyterm 1620 with keyboard (option: Philips P817 Electronic Data Display with keyboard)
OUTPUT: 6 × Diablo Hyterm 1640 (option: Philips P817)
STRIP
  dimensions: 8½ × 1 in (21·6 × 2·5 cm)
  lines: 3 or 4 per strip
  max characters: 72 per line

Status

Production

Manufacturer

Hollandse Signaalapparaten BV, Hengelo, Netherlands

Signaal Bravo/Charlie/Delta Air Traffic Control Systems

Signaal Bravo, Charlie and Delta atc systems are modular units which are designed to be expansible from the basic Bravo system to the more sophisticated Delta configuration. Bravo is an autonomous unit and is also suitable as an initial complement to the strip printing/flight plan processing system, Alpha. It is designed to display raw primary and secondary radar video. The primary radar either D- or E/F-band features digital mti, video correlation, interference suppression and pulse discrimination. The secondary surveillance radar provides active and/or passive decoding and height layer selection by special control modules. Up to 15 working positions can be incorporated with synthetic video maps, picture off-centring and display range scaling. Six ranges are selectable between 4 and 256 nautical miles (7·4 and 475 km) in multiples of 4 nautical miles (7·4 km). Other display features comprise a cursor line with bearing and range displayed, DF lines, and a North marker. The last two are optional and the DF facility requires a display processor.

The Charlie system is an expanded version of Bravo, with the addition of an ssr data processor, flight plan input typewriter, monitor typewriter, and distributed processing with minicomputers for specially assigned functions. Up to 15 operator consoles can be provided with displays of raw primary radar video, ssr plots and synthetic symbols with labels including vectors and alphanumeric data. Autonomous-mixed displays are used with 16- or 23-inch screens, hand-wheels and light pen input devices. The display incorporates tracking with automatic initiation of transponder-equipped aircraft, limited flight plan handling, and automatic track/flight plan correlation of 4-digit ssr tracks. Manual label and hand-over facilities are provided.

Optional features offered comprise analogue video mapping, primary radar tracking with primary/secondary plot correlation, passive ssr video decoding and a DF line display.

Signaal Charlie/Delta atc system operator console

Delta is an expanded version of the Charlie system with up to 25 operator consoles with bright displays of synthetic radar information (symbols with labels), video map and flight data tables. In addition to the Charlie capabilities the system provides tracking of aircraft with a 4-/2-digit ssr or primary responses only. The display is a 23-inch screen with a 55 Hz refresh rate and 50 foot (15·2 metres) lambert brightness.

Optional equipment available includes weather map and DF line displays, and provision for the integration of a computer-controlled strip printing or an electronic data display flight plan handling system. In both Charlie and Delta systems, the hardware can be duplicated and provided with an automatic change-over facility.

Characteristics

SYSTEM	Bravo	Charlie	Delta
WORKING MAX POSITIONS	15	15	25
LABELLED TRACKS		40	100
UNLABELLED PLOTS		80	25
FLIGHT PLANS		40	100

Primary and secondary radar antenna suitable for use with Signaal Bravo/Charlie/Delta atc systems

	Charlie	Delta
DIGITAL MAP	70 lines with 70 symbols each	700 lines with 70 symbols each
TABLES	2	1
NO OF CHARACTERS	300 each	250

Status
Production

Manufacturer
Hollandse Signaalapparaten BV, Hengelo, Netherlands

Signaal LAR II Long-Range Surveillance Radar

LAR II is a high-performance D-band radar suitable for use with sophisticated atc systems. Its operational applications include: long-range surveillance of area traffic, medium-range surveillance for terminal area control, and short-range approach vectoring. The radar system which is also available in a dual (main and standby transmitter/receiver) configuration is suitable for integration with a secondary radar. The antenna is dual beam with a passive horn and two rotational speeds. A crystal-controlled coherent transmitter with pulse compression provides ranges in excess of 220 nautical miles (408 km) on a 5-square metre target, with a minimum range of 0·5 nautical mile (0·93 km) and resolutions in range of 100 metres and 1·5 degrees in azimuth. A moving target indication facility is fitted with digital I and Q cancellers. Other measures against clutter and interference include pulse compression, logarithmic channel with a fast time constant, sensitivity time control and automatic gain control, staggered prf, digital interference suppression unit and continuously variable polarisation. The system incorporates built-in test equipment.

Characteristics

ANTENNA-TYPE: parabolic reflector with dual feed horn
  beam-width
    horizontal: 1·5° main and high beam
    vertical: 3° main and high beam
  gain
    main beam: 35 dB
    high beam: 33·5 dB
  high beam tilt: 3° above main beam
  side-lobe level: –25 dB
ROTATIONAL SPEEDS: 6 or 12 rpm
TRANSMITTER
  type: coherent amplifier with TWT output stage
  frequency: 1250–1350 MHz
  peak power: 150 kW
  pulse length: 24 + 1 μs, 68 + 1 μs
  prf (staggered): 600 Hz, 300 Hz
RECEIVER
  noise figure: 3·5 dB
  receiver channel: pulse compression channel for long pulse, non-pulse compression channel for short pulse
INTERMEDIATE FREQUENCY: 30 and 37·5 MHz
COMPRESSED PULSE LENGTH: 0·6 μs
VIDEO CHANNELS: linear, mti, logarithmic (ftc)

Signaal LAR II primary antenna

Status
Production. LAR II is in service in Singapore as an integral part of the LORADS system installed by Signaal in 1981 for the Singapore atc centre and Changi International Airport, Singapore. This is an integrated civil and military atc system in which both the civil aviation authorities and military atc organisation share the same system using a central data bank. The UK CAA has selected the radar with AEG Telefunken SRE-M5 primary antenna and Cossor monopulse ssr for its later radar systems. Six are on order, the first three are scheduled to be operational in 1983 and two in 1984. The sixth will go to a technical support facility to support the other five modular radars

Manufacturer
Hollandse Signaalapparaten BV, Hengelo, Netherlands

Signaal SARP I and II Atc Automatic Radar Processing Systems

Signaal SARP I atc automatic radar processing system

Signaal SARP II at Amsterdam-Schipol Airport

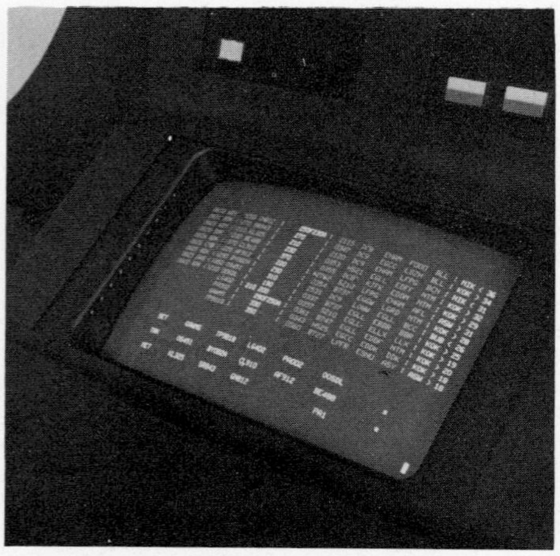

Signaal electronic data display with touch input device

SARP I is an automatic radar data processing and display system for airfield and approach control, designed and manufactured by Signaal. The system provides separated approach and departure routes, early landing sequence planning and a multiple stack concept for optimal routeing and expeditious traffic flow.

The system is controlled through a dual Signaal micromin computer and interface layout, with a software package comprising operational programs, proving programs and support programs. It can process data from up to four radar heads. Seven or 15-inch electronic data displays equipped with a refresh memory and two tone labelled ppi displays, 16- and 21-inch, are used in this system. A light pen, joystick, rolling ball and keyboard provide input facilities.

The system enables tracking of aircraft equipped with 4-digit ssr, 2-digit ssr, or giving a primary response only and automatic initiation of tracking of aircraft equipped with 4-digit ssr. Primary and secondary radar data is processed and the system permits automatic or manual ssr code assignment, flight-path calculation, plan processing and track-plan correlation.

Flight data are displayed in tabular form using electronic data displays, sequenced and distributed by computer. A computer-assisted landing sequencing is provided from the holding beacon to the runway threshold. Other features include a labelled ppi display in two tones, orange for radar data and magenta for labels and additional tables, analogue and digital video mapping and a manual or automatic system reconfiguration facility.

SARP II is an automatic radar data processing system for atc providing a comprehensive and sophisticated area, approach and airfield control system. It incorporates modern techniques for radar and flight data processing and synthetic daylight displays and is delivered as an operational system with all the software required. It can also be supplied as an extension to SARP I.

SARP II matches SARP I in all operational and technical aspects, such as multiple stacks, flow control and flexibility in soft- and hard-ware but is equipped with plan video displays in lieu of the ppis. These have had a 23-inch screen, 50 foot (15·2 metres) lambert brightness and a 55 Hz refresh rate. A separate and independent programming and test system is incorporated which contains a Signaal SMR central processor, an SMR-S display computer, a display unit and a full set of input devices and programming facilities. This system permits program preparation and recording, maintenance, training and monitoring.

In addition to the operational facilities provided by SARP I, the system is equipped with automatic flight updating, a conflict search based on radar criteria, stored flight-plan facility, height filtering and flow control. Other features comprise: departure time calculation, direct data exchange with adjacent centres via computer links, aftn message processing, strip printing for administrative purposes and extensive recording facilities.

Manufacturer

Hollandse Signaalapparaten BV, Hengelo, Netherlands

SWEDEN

SRA Airwatch Series Air Traffic Control Display Systems

SRA Communications produces a family of three modular systems which vary in complexity from the relatively simple Airwatch 1000 to the sophisticated Airwatch 3000 designed to meet the needs of medium- to large-sized atc centres.

Airwatch 1000 has been designed for small atc units which require a limited number of displays. Two typical applications requiring one or more ppis are: small- or medium-sized approach control centres (acc); and control towers operated to control air traffic in the terminal area in the absence of a separate acc unit or used to help control the total flow of air traffic in the immediate vicinity of the airfield, including spacing between departures and arrivals. The system incorporates a number of autonomous operator positions, each including a display unit (ppi), a selector panel, a display processor, a keyboard and a rolling ball unit. If required an optional Alfaskop data terminal can be connected for centralised input of flight data for test purposes. Normally, the system has one primary, secondary or combined radar station, but one or two additional stations can be connected enabling each operator to select the station which best meets his needs at any given time. The system accepts digitised, narrow-band radar data processed by a remote or centralised plot extractor as well as raw video signals. If both types of signals are being used, a sweep compressing unit is added to each display to allow simultaneous presentation of video and synthetic radar data (interscan time is used for the synthetic mode). By connection of a direction finder station, df bearing lines together with numerical data can be displayed for radio channels selected by each individual operator.

The ppi is offered with a 12-, 16- or 23-inch screen. Different types of phosphor are available for operation in different lighting conditions.

The display processor module handles the reception, processing and display of externally generated synthetic data, operator inputs and communication to peripheral devices (inter-console communication is available as an option).

The display processor consists of a micro-processor, a memory system and an input/output system. The controller interfaces with the system via the keyboard and, whenever necessary, uses a joyball, a device which combines the actions of the conventional rolling ball and a joystick. Verification of input is provided on the ppi.

The system displays synthetic plot trails consisting of a position symbol representing the type of target (primary, secondary or combined) and three history plots with unique symbols for spi, garbling and distress codes. Labels include mode C-level, attitude and flight identity (displayed automatically and based on entered code-call sign pairs) or ssr code. The unit is equipped with digital maps and runway extension lines, active and passive decoding and presents a ppi list with entered code-call sign pairs and transient display of current inputs. Label contents are transferred to the ppi list while the aircraft flies in a holding pattern or other high-density area. Other features are a display of df bearing line on the ppi with automatic display of numerical values of qdr and qdm (magnetic tracks to and from an aircraft) and a vector function, controlled by the rolling ball, for displaying a line between two arbitrary points on the ppi and measuring the bearing and distance.

Optional features include inter-ppi co-ordination functions such as pointer symbol communication, semi-automatic handover and display of a signature code identifying the controller responsible for each controlled aircraft. An optional data terminal permits central input of code-call sign pairs to all ppis within the centre.

Airwatch 2000 radar and display processing system has been developed for medium-sized area and approach control centres. It features dual computer processing systems that operate in parallel. Each system carries the full processing load, thus ensuring a smooth changeover if the operational computer should malfunction. Other sub-systems are also duplicated and can be combined to minimise disturbances. The system provides conventional raw video ppi presentation of signals from one psr and extracted synthetic position information from a similarly located ssr. The unit can be connected to existing psr and ssr stations. The information is transmitted over broad-band links to the centre, where the extractor digitises the ssr information for further processing by the Airwatch 2000 computers. Psr information is received by the sweep generation equipment and presented on ppis as conventional raw video. The sweep generation and distribution equipment is duplicated in addition to dual Datasaab SX 1100 ssr extractors and dual display generation systems.

The ppi has a 23 inch screen. Various types of phosphor are available for operations in different lighting conditions. The controller interacts with the system via the keyboard and, whenever necessary, a joyball. Verification of input is provided on the ppi. Each of the two Datasaab Censor 932 computer systems includes a cpu, a core store expandable to 32 K 32-bit words and a console typewriter for operator-computer communication. A paper tape reader is provided for program loading.

The system displays synthetic ssr plots showing type of reply: discrete or non-discrete (group) code, or garbled with a special spi symbol and flashing symbols for distress codes. Labels show: ssr code or call sign (for flights transmitting a code for which a code-call sign pair has been entered); mode C-level (automatically converted to altitude below transition level) or manually entered level; and ground speed. The unit is equipped with active and passive decoding and has code bank for allocation of unique ssr codes based on pre-entered code-call sign pairs and for random allocation. It presents a ppi list containing: transient tellback of current inputs; system data such as transition level, selected range and selected group codes; entered code-call sign pairs; and flights in coast status. This list can be positioned anywhere on the screen. Controller designations indicate the person responsible for each flight under control and identify each pointer symbol. The system incorporates a semi-automatic handover between controllers. Other features comprise a vector function for measuring bearing and distance between two points, digital maps which can be used in any combination, runway extensions lines and continuously variable range setting and off-centring.

Airwatch 1000 atc system

Airwatch 2000 atc system

Moscow area control centre atc system installed by Datasaab

Airwatch 3000 is a newer model designed primarily to meet the needs of medium to large-sized atc centres, its modular architecture permits the system to be tailored to customer requirements. The number of radar stations and operator consoles can be varied, the dual computer feature omitted if considered unnecessary for a particular application and extractors can be provided if desired. The full Airwatch 3000 configuration accepts primary and secondary radar information from two stations. It processes this information for display and tracking and presents it in both synthetic and raw form. The radar picture includes labels containing call sign and other essential flight data, digital maps and supplementary alphanumeric information.

Twenty-three inch ppi screens allow concentration of all relevant information plots: flight data in labels that include call sign, mode C-readout, attitude, cleared level, and speed; lists of expected traffic and flights in holding; and direction finder lines. The controller interacts with the system via the keyboard and, whenever necessary, a joyball. Verification of input is provided on the tell-back line which appears on the ppi. Like tabular data, the tell-back line can be manoeuvred to avoid interference with plots, labels and map information in areas of immediate interest to the controller.

Plots, graphic data and alphanumeric data can be displayed simultaneously on the ppis. In the basic configuration, radar information is synthetic, but supplementary raw radar video is available for mixed presentation. A number of digital maps can be stored. These maps can be selected by each controller in any combination to meet individual needs. A mosaic technique is used for copresentation of information from two or more radar stations. Special routines ensure that continuity is maintained when mosaic boundaries are crossed. Position symbols indicate the type of target: psr, ssr or combined. A plot trail with the previous positions is presented, and a signature code identifies the responsible controller. A leader line connects the label to the track. Labels which maintain the identity of flights carrying a 4096 code ssr transponder are available in many systems. However, tracking is needed to associate a label with primary targets (psr) and with flights transmitting non-discrete (64) codes. Airwatch 3000 tracks all controlled flights, and labels can be associated with all tracks. Irrespective of target-type, each label contains call sign, ground speed and (if entered) cleared level. Mode C-level is displayed when available, and an arrow indicates climb or descent. A label is normally displayed only for the controller responsible for the flight and is transferred to an adjoining sector by means of a semi-automatic handover function. Moreover, a quick-look function is available for inspection of other controllers' labels. Facilities exist to reduce information in labels and to select an alternative label position relative to a track (thus avoiding label conflicts).

Prior to label association, flight data for each expected flight is displayed in a table on the ppi. In this table, flights are arranged on the basis of their estimated entry into the sector. The table presents call-sign, ssr mode (if any), entry point and estimated entry time. Flight data for numerous repetitive flight plans can be stored. A bank of ssr codes for automatic allocation to transponder-equipped aircraft is also stored.

Other features available include selection of high-precision approach lines, pointer symbol communication, vector function for measuring bearing and distance between two points, automatic altitude conversion for mode C-information below transition level and active decoding with retained labels if required.

Characteristics

MODEL: Airwatch 3000
RADAR STATIONS CONNECTED: 1–4
DIRECTION FINDERS CONNECTED: 0–2
RADAR CONTROLLER POSITIONS: 3–15
TRACKED AIRCRAFT: up to 200
MAPS: up to 20
APPROACH LINES: up to 24
SSR CODES IN CODE BANK: up to 315
REPETITIVE FLIGHT PLANS STORED IN LIBRARY
up to 1800

Status

The Airwatch systems were developed by the Interactive Data Systems division of Datasaab. On 1 January 1982 this division was transferred to SRA Communications. Datasaab has installed atc systems in Belgium, France (Eurocontrol Experimental Centre), Netherlands, Norway, Sweden, Switzerland, USSR and Yugoslavia

Manufacturer

SRA Communications, Spanga-Stockholm, Sweden

UNITED KINGDOM

Cossor Adsel/DABS Air Traffic Control System

Adsel (addressing selectively) is a new secondary radar under development by Cossor for the Civil Aviation Authority. It has been designed to be compatible with a similar system being developed in the USA known as DABS.

The system is designed to overcome many of the signal problems experienced with current secondary radars in high density traffic. It does this by a system of selective addressing, only interrogating one aircraft transponder at a time rather than triggering off replies from all transponders within range, as at present.

There is a specially constructed trials site near Harlow, England. The site is fully equipped with antennas, processing equipment and radar and data displays.

The system should be adopted universally in the mid-1980s by international agreement.

Characteristics

TRANSMITTER
frequency: 1090 MHz
Power output: 27 dB W
Modes: SSR: A, B and C
Adsel/DABS:
ssr/DABS all-call
DABS only all-call
Surveillance Comm-A, Comm-C
RECEIVER
sensitivity ssr: 90% at −76 dBM interrogation level
Adsel/DABS: 90% at −72 dBM interrogation level
band-width: ± 3 MHz at 3 dB points
dimensions:
height: 194 mm
width 90 mm
depth: 314 mm
weight: 5·2 kg

Cossor Adsel atc system

power requirements: 115 V ac, 400 Hz, 80 VA
operating temperature: −26°C to +55°C continuous
range: −26°C to +71°C for 30 minutes

Manufacturer

Cossor Electronics Ltd, Harlow, Essex, England

Cossor CEL 850 Secondary Radar System

The CEL 850 is a multi-purpose secondary radar interrogator developed for military and civil aviation use. It is the standard unit used in a wide variety of iff/secondary surveillance radar applications where identification facilities are required in association with long- or medium-range primary radars. These may be fixed or mobile installations.

Four antenna types are available for operation with this equipment. They can be mounted on the primary radar antenna or alternatively equipped with their own turning gear. In the latter configuration, the antenna can be slaved in rotation with an adjacent primary radar or operated in an autonomous, free-running condition. The CRS 390 transmitter/receiver is compact, lightweight and rugged in construction for mobile applications.

Operating features of the equipment include: wide range of acceptable trigger pulses; external computer control facility; external modulation facility; flexible mode programme and video outputs to match processing requirements.

Characteristics

ANTENNA	CRS 320	CRS 379	CRS 381	CRS 389
length	5·65 m	1·16 m	3·35 m	4·35 m
weight	305 kg	13·8 kg	57 kg	75 kg
GAIN (minimum)	23 dB	16 dB	16·5 dB	17·5 dB
half-power beam-width	2·2°	12·5°	6°	4·5°

TRANSMITTER/RECEIVER: CRS 390
TRANSMITTER:
FREQUENCY: 1030 MHz
POWER OUTPUT: 33 dBW
MODES: 1, 2, 3/A, B, C and D external modulation
PRF: 150 to 5000 Hz
MODE INTERLACE: single, dual or triple
RECEIVER SECTION:
CENTRE FREQUENCY: 1090 MHz
TANGENTIAL SENSITIVITY: not less than −85 dBm
BAND-WIDTH: 10 MHz at −3 dB points
VIDEO OUTPUTS: 2–5 V adjustable into 75 Ω
IDENTIFICATION PULSES: pre-trigger; mode pulse pairs
DISPLAY TRIGGER PULSE: +5 V into 75 Ω between P3 and P3 +60 μs

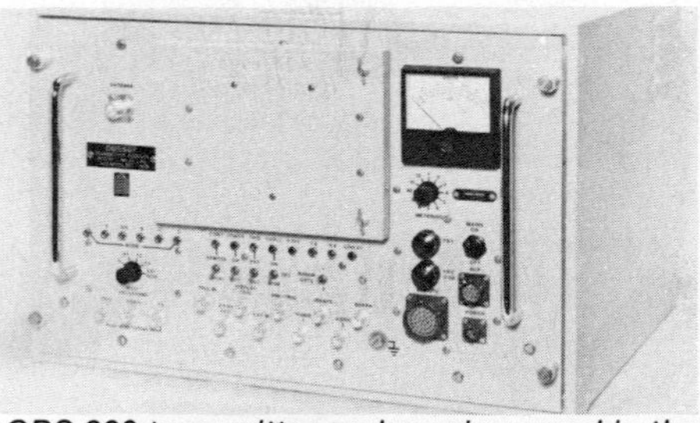

CRS 390 transmitter and receiver used in the CEL 850 radar

Status

The CEL 850 was introduced in 1977. Civil airports equipped with this system include Bergen, Forneby in Norway, Paraguay, six in Nigeria and Barbados. More than 70 systems have been ordered for military applications

Manufacturer

Cossor Electronics Ltd, Harlow, Essex, England

Cossor Compass 9000 Air Traffic Control and Processing System

The Compass 9000 is a flexible atc system which has been designed to be expanded as required to meet more complex requirements without any redundancy of existing equipment.

In its basic form the CRD 2016 raw primary radar display provides the controller with the facility to display broad band radar video, analogue video map signals, stroke lines, coupled with radar head selections, off-set, range, and range mark selection. Ssr data, from manual processing systems, can be mixed with the primary radar presentation. By fitting an additional module to the basic version, vector, character and dot drawing capability can be introduced to the display. This enhanced display is known as the CRDS 2016.

Compass 9000 series radar displays at London Gatwick Airport

With the introduction of radar data processors and refresh control units, to process the signals from the ssr and primary plot extractors for the CRDS 2016 displays, the Compass 9000 system is established. Two radar data sources are simultaneously processed and each data channel can be separately selected and identified to the controller. Targets are correlated on range, bearing and ssr data. Code-call sign and altitude correlation can be provided by the addition of an alphanumeric display. QNH data input from the radar displays is used to correct the ssr altitude data. Site parameter data can be programmed into the system using a teletypewriter.
Display presentation is on 16-inch diameter crts and selected by the operator using single key operations. A function select keyboard and a rolling ball control are provided for display data filtering and numeric data input. The primary radar display is equipped with a separate display control panel.
The Compass 9000 system is capable of handling up to 24 radar displays.

Status
The system is in service with the Civil Aviation Authority at various airports, with the Ministry of Defence in the United Kingdom, and with the Hong Kong CAA at Kai Tak airport

Manufacturer
Cossor Electronics Ltd, Harlow, Essex, England

Cossor CR 62 Precision Approach Radar

The CR 62 par is designed to replace SLA-3 generation equipment in service at Royal Air Force and Ministry of Defence airfields in the United Kingdom, providing runway approach and landing guidance to aircraft equipped with two-way communication in either vhf or uhf.
The CR 62 consists of radar head equipment installed in a transportable cabin together with atc display consoles and associated remote control equipment in the operations room. The radar head equipment is a dual channel system employing solid-state circuit elements except for the main modulator thyratron and coaxial magnetron. Modular construction is used to a maximum extent, the widespread use of small easily replaceable modules being adopted to ensure relatively inexpensive replacement costs. The equipment utilises digital techniques to minimise drift of parameters and warm-up time.
The clear air range of the system on a 1 square metre target has been calculated at 33 km, so that the operational performance of 28 km is comfortably exceeded. The radar will also give an improved performance in adverse weather conditions; it uses selectable digital mti techniques, together with a video correlator to reject uncorrelated clutter returns. In the radar transmitter a five stagger prf is used for eliminating blind speeds between 20 and 600 knots.
In this replacement programme, the existing antennae are retained after being refurbished to interface with the new equipment and to incorporate a simplified drive system together with direct digital angle data take-off for improved stability. Each system comprises an azimuth and an elevation antenna. Both antennae are installed in robust modules that mount directly on to the cabin exterior. The elevation antenna, which measures the position of the aircraft relative to the glide-slope, scans vertically. The azimuth antenna, which measures the position of the aircraft relative to the centre line of the runway, scans horizontally. Resolution is such that two targets at the same range can be distinguished at 0·6 degrees separation in elevation, 0·8 degrees separation in azimuth and in range two targets can be distinguished at a separation of 60 metres.
Two display consoles are supplied for fitting in the operations room. Each with a 16-inch azimuth/elevation display, an electronic map generator with associated controls and a desk unit. Each display shows both azimuth and elevation information from maximum range to touchdown. The displays also contain front panel controls which can individually tilt and pan the antennas to optimise radar returns. The nose of each beam is indicated on the displays. Remote control facilities are provided to transfer data between the radar-head and the operations room over a distance in excess of 3000 metres. The control facilities can select one or other of the two radar channels virtually instantaneously and control the performance of the selected channels while providing continuous status information of both channels. Additionally the system will provide remote control of the antenna and cabin turntable positioning with return status indication.
While employing cross-site coaxial cabling, provision is made for housing extra units to enable the remote control system to be operated over a microwave link for any special site or operational reasons.

Characteristics

ANTENNA	Elevation	Azimuth
BEAM-WIDTH	0·5° vertical	0·5° horizontal
SCANNING ANGLE	−1° to +6°	−10° to +10°
SCANNING CYCLE	1 s (approx)	1 s (approx)

TRANSMITTER
FREQUENCY: 9080 MHz
PEAK POWER OUTPUT
80 kW nominal
65 kW minimum
PRF: 3·3 kHz
STAGGER RATIO: 10 : 12 : 16 : 14 : 10·4
RECEIVER BAND-WIDTH: 8 MHz
VIDEO BAND-WIDTH: 4 MHz (mti and normal)
NOISE FIGURE: 5·2 dB
IF FREQUENCY: 30 MHz
MTI SUB-CLUTTER: −20 dB or better
VISIBILITY
displayed range: 15 nm (maximum)

Status
Cossor is to supply a total of 42 CR 62 par for Royal Air Force and Ministry of Defence airfields in the United Kingdom and overseas. Delivery started in 1981 and is scheduled to be completed by 1985

Manufacturer
Cossor Electronics Ltd, Harlow, Essex, England

Cossor Monopulse Secondary Surveillance Radar

A system of monopulse bearing measurement, this equipment was developed as part of the Adsel project. It overcomes many of the problems of signal interferences encountered in conventional ssr systems, but unlike Adsel, it does not require a special transponder to be fitted to the aircraft.
Monopulse measures the bearing of an aircraft from each received frame and code pulse of its transponder reply. An average of these measurements gives consistent, accurate azimuth data.
This is achieved by determining the bearing of an aircraft with respect to the antenna boresight by comparing the relative amplitudes of transponder replies received through the separate sum and difference patterns of the antenna. Because the bearing of the aircraft can be accurately measured, it is possible to discriminate between the transponder replies of aircraft flying in close proximity.
The system comprises a CRS 370/3 antenna, ssr 950 transmitter/receiver and a CVP 250 plot extractor.
The 31 foot (9·45 metre) monopulse antenna which is a derivative of the ssr antennas, provides sum and difference patterns with an integral control pattern, matched to the same phase centre. A supplementary radiator for four pulse side-lobe suppression is available.
The CRS 370/3 is designed to be installed on a suitable primary radar antenna or to be mounted on its own turning gear.
The SSR 950 interrogator consists of matched logarithmic receivers for monopulse operation and a third receiver for receiver side-lobe suppression.

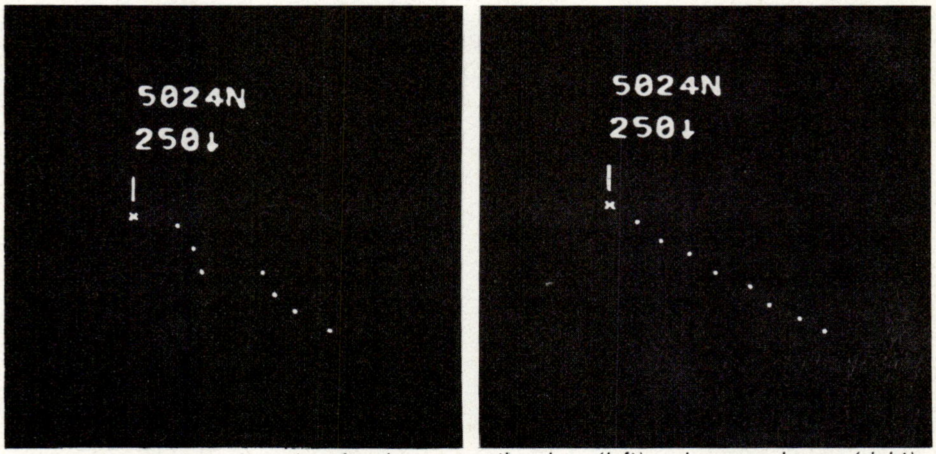

Simultaneous plots of an aircraft using conventional ssr (left) and monopulse ssr (right)

Interrogator side-lobe suppression can be added. The SSR 950 features programmable gain time control in range and azimuth and includes interfaces for the digital plot extractor and for a computerised management system.
The CVP 250 plot extractor uses monopulse data from the SSR 950 to compile target reports and generate tracks, thereby providing a high degree of track accuracy.
A significant feature of the monopulse system is that it can be upgraded to a full selective address ssr (Adsel/DABS) operation.

Status
Currently under development. Two systems have been ordered in the United Kingdom by the CAA

Manufacturer
Cossor Electronics, Harlow, Essex, England

Ferranti Cards Radar System

The Ferranti computer-assisted radar display system (cards) is designed to process primary and secondary radar data and to display it in conjunction with raw primary radar video. It accepts digitised data in any standard format, from any number of sources. Co-ordinate conversion facilities allow remote ssr and processed primary data to be superimposed on raw primary returns from any one of up to four radar heads. The system operates with two computers, both processing all the data all the time. Computer peripherals and power supplies are also duplicated to provide two independent channels. Controllers can choose either channel and existing data is replaced within one eighth of a second when a channel is changed. The system can provide labelled radar display facilities for controllers employed as directors, in approach and departure. A selection of overlay digital maps are available for the displays which are equipped with a mini-tab, electronic bearing and range markers (ebrm) and an inter-console marker (icm). A keyboard enables the controller to perform code selections, and area and height band filtering. The mini-tab, which can be positioned anywhere on the display by means of a rolling ball, provides the controller with details of keyboard entries. Again using the rolling ball, the controller may draw two ebrm reference lines. The line being drawn has its range and bearing displayed in the mini-tab. The icm enables a controller to transmit a symbol either to one selected console or to all consoles in the broadcast mode. Other features include flashing emergency plots, automatic QFE correction on mode C-plots responding below the transition altitude, and an active decode facility which enables the controller to display a specified plot's height and code regardless of code selection or height filtering.
A program development system is offered as an option which comprises magnetic disc backing store, line printer, punch and teletype with their respective peripheral control units. Additional display units are available up to a maximum of 16 autonomous units. Each autonomous unit can drive one or more slave units. Other optional equipment which can be supplied includes choice of phosphors, 23-inch display units, auto tracking facility, code-call sign conversion, minimum safe altitude alert and flight-plan processing.

Characteristics

LOCAL PRIMARY RADARS
MAX NUMBER: 4
OPERATION RANGE: 20–160 nm (37–296·5 km)
PRF: 300–1200 Hz
ROTATION RATE: up to 20 rpm
AERIAL POSITION
synchro: x1, x9, x18, x36 50/60/400 Hz
digital: 4096 increments per revolution
MAX DISTANCE FROM CPU: 2 nm (2·7 km)
RADAR TYPE: any 2d radar
REMOTE SECONDARY RADAR
MAX NUMBER: 4
MAX DISTANCE FROM LOCAL SITE: 80 nm (148 km)
CARDS POWER REQUIREMENTS: 240 V, 50 Hz/115 V, 60 Hz

Ferranti Cards radar display system

Status

Production. Cards is operational at the London Air Traffic Control Centre

Manufacturer

Ferranti Ltd, Bracknell, Berkshire, England

Marconi Air Traffic Control Data Handling Systems

Marconi Radar atc data handling systems cover a range of requirements from the small airfield control centre to the large data handling complex controlling air traffic over large areas of land and sea. They are based on the Marconi Radar Locus 16 processor and use the company's colour and monochrome display units and Digilux touch masks, together with all necessary ancillaries such as specialised consoles, vdus, printers etc.
The data handling systems are the result of extensive experience of air traffic control techniques and place particular emphasis on the reduction of operator stress and workload. The systems allow new installations to use current data handling techniques and permit existing control centres to be modernised and extended.
A typical display and data handling system is the Scottish Air Traffic Control Centre (ScATCC) at Prestwick, which provides a control and advisory service throughout the Scottish Flight Information Region in controlled airspace, middle airspace and upper airspace and handles airways, general aviation and military traffic in an area covering more than 350 000 square km. The ScATCC radar data processing and display system presents the outputs of five plot-extracted radar sites, with an expansion capability of up to ten radar sites. The system incorporates 33 Locus 16 processors in a distributed data processing system.

Marconi Radar data display positions

Manufacturer

Marconi Radar Systems Ltd, Chelmsford, Essex, England

Marconi S 511 Airfield Surveillance Radar

The S 511 is a dual beam 10 cm band airfield surveillance radar with a design capability for operating in severe clutter conditions. The antenna reflector is double curvature, constructed from a carbon fibre composite. A dual horn system is used to produce the main beam (transmit and receive mode) and an auxiliary beam (receive only) with each horn having an integral polariser assembly. Polarisation of the main (lower beam) is continuously variable between circular and linear vertical and fixed circular polarisation is applied to the auxiliary upper beam. The antenna rotates at 15 rpm on its own turning gear. Provision is made for the fitting of a ssr antenna on top of the reflector. A standard self-supporting tower incorporates a main service platform, of which the enclosed upper 2·5 metres houses the turning gear and data take-off.
The transmitter-receiver employs a solid-state line type modulator consisting of eight individually fused modules in parallel, designed so that operation can be maintained despite the failure of a single module. The receiver normally selects and processes the signal from the main beam. When the clutter exceeds the threshold, the auxiliary beam is selected, thus maintaining the level of clutter signal within the dynamic range of the mti. Switching is controlled from the signal processor. The signal processor uses a combination of coherent and non-coherent integration techniques for optimum target detection. The coherent channel employs a digital four pulse filter incorporating time varied weighting for stationary clutter rejection. Both channels use temporal threshold integration techniques to reduce clutter and provide protection against tangential fading. Black/white detections from both channels are combined logically and subjected to sliding window integration before being fed to output buffers producing signals suitable for ppi displays.
The S 511 radar can be controlled and monitored from a display centre located up to two nautical miles from the antenna. Video, trigger and turning data are transmitted to the remote centre where equipment is provided for operational control. Compatible display systems are modularised and range from simple analogue video to ssr labelled raw radar. A monitor is available for installation in the radar head.
Standard packages can be supplied for single and diversity systems providing surveillance cover from 0·5 nautical miles (0·9 km) to a maximum instrumented range of 70 nautical miles (130 km).

Characteristics
FREQUENCY: 2·7–2·9 GHz
TRANSMITTER
peak power: 650 kW
pulse width: 0·85 μs
prf: 1000 pps
RECEIVER
noise factor: 2·7 dB
ANTENNA
rotation speed: 15 rpm
gain: 33·5 dB
HORIZONTAL BEAM-WIDTH: 1·5°
RANGE
max at 80% probability
single transmitter: 61 nm (113 km)
dual transmitter: 75 nm (139 km)

Status
Production

Manufacturer
Marconi Radar Systems Ltd, Chelmsford, Essex, England

Marconi S 511 airfield surveillance radar

Plessey ACR 430 Airfield Control Radar

This I-band radar was designed principally as a simple and inexpensive airfield aid for precision azimuth approach guidance and local area surveillance. Suitable for use with all types of aircraft, it may also be used for monitoring other landing aids, for marshalling traffic and for other surveillance duties. Recently, coastal defence versions were added to the series.

Extensive use is made of fibreglass construction for strength and low weight, and solid-state and integrated circuit design are used throughout.

Installation is facilitated by the breakdown of the radar head equipment into readily handled sub-assemblies.

Two separate coverage beams are radiated and the associated equipment channels use identical parallel electronic units which may be rapidly inter-changed. Reliability is thus further improved by parallel redundancy giving high operational availability.

A double curvature antenna is illuminated by two horns, each of which is coupled to a transmitter/receiver. One horn produces a cosecant pattern; the other produces a pencil beam which is geometrically centred to provide the high degree of accuracy necessary for approach guidance. Antenna rotation rates of either 20 or 40 rpm can be selected from the remote control unit. The two transmitter/receivers, together with their power units, are located in a housing at the back of the antenna. Each transmitter provides a peak power of 65 kW. The signal output at intermediate frequency is passed by cable to the remote radar control unit. The two if receivers housed in this unit provide a wide range of receiver modes for interference and clutter rejection.

Characteristics
FREQUENCIES: 9320 ± 30 MHz and 9460 ± 30 MHz
POLARISATION: circular (fixed)
ANTENNA
ROTATION RATE: 20 ± 10% rpm and 40 ± 10% rpm in wind speeds of up to 55 knots with survival in wind speeds of up to 120 knots
TILT: 0 to +18° adjustable from radar control panel (tilt is referred to nose of low beam)
HORIZONTAL BEAM-WIDTH: 0·55° ± 0·1°
VERTICAL BEAM-WIDTH
high beam: 2·5° cosecant^2 to 25°
low beam: 2·5°
TRANSMITTERS
MAGNETRON PEAK POWER: 55 kW nominal
PULSE DURATION
0·2 μs at 2000 pps
0·5 μs at 1000 pps
RECEIVERS
NOISE FACTOR: not worse than 10 dB
FREQUENCY CONTROL: automatic and manual
CHARACTERISTICS
linear; linear differentiation; logarithmic; logarithmic differentiation; wide-band limiting; pre-set sensitivity time control variable in range and level (all selected at radar control unit)
DISPLAYS
TYPE: autonomous rotating coil, desk-mounted plan position indicators

Plessey ACR 430 airfield control radar antenna assembly

QUANTITY: 2 displays are provided as standard equipment
RANGE SCALE: 4, 8, 16, or 32 nm (7·4–29·3 km) per display radius
RANGE ACCURACY: better than 1% of range selected
RANGE RING ACCURACY: better than ± 0·5% of indicated range down to ½ nm (0·93 km)
AZIMUTH ACCURACY: better than ± 0·5°.
The display system automatically aligns in azimuth

Status
Production. The radar is in service with United Kingdom and other countries

Manufacturer
Plessey Radar, Weybridge, Surrey, England

Plessey AR 5 Air Surveillance Radar

The AR 5 is D-band air surveillance radar in atc systems and provides the solid, long-range, high- and low-altitude cover needed for en route, upper air space, and area and airways operations. The AR 5 complies with the ICAO recommendations for radars serving these categories of traffic, and is designed to handle both current and planned aircraft, including supersonic transport aircraft, from take-off to a range of 370 km on medium-sized targets.
The design concept provides for single and diversity transmitter-receiver configurations for maximum reliability and resistance to ecm, and for use in conjunction with specialised eccm receivers.
Particular attention has been paid to signal processing techniques which provide well defined aircraft responses in a form suitable for easy viewing by controllers using ppi displays and for compatibility with computer controlled data handling in atc systems. Ssr/iff is available and can be fitted. The maximum display range of the AR 5 can be either 370 or 550 km depending upon the requirement. The high level coverage provided is sufficient to enable even the smallest aircraft to be tracked at maximum operational altitude. Low coverage at close range is also good and subject to local siting limitations, it is possible to track aircraft to within 1·8 km of the antenna head. Two horns are used in the antenna feed system to derive the overall coverage pattern. The primary horn is used for both transmission and reception and provides the main pattern. The second horn, used for reception only, is fitted immediately below the primary horn and is adjusted to give a separation in elevation of approximately 5 degrees between the axes of maximum gain of the two horns. The use of the second horn ensures adequate suppression of 'angels' while maintaining full coverage on wanted targets. Appropriate rf-switching between the primary and secondary coverage patterns ensures that there is no loss of wanted targets.
Horizontal antenna beam-width is 1·2 degrees which provides an azimuth discrimination of 1 degree, thus ensuring that aircraft passing at a range of 370 km will be shown as separate echoes when separated by only 7·5 km. The radar has a range resolution of less than 290 metres at maximum range.
The standard antenna system has a rotation speed of 8 rpm. This has been selected to meet the requirements of long-range detection, narrow antenna beam-width and good mti performance.
The AR 5 employs a full transistorised digital mti system which produces an overall sub-clutter visibility of 27 dB with the antenna rotating at 8 rpm. Prf stagger is used to reduce the number of blind speeds. Using a single transmitter, the first blind speed is approximately 1050 knots but with two transmitters operating in frequency diversity, there are no blind speeds within the performance capability of any envisaged aircraft.
Facilities are provided for superimposing raw radar video on to mti video to enable the aircraft controller to select a background of permanent echoes which will not obscure aircraft returns. This facility enables the controller to provide a measure of terrain clearance to aircraft flying in the vicinity of high ground.
Weather returns are suppressed by use of variable linear-to-circular polarisation.

Characteristics

RADIATION: two preset spot frequencies
FREQUENCY: in band 1250–1365 MHz
DIVERSITY GUARD BAND: 35 MHz nominal
POLARISATION: variable linear and variable from linear through elliptical to circular
ANTENNA
GAIN: 34 dB relative to isotropic radiator
HORIZONTAL BEAM-WIDTH: 1·2° at half-power points
VERTICAL BEAM-WIDTH: 4·5° to half-power points with shaping to 40°
ANTENNA ROTATION RATE: 8 rpm other speeds optional
'ANGEL' SUPPRESSION: a second (receive) rf feed is provided

Plessey AR 5 air surveillance radar

TRANSMITTER
OUTPUT VALVE: tunable magnetron
POWER OUTPUT: 2 MW peak
PULSE LENGTH: 2–5 μs
PRF: 270–680 pps
RECEIVER
TYPE: parametric amplifier
OVERALL NOISE FACTOR: not worse than 3·5 dB
INTERMEDIATE FREQUENCY: 30 MHz
CHARACTERISTICS: switchable to — linear, log, Dicke-fix, eccm logical (log with pulse length discrimination gated by Dicke-fix). differentiation can be applied to any of these
DIGITAL MOVING TARGET INDICATOR:
type: double loop cancellation
cancellation: 47 dB

Status

AR 5 has been selected by the British and other governments and meets a wide range of operational requirements

Manufacturer

Plessey Radar, Weybridge, Surrey, England

Plessey AR 15/2C Medium Range Air Surveillance Radar

The Plessey Type AR 15/2C is a medium-range primary air surveillance radar for use in both civil and military atc including terminal area and medium range surveillance, approach control, radar sequencing control, parallel runway approach control, outbound control from take-off and the surveillance element of ground controlled approaches (gca).
The unit uses a two beam antenna system and a sophisticated digital signal processing system. The second or auxiliary beam is used at short range designed to reduce returns from ground clutter, sea clutter and angels when compared with signals from the main beam. The signals from both beams are presented on a conventional ppi display.
The digital signal processing system includes digital mti with simultaneous I and Q processing, two cascaded 8-bit cancellers and a programmable seven-period stagger. It derives its timing from the transmitter, which itself is timed at the system prf by an independent system. A digital monitoring highway is also included for fault location and for routine maintenance and setting up. The signal processing system incorporates a digital background averaging circuit which gives a further improvement in clutter suppression. Other features include a digital video integrator to improve aircraft responses, swept gain and logarithmic signal processing. The antenna system includes a variable linear to circular polarisation facility which is adjustable from the radar control panel.
The radar can be used with all modern display systems and its signals may be presented either at the radar or remotely via cable or radio link. A typical maximum display range is 80 nautical miles (148 km) and it is usual to provide facilities for each controller to select a range scale of 10, 20, 40 or 80 nautical miles (18·5, 37, 74 or 148 km) radius on a 16- or 22-inch display. The main units of the AR 15/2C system comprise the antenna, two solid-state transmitter/receivers (working in frequency diversity), control rack, digital mti equipment and radar control unit. A standard variant of the radar, available if required, uses only one transmitter/receiver. The advantages derived from the use of the two transmitter/receivers version include an increase in coverage of 25 per cent at 80 per cent detection probability over that provided by a single transmitter/receiver and also increased reliability derived from instantaneous standby in the case of a transmitter failure, thus ensuring no break in operation.
Whichever version is chosen, the radio frequency power is transmitted from the main horn and the reflected signals are received both by the main horn and by the auxiliary horn. The auxiliary horn and the antenna reflector together form a small beam tilted up compared with the main beam. Clutter usually occurs at low angles of elevation and hence should not be detected by the auxiliary beam. The changeover from main beam to auxiliary beam is carried out at a range chosen to suit the terrain and the operational requirements. The selected signals are fed to the dmti and then to the ppi. Typically a changeover range of up to 15 nautical miles (28 km) is suitable for most applications.

Plessey AR 15/2C medium range air surveillance radar

Status

The AR 15/2C is a development of AR 15/2B which is operational in the United Kingdom and other countries. The new radar entered production in 1981

Characteristics

FREQUENCY
DUAL OPERATION: 2700–2900 MHz and 3000–3040 MHz (spot frequency in both bands)
SINGLE TX: 2700–2900 MHz or 2900–3100 MHz (spot frequency in either band)
POLARISATION
MAIN BEAM: variable from linear through elliptical to circular
AUXILIARY BEAM: fixed circular
TRANSMITTER
PEAK POWER: 600 kW at 1 μs/700 kW at 0·6 μs
PRF: 700/1000 pps with 7-period stagger
RECEIVER (PARAMP)
NOISE FACTOR: not worse than 3 dB
INTERMEDIATE FREQUENCY: 30 MHz
OVERALL BEAM-WIDTH: 1·2 MHz
MOVING TARGET INDICATOR
CANCELLER IMPROVEMENT FACTOR: 47 dB
SYSTEM IMPROVEMENT FACTOR: 33 dB
ANTENNA
GAIN: 33 dB relative to isotropic radiator (nominal)
HORIZONTAL APERTURE: 4·6 m
HORIZONTAL BEAM-WIDTH: 1·5° to half power points
VERTICAL APERTURE: 2·2 m
HORIZONTAL SIDE-LOBES: 27 dB ± 5°
ROTATION RATE: 15 rpm clockwise (at 50 Hz supply)

Manufacturer

Plessey Radar, Weybridge, Surrey, England

Plessey Mk 8 and Series 9 Radar Displays

Plessey Radar manufactures two types of radar displays, the Mk 8 and the Series 9, which are available in either 16- or 22-inch diameter.

The Mk 8 is a general purpose self-contained radar display which can be used with all atc and surveillance radar signals either with or without ssr labels. A version of the Mk 8 is also available which incorporates a character generator to permit high quality alphanumerics to be written under the command of a computer.

The Series 9 is the latest in the range of Plessey displays. It is a bright display capable of presenting a large amount of alphanumeric and graphical data with high definition and registration accuracy. The Series 9 is available with either monochrome or four colour presentation.

Status

Operational at the London ATC Centre

Manufacturer

Plessey Radar, Weybridge, Surrey, England

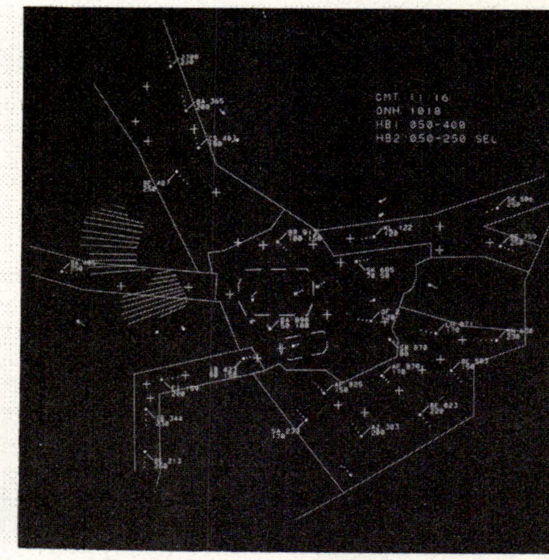

Plessey Series 9 radar data display

Plessey radar display equipment

Plessey Series 200 Secondary Surveillance Radar System

The Plessey Series 200 ssr system provides automatic plot extracted, labelled ssr data for civil and military approach and/or area atc operation. The system is modular in concept and may be configured to meet individual requirements. It can be associated with local primary radar using on-mounted or off-mounted ssr antenna.

The basic equipment comprises an antenna, an interrogator receiver and a plot extractor at the radar head and a processor with up to six displays at the control centre. Each of these modules can be duplicated with automatic changeover facilities to increase the system availability. The system presents data in label form superimposed on the primary radar data on each display showing aircraft call sign or ssr identity code, aircraft destination and a flashing indication of aircraft transmitting emergency codes. An inbound/outbound list of aircraft due to enter the control area is displayed and the system can indicate predicted positions of the aircraft when ssr replies are lost. Each display position is provided with a keyboard and rolling ball facility. Additional features include an electronic range and bearing line and auto-lock facilities with the relative range and bearing displayed at the end of each line, and a silent handover facility between controllers. The series 200 employs the Mk 8 autonomous labelled radar displays. A typical series 200 system with four displays operating with the AR 15/2 radar turning at 15 rpm (the ICAO recommended rotation rate for surveillance in the approach phase) can process up to 200 aircraft plots presenting labels for a maximum of 50 aircraft on each display. Ssr data from a remote radar installation may also be fed into the system.

Status

Plessey developed the first automatic ssr processing system to enter operational service when the London atcc controlled airspace service started in January 1971. The Series 200 ssr entered production in 1975 and is in service in the United Kingdom and other countries

Manufacturer

Plessey Radar, Weybridge, Surrey, England

Racal-Decca Airport Surface Movement Indicator Radar

An airport surface movement indicator (asmi) radar operates in the X-band (with a nominal 3 cm wavelength) giving substantial performance improvements in rain, sand, dust, etc. The systems are produced using modified X-band modules to reduce manufacturing costs. Installation and site selection is facilitated using a light-weight scanner and pedestal giving good mobility to the site-engineer.
The displays use a digital scan conversion process with a high ambient daylight viewing capability incorporating synthetic mapping to enhance realism and sharpen operator response and providing a swept gain law appropriate to each airport and a blanking facility to remove unwanted bright areas.

Status
The radar was announced in December 1981 and is operating on a trial basis at Edinburgh Airport, Scotland

Manufacturer
Racal-Decca Navigator Ltd, Ground Systems Division, New Malden, Surrey, England

Racal-Decca asmi display

UNITED STATES OF AMERICA

Bendix BPS-1000 Surveillance Radar

The BPS-1000 is a long-range D-band surveillance radar for air defence and atc applications, introduced by Bendix in 1976 and available in a variety of configurations to satisfy differing requirements. The dual diversity transmitter is a coherent type using a long-life L3035 power klystron, and can be operated in either simplex or diplex modes. In the former mode, the second channel provides full redundancy except for the common antenna system. Predicted mtbf in this mode is better than 19 000 hours. Solid-state receivers and processors contribute to the high overall reliability.
The coherent transmitter design eliminates second-time-around returns and permits the linear digital receiving system to achieve an mti improvement factor of 45 dB. The receiver processor provides five basic receivers: normal, Dicke-fix, log, mti, and LINAR. With the exception of the normal mode, all provide a cfar output. Anti-jamming and weather countermeasures include: automatic clutter gating, delay-line blanking, leading edge blanking, wide-dynamic range cfar receivers and mti anti-weather system. The LINAR and log receivers are also effective against weather interference and further rejection of rain clutter is provided by circular polarisation.
The log and Dicke-fix receivers provide effective rejection of wideband noise jamming interference, and the Bendix LINAR receiver is effective against wide- and narrow-band noise jamming, as well as CW and swept frequency. The receiver processor provides outputs for manual and automatic systems, simultaneously. Computer-driven displays may be used, in which case the Bendix radar data processor is added to facilitate the target and iff/ssr data transmissions to the display subsystem. The computer-driven graphic displays employed operate in a distributed system without a central data processor. Consoles use identical hardware, and operation is determined by the program loaded and operator keyboard commands.

Bendix BPS-1000 surveillance radar

LEC Air Traffic Control Systems

Lockheed Electronic Company (LEC) designs, manufactures and installs automated atc systems. The company claim to have introduced the first mini-computer-based systems to bring reliable, low cost, data processing technology to atc requirements.

Status
Lockheed atc systems are in service at civil and military installations in Italy, Saudi Arabia and the USA. No technical details on these systems are available

Manufacturer
Lockheed Electronics Company, Plainfield, New Jersey, USA

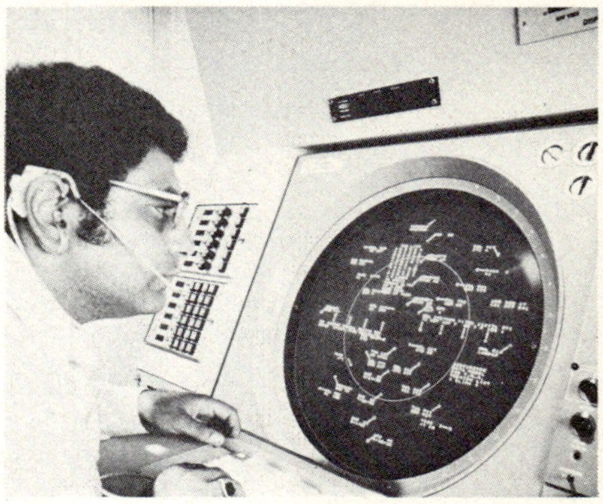

LEC atc display system

Raytheon Air Traffic Control Displays and Systems

Raytheon produces a range of displays for atc including the plan view display, the autonomous display, the Tower bright display unit and the RMD-80 multi-mode display.
The plan view display radar console was designed to meet requirements of the Federal Aviation Administration for en route atc in the USA. Using graphic display technology, Raytheon produced an interactive graphic display system that can display aircraft position, altitude and speed as well as the severity and size of local and distant storm areas. It allows individual flight plans to be entered into the system along with all necessary data for a background map of principal airways, navigational aids, sector boundaries, and geographic reference points. In addition to the all-digital mode of operation, the plan view display can operate in a time-shared fashion to display a mixture of digital data and scan-converted radar video. In this mode the amount of digital data is less than for full digital. In the television time-shared mode, the digital data is displayed during the vertical retrace time of the tv signal. In a normal 945 line system this is about 2 milliseconds every 16·7 milliseconds. The manufacturer recommends the use of a slightly modified television-signal format to provide 2·5 milliseconds display time.
The autonomous display unit is used in the DERD-MC atc system operated in West Germany. In addition to plan view display circuitry, it contains an RDA-500 mini-computer with up to 128 K of 16-bit memory.
The combination of the FAA plan view display and the RDS-500 in one stand alone unit provides fast interactive control action response times with minimum loading of a central, or control, processor. In the DERD-MC system, the autonomous display unit provides ssr target tracking, radar data filtering, and direction finder data processing, in addition to display refreshing control action processing, and display data co-ordinate conversion and control. The stand-alone configuration permits the unit to be connected directly to radar extractor modern links for either local or remote operation.
The Tower bright display unit has been designed for operation in the ambient lighting encountered in an airport control tower. Phosphor and proprietary contrast enhancement filters have been developed to produce a bright display. The unit is capable of operating in a full digital mode, a mixed digital/scan converted television mode, or a television mode. It consists of a display-head containing the crt, high voltage and deflection circuitry, and a remote base assembly with an RDS-500 mini-computer and interfacing circuitry. The RMD-80 multi-mode display is the latest development from Raytheon. It includes a self-contained refresh, built-in display processing capability and a multicolour mode. It can operate in two modes: an all digital mode with either a monochrome or a four-colour penetron phosphor crt; or in a time shared ppi/digital mode with a long-persistence phosphor crt. The console can operate in either mode by changing the crt and selecting an appropriate switch position, all of the necessary hardware being in place. The only system variable is the display to host computer interface which can be modified to match any interface protocol.
The display architecture includes two RP-16 micro-computers, a micro-processor and numerous prom. Each of the three internal processors is assigned a different and specific function; the display processor maintains overall control of all console functions; the ppi processor is used for ppi mode processing, and the third, the micro-processor, interfaces the operator controls to the display processor.
Other features of this radar include: a capability to interface with all terminal atc systems and with all en route surveillance radars, data classes which include graphics, conics, alphanumerics, and special symbols, eight programmable brightness categories (each of the four colours has two brightness categories) and built-in on-line performance monitoring and diagnostics.
A fifth system is installed at the Bundesanstalt für Flugsicherung test centre in Frankfurt.

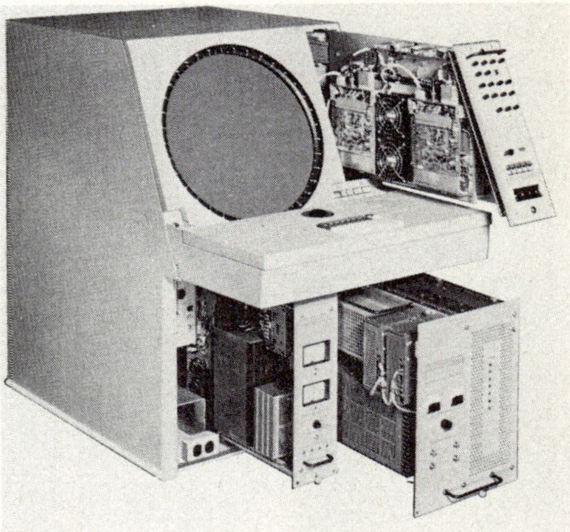

Raytheon multi-mode display showing lower drawers extended for front maintenance

Status
The plan view display was introduced in the early 1970s and more than 1400 have been produced and are in use in the Federal Republic of Germany, Japan and the USA. The majority of the displays are used by the Federal Aviation Administration. The Autonomous Display unit was developed for the DERD-MC system which is an automated atc system installed in the regional control centres of the Federal German Republic at Bremen, Dusseldorf, Frankfurt/Main and Munich, for which Raytheon was the prime contractor

Manufacturer
Raytheon Company, ATC Systems, Sudbury, Massachusetts, USA

Raytheon AN/GPN-22 Precision Approach Radar

The AN/GPN-22 par is intended specifically for use in fixed-base high-density atc system operations. It operates in the radar L-band and is a development of the sophisticated AN/TPN-25 par used in the US Air Force AN/TPN-19 landing control central system. Both these radars employ electronic scanning to perform both scanning and tracking modes over the radar coverage volume without any mechanical movement of the antenna. In the case of the TPN-25, up to six aircraft targets can be tracked simultaneously; in the GPN-22, the same number of targets can be handled but the coverage pattern measures 20 by 8 degrees compared with the 20 by 15 degrees of the earlier equipment. The range is about 35 km in both cases. Provisions are made for the relaying of guidance information from the radar to landing aircraft via an automatic data link.
The antenna feed arrangements are different. The GPN-22 feed assembly is supported by a boom structure projecting over the upper edge of the main reflector, whereas in the TPN-25 it is mounted on the side of the reflector.

Status
Ordered in quantity for US and other air forces. Deliveries began in November 1978 of 39 sets for the US Air Force. Austria has made an initial purchase of the system with options for more to be supplied later. The Netherlands has placed contracts for the provision of 11 sets

Manufacturer
Raytheon Company, Equipment Division, Wayland, Massachusetts, USA

AN/GPN-22 precision approach radar

Raytheon AN/TPN 19 Terminal Air Traffic Control System

AN/TPN-19 is a terminal atc system comprising an airport surveillance radar, a par (AN/TPN-24 and AN/TPN-25), and an operations centre. The radar systems are linked to the operations centre by microwave links, 10 nautical miles (18·5 km). The system is designed for rapid world-wide deployment as well as fixed base installations. Packaging techniques make the system transportable by aircraft, helicopter, or road vehicle.

The AN/TPN-24 airport surveillance radar is a dual-channel, 2 by 500 kW E/F-band mti radar. Its most notable features are a multiple (12) horn antenna feed which produces lobe-free coverage up to 12 000 metres and 60 nautical miles (111 km) on a 1 square metre target; a low-angle pattern which is electrically variable and programmable in range and azimuth to reduce clutter returns; a digital coherent and non-coherent mti system to eliminate weather returns and a staggered prf to eliminate blind speeds below 1100 knots.

The AN/TPN-25 precision approach operates in I/J-band, it has a range of 20 nautical miles (37 km) even with precipitation at 5 cm per hour. This results from the use of the narrow pencil beam (0·75 + 1·4 degrees) produced by the 824-element phased array antenna of 320 kW. A monopulse system, it can track up to six targets simultaneously. For multiple runway coverage the antenna assembly can be slewed through 270 degrees.

In the gca configuration there are three displays but through simple modular extension the system provides seven displays for a full rapcon configuration. All displays are dual mode and can be changed to display either an asr or par presentation simultaneously.

Status

Both the complete TPN-19 and separate examples of the two radar sets incorporated in it have been supplied or ordered for USA and other countries. Austria has one GPN-22 with options on more; Australia has one TPN-19, six TPN-25 pars and one TPN-24 asr; the Federal Republic of Germany is being supplied with 44 TPN-24s for fixed base installation; the Netherlands has fixed installations for six TPN-24 asrs and 11 GPN-22 pars (the latter being an improved TPN-25); the US Air Force has 11 TPN-19s and is taking delivery of 39 GPN-22s

Manufacturer

Raytheon Company, Equipment Division, Wayland, Massachusetts, USA

Raytheon AN/TPN-24 Airport Surveillance Radar

The AN/TPN-24 is the airfield surveillance radar employed in the AN/TPN-19 landing control central system and was originally designed for that system although it has subsequently been used separately on its own or with other equipment. The original version was designed and packaged for use as a highly mobile, tactical radar equipment, but other versions have been configured for different applications, including a fixed-site variant, a tower-mounted model (ASR-910), and semi-static version designated the ASR-909. The antenna and transmitter/receiver units are the same in each version as those of the basic AN/TPN-24.

Status

In addition to those supplied for use in AN/TPN-19 systems (a total of about 12) more than 50 separate AN/TPN-24 sets have been supplied or ordered, including: 44 fixed base installations for the Federal Republic of Germany and six similar installations for the Netherlands. The AN/TPN-24 has also been supplied to Australia and Austria

Manufacturer

Raytheon Company, Equipment Division, Wayland, Massachusetts, USA

ASR-910 airport surveillance radar

Characteristics

RADAR	Dual channel, dual frequency, diversity. Redundant or single channel and spectrum filter options
FREQUENCY	2700–2900 MHz
TRANSMITTER TYPE	Magnetron
PEAK POWER	500 kW (per channel)
PULSE LENGTH	1 μs
PULSE RATE	1050 pps, average
PRF	12 staggers
ANTENNA SIZE	4·26 × 2·43 m
AZIMUTH BEAM-WIDTH	1·6°
ELEVATION BEAM-WIDTH	4°
ROTATION SPEED	13 rpm
BEAM ELEVATION TILT	+15 to −2°
DETECTION RANGE	110 km (clear weather, 1 m^2 target)

Raytheon AN/TPN-25 Precision Approach Radar

The Raytheon AN/TPN-25 is the high-performance precision approach radar (par) developed for the AN/TPN-19 landing control central tactical airfield control system. Designed to fulfil a requirement for a highly mobile system capable of assisting the landing of fixed-wing aircraft and helicopters in adverse weather conditions, the AN/TPN-25 can also be deployed at fixed bases. It employs a unique phased-array antenna that constantly scans the entire service volume of 20 by 15 degrees and 35 km in range, as well as tracking up to six approaching aircraft in elevation and azimuth. No antenna movement is required except when it is necessary to re-orientate it to allow a different runway direction to be used. This can be performed under remote-control to permit any one of four runways on an airfield to be served. The TPN-25 is capable of automatic remote feeding of guidance information to the landing aircraft via a data link.

Operation is in the radar I-band and no mechanical scanning is necessary. A four-lobe monopulse feed horn is directed at the main antenna reflector, radiation having to pass through a digitally-controlled microwave lens consisting of an array of 824 phase-shifting elements. The latter are controlled to produce sharp, digitally-steered, step-scanned pencil beams.

AN/TPN-25 precision approach radar

Characteristics

RADAR: single-channel, dual-function (scan/track)
FREQUENCY: 9000–9200 MHz
TRANSMITTER-TYPE: solid-state oscillator with TWT and CFA amplifiers
PEAK POWER: 320 kW
PULSE LENGTH: 1 μs
PRF: 3500 pps (average)
STAGGER RATIO: 3·2
ANTENNA BEAM-WIDTH: 1·4° azimuth, 0·75° elevation
DETECTION RANGE: 37 km (clear weather, trainer aircraft target)

Status

Production. Apart from about 12 sets supplied as part of the complete US Air Force AN/TPN-19 systems, another six equipments have been supplied to Australia, both as fixed-base systems and in the tactical mobile mode. A development of the AN/TPN-25, with higher performance characteristics, has been developed as the AN/GPN-22 and this has been ordered and supplied in quantity both to the US Air Force and to other countries

Manufacturer

Raytheon Company, Equipment Division, Wayland, Massachusetts, USA

Raytheon DARC Series Air Traffic Control Systems

Raytheon has developed the Direct Access Radar Channel (DARC) series of automated systems for atc radar data processing and display. The modular system has been designed to meet a spectrum of user requirements for en route, terminal or mixed operations, with 1 to 24 input radars, automatic reconfiguration and up to 100 displays. The system has been developed from the RAYDARC system, the DARC contract from the FAA in 1976, the Display of Extracted Radar Data Mini-computer (DERD-MC) contract for the West German BFS in 1977 and the R-2508 award by the FAA in 1979.

The DARC system comprises a radar multiplexer which receives data from up to 15 radars; the RDS-500 display processor, which can handle radar and display processing for up to six plan view displays; quality processor; control processor; system status and control unit; the plan view displays and other peripheral equipment. The latter consists of a card reader, line printer, magnetic tape units, data entry keyboards, typewriters and a printed circuit board tester. The basic DARC system displays the data from a single radar on each plan view display. Selection of the radar to be displayed at each position is under operator control via data entry keyboard entries to the on-line control processor. Appropriate map data displaying routes, sector boundaries, navaids, and georefs is supplied to each display processor for each associated pvd from the disc cartridge. (Maps are generated or modified by the off-line map generation program.) Weather data, if available, is also displayed by DARC, with two weather intensity level displays available.

Plan view display/radar pairing, range scale, off-centre position, geographic filter limits for rejecting unwanted radar data, and other system values and assignments are stored on disc for automatic system initialisation. All such values and assignments may be changed from the data entry keyboard and many may be changed by the controller at the pvd.

For the selected radar, DARC displays the latest target position and up to five histories (target positions from the five previous radar scans); the display symbology identifies each as either a primary or secondary target. Mode 3/A and mode C-beacon codes are displayed for beacon targets, if available. Altitude filtering is available for aircraft equipped with mode C-transponders, allowing inhibiting of the display of these aircraft outside the altitude band of interest at each display.

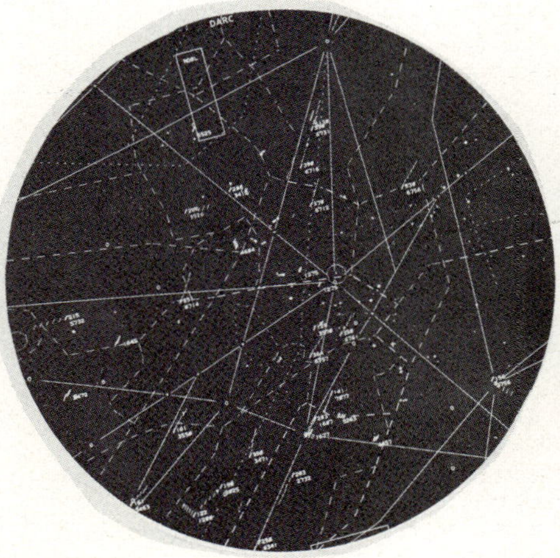
Basic DARC

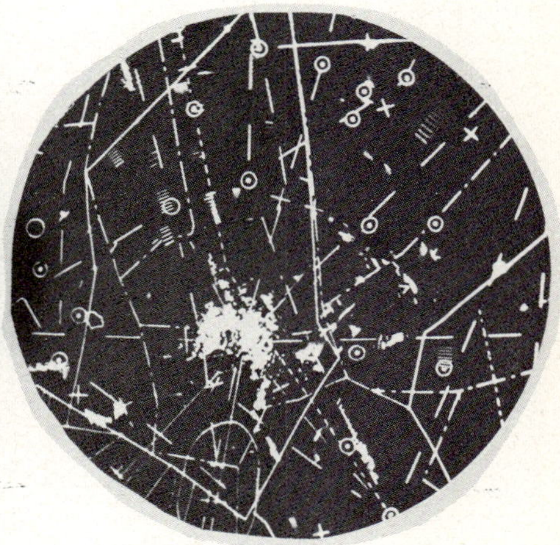
Scan converted display

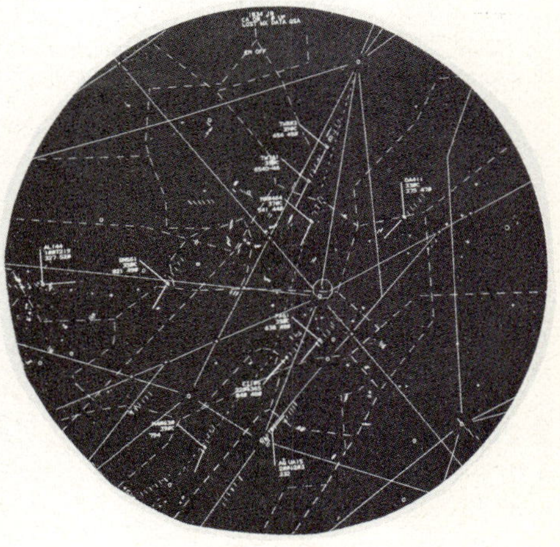
Upgraded DARC

Controllers may enter discrete mode 3/A beacon codes and flight identification (flid), or radio callsign, at the pvd. This code/flid pair will then cause the appropriate flid to be displayed, along with the beacon codes, for an aircraft squawking a particular discrete code.

In addition to the DARC capabilities, R-2508 and RAYDARC provide three capabilities not available in DARC, and a number of additional features. These are 'mosaicking', tracking and realtime quality control. 'Mosaicking' is the technique employed when radar coverage overlaps with multiple radars. It enables signals from the prime radar which is achieving the best results to be selected in preference to the supplementary radar which can be used in the event of a failure in the preferred radar, or missed target reports. R-2508, RAYDARC, and upgraded DARC perform tracking of both primary and secondary (beacon) targets. Real-time quality control is a series of radar input checks which are performed automatically by the system.

Other functions of the upgarded DARC include: intrasystem handoff and jurisdictional control, automatic beacon code allocation, system data recording, six displays per display processor, mode C-establishment, 24 radar inputs and automatic data block offset.

Status

DARC, R-2508 and the DERD-MC atc systems are now in service

Manufacturer

Raytheon Company, ATC Systems, Sudbury, Massachusetts, USA

Texas Instruments ASR-8 Series Airport Surveillance Radars

The ASR-8 series comprises three basic configurations, the K, M and L models. ASR-8K uses a klystron transmitter with quadrature mti and is designed to meet the needs of high density terminals where clutter and environment are particularly severe. Standard features of the system are a crystal oscillator multiplier, frequency diversity with a dual-beam antenna, range azimuth gate and low-pass filter. Processor storage and timing on the standard system is set at 60 nautical miles (111 km).

The ASR-8M operates with a pulse-coherent magnetron system and is suitable for the majority of terminal areas. Apart from the magnetron transmitter it is identical to the ASR-8K. ASR-8L is a single channel version of the ASR-8M

ASR-8K system housed in two portable buildings

designed for lower-density commercial airports and satellite general aviation airports not currently served by radar, which handle large volumes of air traffic including business jets. The ASR-8L is capable of providing all the features contained in ASR-8M except frequency diversity. All systems utilise the same basic antenna receiver and processor sub-systems. The transmitters share common high-voltage power supply and modulator components. Many operational features are designed to be optional to permit each user to select those features which best serve his needs.

Manufacturer

Texas Instruments Incorporated, Dallas, Texas, USA

Characteristics

FREQUENCY: 2700–2900 MHz dual channel operation, 60 MHz separation
TRANSMITTED POWER
1·25 MW (klystron)
500 kW (magnetron)
PULSE WIDTH: 0·6 μs klystron/0·8 μs magnetron
PRF: 1040 pps, normal mode, 4-pulse staggered
RECEIVER
DATA RATE: 12·5 rpm (15 rpm option)
NOISE FIGURE: 40 dB
SYSTEM MTI IMPROVEMENT FACTOR
36 dB (klystron)
32 dB (magnetron)
ANTENNA
TYPE: dual beam, normal and passive (optional)
AZIMUTH BEAM-WIDTH: 1·35°
GAIN
nominal: 33·5 dB
passive: 32·5 dB
REFLECTOR DIMENSIONS
width: 4·9 m
height: 2·7 m

Status

The ASR-8 series has evolved from the ASR-8, which is identical to the ASR-8K. ASR-8 was introduced in 1975 and is in service with the Federal Aviation Administration

Westinghouse AN/TPS 65 Tactical Surveillance Radar

The AN/TPS 65 is a two-dimensional D-band air search radar developed as the primary airfield surveillance radar sensor for the US Marine atc and handling system. This system provides automated and manual control information for aircraft landings at expeditionary airfields. The radar is similar to the AN/TPS 63 designed to provide low altitude coverage for tactical air defence operations but has a dual channel configuration and is equipped with a different antenna. Both radars feature extensive use of micro-electronic and solid-state techniques. An advanced mti facility is claimed to provide a 90 per cent probability of detecting a one metre square target out to the full instrumented range of 80 nautical miles (150 km). The digital mti uses a four-pulse canceller with variable time intervals between pulses to distinguish moving aircraft from ground returns while eliminating mti radar blind speeds. There is an additional digital three-pulse mti weather canceller, and digital cfar processing. Dual diversity operation is normal and there are built-in test facilities. Fixed station configurations of the AN/TPS 65 using sub-systems suitable for commercial airport radar applications are also available.

Characteristics

FREQUENCY: D-band, dual diversity, 51 selectable frequencies between 1250–1350 MHz
RANGE: 80 nm on fighter sized aircraft
DATA RATE: 6, 12, 15 rpm (10 s, 5 s, 4 s)
ELEVATION COVERAGE: 12 000 m shaped coverage to 45°
POWER OUTPUT: 100 kW peak, average 0·3 kW
ANTENNA GAIN: 32·5 dB
ANGULAR RESOLUTION: 2·7°
RANGE RESOLUTION: 152 m
ECCM: coded-pulse anti-clutter, frequency agility, prf stagger
MTBF: 500 h
MTTR: 20 minutes
WEIGHT: 6200 kg

Status

Initial production completed

Manufacturer

Westinghouse Defence and Electronic Systems Centre, Baltimore, Maryland, USA

Westinghouse AN/TPS 65 radar

Westinghouse ARSR-3 Air Traffic Control Surveillance Radar

The ARSR-3 is a two-dimensional atc surveillance radar designed for remote-control operation. The dual configuration includes correlators and target extractors and the unit is equipped with dual weather channels. In the diplex mode both channels operate simultaneously. Other features include an integral primary/secondary antenna with a low beam/upper beam primary pattern, radio frequency sensitivity time control, and solid-state switching and control. Polarisation diversity enables targets and weather to be displayed simultaneously enabling the controller to steer aircraft around the bad weather or see aircraft in it. The equipment incorporates built-in test equipment with fault isolation facilities and automatic reconfiguration.

Characteristics

TRANSMITTER
FREQUENCY: 1·25–1·35 GHz
PEAK POWER: 5000 kW
AVERAGE POWER: 3·5 kW
PRF: 310–365 pulses/s
MAX RANGE ($2m^2$ TARGET)
mode diplex: 240 nm (445 km)
simplex: 193 nm (358 km)
MAX ANGLE ELEVATION: 44°
DIGITAL MTI IMPROVEMENT: 50 dB
ANTENNA TYPE: modified parabola, dual beam, integral beacon/ssr
HORIZONTAL BEAM-WIDTH: 1·25–2°
ROTATION SPEED: 12 rpm
REFLECTOR DIMENSIONS
width: 5 m
height: 2·35 m

Status

Twenty three fixed and four mobile versions (with single channel) have been supplied to the United States Federal Aviation Administration. Nine additional systems have been purchased by Canada and Switzerland

Westinghouse ARSR-3 radar

Manufacturer

Westinghouse Electric Corporation, Baltimore, Maryland, USA

Westinghouse ASR-30 Airport Surveillance Radar

The ASR-30 is a dual-channel airport surveillance radar derived from the ARSR-3 en route radar system. Similar to the latter, the ASR-30 incorporates polarisation diversity, diplex transmitter operation, a dual beam antenna and radio-frequency sensitivity time control. All elements beneath the antenna are duplicated and in the diplex mode both channels operate simultaneously but at different frequencies to provide an improvement in range performance. The transmission frequency is derived from a single crystal source.

Should a fault occur, built-in test equipment detects and indicates the failure and the system is switched automatically to the good channel. The digital mti uses four-pulse cancellation and I and Q processing techniques. Three modes of receiver operation are possible: the logarithmic receiver provides a normal mode for detection in the absence of weather, the weather receiver displays weather information, and the digital mti receiver mode enhances target detection capability in the presence of weather and clutter. The system can be configured to suit individual sites by selections in the range-azimuth gate of dynamic threshold, mti normal, stc curve, low beam/upper beam and variable/fixed interpulse period. Up to 352 selections are available.

The ASR-30 antenna contains two microwave

feed systems which produce an upper and a lower beam. Optimum short-range detection is accomplished by using the upper beam to avoid ground clutter, while the lower beam provides long-range coverage. The antenna feed also incorporates an integrated beacon/ssr system to provide beacon coverage without degrading the primary radar performance. The antenna permits full-weather detection when the system is operating in the circular polarisation mode. The pedestal assembly includes two complete drive systems. Should one fail, the other will continue to rotate the antenna.

Characteristics

TRANSMITTER
frequency: 1·25–1·35 GHz
peak power: 2000 kW
average power: 3·65 kW
prf: 524 pulses/s
max range (2 m^2 target): 120 nm (222 km)
ssr: 200 nm (371 km)

RECEIVER
data rate: 12 rpm
noise figure: 2 dB max

INTEGRATED CANCELLATION RATIO
nominally 21 dB (18 dB minimum)

MTI IMPROVEMENT: 50 dB

ANTENNA TYPE: dual beam, integral beam/ssr
azimuth beam-width: 1·25 to 2°

ANTENNA GAIN
high beam: 33·6 dB
low beam: 34·9 dB

HORIZON RATIO: 16·75 dB

REFLECTOR DIMENSIONS
width: 33 ft (10 m)
height: 31 ft (9·4 m)

POWER REQUIREMENTS: 50/60 Hz, 65 kVA

Status

Production

Manufacturer

Westinghouse Electric Corporation, Baltimore, Maryland, USA

Westinghouse ASR-30 radar antenna

NAVAIDS

AUSTRALIA

AWA Type LDB 101 Distance Measuring Equipment

The Type LDB 101 distance measuring equipment developed by AWA is fully solid-state and designed to meet or exceed ICAO requirements. Compact in design, it is offered in three versions with peak output powers of 100, 500 watt or 1 kilowatt. Either a single- or full-duplicated beacon can be supplied for local or remote-control operation. The system incorporates built-in test equipment.

Characteristics

TRANSMITTER
- operating frequency: 962–1213 MHz
- peak output power: 100 W or 500 W or 1 kW
- spectrum: ICAO Annex 10–DME (N)

RECEIVER
- operating frequency: 1025–1150 MHz
- threshold sensitivity: –90 dBm
- adjacent channel rejection: 75 dB (minimum)

SYSTEM
- transponder capacity: 100 interrogators
- channels: 252 (X and Y)
- reply delay
  - adjustable 35–50 μsec X channels
  - 45–50 μsec Y channels

MONITORING ALARM LIMITS
- reply data: ±1 μsec
- power: -3 dB (adjustable)
- pulse spacing: ±1 μsec
- pulse rate: 900 pulse pairs per second
- identification: if not repeated within 80 s
- reply efficiency: 70%

ANTENNA
- polarisation: vertical
- vswr: 2:1 max
- gain: 9 dB minimum

PRIMARY POWER
- voltage: 21–29 V dc

ENVIRONMENTAL
- temperature: –20°C to +60°C
- relative humidity: up to 90%
- antenna: –50°C to +70°C (100% RH)

Status

The equipment was stated to be in the final stages of development in late 1981

Manufacturer

Amalgamated Wireless (Australasia) Ltd, North Ryde, New South Wales, Australia

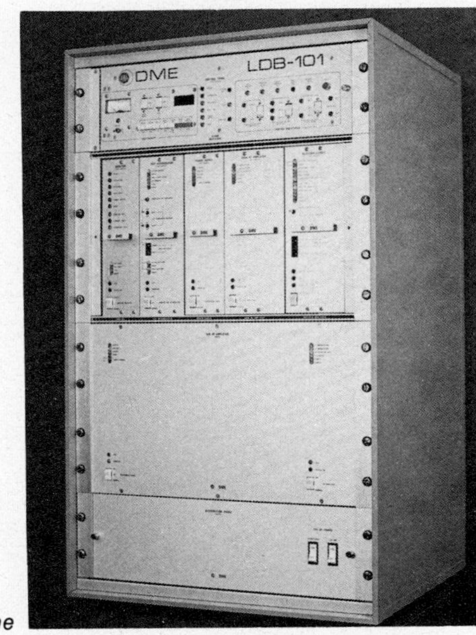

AWA LDB 101 dme

AWA Type VRB-50D Doppler Vhf Omni-range Beacon

The principle of Doppler vor beacons is based on the Doppler change in frequency which results if a radiating antenna is moved around a large diameter circle. If this antenna is rotated 30 times a second, the signal at a distance observation point will be frequency modulated at a 30 hertz rate. The phase of this 30 hertz modulation will be related to the azimuth of the observer.

The effect of a rotating antenna is simulated by using a ring of antennas and switching the rf signal to each in turn. Provided sufficient antennas are used and provided the pulses of rf energy are suitable shaped, the simulation is good.

Two rf signals are radiated from diametrically opposite antennas in the circular array; they are the upper and lower sidebands of the 9960 hertz amplitude modulation which become frequency modulated at 30 hertz. At the vor receiver in the aircraft, this appears as 30 hertz frequency modulation of the 9960 hertz sub-carrier. In Doppler vor the 'variable phase' azimuth dependent information is contained in the 30 hertz fm signal. The 'reference phase' 30 hertz amplitude modulated signal is radiated from a single antenna at the centre of the ring. The wider aperture of the Doppler vor antenna system, when compared with the conventional vor antenna, improves the accuracy and stability of the bearing information by reducing the effects of multi-path propagation. In addition, the frequency modulated signal is demodulated in the aircraft receiver with minimum phase shift error because of the capture effect of an fm system. This capture effect ensures that the stronger, direct path fm signal predominates in the aircraft receiver when received together with weaker, reflected path signals. The output of the fm discriminator in the receiver will then contain principally the required 30 hertz information carried by the stronger signal to the exclusion of the unwanted information contained in the reflected signals. Unlike the conventional vor, siting of this equipment is not critical since it will not be affected by multi-path reflections.

The Type VRB-50D produced by AWA is a double side-band system which conforms to the requirements of ICAO Annex 10. The equipment is modular in design and solid-state including the antenna distribution switch. Forty-eight antennas are used and single crystal clock controls the digital generation of all the bearing information. The double side-band system is designed to eliminate errors which are possible in maladjusted aircraft receivers when single side-band and alternating side-band systems are used.

Monitoring of the system is achieved by the use of a time-shared filtering technique and a zero-crossing detector. A no-break dc operation is provided.

AWA Doppler vor at Lazarus Island, Singapore

Characteristics

SIGNAL OUTPUT
- output power: 50 W minimum
- frequency band: 108–117.95 MHz
- carrier frequency stability: ±0·002%

RADIATION
- antenna system: 1 + 48 Alford loops
- polarisation: horizontal
- bearing accuracy (for all elevation angles from 0° to 40°): typically ±0·5 degree, max error

REFERENCE PHASE MODULATION
- (fm on sub-carrier)
- sub-carrier frequency: 10 000 Hz ±940·01%
- mean depth of modulation
- 28–32% by sub-carrier:
- frequency of modulation: 30 Hz ±0·01%
- deviation ratio: 16 ± 1

SUB-CARRIER MODULATION
- am of sub-carrier measured at least 1600 m from vor: 40% of max

IDENTIFICATION
- modulation tone: 1020 Hz ± 50 Hz
- modulation depth: recommended 10% (adjustable 0–15%)
- code: international Morse, 2/3 letters
- keying speed: 7 words per minute
- repetition rate: 6 times per minute
- distance measuring equipment: keying available

VOICE MODULATION
- frequency range: 300–3000 Hz
- modulation depth: recommended 10% (adjustable 0–30%)
- noise due to signal commutation: negligible

MONITORING ALARM CONDITIONS
- change in transmitted bearing information adjustable in 0·1° steps–max 1°
- depth of modulation of sub-carrier and 30 Hz am: 25·5% (ie 30% modulation depth reduced by 15%)
- code failure: continuous or absent tone

INPUT POWER
- voltage: 24 V dc nominal
- current consumption
  - single system: 18 A
  - duplicated system (standby active): 36 A

ENVIRONMENTAL CONDITIONS
- indoor equipment: 0°C to +55°C
- outdoor equipment: –10°C to +60°C
  - 145 km/h wind in ice and snow

Status

Production. VRB-50D beacons are installed and operating in Afghanistan (1), Australia (10), Malaysia (1), Nepal (2), Philippines (1), Portugal (1), Singapore (2) and Thailand (1)

Manufacturer

Amalgamated Wireless (Australasia) Ltd, North Ryde, New South Wales, Australia

Interscan Microwave Landing System

The time reference scanning system is an air-derived system which provides wide sectors of azimuth and elevation angular guidance as well as low level height information for flare guidance. Range information is provided by distance measuring equipment which relies on the conventional interrogation/reply principle.

The complete system can comprise up to five sub-systems:

- azimuth: approach guidance in the horizontal plane
- missed approach azimuth: missed approach and take-off guidance in the horizontal plane
- elevation: approach guidance in the vertical plane
- flare: low level vertical guidance for aircraft flare manoeuvre before touchdown
- distance measuring equipment; distance information throughout the coverage volume.

All these sub-systems can be provided independently. For example a minimum system could comprise azimuth and elevation only. Other sub-systems could be added later without altering the already installed azimuth and elevation sub-systems.

The Interscan system uses two ground-based antennae near the runway which radiate narrow fan-shaped beams of microwave energy which are scanned to-and-fro at a precisely controlled rate. One beam scans from side-to-side, the other scans up and down.

In a full scan, the side-to-side beam intercepts the aircraft twice, each time producing a pulse in the aircraft receiving equipment. The time interval between the two pulses is a direct measure of the azimuth angle of aircraft relative to the centre line of the runway. This is the time reference scanning beam principle.

The azimuth, elevation and distance information accurately fixes the aircraft's position. This information, which is updated some 40 times a second, enables the pilot to fly any one of a variety of approach paths which may be chosen to suit the requirements of particular airports and aircraft types.

All of the different angle functions, approach azimuth, elevation, and missed approach back azimuth are time multiplexed so that only one airborne receiver channel is required. Each function is identified by a preamble code transmitted immediately before its to/fro scan cycle. The angle guidance sub-systems (azimuth and elevation) operate at C-Band in the range 5031–5090·7 megahertz. The functions, commonly referred to as the angle guidance functions, including approach azimuth, missed approach azimuth and elevation, are radiated sequentially on the same carrier frequency with each function being identified by a data preamble. Additional space is available in the format to incorporate future developments such as missed approach elevation and 360 degree clearance azimuth, as well as auxiliary data. Since each angle function is separately identified, functions can be added or deleted to suit particular airport requirements. The present format requires a bandwidth of only 300 kilohertz for each channel.

Azimuth antenna Interscan mls

Elevation antenna Interscan mls

Characteristics

FREQUENCY	5031–5090·7 MHz 200 channels	
TYPE	basic mls	Small community airport mls
APPLICATION	airports with long runways to achieve threshold window requirements with standard azimuth scan (+40°) or option (+60°)	small airports, special applications.
AZIMUTH BEAM	2°	3°
scan	±40°	±10°
ELEVATION BEAM	1°	1·5°
scan	–1–20°	0–15°
WINDOW RANGE		
width at 10 nm (18·5 km)	16·5 nm (30·6 km)	3·5 nm (6·5 km)
height at 10 nm (18·5 km)	10 000 ft (3048 m)	10 000 ft (3048 m)
WINDOW RANGE		
width at 20 nm (37 km)	33·5 nm (62·1 km)	7 nm (13 km)
height at 20 nm (37 km)	20 000 ft (6096 m)	20 000 ft (6096 m)

The various functions of the new landing system can be understood by considering an approach to landing. An aircraft enters the precision coverage of TRSB mls at ranges up to 20 nautical miles (37 kilometres) in a segment that extends to 40 degrees either side of the runway approach centreline and at angles up to 15 degrees in elevation. At this point, the aircraft obtains guidance from the approach azimuth, and elevation, and distance measuring equipment transmitters. When it reaches final approach, the pilot selects a glide slope suitable to his aircraft (normally 3 degrees) which is provided by elevation guidance and maintains his directional course based on azimuth guidance. Upon reaching the runway threshold, a transition is made to flare-guidance for the 'round-out' manoeuvre leading to touchdown. If the aircraft must abort the attempted landing for any reason it uses the precision guidance provided by the missed approach azimuth transmitter in its departure from the field.

Status

The TRSB system was selected by the member states of ICAO in April 1978 as the new generation landing system. While a further specification for the system known as Standards and Recommended Practices is being promulgated work on the Interscan system is continuing and a prototype system has been produced which meets the predicted ICAO standards and the anticipated engineering requirements of the FAA

Manufacturer

Interscan Australia Pty Ltd, Ryde, New South Wales, Australia, Amalgamated Wireless (Australasia) Ltd, North Ryde, New South Wales, Australia, Hawker de Havilland Pty Ltd, Bankstown, New South Wales, Australia, Wilcox Electric Inc, Kansas City, Missouri, USA

CANADA

Leigh Tacan Beacon

The Leigh Tactical Air Navigation (Tacan) system is a terminal and en-route navigation system. Navigation guidance is provided to suitably-equipped aircraft by means of azimuth and distance information. The beacon is available in fixed ground, vor, tactical mobile, or shipboard configurations.

Any one of 252 frequencies can be used to prevent detection or jamming by an enemy. The mobile system includes: Tacan beacons, antenna system, remote-control indicator, automatic power switching, full monitoring, diesel generator, air conditioner, fuel tank, transportable shelter and trailer.

The system is designed to be carried by helicopter or transported by light cargo aircraft such as the de Havilland Buffalo and can be moved and made operational in 30 minutes. With an aircraft range of 200 miles (322 km) it provides up to 100 aircraft with range information and an unlimited number of aircraft with azimuth information.

The Tacan can be integrated with any vor or Doppler vor system to provide the Tacan azimuth and dme functions of the vortac navigation system.

Characteristics

PEAK POWER: 3 kW (5 kW optional)
SPECTRUM: ICAO Annex 10-MIL STD 291
TRANSPONDER CHANNELS: 252 frequencies (x and y)
WEIGHT
mobile system: 2720 kg
dual beacon: 545 kg
OPERATIONAL TEMPERATURE RANGE: –54 to +65°C
RELATIVE HUMIDITY: 0–95%

Status

Production

Manufacturer
Leigh Instruments Ltd, Pointe Claire, Quebec, Canada

Leigh Mobile Tacan Beacon

Nautel ND 200S Non-Directional Radio Beacon

The ND 200S is a fully solid-state non-directional radio beacon transmitter which has an adjustable carrier from 25 to 50 watts with 10 to 95 per cent modulation and a voice filter and compression. It is available in single or dual configuration and uses a 52 ohm co-axial output allowing transmitter and antenna system separation up to 360 feet (110 metres). The equipment operates with either a suitable automatically tuned Polestar A or D series antenna or via a separate automatic antenna turning unit (model NX 200 TUB) to the operator's existing antenna system. Designed for operating in remote unattended sites with no external heating required, the beacon features an output protection against open- or short-circuits and an automatic transmitter shut-down facility with a remote failure alarm.

Characteristics

TRANSMITTER
frequency: 190–535 kHz
continuous power: 25–50 W adjustable
emission modes
A2 keyed modulating tone
A3 voice
power requirements: 120 V ac, single phase, 60 Hz, 160 VA
standby: 48 V dc, 2·5 A

ENVIRONMENTAL CONDITIONS
temperature: –50 to +70°C
relative humidity: 0–95%

UNIT DIMENSIONS
height: 2 ft (0·61 m)
width: 1 ft (0·305 m)
depth: 1 ft (0·305 m)

WEIGHT: 90 lb (41 kg)

Status

Nautel has over 1500 NDB installations in approximately 45 countries

Manufacturer

Nautical Electronic Laboratories Ltd, Hackett's Cove, Nova Scotia, Canada

Nautel NX series Non-Directional Radio Beacons

Nautel has developed the NX series of non-directional radio beacon transmitters with outputs ranging from 25 watts to 4 kilowatts. They are especially suitable for use at remote, unattended indoor installations. The rf output is fed via 50 ohm co-axial feeder either directly to a suitable automatically tuned polestar antenna or via a separate automatic antenna tuning unit to the operator's existing antenna system. It features an adjustable carrier with 0 to 95 per cent modulation. All are fully solid-state and capable of integral modular reserve by using a number of separate power amplifier modules fed with a common input drive signal and having their outputs added via a passive bridge isolating network known as a hybrid combiner. Each power amplifier module contains its own integral power supply, fusing, rectifying and regulating protection circuits, together with a wide-band linear amplifier circuit. Under normal operating conditions, each of the pa modules in a system produces an equal amplitude coherent signal level, of 62·5 watts, at its output. The passive hybrid combiner system adds these pa module outputs to produce the final transmitter output power. In the event that any of the pa modules should fail, the remaining operational pa modules continue to support output power while the hybrid combiner ensures that the outputs of these pa modules remain matched into the final rf load of 50 ohms. In this way pa module failures result only in a gradual reduction of output power. In transmitters with

Characteristics

TYPE	NX 1000 BD	NX 2000 BD	NX 4000 BD	NX 8000 BD	NX 12000 BD	NX 16000 BD
TRANSMITTER						
FREQUENCY	200–535 kHz	200–535 kHz	200–535 kHz	200–535 kHz	200–535 kHz	200–535 kHz
PEAK POWER	950 W	1900 W	3800 W	7·6 kW	11·4 kW	15·2 kW
CONTINUOUS POWER	250 W	500 W	1000 W	2 kW	3 kW	4 kW
EMISSION MODES	A0, A1 A2, A2/A3 A2 + A3, A3	A0, A1 A2, A2/A3 A2 + A3, A3	A0, A1 A2, A2/A3 A2 + A3, A3	A0, A1 A2, A2/A3 A2 + A3, A3	A0, A1 A2, A2/A3 A2 + A3, A3	A0, A1 A2, A2/A3 A2 + A3, A3
POWER REQUIREMENTS	190–260 V, ac single phase 50/60 Hz 1400 VA	190–260 V, ac single phase 50/60 Hz 2500 VA	190–260 V, ac single phase 50/60 Hz 5 kVA	190–260 V, ac single phase 50/60 Hz 11 kVA	190–260 V, ac 3-phase 50/60 Hz 15 kVA	190–260 V, ac 3-phase 50/60 Hz 20 kVA
ENVIRONMENTAL CONDITIONS						
temperature	–10 to +55°C	–10 to +55°C	–10 to +55°C	–10 to +55°C	–10 to +55°C	–10 to +55°C
relative humidity	0–95%	0–95%	0–95%	0–95%	0–95%	0–95%
UNIT DIMENSIONS						
length	5 ft 1 in (1·55 m)	5 ft 1 in (1·55 m)	6 ft 9 in (2·06 m)	6 ft 9 in (2·06 m)	6 ft 9 in (2·06 m)	6 ft 9 in (2·06 m)
width	1 ft 10½ in (0·57 m)	1 ft 10½ in (0·57 m)	1 ft 10½ in (0·57 m)	3 ft 9 in (1·14 m)	5 ft 7½ in (1·71 m)	7 ft 6 in (2·29 m)
height	2 ft (0·61 m)	2 ft (0·61 m)	2 ft (0·61 m)	2 ft (0·61 m)	2 ft (0·61 m)	2 ft (0·61 m)
WEIGHT	300 lb (136 kg)	332 lb (151 kg)	518 lb (235 kg)	1350 lb (616 kg)	2100 lb (960 kg)	2800 lb (1280 kg)

output levels of 500 watts and greater, the pa modules are arranged in groups of eight for convenient mounting and form a linear amplifier sub-assembly referred to as a 500 watt (8 × 62·5 watt) power-cube.

Status

Production. Nautel has provided NDBs to the Canadian Ministry of Transport and Department of Defence, the FAA and military forces in the USA and to the Governments in Brazil, Burma, Guyana, India, Indonesia, Iran, Libya, Oman, Philippines, Sweden, Trinidad, Yugoslavia and Zaïre. Non-directional radio beacons have been supplied through ICAO to Guinea, Sri Lanka, North and South Yemen and Viet-Nam

Manufacturer

Nautical Electronic Laboratories Ltd, Hackett's Cove, Nova Scotia, Canada

Nautel NX series 500 watt non-directional radio beacon transmitter

Philips Electronics Navigational and Landing Aids

Philips Electronics specialise in the production of ground-based aircraft navigation and landing aids. Equipment manufactured by the company includes vhf omni-directional range equipment, both conventional and using Doppler, distance measuring equipment and instrument landing systems (ils). Philips vor systems are all solid-state, and modular design incorporating fail-safe monitors and integral test facilities. They are available in vor and dvor versions single- or dual-channel with 25, 50 or 100 watt carriers. Most modules are interchangeable between the two versions.
The distance measuring equipment systems are offered in 100 watt and 1000 watt configurations for single- and dual-channel operation. They meet ICAO Annex 10 recommendations and can be co-located with ils or vor equipment. Omni-directional, bi-directional and uni-directional antennae are available.
The Philips ils systems exceed ICAO category I, II and III recommendations in both performance and monitoring. Modular and fail-safe in design, they incorporate built-in test equipment and a choice of antennae systems.

Status

Production. There are over 300 installations of Philips navigational and landing aids in service at civil airports in 32 countries. These include 108 Canadian government installations at Department of Transport locations

Manufacturer

Philips Electronics Ltd, Scarborough, Ontario, Canada

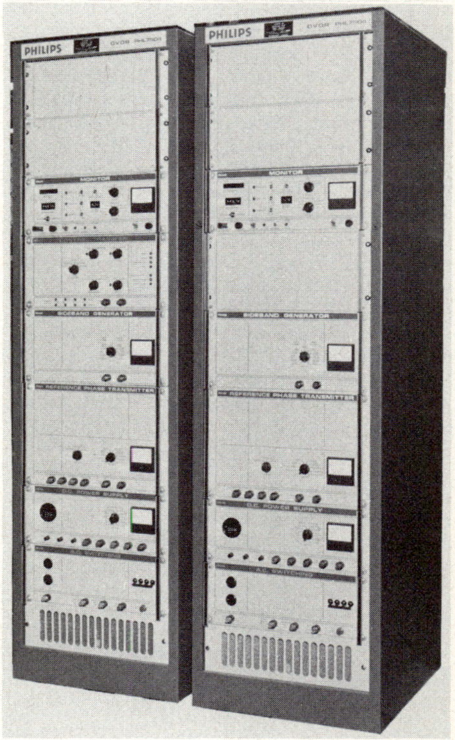

Philips Doppler vor system

FRANCE

Thomson-CSF 721 R/A Distance Measuring Equipment

Thomson-CSF manufacture distance measuring equipment for en-route navigation associated with a vor (Type 721 R) or operating as a landing aid in conjunction with an ils (Type 721 A). The former, using a single or dual 1 kW transmitter with power tube and omni-directional antenna, achieves a radiated power of 36 dBW; the latter type with a solid-state 100 watt transmitter and unidirectional antenna produces 28 dBW. Both types of equipment are housed in a single cabinet. Other features include built-in test equipment and a power supply from no-break mains or a self-contained 48 volt dc generator connected with the associated vor or ils equipment. Remote test and control facilities operate using multi-pair cable, telephone line via a modem or in conjunction with the associated vor or ils systems.

Status

Production. Both types of dme are in service in France and other countries

Manufacturer

Thomson-CSF, Division DRS-TVT, Bagneaux, France

Thomson-CSF vor/dme aerial (511/721R)

Thomson-CSF Doppler/Vhf Omni-range Beacons

Thomson CSF produces both Doppler vhf omni-range beacons and conventional vor equipment. The DVOR 512 is a simultaneous double side-band beacon for difficult sites, which can be co-located with a dme system. Using a solid-state 100 watt broad-band transmitter, the beacon features patented phase- amplitude distribution circuits, proximity monitoring and built-in curve measurement. The antenna has 48 + 1 radiating elements on a 27-metre diameter reduced counterpoise. A no-break power supply operating at 48 volts dc ensures between 3 and 30 hours of autonomy and remote monitoring and control can be carried out using a modem. The equipment can be installed in a prefabricated building or mono-block cabin.

The VOR 511B is offered as a 50 or 100 watt vhf bearing beacon which can be co-located with a dme system. The broad-band single or dual transmitter and the side-band digital generator are housed in a single cabinet. The beacon operates with a high gain antenna and is equipped with built-in error curve measurement and a transmission parameter monitoring system ensuring a bearing accuracy of less than one degree. Remote monitoring and control is available via a multi-pair cable or by modem in association with dme equipment. The no-break operation facility and the available installations are similar to those offered with the dvor beacon.

Status

Production. The beacons are in service in France and other countries

Manufacturer

Thomson-CSF, Division DRS-TVT, Bagneux, France

Thomson-CSF VOR 511B

Thomson-CSF DVOR 512

Thomson-CSF ILS 381 Instrument Landing System

The localiser and glide path electronics of the ILS 381 are each housed in a single cabinet. Modular in construction, the equipment with the availability of four types of localiser antennae and four glide-path antennae offers a wide variety of configurations to enable the system to be adapted to each site and each particular requirement. To ensure operational safety, the antennae have field detectors which are insensitive to over-flying aircraft. The equipment which incorporates monitors, operates with no-break power supplies so permitting autonomous operation for several hours. If space is at a premium, the 75 MHz radio markers can be replaced by a DME 721 A. The equipment forming the ils systems is remote-controlled and remote-tested on a double pair of telephone lines via a modem. It can be supplied in prefitted air-conditioned and sealed cabins. To facilitate installation, adjustment and maintenance, several facilities are available. These include a predicted performance calculation data processing program; remote field monitor (MNL 395) for checking the localiser signal at the level of the intermediate radio marker; mobile field receiver (RME 392) for remote-measurement of the localiser and glide path transmission characteristics; and 90 and 150 hertz precision generators (type GBF 391).

Characteristics

	Localiser	Glide Path	Markers
FREQUENCY	108/122 MHz	328·6/335·4 MHz	75 MHz
FREQUENCY PRECISION AND STABILITY	$\pm 17.10^{-6}$	$\pm 17.10^{-6}$	$\pm 50.10^{-6}$
MODULATION	90–150 Hz	90–150 Hz	400/1300 Hz
IDENTIFICATION	1020 Hz		
MODULATION DEPTH	20% + 20%	40% + 40%	95%
CSB OUTPUT POWER	25 V	5 W	2 W
MINIMUM RANGE AND COVERAGE	25 nm ± 10° 17 nm ± 60°	10 nm ± 8°	ICAO
POWER SUPPLY	48 V cont	48 V cont	24 V cont
CONSUMPTION	5 × 15 A	4 × 10 A	2 A

ILS 381 M array glide-path at Basle

Status

Production. ILS 381 is installed at Innsbruck airport in Austria, and Basle and Zurich airports in Switzerland, the latter as a category III installation

Manufacturer

Thomson-CSF, Division DRS-TVT, Bagneux, France

ILS 381 category III installation at Zurich

MLS 800 azimuth station

Thomson-CSF NFRP 1A Tactical Air Navigation Transmitter

NFRP 1A is a range and bearing radio beacon for en-route navigation, associated with a vor beacon (vortac station) and conforming to military and ICAO standards. Available with single or dual 1 kW solid-state transmitter with power-tube, the equipment has a range of 180 nautical miles (333·5 kilometres). A synthesizer control enables frequencies to be changed in less than 15 minutes. Other features of the equipment are a static electronic-scan antenna and the facility for remote-control by telephone line or radio link up to 20 kilometres. The unit operates from a mains or no-break power supply.

Status

Production. In service in France and other countries

Manufacturer

Thomson-CSF, Division DRS-TVT, Bagneux, France

Thomson-CSF NFRP 1A tactical air navigation transmitter

Thomson-CSF Non-Directional Beacons

Thomson-CSF RBL 10 locator

Intermediate frequency non-directional beacons are available for en-route and terminal operations, with several power ratings. All feature a solid-state transmitter with an automatic test system and high gain daisy-type antennae. The equipment operates either from a mains power supply or using a 48-volt dc power generator. RBL 10 is offered as an autonomous unit or in a transportable configuration. Both are rated at 10 watts.

Characteristics

FREQUENCY RANGE	200–425 k Hz			
TYPE	RBT 20/50	RBT 50/100	RBT 200/400	RBL 10
POWER	50 W or 25 W	100 W or 50 W	400 W or 200 W	10 W
EMISSION MODE	A1 (50 W)	A1 (100 W)	A1 (400 W)	A1
	A0/A2 (25 W)	A0/A2 (50 W)	A0/A2 (200 W)	

Status
Production. The equipment is in service in France and other countries

Manufacturer
Thomson-CSF, Division DRS-TVT, Bagneux, France

Thomson-CSF TRG-741 Vhf Direction Finder Equipment

TRG-741 is a single frequency, two-channel vhf radio direction finder providing the capability of selecting either of the two preset frequencies. The unit is equipped with a solid-state receiver with digital phasemeter, and a local- or remote-digital or compass display providing magnetic headings and true bearings (QDM and QTE). The system operates from a 24-volt dc or mains supply.

Characteristics
FREQUENCY RANGE: 118–136 MHz
RECEIVER SENSITIVITY: 3 μV
ACCURACY: ±2°
MEASUREMENT TIME: 0·33 s
POWER REQUIREMENTS: 24 V dc or mains supply

Status
Production. The beacons are in service in France and other countries

Manufacturer
Thomson-CSF, Division DRS-TVT, Bagneux, France

Thomson-CSF TRG-741 vhf

GERMANY, FEDERAL REPUBLIC

Rohde & Schwartz PA 002 Vhf DF System

The vhf df system PA 002 takes the bearings of aircraft which are equipped for radio communications. The PA 002 consists of the df antenna system AP 002 and the df PP 002.
Besides being used in fixed stations, particularly on small airfields, the PA 002 is suitable for mobile operation.
The received carrier signal is fm resulting from the Doppler effect produced by the simulated antenna rotation. The bearing is obtained by evaluation of the modulated signal. In this way, oblique incidence of the waves causes no bearing error. If the plane of polarisation changes according to the position of the aircraft and in case of reflections, comparatively small errors only result under unfavourable conditions.
The antenna is comprised of six dipoles arranged in a circle with a diameter of 1·46 metres. The dipoles are successively connected to the df receiver (simulated antenna rotation of 170 rps) and lighting protection is provided. The rf section designed as a double-heterodyne receiver has an fm branch for bearing determination and an am branch for processing the information. Any two

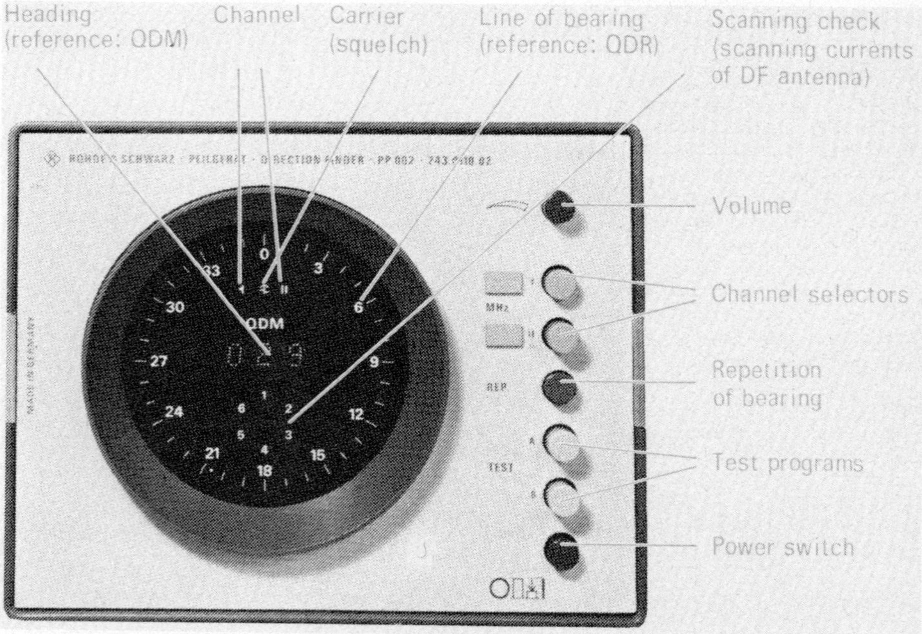

Front panel controls, direction finder PP 002

frequencies in the range 117·5 to 136·5 (144) MHz can be selected by means of push-buttons.
The incoming signals pass through separate input sections, each comprising a preliminary stage, crystal oscillator and first mixer. The receiver frequencies can be changed by replacing the input modules.
The df filter clamps the indication when the bearing is processed. Fluctuations of the bearing signal and modulation effects are largely suppressed, whereas the bearing variations caused by motions of the aircraft are taken into account. When a signal is received the display section lights up and the monitoring channel is switched on. After approximately ½ a second the bearing is determined and as long as a signal comes in, the bearing is updated every second. The last bearing is stored until a new signal is received.
There are two bearing indications available simultaneously. The first is a three-digit indication (1 degree resolution) for the heading of the aircraft to the dc site referred to magnetic north (QDM). The second is a bright dot on the compass with 10 degree divisions, indicating the direction of the aircraft (reference direction QDR). Both indications are updated in intervals of 1 second. The illumination goes out 1½ seconds after the radio communication has been terminated. The stored bearing can again be indicated by pressing the repeat-button. Reception of a new signal does not change the indication as long as the repetition mode is selected.

AP 002 df antenna system

Status
Production

Manufacturer
Rohde & Schwartz, Munich, Federal Republic of Germany

Rohde & Schwarz Vhf/Uhf Doppler Direction Finders

Rohde & Schwartz have developed a new range of Doppler direction finders, designated PA 007, PA 008, PA 009.
The principal differences between the systems PA 007, PA 008 and PA 009 of the new Doppler df range are in their frequency ranges and in the antennae used. The basic equipment of any system consists of df antenna system, front-end equipment and bearing-processing equipment. Through the addition of further units, such as displays, radar combiners and data-transmission equipment, each system can be matched to the user's requirements. Particular flexibility is provided with regard to the number of df channels to be handled by the system. The direction finders of the PA 007 series are characterised by the use of the wide-aperture vhf antenna AP 007. This operates from 100 to 162 megahertz, 32 dipoles arranged in a circle of 6 metres in diameter. The fact that the antenna diameter is significantly greater than the wavelength of the signals ensures that errors due to reflections are kept relatively small.
Along with the PA 007, the PA 008 with the wide-aperture antenna system AP 008 (100 to 162 MHz, 16 dipoles on 3·3 metres circle) is also available for the vhf range. The PA 008 is particularly suited for use as a compact installation in environments presenting few difficulties for df, or as a mobile unit. The basic configuration provides for only one df channel.
Since it presents a relatively low wind resistance the antenna can be installed on the roof of a control tower. The only difference between the uhf system PA 009 and the vhf system PA 007 is in the frequency-determining sections so different receivers and another antenna (the uhf wide-aperture antenna AP 009, 225 to 400 megahertz, 32 dipoles on a 2·4 metre circle) are used.

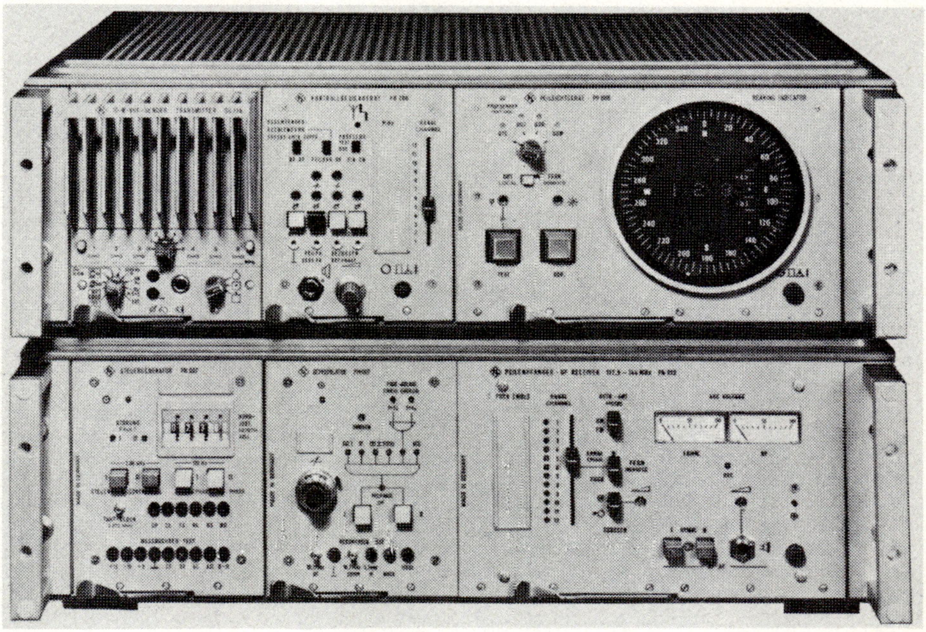

Rohde & Schwartz PA 007 df system

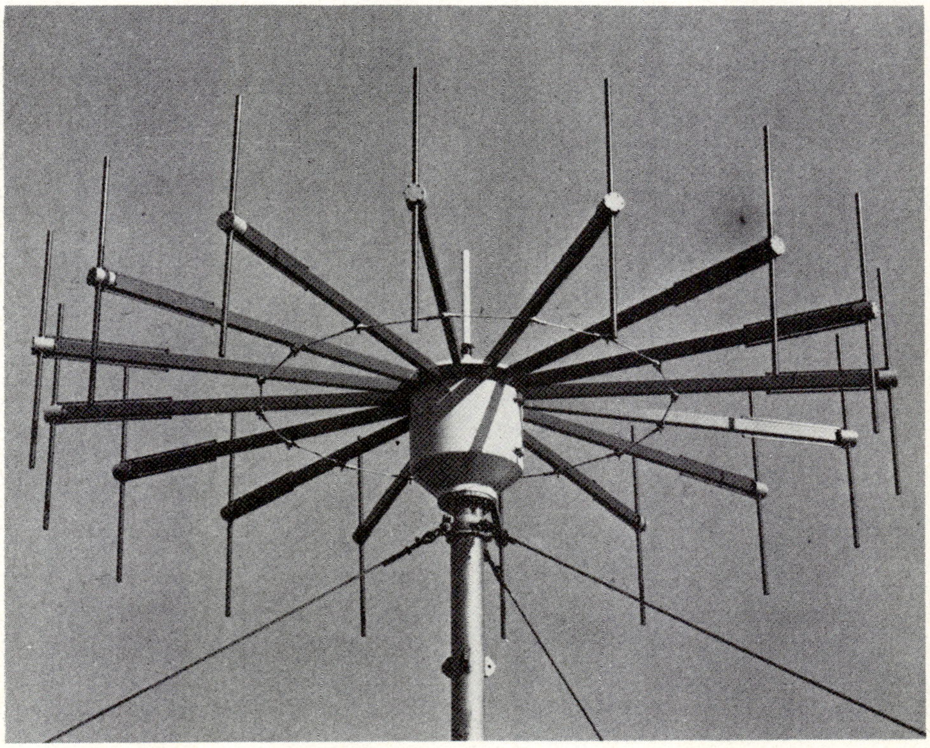

Rohde & Schwartz AP 008 antenna system

Characteristics

MODEL	PA 007	PA 008	PA 009
FREQUENCY RANGE	117·5–144 MHz/ 100–162 MHz/ 146–174 MHz	117·5–144 MHz/ 100–162 MHz	225–400 MHz
CHANNEL SPACING	25 kHz	25 kHz	25 kHz
SYSTEM ERROR	± 0·5°	± 0·5°	± 0·5°
POLARISATION ERROR	± 1°	± 1°	± 1°
SYSTEM SENSITIVITY FOR BEARING FLUCTUATION ± 1°			
vhf	3 μV/m	3 μV/m	3 μV/m
uhf	6 μV/m	6 μV/m	6 μV/m
VERTICAL COVERAGE	Up to 65°	Up to 65°	Up to 65°
ANTENNAE	AP 007	AP 008	AP 009
NUMBER OF DIPOLES	32	16	32
APERTURE DIAMETER	6 m	3·3 m	2·4 m
POWER REQUIREMENTS	110/127/203/220/237 V, ac, 47–63 Hz		

Status
Production

Manufacturer
Rohde & Schwartz, Munich, Federal Republic of Germany

SEL 3100 series Doppler/Vhf Omni-range Equipment

SEL produces modular equipment for both vhf omni-range beacons and Doppler based vhf omni-range systems.

The VOR 3100 can be supplied with 50 or 100 watts output power in a single or dual installation. The ground beacon consists of a cabinet housing the transmitter and the associated drive units, an electronic goniometer, monitoring and testing devices and the power supply. The cabinet can be installed in a shelter whose roof constitutes the counterpoise and carries the antenna. A field detector whose function is to monitor the radiated bearing information is erected on a small mast on the transmitter shelter. All sub-units are of the plug-in type. The carrier transmitter has wide-band characteristics so that no alignment is needed within the limits of the vor frequency range. The modulator for amplitude modulation of the transmitter with a 9960 hertz sub-carrier, the 1020 hertz identity tone, and voice works on the principle of a pulse-width modulated dc converter.

The earlier vor equipment used the mechanical rotation of a dipole or goniometer at 30 revolutions per second to produce the azimuth-dependent (variable) signal. Disadvantages of this method are short service life and high driving power. To meet the requirements of reduced maintenance and high reliability, SEL developed an electronic goniometer, enabling a non-rotating antenna to radiate a field having a rotating directional characteristic.

The antenna is designed as a printed-circuit board. The board comprises the omni-directional radiator, which radiates the carrier frequency amplitude-modulated at 9960 hertz, and the two 90 degrees crossed dipoles, which are fed from the electronic goniometer and produce the rotating figure-of-eight pattern. The use of a two-element antenna is stated to increase the field strength by up to 5 dB at low elevation angles.

The beacon is fed from a float-charged battery and a built-in battery charger is provided for that purpose. If the equipment fails, alarm is given and in dual installations the standby transmitter is put into service.

In the shelter the emergency power supply is housed in a compartment isolated from the main equipment by gas-tight seals. Sufficient space is provided in the equipment room to house the additional racks required to extend the system at a later date by the installation of a distance measurement equipment or tactical air navigation transmitter.

SEL has developed a mobile vor for use in cases where a vor ground station is needed quickly or at a temporary basis, or where the erection of a permanent installation cannot be justified. It comprises a standard vor system (single or dual installation), an extensible standard single element antenna and fold-out antenna counterpoise.

The mobile vor is housed in a special cabin mounted on a standard trailer chassis. A diesel power unit and battery ensure uninterrupted operation when no connection to mains supply is available or in the event of mains power failure. The battery and diesel generator will provide an emergency power supply for up to 52 hours without refuelling.

The system offers all the facilities of the permanent vor beacon and complies with Annex 10 of the ICAO recommendations.

In cases where (because of unfavourable siting conditions) a conventional vor cannot provide the required degree of accuracy in navigational information, it is possible that the use of a dvor system will provide a solution.

Large topographical obstacles such as buildings and high terrain can reflect the transmitted signal. Dependent on the direction of propagation, the result is a reflected signal superimposed on the transmitted signal. The resultant error in displayed information can be almost completely eliminated by using the Doppler principle. The information supplied by dvor installations has a high degree of accuracy in difficult terrain and using a dvor installation the effect of reflected signals is reduced by a factor of about 10.

The DVOR 3100 is a medium-range navigational aid recommended by ICAO. The dual system consists of one cabinet with transmitting equipment of 50 watts output power each, one cabinet accommodating the antenna switching unit, the antenna system including the counterpoise and a remote control facility. The latter allows remote control of the transmitter from the control tower of the airport via a two-wire line. All remote on, off and changeover switching is initiated by depressing the appropriate keys. All the essential DVOR 3100 functions are checked on the monitoring facilities. A switchable microphone serves for transmitter modulation. Communication between control tower and maintenance staff can be established by means of an additional telephone set.

The carrier frequency is amplitude modulated with 30 hertz, identity tone and voice, and the identity keying is electronic. A monitor in the transmitter cabinet effects automatic changeover to the standby transmitter in the event of a failure of the operating transmitter, whereas single systems are simply switched off. The radiated information is received by a field detector (Yagi antenna) positioned at a distance of at least 200 metres from the centre antenna and fed to the monitor.

Electronic switches for the antenna system and their control circuits are accommodated in the antenna switching unit.

The antenna system comprises 40 omni-directional antennas with horizontally polarised fields: one centre antenna radiating the carrier (reference signal) and 39 antennae equally spaced on a circle of 13·5 metres diameter radiating the side-bands (azimuth-dependent-signal). All antennae have a height of approximately 1·40 metres and they are mounted on a counterpoise having a diameter of 30 metres. The height of the counterpoise above ground can be 3, 5, 7 or 10 metres. In special cases the antenna system can be installed on top of a higher steel structure (tower DVOR 3100).

The power consumption of the system allows feeding from a float-charged battery. A power supply is provided for that purpose. All transmitting equipment including the antenna switching unit is accommodated in a shelter underneath the counterpoise.

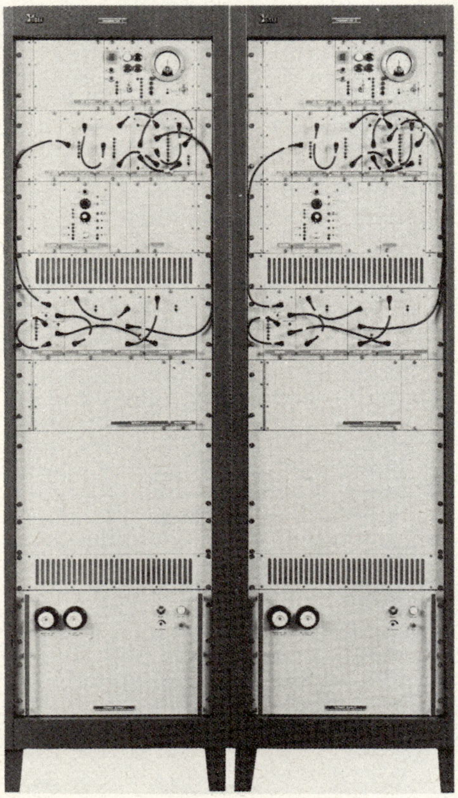

SEL VOR 3100 dual 50-watt installation

SEL DVOR 3100 antennae with tower installation

SEL mobile VOR in transit

Characteristics

TYPE	VOR 3100	DVOR 3100
TRANSMITTER		
FREQUENCY RANGE	108–118 MHz ± 0·002%	108–118 MHz ± 0·002%
OUTPUT POWER	50 W	50 W
MODULATION AM		
reference	30 Hz ± 1%	30 Hz ± 0·2%
identification	1020 Hz ± 5%	1020 Hz ± 5%
voice	300–3000 Hz ± 3 dB	300–3000 Hz ± 3 dB
HARMONIC RADIATION	>m 65 dB	>m 65 dB
BEACON ACCURACY	Better than ± 2°	Better than ± 1°
BEARING STABILITY	± 1°	± 0·5°
ANTENNA	Loop antenna with crossed dipoles	Alford loop antenna
POLARISATION	Horizontal	Horizontal
COUNTERPOISE		
diameter		30 m
height		3, 5, 7, or 10 m
POWER REQUIREMENTS	230/115 V ac, single phase 45–62 Hz or 40 V dc, 26A	230/115 V ac, single phase 48–62 Hz or 40 V dc, 25A
ENVIRONMENTAL CONDITIONS		
external temperature	–40 to +70°C	–40 to +70°C
windspeed	100 mph (160 km/h)	
ice/snow loading	1·2 m (3 cm) causes negligible errors	

Status

Production. SEL has supplied over 350 vor installations to 62 countries and over 120 dvor systems to 26 countries

Manufacturer

Standard Elektrik Lorenz AG, Stuttgart, Federal Republic of Germany

SEL 3000 Series Instrument Landing Systems

SEL produce a family of instrument landing systems which meet the operating requirements for landings under ICAO category I, II and III conditions.

An ils comprises localiser, glide-slope, middle marker and outer marker beacons. Tower system equipment enables the system to be remotely-controlled and monitored from the control tower.

In the localiser and glide slope ils transmitters, course position, course width, modulation level, field strength and identification signal are continuously monitored. All monitors are integrated with the system and provide an alarm in the event of a malfunction. The monitors operate on the majority decision principle. The alarm tolerances are adjustable and comply with the requirements of category I, II or III operations as defined in ICAO, Annex 10. Localiser, glide slope, and marker beacon are supplied as a dual installation. One transmitter is always in operation, the second serving as a standby which automatically comes into operation if the main transmitter fails. For maintenance purposes manual switch-over can be carried out at the transmitter rack.

The localiser antenna is normally installed at the extended centreline within the obstacle clearance regulated by ICAO.

The distance of the glide-slope antenna position from threshold depends on the glide angle while the lateral spacing is regulated by the obstacle clearance. A Yagi antenna comprising four elements and a monitoring pick-up is used for each marker beacon transmitter. The middle and outer marker beacons complete with antenna are positioned approximately 1000 and 7000 metres from the runway threshold on the extended centreline. An additional beacon can be installed and used as an airfield boundary marker positioned approximately 60 metres from the threshold on runway centre-line. The marker antennae are intended to be installed on top of the roof of the transmitter building.

In the event of a mains failure, an emergency power supply (batteries) provides uninterrupted operation for seven to ten hours.

SEL supplies a prefabricated shelter to house the equipment, either a single or dual installation, including batteries and provides adequate space for other equipment.

To solve problems with severe sitting conditions a two-frequency localiser (capture effect) is available. For glide slope installations with irregular reflection areas and positive slope a capture-effect system (M type) can be supplied. Airports with a limited reflection zone in front of the glide slope array and negative slope can be served with a side-band reference (B type) installation.

To convert from a single-frequency into a dual-frequency the LOC 3000 localiser transmitter is complemented by the dual-frequency auxiliary rack LOC 2F and the 12 element antenna is extended to 13 elements.

SEL GS 3000-M glide slope antenna

SEL LOC 3000-2F localiser antenna

SEL LOC 3000 localiser antenna

The glide-slope transmitter GS 3000 is extended using the auxiliary rack M-kit and an antenna comprising three elements.

The localiser LOC 3000-2F and glide-slope transmitter GS 3000-M radiate a highly directional beam thereby eliminating interference caused by reflected signals. In addition, each transmitter radiates a clearance signal which provides the flight crew with adequate information if the aircraft is outside the approach area or below the glide-slope.

The carrier frequencies of the clearance signal and the track information signal are separated by several kilohertz but remain well within the bandwidth of the aircraft receiver. Using the capture-effect, the receiver discriminates the two signals, accepting the stronger and suppressing the weaker signal.

The SEL GS 3000 glide-slope installation type B is recommended where an airport has local topographical conditions which do not comply with the recommendations governing the installation of a standard ils glide-slope transmitter (eg antenna reflection areas which are too small and those airports with inclines).

Whereas the standard glide-slope installation operates on the null reference principle, the type B ils utilises a side-band reference system. A conversion kit for changing a standard glide-slope into a type B is available which incorporates a feed network for the side-band referencing process in the Number 2 transmitter rack of the dual installation. Except for the arrangement of the dipole arrays, the type B unit antennae do not differ from those of the ils GS 3000 null-reference. The application of the side-band reference principle can achieve a satisfactory glide-slope alignment in difficult terrain.

The category I ils equipment is suitable for regional airports with a high traffic frequency but whose use of an ils suitable for all-weather landings in ICAO category II or III is unnecessary or precluded by operational or financial considerations. Ils (category I) is a fully-solid-state navaid and basically comprises the same units as the category II/III installations but with a reduced number of plug-in units and a smaller antenna system. The basic category I ils system can be extended to a higher category system, and if necessary, to the high performance category III system.

Characteristics

LOCALISER: LOC 3000
FREQUENCY: 108–112 MHz ± 0·002%
OUTPUT: 40 W
MODULATION FREQUENCIES: 90 and 150 Hz
COURSE SECTOR WIDTH: 2–6° adjustable
ANTENNA TYPE: 12 dipoles in front of reflector
POLARISATION: horizontal
GLIDE SLOPE GS 300: > 0·02 of path angle θ
FREQUENCY: 328–336 MHz
OUTPUT: 15 W adjustable
MODULATION FREQUENCIES: 90 and 150 Hz
GLIDE PATH ANGLE (θ): 2–4° adjustable
ANTENNA TYPE: 2 × 2 rows of 4 dipoles
MARKER BEACON: M 3000
FREQUENCY: 75 MHz ± 0·005%
OUTPUT: 0·5–3 W adjustable
IDENTIFICATION FREQUENCIES: 400 Hz, 1300 Hz, 3000 Hz ± 2·5%
ANTENNA TYPE: four element Yagi
POLARISATION: horizontal

Status

Production. SEL has installed over 200 ils systems in 26 countries

Manufacturer

Standard Elektrik Lorenz AG, Stuttgart, Federal Republic of Germany

ITALY

Face Standard FSD-10/-15 Distance Measuring Equipment

Two models of dme are offered by Face Standard which combined with the two types of Tacan beacons form a family of equipment with standard modules. The cabinet, rf change-over control unit, monitors, pulse processor, low-level transmitter and power supplies are the same on all beacons and distinction exists only for rf high power amplifiers and associated high voltage power supplies.

The FSD-10 is a dme ground beacon, particularly designed for combined operation with any type of ils and mls facility or with vor or dvor equipment for limited area coverage. Its rf output power is 150 watts, and its accuracy is ± 30 metres. For ils combined operations a sectorial antenna (FAN-88) is available having the same ils antenna coverage. The FSD-10 can be installed inside an existing glide-slope or localiser shelter.

The FSD-15 is a 1 kilowatt dme ground beacon, suitable for en route navigation. It is basically an FSD-10 with a fully solid-state 1·6 kilowatt amplifier. This is designed in accordance with a fail-soft technique since it consists of eight amplifiers in parallel. No blower or other moving parts are used. The FSD-15 can be co-located with any type of existing vor or dvor equipment to give distance and azimuth information to the aircraft in accordance with ICAO specifications.

FAN-86 is a 9 dB gain dme antenna. The horizontal radiation pattern is circular and in the vertical plane the maximum radiation is at 3 ± 0·5 degrees above the horizon in free space. The beamwidth in the vertical plane is 9 degrees at half-power points.

The antenna covers the whole dme frequency band. With 1 dB antenna feeder loss, the effective radiated power is 28 dBW in case of operation with FSD-10 and 38 dBW with FSD-15.

FAN-87 is a 12 dB gain dme omni-directional antenna, which provides 31 dBW effective radiated power in case of operation with FSD-10 and 41 dBW with FSD-15. The antenna covers the whole dme frequency band. Main radiation is 3 ± 0·5 degrees over the horizon in free space. The vertical radiation pattern has a lobe of energy not less than 5 degrees at half-power points.

FAN-88 is a 15 dB gain sectorial dme antenna. The coverage sector is ± 35 degrees, and can be centred on the runway, reducing the effects of obstacles and increasing signal-to-noise and signal-to-echo ratio. It is suitable for heavy traffic airports and when operating with FSD-10 the effective radiated power is 33 dBW and 44 dBW with FSD-15.

Where remote unattended operation is needed the remote-control model FRC-4/7 is offered with four controls, four independent indications, and telephone and acoustical check of the identification signal. An alarm buzzer is provided to indicate a fault. This ancillary consists of a local plug-in module (FRC-4L) plus a remote unit (FRC-4R). An auxiliary indicator module (FRC-4RA) is optionally available for control installation. The Tx signal to and from the remote station is by a radio link or telephone.

Face Standard FSD-10 dme equipment

Face Standard FSD-15 dme equipment

Status

Both the dme and the Tacan beacons are new developments from Face Standard who have been producing beacon systems for more than 20 years and have installed 50 Tacan and 300 dme beacons throughout the world

Manufacturer

Industrie Face Standard SpA, Milan, Italy

Characteristics

TRANSMITTER	FSD-15	FSD-10
PEAK POWER	1 kW	100 W
SPECTRUM	ICAO Annex 10 — dme (n)	
RECEIVER		
sensitivity	-89 dBm	-79 dBm
adjacent channel rejection	80 dB	80 dB
SYSTEM		
TRANSPONDER CAPACITY	200 interrogators	200 interrogators
CHANNELS	126X, 126Y	126X, 126Y
REPLY DELAY	50 μs (X mode) 56 μs (Y mode)	50 μs (X mode) 56 μs (Y mode)
BEACON ACCURACY	30–80 m	30 m
ANTENNA		
EFFECTIVE RADIATED POWER		
FAN-86	38 dBW	28 dBW
FAN-87	41 dBW	31 dBW
FAN-88	44 dBW	34 dBW
PRIMARY POWER	220 V (± 15%) and no-break battery operation	
ENVIRONMENTAL CONDITIONS		
temperature (outside)	-30 to +60°C	-30 to +60°C
relative humidity	up to 100%	up to 100%
ice load	12 mm	12 mm
wind velocity	150 km/h	150 km/h

Face Standard FTA-10/-10S Tacan Beacon

Two models of Tacan ground equipment are available, FRA-10 and FTA-10S using similar components to the dme beacons except for rp high-power amplifiers and associated high voltage power supplies.

The FTA-10 Tacan is a 3·5 kilowatt ground beacon, solid-state with the exception of the rf power amplifier using four low-cost, long-life planar triodes. The beacon which provides distance and azimuth information to aircraft equipped with airborne Tacan, can be associated with any of the existing vor or dvor when rho-theta navigation service has to be provided to both commercial and military aircraft. Monitor interrogation rate has been reduced in order to route maximum reply percentage to aircraft interrogation and to permit servicing up to 200 aircraft. The beacon is normally associated with the electronic antenna model FAN-29.

The FTA-10S is a 1 kilowatt fully solid-state Tacan ground beacon. The 1 kilowatt output is achieved by eight amplifiers in a fail-soft configuration. The beacon can be accommodated in a transportable shelter. Measuring 2 metres wide, 2 metres deep and 1·8 metres high, it can be mounted on a truck and after arrival on site, the equipment can be put into operation after carrying out minor adjustments. The FTA-10S is offered in a special version which can be installed in a jeep when a rapid change of sites is needed. The Tacan antenna model FAN-29 is an electronically-scanned antenna with a 4 dB gain. A number of parasite dipoles, concentrically disposed around a central radiator, are continuously commutated simulating the rotation of a rotating antenna. Solid-state in construction and light and small in size, the FAN-29 antenna is suitable for installation on fixed and portable stations and incorporates performance monitoring equipment. The Tacan mechanical antenna model FAN-4 is suitable for fixed Tacan stations and because of its high gain, for en route navigation. It can operate with any Tacan beacon. A typical combination is antenna model FAN-4 and Tacan beacon model FTA-10. In order to give satisfactory long-range coverage to aircraft flying at high altitudes, the maximum energy is radiated at about 5 degrees above the horizon. Below the horizon there is a sharp energy fall-off which avoids reflections. The antenna system model FAN-4 includes a radiation unit, a separate control cabinet and a monitor pick-up probe. Where remote unattended operation is needed the remote-control model FRC-4/7 is offered with four controls, four independent indications and telephone and acoustical check of the identification signal. An alarm buzzer is provided to indicate faults. This ancillary consists of a local plug-in module (FRC-4L) plus a remote unit (FRC-4R). An auxiliary indicator, (model FRC-4RA), is also optionally available for control installation. The Tx signal to and from the remote station is by a radio link or telephone.

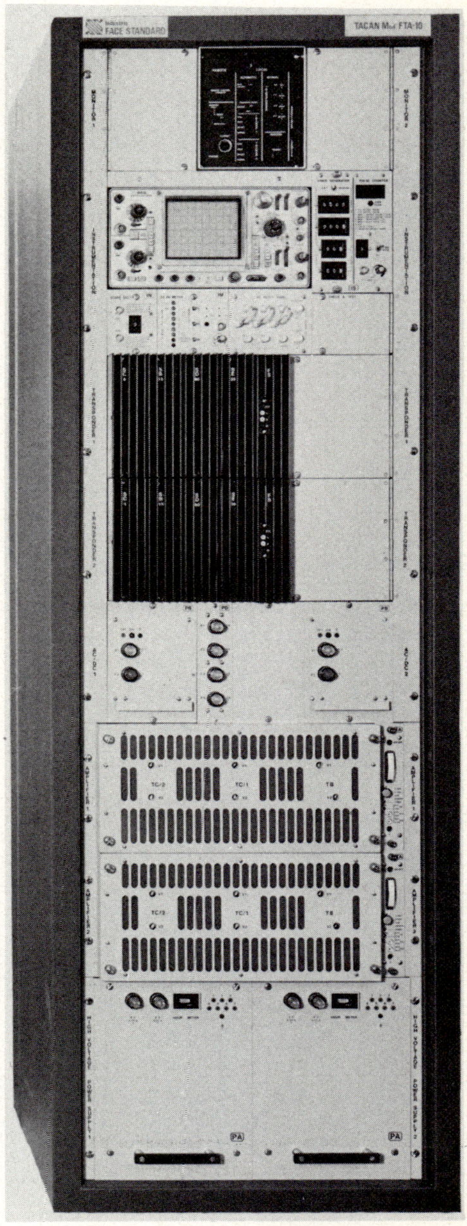

Face Standard FTA-10 Tacan Beacon

Characteristics

TRANSMITTER	FTA-10	FTA-10S
PEAK POWER	3·5 kW	1 kW (typical) 800 W (minimum)
SPECTRUM	Stanag 5034, MIL STD 291-B	
RECEIVER		
sensitivity	-89 dBm	-92 dBm
ADJACENT CHANNELS REJECTION	80 dB	80 dB
SYSTEM		
TRANSPONDER CAPACITY	Bearing to unlimited number of aircraft	
CHANNELS	126 X, 126 Y	126 X, 126 Y
REPLY DELAY	50 μs (X mode) 56 μs (Y mode)	50 μs (X mode) 56 μs (Y mode)
BEACON ACCURACY		
bearing	0·67°	0·67°
range	60–80 m	30–80 m
ANTENNA		
EFFECTIVE RADIATED POWER	38 dBW (minimum)	33 dBW (typical)
PRIMARY POWER	220 V (± 15% with voltage regulator)	220 V (± 15% with voltage regulator)
ENVIRONMENTAL CONDITIONS		
temperature (outside)	-30° to +60°C	-30° to +60°C
relative humidity	Up to 100%	Up to 100%
ice load	12 mm	12 mm
wind velocity	150 km/h	150 km/h

Status

Both the dme and the Tacan beacons are recent developments from Face Standard who have installed 50 Tacan and 300 dme beacons world-wide

Manufacturer

Industrie Face Standard SpA, Milan, Italy

JAPAN

NEC Distance Measuring Equipment

Two systems are available from NEC, NNG-560 rated at 100 watts and designed for terminal operations, and NNG-570 producing 1 kilowatt power for en route navigation. Both are fully solid-state and use 1·6 kilowatt power amplifier which covers the whole assigned dme frequency band without any adjustment. Other features of this equipment are: modulation pulse shaping for minimising the spectrum spread, time-delay circuitry, modular construction with plug-in card modules, and built-in test equipment with a digital display and self-diagnostic capability. Digital remote-control is available via a pair of telephone lines. The antenna is omni-directional for en route navigation and uni-directional for terminal use. Optional equipment available includes: improved echo suppression facility and co-location installations with vor or ils systems.

Status
Production. The equipment was developed in 1978 and meets ICAO standards. NEC has installed the majority of dme systems in Japan and has equipment in operation in other countries in the Far East, South America and Africa

Manufacturer
Nippon Electric Co Ltd, Tokyo, Japan

NEC NNG-500 series Vhf Omni-directional Range Beacon

NEC produce three vor beacons in this series, NNG-500 a 50 watt unit, NNG-501 rated at 100 watts and NNG-502 producing 200 watts.
Each vor beacon is of similar modular construction and design and incorporates a micro-processor and other micro-electronics controlling the goniometer and fm reference generator. The beacon is fitted in a single cabinet for dual configuration.

Status
Production. The beacon meets the latest ICAO standards for 50 Hertz separation. NEC has installed a number of vor/dme and vortac systems internationally

Manufacturer
Nippon Electric Co Ltd, Tokyo, Japan

NEC vor/dme antennae with shelters for equipment (left) and for engine generator (right)

NEC NNG-690 Instrument Landing System

NEC have developed ils in various configurations to suit different airports. In cases of airports with difficult terrain, NEC use a dual-frequency localiser with an M-array/ capture glide-path antenna. The standard configuration for the NNG-690 ils comprises localiser, glide-path, marker beacon or terminal dme and remote-control and monitoring equipment. The system is completely solid-state and features a digital modulator, directional antenna, fail-safe monitoring system with remote-control, built-in test equipment and uninterrupted operation with a no-break power supply.

Status
Production. The NNG-690 satisfies the category III requirements of ICAO. NEC claim to have installed the majority of ils systems in Japan and some overseas

Manufacturer
Nippon Electric Co Ltd, Tokyo, Japan

NEC ils dual-frequency localiser antenna

NEC ils null reference glide-path antenna

NEC NNG-860/-970 Tacan System

The NNG-970 Tacan developed by NEC incorporates a 5 kilowatt ground beacon which is solid-state with the exception of the final power amplifier tube. It performs over the full Tacan assigned frequency range in X and Y modes, and uses advanced digital micro-electronic technology including an echo suppressor, time-delay circuitry, identification generator and keyer and a frequency synthesizer for both transponder and monitor signal generator which can be preset. A micro-processor provides overall monitoring of the Tacan system with automatic data processing for measurement of receiver sensitivity, system delay and other characteristics and an alpha-numeric plasma display to facilitate maintenance. Digital remote-control is achieved with one pair of telephone lines. The system can also be used as a vortac facility.

Status
NEC began production of Tacan equipment in 1960. A number of conventional Tacan ground beacon systems have been supplied to the Japan Civil Aviation Bureau and the Japan Defence Agency. In 1978 NEC developed the NNG-970 system designed to comply with the ICAO standards and military specifications

Manufacturer
Nippon Electric Co. Ltd, Tokyo, Japan

NEC vortac station at Kyusha Island

UNITED KINGDOM

Fernau Avionics Vhf/uhf Direction Finder System

Fernau Avionics direction finder system is a solid-state wide-aperture df system, operating in either or both the vhf (100–162 MHz) and uhf (225–400 MHz) bands, designed to meet present and future requirements of atc for military and civil aviation. The system employs an electronically-scanned ring of 18 aerials giving a pseudo-Doppler-modulated output which is compared with an unmodulated output from an auxiliary aerial. The signals are frequency-shifted and fed to a standard communications receiver then passed over telephone lines to the resolver. Here the bearing is measured by a hybrid analogue-digital computer consisting of three small printed circuit cards. The bearing information is produced in the form of dc voltages proportional to the sine and cosine of the angle for presentation on radar ppi or crt displays; and in a digital form for numerical displays, computer interface and transmission over telegraph lines. The modular construction gives high system flexibility permitting expansion from simple vhf systems to complex vhf/uhf systems with 16 or more independent display channels and 20 or more separate receivers. The equipment has an automatic storage facility in which the last displayed bearing is stored automatically and may be recalled by the operation. A two oscillator system gives the operator a full integrity check. A calibrator, multi-channel test oscillator and other test facilities are supplied with the equipment.

Characteristics

FREQUENCY RANGES
vhf: 100–162 MHz
uhf: 225–400 MHz
CHANNEL SPACING
vhf: 50 or 25 k Hz
uhf: 50 k Hz
INSTRUMENTAL ERROR: ± 0·5°, plus additional ± 0·5° for ± 25°C indoor temperature change
POLARISATION ERROR
vhf: ± 1°
uhf: ± 0·5°
SYSTEM SENSITIVITY: bearing ± 3° random fluctuation with field strength 5 μV/m
RESPONSE RATE: less than 0·5 s with bearing updated every 1·1 s
VERTICAL COVERAGE: up to 60°
POWER REQUIREMENTS: 200–240 V, 45–65 Hz
ENVIRONMENTAL CONDITIONS
external temperature: –40 to +55°C
aerial wind loading: 200 km/h
internal temperature: 0 to +55°C
relative humidity: up to 95%

Fernau Avionics df system

Status

Production. The unit is installed at Manchester International Airport, England

Manufacturer

Fernau Avionics Ltd, St Albans, Hertfordshire, England

IAL ABX 400 Non-directional Beacon

IAL design, manufacture and supply the ABX 400, a compact, medium-range aeronautical navigation beacon, capable of continuous service at remote sites. The unit is fully solid-state and has a power output of 100 watts (400 watts peak envelope power). An automatic antenna tuning unit complements the beacon, compensating for climatic changes in antenna characteristics and ensuring optimum transmission. Also available is a beacon monitor receiver, fully equipped with fail-safe alarm facilities.

Status

Production

Manufacturer

IAL, Southall, Middlesex, England

MEL MADGE Microwave Guidance Equipment

MADGE is a microwave landing system which has been developed to meet the all weather requirements of military and civil aviation operations. The modular system has been designed to operate in all classes of aircraft at a variety of permanent airfields and heliports and temporary landing sites. Originally designed to the specific requirements of NATO and the United Kingdom Ministry of Defence the system has been further adapted to encompass use on board ships, oil exploration rigs and oil production platforms. Based on a high speed, low-duty cycle air/ground, ground/air digital data link, the system provides landing guidance, identification, navigation and atc capabilities including talk down.

The portable approach-aid consists of co-operating aircraft and ground elements. It operates in the frequency band 5 to 5·25 GHz, and provides guidance on a number of selectable approach paths down to a decision height of 30 metres when the runway visual range is 400 metres. Lower limits are possible for VTOL operations. The military system is being certificated for civil use and meets the British Helicopter Advisory Board's Helicopter Approach Aid Requirement.

The aircraft equipment consists of three units, a controller (interrogator set) and transmitter/receiver (interrogator set) coupled to an antenna system. Normally two antennae are used with automatic selection.

The logic unit interfaces with standard ils displays. Simple interface can also be made with auto pilots and flight control systems and displays. The aircraft equipment operates from primary 28 volt dc supplies.

The ground equipment consists of an antenna elevation, an antenna azimuth and a transponder set. A second antenna azimuth is required for overshoot/missed approach guidance. All units operate from 24 volt dc supplies and are man portable and suitable for an airdrop. The ground equipment may be interfaced with atc and air defence systems to provide identification, position, track speed and height of all aircraft using the system. The data provided by the co-operating aircraft and ground elements has application to navigation and communications techniques.

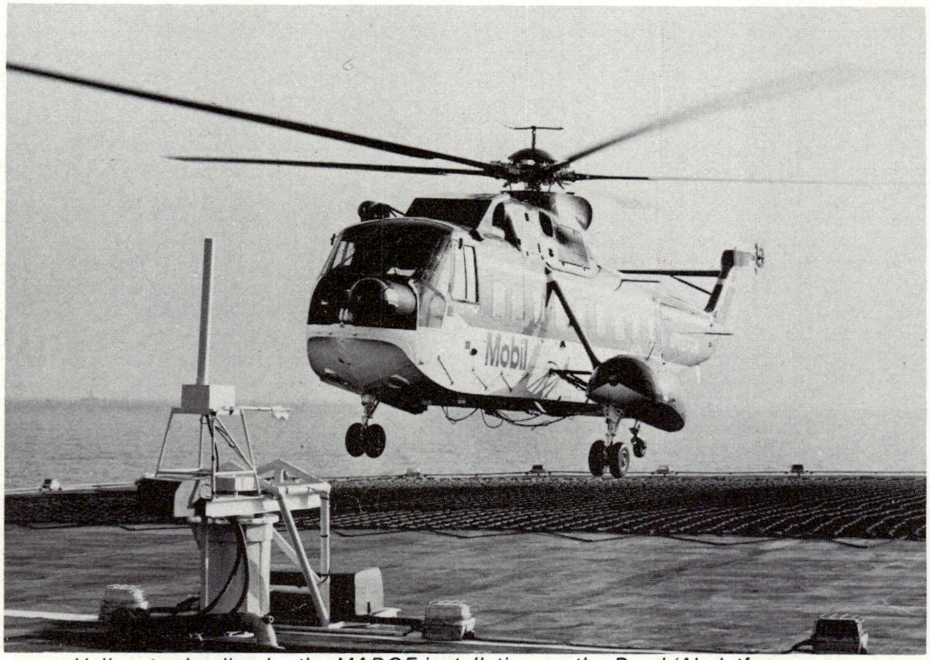

Helicopter landing by the MADGE installation on the Beryl 'A' platform

MADGE is suitable for helicopters and all types of fixed-wing aircraft. It provides azimuth coverage over 90 degrees both on the approach and overshoot. Elevation (glide-slope) coverage is provided up to a 25 degree range measurement through 360 degrees of azimuth. Selection of the approach path in azimuth (localiser) and elevation (glide-slope) is made by the pilot. The system will handle over 100 aircraft simultaneously. In addition ground equipments can operate within close proximity on the same frequency without causing mutual interference.

The ground equipment can be set up by two people in 15 minutes and the equipment can be carried in a light vehicle (eg Land-Rover) or in a single container which can be air-dropped. Built-in automatic test equipment and tilt monitoring mechanisms permit the unattended operation of the ground equipment with monitoring at a remote central position.
In the aircraft the pilot is given a warning when the aircraft is too low on the approach or when the aircraft flies to the overshoot sector on a missed approach. Individual failure warning flags are incorporated in the aircraft for azimuth, elevation and range data and the pilot is supplied with a serviceability state of the ground station when the aircraft is within dme range.
The system is resistant to electronic countermeasures and jamming will not result in erroneous information being used in the aircraft guidance system, since MADGE is passive except for the short periods of actual interrogation by an aircraft which causes the transponder to switch the ground station from the standby to the operational mode.
For area navigation, display facilities permit the MADGE range and azimuth outputs to be used in the vicinity of airfields and for tac/nav. The outputs include range rate/velocity information. This feature permits interface with automatic flight control systems, tac/nav systems and provides key data for pilot interpreted, and automatic operation of transition from horizontal to vertical flight in VTOL aircraft.
MADGE can provide lateral offset approach paths as well as angular offset approach paths, for an approach to a ship, offshore rig or platform and other military applications such as airdrop supplying.

Characteristics

EQUIPMENT: transmitter
FREQUENCY: 5·0–5·25 GHz
PEAK POWER: 200 W
RECEIVER SENSITIVITY: –86 dBM
DYNAMIC RANGE: 65 dB
BAND-WIDTH: 3·0 MHz
EQUIPMENT: system
FREQUENCY CHANNELS
4 air/ground (receiver)
3 air/ground (transmitter)
AZIMUTH: over 90° on approach and overshoot
ELEVATION (GLIDE SLOPE): up to 25°
RANGE MEASUREMENT: through 360°
RANGE: 30 nm (55·5 km)
PRIMARY POWER: 24 V dc
ENVIRONMENTAL CONDITIONS
ground equipment temperature: –46 to + 44°C
relative humidity: 20–100%

Status

Production. MADGE has been selected by NATO as the recommended Tactical Portable Approach Aid and is installed in Royal Naval ships operating Sea Harrier aircraft. In the commercial application, MADGE is operating in the North Sea on the Mobil Beryl 'A' offshore oil platform

Manufacturer

MEL, Crawley, West Sussex, England

Plessey Doppler Vhf Omni-Range System

Plessey Doppler vor antenna

The Plessey PLAN 50 Doppler vhf Omni-Range equipment enables all aircraft fitted with vor receivers to obtain bearing and homing information for en route and terminal navigation. The service supplied by this equipment fully meets the requirements of ICAO.
PLAN 50 operates in the 108 to 118 MHz vor-band and will co-locate and co-operate with dme. The dvor is capable of remote unattended operation. It is a dual system with automatic switch-over and battery stand-by, modular in construction and can be supplied in a number of configurations depending upon user requirements. It is supplied with remote- and local-control and monitoring, the amount of which depends on the system configuration and operational requirements.
The PLAN 50 comprises an all solid-state transmitter system feeding an electronic distributor, which energises the aerial system to provide the vor signal. The aerial consists of a ring of 50 omni-directional Alford loops, mounted on a counterpoise with a single Alford loop mounted in the centre for carrier transmission.
The distributor feeds the output from the side-band transmitters sequentially to the ring of aerials to produce the side-band components of the vor signal. Switching is by logic units which are driven from a high-accuracy crystal clock. The switching elements are high-frequency PIN diodes.
The transmitters are wide-band, convection-cooled units of modular design. The carrier and two-sideband transmitters are crystal frequency-controlled and phase-locked together to achieve the high stability required to produce the correct signal.
The carrier transmitter is modulated with identification tones and with voice modulation. A 30 Hertz reference signal is frequency modulated directly on to the carrier. Both external and internal monitoring are provided. The external monitoring checks the bearing accuracy and modulation content and the internal monitoring checks performance parameters as required.

Characteristics

FREQUENCY: 108–118 MHz ± 0·002%
OUTPUT POWER: 25/100 W
MODULATION-AM
reference: 30 Hz ± 0·5%
identification: 1020 Hz ± 50 Hz
voice: 300–3000 Hz
MONITORS: external field and internal parameter
AERIALS: 51 Alford loops
POLARISATION: horizontal
COUNTERPOISE DIMENSIONS
diameter: 30 m
approximate height: 4 m
POWER REQUIREMENTS
100–125 V ac/200–250 V ac
45–60 Hz 2 kVA
UNIT DIMENSIONS
width: 1·11 m
depth: 0·72 m
height: 1·83 m

Status

Production

Manufacturer

Plessey Radar, Weybridge, Surrey, England

Plessey Model 1020 Distance Measuring Equipment

The Model 1020 dme has been designed to meet FAA and ICAO Annex 10 requirements and has been engineered to minimise the effect of reflections at difficult sites. Except for the 1000 watt amplifiers, the 1020 is fully solid-state. It is available in 100 and 1000 watt versions with various options, such as single or dual transponders and monitors, the provision of single or dual keyers and of a specially designed test set. A monitor, which serves as a test signal generator for aligning and testing the transponder, is also provided. Whatever equipment configuration is selected, it can be housed in a single standard cabinet.
Omni-directional and uni-directional antennae are available and these can be co-located with vor and ils installations.

Characteristics

EQUIPMENT: transmitter
FREQUENCY: 962–1213 MHz
OUTPUT POWER: 100/1000 W
SPECTRUM: ICAO annex 10–dme(n)
EQUIPMENT: receiver
FREQUENCY STABILITY: ± 60 kHz
SENSITIVITY: 70% at –126 dBW
ADJACENT CHANNEL REJECTION: 80 dB
EQUIPMENT: system
CHANNELS: 252 (X and Y)
TIME DELAY: 50 ± 0·25 μs
EQUIPMENT: monitoring system
SIGNAL ALARM CONDITIONS
reply delay error exceeds 1 μs
reply efficiency less than 65 ± 5%
effective radiated power decreased by 3 dB
transmit pulse count falls below 850 pps
transmit pulse spacing error exceeds 1 μs
identification signal is not transmitted every 75 s
POWER REQUIREMENTS: 105–130 V or 210–260 V ac, 50–60 Hz
UNIT DIMENSIONS
width: 0·53 m
depth: 0·65 m
height: 1·7 m
WEIGHT: 125 kg

Status
Production

Manufacturer
Plessey Radar, Weybridge, Surrey, England

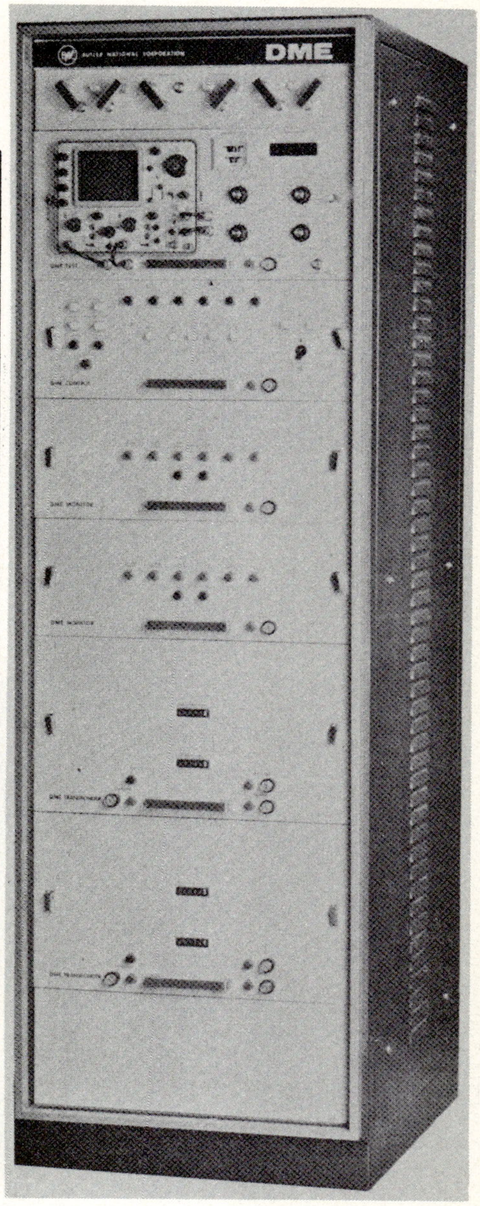

Plessey dme cabinet fitted with dual transponders, dual monitors and test set

Plessey PLAN 17/18 Instrument Landing System

The Plessey PLAN 17/18 is a system for category I and category II ils in accordance with ICAO standards. The system provides guidance information for suitably equipped aircraft in the landing approach phase, using in the azimuth plane a localiser Plessey PLAN 17, and in the vertical plane a glide-path Plessey PLAN 18. Marker beacons for indicating discrete ICAO promulgated ranges are also available.

The PLAN 17 localiser is a fully solid-state equipment which can be supplied in either single or dual-channel configurations. The common path electronic modulator gives high stability due to the locked phase relationship of the 90 and 150 Hertz signals. A choice of aerial systems is available, either in single or double aperture forms and the Plessey quadrature clearance technique can be incorporated to minimise effects of siting limitations.

The PLAN 18 glide-path features signal generation circuits similar to those used in the PLAN 17 localiser, and electronic modulation and strip-line aerial distribution. Three types of aerial arrays are available: null reference, side-band reference and the type M. The Plessey switched clearance technique is incorporated in type M arrays to provide below-path clearance.

The system performs integral monitoring of the radiated signals. Field monitors are available if desired. Remote-control units are provided to give information and control of the mode of operation at a control tower or other remote point.

Plessey PLAN 17 localiser array

Plessey PLAN 18 glide-path antenna

Characteristics

EQUIPMENT	PLAN 17	PLAN 18
RANGE	25 nm (46·25 km) within ±10° from front course line 17 nm (31·45 km) between 10 and 25° from front course line	10 nm (18·5 km) between 0·45 θ and 1·75 θ within ± 8° front course sector
POWER OUTPUT	25 W	8 W
SENSITIVITY	0·00145 ddm/m at threshold for aerial/threshold separation up to 4349 m	0·0875 ddm at 0·12 θ ± 0·02 θ
POWER REQUIREMENTS	30 V dc, batteries float charged from 100–125 V or 200–250 V ac, at 45–60 Hz	30 V dc, battery float charged from 100–125 V or 200–250 V ac, at 45–60 Hz
ENVIRONMENTAL CONDITIONS		
temperature — internal	-10° to +50°C	
external	-40° to +60°C	
max winds (survival)	100 knots	

Status
Production

Manufacturer
Plessey Radar, Weybridge, Surrey, England

Plessey STAN 37/38 Instrument Landing System

The Plessey STAN 37/38 system comprises the Type 37 localiser and the Type 38 glide-path designed to meet the standards required for categories I, II and III landing systems in accordance with the requirements of ICAO Annex 10. The system provides guidance information in azimuth and elevation for suitably equipped aircraft in the landing approach phase. Two separate system elements are employed, a localiser Plessey STAN 37 for azimuth information and a glide-path Plessey STAN 38 for elevation angle. Marker beacons for indicating discrete ICAO promulgated ranges can be provided.

The localiser Type 37 equipment can be supplied to meet all ICAO requirements from the simpler category I application where single-channel systems are supplied to the higher category II and III applications with fully duplicated equipments. Simple initial installations can subsequently be upgraded to the higher category requirements. Different equipment configurations allow various monitoring and stand-by/change-over arrangements to suit specific operational requirements.

In configuration A the system is a single channel equipment with a single set of monitors, and an associated alarm unit, to monitor the position, width and clearance in the field. Configuration B is offered in two versions. In the first, dual equipment is fitted with two sets of monitor units connected to field monitor aerials. The stand-by equipment is not running and in the event of a failure of the main equipment there is a delay of approximately 20 seconds, to allow the modulator to run up to speed before the stand-by transmitter starts to radiate. The field monitors initiate change-over on a two-out-of-two agreement basis. In the second version, dual equipment is fitted with two sets of field monitor units as above, and a third monitor set is connected to the stand-by system which is running with dummy loads. In the event of failure of the main transmitter, change-over to the stand-by transmitter will be completed within one second. Again, the field monitors initiate change-over on a two-out-of-two agreement basis.

In configuration C dual equipment is fitted with two sets of field monitors, a third set connected to the running stand-by transmitter as in the second version of configuration B. A fourth set of monitors measures the outputs of the operational transmitter within the cabinet which is insensitive to disturbances in the near field caused by overflying aircraft. The two external and one internal monitor sets connected to the working transmitter are programmed to initiate change-over on a two-out-of-three agreement principle.

The glide-path Type 38 configurations are basically similar to the localiser configurations. Only position and width are monitored, there being no clearance signal except on Type M arrays.

Status

Production. In service in the United Kingdom and in many other countries

Manufacturer

Plessey Radar, Addlestone, Weybridge, Surrey, England

Plessey STAN 37/38 ils

Plessey STAN 37/38 ils

Racal-Decca Navigation and Landing Aids

Racal-Decca produce Doppler vhf omni-range beacons, non-directional beacons and instrument landing systems.

The Racal-Decca DN 811 ils comprises the following elements: a localiser antenna system and electronics sub-system; a glide-path antenna system and electronics sub-system (similar to the localiser assembly but operating at uhf), marker beacons electronic sub-system housed in a building with a pole- or mast-mounted antenna array and an optional remote-control unit. This system conforms in all respects to the requirements of ICAO Annex 10. The electronic sub-systems are designed in a modular format, permitting the different ICAO performance categories to be met and providing for easy maintenance.

The Racal-Decca non-directional beacons conform to ICAO Annex 10 and the appropriate sections of ITU regulations. They are fully approved to relevant safety regulations, including those of the United Kingdom's Civil Aviation Authority and the Federal Aviation Authority. They provide an omni-directional radiation pattern carrying a specific identification in morse symbols in the low and medium frequency range in either icw or mcw modes to permit suitably equipped aircraft to determine their bearing with respect to the receiving aircraft's position.

The Decca type DVB-50 dvor beacon conforms in all respects to the requirements of ICAO Annex 10. It is fully approved by the United Kingdom Civil Aviation Authority.

The beacon is a fully double sideband system, available in single or dual configurations with all the accessories necessary to complete a dvor station and is compatible with all types of airborne vor receivers.

Status

Production. The ils is manufactured in the UK under licence from Wilcox Electric (USA). It is in service at Cranfield and Exeter airports and 37 systems are being installed at RAF airfields for the UK Ministry of Defence. The ndb's are operational in 20 countries

Manufacturer

Racal-Decca Navigator Ltd, Ground Systems Division, New Malden, Surrey, England

Racal-Decca Doppler vor beacons

Racal-Decca non-directional beacons

UNITED STATES OF AMERICA

Bendix Microwave Landing System

The mls is an all-weather precision approach and landing guidance system.

The ground sub-system consists of azimuth (az) and elevation (el) angle guidance elements which provide angle information over volumetric coverage about the runway centre line. The mls ground sub-system may also use a dme transponder for distance information.

The mls ground equipment provided by Bendix is fully compliant with the ICAO Standards and Recommended Practices (SARPs) and the FAR-171, Sub-part J. Azimuth and elevation ground antennae provide the TRSB signal-in-space over the required volume of airspace. The azimuth is typically on the runway/approach centreline; the elevation coverage sector originates at a point on the runway centreline opposite the el site. (This point is called the mls datum point in the ICAO SARPs.) Any aircraft within this volume of airspace can receive the guidance information and establish its angular position. A time division multiplex (tdm) signal format is employed. Various basic data items, used to report parameters, such as siting constants and equipment status, are an integral part of the mls format.

The proposed system for most non-Federal users is the B 32-40S which employs 3 degree azimuth and 2 degree elevation antennae, providing the full ±40 degree proportional coverage about runway centreline. Larger antennae are offered for longer runways and difficult siting geometries. The ground system consists of an azimuth sub-system (az-3-40S), elevation sub-system (el-2-40S), a remote-control/status unit, and a remote monitor terminal (printer keyboard, and telephone modem). The az and el sub-systems each consist of an array antenna, a field monitor antenna, and the electronics assembly containing the transmitter, power supplies, timing and monitor electronics, and the associated enclosures/shelters. All solid-state electronics are used in the basic mls equipment.

The ICAO document allows two data rates on the azimuth sub-system; a low rate at 13 hertz and a high rate at 39 hertz, and recommends that high rate azimuth be used on all systems not incorporating the flare guidance and the ±60 degree azimuth coverage. The Bendix basic system incorporates the high rate azimuth function.

The antennae and electronics can be housed in waterproof metal enclosures, or they can be installed in personnel shelters as required by the environment. The equipment is of modular design with extensive sharing of modules between the azimuth and elevation hardware. Split-size azimuth and elevation equipment, each having a single channel of electronics, is the standard configuration.

System integrity is assured by using both integral and field monitors. Executive monitors continuously check critical parameters and automatically initiate equipment shut down if alarm limits are exceeded. Integral maintenance monitors indicate failure trends. The timing/control and monitor electronics contain microprocessors which continuously perform internal self-diagnostics to verify performance and identify defective modules. All transmitter modules, power supplies, array phase shifters, system timing and control elements are checked by the monitor electronics and out-of-tolerance parameters identified. A remote-monitor system is also provided. Module and system parameter data is received by a remote data terminal via standard telephone lines. Data available to this terminal includes measured system parameters such as power supply voltages, transmitter power, guidance error, system shut down history data giving time and cause of shut downs, phase shifter failure data, and various other maintenance information. The data terminal furnished by Bendix is a commercial unit with printer.

Bendix mls azimuth sub-system Washington DC National Airport

Bendix mls azimuth sub-system Washington DC National Airport

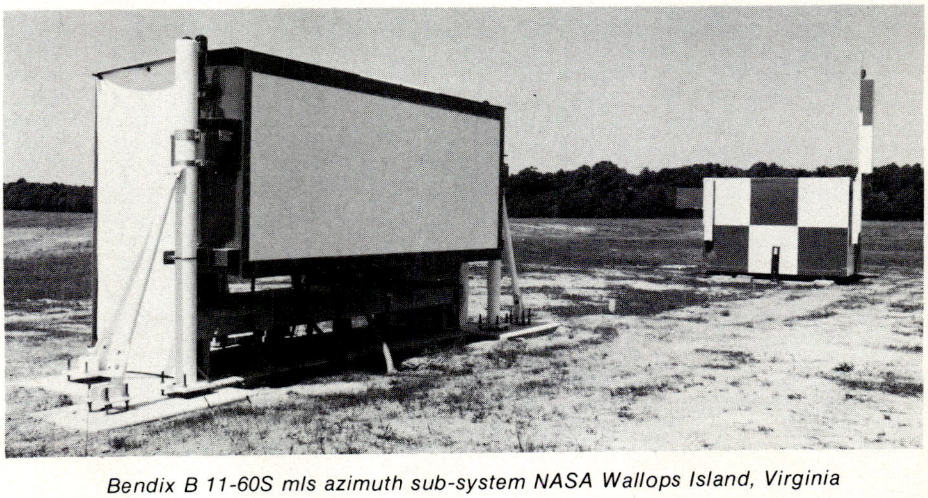

Bendix B 11-60S mls azimuth sub-system NASA Wallops Island, Virginia

A remote-control and status unit, normally positioned in the airport control tower enables personnel to remotely turn the mls equipment on and off and to attempt restart following automatic shut down. The panel also provides indicators which show system status (on or off) and monitor status (existence of a maintenance problem such as out-of-tolerance power supply voltage or low transmitter power). The remote-control and status panel is connected to the mls sites via a multi-conductor cable. Primary power to the mls equipment is 120/240 volts ac, ±10 per cent, single phase, 47 to 63 hertz. An uninterruptible power source is also furnished with the equipment. This system provides two hours of system operations, using batteries in the event of primary power failure. The batteries 'float' on the primary dc input to the electronics providing transient protection and uninterrupted operation at times of power failure.

Characteristics

MODEL	B 32-40S	B 21-40S	B 11-60S
APPLICATION	Compliant: ICAO SARPs FAR-171 J for heliports and airports with 9000 ft (2727 m) length runways	Designed to meet FAA-G-2100 with high multi-path possibilities and 12 000 ft (3658 m) length runways	Meets ICAO requirements for 15 000 ft (4573 m) length runways
AZIMUTH BEAM (az)	3°	2°	1°
SCAN	±40°	±40°	±60°
ELEVATION BEAM (el)	2°	1°	1°
SCAN	0·9–7·5°, extensible to 10°	0–20°	0–20°
ACCURACY			
at threshold az	±20 ft	az ±13 ft	az ±13 ft
el	±2 ft	el ±2 ft	el ±2 ft
at 20 nm az	±0·29°		
el	±0·2°		
RANGE		20 nm	20 nm

Status

Bendix mls is in production for non-Federal navigation facilities. Sales have been made to Golden West Airlines for an installation at Lake Tahoe, California, and to the city of Valdez in Alaska, USA. These are scheduled for acceptance for public ifr use in late 1982. Licensed production has been agreed with Philips Electronics Ltd, Canada, Standard Elektrik Lorenz AG, Federal Republic of Germany and Toshiba Corporation of Japan

Manufacturer

Bendix Communications Division, Baltimore, Maryland, USA

Cardion Distance Measuring Equipment

The dme ground station is an important component of en route navigation and landing approach systems. When operating in conjunction with an airborne interrogator, the dme ground station provides a read-out in the aircraft cockpit of the slant-range distance between the aircraft and the ground station. The dme can be used in conjunction with an ils for landing approaches, or as a replacement for ils outer- or middle-marker beacons. Another primary use of dme is for en route navigation which, when co-located with a vor system, provides polar co-ordinate information to the pilot. Cardion produce the 9639B and the 9783 models.
The 9639B system is available in 100 and 1000 watt versions for terminal area or for use en route, respectively. Coverage of up to 200 nautical miles can be obtained with the 1000 watt system. The 9639B dme will provide a typical accuracy of ±100 feet (30·5 metres).
To meet an operator's particular requirements, such as the mls and category III ils, dme modifications can be made to provide high accuracy, including close-in multiple effects. The dme system has been designed for a traffic handling capability of 100 aircraft. During saturation the ground station reduces sensitivity on the basis of signal strength and reduces replies to those aircraft furthest from the station. The dme uses pulses in the 962 to 1213 MHz frequency range with the frequency selection determined by the application (ils, mls, dvor or Tacan).
The main components of the 9639B system are: transponder, monitor, transfer/communications unit, test unit, system power supply, antenna and radome. The transponder, monitor, transfer/communications unit, test unit and system power supply are contained in standard 19-inch slide-mounted drawers in a 76-inch high equipment cabinet. The system is designed for expansion, from single to completely dual operation. A variety of configurations can be chosen such as; single monitor/single transponder, dual transponder/single monitor or dual transponder/dual monitor. The maximum dual configuration is supplied in a single equipment cabinet.
The transponder is a 100 watt/1 kilowatt expansible solid-state transmitter with a low noise rf preamplifier, wide dynamic range log receiver, echo-suppression circuits and a dc-operated switch regulated power supply.
The monitor is an 80 dB dynamic range signal generator with on-line and maintenance test modes and a monitor test capability. The transfer/communications unit controls the dme system in both manual and automatic modes. For dual dme configurations, automatic transfer operation is provided. This unit also provides a communications network capable of transmitting and receiving digital information to and from the remote monitoring point. The method of digital information transfer between the shelter and the remote-control site consists of fsk modulation transmitted and received over a two-wire telephone line between the two places providing both remote-control and status of the dme system. The test unit contains a video test generator, a pulse counter with led display, a variable prf generator, crystal controlled markers, and an enclosure for a Tektronix type 465 M oscilloscope. Power to the system power supply is 120 or 240 volts ac, 60/60 hertz.
Either an uni- or omni-directional antenna can be used with the dme. Each antenna uses broad band elements for the required performance over the 962 to 1213 MHz frequency band without recourse to tuning. In addition, each antenna contains two probes for monitoring reply pulses. The antenna is enclosed in a sealed fibreglass radome for protection from adverse weather conditions.
The 9783 is similar to the 9639B and is expansible from 100 to 1000 watts and from single to dual transponder configurations. It features a remote maintenance monitoring system that allows certification of the ground station at a central, remote, attended control point by providing quantitative data over standard commercial telephone lines. Remote certification eliminates the need for maintenance technicians to make periodic visits to the ground station shelter to verify system performance. Monitored data retention is provided by a non-volatile microprocessor memory. The data are transmitted remotely by means of an interface and can be displayed by means of an input/output crt terminal. Hard copies of the transmitted data can be obtained by connection to a printer.

Characteristics

TYPE	9639B	9783
TRANSMITTER FREQUENCY	962–1213 MHz	960–1215 MHz
MINIMUM POWER	100/1000 W	100/1000 W
SPECTRUM		
at ± 0·8 MHz	−37 dB/−47 dB	−37 dB/−47 dB
at ± 2 MHz	−57 dB/−67 dB	−57 dB/−67 dB
RECEIVER		
frequency	1025–1150 MHz	1025–1150 MHz
sensitivity	−90 to −93 dBM	−90 to −93 dBM
ADJACENT CHANNEL REJECTION	75 dB	75 dB
SYSTEM		
pulse spacing		
X channel	X channel 12 ± 0·25 μs	12 ± 0·25 μs
Y channel	Y channel 30 ± 0·25 μs	30 ± 0·25 μs
reply delay	± 0·2 μs	± 0·2 μs
beacon accuracy	± 100 to 200 ft (± 30·5 to 61 m)	± 50 to 100 ft (± 15·2 to 30·5 m)
ANTENNA		
gain	8–9·5 dB	6 dB
effected radiated power	550 W	550 W (100 W dme) 6310 W (1000 W dme)
polarisation	vertical	vertical
primary power	120 ± 18 V ac, or 240 ± 36 V ac, 50/60 Hz or 24 V dc	115/230 V ac, single phase 50/60 Hz
ENVIRONMENTAL CONDITIONS		
exterior antenna	−50 to +70°C 100% humidity	−50 to +70°C 100% humidity
remote control unit	−10 to +50°C 90% humidity	
interior	−10 to +50°C 90% humidity	−10 to +50°C 90% humidity

Status

Production. Both systems have FAA approval

Manufacturer

Cardion Electronics, Woodbury, New York, USA

Cardion Vhf Omni-Range System

The vor system provides bearing and azimuth information for the pilot. The vor is an extension of the low-frequency range navigational aids commonly referred to as the nondirectional beacon. The vor operates in the vhf band (108–118 MHz) and the signals are radiated in a 360 degree pattern with magnetic north set at 0. Each degree represents a different radial for the pilot for a total of 360 separate and distinct courses.

The main components of the vor system are: shelter, antenna and radome and the transmitter group. The Cardion vor can be supplied with a basic 8 by 12 by 8 feet (2·44 by 3·66 by 2·44 metres) shelter with an integral hinged counterpoise. The shelter is supplied with necessary conduit, distribution panels, junction boxes, power wiring and hardware for complete site installation and antenna interface. The antenna is a cylindrical slot type with four vertical slots equally spaced on the circumference. It is mounted on top of the shelter which is supplied with a 15-foot (4·6-metres) nominal counterpoise. Protection for the antenna, from wind and ice loading, is provided by a fibreglass radome supplied with the antenna. Construction of the antenna/radome assembly allows for the addition of a dme antenna when vor/dme operation is required. Dual obstruction lights are provided in both cases described.

The Cardion Model 9467 vhf transmitter consists of a modulator/power supply and an rf transmitter. Both units are housed in standard 19-inch drawers. The modulator/power supply accepts 240 volts ac, 50/60 hertz power and contains circuitry for either local or remote-control. This drawer also provides an identification oscillator and keyer that can be programmed to provide a morse code of any three or four letters with interlacing of dme and Tacan keying.

A voice processor uses compression amplifier techniques to limit voice peaks.

The rf transmitter drawer contains an ultra-stable crystal oscillator/multiplier that provides crystal selectable output frequencies from 108 to 118 MHz. The low level signal is amplified up to a level of 100 watts minimum carrier power at the end of a 30 foot (9·14 metres) cable.

A goniometer is housed in a 7-inch (17·8-cm) high enclosure and is suitable for mounting in a standard 19-inch cabinet. The system incorporates a transfer/communications unit which provides the following functions: power measurement of on-line equipment, automatic transfer of equipment under alarm conditions and system input/output interface at remote locations. A monitor unit is used in the vor station and ensures that the vor transmitted signal does not vary from normal. Variable alarm limits are provided so that monitored parameter limits can be set according to user preference.

A field detector which consists of a dipole antenna and a detector is used to demodulate a sample of the radiated signal. The detector is placed in a fixed place where it intercepts a known radial chosen to be representative of the radiated signal. The detected output is sent to the vor monitor to evaluate system accuracy during normal monitoring.

For unattended or remote vor station, microwave, land-lines or rf communication links can monitor the control and status of the vor with audio and visual alarms. Recorded real-time weather information can be transmitted over the vor voice link or sent to the controller in a nearby airport upon command. The system is compatible and can be expanded to include Tacan.

Cardion vor system

Optional equipment available includes a remote-control and status unit, fail-safe default mode for light traffic airports using a single vor system, and an emergency-battery reserve-power system.

Characteristics

FREQUENCY RANGE: 108–118 MHz
DIRECTION: omni-range, 360°
RANGE: 40–200 miles (64–322 km)
POWER: adjustable 25–100 W
STABILITY: ± 0·5 dB
SPECTRUM: ICAO Annex 10
IDENTIFICATION: code and/or voice
POWER REQUIREMENTS: 240 V ac, 60/60 Hz

Status

Production. Installed at airports in the USA

Manufacturer

Cardion Electronics, Woodbury, New York, USA

Edmac Model 9500 Tacan Transponder

The Edmac Model 9500 radio set, Tacan transponder is a fully solid-state unit designed to replace AN/GRN-9, RTB-2, and other older Tacan transponders. It is compatible with the antenna, test, monitor and control equipment commonly associated with these transponders. The equipment which has 252 channels is capable of accepting either X-channel (12 micro-second) or Y-channel (36 micro-second) interrogations and in either mode synchronises the response to the first pulse of the interrogation. A combination of interpulse and post-interrogation blanking ensures minimal response to multi-path interrogations, and the equipment allows the blanking to be adjusted to the optimum for each site. The unit features a logarithmic if amplifier and a digitally-operated adjacent channel rejector which blocks off-frequency interrogations.

The adjustable replay rate control can be set for the standard 2700 per second, or at higher limits up to 4000 per second for operation at heavy traffic locations. The unit is capable of transmitting either X-channel or Y-channel pulse codes. The selection of X- or Y-channel decoding and encoding as well as selection of the channel number, is accomplished with a single four-section thumb-wheel switch. Tuning the receiver requires only one control, while the transmitter has no tuning adjustment.

The Model 9500 includes a solid-state keyer which transmits either a three- or four-character identification signal. Selection of the call letters is by a four-section thumb-wheel switch. As an alternative the Model 9500 can be connected to an external identification keyer.

Characteristics

TRANSMITTER
frequency: 962–1215 MHz
peak power: 3·1 kW
RECEIVER
sensitivity: −93 dBm
ADJACENT CHANNEL REJECTION: independent of interrogation strength
SYSTEM
channels: 252 (X and Y)
reply delay: adjustable 1350 pps (dme service)
2700 pps (standard Tacan service)
4000 pps (increased traffic Tacan service)
PRIMARY POWER: 230/115 V, 50/60 Hz, single phase 745 VA
OPERATING TEMPERATURE RANGE: −20 to +55°C
UNIT DIMENSIONS
width: 2 ft (0·6 m)
depth: 1 ft 10 in (0·55 m)
height: 2 ft 5⅝ in (0·75 m)
WEIGHT: 198 lb (90 kg)

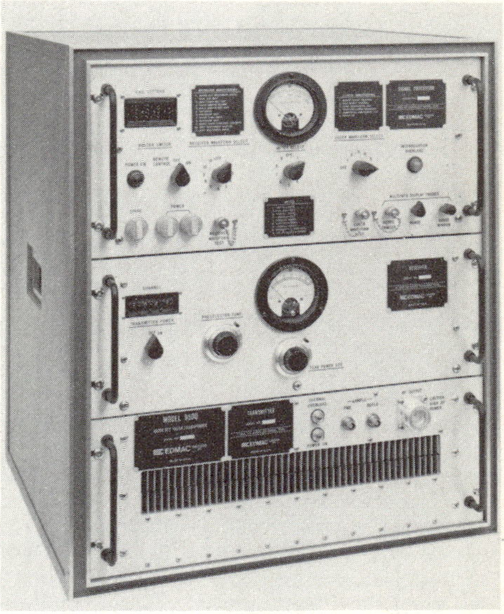

Edmac Model 9500 Tacan Transponder

Manufacturer

Edmac Associates Inc, East Rochester, New York, USA

Hazeltine Model 2500 Series Microwave Landing System

Hazeltine has developed a family of mls using a time reference scanning beam. The mls was adopted by ICAO during 1980 as the next generation landing system. The Hazeltine equipment uses a flexible core modular system concept and incorporates a patented phased array circuit technique.

Status

Hazeltine has been developing the equipment for ten years. The system has been selected by the Ministry of Transportation in Canada. The contract award, announced in December 1981, represented the Canadian Government's initial procurement of mls equipment prior to national implementation which is presently planned for the mid-1980s. In addition, two systems have been procured by the State of Michigan Aeronautics Commission for its state precision approach programme.

Manufacturer

Hazeltine Corporation, Commack, New York, USA

ITT/Wilcox System 21 Vortac

System 21 Vortac is an integrated system offering from one fixed station vor azimuth, Tacan azimuth and Tacan (dme) distance information. It is a result of collaboration between ITT Avionics and Wilcox Electric Inc. The system which is fully solid-state including the 5 kilowatt Tacan equipment, incorporates computer-assisted remote maintenance monitoring and provides remote certification with a 16-point vor ground-check programme.

A telephone line connects the remote computer terminal and hard-copy printer with the computer unit in the Vortac shelter. This line provides for both digital information transfer and voice modulation of the vor transmitter. Routine maintenance schedules are programmed into the computer enabling interrogation of all critical operating parameters. A collaboration of four internal monitors and the central remote computer reports any out-of-tolerance conditions with positive identification of faults down to the particular circuit board. Separate vor and dme/Tacan files are available via local- or remote-crt display or printer. The Tacan final amplifier provides fail-soft modes, permitting continued system operation with up to 25 per cent of its output amplifiers failed. The system features a floating battery back-up power system to ensure uninterrupted service despite power failures. In cases where a separate dme antenna is installed, the Vortac automatically converts to a vor/dme operating mode while on back-up battery power.

Tacan antennae are available from ITT Avionics. The YNI-103A (low-band) or YNI-104A (high-band) antennae are recommended for new Vortac installations. These antennae employ a wide aperture for sharp horizon cut-off and to suppress lobes below the horizon and internal phase shifters for adjusting elevation patterns to tailor the operation to a given site.

Vor antennae are available from Wilcox. The cylindrical slot antenna is claimed to produce a signal with a low radiation angle above the horizon, and a minimal cone of silence above the station. The Wilcox dme antenna consists of nine biconical main radiation elements and secondary monitor pick-up elements. The antenna radiates a 6 degree beam width with slight lobe tilt which permits a balance between undesired null depths and power on the horizon to minimise echo generation.

Characteristics

EQUIPMENT: vor
FREQUENCY: 108–118 MHz
POWER OUTPUT: adjustable to 150 W
RANGE: 174 nm (322 km)
ACCURACY: ±0·75°
EQUIPMENT: Tacan/dme
FREQUENCY: 962–1213 MHz
POWER OUTPUT: 5000 W (Tacan) 1000 W (dme)
SPECTRUM: FAA-E-2678 a/ICAO Annex 10/Mil standards
EQUIPMENT: receiver
SENSITIVITY: −94 dBM for 70% replies
SELECTIVITY: 84 dB adjacent channel rejection
EQUIPMENT: system
CAPACITY
Distance service to 200 aircraft
Bearing service unlimited
CHANNELS: 126 (X and Y channels)
REPLY DELAY
−10 to −79 dBM 50 ± 0·1 μs (X mode)
56 ± 0·1 μs (Y mode)
−79 to −94 dBM 50 ± 0·25 μs (X mode)
56 ± 0·25 μs (Y mode)
PRIMARY POWER
Vortac: 120/208 V ac, 3-phase, 60 Hz or 220/380 V ac, 3-phase, 50 Hz
vor/dme: 120/240 V ac, 1-phase, 60 Hz or 110/220 V ac, 1-phase or 36 V dc
CONSUMPTION: Vortac 4200 W, vor/dme 2400 W
EQUIPMENT: station equipment
DIMENSIONS
width: 6 ft (1·83 m)
depth: 2 ft 6 in (0·76 m)
height: 6 ft 4 in (1·92 m)
WEIGHT: 2000 lb (910 kg)
OPERATING TEMPERATURE RANGE: −40 to +50°C

Status

Production. Vortac System 21 was selected by the FAA to meet the revised US National Standards for vor, dme and Tacan systems and will be installed in 950 places in the USA. The system complies with ICAO vor/dme and Military Tacan standards.

Manufacturers

Avionics Division, ITT, Nutley, New Jersey, USA
Wilcox Electric Inc, Northrop Corporation, Kansas City, Missouri, USA

ITT/Wilcox System 21 Vortac

Raytheon Distance Measuring Equipment

Raytheon dme is a complete ground station consisting of a dual transponder, dual monitors, antenna, and if required, shelter and stand-by power supplies. The equipment which is solid-state except for the high-power output stage, provides an effective radiated power of 39 dBW. The output stage uses planar triodes. Other features of the system are high accuracy pulse timing and amplitude discrimination and a high gain antenna to minimise nulls. The beacon is compatible with vor and ils equipment and has a mtbf of 3500 hours for each channel.

Status

Production. The system meets and exceeds the requirements of ICAO Annex 10

Manufacturer

Air Traffic Control Division, Raytheon Co, Sudbury, Massachusetts, USA

Raytheon dme

Raytheon Vhf Omni-range Beacons

Raytheon manufactures vor equipment in both single and dual channel configurations complete with antenna and self-contained performance monitoring facilities. The equipment, which is modularised and fully solid-state, provides 100 watts carrier power to the antenna. It can be adapted to power from any ac source or batteries and incorporates a self-calibrating monitor which employs digital techniques indicating azimuth to the nearest 0·1 degree.
Two types of vor are manufactured, Standard and Doppler. Most modules are interchangeable between the two types. The Doppler vor is recommended for sites in hilly terrain or with other obstructions. The increased accuracy usually overcomes the distortion caused by the environment and results in much improved performance.
The equipment is designed for a single channel mtbf of 5000 hours at an ambient temperature of 35°C.

Status
Production. The system meets or exceeds the requirements of ICAO Annex 10

Manufacturer
Air Traffic Control Division, Raytheon Co, Sudbury, Massachusetts, USA

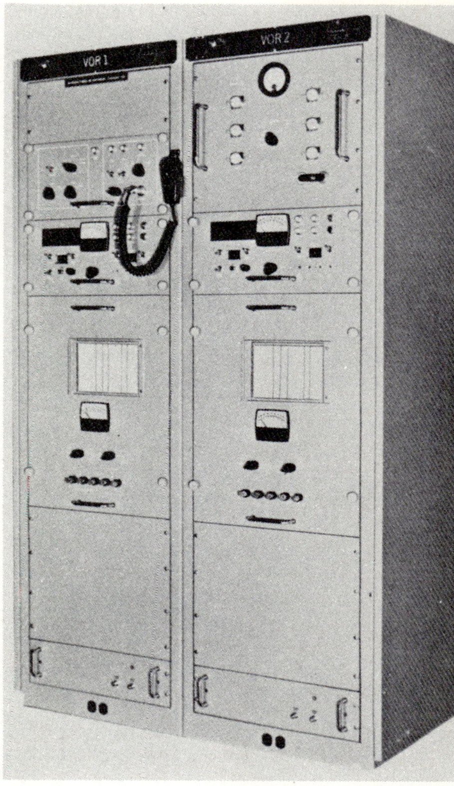

Raytheon vor beacon

Servoflight Doppler Vhf Direction Finder System

Servo produce several models of Doppler df systems for aviation applications. The latest development is the model 7050.
Frequency range is 118 to 135·975 MHz, with 720 operating channels, and 25 kHz spacing. Basic components comprise a fully synthesised solid-state receiver/bearing processor, a wide-aperture 16 element quasi-Doppler antenna array with electronic commutation and a remotely-located bearing display/control.
The equipment is designed for local or remote-control operation and features a led 'ring of lights' vector display, led digital presentations of bearing and frequency, and a built-in test facility. The model is capable of a bearing accuracy of ±3 degrees on most sites with an instrumental accuracy of ±1 degree.
Options and modifications available include: other display configurations, eight-element lightweight antenna array, computer interface, expanded frequency range, remote maintenance monitoring, and a single unit configuration.

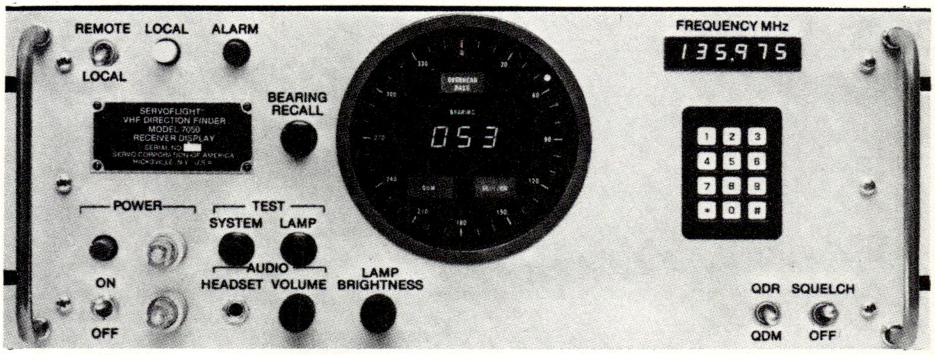

Servoflight model 7050 vhf df system

Others models produced include the 7010–11, both vhf units with up to 12 channels; the 7020 uhf equipment, fully synthesised with 7000 channels and the 7030, a vhf/uhf model with dual antenna and simultaneous operation.

Status
Production. The model 7050 is the basic element of the df system selected by the US Federal Aviation Administration for installation at flight service stations throughout the USA

Manufacturer
Servo Corporation of America, Hicksville, New York, USA

Wilcox Instrument Landing System

Wilcox ils supply approach and landing information to pilots through three major sub-systems: the localiser, the glide-slope and the markers (outer, middle and inner). Additions to the basic system configuration include a remote monitor/control unit and a compass locator, a non-directional beacon co-located with the outer marker.
The localiser radiates a signal which provides azimuth (lateral) guidance for a landing aircraft. The localiser transmits information on to a visual display through the standard vor-loc receivers in the aircraft, and depicts the aircraft's relative position to the central line of the runway. In the case of a coupled approach, the autopilot works in conjunction with the vor-loc receiver to accomplish these corrections automatically. The localiser normally placed 330 to 2000 feet (92 to 610 metres) on an extended centre-line beyond the end of the primary instrument runway uses log periodic dipole (lpd) radiating elements and is available in several antenna configurations.

Wilcox end-fire, non-imaging glide-slope system

Many airports use an eight-element narrow aperture lpd but, in restricted space, a 14-element single or dual frequency array can be substituted. An environmentally-controlled shelter, generally about 250 feet (76 metres) from the end of the antenna array, houses the localiser electronics.

The glide-slope provides vertical guidance to an aircraft on final approach. The signal radiated is translated into a visual display in the cockpit which indicates whether the aircraft is above, below or on the correct angle of approach during descent.

The glide-slope antenna tower, normally positioned approximately 1100 feet (335 metres) from the runway threshold and 400 feet (122 metres) to the side of the runway, is available in any one of four signal formats.

The null reference signal is standard for most installations. Where tighter tolerances are required, or where siting problems prohibit the null reference, several others are available. One system, the modified side-band reference, would be installed at sites where level terrain is available in front of the antenna and where the terrain drops off shortly beyond the end of the runway. Another system, the capture effect, is recommended where the terrain beyond the end of the runway rises slightly.

Where neither format is appropriate end-fire non-imaging is used. This system employs a slotted-cable transmitting antenna.

The marker beacon sub-systems consist of an outer, middle and, in some cases, an inner marker. These beacons provide a pilot on final approach a series of visual and audio indications of precise position reference points.

With the information supplied by the marker beacons, a pilot knows when to execute the procedure turn (or maintain minimum holding altitude) when he has reached the decision height and when the runway threshold is imminent. For non-precision approaches, the outer marker may be used as an airway or fan marker. In addition, it is possible to co-locate a non-directional beacon with the outer marker to serve as a compass locator.

In all three categories of ils installations, the markers supply pilots with vital location references independent of weather conditions. Marker beacon electronics are housed in pole-mounted shelters or optional environmentally-controlled walk-in shelters.

Wilcox ils localiser at Brasilia Airport

The critical operating parameters of any category can be monitored from a remote location with one of several optional remote monitoring systems. The status of individual components in a category III installation can be monitored from a remote location via a digital data link or land-lines between remote maintenance monitors and on-site, computer-controlled executive monitor units. These on-site monitors, one each for glide-slope and localiser sub-systems, feature full redundancy with micro-processor control of pre-set digital alarm limits. In addition to indicating when a sub-system's operation falls outside the tolerance levels for category III performance, internal self-checking enables the micro-processors to locate the source of the trouble down to the individual solid-state circuit board.

In a category II installation, a remote status/control unit, connected to the major sub-systems via land-lines or a data link, gives a visual indication of the status of each facility and an audio alarm if conditions change.

A remote alarm receiver may be used in category I installations if land-lines are impractical.

Status

Production. The first Wilcox category III system was installed at Kansas City in September 1981. Other airports include Houston, Chicago O'Hare, New York-Kennedy, Los Angeles, Portland, Seattle Seatac and Washington Dulles, USA

Wilcox ils glide-slope antenna

Manufacturer

Wilcox Electric Inc, Northrop Corporation, Kansas City, Missouri, USA

AIRFIELD LIGHTING

AUSTRALIA

AWA T Visual Approach System

AWA T-Vasis Runway 16, Sydney Airport

The T visual approach system (T-Vasis) which was designed and developed in Australia, has been adopted by ICAO and is specified as a standard in Annex 14. With seven distinct zones of information, it provides an accurate visual indication to assist the pilot to correct and maintain the desired approach slope. Corrective information is in the form of T or inverted-T groups of lights with one, two or three white lights in the vertical legs to indicate the extent of divergence from the standard slope. An all-red condition indicates gross undershoot.

The T-Vasis consists of 20 light-sources in the form of light-projection units which produce closely controlled beams of light. These light-projection units are alongside the runway. A smaller model, the AT-Vasis, consists of ten light-projection units along one side of the runway.

The T-Vasis employs three types of light-projection unit which are of the same basic construction and vary only in detail; each unit is constructed of lightweight grp. A unit is supported near each of its four corners by threaded support rods. These rods, which are used in adjusting the unit, are rigidly fixed to the foundation by frangible mountings. In some cases, where sloping terrain makes it necessary, pillar assemblies are inserted between the unit support rods and the frangible mountings. The lamp assembly is fixed into the rear of the unit and carries the cable-terminal block. Blades are fitted within the light-projection units to screen and filter the light. Just in front of the lamp assembly is the rear blade; the front blade is at the other end of the box. Fabricated from light gauge metal, the blades have a small post at each end on which a special precision level may be located to permit horizontal adjustment.

The three types of T-Vasis light-projection units are: the fly-down, bar and fly-up unit. The fly-down unit is positioned in the leg of the inverted-T and carries a rear cut-off blade fitted above the beam and a front cut-off blade below the beam. It provides a beam extending from an elevation of 6 degrees down to approximately the approach slope where it has a sharp cut-off.

The bar unit is in the horizontal bar of the T. It features a rear cut-off blade fitted above the beam and a red filter at the lower part of the front. This light unit produces a beam from ground level up to 6 degrees, the low part up to 1 degree 54 minutes being red.

The fly-up unit is in the leg of the upright T and has rear cut-off blade below the beam which, together with a front blade above the beam, provides a sharp cut-off at the top of the beam. A red sector of light is formed by a rear red filter fitted above the beam and a front blade fitted below the beam. This light unit produces a beam from approximately the approach slope down to ground level, the lower part below 1 degree 54 minutes being red.

All units are fitted with service covers which provide weather protection for the lamp assembly and cable-termination block. The lamp assembly consists of a light sheet-metal channel section fitted with lamp-mounting plates which are independently adjustable in azimuth and elevation. The four day lamps and two night lamps are operated on separate chassis.

Control equipment permits the selection of a variety of day- and night-time intensities with a choice of local only, or local and remote-control monitoring.

Characteristics

TYPE: T-Vasis/AT-Vasis
LIGHT UNITS: 20/10
LAMPS
type: PAR 36
day: 4 × 6·6 A, 100 W quartz-halogen spot
night: 2 × 6 V, 35 W tungsten filament, spread beam

Status

Production. More than 100 systems have been commissioned since 1964

Manufacturer

Amalgamated Wireless (Australasia) Ltd, North Ryde, New South Wales, Australia

BELGIUM

ADB Airfield Lighting System

ADB produce a comprehensive range of airfield lighting equipment for military and civil operators.

Approach lighting systems comprise elevated or inset lights, uni- or omni-directional, low or high intensity for non-instrument, instrument approach or precision approach runways, category I, II or III. These include capacitor discharge lights, uni- and omni-directional for a sequence flashing lights system for runway threshold identification (reil system).

Visual landing aids available consist of two or three bar, two colour Vasi, and Avasi and one or two bar, two colour Papi system. The Vasi informs the pilot whether the approach path is correct, too high or too low.

A new system to give a visual indication of the correct glide-slope has been developed. The system, called the Papi, gives precise guidance right down to the runway threshold.

A pilot using the Papi system sees a bar of lights made up of four units on either side of the runway adjacent to the glide-path origin. Each unit contains two lamps, two red filter glasses and two lenses aligned to produce a high intensity beam of light, the upper part being white and the lower red. The four units are set up at different angles so that there is 1 degree between the highest and lowest settings with the highest setting being nearer the runway. If the aircraft is on the correct glide-slope the pilot sees two red and two white lights. If it goes high progressively more white lights appear, if low, more red lights are seen.

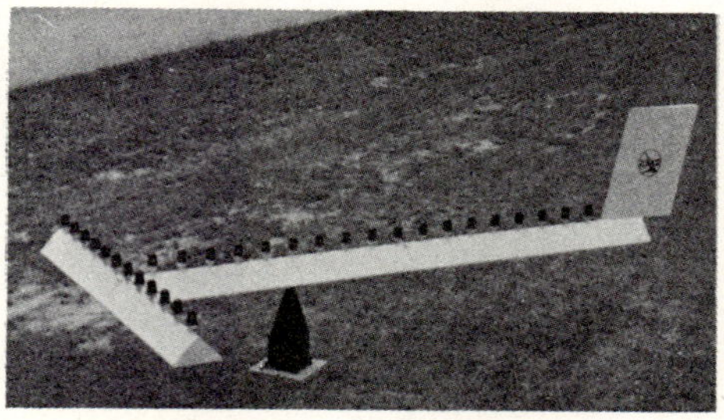

ADB illuminated landing-T

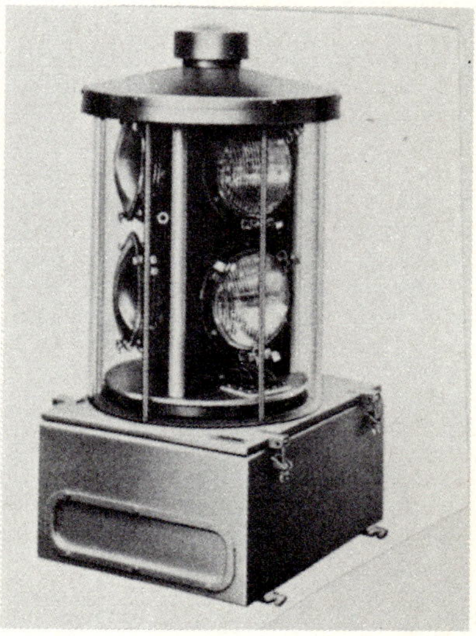

ADB airfield identification beacon

A range of lights, elevated and inset, high- or low-intensity are offered for the runway edge and end, threshold and wing-bar, centre-line, touchdown zone, taxiway edge, centre-line, stop bars and clearance bars.

Obstruction lighting produced for marking high voltage overhead lines, towers and other obstructions to air traffic equipment includes single or double lights, flashing beacons, autonomous flashing beacons supplied by batteries recharged by solar cells, and long life obstruction lights using cold cathode neon lamps.

Accessory lighting equipment offered includes airfield and heliport identification beacons, illuminated landing tee to show the runway in use by day or night and an illuminated wind cone. Ground guidance and signalling equipment comprises traffic lights, distance panels and taxiway guidance signs.

Power control and distribution systems for airfield lighting consist of isolating transformers and connecting units, constant current regulators, remote-control desks and emergency supply systems. The isolating transformers insulate the lighting fitting from the high voltage primary series circuit and ensure the continuity of the primary loop in case of lamp failure. Also, the value of the primary current being given, a lamp rated at any other current can be fed through an isolating transformer. These transformers are fully protected in a polychloroprene moulded sheath. The primary and secondary cables are an integral part of the transformer body.

Constant current regulators for the supply and control of brilliancy of series circuit airfield lighting are available to provide step-by-step brightness control (Type NBF), continuous brightness control (Type SCF) and brightness controlled by thyristors (Type TCC). The latter is a special model designed for feeding small circuits up to 10 kW (ie heliports and portable lighting). Circuit selectors are offered for the alternate or simultaneous supply of one or more series circuits using only one constant current regulator. The earth fault detector for series circuits indicates the existence of an earth fault and shows it approximate position. A burnt-out lamp detector alarm indicates when a certain percentage of lamps in a series circuit have failed.

The remote-control desks, panels and switchboards are generally constructed to meet the particular requirements of each airport. These can include remote control with an electronic memory and ground traffic control by loop detectors and computer. The remote control of the lighting can also be performed by means of a transmitter/code generator installed in the aircraft and a vhf receiver/decoder which will respond to correctly coded modulations.

ADB P702T capacitor discharge light

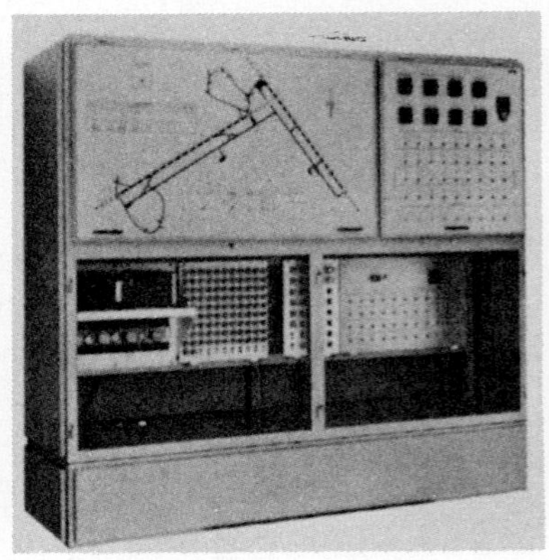

ADB remote-control panel

ADB has a range of emergency power supply units comprising regulated battery charger-rectifiers for automatic, monitored recharging of stationary batteries; solid-state, single-phase inverters type Invermatic (power rating range 250 volts amperes to 10 kilovolt amperes); and solid-state uninterruptible power systems, single- or three-phase.

Status
ADB equipment complies with FAA and ICAO specifications for civil airfield lighting systems and with NATO, STANAG and MIL specifications for military bases. A total of 48 NATO bases have been equipped with ADB lighting, and the company has supplied systems in 21 countries. Civil systems are in operation at airports world-wide

ADB FTS 25 taxiway edge light

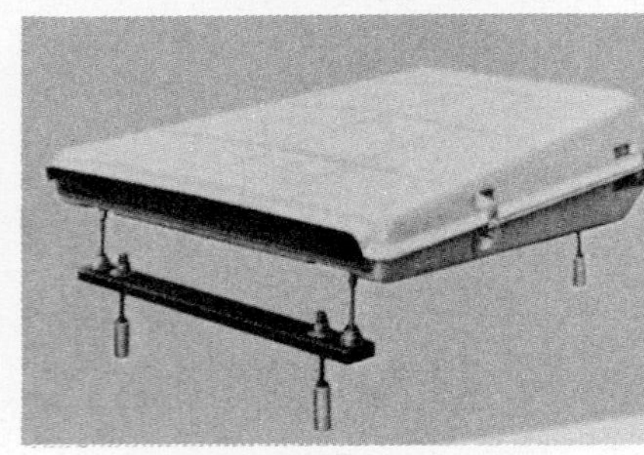

ADB GP 600/3 Vasi unit

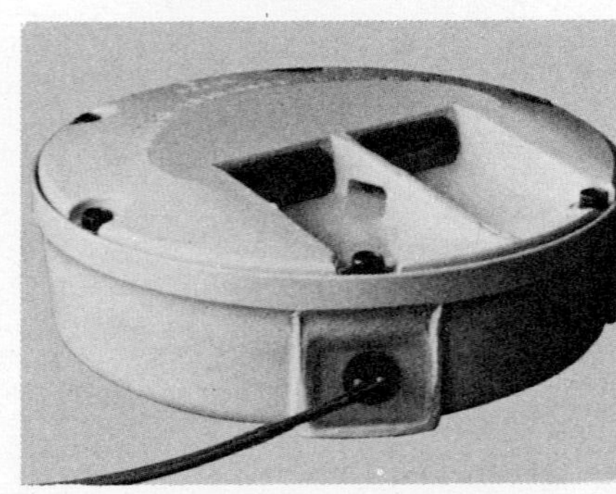

ADB SQ 1200 runway touchdown zone unit

Manufacturer
ADB SA, Zaventem, Belgium

CANADA

Amerace Ground Lighting Transformers

Amerace manufacture a series of isolating transformers for use in the series circuits that light runway, taxiway and approach lighting systems on airfields.

The transformers can be supplied either encapsulated in epoxy as specified by Canada's Ministry of Transport or in rubber as specified by the FAA in the USA.

The manufacturers claim that epoxy encapsulation material offers greater resistance to petrol, oil, acid and alkaline soils, is rodent- and termite-proof and capable of operating at higher ambient temperatures. In conjunction with the kits specifically designed for the primary and the secondary connections to the transformer, the supply and output connections can be made to form a waterproof circuit.

In addition to stringent physical and electrical tests and inspections, each epoxy encapsulated transformer is immersed in water while hot and allowed to cool and soak. At the end of the soaking period, a high voltage direct current insulation resistance test is performed, along with the current ratio tests.

Characteristics
FREQUENCY: 50/60 Hz

RANGE

30/45 W	6·6/66 A
65 W	6·6/66 A
100 W	6·6/66 A
200 W	6·6/66 A
250 W	6·6/20 A
300 W	6·6/20 A
500 W	6·6/20 A

Status
Amerace transformers are installed at airports in Burma, Canada, Dubai, Ecuador, Federal Republic of Germany, Hong Kong, Iraq, Israel, Mexico, New Zealand, Saudi Arabia, Singapore, Trinidad and Tobago, United Kingdom, USA and Yugoslavia

Manufacturer
Amerace Ltd, Markham, Ontario, Canada

CZECHOSLOVAKIA

Elektrosignal Airfield Lighting

Elektrosignal produce a range of airfield lighting equipment. The control and distribution systems include power switchboards, control panels and control racks, constant current power supplies, short-circuiting relays and short-circuiting gear, single-core airport cables, low-voltage cables and insulating transformers.

Lights and signs available comprise active and passive airport signal boards, obstacle warning lights, signalling floodlights, elevated lights, flush mounted lights, incandescent bulbs and discharge tubes. A system assembled from the above components permits airport lighting facilities to be installed on all categories of airports handling all kinds of air traffic.

Status
Production. Elektrosignal equipment is installed in over 50 international airports and several hundred domestic airports in Europe and Asia

Manufacturer
Elektrosignal, Omnipol Foreign Trade Corporation, Prague, Czechoslovakia

Characteristics
STANDARD SETS

MODEL	D3	D2-1	M2-2	MLOK C	MLOK 2	MLOK P	SE DLO	SVETKUSKA 2
TYPE OF INSTALLATION	Fixed	Fixed	Fixed	Mobile	Mobile	Mobile	Mobile	Portable
APPLICATION	Cat I, II and III	Domestic 1 direction of approach	Domestic 2 directions of approach	Domestic 1 direction of approach	Domestic 2 directions of approach	Taxiway lights	Glide-path beacons	Heliports

FINLAND

Idman IDM Series Approach Lights

The IDM-2982 is a high intensity light intended for category I, II and III operations at the approach, threshold and end of the runway. The optical system consists of an anodised aluminium reflector and front reflector. Colour filters can be mounted between the reflector and front glass and the beam can be aligned using two graduated scales.
The light can be mounted through a breakable coupling on the top of a transformer case or a safety pole with specified height. Minimum luminaire height from the ground when installed is 270 mm.

Characteristics
MODEL: IDM-2582
LAMP
single-ended halogen lamp 200 W, 6·6 A
rated life 1000 h
AVERAGE INTENSITY WHITE: 42 Kcd
PEAK INTENSITY: 56·4 Kcd
RED/GREEN FILTERS TRANSMISSIVITY: 23–24%

Status
Production

Manufacturer
Idman Ltd, Helsinki, Finland

Idman IDM Inset Runway Lights

The IDM inset-type lights for runways are all semi-flush types with different protrusion heights using mainly the same mechanical and optical construction.
The lights consist of the projector assembly with lamps and anodised aluminium reflectors, the top cover with the refractive prism and the base receptacle. The cover is constructed of cadmium-plated cast iron with a silicone aluminium alloy projector body and base receptacle. The gaskets are heat and weather resistant. The base receptacle has two glands for input cables at the right angles. Turning it permits cable installation from four different directions. A watertight plug is fitted between the input cable and the projector cable.
The inset lights are fastened to the pavement with special two-compound plastics, strengthened by grooves on the outer surface of the base receptacle. The base receptacle has a 300 mm diameter with an inset depth of 103 mm (190·5 km). The lights are each rated for a 410 kN static load.

Status
Production

Manufacturer
Idman Ltd, Helsinki, Finland

Characteristics

TYPE	2885	2885	2885	2885	2495	5067	2807	2987	2819	2984
APPLICATION	Approach cross-bars	Approach side row	Threshold	Threshold wing bar	TDZ	TDZ	Runway	Runway	Runway end	Runway edge
COLOUR	White	Red	Green	Green	White	White	White	White	Red	White
LAMP	2 × 200 W	2 × 200 W	2 × 200 W	2 × 200 W	200 W	100 W	200 W	2 × 100 W	200 W	2 × 200 W
INTENSITY ICAO rated average	27 000 Kcd	7400 Kcd	21 000 Kcd	17 500 Kcd	17 000 Kcd	6500 Kcd	6000 Kcd	6500 Kcd	2500 Kcd	11 300 Kcd
peak	36 500 Kcd	10 500 Kcd	27 100 Kcd	27 100 Kcd	21 500 Kcd	8000 Kcd	9300 Kcd	8000 Kcd	3500 Kcd	17 400 Kcd
PROJECTED HEIGHT	25 mm	25 mm	25 mm	25 mm	15 mm	13 mm	9 mm	13 mm	25 mm	25 mm

Idman Inset Taxiway Lights

The IDM inset-type lights for taxiways are of semi-flush type with 9 mm protrusion height using similar mechanical and optical constructions except for the omni-directional edge light.
The lights consist of a projector assembly with lamp, top cover with lenses and a base receptacle. The cover, projector body and base receptacle are of silicone aluminium alloy, the gaskets are heat and weather resistant and a watertight plug is fitted between the input cable and the projector cable. The lights are fastened to the pavement with special two-compound plastics, strengthened by grooves on the outer surface of the base receptacle. The base receptacle has a 200 mm diameter and an inset depth of 75 mm. Narrow-beam lights and lights with 100 watt lamps are designed for category III operations.

Characteristics

TYPE	5504	5505	5506	5507	5508	5507	5468
APPLICATION	Bi-directional CL narrow beam	Uni-directional CL narrow beam	Bi-directional CL wide beam	Uni-directional CL wide beam	Bi-directional CL curved sections	Stop-bar	Omni-directional edge
COLOUR	Green	Green	Green	Green	Green	Red	Blue
LAMP	45 W	45 W	45, 100 W	45, 100 W	45, 100 W	45, 100 W	45 W
INTENSITY							
ICAO rated							
average	400 Kcd	440 Kcd	105, 370 Kcd	115, 405 Kcd	85, 320 Kcd	65, 220 Kcd	
peak	530 Kcd	585 Kcd	125, 415 Kcd	140, 455 Kcd	125, 415 Kcd	80, 250 Kcd	
PROJECTED HEIGHT	9 mm	9 mm	9 mm	9 mm	9 mm	9 mm	25 mm

Status
Production

Manufacturer
Idman Ltd, Helsinki, Finland

FRANCE

Augier Ground Lighting Transformers

Augier manufactures watertight non-corrosive transformers housed in epoxy-resin mouldings. The transformers can be fitted on both primary and secondary connections with cables incorporating hypalon plugs and sockets complying with FAA specifications or specially designed leads and connectors which can be assembled on site. The transformers which are equipped with metallic handle for carrying and earthing the unit, supply one or more lighting fixtures depending on their capacity.

Characteristics
RATINGS: 45–550 VA
PRIMARY CURRENT: 6·6/20 A
SECONDARY CURRENT: 6·6/20 A

Status
Augier transformers are installed at Charles de Gaulle Airport, Paris, France

Manufacturer
Augier SA, Carros, France

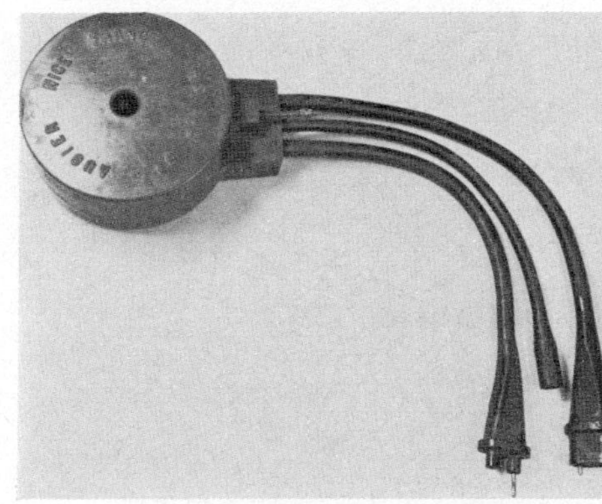

Augier insulated transformer

BBT Airfield Lighting Equipment

BBT produce equipment for visual landing aids, approach runway and taxiway lighting, obstruction warning lights, apron and general purpose lighting and control and distribution systems.

Visual landing aids available comprise a Vasi and a portable slope angle indicator. Approach and runway lights consist of high and medium intensity, uni- and bi-directional elevated lights, a condensed discharge light, threshold/approach inset lights, and high intensity uni- and bi-directional runway inset lights. For taxiways an elevated medium- and a low-intensity light is offered. Inset units for taxiway centre lines include a 45 watt symmetrical light, 100 watt symmetrical or asymmetrical lights and a high intensity 200 watt symmetrical or asymmetrical light with a wide-angle beam.

Requirements for apron and general purpose lighting are met by the Horizon, Zirkon and Hexaline floodlights. The Horizon and Zirkon models use 250/400 watt fluorescent high pressure sodium lamps; the Hexaline models are fitted with 250/400/1000 watt metal halide and high pressure sodium lamps.

For distribution and control BBT produce series transformers with cast metal housings, or sheathed to meet FAA or STNAe specifications, static resonant circuit and magnetic amplifier regulators and airfield lighting control desks. Accessory lighting equipment offered includes revolving airport beacons, producing 30 000 or 100 000 candelas, single and double obstruction lights and an illuminated tee.

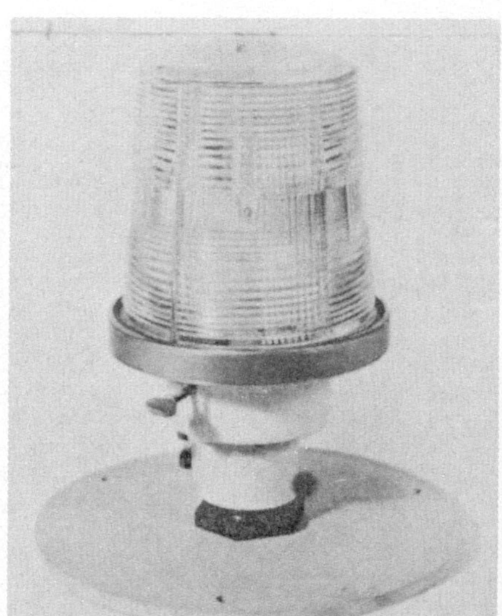

BBT elevated high intensity bi-directional light

Status
Lighting equipment installed at Paris, Charles de Gaulle and Lyon Satolas airports in France and at airports in Australia, Algeria, Denmark (Billund), Greece (Athens), Luxembourg, Morocco, Saudi Arabia, Switzerland (Zurich), Venezuela (Caracas) and Yugoslavia (Zagreb)

Manufacturer
Barbier, Benard and Turenne, Paris, France

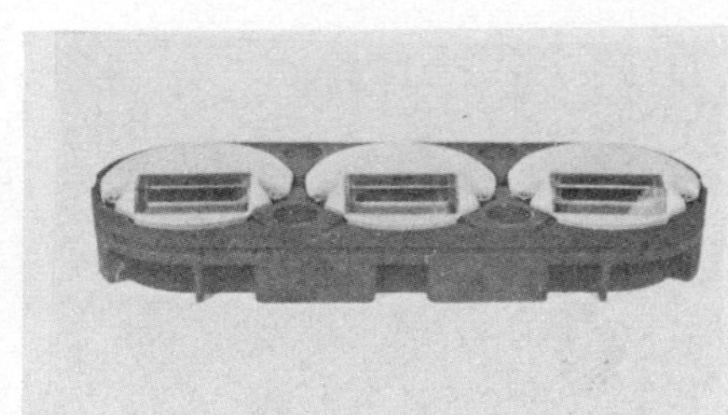

BBT 100 000-cd revolving airport beacon

Claude Obstacle Warning Lights

Claude produce three types of obstacle warning lights, the Obsta, an autonomous solar energy powered light and a strobe.

The Obsta is an explosion-proof flashing light which uses a xenon discharge-tube powered by an electronic converter. Its energy is 20 joules per flash. Power supply is 24 or 48 volts dc or 230 volts ac on request.

The autonomous warning station is equipped with silicon solar panels, storage batteries, charging regulator, dc/ac converter and a standby set.

The lights, one or two working together or with one in a standby mode are powered through the 24 or 48 volt dc batteries which are recharged by solar energy. The unit is designed for providing visual warning of isolated obstructions which are without electrical power supplies.

The strobe light, fitted with a printed-circuit board, provides 60 flashes per minute at 20 joules per flash. Power supply is from 48 volts dc to 220 volts ac.

Status

Production. The Obsta and the autonomous warning station were developed in 1981

Manufacturer

Societe Claude, Division Protection and Systems, Reims, France

Claude Obsta flashing obstruction light

Claude obstacle warning station

CSEE Essential Power Supplies

CSEE produce emergency power plants for airports. The plant installed at Paris-Orly consists of four 1500 kVA, 5·5 kV permanently running generators and one 800 kVA, 380 volt stand-by generator starting automatically.

Status

Production. CSEE emergency power plants have been supplied to the following civil airports: Toussus, Pontoise, Orly, Brest, Bordeaux, Poitiers, Nîmes, Perpignan, Lyons, France; Nouasseur (Morocco) with outputs ranging from 80 to 1600 kVA. At military air bases 57 automatic starting plants (40 to 600 kVA), 26 generators with permanently coupled alternators (135 to 300 kVA), six generators with instantaneous alternator coupling (120 to 500 kVA) have been installed

Manufacturer

Compagnie de Signaux et d'Entreprises Electrique, Paris, France

Holophane Visual Landing Aids

Visual landing aids offered by Holophane comprise a Vasi, a T-Vasi and a Papi system for helicopter platforms.

The Vasi is a standard ICAO landing aid which usually includes two pairs of wing bars with each bar consisting of three F19 indicators. Each indicator produces two light beams, the upper beam being white and the lower beam red, and the colour transition defines an upper or lower approach zone limit. The pilot manoeuvres so as to bring and maintain his aircraft on the correct glide-slope when the downwind bars of the Vasi are white and the upwind bars red. When the system has to be used by large aircraft on which the height between the pilot's eyes and the wheels is between 4·5 and 16 metres, a three-bar Vasi is installed. The pilot of an aircraft of normal height uses the information from the downwind and intermediate bars; on large aircraft he uses the information from the intermediate and upwind bars.

The F19 indicator complies with the latest ICAO recommendations with a visual range greater than 7·4 kilometres. It is equipped with two quartz halogen 100 watt lamps producing 60 000 candelas or alternatively two quartz halogen 200 watt lamps producing 100 000 candelas. The unit is installed on a frangible mounting designed to withstand jet blast but break off on impact with an aircraft.

The T-Vasi is also a standard ICAO landing aid. It enables the pilot to be given the necessary information for placing his aircraft on the ideal approach slope, and can be used by day and by night in good and average visibility. The T-Vasi system has the following advantages over the Vasi device. The system provides progressive information on vertical deviation from the glide-slope and homochromatic indications avoiding errors in the identification of Vasi colours in reduced visibility. On each side of the runway the system comprises: a wing-bar consisting of four F21 indicators with a constantly visible white and red beam without a cut-off; three indicators aligned upwind from this bar, with a white beam and a low cut-off making them practically invisible when the aircraft is too low; and three indicators aligned downwind from the bar, with a white and red beam and a high cut-off making them practically invisible when the aircraft is too high. Similar to the wing-bar indicators the low part of the beam is red so as to warn the pilot coming in too low. When the aircraft is on the glide-path the pilot sees only the wing-bar with the lights just before and just after slightly lit.

The F21 indicator complies with the latest ICAO recommendations and has two types of lighting using the same optical system: day lighting by means of two quartz halogen 200 watt focused lamps in the reflectors, and night lighting at reduced intensity, with a wider horizontal beam spread by means of two defocused quartz halogen 100 watt lamps in the same reflectors. In a similar manner to the Vasi, this unit is installed on frangible mountings.

The Papi system for helicopter platforms comprises two wing bars each consisting of two F19 indicators. Like the F19, the F19 H produces two beams, white and red, and their transition fixes an approach zone limit. Generally the approach is made at a descent angle of 8 degrees. The transition angle of the inner indicator will be set at a degree above the ideal approach angle (9 degrees) and that of the outer indicator a degree below (7 degrees). If the helicopter is at the correct angle of approach, the pilot will see the inner indicators on either side of the platform indicating red and the outer indicators showing a white light. Above the glide-path, all indicators will be white and below the glide-path, all will be red. The F19 H indicator will normally be equipped with two quartz halogen 45 watt lamps producing a visual range greater than 5 kilometres.

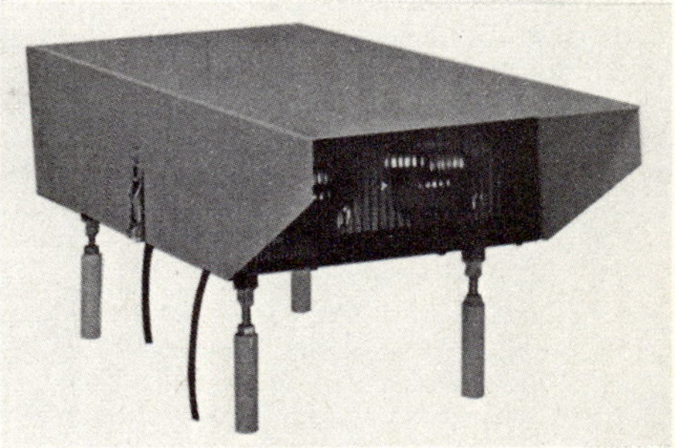

Holophane F19 Vasi unit

Status

Production. Holophane airfield lighting systems are installed at airports in Europe, the Middle East, Africa, Australia and Malaysia

Manufacturer

Holophane, Paris, France

Holophane Inset Airfield Lights

Holophane F9-4 approach and runway flush lights

Inset lights are available for the range of category I, II and III requirements. Runway centre-line lighting is standardised on runways which can be used for category II and III operations and is recommended by ICAO for category I operations. Like the F7-4, its uni-directional counterpart, the flat, bi-directional, F10-4 light is specially intended for these runways. It is small in size and is designed for sealing into the runway pavement by means of organic cement. Its active part is removable. Power supply is separated for the two lamps so as to permit operation in one direction at a time when required. Similar to the runway centre-line lighting, touchdown zone lighting is standardised on runways which can be used for category II and III operation. The F7-4 flat, uni-directional light is specially designed for these runways. The top assembly is identical, and therefore interchangeable, with those used in threes on the series F9-4 flat approach or threshold light.

In most cases flush lights should be used for threshold and end lighting on categories I, II and III runways, since ICAO recommends that these runways should be equipped with paved overrun areas beyond the threshold/end and only flush lights can be fully satisfactory in these zones.

Series F9-4 lights are designed for these areas and for lighting the end of the approach system on a runway with a displaced threshold. These lights enable aircraft to pass easily owing to their limited projection above the ground.

The normal complement to runway centre-line lighting, taxiway centre-line lighting is standardised by ICAO on taxiways associated with a category III runway. This lighting is also recommended for taxiways associated with lower category runways, such as categories I and II, when traffic is heavy. Like its F13-4 asymmetric counterpart, the flat, symmetrical, bi-directional F6-4 light is intended for taxiway centre-line marking. Owing to its photometric characteristics, it is suitable as standard equipment for sections of straight taxiways. With a 100 watt lamp, it also meets the specifications as a runway centre-line light when it is spaced at 7·5 or 15 metres on category I and II runways.

The asymmetric, bi-directional, flat F13-4 light is also designed for taxiway centre-line lighting. The width of its light beams makes it suitable for curved taxiways, even when the curve radius is very low and permits good visibility of the taxiway configuration.

The F3 series of inset lights consists of two models which are identical in design but protrude to different heights above the ground. The light is immersible and its glass is claimed to be capable of withstanding the weight of aircraft running at moderate speeds and road vehicles equipped with tyres. It can be used in all areas over which aircraft pass outside the runway. Owing to its low projection, the F36is light is reserved especially for lighting the touchdown area of helicopter platforms and for use on roads. For other purposes the F3 has a better light distribution at low incidence.

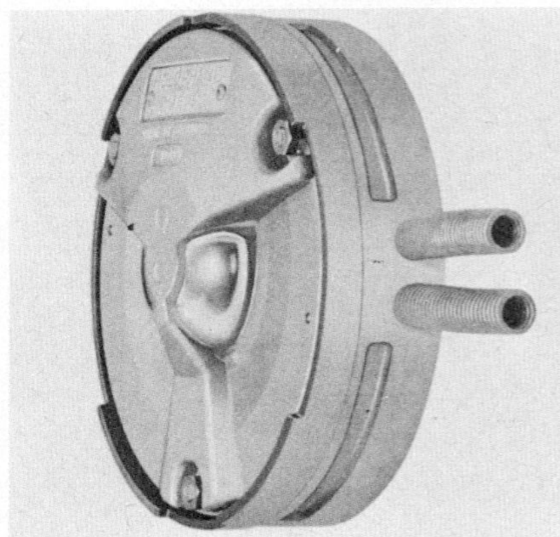

Holophane F13-4 taxiway light

Holophane F3 low intensity flush light

Characteristics

SERIES	F10-4	F7-4	F9-4	F6-4	F13-4	F3
TYPE	Bi-directional flush	Uni-directional flush	Uni-directional flush	Bi-directional flush	Bi-directional flush	Omni-directional low intensity
COLOUR	Clear, green/red	Clear, green, red yellow	Clear, red, green	Clear, green/red	Green, blue	Blue, red clear, yellow, green/red
LAMPS	Quartz halogen 2 × 100 W 6·6 A	Quartz halogen 2 × 45/100 W 6·6 A	Quartz halogen 3 × 45/100 W 6·6 A	Quartz halogen 1 × 45/100 W 6·6 A	Quartz halogen 1 × 45/100 W 6·6 A	F3/E27 1 × 25–65 W 6·6 A

Status

Production. Holophane airfield lighting systems are installed at airports in Europe, the Middle East, Africa, Australia and Malaysia

Manufacturer

Holophane, Paris, France

Holophane Elevated Airfield Lights

A range of elevated lights are produced for a wide variety of applications.

The F12 and F15 series lights, when used in the approach pattern, can be mounted on a pole permitting heights of up to 17 metres or a safety mast with frangible coupling. The F15 operated in a tracer-ball approach pattern is supplied with a control box incorporating a timer which operates the 20 lights in the system. When used for runway identification, the light is synchronised by means of a clocking card inserted in one of the two lights. It becomes the master light, the other a slave. The F20 is a new model for marking runway edges, replacing the F17.

As a complement to high-intensity runway lights the F2-1 light, like its counterpart the F3 light, is mainly intended for marking the boundaries of aprons and taxiways. It is also used for the medium or low intensity lighting of runways and secondary approach systems. It is equipped with a P28S socket for a 6·6 ampere lamp and is supplied in a series circuit. Owing to the simplicity of the F2-1 light with an E27 socket and a standard lamp, this model is suitable for general use in lighting obstructions and lighting helicopter platforms. The F1 model is a variant of the F2-1 light; the only difference is its glass which is smooth on the outside. It is more economical but has a lower light intensity and is suitable for lighting close-lying obstructions, such as anti-blast fences.

The F18 transformer light is designed for simplified aerodrome or helicopter platform lighting without the addition of an insulating transformer. It is particularly suitable for fixed or field military aerodromes, helicopter operations and small aerodromes intended for seasonal civil transport. The F18 can be placed on the ground, stake-mounted or sealed and can be used on earth, paved runways and platforms for boundary marking of runway and manoeuvring areas, approach, etc.

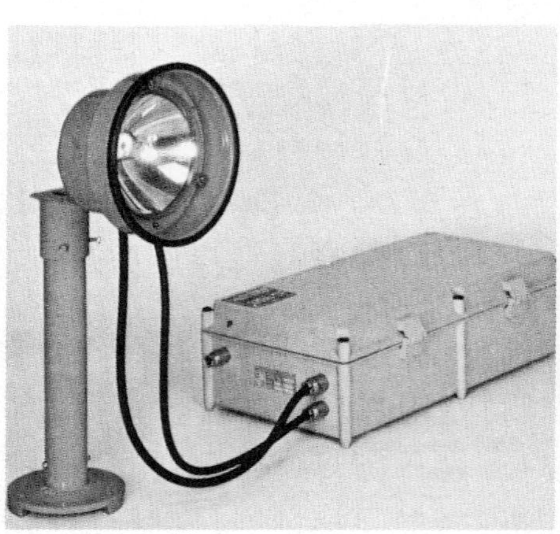

Holophane F15 condenser discharge light

Holophane F18 transformer light

Characteristics

SERIES	F12	F20	F15	F2-1	F18
TYPE	Uni-directional high intensity	Bi-directional high intensity	Uni-directional condenser discharge flashing	Omni-directional medium/low intensity	Omni-directional low intensity portable transformed
COLOURS	Clear/red/green	Clear/green/red/yellow	Clear	Clear/red/blue/green/red	Clear/red/blue/green/yellow
LAMPS	119/200/300 W 6·6 A	200 W 6·6 A	Xenon PAR 56 FT 34/HP2 × 60 J/s	40–100 W 6·6 A	30/45 W 6·6 A

Status
Production. Holophane airfield lighting systems are installed at airports in Europe, the Middle East, Africa, Australia and Malaysia

Manufacturer
Holophane, Paris, France

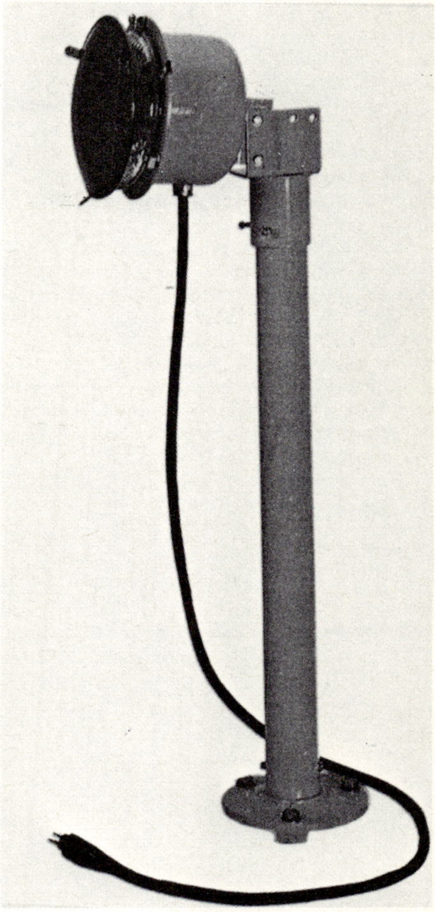

Holophane F12 elevated approach light

Holophane Apron Obstruction and Accessory Lighting

The series PRT 25, 26, 35 and 36 lights operate by the principle of lamp defocusing and the use of an internal deflector to obtain a beam suitable for large sized apron lighting with a reduced number of projectors. Generally, double envelope-type halogen lamps are used in PRT 25 and 35 projectors. The variants PRT 26 and 36 are equipped with halogen quartzine lamps. The lights can be mounted on masts supplied on request, or on the elevated structures of the terminal building.

Also available for apron lighting are the series PRTA 25 and 35 light which use high pressure sodium lamps, 400 or 1000 watt, for installations where high illuminance is requested.

Obstruction lights offered by Holophane comprise F2-1 single light, F22 twin configuration, Obsta light and Balisor line lights.

The F2-1 is an omni-directional red light with Fresnel prisms which uses a 55 watt, 230 volt lamp. The twin configuration F22 incorporates two F2-1 units supported in a box in which there are two fuses and a reversing relay transferring the supply from one lamp to another when it is out of service and vice versa after a change of lamp. Remote signalling can be provided.

The Obsta lights, which are more expensive, are equipped with a neon lamp with a service life exceeding 20 000 hours. They are used on obstructions which are difficult to reach.

Balisor line lights are intended for lighting high-tension lines exceeding 60 kV. They are fixed to the conductor itself and draw the necessary energy from its electrostatic field.

Accessory lighting includes signal panels for large airports with complex taxiways, illuminated wind sock, illuminated landing T, aerodrome beacon, and an identification beacon. The landing T consists of two perpendicular segments which are each 4 metres long and 0·5 metres wide. They are white and support the 18 lights regularly distributed over the structures. The system is mounted on a cabin with a vertical-rotation axis driven by a steering servomechanism.

The standard model of aerodrome beacon gives

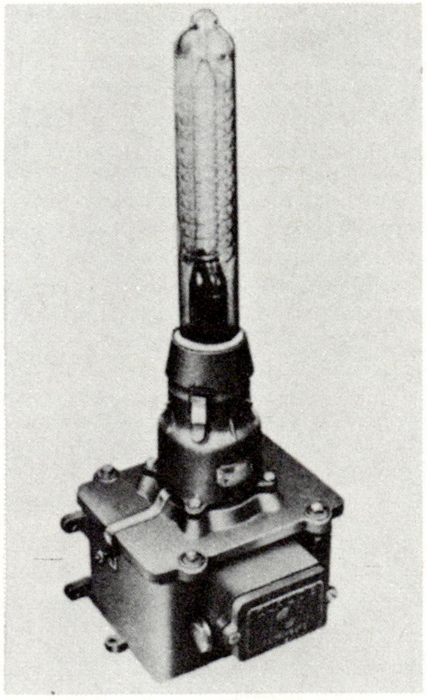

Holophane PRT 25 apron light

Holophane Obsta light

Holophane apron lighting

alternately one white flash and one green flash at a regular rate of 12 flashes a minute, but other colours and other rates can be provided. The device consists of: a revolving lantern with four sides using optical reflecting lamps. The other two sides are used for stand-by beams automatically activated as soon as a normal beam lamp fails. The housing contains the electrical control and protection components and the motor turning the lantern.
An identification beacon is installed on an aerodrome or heliport when, if it is used by night, it cannot be easily identified in flight by other visual means. The beacon is installed so that it is visible in all azimuthal directions of observation. It is fixed by means of four bolts and the device must be carefully levelled. This beacon gives off white or green flashes at ground aerodromes and heliports, and yellow flashes at water aerodromes; these flashes are Morse coded using the identification letter allocated to the base, and controlled by an electrical eclipsing device. It can also be used as a hazard beacon, providing an average of 30 red flashes a minute.

Status
Production. Holophane airfield lighting systems are installed at airports in Europe, the Middle East, Africa, Australia and Malaysia

Manufacturer
Holophane, Paris, France

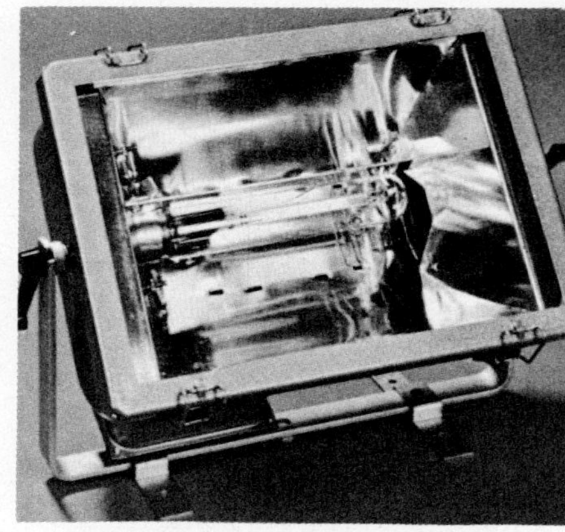

Holophane aerodrome beacon

Characteristics

MODEL	PRT 25/ PRT 26	PRT 35/ PRT 36	PRTA 25/ PRT 35	F2-1	F22	OBST	Balisor line	25219	25220	25188
TYPE	Floodlight	Floodlight	Floodlight	Omni-directional	Omni-directional	Omni-directional	Omni-directional	Omni-directional	Flashing beacon	Flashing beacon
APPLICATION	Apron terminal areas	Apron terminal areas	Apron terminal areas	Obstruction	Obstruction	Obstruction lighting for inaccessible places	Obstruction lighting for high tension lines	Illuminated landing T	Aerodrome light	Identification /hazard
COLOUR	Clear	Clear	Clear	Red	Red	Red	Red	Red	Green/white 12 flashes/ minute	Green/white 30 flashes/ minute or red
LAMP	Double envelope 1 × 500/ 1000/2000 W 240 V	Quartz halogen 1 × 1000/ 2000 W 240 V	HP sodium 1 × 400/ 1000 W	1 × 55 W 230 V	2 × 55 W 230 V	1 × neon 230 V		18 × 40 W 240 V	4 × 500 W 240 V	1 × 1000 W 1 × 500 W 230 V

Holophane Control and Distribution Systems

There are two types of feed for lighting fixtures along a runway or a taxiway; parallel or series feed. Holophane claim that with the parallel feed, since circuits of a runway are several kilometres long, a progressive voltage drop occurs from one light to the other, even with wide cable sections: this results in expensive cable and a heterogeneous luminous intensity of lights and low voltage prefocus lamps still require a step-down transformer. With a series feed the same current runs through all the lamps, providing the possibility of a smaller cable section, with a subsequent reduction in the voltage drop. A series feed requires a constant current regulator, insulating transformers and lamps specially for a rated current of 6·6 or 20 amperes. All lighting fixtures are designed for use of these lamps.
Three types of insulating transformers are produced. The INT transformer is coated with a layer of vulcanised polychloroprene which covers the two primary uni-polar conductors and the secondary bi-polar conductor. The FR transformer is also coated with a layer of polychloroprene but differs in that it permits direct supply by a shielded cable. The GR-type transformer is moulded in epoxy resin. It is also immersible. Although it has no outer conductors, it may be disconnected by means of special connectors inside the overmoulding. The transformers are available with ratings from 30 to 500 watts.
Holophane produces ferro-resonant regulators for step regulation of brightness, and magnetic amplifier regulators for progressive brightness control. The lighting control system is almost always installed in the control tower. Control panels can be integrated with other systems custom-built to suit individual operators. They can incorporate: general key-operated switch, brightness step switches or contactors, potentiometers for continuous brightness regulation, direction reversers, pre-selection contactors for grouped devices, auxiliary switches for an illuminated landing T, windsock, etc and positive or alarm repetition indicators. A representation of the aerodrome runways and taxiways can be provided, either on the same panel or on a separate panel.
When a complete console is required, a control desk, made of steel sheet, can be supplied, according to dimensions given by the customer. Control desks incorporate a 48 volt dc mains for remote control, and controls of the transformers.
Optional features offered comprise switches and reversers on a keyboard, with push-buttons, luminous; control potentiometers, for progressive control of regulators (am type) and provision for adjusting the intensity of the tell-tale lamps at night.

Status
Production. Holophane airfield lighting systems are installed at airports in Europe, the Middle East, Africa, Australia and Malaysia

Manufacturer
Holophane, Paris, France

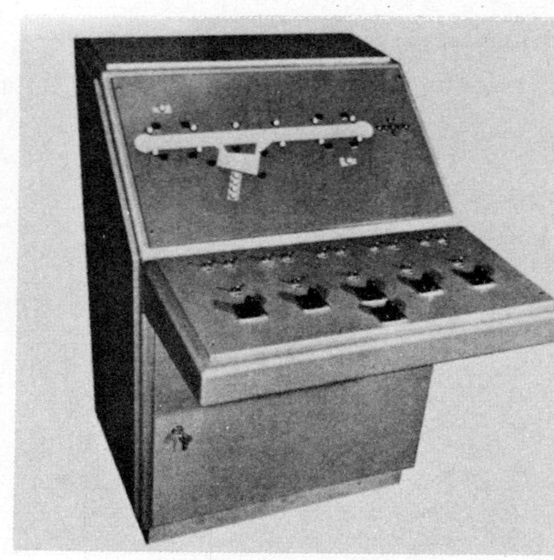

Holophane remote-control desk

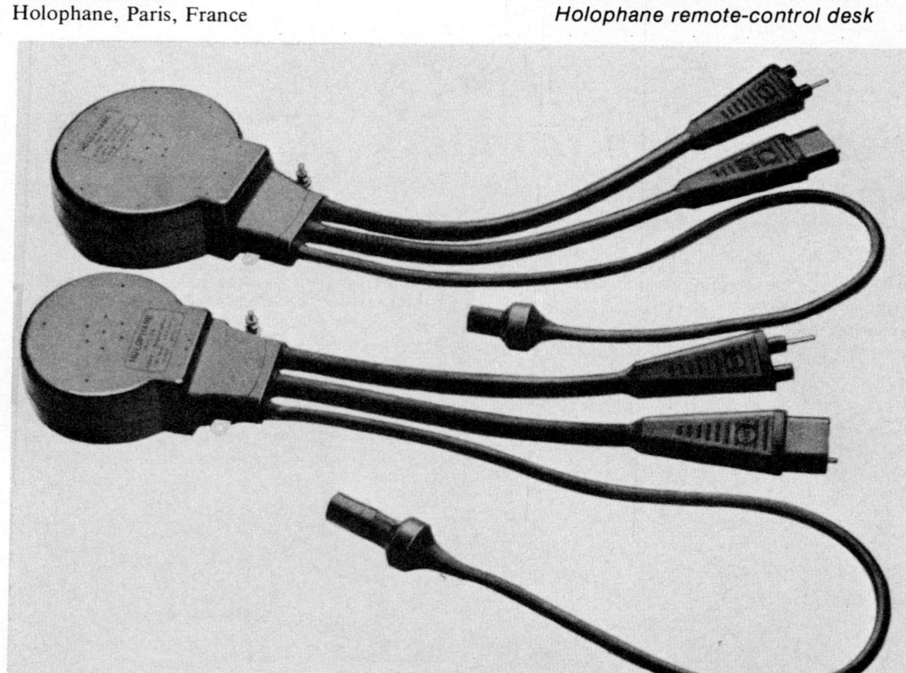

Holophane FR series isolating transformer

Santerne Airfield Lighting

Santerne produce airfield lighting systems which include elevated approach lights, runway and taxiway lights, obstruction and marker lights and environmental lighting. Control and distribution equipment is available and the company can supply emergency generator stations, individual generator units and uninterruptable power supplies.

Status
Production. Civil installations of Santerne equipment include external lighting of bridges and roadways at Paris Charles de Gaulle Airport, France; runway, marker lights and luminous indicators at Paris Orly, marker lights at Toulouse Blagnac, France, marker lights and external lighting at Lyons Satolas, France, and the airfield lighting at Strasbourg including the control and distribution network

Manufacturer
Santerne, Arras Cedex, France

Santerne airfield lighting

Ulmer Navigation and Elevated Approach Lights

Ulmer produce a Vasi, condenser discharge light for use as an approach marker and elevated approach lights. The BP 5862 Vasi can be used with two- and three-bar systems, indicating all white bars to the pilot that the aircraft is above the glide-slope, red and white on the glide-path, and all red lights that the aircraft is below the glide-slope. The unit uses 200 watt halogen lamps to produce a beam with a luminous intensity exceeding 100 000 candelas in white and 15 000 in red, extending 20 degrees in azimuth and 8 degrees in elevation. The width of the pink zone in the transition from red to white or vice versa is an arc of approximately 15 minutes. The unit is housed in an aluminium structure on a frangible mounting designed to withstand jet blasts but collapse on impact with an aircraft.

BP 5761 is a condenser discharge light designed for use as a marker light in an approach path. A typical system might consist of 20 lights on the extended centre-line of the runway at 30 metre intervals from 900 to 300 metres from the touch-down end of the runway facing up the glide-path. The lamps are flashed one after the other from the outer end of the path to the inner end in about ⅓ of a second. The cycle is repeated at ½ second intervals. The optical system consists of a PAR 56 xenon flashing lamp with a control box in each lamp which includes a timer providing the required flashing sequence. The cast aluminium body contains the ignitor and the terminal strip. A safety lock prevents contact from high voltage, when the lamp is disconnected. The unit is mounted on a frangible coupling. The elevated approach lights available comprise the UL23 for category I, II or III runways and the UL18 and Type 233 for a simplified approach pattern. The UL 23 is designed for installation on a tripod mast or a frangible post with a maximum height of two metres. The sealed beam PAR 56 lamp is rated at 200, 300 or 500 watts. Coloured filters and an obstacle beacon light or a low-intensity approach light can be added if required. The UL18 is an omni-directional light suitable for low voltage approach operations. It uses a 45 watt halogen lamp, with a prefocus system, fresnel lens, super-polyamide base and an aluminium alloy mounting. The type 233 is a dual-purpose uni-directional light, which can be used with a 100 watt, 6 ampere halogen lamp to provide a simplified approach pattern. Alternatively, fitted with a 45 watt, 6·6 ampere halogen lamp, the unit is suitable for marking runway edges.

Characteristics

MODEL	UL23	UL18	Type 233	BP 5761
TYPE	Uni-directional high intensity	Omni-directional medium intensity	Uni-directional medium intensity	Uni-directional flashing
APPLICATION	Approach Cat I, II, III	Simplified approach	Simplified approach/runway edge	Runway threshold identification (Reils)
LAMP	PAR 56 200/300/500 W	1 × 45 W P 18S 6·6 A	1 × 100/1 × 45 W P 18S 6·6 A	1 × PAR 56 xenon 60 W

VASI	
AZIMUTH	20°
ELEVATION	white 4°, red 4°, pink 0·5° max
INTENSITY	white: 100 000 cd red: 15 000 cd
LAMP	2 × 200 W, 6·6 A
UNIT DIMENSIONS	0·954 × 0·68 m
WEIGHT	16·5 kg

Status
Production. All lights are manufactured to meet ICAO requirements

Manufacturer
Ulmer Aeronautique, Clichy, France

Ulmer Runway and Taxiway Elevated Lights

A variety of lights are offered in this category for marking runways and taxiways. The UL22 light is designed to supply high- and low-intensity lateral lighting for runways, with category I, II and III approach precision. It can be used for lighting the thresholds and ends of runways.

The BP 5732 is a high intensity lamp. Multi-directional with two preferential beams, it can be used for marking the sides or thresholds of runways equipped for instrument landings.

The UL2 is a low intensity omni-directional light for marking the edges of runways and roadways and the thresholds and extremities of runways with low traffic intensity. A second configuration of this light, the UL2H, is designed for use at heliports.

Characteristics

MODEL	UL22	BP 5372	UL2	UL2H
TYPE	Omni-directional high intensity	Multi-directional high intensity, 2 preferential beams	Omni-directional low intensity	Omni-directional low intensity
APPLICATION	Runway edges	Runway edges ifr runways	Runway edges threshold/edges–low traffic runways	heliports
LAMP	1 × 100/200 W, 6·6 A	1 × 150/200 W, 6·6 A	1 × 30/45 W, 6·6 A	1 × 60 W, 24–220 V

Status
Production

Manufacturer
Ulmer Aeronautique, Clichy, France

Ulmer Runway and Taxiway Inset Lights

Ulmer UL7-1 inset light

Ulmer UL9-1 inset light

Inset light fittings produced by Ulmer are suitable for use in the touchdown area, runway end, fast runway clearing, turning points and taxiways.
All comprise a mounting base which is set into the pavement, an optical system with appropriate support device and the cover. The supporting chassis allows the uni- or bi-directional lights to be adjusted in elevation between 0 and 8 degrees. Other inset lights are the UL3, an omni-directional light for marking taxiways, parking areas and the runway threshold; the 438 Al, a bi-directional light of high- and low-intensity for runway edge marking and the UL3H, an omni-directional light with fresnel lens cover intended for use at heliports.

Characteristics

MODEL	UL6	UL7-1	UL9-1	UL10-1	UL13	UL3	438 Al	UL3H
TYPE	Bi-directional symmetrical	Uni-directional	Uni-directional lights outer lights angled towards centre by 2°/6°	Bi-directional	Bi-directional asymmetrical	Omni-directional	Omni-directional high/low intensity	Omni-directional
APPLICATION	Taxiway axis runway axis	Touchdown area fast runway clearing runway end	Runway threshold runway end	Runway axis	Taxiway turns	Taxiway edge parking area runway end/ threshold low intensity	Runway edge	Heliports
COLOURS	Clear/green	Clear/red/ green	Clear/red	Clear/red	Green	Clear/red/ green	Clear/red blue/green/ yellow	Clear/yellow
LAMP	1 × 100/45 W, 6·6 A	1 × 100 W, 6·6 A	3 × 100 W, 6·6 A	2 × 100 W, 6·6 A	1 × 100 W, 6·6 A	1 × 45/100 W, 6·6 A	1 × 100/40 W, 6·6 A	1 × 45 W, 6·6 A

Status
Production. The UL series lights are designed in accordance with ICAO recommendations

Manufacturer
Ulmer Aeronautique, Clichy, France

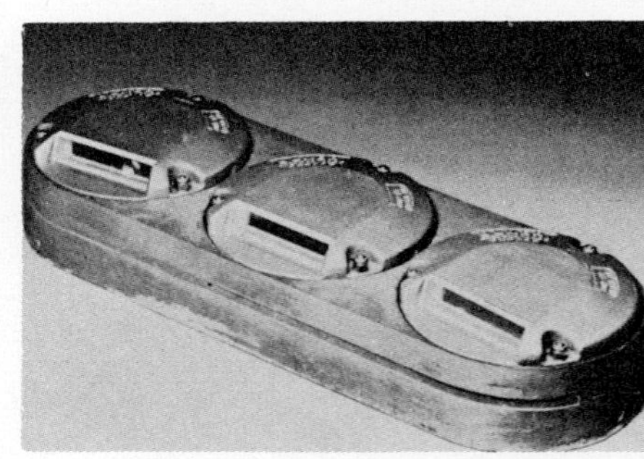

Ulmer UL10-1 inset light

Ulmer Obstacle and Accessory Lighting

In this category Ulmer produce a range of lights suitable for marking obstacles, airfield beacons, portable lights, direction indicators, landing T and wind sock. The signs, T and wind sock are all designed to be illuminated for night operations. Single and double omni-directional obstacle lights are available with single or dual 67 or 111 watt lamps. For use in a series installation 65 watt lamps are offered optionally. A portable obstacle beacon light is produced which projects a clear beam. Red, green, yellow and blue filters can be fitted and the flash frequency regulated between 12 and 170 flashes per minute. The airfield beacon designed for airfield identification from the air, incorporates four optical systems. The two main systems producing a white and a green beam, are replaced by the standby systems in the event of a lamp failure. The light rotates at 6 revolutions per minute producing 12 flashes per minute.

Characteristics

MODEL	BP 5272	BP 5273	BP 4910	BP 5138	BP 5412	BP 5149	BP 5279/ 5280
TYPE	Single omni-directional low intensity	Double omni-directional low intensity	Portable omni-directional flashing	Portable omni-directional trans-former lights	Green/ white rotating beams	Landing T	Wind-sock unit
APPLICATION	Obstruc-tion marking	Obstruc-tion marking	Obstruc-tion marking	General purpose marking	Airfield location	Day/night	Day 5279 day/night 5280
LAMP	1 × 67/ 111 W 110/220 V or 1 × 65 W 6·6 A	2 × 67/ 111 W 110/220 V or 2 × 65 W 6·6 A	1 × 7 V	1 × 30/ 45 W 6·6 A	1 × 500 W 24 V	2 × 16 × 15–40 W 220 V	1 × 150 W

Status
Production

Manufacturer
Ulmer Aeronautique, Clichy, France

Ulmer Power Control and Distribution Systems

Ulmer manufacture a range of equipment for airfield lighting control and power distribution. These include remote-control desks, constant current regulators, isolating transformers and sealed connectors.
The remote-control desks can be supplied as a standard installation in desk form, for mounting on a table or stand, or mounted on a wall. The standard models comprise a diagrammatic representation of the airport and the lighting system, and change-over and step switches for controlling intensity and landing direction. When continuous brilliancy control regulators are used for series circuits, step switches are replaced by potentiometers, either with or without spring catches. Alternatively, a custom-built desk can be provided to meet operator requirements.
Several types of constant current regulator (ccr) are offered. The Ferroresonant oil-cooled ccr is produced with ratings from 2–30 kVA, and air-cooled Ferroresonant units are rated from 2–7·5

kVA. Magnetic amplification ccr equipment which is oil-cooled is available rated at 2–30 kVA and air cooled versions are offered at 2–7·5 kVA. A further type of ccr is produced, the THA-type with a bi-directional thyristor rated at 2·5 to 20 kVA.

Two types of isolating transformers are manufactured. The transformer insulates the primary series high voltage circuit from the secondary circuit supplying current to the beacon light and ensures the continuity of the primary circuit in the event of failure of a bulb.

The BP 5371 series are capable of handling 30–600 watts in a wide variety of primary and secondary current configurations. These units are sealed in a polychloroprene body. The second version BP 5747 available with power ratings from 45 to 550 watts at 6·6 ampere are designed for supplying power to visual aids. The sealed transformers are coated with an epoxy resin.

Status
Production

Manufacturer
Ulmer Aeronautique, Clichy, France

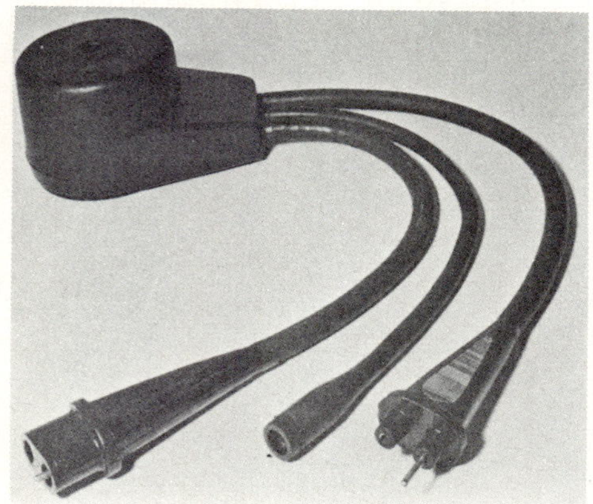

Ulmer type BP 5371 isolating transformer

Ulmer constant current regulator

GERMANY, FEDERAL REPUBLIC

AEG Telefunken Approach and Navigation Lights

High- and low-intensity approach lights, steady and flashing and an approach-angle light to form a Vasis, are manufactured by AEG Telefunken. The lights conform to ICAO specifications.

ATR 1.1 is a high- and low-intensity light. It serves as a marker for the approach, the threshold, the overrun strip and the touchdown zone.

ATR 1.2 has a similar function but is not equipped with the low-intensity light. In both versions the elliptic-shaped light-metal housing is fixed by a support column. Column and column support are mounted on an interchangeable breaking coupling which is screwed to the bottom flange. For elevated mountings on poles or on structures the lights are normally supplied without breaking coupling. The lights are set level by means of a bubble which can also be used to adjust the required horizontal angle towards the runway. The graduated scale 0 to 15 degrees for the required angle is placed at the lamp support. The horizontal range is ±180 degrees. The lamps can be switched and adjusted independently from each other and are connected to the isolating transformers by a vulcanised type two- or four-pole plug device, an electric breaking coupling, which passes through the frangible tube. The isolating transformers are installed in a water-tight cast iron container or in a resin moulding.

The approach flashing light BL 357 LK serves as marker for the approach system in addition to the fixed lights. Flash lights are installed along the approach centre-line and at the threshold outer points. The periodic flash cycles with extreme high-light intensity enable the pilot to recognise the approach even at poor visibility when the normal approach lighting has become insufficient. The light which is contained in a waterproof housing, uses a xenon discharge tube producing 60 watts with two flashes per second at 1·8 kV discharge flash voltage. Mounting is via a frangible coupling with a bottom flange for a concrete foundation or an adaptor for elevated mounting on a pole or structure.

The approach angle light AWF 51 forms the Vasis visual landing aid. The light has been designed in accordance with the standards of the ICAO and the Federal German Ministry of Transport and is equipped with two optical systems. If the 200 watt halogen lamp fails, the light is automatically switched to the second system. With a two bar Vasis, three lights in two rows either side of the runway indicate to the pilot whether the aircraft is on the glide-path, above or below during the approach phase. Above the glide-path, all lights are white, below the glide-path indicating to the pilot to reduce the rate of descent, all lights are red. One row of red and one of white show the pilot to be on the glide-path.

AEG AWF 51 approach angle light

AEG BL 357 LK approach flashlight

Status
Production. Installed at airports world-wide, all lights meet ICAO specifications

Manufacturer
AEG Telefunken, Wedel (Holst), Federal Republic of Germany

Characteristics

MODEL	ATR 1.1/ATR 1.2	BL 357 LK	AWF 51
TYPE	Uni-directional/ omni-directional high/low intensity (ATR 1.2 high intensity only)	Flashing uni-directional	Uni-directional
APPLICATION	Approach	Approach	Vasis
COLOUR	Clear/red/green/ yellow	Clear	Clear/red
LAMP(S)			
high intensity	200 W, 6·6 A or 300 W, 20 A	Xenon 60 W, 220 V	2 × 200 W, 6·6 A
low intensity	45 W, 6·6 A		
UNIT DIMENSIONS			915 × 450 mm
projected height	530 mm	630 mm	411 (minimum)
light diameter			
low intensity	98 mm		
high intensity	204 mm	445 mm	

AEG Telefunken Elevated Runway Lights

AEG Cl/M high intensity runway light

AEG ATR 2.1 high/low intensity lights

Elevated runway lights produced by AEG Telefunken comprise the ATR 2.1, ATR 2.2 and Cl/M high intensity units. The ATR 2.1 is a high-and low-intensity light. The ATR 2.2 is similar but without the low intensity feature. The lights serve as a runway marker and can also be used as a threshold and overrun-strip light. The elliptic-shaped light-metal housings are fixed by a support to the vertical column. Column and column support are mounted on an interchangeable frangible coupling which is screwed to the bottom flange. For elevated mountings on poles or on structures an adaptor is available. The lights are set level by means of a bubble which can also be used to adjust the required horizontal angle towards the runway. The graduated scale 0 to 15 degrees for the required vertical angle is placed at the lamp support. The horizontal adjustable range is ±180 degrees. The lamps can be switched and adjusted independently from each other and are connected with the isolating transformers by a vulcanised-type four- or six-pole plug device, an electric breaking coupling which passes through the breaking tube.

The isolating transformers are installed in a waterproof cast iron container or in a resin sheathing. The Cl/M is designed to serve as a runway marker. It can also be used as threshold and overrun-strip light. The refractor support and lamp socket with incandescent lamp are mounted on a ball-type flange which can be adjusted on a socket by four screws. The refractor accommodates a concave mirror which in combination with the lenses of the fresnel lens effect a fixed main beam direction of 4 degrees towards the runway centre-line and also of 4 degrees in an upward direction. The lamp is connected to the isolating transformer by a vulcanised-type two-pole plug device.

Status
Production. Installed at airports world-wide

Manufacturer
AEG Telefunken, Wedel (Holst), Federal Republic of Germany

Characteristics

MODEL	ATR 2.1/ATR 2.2	Cl/M
TYPE	ATR 2.1 high and low intensity ATR 2.2 high intensity only	High intensity
APPLICATION	Runway light	Runway light
COLOURS		
high intensity	Clear/red/green/yellow	Clear/red/green/yellow
low intensity	red/green/yellow or clear protective glass with red/green half filters	
LAMP		
high intensity	1 × 200 W, 6·6 A or 1 × 300 W, 20 A	1 × 200 W, 6·6 A
low intensity	1 × 45 W, 6·6 A	
UNIT MEASUREMENTS		
projected height		
ATR 2.1	485 mm	
ATR 2.2	362 mm	380 mm
light diameter		
high intensity	204 mm	190 mm
low intensity	98 mm	

AEG Telefunken Inset Light Fittings

Inset light fittings available are suitable for runway, taxiway and apron applications. Included in this category are BOF 41 to 45, BOF 36 QJ and BOF-L8.

The inset light BOF 41 to 45 models are designed to serve as markers for runway and taxiway centre-lines and also as threshold or touchdown zone lights for category I, II and III operating conditions. The flush-mounted inset lights have a modular design incorporating an optical system with parabolic mirror and glass cover, and all lights have the same dimensions. By changing the upper part of the unit, the different versions of the BOF light family are created.

The BOF 36 QJ is a medium-intensity light suitable for use as a marker for taxiway edge and centre-line, apron edge and guidance lines and for all kinds of heliports. In special cases it can also be used as runway edge and centre-line light. It is used where aircraft are likely to overrun the light since it is embedded in the runway so that it is level with the surface. The housing is made of cast aluminium, and comprises an upper and a lower part. The elastically supported omni-directional lens seals it to render it dustproof and watertight. The interior of the light accommodates the removable lamp socket-holder with lamp socket and halogen lamp. Colour filters can be fitted to produce any required colour.

The BOF-L8 is a low-intensity light of similar construction to the BOF 36 QJ being a flush mounted light with omni-directional lens in an aluminium housing. All aviation colours can be obtained by placing the corresponding filters under the omni-directional lens. A water tight two-pole plug device connects the lamp with the vulcanised isolating transformer.

Characteristics

MODEL	BOF 41	BOF 42	BOF 43	BOF 44	BOF 45	BOF 36 QJ	BOF-L8
APPLICATION	Runway centre-line, stop bar touchdown zone	Runway centre-line threshold	Taxiway centre-line straight	Taxiway centre-line straight Cat I, II, before/after curves Cat III	Taxiway centre-line curves	Taxiway edge centre-line apron edge and guidance lines	Taxiway edge centre-line apron edge and guidance lines
OPTICAL SYSTEM	Parabolic mirror smooth glass cover	Parabolic mirror smooth glass cover	Parabolic mirror smooth glass cover	Parabolic mirror glass cover with smooth front and diffusing rear	Parabolic mirror glass cover with smooth front and diffusing rear	Omni-directional lens	Omni-directional lens
FILTER	Red	Red/green	Green	Green	Green	Blue/red/green/yellow	Blue/red/green/yellow
LAMP	Halogen 1 × 100 W, 6·6 A	Halogen 2 × 100 W, 6·6 A	Halogen 2 × 45 W, 6·6 A	Halogen 2 × 100 W, 6·6 A	Halogen 2 × 100 W, 6·6 A	Halogen 1 × 45 W or 1 × 100 W, 6·6 A	Halogen 1 × 45 W or 1 × 100 W, 6·6 A
UNIT DIMENSIONS							
projected height	63 mm	63 mm	63 mm	63 mm	63 mm	Flush	Flush
overall height	970 mm	970 mm	970 mm	970 mm	970 mm	100 mm	170 mm
diameter	290 mm	290 mm	290 mm	290 mm	290 mm	310 mm	200 mm
WEIGHT	10 kg	10 kg	10 kg	10 kg	10 kg	13 kg	9 kg

Status

Production. Installed at airports world-wide. The BOF 41–45 lights were designed in accordance with ICAO and FAA specifications

Manufacturer

AEF Telefunken, Wedel (Holst), Federal Republic of Germany

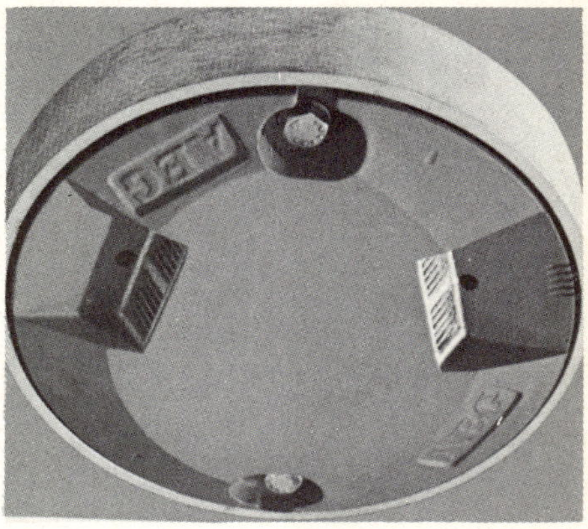

AEG BOF 45 inset light

AEG Telefunken Obstruction and Accessory Lights

The lights included in this category are hazard warning, rotating beacon and landing and wind-direction indicators. Both of the indicators can be illuminated at night.

KHF 51 and DKHF 51 are single and twin omni-directional fixed lights respectively used for marking obstructions likely to endanger air traffic. The interchangeable glass domes are fixed on a cast aluminium base by means of a hinged holding ring providing dustproof and watertight seal. The lamp socket is accommodated in the lower part. The glass domes are designed as drum lenses. The type KHF 51 is the standard version, and the type DKHF 51 is mainly installed on objects which are not easily accessible. The lights are fitted in a side-by-side arrangement. The lower parts of the lights are mounted on pipe bends which are attached to a cast iron box that accommodates the change-over relay which switches on the spare lamp in case the operating lamp fails.

The GF 2000 is an omni-directional intermittent light for night-marking of obstructions which are particularly dangerous to air traffic. It is mainly installed on very high danger points, and on obstructions that are not easily accessible. The cast aluminium housing incorporates two omni-directional clear glass refractors, one for the operating lamp and the other for the spare lamp, which are arranged one above the other. An aviation-red cylindrical coloured filter is fixed externally. The flashing unit and change-over relay are installed in a separate cast housing which can be placed separately. In case the operating lamp fails, the change-over relay switches on the spare lamp. The failure is indicated by a lamp on the cast housing.

The DDS 20 is a twin rotating beacon by which the position of an airfield can be determined at night. It is suitable for use on small airfields and heliports. The two lights with parabolic mirrors, are mounted on a turntable driven by an electric motor with gears and slipring assembly. The motor, together with the transformer and the terminals for the fixed and moving parts of the beacon is fastened on the cover of the cast light-alloy base. The lamps are adjustable and the diffusion of the single beam is about 12 degrees.

AEG DDS 20 twin rotating beacons

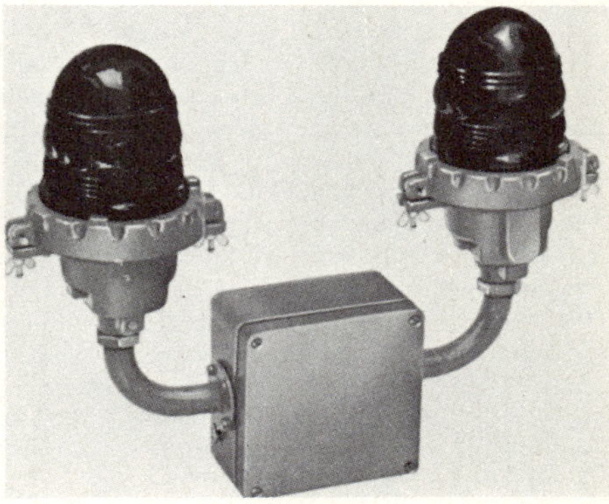

AEG DKHF 51 obstruction lights

Characteristics

MODEL	KHF 51	DKHF 51	GF 2000	DDS 20
TYPE	Omni-directional fixed light	Twin omni-directional fixed lights	Omni-directional intermittent	Twin rotating beacons
APPLICATION	Obstruction	Obstruction	Obstruction	Airfield location
OPTICAL SYSTEM	Drum lens	Drum lens	2 omni-directional refractors	2 parabolic mirrors
FILTER	Red	Red	Red	Clear
SPEED OF ROTATION				6 rpm
FLASHING UNIT			40–60/minute	
LAMP	1 × 100 W, 220 V	2 × 100 W, 220 V	2 × 1000 W/ 2 × 500 W, 220 V	2 × 200 W, 24 V
UNIT DIMENSIONS				
height	240 mm	385 mm	699 mm	710 mm
diameter/ length	180 mm	515 mm	300 mm	645 mm
width				380 mm
WEIGHT	3 kg	15 kg	50 kg	40 kg

AEG KHF 51 obstruction light

Status

Production. The obstruction lights/beacon conform to ICAO specifications and the twin rotating beacon to the recommendations of the Bundesanstalt für Furgsicherung in the Federal Republic of Germany

Manufacturer

AEG Telefunken, Wedel (Holst), Federal Republic of Germany

AEG Telefunken Power Control and Distribution Systems

AEG Telefunken offer a comprehensive range of equipment for power control and distribution systems. The switchgear is arranged as a standard cabinet system and comprises the control cabinet and transformer cabinet. The control cabinet contains the regulator, pulse-generator and switching units as plug-in systems. The transformer cabinet accommodates the transformer modules required for the operation of the constant-current regulators in a series circuit. All 5 different modules with ratings of 2·64, 6·27, 10·56, 15 and 25 kVA can be installed in the same type of cabinet.
The CR 400/25 constant current regulator supplies a series circuit for airport lighting systems via an interstage transformer. The current in the series circuit is maintained at a constant level corresponding to the set-point by a thyristor final control element and via a closed-loop circuit.
Operation is possible at five different and permanently set current-levels. The set-points are set at the factory and ensure that the following five light-intensities are obtained with the output currents at the lamps of the lighting system: 100, 30, 10, 3, and 1 per cent.
The constant current regulator is switched on either via remote-control inputs (normal operation) or by a function switch mounted on the front panel of the regulator.
Other control equipment offered includes the SFU 400/32 sequence flash control unit which is used for activating and monitoring the approach flash light system.

Status
Production. Installed in airports world-wide

Manufacturer
AEG Telefunken, Frankfurt, Federal Republic of Germany

AEG Telefunken Essential Power Supplies

AEG Telefunken produce equipment for emergency and uninterrupted power supplies. These range from automatic emergency generating sets, short-break sets to no-break sets. With the no-break sets, the operators are always supplied by the generator. They are connected to a separate bus-bar which is isolated from the mains and thus remains uninfluenced by mains faults.
The generator and flywheel are continuously driven by an electric motor which is normally laid out as a three-phase induction motor. In special cases a synchronous motor or a controlled dc motor is used instead of the induction motor to avoid the slip.
The slip of the induction motor amounts, at full load, to approximately 2 per cent, ie with mains operation a generator frequency of 49 or 59 Hz results. The magnetic coupling is open (off-circuit), as in the case of the short-break generating system. The diesel engine is at a standstill. If the mains fail, the operators continue to be supplied without interruption. The electric motor is switched off and the magnetic coupling excited. As with the short-break generating system, the diesel engine is started by the flywheel energy via the magnetic coupling. In this short space of time up to the load take-over, the flywheel also supplies the driving power for the generator. During the delivery of power by the flywheel mass the generator frequency, with a standard system design, drops by approximately 4 per cent compared to the initial value. The special system can ensure ±1 per cent frequency accuracy for 50 or 60 Hz supplies. In the case of systems with a dc motor drive, the supply of the rectifier takes place via a three-phase rectifier bridge from the mains. The diesel engine takes over the drive of the set and compensates the frequency drop after a few minutes.
With the no-break generating system, the coupling switches off after a preset delay time and the electric motor is switched on. The diesel engine is not connected with the flywheel during the cooling off run. The mains voltage, lubricating oil pressure, cooling of the diesel-engine as well as loading of the generator are monitored by the automatic control. The winding of the magnetic coupling and the operation of the electric motor are also monitored. In the case of a single- or two-phase mains failure, there is an automatic switchover to diesel operation.

Status
Production

Manufacturer
AEG Telefunken, Frankfurt, Federal Republic of Germany

INDIA

AMA Approach and Navigation Lights

AMA produce elevated approach lights and a Vasi. The Vasi systems are supplied to suit specific requirements of all types of airports. For example, the three-bar system, conforming to ICAO standards, comprises 16 units for use at large international airports. Each unit projects high intensity white and red beams of light and consists of three sealed beam lamps of 200 watt, 6·6 ampere each, producing a beam which covers 20 degrees in azimuth.
Other features of the system include a red filter of heat resistant material, protected by a sturdy frame. The housing provides protection for the optical system against rain water and all parts are constructed from frangible aluminium alloy. Adjustable lengths of mounting columns and a jig enable accurate positioning of equipment on the concrete base. A tool kit has been designed for alignment of the equipment by non-specialised personnel.

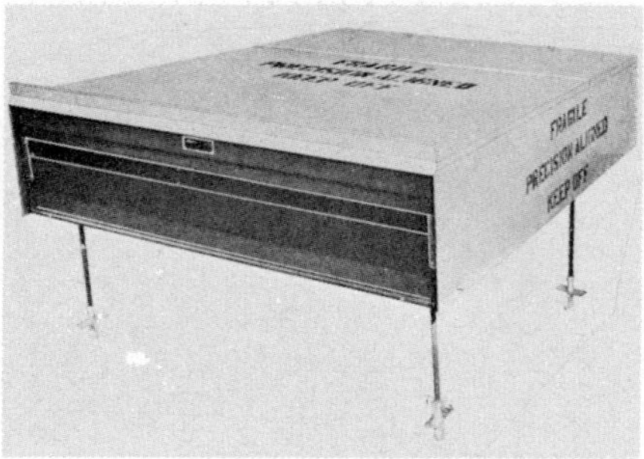

AMA Vasi unit

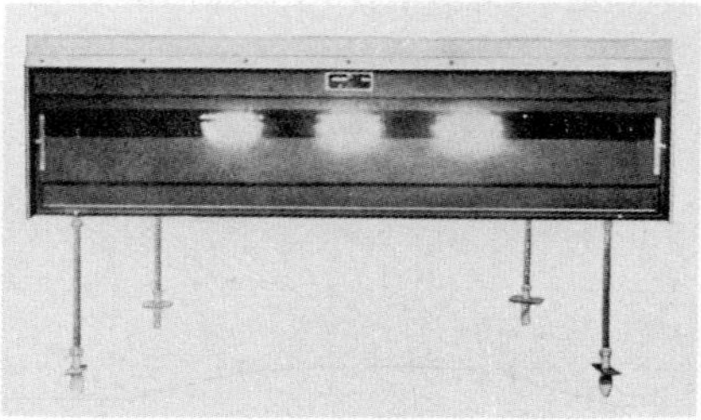

AMA Vasi unit

Status
Production. In service at civil and military airports in India

Manufacturer
AMA Private Ltd, Bombay, India

AMA Elevated Runway and Taxiway Lights

Three types of lights are available from AMA for runway and taxiway lighting. The bi-directional light is designed for use as a high-intensity unit for larger airports which can be adapted as a medium-intensity unit for smaller airfields with a 45 watt lamp.
The medium/low-intensity runway light is for use at smaller aerodromes. It features a clear, asymmetric lens with a 30 watt lamp.
The taxiway light is for use at all types of airfield. It is fitted with a blue, symmetric lens and a 45 watt lamp.
All units feature a breakable coupling which is designed to withstand high velocity jet blasts. As required by FAA specifications, the height above ground level does not exceed 1 foot 2 inches (35·56 cm). A two-pin plug connector is available

AMA elevated runway light

AMA elevated runway light

for coupling with the secondary socket of the isolating transformer, and its leads meet FAA specifications.
A cast-iron base plate conforms to FAA specifications and all other parts are constructed of non-corrosive aluminium alloy. The light fittings are also offered with a mounting stake instead of base plate.

Status
Production. In service at civil and military airports in India

Manufacturer
AMA Private Ltd, Bombay, India

AMA Accessory Light Fittings

Included in this category are an illuminated 'H' indicator for landing direction and an illuminated wind sock.
The 'H' indicator provides day and night indication of the correct landing direction of runways. The lights are arranged in three rows to form an 'H' shape. Illuminated rows of the indicator, viewed from above in the landing direction, form a 'T', as lights on one of the parallel arms are switched on, along with those of the central arm. The indicator consists of three rectangular limbs of eleven lights each. Its main body is of welded mild steel sheets supported on galvanised pipe legs. Cast aluminium fixtures, fastened to the limbs, support screw-type lamp holders in ceramic housings. All the 33 lamps are clear 230/250 volt, 60 watt and each is covered with clear glass dome with watertight fastenings. The unit is connected to the remote-control desk in the tower.
The AMA wind socks indicate true direction of the wind at velocities above 3 miles per hour (5 km/h). The unit consists of a nylon wind sock, support basket, set of two bearing assemblies, main pole and shaft, four external lights and single obstruction marker light. The obstruction marker light is mounted at the top of the support pipe. Its housing is of cast aluminium with red fresnel lens. The wind socks are supplied in two sizes, 12-foot (3·66-metre) wind cones for large commercial and military airports and 8-foot (2·44-metre) wind cones for small airports.

Status
Production. In service at civil and military airports in India

Manufacturer
AMA Private Ltd, Bombay, India

AMA Power Control and Distribution Equipment

Equipment for power control and distribution includes a remote-control desk, static constant current regulator, power distribution panel, circuit indication equipment and a transformer housing box.
The remote-control desk regulates all lighting equipments on airports from the control tower. Lights for runways, taxiways, Vasi systems and auxiliary circuits are controlled by energising selected circuits which are shown on the control desk by indicator lamps. Brightness of runway and Vasi lights is controlled by rotary selector switches.
The control desk for each airport is designed to meet its specific requirements and the size of the desk depends on the number of circuits to be operated. A mimic diagram of the circuits, made to scale, is shown on the panel.
The static constant current regulators maintain constant current through the various circuits regulating the intensity of the airport lighting systems. Current is maintained in one of the five steps for selected brightness, depending on visibility conditions.
The equipment consists of induction coils and capacitors, mounted in a two-compartment ventilated sheet steel cabinet. Main input contactors, brightness contactors, protective relays, input and output terminals, control terminals, lightning protection units and fuses are in the front compartment mounted on a hylam sheet. Main transformer components, reactors and capacitors are contained in the rear compartment.
Standard equipment operates on single phase, 230 volts, 50 cycles, ac supply; single phase 415 volt regulators are available. The input circuit is disconnected automatically when an open circuit occurs in an output system, and thermal overload protection is incorporated. The equipment is air cooled up to 7·5 kW.
Circuit indication equipment shows positively whether an airport lighting circuit is energised through its constant current regulator. Each of the runway, taxiway and Vasi system circuits is connected via a circuit indicator coupled to the constant current regulator. The equipment consists of matched transformer and relay, housed in a steel case. Red and yellow lights of the equipment indicate energising of the main supply and of the lighting circuit respectively. An ammeter is provided to indicate current in the lighting circuit and the relay can be seen through a glass panel. A circuit diagram is displayed on the equipment.
The transformer housing box provides underground protection for up to three standard

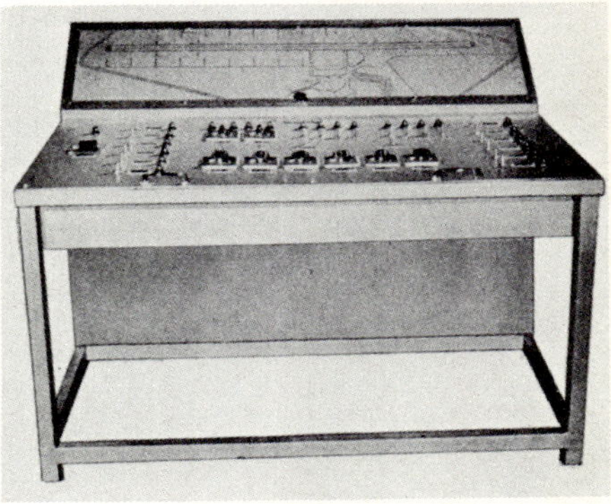

AMA remote-control desk

isolating transformers required for runway/taxiway lights or for Vasi units. The box has a cast-iron cylindrical body, with top flange and cable box. Rubber gaskets are used for watertight integrity. The circuit cable enters through glands and is connected by splicing leads.

Status
Production. In service at civil and military airports in India

Manufacturer
AMA Private Ltd, Bombay, India

ITALY

Panerai SAP 301 Portable Landing Lights System

The SAP 301 is a portable landing lights system developed by Officine Panerai. The system is suitable for tactical deployment in battle areas or for use at emergency landing sites. It comprises a gpi, runway lighting system and an independent power supply. The gpi has three beams. The top with an 8 degree beam width provides a yellow light to indicate to the pilot that the aircraft is above the glide-path.
The centre green beam which has a sector width of 3 degrees indicates the glide-path. The lower red sector is 4 degrees wide, indicating that the aircraft is below the glide-path. The glide-path angle can be adjusted to meet operational and site requirements.
The runway lighting system consists of omni-directional flashing and steady lights with different coloured filters connected by cable to the power supply. The lights can be arranged to form runway and taxiway edge lights or to indicate helicopter landing sites and parking areas.

Panerai SAP 301 portable lamp

Panerai SAP 301 glide-path indicator

Status

Production. The system has been developed to meet the requirements of the Italian Army Helicopter and Light Aircraft Wing. Thirty systems were procured in 1981

Manufacturer

Officine Panerai Srl, Florence, Italy

Panerai SAP 301 cable and reels

Siderpali Apron and General Purpose Lighting

Siderpali produce a range of lights, masts and columns suitable for apron and general purpose lighting at airports, including the Siderlux model, with a 30-metre high mast and a mobile crown. The crown on which 12 1000-watt floodlights are mounted, can be lowered for maintenance using a transportable power unit at the base which operates the internal pulley system. The crown is raised and lowered at approximately 8 metres a minute. A remote-control permits the operator to control the system up to 5 metres away from the base. One power unit can service all the high masts at the airport. Other features of the lighting unit include galvanised steel head with polyamide pulleys, mobile crown constructed in steel and aluminium and a distribution block housing power supply cables, connectors and a chain/rope coupler.

Manufacturer

Siderpali SpA, Crespellano, Bologna, Italy

Siderlux lighting unit lowered and transportable power unit

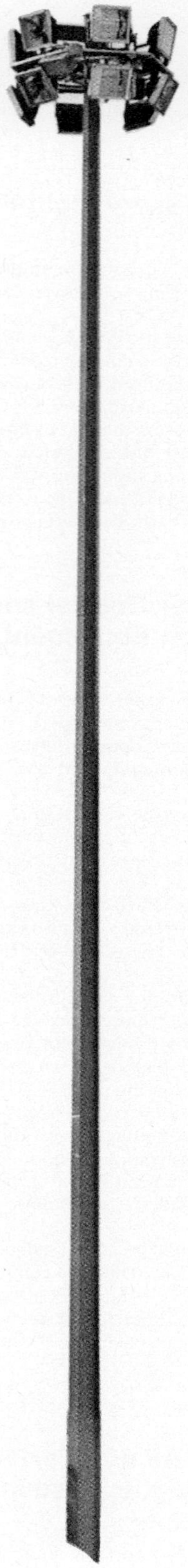

Siderlux lighting unit, crown extended

Buini and Grandi Visual Landing Aids

Two types of visual landing aids are available, the QSL-600 Vasi and the BG 800/T type T-Vasi.

The QSL-600 is a two- and three-bar Vasi designed to provide glide-slope information. The unit comprises three identical optical systems, each consisting of an ellipsoidal aluminium mirror, with a halogen lamp in the centre, focused on a front lens. The upper part of the elliptical reflector, the flux of which corresponds to the lower part of the beams, is covered by an aviation-red filter. The resulting appearance of the indicator is red in the lower zone, below the glide-path, and white in the upper zone. The pink zone width, between the upper and lower zones, is not more than 0·5 degrees. The width of the luminous beam is 8 degrees vertically and 20 degrees in azimuth.

The unit is mounted on a frangible structure designed to withstand the force of jet blasts and strong winds but which would collapse instantly if struck by an aircraft.

The BG 800/T is a T-Vasi system designed to provide glide-path information to the pilots of large commercial jets and suitable for any other type of aircraft. The system meets ICAO specifications-Annex 14 as an alternative to the three-bar Vasi system. The T-Vasi consists of three modules containing 'fly-down', bar and 'fly-up' indicators. The 'fly-down' is an inverted-'T' in white lights which indicates to the pilot that he is above the glide-slope and should increase his rate of descent. The bar indication, a horizontal line of white lights signifies the correct glide-slope. The 'fly-up' unit is a white 'T' which shows the pilot that he is below the glide-path and should decrease his rate of descent. This indication changes to red if the pilot is dangerously low. The visual range exceeds 7·5

kilometres and complies with ICAO requirements for visual range with a background luminance of 4–50 candelas per square metre by night and 1000–12 000 candelas per square metre by day. It covers a wide approach zone and the vertical plan can be adjusted using a clinometer, to an accuracy of one minute.
The optical system consists of eight front lenses and eight lamps: six for day and two for night operation. At dusk and dawn, day lamps can be used with the brilliance dimmed.
The unit is mounted on four aluminium supports with breakable couplings and powered by two 200 watt and two 65 watt series transformers connected to different series circuits.

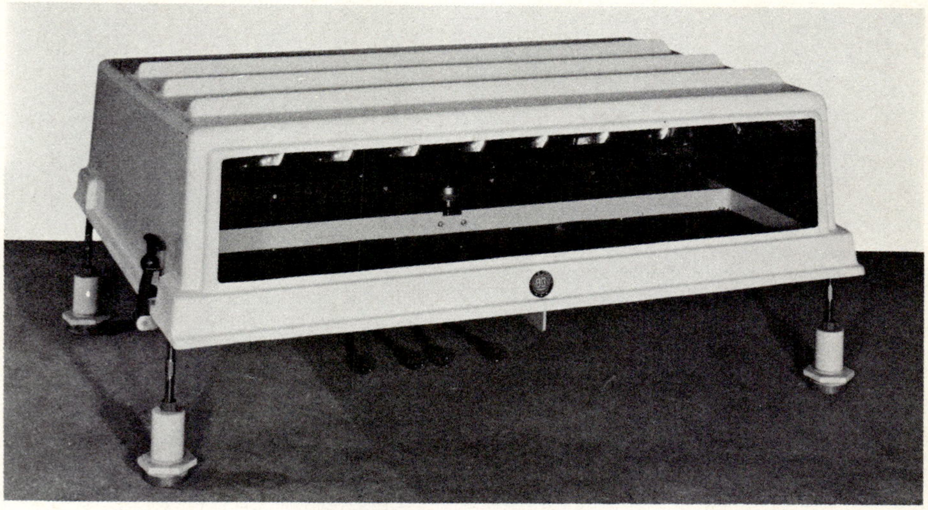

Buini and Grandi BG 800/T Vasi

Characteristics

MODEL	QSL-600	BG 800/T
TYPE	Vasi	T-Vasi
APPLICATION	Visual landing aid	Visual landing aid
BEAM WIDTH		
azimuth	20°	
elevation	8°	
fly-down unit:		white 3°8′
bar:		white 4°6′, red 1°54′
fly-up unit:		white 0°58′, red 1°54′
MAX INTENSITY		
white	100 000 cd	95 000 cd
red	15 000 cd	20 000 cd
LAMP	3 × quartz halogen 200 W 6·6 A	1 × quartz halogen (per unit) 200 W 6·6 A

Status
Production

Manufacturer
Buini and Grandi SpA, Bologna, Italy

Buini and Grandi Approach Lights

Buini and Grandi approach and runway lighting, Trieste Airport, Italy

The range of lights for use in the approach pattern comprises FA-200, BG60-X, 2LP-400/2LP-400/BIS, and the ILP-200. The last two units which are uni-directional inset lights, can be used either as an approach or runway light.
The FA-200 is a high-intensity marker light for the approach, threshold and runway end. It comprises a cast aluminium body, PAR 56 lamp and support, complete with two graduated scales, for horizontal and vertical alignment of the beam. The mounting pole consists of a 2-inch (50·8 mm) diameter steel pipe of a length to be specified. The unit incorporates an aluminium breakable coupling meeting FAA specifications. A colour filter can be mounted in front of the lamp, by means of three stainless-steel supports.
The BG60-X marker light is used in approach paths in addition to normal lights, when required by the characteristics of the approach system and by the nature of the meteorological conditions. A typical system might consist of 21 lights on the extended centre-line of the runway at 30 metres intervals from 900 metres to 300 metres before the touchdown end of the runway facing up the glide-path. The lamps are flashed sequentially from the outer end of the path to the inner end in about one third of a second. The cycle is repeated at half second intervals. The control box of each lamp includes a timer providing the required flashing sequence.
The light fitting comprises a cast aluminium body, graduated support for vertical alignment, slip fitter, frangible coupling and optical system and control box. The body includes the ignitor and the terminal strip. A safety lock prevents contact from high voltage when the lamp is disconnected. The slipfitter permits free rotation on the horizontal plane and is complete with graduated scale. The optical system consists of a PAR 56 xenon flashing lamp producing a maximum intensity of 13 700 candelas.
The 2LP-400 marker lights have been specifically designed for use in connection with category I, II and III approach paths and in accordance with the requirements of ICAO Annex 14. In the case of a displaced threshold the lights have to be fixed type semi-flush, and their construction permits a safe roll-over of the heaviest type of aircraft. The lights can be used of category I and II axial barrettes as white lights; category II side row barrettes (red), runway end lights (red) or runway edge lights in particular cases (white or yellow). If the air traffic is heavy or consists of light aircraft equipped with thin tyres, the type 2LP-400/BIS fitted with central protection ribs can be used.
The light fitting consists of a mounting frame, base and two separate domes each containing an optical system. The optical units are interchangeable with SLP-600 threshold lights. A coloured filter can be mounted in the system.
The ILP-200 inset marker light is uni-directional. It comprises cast-iron cover, optical assembly with filter frame and one or two prismatic lenses, PAR 56 lamp and mounting.

Status
Production

Manufacturer
Buini and Grandi, Bologna, Italy

Characteristics

MODEL	FA200	BG60-X	2LP-400/2LP-400/BIS	ILP-200
TYPE	Elevated high-intensity, uni-directional	Elevated uni-directional flashing	Inset semi-flush uni-directional	Inset semi-flush bi-directional
APPLICATION	Approach, threshold, runway end	Flashing approach (Reils)	Approach Category I, II, III	Approach/runway
FILTERS	Clear, red, green	Clear	Red, yellow	Red, green
LAMP	PAR 56 199/200 W, 6·6 A or 300 W, 20 A	PAR 56 xenon 60 W/s flash	2 × halogen 200 W, 6·6 A	1 PAR 56 200 W, 6·6 A/300 W 20 A

Buini and Grandi Runway Lights

Buini and Grandi PK 2 × 100 bi-directional inset light

Buini and Grandi 3LP-600 high intensity, semi-flush lights

In addition to the 2LP-400/2LP-400 BIS and ILP-200 units which can be used for runway and approach lighting and are covered under approach lights, the company produce elevated and inset lights for runway lighting. These are the FP-200, HL-2, TL-2 × 100, 2LP-200, PK 1 × 100, PK 2 × 100 and 3LP-600/3LP-600 BIS.

The FP-200 is an elevated high-intensity bi-directional light. It features a fresnel glass lens which produces two opposite main beams on an aluminium body with adjusting plate, medium pre-focus lamp holder and lamp, slipfitter and a frangible coupling. An aluminium daymarker cone also made of aluminium, and painted yellow, is offered as an option.

The HL-2 × 100/45 has an optical system which consists of two 100 watt lamps for the high intensity bi-directional beams and of one upper omni-directional lens with a 45 watt low intensity lamp. Luminous uniformity for the bi-directional beams is obtained by means of front filters. When equipped with colour filters, the light can be used for displaced threshold and runway end zones. The lampholders are complete with a focusing device and each high-intensity side is equipped with two graduated scales for precise horizontal and vertical alignment of the beams.

Other standard features include aluminium body, frangible coupling, and daymarker cone. Two models are available, the HL2 × 100/45, which is equipped with both high- and low-intensity lights, and the HL-2 × 100 with the high-intensity lights only.

The TL-2 × 100 semi-flush marker lights have been specially designed to be installed along the edges, when the runway is used for very large aircraft. These lights meet the latest requirements of Stanag and ICAO specifications. The light gives two opposite high intensity beams, produced by two lamps, that can be lit separately.

A second version, the TLK-2 × 100/45 marker light consists of the same body, with the addition of an upper, medium-intensity light, with symmetric distribution.

Type 2LP-200 marker light is a bi-directional inset unit of similar construction to the uni-directional 1 LP-200.

PK 1 × 100 is a high intensity, uni-directional inset light with cast aluminium base and cover secured by four stainless-steel captive screws and nuts. The optical system consists of a prismatic lens, aluminium reflector and halogen lamp. A coloured filter when required is fixed inside the lens by means of stainless-steel springs.

The PK 2 × 100 is a bi-directional version of the PK 1 × 100 light.

The 3LP-600 and 3LP-600 BIS are high intensity semi-flush lights for use as threshold lights in the case of a displaced threshold where the construction permits a safe roll-over of the heaviest type of aircraft. The lights can be used for threshold lights (green), threshold side-row barrettes (green), and pre-threshold lights (red) in the ALPA-IATA system. If the air traffic is heavy or consists of light aircraft equipped with thin tyres, the type 3LP-600 BIS, reinforced with central protection ribs, can be used. The body of the unit consists of a base and three separate domes, each containing an optical system which parabolic reflector, halogen lamp and a lens. A colour filter can be added. The base is galvanised sheet steel with reinforcing ribs and upper steel flange.

Status
Production

Manufacturer
Buini and Grandi, Bologna, Italy

Characteristics

MODEL	FP-200	HL-2 × 100/45 HL-2 × 100	TL-3 × 100 TL-2 × 100/45	2LP-200	PK 1 × 100	PK 2 × 200	3LP-600/3LP-600 BIS
TYPE	Elevated high-intensity bi-directional	Elevated high-intensity 2 main beams at 180° low-intensity omni-directional	Semi-flush high-intensity bi-directional low-intensity omni-directional	Semi-flush bi-directional high-intensity	Inset uni-directional high-intensity	Inset bi-directional high-intensity	Semi-flush uni-directional high-intensity
APPLICATION	Runway edge displaced threshold runway end	Runway edge displaced threshold runway end	Runway edge	Runway	Runway	Runway	Threshold, threshold side-row barrettes pre-threshold lights (ALPHA-IATA)
FILTERS	Red, green, yellow, green/red	Red, green, yellow	Red, green, yellow	Red, green	Red, green, yellow	Red, green, yellow	Green, red
LAMP	1 × 200/210 W 6·6 A	2 × Q-100 W 1 × 45 W, 6·6 A PAR-36, 6·6 A	2 × Q-100 W PAR 36 6·6 A 1 × 45 W, 6·6 A halogen	PAR-56 1 × 200 W, 6·6 A/ 1 × 300 W, 20 A	Quartz halogen 1 × 100 W, 6·6 A	Quartz halogen 2 × 100 W, 6·6 A	Quartz halogen 3 × 200 W, 6·6 A

Buini and Grandi Taxiway, Apron and Heliport Ground Lighting

The lights offered in this category comprise VC-35, PK 2 × 45, PK 2 × 65, TL-35, and the TLP-60.

The VC-35 is a medium intensity elevated light for taxiways. It can also be used for non-instrument runway lighting. The fresnel lens is of heat-resistant glass, available with symmetric distribution, clear or in aviation blue, and asymmetric distribution, clear or in aviation green. Other colours are obtainable by the addition of a filter to the clear lens. The body is aluminium and is complete with breakable coupling, medium pre-focus lampholder, two-pole lead and plug, meeting FAA specifications. The colour filter when used, is secured by means of a neoprene ring and a neoprene block on the top. An optional daymarker cone is available in aluminium, and painted aviation yellow. A spun aluminium screen with adjustable aperture can be supplied in order to obtain a uni-directional beam. The marker light can be mounted on a pipe elbow, or on a universal base with base plate.

The PK 2 × 45 light is used on straight taxiways with a beam width of 14 degrees.

The PK 2 × 65 light has a wider beam of 60 degrees and is used on curves.

The markers are also available in uni-directional versions (PK 1 × 45 and PK 1 × 65) as centre-line lights on exit taxiways.

The TL-35 is an omni-directional semi-flush light for use on airport aprons. It comprises a cast-iron base with medium pre-focus lamp holder and two-pole cable lead heavy duty cover with a six beam optical system, spherical aluminium reflector, circular symmetrical fresnel lens and 35 watt lamp.

The TLP-60 is an omni-directional medium intensity flush light for marking landing areas for helicopter operations.

Status
Production

Manufacturer
Buini and Grandi, Bologna, Italy

Characteristics

MODEL	VC-35	PK 2 × 45 (PK 1 × 45) PK 2 × 65 (PK 1 × 65)	TL-35	TPL-60
TYPE	Elevated medium-intensity omni-directional	Inset medium-intensity bi-directional	Semi-flush low-intensity omni-directional	Flush medium-intensity omni-directional
APPLICATION	Taxiway non-IFR runway	Taxiway	Apron	Heliport
FILTERS	Blue, red, green, yellow red/green	Green	Green	Yellow, green, red
LAMP	30/45 W, 6·6 A	2 or 1 × 45 W, 6·6 A/ 2 or 1 × 65 W, 6·6 A	1 × 35 W, 6·6 A	1 × 30/35/45 W, 6·6 A

Buini and Grandi Beacons, Obstruction, Apron and Accessory Lighting

The light fittings in this category comprise airport hazard and identification beacons, obstruction marker lights, apron floodlights, illuminated wind-sock and taxiway guidance signs.
BG 1000/RO is an airport beacon which features a rotating assembly with four optical systems and each contains an aluminium reflector, 500 watt lamp and support for six glass prisms. Two of the four systems are equipped with green glass filters having 20 per cent transmittance, and the revolving light produces green and white beams with a frequency of 12 flashes per minute. Two lamps are normally operating, the other two on stand-by are illuminated if a main lamp fails.
BG/2R-420 and BG/IV-420 are hazard and identification beacons respectively. The hazard beacon is equipped with two main fresnel lenses provided with a concentric red filter and an upper circular fresnel lens. Two 420 watt lamps each give a symmetric distribution having an intensity of over 2000 candelas. When one lamp is used for continuous service, a current relay automatically transfers the power to the second lamp if the main lamp burns out. Flashing equipment provides 20 to 60 flashes per minute, with a ⅔ on and ⅓ off cycle. A transformer rated at 500 watt provides 24 volt to the lamp, from the 220 volt network. The transfer relay, flasher and the transformer are contained in a separate weatherproof box mounted at the base of the obstruction.
The identification beacon is similar to the hazard beacon, but is equipped with one lamp, inner green filter and different upper lens. Light is emitted through 360 degrees in azimuth and up to 60 degrees above the horizontal. The identification characters are transmitted in the International Morse Code. The electronic gear and the 500 watt 220/24 volt transformer are supplied in a waterproof box installed near the light.
BG/IR-100 and BG/2R-100 are single and dual obstruction lights with one or two 40 or 60 watt lamps and a red symmetric lens. A flashing system, transfer relay and photo-electric switch are optionally available.
Apron floodlights offered comprise filament lights using a 1500 watt halogen lamp and Multilux gas discharge lights with 400 or 2000 watts discharge lamp. All feature aluminium reflector, shatter-proof front glass and metal shield to avoid dazzling pilots of arriving aircraft.
The MV-100 is a red and white wind-sock, equipped obstruction light and four 200 watt overhead floodlights.
TVC-30 taxiway guidance signs are mounted on frangible couplings and conform to FAA requirements. Each panel is lit from the interior by a 25 or 30 watt lamp.

Status
Production

Manufacturer
Buini and Grandi, Bologna, Italy

Characteristics

MODEL	BG 1000/RO	BG/2R-420	BG/IV-420	BG/IR-100 BG/2R-100	BG Quartzlux 1501 Multilux 2000 Multilux 400
TYPE	Rotating green/white beams	Flashing omni-directional red beam	Flashing omni-directional green beam	Omni-directional red beam	Floodlighting
APPLICATION	Airport beacon	Hazards beacon	Identification beacon	Obstruction	Apron
FILTER	Green	Red	Green	Red	Clear
LAMP	4 × 500 W 24 V	2 × 420 W 24 V	1 × 420 W 24 V	1/2 × 40/ 60 W, 6·6 A	1 × 1500 W 220 V halogen/ 1 × 400 W Multilux/ 1 × 2000 W Multilux

Buini and Grandi Power Distribution and Control Systems

Power distribution and control systems for airfield lighting comprise constant current regulators and isolating transformers.
The constant current regulators are specially designed to meet the requirements of FAA specifications for airport lighting circuits. The regulators are available in eight different power ratings from 2·5 to 40 kilowatts, with a primary voltage of 200 or 380 volts and frequency of 50 or 60 hertz. The secondary current is 6·6 amperes. Regulators rated at 15 kilowatts and over can be supplied for 20 amperes secondary current on request.
The regulators are manufactured without brilliancy control; with continuous brilliancy control; with three step control, corresponding to 100, 30 and 10 per cent of full brilliancy, or with five steps control, corresponding to 100, 25, 5, 1 and 0·2 per cent of full brilliancy.
The series to series transformers for airport lighting circuits perform the double function of maintaining the continuity of the primary circuit and isolating the secondary circuit from the primary voltage for safe maintenance.
The transformers are moulded in a thick neoprene sheath and they can be directly buried or mounted inside the light bases. They are equipped with two single-pole primary cable leads with a neoprene sheathed plug and receptacle, and a two-pole secondary cable lead, with receptacle. They are available with ratings ranging from 30/45 to 600 watts.

Status
Production

Manufacturer
Buini and Grandi, Bologna, Italy

NETHERLANDS

Alreso CARIN Runway Lights Controller

In conjunction with the Schiphol Authority the CARIN system has been designed to control the runway and taxiway lighting at Schiphol Airport, Amsterdam, Netherlands. It includes a monitoring system using fibre optics. The system requirement is to minimise human and machine error in controlling the airport runway lights. Each function is verified by the computer three times before change of status is accepted and executed. All this takes place within 200 milliseconds. The computer is sub-divided into two identical units, one operational and the other stand-by. The stand-by unit follows the operational system in all functions, especially the memory register, so that in the event of malfunction, it can take over immediately and automatically assume operational responsibility.

Manufacturer
Alreso Elektronika, Zaandam, Netherlands

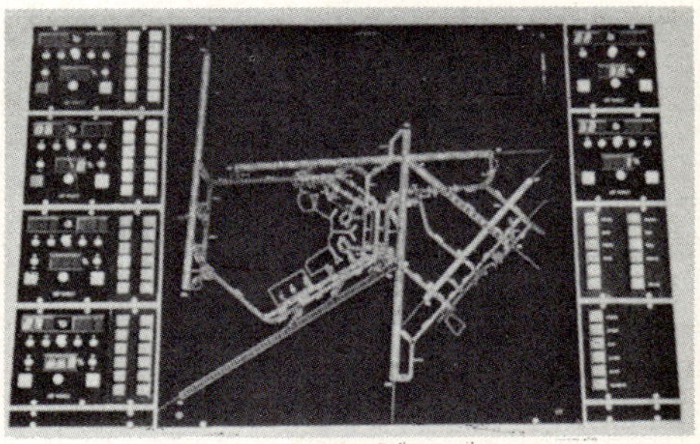

Alreso CARIN control panel for runway and taxiway lights

Nolte Airfield Lighting

Nolte design, manufacture and install a range of airfield lighting including approach, runway and taxiway lights, floodlighting for apron, freight centres and service areas, and power control and distribution systems.

Status
Nolte installed the airfield lighting systems at Schiphol Airport, Amsterdam, Netherlands

Manufacturer
Nolte Mastenfabriek BV, Maarheeze, Netherlands

Nolte lighting at freight terminal Schiphol Airport

Philips Airfield Lighting

Philips produce a comprehensive range of airfield lighting including Vasis and T-Vasis visual landing aids, approach, runway and taxiway lights, apron, and general purpose lighting, and control and distribution systems.

Philips PS 43 T-Vasis consists of ten high-intensity light units at each side of the runway; four at a right angle so that they form a wing-bar. Slope deviation is indicated by the number of lights visible in the leg of a T-arrangement, the wing-bar being continually visible. An upright-T indicates to the pilot that the aircraft is below the glide-path, and rate of descent should be decreased. An inverted-T signifies that the aircraft is above the glide-path and rate of descent should be increased. All lights in line indicate to the pilot that aircraft is on the glide-path. Thus, with the exception of gross undershoot, indicated by an all-red upright-T, T-Vasis guidance relies on the arrangement of clear high-intensity light units rather than colour differential as used in the conventional Vasi system.

Status
Installed at major airports world-wide

Manufacturer
NV Philips Gloeilampenfabrieken, Light Division, Eindhoven, Netherlands

Philips T-Vasis installed at Vienna Schwechat Airport, Austria

SWEDEN

Desmi RL 30 Obstruction Light

The Desmi RL 30 is an obstruction light which meets Swedish and ICAO regulations for marking obstacles to aviation inside and outside airfield area. The light consists of a stainless-steel housing in which is mounted a ring-core transformer. On top of the transformer is a halogen bulb surrounded by a cover of Makrolon. The halogen bulb is protected against vibration and shocks, due to, for example, falling ice and gusts of wind, since it is mounted on the transformer which is connected to the housing via a vibration absorber.

Characteristics
TYPE: omni-directional low-intensity red obstacle light
MOUNTING: vertical
LAMP: GE type Q 6·6 A, 200 W, halogen
OPERATING VOLTAGE: 21 V
POWER REQUIREMENTS: 220 V, ac, single phase
UNIT DIMENSIONS
height: 0·33 m
diameter: 0·15 m
WEIGHT: 4·5 kg

Status
In service in Sweden

Manufacturer
Desmi, Tyreso, Sweden

Desmi RL 30 obstruction light

UNITED KINGDOM

Abacus Apron and General Purpose Lighting

Abacus Grandelume and Multilume lights are suitable for apron and airfield general-purpose lighting. They incorporate a unique raising and lowering mechanism and cover a range of heights from 5 to 35 metres. The raising and lowering system is operated by an electrically-powered hydraulic unit. A double-acting ram is employed except on the 30- and 35-metre masts where twin double-acting rams are used. The masts can be supplied with a wide selection of lights.

Characteristics

MODEL	Grandelume	Multilume
HEIGHT RANGE	20–35 m	5–35 m
RAISING AND LOWERING SYSTEM		
HYDRAULIC PRESSURE	173 bar (2500 psi)	173 bar (2500 psi)
POWER REQUIREMENTS	220–240 V or 380–415 V, 3 phase	220–240 V or 380–415 V, 3 phase
TIME		
lowering	4 minutes	4 minutes
raising	8 minutes	8 minutes

Status
Production. Installed at East Midlands and Jersey Airports, England, at Royal Air Force airfields and Piarco Airport, Trinidad

Abacus Grandelume 25-metre lighting masts at East Midlands Airport

Manufacturer
Abacus Municipal Ltd, Sutton-in-Ashfield, Nottinghamshire, England

Auriga Mobile and Portable Floodlighting Equipment

Auriga produce a range of portable and mobile floodlighting equipment. The standard products form the basis of a floodlighting system to provide apron and other area lighting for airport use.
The M range of mobile floodlighting equipment offers towable units in two product families: the M8 and M18.
The M8 lighting unit incorporates a generating set and an 8-metre mast to achieve mobility for continual on-site movement of the light source.
The M18 lighting unit is equipped with 18 metre mast which illuminates areas up to 18 000 square metres for lighting large site areas.
The PM range of hand-portable lighting equipment offers small lightweight self-contained units designed for single operator use.
The PMG5 is a self-contained lighting unit incorporating a lightweight 1·5 kW generating set and a 5-metre mast.
The PM3 model has a 3-metre mast which is mounted on a secure solid base with optional power supplied by electricity mains or site generating plant.
The PM7 is a stand-alone 7-metre mast unit for lighting larger areas. The floodlights are powered by the same options as the PM3 model.

Status
Production

Manufacturer
Auriga Lighting Systems Ltd, Hawkhurst, Kent, England

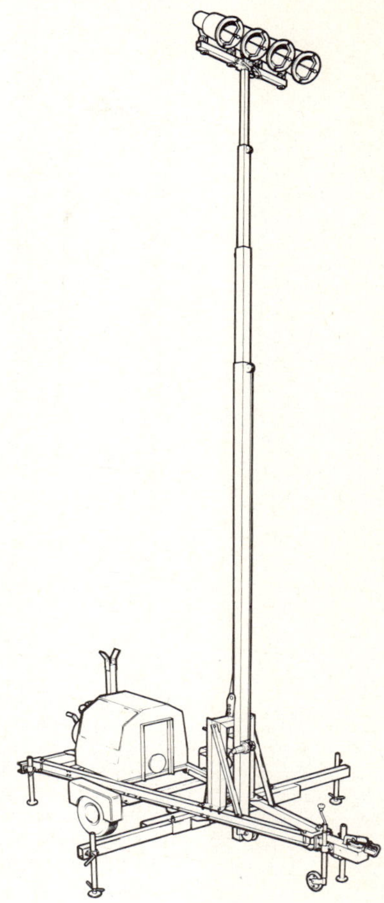

Auriga M18 mobile floodlighting unit

Auto-Diesels Braby Essential Power Supplies

Auto-Diesels Braby offer a range of diesel driven generating sets designed for no-break power supplies up to 200 kVA for essential airfield services.

Status
Production

Manufacturer
Auto Diesels Braby Ltd, Uxbridge, Middlesex, England

Auto-Diesels Braby no-break generating set

Barrel Airfield Lighting

Barrel produce airfield lighting equipment with the associated control gear. Visual landing aids available comprise two and three lamp papi systems, both tactical and portable models and Vasi. Signs and indicators offered include internally illuminated taxiway guidance signs and reflective taxiway markers. Other lighting equipment manufactured by Barrel comprises mobile floodlights, constant current regulators, 30 to 500 watt transformers and all types of filament lamps and par lamps.

Status
Production. Two lamp papi systems have been supplied to the Swedish Air Force and three to NASA, installed at the three landing sites for the space shuttle

Manufacturer
Barrel Lighting Co Ltd, Stansted, Essex, England

Barrel papi visual landing aid

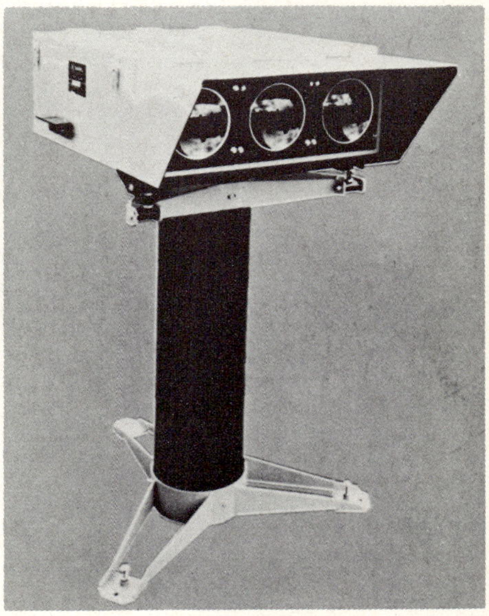

Barrel taxiway guidance sign

Cooch Mobile Floodlighting Equipment

Henry Cooch manufacture a range of mobile floodlighting equipment.
The Skylite 60 Mk3 has a four-section telescopic mast made from box-section steel tube. The mast is fitted with automatic locks to retain it in the vertical position and to lock the sections together when folded down. A hand-operated, self-locking safety winch is used to raise the mast from the horizontal and then to extend it to the full height of 18·3 metres. The lighting array consists of four 1500 watt tungsten/halogen floodlamps. Each lamp can be adjusted individually for direction. Power for the lamps is provided by a 7 kilowatt Lister diesel-powered generating set mounted on the trailer.
The Skylite 30 is a smaller version of the Skylite 60 Mk3. It has a 9 metre three-section telescopic mast made from box-section steel tube. The lighting array consists of either four 750 watt tungsten/halogen floodlamps powered by a 3·5 kilowatt Lister diesel-powered generating set or four 1500 watt tungsten/halogen floodlamps powered by a 7 kilowatt Lister diesel-powered generating set.
As an alternative, these units can be fitted with high-pressure sodium lamps or other types of discharge lamp.
The Skylites are mounted in purpose-built, two-wheel trailers fitted with overrun and parking brakes and jockey wheel. Telescopic outriggers fitted with jacks are provided to make a stable platform for the mast when erected.

Status
Skylite is in service with the British Army where its primary function is illumination for rapid airfield repair work

Manufacturer
Henry Cooch Ltd, Sevenoaks, Kent, England

CU Apron and General Purpose Lighting

The CU range of floodlights and masts has been designed to provide lighting for airports. They can be used to light airport aprons, car parks, access roads and freight areas. The CU masts can be provided in heights from 15 to 50 metres. The floodlights at the top of the mast can be lowered to the ground by a special patented double-drum winch. Alternatively the masts can be supplied with ladders to allow access to the headframe. The masts are supplied with all the necessary cables, ropes, foundation bolts and accessories and have a hot-dip-galvanised finish which may be painted if required. The mast is manufactured from mild steel plate to British Standard BS 4360, cut and folded to form a telescopic-jointed 20-sided section. All joints are welded by semi-automatic carbon-dioxide process or low hydrogen electrodes. The base is flanged and reinforced to take ten securing bolts set in a concrete foundation; the size and number of bolts used depends on the height of the mast. The base compartment has a reinforced opening fitted with a steel door and lock. A removable resin-bonded plywood backboard is installed to carry electrical equipment. Raising and lowering gear is operated by a self-sustaining winch of single- or patented double-drum pattern fitted in each mast. The winch is operated by a power tool or by hand with a special torque-limiting arrangement.
The head-gear comprises a frame within which are mounted large diameter pulleys. The pulleys support stainless steel wire ropes and multi-core electric cable. A temporary extension or trailing cable is used to maintain the electrical circuit when the lanterns are lowered. The lamps can be energised at ground level for testing.
CU Phosco floodlights use high-pressure sodium lamps. The FL300 floodlight is designed for use with a 400 watt high-pressure sodium lamp. It provides a controlled light beam. This floodlight is suitable for apron lighting.
The FL310 floodlight uses a 1000 watt high-pressure sodium lamp. It can be used in car parks and freight areas where beam control is not as important as in apron lighting, to provide a lighting scheme using a small number of floodlights and masts.

Status
CU equipment is in service at airports in Australia, Dubai, Jamaica, Oman, Philippines, Singapore and the United Kingdom

Manufacturer
Concrete Utilities Ltd, Ware, Hertfordshire, England

CU apron lighting at Dubai Airport

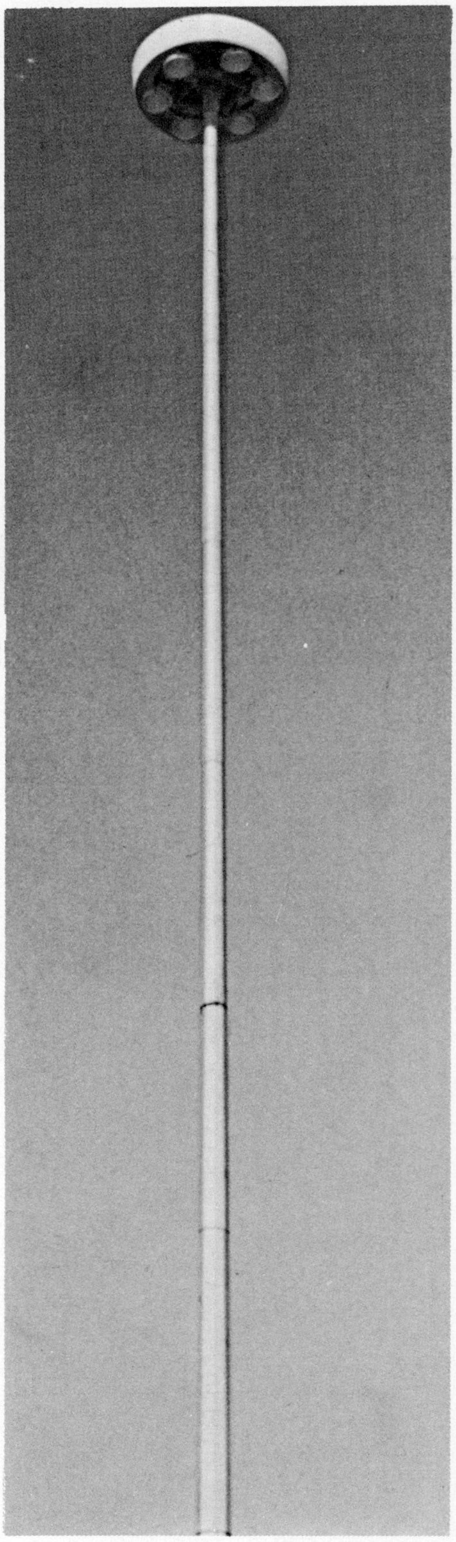

CU 25-metre octagonal mast with six lanterns

Dale Essential Power Supplies

Dale produce a comprehensive range of generating equipment for no-break power supplies. There are two types of no-break systems. The first system uses a battery giving a stand-by period of three minutes used in conjunction with a mains failure generating set. This system allows the battery to provide the uninterrupted power supply while the mains failure generating set starts automatically and runs up to speed. As the generating set is able to take over the load within 15 seconds the three minute stand-by period of the battery allows a generous safety margin.

The second system uses a battery providing a larger stand-by period, eliminating the need for a mains failure generating set.
The equipment consists of: control cubicle, constant potential charger, main battery and dc motor alternator set. The no-break plant input is connected to the ac input of the constant potential charger. The dc output of the charger is connected to the battery and the dc motor, the charger having sufficient output to simultaneously drive the dc motor and float charge the battery. The dc motor drives the alternator to provide the no-break plant output. On mains failure the battery provides the driving power for the dc motor. On restoration of the mains supply the charger again takes over the drive and slowly recharges the battery to its original capacity.

Status
Production

Manufacturer
Dale Electric of Great Britain Ltd, Filey, North Yorkshire, England

---

Dorman Smith Portable Airfield Lighting System

The Dorman Smith portable airfield lighting system is based on the range of standard hazard lighting equipment manufactured principally for road use by Dorman Smith Traffic Products Ltd.
The system consists of two types of standard portable battery-operated lamps, both of robust construction, weatherproof and insensitive to normal temperature variations.
The runway lamp is fitted with a clear lens and is supplied with rechargeable lead/acid cells, with integral charger. The batteries are recharged by connecting the charger to the mains supply, for approximately 12 hours. The lens is manufactured from high impact plastic and is mounted on the battery case, which is of metal construction, finished in yellow stove enamel and fitted with a carrying strap. Alternatively this unit can be supplied with a green or red lens for threshold use.
The TL2 for use on taxiways has a similar high impact lens mounted on a battery case made from plastic material. The unit is powered by standard, dry batteries.

Dorman Smith portable airfield light

Characteristics

TYPE	Runway	Taxiway
LENS COLOURS	Red/blue/green	Clear, green, red, blue
BULB	12 V 12 W	6 V 0·36–6 W
LIGHT INTENSITY	Approx 200 cd	1·5–60 cd
BATTERY CAPACITY	10–12 hours	Up to 150 hours
UNIT DIMENSIONS		
width	$9\frac{3}{8}$ in (238 mm)	$7\frac{1}{2}$ in (190 mm)
depth	$6\frac{3}{4}$ in (172 mm)	$3\frac{3}{4}$ in (95 mm)
height	1 ft $3\frac{1}{2}$ in (387 mm)	1 ft $1\frac{1}{2}$ in (343 mm)
WEIGHT	21 lb (9·5 kg)	$2\frac{1}{4}$ lb (1 kg)

Status
Production

Manufacturer
Dorman Smith Traffic Products Ltd, Southport, Lancashire, England

---

Forest City Lightowers Floodlighting Units

Forest City Lightowers are self-contained lighting units with an integral generator which is independent of any external source. The range extends from a static 14-foot (4·3-metre) unit to a 60-foot (18·3-metre) mast mounted on a fast tow road trailer. Four types of base are used depending on the weight of the set and payload. These are the portable base, two-wheel barrow, four-wheel trolley and fast tow trailer. The generator is in all cases dual voltage and can therefore be used to power other equipment when the lights are not being used. The pneumatic masts are erected by hand- or electric-pump depending on size. The lattice masts can be erected by hand or electric winch, and can be locked at any desired height. All units can be erected by one person.

Forest 60L/LA4/RT Lightower mobile floodlighting unit

Forest 14P/LA2/B Lightower mobile floodlighting unit

Characteristics

MODELS	14P/B	20P	30P/L/T	40P	60L
MAST TYPE(S)	Pneumatic	Pneumatic	(P) Pneumatic/ (L) Lattice/ (T) Telescopic	Electro-pneumatic	Lattice
HEIGHT	14 ft (4·3 m)	20 ft (6·1 m)	30 ft (9·1 m)	40 ft (12·2 m)	60 ft (18·3 m)
MOUNTING	Portable tubular steel frame/2-wheel barrow	4-wheel trolley/road trailer	Road trailer	Road trailer	Road trailer
LIGHTS	2 × 500 W or 2 × 1000 W	4 × 750 W or 4 × 1000 W	4 × 1000 W (P) or 3 × 1000 W (L) or 4 × 1500 W (T)	4 × 1500 W	4 × 1500 W

Status
Production

Manufacturer
Forest City Electric Co Ltd, Walkden, Manchester, England

GECOS Visual Landing Aids

GECOS produce three types of visual landing aids designed to assist the pilots of rotary and fixed wing aircraft to maintain the correct approach slope from the beginning of the final approach to touchdown.

These are the Vasi, Chapi and gpi.

The Vasi is a sharp transition unit suitable for use in two- or three-bar Vasi systems or in a Papi system. Both units feature an identical transition performance, the choice of unit being determined by the operational requirements. Glide-path angles from −1 to +10 degrees can be accommodated.

The Chapi is available as a three or two lamp unit and is based on the Papi principle. An approaching helicopter can acquire the approach slope indicated by a red/white beam projected by the equipment at ranges from 2 miles (3·2 km) to 5 miles (8 km), depending on the visibility. At closer ranges a green signal appears in the white sector to define the desired approach slope more accurately. By remaining on the green approach slope (normally 7 degrees) the helicopter will reach the touchdown area (hover point) with an eye/ground height of approximately 10 feet (3 metres).

The unit can be installed on any type of helipad, either on or off shore. In a two unit configuration roll guidance is provided as well as additional rate of descent information.

The gpi is used as part of a visual approach system for helicopter deck landings. The unit projects a beam of light split into three coloured zones: red, green and yellow. The red zone indicates to the pilot that the helicopter is below the glide-path, green zone on glide-path, yellow zone above glide-path. The beam is stabilised to correct for pitch and roll of the vessel.

Status
Production. ZA 757 is installed and in operation at Royal Air Force Farnborough, England. Chapi is in service at Aberdeen Dyce Airport, Scotland and gpi in service with the Royal Navy

Manufacturer
GEC Electrical Projects Ltd, Aviation Division, Rugby, Warwickshire, England

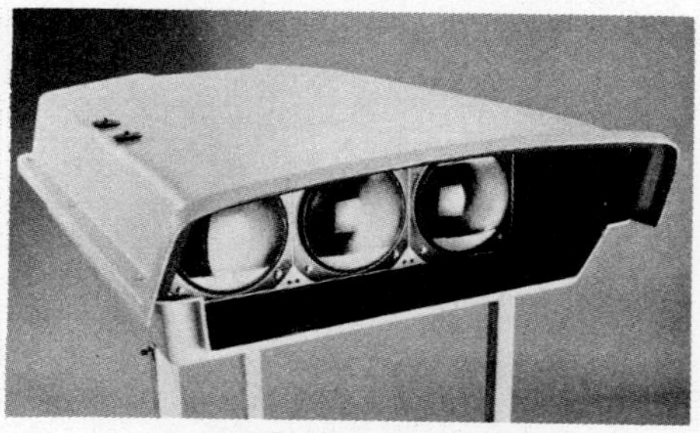

GECOS ZA 757 Vasi system

Characteristics

UNIT	ZA737/ZA757	ZA730/ZA720	ZA710
APPLICATION	Vasi/Papi	Chapi	gpi
LAMP			
NUMBER	3 (ZA737) 2 (ZA757)	3 (ZA730) 2 (ZA720)	1
TYPE	Tungsten halogen 200 W J1/39 or 100 W J1/58 or 45 W J1/57		Tungsten halogen GECOS 26368

GECOS Approach Lighting

The approach lighting comprises uni- and omni-directional lights of low- and high-intensity, with pole-mounted elevated and inset fittings designed to form part of an all weather approach lighting system.

ZA 401 is a low-intensity omni-directional approach light which can be used as a lead-in light to runways equipped to low-intensity standards, or as part of an all-weather approach lighting system. Alternatively, the unit is suitable as a medium-intensity obstruction light in areas where frangible equipment is needed. The light is designed for pole or ground mounting and is normally fitted with a red refractor.

ZA 409 is a high-intensity elevated approach light suitable for use as a white centre-line or cross-bar light, or with appropriate filters, as a threshold, wing-bar or runway end light.

ZA 410 and ZA 411 are high-intensity uni-directional inset approach lights. ZA 410 can be installed as a white centre-line approach light. ZA 411 units, with appropriate filters, are designed for use as side-row barrettes, threshold lights, wing-bars or runway end lights.

ZA 413 are high intensity uni-directional inset approach lights suitable for use as a centre-line and cross-bar lights for categories I, II and III approach lighting systems. A modified version of the same unit may be used for side-row barrettes or threshold and wing-bar lights.

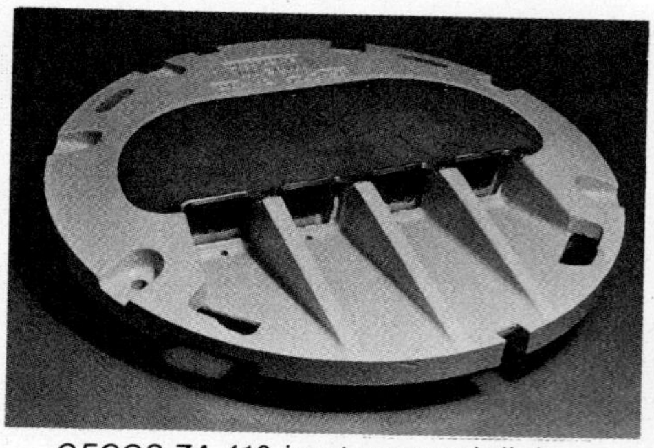

GECOS ZA 410 inset approach light

GECOS ZA 409 elevated approach light

Characteristics

MODEL	ZA 401	ZA 409	ZA 410/ZA 411	ZA 413
TYPE	Low intensity, omni-directional elevated	High intensity, uni-directional elevated	High intensity, uni-directional	High intensity uni-directional elevated
MOUNTING	Pole/ground	Pole/ground	Inset	Pole/ground
LAMP	Tungsten halogen 200 W J1/39 or J1/48, J1/50, J1/52	Tungsten halogen 200 W J1/39	2 × tungsten halogen 200 W J1/39	Tungsten halogen 200 W J1/39

Status

Production. In service in more than 300 civil and military airfields. Major systems are operating in Bangladesh; Luxor, Egypt; Hong Kong International Airport, Hong Kong; Iran; Amman, Jordan; Tripoli International Airport, Libya; Kuala Lumpur International Airport, Kuching Airport, Malaysia; Luqa Airport, Malta; Abu Dhabi International Airport, United Arab Emirates; Lagos Mutalla Mohammed International Airport, Nigeria; Karachi International Airport, Pakistan; West Indies and Belgrade, Yugoslavia

Manufacturer

GEC Electrical Projects Ltd, Aviation Division, Rugby, Warwickshire, England

---

GECOS Runway and Taxiway Lighting

Fittings designed for use as runway and taxiway lights to meet the ICAO requirements for all-weather operation include high- and low-intensity, omni, bi- and uni-directional, elevated and inset lights.

ZA-107 is a high-intensity bi-directional elevated runway edge light for use on instrument runways in all weather conditions. An omni-directional component provides orbiting guidance in good visibility.

ZA-117 is a high-intensity uni-directional inset touchdown zone light which when fitted with red filters can be used as a runway end light.

ZA-118, a high-intensity bi-directional inset light is designed as an inset runway edge light. It is suitable for use on airfields equipped for all weather operation.

ZA-127 is a high intensity bi-directional inset light used as a runway centre-line light. It has a uni-directional capability for use in mist or sandstorm conditions.

The ZA-201 and ZA-209 fittings are low-intensity omni-directional inset lights suitable for installation as taxiway centre-line, stop-bar, runway end lights or fast turn-off markers. Units can be supplied with a lens system to give symmetric or asymmetric bi-directional medium-intensity elements.

ZA-202 low-intensity omni-directional inset lights can be used as a taxiway light or as a low-intensity runway light.

ZA-206 is a low-intensity omni-directional elevated runway or taxiway light, designed for use on runways equipped to low-intensity standards, or as a taxiway edge light. The unit has a medium-intensity capability when fitted with an asymmetric refractor.

ZA-210 and ZA-212 are high-intensity bi-directional inset taxiway centre-line and stop-bar lights, designed to assist aircraft ground movements in bad weather conditions. The ZA-210 beam is non-axial for use on curves, and the ZA-212 beam is axial.

ZA-214 is a low-intensity omni-directional elevated multi-purpose light. This miniature fitting may be used as taxiway and runway edge lights, or an obstruction light. It is also useful for installation on temporary pavement surfaces.

Status

Production

Manufacturer

GEC Electrical Projects Ltd, Aviation Division, Rugby, Warwickshire, England

Characteristics

MODEL	ZA 107	ZA 117	ZA 118	ZA 127	ZA 201/	ZA 202 ZA 209	ZA 206	ZA 210/	ZA 214 ZA 212
TYPE	High-intensity, bi-directional	High-intensity, uni-directional	High-intensity, bi-directional	High-intensity, bi-directional	Low-intensity, omni-directional	Low-intensity, omni-directional	Low-intensity, omni-directional	High-intensity, bi-directional	Low-intensity, omni-directional
MOUNTING	Elevated	Inset	Inset	Inset	Inset	Inset	Elevated	Inset	Elevated
LAMP	Tungsten halogen 200 W J1/39 or J1/43, J1/53, J1/53, J1/42, J1/50	Tungsten halogen 100 W J1/58 or 200 W J1/39	Tungsten halogen 200 W J1/39	Tungsten halogen 2 × 100 W J1/58 or 2 × 200 W J1/39	Tungsten halogen 45 W J1/57 ZA 201, J1/7	Tungsten halogen 45 W J1/57 or J1/7, J1/14	Tungsten halogen 100 W J1/58 or J1/7, J1/57, J1/14, J1/3, J1/13, J1/24	Tungsten halogen 100 W J1/58	Tungsten halogen 45 W J1/57

GECOS Apron and General Purpose Lighting

In this category GECOS offer floodlights, airport location beacons, obstruction lights and markers.

ZA 352 is an apron floodlight suitable for wall or pole mounting.

The airport location beacon, ZA 503 may be used as a special hazard beacon when equipped with red filters. It is fitted with a stand-by lamp with automatic lamp changer.

ZA 743 and ZA 753 are obstruction lights. Normally within the perimeter of an airport or airfield, any marked obstruction is lit, and for this purpose low-intensity obstruction lights are used such as the ZA 743, a single lamp unit, or the ZA 753, a double lamp unit which can be fitted with a lamp change-over relay.

ZA 772 is approved for use as a beacon in category II areas. Alternatively, it can be installed as an obstruction light or as a single light marker unit on a helipad.

ZA 763, ZA 764 are medium-intensity obstruction lights suitable for use on high obstructions. ZA 763 is a single lamp unit and ZA 764 is a double lamp unit which can be fitted with a lamp change-over relay.

Three models of general-purpose neon obstruction lights are produced, the ZA 750 and ZA 756 which are low-intensity units and the ZA 758, a medium-intensity light. All use long-life sources. Flasher and photo cell units can be supplied for these fittings as optional equipment, and different coloured light sources are available.

Other fittings produced by GECOS include an illuminated wind direction indicator, azimuth guidance, nose-in stand for aircraft alignment in terminal areas, signalling lamp, portable ground light, taxiway guidance sign and distance-to-go marker.

Characteristics

MODEL	ZA 352	ZA 503	ZA 743/ZA 753	ZA 772	ZA 763/ZA 764	ZA 750/ZA 756	ZA 758
TYPE	Apron floodlight	Airport location beacon	Low-intensity omni-directional	Omni-directional	Medium-intensity	Low-intensity omni-directional	Medium-intensity omni-directional
MOUNTING	Wall/pole	As required	Obstruction	As required	Obstruction	As required	As required
LAMP	400 W Son L or tungsten halogen 1500 W	100 V, 2·25 kW bi-post G 38	ZA 743: 1 × 100 W or 75 W GLS ZA 753: 2 × 100 W or 75 W GLS	100 W ES GLS	250 W B1/7 or 200 W J1/48, J1/52 GLS	Neon	Neon

Status
Production

Manufacturer
GEC Electrical Products Ltd, Aviation Division, Rugby, Warwickshire, England

GECOS Lighting Control and Distribution

This category includes electrical power distribution, cabling, transformers, switchgear, substations, mains and stand-by generation equipment using no-break sets where necessary, all associated lighting control switchboards and regulators together with remote-control consoles and mimic panels.

A range of constant current regulators (ccr) has been designed to control the brightness of airport series circuits. These ccrs are capable of supplying constant current with variable load. By remotely varying the controlled series current, the overall light intensity of the connected circuit can be adjusted to the required level of brilliancy. Standard ratings are from 4–30 kW. Regulators available are the XG 200 and XG 300 series which are solid state assemblies, complete with cabinet and tray for use on 380/440 volt, 50/60 Hz supply and the XT 120–130 series. The latter are resonant bridge, static oil-filled units for standard voltages and frequencies.

The switchboards are built on the extensible cubicle system. Each cubicle consists of a completely enclosed unit housing the various items of equipment which are of the panel mounted type. This arrangement enables a variety of equipment required for airport services, eg remote-controlled contactor gear, air and oil circuit breakers, manually operated fuse swichgear and standby generator panels, to be operated in a switchboard of uniform appearance. Purpose-built lighting control panels are available to provide remote-control of all lighting circuits, and to indicate the state of these circuits on a mimic diagram.

Status
Production

Manufacturer
GEC Electrical Products Ltd, Aviation Division, Rugby, Warwickshire, England

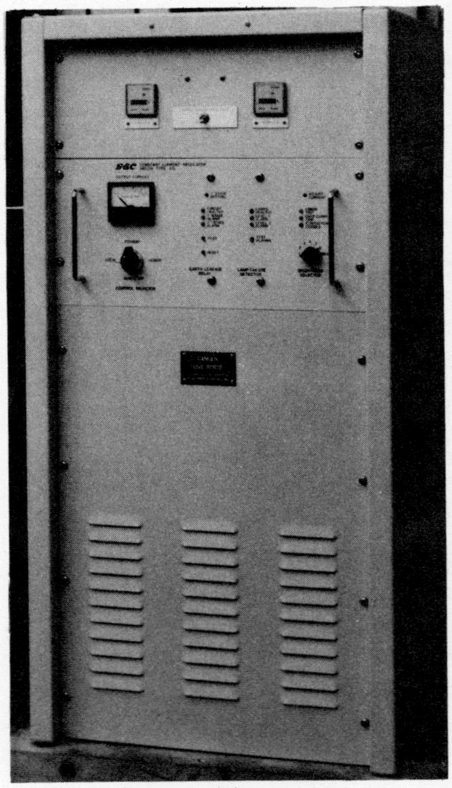

GECOS constant current regulator

Metalite LE/58 Portable Airfield Ground Lights

The Metalite is a battery-operated portable cold cathode light, designed and developed for use as standard, auxiliary and emergency runway and airfield lighting, and for general purpose taxiway, obstruction, threshold, end of runway services.

The lamp is a sealed unit containing a cold cathode lighting tube in any one of five colours, white, red, green, blue, yellow. The lighting tube is mounted through a shock absorbent carrier on a reflectorised base within a clear acrylic lens giving omni-directional light in azimuth. The lamp can be unclipped by toggle fasteners and replaced by alternative colours for immediate use in other applications and emergencies.

The control case is moulded in frangible glass fibre reinforced polyester resin and comprises recessed manual control lever switch. The plastic coated carrying handle hinges down flat to case. A sealed compartment within the control case contains the solid-state transistorised inverter, panel type fuse holder for cartridge fuse, polarised battery plug and other optional controls. A ventilated compartment also in glass fibre, houses the preferred power pack of primary or secondary cells. Model type B designed for low clearance has a battery compartment of more limited size.

Toggle fasteners provide access to the battery compartment for recharging or replacement. Anchor flanges allow for ground pinning with 6 mm diameter steel plastic coated pins.

Optional facilities offered comprise photocell, flashing, radio, and linked controls. The photocell control can be used for dusk to dawn automatic switching, or as runway mains failure standby lighting. When placed adjacent to mains runway lights the LE/58 will be held 'off' with minimum current drain during normal operation of mains lightings, but will switch on automatically in the event of mains failure. An electronic transistorised switching unit provides a flashing light with any colour of lamp and extends battery life. Radio control is achieved by a radio receiver within the LE/58 which operates an on/off relay through a signal transmitted from an aircraft, or from ground control. Cable linking the LE/58 enables up to 100 lights to be switched on/off from one master unit, which may be manually operated or radio remote-controlled.

Status
Production. In service with the Royal Air Force

Manufacturer
Metalite Aviation Lighting, Winster Grove, Birmingham, England

Metalite LE/58 portable ground light

Characteristics

MODEL	LE/58 A	LE/58 B
TYPE	Portable omni-directional – azimuth 0–20° elevation	Omni-directional – azimuth 0–20° elevation
LIGHTING INTENSITY	45 cd	45 cd
MAX RANGE	10 nm (18·5 km)	10 nm (18·5 km)
BATTERY CONSUMPTION	1·5 A – 12 V	1·5 A – 12 V
OPERATING TEMPERATURE RANGE	−20 to +50°C	−20 to +50°C
UNIT HEIGHT	1 ft 1½ in (0·345 m)	9 in (0·23 m)
MAX DIAMETER	10 in (0·25 m)	9½ in (0·242 m)
OPERATING WEIGHT	25 lb (11·4 kg)	9 lb (4·1 kg)

Parmeko Ground Lighting Controls

Parmeko produce isolating transformers and series circuit regulators for airfield lighting. The range of isolating transformers meets all relevant specifications of the British CAA and Department of the Environment and the FAA in the USA.

Each unit is tested for electrical and mechanical characteristics and random checks are made for water immersion insulation resistance. The primary and secondary cables are terminated with plugs and sockets to FAA specifications, to facilitate connection to the lamp and series circuits.

Parmeko isolating transformers can be installed on the surface, in concrete pits or buried underground. These locations may be waterlogged or contaminated by oils, kerosene, other aircraft fuel, soil acids or alkalis. Operating ambient temperatures range from −10°C to +50°C.

The series circuit regulators ensure that a constant current is fed to the lighting circuit irrespective of power supply fluctuations or variations in loop impedance due to lamp failure.

The saturable reactor and supply-isolating transformer is designed to withstand a prolonged short circuit at the regulator output, even if the earlier stages of the constant-current control system should fail.

Optional features available include local/remote-control, earth leakage fault indicator, open-circuit protection equipment and back indication of pre-set value of output current.

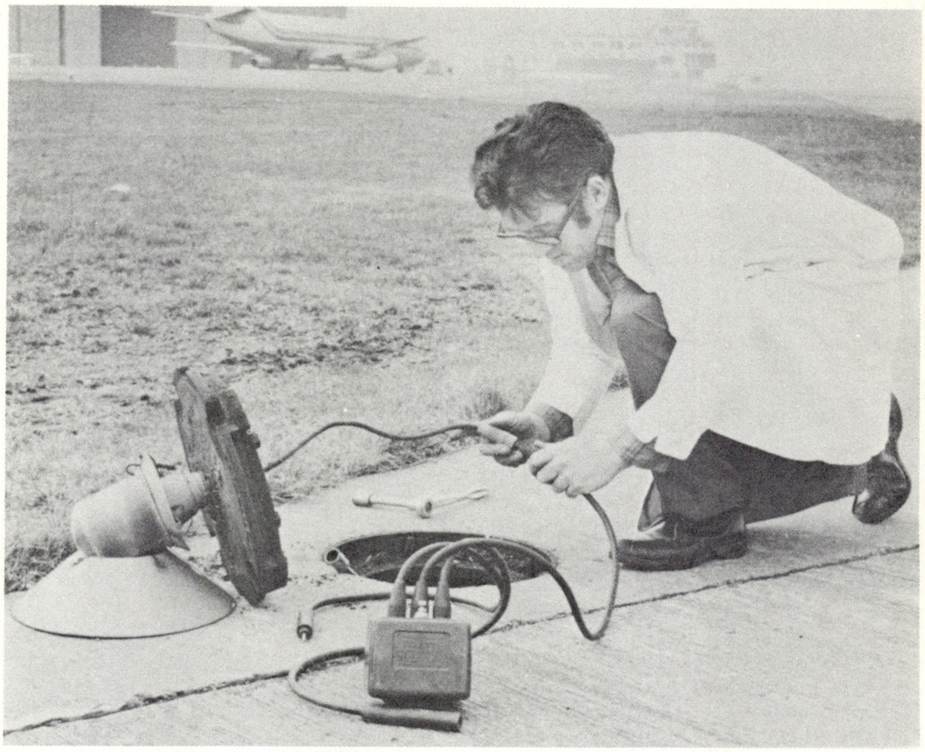

Parmeko series isolated transformer

Status
In service at civil and military airports world-wide

Manufacturer
Parmeko Ltd, Leicester, England

Paxman Essential Power Supplies

Paxman Diesels design and manufacture diesel power generating sets to provide stand-by or emergency power for essential power supplies in airports.

These take the form of conventional generating sets or of the more sophisticated no-break stand-by sets incorporating high-energy flywheels.

The company can supply conventional short-break stand-by generator sets ranging from 600 to 2500 kW, at speeds up to 1500 rpm. Generating sets are normally custom-built to specific requirements.

Status
Production. Installed in airports in Darwin, Melbourne, Sydney, Australia; Al Muharraq, Bahrain; Guernsey, Channel Islands; London Gatwick, London Heathrow, England; Bombay, Mahe, India; Dublin, Ireland; Mexico; Wellington, New Zealand; Doha, Qatar; Glasgow, Scotland; Johannesburg, Windhoek, South Africa; Abu Dhabi, United Arab Emirates

Manufacturer
Paxman Diesels Ltd, Colchester, Essex, England

Piller Essential Power Supplies

Piller specialise in the production of power supply equipment.

The DH system is used for no-break power supplies, incorporating the Uniblock converter, which combines the windings of both synchronous motor and generator in one housing. In the DM system electrical power is fed via a static switch and choke to the Uniblock converter and then to the load. The Uniblock consists essentially of a stator in which alternate slots are wound with input and output windings. When fed with a three-phase supply, a rotating magnetic field is created, causing the dc excited rotor to spin at synchronous speed. By controlling the rotor, control of the secondary stator windings is effected, and the system allows fine control of power conversion. Relatively large mains voltage variations on the input windings are not passed on to the secondary winding, and in practice even a complete loss of ac mains does not cause loss of ac output. High voltage spikes, and frequency variations are similarly attenuated.

The Uniblock forms one of two current paths through the system, the other path being via a simple diode rectifier stand-by battery and static inverter. Under normal operation the two circuit paths are used together, but 95 per cent of power passes through the Uniblock. The second path is used for only 5 per cent of the power, and this is done to maintain the rectifier/inverter in an active state. In practice there is also a further system by-pass circuit which comes into operation if the Uniblock path fails. This would ensure no break in power supply to the load if a machine breakdown occurred.

Piller DH-type

Characteristics

UNIBLOCK	300 kVA	50 kVA
DIMENSIONS		
length	2·3 m	2 m
depth	1·2 m	0·8 m
height	2 m	1·8 m

Status
Production

Manufacturer
Anton Piller (UK) Ltd, Cirencester, Gloucestershire, England

Petbow Essential Power Supplies

Petbow manufacture generating sets in the range from 25 to 6000 kVA for stand-by and emergency power supplies. The range offers a wide choice of diesel engines, generators, alternators and control systems.

Status

Petbow generators were chosen by British Aerospace for continuous use throughout the Concord programme and are installed at airports worldwide

Manufacturer

Petbow Ltd, Sandwich, Kent, England

Petitjean Apron and General Purpose Lighting

Petitjean manufacture a wide range of lighting columns and masts for airport environments. The masts for apron and open-space lighting are offered from 3·5 to 50 metres mounting height. Standard columns are available from 5 to 15 metres high. A typical example is the Leven column, a hinged 5-metre pole which can be lowered for maintenance. The columns manufactured from galvanised sheet steel to British Standard BS 729, are of octagonal cross-section and are tapered.

Status

Petitjean masts are installed at airports worldwide

Manufacturer

Petitjean and Company Ltd, Thornaby, Cleveland, England

Research Engineers Pasi Visual Landing Aid

The Research Engineers Pasi unit has been developed specifically for precision glide-slope indicator systems such as Papi.

The Papi system uses a two-colour light projector. Each unit usually consists of three optical projectors placed side-by-side. They produce a beam of light, the upper part of which is white and the lower part red. In translating vertically through the beam the transition from one colour to the other is almost instantaneous and very obvious. In a basic configuration the Papi system consists of four of these sharp transition units at the side of the runway adjacent to the glide-path origin. The setting angles of the red/white interface of the four units are graded, the differences in angle between the units being 20 minutes of arc. The nominal glide-slope angle is midway between the angular settings of the centre pair of the four units. The 'on course' signal is thus two red and two white lights in the bar. If the aircraft goes below the glide-slope the pilot sees a progressively increasing number of red lights. Conversely, if the aircraft goes above the glide-slope the number of white lights seen is increased. The company claims that it is possible to follow the indications down to ranges as short as 1000 feet from touchdown (50 feet above the runway threshold).

This basic system, which provides multi-position information from which the pilot can derive rate information, is all that is needed to support landing operations provided that all the aircraft using the glide-slope indicator are of a similar size. If a wide range of aircraft sizes is to be catered for, a second bar of four units is sited approximately 500 feet upwind of the first bar and the units are set to the same angles. This bar is used by aircraft with large ils aerial to pilot eye heights or large pilot eye to wheel heights.

While the above description is based on one or two wing-bars, (one up- and one down-wind) for larger aircraft, other configurations are possible, for example a bilateral array (ie a symmetrical system on the opposite side of the runway) or simplified arrays of two units only, for smaller airfields where minimal glide-slope indications are acceptable.

The distinguishing feature of the Pasi unit is its modular design which enables the pre-aligned projectors to be replaced if faulty and permits two or three lamp versions to be assembled on the same basic structure.

Research Engineers Pasi visual landing aid in situ

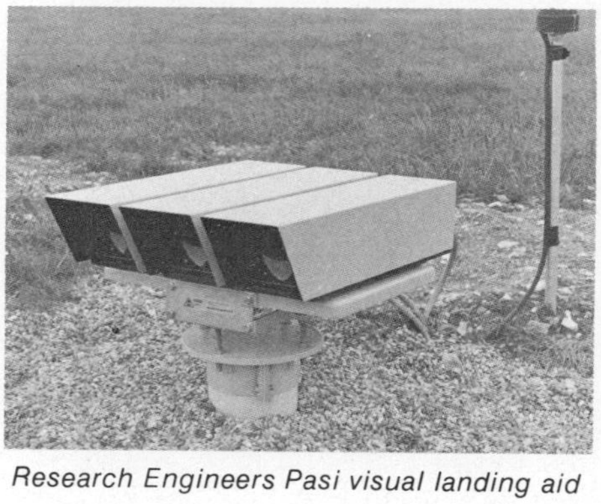

Research Engineers Pasi visual landing aid

Status

Production. Accepted by the British CAA for use in Papi installations and as a sharp transition Vasi. It is installed at London Heathrow Airport, England.

Crouse Hinds has obtained manufacturing and marketing rights for the Pasi unit, from Research Engineers Ltd

Manufacturer

Research Engineers Ltd, Shoreditch, London, England

Characteristics

TYPE	(1)	(2)	(3)
PEAK INTENSITY	200 W	45 W	55 W
WHITE	130 000 cd	49 000 cd	39 000 cd
RED	26 000 cd	10 000 cd	8000 cd
HORIZONTAL SPREAD	± 15°	± 15°	± 15°
TRANSITION (TYPICAL)	2′	2′	2′
DIMENSIONS (excluding pillar)	560 × 430 × 215 mm	560 × 430 × 215 mm	560 × 430 × 215 mm
WEIGHT	14·5 kg	14·5 kg	14·5 kg

Thorn Visual Landing Aids

Two types of visual landing aids are produced by Thorn, the conventional Vasi and the later development, the Papi. Vasi uses two parallel rows of lights to provide glide-path indications to the pilot, a system which cannot be used accurately at heights below 200 feet. Papi presents its information from a single horizontal row of lights and can be operated accurately down to the runway threshold.

It was because a bulky unit can be vulnerable to jet blast and a danger to landing aircraft that Thorn originally developed the mini-Vasi unit, the smallest of its type available, for Vasi installations. The unit has been further developed and now it can either be used in existing Vasi systems or set up in the Papi configuration.

Papi has a single horizontal row of lights instead of two parallel lines and, if there are other visual aids sufficient to provide roll guidance, the Papi units need only be placed on one side of the runway. However, these can be supplemented if necessary by a second bar on the opposite side of the runway.

In each bar there are four groups of lights, each group set at a slightly different angle to the others. The glide-path indication to the pilot comprises two whites and two reds in each bar. A slightly low signal is one white and three reds, and well below the glide-path is indicated by four red lights. Slightly high consists of three white lights and one red and a high signal is seen as four whites.

For a single-sided installation it is recommended that each group of lights is fed from two separate circuits in order that the signal is maintained in the event of power failure. This is achieved by positioning two units together in each group, each unit in the pair fed from a separate circuit. In the event of a circuit failure an adequate signal is provided by the two remaining lamps in each group.

The fitting consists of a cast aluminium base with frangible coupling for taking various mounting arrangements. The optical system, which provides the necessary sharp transition, consists of a dichroic-coated red filter glass and double lenses. The system is protected from jet blast and dirt by a clear flat glass. Two AF7/2 tungsten halogen lamps of 200 watts, 6·6 amperes are fitted as used in Thorn inset and elevated airfield lighting fittings.

Under most normal operating conditions the heat generated by the lamps at full intensity disperses any ice originating from water condensation when the unit was unlit. Where extreme condensation is likely to occur an optional heating system is available.

The Vasi unit produced by Thorn comprises a base frame, an optical bench unit and a cover. The base frame is of welded-tubular construction, and is fixed to support pillars. The welded-tubular optical bench unit, carries the three pre-focus sealed beam projector lamps in quick-change housings and three filter spreader glasses. The fibreglass cover has a hinged back section to permit access to the lamps and filters.

Characteristics

TYPE	Vasi	Mini Papi
LAMP	3 × 200 W, 30 V, 6·6 A or 16·6 A	2 × AF7/2 tungsten halogen, 200 W, 6·6 A
INTENSITY		
red	15 000 cd	4860 cd
white	60 000 cd	39 000 cd
BEAM		
width		± 10°
depth		± 3·5°
RANGE		
day	4–5 miles (6·4–8 km)	
night	> 15 miles (24 km)	
ANGLE OF APPROACH	2–4·5°	
DEPTH OF APPROACH	25–50 ft (7·6–15·2 m)	
UNIT DIMENSIONS	1·45 × 1·45 × 0·406 m	0·42 × 0·285 × 0·195 m
WEIGHT	36·3 kg	

Status
Production. In service at London Gatwick Airport, England

Manufacturer
Thorn EMI Lighting Ltd, London, England

Thorn mini Papi

Thorn Elevated Light Fittings

Elevated light fittings are available for use as approach lights, threshold lights and as edge lights for the runway, taxiway and apron. All versions embody a frangible base designed to withstand jet blast but break off readily on impact to avoid damage to an aircraft.

Thorn HIA6/1 elevated light

Thorn LIA3/3 elevated light

Characteristics

MODEL	HIA6/1	LIA3/3	HIR4/3	LIR 4/4
TYPE	Uni-directional high-intensity	Omni-directional low/medium intensity	Omni-directional high-intensity two main beams	Omni-directional low/medium intensity
APPLICATION	Approach	Approach	Runway edge threshold stopway runway end	Runway edge threshold taxiway/ apron edge
LAMP	AF7/2 tungsten halogen 200 W	AF7/2 tungsten halogen 200 W	Mainly AF7/2 halogen 200 W	Various including AF4/2 45 W
LIGHTING CATEGORY	All including Cat III (L1)	L3	All including Cat III (L1)	Cat I (L2)
UNIT DIMENSIONS				
height	270 mm	290 mm	240 mm	200 mm
diameter	200 mm	178 mm	203 mm	127 mm
depth	115 mm			

Status
In service world-wide

Manufacturer
Thorn EMI Lighting Ltd, London, England

Thorn Inset Light Fittings

Inset light fittings are produced for the approach, threshold touch-down zone, runway centre-line and edge and taxiways. Standard features of all the lights are a fully sealed plug-in light module, tungsten halogen lamps and a common base to facilitate installation in new pavements (the HIA B1 base is slightly larger than the standard). Two bolts which are spring-loaded hold the cover in position. Alternative size bases are available for direct mounting on existing or new pits.

Status
In use throughout the United Kingdom, Africa, Middle East, Far East and Australasia

Manufacturer
Thorn EMI Lighting Ltd, London, England

Thorn inset light fitting being installed

Thorn CRH B2 inset light

Characteristics

TYPE	HIA 81 approach	HIA 81 Threshold	CRH C1	CRH B1
APPLICATION	Approach	Threshold	Touch down zone	Touch down zone
LAMP	2 × AF7/2	2 × AF7/2	AF7/2	AF7/2
LIGHTING CATEGORY	Cat III (L1)	Cat III (L1)	Cat III (L1)	Cat I (L2)
MAX PROJECTION	25·5 mm	25·5 mm	12 mm	17 mm

TYPE	CRH C2	CRH B2	ERH-B2T toe in	ERH-B2T non-toe in
APPLICATION	Runway centre-line	Runway centre-line	Runway edge	Runway edge
LAMP	AF7/2	AF7/2	AF7/2	AF7/2
LIGHTING CATEGORY	Cat III (L1)	Cat I (L2)	Cat III (L1)	Special Application
MAX PROJECTION	12 mm	17 mm	38 mm	38 mm

TYPE	CTHC-B2	CTHCS-B2	CTHS-B2	OLTF-1
APPLICATION	Taxiway-curve	Taxiway entering curve	Taxiway straight	Taxiway, omni-directional edge/ centre-line
LAMP	AF6/5 100 W	AF6/5 100 W	AF6/2 45 W	AF6/2 45 W
LIGHTING CATEGORY	Cat III (L1)	Cat III (L1)	Cat III (L1)	Various
MAX PROJECTION	25 mm	25 mm	25 mm	25 mm

Thorn Apron, Obstruction and General Purpose Lighting

Thorn produce a comprehensive range of lights for apron, terminals, freight centres, maintenance areas, car parks and approach roads. Two types of hazard warning lights are offered, a high-intensity tower and obstruction light fitting, and a low-intensity light. The high intensity version has a 160 watt neon light source. It is housed in a resin bonded glass fibre moulding with an acrylic front screen. The unit has been developed to give a 2000 candela beam of red light in the angular zone where aircraft would be in danger. It is generally used on high structures such as tall chimneys, towers, etc, typically in a four-unit arrangement. A flashing facility is optional. The low intensity tower and obstruction light is offered for those locations, such as control towers, hangars, passenger terminals etc, which constitute a hazard to aircraft movements, but which do not require the higher light output of the high-intensity unit.

Status
In service at airports world-wide

Manufacturer
Thorn EMI Lighting Ltd, London, England

Thorn low-intensity obstruction light

Characteristics

TYPE	High-intensity	Low-intensity	
APPLICATION	Obstruction light	Obstruction light	
LAMP	1 × 160 W neon	1 or 2 × 240/250 V 65 W ES lamp	
UNIT DIMENSIONS		Single fitting	Double fitting
height	508 mm	229 mm	232 mm
width	1073 mm	118 mm	245 mm
depth	448 mm	137 mm	137 mm
WEIGHT	47 kg	3 kg	6 kg

Thorn Control and Distribution Systems

In order to obtain a consistent lighting intensity, airfield systems operate by energising the lamps from a constant current series circuit, each lamp being supplied from its own isolation transformer to protect the rest of the circuit in the event of a lamp failure. The current is controlled by a constant-current regulator which allows different levels of current to be selected, providing controlled stages of intensity.

In the constant-current regulators current control is obtained by means of a transductor driven by a semi-conductor sensing circuit. In the smaller sizes both the power circuit and the control gear are air-cooled and mounted in a free-standing ventilated case. Units of 20 kVA and above will normally have the main transformer and the transductor in an oil filled tank with the control gear mounted on

the front of the unit in a separate compartment. All are based on a modular design. Facilities are provided for local and remote operation. Units can be supplied for various input voltages at 50 or 60 Hz. Standard designs have ratings of 2·5, 4, 6, 7·5, 10, 12, 15, 18, 20, 25 and 30 kVA. Output current is 6 or 20 amperes and alternative ratings and output current are available on request.
The remote control of constant current regulators is usually required so that visual guidance systems can be operated from the air traffic control tower. Because requirements tend to differ widely from airport to airport, these are usually made to suit the operator's individual specifications. A range of units are offered by Thorn, from a desk top panel to the more complex control and mimic diagram for installation in a controller's desk unit.
The isolating transformers isolate the primary series circuit from the secondary connections to the light fittings. This ensures the continuity of the primary series circuit in the case of a lamp failure. The most generally used current rating is 6·6 ampere primary and secondary. Performances conform to the requirements of the relevant FAA specifications. Several types are available.
In the type OC and AC series the transformer and its cable terminals are enclosed in a cast-iron case. This has three watertight compression glands for the primary and secondary cable entries. The top cover has a watertight screwed plug for final topping up with the insulation compound. Provision is made for an earth continuity link to earth and bond the brass tape screen of the primary cable.
The type OC series is designed for standard airfield lighting cable, with or without brass tape protection.
The AC-type series is the same as the OC series but with cable entry glands suitable for use with steel wire armoured cable.
In the NPE-type series, the transformer is encased in synthetic rubber, and is designed for direct burial in the ground. Or it can be fitted into either a metal mounting base or a concrete pit.
The primary and secondary cables are moulded into the body of the transformer and are fitted with connectors to FAA specifications.
Also available are transformers encased in epoxy resin (EPR series). The choice of type (OC or AC, NPE or EPR series) is often dependent on soil conditions. Plug and socket connector kits are produced to facilitate connection in the field.

Status
Thorn control and distribution systems are installed in airfield lighting systems world-wide

Manufacturer
Thorn EMI Lighting Ltd, London, England

Wysepower Mobile Lighting Units

Wysepower has developed a range of heavy-duty, self-contained mobile lighting units suitable for illuminating hazard or accident areas, aircraft aprons and service areas. With mast heights of from 5 to 18 metres, the units can provide illumination at levels up to 10 lux over an area of 1 hectare, using 4 kilowatts of power.
The Juniorlite Type JL is a lightweight trolley-mounted lighting set powered by a 2 kVA petrol engine. A hand-extended mast raises to a height of 5·2 metres and accommodates 2 × 500 watt tungsten halogen floodlights. Sockets outlets provide 110/240 volt power for auxiliary tools and lighting.
The Seniorlite type SL is also a lightweight trolley-mounted lighting and generating set. It is powerec by a Petter diesel engine delivering 3·2 kilowatt of power at 110/240 volt which can provide auxiliary supplies for additional lighting. Standard equipment is 2 × 500 watt tungsten halogen floodlights mounted on a 5·2 metre telescopic mast.
The Powerlite 9 type PL9 is a heavy-duty mobile floodlighting unit mounted on a road-towing trailer. It is available with engine-driven power generating units with outputs from 4 to 7 kilowatts and a selection of lighting heads ranging from tungsten halogen with an output of 20 000 lumens to high-pressure sodium with 375 000 lumens. The trailer is of heavy-duty welded steel construction. The three-section tubular mast is raised manually by an autobrake winch to a full height of 9 metres. The chassis, mast and stabilising jacks are all fully galvanised for weather protection.
The Powerlite 18 type PL18 is designed to floodlight larger areas. It has a lattice mast in four sections which may be raised to a full height of 18 metres. The PL18 has all the main features of the PL9 but is available only with the Lister 7 kVA power unit.
A full range of lighting heads is offered plus mercury vapour units with 200 000 lumens output. Outlets for power tools are available on all models at 110 volts.

Wysepower Juniorlite type JL mobile lighting unit

Wysepower Powerlite 9 type PL9 mobile lighting unit

Wysepower Powerlite 18 type PL18 mobile lighting unit

Status
Production

Manufacturer
Wysepower Ltd, Sandy, Bedfordshire, England

UNITED STATES OF AMERICA

Crouse-Hinds Approach and Navigational Lights

Approach lights comprise elevated and inset units. The EAL is an elevated approach light constructed in aluminium. This uni-directional PAR 56 lamp holder is suitable for mounting on frangible towers. It is also available with a bypass relay to allow more than one light to be operated from a transformer. Colour screens can be added for threshold or red side bar application. Units are available with 200, 300 or 500 watt lamps.
The HTL fixed focus MB-1 threshold light is designed and constructed to meet military requirements. The light has an aluminium housing with an internal levelling device. The optical system is a prismatic globe, which covers the 180 degree green and red colour filters. The fixture is mounted on a frangible coupling and uses a 500 watt, 20 ampere lamp.
The HAP is a 500 watt, 20 ampere inset high intensity approach or threshold light used where the approach light system must be placed in a paved area. It consists of a mounting ring and a sealed optical assembly. The sealed optical assembly uses a compound ellipsoidal reflector to produce a light compatible with high-intensity elevated approach lights. Components are designed to withstand take off and landing loads from all types of aircraft.
The HTL is a high-intensity threshold light that incorporates a red runway end light. The two lights may be connected to separate circuits. The 65 watt, 6·6 ampere end light can be operated from the edge light circuit and the 500 watt green light can be operated with the approach light system. This unit is similar to the HAL light unit in construction.

Crouse-Hinds EAL elevated approach light

The RAL is an inset high-intensity runway approach light designed for runway overrun areas that have category I approach lights. This unidirectional light has a sealed optical assembly and is supplied with a 100 watt, 6·6 ampere or 200 watt, 6·6 ampere quartz lamp. This unit is not recommended for airports with heavy snowplough activity.

Visual landing aids produced by Crouse-Hinds include a Vasi, an abbreviated Vasi and a Papi.

The Vasi system provides the pilot with approach slope information to provide safe descent guidance. The unit, in a fibreglass corrosion-resistant housing, projects a split red and white beam to form a descent path to the runway.

The Vasi system is available in several configurations, depending on the day vfr range required and category of aircraft using the runway.

The abbreviated Vasi is usually supplied in the Vasi-2, Vasi-4 or Vasi-6 configuration with a power adaptor unit. This is an economical system with photocell intensity control or optional remote-control. Tilt switches are available to turn the system off in case of misalignment. This system provides visibility of up to 4 nautical miles during daylight operation.

Papi is a visual landing aid that gives pilots specific glide-path information on approach. The system indicates with a pattern of red and white lights whether or not the aircraft is at the best angle for safe touchdown. A pattern of two white and two red lights signifies the correct glide-path while three red and one white signify below glide-path and three white and one red above glide-path.

Other navigational lights offered comprise a rotating beacon, runway end identifier light (Reil), and two condenser discharge lights, one inset, the other elevated.

The high intensity rotating beacon emits an alternate green/white beam. The unit is constructed of a cast aluminium housing, central 1000 watt lamp, and automatic lamp changer. The beacon can be supplied with a lamp failure indication.

Crouse-Hinds abbreviated Vasi system

The Reil system consists of two uni-directional strobe lights flashing simultaneously. The twice-per-second flashing strobe is a positive means for the pilot to identify the runway. The Reil lights can also be controlled directly from an approaching aircraft by means of an optional radio control unit. This is a suitable aid for runways that have no approach light system.

The condenser discharge system is a series of sequentially flashing lights designed to attract the pilot's attention. These lights can be easily seen even when the background brightness is very high. Variable intensities permit the control tower to vary the light output to suit local conditions. These units are lightweight and can be mounted on frangible towers.

The inset version is designed for runway overrun or displaced threshold areas where approach light systems must be set into the pavement. This unit is compatible with elevated flashers.

Crouse-Hinds RB rotating beacon

Status

In service at civil and military airfields

Manufacturer

Crouse-Hinds Co, Syracuse, New York, USA

Crouse-Hinds Runway and Taxiway Elevated Lights

These are elevated lights designed to indicate the edge of the runway and taxiway to the pilot. High-intensity lights comprise the HRL and HRLQ units. The HRL runway edge light consists of a fresnel globe and lamp holder assembly, which can be mounted to a frangible coupling and base plate. Colour filters are offered for threshold applications. This unit is available for civil or military applications and is normally supplied with a 200 watt, 6·6 ampere lamp.

The HRLQ is a fixed focus quartz halogen light designed to withstand high velocity jet blast. The light comes complete with lamp, lens, colour filters and frangible coupling. The unit is normally supplied with a 115 watt lamp for runway edge and a 175 watt lamp for threshold. Other lamps are available on request.

Other elevated lights in this category are the ERLQ, ERLP, ELM, LRL and ETLS models. The ERLQ is a quartz light designed for medium-intensity runway light applications and for the traditional blue taxiway edge light. The unit incorporates a twist-lock feature, which simplifies changing the lamp. It is available with 30 or 45 watt lamps and coloured domes to meet a variety of applications. Extension kits are produced for increased height in snow regions. The ERLP medium-intensity light is designed to mark medium intensity thresholds, runway edge, taxiways and heliports. The light features a twist-on dome, non-metallic construction and miniature quartz lamp. Optional anti-tamper hardware is offered. The unit can be mounted on a stake assembly or base plate. Available in 30 watt and 45 watt, 6·6 ampere lamps with all aviation filter colours.

The ELM taxiway edge light provides low profile lighting for military airfields. The low profile minimises damage due to aircraft drag chutes, which are used to slow landing aircraft. The ELM features a stamped steel housing, medium pre-focused socket, prismatic globe and frangible coupling. Coloured lenses and a 30 or 45 watt lamp are available. Both class A and B type lights are produced.

Crouse-Hinds ELM elevated light for military use

The LRL is designed for lighting small general aviation runways and taxiways. The unit is available with a 15 watt lamp in a one-piece cast aluminium housing. The prismatic globe is held by a spring clamp for easy lamp replacement. The light may be stake-mounted.

The ETLS is a wide beam threshold light designed to give higher definition to medium intensity runways without approach lights or runway end identifiers. The unit features 115 watt quartz lamp, cast aluminium construction, and smooth dome. Other lamp sizes are available.

Status

In service at civil and military airports

Crouse-Hinds HRL high-intensity runway light

Crouse-Hinds ETLS elevated threshold light/special

Manufacturer

Crouse-Hinds, Syracuse, New York, USA

Crouse-Hinds Runway and Taxiway Inset Lights

A range of inset lights are produced for the runway and taxiways.
The ATL and FMU units are suitable for approach and threshold lighting.
The ATL runway threshold light is uni-directional and is designed to be used for category I threshold applications. The light uses one 200 watt lamp and a dichroic green filter. The optical assembly is housed in a sealed iron casting ready for installation with its mounting ring. The mounting ring may be used with either a L-857 Type I or Type II light base.
The FMU semi-flush, uni-directional runway light operates with sealed beam lamps for runway approach and threshold applications. A variety of lamps may be employed to meet civil and military airport requirements. A filter bracket arrangement allows for the addition of single colour filters when needed. These fixtures are normally installed on L-857 Type I light bases. The assembly is supplied with plug assembly, gaskets, bolts and clamp ring. For runway edge, end and threshold applications the FMB, REL and RTEL models are produced. The FMB is similar to the FMU model and uses a sealed beam lamp. Single- or dual-colour glass filters may be fitted. The REL is a runway edge light designed specifically for use where taxiways intersect runways. The light beams are optically matched to adjacent elevated runway light fixtures. The optical assembly contains two lamps housed in a sealed iron casting ready for installation with its mounting ring. The mounting ring may be used with either an L-857 Type I or Type II light base. Two lamps (115 and 200 watt) and a variety of filters are available.
The RTEL is a high-intensity runway threshold bi-directional light for use as a threshold light with green and red beams. The light uses dichroic filters. The optical assembly contains two 200 watt lamps in a sealed iron casting ready for installation with its mounting ring, which may be used with either an L-857 Type I or Type II light base.
The TDZ is a high-intensity runway touchdown zone light. It is uni-directional with a sealed optical assembly ready for installation on either shallow or deep L-857 Type II light bases. One 6·6 ampere, 115 or 200 watt quartz lamp may be used for all applications.
For runway centre-line applications the RCL and RCLN are available. The RCL is a high-intensity bi-directional light. Fixtures may be installed on shallow inset or L-857 deep light bases. The optical assembly consists of two sealed iron castings. One 115 watt, 6·6 ampere or 200 watt, 6·6 ampere quartz lamp is used for most applications. The RCLN was designed for the US Navy to withstand the impact from aircraft arresting hooks. It features a high strength optical cover assembly mounted in a hardened stainless steel base casting. These units are normally mounted on 25-foot (8-metre) centre-lines with a 45 watt, 6·6 ampere lamp. The lights are supplied in four styles for shallow or deep can mounting in the runway.
Taxiway inset lights available comprise the TOL, TOLF, TCL and FMT models.
The TOL omni-directional light assembly serves as a general purpose inset light for taxiway intersections, apron edge, and heliport applications. This light can be installed on L-850 or L-857 Type II bases and is normally supplied with a 45 watt lamp. Coloured lenses are available as required.
The TOLF omni-directional flush light assembly is a general purpose inset light designed for airport areas with snowploughs. The flush profile is designed to allow snowplough blades to pass over the top of the unit without damage. The unit can be provided with colour filters as required. Fitted with a 45 watt, 6·6 ampere lamp, it can be supplied to fit on L-857 Type I or Type II base.
The TCL taxiway centre-line light is the cartridge type with a rectangular optical assembly. The Type I and Type II units are designed to be core-drilled and bonded into existing pavements. These units are suitable for taxiway systems where low-visibility operations will be conducted, and are available in wide- or narrow-beams with 30, 45 or 60 watt lamps.
The FMT provides semi-flush lighting for operational areas where elevated lights may constitute a hazard. The units consist of an iron casting with lens, a holder for the lamp and colour filter, plug assembly. This light is designed for installation in the L-857 Type I deep base style cans. Lamps are available in either 45 or 100 watts.

Crouse-Hinds REL runway edge light

Crouse-Hinds TDZ runway touchdown zone light

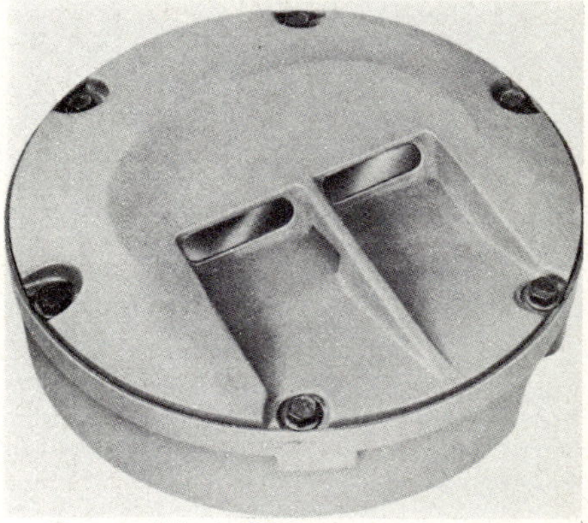

Crouse-Hinds TDZ runway touchdown zone light

Status
In service at civil and military airfields

Manufacturer
Crouse-Hinds, Syracuse, New York, USA

Crouse-Hinds Obstruction, Apron and General Purpose Lighting

Hazard warning lights produced by Crouse-Hinds include FBC-12, EOL and VAW models.
The FBC-12 is a flashing red hazard beacon for marking towers and other obstructions to air navigation. It must be used in conjunction with TSS flashing switch. The unit features a watertight, non-ventilated, cast aluminium base and frame, four clear fresnel lenses and two internal aviation red colour screens to emit a 360 degree concentrated beam. Top and bottom sections are hinged at the centre to allow for changing lamps.
The EOL and VAW units are steady burning red lights. Both have aluminium housings, red fresnel globes, and medium screw lamp receptacles. FAA units have approved international orange finish. Military units are in aluminium. All units are available in a double housing.
Apron and ramp lighting requirements are met by the CLX, GALR and SFL units. The CLX is available rated at 400–1000 watts and features an aluminium reflector designed to ensure a sharp cut-off point both in front and to each side of the luminaire, with virtually no light being emitted behind the unit. The reflector produces the peak candlepower at an angle of 65 degrees with cut-off

Crouse-Hinds FBC-12 hazard beacon

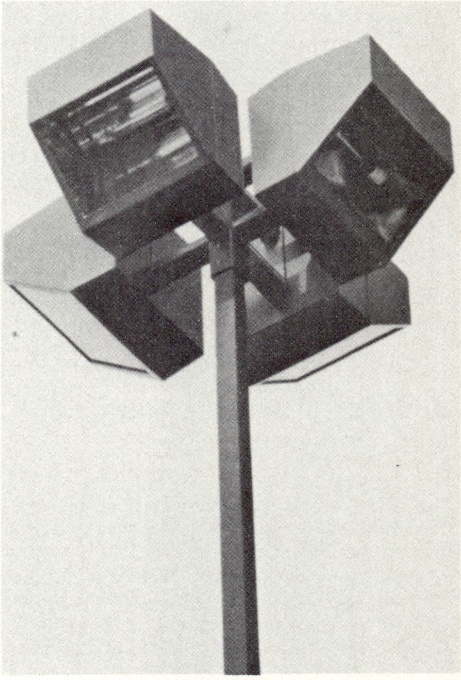

Crouse-Hinds CLX lights

at 70 degrees and is designed for maximum output of high pressure sodium sources.
The GALR uses an asymmetrical reflector which produces a rectangular pattern of light rated at 18–1000 watts.
The SFL is suitable for general floodlighting and features a variety of light shielding and cutoff combinations, and spun aluminium housing with louvres. Other models produced are suitable for lighting access roads, parking lots and terminal buildings.
Accessory equipment available includes illuminated wind socks, taxiway guidance signs and distance markers.

Status
In service at civil and military airports

Manufacturer
Crouse-Hinds, Syracuse, New York, USA

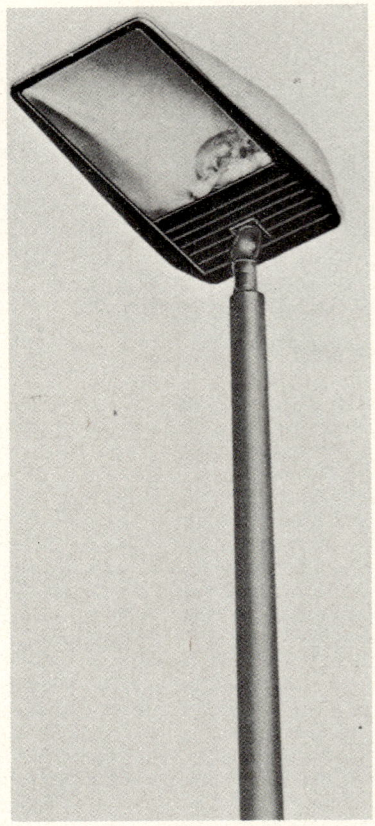

Crouse-Hinds GALR light

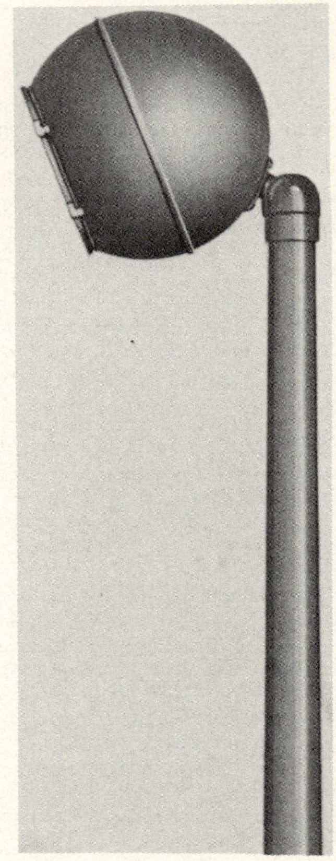

Crouse-Hinds SFL light

Crouse-Hinds Power Distribution and Control Equipment

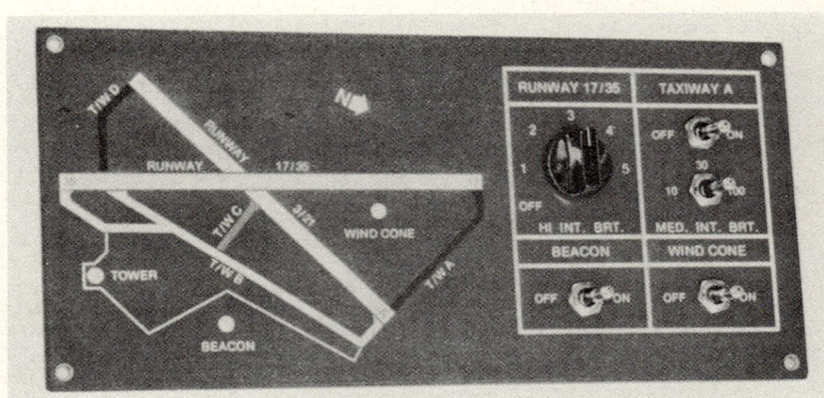

Crouse-Hinds remote-control panel

In this category Crouse-Hinds supply control panels, constant current regulators, switching gear, radio and photo-electric control equipment and series isolation transformers.
Series isolating transformers are designed for direct burial and insulate the runway lights from high voltage series lighting circuits. The control of airfield lighting is accomplished by means of a control panel located in the control tower. These panels may be obtained in several configurations. Some are supplied with on-off switches and intensity selector switches and some panels are provided with a mimic display to give a visual presentation of the lighting status on the airport. Control panels are usually custom-built for a particular airport.
Two types of constant current regulators are available. The REGW constant current oil-filled regulators are designed for all series lighting. These units can be directly controlled from up to 3 kilometres without the use of auxiliary relays. Other special features, such as positive feedback circuit status, are optionally available on special order. Models range from 10 to 70 kW for 50 and 60 Hz frequencies.
The REGD dry-type constant current regulators are available in 4, 7·5 and 10 kW models. The features of this unit are similar to the REGW. The unit which may be wall mounted, is fitted with load protection devices.

A variety of circuit selector switches and pilot relay cabinets are offered for airport remote control functions. Circuit selector switches allow tower personnel to select one or more circuits to be operated from a single regulator and auxiliary relay cabinets may be used to control several regulators from a single switch. Units can be provided to meet civilian or military applications for regulators up to 50 kW.
The radio controller is used by the pilot to activate approach lights, runway edge lights and Vasi by means of a series of clicks on the radio microphone in the aircraft. This aid is especially useful on smaller airports that do not always have manned-control towers.
The PEC-4 is an automatic photo-electric control unit with a photocell assembly, 15 ampere relay and surge protection. Factory set per FAA requirements, the unit can be used to control obstruction lights or other airport lighting equipment.

Status
Installed at civil and military airports

Manufacturer
Crouse-Hinds, Syracuse, New York, USA

DeVore Pulse Light Approach Slope Indicator

The DeVore pulse light approach slope indicator (Plasi) is a ground-installed, self-contained device which visually provides vertical glide-path information including correct or acceptable position, direction, and degree of deviation from glide-path. The equipment has a range of at least four miles (6·5 kilometres) in bright sunlight and can be adjusted to any approach angle.

The device generates and projects three horizontal bands of light, only one of which can be seen by the pilot at a given instant. The centre band is a steady white light projected as a 0·35 degree (±0·02 degree) high angular wedge, 16 degrees wide with the apex at the approach aid. This centre band defines the correct glide-path. An upper band of white light pulsing at approximately 2·25 pulses per second is a wedge approximately 2·5 degrees high by 16 degrees wide which gives 'above glide-path' indication. The pulses of the white and red 'above' and 'below' lights vary in length from continuous at the edge of the glide-path to zero length at the limit of visual contact off the glide-path. This variation in light pulse length, long near the path, shorter and shorter as deviation from

glide-path increases, gives the pilot quantitative deviation information. Rate of change of pulse length provides rate of deviation from or closure with the glide-path.
The visual presentation is accomplished through the use of optical components, a movable shutter, and red filter. One tungsten halogen lamp is positioned behind the condenser lens.
The system is equipped with an automatic lamp changing system with five lamps in the unit. When one burns out, the next one is automatically moved into place. When the final lamp is in operation, a red warning light is illuminated on top of the unit to alert maintenance personnel that lamp replacement is necessary.

Status
Production. The system received FAA approval for use by fixed-wing aircraft and helicopters in January 1982. In service at a number of airports and heliports in the USA and elsewhere

Manufacturer
DeVore Aviation Corporation, Albuquerque, New Mexico, USA

DeVore Plasi visual landing aid

Manairco Elevated Runway/Taxiway Lights

Manairco produce medium- and low-intensity elevated lights available for smaller airfields and heliports. The low intensity 1000 and 2000 series, for runway and taxiway edge, runway threshold and heliport use, features a tempered glass lens incorporating magnifying prisms, with a yellow marker cone. The light is available with either a 2 foot 6 inch (0·76 metre) metal stake or a 1 foot 4 inch (0·41 metre) conduit pole with breakable coupling. The medium-intensity 6000, 7000 and 8000 series for runway and taxiway edge and threshold lighting comprise a lamp unit, glass lens with securing ring, adjustable base with pipe adaptor, 5 or 16 inch (0·13 or 0·41 metre) conduit pole and breakable coupling. Two mountings are offered, a 2 foot 6 inch (0·76 metre) metal stake or a low-profile pad mounting. The model 6300 is available with a red lens for use as an obstruction light.
Model R-200 Reil are omni-directional units with a xenon flasher tube and a 6 inch diameter (0·15 metre) fresnel lens on an aluminium base. The unit can be mounted on either a 2 foot 6 inch (0·76 metre) metal stake or a 1 foot 4 inch (0·41 metre) conduit pole with breakable coupling.

Characteristics

SERIES/MODEL	1000	2000	6000	7000	8000	R-200
TYPE	Low-intensity	Low-intensity	Medium-intensity	Medium-intensity	Medium-intensity	Medium-intensity
LAMP	15 W, 117 V	15–40 W, 117 V	30–45 W, 117 V	30–45 W, 6·6 A	25–100 W, 117 V	Xenon 30 W, 117 V

Status
Production. The series 2000, 7000 and 8000 are approved by the FAA

Manufacturer
Manairco Inc, Mansfield, Ohio, USA

Manairco Navigation Lights

Manairco offer a range of rotating beacons for airfield location and a Vasi as a visual landing aid. Seven models of the beacon are produced ranging from 75 000 candelas from two 100 watt lamps visible at 10 miles (16 km) to 400 000 candelas from four 1000 watt lamps giving a range of 40 miles (64 km). Optional equipment available includes photo-electric control to switch the lights on at sunset and off at sunrise, 24 hour time clock to activate the lights at preset times, radio controlled operation facility and an arctic kit for cold weather environments.
The Richter Vasi projects three beams which are visible at distances up to 4 miles (6·4 km). The middle green light signifies the glide-path to the pilot. The yellow upper beam indicates that the aircraft is above the glide-path and the red lower beam that the aircraft is below the glide-path. The green glide-slope reticle is 2 degrees high and 6 degrees wide. The unit is equipped with two 300 watt quartz lamps. If the primary lamp fails, the stand-by lamp illuminates automatically and a small tell-tale light outside the housing indicates the failure. Other features include weatherproof steel housing, scope sight for visual alignment, front levelling device to ensure horizontal beam pattern and automatic shut-down if the unit is tilted. In the latter facility, a goniometer arm controls two balanced mercury switches to extinguish the projected beam.

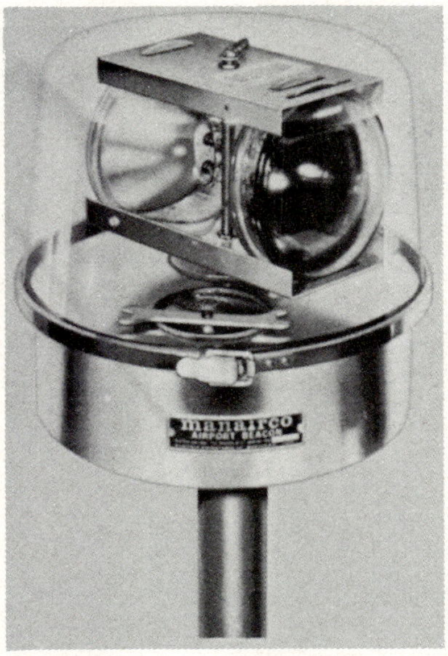

Manairco AB-100 airport becon

Manairco AB-500H airport beacon

Characteristics

MODEL	AB 100	AB 500	AB 1000	AB 1000D	AB 500H	AB 500F	AB 1000F
INTENSITY	75 000 cd	190 000 cd	400 000 cd	400 000 cd	285 000 cd	190 000 cd	400 000 cd
RANGE	10 miles (16 km)	30 miles (48 km)	40 miles (64 km)	40 miles (64 km)	30 miles (48 km)	30 miles (48 km)	40 miles (64 km)
LAMPS	2 × 100 W	2 × 500 W	2 × 1000 W	2 × 1000 W	3 × 500 W 3 × 500 W	4 × 500 W 4 × 500 W	4 × 1000 W 4 × 1000 W

Vasi
BEAM PROJECTION
ELEVATION: yellow and red 4°, green 2°
AZIMUTH: 6°
LAMP: 2 × 300 W
POWER REQUIREMENTS: 117 V ac, 2·6 A
UNIT DIMENSIONS
length: 16 in (0·4 m)
width: 6¼ in (0·16 m)
height: 12½ in (0·32 m)
WEIGHT: 8 lb (3·6 kg)

Status
Production

Manufacturer
Manairco Inc, Mansfield, Ohio, USA

Over-Lowe Airliner Mobile Floodlighting Unit

The Airliner is a mobile floodlighting unit designed specifically for use in airport maintenance and emergencies. The unit is equipped with an adjustable tower and T bar on which either one or two 1000 watt metal halide floodlights are mounted. The tower is attached to a frame fitted with rubber-tyred wheels. Other features include a double-pole switch and a 50 foot (15·2 metre) length heavy-duty extension cord.
A 14 foot (4·3 metre) telescopic tower and a detachable Over-Lowe KP 3000 generator are optional equipment.

Status
Production. Supplied to several airlines and airports

Manufacturer
Over-Lowe Co, Englewood, Colorado, USA

Over-Lowe Airliner mobile floodlighting unit

COMMUNICATIONS

CANADA

Garrett G 2000 Series Vhf Communications Equipment

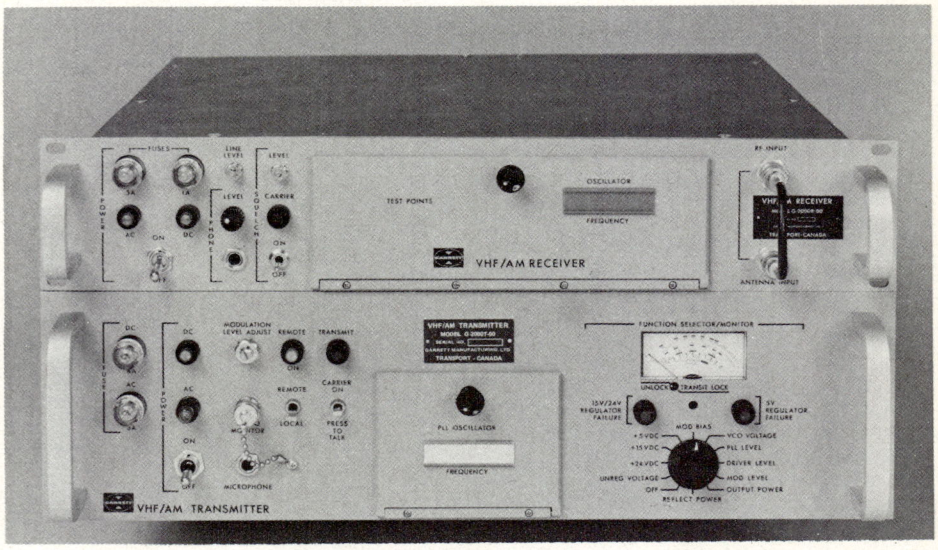

Garrett G 2000 series vhf transmitter/receiver units

Garrett has developed a family of solid-state vhf/am single channel transmitters and receivers for fixed station ground to air communications.
The Model G 2000T is a single channel transmitter operating in the 117·5 to 137 MHz frequency band with an adjustable power output from 10 to 25 watts. The unit incorporates an audio compressor to reduce the modulation depth variation and a built-in transmit/receive relay. The equipment features fault isolation and monitoring with protection for excess voltage, current and temperature.
The Model G 2000R is a single-channel receiver of plug-in modular construction. An audio compressor is incorporated to ensure constant audio output. The unit is fitted with a front panel test matrix for fault isolation and protection for excess current.
Both transmitter and receiver are equipped with an automatic change-over facility to 24 volt dc stand-by power.

Characteristics

FREQUENCY RANGE: 117·5–137 MHz
CHANNEL SPACING: 25 kHz
POWER OUTPUT: adjustable 10–25 W
POWER REQUIREMENTS: 110/120 V, 50/60 Hz
(220/240 V, 50/60 Hz optional)
stand-by: 24 V dc

UNIT DIMENSIONS	G 2000T	G 2000R
height	5¼ in (133 mm)	3½ in (89 mm)
width	19 in (483 mm)	19 in (483 mm)
depth	17 in (432 mm)	17 in (432 mm)
WEIGHT	33 lb (15 kg)	15 lb (6·8 kg)

Status
Production

Manufacturer
Garrett Manufacturing Ltd, Rexdale, Ontario, Canada

Leigh ANT-57MT Mobile Air Traffic Control Tower

Leigh ANT-57MT mobile atc tower

The mobile atc tower ANT-57MT is designed to meet the needs of civil or military atc requirements at secondary airports, or to provide emergency or supplemental facilities for primary airfields.
The control tower is relatively lightweight for airlifting via cargo aircraft or helicopter. On the land, the unit can be towed around on its own castor wheels and then positioned in the optimum location by means of four levelling jacks. The ANT-57 MT is also available mounted on a standard heavy-duty truck chassis. Coupled to a mobile generator set, the control tower is ready for operation, wherever required.
The standard tower provides a complete atc capability under vfr and ifr flight conditions and will serve in all atc density levels. There is accommodation for up to three controllers and the necessary equipment: hf (ssb) vhf, uhf, vhf-fm, atc logging recorder, radio switching system, meteorological equipment, air conditioning/heating, and emergency ground power unit. Options are available for remote-control/monitoring to other terminal landing facilities such as lighting systems, navaids, vhf-df, par and acquisition radar. Standard signalling facilities are provided for conveying visual advisories to the operational area to at least 1200 feet above field elevation and a minimum unobstructed radius of 3 miles (4·8 km), permitting aircraft control under vfr conditions.
The ANT-57MT is available mounted on a vehicle fitted with a scissors-lift hydraulic platform, which can be elevated to a height of 25 feet (7·62 metres).

Characteristics

MODEL	Standard tower	High lift tower
BASE AREA	120 ft² (11·25 m²)	
GROSS FLOOR AREA	100 ft² (9·35 m²)	
NET FLOOR AREA	44 ft² (4·12 m²)	
ROOF AREA	120 ft² (11·25 m²)	
INTERIOR HEIGHT	6 ft 9 in (2·08 m)	
OVERALL DIMENSIONS		
length	15 ft 5 in (4·7 m)	31 ft 5 in (9·69 m)
width	8 ft (2·44 m)	8 ft (2·44 m)
height	7 ft 8 in (2·34 m)	13 ft 6 in (4·11 m)
elevated height		26 ft 1 in (7·95 m)
WEIGHT	7000 lb (3270 kg)	37 310 lb (15 929 kg)

Status
Production

Manufacturer
Leigh Instruments Ltd, Pointe Claire, Quebec, Canada

Leigh Fixed and Transportable Control Towers

Leigh Instruments produces fixed and transportable control towers. Atc tower cab models CT-250, CT-350 and CT-450 have been developed in accordance with present and anticipated airport needs. These atc tower cabs are pentagonal structures meeting specific structural and operational requirements to cope with low-medium, medium-high, and high activity airport atc operation levels. They are equipped with: lightning conductor air terminals, obstruction lights and photo-electric control, antenna/mounts, roof hatch, roof exhaust fan, suspended antivibration acoustic ceiling, and directional and dimming ceiling lights.

Control tower consoles are designed and manufactured to the customer's requirements, with the amount and type of equipment specified by the customer.

The Leigh model 909E communications switching system is normally installed in each controller position. The system is a modular integrated solid-state communications control system, specifically designed for atc providing control of radios and intercommunication channels. The company provides air ground vhf, uhf and hf communications equipment, analogue or digital meteorological systems, logging recorder, airfield lighting indicator and control panel, navigation aid monitors, and all essential atc operating accessories.

The transportable atc tower model TT-070 is a self-contained, lightweight unit which provides positive control of aircraft under vfr and ifr flight conditions, and can be operated in low-medium activity airport atc operation levels. Hexagonal in shape with window sections sloping out at 15 degrees, model TT-070 is manufactured of aluminium alloy, insulated and waterproof. The tower is modularly constructed. Assemblies are sized for shipping and erection on site and all connections are bolted.

Status
Production

Manufacturer
Leigh Instruments Ltd, Pointe Claire, Quebec, Canada

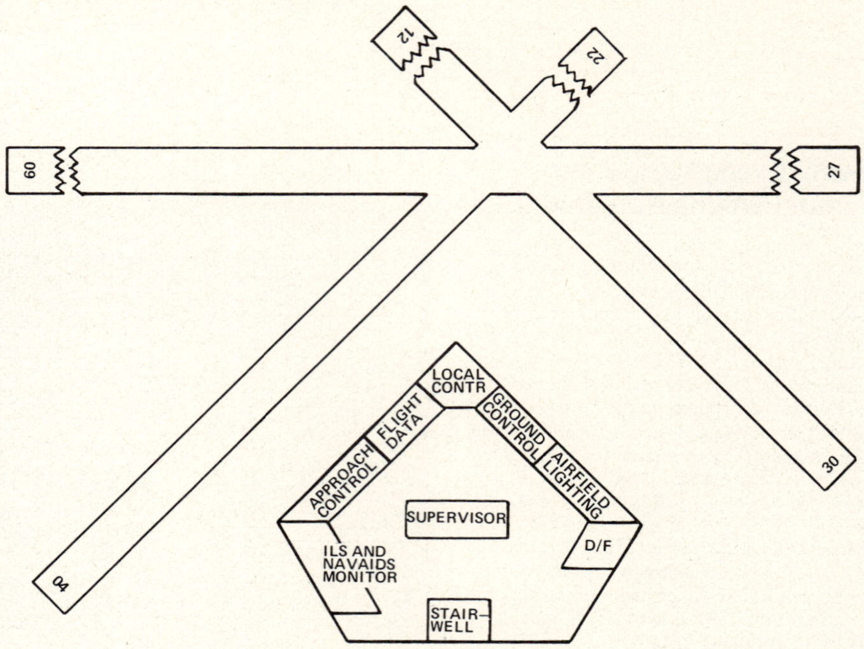

Interior design of Leigh control tower

Leigh TT-070 transportable control tower

Characteristics

MODEL	CT-250	CT-350	CT-450	TT-070
TYPE	Fixed base	Fixed base	Fixed base	
FLOOR AREA				
total	261 ft² (24·3 m²)	358 ft² (33·25 m²)	468 ft² (43·5 m²)	66 ft² (6·18 m²)
with consoles	182 ft² (16·88 m²)	301 ft² (28 m²)	398 ft² (37 m²)	
HEIGHT				
base to coping	14 ft 1½ in (4·31 m²)	14 ft 1½ in (4·31 m)	14 ft 1½ in (4·31 m)	9 ft 3 in (2·82 m)
floor to ceiling	9 ft 4½ in (2·86 m)	9 ft 4½ in (2·86 m)	9 ft 4½ in (2·86 m)	7 ft 4 in (2·23 m)
GLASS HEIGHT AT 15°	6 ft 6 in (1·98 m)	6 ft 6 in (1·98 m)	6 ft 6 in (1·98 m)	4 ft (1·22 m)
GLASS AREA (TOTAL)	484 ft² (45 m²)	564 ft² (52·3 m²)	678 ft² (63 m²)	
VISION				360°
ROOF AREA	468 ft² (43·5 m²)	616 ft² (57·2 m²)	872 ft² (81 m²)	
VOLUME OF CAB (INTERIOR)	3263 ft³ (92·4 m³)	4374 ft³ (124 m³)	6066 ft³ (171·75 m³)	

FRANCE

CIT-Alcatel KLB Series Automatic Message Switching Centres

The KLB series is an automatic message and data switching system, controlled by recorded programmes and integrated microprogrammes. Total traffic protection, with no loss of message or character is ensured by the use of two independent processing chains. Messages are stored while awaiting dispatch and for a period after reception in order to meet repetition requests. The system provides automatic surveillance of both network and centre status. The range of message and data switching requirements is covered by two sizes of equipment, to which may also be added line-switching capability. The KLB 3 offers up to 64 low-speed and 2 medium-speed lines, and the KLB 5 up to 256 low-speed or 32 medium-speed lines.

Status
Production. In service with civil and military operators in Europe and Africa

Manufacturer
CIT-Alcatel DTM Division, La Verriere, France

CSEE Sigphone Communications Switching Systems and Air Traffic Control Consoles

CSEE atc console

CSEE offers a range of products for communications switching operations including atc consoles. The message switching systems include the Sigphone models S, ACD and C.
Sigphone S is a space-controlled pabx system functioning with a pre-recorded programme. It provides telephone communications between operators in the same centre (internal links) or remote centres (external links). The external links comprise special lines and the trunk network.
The system consists of an operator's desk for the various controls, coupled to a central computer. The operator's desk contains a set of 30 individual call keys for direct contact with all internal or external correspondents, a dialling keyboard with a call transfer key, a function keyboard and a telephone handset.
The central computer controls and monitors the telephone functions. It consists of two programmed processors, one of which is on stand-by. It systematically scans the line couplers and operator stations and processes all the messages transmitted by them. The system equipment is monitored and any faults detected are indicated by an alarm, or, in certain cases, the equipment, switches over to the stand-by processor without disturbing communications.
The system can simultaneously distribute calls to a large number of operators' desks. The degree of distribution varies throughout the day, depending on the density of traffic. The call control mode is modified by the supervisor by means of an order from the desk, indicating the choice of operation.
The ACD Sigphone is a computer-controlled pabx which operates a range of recorded messages, for example 'welcome functions', information, bookings etc.
The C Sigphone is intended for civil and/or military security or command centres. This system is in operation, in particular, at the atc centres at Brest and Athis-Mons. Air traffic control consoles produced by CSEE can include the Sigphone switching systems.

Manufacturer

Compagnie de Signaux et d'Entreprises Electriques, Paris, France

---

Nardeux Hf Communications Equipment

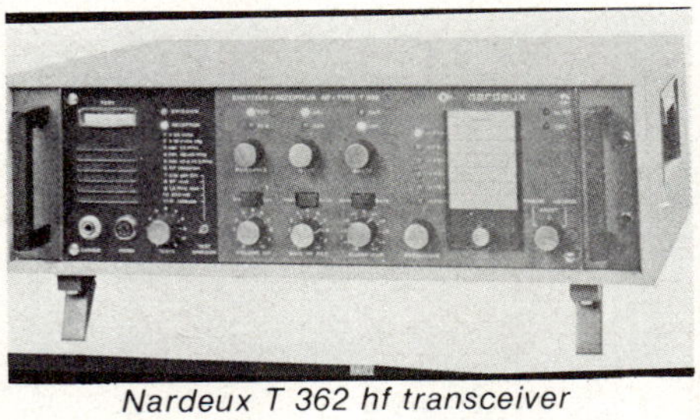

Nardeux T 362 hf transceiver

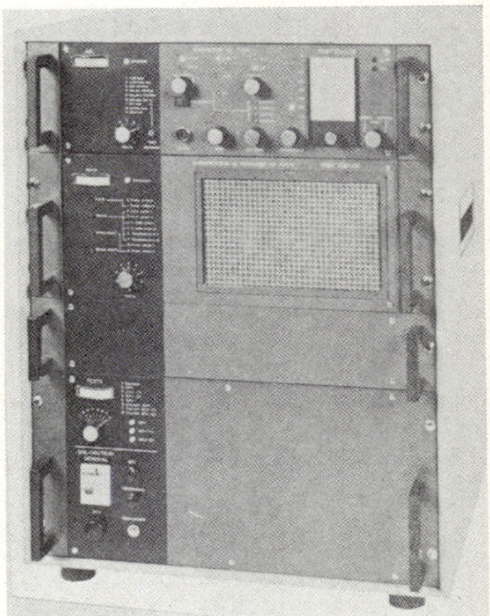

Nardeux T 164 hf transmitter

Hf communications systems offered by Nardeux include the T 150 hf transmitter which operates in the frequency band 1·5 to 30 MHz with a power output of 1·5 kW. Up to six frequencies can be preselected. The equipment is housed in standard 19-inch racks which include a removable control panel that can be placed either on the upper part of the cabinet or on a desk; switchable antenna filter, exciter, power amplifier rack which amplifies the signal coming from the exciter to obtain 200 watts pep; power supply rack and 24 volt converter. Optional racks offered contain modulation F1-A3J (low or high upon request), modulation A3H-A2J-A3J (low or high upon request) and modulation A1-A2-A3.
Each receiver in the Nardeux T 250 series is designed for a specific mode of reception. All have a common chassis. Only the interchangeable front panel and the plug-in intermediate frequency circuit cards vary with each reception mode. The selectivity and the type of demodulation is determined by each of the various optional intermediate frequency cards. The receiver may be mounted on a table top or in a standard-rack cabinet.
T 362 is a 70 watt wide-band solid-state transceiver. It operates between 1·5 and 30 MHz with six preselected frequencies.
T 164 is a 400 watt wide-band solid-state transmitter operating in 2 to 24 MHz frequency or optionally 3 to 30 MHz, with six preselected frequencies.

Characteristics

MODEL	T 150	T 250	T 362	T 164
TYPE	Hf transmitter	Hf receiver	Hf transceiver	Hf transmitter
FREQUENCY	1·5–30 MHz	1·5–30 MHz	1·5–30 MHz	2–24 MHz 3–30 MHz optional
NO OF CHANNELS	6	8	6	6
MODES	A1, A2, A3, A2J, A3H, A3J, F1	A1, A3, F1, F4	A3J (LSB/USB) A3H	A1, F1, A3J (LSB or USB), A3H
POWER OUTPUT	6 kW pep A3J–A3B 3 kW average A1–F1 1·5 kW carrier A2–A3		70 W pep	400 W pep
POWER REQUIREMENTS	220/380 V ac, 40–60 Hz	110/220 V ac, 50–60 Hz	24 V dc, optional 12 V dc 110/220 V ac, 50–60 Hz	110–120/ 220–240 V ac, 50–60 Hz

Status

Production. In service with CAAs in France, Greece and Sweden

Manufacturer

Nardeux SA, Loches, France

Nardeux Vhf Communications Equipment

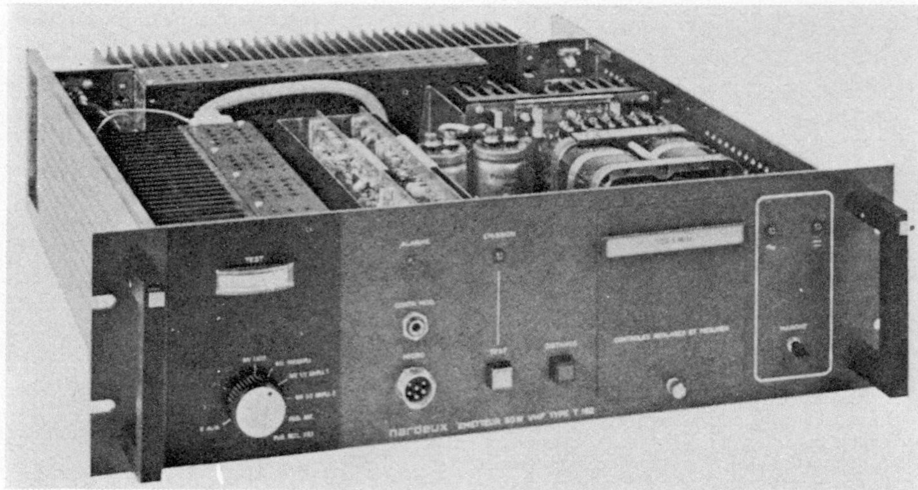

Nardeux T 180/T 182 vhf transmitters

The T 122 with two preselected frequencies and the T 128 with ten preselected frequencies are vhf transceivers designed for aviation use. They both operate in the 118 to 136 MHz frequency band and are available as fixed base or mobile installations.
T 180 and T 182 are wide-band transmitters, T 180 operating between 118 to 136 MHz with a single channel.
T 280 and T 282 are wide-band receivers designed to be used in conjunction with the T 180 and T 182 transmitters.

Status
Production. In service with CAAs in France, Greece and Sweden

Manufacturer
Nardeux SA, Loches, France

Nardeux T 280/T 282 vhf receivers

Characteristics

MODELS	T 122	T 128	T 180	T 182	T 280	T 282
TYPE	Vhf transceiver	Vhf transceiver	Vhf transmitter	Vhf transmitter	Vhf receiver	Vhf receiver
FREQUENCY	118–136 MHz	118–136 MHz	118–156 MHz	118–136 MHz	118–156 MHz	118–136 MHz
NO OF CHANNELS	2	10	4	1	4	1
POWER OUTPUT			50 W	50 W		
POWER REQUIREMENTS	6, 12, 24 V dc, 110–220 V ac, 50–60 Hz	6, 12, 24 V dc, 110–220 V ac, 50–60 Hz	24 V dc, 110–220 V ac, 50–60 Hz	24 V dc, 110–220 V ac, 50–60 Hz	24 V dc, 110–220 V ac, 50–60 Hz	24 V dc, 110–220 V ac, 50–60 Hz

SAF 79 STNA Standard Air Traffic Control Console

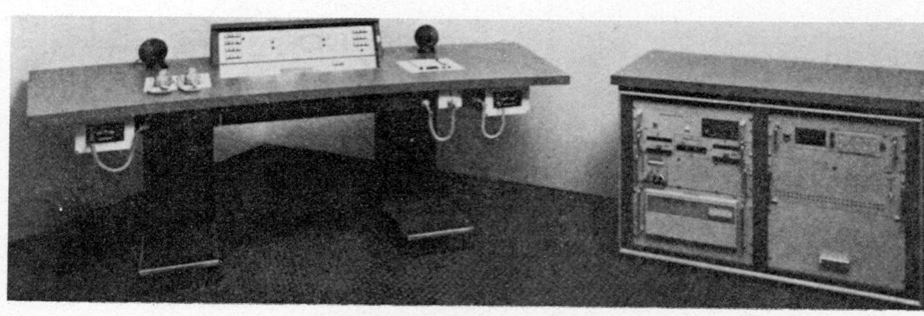

SAF 79 STNA atc console

The 70 STNA standard unit is designed for use in Category III airport control towers. It is divided into two separate elements, the control console and the technical cabinet, combining all equipment required for atc operations.
The control console contains: two Nardeux T 122 dual frequency vhf modules; a beacon and QFU control unit; anemometer and wind-vane data indicator; a df unit; clock/barometer; Avasis control unit; sound system control; power supply and alarm control unit with line generator supply selector. On the work top an intercom module, two telephone units and two speakers are positioned.
The technical cabinet contains a time modulation coder, automatic four track recorder, compartment to house a vhf df and OM Nardeux T 228/1 locator monitor (options), 220 volt 50 Hz ac and 24 volt dc power supply module, 25 watt amplifier driving two external and one internal speaker, intercom and 24 volt 16 ampere/hour nickel cadmium battery.

Status
Production. Supplied to civil and military operators in France

Manufacturer
SAF, Paris, France

Safare-Crouzet Air Traffic Control Communications Equipment

French Air Force control tower console

The Safare-Crouzet range of equipment enables maximum advantage to be taken of available communications facilities by remote-control of transmitters and receivers, electronic switching of radio channels for the operators, coded instruction and data transmission to reduce local traffic.
Increasing aircraft numbers and speeds created a need for highly reliable radio-telephone communications systems capable of high-speed radio channel switching. Electronic switching in the operational systems designed by SC allows several operators to maintain a watch on a large number

of frequencies. At any moment the operators can choose the necessary frequencies for ground-to-air, air-to-ground or ground-to-ground traffic from the entire range of hf, vhf and uhf transmitters and receivers of the associated transmission and reception centres.
This equipment is ground-based, permanently installed and designed for continuous operation. According to the number of radio channels used, the mixing and switching modules can be independent or combined in one bay. In small installations the whole apparatus is integrated into a single operator's console.

Status
Developed for the French Air Force Telecommunications Technical Services for military atc purposes. In service

Manufacturer
Safare-Crouzet, Nice, France

Secre Meteor Communications Switching System

The Meteor communication switching system offers: remote-control of transmitters and receivers; transmission of radio messages; selection of frequency; setting up of radio channels (maximum 64); distribution of these channels to a maximum of 48 operators; a facility for monitoring and operating on five channels simultaneously and management of all elements of the system at the level of an operational manager and a technical supervisor.
Additional functions of the modular system include telephone line switching and transfer of radio channels to another location. Other possible functions include switching facilities for multi-frequency transmitters.
The modular design permits the system to be used in medium and small size configurations.
The local airfield control-type CLA provides facilities for four operators using four multi-frequency units. The FATAC configuration is suitable for mobile systems used in surveillance or tactical support. It consists of six operators using six single frequency channels and four multi-frequency channels.

Status
Production. In service with civil and military operators in France

Manufacturer
Secre-Societe d'Etudes et de Constructions Electroniques, Paris, France

Socrat Vhf/uhf Communications Equipment

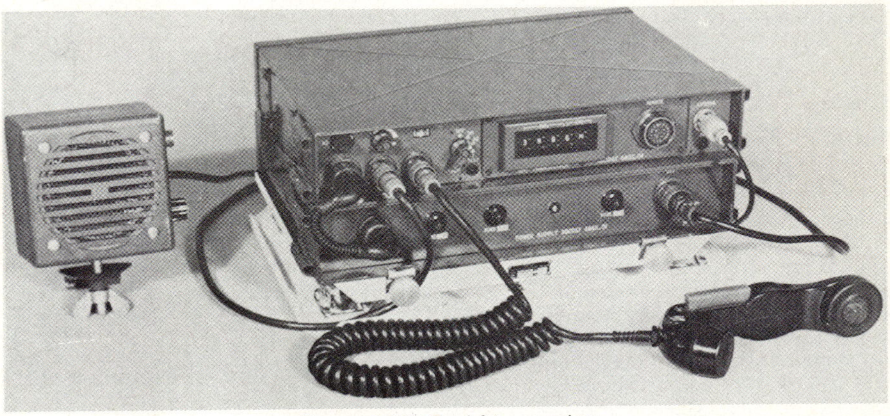
Socrat 6803-G uhf transceiver

Vhf and uhf communications systems are produced by Socrat suitable for aviation operations. The 4603 is a vhf/am ground station intended to provide ground-to-air communications in 118 to 144 MHz range with a channel spacing of 25 kHz. The unit can be used as a fixed station, in a workshop on a vehicle with mounting on a standard table, or as a movable station. In the latter version, retractable legs are supplied. The unit can be operated with the H 33 EPT headset or the AN/GRA-6 remote-control unit. In the basic version, the station is delivered with whip antenna and antenna with ground plane secured on a small mast; both antennae can be folded and stored in a special stowage compartment. The unit is supplied with 28 volt dc but can be made available with 220 volt 50 Hz power supply.
The 6803-C uhf transmitter-receiver assembly is designed for military communications in am mode within the 225 MHz to 399·975 MHz frequency. The equipment delivers a carrier with 15 watt output power. It is provided with a frequency synthesiser which permits operation on any of the 7000 channels spaced by 25 kHz. It can be installed as a fixed station, on a light, heavy or armoured vehicle or on a ship and can be fed from 28 volt dc main system. A monitor system is used to check the correct operation during the transmission and reception. The unit, designated 6803-G is available with a facility which permits remote-control operations.

Status
Production

Manufacturer
Socrat, Paris, France

Characteristics

MODEL	4603	6803-C/-G
TYPE	Vhf	Uhf
FREQUENCY	118–144 MHz	225–399·975 MHz
OUTPUT POWER	20 W	15 W
NO OF CHANNELS	1040	7000
SPACING	25 kHz	25 kHz
POWER REQUIREMENTS	28 V dc Optional: 220 V, 50 Hz	28 V dc 220 V, 50 Hz
UNIT DIMENSIONS		
height	260 mm	140 mm
width	300 mm	380 mm
depth	550 mm	270 mm
WEIGHT	19 kg	6803-C: 13·5 kg 6803-G: 12 kg

Socrat 4603 vhf ground station

Systelec ST 4000 Series Multi-track Communications Recorder

The ST 4000 multi-track communications recorders record on magnetic tape, from 4 to 16 telephone or radio-telephone communications simultaneously, together with the time at which these communications were exchanged. The modular system consists of several two- or four-track decks. When the tapes are played back the time at which each event took place is shown on a digital display unit. A standard DIN rack, height 2 metres, contains a maximum of four two- or four-track decks; one operating and monitoring unit, and one ST 304 digital clock. The front face is protected by a smoked plexiglass door. The ST 4000 series multi-track communications recorders are available in two versions: ST 402 equipped with two-track decks and ST 404 with four-track decks.

Characteristics

MODEL	ST 402	ST 404
NO OF TRACKS	2, 4, 6 or 8	4, 8, 12 or 16

RECORDING: triggered on modulation
DURATION: 6 hours of communications
POWER REQUIREMENTS: 220 V, 50 Hz, ac

Status
Production. In service at airports in France

Manufacturer
Systelec, Montrouge, France

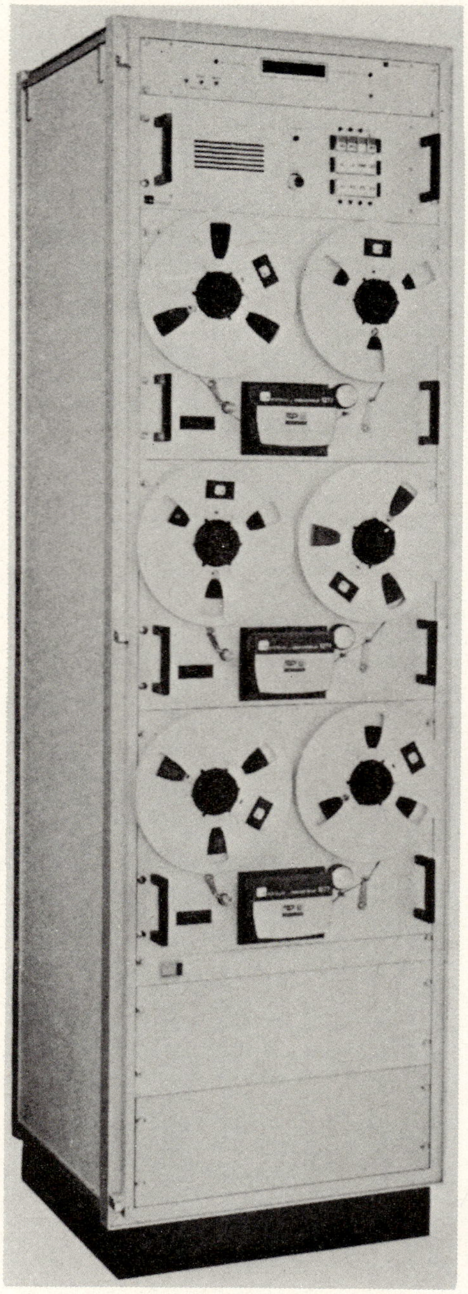

Systelec ST 4000 series multi-track communications recorder

Telerad Uhf Communications Equipment

Telerad produces a range of uhf equipment for aviation operations operating in the 225–400 MHz frequency band. The 100-watt uhf transmitter assembly includes a uhf 30-watt transmitter, type EL 745, associated with a 100-watt amplifier, type PA 746. With its mains power supply unit, type AL 743, the assembly occupies a ten unit space in a 19-inch rack. The equipment operates either on a single frequency with an internal oscillator or as a multi-channel transmitter with a SY 744 synthesizer.
The LB 760 is a wide-band superheterodyne receiver with a double frequency change, designed to be driven by the SY 744 synthesizer.
The SY 744 synthesizer is designed to operate as a driver for the 30- and 100-watt transmitters or as a local oscillator for the LB 760 receiver, it can operate in an integrated transmission-reception assembly. It is controlled by thumbnail selectors situated on the front panel or by means of a programmer or a remote-control device. It features 7000 channels with 25 kHz spacing.
The FC 710 is a single frequency superheterodyne receiver with a single frequency change which is crystal controlled.

Status
Production. Supplied to civil and military authorities in France and other countries

Manufacturer
Telerad, Anglet, France

Telerad Vhf Communications Equipment

The E 667 series transmitters are crystal controlled single channel transmitters operating in the 108 to 136 MHz frequency band with an optional band offered between 118 and 144 MHz. Three configurations are available.
FC 680 and FC 730 are crystal-controlled, single frequency change superheterodyne receivers operating in the 108 to 144 MHz band. Compared to the FC 680, the FC 730 receiver features improved characteristics as regards intermodulation, cross-modulation and protection against spurious signals.
EL 750 rated at 50 watts and EL 752 producing 100 watts are wide-band transmitters which feature linear amplification, and single frequency operation with a plug-in oscillator or multi-channel operation using an SY 7320A synthesizer.
LB 740 is a single frequency change superheterodyne receiver. Three versions are available: LB 740 A, designed to be driven by synthesizer; LB 740 B, equipped with a four frequency oscillator and LB 740 C, using a single frequency oscillator.

Status
Production.Supplied to civil and military authorities in France and other countries

Manufacturer
Telerad, Anglet, France

Characteristics

MODEL	E 667 series	FC 680/FC 730	EL 750/EL 752	LB 740 series
TYPE	Vhf transmitters	Vhf receivers	Vhf transmitters	Vhf receivers
FREQUENCY	108–136 MHz 118–144 MHz optional	108–144 MHz	118–144 MHz	118–144 MHz
POWER OUTPUT	T/1 - 50 W T/2 - 15 W T/3 - 15 W		EL 750 - 50 W EL 752 - 100 W	
POWER REQUIREMENTS	T/1 - 24 V dc T/2 - 24 V dc T/3 - 127/220 ac 50/60 Hz			

Thomson-CSF Automatic Message Switching Centres

Automatic message switching centres (amsc) controlled by computers are gradually replacing tape systems. The amsc is placed as a buffer between the subscribers to ensure automatic reception, verification, recording and priority arrangement of messages as well as their queuing, retransmission and distribution.
The two amsc ranges proposed by Thomson-CSF have basic characteristics in common. They comply with ICAO specifications, utilise standard low- or medium-speed transmission lines (for duplex links) and have short/long term storage facilities. The searching, checking and correction of messages takes place on interactive screen terminals. The system can be duplicated with testing and automatic switchover to stand-by equipment without any loss of data.

Status
Production. Installed at airports in France and other countries

Manufacturer
Thomson-CSF, Division DRS-TVT, Bagneaux, France

Thomson-CSF SECAD 65 amsc installed in China

Characteristics

MODEL	DS/200	DS6/400 A	DS6/400 B	SECAD 40	SECAD 65	SECAD 80
STANDARD LINES	128	256	256	64	256	512
MEDIUM SPEED LINES	8	32	64	4	16	32
COMPUTER	GA 16440	GA 16440	GA 16440	SOLAR 16/40	SOLAR 16/65	SOLAR 16/65

Thomson-CSF Hf Communications Equipment

Thomson-CSF hf antennae field in Tahiti

Air-ground communications are still provided in hf (A3) in many areas when the telephone infrastructure does not allow a vhf station network to be formed.
Thomson-CSF experience in the study, installation and utilisation of hf radio communications networks for air-ground or point-to-point links has resulted in the development of a range of hf, independent side-band (isb) or single side-band (ssb) transmitters and receivers as well as auxiliary equipment; antennae, switching matrixes, long-range remote-control and remote-testing and multi-couplers for the connection of several receivers to the same antenna.

Status
Production. Installed world-wide

Manufacturer
Thomson-CSF, Division DRS-TVT, Bagneaux, France

Characteristics

TRANSMITTERS

model	TRC 222	TRC 393	TRC 395
	Remote-controlled and remote-tested transmitters for air-ground and point-to-point links		
TYPE	Solid-state	Solid-state + power tube	Solid-state + power tube
PEAK POWER	1 kW	1 kW	6 kW
FREQUENCIES	1·6–30 MHz	1·6–24 MHz	1·6–28 MHz
NO OF CHANNELS	284 000	224 000	264 000
PRESET CHANNELS	14	6	6
MODES	ISB, SSB (A3A, A3J, A3H, 6A3B, A1, F1)		

RECEIVERS

model	TRC 172	TRC 394 A	TRC 394 B	TRC 394 C
APPLICATION	Air-ground	Air-ground and point-to-point	Air-ground and point-to-point	Air-ground and point-to-point
FREQUENCIES	2–30 MHz	400 kHz–30 MHz	400 kHz–30 MHz	10 kHz–30 MHz
CHANNELS	Mono-frequency	10–1000 Hz around selected frequency	1–12 set frequencies + interpolator	20 frequencies in memory + scanning
MODES	A1 or A2J or A3J	A1, F1, A2, 6A3B, A3J	A1, F1, A2, 6A3B, A3J	

SSB TRANSMITTERS RECEIVERS

model	TRC 492	TRC 493	TRC 382
APPLICATION	Point-to-point links		(mobile)
PEAK POWER	70 W	400 W	100 W
FREQUENCIES	6 preset 2–12 MHz	4 preset 2–12 MHz	160 000 channels by 2–18 MHz synthesiser
MODEL	TRC 1804	TRC 706	TRC 523
	1-channel F1 mono-frequency or diversity demodulator associable with TRC 394	200-baud mono-channel TOR/ARQ conforming with CCITT recommendations	4-channel 50-baud multiplex TOR/ARQ conforming with CCITT recommendations

Thomson-CSF Vhf/uhf Communications Equipment

Thomson-CSF offers a complete range of vhf and uhf transmitters and receivers, either mono- or multi-frequency (by synthesizer), meeting all ground-air aeronautical communications requirements.
The development of stations combining transmission and reception with a set of wide-band antennas, co-located on the same mast, has simplified the installation, supply and remote operation of extended range stations used for links between pilots and controllers.
By using cavity multi-couplers, combined with the vertical alignment of radiating elements and receivers, good channel separation (at least 60 dB) is obtained in spite of transmitters and receivers working on the same antennas, simultaneously on 8, 16 or more different frequencies. Omni-directional radiation is achieved by distributing dipoles and panels to form a ring-array around the mast.
Integrated stations are available for vhf, uhf or combined vhf and uhf, with duplicated or non-duplicated transmitters and receivers, mono- or multi-frequency, and at power ratings from 12 to 100 watts. They can be supplied in prefitted shelters. The stations can be controlled and tested remotely using two pairs of telephone lines connected to the TXM 1210 system. This system reduces links between the operations centre and the station to a single duplex voice communication line for each channel being used.
The TXM 1210 sends the following to the station transmitters: modulation; push-to-talk signal; 20 or 40 commands (FSK, 100 bauds), and receives from the station receivers: modulation; squelch and 20 or 40 status signals (test reports).

Thomson-CSF integrated vhf station

Characteristics

VHF

TRANSMITTER	E 764	E 667	E 750	EL 752
FREQUENCY	Mono- or multi-frequency	Mono-frequency	Mono- or multi-frequency	Mono- or multi-frequency
POWER	12 W	15/50 W	50 W	100 W
APPLICATION	Control towers	Terminal area, en route control	Terminal area, en route control	Terminal area, en route control
RECEIVER	FC 680 Mono-frequency	FC 730 Mono-frequency	740/A Multi-frequency; 740/B 4 preset	740/C Mono-frequency

UHF

TRANSMITTER	TRC 261	TRC 262	EL 745	EL 746
FREQUENCY	Mono-frequency	Multi-frequency	Mono/multi-frequency and 16 preset frequencies	Mono/multi-frequency and 16 preset frequencies
POWER	20 W	20 W	30 W	100 W
RECEIVER	TRC 271	TRC 272	FC 710	LB 760
FREQUENCY	Mono-frequency	Multi-frequency	Mono-frequency	Multi-frequency (16 preset frequencies)

Status

Production. Installed in France and other countries

Manufacturer

Thomson-CSF, Division DRS-TVT, Bagneaux, France

GERMANY, FEDERAL REPUBLIC

Assmann Communications Recording Equipment

Assmann MS 300 documentation recorder

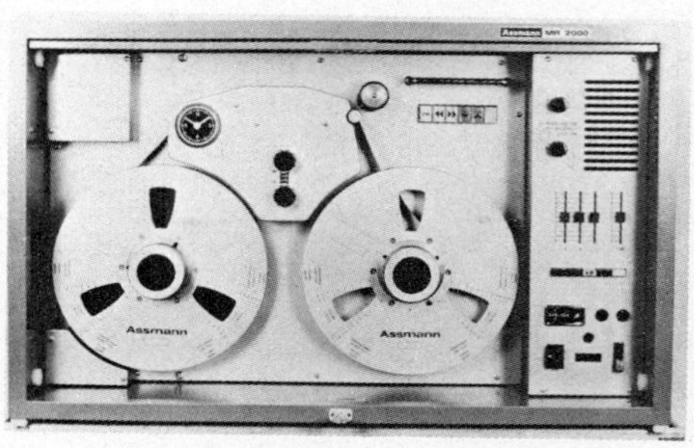

Assmann MR 200 tape reproducer

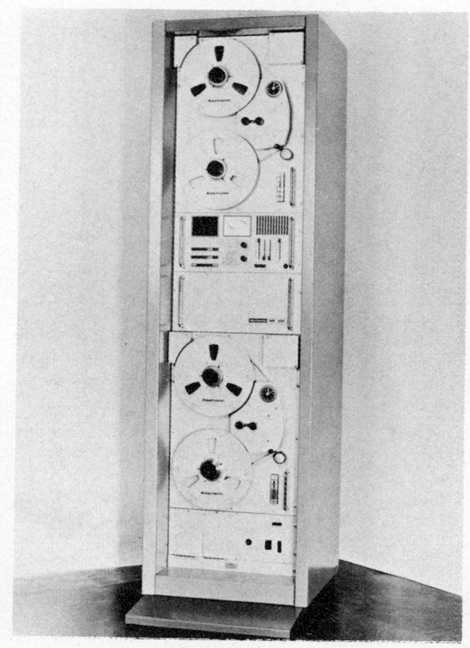

Assmann MS 200 multi-channel recorder

Two types of recorders are produced by Assmann, the multi-channel MS 200 and the MS 300 universal documentation recorder. The MS 200 is designed for operation in medium- and large-size international airports. Each multi-channel tape-recorder is equipped with two identical tape-decks providing a recording capacity for each tape-deck of 24-hours continuous operation. During normal operation one of the tape-decks is running while the second one is on stand-by. This concept gives up to 48-hours unattended operation. The equipment is designed in a modular building-block system in order to meet the varying specifications and requirements of different customers. Each recorder is wired for the maximum capacity of 36 tracks so that each system used for a lower number of channels may be extended to the maximum capacity at any later date by inserting additional printed circuit boards.

Any vital part of the equipment can be duplicated with an integrated switch-over-facility and an all-track overtape monitoring system may be added. This ASCAM-PR system simultaneously and continuously monitors all tracks overtape and includes automatic switch-over of a defective track to a separate spare track without interrupting the continuity of recording. During low density traffic the equipment may be operated in a VOX control mode in order to economise tape. All operational conditions of the MS 200 system can be displayed at a remote indicator board.

For immediate evaluation of tapes a separate reproducer MR 2000 is available. It is also possible to use one of the recorder's duplicate tape-decks for playback while the other tape-deck continues in recording mode.

The MS 300 universal documentation recorder is intended to fit the specific requirements of smaller airfields with less air traffic and a smaller number of voice channels that need to be recorded simultaneously. The unit is designed with a single tape-deck having a continuous recording capacity of 24 hours where a maximum of eight channels can be recorded simultaneously. The equipment offers the option of operating in auto-reverse mode (logger) with alternating tape direction. In this mode it is possible to record a single information channel on one magnetic tape for a period of 32 days continuously or two tracks for 16 days, four tracks for eight days or eight tracks for four days respectively. In order to fit existing power supplies the equipment may be directly operated either from an ac or dc power source.

Status

Production. Currently installed at airports in Bahrain; Riyadh, Saudi Arabia and at Abu Dhabi and Sharjah Airports, United Arab Emirates

Manufacturer

Assmann GmbH, Bad Homburg, Federal Republic of Germany

Becker GS 2000 Vhf Communications System

GS 2000 is a range of ground-to-air vhf equipment available as a fixed base or transportable station or in a portable configuration. The T264 fixed base station is offered with a choice of 8 or 720 channels (360 channels optional) with transmitter powers of 5, 10 or 16 watts. The design allows mounting in a desk cabinet as well as in standard 19 inch racks. Optional features provided comprise remote-control and tape-recorder interface and built-in batteries with automatic power transfer for emergency operation. The transportable version is designed for vehicle installation. It is available with a choice of 8, 360 or 720 channels with transmitter powers of 5 or 10 watts. A remote-control console is optional.

The portable vhf transceiver operates on 720 channels in the 118–135·975 MHz band in 25 kHz steps with a 2–3 watt transmitter using rechargeable batteries. The unit is housed in a leather case with shoulder-strap and the antenna and microphone can be stored in the lid of the case. Transmitter output of 6–8 watts is offered as an optional feature.

Characteristics

	Fixed base		Transportable	Portable
CONFIGURATION	Fixed base		Transportable	Portable
FREQUENCY RANGE	118–135·975 MHz		118–135·975 MHz	118–135·975 MHz
NO OF CHANNELS	8/720 (360 option)		8/360/720	720
CHANNEL SPACING	25 kHz (50 kHz option)		25/50 kHz	25 kHz
OUTPUT POWER	5/10/16 W		5/10 W	2–3 W/6–8 W
POWER REQUIREMENTS	110/220 V ac, 50/60 Hz 12/24 dc		110/220 V ac, 50/60 Hz 12/24 dc	12 V dc
	Emergency battery – 12 V			
UNIT DIMENSIONS	desk cabinet	19-inch module		
height	111 mm	88·1 mm	168 mm	303 mm
width	505 mm	483 mm	345 mm	270 mm
depth	395 mm	376 mm	190 mm	85 mm
WEIGHT	13·5 kg	13·5 kg		5 kg

Status
Production

Manufacturer
Becker Flugfunkwerk GmbH, Rastatt, Federal Republic of Germany

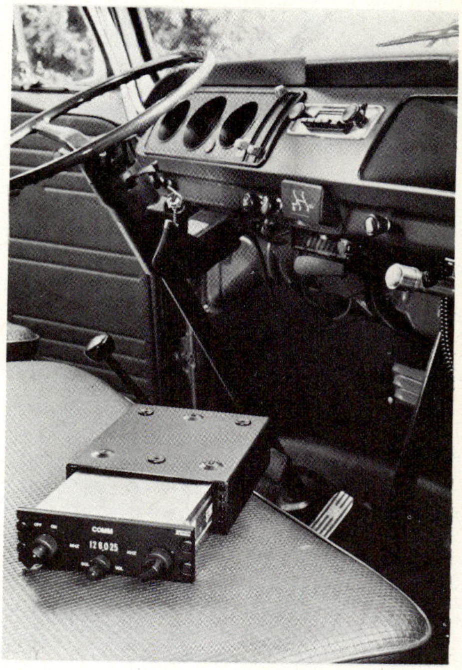

Becker GS 2000 Becker vhf transportable station

Rohde and Schwarz Hf Communications Equipment

Hf communications equipment produced by Rohde and Schwarz includes the SK1 transmitter, EK 070 receiver, telegraphy demodulator, antenna signal distribution and hf antennae.

The SK1 transmitter rated at 1 kW operates between 1·5 and 30 MHz with automatic tuning. It is suitable for aeronautical and broadcasting services and features a central control unit for the selection of frequency, class of emission, power steps and stand-by/on/off modes. Remote-control is possible with an optional facility to provide control via a radio link. The unit is solid-state except for the power stages.

EK 070 is a vlf-hf receiver operating between 10 kHz and 30 MHz. Micro-processor controlled, the unit is equipped with a memory for 30 receiver settings, automatic memory recall and a single row, liquid crystal display showing operating status. Frequency setting is via a rotary switch or keyboard and the unit includes a standardised data interface for remote-control. The receiver is fitted for all usual classes of emission.

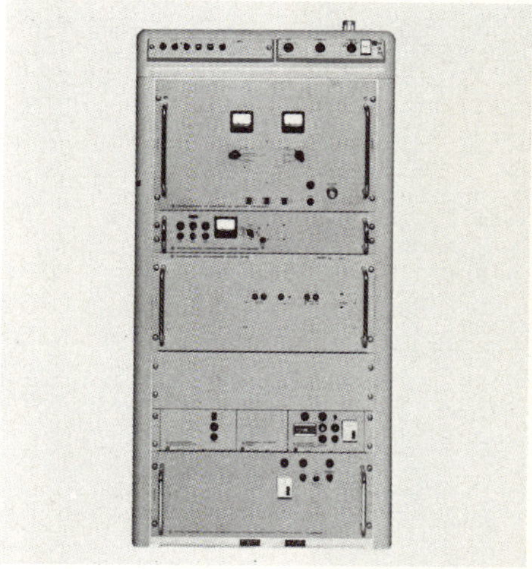

Rohde and Schwarz SK1 hf transmitter

Characteristics

MODEL	SK1	EK 070	
TYPE	Hf transmitter	Vlf-hf receiver	
FREQUENCY RANGE	1·5–30 MHz	10 kHz–30 MHz	
INCREMENTS/RESOLUTION	100 Hz	10 Hz	
POWER OUTPUT	1 kW Adjustable 200–800 W		
POWER REQUIREMENTS	117/220 V, 50/60 Hz ac	115/125/220/225 V, 47–420 Hz ac 12/24 V dc	
UNIT DIMENSIONS		Cabinet model	Rack-mount
height	1315 mm	154 mm	132 mm
width	613 mm	470 mm	483 mm
depth	850 mm	487 mm	515 mm
WEIGHT	327 kg	23 kg	23 kg

Status
Production

Manufacturer
Rohde and Schwarz GmbH, Munich, Federal Republic of Germany

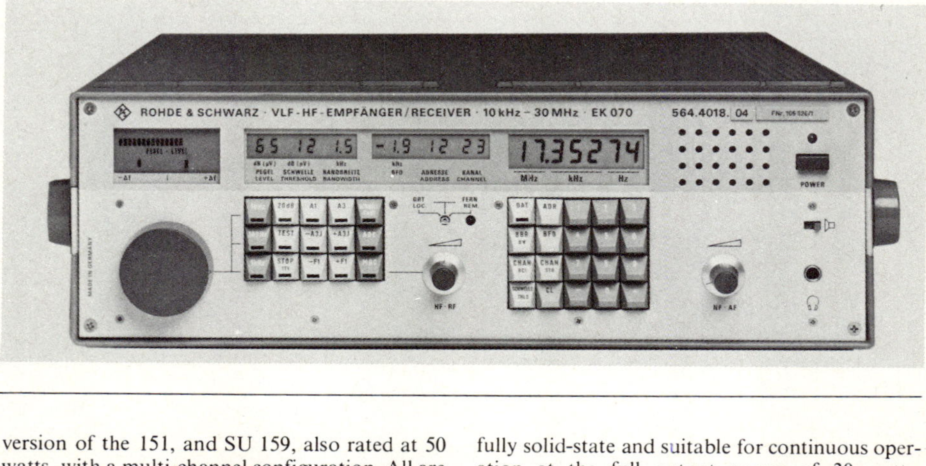

Rohde and Schwarz EK 070 vlf-hf receiver

Rohde and Schwarz Vhf/uhf Communications Equipment

A range of equipment is available from Rohde and Schwarz for ground-to-air vhf/uhf communications. Four vhf transmitters are produced: SU 116, a 10 watt transmitter, with six selectable channels; SU 151, producing 50 watts, SU 156, a six channel version of the 151, and SU 159, also rated at 50 watts, with a multi-channel configuration. All are designed for atc operations and are particularly suitable for complex transmitting stations operating simultaneously on different frequencies or for combined transmitting/receiving systems.

SD 131 and SD 139 are uhf transmitters, the former operating on a single channel, the latter designed for multi-channel operation. Both are fully solid-state and suitable for continuous operation at the full output power of 30 watts. Constant output power and modulation depth is ensured by control circuits.

The single-channel receivers types EU 25002/2 for vhf and ED 21002/2 for uhf and the vhf 6-channel receiver type EU 25004/2 are used for the reception of am signals in the vhf range from 117·5 to 136·5 MHz and in the uhf range from 225 to 400

MHz. They can be tuned to fixed frequencies. The six-channel receiver enables one out of six preset, fixed channels to be selected. The 50 kHz channel spacing of the standard receivers can be altered to 25 kHz, if required, by changing the crystal filter. Adjacent channel interference due to cross modulation, intermodulation and blocking has been minimised. A selective preamplifier to be inserted into the channel receiver or, in the case of simultaneous operation of several channel receivers from one antenna, a broad-band multi-coupler provides the necessary rf gain. The various models of the channel receiver for vhf (single- or six-channel) and uhf (single-channel only) differ only in the rf section and in the oscillator. They operate with the same intermediate frequency of 10·7 MHz. This allows the use of identical sub-assemblies, such as crystal filter, if-af amplifier and power supply.

Series 400 is a further development in vhf/uhf communications from Rohde and Schwarz. The modular design permits single- or multi-channel configurations of vhf/uhf receiving, df or transmitting systems and vhf/uhf transmitting/receiving systems for ground-to-air and ground-to-ground communications. The systems operate in the vhf range 118 to 146 MHz or optionally 100 to 162 MHz and in the uhf range between 225 and 400 MHz.

A comprehensive range of preamplifiers, multi-couplers, filters and antennae are produced by the company.

Rohde and Schwarz SU 116 vnt transmitter

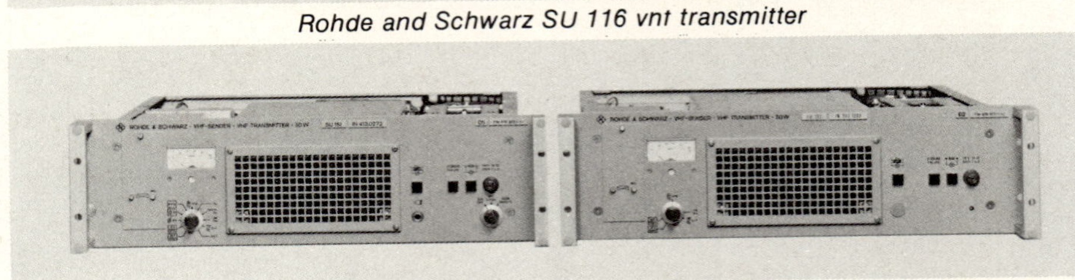

Rohde and Schwarz SU 151 vhf transmitter

Characteristics

MODEL	SU 116	SU 151	SU 156	SU 159	SD 131	SD 139	EU 25 series	ED 21	400 series
TYPE	Vhf am transmitter	Vhf am transmitter	Vhf am transmitter	Vhf am transmitter	Uhf am transmitter	Uhf am transmitter	Vhf am receivers	Vhf am receivers	Vhf/uhf am and fm transmitters/ receivers
FREQUENCY RANGE	117–144 MHz	117–144 MHz	117–144 MHz	117–144 MHz	225–400 MHz	225–399·975 MHz	117·5–136·5 MHz	225–400 MHz	118–146 MHz (option: 100–162 MHz) 225–400 MHz
NO OF CHANNELS	6		6	30/6	1	30/6	1/6	1	1/12
CHANNEL SPACING				25 kHz		25 kHz	50 kHz	50 kHz	25 kHz
POWER OUTPUT	10 W	50 W	50 W	50 W	30 W	30 W			10 W
POWER REQUIRE-MENTS	115/125/220/ 235 V 47–63 Hz ac 22–30 V dc	115/125/220/ 235 V 47–63 Hz ac 22–30 V dc	115/125/220/ 235 V 47–63 Hz ac 22–30 V dc	115/220 V 47–63 Hz ac 22–28 V dc	115/125/220/ 235 V 47–63 Hz ac 22–30 V dc	115/220 V 47–63 Hz ac 22–28 V dc	115/125/220/ 235 V 47–63 Hz ac	115/125/220/ 235 V 47–63 Hz ac	93/110/127/ 203/220/237 V 47–63 Hz ac
UNIT DIMENSIONS									
cabinet + mount									
height		150 mm	150 mm		150 mm		132·5 mm	132·5 mm	
width		484 mm	484 mm		484 mm		106 mm	106 mm	
depth		436 mm	436 mm		436 mm		307 mm	307 mm	
WEIGHT		20 kg	20 kg		20 kg		4·5 kg	4·5 kg	
rack-mounts									
height	133 mm	133 mm	133 mm	133 mm	133 mm	133 mm			
width	106 mm	483 mm	483 mm	483 mm	483 mm	483 mm			
depth	345 mm	371 mm	371 mm	371 mm	371 mm	371 mm			
WEIGHT	5·2 kg	19·5 kg	19·5 kg	30 kg	19·5 kg	30 kg			

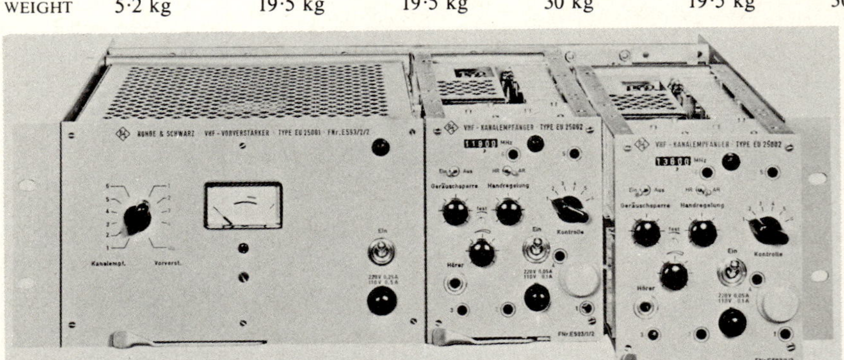

Rohde and Schwarz EU 25 series vhf receiver

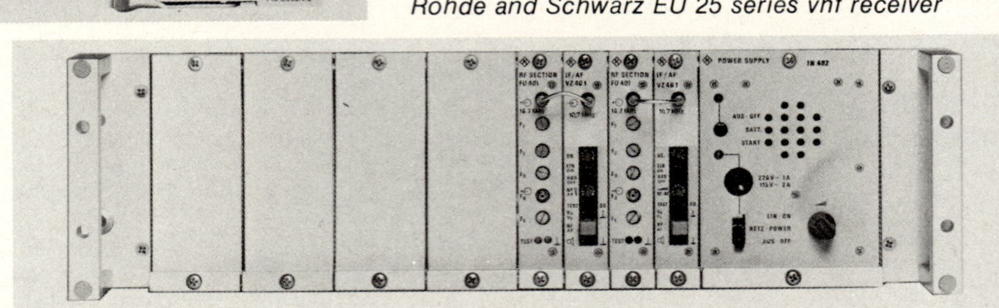

Rohde and Schwarz 400 series vhf/uhf equipment

Status
Production, except for Series 400 which is in final stages of development, and in service at German airports and in other countries

Manufacturer
Rohde and Schwarz GmbH, Munich, Federal Republic of Germany

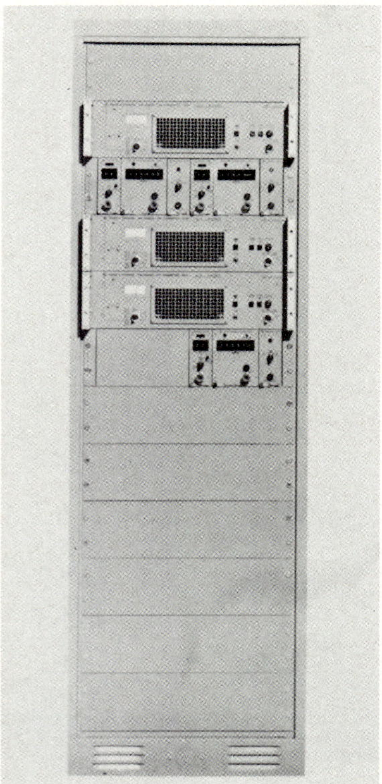

Rohde and Schwarz SU 159 vhf transmitter

Siemens S 307/S 305 Transmitters 1·5 to 30 MHz

Siemens produce the S 307 transmitter operating at 1 kW and the S 305 rated at 5–10 kW.
The S 307 operates in the frequency range 1·5 to 30 MHz offering all standard telegraphy and telephony transmission types, including single-band operation or two independently operating sidebands. Of particular interest is its application for ARQ Simplex operation. The continuous hf-power for telegraphy, or peak power for telephony, is 1 kW in each case. The antenna matching unit automatically adapts antennas so that the full 1 kW hf-power is passed to the antenna. The mean tuning time is seven seconds. For applications requiring a high-speed frequency change a low-pass filter is provided. The transmitter can be entirely remote-controlled. Ten complete transmitter settings can be stored using an accessory program module. Faults can be identified with the aid of a built-in diagnostic system. For mobile applications in radio vehicles the transmitter is mounted in a shock-absorbing frame.
The S 305 is designed for all standard telegraph and telephone transmission types. Its maximum continuous hf power for telegraphy operation is 5 kW, its peak power for telephony 10 kW. When the send frequency has been adjusted within the 1·5 to 30 MHz range, tuning is automatic. The transmitter can be remote-controlled and from a distant location it is possible to switch on and off, select the frequency and type of transmission and add or suppress the carrier. The transmitter is suitable for use in stationary and mobile applications and for installation in vehicles.

Status
Production

Manufacturer
Siemens Aktiengesellschaft, Munich, Federal Republic of Germany

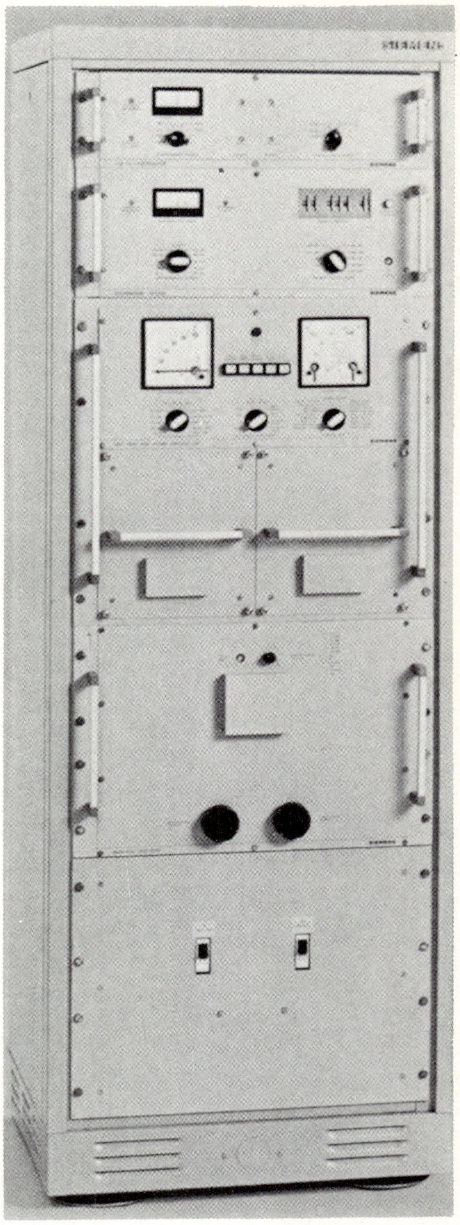

Siemens S 307 radio transmitter

Siemens E 530 Series Communications Receiver 100 kHz to 30 MHz

The E 530 series receiver operates in the band from 100 kHz to 30 MHz. Preselection, a crystal filter and a high first intermediate frequency result in optimal suppression of all internal and external interference factors. Modular in design, the receivers can be adapted to the requirements of various radio services. An F1 demodulator ensures that teleprinters can be directly connected, while a special electronic switch makes operation with the ARQ Simplex data protection device possible. A built-in testing facility permits the checking of all functions of the receiving section. There are several receiver versions which differ as follows:
E 530 is a manually-operated receiver with decadic frequency adjustment, adjusted receive frequency output in BCD code.
E 531 is a manually-operated receiver with quasi-continuous and decadic frequency adjustment; frequency storage for storing twelve receive frequencies independent of network failures; selected receive frequency output in BCD code.
E 532 is a manually and remote-controlled receiver with address switches, data interface V28 for remote-control and status inquiry. In addition the complete receiver setting can be transmitted to other receivers (E 532/E 533). Program storage for 40 frequency and operational adjustments. Input and call down of programs is possible without altering the receiver setting.
E 533 is a remote-controlled receiver without manual controls but with address switch, data interface V28 for remote-control and status inquiry.
The control device N 530, which can also store 40 adjustment programs, enables the remote-control of receivers E 532 and E 533. Receiver adjustment can also be carried out by an external computer.

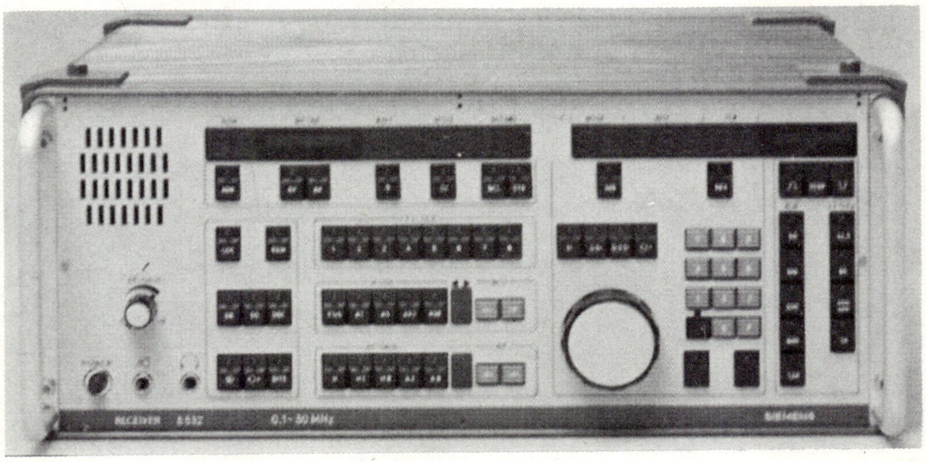

Siemens E 532 series communications receiver

Status
Production

Manufacturer
Siemens Aktiengesellschaft, Munich, Federal Republic of Germany

ISRAEL

ECI ATC/4950-AM Air Traffic Control System

ATC/4950-AM is a mobile/airborne atc system in the form of a jeep-sized unit. Mounted on an all-terrain vehicle, the system can travel on almost any kind of surface and incorporates within its logistic facilities all the equipment and accessories necessary for instantaneous operation as a fully-equipped atc tower.

Status

In production for the Israeli armed forces

Manufacturer

Electronic Corporation of Israel, ECI, Tel Aviv, Israel

ATC/4950-AM in operation

ATC/4950-AM in transit

ECI ATC-4952/M Mobile Air Traffic Control Tower

The ATC-4952/M is a completely independent mobile atc system capable of continuous operation in the field under extreme environmental conditions.
The mobile unit, including the vehicle transporter, tower and trailed generator, can be put into operation in minutes and contains communications and control equipment needed for a total control-tower operation. The vehicle used is a D-500 Chrysler truck, equipped with a diesel engine and with modifications in the truck's frame and electrical system.

Status

In production for the Israeli armed forces

Manufacturer

Electronic Corporation of Israeli, ECI, Tel Aviv, Israel

ATC-4952/M in operation

ATC-4952/M in transit

ECI ATC-4954/AM Air Traffic Control System

The ATC-4954/AM is a mobile atc system which incorporates a total crew protection environment with air-conditioning and bevelled, double-paned and tinted observation windows for all-weather operation under all field conditions.
The system uses the pop-top concept and literally folds within itself to protect the windows and tower equipment. All the equipment and supplies needed for total operational independence are placed inside the tower and the system can be set up and in operation within minutes.
During flight transportation and when travelling under its own power, the ATC-4954/AM trails a generator to operate the tower equipment in the field. The vehicle used is a modified WM-325 command car.

Status
Announced in 1980

Manufacturer
Electronic Corporation of Israel, ECI, Tel Aviv, Israel

Model of ATC-4954/AM in transit

Model of ATC-4954/AM in operation

Tadiran Mobile Communication Shelters

The family of Tadiran mobile shelters are for use in mobile tactical communication networks. The various shelter configurations are completely self-contained, watertight units, manufactured to the operational performance and testing procedures of military specifications MIL-S-55541B (EL), MIL-S-52059B (EL), and MIL-S-55286 (EL). They are built of lightweight aluminium sandwich panels with internal thermal insulation, to give the optimum volume to weight-to-payload ratios. They can be transported in a variety of vehicles, aircraft and helicopters.
Shelters are of three basic types: S-318, S-250, and S-280. Types S-318 and S-250 have a stepped-bottom construction, especially suitable for mounting on a four- by four-wheel drive power wagon. Type S-280 is flat-bottomed and larger, and is mounted on a 2½-ton six- by six-wheel drive power wagon. It is also transportable by other military vehicles as well as by helicopters.

Characteristics

SHELTER TYPE	S-318	S-250	S-250/G modified	S-280
INTERIOR				
length	5 ft 10 in (1·78 m)	6 ft 6 in (1·98 m)	6 ft 4½ in (1·94 m)	11 ft 6 in (3·51 m)
width	5 ft 9 in (1·75 m)	6 ft 3 in (1·91 m)	5 ft 7 in (1·7 m)	6 ft 9½ in (2·07 m)
height	5 ft 5 in (1·65 m)	5 ft 4 in (1·63 m)	5 ft 3 in (1·6 m)	6 ft 2½ in (1·9 m)
EXTERIOR				
length	6 ft 3⅝ in (1·91 m)	7 ft 1 in (2·16 m)	7 ft 1 in (2·16 m)	12 ft 2 in (3·71 m)
width	6 ft (1·83 m)	6 ft 7 in (2·01 m)	5 ft 11 in (1·8 m)	7 ft 3 in (2·21 m)
height	6 ft 7⅞ in (2·02 m)	5 ft 10 in (1·78 m)	5 ft 9 in (1·75 m)	6 ft 11 in (2·11 m)
WEIGHT	400 lb (181 kg)	630 lb (286 kg)	580 lb (263 kg)	1200 lb (544 kg)
CAPACITY	1100 lb (499 kg)	1900 lb (862 kg)	1900 lb (862 kg)	1500 lb (680 kg)
TRUCK SIZE	¾ ton (680 kg)	1 ton (907 kg)	1 ton (907 kg)	2½ ton (2268 kg)

Status
Production

Manufacturer
Tadiran Israel Electronics Industries Ltd, Tel Aviv, Israel

Tadiran URC-240T Uhf/am Ground-to-air Radio System

The URC-240T system is a series of military communication equipment designed for ground-to-air uhf/am service at fixed stations. The fully solid-state equipment is adaptable for operation from a mobile communication van. All major components of the system have independent built-in power supplies; thus each unit can be operated either individually or as part of a complex configuration. All major units fit into a standard 19-inch rack, whereas the control and audio boxes fit into standard atc consoles.
Receiver R-241/URC-240T and transmitter T-242/URC-240T each incorporate a plug-in digital synthesizer, accessible through the front panel, which allows them to cover the 225 to 400 MHz range with 3500 channels at 50 kHz increments. A 7000-channel option with 25 kHz increments is also available. Receiver R-241 SC/URC-240T and transmitter T-242SC/URC-240T are single frequency versions, each incorporating a plug-in crystal oscillator unit. Conversion for full channel coverage is readily made by replacing the crystal oscillator with a synthesizer unit. The transmitters provide 35 watts of unmodulated rf output (140 watts pep when modulated).
The transmitter power can be increased to 100 watts unmodulated output (400 watts pep when modulated) by the addition of broadband rf amplifier AM-245/URC-240T. This unit provides linear amplification of the am signals.
For additional protection against co-location interference, tunable filter F-244A/URC-240T is available as a separate unit which can be added to any system configuration.
Control box C-243/URC-240T permits remote-control with back indications of all transmitter and receiver functions, including frequency selection and squelch operation. A built-in non-destructive core memory enables remote presetting of up to 20 frequency channels. Control is possible over four telephone line pairs from virtually any distance, using serial data transmission, by fsk. Audio box C-246/URC-240T provides audio and ptt functions.
Besides conventional voice communication, provision has been made in the system for the wide-band audio signals used in data communication and/or secure mode. Audio mode changeover is effected by means of a single-control wire.

Status
In service with Israeli armed forces

Manufacturer
Tadiran Israel Electronics Industries Ltd, Tel Aviv, Israel

ITALY

OTE Integrated Radio-telephone Communication System

OTE has developed an integrated communication system which allows all radio subscribers with either a fixed, vehicle-mounted or personal set, to contact any person equipped with another radio or standard telephone, immediately and automatically by dialling the appropriate number. Similarly, all the telephone subscribers can contact any radio user anywhere within the confines of the airport. The system consists of three main sections: a radio central station; radio network implemented with fixed, mobile and portable equipment and a telephone exchange. The radio central station allows the peripheral equipment to link with an existing telephone exchange which can be a pabx, a private intercom or a central office. The only condition is that the exchange must have trunk connection facilities with the radio central station.

The radio central station can be supplied with different channel capacities according to the operating needs. Eight radio channels are typical. When fully equipped, an eight-channel capacity radio central station consists of eight single-channel duplex transceivers, each one connected through a trunk to a telephone exchange by a logic unit with a telephone interface and a channel control unit. All the transmitters can operate on both their own channel frequency and a second standard frequency (service channel) used for calls to the peripheral radio users. All channels include the same basic equipment and modular units and each can operate independently as the relevant local unit controls the radio equipment dynamically, handles the outgoing and incoming calls, and provides for the signalling transmission. Therefore, each radio central station can be built up in a modular way from one channel up to the maximum number of radio channels required. Several radio central stations can be linked to a telephone exchange, each one including the required number of channels, to cover different areas or services.

The radio equipment may be selected in any frequency band, either vhf or uhf. However, experience shows that the uhf band has advantages over vhf when medium size areas are to be covered, because of reduced interference probability and absence of radio noise.

The equipment at the radio central station is powered from both an ac mains supply (220 volt, 50 Hz) and 12 volt dc floating battery. Each channel control unit is powered from the relevant radio equipment and each telephone interface is powered from the telephone exchange with which it connects.

Many kinds of peripheral radio equipment can be used, depending on operating requirements. OTE proposes a single interchangeable radio equipment suitable for fixed, vehicular and portable installations. This equipment is based on a portable set with a special control unit provided with a touch-tone pad, and it can be also used inside a vehicle inserted in a housing providing inductive coupling between the portable helical antenna and the omni-directional one mounted on the vehicle. The af controls and sockets for microphone and loudspeaker are available on the housing. The same housing allows the equipment to be used in fixed installations (for example, on a desk, a counter, etc) with a suitable antenna. Should the operator move away from the vehicle, he may carry the unit with his own selection number inserted. Due to the use of a portable equipment, the radio system can only operate in two frequency simplex mode and the user can only talk or listen alternatively. This limitation does not apply to the telephone subscribers.

Status

Production. This radio-telephone integrated communication system has been adopted by Alitalia, to improve the ground facilities of Rome Fiumicino Airport, Italy. The system connects several hundred intercom telephones with 225 vehicular, fixed and portable radio units

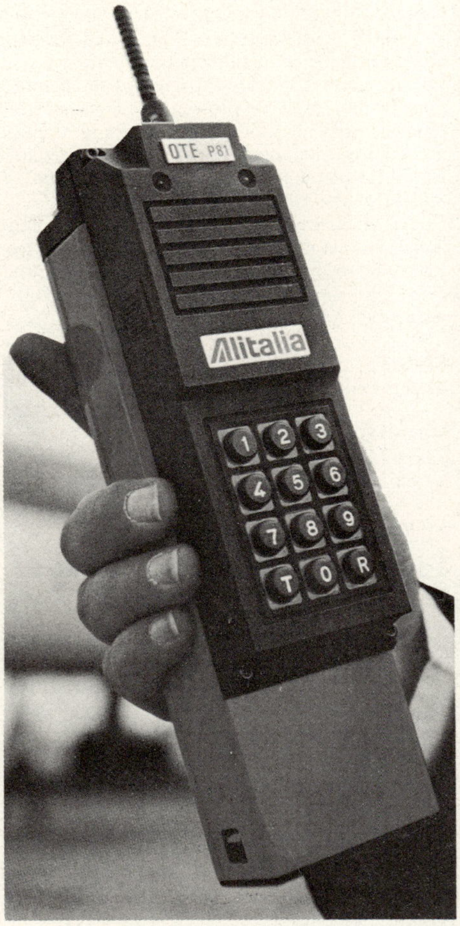

OTE integrated radio-telephone communication system

Manufacturer

OTE, Florence, Italy

OTE Vhf/uhf Communications System

This ground to air telecommunications system developed by OTE has been designed to meet civil and military requirements for a highly flexible integrated vhf/uhf communications network. A typical integrated station is situated in one building and comprises vhf/uhf antennas, multicouplers, transmitting and receiving equipment, power amplifiers where it is necessary to increase the station's coverage, and remote-control units. Alternatively, the system can consist of vhf and/or uhf stations where transmitters and receivers can operate and work as transceivers, or vhf and/or uhf mobile stations for military use.

The equipment is modular in construction, solid-state and available in either single- or multi-channel versions. The latter can be obtained by substitution of some of the modules of the single channel version.

The operating frequencies are between 110 and 150 MHz (civil band) and 225 and 400 MHz (military band). The main equipment is universal and changes between vhf and uhf are effected by substitution of plug-in modules.

The frequency of the operating channel which can be selected by the operator, between the wavebands indicated above, is generated by a synthesizer, phase-stabilised by a crystal oscillator. The oscillator is used in both the single- and multi-channel versions.

The power amplifier is fully solid-state and modular in construction. It features built-in test and diagnostic equipment to isolate a fault to module level and a by-pass facility in case of malfunction of part of the equipment.

The remote-control equipment enables the operator to transmit and receive on any one of 20 preset channels. The selection of the channel is made by the operator and the frequency in use is displayed on the console.

OTE GT 80-MU multi-channel transmitter

Characteristics

TRANSMITTER
- single-channel: GT 80-V (vhf)/GT 80-U (uhf)
- multi-channel: GT 80-MV (vhf)/GT 80-MU (uhf)

FREQUENCY RANGE: 110–150 MHz/225–400 MHz
NO OF CHANNELS: 1600/7000
CHANNEL SPACING: 25 kHz
UNIT DIMENSIONS
- length 7 in (180 mm)
- width 1 ft 7 in (483 mm)
- height 1 ft 9 in (540 mm)

WEIGHT: 35 kg

RECEIVER
- single-channel: GR 80-V/GR 80-U
- multi-channel: GR 80-MV/GR 80-MU

FREQUENCY RANGE: 110–150 MHz/225–400 MHz
NO OF CHANNELS: 1600/7000
CHANNEL SPACING: 25 kHz
MODULATION TYPE: amplitude (A3)
UNIT DIMENSIONS
- length $3\frac{1}{2}$ in (88 mm)
- width 1 ft 7 in (483 mm)
- height 1 ft 9 in (450 mm)

WEIGHT: 7 kg (single), 14 kg (multi)

POWER AMPLIFIER
TYPE: GPA 80-V (vhf)/GPA 80-U (uhf)
FREQUENCY RANGE: 110–150 MHz/225–400 MHz
POWER OUTPUT: 100 W nominal
UNIT DIMENSIONS
- length 8 in (222 mm)
- width 1 ft 7 in (483 mm)
- height 1 ft 9 in (540 mm)

WEIGHT: 50 kg

OTE GR 80-MU multi-channel receiver

POWER REQUIREMENTS: 110/220 V ac, ± 15%, 47–53 Hz
OPERATING TEMPERATURE RANGE: –20 to +60°C

Status
Production. Supplied to civil and military operations in Italy

Manufacturer
OTE, Florence, Italy

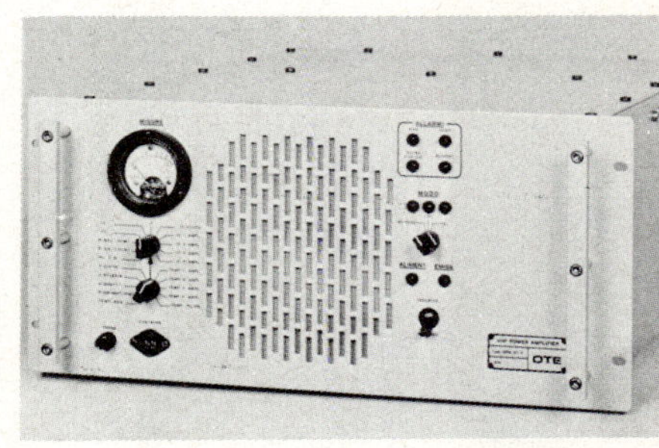

OTE GPA 80-U power amplifier

SITTI MULTIFONO Communications Switching Equipment

MULTIFONO communications switching equipment has been designed for atc operations and is suitable for use in an area control centre, approach control and the airfield control tower. A modular system, it can be used for simple and complex operations; from a control tower with one console to an area control centre with 80 operators. The system can be equipped for radio-telephone and interconsole communications. Radio communications comprise hf, vhf and uhf ground/air/ground services, vhf/uhf integrated configurations and conference facilities. Telephone services consist of local battery lines, central battery lines as a branch exchange or as a user, and automatic branch exchange lines (pabx). Interconsole facilities provide point-to-point or conference communications for the operators.
The system can be incorporated in consoles produced by SITTI or by other manufacturers. For area control centres and approach control, the modular equipment on the console comprises radio keyboard, satellite time display of master clock, loudspeakers, telephone and interconsole keyboards. For tower operations the console equipment includes radio and telephone keyboards, loudspeaker equipment connection points for headsets and handsets and satellite time display of master clock.

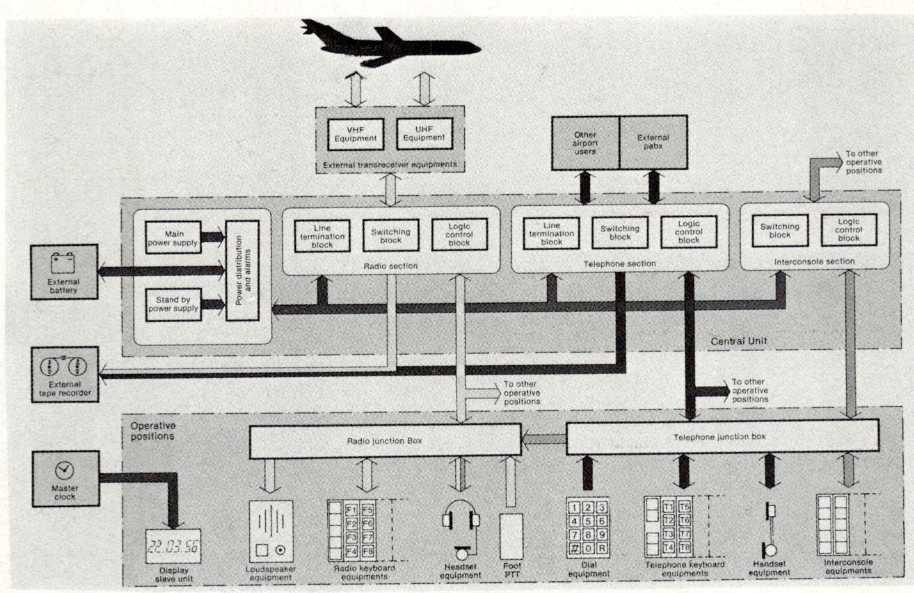

SITTI MULTIFONO switching equipment network

Status
Used for atc operations in Italy

Manufacturer
Speciali Impianti Telescriventi Telefonici Interni SaS, Milan, Italy

JAPAN

NEC Vhf Ground-to-air Communications

Vhf communications equipment offered by NEC for atc includes the NRG-628, vhf transmitters/receivers rated at 50 and 25 watts, NRG-629, a vhf 10 watt transceiver, NRV-722 a mobile 10 watt transceiver and NEP-719 vhf 1·5 watt portable transceiver. All equipments are fully solid-state and include built-in automatic channel transfer facilities, and channel separation of 25 kHz compatible with ICAO requirements. An off-set carrier system is available for en route services.

Status
Production. NEC communications equipment is in service in Japan and other countries

Manufacturer
Nippon Electric Co Ltd, Tokyo, Japan

NEC vhf antenna installation, Mount Olympus, Cyprus

NEC NRG-628 vhf transmitters/receivers

NETHERLANDS

Philips Communications Recording Equipment

Philips has been producing voice-logging equipment for atc purposes for over 25 years and the company's equipment is installed at over 120 airports world-wide. During this period, Philips has developed four distinct generations of voice loggers specifically to meet the communications recording needs of the air transportation industry. Successive generations of the equipment have shown increases in both capacity for simultaneous recording on multiple channels and continuous recording time.

For example, the fourth-generation voice loggers have a simultaneous recording capacity of up to 44 channels (although smaller configurations are also available with 11, 22 or 33 channels) to meet the needs of centres where large numbers of aircraft may be under control simultaneously. Continuous recording times have been steadily extended so that equipment can operate non-stop for 24 hours. After this period, change-over to a stand-by tape-deck carrying a fresh tape reel takes place automatically, with a 30-minute overlap period, so there is no danger of any loss of recorded information. A warning is automatically given of system malfunction and the built-in monitor panel contains indicators to help pinpoint and correct any fault in the minimum tape. Automatic switch-over to the stand-by tape deck takes place when the fault is such as to endanger the security of recording.

The XMN 6 is a compact voice logger for recording low- to medium-density communications traffic on a continuous 24 hour basis. The unit is a stand-alone system with six-channel capacity. Table top or rack mounted, up to four XMN 6s may be linked together to provide 18-channel capacity plus a further six on emergency stand-by. Continuous monitoring, visual and audible alarm systems, lockable operation and long-life tape heads are features built-into the XMN 6 to guard against information loss caused by equipment failure. The voice logger is backed by a full range of system accessories such as automatic search, voice operated start/stop and a time code generator/reader.

Philips fourth-generation multi-channel voice-logging system

Status
Production and in service world-wide. Installations in Europe, Malaysia, Pakistan and Sudan. Philips announced an order at the Airport 81 exhibition in late 1981 for eight multi-channel

voice logging systems from the Hellenic Civil Aviation Authority.
The multi-channel documentation recorders selected by the HCAA are from the 'fourth generation' range of equipment. Two 44-channel recorders are to be installed at Athens and Salonika Airports, Greece. The other six units, each with 11-channels, are to be installed on the islands of Myconos, Skiathos and Karpathos, and in the towns of Kozani, Kastoria and Sparta in Greece. All eight systems will be used for recording communications between aircraft and air traffic controllers. The HCAA order brought the total of fourth-generation vls systems sold, since their introduction in 1979 to over 150

Manufacturer
NV Philips Gloeilampen fabrieken, Eindhoven, Netherlands

Philips XMN 6 voice-logging system

NEW ZEALAND

AWA Vhf Communications Equipment

AWA produces mobile and fixed station vhf/am communications equipment.
TR 235 is a mobile unit designed for mounting in a vehicle providing am operation on up to ten channels within a 4 MHz band-width in five vhf bands. The unit has a power output of 10 watts in the aeronautical waveband.
The TR 2280 series is a modular fixed station designed for vhf ground-to-air communications. Simplex or Duplex operation is possible on specific bands in the frequency range 72–136 MHz, on 12·5 kHz, 25 kHz or 50 kHz carrier separation systems, with a transmitter power output of up to 35 watts. Line-pad switching is incorporated to allow remote operation over long, two-wire, 600 ohm lines.

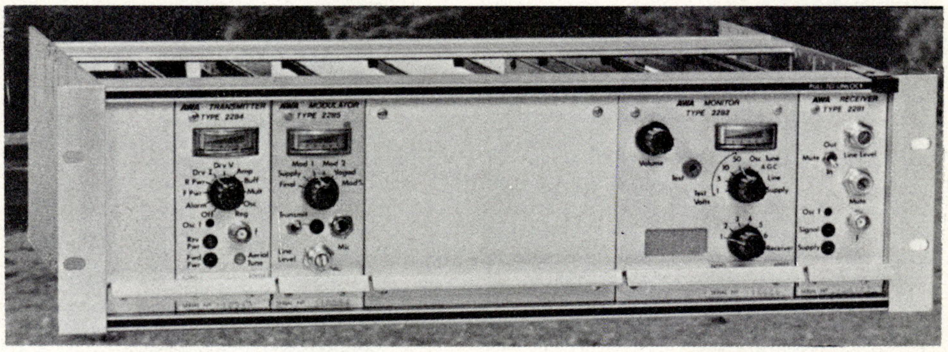

AWA TR 2280 vhf am fixed station communications equipment

Characteristics

MODEL	TR 235	TR 2280
APPLICATION	Mobile vhf am	Fixed station vhf am
FREQUENCY RANGE	68–88 MHz 88–108 MHz 118–136 MHz 105–108/138–141 MHz 148–174 MHz	72–88 MHz 88–108 MHz 118–136 MHz
NO OF CHANNELS	Up to 10 with 4 MHz band-width	Single or multi-channel
CHANNEL SPACING	12·5 or 25 kHz (30 kHz option)	12·5 kHz, 25 or 30 kHz, 50 or 60 kHz
POWER OUTPUT	10 W - 118–174 MHz 20 W - 68–108 MHz	25–35 W
POWER REQUIREMENTS	13·6 V dc	12–14 V dc, 110–120/230–250 V ac, 50/60 Hz
UNIT DIMENSIONS		
height	50 mm	133 mm
width	150 mm	483 mm
depth	250 mm	402 mm
WEIGHT	1·3 kg	12·5 kg

Status
Production. TR 2280 has been supplied to the New Zealand Civil Aviation Division of the Ministry of Transport for their entire vhf network

Manufacturer
AWA New Zealand Ltd, Communications Division, Porirua, New Zealand

NORWAY

Nerion Garex Communications Control System

The Garex system integrates the communications control of radio channels, external telephone lines and intercom between air traffic controllers, giving each controller access to all facilities by means of console-mounted control panels. The Garex programme includes systems adaptable to requirements ranging from those of the smaller domestic airfield to the most sophisticated systems needed at large international airports.
A modular system, each standard panel comprises a radio panel with 10 radio channels; telephone panel with 20 telephone lines and an intercom panel with 12 intercom lines between controllers. The number of control positions, radio channels, telephone and intercom lines can be adapted to suit individual airport requirements.
The smaller Garex-21 system completes the family of software controlled systems. It is based on the same units as the Garex-14 system.
Prototypes of the Garex-6 system have been delivered to the US Air Force as the communications control system in the AN/TPN-19 mobile landing system. Garex-6 is designed, produced and documented in accordance with military specifications.

Manufacturer
Nerion A/S, Oslo, Norway

Status
The standard Garex-5 system is installed at all civil airports in Norway and at several international and domestic airports in Europe, Asia and South America.
Amsterdam Schiphol Airport, Netherlands is equipped with the Garex-14 computer controlled communications control system. The same system concept has been procured by Geneva Airport, Switzerland, Changi Airport, Singapore and by the Swedish Authorities for the Malmo atc centre.
Fifty-eight stationary land-line switching systems, Garex-20 (AN/G TC-28), have also been delivered to the US Air Force for installation at bases in the United States and other countries

SWEDEN

Magnetic MARA Communications Switching System

MARA is a modular system designed for use in airport control towers for the distribution and selection of radio channels. Up to 20 radio channels can be distributed to one or several operators. At each operator position two small units are placed, a control unit which contains all the audio circuits and a selector unit with illuminated push-buttons for selection of the radio channels. With one selector unit the operator has a choice of five channels. If the operator requires more than five channels, a second selector unit can be added. These units are attached to a control terminal where the lines (600 ohms balanced) from the transmitters and receivers are connected. The central equipment contains amplifiers to compensate for line attenuation, a distribution unit with an x-y matrix for pre-selection of the radio channels to the various operators, a test unit and a power supply.

Status
Production

Manufacturer
Magnetic AB, Stockholm-Bromma, Sweden

Magnetic MARA communications switching equipment

SWITZERLAND

Patek Philippe TDA Communications Recording Equipment

The TDA system automatically records information from the telephone, radio and intercom, and inserts the time date code in a manner which makes it impossible to modify or to falsify the recordings. The TDA equipment is composed of professional tape-recorders and Z4 modules. Its basic components comprise ZG4 time and data code generator, ZL4 time code reader, SE 4/4 automatic command module for four-channels, USG fail safe and control module, AHG monitor and play back module, two TC 428 tape-recorders and one 19-inch rack.

The two tape-recorders TG 428 are used in turn with one operational and one on stand-by. Without need to erase or to change tapes they can store an uninterrupted recording period of 24 hours. If the automatic command module ZE 4/4 is used, the recorder is turned on only when something needs to be recorded. The recording capacity is therefore increased. In this configuration about 5 minutes before the operational recorder reaches the end of the tape, an alarm is activated and, at the same time, the stand-by recorder is switched on. The two recorders then work simultaneously until the first one has reached its tape end. It then makes a fast rewind operation (approximately 140 seconds duration), stops and is ready for stand-by service, while the other recorder operates as the operational recorder for the next 12 hours.

This alternating system allows two different service modes to be used: with a new tape at each recorder change-over or without a tape change. In the latter mode the messages of the last 24 hours are retained on the two tapes which remain permanently on the TDA.

It is possible at any time to make a manual change over from one recorder to the other and therefore to find and to play back a certain message on the tape without having to wait until the tape end.

Status
Production

Manufacturer
Patek Philippe Electronic, Geneva, Switzerland

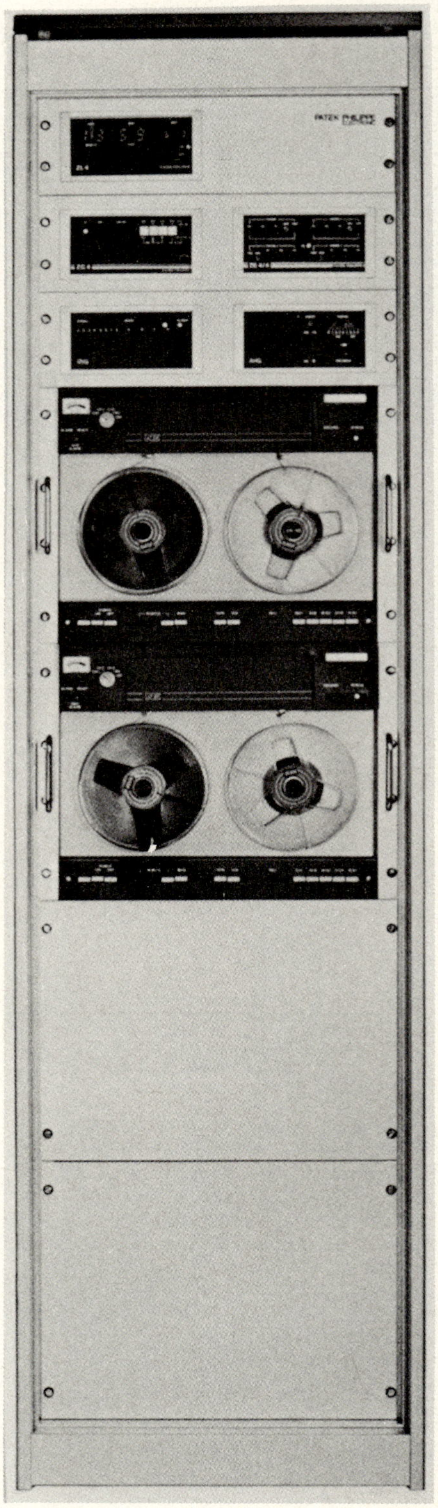

Patek Philippe TDA communications recording equipment

Schmid Communication Switching and Control System

Schmid communications switching and control system

Schmid radio and intercom control is a special switching installation for atc centres. It links co-ordination talks between traffic controllers, radio conversations between aircraft and control centre and talks between pilots, who communicate with the same controller on different frequencies. The system capacity allows 42 air traffic controllers at seven different transmitter places to control 25 transmitters tuned to different frequencies or to call each other. For this control function the controllers have operating units which exchange data with the central control station via remote-control systems. Remote air traffic controllers at neighbouring airports for instance, can conduct radio and intercom conversations through this system. Micro-processors fitted in the operating units permit simple operation and verify that input commands are properly executed. Audio frequency amplifiers equalise both the incoming lines from outlying air traffic controllers and the lines to the various transmitter locations. Signalling sets fitted in the af amplifiers transmit the keying criteria for the transmitters and a signal for monitoring the transmission paths simultaneously. If the controller has selected more than one transmitter, the system switches over automatically to the one whose reception signal has the best signal-to-noise ratio.

The equipment is duplicated and provided with monitoring facilities. Faults are signalled to the air traffic controllers concerned on the operating units and displayed to the service technicians on a monitor; they are also logged with a printer. The alarms from the transmitters are sent to the central control station via leased and dialling lines, by remote-control systems. For fault diagnosis the switching or memory status of every system component can be displayed on the monitor. An integral measuring module automatically checks the transmission quality of all possible paths, including the amplifiers.

Status
Production. The equipment is installed at atc centres in Switzerland

Manufacturer
Schmid Telecommunication Zurich, Zurich, Switzerland

UNITED KINGDOM

Alifabs Visual Control Rooms for Air Traffic Control

Alifabs visual control rooms are available in a range of standard designs and factory-produced to cover a wide range of applications. Designed in accordance with the British Civil Aviation Authority requirements or US Federal Aviation Administration specifications, Alifabs visual control rooms are suitable for installation worldwide.

The structure provides a weatherproof shell in an ergonomic layout and environmentally-controlled interior working conditions for personnel and equipment. If required, a service module can be supplied below the control room, providing space for equipment and racks, battery rooms and air-conditioning equipment.

A temporary atc facility (Tatcof) has been developed by Alifabs. Tatcof is an auto nomous unit complete with its own transmitting/receiving and ancillary equipment, power supply and fittings. Built to the iso dimensions of a 20-foot equivalent container, it can maintain operational requirements in the event of a permanent control system being temporarily out of action or prior to completion of the permanent facility.

All visual control rooms can be fitted with Alifabs Sonarview, an automatic electrically-powered anti-condensation, anti-icing window system. Designed to meet sound attenuation standards of the Civil Aviation Authority, the Sonarview window incorporates solar retardant glass, laminated with tungsten wire webbing which acts as the heating element.

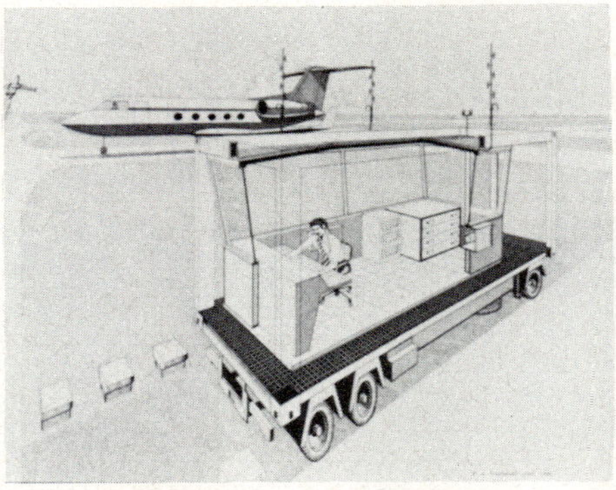

Alifabs Tatcof system

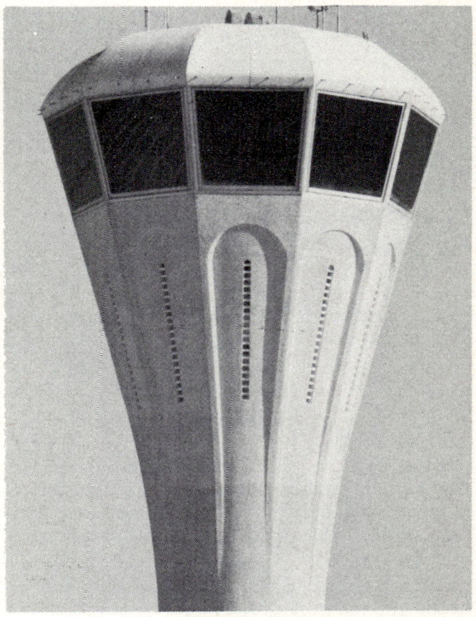

Alifabs visual control room Sharjah Airport

Status
Production. Installed world-wide

Manufacturer
Aluminium Alloy Fabrications (Contracts) Ltd, Woking, Surrey, England

Characteristics

TYPE	A		B		C	D	E
APPLICATION	International		Regional		Local	Transportable	Tatcof
SHAPE	Hexagonal	Octagonal	Hexagonal	Octagonal	Square	Rectangular	Rectangular
LENGTH (OF EACH SIDE)	4·4 m	3·3 m	3·6 m	2·8 m	3·6 m	4–10 m (2 m modules)	6·1 m
INTERNAL FLOOR AREA	50 m²	50 m²	37 m²	37 m²	13 m²	17–25 m²	14·9 m²
FLOOR TO CEILING HEIGHT	3 m	3 m	2·7 m	2·7 m	2·7 m	2·3 m	2·4 m
UNIT WIDTH	8·1 m	7·4 m	6·8 m	6 m	3·6 m	2–6 m (2 m modules)	2·4 m

Burndept Communications Equipment

Communications equipment produced by Burndept suitable for use at airports includes uhf hand-held and fixed station transceivers and a communications switching system.

The BE 471 is a uhf hand-held transceiver with a miniature helical antenna providing three channels in the 420–470 MHz range. It has a power output of 500 milliwatts.

Type BE 454 is a solid-state uhf fixed station of modular construction. It is designed to be operated either locally or remotely in a communications network. The unit can be remotely-controlled by tone or dc signalling over two- or four-wire systems. In addition two units can be interfaced with a change-over unit providing hot stand-by.

Consort 2 is a computer-based switching system designed specifically for the control of line and vhf/uhf radio channels. It is suitable for existing and planned multi-channel, multi-operator configurations with a capability of 99 channels and up to 24 operators.

Burndept Consort 2 network control system

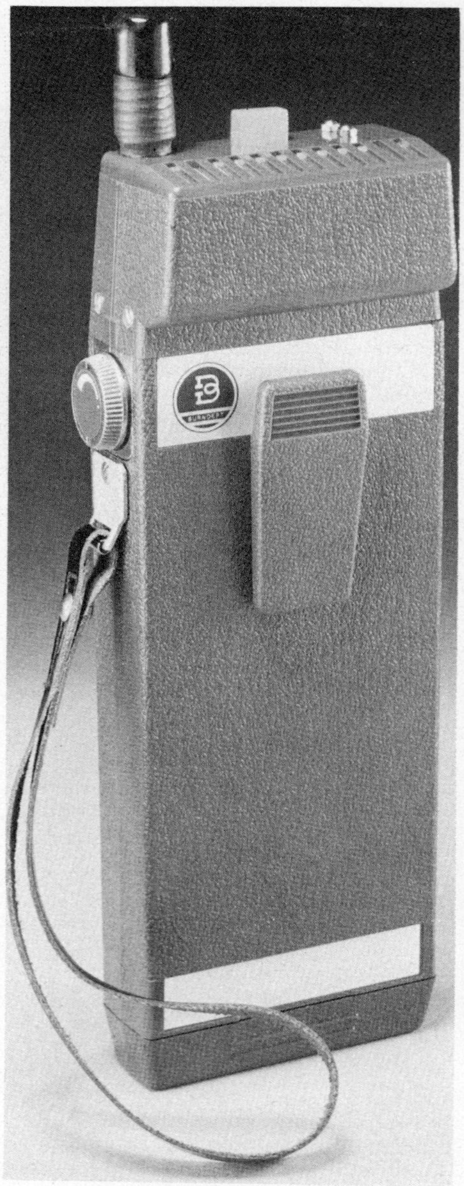

Burndept BE 471 hand-held uhf transceiver

Characteristics

MODEL	BE 471	BE 454	
APPLICATION	Uhf fm hand-held	Uhf fm fixed station	
FREQUENCY	420–470 MHz	420–470 MHz	
NO OF CHANNELS	3	1	
CHANNEL SPACING	25 kHz	25 kHz	
OUTPUT POWER	500 mW	5 W	
POWER REQUIREMENTS	2 × 100 m A/h Ni-Cd batteries	105/115 V or 210/245 V, 50/60 Hz or 24 V dc	
UNIT DIMENSIONS		Single unit	Double unit
height	8 in (203 mm)	5·9 in (150 mm)	11·2 in (285 mm)
width	3·125 in (79 mm)	19 in (483 mm)	19 in (483 mm)
depth	1·42 in (36 mm)	15·5 in (390 mm)	15·5 in (390 mm)
WEIGHT	20 oz (570 g)	26 lb (118 kg)	45·2 lb (20·5 kg)

Status
Production

Manufacturer
Burndept Electronics Ltd, Erith, Kent, England

Cable and Wireless Space Communications Control Equipment

Cable and Wireless Space

The speech processing and control equipment (Space) provides centralised radio, intercom and telephone control. The equipment is modular and can be tailored to meet the requirements of individual airports. The consoles may be equipped so that the operators have access to all available services such as radio, intercom and telephone. If required, the radio control units may be duplicated providing two independent systems. The central processing equipment contains the main switching and control circuits which are mounted on plug-in printed wiring boards for ease of system expansion and maintenance.

Modules available in the system comprise frequency intercom and telephone selection units, telephone keying unit, loudspeaker monitor, headset and auxiliary headset control units, lamp glow and console illumination units, auxiliary control unit to accommodate certain auxiliary functions required by the user (for example, lamp test push-button), runway lighting control panel and navigational aids monitor panel. A full navigational aids switching panel can also be supplied if this is appropriate.

The central processing equipment consists of transmit/receive units, frequency selector switching units, combiner units, voice operated switches and telephone interface units. The transmit/receive units operate at voice frequencies and are situated, electrically, between the transmitters/receivers and the frequency selector switching units. One transmit/receive unit is associated with each discrete radio frequency. The outputs from the receive side of all transmit/receive units are bus-wired via a matrix board to any number of frequency selector matching units.

Frequency selector switching units are situated, electrically, between the transmit/receive units and the control positions. Each unit contains remote-controlled switches which are used to select the required radio frequency (or frequencies) for monitoring and transmitting purposes.

Combiner units join a number of voice frequency inputs on a single line. The combiners are used to route the selected receive frequencies to a controller's console and to connect the transmit and receive lines and/or the intercom and telephone

lines to multi-channel recording equipment.
Voice-operated switches, associated with the receive side of a transmit/receive unit, provide visual indication (on the control console) of an incoming call on the associated radio frequency.
Telephone interface units connect the control console to external and/or internal telephone lines.

Status

Production. In operation at airports in Europe, the Middle East and the Far East

Manufacturer

Cable and Wireless Ltd, London, England

Cescom Communications Switching Equipment

Cescom specialises in the design and manufacture of communications switching equipment.
The Cescom 10273 store-and-forward message switch, when connected to a private terminal network, allows the user to send messages into the switching system, knowing that they will be delivered by the system as soon as practical. Once a message has been sent into the system no further action is required by the originator who is therefore able to originate another message. If the called terminal is busy then the message is stored until delivery can be made.
The 10273 can switch sixteen full duplex channels (sixteen send and sixteen receiver). It can be operated as a stand alone switch or can interface into a switching network for users who have insufficient switching capacity on their existing system and require a satellite to their system. Messages are entered into the system in a simple format which consists of three main elements: a header, giving details of the destination and priority of the message, the message text and the end of message sequence.
The system checks the channel identity and serial number of the incoming message and reports any errors to the supervisor. The system then checks the address and priority of the message and any error will result in the message being sent to the supervisor for correction. Correctly received messages will automatically be forwarded by the unit to the appropriate destination as soon as the outgoing line is free. Messages are delivered in order of entry to the system within the three priority levels.
The switcher, which is housed in a standard 19-inch cabinet, has a maximum memory capacity of 64 K characters. The IATA, AFTN, ICAO, SITA and OFTS message formats are all available as standard packages with the message switcher.

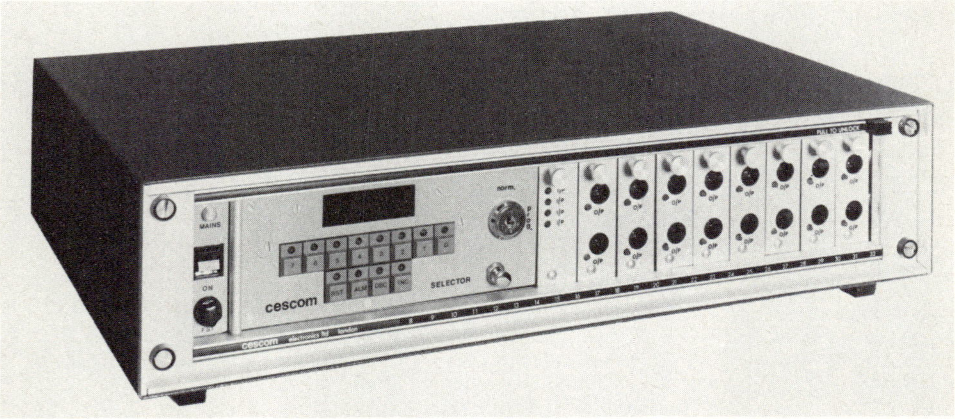

Cescom 10312 multi-way selector unit

The 10312 micro-processor based multi-way selector unit monitors the input line and in response to predetermined codes, controls the operation of up to 16 output lines. The standard unit can be controlled by up to six types of control code, each of which may consist of up to five characters: start of message, end of heading, address code, end of address, end of text and end of message.

Status

Production. Both units have been procured by IAL. The 10273 is installed at the IAL training centre Bailbrook College, Bath, UK and the 10312 in service in Riyadh, Saudi Arabia

Manufacturer

Cescom Electronics, Leytonstone, London E11, England

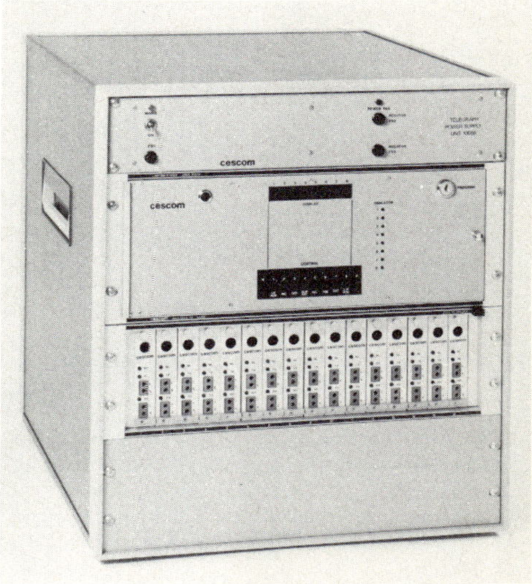

Cescom 10273 message switcher

Cossor CGR 1020/1020M Vhf/uhf Communications Systems

The CGR 1020 series has been designed to meet both military and civil requirements for single-channel, crystal-controlled, ground-to-air equipment operating in the vhf and uhf frequency bands.
The equipment range comprises a receiver, 20-watt transmitter and 100-watt power amplifier. The three chassis contain all the non-frequency sensitive elements of their respective units and are therefore common to both the vhf and uhf frequency bands. Discrete plug-in modules, containing the frequency sensitive circuit elements, are inserted through the front panel of the appropriate chassis as required, to select vhf and uhf frequency bands. The receiver chassis will accept only one plug-in module at a time, and the transmitter and power amplifier chassis will accept up to two, one vhf and/or one uhf, although for non-simultaneous operation.
The universal chassis solution ensures that a high proportion of components, sub-assemblies, and pcbs are common.
The transmitter and power amplifier employ a power tetrode in the final output stages of the respective units. The cooling system has been designed to be as quiet as possible by using high-volume low-pressure air, minimising noise and removing the high-frequency element from the spectrum.
A new generation multi-channel version, the CGR

CGR 1020 in service at Westland's airfield, Yeovil, England

1020M, is now available. The new equipment makes use of designs from existing equipments but, unlike the CGR 1020 which uses valves in the transmitter output stages, the multi-channel equipment is all solid-state. Using two basic equipment cases, one for the receiver and one for the transmitter, it is possible to achieve a variety of vhf and uhf configurations for simultaneous operation. As with the single-channel equipment, the transmitter output power is either 20 or 100 watts with the addition of the power amplifier.

Status

Developed in the USA by E Systems, the CGR 1020 is manufactured in the UK by Cossor, and supplied to the Royal Air Force as standard equipment for ground-to-air communications. It is operational at civil airports world-wide.
The CGR 1020M was developed by Cossor in conjunction with E Systems. First production equipments were delivered to the RAF in 1980

Manufacturer

Cossor Electronics Ltd, Harlow, Essex, England

Characteristics

TYPE	Receiver	Transmitter	Power amplifier
SERVICES	A3am	A3am	A3am
MODE		Continuous at max mod depth	Continuous at max mod depth
FREQUENCY RANGE	(Vhf) 117–137 MHz (Uhf) 225–400 MHz	(Vhf) 117–137 MHz (Uhf) 225–400 MHz	(Vhf) 117–137 MHz (Uhf) 224–400 MHz
AMPLIFIER DRIVE POWER			(Vhf) 7·5–15 W (input impedance nominally 50Ω) (Uhf) 7·5–20 W (input impedance nominally 50Ω)
CHANNEL SPACING	25 or 50 kHz	25 or 50 kHz	
FREQUENCY STABILITY AND ACCURACY	Equipment drift 12 ppm over temperature change	Drift 12 ppm over any 30°C any 30°C temperature change	
SENSITIVITY	(s + n)/n ratio with input of 10 dB μV emf (M=30% at 1 kHz)		
SELECTIVITY	Reference centre frequency -6dB ±15 kHz ± 25 kHz -60 dB		
POWER OUTPUT		20 W into 50Ω	100 W into 50Ω
TEMPERATURE RANGE	-10 to +55°C	-10 to +55°C	-10 to +55°C
RELATIVE HUMIDITY	95% from -10 to +40°C	95% from -10 to +40°C	95% from -10 to +40°C
UNIT DIMENSIONS			
height	3½ in (8·8 cm)	10½ in (26·6 cm)	10½ in (26·6 cm)
width	19 in (48·2 cm)	19 in (48·2 cm)	19 in (48·2 cm)
depth	12 1/10 in (30·6 cm)	17 in (43·1 cm)	17 in (43·1 cm)
WEIGHT (RECEIVER OVERALL WITH UHF MODULE FITTED)	15 lb 12 oz (7·1 kg)	48·3 lb (21·8 kg)	65·8 lb (29·7 kg)

Feedback Cygnet Telewriters

Telewriters are a system for transmitting handwriting or drawings almost any distance without delay or error.
Cygnet is offered as a basic hard-wired system between two stations. This would normally consist of a transmitter directly linked with a receiver. This initial link can be extended to form a multiple-selective system, consisting of a single transmitter to several receiver out stations, enabling a central-control point to issue instructions to a number of departments.
The multiple selective system can also function in reverse, so that a number of transmitters can report to a single receiving point.
A more sophisticated system is the hard-wired network selector system, which is capable of handling up to ten stations, all inter-communicating. The telewriters are all transceivers connected through a central control unit providing in-house-two-way communication between several departments. It is possible to extend this system to up to 30 stations by the addition of further control units, together with the necessary out stations.
Telewriter communication over longer distances can be provided by use of a conventional telephone system. The telewriter is connected to a standard GPO private wire that has been dedicated specifically for this use. As such it may be used in the same way as the normal hard-wired system.
Telephone authorities require that some form of spoken announcement should be made on receipt of all calls at the receiving end. The telewriter is, therefore, linked to a standard automatic answering machine and operates after a recorded message has been given.
Airport applications include ground communications for atc passenger cargo and baggage handling, aircraft maintenance and fault reporting.

Feedback Cygnet telewriter in terminal control room, London Heathrow Airport

Status

Production. British Airports Authority use Cygnet machines at London Heathrow, Gatwick Airports, England, Aberdeen and Prestwick Airports, Scotland. Manchester Airport, England operate the telewriters for their information desks

Manufacturer

Feedback Instruments Ltd, Crowborough, Sussex, England

Feedback Cygnet telewriter in operation at Prestwick Airport's apron control

IAL Stratus Communications Switching System

Stratus is a digital voice communications switching system. Modular in concept, the system can be expanded to cater for all atc communications control requirements. Operators are given flexibility in the selection of radio, telephone and intercom channels. The audio and control channels are separated and micro-processor technology is used to provide distributed intelligence. The system is designed for 100 per cent availability so that any failure that does occur can be isolated.

For systems where channel access is fixed, as in the majority of small and medium-sized systems, switching between the audio channels from each console position and the external radio and telephone links, is achieved with a solid-state, space-division, multiplex arrangement of cross-point switches linking into the system's shared audio hiways. The control consoles connect with these multiplex cards and the quantity of each one depends on the number of audio hiways demanding access.

In larger systems, particularly where there is a limit to the number of channels an operator can handle and where the possibility of reconfiguring these channels is desirable, time-division multiplexing techniques are used. By digitising speech into the form of a pulse code modulated signal, the speech can be time-division multiplexed on to duplicated communicatioans buses, allowing up to 256 links to interface directly, for example, it is possible for 128 operators to talk on 128 channels. Larger systems are arranged through interlinks to provide 1000 crosspoints or more.

To access a particular channel, an operator programmes a decoder card to select the time-slot carrying the channel required. Programming involves typing in a predetermined alphanumeric channel code on a small console switch module, then transferring it on to the operator's control module where the usual range of monitor and transmit functions can be performed. Any five or ten channels may be accessed at any one time, depending on whether one or two control modules are used. The heart of Stratus is a set of independent, dedicated microcomputers that control the access of each console to the mix of external and internal audio channels. The microcomputers communicate through a high speed parallel data bus. In this way, each microcomputer keeps track of transactions across the entire system, enabling decisions to be made with reference to overall system status. Illegal or contentious situations, such as two controllers attempting simultaneous transmission on one channel, are inhibited. Standard software is available to locate faults at module level, and in some cases right down to the individual printed circuit boards. Diagnostic tests are carried out continually and automatically by the system and a special micro-computer acts as the supervisory controller for these diagnostic functions, alerting the operator to any faults through coded error messages on a display module.

Any changes in operating layout or procedure can be made by changing the software. New system configurations can be prepared off-line and then stored on disc, ready for implementation in case of a console failure, without the need for hardware modifications.

Status

Stratus was selected by the British CAA for the London atc centre at West Drayton in late 1981. The installation will provide 96 executive and production support positions with access to 144 radio telephone channels. It is designed to be expanded to 175 executive and support positions with 300 channels if necessary.

A smaller Stratus system has been installed at Abu Dhabi International Airport, United Arab Emirates

Manufacturer

IAL, Southall, Middlesex, England

IBM Message Switching Equipment

The IBM 3750 voice- and data-switching system is a computer-based telephone exchange with elaborate facilities for storing and re-routeing incoming calls; setting up three-way calls when necessary; providing automated dialling procedures for std and overseas calls; and various links to word-processing systems.

Status

Production. The system was procured by the British Airports Authority to control the internal telephone network at London Gatwick Airport, England

Manufacturer

IBM (United Kingdom) Ltd, Feltham, Middlesex, England

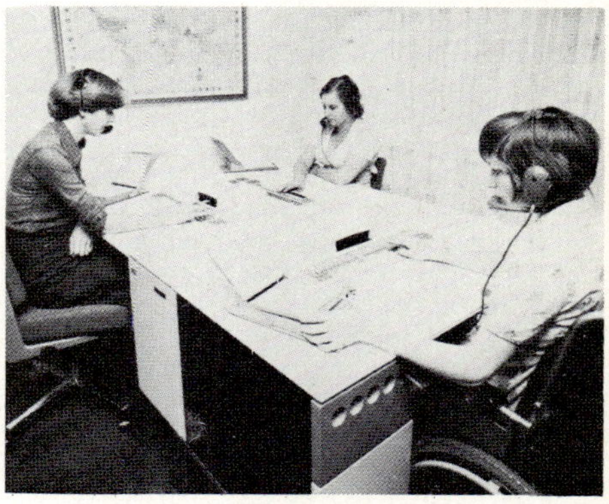

Operator Desks of the IBM 3750 voice- and data-switching system at London Gatwick Airport

---

ITT Automatic Message Switching Equipment

ITT has produced a message switching system with a software package and facilities specifically designed for use in the Aeronautical Fixed Telecommunications Network (AFTN). Normally supplied in a dual hardware configuration, the ITT 6405 ADX offers a short term file on magnetic disc for message filing and retrieval with a minimum capacity of one hour in accordance with ICAO requirements. Other features of the equipment include 64, 96 or more low speed lines (50–200 baud); telegraph or V28 line interfaces; and ITA 2 and ITA 5 character codes. The system can be interfaced with other networks if required, for example, MOTNE, SITA, Telex. The unit is available in a dual configuration with a hot standby mode and automatic switches for continuity of operation. The hardware of the 6405 ADX is contained in a suite of cubicles. One or more cubicles are used, depending on the configuration, each cubicle being 600 mm wide, 660 mm deep and 1625 mm high. Access to the cubicles is via front and rear doors. Each cubicle is fitted with cooling fans and power supplies as appropriate. Inter-cubicle connections are made within the suite.

Characteristics

NO OF LINES: 224 input, 224 output
MEDIUM SPEED LINES: up to 16
LINE SPEEDS: all standard speeds from 50–9600 baud
DATA CODES: CCITT No 2/CCITT No 5 or equivalent
LINE TYPES: full and half duplex, simplex
LINE INTERFACE: CCITT V28/RS232C Telegraph, single or double current
POWER REQUIREMENTS: 240 V ± 10%, 50 Hz/60 Hz (optional: 115 V ± 10%)
OPERATING TEMPERATURE RANGE: −15 to +32°C

Status

Production. Installed in a number of major international ICAO switching centres including Copenhagen, Denmark; Shannon, Ireland; Johannesburg, South Africa; Madrid, Spain and Caracas, Venezuela

Manufacturer

ITT Business Systems, Brighton, East Sussex, England

ITT 6405 message switching equipment

Park Air Electronics Air Traffic Control Console

Consoles for atc operations produced by Park Air Electronics include the mini-desk and the series A desk.

The mini-desk has been developed for use at smaller airports, airfields and airstrips. Intended to stand on top of an existing desk or table, the mini-desk provides the airfield with wind velocity (speed and direction) on two 5-inch (127 mm) dial lit instruments. A digital read-out altimeter provides local pressure information and a quartz-crystal referenced battery operated clock with liquid-crystal display provides time information. All of these facilities are incorporated in a single panel of the two panel format of the desk superstructure. The second panel provides radio communications facilities for the allocated air/ground service. Using the 1300 transmitter/receiver, the radio equipment matches ICAO's latest 25 kHz channelling recommendations and at 10 watts output power matches the UK CAA specification for local operation and airfield service. Power supplies may be 110 volt or 240 volt standard ac mains supplies (50–60 Hz) or 24 volt dc batteries where no ac supplies are available. The digital clock is independently powered by integral batteries which permit the main supply to be interrupted without loss of time information.

Intended for use at the smaller airport, the series A desk is designed for single controller operation. The standard desk unit is supplied with a vhf radio communication channel PAE-type 1400 transmitter and PAE-type 1901 receiver, desk microphone with press to transmit facility, and sockets for headsets, microphones and push-to-talk key switches.

The centre superstructure contains meteorological instruments comprising wind direction, wind speed in knots, barometer calibrated in millibars and inches of mercury and loudspeakers in two positions.

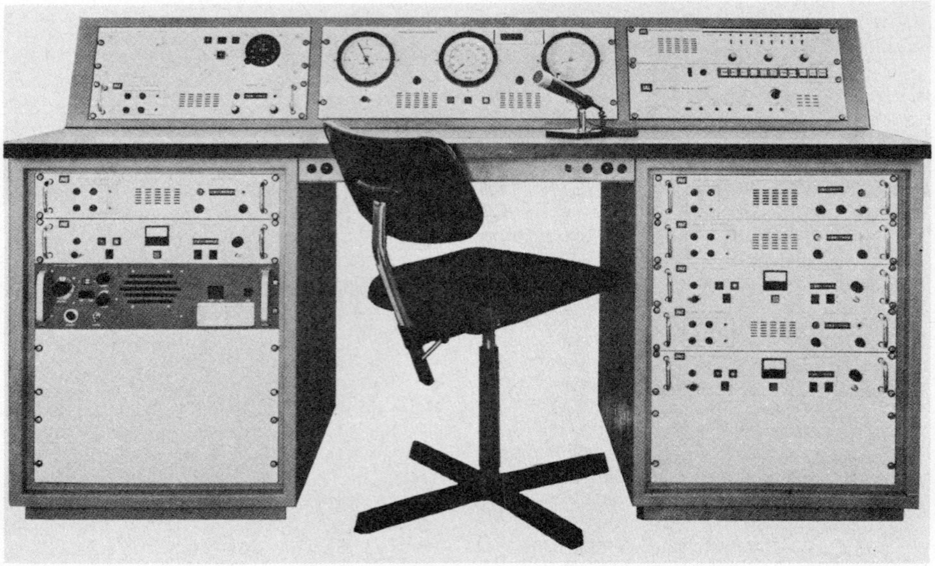

PAE-type 2200 uhf am receiver

Characteristics

SERIES: A desk

DIMENSIONS

width:	1795 mm
height:	933 mm
depth:	725 mm
kneehole width:	625 mm
depth:	725 mm
working surface	
width:	1795 mm
height:	765 mm
depth:	380 mm

Status

Production

Manufacturer

Park Air Electronics Ltd, Peterborough, Cambridgeshire, England

Park Air Electronics Uhf Communications Equipment

Uhf communications equipment offered includes the type 2200 receiver, 2300 and 2400 transmitters and the combined vhf/uhf receiver type 2500.

The type 2200 receiver is a single channel equipment designed for airport use in the 225–400 MHz uhf am band. The receiver is a single conversion super-heterodyne design, and makes extensive use of integrated circuits. Modular construction has been used throughout. Each module (the rf converter, if amplifier, audio-processing and output) plugs-in to the universal mother-board. The mother-board is common to both 2200 uhf and 2100 vhf receiver types, similar module types are interchangeable, and different rf modules determine uhf or vhf band operation.

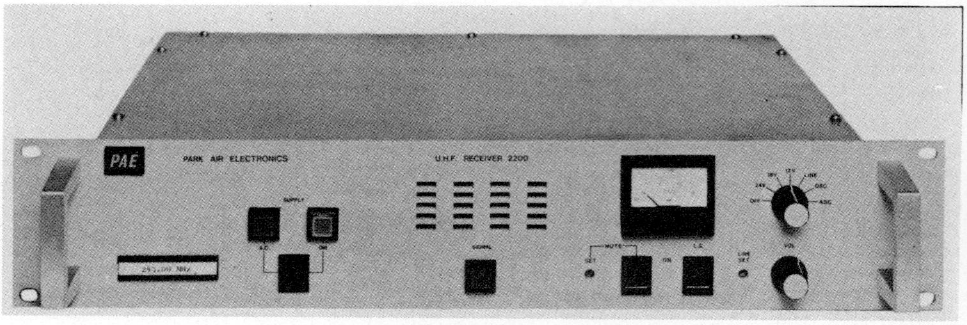

PAE-type 2200 uhf am receiver

The receiver utilises a balanced mixer and helical filters have been used in the rf amplifier front end design. Both rf and if modules are fully screened, to ensure not only very direct radiation of local oscillator signals from the receiver, but also to provide the receiver with a high degree of immunity to extraneous rf fields. Mute circuits are s/n operated, thereby eliminating false triggering under noisy conditions. A noise blanker has been included to improve reception when high levels of impulse interference are present. The 2200 is suitable for local or remote operation, and has 600 ohm line output circuits fitted as standard. Power may be obtained from ac mains or 24 volt dc supplies with automatic changeover to dc operation in the event of ac mains failure. The unit offers front panel metering, loudspeaker volume control, supply switches, mute and line level adjustment and an integral loudspeaker.

The type 2300 transmitter is a single channel equipment intended for airport use on the 225–400 MHz uhf am air-band. The units features rf amplifier modules which are wide-band, and fully interchangeable, no retuning being necessary. The transmitter can be driven by a synthesized drive unit, allowing full selection of channels throughout the band. Overcurrent and overvoltage protection is included as standard and thermal shutdown switching has been incorporated. Continuous operation is possible in temperatures up to +60°C. The transmitter has a lower power modulated rf stage which feeds a wide-band amplifier. A sample of the output signal is fed back to the modulator stage via a directional coupler and detector. This technique provides linear modulation and stable output even with large supply voltage variations.

The af processing circuits include Vogad, muting during gaps in speech periods and rf clipping. All may be switched in or out of circuit as required. The rf clipper provides immunity to over modulation and an effective increase in average modulation depth for longer range communications. The transmitter can be controlled locally or remotely via telephone lines. Automatic local/remote switching is incorporated giving priority to the local operator.

The type 2400 transmitter is a new development designed for civil/military ground station use in the frequency range 225–400 MHz. The transmitter has a maximum power output of 100 watts carrier wave (400 watt pep) and incorporates comprehensive amplitude modulation facilities. For single channel operation a crystal oscillator is included within the transmitter, all necessary tuning adjustments are accessible from the front panel. For multichannel use the transmitter may be driven from a separate synthesizer drive source (PAE-type 2600).

The transmitter operates from 28 volt dc which is provided from the standard ac mains supply via a separate power unit type 2401. Emergency 28 volt dc supplies may be combined with the 2401 PUS which then provides automatic changeover to the emergency supply in the event of an ac supply failure. The transmitter rf power output stages are all wide-band and have no tuning controls. The output stages and the driver amplifier are monitored and the parameters include voltage, current, vswr and temperature. The condition of the monitoring circuits is shown by internal leds. A thermostatically-controlled cooling fan is employed to maintain a stable thermal environment for the amplifiers. Audio processing circuits include agc, muting during no speech periods and rf clipping to provide immunity to overmodulation and an effective increase in average modulation depth. A switched time-out circuit is also provided to automatically unkey the transmitter after approximately three minutes. The transmitter can be controlled locally or remotely via telephone lines. Automatic local/remote switching is incorporated.

The type 2500 is another new development. It is designed for operation in conjunction with a PAE synthesised drive-source type 2600. The equipment is suitable for civil/military vhf/uhf ground station use in the frequency range 100–156 MHz or 225–400 MHz. Both the 2500 receiver and the 2600 drive unit are designed as plug-in modules which may be combined together in a single 3-unit high 19-inch rack mounting main frame. The 2600 drive unit includes a channel memory which enables up to sixteen frequencies to be programmed via thumbwheel switches. Programmed channels may be recalled by pressing one of 16 recall buttons. An led display of frequency and channel number is provided. To prevent unauthorised corruption of pre-programmed frequencies a key operated lock-switch is provided. When locked (with the key in the lock position) pre-programmed channels can be recalled but re-programming is not possible.

The 2600 drive unit may be used with transmitters of various power outputs from the PAE range of build multi-channel transmitters/receivers or base stations as required. Remote-control of a main/stand-by base station with selection of the 16 pre-programmed channels may be achieved via two telephone line pairs. Extended control of all facilities is provided via a separate control head connected to the main equipment via a 20-way ribbon cable.

The equipment is designed to operate from standard ac mains supplies or a 24/28 volt dc supply. Where both supplies are available the equipments will automatically switch to the dc supply in the event of an ac supply failure.

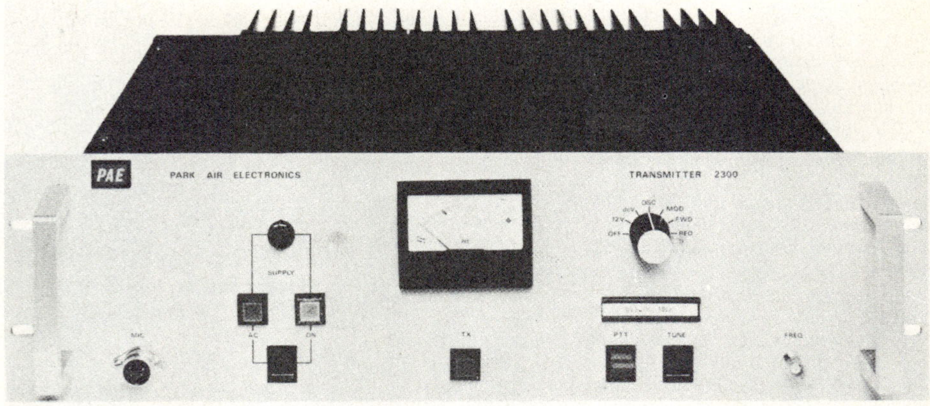

PAE type 2300 uhf am transmitter

Characteristics

MODEL	2200	2300	2400		2500
TYPE	Uhf am receiver	Uhf am transmitter	Uhf am transmitter		Vhf/uhf am receiver
FREQUENCY	225–400 MHz	225–400 MHz	225–400 MHz		110–156 MHz/ 225–400 MHz
NO OF CHANNELS	1	1	1/7000		Vhf- 2240 Uhf- 7000
CHANNEL SPACING	50 kHz (option 25 kHz)				25 kHz
POWER OUTPUT		25 W	100 W max adjustable		
POWER REQUIREMENTS	100–120/200–240 V 40–60 Hz ac, 24 V dc	100–120/220–240 V 40–60 Hz ac, 24 V dc	28 V dc, 2401 power supply unit 110–120/220–240 V 40–60 Hz ac		110–120/200–240 V 40–60 Hz ac, 22–30 V dc
UNIT DIMENSIONS			2400	2401	
width	483 mm	483 mm	483 mm	483 mm	483 mm
depth	270 mm	365 mm	425 mm	450 mm	400 mm
height	89 mm	133 mm	177 mm	133 mm	133 mm
WEIGHT	4·1 kg	15 kg	11·5 kg	21·5 kg	

Status

Production. In service with civil and military operators world-wide

Manufacturer

Park Air Electronics Ltd, Peterborough, Cambridgeshire, England

---

Park Air Electronics Vhf Communications Equipment

Park Air Electronics produces a wide range of vhf communications equipment for airport and en route stations.

The 1201 series is an all solid-state am receiver with a single channel, full metering facilities and built-in monitor loudspeaker. It may be operated from standard ac supplies or 24 volts dc. An auto no-break change over facility is incorporated which will cause the receiver to switch to dc supplies if the ac mains fail.

The type 1250 am transmitter operates in the 118 to 136 MHz frequency range producing 25 watts carrier power, 100 watts pep, from a mains- or battery-power source with automatic change over. The equipment is fully solid-state and meets ICAO recommendations for operation in 25 kHz channelling environments. The af processing circuits include vogad, muting during no-signal periods and rf clipping. All may be switched in or out of circuit as required. The rf clipper provides immunity to over modulation and an increase in average modulation depth for longer range communication. An external socket is provided for a multichannel or high stability drive source. The transmitter can be controlled locally or remotely via telephone lines. Automatic local/remote switching is incorporated giving priority to a local operator.

The type 1300 is an equipment of modular construction for use at the smaller airfield or airport, oil rig or as a company vhf ground-to-air transmitter/receiver and designed to meet the British Civil Aviation Authority requirements for an effective radiated power of 5 watts. The 1300 may be used for ground-to-air communications at ranges of up to 200 miles (322 km). Optional accessories are available for remote-control operation.

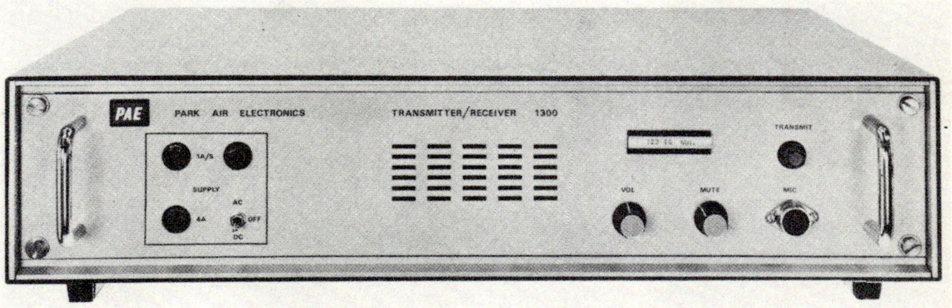

PAE type 1300 vhf am transmitter/receiver

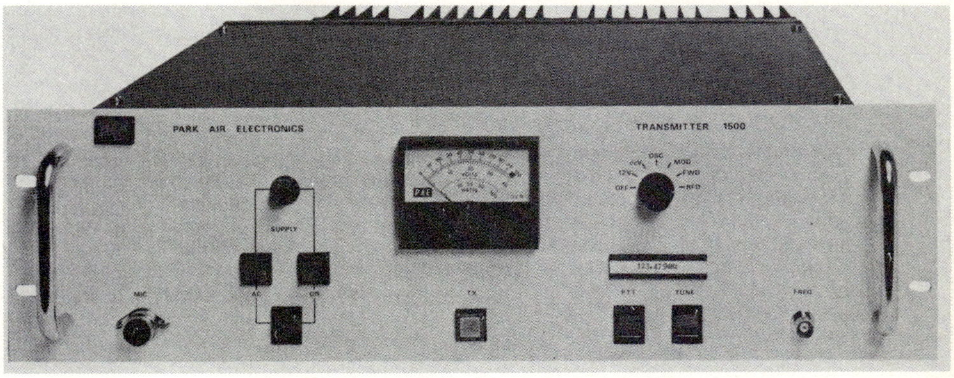

PAE type 1500 vhf am transmitter

The type 1400 transmitter is an amplitude modulated equipment intended for airport service in the vhf band 118–136 MHz. The equipment is fully solid-state and meets ICAO recommendations for operation in 25 kHz channelling environments. The transmitter has a low power modulated stage which feeds a wide-band power amplifer, part of the output signal is fed back to the modulated stage via a directional coupler. This technique provides linear modulation and stable output power even with large supply voltage variations.

Provision is made for simple adjustment of output power to suit station requirements. The af processing circuits include vogad, muting during no signal periods and rf clipping. All may be switched in or out of circuit as required. The rf clipper provides immunity to over modulation and an effective increase in average modulation depth for longer range communication. An external socket is provided for multichannel or high stability drive source. The transmitter can be controlled locally or remotely via telephone lines. Automatic local/remote switching is incorporated giving priority to a local operator.

The type 1500 transmitter rated at 50 watts has a low power modulated rf stage which feeds a wide-band driver amplifer. The output from the driver is fed to two power amplifiers via an equi-phase power splitter. The outputs from the two power amplifiers are then summed via a power combiner. Part of the output signal is fed back to the modulated stage via a directional coupler. All other features are similar to the type 1400.

The type 1600 mobile radio-telephone has been designed to fulfil the requirements which exist for mobile radios within modern airport environments.

The equipment can operate on any four channels within the band 118–136 MHz with no band-width restrictions. Channel selection is achieved via push-button switches mounted on a small remote-control head. The remote-control head is small enough to be mounted in any position within the vehicle. An optional plug-in priority receiver circuit board may be fitted which provides priority reception of a particular channel. If it is required to reply to a priority call then this must be one of the four operating frequencies. If reception only of the priority channel is required, then this may be on any other specified frequency. The priority receiver circuits are completely independent of the main receiver, and its performance is unaffected by any signals being received by the main receiver. In operation, if a priority signal is received then a lamp is illuminated on the control head and the signal is passed to the loudspeaker; if at the same time, a signal is being received by the main receiver, then both signals are passed to the loudspeaker with priority signal at a higher level.

The public address amplifier is designed as an optional add-on unit which matches the remote-control head. It may be mounted independently or in association with the control head using a double-mounting bracket. A three-position switch on the remote head provides for selection of either the receiver output, or the microphone to the amplifier.

The type 1700 channel scanning monitor is a multipurpose crystal-controlled eight-channel receiver operating in the vhf air mobile band 118–136 MHz am. The signal searching technique of the monitor is accomplished by the use of low current drain logic circuits which permit operation from internal rechargeable batteries (optional). An integral telescopic antenna is included which allows operation in both desk top or hand-carried modes. Provision is made for the deselection of unwanted channels and the automatic mode may be inhibited to permit manual operation of any one channel.

A separate socket on the rear panel allows the equipment to be operated from the dc supply of an automobile. The negative pole of the supply is grounded. Extension loudspeaker facilities are provided and also a tape-recorder socket. Indication of scanning or activated channels is by light emitting diodes built into the finger-tip control micro switches. A led flashes when the internal batteries require charging. Charging may also be carried out at the same time as the set is operating from the appropriate ac supply. A feature of the equipment is the special protection of the set from interference by nearby fm broadcast transmitters and the individual 720 channel capability of the equipment by the use of crystal lattice filters.

The type 1800 receiver has been specially developed for use in conjunction with the 1210 drive unit and an associated transmitter type 1400/1250 or 1500. The receiver is crystal controlled with up to eight channels which are selected via the 1210 switching circuits. Dc power for the receiver is obtained from the associated transmitter.

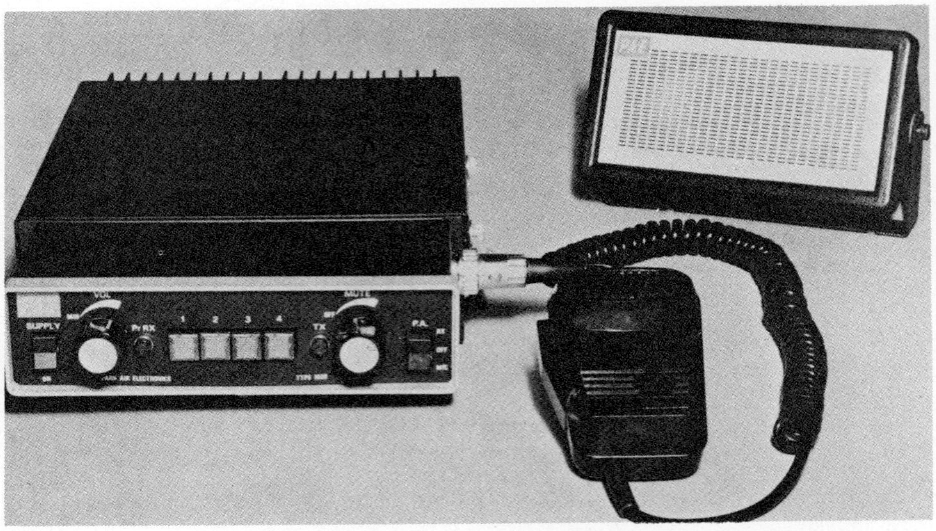

PAE-type 1600 vhf am mobile radio-telephone

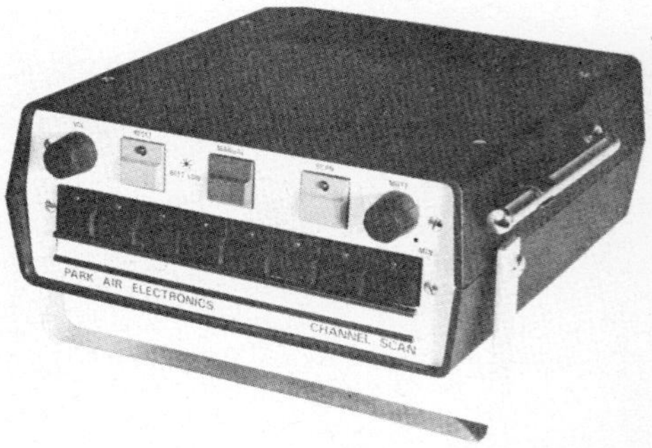

PAE-type 1700 channel scanning monitor receiver

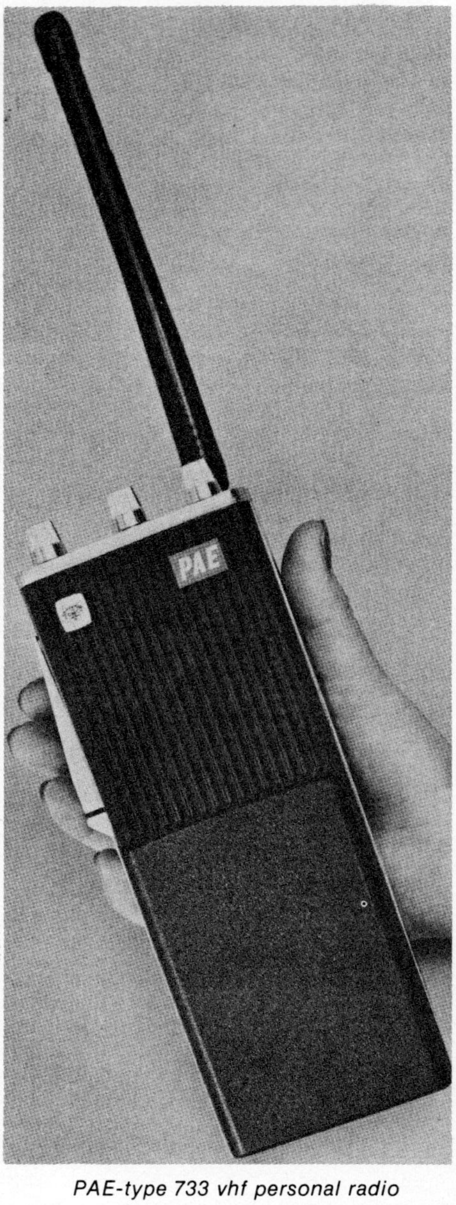

PAE-type 733 vhf personal radio

The 1210 drive unit is one of three offered by Park Air Electronics for use in conjunction with their transmitters. The 1210 provides a multichannel drive source. It features an eight channel capacity, all solid-state rf switching, local or extended channel selection and is powered from the transmitter unit. The 1230 provides a fully synthesised 720 channel drive source, and the 1240 is designed for single frequency operation.

The type 1901 receiver is a single conversion super-heterodyne design and makes extensive use of integrated circuits. The mute circuit is s/n-operated, thereby eliminating false triggering under noisy conditions and a blanker circuit is included to improve reception when high levels of impulse interference are present. The 1901 is suitable for local or remote operation and has 600 ohm line output circuits fitted as standard. Power may be obtained from ac mains or 24 V dc supplies with automatic change-over to dc operation in the event of an ac supply failure.

The type 2100 receiver is a single-channel equipment designed for airport use in the 110–156 MHz vhf am band. The receiver is a single conversion super-heterodyne design, and makes extensive use of integrated circuits. Modular construction has been used throughout. Each module, the rf converter, if amplifier, audio processing and output plugs-in to the universal mother-board. This mother-board is common to both 2200 uhf and 2100 vhf receiver types, similar module types are interchangeable, and different rf modules determine uhf or vhf band operation. The receiver uses a balanced mixer and helical filters have been used in the rf amplifier front end design. Both rf and if modules are fully screened to ensure not only very direct radiation of local oscillator signals from the receiver, but also to provide the receiver with a high degree of immunity to extraneous rf fields. Mute circuits are s/n-operated, thereby eliminating false triggering under noise conditions. A noise blanker has been included to improve reception when high levels of impulse interference are present. The 2100 is suitable for local or remote operation, and has 6000 line output circuits fitted as standard. Power may be obtained from ac mains or 24 volt dc supplies with automatic changeover to dc operation in the event of ac mains failure. The unit offers front panel metering, loudspeaker volume control, supply switches, mute and line-level adjustment and an integral loudspeaker.

The type 2500 receiver is a high performance multichannel equipment designed for operation in conjunction with a PAE synthesized drive source type 2600. The equipment is suitable for civil/military vhf/uhf ground station use in the frequency range 100–156 MHz or 225–400 MHz. A full description of these units is given in the Park Air Electronics uhf communications entry above.

The type 2700 transmitter is one of the latest developments. It is designed for civil/military ground station use in the frequency range 108–156 MHz. The transmitter has a maximum power output of 250 watt carrier wave (1000 watt pep) and incorporates comprehensive amplitude modulation facilities. For single channel operation a crystal oscillator is included within the transmitter, all necessary tuning adjustments are accessible from the front panel. For multichannel use the transmitter may be driven from a separate synthesizer drive source (PAE type 2600 or 1230). The transmitter operates from 28 volt dc which is provided from the standard ac mains supply via two separate power units type 2401. Emergency 28 volt dc supplies may be combined with the 2401 power supply units which then provide automatic changeover to the emergency supply in the event of an interruption in the ac supply. Failure of a single 2401 unit causes a 6 dB reduction in the transmitter output power.

The transmitter rf power output stages are all wide-band and have no tuning controls. The final output stages consists of four 300 watt (pep) push-pull modules interconnected via a splitter/combiner system. Parameters which are monitored include voltage, current, vswr and temperature, the condition of the monitoring circuits is shown by internal leds which used in conjunction with the front panel meter and alarm lamp provide for fault isolation. A thermostatically-controlled cooling fan is employed to maintain a stable thermal environment for the amplifiers.

Audio-processing circuits include agc, muting during no speech periods and rf clipping to provide immunity to over-modulation and an effective increase in average modulation depth. A switched time-out circuit is also provided to automatically unkey the transmitter after approximately three minutes. The transmitter can be controlled locally or remotely via telephone lines. Automatic local/remote switching is incorporated.

The type 551/661 is a vhf portable air-mobile band 360 channel am transceiver designed for ground or airborne use as an emergency or occasional use station. It is available with 360 channel capacity in 50 kHz steps of the full air-mobile communications band or has 720 channel capacity over the same band in 25 kHz steps. The internal power source is a 4·5 ampere hour gelled electrolyte battery and a charger unit for use on either 110 or 230 volt ac standard supplies is provided. Optionally an ac power supply is available for continuous ac fixed site operation using either 110 volt 60 Hz or 240 V Hz ac supplies. Frequency selection is by two front panel mounted knobs. Also mounted on the front panel are controls for squelch, volume, dial lamps, battery state indicator with jacks for headset and microphone. A carbon hand microphone is included as part of the 551/661 station.

The type 733 hand-held vhf am transceiver is a four-channel portable, operating in the vhf 118–136 MHz band. A special feature is the broad-band design, enabling four channels to be utilised anywhere within the band. Lightweight aluminium construction protects the internal circuitry. The controls include adjustable mute, volume and channel selection. A led transmit indicator is provided, along with accessory antenna and speaker jacks. The rechargeable battery is integral with the rear panel. Along with a low stand-by current drain (15 mA), the unit will operate for over eight hours at the specified duty cycle before recharging is required. This is achieved by simply plugging the unit into a desk-top battery charger.

Status

Production. In service with civil and military operations world-wide

Manufacturer

Park Air Electronics, Peterborough, Cambridgeshire, England

Characteristics

MODEL	1201	1250	1300	1400	1500	1600	
TYPE	Vhf am receiver	Vhf am transmitter	Vhf am transmitter/ receiver	Vhf am transmitter	Vhf am transmitter	Vhf am mobile radio-telephone	
FREQUENCY	118–136 MHz	118–136 MHz	118–136 MHz	118–136 MHz	118–136 MHz optional 108–136 MHz	118–136 MHz	
NO OF CHANNELS	1	1	1	1	1	4	
CHANNEL SPACING						25 kHz	
POWER OUTPUT		25 W	10 W	2–10 W adjustable	50 W	10 W	
POWER REQUIREMENTS	100–125/ 210–250 V, 40–60 Hz ac, 24 V dc	110/120–220/ 240 V, 40–60 Hz ac, 24 V dc	100/120–220/ 240 V, 40–60 Hz ac, 24 V dc	100/120–220/ 240 V, 40–60 Hz ac, 24 V dc	110/120–220/ 240 V, 40–60 Hz ac, 24 V dc	12/24 V dc	12/24 V dc
UNIT DIMENSIONS						control head	main assembly
width	483 mm	483 mm	500 mm	483 mm	483 mm	175 mm	245 mm
depth	254 mm	365 mm	310 mm	300 mm	410 mm	75 mm	190 mm
height	89 mm	132 mm	130 mm	89 mm	133 mm	50 mm	70 mm
WEIGHT	4·8 kg	14 kg	10 kg	6 kg	16 kg		

MODEL	1600	1700	1800	1900	1901	2100
TYPE	Vhf am ground station	Vhf am channel scanning monitor receiver	Vhf am receiver	Vhf am receiver	Vhf am receiver	Vhf am receiver
FREQUENCY	118–136 MHz	118–136 MHz	118–136 MHz	118–136 MHz	118–136 MHz (108–118 or 126–144 MHz special order)	110–156 MHz
NO OF CHANNELS	4	8	8	720	1	1
CHANNEL SPACING	25 kHz		25 kHz/50 kHz	25 kHz	25 kHz	25 kHz
POWER OUTPUT	10 W					
POWER REQUIREMENTS	110/120–220/ 240 V, 40–60 Hz ac, 10·5–16 V dc	100–120/210/ 250 V, 40–60 Hz ac, 10·5–16 V dc	24 V dc	100/120/ 200–240 V, 40–60 Hz ac, 20/30 V dc	100–120/ 200–240 V, 40–60 Hz ac, 20/30 V dc	100–120/ 200–240 V, 40–60 Hz ac, 24 V dc
UNIT DIMENSIONS						
width	288 mm	210 mm	483 mm	483 mm	483 mm	483 mm
depth	250 mm	180 mm	190 mm	180 mm	180 mm	270 mm
height	120 mm	80 mm	89 mm	89 mm	89 mm	89 mm
WEIGHT	7 kg	1·6 kg	3·5 kg	5·5 kg	5·5 kg	4·1 kg

MODEL	2700		551/661	733
TYPE	Vhf am transmitter		Vhf am portable	Vhf am hand-held portable
FREQUENCY	108–156 MHz		118–135·95 MHz	118–136 MHz
NO OF CHANNELS	1/1920		360/720	4
CHANNEL SPACING			50/25 kHz	25 kHz
POWER OUTPUT	250 W		6 W	2 W
POWER REQUIREMENTS	28 V dc, Type 2401 power supply 110/120–220/240 V ac, 40–60 Hz		12–1375 V dc, option 10 V 60 Hz/ 240 V 50 Hz	rechargeable Ni-Cd battery
UNIT DIMENSIONS	2700	2401		
width	483 mm	483 mm	222 mm	88·9 mm
depth	425 mm	450 mm	337 mm	44 mm
height	267 mm	133 mm	89 mm	225 mm
WEIGHT		21·5 kg	6·4 kg	0·9 kg

Racal Communications Recording Equipment

Racal produces a range of recording equipment suitable for atc communications. These include the International Communications Recorder (ICR) series, Autostore and Callstore. ICR is a modular system available in 8, 16, 32 or 40 channel models. After determination of maximum required recording channel capacity, allowing for a time indexing channel and future channel expansion, selection of optional systems can be made. Single or dual tape transports may be fitted into a system. Single transports provide 24 hours of continuous recording. Dual transports may be used to provide continuous recording with standby facilities. This ensures continuity during tape change or recording failure and the capability of maintenance without interruption of recording.

Pilot tone monitoring is a method of continuously and automatically checking the correct operation of each channel. A sub-audio tone is injected at each recording amplifier input and subsequently detected from the tape by the individual channel monitoring amplifiers. In the event of a drop in pilot tone level on a particular channel, the ICR will take corrective action by automatically activating the pre-determined spare track guard channel and indicating this condition. Coupled with integral alarm systems, pilot tone monitoring also provides a high level of security, instantly detecting any unauthorised tampering with the ICR. The guard channel must be included in the maximum required channel recording capacity.

Service modules which are common to all channels in the system operation can be duplicated. These include the electronics, power supply modules, bias oscillator and pilot tone oscillator modules.

Equipments may be operated in a voice-operated record start mode by the inclusion of a voice-operation module. This unit also allows the transports to be started from a squelch or remote-contact closure. This facility is provided as standard with automatic pilot tone systems.

Time signals in the form of any-language voice idents or digitally encoded signals can be recorded on any ICR. One channel is allocated to the time signal when this format is selected.

Other ICR optional equipment offered includes a remote facilities interface providing alarm and status indications and control. A separate reproducer can be provided for transcription and replay remote from the recording site. The standard version is fitted with replay only heads, but an optional version fitted with both record and replay heads provides a spare head-plate assembly for the recorder. The tape transport is identical to that used in the recorder which enables it to be used as an emergency substitute. An integral loudspeaker is included for playback of four channels simultaneously; any two selected channels, the time channel and the guard channel. The signal is available as a line output and on two jack sockets and a DIN socket for monitoring by headphones or external loudspeaker. The pilot tone signal is normally filtered out on replay but the filter may be switched out as a means of checking that the recorder was operating, even if no audio signal is present.

Retrieval of information is assisted by the standard clock indicator or, alternatively, the automatic Timesearch accessory.

Timesearch is a digital date and time clock system which provides accurate time reference to all recordings, and can automatically locate recordings with reference to time. Timesearch is supplied either as a clock system for recording only or as a complete clock and automatic search facility for recording and play out. The unit uses a microprocessor both to keep time and make the necessary decisions in its search for recorded information. Its recorded time code contains a day number, hours, minutes and seconds and an identity code for the recorder. Real time or time read from the tape is displayed on a nine-digit fluorescent display. Time selection for locating a message, or setting in real time is made by a set of leverwheel selection switches. When deck selection and message time have been entered, operation of the search-button commands the microprocessor to find the correct portion of tape. After precise location, the relevant passage is automatically replayed.

Racal Autostore recording equipment

Racal Callstore

Racal International Communications Recorder

Racal Timesearch

Autostore is a multichannel cassette-loaded communications recorder suitable for smaller airfields, which provides a 24-hour log of communications within the voice band-width of 300 to 3400 Hz. Four versions are available, giving four or eight channels of recording capacity on one or two tape transport decks. Each deck operates for approximately 12 hours with, in the case of dual decks, automatic change-over of decks as the end of a tape is reached. Autostore machines can be supplied for ac only operation or for ac/low voltage dc supplies. The input circuits accept audio signals from line or radio systems and a voice-operated start mode is incorporated for use in low-density traffic applications. This feature substantially adds to the period of time between tape changes as the tape deck is then only operating when an input signal is present on one or more channels. Continuous operation mode provides approximately 12 hours per deck operation. A Timesearch time-indexing module can be incorporated within Autostore. In multiple Autostore

installations, a single Timesearch can provide sufficient signal output to operate up to ten machines.
Callstore is a cassette recorder/reproducer designed to provide immediate access to recently recorded information. Principally designed for use in the voice operation or contact start mode, Callstore is available with up to three audio channels plus a search control track. The option of three channels allows the operator to segregate incoming and outgoing signals (eg duplex radio system) or signals from different sources, such as telephone and radio, and at the same time record time signals for reference. Callstore can be mounted in a standard 19-inch rack capable of carrying up to four recorders. Each recorder is operated from a separate console position by its own remote control unit, up to 100 metres away from the tape transports. The major feature of Callstore is the instant recall facility which is achieved by use of a cue signal on the search control track.
Other features of the Callstore include the use of standard C60 cassettes, replayable on any commercial recorder (audio tracks only), and a five minute end-of-tape warning light.

Status
Production. In service in over 120 countries. The ICR 40 channel variant was added to the range in September 1981

Manufacturer
Racal Recorders Ltd, Southampton, Hampshire, England

Racal Messenger Hf/ssb Communications Equipment

The Racal Messenger hf ssb range operates in the 2–16 MHz frequency range to fulfil the mobile and fixed requirement for long distance communications. Solid-state techniques are used throughout and a common transceiver operates for both mobile and base stations. A range of ancillaries is available and these include whips, dipoles and long-wire antennas, audio ancillaries, antenna tuning units, and special vehicle mounting frames.
The VRM 4555A is an all solid-state 100 watt transceiver contained in one cabinet. All operating controls are on the front panel and a fist microphone and built-in loudspeaker provide the audio interfaces.
The VRM 4555A hf ssb transceiver and the MA 4557 A ac power supply together form a compact base station.

Status
Production

Manufacturer
Racal Messenger Ltd, Reading, Berkshire, England

Racal VRM 4555A hf ssb transceiver

Characteristics

MODEL	VRM 4555A	VRM 4555A/MA 4557A
TYPE	Hf ssb mobile	Hf ssb base station
FREQUENCY RANGE	2–16 MHz	2–16 MHz
NO OF CHANNELS		
single frequency simplex	1–6	1–6
two frequency simplex	1–5	1–5
POWER OUTPUT	100 W	100 W
POWER REQUIREMENTS	12 V dc	240 V ac, 50 Hz

Racal Messenger Uhf/vhf Communication Equipment

Racal Messenger vhf and uhf equipment operates in four frequency bands; 25 to 50, 66 to 88, 138 to 174 and 406 to 512 MHz.
The personal hand-held radios provide frequency modulation on a single or two frequency simplex basis with the option of a duplex capability on mobile and base stations.
A repeater/talk-through function is also available as an option in the mobile and base stations, providing greater flexibility within a radio network. Various tone signalling facilities and channel capacities are offered to meet individual requirements. Power outputs range from 2 watts in the personal radios to 110 watts in the mobile and base stations.
Two personal radios are available, the PRM 4510 with two frequency ranges 29·7 to 50 MHz and 66 to 88 MHz, and the PRM 4500 operating in two frequency bands 138 to 174 MHz (vhf) and 406 to 512 MHz (uhf).
The FRM 4560 base station also provides a choice of four frequency ranges, up to four channels and options for power outputs from 20 to 100 watts. The FRM 4561 variant is equipped with a repeater/talk through facility.

Status
Production

Manufacturer
Racal Messenger Ltd, Reading, Berkshire, England

Racal PRM 4500 personal radio

Characteristics

MODEL	PRM 4510	PRM 4500	VRM 530	FRM 4560
TYPE	Vhf fm personal hand-held	Vhf/uhf fm personal hand-held	Vhf/uhf fm mobile	Vhf/uhf fm base station
FREQUENCY RANGES	29·7–50 MHz 66–88 MHz	138–174 MHz 406–512 MHz	25–50 MHz 66–88 MHz 138–174 MHz 406–512 MHz	25–50 MHz 66–88 MHz 138–174 MHz 406–512 MHz
NO OF CHANNELS	1–8	1–10	1–8	1–4
POWER OUTPUT	2 or 5 W/5 W	2·5 or 6 W/2 or 5 W	20–110 W	20–100 W
POWER REQUIREMENTS	700 m Ah battery (500 m Ah option)	750 m Ah/ 1200 m Ah battery	± 12 V dc	120/240 V ac 50/60 Hz 12 V dc (option)

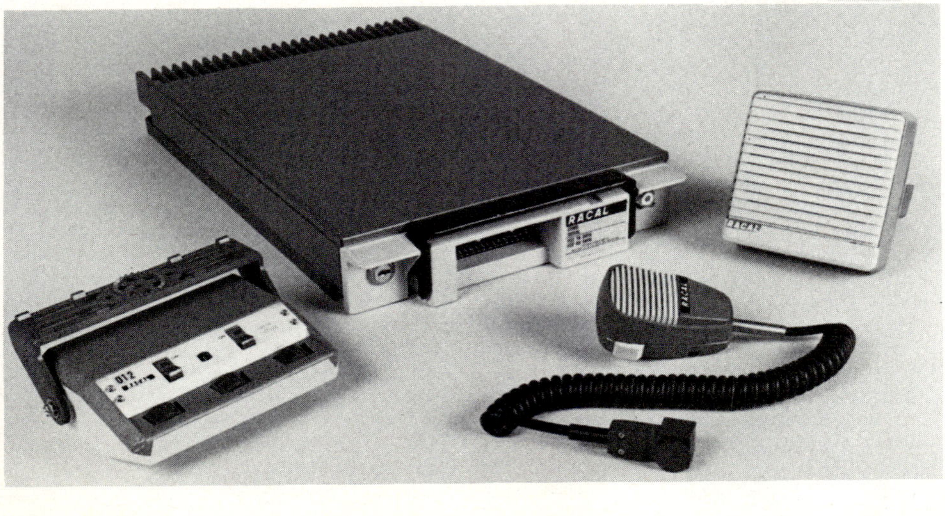

Racal VRM 530 mobile radio

Rediffusion Air Traffic Control Communications Systems

Rediffusion produces a comprehensive range of communications for atc operations with hf, vhf and uhf radio equipment.

A mobile self-contained atc tower has been developed by the company. The atc cabin is mounted on a Leyland Clydesdale vehicle chassis with scissor lift. The unit is mobile and tows its own power generator. The cabin is fully-equipped with a comprehensive range of hf, vhf and uhf radio equipment and a selection of meteorological instruments, according to the requirements of the operator. The cabin is rigidly constructed and is designed to meet the stringent atc regulations. It has full air-conditioning and the tinted windows are made from a special glass which has heated elements to control misting. Other features include an observation dome in the roof and an observation platform outside the cabin door which also supports the ladder giving access to the ground. The tower is raised to its maximum operating height of approximately 8 metres by means of a hydraulically-operated scissor lift which can be operated either from the vehicle cab or the atc cabin itself. Four support legs are also provided to enable the vehicle to be removed to leave the cabin free-standing if required.

The FIC system installed at Kotoka International Airport, Accra, Ghana has remote-control facilities and is computer-assisted featuring an engineering console with a programmable vdu. This enables an operator to pre-program the system operating requirements, with a visual record of all instructions, and allows the air traffic controllers to control the relevant remotely-located transmitters and receivers by push-button operations.

Three ground-to-air communications systems have been custom-built by Rediffusion for installation in the Far East. Each system contains 12 single-channel hf receivers and two 1 kW transmitters together with a range of ancillaries and makes extensive use of remote-control techniques. The control desks are equipped with selective calling facilities plus all necessary terminal equipment designed to give the operator control of the remote receivers and transmitters over ordinary land-lines. The systems are used to alert and provide flight information to overflying aircraft outside controlled air space.

Status

Production. The first mobile atc tower was delivered to a Middle East operator in early 1982. Equipment is being supplied to the flight information centre at Kotoka International Airport, Accra, Ghana and this is complemented by a number of aeronautical fixed telecommunications networks at the provincial airports of Kumasi, Sunyani and Tamale, Ghana

Manufacturer

Rediffusion Radio Systems Ltd, Wandsworth, London, England

Rediffusion mobile atc tower

Rediffusion atc communications system

UAS Matcon Mobile Air Traffic Control Tower

The Matcon mobile and trailer-mounted range of atc units has been designed by Universal Aviation Support (UAS) to provide atc on satellite airfields and remote landing strips, as well as for the relief of main atc towers on major airfields during periods of heavy congestion. These units offer a choice of interior configurations for either single or dual controller observers. Approved communications systems to customers' specific requirements and the choice of other comprehensive atc equipment can be supplied and fitted where necessary.

Both the mobile and trailer versions can be supplied for tropical conditions and are fitted with double-glazed tinted glass, insulated bodies, interior heating and air-conditioning. A choice of alternative chassis and cabs is available for the mobile units, such as the 8-ton Bedford or other suitable 4 by 4 vehicles when required to operate across country or in desert terrain.

Status

Production

Manufacturer

Universal Aviation Support Co Ltd, Tolworth, Surrey, England

UNITED STATES OF AMERICA

Aydin ACS-1000 Airport Communication System

The Airport Communication System, ACS-1000, has been developed to provide communications at military airfields. The system provides reliable telephone communications between all major facilities at an airport, push-button command of aircraft launch operations from the operations centre to hangars, and connection to a public address system from operations centre to all key facilities. Connection of this system with other airports is provided either by public telephone lines or by use of troposcatter or line-of-sight microwave radio links in frequency bands from 800 MHz to 8 GHz.

The system is flexible, and can be implemented in any of a variety of configurations and system capabilities. Two-way telephone communication is provided between all major facilities at the airport, using an automatic telephone exchange. Interconnection for short runs is by direct-burial multi-pair telephone cable, and for long runs, or over adverse terrain by microwave radio link employing voice channel multiplex equipment.

Connection to most subscribers is by a two-way ring of telephone cables (or radio link and telephone cable), terminating in distribution boxes at each position. This permits isolation of damaged lengths of cable, and tie-in of any subscriber in either of the two directions around the ring in the event of a cable break. Connections to subscribers of lower priority are made by buried cable branching off from the ring at the distribution boxes.

The automatic telephone exchange is equipped for, typically, 60 subscribers and ten external trunk lines, with expansion capability in the order of 180 subscribers and 30 trunk lines. Conference facilities are supplied for up to 20 subscribers, and an emergency conference feature is provided for the operations centre enabling it to call all of the 20 subscribers at once with a single pushbutton, in the event of an emergency.

The telephone system is provided with a back-up battery bank permitting a minimum of eight hours operation in the event of failure of the prime power mains or rectifier.

Conventional telephone instruments are used in

protected places; but, for buildings such as aircraft hangars, rugged outdoor-type instruments are used, equipped with external loud signalling horns and brightly lit call indicators. The radio link consists of pulse code modulation (pcm) multiplex equipment and microwave transceiver configured for, typically, 18 channels, with expansion capability of 120 channels. This can be backed up by a stand-by link of buried cable or by stand-by equipment, with automatic fault detection and switch over in the event of failure of the primary link.

The public address system provides one-way communication from the operations centre to all key locations within the airport. Provisions are included for control of this system from an alternative place in the event of transfer of control because of damage or other reason. In addition to voice messages, the public address system is used to transmit klaxon-type warning signals, using a burst oscillator provided at the operations centre. The public address system can be fitted for either of two optional types of control. The standard model transmits messages simultaneously to all locations. The selective calling model employs tone coding for selective calling of groups, such as groups of aircraft hangars. Both models use only a single pair of the multi-pair telephone cable (or a single channel of the multiplex/ radio link) between the operations centre and other locations. Aircraft hangars and similar noisy areas are equipped with high-power amplifiers and weatherproof trumpet speakers, capable of producing a sound pressure level of about 125 dB four feet on axis, at 30 watts. Other locations usually employ wall to ceiling mounted speakers, rated at 8 watts continuous and 15 watts peak power.

The Launch Light System provides push-button command of aircraft launch signals from the operations centre to aircraft hangars, with automatic acknowledgement of message received and command displayed. As with the public address system, control of this system may be taken over at an alternative site in the event of transfer of control because of damage or other reason. By use of time-position and tone-coded commands and acknowledgements, the system is capable of providing any of three commands to any of up to 15 discrete addresses, using one pair of the multi-pair telephone cable or one channel of the multiplex/ radio link. The control panel at the operations centre contains individual push-buttons for control of launch light indicators at each aircraft hangar, as well as indicator lamps showing the status of the launch lights at each hangar.

Each aircraft hangar contains a three-state launch light indicator, consisting of red, white, and green lamps, an audible alarm with timer, and a receiver which decodes commands from the control centre and returns status information. The audible alarm horn is turned on automatically for a preset interval whenever the launch light status is changed. The launch light system includes several features to ensure message reliability. The system employs automatic continuous updating of all commands and status messages approximately every 2·5 seconds. In addition, a message must be received three times consecutively, before the launch light circuitry will respond. The system contains built-in fault monitoring circuits, which provide automatic continuous detection of faults in cable, multiplex-radio channel or launch light equipment. In the event of an uncorrected cable break or link failure, the system ensures that the maximum number of aircraft hangars is always reached, since all command messages and return acknowledgements are sent in both directions around the ring, alternately every 2½ seconds.

Status

Production. The equipment is installed at 28 NATO airfields in Turkey

Manufacturer

Aydin Monitor Systems, Fort Washington, Pennsylvania, USA

Craig Transportable Air Traffic Control Tower

Craig produces a transportable atc tower and equipment pallet-mounted on levelling jacks. The shelter features include tinted and heated glass, acoustic ceiling tiles, lighting, cable and power entry panels, Elasto-Flex shock absorbing skids, access ladders, protective screens, and provisions for air-conditioning. This shelter was developed for use by the US Department of Defense.

Manufacturer

Craig Systems, Lawrence, Massachusetts, USA

Craig transportable atc tower

Dictaphone Communications Recording System

Dictaphone produces equipment suitable for recording of atc communications. The models available include the Veritrac, Veritrac SL and the Series 5600. Veritrac is a solid-state system which provides multichannel recording of messages to preserve, locate and replay the time and content of the request. A time/date encoding system with light emitting diodes automatically records the precise time and date of the response on the tape. Multiplex engineering eliminates the need for a dedicated time/date track. Several formats of time/date display are available. The recording system employs two magnetic tape decks, each with a maximum of 25 hours of recording capacity. All channel Safe-Scan is a monitoring function that instantly detects any recording failure and an auto transfer facility permits the automatic switching of recording to the second deck on a pre-set schedule or in the event of malfunction. Visual and audible alarms inform the operator of recording interruptions. The system also offers an optional remote-control capability that permits complete system control including automatic search of recorded material from a remote place. Veritrac is available in 4-, 8-, 10-, 20-, 30- or 40-channel modes.

The Veritrac SL system has a one-channel recording capacity of 192 hours, two channel capacity of 96 hours and four-channel capacity of 48 hours. It is designed for applications where recording capacity is more important than channel capacity. The unit features solid-state modular construction, a Uniband braking system and a fail-safe test system to monitor individual tape channels. A tape sensing system permits a change of function without stopping the machine. An optional second magnetic tape-deck is available which, in the case of malfunction, operates automatically to provide continuous recording. Veritrac SL can be equipped with an optional time code system which records the time of day simultaneously with audio recording, and eliminates the need for a dedicated time channel.

The Series 5600 is a modular system that can be ordered initially as a basic single recorder and then expanded, via preplanned options, into a dual transport system with auto transfer, time coding and fail-safe operation. The choice of recording configurations includes four channel/one pass, two channel/two pass, one channel/four pass. The minimum capacity of the single transport model is 24 hours per reel with a maximum of 96 hours on the four pass unit. Addition of an optional second transport extends the uninterrupted recording time to a maximum of 192 hours. This system, specifically designed for extended periods of totally unattended operation, can also be equipped with a time code generator/reader, fail-safe alarms and automatic transport transfer in the dual transport models. All Series 5600 modules are mounted in standard 19-inch racks.

Dictaphone Veritrac communications recording unit

Status
Production. Veritrac units were introduced in 1981

Manufacturer
Dictaphone Corporation, Rye, New York, USA

Dictaphone Series 5600 communications recording unit

E Systems 1020M Single-channel Vhf/uhf Voice Communications Equipment

1020M is a multichannel ground-to-air communication system developed from the 1020 single-channel system. The new equipment, like its predecessor, will be the standard fit for the Royal Air Force ground-to-air requirements, manufactured as the CGR 1020M by Cossor.

Two basic equipment cases, one each for the receiver and transmitter, provides a choice of the following configurations: vhf and uhf receive by module substitution in the receiver case; vhf or uhf 20-watt transmitter by the insertion of the appropriate module in the transmitter case; vhf and uhf simultaneous 20-watt transmitters by insertion of both 20-watt modules into the transmitter case; two uhf simultaneous 20-watt transmitters by insertion of both 20-watt modules into the transmitter case; and uhf 100-watt transmitter, achieved by using the uhf 20-watt transmitter module as a driver for the uhf 100-watt power amplifier module with both inserted in the basic transmitter case.

The equipment's control unit will operate digitally and will afford frequency/channel control selected by dial with up to 30 channels.

As a 20-watt vhf/uhf simultaneous operating transmitter the equipment is known as the 1021M and as a 100 watt uhf transmitter, as the 1022M.

Characteristics
FREQUENCY RANGE
(vhf) 117–136·975 MHz
(uhf) 225–399·975 MHz
CHANNEL SPACING: 25 kHz
SENSITIVITY: better than 10 dB (s + n)/n with +12 dB input μV, emf = 30% 1000 Hz
IMAGE REJECTION: better than 65 dB
IF REJECTION: better than 80 dB

Status
In production by Cossor for the Royal Air Force and E Systems for a number of other air forces

Manufacturer
E Systems, ECI Division, St Petersburg, Florida, USA

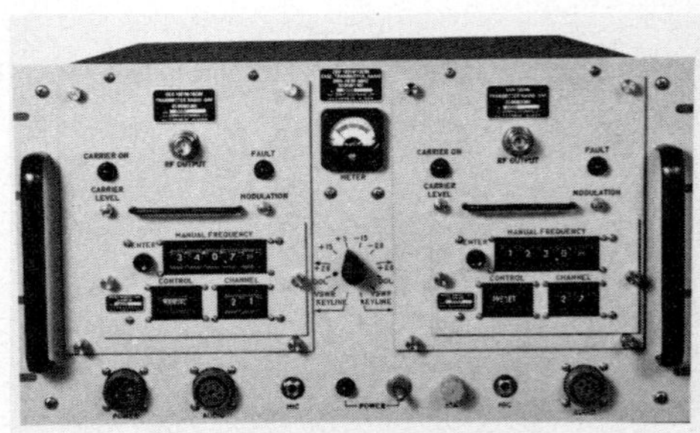

E Systems 1021M/1023M transmitter

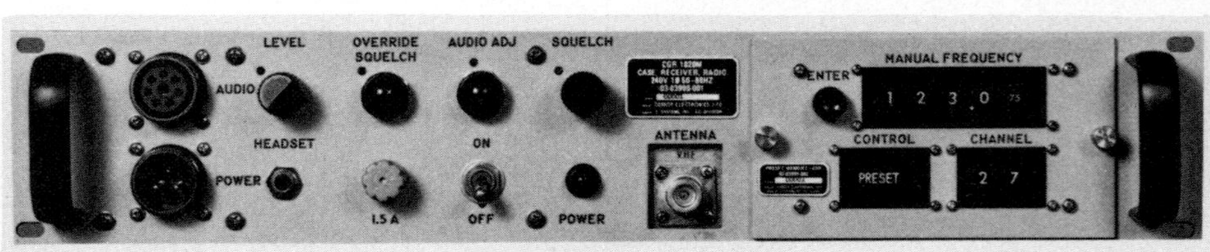

E Systems 1020M receiver

Genave Vhf Communications Equipment

Vhf communications equipment produced by General Aviation Electronics includes hand-held transceivers and units for portable, vehicle and ground station applications.

The two hand-held transceivers are the Alpha-Six and the AirCom.

The Alpha-Six provides six channels selected from the 720 channel coverage with an 18 MHz frequency spread, electronic tuning and diode antenna switching. The AirCom offers four channels with an output power of 3 watts pep.

Alpha 12 is a vhf transceiver which can be supplied in separate configurations for vehicle, ground station or portable application, with transmitter output power of 16 watts pep, 4-watts carrier and a 12 channel capability.

The Alpha 100 is a vhf communications transceiver designed for portable, vehicular or fixed application. It has a 100 channel capability with an output power of 8 watts pep, 2 to 3 watts carrier. Fully solid-state the Alpha 100 has a two knob frequency selector with digital readout for channel selection and a built-in microphone jack on the front panel.

Alpha/720 uses one crystal for the transmission and reception of all 720 channels in the communication network (118·00–135·975 MHz). A digital-phase locked loop synthesizer provides access to any of the 720 frequencies. The unit features 4 watts transmitter output power nominally and 16 watts peak power, with a transformerless series modulator, single conversion receiver, FET front end and mixer, an automatic squelch, and an active impulse noise limiter. The unit is fitted with a dual control frequency selector and a built-in microphone jack on the front panel. The Alpha/720 can be adapted for aircraft, vehicle, ground station or portable application.

Genave Alpha/720 vhf transceiver

Characteristics

MODEL	Alpha-Six	AirCom	Alpha 12	Alpha 100	Alpha/720
APPLICATION	Hand-held	Hand-held	Portable/vehicle/ ground station	Portable/vehicle/ ground station	Portable/vehicle/ ground station
FREQUENCY	118–136 MHz	118–135·975 MHz	118–135·975 MHz	118–127·9 MHz	118–135·975 MHz
NO OF CHANNELS	6	4	12	100	720
FREQUENCY SPREAD	18 MHz	10 MHz			
CHANNEL SEPARATION		25 kHz	25 kHz	100 kHz	25 kHz
OUTPUT POWER	2 W	3 W	4 W	2–3 W	4 W
POWER REQUIREMENTS	8 AA Ni-Cd	8 Ni-Cd	13·7 V dc	13·75 V dc/120 V ac	13·75 V dc
UNIT DIMENSIONS					
height	8 in (203 mm)	8·1 in (206 mm)	2½ in (63·5 mm)	2½ in (63·5 mm)	2½ in (63·5 mm)
width	2⅔ in (67 mm)	2·9 in (74·6 mm)	6½ in (165 mm)	6½ in (165 mm)	6½ in (165 mm)
depth	1½ in (32·5 mm)	1·8 in (46·5 mm)	10 in (254 mm)	9 in (229 mm)	10 in (254 mm)
WEIGHT	2¼ lb (1·02 kg)	1 lb (0·45 kg)	4 lb (1·8 kg)	5–6 lb (2·3–2·7 kg)	4 lb (1·8 kg)

Status
Production

Manufacturer
General Aviation Electronics Inc, Indianapolis, Indiana, USA

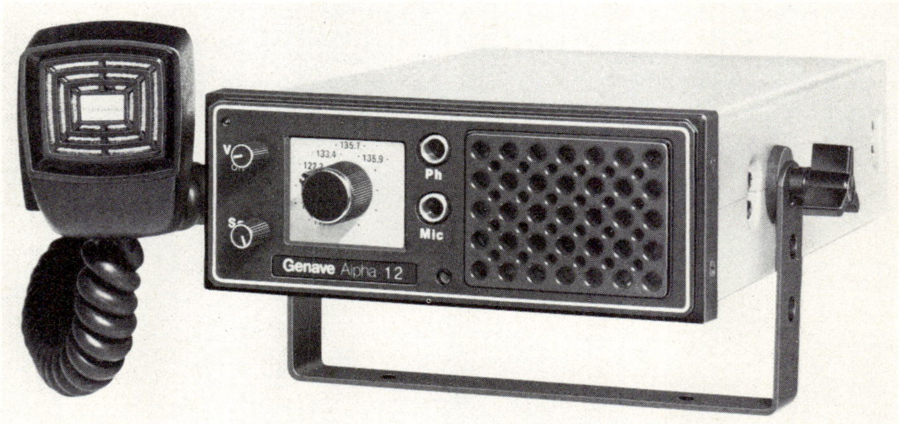

Genave Alpha 12 vhf transceiver

Genave Alpha-Six hand-held transceiver

Genave AirCom hand-held transceiver

Hamilton Standard AN/MRN-20 Air Traffic Control Set

AN/MRN-20 atc set is designed for the airfield guidance of aircraft during landings and departures, providing mobility and operation under various climatic conditions. It is also capable of being towed over roads and rough terrain on its own detachable trailer. A retractable astrodome is built into the top of the enclosure. The dome is raised when in operation to provide head space for the operator. When not in operation, the dome is retracted into the enclosure. The enclosure contains a radio control set, Hamilton Standard GS1331, capable of remotely-controlling transmitters and receivers for audio and continuous wave communications. This set has two transistorised power supply modules, four identical audio-recorder amplifier-squelch alarm modules and one microphone preamplifier-side tone oscillator module. Each of the set's independent channels can be operated singularly or in any combination up to four.

Hamilton Standard AN/MRN-20 atc set

Characteristics
DIMENSIONS
length: 6 ft (1·83 m)
width: 5 ft (1·52 m)
height: 5 ft (1·52 m)
WEIGHT: 3100 lb (1406 kg)

Status
Production. About 150 sets have been supplied to the United States Air Force Tactical Air Command

Manufacturer
Hamilton Standard Division, United Technologies, Windsor Locks, Connecticut, USA

Harris RF-727 Hf Transmitters

The RF-727 transmitter is a 5 kW rated transmitter designed for ground-to-air service covering 1·6 to 30 MHz. Suitable for unattended remote-controlled operation on up to 10 preset frequencies the unit switches channels in 0·2 seconds. Rated for continuous duty, key down, fsk operation at 5 kW, it is fully solid-state up to the output stage. Remote-control of channels, modes and functions is possible either over multi-wire cable up to 300 feet; or over a voice-grade telephone line using the RG-795 remote-control unit, for unlimited distances.

Characteristics

FREQUENCY RANGE: 1·6–30 MHz
NO OF CHANNELS: 10
MODES: A3J, A3A, A3H, A1, F1 (with external modem)
POWER OUTPUT: 5 kW
POWER REQUIREMENTS: 208 V, 50/60 Hz, 3-phase, 12 kVA
UNIT DIMENSIONS
height: 6 ft 6½ in (2 m)
width: 2 ft 4½ in (0·72 m)
depth: 2 ft 6½ in (0·78 m)
WEIGHT: 1050 lb (477 kg)

Status

Production. The Canadian government purchased 50 units for ground-to-air communications. Two units are installed at Bodo, Norway to provide commercial aircraft communications between Iceland, Greenland and Norway

Manufacturer

Harris Corporation, RF Communications Division, Rochester, New York, USA

Harris RF-727 hf transmitter

ITT 3650/3660/3670 Remote-tuned Receivers/Transmitters

Designed for fixed-station or transportable ground-to-air radio communications, these multi-channel units are the result of modifications to ITT's AN/GRR-23, AN/GRR-24, AN/GRT-21 and AN/GRT-22 single-channel radios. Remote-frequency control employs existing interconnecting audio circuits and requires no additional interface wiring. The remote tuned receivers/transmitters are also adaptable to radio link interface, and the remote-control unit can be used for a single system or expanded to control multi-systems from a central control point. Functions controllable remotely include power on-off keying, and modulation of transmitters, monitoring of receiver output, selection of high or low power output, modulation mode, preset channel selection and changing frequency of preset channels.

The extensive sharing of modules and parts with the basic single-channel radios, facilitates logistics and maintenance support of the multi-channel, remote-tuned systems when the operator uses both types.

ITT 3660 series transmitter

Typical characteristics retained in the multi-channel radios are the capability to operate without interference within the most dense atc environment, a specified reliability of 10 000 hours mean-time-between-failure, and a low mean-time-to-repair.

Manufacturer

ITT Aerospace/Optical Division, Fort Wayne, Indiana, USA

Characteristics

TYPE	Vhf/uhf receiver 3650	Vhf 3660 and uhf 3670 transmitters
FREQUENCY RANGE	(vhf) 100–156 MHz (uhf) 225–400 MHz	(vhf) 100–156 MHz (uhf) 225–400 MHz
FREQUENCY STABILITY	0·001%	±0·001%
SENSITIVITY	(am) 10 dB (s + n) at 2 μV, 30% mod (fm) 10 dB (s + n)/n at 2 μV, +5 kHz deviation	
RF POWER OUTPUT		20 W am carrier (optional 100 W linear power amplifier available)
MODULATION CAPABILITY		(am) 95% am, −35 to +10 dBm input (fm) +4 kHz deviation, −35 to +10 dBm input
AUDIO FREQUENCY RANGE	(voice) 0·3–3 kHz (wb) 0·3–25 kHz	(voice) 0·3–6 kHz (wb data) 0·3–25 kHz
AUDIO HUM AND NOISE	50 dB below 100 mW	
IMAGE REJECTION	70 dB minimum	
SPURIOUS RESPONSE	−70 dB minimum	
POWER SUPPLY	105/120, 210/240 V ac, ±10% at 47–420 Hz, 22–30 V dc, with automatic switchover to dc in ac failure	105/120, 210/240 V ac, ±10% at 47–420 Hz, 22–30 V ac, with automatic switchover in ac failure
TEMPERATURE RANGE	−10 to +55°C	−10 to +55°C
WEIGHT	22 lb (10 kg) max	43 lb (19·5 kg) max

ITT 3701 Vhf/uhf Portable Transceiver

Model 3701 vhf/uhf portable transceiver offers am coverage of the vhf and uhf ranges without need for changing any component.
The transmitter can be operated in the high power mode of 5 watts carrier power or switched to the low-power mode of 1 watt to extend battery life. Wide-band power amplifiers permit transmitter operation without manual adjustments for tuning or loading.
System design employing up-conversion in the receiver permits low interference reception without the need for variable tuned filters.
The special multi-band frequency synthesiser develops all signals required to operate the transceiver, providing 25 kHz tuning increments in both vhf and uhf by selection from front-panel switches. In addition, four preselected channels are set into the synthesiser for simple call-up from the front panel without the need to dial the exact frequency.
Optional accessories are available to provide operation from 12·5 to 15 volt dc supply or 115/230-volt ac supply.
The 3701 vhf/uhf portable transceiver provides frequency flexibility for dual band availability, made possible by advanced circuit designs using the latest in solid-state components and assembly techniques.

Characteristics

MODES: dsb-am, push-to-talk voice
FREQUENCY RANGE: 116–149·975 MHz and 225–399·975 MHz
STABILITY: 5×10^6
CHANNELS AVAILABLE: (uhf) 7000; (vhf) 1360
PRESELECTED CHANNELS: 4 total vhf and uhf
TUNING STEPS: 25 kHz vhf and uhf
TRANSMITTER OUTPUT: (high power) 5 W cw + 2 dB; (low power) 1 W cw + 2 dB
AUDIO RANGE: 300–2500 Hz
BATTERY LIFE (RECHARGEABLE): 16 h; 10% transmit in high power
TEMPERATURE RANGE: −10 to +50°C
UNIT
height: 5 in (127 mm)
width: 11 in (279 mm)
depth: 16 in (406 mm)
WEIGHT
16 lb 8 oz (7·5 kg) excluding battery
25 lb (11·5 kg) including battery

Manufacturer

ITT Aerospace/Optical Division, Fort Wayne, Indiana, USA

ITT AN/GRR-23 and AN/GRR-24 Vhf/uhf Receivers (Model 3101 and 3102)

In operation with the US Air Force with their associated transmitters (AN/GRT-21 and AN/GRT-22) for atc, these solid-state single-channel receivers offer a choice of 7000 uhf or 1360 vhf channels.

Characteristics

FREQUENCY RANGE: 225–399·975 MHz, 116–149·975 MHz
CHANNEL SPACING: 50 kHz (optional 25 kHz)
NO OF CHANNELS
(uhf) 7000
(vhf) 1360
FREQUENCY STABILITY: ± 0·0005% with synthesiser; 0·001% with crystal oscillator
SENSITIVITY: 10 dB (s + n)/n at 3 μV, 30% mod
CHANNELS
(uhf) 3500 (50 kHz), 7000 (25 kHz)
(vhf) 680 (50 kHz), 1360 (25 kHz)
STABILITY: ± 0·001%
AUDIO FREQUENCY RANGE: 0·3–3 kHz
AUDIO OUTPUT (2): 100 mW (adjustable) (600 Ω)
SELECTIVITY
(50 kHz spacing) ± 18 kHz (6 dB)
(25 kHz spacing) ± 10 kHz (6 dB)
MTBF: 10 000 h (demonstrated)
MTTR: 15 minutes
POWER SUPPLY: 105/210, 120/250 V ac, ± 10% (47–420 Hz), 22–30 V dc
TEMPERATURE RANGE:
(operating) −29 to +60°C
(storage) −62 to +71°C
HEIGHT: 3½ in (89 mm)
WIDTH: 19 in (483 mm)
DEPTH: 12 in (305 mm)
WEIGHT: 22 lb (10 kg)

Status

In service with US military departments and corresponding agencies in 40 other countries

Manufacturer

ITT Aerospace/Optical Division, Fort Wayne, Indiana, USA

ITT AN/GRT-21 and AN/GRT-22 Vhf/uhf Transmitters

In operation with the US Air Force with their associated receivers (AN/GRR-23 and AN/GRR-24) for atc these solid-state single channel transmitters offer a choice of 7000 uhf and 1360 vhf channels. The vhf exciter and power amplifier are designated models 3201 and 3211 respectively, and the uhf exciter and power amplifier models 3202 and 3212 respectively.

Characteristics

FREQUENCY RANGE: 225–399·975 MHz; 116–149·975 MHz
CHANNEL SPACING: 25 kHz
NO OF CHANNELS: (uhf) 7000; (vhf) 1360
FREQUENCY STABILITY: ± 0·0005% with synthesiser; 0·001% with crystal oscillator
RF POWER OUTPUT
(solid-state vhf exciter 3201)
10 W carrier minimum
(solid-state uhf exciter 3202)
10 W carrier minimum
(linear vhf amplifier 3211)
50 W carrier minimum
(linear uhf amplifier 3212)
50 W carrier minimum
MTBF: 10 000 h (demonstrated)
MTTR: 15 minutes
KEYING MODES: carbon and dynamic microphones, optional keying by current voltage or tone
POWER SUPPLY: 105/120, 210/240 V ac, ± 10% (47–420 Hz)
(exciter only) 24–30 V dc
TEMPERATURE RANGE
(operating) −29 to +60°C
(storage) −62 to +71°C
SOLID-STATE EXCITER
height: 5¼ in (13·3 cm)
width: 19 in (48·3 cm)
depth: 15⅝ in (42·1 cm)
weight: 45 lb (20·4 kg)
LINEAR AMPLIFIER
height: 7 in (17·7 cm)
width: 19 in (48·3 cm)
depth: 18⅜ in (46·6 cm)
weight: 75 lb (34 kg)

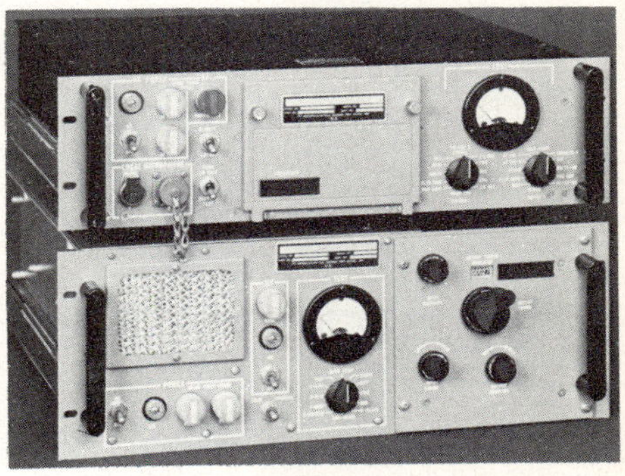

ITT AN/GRT-21/22 vhf/uhf transmitter

Status

In service with US military departments and corresponding agencies in 40 other countries

Manufacturer

ITT Aerospace/Optical Division, Fort Wayne, Indiana, USA

Litton Amecom 3080 Integrated Communications Switching System

The 3080 integrated communications switching system combines the decentralised control approach with time division multiplexing/pulse code modulation (tdm/pcm) techniques, plus digital technology, in production equipment. It can provide automated management and routeing of all voice communications in a facility. In a typical atc application, intercom connections as well as access to ground-air radios and to telephone land-lines are available in a simple and integrated communications interface with the operator.
Composed of independent modules at each operator position, the 3080 is a distributed system with microprocessing distributed among the modules of the radio, intercom and interphone subsystems. It provides a flexible, reliable, non-blocking communications management system for the intercoms, phones and radios necessary for atc operations.
It can also fulfil numerous communications roles within C^3 areas, providing integrated communications support for the command and control functions of military forces. The 3080 enables the rapid reconfiguration of communications access and resources allocations from normal operation to alert conditions, providing optimum communications management during escalating phases of conflict. Typical C^3 applications would include strategic and tactical communications systems, information systems, war headquarters, warning systems, and air command and control systems.
For integrated communications between sites, the 3080 can be combined with other series 3000 telecommunications units to form a global (long-haul) network, using a variety of transmission

media, to transmit voice, data, video and facsimile.

Status

Currently in production for the FAA, US Air Force and Singapore, and installed at Dallas/Fort Worth Airport, USA, and Riyadh, Jeddah and Dhahran Airports in Saudi Arabia

Manufacturer

Litton Amecom, College Park, Maryland, USA

Amecom 3080 integrated communications switching system

Magnasync Radar and Communications Recording Equipment

Automatic logging of voice communications between aircraft and atc operators has become a standard and essential aid to safe and effective control operations. Equipment is now available for recording data from communications and the radar display.

The Magnasync 2000 Series radar and beacon records the voice communications plus the screen display complete with beacon (transponder code) data, primary data and operational data. The equipment processes computer data so it can be recorded along with the voice communications on a standard magnetic tape. A single machine records up to 30 channels of visual data and/or voice communications on a standard ½- or 1-inch magnetic tape. A simultaneously-recorded time code signal allows the exact time of an incident to be displayed and an automatic search feature enables a recording to be located at any time.

The system is based on the Magnasync 2000 Series logging recorder. This machine continuously records and detects a low-level control tone, monitoring its own function and automatically switching to a standby transport if necessary. The equipment records 16 hours of activity on each reel of tape and automatically switches to the second transport, allowing 32 hours of unattended operation.

The 2000 Series equipment can be interfaced with most radar scope beacon systems. The equipment records the processed sif information from a video signal processor, stores it and plays back the data to an indicator processor for reproduction and display.

Other features of the equipment are selected multi-channel monitoring during recording or playback, visual monitoring of each channel via led indicators, and time signal multiplexing.

Characteristics

MODEL: TR 2000 Series
APPLICATION: radar and voice communications recording
NO OF CHANNELS: 10, 20 or 30
RECORDING TIME: 16 h per transport
REEL LENGTH: 4600 ft
PRIMARY POWER REQUIREMENT: 117 V ac, 60 Hz, optional: 220 V ac, 50 Hz
UNIT DIMENSIONS
height: 6 ft 3 in (1·91 m)
width: 1 ft 10 in (0·56 m)
depth: 1 ft 8 in (0·51 m)
WEIGHT: 295 lb (135 kg)

Status

The 2000 Series was introduced at the Airport 81 exhibition in Manchester, England October 1981

Manufacturer

Magnasync/Moviola Corporation, North Hollywood, California, USA

Magnasync TR 2000 Series radar display/ voice communications recorder

RCA AN/TSW-7A Transportable Air Traffic Control Tower

The AN/TSW-7A is a tactical mobile atc system designed to convert an unprepared clearing into an instant airfield.

The AN/TSW-7A can be operational within half an hour of arrival on site when operated from the back of standard military 2½-ton trucks. It can also be transported by cargo aircraft, helicopter, ship or rail. The tower accommodates three controllers and a supervisor. Its equipment complement includes ten transceivers in the hf, vhf (am and fm) and uhf bands, telephone system, meteorological instrumentation and signal light guns.

Status

In production for the US Army. Other configurations of the basic AN/TSW-7A system are currently operational with the US Air Force, Air National Guard and in several other countries

Manufacturer

RCA Automated Systems, Burlington, Massachusetts, USA

Ritron Portable Communications Equipment

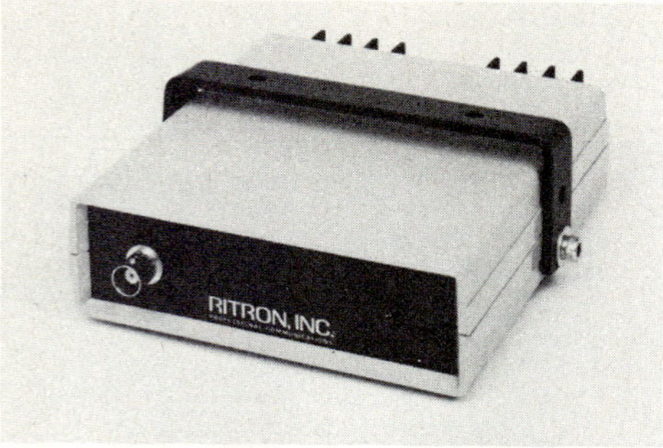

Ritron mobile amplifier charger

Ritron RT-155 portable vhf transceiver

Ritron produces portable transceivers and mobile amplifier chargers suitable for the airport environment ranging from the RT-50 operating in the vhf between 25–50 MHz to the RT-450 uhf set. The RT-125 operating in the vhf aircraft band with four channels is equipped with rechargeable nickel cadmium batteries. Optional equipment available to all transceivers includes sub-audible tone options, paging using appropriate decoders for converting the transceiver, touchpads to enable the unit to be used for selective calling or telephone interconnections and battery chargers.
The mobile amplifier chargers are available for vhf high band and uhf. These units convert the portable unit for use in a vehicle and incorporate an automatic battery charging facility.

Characteristics

TRANSCEIVERS

MODEL	RT-50	RT-125	RT-151/ RT-152/ RT-153	RT-154	RT-155	RT-450
TYPE	Vhf low band	Vhf aircraft band	Vhf high band	Vhf high band	Vhf high band	Uhf
FREQUENCY	25–50 MHz	118–136 MHz	144–174 MHz	144–174 MHz	144–174 MHz	450–512 MHz
POWER	4 W	3 W	1 W/2 W/ 3 W	4 W	4 W	1 W/2 W
CHANNELS	2	4	2	4	10	2

MOBILE AMPLIFIERS

MODEL	RA-150	RA-450
TYPE	Vhf high band	Uhf
FREQUENCY	144–174 MHz	450–512 MHz
POWER	25 W	15 W

Status
Production. In service in the USA

Manufacturer
Ritron Inc, Carmel, Indiana, USA

Rockwell-Collins AN/GRC-211 VHF-400 Vhf/am Transceiver

The VHF-400 high-performance vhf transceiver is designed for reliable atc communications at co-located vhf/uhf transmitter/receiver sites. The equipment is completely solid-state, providing am communications on any one of 1440 channels in the 116 to 151·975 MHz band with 25 watts am carrier output. The solid-state power amplifier is ruggedised for operation into high antenna vswrs under severe environmental conditions. The manufacturers claim that interference problems at co-located sites are eliminated by an integral rf filter and that selected module commonality with the AN/GRC-171 can reduce operational spares requirement substantially.

Status
In production for the US Air Force

Manufacturer
Rockwell-Collins, Cedar Rapids, Iowa, USA

Terra TPX 10/TPX 720 Portable Transceivers

Terra produces two portable transceivers suitable for airport operations.
The TPX 10 is a 2-watt unit with up to ten channels in vhf frequencies 118–136 MHz. Features of the equipment are single crystal operation, noise cancelling circuit to reduce ignition interference, jacks for standard aircraft headphones and microphone, and a high/low transmitter power switch.
The TPX 720 is a self-contained portable hand-held transceiver with a flexible antenna, rechargeable nickel cadmium batteries and ac battery charger. Any frequency in the aviation band can be selected by turning a miniature digital thumbwheel switch on the top panel. If desired, standard microphones or headsets can be plugged into the bottom of the case.

Characteristics

MODEL	TPX 10	TPX 720
FREQUENCY	118–136 MHz	118–136 MHz
PEAK POWER	2 W	8 W
CHANNELS	10	720
UNIT DIMENSIONS		
width	3 in (76 mm)	$3\frac{1}{4}$ in (83 mm)
depth	$1\frac{3}{4}$ in (44 mm)	2 in (48 mm)
height	10 in (254 mm)	$9\frac{1}{2}$ in (241 mm)
ANTENNA LENGTH	$9\frac{1}{4}$ in (235 mm)	$9\frac{1}{4}$ in (235 mm)
WEIGHT	2 lb (0·91 kg)	2·1 lb (0·95 kg)

Terra TPX 10 portable transceiver

Status
Production. In service in the USA

Manufacturer
Terra Corporation, Albuquerque, New Mexico, USA

Terra TPX 720 portable transceiver

Wulfsberg WCS-100/-200/-300 Vhf Communication Systems

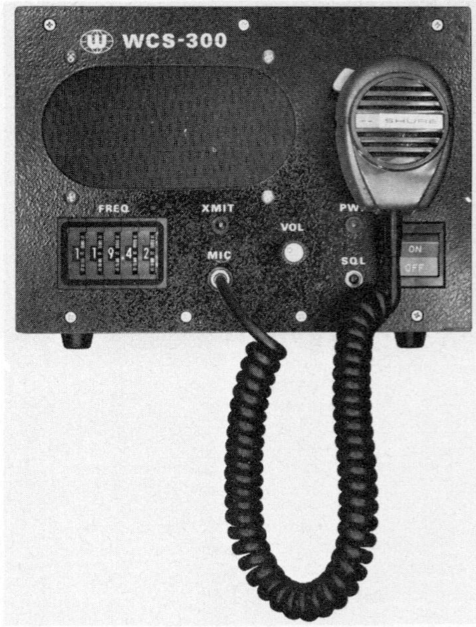

Wulfsberg WCS-300 vhf communication system

The WCS-100 remote and WCS-200 desk ground-to-air vhf communications systems provide a single channel of vhf am voice or data ground station communication service in the 118 to 136 MHz range. The remote system consists of a WR-100 receiver and a WT-100 transmitter mounted beside each other on a RM-100 19-inch relay rack. For the desk system, the receiver and transmitter are designed to stack vertically. The WR-100 and WT-100 are self contained and may also be used separately.
The WCS-300 desk ground-to-air vhf communication system provides 720 channel am voice ground station communication service in the 118 to 136 MHz range.
The WCS-300 is a modified version of the WT-200 synthesised transceiver. Any desired frequency in the 118–136 MHz band is available with a thumb-wheel frequency. The WCS-300 may be used for local or extended local installations. A transfer capability is provided so that a frequency selector may be used in extended local installations to control the transceiver's desired frequency.

Characteristics

MODEL	WCS-100	WCS-200	WCS-300
APPLICATION	Remote vhf	Desk vhf	Desk vhf
FREQUENCY RANGE	118–136 MHz	118–136 MHz	118–136 MHz
NO OF CHANNELS	1	1	720
CHANNEL SPACING			25 kHz
POWER OUTPUT	Adjustable 5–20 W	Adjustable 5–20 W	Adjustable 5–20 W
POWER REQUIREMENTS	115/230 V ac 50–400 Hz or 28 V dc	115/230 V ac 50–400 Hz or 28 V dc	115/230 V ac 50–400 Hz or 28 V dc
UNIT DIMENSIONS			
height	$3\frac{1}{2}$ in (88 mm)	7 in (178 mm)	6 in (152 mm)
width	19 in (483 mm)	8 in (203 mm)	$8\frac{1}{4}$ in (210 mm)
depth	$13\frac{1}{4}$ in (338 mm)	$13\frac{1}{4}$ in (338 mm)	$15\frac{1}{4}$ in (386 mm)
WEIGHT	17 lb (7·7 kg)	17 lb (7·7 kg)	15 lb (6·8 kg)

Status
Production

Manufacturer
Wulfsberg Electronics Inc, Overland Park, Kansas, USA

Wulfsberg WCS-100 vhf desk unit

METEOROLOGY

BRAZIL

IPD-PEA S-Band Meteorological Radar

This radar operates in the 2700 MHz frequency band and has preset ranges of 25, 50, 100, 200 and 400 km. A feature of this equipment is the large circular reflector which can be steered in azimuth and elevation on its tower-mounted pedestal. This presumably allows the radar to be employed for both radiosonde tracking or weather/precipitation surveillance purposes. Digital techniques are employed in signal processing with presentation as raw radar video. Other facilities include azimuth and range integration, iso-echo and iso-contour presentation with height density levels, colour television screen display and remote facsimile presentation. The normal display arrangements provide for sectorial covering, ppi, rhi displays, and the antenna can be adjusted in elevation from the operating console.

Status

The PEA-built prototype has been fully tested and approved, and transfer to private industry for commercial production is now in progress. Full scale production is anticipated in 1983. Companies involved in this include Tecnasa for system integration and electronics, Motoradio for the microwave elements, and Jamy for tower, pedestal and antenna. Thirty-two of these radars are due to be procured, initially as part of the Brazilian Ministry of Aeronautics' Plano de Desenvolvimento do Sistema de Proteção ao Vôo (Flight Protection System Development Plan)

Antenna assembly of the Brazilian S-band meteorological radar

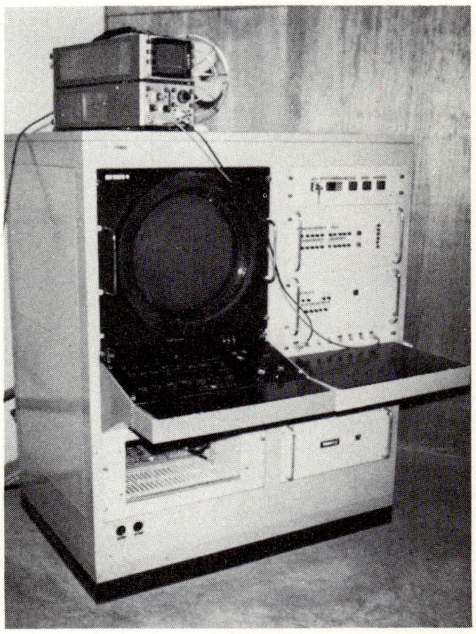

Brazilian meteorological radar console

Manufacturer

IPD/PEA - Divisãe de Electronica, São José dos Campos, Brazil

CANADA

Goodwood Weather Radar and Processing System

Goodwood has developed a weather radar and processing system for use in air traffic control. The display systems offer complete three-dimensional pictures of the weather as it exists and how it will appear in three hours time. The three-hour advance warning allows air traffic control and flight crews to plan aircraft movements in great detail. The point forecast map shows how a weather system will affect an area containing a number of airports and other critical locations. In particular, this map permits air traffic control to change runways and decide on diversion airports extremely rapidly. In the past, the company claims, weather has been treated as an area phenomena, whereas this concept treats weather as a combination of area, lines and points. The line forecast allows air traffic control to determine the weather in the approach path of an aircraft.

In a companion development program, called Raisat, Goodwood has developed a technique of integrating weather radar and weather satellite data to provide detailed weather information and warnings which would be available over a large area. The modules of coverage are 2000/1600 km with rain delineation to 4 or 8 km.

Status

This system has been supplied to the Royal Observatory in Hong Kong to supply the meteorological department and air traffic control with accurate forecasts of typhoon and severe weather movements. The system comprises a radar installed north of Kowloon, processing equipment at the Royal Observatory and display equipment at Kai Tak Airport

Manufacturer

Goodwood Data Systems, Carleton Place, Ontario, Canada

FINLAND

Vaisala MicroCORA Upper-air Observation System

The MicroCORA system provides data on upper-air pressure, humidity, temperature and winds and automatically edits the information for standard meteorological messages.

The data is displayed on the system console and printed on perforated paper-tape. It can be connected directly to a telecommunication network. Alternative output formats, windfinding systems and a selection of data analysis programs can be provided.

The system is designed for use with Vaisala radiosondes transmitting on the 403 MHz frequency band. Windfinding is based on the use of the Omega navigation network. Alternatively a Loran C windfinding module can be fitted.

Characteristics

ANTENNA: omni-directional
FREQUENCY: 403 MHz
RANGE: 100 km

Status

In production for several years, the system is a development of the CORA upper-air system which was introduced in 1973. Units are operating in 20 countries

Manufacturer

Vaisala Oy, Helsinki, Finland

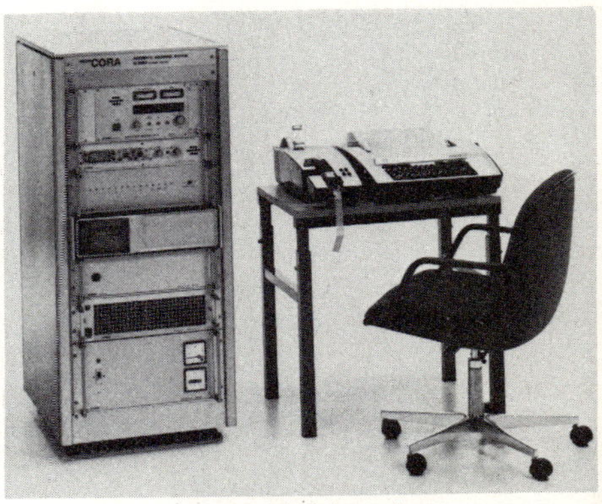

MicroCORA system

Vaisala MIDAS Weather Observation System

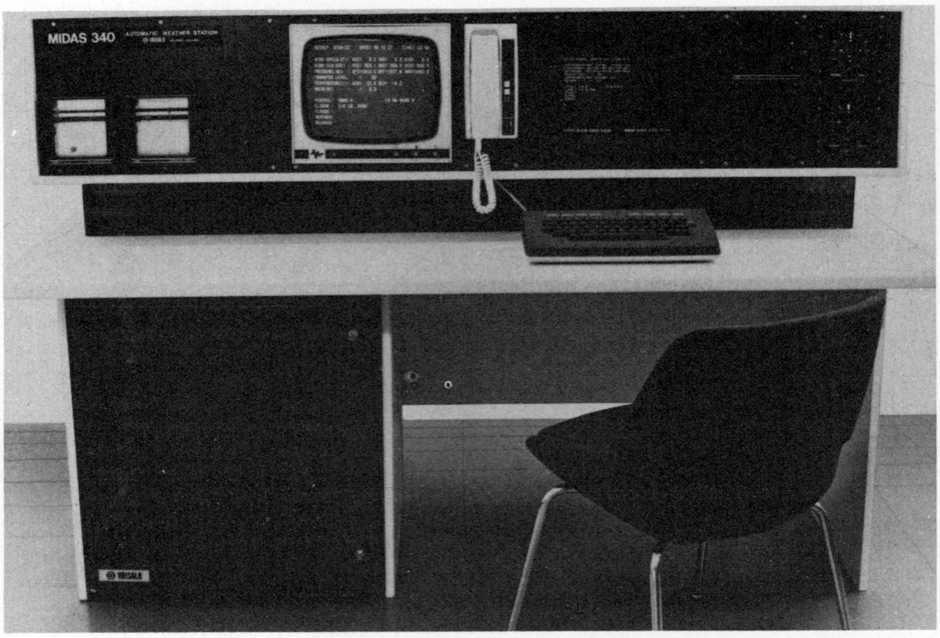
MIDAS operator's console

MIDAS is an automatic integrated system for surface weather observations, designed for medium and large airports. The system performs the meteorological data collection and processing from one or more runways in real time. A variety of sensors can be incorporated in the system to measure wind speed and direction, air and runway temperature, atmospheric pressure and humidity, runway visual range, cloudbase and also the MILOS system supplying data from remote stations. The central unit collects data from the sensors and generates standard messages at prefixed times for distribution via a closed-circuit television network. Separate displays are available for wind information. The system provides automatic monitoring of weather parameters and alerts the observer when significant changes occur. The MIDAS system can be used to prepare meteorological reports and send them to the Aeronautical Fixed Telecommunications Network (AFTN) with a minimum of manual work. The system can be expanded to receive and automatically distribute around the airport, meteorological reports and terminal area forecasts from the AFTN.

Characteristics

TYPE OF SENSOR	MEASURED PARAMETER	SAMPLES/ MINUTE	MEASURING RANGE	SYSTEM RESOLUTION	ACCURACY
Anemometer	Wind speed	15	0·8–150 knots 0·4–75 m/s	1 knot 0·1 m/s	± 0·4 m/s
Wind vane	Wind direction	15	0–360°	1°	± 3°
Pt 100 platinum probe with copper shield	Air temperature	1	−40 to +60°C	0·1°C	± 0·2°C
Pt 100	Runway temperature	1	−30 to +70°C	0·1°C	± 0·2°C
Aneroid capsule	Atmospheric pressure	1	940–1060 mb	0·1 mb	± 0·3 mb
Thin film hygrometer	Humidity	1	0–100%	1%	± 3%
Fixed base line	Runway visual range 2 units	4	50–1500 m	10 m	± 15 m
Laser	Cloud-base height	1	10–1500 m	10 m	± 10 m

OPERATING TEMPERATURES (SENSORS): −40 to +55°C
SIZE: basic system can be housed in standard 19-inch cabinet
POWER: 110 or 220 V ac, 50/60 Hz

Status
MIDAS systems have been in production since 1977. MIDAS 310 entered production in 1979 and 40 systems are in service in eight countries. A network of 20 stations based on the MIDAS 330, a synoptic automatic weather station is being delivered to the British Meteorological Office

Manufacturer
Vaisala Oy, Helsinki, Finland

Vaisala MILOS Automatic Weather Station

The MILOS station is designed for unmanned or small manned airfields. The station has sensors for nine basic meteorological parameters. These are temperature, wind speed and direction, relative humidity, atmospheric pressure, solar radiation, rainfall, precipitation and sunshine durations. The system computes averages and extremes and composes messages on-site. The information is displayed using a printer. The data can be transmitted automatically via a radio link or the public telephone network.

Status
In production since 1978, 50 systems have been delivered to 14 countries

Manufacturer
Vaisala Oy, Helsinki, Finland

Characteristics
SENSORS

TYPE OF SENSOR	MEASURED PARAMETER	DATA
Platinum probe with copper shield	Air or ground temperature	Instantaneous and extreme values −40 to +60°C
3-cup anemometer	Wind speed	Every 4 s taking 10 minutes running average
Wind vane	Wind direction	Every 4 s taking 10 minutes running average
Thin film hygrometer	Relative humidity	Every minute, instantaneous and extreme values
Aneroid barometer	Atmospheric pressure	Every minute, max, minimum and average pressures
Pyranometer	Solar radiation	Every minute, average radiation over given period
Tipping bucket pluviometer	Rainfall	Accumulations over given period
Precipitation detector	Precipitation	Every minute, records minutes precipitation over given period
Bi-metal strip	Sunshine	Every minute, records minutes sunshine over given period

SIZE (excluding sensors): 300 × 250 × 280 mm
WEIGHT: 7·5 kg
POWER: 500 mw excluding sensors 1 w with sensors connected

Vaisala WA 21 Wind Monitoring System

The WA 21 is an intelligent wind data-processing, distribution and display system designed for airports. The system features a two-wire connection between the measurement and display points for data transmission. Power to the sensors is supplied via the same line. The display unit provides indication of instant values, two and ten minute averages and maximum and minimum for wind speed and direction. It can be equipped to display wind data from several different sensor locations and a three-channel recorder can be incorporated in the system.

Characteristics

SENSORS
WAA 12 anemometer
WAV 12 wind vane

MEASURING RANGE
Speed: 0–150 knots
Direction: 0–360°

ACCURACY
Speed: ±1% 0–40 knots
±2% 40–150 knots
Direction: ±2·7°

POWER REQUIREMENTS
WAD 21 averaging display unit/WAR 13/3 recorder:
220 V ac, 50 Hz or 110 V ac, 60 Hz ±20%

OPERATING TEMPERATURE RANGE
Sensors: –40 to +55°C
WAD 21 +5 to +55°C

Status

Production started in 1980. Fifty systems are in operation in six countries

Manufacturer

Vaisala Oy, Helsinki, Finland

WA 21 system display unit

FRANCE

Bertin Sodar Wind Measurement System

The Sodar system uses Doppler-acoustic techniques to obtain data on wind speed and direction from the surface up to 1300 metres. It is operated at airports for low level wind shear detection.

The system which functions unattended, comprises three antennas, power amplifier, transceiver, micro-computer with a keyboard, and data display terminals. The antennas, mounted on trailers for mobility, operate in three different axes to obtain the three components of the wind direction. The data is processed in real-time and displayed using a crt, a printer or by the use of colour graphics. For each sampled layer, information is available on the speed and direction of the horizontal wind, the vertical velocity of the wind and its standard deviation, and the intensity of acoustic signals received and the standard deviation. The data can be stored on cassettes, floppy discs or nine-track magnetic tapes.

Status

The system entered production in mid-1980 and is in operation at several locations including Lyons Satolas and Nice Airports

Manufacturer

Bertin & Cie, Plaisir, France

Sodar system

CIT-ALCATEL Meteorological Equipment

CIT-ALCATEL produces ground stations for the reception of data from meteorological satellites and communications equipment for transmitting meteorological information to aircraft.

The Satellite ground stations offered can receive information from either polar orbiting satellites such as the TIROS N family and Meteor, or from the geostationary satellites for example METEOSTAT, which are placed around the equator. Every polar orbiting satellite encompasses the entire globe twice a day whereas the geostationary satellites cover specific areas on a continual basis. The ground stations operate with the STRI 2000 laser image recorder which uses a low-power laser beam illumination to reproduce the received signal. Dry silver photographic paper is developed by heating the paper after it has been exposed to the laser beam. The unit permits full-definition pictures of the polar orbiting satellites as well as high-definition pictures of the geostationary satellites in sector format.

The Declam system receives, processes and transmits audio meteorological messages to aircraft. The unit is automatic, operating under the control of recorded programmes. It can handle up to ten telephone lines and is capable of editing the VOLMET bulletin by analysing and selecting key meteorological data and broadcasting the information in vhf in both French and English.

Status

The satellite ground stations are operating with the French Meteorological Service, European Space Agency, French Navy, and in New Zealand. Declam units have been supplied to the French Civil Aviation Authorities

Manufacturer

CIT-ALCATEL, Paris Cedex 15, France

ELECMA Delta Automatic Weather Station

The Delta system is designed for automatic data collection at the point of measurement. Sensors are interrogated by the out-station and the data is transmitted in real-time to the central station. The system can handle up to eight sensors and this is extensible in modules of eight channels. Each module comprises three resistance inputs suitable for platinum probe sensors; two voltage inputs for pressure, humidity and radiation sensors; one rheostatic input for a wind vane; one counting input for an anemometer or a rain gauge; and a switching input for a rain gauge. The data can be transmitted by radio or using a telephone line.

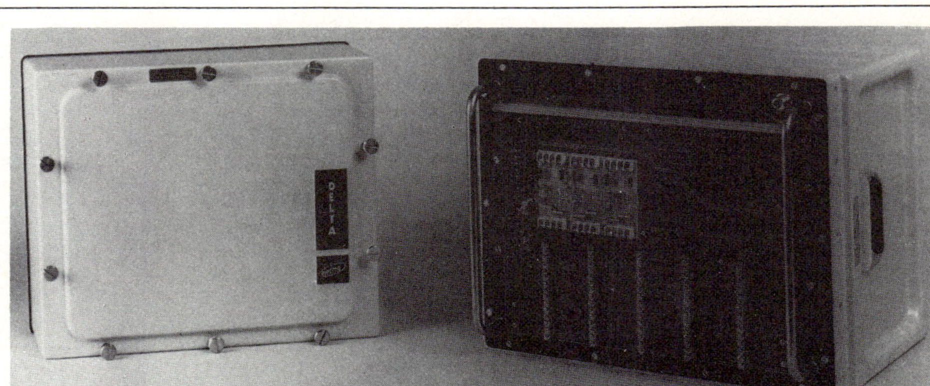

Delta automatic weather station

Alternatively, the information can be stored on a mini-cassette. The transmitter is powered by a 12 volts battery which can be recharged using solar panels.

Characteristics

SENSORS: 8 per module
POWER REQUIREMENTS: 12 V battery
POWER CONSUMPTION: 2 W
OPERATING TEMPERATURE RANGE
-20 to +60°C standard
-40 to +70°C optional

Status

Production was started in 1980 and over 60 stations are now in service

Manufacturer

ELECMA - Electronic Division of SNECMA, Suresnes, France

ELECMA Lynx 2 Transmissometer Luminance Sensor

Two sensors which, combined, are used to determine runway visual range are produced by ELECMA. These are the Lynx 2 transmissometer, and the luminance sensor.

The transmissometer measures the mean atmospheric transmissivity in a horizontal plane along a given path. It consists of a modulated light transmitter covering the visible spectrum and a light-sensitive receiver that supplies a direct current proportional to luminous flux. The distance between the transmitter and the receiver is the measurement base.

The transmitter consists of a halogen lamp and an optical unit. The optical unit ensures the projection of a homogeneous beam and the transfer to an optical fibre, of a flux proportional to the one transmitted to the receiver. The receiver comprises two cells; one receives a part of the flux reduced by the atmospheric transmissivity and the other receives the flux transmitted through the optical fibre.

The luminance sensor is an optical and electronic device which provides a continuous measurement of the atmospheric background luminance relative to four thresholds; night, twilight or dawn, daylight and bright sunlight.

Characteristics

TRANSMISSOMETER
Length of measurement base: 30 or 50 m
Accuracy: 0·5%
Operating temperature range: -30 to +50°C
LUMINANCE SENSOR
Measurement range: 0-12 000 cd/m²
Accuracy: ±10%
Operating temperature range: -30 to +50°C

Status

In production. The visibility measuring equipment is installed at airports in over 20 countries

Manufacturer

ELECMA - Electronic Division of SNECMA, Suresnes, France

ELECMA Lynx 3 Weather Observation System

Lynx is a modular integrated system for surface weather observations. The system performs acquisition of meteorological data processing and dissemination. A large number of sensors can be incorporated in the system which can include measurements of wind speed and direction, air and runway temperature, atmospheric pressure, humidity, runway visual range, cloud base, precipitation quantity and sunshine duration. The operator in the central station interrogates the out-stations through a telephone line or radio using a crt console, keyboard or a teleprinter. The information received from the out-stations can be displayed using plasma indicators or crt screens. Data is stored in the system by means of graphic recorders, cassette recorders, alphanumeric printers or by the use of punched tape with a teleprinter.

The Lynx 3 system can be used to prepare meteorological reports and can be adapted for the dissemination of meteorological and aviation data, meteorological radar images and the acquisition, monitoring and validation of meteorological messages concerning diversion airports.

Characteristics

SENSORS:

TYPE OF SENSOR	MEASURED PARAMETER	MEASURING RANGE	ACCURACY	PROCESSING
Wind vane with 18-sector rheostatic track	Wind direction	0-360°	±10°	Calculation of sliding mean value over 2 or 10 minutes
3-cup anemometer	Wind speed	0·5-60 m/s (1-117 kt)	±5%	Calculation of sliding mean value over 2 or 10 minutes Calculation of wind gust speed during last 10 minutes
100-Ω platinum probe	Air or soil temperature	-40 to +60°C	±0·3°C	Linearisation of measurement
Hair-bundle hygrometer	Relative humidity	10 to 100% (-40 to +60°C)	±2%	Scaling of measurement
LiCl probe and 100-Ω platinum resistance	Absolute humidity	-30 to +60°C	±0·5°C	Conversion of probe temperature to dew-point temperature
Barometric capsule	Absolute atmospheric pressure	150 mb (setting between 650 and 1100 mb abs)	±0·5 mb	Scaling of measurement calculation of qnh and qfe values
Laser-diode telemeter	Cloud-base height	10-1500 m or 15-1500 m	±5 m or ±7·5 m	Time range conversion
Transmissometer in visible spectrum and background luminancemeter	Runway visual range	75-1500 m or 150-2000 m 30-30 000 cd/m²	±0·5% ±10%	Determination of visibility by contrast determination of visibility of light sources and determination of best of these 2
Tipping bucket pluviometer	Precipitation		±3% at 4·5 mm per minute rainfall	Counting over given period: 1-6 or 12 h conversion into mm
Heliometer (direct solar radiation)	Sunshine radiation	A day		Comparison with reference threshold counting
Pyranometer (sun and sky radiation)	Solar radiation	Hourly and/or daily value expressed in J/cm²	± 1%	Integration of instantaneous value of global radiation over 1 h and/or day

OPERATING TEMPERATURES (SENSORS)
-30 to +60°C
SIZE
out-stations: 650 × 450 × 290 mm
central station: can be housed in standard 19-inch cabinet
POWER: 220 V ac, 50 Hz

Status
The system is installed and operating at airports in 26 countries

Manufacturer
ELECMA - Electronic Division of SNECMA, Suresnes, France

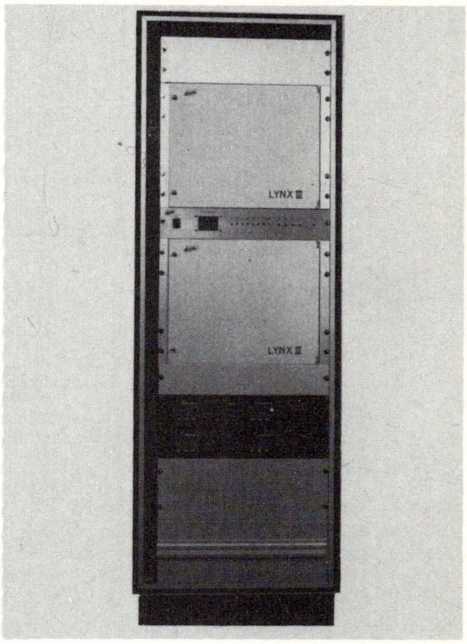

Lynx 3 Central Processing Unit

Enertec CZ 9140 Barometric Pressure Transducer

The CZ 9140 is an absolute transducer, designed for measurement of atmospheric pressure in a fixed meteorological station or mobile station mounted on a vehicle. It incorporates a piezo-resistive absolute pressure transducer with a power supply, amplifier and linearizer temperature corrections. The sensing element is a monocrystalline silicon diaphragm, incorporating four strain-gauges diffused into the material. The transducer is compatible with the Simoun, Caviar and Eole automatic meteorological stations.

Characteristics
RANGE: 150 mb
SELECTABLE: Between 650 and 1100 mb in 50 mb steps
TEMPERATURE FACTOR: 0 to +20°C; 0·5 mb
POWER SUPPLY: 12 V dc <20 mA
DIMENSIONS: 50 mm length, 80 mm diameter
WEIGHT: 0·5 kg

Status
Production. Officially approved by French Meteorological Authority

Manuafacturer
Enertec Schlumberger, Enertec Division, Velizy-Villacoublay, France

Enertec Eole Remote Weather Station

The Eole is designed to operate as an autonomous recording unit or as a remote runway station connected to a Simoun or Caviar central station. It can accommodate a maximum of 15 parameters, providing continuous information on those sensors with data processed on-site, and at intervals of 2, 12 and 60 minutes for the other parameters. Information can be obtained on temperature, wind speed and direction, relative humidity, atmospheric pressure, rainfall, sunshine duration, radiations: global, diffuse, direct, net, water level, visibility, cloud base height and dew point temperature. The last three parameters require an independent power source. Power can be provided using a mains supply, windmill generator, solar panels, or 12 volt dc batteries.

Characteristics
PARAMETERS: Up to 15
POWER SUPPLY: 220 V 50 Hz, 117 V 60 Hz, 12 V dc
OPERATING TEMPERATURE RANGE: −35 to +60°C

Status
In service in France, Jersey, Channel Islands and Egypt

Manufacturer
Enertec Schlumberger, Enertec Division, Velizy-Villacoublay, France

Enertec Eole remote weather station

Enertec EV 1000 Visibility Meter

The EV 1000 visibility meter determines horizontal visibility range in a fog sample of known, constant volume. It measures the luminous flux diffused by the water droplets forming the fog, at an angle of 20 degree with the line of incident light. It also measures the luminous flux transmitted, calculating the ratio of transmitted flux to diffused flux. This ratio is proportional to visibility and corresponds to the definition of meteorological visibility.
A parallel beam of visible light is produced by a flash lamp, delivering 6 μs light pulses. The receiver is a photodiode. Two mirrors, one of which is fixed and the other mobile, are used to measure transmitted and diffused flux. The mobile mirror is set at 40 degrees to the first mirror for transmission, and at 10 degrees for diffusion. In the transmission position, the receiver measures part of the luminous flux transmitted, after reflection by both mirrors. In the diffusion position, it measures the luminous flux diffused by the volume of fog, after reflection by the mobile mirror. A number of successive transmission measurements are made, followed by a similar number of diffusion measurements. The micro-processor calculates the ratio transmission measurement/diffusion measurement, deducing visibility range from this ratio.
The measurement range of the unit is 10 to 10 000 metres. Visibility values are converted to analogue current form (10 m = 0 mA, 10 000 m = 5 mA). Data is available at a repetition rate of 1 minute for values below 5000 metres, or 3 minutes over 5000 metres. In addition when visibility drops below a predetermined threshold an alarm signal is given. A fault indication used to detect an abnormal decrease in transmission power, or excessive dirt build-up on the meter windows, is also available. The EV 1000 visibility meter is compatible with the Simoun and Eole automatic meteorological stations.

Characteristics
MEASUREMENT RANGE: 10–1000 m
ACCURACY
 at 100 m visibility (v): 0·3%
 at 1000 m visibility (v): 3·3%
SAMPLE RATE
 1 measurement/minute for V < 5000 m
 1 measurement/minute for V > 5000 m
TRANSMITTER: Xenon lamp
 spectral range: 0·32–1·1 μm
RECEIVER: Pin 10 D photodiode
 sensitivity: 2400 μA/lumen
POWER SUPPLY: 220 V ±15%, 50 Hz ±2%, 88 VA
DIMENSIONS: 1·11 × 0·27 × 11·36 m
WEIGHT: 3 kg (including stand)
RECORDER: Europont × 2100
 accuracy: ±0·5%
 chart: 100 mm scale width, 20 mm/h speed
 response: 0·6 s
 duration: 30 days

Manufacturer
Enertec Schlumberger, Enertec Division, Velizy-Villacoublay, France

Enertec EV 1000 visibility meter

Enertec Mistral Automatic Meteorological Station

The Mistral is designed to function with a range of meteorological sensors accommodating a maximum of 23 parameters with either continuous or cyclical inputs. The unit includes a processor which, for each acquired measurement provides the true significant value as well as the instantaneous, mean or maximum value, depending on the parameter. An internal clock, with back-up power supply provides dating and the calculation of true solar time.
Depending on operating conditions, the user can choose immediate print-out of the data, recording on a cassette recorder or data transmission to a central station or computer. The printer can be used in all configurations. Runway visual range is obtained from a transmissometer and luminance meter in the system, calculated using Allard's and Koschmieder's laws.
The selections available are with the transmissometer base at 30 or 50 metres and a maximum runway light intensity of 63 500 candelas adjustable in steps of 500 candelas. A 5 per cent eye contrast threshold value is used as recommended by ICAO.
The unit is housed in a weather-proof grp casing.

Status
Production. Approved by the French Meteorological Department

Manufacturer
Enertec Schlumberger, Enertec Division, Velizy-Villacoublay, France

Enertec Mistral automatic meteorological station

Characteristics

SENSOR	Measured parameter	Max quantity	Range	Resolution
ANEMOMETER	Wind speed	1	0–60 m/s	1/10 m/s
WIND VANE	Wind direction	1	360°	1°
BAROMETER	Atmospheric pressure	1	900–1050 mb	1/10 mb
PT RESISTOR	Air temperature	1	−40 to +70°C	1/10°C
HYGROMETER	Relative humidity	1	20–100%	1%
PLUVIOMETER	Precipitation	1		0·2 mm
CEILOMETER	Cloud base height	2	0–500 m	10 m
TRANSMISSOMETER	Atmospheric transmissivity	3	0–100%	1%
LUMINANCE METER	Luminance	1	0–1200 cd/cm^2	
	Runway light intensity			

Enertec Simoun Automatic Meteorological Station

The Simoun automatic station was designed by Enertec in collaboration with the French National Meteorological Authority. The station can handle the following parameters: wind speed and direction, dry bulb temperature, wet bulb temperature, relative humidity, atmospheric pressure, dew point temperature, radiation: direct, diffused, global or total, cloudbase altitude, meteorological visibility, atmospheric transmissivity, general luminosity of the air, rainfall and runway light intensivity.
The basic element of the station is a microprocessor. The station comprises two main parts: a functional unit and a power supply rack. The functional unit contains the acquisition, dating and processing cards for parametric data. The number of cards varies according to the number of transducers connected to the station. A maximum of 21 cards can be included in the unit. The power supply rack delivers the voltages required for operation of the station. These voltages are obtained from 110/220 volts, 50 or 60 Hz line supply, an uninterruptible line source, or a 24 volt dc supply. The unit can be connected to a printer, crt display, magnetic tape cassette recorder, chart recorder or to public telephone or telegraph lines. The Simoun station when used for the measurement and calculation of runway visual range is designated Caviar. The Caviar station is equipped with a display, printout and recording devices. One station can handle between two and ten runway visual range values.

Status
In service with the French Meteorological Service and Egypt, Jersey, Channel Islands, Mexico, Venezuela, Yugoslavia and Zaire

Manufacturer
Enertec Schlumberger, Enertec Division, Velizy-Villacoublay, France

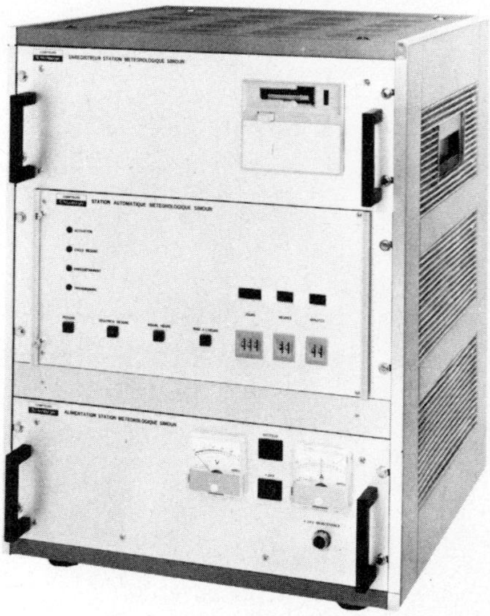

Enertec Simoun automatic meteorological station

Enertec TNL 1500 Cloud Ceilometer

The TNL 1500 cloud ceilometer provides continuous measurement of cloud base altitude up to 1500 metres, and conforms to ICAO recommendations.

The unit uses a low-power laser diode with an emission in the infra-red spectrum of a short pulse duration. The ceilometer measures the time taken by a pulse to cover the path from the ground to the cloud base and back. Operation is controlled by a clock which regulates two separate signals: laser-diode control pulses and a scan signal. The echo obtained by the receiver is analysed using the scan signal to provide echo validation. The time interval between the start of the analysis and the echo signal is proportional to the measured height. Analysis duration is between 2 and 33 seconds. The digital data stored in the register, is available at the ceilometer output after conversion into analogue form.

All sub-assemblies of the TNL 1500 are fitted to a chassis in a single fibreglass housing. There are two main sub-assemblies: the optical unit comprising the laser-diode, the receiver diode and their associated lenses, and the modular logic unit, performing the measurement timing, the measurements themselves and the associated processing. The unit is equipped with internal heaters which maintain the temperature above the external ambient temperature, vent holes to prevent the formation of internal condensation, and for use in tropical zones, solar shutters can be fitted to protect the laser and receiving diodes.

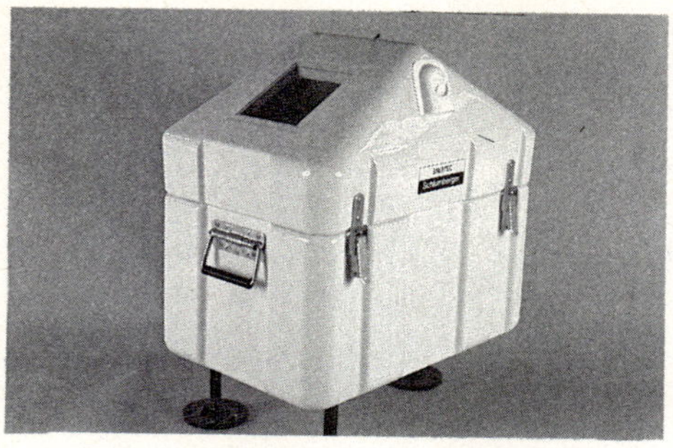

Enertec TNL 1500 cloud ceilometer

The circuitry of the TNL 1500 is shielded from airport radar and radio-navigation equipment, and the equipment features a narrow-band laser emission permitting the use of interferential filters; high amplification, low noise avalanche diode reception and sector analysis. The output signal of the TNL 1500 is processed by a remote device, either: an Enertec automatic station type Simoun or Mistral, the recorder unit ET 1503/0 or the recorder and display unit ET 1504/0.

The ET 1504/0 unit incorporates a seven-segment led digital display for indicating the cloud base height and a strip chart recorder. When cloud base height is above 1500 metres the chart recorder shows a value equivalent to 1590 metres. The digital display provides similar indication of this situation.

The ET 1503/0 unit performs graphic recordings of cloud base height.

Characteristics

TNL 1500
RANGE: 15–1530 m
ACCURACY
up to 5 m: −15 to +0 m
above 510 m: −30 to +0 m
POWER SUPPLY: 220 V ±15%, 50 Hz ±2%, 60 VA
DIMENSIONS: 1 × 0·9 × 0·71 m
OPERATING TEMPERATURE RANGE: −50 to +50°C

ET 1504/0 recorder and display/ET 1503 recorder
RECORDER: 100 mm width scale, 0·5% accuracy
CHART SPEED: 20 mm/h
duration: 1 month
POWER SUPPLY: 220 V ±10%, 50 Hz ±2%,
24 VA (1504)
20 VA (1503)
DIMENSIONS: 491 × 203 × 510 mm
WEIGHT
15·5 kg (1504)
15·1 kg (1503)
OPERATING TEMPERATURE RANGE: 0–50°C

Status

The TNA 1500 has been supplied to the French meteorological service, and to airports in Algeria, Belgium, Cameroon, Iran, Ivory Coast, Libya, Luxembourg, Morocco, Democratic People's Republic of Korea, Sri Lanka, Turkey and Zaire

Manufacturer

Enertec Schlumberger, Enertec Division, Velizy-Villacoublay, France

---

LCT Ronsard Weather Radar

Ronsard is a weather radar system for studying atmospheric convection developed in clouds, storms and frontal precipitations. It consists of two identical pulsed Doppler radar sets which measure the reflectivity of atmospheric particles and their radial velocities. A true speed map is produced, off-line, by processing the measurements made simultaneously by both radars. Each set comprises a trailer-borne antenna and a transportable shelter which contains the signal processing equipment and a computer. The system has three operating modes. In the first, both radars operate simultaneously on two different frequencies. In the second, the bi-static mode, one radar transmits and receives while the other one receives the signals scattered by the clouds. In the third, the radiometry mode, both radars operate as receivers.

Characteristics

ANTENNA: 4 m diameter, trailer-borne
FREQUENCY: 5·5 GHz
PEAK POWER: 250 kW
NOISE FACTOR: 1·2 dB
RANGE: 200 km
MAX DISTANCE BETWEEN RADARS: 50 km
WEIGHT
antenna unit and trailer: 4500 kg
shelter: 3500 kg
OPERATING TEMPERATURE RANGE: −20 to +50°C

Manufacturer

Laboratoire Central de Telecommunications (LCT), Velizy-Villacoublay, France

---

LCT Zephyr Windfinding Radar

Zephyr is a lightweight, self-contained mobile radar designed for windfinding, which tracks automatically, a balloon-borne passive reflector. The radar is equipped with a sighting telescope, to enable the operator to acquire the reflector visually, as the balloon is released near the radar. When the target is centred in the collimator, the operator switches to the automatic tracking mode. The equipment then provides constant indications in digital form of the polar co-ordinates (azimuth, elevation and slant range) of the balloon. This data is processed, by an on-line computer, to produce real- (or delayed-) time calculation of the wind vector parameters. Changes to the computer program can be effected from the keyboard.

The Zephyr system is available in two versions, Zephyr A which has a maximum range of 40 km and Zephyr B which has a maximum range of 100 km.

Characteristics

RANGE
Zephyr A: 0·1–40 km
Zephyr B: 0·1–100 km
ACCURACY
angular 0·1° at 40 km
range 10 m at 40 km
ANGLE TRACKING LIMITS
continuous in azimuth
−5 to +85° elevation
WEIGHT: 95 kg
OPERATING TEMPERATURE RANGE
−15 to +55°C

Manufacturer

Laboratoire Central de Telecommunications (LCT), Velizy-Villacoublay, France

---

Omera Meteorological Radars

Omera-Segid produces a range of weather radars designed to measure wind speed and detect, measure and survey precipitation. The models are ORV 31 for wind speed measurement, the ORP 310, 330 and 340 series and the ORP 105.

The ORV-31 radar is designed to measure wind speed at different levels of the atmosphere by automatically tracking a balloon-borne radar reflector. It is produced in both fixed installation and mobile versions. Either manual or automatic computation of windspeed is possible, the latter being by means of a Type 316 computer. The 80 kW transmitter and a 2 metre diameter parabolic reflector give a minimum tracking range against a 60 cm tetrahedron reflector target of 150 metres, and maxima of 180 km or 150 km with 4 mm per hour rainfall. Manual or automatic acquisition of the target are possible; in the latter case either a helical scan or a high-speed limited-scan sectorial search pattern can be used depending upon the operational circumstances. An optional recorder provides a continuous print-out at 30 second intervals of the measured and computed parameters.

The ORP 310, 330 and 340 are I-band weather radars used for detecting and locating meteorological phenomena, the intensity of which can also be assessed. Detection takes the form of a 'paint' on the ppi which indicates the strength of the echo. The echo is located in range by regularly-spaced range markers on the ppi and in bearing by an alidade graduated in degrees. The Type 310 is designed to detect precipitation within a radius of 200 km. The maximum transmitted peak power of 35 kW for the Types 330 and 340 provides detection ranges of up to 400 km. The Type 340 also incorporates a 15-level iso-echo system and the rhi function.

The ORP 105 (DLM 10E) is a large E-band radar used for fixed installations. The operating frequency is between 2700 and 2900 MHz and the transmitter peak power is 800 kW. Two pulse widths and two prfs are provided, 0·5 and 2 microseconds and 1000 and 250 Hz, respectively. A dish antenna of 4·12 metres diameter is used and the operating range is up to 400 km. Sector or continuous scan in azimuth at variable speed, 0 to 10 rpm are provided and in conjunction with various optional elevation scan patterns a number of automatic and manual surveillance programmes are available. A ppi/rhi display is employed and a full range of signal processing and presentation facilities is provided.

Characteristics

MODEL	ORV-31	ORP-310, 330, 340	ORP-105 (DLM 10E)
FREQUENCY	9376 MHz	9375 MHz	2700–2900 MHz
PEAK POWER	80 kW	8–35 kW	800 kW
PULSE LENGTH(S)	0·4 and 1 μ/s	2–4 μ/s	0·5 and 2 μ/s
PRF	800 Hz	300–500 Hz	250/1000 Hz
RANGE	150 km with rainfall 4 mm/h, 180 km clear weather	200–400 km	400 km

Status
The ORV-31 is in service with the French Meterological Service and in Zaire. The ORP-310, 330 and 340 series are also used by the French Meterological Service and have been supplied to Air Alpes, Royal Thai Air Force and the Tunisian Meteorological Institute

Manufacturer
Omera-Segid, Argenteuil Cedex, France

Systelec ST 400/A Meteorological Station

The ST 400/A meteorological station processes data from four sensors and displays the information on crt and line printers. The sensors which can be sited several kilometres from the centre, are connected to the processing unit by a pair of telephone cables. They comprise a temperature probe, absolute pressure transducer, anemometer and a wind vane to produce data on temperature, barometric pressure, wind speed and direction. The rack-mounted processing system consists of three units. The first receives and compiles the sensor data. The second contains the transmission unit for interrogating the sensors and the third incorporates the power supply unit.

Characteristics

SENSORS	Measured parameters	Range	Accuracy
PROBE	Temperature	–40 to +60°C	±0·5°C
TRANSDUCER	Pressure	QFE or QNH in mB	
ANEMOMETER	Wind speed	0 to 99 knots	±1 knot
VANE	Wind direction	0 to 360°	

Manufacturer
Systelec, Montrouge, France

Thomson MT/MR 750 Meteorological Data Display System

The MT/MR 750 is a digital system for weather radar data extraction. The system displays a map of areas of precipitation differentiating up to 14 activity levels by a range of colours. Cloud contours can be stored and up to 14 recordings superimposed on the scene in order to follow the development of the phenomenon. The system provides an analysis of cloud-formation structure by means of vertical and horizontal sections. It can be used for labelling the image and transmitting in real-time a detailed situation using conventional telephone lines. Pictures can be stored on a video-recorder or in hard-copy form.

Manufacturer
Thomson CSF, Division DRS-TVT, Bagneux, France

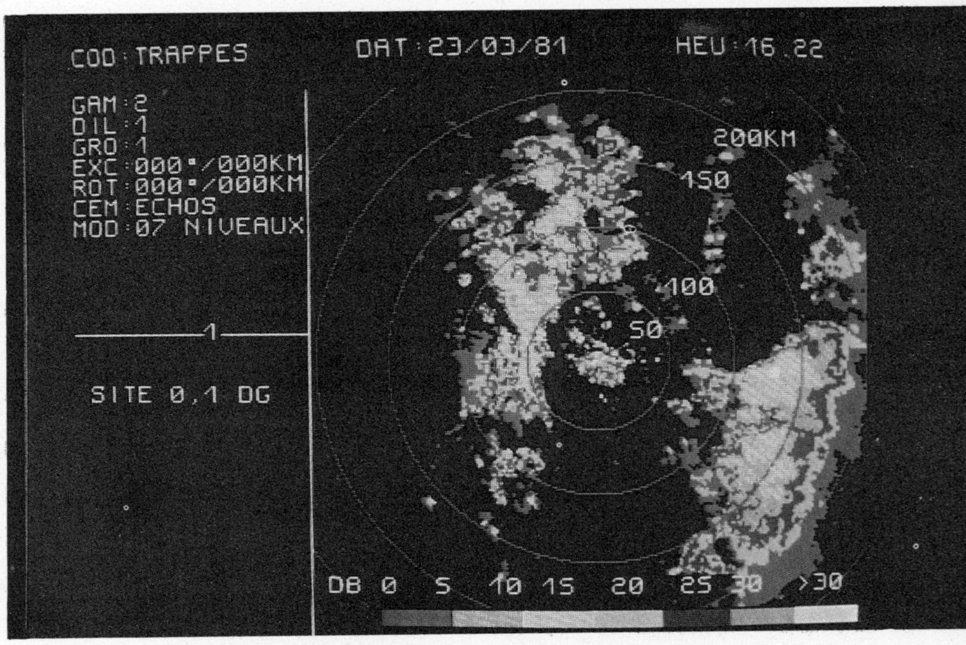

Meteorological picture from MT 750 display system

Thomson TRS 2730 Weather Radar

The TRS 2730 is a C-band long-range surveillance radar for the detection, analysis and tracking of cloud formation. The radar will operate to range of 400 km and enable manual tracking of meteorological balloons. Automatic tracking is an optional feature. The equipment comprises a parabolic antenna, a servo-cabinet, a transmitter/receiver cabinet and switchable display (ppi/rhi) on a colour crt display. The antenna 3·05 metres in diameter is mounted on a turret which provides mobility in both elevation and azimuth.

Characteristics
FREQUENCY: Tunable 5·6–5·8 GHz
PEAK POWER: 250 kW
PULSE DURATION: 2 μs
REPETITION FREQUENCY: 330 Hz
ANTENNA
beam width: 1·25°
gain: 40 dB
polarisation: horizontal
scan: elevation: –2 to +90° at 5 and 2·5°/s
azimuth: 360° at 5 and 2·5°/s

Status
First set installed in 1976 and the radar now in production

Manufacturer
Thomson CSF, Division DRS-TVT, Bagneux, France

Thomson TXM 1110 Airport Weather Station

The TXM 1110 is a weather data-processing system which co-ordinates, files and presents information collected from a variety of sensors located on the airport. These include transmissometers, radiance sensors, cloud-measuring equipment, wind speed and direction indicators, barometers, hygrometers and thermometers.

The processing system, controlled by a microprocessor, indicates any operating faults in the sensors while calculating and distributing the data required for airport operations.

The information displayed is: the runway visual range according to data received from the transmissometers, background radiance, the approach lighting system characteristics and test and lighting thresholds; the wind parameters with average speed, direction and peak speed determined over a programmed time; instantaneous direction and speed; the cloud ceiling with ambient temperature and runway temperature, dew-point, qfe and qnh.

The parameters are recorded and displayed as a block diagram presented to the operator on a crt display or through a network of alphanumeric consoles.

Manufacturer
Thomson CSF, Division DRS-TVT, Bagneux, France

GERMANY, FEDERAL REPUBLIC

Assmann MET 100 VOLMET Weather Broadcast

The Assmann MET 100 system offers a continuous distribution of the actual airport weather and landing conditions of the destination airport as well as those of the alternative and auxiliary airports.
In accordance with ICAO recommendations an internationally agreed selection of meteorological data is transmitted through ultra-short-wave transmitter stations by the national air-traffic authorities.
The MET 100 system provides the capability of transmitting up to 18 different messages in a fixed or variable sequence. Partial announcements may be altered at any time without disturbing the set sequence.
Features of the equipment are: magnetic disc recording, remote-control and recording capability, automatic monitoring and indication of all operational conditions, and a system which permits access to any equipment at any time. The unit provides continuous operation and frequent repetition of messages, with brief time intervals between individual announcements, independent of message length.

Status
Production

Manufacturer
Assmann GmbH, Bad Homburg, Federal Republic of Germany

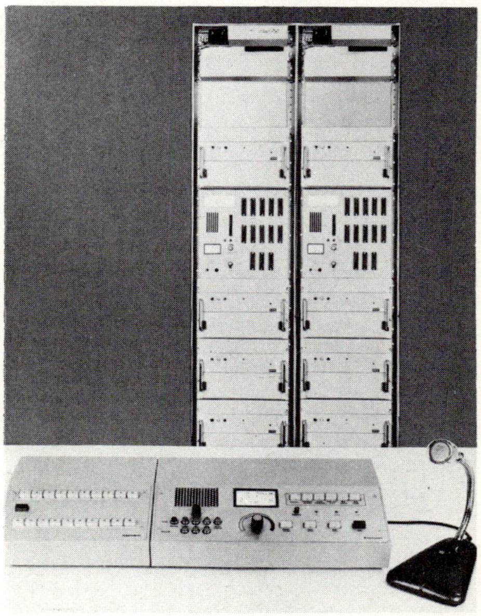

Assmann MET 100 VOLMET weather broadcast

Gematronic Meteor 300 Weather and Windfinding Radar

The Meteor 300 system is designed for operation as an analysis radar, and for windfinding. It can be operated in X-, C-, or S-band.
In the 300A configuration, as an analysis radar, the equipment is used for detection and evaluation of precipitation areas in the atmosphere. These can be measured to give position, height and intensity enabling direction and speed to be calculated up to a distance of 400 km. Employing a video digital converter, a process control computer can provide presentation of quantised precipitation intensities in six colours. A modem enables data and pictures to be transmitted via telephone lines.
In the 300-watt configuration the equipment is used for windfinding, operating in X-band with conical scanning. In the balloon tracking mode the system provides constant indications of elevation, azimuth and slant range. The data is processed by an on-line computer to produce a profile of the upper winds giving horizontal wind, direction and speed.
The Meteor 300 is available as a fixed installation. Alternatively, it can be mounted on a vehicle or trailer, or installed in a transportable shelter.

Characteristics
ANALYSIS X-BAND RADAR ANTENNA
parabolic, 2 m diameter
FREQUENCY: 9375 MHz ±30 MHz
PEAK POWER: 180 kW
PULSE WIDTH: 0·5 and 3 μs
PRF: 1200 and 200 Hz
NOISE FIGURE: ⩽ 10 dB
RANGE: 200 km
POWER REQUIREMENTS: 380/220 V, 50 Hz, 6 kVA

Manufacturer
Gematronic GmbH, Neuss, Federal Republic of Germany

Impulsphysik Meteorological Sensors

Impulsphysik produce a range of sensors for the observation of weather and visibility conditions. These include the Skopograph, Skopolog and Videograph units for the measurement of visibility and the Ceilograph II and Laser Ceilograph for cloud-height evaluation. The Skopograph transmissometer provides data on visibility and when coupled to the Skopolog processor, information on the runway visual range is obtained. The processor which can be connected to any transmissometer using a Digilog computer with microprocessors. Digital display of the runways visual range with data recording can be presented using conventional recorders and printers.
The Videograph is a visibility meter which operates on the backscatter principle with the projector and receiver in the same housing. The unit is equipped with a relay for controlling fog warning devices.
The Ceilograph II is a pulsed light, cloud-height meter for day or night operation with continuous recording, analogue and digital display, and alarm-tripping device.
A further development is the Laser Ceilograph which has a semi-conductor laser diode transmitter. The unit is capable of providing cloud-base information up to a height of 1500 metres.
The sensors can be connected to an automatic weather station, a digital data processor with a central micro-processor, containing plug-in converters for the majority of meteorological sensors. This equipment can supply serial data output for remote transmission.
Sensor information can be coupled to a Weather Data Monitor, which is a micro-processor-based unit for data reception, processing and transmission. Output of the meteorological data is in ordered groups on video monitors. An alphanumeric keyboard is provided for manual text input.

Status
Over 200 Skopographs, ceilographs and videographs are in operation in the Federal Republic of Germany. On European airports and waterways, 210 Skopographs, 60 Skopolog runway visual range systems, 190 ceilographs and 100 videographs are in service with approximately the same number spread throughout 70 other countries. A further 550 videographs have been built under licence in Canada and the USA

Manufacturer
Impulsphysik GmbH, Hamburg, Federal Republic of Germany

Rosenhagen Doppler Sodar DS 100 Measuring System

The Doppler Sodar DS 100 is a remote measuring system for obtaining profiles of the wind speed and direction up to 500 metres above ground level as well as producing data on turbulence and temperature differentials. It can be used at airports to detect wind shears and wake turbulence.
The DS 100 works on the Dopper principle, determining wind speed from the time delay and frequency shift of sound pulses. The equipment comprises up to three antennas each mounted on a trailer. The antenna consists of a bell-shaped sound screen made of grp with an array of seven horn-shaped pressure-chamber loudspeakers which also serve as receivers. Using an array of loudspeakers of equal phases produces a higher transmitting capacity and shorter reverberation time thereby increasing the height range over which the equipment can operate.
The electronic element includes transmitter/receiver switches with power amplifiers and a low-noise pre-amplifier; switched-capacitor filters for the spectral analysis of the receiver signal and a mini-computer for process control and signal analysis. A special signal analysing procedure has been developed to produce wind data at a low signal-to-noise ratio enabling the equipment to be used in an area with a high level of environmental noise.

Characteristics

TRANSMITTER
frequency: 1675 Hz
pulse length: 50, 100, 200 m/s
POWER REQUIREMENT: approx 50 W
RANGE
wind speed: 0–30 m/s
height: 30–500 m
PULSE REPETITION FREQUENCY
selectable in seconds
0–255 s
TIME AVERAGE
selectable in seconds
0–9999 s
HEIGHT AVERAGE
selectable 5–255 m
DISPLAYED HEIGHT INTERVALS
selectable 1–24
PROFILE DATA
wind speed, direction, N-S component, E-W component, radial components

Manufacturer

Dr Justus Rosenhagen GmbH, Norderstedt, Federal Republic of Germany

---

Wimera Tracking Radar

The Wimera tracking radar is a development of the MBVR 120 mobile balloon and drone tracking radar currently in use with the Federal German Armed Forces. Wimera's wide operational scope necessitated modular construction and high mobility. Two versions have been designed, one for balloon tracking and the other for drone tracking.

The balloon tracking version provides all the meteorological data necessary to compile complete weather reports. In this configuration, Wimera has three components: radar, radio-sonde system and computer.

All modules of each version are housed in a compact shelter which enables operation in any terrain independent of the weather. The system requires only one operator. After a change of position the Wimera can be fully operational again in 15 minutes with two people to set up and level the radar units.

Target acquisition incorporates pre-selection facilities for: optical target assignment; spiral scanning; sector scanning and manual target assignment.

After the target has been acquired the Wimera system switches to automatic target tracking. The radar data are indicated in digital form and printed out. Together with the data received from the radio-sonde, eg (temperature, air pressure, humidity) a complete weather report can be transmitted by wire or radio.

Characteristics

FREQUENCY
fixed: 9375 MHz
adjustable: 8500–9500 MHz
PEAK POWER: 70 kW
PULSE LENGTH: 0·05 μs/1 μs selectable
PRF: 1000 Hz
IF: 30 MHz
ANTENNA DIAMETER: 1220 mm; (parabolic optional) 2100 mm
ANTENNA GAIN: 38·5 dB; 44 dB
BEAMWIDTH: 1·8° or 1·1°
POLARISATION: vertical
RANGE: ⩾60 km (standard reflector 92 × 92 cm); optional ⩾120 km
ACQUISITION RANGE: ⩾40 m minimum
RESOLUTION: 150 m
SCAN FREQUENCY: 30 Hz

Status

In production

Manufacturer

Elektro Spezial, Philips GmbH, Bremen, Federal Republic of Germany

---

SWITZERLAND

Tecnavia SKYCEIVER Weather Satellite Data Processor

The SKYCEIVER is a modular system designed to produce a variety of configurations for meteorological use. The SKYCEIVER VI series are designed to receive data from the global system of equidistant geostationary satellites around the equator: GOES A & B (USA), METEOSAT (Europe), GOMS (USSR) and GMS (Japan). The system comprises a parabolic antenna, S-band pre-amplifier, receiver, video-processor and a Kudelski Nagrafax picture recorder. The station can be left unattended. A micro-processor permits reception of selected pictures automatically, in a visible infra-red, or side-by-side format. Different gamma curves for visible and infra-red pictures can be programmed.

For the reception of polar orbiting satellites such as the TIROS-N family and METEOR, the SKYCEIVER VII series are suitable with VHF- and S-band antennas. The image processing is carried out by a modified Harris Laserfax 850 image recorder with a low-power laser beam illumination to reproduce the received signals. Dry silver photographic paper is developed by heating the paper after it has been exposed to a laser beam, resulting in a high definition picture produced approximately 30 seconds after receiving the data.

Status

SKYCEIVER is in service in 14 countries in four continents and the Antarctic. The equipment has also been procured by the United Nations' World Meteorological Organisation

Manufacturer

Tecnavia SA, Agno, Switzerland

UNITED KINGDOM

Marconi IVR Meteorological Weather Station

The Marconi Instrumented Visual Range (IVR) system, extensible to accept and process data from other meteorological sensors, and designated IVR-Met, can provide a complete airfield surface meteorological system.

The system is based around the MET-1 transmissometer which measures and displays runway visual range in all three categories of flying conditions. MET-1 calculates visibility by the measurement of atmospheric transmission over a baseline, both ends of which are mounted on a common boom. The base-line measures 2 metres by 2 or 3 metres, the difference depending on the application.

The IVR-Met system consists of between one and eight field sites installed alongside the airfield runway(s). Each field site comprises up to eight sensors feeding into a common data-transfer unit. Parameters which may be monitored are: atmospheric transmission, background luminance, wind speed and direction, air temperature and humidity, atmospheric pressure and cloud base. The data is transmitted from the field sites(s) over speech-grade telephone circuits to the central processing unit. All incoming data is read and

Characteristics

SENSORS	TYPE	DYNAMIC RANGE	TYPICAL ACCURACY
Wind speed and direction	Anemometer and windvane	0·25 to 50 ms^{-1}	±2% of reading
Air temperature	Thermometer*	−50 to +50°C	±0·4°C
Humidity	Psychrometer*	0–100%	±2·5%
Air pressure	Barometer*	900–1100 mb	±1·2 mb
Cloud base	ASEA infra-red laser recorder	10–1500 m	±5 m ±0·5%
Visual range	MET-1 Transmissometer	50–1500 m indication out to 6 km	between ±5 and ±50 m at 1200 m

*housed in a Stevenson screen

checked by the processor before computing the information and providing the final data. Averaging periods and updating intervals for the data packages can be changed as required by the operator via the keyboard.

A variety of displays are available to present the data. Display panels which are solid-state incorporating a five- by seven-dot matrix display are suitable for air traffic controllers in radar and visual control rooms. The visual display unit provides a tabular display in television-format which is suitable for presenting the overall meteorological status to meteorological and airline users. Optional keyboard facilities can be used to call for more detailed data.

Recording the information can be done by a DEC writer, which produces hard copy in a tabulated format; a teletype with reader punch; a DEC cassette for program loading and data recording, or pen recorders in various colours.

Status

MET-1 has been procured by the Royal Aircraft Establishment and the Meteorological Office at Bracknell in the United Kingdom and the FAA in Washington in the USA

Manufacturer

Marconi Radar Systems Ltd, Leicester, England

Munro Type D1/119/617 Series Digital Wind Measurement System

The digital wind measurement system provides a display of instantaneous wind readings at two second intervals with push-button selection for two and ten minute averaged displays. Maximum and minimum gust values are indicated, collected over ten minute periods and all readouts are updated every two seconds.

The data is supplied by a Type IM 146 combined wind speed and direction sensor which comprises a three-cup anemometer and a wind direction vane designed to the British Meteorological Office specification.

For two-wire data transmission over dedicated lines, a Type 1000 S single card Transmit modem and a Master station receiver modem are included in the system. The transmit modem is interfaced to the wind sensors by a micro-processor based signal-conditioning unit.

Characteristics

SENSORS
cup Anemometer Type IM 124 Mk II
wind direction vane Type IM 144
dimensions (combined sensors):
height: 828·67 mm
vane length: 796·92 mm
vane balance arm length: 355·6 mm
weight: 17 kg
power requirements:
50 V/50 Hz or 60 V/60 Hz

ELECTRONIC UNIT
Wind speed: ±1 knot from 5 to 40 knots
±2 knots from 41 to 90 knots
Wind direction: ±2° resolution 1°

DIMENSIONS
width: 330 mm
height: 133 mm
depth: 260 mm

POWER REQUIREMENTS: 115/240 V, 50/60 Hz
OPERATING TEMPERATURE RANGE: +10 to +40°C

Status

This development was launched by Munro at the ICAA exhibition in Manchester, England in October 1981

Manufacturer

R W Munro, London, England

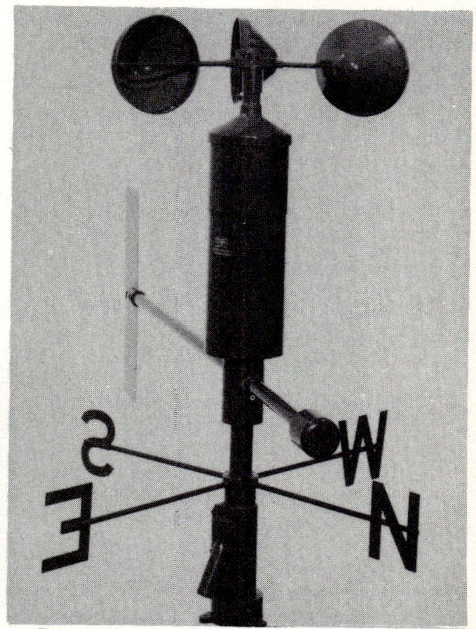

Type IM 144 combined wind speed and direction sensors

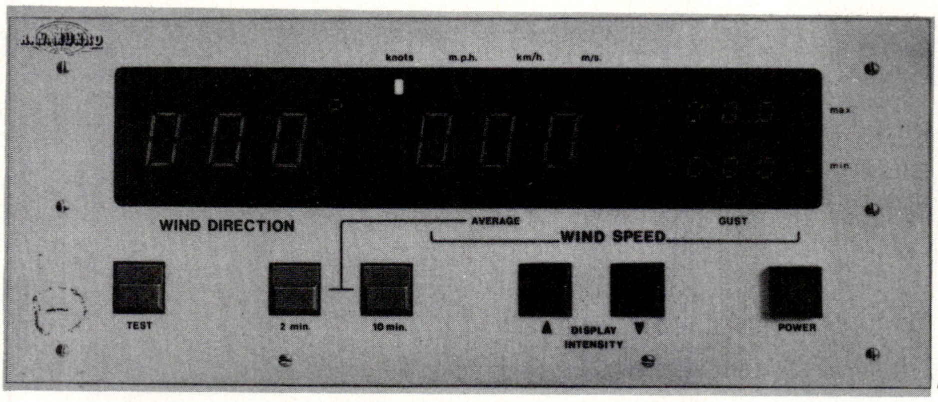

Munro digital wind measurement display

Negretti and Zambra Precision Aneroid Barometer Mk 2

This barometer designed by Negretti and Zambra uses an aneroid displacement measuring technique. A contact arm and micrometer lead-screw form the magnifying system. One end of the contact arm is held against the capsule contact by a hairspring and when the reading is taken the micrometer lead-screw is brought into contact with the other end by the manual operation of an external driving knob. Contact between the lead-screw and contact arm is detected by an electronic circuit which provides a visual indication on a digital counter to the operator.

The aneroid capsule is enclosed in a sealed pressure chamber which is normally vented to atmosphere. An adaptor can be used to connect the pressure chamber to a closed-circuit system when it is required to take comparative readings under stable conditions. Where ambient conditions are likely to produce rapid pressure fluctuations a pressure choke can be fitted. This decreases the rate of response reducing fluctuations of 1 mb over 12 seconds to less than 0·1 mb. The Baromec instrument uses the same measuring techniques but is smaller and designed for use in the field.

Manufacturer

Negretti and Zambra (Aviation) Ltd, Croydon, Surrey, England

Precision Aneroid Barometer Mk 2

Plessey Ceilometer Cloud Measuring Device

The Ceilometer is a device which measures the height of the cloud base automatically. It consists of two units: the field unit, which includes the digital data-processing circuitry and the display unit, connected by cable.

The height of the cloud base is obtained by timing a light pulse from the transmitter to the base of the cloud and back to the receiver. Identical, adjacent, Newtonian optical systems are used for both the transmitter and receiver. The gallium arsenide laser source produces 130 ns pulses of infra-red radiation at a repetition frequency of 600 Hz, with a peak pulse power output of 70 watts. The optical detector is a silicon pin photodiode. To maintain as high an optical signal-to-noise ratio as possible, the level of sky radiation which falls on the photodiode is kept to a minimum by the use of a narrow band optical filter and an optical baffle. Gold-surfaced mirrors are used and lenses and windows are coated, to have a high transmission at the laser wavelength. In this way considerable window fouling can be tolerated.

The return signals are processed within the field unit into range gates spaced sequentially and each 100 ns long (corresponding to 15 metres). The results of a number of laser firings are integrated in the stores associated with each range gate. The contents of the store indicate the presence of cloud at the various ranges. The data is transmitted as digital words to the display unit by cable. At the display unit the message is decoded and the chart recorder marked at the appropriate point, the chart moving horizontally and the cloud height being displayed vertically. The chart is marked only when cloud is detected at the height appropriate to the cloud base. This information is updated every 60 seconds. The field unit output can be used to drive a digital display.

Status

Production

Manufacturer

Plessey Radar, Weybridge, Surrey, England

Plessey Type 45 C Weather Radar

The Type 45 C is a C-band weather radar for rainfall measurement and general airport use where there are low to medium precipitation levels.
The antenna is a parabolic reflector front-fed by a horn in a metal space-frame radome allowing operation in wind speeds up to 240 km/h. It is capable of automatic scan in elevation at a rate of 10 degrees per second and the upper limit of the scan can be varied. Alternatively the elevation can be set manually. The antenna rotation rate is variable.
The transmitter/receiver operates in the band from 5450 to 5825 MHz. The receiver has a wide dynamic range and swept gain to a fixed $1/R^2$ law and automatic fade control are incorporated. Presentation of data with linear or two logarithmic characteristics can be selected. The combined plan position indication (ppi) and relative height indication (rhi) is presented on a Series 5 fixed-coil autonomous display using a 305 mm crt. The ppi mode has range scales from 25 to 200 km per radius and a X2 switch providing a 400 km scale. Full off-centring and electronic range and bearing lines are provided. The rhi mode has four horizontal scales from 25 to 200 km displaying ground range and two vertical scales of 12·5 and 25 km displaying height above the radar. There is an electronic range and height line and full off-centring to 200 km. The main scan is corrected for earth's curvature and standard atmospheric refraction.
The master control unit carries all necessary controls and indicators operating the radar system with the exception of specific display functions. The antenna and operating positions can be separated by up to 2 km to meet specific siting requirements.

Characteristics
ANTENNA: 3·66 m diameter
MANUAL ELEVATION: –2 to +90°
ROTATION RATE: 0·1–6 rpm
FREQUENCY: 5450–5825 MHz
BEAM WIDTH: 1°
PEAK POWER: 250 kW
PULSE LENGTH: 2 μs
NOISE FIGURE: ≤8·5 dB

Status
Production

Manufacturer
Plessey Radar, Weybridge, Surrey, England

Plessey Type 45 S Weather Radar

The Type 45 S is an S-band weather radar developed from the Type 45 C, designed for airports in regions where heavy precipitation can be expected. Because of its frequency, the radar is free from the attenuation effects of heavy rainfall and is of particular value in detecting and measuring area rainfall and detecting and tracking severe cyclones.
The basic equipment incorporates a combined ppi/rhi display and ancillary equipment includes an iso echo unit, so that the echo pattern can be displayed in up to four levels of brightness corresponding to preset values of precipitation intensity and camera equipment.
The radar has been designed to simplify siting the antenna apart from the display position. The standard equipment allows the two positions to be separated by up to 2 km, while the addition of the appropriate modules permits the interconnections to be reduced to a single way, wide-band link.
The transmitter/receiver operates in the band from 2700 to 3050 MHz with a minimum peak pulse power output of 650 kW at a pulse repetition frequency of 300 pulses per second and a pulse length of 2 seconds. A balanced crystal mixer incorporates a low noise, solid state pre-amplifier for optimum noise stability and dynamic range. In all other respects, the Type 45 S is similar to the Type 45 C above.

Status
Production

Manufacturer
Plessey Radar, Weybridge, Surrey, England

Plessey WF 3 (Mk 2) Windfinding Radar

The WF 3 (Mk 2) is a windfinding radar designed to have a maximum range of 200 km on a simple corner reflecting target, carried aloft by a balloon. Accurate wind data can be obtained up to heights of between 20 and 30 km. The radar is suitable for a permanent roof-mounted installation. In common with the WF 3 the radar consists of two small, lightweight units: the antenna unit and the operations control rack. These units are interconnected by cable and may be sited up to 30 m apart. A kW of prime power is required to drive the equipment which may be obtained from either a mains source or a portable generator.
After initial acquisition the radar tracks the target automatically and displays the target positional data as slant range, azimuth angle and elevation angle using digital indicators. These indicators may be frozen for manual recording of the data. Alternatively an automatic print-out system can be used to record the data together with elapsed flight time.
The antenna consists of a 1·1 m diameter parabolic dish with offset rotary sub-reflector. The radar transmitter and receiver are mounted immediately behind the dish. The complete antenna system is housed in a hemispherical front radome and rear cover.
The operator's control rack is a small unit which may be desk-mounted and connected by cable to the antenna unit. It contains the operator's controls and displays and houses the electronic circuits which control the automatic tracking system.

Status
Production. This system has been developed from the WF 3 of which more than 70 installations are in operation throughout the world

Manufacturer
Plessey Radar, Weybridge, Surrey, England

Rosemount Ice Warning System — Antennas

The ice warning system consisting of an ice detector and a controller, is designed specifically to automate antenna de-icing systems. It detects ice and activates antenna heaters, when ice is present, to prevent ice accumulation which can reduce antenna performance and cause structural damage. The ice detector is a sensing element at the top of the mast. The element vibrates axially at its resonant frequency of 40 kHz. When ice forms on the sensing element the frequency of the vibration decreases and at a preset level, which can vary from ice thickness of 0·02 to 0·2 inches (0·5 to 5 mm), an icing signal is given. This signal is received by antenna heaters which switch on automatically. In this configuration, the heaters have to be switched off manually. Alternatively a controller can be used to automate the system activating the heaters through timers. The duration of heating can be adjusted according to geographical area and individual antenna characteristics. The sensing element will continue to give icing signals as long as icing persists, reactivating the heaters as required.

Characteristics
Ice detector
DE-ICE TIME: 90 s nominal
POWER REQUIREMENTS: 115 V ac, 60 Hz
WEIGHT: 1·6 kg
OPERATING TEMPERATURE RANGE: –40 to +60°C

Manufacturer
Rosemount Engineering Company Ltd, Bognor Regis, West Sussex, England

Rosemount Model 1201F Pressure Transducer

The model 1201F is a pressure transducer for precision pressure measurement in unmanned environments. The transducer is accurate to 0·1 per cent full scale including linearity, hysteresis, repeatability and resolution errors and uses a high-level signal (0–10 V dc over the pressure range 800 to 1100 mb).
Rosemount also produce a range of platinum resistance thermometers suitable for meteorological applications.

Manufacturer
Rosemount Engineering Company Ltd, Bognor Regis, West Sussex, England

Rosemount Orthogonal Airspeed Sensor

The orthogonal airspeed sensor was developed for airborne use but is suitable for measuring wind speed and direction. The basic sensor measures two orthogonal airspeed vectors of airflow. Pressure sensing ports along the sensor's cylindrical body develop differential pressures proportional to the wind speed and direction. The sensor can be modified with additional sensing ports to provide static pressure in addition to wind data.

Manufacturer

Rosemount Engineering Company Ltd, Bognor Regis, West Sussex, England

UNITED STATES OF AMERICA

Belfort Automatic Meteorological Data Acquisition System

Belfort designs and manufactures meteorological instruments which can be supplied individually or as a complete system known as AMDAS, Automatic Meteorological Data Acquisition System.
The AMDAS system consists of a central station, out-stations and sensors.
The prime functions of the central station are data collection, computation, formatting, recording, display and operator input. The equipment comprises a meteorological data processor, altimeter setting indicator, runway data converter/monitor, keyboard recording devices and displays. The data processor, equipped with a keyboard, controls the computational functions, monitors the system and formulates the data for display, magnetic tape and printer messages. QNH information is provided by the altimeter setting indicator. The runway data converter/monitor acts as an interface with the out-stations and generates signals to drive air traffic control displays. All data is recorded and a hard copy printout is available.
The out-station structure is designed to meet individual operator's requirements. A complete system measures wind, cloud height and visibility at both ends of the runway and those parameters that may be taken as representative of the airport, at one end only.
A two wire current loop system is employed for transmission of data to the central station.

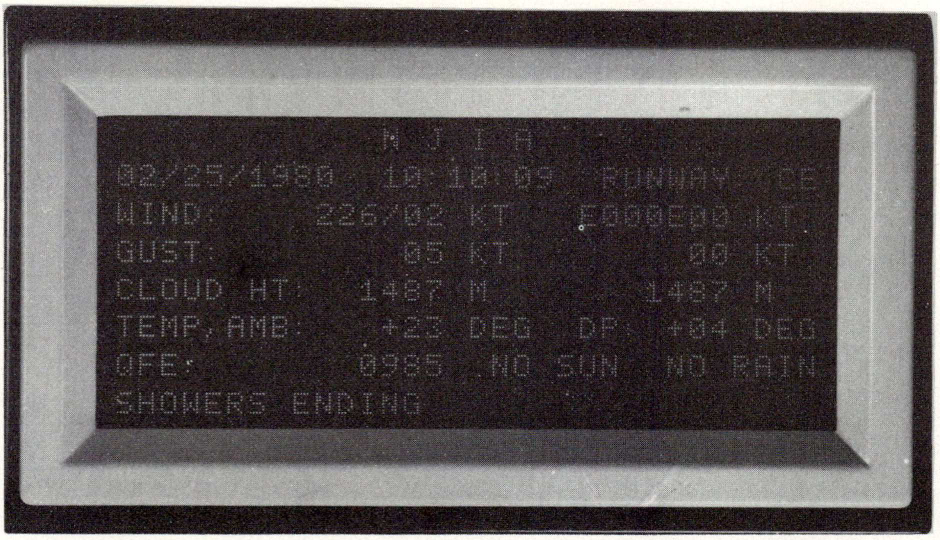

Visual display unit in Belfort's AMDAS installed in the King Abdul Aziz International Airport, Jeddah, Saudi Arabia

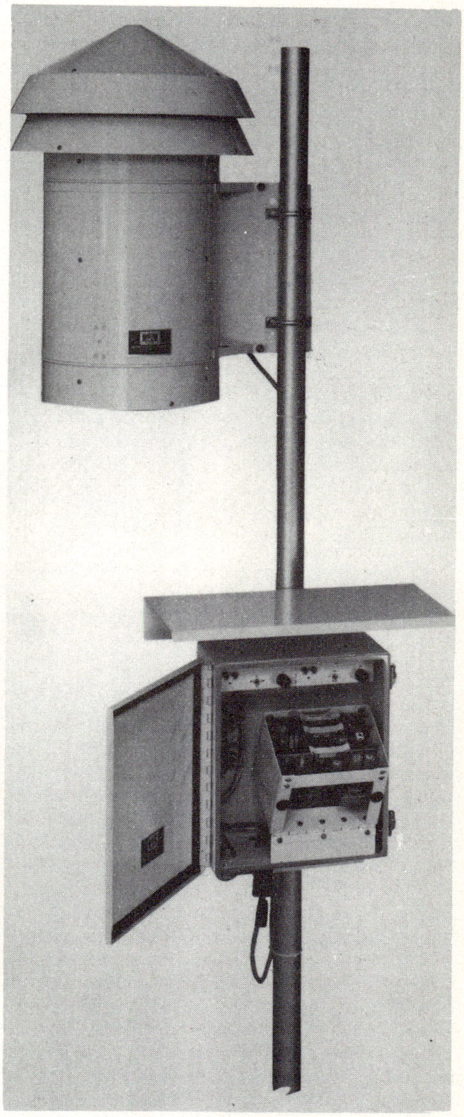

Belfort aspirated temperature/dew point system with data transmitter

Characteristics

SENSORS

TYPE OF SENSOR	MEASURED PARAMETER	MEASURING RANGE	ACCURACY
5-120 Wind transmitter	Wind speed and wind direction	0–120 knots 0–359°	±2% ±3%
7002 Temperature/dew-point system	Temperature and dew-point	−40 to +60°C −40 to +60°C	±0·5°C ±1·8°C
7003-1 Sun sensor/pyranometer	Solar radiation	Threshold detection/ 0 to 140 mWcm^{-2}	±5 to ±1%
5-405A Tipping bucket rain gauge	Rainfall	0·2 mm sensitivity	±1% at 25 mm per h
6007 Altimeter setting indicator	Barometric pressure	800–1100 mb QFE, QNH	±0·2 mb
7013 Ceilometer	Cloud height	30–1440 m	±15 m
7014 Transmissometer	Runway visual range	½ to 20 times the base line	±25–150 m ±50 m between 150–500 m ±100 m between 500–1000 m

OPERATIONAL TEMPERATURE RANGE
standard: −20 to +65°C
optional: −40 to +65°C

POWER REQUIREMENTS (OUT-STATION)
standard: 220 V, 50/60 Hz
optional: 110 V, 50/60 Hz

Status

Belfort has supplied meteorological sensors to military and civil airports in the USA and other countries. AMDAS is installed at the King Abdul Aziz International Airport, Jeddah, Saudi Arabia

Manufacturer

Belfort Instrument Company, Baltimore, Maryland, USA

Cardion Automatic Surface Weather System

The automatic surface weather system is designed to collect, process, disseminate and display meteorological data at airports in real-time. The system is modular so that any combination of sensors can be used and integrated into the system. Typically, the system includes sensors for wind speed and direction, peak gusts, barometric pressure, air temperature, dew point temperature, runway visual range and cloud height. The data is processed and can be shown on remote digital displays, crt terminals or using a teletype presentation.

Manufacturer
Cardion Electronics, Woodbury, New York, USA

Characteristics
All average readings are obtained over a one or two minute period.
The following parameters are displayed digitally:

PARAMETERS	DATA	RANGE (INCREMENTS)
Average cloud height	Metres	as 20, 30, 40, 50, 70, 90, 120, 180, 210, 250, 300, 400, 500, 600, 800, 1000, 1200 or 1500 m
Altimeter setting	Millibars	900–1100 millibars (0·1 mb)
Average air temperature	Degrees Celsius	–30 to +55°C (1°C)
Average dew-point temperature	Degrees Celsius	–30 to +50°C (1°C)
Average wind direction	Knots	0–99 knots (1 knot)
Max wind speed (over 1 s in last 10 minutes)	Knots	0–99 knots (1 knot)
Runway visual range (from up to 3 locations per runway)	Metres	150–800 (50 m) 900–1800 (100 m)

Cardion DRWS-20A Automatic Weather Station

The DRWS-20A is an automatic digital weather reporting system developed for use at smaller airports. The system, which is self-contained, automatically measures wind speed and direction, temperature, dew-point and barometric pressure (altimeter setting) and transmits the processed data with a synthesised voice output on the radio, vor, or non-directional beacon-voice communication link with the pilot in real-time. Synthesised voice input also allows other weather messages or runway operational activity to be included in the message format.

Characteristics
SENSORS

PARAMETER	RANGE	ACCURACY
Wind speed	0–99 knots	±1 knot from 1–25 knots ±2 knots from 26–99 knots
Wind direction	0–360°	±3°
Air temperature	–34 to +55°C	±0·2°C
Dew-point	–30 to +50°C	±0·5 to +40°C ±1 to +50°C
Barometric pressure	800–1100 mb	±0·2 mb

Status
Systems have been built for installations in Brazil, Jordan, South Korea, Abu Dhabi and Oman

Manufacturer
Cardion Electronics, Woodbury, New York, USA

DSD SWIMS Surface Wind Monitoring System

SWIMS is a surface wind monitoring system developed under the FAA programme to produce a low level wind shear alert system. It is a real-time, data acquisition, analysis and display system. It collects wind direction and wind speed data from six anemometers. One instrument is near the centre of the airport and five boundary anemometers are close to the approach and take-off areas of the airport.

Studies of gust-front propagation indicate that the five boundary anemometers should be about 1200 metres from the runway threshold, beyond the instrument landing system middle-marker location and between 9 and 15 metres above ground level. They are positioned more or less symmetrically around the airport, each approximately 1200 metres from the nearest runway.

Each anemometer supplies vector component values of wind data in analogue form to associated remote station electronic equipment at the anemometer site. The analogue vector components are conditioned and filtered to remove microscale wind variations and retain mesoscale wind variations associated with wind shear. The remote station equipment converts the data from analogue to digital data for transmission back to a central station at the control tower. Transmission of data from the anemometer sites to the central station is by radio, operating in the vhf range. Data can be transmitted by telephone lines if these are available.

The SWIMS micro-computer at the central station controls the data acquisition process, analyses the incoming wind information and updates wind displays for the air traffic controllers. The system incorporates performance analysis routines permitting continuous, on-line review of system performance.

SWIMS determines wind shear by calculating the difference between the vector component values of the wind at a boundary and the centre-field location. The threshold for wind shear is 15 knots. When the vector difference is 15 knots or greater a hazardous wind shear exists and the system automatically alerts the tower controllers by visual and audible alarms. The display identifies each boundary anemometer where a wind shear is occurring and shows the wind direction and speed at the boundary site.

To compute wind gusts SWIMS uses a gust threshold of 9 knots. Wind gusts appear on the displays only when the threshold is exceeded. A separate peak-detecting measurement of wind is compared with the average of the last 15 vector component measurements from the centre-field. When the difference is greater than 9 knots, the actual wind-gust value appears on the wind displays.

Status
The system started full scale operational tests at six airports in the USA in 1977. As a result of these tests SWIMS production was initiated and to date 58 systems are in operation at US airports

Manufacturer
Data Systems Division, Sangamo Weston, Sarasota, Florida, USA

MRI Model 1070/1080 Mechanical Weather Station

Meteorology Research produce a self-contained recording weather station which supplies data in either Imperial (Model 1070) or metric measurements (Model 1080). The system provides a strip chart recording of four sensors, wind speed and direction, temperature and precipitation. The temperature range is selectable in the model 1070. The unit is powered by battery.

Manufacturer
Meteorology Research Inc, Altadena, California, USA

Characteristics

MEASURED PARAMETER	Range	Accuracy
AVERAGE WIND SPEED	0–105 mph	±2%
	0–180 km/h	±2%
WIND DIRECTION	0–360°	±1%
TEMPERATURE RANGE		
low	–90 to +60°F	±3°F
high	–30 to +120°F (–32 to +50°C)	±3°F (±2°C)
PRECIPITATION	0·01 in per event (0·2 mm per event)	±1% at 3 in/h (±1% at 75 mm/h)
CHART SPEED	10 mm/h or 20 mm/h	
MAX RECORDING TIME	31/63 days	

MRI Meteorological Sensors

A range of sensors are produced by Meteorology Research, suitable for airport applications. These provide data on visibility, wind speed and direction, temperature, relative humidity and precipitation.

Transmitters are available to provide signal amplification and conditioning of these sensors to produce an analogue output suitable for recording, display or for use by other peripheral equipment. All are capable of 115/230 volt ac or 12 volt dc operation. Four models are produced accommodating up to a maximum of 16 meteorological variables.

Analogue chart recorders, digital recorders, readers and line printers are also offered by MRI.

Manufacturer

Meteorology Research Inc, Altadena, California, USA

Characteristics

MODEL	1580	1022S	1022D	1074	800 series	302/382
TYPE	Visiometer	Anemometer	Vane	Anemometer/vane	Natural or power operated sensors	Pluviometer
MEASURED PARAMETER	Visibility	Wind speed	Wind direction	Wind speed and direction	Temperature at relative humidity temperature, dew-point/temperature diffuse	Precipitation
RATE/RANGE	78–7800 m (250 ft–4·8 miles)	0·22–50 m/s (0·5–105 mph)	001–360°	0·33–50 m/s/001 to 360° (0·75–105 mph)	−50 to +50°C/up to ±15°C (±27°F) (−58 to +122°F)	0·01–10 in/h (302) 0·2–250 mm/h (382)
ACCURACY	±15%	±1%	±1%	±1%	±1°C (±1·8°F) temperature ±0·1°C (±0·18°F) temperature difference	±1% at 3 in/h (302) ±1% at 75 mm/h (382)
POWER REQUIREMENTS	115 V ac 50 watts nominal 150 watts with heater					

MRI Model 5000 Digital Weather Station

The Model 5000 is a micro-processor-based meteorological monitoring system that measures, processes and records data for up to seven meteorological variables and six auxiliary inputs. Data records are on a single magnetic cassette tape for a line print-out presentation or for processing by machine for evaluation. The meteorological variables are converted into engineering units, auxiliary inputs are recorded as a voltage in a 0 to 2 volt range.

The system is modular in construction consisting of three major components. They are: the electronics enclosure assembly, sensor array unit, and powr supply module. The electronics enclosure assembly includes an environmentally-sealed metal enclosure which houses a micro-processor unit, signal conditioning electronics for the various sensors employed with the system, local controls and displays and the digital cassette recorder.

Operator controls and indicators are all in the electronics enclosure. Controls and indicators include power on/off, station identifier, recording interval, sensor select and a display unit, mounted on the front of the computer assembly. The latter is used to set the station in operation and to select and read the last recorded data for each measurement made by the station.

The sensor array is a single assembly providing four of the meteorological measurements. On a single coaxial column, 4·5 inch (114·3 mm) diameter cups measures wind flow while an aluminium vane senses direction. A short boom attached to the wind sensor housing supports the radiation shield for the air temperature sensor and relative humidity sensor. The air temperature sensor is placed furthest away from the boom with two shields on top and a reflective shield underneath. A resistive sensor and oscillator circuit for the relative humidity sensor are located under the top, larger shield. The power supply uses 12 volts dc as a standard power source which can be provided by two 6-volt carbon-zinc batteries, two 6-volt batteries or two 6 volt sealed lead-calcium rechargeable batteries. The latter are used with an optional solar panel charger. A 115 volt, 60 Hz single phase ac power supply with an ac/dc converter and trickle-charged battery to operate the unit in the event of a power failure are available as an option.

Characteristics

SENSOR TYPE	Measured parameter	Measurement range	Resolution /	Accuracy	Data
ANEMOMETER	Wind speed	0–50 m/s	0·1 m/s	±1%	Pulse every 2·4 m of wind run – max gust = max 15 s interval
VANE	Wind direction	001°–360°	1°	±1%	Samples every 10 s
LINEAR THERMISTOR NATURALLY ASPIRATED	Temperature	−30 to +50°C	0·1%	±1%	Sample every 60 s
POWER ASPIRATED (OPTION)	Temperature	−30 to +50°C	0·1%	0·15%	Minimum/max every 24 hours
LINEAR THERMISTOR	Relative humidity	10–95%	1%	±6%	Sample every 60 s
TIPPING BUCKET	Precipitation	0–99·8 mm	0·2 mm	+1% up to 76·2 +5% from 76·2 to 254 mm/h	Variable period each bucket tip represents 0·2 m of precipitation
THERMOCOUPLE	Solar radiation precipitation	0–140 mw/cm^{-2}	1 mw/cm^{-2}	±5 mw/cm^{-2}	Sample every 60 s
ANEROID CAPSULE (OPTIONAL)	Barometric pressure	65–85 kPa/ 75–95 kPa/ 85–105 kPa	0·1 kPa	±0·2 kPa	Sample every 60 s

POWER REQUIREMENTS: 12 V dc, 3mA

OPERATING TEMPERATURE RANGE: −30 to ±50°C

Manufacturer

Meteorology Research Inc, Altadena, California, USA

Silver Instruments MOS-8000 Meteorological Observation Station

The MOS-8000 is a real-time data acquisition system, which, in its basic form, provides information on wind speed and direction, barometric pressure and temperature. Additional sensors for dew-point, rainfall, wind gusts and density altitude are available. The standard system consists of a master electronic unit and one slaved unit. An optional field digitiser can be used to enable the sensors to be located near the runway threshold or any other remote location. Data is transmitted by wire or using a single-channel radio link.

Each master unit can drive as many remote displays as required, using a serial-bit multiplex encoder-decoder technique. The master unit has a single display which can be switched to each of the measured functions; display units present all information simultaneously. A clock with a 24-hour master unit that can drive up to eight-slaved clock displays is available.

Characteristics

TYPE OF SENSOR	MEASURED PARAMETER	DATA	ACCURACY
Wind vane with opto-electronic mechanical position transducer	Wind direction	0–360°	±2°
3-cup anemometer with opto-electronic speed transducer	Wind speed	1–99 knots	±0·3%
Barometer: aneroid drive with linear variable differential transformer	Atmospheric pressure	27–31 inches Hg	0·01 inch
Linear thermistor network in radiation shield	Temperature	−16 to +52°C	±0·5°
Quartz crystal	Time	24 h in 1 s increments	±1 s/week

DIMENSIONS	MASTER UNIT	DISPLAY UNIT
length	482·6 mm	482·6 mm
height	88·9 mm	88·9 mm
depth	165·1 mm	104·8 mm

POWER REQUIREMENTS: 105–125 V ac, 50–400 Hz, 0·8 A or 208–240 V ac, 50–400 Hz, 0·4 A
OPERATING TEMPERATURE RANGE: 0–60°C

Manufacturer
Silver Instruments Inc, Concord, California, USA

WeatherMeasure Midas III Remote Weather Station

The Midas III is a micro-processor-based remote weather station which samples the environment at programmed intervals from up to nine sensors and converts the information into engineering units. It calculates averages, maximums, minimums, standard deviation, time of event, and other selected output data programmes and records and stores these measurements on either a cassette tape recorder or a printer. The computer can display the stored measurements directly on the digital readout of the control panel or transfer stored data at selected intervals to the tape recorder or printer. The instrument can also send data through a telephone modem. The sensors available allow the measurement and on-site analysis of wind speed and direction, temperature, relative humidity, solar radiation and precipitation. The system is offered in two basic configurations, the Midas IIIA and IIIB. Both use the same tripod and offer the same micro-computer control panel, and cassette tape recorder. Differences are in the type and number of sensors. For data computing and processing capability that goes beyond the Midas III and Meteorite micro-processor system, the company has a variety of remote data acquisition systems. These include the Meteor micro-computer. It has the capability to operate as a process monitoring system, measuring, processing and storing data, and in addition, can make automatic adjustments or perform control functions to specified limits, providing signals to activate hardware such as valves and pumps, compute efficiency factors and perform system tests.

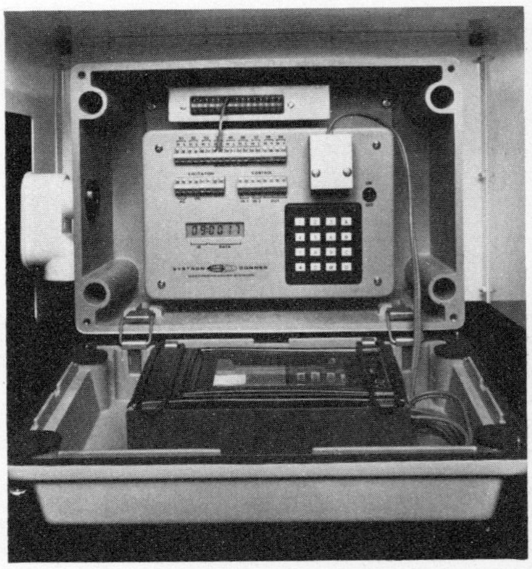

Midas III processor

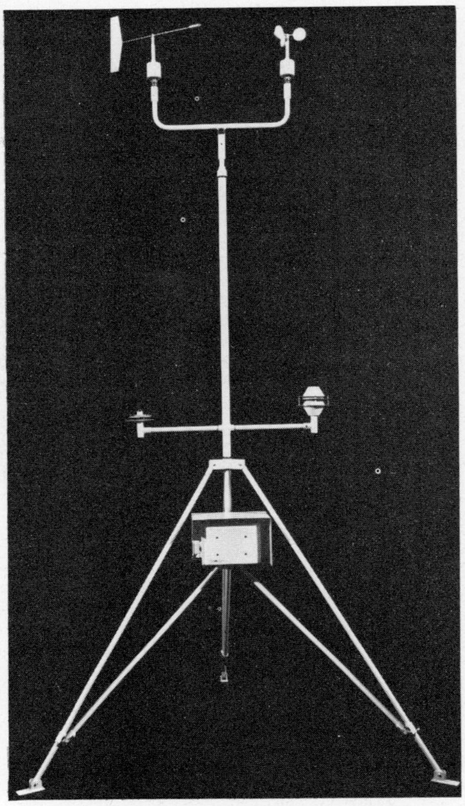

Midas III weather station sensors

Characteristics

SYSTEM	MEASURED PARAMETER	TYPE OF SENSOR	MEASUREMENT RANGE	OVERALL ACCURACY
MIDAS IIIA	Wind speed	W203 Anemometer	0·5 m/3–50 m/s	±1%
MIDAS IIIA	Wind direction	W204 wind vane	0–360°/0–540°	±0·5%
MIDAS IIIA/B	Relative humidity and temperature	M800-03-01 electronic hygrometer	0–100% −35 to +55°C	3·5% ±0·2°C
MIDAS IIIB	Wind speed and direction	W200 combined anemometer/wind vane	Up to 45 m/s	
Optional	Solar radiation	M800-05-01 silicone detector in a cosine-corrected head	In W/m²	
Optional	Precipitation	P501-1 tipping bucket pluviometer	From 0·01 in (0·25 mm)	0·5%

METEORITE DATA PROCESSOR
SCAN INTERVAL: 10 s
DIMENSIONS: 2·7 × 5·7 × 8·2 in (68·6 × 144·8 × 208·2 mm)
WEIGHT: 5 lb (2·27 kg)
OPERATING TEMPERATURE RANGE: −25 to +50°C

Manufacturer
WeatherMeasure Corporation, Sacramento, California, USA

WeatherMeasure Skyvane Wind Speed and Direction Measuring System

The Skyvane system consists of a Skyvane wind sensor and a remote indicator that provides continuous indication of wind speed and direction. A cable connects both units and the separation distance between sensor and indicator can be up to 500 feet (152·4 metres). This distance can be extended through use of a specially shielded cable. The Skyvane sensor uses a four-bladed grp propeller which has a starting threshold of 1 mile per hour (16 km per hour). This model has an ac generator as the speed sensor and a selsyn motor as the wind director sensor. The display comprises two dials, one for wind speed and the other for wind direction, both 6 inches (152 mm) in diameter. The standard wind speed indicator plate is scaled in miles per hour with optional versions offering scales in metres per second, knots and km per hour.

Manufacturer

WeatherMeasure Corporation, Sacramento, California, USA

Skyvane wind speed and direction measuring system

Characteristics

WIND SENSOR: Model W102
MEASUREMENT RANGE: 1–200 mph (0·45–91 m/s)
ACCURACY: ±1% below 25 mph (40 km/h) ±3% above 25 mph (40 km/h)
DIMENSIONS: 2 ft 5¾ in long × 2 ft 6 in high (0·756 × 0·762 m)
DISPLAY: Model W112
SCALE
WIND DIRECTION: 16 cardinal points +10° sections
WIND SPEED: 0–25 and 0–50 mph/0–50 and 0–100 mph (m/s/knots/km/h optional)
DIMENSIONS: 1 ft 2¾ in × 6¾ in × 8 in (0·37 × 0·17 × 0·2 m)
WEIGHT: 7 lb (3·18 kg)
POWER REQUIREMENTS: 115/230 V ac, 50/60 Hz

DESIGN, CONSTRUCTION, OPERATIONAL AND TECHNICAL SUPPORT SERVICES

Acculab
3201 Ridgewood Drive, Columbus, Ohio 43220, USA

TELEPHONE: (614) 876 5108

Chief executive: A J Campanella, PE, PhD

Acculab provides acoustical consulting services. Chief aviation services include computerised calculation of airport noise contours with the FAA integrated noise model; noise reduction design of small air-frames and power-plants; and surface structure design to attenuate aerospace noises.

Aeroport de Paris
Orly Sud 103, 94396 Orly Aerogare Cedex, France
TELEPHONE: (33.1) 884 52 43
TELEX: 200376

President: Raoul Moreau
Executive director: Jean Costet
Deputy executive director: Gerard Franck
Director projects and consultancy: Paul Andreu

Aeroport de Paris is a self-supporting French public authority financially autonomous formed in 1945 by the French Government to plan, design, operate and manage the civil airports and airfields in the Paris area: Charles de Gaulle, Orly and Le Bourget.

As operator, Aeroport de Paris are responsible for the technical and commercial management of the installations, maintenance, airport handling for air carriers, supply of utilities and services such as production and distribution of the principal fluids, data processing, fire fighting, accident and medical assistance, as well as air traffic and aircraft noise control. Shortly after it was formed Aeroport de Paris was called on to bring technical assistance to airport operators, to aviation or government authorities in several countries, first on an advisory basis then as fully qualified consultant. This part of its activity has developed considerably since 1960 and Aeroport de Paris is believed to be one of the top consultants in the field of airports and aeronautical problems.

The company has an engineering service composed of a permanent team of 700 engineers, architects and technicians out of a total of over 5000 people. The company's work include studies of the following areas: aspects of airport planning including traffic forecast and market studies, air traffic and flight control organisation, construction programmes, training, financing and overall co-ordination; infrastructure engineering: runways, taxiways, bridges, tunnels etc; building engineering: terminal buildings, air traffic control towers etc, air traffic control and emergency services; electromechanical equipment: power plants, telescoping air bridges, baggage sorting and handling etc; electricity, water, fuel, lighting, telephone; operational and maintenance support.

Aeroport de Paris achieved turnovers of 1399 million francs (1978), 1586 million francs (1979) and 1795 million francs (1980) after tax.

For consultancy services only, the figures were 39 million, 48 million, and 47 million francs respectively.

Roissy II Paris Charles de Gaulle Airport

Aeroport de Paris is a shareholder in the following companies whose activities relate to airport operation: Roissy-Fret, a service company working for air carriers in warehouse handling of freight; SOGECEBA, a company which operates computer terminals on the SOFIA system for common use by customs personnel; and SARESCO an airport restaurant, service and commerce company which studies, sets up and manages terminal bars and restaurants.

Aeroport de Paris has completed the following projects:

Roissy-Charles-de-Gaulle Airport, France
West Terminal Building at Orly Airport, France
New Abu Dhabi International Airport, Abu Dhabi
Damascus International Airport, Syria
New Dacca International Airport, Kurmitola, Bangladesh
Athens Airport site selection, Greece
Development of Goma Airport, Zaire
New passenger terminal zone of Cairo Airport, Egypt
New Riberalta Airport, development of Santa Ana de Yacuma Airport, Bolivia
Mexico City International Airport, Mexico – technical and social-economical study, Mexico
New Jakarta International Airport, Indonesia
Sibu, Limbang and Bintulu Airports, Malaysia
Seven domestic airports, Indonesia
Canlubang airport, Philippines
Dar-es-Salaam International Airport, Tanzania

Airways Engineering Corporation
1700 North Moore Street, Suite 810, Arlington, Virginia 22209, USA

TELEPHONE: (703) 522-4050
TELEX: 440097

President: Rufus C Phillips
Executive vice president: David Jaffe
Director, airport systems planning: Bobby D Ryan
Director, atc planning: Tirey K Vickers
Assistant chief engineer: Eugene K Fox
Chief architect: Harry A Morris

Airways Engineering Corporation was founded in 1944 as an engineering and architectural consulting firm specialising entirely in airport planning and design. Based at Arlington, Virginia, USA, the Corporation's first projects were airport and airlines studies in Washington DC, Newport, Houston, Baltimore, Birmingham, Memphis and St. Louis. In the late 1940s and early 1950s the firm undertook initial studies of such airports as Detroit, Philadelphia, Toledo, Niagara Falls and Mexico City. The Philadelphia project included detailed design of all airfield facilities. During this period the firm also did work in the design of loading bridges, mechanical baggage-handling and aircraft ramp handling devices.

In the mid-1950s Airways contracted for over 200 airport master plans for the US Air Force and US Navy. At the same time the firm began to undertake international airport work and since that time has completed many foreign projects.

Airways has undertaken the master planning and/or feasibility study of more than 250 civil and military airports in the USA and in 30 other countries. This includes the feasibility study and master plans of over 30 international airports and the detailed design of more than 15 major airports. Within the USA, much of this work has involved the development of airports using FAA Airport Development Aid Programme funding and prior grant programmes.

In 1968 the aviation consulting and airport planning firm of James C. Buckley Inc became an affiliate of Airways. Airways and Buckley are subsidiaries of Intercontinental Consultants, Inc. Their total experience extends over 31 countries in North, Central and South America; and in Europe, Africa, Asia and the Middle East.

Airwork Limited
Bournemouth-Hurn Airport, Christchurch, Dorset BH23 6EB, England

TELEPHONE: (020 16) 2271
TELEX: 41282

Airwork Ltd, a member of the British and Commonwealth Shipping Group, has been developing, managing and staffing civil and military airfields and air navigation facilities in the United Kingdom and other countries since 1928. Airwork Ltd is currently a contractor to the British Ministry of Defence, the Civil Aviation Authority, governments, local authorities, and air forces of other countries and private companies. It can provide services for development, management and staffing, navigation and communications, air traffic control and rescue and airfield vehicles. The company also offers language tuition and aviation training and procurement services.

The company can advise on the development of airfields and act as a consultant to civil, mechanical and electrical engineers on the layout and provision of runways, lighting, building, hangars, air traffic control, navaids, passenger handling, crash and rescue services.

Airwork operates five civil airports in the United Kingdom and can provide management and staff for all airport services. The company operates and services all forms of air traffic control systems and air navigation aids including airfield radars, distance measuring equipment (dme), vhf omni-directional ranging (vor), non-directional beacons (ndb) and tactical air navigation (Tacan) systems as well as land-line, microwave, radio and satellite communications systems. Its staff of licensed air traffic controllers provide and maintain air traffic control, and crash and rescue services at civil and military airfields.

Airwork specialises in the layout, design, installation and commissioning of air traffic control facilities and can also provide experienced technical staff to assist the repair, maintenance and inspection of electronic and mechanical systems.

The company operates and services the complete spectrum of airfield vehicles, including fire tenders, ambulances, refuellers and aircraft tugs, and associated mechanical equipment such as runway ice measurement trailers, glycol spray equipment and baggage loaders.

Airwork can provide specialised aviation English language tuition at the Air Service Training School at Perth, Scotland for airfield management, administration, passenger handling and air traffic control staff. Air Service Training also offers lead-in training including flying experience for international students taking air traffic control qualifications and periodic courses for flight information safety and airline operations officers.

Airwork has established, managed and staffed aviation training schools in the Middle and Far East and in East, Central and West Africa.

The company acts as a purchasing agency for British and other aviation authorities and airfield operators. The company has its own warehouses and a bonded store for the handling of all forms of equipment.

Amey Roadstone Construction Ltd

Sutton Courtenay, Abingdon, Oxfordshire OX14 4PP, England

TELEPHONE: (0235) 82811
TELEX: 83496

Amey Roadstone Construction Ltd has more than 30 years' experience of major airfield construction projects in the United Kingdom and other countries. Its knowledge of the specialised problems and requirements of airfield contracts is supported by technical backing and testing of materials to meeet the requirements for this type of construction.

Major airfield projects undertaken by the company cover both civil and military airports. In the United Kingdom these contracts have included work for the British Airports Authority at London Heathrow and Gatwick, England and at Aldergrove Airport, Belfast, Northern Ireland. Major contracts have also been carried out at many Royal Air Force stations.

Other contracts have included projects in Nigeria, the Middle East and Gibraltar where local materials have been used to produce products of the required standards under varying climatic conditions.

Association of Consulting Engineers

Alliance House, 12 Caxton Street, London SW1H 0QL, England

TELEPHONE: 01-222 6557
TELEX: 268312

The following is a summary of the projects by members of the Association in hand on 1st January 1982 or completed in 1981.

COUNTRY	NATURE OF SCHEME	CONSULTING ENGINEERS	MAIN CONTRACTORS
Bahrain	Bahrain International Airport: second phase extensions to terminal building,	Gibb Petermuller and Partners	Zachariades-Shaheen WLL
	extensions to aprons	Gibb Petermuller and Partners and Sir Alexander Gibb and Partners	Not yet awarded
	Reclamation and construction of earth fill platform for new radio receiver station	Sir Alexander Gibb and Partners	Design in progress
Botswana	Jwaneng Airfield: 2000-metres flexible runway and terminal area (supervision only)	John Burrow and Partners	Basil Read (Botswana) Pty Ltd
	Runway for Jwaneng Airfield Design and supervision in part	Brian Colquhoun, Hugh O'Donnell and Partners	Basil Read (Botswana) Pty Ltd
	Gaborone New International Airport	Sir Alexander Gibb and Partners (Botswana) with Gibb Petermuller and Partners	Design completed
Ecuador	Feasibility studies and master planning of 4 regional airports and complete design of 2	Sir Frederick Snow (International) Ltd in association with CIC	Not yet appointed
Greece	Athens New International Airport	British Airports Authority in association with Sir William Halcrow and Partners	Not yet appointed
Hong Kong	Specialised taxiway lighting	Kennedy and Donkin International in association with Scott Wilson Kirkpatrick and Partners	China Engineers
	Hong Kong Airport: aerodrome lighting control equipment	Kennedy and Donkin	Westinghouse Brake and Signal Company, England
	Hong Kong Replacement Airport at Chek Lap Kok: design and supervision for land development and major reclamation works	Mott, Hay and Anderson, Far East as sub-consultants to Ralph M. Parsons	Not yet appointed
Iraq	Baghdad International Airport	Scott Wilson Kirkpatrick and Partners International in association with Preece, Cardew and Rider, Scott Houghton and Co and Widnell and Trolloppe	Secondment of staff to ICAO
	3 air bases	Scott Wilson Kirkpatrick and Partners in association with George Trew Dunn Beckles Wilson Bowes; Ewbank Design Partnership and DG Jones and Partners	Design in progress
	Baghdad International Airport	Maunsell Consultants Ltd in association with Dunwoody Consultants Ltd	Fougerolle/Spie Batignolles
	Basrah International Airport	Maunsell Consultants Ltd in association with Dunwoody Consultants Ltd	Joint Venture Stabag Bau Ag

Ivory Coast	Abidjan International Airport: planning and detailed design of first stage of development including 4 000-metres runway and taxiway system, new international passenger terminal, cargo terminal, ancillary buildings and infra-structure	Plessey Radar Ltd and Sir Alexander Gibb and Partners	Not yet appointed
Jordan	Queen Alia International Airport, Amman: supervision of construction and review of designs. Work includes runways, taxiways and aprons and terminal, control and ancillary buildings	Sir Frederick Snow (International) Ltd in joint venture with Dar Al Handasah	Geneco, Amman
Kenya	Nairobi Airport development and extensions to infra-structure	Sir Alexander Gibb and Partners (Africa) with Gibb Petermuller and Partners	Sogene and Laxmanbhai and Co Ltd
	Wajir Airfield: development of existing airfield including new runway, taxiways, parking apron	Sir Alexander Gibb and Partners (Africa)	HZ and Company Ltd
	buildings Nyahururu Airfield	Sir Alexander Gibb and Partners (Africa)	HZ and Company Ltd
	Complete airport lighting for Mombasa Airport	Kennedy and Donkin	Mitsubishi
Libya	Tripoli International Airport: air freight terminal, aircraft maintenance facilities	Wallace Evans and Partners	Design stage
		Wallace Evans and Partners	Not yet appointed
	Tripoli International Airport: extensions to passenger terminal aprons	Sir Alexander Gibb and Partners and Gibb Petermuller and Partners	Samsung Construction
	Provision of General Aviation Facilities		Design in progress
	Provision of new taxiways		Design in progress
Malawi	Zomba Airfield Kasungu Airfield	Sir Alexander Gibb and Partners (Malawi)	Dowsett Engineering BRL Construction Ltd
	Lilongwe International Airport 3540-metre runway, parallel taxiway, passenger terminal buildings, operations block, control tower, aircraft parking apron, air cargo terminal, flight catering building, airfield operational lighting and navigational aids, water and sewage reticulation, main access road and various ancillary works	Scott Wilson Kirkpatrick and Partners, Malawi, in association with Murray Ward and Partners (Architects), Scott Houghton and Company (Building Services) Preece, Cardew and Rider (airfield operational lighting and navigational aids)	Nello L. Teer Co Civil Works Taehwa Construction Co Ltd Cargo Terminal and various ancillary buildings Mitsubishi Corporation, Passenger terminal operations block and control tower. Electrical works and navigational aids W and C French Flight Catering building, ancillary airline buildings
Malaysia	Kota Kinabulu International Airport: phase I airfield lighting	Preece Cardew and Rider	GEC of Malaysia Sdn Bhd
	Phase II: airfield lighting and electrical distribution	Preece Cardew and Rider	AEG – Telefunken
	Phase III: terminal development	Scott Wilson Kirkpatrick and Partners in association with Robert Matthew Johnson-Marshall and Partners (Architects), Architects Team 3, Preece Cardew and Rider and Yusoff Ibrahim Sehu Berakan (Consulting Engineers)	Civil Works Contract: Syarikat Setia Jaya, Goh and Lim Building works: Lim Kah Ngam
	Sandakan Airport Development	Scott Wilson Kirkpatrick and Partners in technical association with Malaysia International Consultants Sdn Bhd	Design in progress
Nigeria	Murtala Muhammed Airport, Lagos: survey, design and preparation of 3-year mechanical and electrical services maintenance contract and detailed appraisal of tenders	Sir Frederick Snow (International) Ltd	Not yet appointed
Qatar	Doha Heliport: helipad services (mechanical and electrical) plus new hangar	Trevor Crocker and Partners	Gulf Housing Ltd
	Doha International Airport: alteration and expansion to terminal building	Scott Wilson Kirkpatrick and Partners in association with Murray Ward and Partners, Scott Houghton and Co and Langdon and Every	Phase I Al Kaabi Contracting and Trading Company Phase 2 Tryad Contracting Phase 3 and Trading Company
Sarawak	Kuching International Airport: replanning and specialist site supervision	Sir Frederick Snow (International) Ltd in association with Malaysian Associate Architects	Buildings, Hoo Sheng Construction Piling Maltimor Piling Runway and Apron Civil Works; Public Works Department, Kuching
Saudi Arabia	Jubail Airport	Sir Alexander Gibb and Partners Petermuller and Partners Saudconsult	Various contractors
	Design for air base extension, including new aircraft apron, hangars and other associated buildings	Mott, Hay and Anderson International as sub-consultants to Dooh Consultant Engineers	Not yet appointed
Seychelles	Seychelles International Airport Phase II: Extensions to terminal and aprons	John Burrow and Partners	Allied Builders (Sey) Pty Ltd
Singapore	Singapore Airlines Ltd Changi engineering hangar	Ove Arup and Partners Singapore	Low Keng Haut
South Africa	Hangar 8, Jan Smuts Airport, Johannesburg. Partial structural design and mechanical and electrical services for hangar	Ove Arup and Partners, South Africa	L and F Metter (Pty) Ltd

Sudan	Construction of new airport at Port Sudan as international alternative to Khartoum. Provision for navigation aids, communications, fire/crash tenders and maintenance equipment	Brian Colquohoun and Partners in association with Cable and Wireless Ltd.	Tenders for Port Sudan re-submitted July 1980. Contractor not yet appointed
Trinidad and Tobago	Infrastructure Works at Piarco International Airport	Lee Young and Partners	Coosal's Quarry Ltd
	Coast Guard Hangar at Piarco International Airport	Lee Young and Partners	Shali Construction and Contractors Ltd
Turks and Caicos Islands	Expansion of existing airport facilities, Providenciales Island	Wallace Evans and Partners	Johnston Construction Ltd
United Arab Emirates	Al Ain International Airport development, Abu Dhabi	Scott Wilson Kirkpatrick and Partners in association with Murray Ward and Partners (Architectural), G.H. Buckle (Consulting Engineers), Northcroft Neighbour and Nicholson (Quantity Surveyors), British Airports International (Airport Management Consultants), T.W. Welch and Partners (Consultants in Radar and Navaids) and Minister Agriculture (Agronomists)	Stage 4 AKC Contracting Design in progress
	Four heliports	A.C. Ross and Partners (Middle East)	Out to tender
Western Samoa	Development of Airports and Urgent Upgradings to accept larger aircraft	Sir Alexander Gibb and Partners Australia	Direct labour

The following is a list of studies by members of the Association in hand on 1st January 1982 or completed in 1981.

COUNTRY	NATURE OF SCHEME	CONSULTING ENGINEERS	PROGRESS
Australia	Darwin Airport users survey	Maunsell and Partners Pty Ltd	Report submitted
Ivory Coast	Abidjan International Airport. Preparation of Master Plan and preliminary designs for future development of the airport, new terminal and ancillary buildings	Plessey Radar Ltd Sir Alexander Gibb and Partners with Gibb Petermuller and Partners	Report completed
Libya	Tripoli International Airport: survey of requirements to meet traffic forecast to 2000 and to report on further development	Sir Alexander Gibb and Partners with Gibb Petermuller and Partners	Report in hand
Malaysia	Master plan for airport development	British Airports International in association with Sir William Halcrow and Partners and MINCO	Completed June 1981
Middle East	Military Airport siting study for self-contained military airport. Master planning the staged development of the airport	Gibb Petermuller and Partners with Sir Alexander Gibb and Partners	
Tanzania	Pemba Airport: extension and rehabilitation of	Norman and Dawbarn Associates	Report submiotted
United Arab Emirates	Abu Dhabi New international airport: structural checks on airport buildings	Mott, Hay and Anderson International Ltd	Report submitted

The Austin Company

3650 Mayfield Road, Cleveland, Ohio 44121, USA

TELEPHONE: (216) 382 6600

CABLE: Austinclegen

Executive: Marvin M Epstein

The Austin Company specialises in the design, engineering and construction of airport facilities. The company has experience in aircraft manufacturing, aircraft maintenance, terminal buildings, cargo handling, flight kitchens, administrative buildings, hangars, computer operations, reservation centres, engineering development facilities, basic aerospace research centres, wind tunnels and passenger facilities. These have been developed in Australia, Belgium, Canada, China, Greece, Italy, the United Kingdom and the USA.

The company has designed and built over 100 projects for The Boeing Company. Among them is the manufacturing complex for the Boeing 747, 757 and 767 aircraft.

Recent projects in this field include the Hellenic Aerospace Industry (HAI) complex at Tanagra, Greece.

Austin was responsible for the master planning, design, engineering and management of construction of HAI.

The Italian subsidiary, Austin Italia SpA, is currently engaged in remodelling, modernising, refurbishing and expanding Leonardo da Vinci Airport, Rome,

In the USA Austin has been responsible for major improvements to a number of the larger regional airports.

The company operates 52 offices in 15 countries.

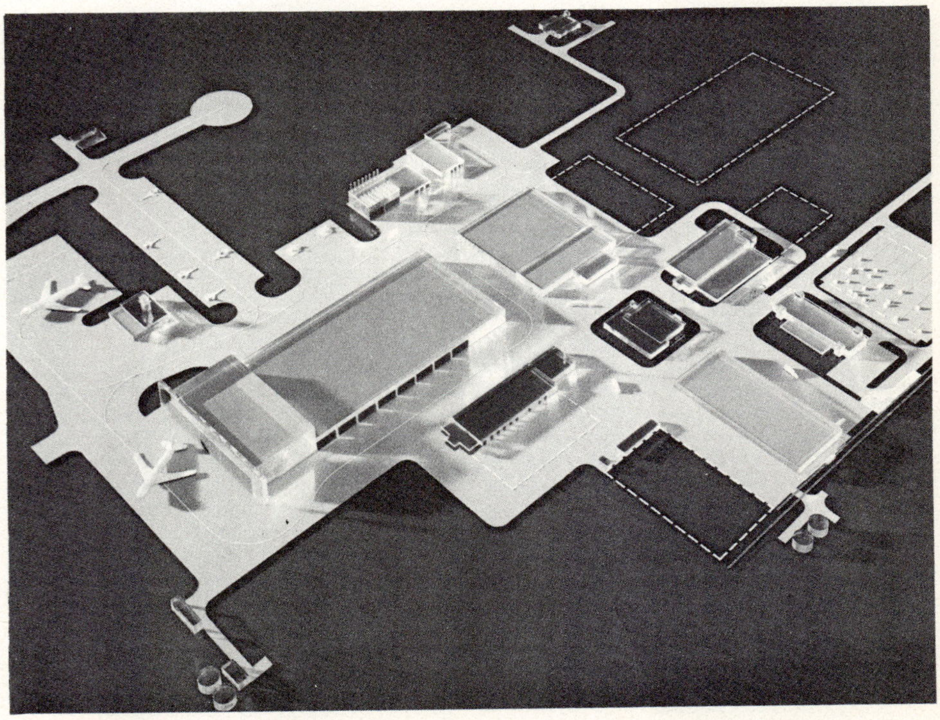

Model of the Hellenic Aerospace Industries complex, Tanagra, Greece

Aviation Planning Services Ltd

800 Dorchester West, Montreal, Quebec H3B 1X9, Canada

TELEPHONE: (514) 878 4388
TELEX: 055 60725

President: C Eric B McConachie
Vice president systems planning: M Stade
Vice president transport planning and special projects: R S Fisher

Aviation Planning Services (APS) provides professional consulting services to all sectors of the aviation industry. The major activities of the firm are concerned with commercial air transport, airport planning, general aviation analyses, product analyses for aerospace manufacturers, aircraft and engine maintenance and overhaul facility planning, as well as aviation systems planning and development programmes for all levels of government.

The aviation facilities planning services range from airport traffic demand forecasts to complete airport master planning. Typical services are: traffic forecasts: passengers and cargo/mail, annual and peak hour passenger and aircraft movements, itinerant and local movements, aircraft capacities; general aviation activities. Site selection: meteorological observations and analyses; topographical, soils, access, drainage and environmental assessments; obstacle clearance and airspace analyses with approach and missed approach collision risk estimates; navigational aids. Master plans: runway and taxiway layout and demand/capacity analyses; air traffic control tower; terminal area and gate requirements; lighting, landing aids and telecommunications; passenger and cargo terminal sizing and location, ground access and parking; utilities requirements, crash fire and rescue services; land use; construction phasing. Aircraft performance analysis: runway length/strength requirements; aircraft operating characteristics; weight limitations; government regulations; airline operating standards; environmental considerations. Airport facilities: passenger terminal concepts and operations; passenger flow, baggage handling and concessions requirements; cargo facilities and equipment; ramp layout and servicing requirements; maintenance hangars and equipment planning: auxiliary service demands; parking areas, access and internal roads. General aviation planning; equipment types; aircraft operations forecasts; hangar facilities; service facilities requirements; analyses of based and transient aircraft. Regional airport systems planning: airport inventory and analysis; air traffic analyses and forecasts; capacity of airport systems in metropolitan areas; long-range regional airport system plans and integration of airport system plans into overall transport programmes. Equipment procurement programmes: to procure the desired mechanical, electrical and electronic or other equipment, arrange the financing requirements, oversee the final installation and check-out of the equipment on a package or turn-key basis. Economical and environmental analyses: relationship of airport plans to the dispersal of atmospheric pollutants from aircraft and fixed ground sources; assessment of overflight noise on communities; noise exposure forecasts and recommendations for alleviation through compatible land use and operational techniques. Computer programs: APS has developed or can access computer programs for runway optimisation for earthwork; airspace and airport movements simulation; passenger terminal movements simulation; airport and environment geometery and lighting for use in flight simulators for evaluation of new airport designs; ground access comparison for site selection; noise exposure forecasts. APS has provided services to airlines in Canada, France, Iceland, Indonesia, Nigeria and USA; to ICAO and other associations; government agencies in Brazil, Dominican Republic, Guinea-Bissau, Indonesia, Malaysia, Peru, Portugal, Trinidad and Tobago, United Arab Emirates and completed projects in 12 other countries.

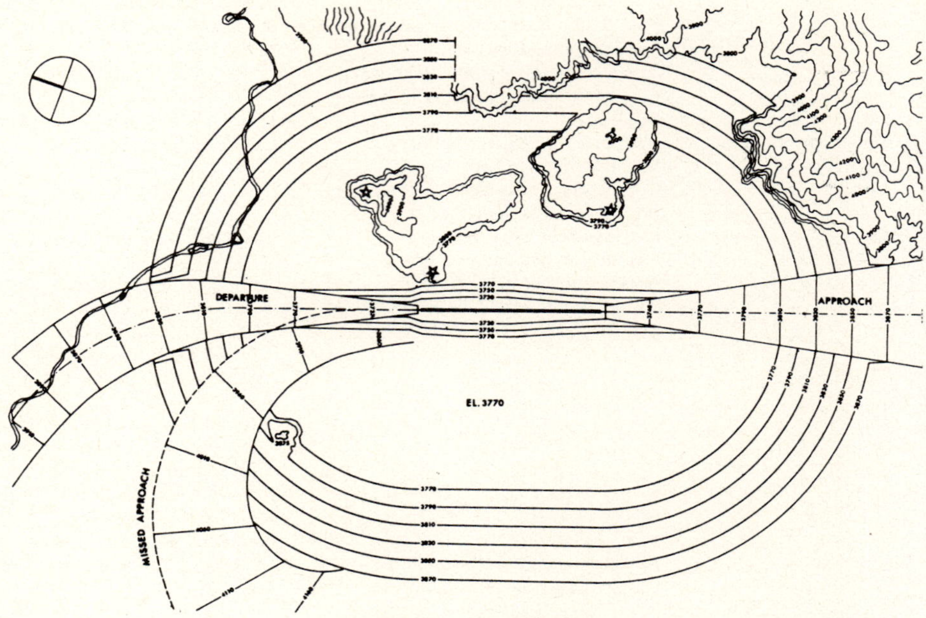

APS airspace analysis

R.L. Banks and Associates Incorporated

900 17th Street, NW, Washington DC 20006, USA

TELEPHONE: (202) 296 6700

Executive: Richard B Blackwell

R.L. Banks and Associates are consultants in transport economics, including air transport. Airport experience has included evaluation of city-pair travel markets, air carrier service evaluation, demand analysis and traffic forecasting. The company has completed air transport assignments for states, local communities, airlines and trade groups.

Barton ATC Incorporated

4986 Gateway Drive, Medina, Ohio 44256, USA

TELEPHONE: (216) 725 6776

Barton ATC Inc is experienced in the fields of air traffic control and meteorology, primarily the operation of control towers and providing weather services at airports where traffic demands call for a control system but do not as yet meet FAA criteria.

Barton ATC Inc started in 1968 with a contract to the Latrobe Airport, Westmoreland County, Pennsylvania, USA, and has expanded, with tower operations for city, county and state-owned US airports including Martin Tower, Glenn L. Martin State Airport, Maryland: Bolton Tower, Bolton Field, Columbus, Ohio; Anderson Tower, Anderson Municipal Airport, Anderson, Indiana; Smyrna Tower, Smyrna, Tennessee; Waukesha Tower, Waukesha County Airport, Wisconsin; Gillette Tower, Gillette-Campbell County Airport, Gillette, Wyoming; Falcon Field Airport, Mesa, Arizona.

The company offers an aeronautical consulting service providing airport planning, site surveys and air traffic surveys, and a turn-key service for the supply of air traffic control equipment.

Air BP, BP Trading Limited

Britannic House, Moore Lane, London EC2Y 9BU, England

TELEPHONE: 01-920 6286
TELEX: 888811

Air BP, the international aviation refuelling organisation of The British Petroleum Company Ltd, through its network of associated companies, has operated an international refuelling service for more than 50 years. The company's operations range from the off-shore drilling platforms to the world's largest airports.

Air BP offers extensive experience to those involved in the development of airport refuelling facilities on a consultancy basis, tailored to meet the requirements of individual operations. Such technical service include operational inspections and specialist advice on organisation, training, documentation and maintenance.

Currently Air BP is acting as consultant in Malaysia, Spain and Lebanon for the management of aircraft refuelling systems and aircraft services. Air BP's operating and technical liaison with airlines and aircraft manufacturers provides a basis for consultancy in the design and construction of all types of aircraft refuelling facility. Services can be either direct or in conjunction with airport consultants and include basic studies of future fuel demand at new and existing airports; studies to determine the optimum types of refuelling system; detailed design of selected refuelling systems; a full materials and equipment purchasing service based on BP's operating experience; expert construction, supervision and commissioning teams and advice in the design and construction of mobile fuelling vehicles.

Air BP has designed the fuelling systems now in operation at many major airports including Bahrain; Dacca, Bangladesh; London Heathrow and Gatwick, England; Munich, Federal Republic of Germany; Kuala Lumpur, Malaysia; Singapore and Zurich, Geneva, Switzerland.

British Airports International

8 Little College Street, London SW1P 3SH, England

TELEPHONE: 01-222 8592
TELEX: 912266

Chairman: J Mulkern
Acting general manager: J Cousins

British Airports International (BAI) was formed in April 1978 as a joint partnership between the British Airports Authority, owner and operator of seven United Kingdom airports and IAL, the international aviation, computers, communications and medical services company. BAI provides consultancy on airport design and construction projects, including the forecasting and planning of cargo and passenger handling requirements, air traffic control facilities, and services such as security and meteorology. The company can arrange for the design and construction of new, or the redevelopment of existing, airports in association with consulting engineers and contractors. Past and current projects include: detail study of requirements for second terminal building at Portela Airport, Lisbon, Portugal; Malaysia National Airport plan and development of airport master plans at Ipoh, Pulau, Langkan and Tawau, Malaysia; preparation of master plan and feasibility study for development of Bahrain International Airport; terminal master plan, Aqaba Airport, Jordan; master plan and project consultancy for new Athens International Airport, Greece; provision of personnel at Kuwait Airport to undertake air traffic control, terminal management and training; marshaller training, Amman Airport, Jordan; provision of training for Iraq nationals in airport electronics and communications engineering and operation for state organisation of Iraqi Civil Aviation.

The company undertakes short- or long-term secondment of fully qualified, experienced staff to provide services to the client's airport. Specialists are available in fields such as airport management, atc, meteorology and fire and rescue. Engineers and technicians can also be supplied. BAI offers a range of courses in the United Kingdom or overseas, covering all aspects of airport management and operation.

Technical and academic facilities are provided at IAL's training centre at Bailbrook College in England, while on-the-job training is available at the airports operated by BAI's parent companies.

Past and current projects include: provision of specialist consultants for the management and operation of Murtala Muhammed International Airport, Lagos, Nigeria; provision of air traffic control services and operation and maintenance of technical services at Abu Dhabi's New International Airport; provision of air traffic control services at Kuala Lumpur International Airport, Malaysia; New Sharjah International Airport, United Arab Emirates: technical services management and operation.

Establishment and operation of training centre in Jeddah for Saudi Arabians.

Provision of aviation technical services in Mozambique at Maputo and Beira International Airports; advice on management and operation of Maldives International Airport; contract with Saudi Arabia's Meteorological and Environmental Protection Administration to provide scientific, operations, maintenance training, consultancy and supplies services; management and technical services contract at Ras Al Khaimah International Airport, United Arab Emirates; technical services at Dubai International Airport, United Arab Emirates.

Operation of East Shetland Basin Flight Information Service and provision of air traffic services in Ninian Field, Scotland; technical services management and operation at United Kingdom regional airports including Liverpool, Humberside, and Hurn Airport, Bournemouth, England; operation of Battersea Heliport, London, England.

John F. Brown Company Incorporated

3057 Erie Avenue, Cincinnati, Ohio 45208, USA

TELEPHONE: (513) 321 6080

President: John F Brown

Airport management and financial consultants.

Conder Buildings Overseas Limited

Montfort House, Romsey, Hampshire SO5 8ZR, England

TELEPHONE: (0794) 517000
TELEX: 477275/477283

Directors: G Kiley, B L Chaffey, K J Churchill, J Druce, A B Paxford, R M T Mitchell

Conder Buildings Overseas specialises in the design and construction of large buildings, using the Kingsworthy concept.

The company's 'dry envelope' is a shell formed at a very early stage in the contract which allows internal work to proceed on plan unhampered by weather conditions. The Kingsworthy method can be adapted for special conditions such as buildings in hazardous areas subject to blast or ground subsidence.

A typical example is the Wils Ness Development at Sumburgh Airport in the Shetland Islands, Scotland. The 6500-square metre terminal building was constructed in 36 weeks.

Sumburgh Airport, Shetland Islands

Costain International Limited

111 Westminster Bridge Road, London SW1 7UE, England

TELEPHONE: 01-928 4977
TELEX: 8811804 Cosdon G

Costain offers an airport building and civil engineering capability.

Costain has built substantial parts of the airports in Melbourne, Sydney, Australia; Brunei; Bradford, East Midlands, Leeds, London Heathrow and Gatwick, Manchester, England; Maldive Islands; Glasgow, Outer Hebrides Islands, Prestwick, Shetland Islands, Scotland; Seychelles Islands; Dubai, United Arab Emirates; Hodeidah, Yemen Arab Republic; Lusaka, Zambia.

The work has included terminal buildings, control towers, hangars as well as runways, taxiways, aprons and all services.

Costain completed, as a joint venture with Amey Roadstone Corporation Ltd, the new Hodeidah Airport in Yemen Arab Republic. Work is also currently in progress at both London Heathrow and Gatwick Airports, England.

Dubai International Airport, United Arab Emirates was built by Costain International under a £4.1 million design and construction contract and was completed in 1971. The contract included all structural, mechanical, electrical and civil engineering, building the control tower, a flyover, car park and elevated road to the terminal. The building clad with grp panels, is air-conditioned and contains administrative offices, shops, a restaurant and all passenger facilities. Access from the three-storey building and adjoining two-storey annex containing the VIP lounge to the aircraft is by four covered bridges projecting on to the apron. Dubai airport was the first in the Middle East designed and built for operating wide-body aircraft. Since 1971 Costain has extended the runways and aircraft parking spaces so that four wide-body aircraft can be parked together.

Dewberry and Davis
8401 Arlington Boulevard, Fairfax, Virginia 22081, USA

TELEPHONE: (703) 560 1100

Executive: Vernon R Gingell, PE

Dewberry and Davis are airport consultants who offer the following services: airport facilities design, sub-surface evaluations, grading, drainage and pavement evaluations design, mechanical and electrical design, specification preparation, construction supervision, management studies, site selection, architectural design, financial planning, structural engineering, land use planning, airfield lighting systems, industrial airpark planning, navaid siting and installation, specialised airspace and flight pattern analyses, air service consultation, airport master plans, airport system plans, feasibility studies, demand forecasting, environmental impact assessment and noise surveys.
Airport projects completed by the company include: Monrovia, Liberia; Sussex County, Delaware; Morgantown, Raleigh County, West Virginia; Cumberland, Maryland; Shenandoah Valley, Washington DC, Dulles International and National airports, Norfolk, Chesterfield County, Petersburg, Virginia, USA.

Washington DC National Airport

Diemler and Diekemper Incorporated
715 Loda Drive, Cincinnati, Ohio 45245, USA

TELEPHONE: (513) 752 6200

Executive: H E Diemler

and

1435 Powhatan Street, Alexandria, Virginia 22314, USA

TELEPHONE: (703) 549 5727

Executive: B H Diekemper

Diemler and Diekemper Inc is an airport consulting firm providing a broad scope of studies and advisory services to the operators and managements of airports world-wide. The types of airport studies and services provided include: site selection and development studies, airport master plans, airport system plans, leasing policies and lease documents, tenant rates and charges programmes. Concession development programme, economic feasibility studies, environmental protection plans, safety and security programmes, organisation, staffing and compensation plans; management information systems, operating policies and procedures, preventive maintenance and repair programmes, property and liability insurance programmes and special advisory services and studies designed to meet the needs of individual clients.

RM Douglas Construction Limited
396 George Road, Birmingham B23 7RZ, England

TELEPHONE: (021) 356 4888
TELEX: 338399

RM Douglas Construction has constructed many airfields for the Royal Air Force and for the United States Air Force in the United Kingdom, including runways and airfield paving with rigid or flexible surfaces.
The company has built terminal buildings for aircraft passenger, baggage, and cargo facilities and surface and multi-storey car parks.

The company and its associated specialist companies can provide in all parts of the world a complete airport design and construction service including soil mechanics and geophysical test facilities, building, civil and structural engineering design.
The specialist companies include Douglas Technical Services Limited (who provide civil and structural engineering and building design, soil mechanics and foundation engineering); RM Douglas Roofing Ltd (to fit insulated metal roof and wall cladding); RM Douglas Asphalt and Paving Ltd (to carry out mastic asphalting, tar paving and surfacing); Marshall David and Company Limited (for decorative and wood block flooring, tiling and suspended ceilings); Douglas Environmental Engineering Limited (who specialise in heating, plumbing, electrical engineering and painting); and British Lift Slab Limited (who offer the Lift Slab method of multi-storey design and construction and the Siemcrete slipform system).

John W Drake
1815 Woodland Avenue, West Lafayette, Indiana 47906, USA

TELEPHONE: (317) 463 5882

John Drake is a transport consultant specialising in forecasts of airline passenger, freight and mail, forecasts of air traffic activity, and civil aviation master plans, cost/benefit analysis of aircraft, traffic control and other investments, fare studies, merger analysis, public policy analysis of air transport issues and logistics studies. Primary clients are US and foreign airlines, manufacturers, governments and airport authorities.

Drover, Welch and Lindlan
Suite 1200, 3225 North Central Avenue, Phoenix, Arizona 85012, USA

TELEPHONE: (602) 264 9731

Executives: Wallace E Welch, AIA; James F Lindlan, AIA; Carleton W Van Deman, AIA; Lawrence E Metcalf, AIA

Drover, Welch and Lindlan are a firm of architects. Professional services offered by the company comprise pre-design services including project budgeting and liaison with regulatory agencies; site analysis; schematic design; design development including structural, mechanical and electrical systems design and project costing; construction document services; bidding and negotiating phase services; construction contract administration; post-construction and supplemental services.
The company's projects include the sky harbour terminal and air traffic control tower, Phoenix Airport, Arizona, USA.

Edwards and Kelcey Incorporated

70 South Orange Avenue, Livingston, New Jersey 07039, USA

TELEPHONE: (201) 994 4520

Partner, president and chairman of board: Gordon L Kirjassoff
Partner and senior vice president: Wesley La Baugh, PE
Partner and senior vice president: Satoshi Oishi, PE, RA
Partner, senior vice president and secretary-treasurer: Norman C Possiel, PE

Branch offices in Boston, Minneapolis, New York, Philadelphia, Washington DC and Chicago, USA

Edwards and Kelcey International Incorporated

One World Trade Center, Suite 8721, New York, NY 10048, USA

TELEPHONE: (212) 466 1383

Edwards and Kelcey is an interdisciplinary consulting firm with a professional and technical staff of some 300 members specialising in engineering, landscape architecture, planning and environmental services. The organisation includes an airport division offering capabilities for the study, planning, design, and construction supervision/inspection of airport facilities.

The airport division provides such services as regional and local master planning, feasibility reports, site selection, pavement evaluation, air and vehicular traffic studies; design of approach systems, airfield facilities, and terminals; and construction supervision and/or inspection.

Airport projects undertaken by the company in USA include: Logan International Airport, Boston, Massachusetts; air carrier/small hub airports at Hyannis and Worcester, Massachusetts; air carrier/air taxi airport, Eau Claire, Wisconsin; general aviation airports at Beverly, Chatham, Fitchburg, Lawrence North Adams, Provincetown, Orange Southbridge and Taunton in Massachusetts; Rockland, Maine; Concord and Wolfeboro, New Hampshire and New Castle, Delaware.

Military bases include: USAF, Keflavik, Iceland; USAF Falmouth, USN South Weymouth, Massachusetts USA.

ESG Elektronik System-Gesellschaft mbH

Vogel weide platz 9, D 8000 Munich 80, Federal Republic of Germany

TELEPHONE: (089) 92 16 1
TELEX: 522594

Chairman of the supervisory board; Hans-Joachim von Ludwig
Executives: Dipl-Ing Horst Klassen, Dr jur Norbert Servatius, Dipl-Ing Reinhart Trautwein

ESG specialises in systems engineering and project management. Aviation related subjects include air traffic control, airport planning and airport equipment.

ESG can offer solutions to problems in all the sub-areas of air traffic control. It undertakes complete planning, beginning with basic technical and operational studies and continuing through to the preparation of tender documents and supervision of installation and commissioning. Software development and system integration take place in ESG's laboratories. The services offered by ESG within the above framework extend to: radar installations; navigation aids; data processing systems to process flight schedule and radar data and condition information on visual display units; communication facilities such as radio or telephone systems, computer-to-computer data links; operational planning of air-space regions and sectorisation; planning of approach and departure procedures for airports and large-scale computer simulation for the study of traffic patterns.

Under contract to the Bundesamt für Flugsicherung, ESG participated in the study for extending the secondary radar system to vfr air traffic.

ESG has experience in the operational and technical planning of airport operations and airport equipment, such as: technical equipment for air traffic control; radio navigation systems; voice and digital communication facilities; meteorological equipment; infrastructure for approach aids; information and display systems for airports; protection and security systems for airports; vehicle allocation for flight operations; automatic check-out systems for aircraft maintenance.

The range of services offered by ESG in this area covers: basic surveys; devising of concepts; preliminary and design planning; provision of tender documentation; systems integration and development of software; supervision of installation; acceptance testing of equipment after installation and training, documentation and logistic support.

ESG has developed a model for the simulation of the rolling traffic on the apron in connection with the planning of Munich II Airport.

Federal Electric Corporation

621 Industrial Avenue, Paramus, New Jersey 07652, USA

TELEPHONE: (201) 967 0123
TELEX: 134458

Federal Electric Corporation, a subsidiary of ITT, is a service company specialising in providing aviation ground electronics systems on a turn-key basis for government, industrial, commercial and military users.

The service covers air traffic control towers: towers, cabs, consoles and attendant electronic facilities; navigational aids: terminal and en route electronic and visual systems; communications: ground-to-air, point-to-pont, and mobile systems; airport radar; primary, secondary and surface detection systems; meteorological systems; weather measuring and recording and display systems; general support facilities; telephone, lighting, public address, cctv and flight information systems.

The service comprises: programme management: on-site management teams dedicated solely to the organisation and direction of individual turn-key programmes; planning: evaluation of facility requirements and the establishment of design criteria and system specifications; engineering: working from design criteria by integrating specific components into a functional system; procurement: selection of equipment and associated components; installation: installation of information transfer systems; testing and commissioning: to align, test and commission electronics facilities; civil works: to design and construct civil works associated with the installation of information transfer systems; maintenance and operation: system maintenance, operation and technical support; documentation: publication department provides comprehensive installation, operations and maintenance manuals for equipment and systems; training: on-the-job and classroom training for customer personnel; quality assurance; logistics support: complete logistics services for turn-key programmes including vendor surveillance, spare parts provisioning and documentation, transport, customs clearance and export licensing, warehousing and inventory control. In Saudi Arabia, Federal Electric provided Air Traffic Control facilities which included Vortac, Dvortac, ils, telecommunications and ground to air communications.

Other examples are projects for the Federal Aviation Administration, United States Air Force/Navy, Manila International Airport, Philippines, Santiago International Airport, Chile, Madrid air traffic control centre, Spain and the Roberts International Airport, Monrovia, Liberia.

The company can provide operation and maintenance support service for communications and radar systems, for which the company believes it is the largest contractor in the USA. The company maintains a wide variety of projects in USA. Outside the USA contracts range from radio transmission in Ecuador to the Area Maintenance and Supply Facility that supports the United States Army's 5th Signal Command in Federal Germany.

Flughafen Frankfurt/Main AG

Airconsult, D6000 Frankfurt am Main 75, Federal Republic of Germany
TELEPHONE: (49/611) 690 2038
TELEX: 413384

Chairman: Erich Becker
Deputy chairman, executive director, operations: Walter Schroder
Executive director, commercial matters: Hans Gunter Beck
Executive director, technical matters: Gottfried Mucke
Director, airport data processing and planning: Klaus Brendlin
Director, project management air freight: Hans-Jurgen Fischer
Director, passenger handling: Karlheinz Heck
Director, construction and facilities: Gottfried Meinicke
Director, ground handling services: Peter Pflugbeil
Director, air freight handling: Walter Reuther
Director, airport security services: Otto Zeller

Flughafen Frankfurt/Main AG (FAG), also known as Airconsult, is a publicly-owned company as part of Frankfurt Airport Authority. Airconsult offers a wide variety of services including: market research and traffic forecast, development of air transport systems; infrastructure planning and environment protection measurements; planning of integrated transport systems; air space studies; feasibility and site studies for airports and airfields; master plans and capacity calculations; systems analyses and procedure studies; functional and design plans; tender documents; construction supervision; construction management; organisation and management studies; edp systems and software; operational support; maintenance and upkeep; personnel training.

Projects completed by FAG include: Saudi Arabia: air traffic plan 1985–2000; Mauritania: air transport plan; Bangui, Central African Republic: airport development; New Athens International Airport, Greece: new airport; Hulule, Maldives: airport extension; St. Maarten, Netherlands Antilles: airport extension, apron planning; Nema, Mauritania: new airport; Rujan, People's Democratic Republic of Yemen: aeronautical equipment and new passenger terminal; Caracas, Venezuela: organisation and management study; Sharjah, United Arab Emirates: airport management.

Hulule Airport, Maldives

The Foundation Company of Canada Limited

1 Yonge Street, Toronto, Ontario M5E 1E8, Canada

TELEPHONE: (416) 361 4611
TELEX: 065 24213

FCC provides a full range of construction capabilities in the building, marine, heavy and industrial fields. The company will handle jobs on a project management, construction management or on a unit price or lump sum basis. FCC will take on projects either on its own or on a joint venture basis depending on the size of the job or other circumstances.

The project management service offered comprises: preliminary study: assistance on feasibility studies, preliminary design, preliminary cost estimates, site investigations; co-ordination of design: work with client, architects and consultants, select technologies, determine most economical design, prepare estimates, budgets and cash flow, develop technical specifications, apply value engineering, determine permits required; preparation and evaluation of tenders; prepare tender documents, call and evaluate tenders, draft contract documents; schedules and procurement: establish schedules, budgets, control procedures, prepare pert or cpm programmes, carry out procurement expediting, inspection, arrange for shipping and insurance; construction and site management: carry out construction as required, control work's progress, inspect and accept work, establish quality control, arrange, supervise and implement labour agreements, settlement of accounts; delivery and follow-up: assure project meets client's specifications, hand-over on time and within budget, control on warranties.

Airport projects undertaken in Canada include: Toronto International aeroquay; Calgary International terminal building; Montreal Dorval main terminal building, loading area, satellite aeroquay and interior connecting tunnels; Vancouver International terminal building.

Vancouver International air terminal

Hanscomb Associates Incorporated

La Salle-Wacker Building, 221 North La Salle Street, Chicago, Illinois 60601, USA
TELEPHONE: (312) 984 1750
TELEX: 25 3600

with offices in Anchorage, Atlanta, New York, Newport Beach, Seattle and Alexandria, Virginia, USA.

Overseas: Construction Management Services
36 Strachan Street, PO Box 7441, Lagos, Nigeria.
TELEPHONE: 57197

International affiliation with the Hanscomb Partnership with offices in London, England; Rome, Italy; Athens, Greece; Glarus and Zurich, Switzerland; Manama, Bahrain, Lantyre, Lilongwe, Malawi; and with the Hanscomb Roy Associates with offices in Montreal, Toronto, Calgary, Edmonton, Hamilton, Ottawa, Quebec City, Regina, Vancouver and Winnipeg, in Canada and in Bermuda.

The name Hanscomb is the collective identity for a number of individual professional practices concerned with the management and financial control of all aspects of design and construction activity. All the practices have developed from the founding practice Hanscomb Partnership, quantity surveyors and cost consultants. There are now three associated major firms working worldwide — Hanscomb Partnership operating from Europe; Hanscomb Roy Associates formed in 1968 and based in Canada; and Hanscomb Associates Inc. formed in 1974 and based in the United States of America. Certain ownership interests within the major practices are interlinked and there is a high degree of interface and collaboration at both the management and the working levels of operation. Each of the practices also owns subsidiary firms working in specific fields of activity.

The basic expertise of the practices examines the management and financial aspects of development

Malton I and II International Airport

and construction on many levels.
No design consultants' work is undertaken but the analysis of design and the rationalisation of choice, considering time and economic consequences, are very much a part of the activities.
In broad terms three distinct types of activity are carried out:
Management services under which emphasis is upon the direction and co-ordination of the work of other parties.
Traditional services which concern financial management and control from the pre-design phase and throughout the design and construction phases of a contract.
Specialist services which include life cycle costing, energy studies, research, information systems and cost indices, compilation of computer based estimating data and other such activities.
Airport projects undertaken by the group started in 1961 and since 1978 they have been involved in the following facilities:

1979	New Airport Facility	Louisville, Kentucky	Cost Estimating Scheduling
1979	Eastern Airlines National Airport	Washington D.C.	Cost Estimating
1979	Royal Saudi Air Force, Operations Centres	Saudi Arabia	Cost Estimating
1979	Anchorage Intl. Airport	Anchorage, Alaska	Cost Control
1979	Chugiak Airport, Vehicle Shop	Chugiak, Alaska	Cost Estimating
1979	Wrangell Airport, Snow Removal Shop	Wrangell, Alaska	Cost Estimating
1980	Cincinnati Airport Parking Canopies	Cincinnati, Ohio	Review Final Estimate
1980	Addition at Naval Air Station	Atlanta, Georgia	Concept Estimate
1980	Dutch Harbour Airport Terminal	Dutch Harbour Terminal	Cost Estimating
1980	Federal Aviation Administration Air Route Traffic Control Centre Alternations and Remodelling, Elmendorf Air Force Base	Anchorage, Alaska	Cost Estimating

Heliport Systems Incorporated
Airport Park, Morristown, New Jersey 07960, USA
TELEPHONE: (201) 540 0011
TELEX: 220883T AUR

President: William E Davis

Heliport Systems offers a specialised service to communities or private companies seeking to establish a heliport. These services extend from basic site selection, engineering, and regulatory compliance to heliport management. The programme includes a plan to acquire all necessary approvals with special emphasis on noise study documentation, and representation at public hearings. For eligible public-use heliports in USA the company assists in acquiring federal funding under the Airport Development Aid Programme and they undertake full management responsibility on a contract basis.
Consulting services in helicopter operations are also available. These include the selection of a rotary-wing vehicle to suit specific operational requirements; avionics equipment; review of lease, purchase, and financing options; insurance; pilot and mechanic selection; and operations and maintenance under private (FAR Part 91) or commercial (FAR Part 135) operating certification.

CH2M Hill
1301 Dove Street, Suite 800, Newport Beach, California 92660, USA

TELEPHONE: (714) 975 1000 and (213) 628 5764

Director transport projects: Vaughan G Sterling

Consulting engineers and planners with experience in airport projects.

Herbert H. Howell
232 Northwood Avenue, St Louis, Missouri 63105, USA

TELEPHONE: (314) 725 8810

Herbert Howell provides airport consultation services to airport owners and operators, their engineers, architects and planners.

Hoyle, Tanner and Associates Incorporated
One Technology Park, Londonderry, New Hampshire 03053, USA

TELEPHONE: (603) 669 5420

President: John W Tanner, PE
Senior vice president: Armen Der Hohannesian, ASLA
Senior vice president: James D Boothroyd, PE
Director of aviation planning: J Richard Ludders

Hoyle Tanner and Associates (HTA) are a firm of engineers, architects and planners who specialise in providing consulting services to airport and heliport operators.
The professional capabilities of the company comprise: planning services: airport and heliport master plans, aviation system studies; site selection studies, airport and heliport facilities, airfield capacity analyses, feasibility studies and traffic studies; design and construction phase services: airfield pavements, airport and heliport lighting, navaids/visual aids, drainage and utilities, terminal buildings, air cargo facilities, hangars and other airport buildings, access and parking facilities, industrial airparks, land acquisitions, obstruction removal and landscaping; environmental services: environmental assessments, environmental impact statements/reports, aircraft noise studies, land use compatibility studies; financial/ management consultation: grant administration, EEO/MBE programme assistance, minimum standards, operating rights agreements, tenant leases, rates and user charges.
The company has undertaken projects for 21 air carrier and commuter airports and 19 general aviation airports and heliports in the USA.

HTB Incorporated
1411 Classen Blvd, Box 1845, Oklahoma City, Oklahoma 73101, USA

TELEPHONE: (405) 525 7451

Executives: Rex M Ball, P G Wilson Jr, James L Slade, Keith Hinchey, Betty Jackson, Frank Chitwood, Robert Leise, Leon Harris, Leonard F Ball, Jim Freehof, Domby L Zinn, Bill J Blair

The company specialises in engineering, architectural, landscape and planning producing feasibility studies and engineering reports, master planning and design for commercial and industrial facilities including airports and terminals.

IAL

Aeradio House, Hayes Road, Southall, Middlesex UB2 5NJ, England

TELEPHONE: 01-843 2411
TELEX: 24114

Chairman: S Wheatcroft
Deputy chairman/managing director: J P Utterson
GM aviation marketing: E C Carpenter

IAL provides aviation, communications, computer and medical services world-wide, with 30 subsidiary and associate companies and a network of agencies operating in over 50 countries. The company currently provides technical, operational and management personnel for the following: aviation consultancy; air traffic services; aeronautical telecommunications project; radio and radar systems; meteorological services; security systems; airport fire and rescue services; motor transport services; training.

At Bailbrook College, Bath, England IAL has its own training establishment which, as well as housing the company's air traffic control radar simulator, provides a wide range of training courses in other aviation and communications-related subjects. Courses are tailored to meet customers' individual requirements. The company has been involved in the establishment and operation of training schools in many parts of the world.

John Laing Limited

Page Street, Mill Hill, London NW7 2ER, England

TELEPHONE: 01-959 3636
TELEX: 263271

Group marketing director: J M K Laing

John Laing Limited is the parent company of a major international organisation engaged in building, civil, mechanical and industrial engineering, construction, research and development as well as the manufacture of products for the construction industry.
The Group operates from headquarters at Mill Hill, London, England and permanent offices throughout the United Kingdom and overseas.
The Group's international construction activities are supported by a wide range of specialist and technical services covering the construction process: building, civil, mechanical and industrial engineering; construction, management and project consultancy; multi discipline design and land surveying; research and development; computing facilities; water and waste treatment; ground engineering; fire protection; manufacture of construction materials. The company provide a total construction capability for airport development and ancillary facilities: terminal buildings; aprons and runways; hangars; cargo handling facilities; control towers; pipework and storage tanks; infra-structure.
Projects have included: operations facilities at RAF Boscombe Down, Amesbury, new runway and terminal buildings in Birmingham, motor transport operation base at London Heathrow Airport, apron paving in Manchester, England; additional works to the Queen Alia International Airport, Amman, Jordan; airport flyover and extensions, Abha, Saudi Arabia; expansion to the International Airport, Dubai, United Arab Emirates.

Abha Airport, Saudi Arabia

Hugo Marom Aviation Consultants Limited

50 Levi Eshkol Street, Tel-Aviv, Israel

TELEPHONE: 03-411224, 03-415141
TELEX: 35770

Hugo Marom Aviation Consultants International Incorporated

28903 Hollowbrook Avenue, Agoura, California 91301, USA

TELEPHONE: (213) 347 7207

Hugo Marom Consultants specialises in aviation consultancy including the production of master plans with an economic analysis, airfield security, forecasting and accident investigation. There are two subsidiaries, Airport Engineering Hugo Marom and Aerospace Design Hugo Marom. The Airport Engineering subsidiary is experienced in the preparation of airfield concepts: total master planning of military and civil airports; architectural detail design of complete airport buildings, hangars and underground aircraft shelters; hardened fighter maintenance facilities, fighter aircraft run-up stands and engine test cells with silencers; specialists in offshore airports and airport automated information and recording systems.
The Aerospace Design company specialises in the engineering of all aircraft and airport ground equipment, aircraft test equipment, engine test cells, sound control equipment, airborne equipment and systems, aircraft installations modifications. Updating and improving aircraft performance; design of aircraft reconnaissance installations, collimator facilities; equipment standarisation studies, specification preparation, cost and programme management.
Clients of the Hugo Marom organisation comprise aircraft and engine manufacturers, airlines and operators, business and financial organisations and government bodies.

C Maurice Company Limited

Southmill Road, Bishop's Stortford, Hertfordshire CM23 3DH, England

TELEPHONE: (0279) 52231
TELEX: 817130

Director: N R Macknelly, CEng, MIEE, MBIM

The company specialises in designing and installing airfield lighting systems. They are members of the Aerodrome Owners Association and independent manufacturers of airfield lighting equipment. The service offered by Maurice includes surveying and preparation of approach profiles, setting out of schemes, installation of ancillary equipment such as cloud height recording apparatus, with modern mechanical aids and experienced technical supervision. The company also designs and prepares schemes and installs systems for base-load and stand-by power supplies.
Airport projects undertaken by the company include: Benina Airport, Benghazi; Cairo, Egypt; Yundum, Gambia; Amman, Jordan; Kuala Lumpur International Airport, Malaysia; Jeddah, Saudi Arabia; Grand Caymen and Antigua, West Indies; Guernsey and Jersey, Channel Islands and 11 other airports in the United Kingdom.

The Mitre Corporation

Metrek Division, 1820 Dolley Madison Boulevard, McLean, Virginia 22102, USA

TELEPHONE: (703) 827 6000
TELEX: 248923

Vice president and general manager: S William Gouse
Vice president: Charles C Grandy
Technical and associate directors (NL) systems development: William F Mason, Frederick C Holland, Lawrence L Holmes
Air transport systems: Howard J Kirshner, David L Bailey, J Paul Locher III
Environment: Robert P Ouellette, Lydia W Thomas
Energy and resources: Edward G Sharp, Martin M Scholl

The Metrek Division of the Mitre Corporation specialises in air traffic system engineering. Since the late 1950s, Mitre has assisted the FAA in systems engineering for: design, development, test and integration of air traffic control systems; telecommunications system design and integration; navigation aids and landing systems; airport and airspace planning, environmental and fuel

conservation analysis; economic analysis of transport alternatives.
These Mitre capabilities were applied, for example, in the modernisation of more than 150 air traffic control facilities in the USA. The same capabilities have been used and are available to other civil aviation administrations which require similar assistance.
Mitre played a part in the acquisition of the present US semi-automated air traffic control system which replaced the earlier non-radar and radar systems.
The nation-wide system includes computers, displays, digital telecommunications and primary and secondary radar processors. Services provided by Mitre included the preparation of designs and specifications for the major hardware and software sub-systems, technical evaluation of proposals, review of contractor technical design data, review and control of system interfaces, technical monitoring and direction, implementation planning, reliability and maintainability analysis and system testing.
Examples of the systems for which Mitre has written or helped write System Programme Plans include: the current air traffic control system in the USA; a major new training system for the FAA academy in the USA; the Next Generation Weather Radar Programme which is jointly sponsored by the Department of Defence, Department of Transportation and Department of Commerce in the USA; a system to process flight plans for air traffic control in the Federal Republic of Germany; an area air traffic control system, including primary and secondary radar, radar and flight data processing, buildings and roads, for Venezuela; a master plan for a ten-year programme for air traffic control improvements in China.

Monenco Limited

2045 Stanley Street, PO Box 6088, Station 'A', Montreal, Quebec H3C 3Z8, Canada

TELEPHONE: (514) 286 3636
TELEX: 05560735

The Consortium of Monenco Limited which comprises Montreal Engineering Limited, Peat Marwick and Partners, Parkin Partnership Architects and Planners and IBI Group provides comprehensive consulting services for overseas airports. The four firms in combination have the full range of airport planning and design services necessary for a range of airport systems.
Professional services offered by the consortium comprise: investigations: airspace analysis, traffic forecasts, road user surveys, passenger forecasts, passenger terminal operations, environmental impact, cargo demand analysis, economic studies, operations research, telecommunications, computer simulation models, route analysis; management: organisational planning, financial analysis, strategic planning, risk analysis, airport operations, land management and procurement, policy analysis, corporate planning, cost effectiveness analysis, airline operations, public participation, staff procurement; planning: airport master plans, ground transport systems, rapid transit, bus systems, passenger terminal concepts, control tower siting, navaids siting, cargo systems, commercial/industrial facilities, community design; design: passenger terminals, runway systems, aprons, hangars, parking structures, refuelling systems, roads, utilities, cargo facilities, navigation facilities; construction: contract documents, bills of quantities, selection of contractors, resident supervision, equipment procurement, progress programming, project management, operating manuals, inspection and commissioning, testing.
Major airport projects undertaken by the consortium include: Logar International Airport, Kabul, Afghanistan: master plan; Nassau International, Bahamas: expansion and feasibility study; Brazil International Airport, Brazil: master plan; Marabel International Airport, Montreal, Canada: master plan; Toronto International Airport, Canada: master plan; Second Toronto International Airport, Canada: master plan; Winnipeg Airport, Canada: master plan for terminal facilities; Copenhagen International Airport, Denmark; master plan; Leonardo da Vinci Airport, Rome, Italy: master plan; Sangster International, Kingston, Jamaica: master plan; Norman Manley International, Montego Bay, Jamaica: master plan; Crown Point, Tobago: master plan; Piarco International Airport, Trinidad: master plan; United States Naval Facility, Miramar, San Diego, USA: updating of master plan; Caracas International Airport, Venezuela: review of master plan; Golden Rock, St Kitts, West Indies: master plan.

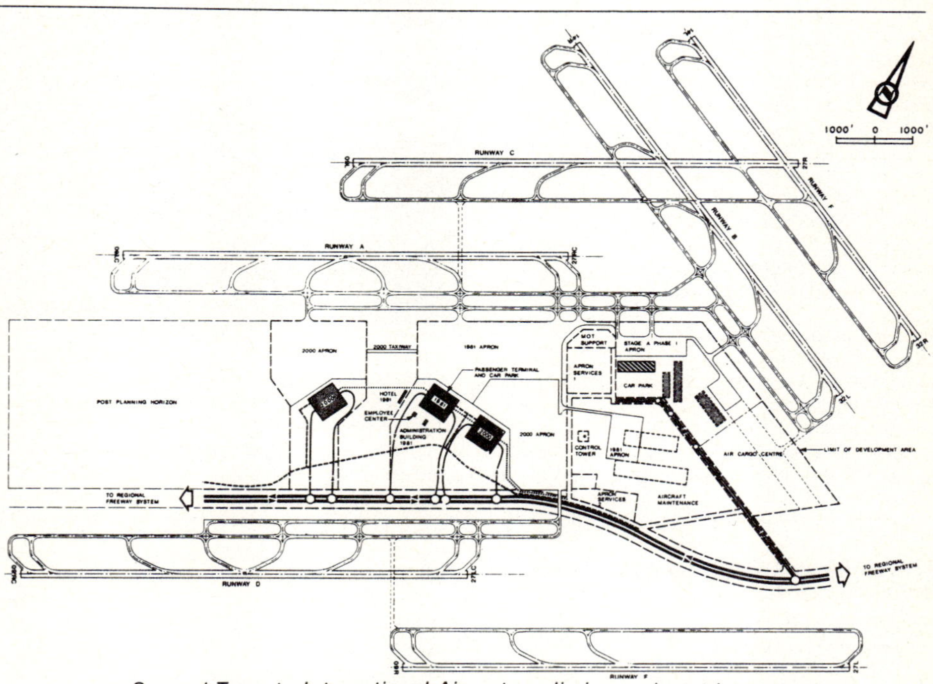

Second Toronto International Airport: preliminary airport layout

---

Morrison-Maierle Incorporated

910 Helena Avenue, PO Box 6147, Helena, Montana 59604, USA

TELEPHONE: (406) 442 3050
CABLE: Mmhelenamt

President: John H Morrison Jr
Senior vice presidents, Engineering: Harold L Eagle; Willis Wetstein
Chief transport engineer: Walter Scott

Morrison-Maierle is an international engineering corporation offering services in planning, feasibility studies and civil engineering. The firm has completed a broad variety of assignments in western USA and other countries.
Airport services offered comprise: feasibility studies, master plans, site selection, environmental impact studies and the design of control towers, runways, taxiways, aprons, helicopter pads, hangars, maintenance facilities, air cargo buildings, aircraft fuelling systems, airfield lighting, instrument landing systems, access roads, potable water systems, waste water systems, solid-waste disposal and car parking.
Typical projects undertaken by the company are at: Malmstrom Air Force Base: runway, taxiway and apron construction, base utilities and waste water treatment, housing project (Strategic Air Command US Air Force).
Glasgow Air Force Base: runway, taxiway and apron design, base utilities and waste water treatment (Strategic Air Command US Air Force).
Helena Airport: passenger terminal, crash fire rescue/maintenance.
Building, runway taxiway and apron construction for air carrier non-hub airport.
Billings Logan Field: runway, taxiway and apron rehabilitation for small hub air carrier airport.
Over 90 separate projects have been completed by Morrison-Maierle for seven air carrier airports, two military bases and 16 general aviation airports.

Helena Airport terminal complex

Netherlands Airport Consultants BV (NACO)

115 Jan Van Nassaustraat, 2596 BS The Hague, Netherlands

TELEPHONE: (070) 264131
TELEX: 33511

NACO is an independent consulting firm of airport consultants, civil, mechanical and electrical engineers, architects and economists. For over 30 years the firm has been involved with about 250 airport projects world-wide. NACO's range of services extends to all aspects of airport work from project management, initial technical and economic feasibility, financial analysis of revenue and costs through site selection, preliminary and final design, tender document preparation, selection of contractors and construction supervision. These services include determination of airport capacities based on forecast, site selection, runway and taxiway patterns, parking apron, road systems, layout of terminal buildings, systems for passengers and luggage handling, cargo buildings, maintenance buildings, power-plants, hangars, airport lighting, landing and take-off patterns and navigational aids.

Many projects are executed in collaboration with local offices to take advantage of local experience. NACO is involved on assignments for high density international airports, domestic airports, as well as national airport planning.

Amsterdam Schiphol Airport

NACO has been involved with major airports including Schiphol, Amsterdam, Frankfurt, Dusseldorf, Munich, Bogota, Caracas, Singapore, Istanbul, Lagos, Athens, Geneva, Kuwait, Colombo and Dhahran; domestic airports world-wide and with national airport planning in Saudi Arabia (22), Nigeria (18), Sudan (14), Indonesia (8), Iran (23) and Colombia (5).

The firm is currently active on more than 60 airport projects in tropical and arctic climates.

NORR Airport Planning Associates

40 University Avenue, Toronto M5J 2G3, Canada

TELEPHONE: (416) 977 1266
TELEX: 065 23255

Chairman: Hedley E H Roy
Vice chairman: Victor S Marko
President: T Carnahoff
Vice-president: P J Manson-Smith

NORR Airport Planning Associates Limited (NAPA) is a wholly-owned subsidiary of Neish Owen Rowland & Roy (NORR), a Canadian architectural and engineering consulting organisation.

NAPA was formed to concentrate on airport planning and design work which has been the focus of NORR's planning activities for the past 20 years. Airport planning and design services have been provided to public and private clients in Canada and abroad in Sri Lanka, Nepal, Singapore, Indonesia, Malaysia, Denmark, Italy and Brazil. NAPA's experience has been gained both on large complex airport development programmes such as Toronto International Airport which now handles over 12 million annual passengers, where the firm has been continuously involved for over 20 years, as well as on smaller airports both international and domestic, particularly in the developing countries.

Special airport consulting services provided by NAPA are: airport systems analyses and planning; traffic forecasts and route analyses; multiple airport system traffic assignments; ground movement system analyses; terminal area systems analyses; airport facility programming; information systems analyses; servicing facilities systems analyses.

Comprehensive airport planning: national airports system planning; regional airports evaluation and selection; master planning; terminal area master planning; facility site evaluation and selection; airport development financial and feasibility analyses.

Airport facilities architectural and engineering design: passenger terminal buildings; parking facilities; cargo terminal buildings; control towers; airport hangars; roads and bridges; airport information centres; power plants; hydrant refuelling systems.

Rio de Janeiro International Airport

Pan Am World Services Incorporated

666 Steamboat Road, Greenwich, Connecticut 06830, USA

TELEPHONE: (203) 661 1637
TELEX: 643536

In 1979 Pan Am consolidated its various service businesses within a new subsidiary Pan Am World Services Incorporated. Three operating divisions provide management and technical services to particular client interests. Additionally, a consultant engineering and design group provides support to the divisions and to special high technology customers.

Principal divisional fields of interest are: International services division: international services for management; operation and maintenance of foreign airlines; airports; communications; military and other complex programmes. Aerospace and base support services division: management; operations and maintenance of US Government owned technical facilities and military bases. General aviation services division: domestic airport and heliport management and development ADAP funding applications and programme implementation. Engineering and design department: consultant engineering services for aviation and space facilities design and development, computer applications and documentation.

Airport Management and Operations, Oman

Airport Management, Teterboro, New Jersey

The International Services Division has provided technical and managerial assistance to more than 26 countries as they developed their own air transport systems. Among them are: Pakistan International Airlines, Turkish Airlines, Ariana Afghan Airlines, Yemen Airways, Alia (The Royal Jordanian Airline), Iran Air, Avianca, Bahamasair.

Today ISD provides management, operations and services to international airports in Oman, Liberia, and Saudi Arabia. Services include:

Management operations and maintenance services for overseas airlines and airports; consultation and training in aircraft maintenance services; design and operation of air traffic control and communications systems; management of harbour and seaport facilities; and logistics planning.

The Aerospace and Base Support Services Divisions (ABSSD) were started when Pan Am became prime contractor for the United States Government in the construction of the 10 000 mile (16 100 km) long Eastern Test Range (ETR) based near Cape Canaveral, Florida.

At each island tracking station Pan Am installed complex radar, telemetry, and communications systems and constructed housing for operations personnel. ABSSD today has complete responsibility for providing the following services at the US Air Force's $1 000 million ETR facility: Base support, engineering, programme management, test operations.

Pan Am World Services has been selected to provide management operations and maintenance services at many US Government installations. Among them are: Kennedy Space Centre, Florida; Fort Gordon, Georgia, Trident Naval Submarine Base, Washington; National Space Technology Laboratories, Mississippi; Johnson Space Centre, Texas; Arnold Engineering Development Centre, Tennessee, USA.

The General Aviation Services Division (GASD) provides professional management and management consulting services for general aviation airports and heliports in the USA, including Teterboro, New Jersey; Westchester, New York; and Lunken Airport, Ohio, USA.

Services available through the company are master plans, fiscal management, airport management, operations and maintenance and tenant lease negotiations. Specialised consulting services on site include: computerised accounting; 'Compuland', a computerised landing fee system, and contract administration.

The Engineering and Design Department (EDD) produces airport and airline master plans and systems engineering for: airport terminals, baggage systems, cargo buidings, air traffic control centres, navigation aids, hangars, engine and GSE shops and military and civil airport approach lighting systems. The group also designs data base management systems for: financial analysis and development planning, automatic documentation, resources scheduling, control of management information and accounting.

Pavement Consultancy Services BV

Kleine Tocht 23, 1507 CB Zaandam, Netherlands

TELEPHONE: (075) 31 1321
TELEX: 19331

Pavement Consultancy Services (PCS) is a company of the Royal Dutch/Shell Group. Most of the activities of PCS for airports are concerned with the determination of the bearing capacity of existing pavements. For this it is essential to carry out measurements in situ and PCS normally applies two different non-destructive techniques: vibration measurements with a heavy vibrator whereby the speed of the waves generated in the various pavements is a measure for the properties of these layers and deflection measurements with a heavy-falling weight deflectometer whereby deflection and the deflection bowl are determined under a pulse load of approximately 11 tons. The outcome of these measurements is transferred into properties of the various pavement layers.

These properties are used in the BISAR programme to calculate stresses and strains in the pavement layers under the effect of the actual wheel-load configurations of various aircraft. In this way it is possible to work out the maximum allowable aircraft load and the permitted number of load applications. The latter is related to the fatigue properties of the pavement layers.

The methods described above are based on an analytical approach and are therefore tailor-made for each aircraft pavement. It is also possible to use the outcome of the measurements to determine an lcn-value, or more recently a pcn-value. The latter is important in view of the new acn/pcn method which is currently being proposed by ICAO.

PCS has used vibration measurements to evaluate the pavements of the airports of New York Kennedy, La Guardia and Newark, USA, Brussels, Belgium, Zurich, Switzerland and Rotterdam, Netherlands. More recently the modern heavy falling weight deflectometer (fwd) has been applied at airports in Brussels, Belgium; Amsterdam, Beek and Rotterdam, Netherlands; Bremen, Frankfurt, Hamburg, Memmingen, Nürnberg and Stuttgart, Federal Republic of Germany; Kuwait; Colombo, Sri Lanka and about 15 military airports in the Netherlands, Federal Germany and Jordan.

The advantages of the fwd are that it is mobile, quick with 200 measuring points per day or night, and that the load level is representative of the wheel load of a heavy aircraft.

In many cases the services of PCS establish the current bearing capacity of an aircraft pavement, assess the residual life and determine the thickness of an overlay. The latter depends on the subgrade properties, the strength of the existing pavement, the climate, the future traffic loads and the materials used for the overlay.

In view of the company's experience in asphalt mixes and their application PCS are often engaged in projects where maximum use has to be made of locally available aggregates. For such projects it is necessary to design suitable asphalt mixes with local materials and to adapt the thickness of the pavement for the use of such asphalt mixes so that a structurally-sound solution is found which is economical.

Examples of projects where asphalt technology played an important role are two airfields in Libya (maximum use of sand) and the airports of Dhahran and Tabuk in Saudi Arabia.

Pavement Specialists Incorporated

4405 Reese Drive, SR 114 Esters Road South, PO Box 61327, Dallas/Fort Worth Airport, Texas 75261, USA

TELEPHONE: (214) 258 0660

Vice president: Joseph A Kurtz

Pavement Specialists Incorporated, a subsidiary of Reese Maintenance Corporation of Napoleon, Ohio, USA are specialist sub-contractors to perform: grooving, grinding, concrete and epoxy repairs, joint cleaning and resealing, rubber removal and pavement marking.

Pavement grooving is a method of combating wet-weather skidding accidents on airport runways. Dynamic hydroplaning, one of the major causes of aircraft skidding, occurs when the landing-gear tyres ride on a cushioning film of water on the runway surface. Viscous hydroplaning occurs if a thin film of oil dirt or rubber particles mixes with water and prevents the tyres from making good contact with the pavement. A pattern of shallow, parallel Safety Grooves sawed transversely across a runway provides better steering stability and increase braking traction.

The grooves are designed to allow water to flow from underneath a tyre, and break up a slippery viscous film of oil and water.

Pavement grooving has been carried out at Dallas/Fort Worth Airport and 13 other civil airports in USA and at US Air Force bases in Okinawa, Japan, Republic of Korea and the USA.

Plessey Airports

The Plessey Company Ltd, Addlestone, Weybridge, Surrey KT15 2PW, England

TELEPHONE: (0932) 47282
TELEX: 262329

Plessey Airports provides a complete airport package service to civil aviation authorities world-wide. Under a single contract the company undertakes the creation of new airports or the expansion or modernisation of existing ones, from initial planning through to hand-over in a fully-operational state. Through connections with the financial institutions in London and world-wide Plessey can also assist in the provision of government-backed preferential credit to finance airport projects.

Plessey provides a total consultative service throughout the planning stage. Planning criteria are in accordance with local requirements using as a reference the standards and recommendations of the ICAO as applicable.

Initial requirements are defined in discussions with the authority concerned. Detailed studies then determine the full operational requirement and the extent of the civil works design, systems engineering, construction works, and technical facilities and equipment necessary for its implementation. Operational and siting appreciation include traffic studies to determine airport parameters and budget requirements. From all these studies a detailed design, specification and programme are prepared.

Design studies are carried out in two stages. Initially, the basic concepts are established and the comprehensive plan developed in terms of construction works, including aircraft pavements, terminal buildings, operational and technical buildings and the airport equipments to be provided.

At this stage systems are planned and the constructional aspects are tailored to fit in with the overall systems design. In the second stage detailed designs for the construction works are produced and specifications prepared for all equipments and sub-systems. In this phase Plessey employs the services of consultants.

For construction, Plessey provides an overall management service, the civil engineering normally being undertaken by suitably qualified contractors, preferably already established in the territory, using local labour where possible. Plessey prepares tender documents for the civil engineering aspects and will make recommendations on the submissions. The company, through its civil engineering management, provides an overseeing service during the construction phase.

An airport package encompasses the procurement and supply of all the equipment and services necessary for a functioning modern airport. These are likely to include: passenger and freight terminal facilities; control tower and technical block facilities; radar, data and display equipment; radio, telephone and telex systems; navigational aids; lighting; power generation and distribution; meteorological equipment; runway and apron services; fire and emergency services; cargo handling and test equipment and tools.

The company will select and train qualified personnel both overseas and on site during the construction stages and will maintain a complete operational service until the new staff are ready to take over.

Plessey provides the logistic support on completion of the project to ensure that the airport continues to operate efficiently. A comprehensive service is offered including maintenance and operational services, spares management and post design services.

Airport projects undertaken by the company include development contracts for 25 airports as follows:

development of Garoua International Airport, Cameroon; development of Cairo International Airport, Egypt; supply, installation and commissioning of equipment and facilities at 17 airports in Gabon; development of Abidjan Port Bouët International Airport, Ivory Coast; supply and installation of navigation aids and communications systems of Sebha International Airport, Libya; upgrading of runways to international standards at six airports in Zaire.

Projects announced in May 1982 include a new international airport at Port Salines, Granada and a new airport at Mmabatho, South Africa.

Prismo Universal Limited

Rowfant, Crawley, West Sussex RH10 4NF, England

TELEPHONE: (0342) 714949
TELEX: 95190

Prismo Universal is believed to be the world's largest company engaged in highway and airfield marking. In association with local licensees and joint-venture companies, it has operational facilities in the UK and throughout Europe, the USA, the Middle and Far East, Africa and Australia.
Products and services offered include Sprayplastic runway marking, Slurritex runway and taxiway surfacing and Fuelseal surfacing for refuelling areas.
Sprayplastic is a cost-effective system of applying durable, fast-setting thermoplastic markings to runways from automatic truck-mounted equipment. Sprayplastic is applied at a thickness of 1.5 mm which is half the thickness of normal thermoplastic marking material, yet has the same service life. Glass beads can be sprayed on to the surface to produce an instantly reflective line. Sophisticated machines with skilled crews can apply up to 2500 square metres of marking per day, so that airport runways can be remarked with the minimum delay to operations.
Slurritex is a thin veneer coat of slurry designed to rejuvenate a runway surface which is becoming spalled and oxidised. The coating extends the life of surfaces by up to five years. Slurritex has been well tried and tested on many airfields, both military and civil in the United Kingdom, the USA, the Middle East, Gibraltar and Africa. Slurritex is a mixture of bitumen emulsion, blended aggregate, water and cement. Two types of emulsions are used: cationic emulsion is used in cooler climates because of its advantage of a quicker setting time. Ationic emulsion is used in warmer climates, because of its flexibility in extreme climatic conditions.
Slurritex is mixed and applied by one vehicle. Sufficient material is carried to lay 3000 square metres of 1.5 mm aggregate surfacing and 1500 square metres of 3 mm aggregate surfacing from one load. Replenishment of the vehicle is from a central stockpile usually located on the airfield. On some contracts over 10 000 square metres of surfacing can be carried out in a single day.
Slurritex meets the requirements of the British Department of the Environment Bridges and Highways Specification, of the FAA specification in the USA and of the International Slurry Seal Association Specification. Fuelseal is a hand-applied slurry designed to seal and protect bitumen-based surfaces from oil spillage in refuelling areas and on parking aprons.
It is a mixture of tar emulsion, blended aggregates, fine filler and water, which is applied in two coats with squeegees. When cured, it provides a hard-wearing veneer.

Reynolds, Smith and Hills Incorporated

4019 Boulevard Centre Drive, PO Box 4850, Jacksonville, Florida 32201, USA

TELEPHONE: (904) 396 2011

Associate vice president, airport planning: James A Meehan, AIAA
Executive vice president, architecture and planning: Robert F Darby, AIA
Senior vice president, corporate marketing: K N Henderson, PE, NSPE

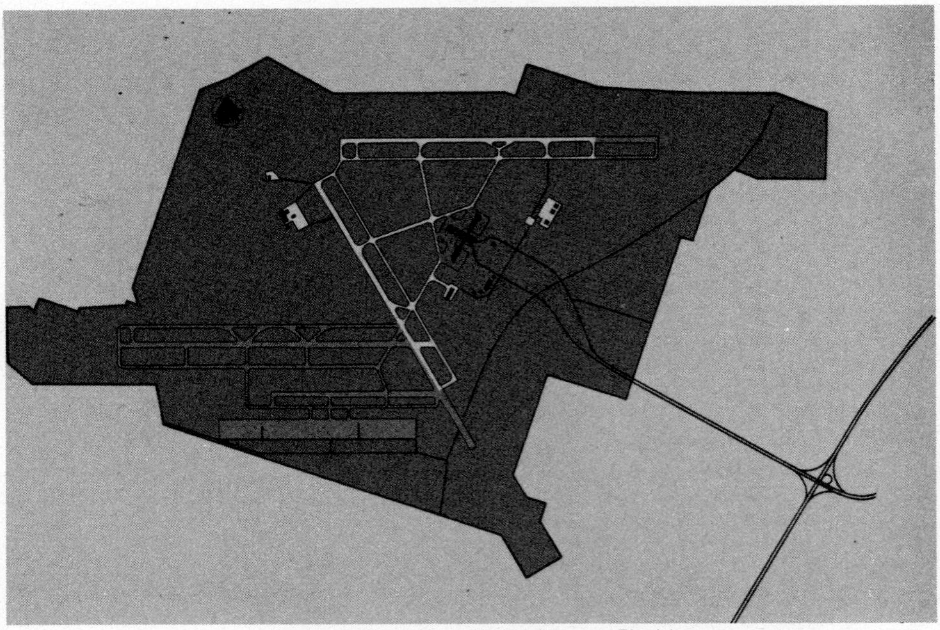

Jacksonville International Airport, Florida

Reynolds, Smith and Hills are a firm of architects, engineers and planners. The company's approach to airport design involves development of alternative airport expansion designs suited to each viable airfield and access configuration. These concepts are judged against a set of evaluation criteria jointly established with the owner.
A two-level evaluation is used to minimise the time required for analysis of all possible terminal/airfield/access combinations. The first level provides a preliminary analysis and cost estimate for each combination, and leads to a selection process which reduces the number of viable ideas to three or four. The second level of effort involves a much more comprehensive analysis of these and results in the selection and recommendation of the one which best meets all requirements.
This two-level analysis calls on data previously developed from the airport master plan and the environmental assessment. The preliminary terminal analysis employs the preliminary design and space programme whereas the concepts selected for final evaluation are developed with the use of more refined data which include input from the air carriers.
Approval by the owner of the airport expansion concept marks the completion of Phase I; concept development. Authorisation to begin basic architectural-engineering services marks the opening of Phase II.
During the initial stages of Phase II, the selected terminal design is developed. Structural, mechanical and electrical engineering systems are selected and integrated. Refinements are added to provide visual and audio communications, interior design and landscaping. And the special requirements of airports are met: car rental facilities, food services, concessionaires, baggage handling, fuelling, security and others. At the conclusion of this phase the scope and cost of the project are well defined and airlines are included. The approved design is converted into working drawings and specifications suitable for bidding and construction.
In the final phase, post operational services, the project team assists the users in start-up, shake-down and other tasks associated with commissioning a new airport project.
Reynolds, Smith and Hills have been involved in the following projects in the USA. Tampa International Airport, Florida; Fort Lauderdale Hollywood International, Florida; Greensboro/High Point/Winston-Salem Regional, Greensboro, North Carolina; Jacksonville International, Florida; Great Pittsburgh International, Pennsylvania; Metropolitan Nashville, Tennessee; Bradley International, Windsor Locks, Connecticut; Port Columbus International, Columbus, Ohio; Shrewsport Regional, Louisiana; Daytona Beach Regional, Florida; Raleigh Durham, Raleigh, North Carolina; Mobile Municipal, Alabama; Palm Beach International, Florida; Pensacola Municipal, Florida.
Overseas airport projects comprise: Freeport International, Grand Bahama Island, The Bahamas; Piarco International, Port of Spain, Trinidad; Geneva Intercontinental, Geneva, Switzerland

Royal Volker Stevin NV

PO Box 2695, Oostmaaslaan 71, 3000 CR Rotterdam, Netherlands

TELEPHONE: (010) 331533
TELEX: 23486

Royal Volker Stevin is a European-based international contractor with world-wide activities. The roads and asphalt division within the group is responsible for airfield and airport projects. Volker Stevin Roads within this division is responsible for all activities outside the Netherlands.
The company have been active in airfield construction, reconstruction, and maintenance, in the Netherlands and abroad for many decades.
If required the company calls on other divisions within the group for all types of building, civil engineering works, dredging, pipelines and any other discipline for such projects.
The company works with consultants and suppliers of all types of systems and equipment when projects call for a turn-key approach and therefore offer services for construction of any type of airfield under any conditions.
Projects undertaken by the company include: Schiphol Airport, Amsterdam, Netherlands: construction and development; Rotterdam Airport, Netherlands: development; Jos, Sokoto, Ilorin Airports, Nigeria: construction; Zanderij International, Suriname: construction; Waginengen airfield, Suriname: development; Dr. Albert Plesman International airport: maintenance and development; St. Eustatius, Flamingo airfield, Netherlands Antilles: development; Bonaire,

Juliana International Airport, Netherlands Antilles: development; St. Maarten, Netherlands Antilles.
In addition Royal Volker Stevin have constructed 15 military airfields in the Netherlands, Belgium and outside Europe. Two are also used for civil aviation.

Amsterdam Schiphol Airport

Sar-Assist Incorporated
One Island Lane, Greenwich, Connecticut 06830, USA

TELEPHONE: (203) 869 1322

Principal director: G W Heath

The company provides consultancy on rescue, recovery, survival and safety in aviation.

Schiphol Airport Authority
PO Box 7501, 1118 ZG Schiphol Airport, Netherlands

TELEPHONE: (020) 5172582
TELEX: 14653

Chief executive: Martin C van Gent

Airport Management Consultancy is a branch created by the Schiphol Airport Authority to advise on planning of facilities at airports, the organisation of airport-operating companies and to provide training programmes and courses, both on site and at Amsterdam Airport, Schiphol.
The branch is able to provide: commercial management studies, scientific investigations, market research studies, traffic forecasts and analysis of costs and revenues. It can give advice regarding: bid documents, financing, concession agreements, operational support, maintenance and upkeep, terminal accommodation, car parking, information systems, signposting and airport facilities.
The requirements are carefully studied and evaluated. This results in a working programme. During the preparation and execution of this programme there is close co-operation within the Airport Authority and, if necessary, with external advisers.
Interdisciplinary teams are formed for each project and consideration given to ways in which the costs related to the studies, reports, educational programmes, etc can be financed through appropriate funds.

Scott Wilson Kirkpatrick and Partners
Scott House, Basingstoke, Hampshire RG21 2JG, England

TELEPHONE: (0256) 661161
TELEX: 858805

Clifton House, Euston Road, London NW1 2RA, England

TELEPHONE: 01-388 6621

Executives: G M J Williams, MA, MICE, FIStructE; K C W James, MBE, BSc, MICE, FIHE; W A D Sterling, BSc, MICE

Scott Wilson Kirkpatrick and Partners are an international group of consulting engineers with offices world-wide formed by amalgamation of Sir Cyril Kirkpatrick and Partners and Scott and Wilson.
The company's activities comprise transport plans etc, airports, roads, bridges and tunnels, maritime, water and waste water engineering, buildings, geotechnical and appropriate technology in developing countries.
The airport services include master plans; feasibility studies; traffic forecasts, economic analyses; site investigations; runways, taxiways, aprons; terminal buildings, ancillary buildings; airfield lighting, landing aids; maintenance bases, hangars, workshops; detailed designs and supervision of construction.
Airport projects have been undertaken by the company at Hong Kong; Kota Kinabalu, Sabah, Malaysia; Brunei; Mombasa, Kenya; Lilongwe, Malawi; Doha, Qatar.
Engineering bases at British Airways, London Heathrow, England, Glasgow, Aberdeen and Sumburgh, Shetland Islands, Scotland. Master plans have been completed for Khartoum, Sudan, Singapore and Munich II airports and SWK have been engaged on airport studies for the UK Crown Agents at Antigua and St Kitts in the West Indies and at Seiyun in the People's Democratic Republic of Yemen; for the British Ministry of Overseas Development at Grenada in the West Indies; for British Petroleum in Libya and Abu Dhabi, United Arab Emirates; and for three different local authorities in England.

British Airways Engineering Base at London Heathrow Airport

Seri Renault Ingenierie
2 avenue du Vieil-Etang, 78349 Bois d'Arcy, France

TELEPHONE: (1) 043 81 34
TELEX: 695426 F

Airports sales manager: J M Bas

Seri Renault Ingenierie, a subsidiary of the Renault group, specialises in the design, engineering, construction and installation of air freight terminals. The company have produced facilities for: Aeroport de Paris: Paris Orly, France; Air France: Paris Orly and Charles de Gaulle, France; New York JF Kennedy, USA; Sabena: Brussels, Belgium; UTA: Paris Le Bourget and Charles de Gaulle, France; Air Afrique: Paris Charles de Gaulle, France; Pakistan International Airlines: Karachi, Pakistan; London Heathrow, England; Egyptair: Cairo, Egypt.

Shorgard Limited
22A Manor Place, Edinburgh EH3 7DS, Scotland

TELEPHONE: (031) 226 3901
TELEX: 95452

Director: H D Gardiner

Shorgard specialises in the design, supply and installation of airfield lighting systems.
The company has completed projects at the following airports: Edinburgh, Glenrothes, Flotta, Kirkwall, Orkney Islands, Benbecula, Hebrides Islands, Sumburgh, Shetland Islands, Scotland; Eglinton, Londonderry, Northern Ireland

Skanska

International Division, S 21102 Malmo, Sweden

TELEPHONE: (46) 401 440 00
TELEX: 32247

Skanska is a civil engineering group of companies. The group has a number of divisions and subsidiaries specialising in every phase of construction: design, engineering, component fabrication, erection, management, administration, finance and economy. The company's senior management group consists of specialists in construction, management and finance. The company uses data processing systems to handle construction management from planning and design to administration of a completed facility.
Skanska has some 60 regional and local offices in Sweden and subsidiaries in ten countries abroad. Skanska carried out the construction of the Gothenburg Landvetter International Airport, Sweden which required the blasting and excavating of 4 million cubic metres of rock, moving 1·5 million cubic metres of earth and draining three lakes.

Gothenburg Landvetter International Airport

Skidmore, Owings and Merrill

33 West Monroe Street, Chicago, Illinois 60603, USA.

TELEPHONE: (312) 641 5959
TELEX: 25 4337

General partner: Albert Lockett

Skidmore, Owings and Merrill provides complete services in the fields of architectural design, planning and engineering. They have undertaken a variety of projects in the USA and in more than 40 countries world-wide. Over 1800 architects, engineers, urban planners, landscape architects, interior architects, lawyers, economists, social scientists and management specialists work in nine major offices.
The firm's work includes large- and small-scale projects including offices, banks, commercial, industrial and educational buildings, hotels, hospitals, airports, recreational, religious and residential facilities. A broad range of urban planning projects includes central business district redevelopment and urban design, environmental impact and regional transport studies.

Smith, Hinchman and Grylls Associates

455 West Fort Street, Detroit, Michigan 48226, USA

TELEPHONE: (313) 964 3000

Smith, Hinchman and Grylls, a member of The Smith Group, are a firm of architects, engineers and planners.
Airports designed by the company include: William B Hartsfield International Airport, Atlanta, Georgia, USA; Metropolitan Wayne County International Airport, Detroit, Michigan, USA

Societe Internationale de Telecommunications Aeronautiques

Administrative headquarters:
112 avenue Charles de Gaulle, 9522 Neuilly-sur-Seine, France

TELEPHONE: 758 12 22

SITA Management

Director general: C Lalanne
Assistant director general: J Glories, K Molenaar
Secretary general: S Dwek
Assistant to the director general: M Thwaites
Technical: A Puerto
Financial: C Deroual
Personnel: B Hussenot-Desenonges
Regions co-ordination: R Bebie
Services development: D Andorka
Operations: C Avignon
General studies and contracts: J C Barat
Software development: J Brandt
Information and public relations: R Duport
Technical studies and development: G Giraudbit
Organisation and methods: B Jaureguiberry
Projects implementation: I Michelson
Data processing and statistics: J Rondeaux
Corporate planning: G Strigari
Supplies and services: P Zerdoun

The Societe Internationale de Telecommunications Aeronautiques (SITA) is a non-profit-making, co-operative organisation registered in Belgium.
SITA was founded in 1949 by 11 airlines: Air France, three British companies, which form the present-day British Airways, KLM, Sabena, three Scandinavian companies which today are merged in SAS, Swissair and TWA.
SITA serves 240 member airlines in 152 countries. There are over 16 000 teleprinter and computer terminals in 12 000 airline offices in 800 cities connected to the SITA network. Included are approximately 5 000 terminals for 35 different seat reservation systems used by 85 member airlines. In addition to its data network, SITA operates a large reservation system in Atlanta, USA which is shared by 27 airlines.
Services provided by SITA fall into one of two broad categories: telecommunications services, and information, handling and supplementary services.
Telecommunications services are provided by switching centres and teletype and data circuits, the circuits being meshed so as to provide a fail-safe alternative in case of failures on a particular route. The 163 switching centres perform a message and data switching function under the store-and-forward transmission principle.
The backbone of the network is the packet-switched, integrated high-level network currently comprised of ten large switching nodes in Amsterdam, Beirut, Frankfurt, Hong Kong, London, Madrid, New York, Paris, Rome and Tokyo which are linked by alternate voice-data (avd) circuits, operated in most cases at 9600 bits per second (bps). It controls a second level of computerised nodes known as medium level centres in Bangkok, Manila and Sydney and 48 satellite processors (programmed data concentrators) serving as interface devices for the connection of data terminals and teleprinters. A third level of the network is formed by time division multiplexers and other types of data concentrators and by manual telegraph switching centres.
The SITA network is interconnected by over 500 medium-speed and 5000 low-speed circuits, all leased from the post and telecommunications authorities, or recognised private operating agencies (common carriers). The SITA network transmits conversational and teletype messages.
The heavy demand for transmitting conversational messages led SITA to develop the Advanced Network.
Intended to meet future airline requirements and changes in the flow and characteristics of traffic the first phase of the development will be the Network Control System, intended to make optimum use of present network resources by improving supervisory and reporting facilities, allowing the cut-over of new systems without impairing existing ones and ensuring that facilities evolved for control functions conform with the long-range requirements of the network. It will provide overall network supervision from a few regional centres and one world-wide centre and have operator communication between the various centres. It will also relieve the High Level Network of a significant proportion of the overall task of network control.
The second phase of the Advanced Network implementation is the Data Transport Network, which will assume some of those functions of the High Level Network associated with transmitting data from one node to another and from nodes to the satellite processors.
With the Advanced Network, SITA will introduce aeronautical telecommunications services such as multi-access terminals enabling terminals to access different reservations computers (or other applications computers) via medium-speed circuits using mediumspeed procedures and to generate and receive teletype traffic; liaison services which will provide for the computer-to-computer transfer of large amounts of data (bulk traffic); combined services, by which crts used for conversational traffic can also generate and receive teletype traffic via associated printers; switched data for query/response and message applications between airline computer systems.
SITA offers the following information handling and supplementary services: GABRIEL reservation system which is an airline reservation system based on a triple UNIVAC configuration in Atlanta, Georgia, USA and provides a comprehensive reservation and management information service to those airlines which, for one reason or another, do not choose a privately-owned or leased system. There are currently 27 carriers using the system; departure control services are based on a stand-alone system aimed at catering for the requirements of a group of airlines or individual carriers whose passenger volumes do not exceed 1000 per peak hour. The system provides for three basic functions, namely check-in, weight-and-balance and boarding pass printing. There are installations in Budapest, Hungary (1978); Sofia, Bulgaria (1979); and Abidjan, Ivory Coast (1980).
The IATA/SITA baggage tracing system is a world-wide automated baggage tracing service, jointly developed by IATA and SITA.
Credit authorisation and document verification

service is available to all SITA member airlines which have crt connections to the SITA network and can develop their own reservations system.
The personnel training service is designed to help the airlines to train their personnel in SITA procedures, data processing and telecommunications. Training courses cover telecommunications, transmission and multiplexing techniques, networks, communications procedures, computer and terminal equipment, data processing techniques etc. The courses are normally given in Paris, in French, English or Spanish, using audio-visual aids. Arrangements can also be made to hold courses at customers' premises, if so required.
The meteorological services provided to several airlines, consist of the selecting of global meteorological data obtained from the Suitland, USA, National Weather Bureau.
According to specific user requirements, data is transmitted twice daily to the service participants for flight planning purposes.
Consulting services are available to all member airlines and include the performance of feasibility studies, economic evaluations, system evaluations, network design and project management.
Maintenance services based on experience of electromechanical and electronic equipment SITA offers field maintenance services to member airlines through its subsidiary company ITS (International Telecommunication Services).

Sofreavia
75 rue de la Boétie, 75008 Paris, France

TELEPHONE: 359 22 93/359 00 56 57
TELEX: 641360 Sofavia/640288 Sofaser

Sofreavia and its subsidiary Sofreavia Service specialise in airport design, planning, project management and technical support.
The company can undertake air transport studies, master plans and architectural design, financial estimates, project management including contract negotiations, technical support and training.
Projects include: study of the aeronautical telecommunications of different African countries, on behalf of the International Civil Aviation Organisation; study and construction of a complex for the technical facilities on the Boufarik air base in Algeria; studies and detail for adapting Bujumbura Airport, Burundi, Kigali airport in Rwanda, and the Algerian airports of Annaba, Bejaia and El Oued Guemar to wide-body aircraft; design and detailed studies of the Indonesian airports of Balik-Pan, Jayapura, Medan, Palembang, Pontianak, Semarang, Sorong on behalf of the Indonesian government and the Asiatic Development Bank, in collaboration with Aeroport de Paris; study and definition of the equipment of St Kitts Airport, West Indies; study of technical equipment in the modernisation of Dar Es Salaam Airport, Tanzania; updating the meteorological equipment of Kenya; supervision of the construction of the airport of Abuja, Nigeria; the preliminary design and working studies for the airports of Owerri, Akure and Bauchi, Nigeria; consulting engineering services on the airport of Niamtougou, Togo; updating the detailed studies for the passenger air terminal in Hodeidah Airport, North Yemen; supply and installation of aeronautical and meteorological equipment in Malawi.

Model of Kigali Airport, Rwanda

Software Sciences Limited
Abbey House, 282/292 Farnborough Road, Farnborough, Hampshire GU14 7NB, England

TELEPHONE: (0252) 44321
TELEX: 858228

Software Sciences, a subsidiary of BOC International, has a Special Systems division which concentrates its activities on real time operational computing, communications and information systems for a variety of aviation applications including air traffic control.
Project involvement may range from consultancy, feasibility analysis, system design and evaluation, and technical audits, to software production, joint venture developments, system procurement and trials management. This is reflected in such projects as: design of flight information airline scheduling and resource allocation systems; software development for radar data processing in aviation environments; airport and airspace system studies world-wide.
The company can assume responsibility for turnkey implementation of major projects in these areas. A current example of these include the supply and installation, for the British CAA, of a flight data processing system for the Oceanic Air Traffic Control.
Software Sciences operates overseas, both on a divisional basis and through its overseas companies, Software Sciences Nederland BV and Software Sciences Inc.

Alan Stratford and Associates Limited
Fairoaks Airport, Chobham, Woking, Surrey, England

TELEPHONE: (09905) 7822
TELEX: 858348

Managing director: D M Lumb

Alan Stratford and Associates is an independent aviation consulting organisation specialising in the airport and airline fields. During the last 16 years, it has carried out studies at more than 60 airports in 40 countries. It covers most of the range of services required in airport planning including passenger, freight and type mix forecasting, locational and cost-benefit analysis, feasibility studies, master planning, runway, apron and taxiway layouts, terminal building sizing and flow-charting, facility planning and air traffic control studies. The firm has carried out airport regional planning studies in the Caribbean, Guyana and Malaysia. Its clients include a number of overseas governments, the British CAA and Ministry of Defence, the US Department of Commerce, the Japanese Ministry of International Trade and the World Bank.

Tarmac Group
PO Box 8, Ettingshall, Wolverhampton WV4 6JP, England

TELEPHONE: (0902) 41101
TELEX: 338544

Tarmac is an international group of construction and building materials supply companies.
The group's capability for airport construction covers most aspects from the initial civil engineering work to the surfacing of runways, taxiways and parking aprons, as well as the design and construction of administration, maintenance and control buildings.
Tarmac co-ordinate projects throughout the company: design, building and civil engineering from the Construction and International Divisions; roofing, cladding and insulation from its Building Products Division; and surfacing technology from the Roadstone Division.
Contracts include work in Bahrain; London Heathrow Airport, England; Paris Charles de Gaulle Airport, France; Hong Kong; Edinburgh, Prestwick Airport, Scotland; Yesilkoy Airport, Turkey; Sharjah Airport, United Arab Emirates; Rhoose Airport, Wales; Sana'a Airport, Yemen Arab Republic as well as at many military installations.
Tarmac specialises in surfacing developments. In addition to the traditional surfacing applications, the group specialises in the laying of Salviacim, a high technology surface for taxiways and parking areas where a strong, fuel- and heat-resistant surface is essential.

Taylor Woodrow International Limited
Western House, Western Avenue, London W5 1EU, England

TELEPHONE: 01-997 6641
TELEX: 23503

Taylor Woodrow is involved in the construction of airfields and airports in the United Kingdom and other countries. The company has wide experience of all aspects of the construction industry and of contract management.
Work undertaken includes runways and aircraft parking aprons, passenger and cargo terminals, cargo handling facilities, hangars and maintenance facilities, hotels, restaurants and ancillary electrical and mechanical services world-wide.
Various other types of civil construction have also been carried out, as well as specialised installations such as hydrant systems and security systems. The company offer full management services for all such projects.

Teleplan AB

PO Box 1310, S-171 25 Solna, Sweden

TELEPHONE: 08-98 10 00
TELEX: 175 87 Telplan S

The company is owned by Bofors, Saab-Scania, Philips, SRA Communications and LM Ericsson. It covers the field of electronics for different applications: computers, telecommunications, surveillance systems, command and control systems and security systems. Teleplan also has specialists in office automation, process control, project management and information processing. Within these fields the company are prepared to take total responsibility for a project.

Professional services offered cover requirement analyses, functional specifications, system designs, technical specifications, RFQ documents, tender evaluation, contract negotiations and follow-up, acceptance testing, site planning, installation planning, and supervision, start of operations, training, maintenance planning and project management.

Assignments within the airport field include: airport security systems: Malmo Sturup, Gothenburg Landvetter, Stockholm, Aralanda, Sweden; power supply and general electrical installations, airport lighting systems, air traffic control centres, navigation/landing aids and communication systems in civil and military airports in Sweden.

Dala Airport, Borlange, Sweden

Arnold W Thompson PC

316 Elwood Avenue, PO Box 221, Hawthorne, New York 10532, USA

TELEPHONE: (914) 769 4456

President: Arnold W Thompson

The company specialises in the planning of airport passenger terminals and associated facilities. Services include: forecasting, space programming, functional and operations studies, monitoring of design documents and preparation of financial feasibility reports. Clients include government agencies, airport authorities, architects, engineers, car rental companies, airlines and other airport concessionaires.

Tippetts-Abbett-McCarthy-Stratton

1101 15th Street, NW Washington, DC 20005, USA

TELEPHONE: (202) 296 4371
TELEX: RCA 248677

Partners: John Lowe, III, PE; Wilson V Binger, PE; Raymond J Hodge, PE; Austin E Brant, Jr PE; John E Bardes, PE; Robert F Heins, PE; Dana E Low, PE; Eugene O'Brien, PE; Donald R Peirce, RA; Patrick J McAward, Jr PE; Philip Perdichizzi, PE

Tippetts-Abbett-McCarthy-Stratton (TAMS) has provided engineering-architectural services for airport and terminal facilities for over 75 airport projects world-wide.

Since 1943 TAMS has been providing comprehensive analysis, planning, design, and engineering management services on a world-wide basis.

TAMS maintains more than 20 offices in the USA, the Middle East, Africa, Asia and Latin America.

TAMS provides professional services for: airport-related studies: economic and demographic, transport market demand, air transport demand, regional planning, airport traffic forecasts, financial and revenue, computerised aviation programs; airport design criteria: aircraft performance capacity, airfield design standards, air space and traffic, ground transportation, passenger facilities, air cargo facilities, general aviation, air traffic control – navaids, aircraft noise, mechanical systems, airport and airfield lighting, baggage handling, aircraft refuelling, airport systems planning, people-mover systems; master planning: airport layout planning, land use planning, terminal facilities, environmental analysis, regional impact of airport, comparative analysis of alternatives, project cost analysis, economic and technical feasibility; project design: conceptual and preliminary design, engineering analyses, design drawings, specifications and contract documents, detailed design and construction drawings, computer analysis for civil and structural engineering design, airport and terminal simulation studies; engineering management: development of project criteria, supervision and review of designs by other consultants, development of construction procedures, estimates of construction costs, cost and budget control, construction schedules, using critical path method, technical inspection, supervision of construction and equipment installation, processing of contractors' payment vouchers, management of construction materials and equipment purchasing, construction housing, training of personnel for project implementation and management.

International passenger terminal, Kimpo International Airport

Airport projects carried out by TAMS include: Aryamehr International Airport, Iran; Queen Alia International Airport, Amman, Jordan; Kimpo International Airport, Seoul, Republic of Korea; Bangkok International Airport, Thailand; Palm Beach International Airport, Florida, San Juan International Airport, Puerto Rico, Portland International Airport, Maine, Bangor International Airport, Maine, Teterboro Airport, New Jersey, Dallas/Fort Worth Airport, Texas, Pan Am World Terminal, New York, JF Kennedy Airport, Greater Pittsburgh International Airport, Pennsylvania, USA

TRA Airport Consultants
215 Columbia Street, Seattle, Washington 98104, USA

TELEPHONE: (206) 682 1133
TELEX: 321 298 Tra sea

Partner, airport consulting director: Edward K McCagg II, AIA
Airport planning director: Carl V Robart, PhD

TRA Airport Consultants provides a wide range of services, including: airport master planning; land use and environmental planning; facilities programming; terminal planning and design; central monitoring and control systems; baggage handling systems; interior design; signing and graphics; airline tenant facilities; and architecture, engineering and planning.
Projects range from large international airports (both within the USA and in other countries) to medium- and small-hub and general aviation airports.
Projects include: Doha International Airport, Qatar; Mexico City International Airport at Zumpango, Mexico; Guam International Airport, Guam; Mactan International Airport, the Philippines; Pago Pago International Airport, American Samoa; Salt Lake City International Airport, Salt Lake City, Fairbanks, Anchorage and Juneau International Airports, Alaska, Sea-Tac International Airport, Seattle, McCarran International Airport, Las Vegas, USA

Alistair Tucker Associates
39 Charing Cross Road, London WC2H 0AW, England

TELEPHONE: 01-439 3631
TELEX: 267678 Aircon G

Alistair Tucker Associates specialises as consultants in management economics and the regulation of the air transport industry. Studies concerning airport siting, expansion and development; airline corporate strategy and planning, organisational and management structure and systems, aircraft evaluation and selection, route analyses, marketing research, tariff policies; international and national regulatory objectives and control procedures; and considerations of international relations are all examples of work which have been undertaken.
Such programmes may include examinations of overall national or corporate policy and objectives, reviews of the overall feasibility of projects and evaluation in terms of financial returns, financial systems and control, and implementation and monitoring procedures. Economic studies include techniques such as social cost/benefit analysis. An additional area of air traffic control experience concerns the setting up of objectives for, and advice on the negotiation of, international air services agreements.
Clients include international aid and development agencies, governments and regional authorities, regulatory bodies, aircraft manufacturers, airport authorities and leading airlines. Projects have been undertaken in the Caribbean, Europe, Indian Ocean Region, Mediterranean Region, Near, Middle and Far East, North America, Pacific Region, and West, Central and Southern Africa.

Wainwright Planning
PO Box 11569, Montgomery AL 3611, USA

TELEPHONE: (205) 284 1797

Director of planning: Allen Q Thames, Jr

Wainwright Planning is a Celtech company which performs professional engineering and planning assignments for both governmental and private clients. Fields of specialisation include airport design, runway/taxiway design, environmental, civil, sanitary, mechanical, electrical engineering, airport and industrial complex systems and a range of planning services.

George Wimpey Limited
International and Engineering Division,
Hammersmith Grove, London W6 7EN, England

TELEPHONE: 01-748 2000
TELEX: 22436

The Wimpey Group is engaged on contracts in over 30 countries world-wide.
George Wimpey Ltd is the holding company for a number of subsidiary operating companies which are organised into four main divisions: Wimpey Construction UK, International Engineering, Specialist Holdings, Wimpey Group Services.
The activities of Wimpey Construction UK cover every aspect of commercial, industrial and community facilities including shopping centres, offices, factories, warehouses, education facilities and housing.
George Wimpey International undertakes building and civil engineering projects world-wide. George Wimpey Mechanical Electrical and Chemical Engineering specialises in design, engineering, construction and project management of oil refining, chemical and industrial plants in the United Kingdom and other countries. The activities of other subsidiary and associate companies in this division include heating, ventilation and electrical engineering, offshore maintenance, offshore structure fabrication, smelter construction and sewage treatment. In addition Wimpey Marine owns and operates a fleet of tug and supply ships which service offshore oil installations.
The Specialist Holdings division includes Wimpey Asphalt which is principally engaged in road and airport surfacing and quarry management. Wimpey Waste Management offers specialist cleaning and maintenance services to the primary processing industries as well as solid and liquid waste collection and disposal. The group's interests in mining and minerals, builders merchants and in plant and transport are also included in the Specialist Holdings Division.
Wimpey Group Services co-ordinates the activities of technical and other service departments, architecture, estimating, surveying, planning and structural design, which provide support to all the Group's operations. In addition Wimpey Laboratories undertake land and offshore investigations and surveys for building and civil engineering work. They also provide technical services for quality control on company products and research into building materials and techniques.
The group also has major interests in commercial property development and investment through Wimpey Property Holdings. Services offered by the group comprise: turn-key projects; technical surveys and feasibility studies; arranging funding packages; fast contract mobilisation world-wide; locating, extracting and handling materials; overslabbing runways including design of material mines; constructing passenger and freight handling buildings and installing systems; installing runway mechanical and electrical systems.
The company has been involved in building, modernising or resurfacing 150 airports world-wide. Projects include: construction, maintenance and development at London Heathrow Airport, England; airport development at London Gatwick, England; six airports in Zaire; Abha, Saudi Arabia; Port Harcourt, Nigeria; Port of Spain, Piarco, Trinidad and Tobago; Lagos, Murtala Muhammed, Nigeria; upgrading operational runways: RAF Mildenhall and RNAS Culdrose, Manchester International Airport, England; Hong Kong Kai Tak; Singapore; Paya Lebar, Doha International Airport, Qatar.

Manchester International Airport

EQUIPMENT GROUPS

Airport Ground Equipment
54 rue Pasteur, 94120 Fontenay sous Bois, France

TELEPHONE: (1) 875 10 40
TELEX: 230 445 F

Airport Ground Equipment is a consortium providing equipment for aircraft handling and servicing and for cargo terminal operations:

Société Nouvelle d'Albret: passenger steps, working and loading platforms
Guinault: ground power units and frequency converters, jet air starters, cabin air heaters
Établissements Pinon: pallet/container dollies, racks, baggage trolleys, portable floodlights
Sovam: self-propelled transporters, loaders, catering vehicles, apron buses, elevating work platforms
Saxby Tracma: tractors

British Airport Equipment Group
Stowe House, 5 Bishopgate Street, Leeds LS1 5DY, England

TELEPHONE: (0532) 441911/433460
TELEX: 55293 Chacom G

Managing director: Alan Buckley
General sales manager: Richard Stott
Technical managers: Bob MacDonald, Trevor Cressey
Executive secretary: Lesley Kay
Sales administrator: Sandra Vant

Planning and Construction:
British Airways Associated Companies Ltd: airport planning, forecasting and design
British Airports International: general airport planning
Fletcher, Ross and Hickling: architectural and quantity surveyors
Terminal and Apron Planning Associates: airport consultancy service
White Young Consulting Engineers: civil engineering consultants

Electrical
Turner Ross: consultancy and electrical installation

Maintenance and service
Briton Handling Systems Ltd: spare parts, mechanical handling equipment, overseas engineers
Rovert Controls (Cheltenham) Ltd: hydraulic systems, test rigs

Aircraft sales and lease
DK Aviation Ltd: civil and military aircraft sales, leases and training

Refurbished equipment
Aircraft Handling Factors Ltd: refitted equipment of all types

Ground support equipment
Access Equipment Ltd: hydraulic platforms
Atol International Ltd: baggage trailers, dollies, tow bars, passenger stairs
Douglas Equipment Ltd: aircraft towing tractors
Gloster Saro Ltd: crash tenders, fire vehicles, refuellers
Montgomerie Reid Ltd: baggage towing tractors
Raynar Power Ltd: gpu, air starters, air conditioning units, generators
Ross Auto Engineering Ltd: electric tractors, platform trucks
Static Systems Group Ltd: electronic monitoring and emergency systems
Universal Conveyor Co Ltd: freight handling and check-in conveyor systems
Vaughan Systems and Programming Ltd: flight information systems
VJ Engineering Ltd: toilet and water servicing vehicles

British Association of Airport Equipment Manufacturers and Services (BAAEMS)
17 Castle Street, High Wycombe, Buckinghamshire HP13 6RU, England

TELEPHONE: (0494) 30430

Secretary: A Whittle

BAAEMS is the official trade organisation representing the British ground support industry. It represents companies that provide services and equipment connected with the planning, construction, operation and maintenance of airports irrespective of size and location.
BAAEMS offers a single contact point from which the airport developer can draw information on what British industry can provide to meet the specific requirements.

Finnaviation Group
PO Box 908, SF-00101 Helsinki 10, Finland

TELEPHONE: (358) 0 694 11 22
TELEX: 121696 trade SF

Member companies provide the following products:

Aspo Electronics: airport information systems, video-monitors
Erikoiskori: intervention vehicles, major foam and water tank vehicles
Finnair Technical Department: airframe and component overhaul
Idman: airport lighting systems, visual approach slope indicator systems
Kone Instrument Division: traffic operations information systems
Outokumpu Engineering Division: walk-through metal detectors
Perusyhtymä Ara and Rahtikone: runway sweepers, snow-blowers
Starckjohann: fire protection systems
Vaisala: ceilometers included in total meteorological systems
Valmet: aircraft and towing vehicles, passenger bridges

German Airport Equipment
PO Box 2049, 2120 Lueneburg, Federal Republic of Germany

TELEPHONE: (04131) 32011
TELEX: 02 182 101

Member companies provide the following products:

Binz GmbH and Co: airport emergency ambulances
BMA-Braunschweigische: cargo handling systems, pallet/container
Maschinenbauanstalt AG: loader and transporter
Bruggemann and Brand GmbH and Co KG: dollies, parachutes, air containers, cargo handling systems, pallets, nets
Dahms and Co: passenger steps, lavatory and potable water carts
Eisenwerk Schroder: catering/commissary trucks, lifting platforms
Faudi Feinbau GmbH: filter/water separators, filter elements, microfilters
Faun-Werke: chassis for crash trucks, runway sweepers
Garrett GmbH: aircraft ground support equipment, jet air starter, gse, gpu
Hydro-Geratebau GmbH and Co KG: aircraft jacks, platforms, tow bars, servicing carts
Mero-Raumstruktur GmbH and Co Wurzburg: maintenance platforms, docks, hangars
Messerschmitt-Bolkow-Blohm GmbH: intra-airport transit systems, airport vehicle control systems
Neoplan Gottiob Auwarter KG: apron coaches
Schoff Maschinenbau GmbH: aircraft towing tractors, towbars, tow tractors
Siemag Rosenkaimer GmbH: baggage handling systems
Ad. Struver KG (GmbH and Co): ground power units, diesel mains, failure sets, aircraft refueller
Telefonbau und Normalzeit: communication and information systems
Thyssen Henschel Flughafentechnik: passenger loading bridges, cargo handling equipment
Trepel AG: lift lounge, catering trucks, cargo handling systems
Velte and Co GmbH: development and installation of electric energy systems for airports
Konrad Wiedemann: tractors, conveyor belts, baggage loaders, trailers

International Airport Equipment Group
Taywood Road, Northolt Road, Middlesex UB5 6RL, England

TELEPHONE: 01-578 2366
TELEX: 24428

Chairman: D J W Wills, M.InstM
Treasurer: A H Marshall, BSc
Publicity director: M Schindler, BA, MBIM

Member companies provide the following products:
Armfield Engineering: elevating work platforms
Carmichael Fire and Bulk Ltd: foam crash tenders, rapid intervention, domestic and high rise fire appliances
Edbro Ltd: catering vehicles, maintenance platforms, invalid lifts, main deck baggage loaders, recovery vehicles
Gecos Aviation Division of GEC Electrical Products: airport electrical and mechanical services
Houchin Ltd: ground power equipment
IAL: aviation and communications technical services, equipment and systems
Reeve Burgess Ltd: coaches
Reliance Mercury Ltd: tractors and technical handling equipment for aviation industry and ports
Taymech Aviation Equipment Ltd: aircraft refuelling bowsers and hydrant dispensers

Italian Airport Equipment
Via Conca del Naviglio, 37 20123 Milan, Italy

TELEPHONE: 02 832 58 51
TELEX: 320595

Member companies provide the following products:
Aviogei: passenger and cargo handling equipment
Buini and Grandi: airfield lighting
Caproni Vizzola: baggage/cargo trailers
Cella: hydraulic elevating equipment, aircraft servicing vehicles
Eurocar: tractors
Gilardoni: baggage screening equipment
Industrie Pirelli: warehouse mechanisation
Irvin Manifatture Industriale: containers
Mobil Italiana Spazzatrici Stradali: airport sweeper
Officiene Fresia: aircraft towing tractors, snowploughs
Officine Viberti: refuellers
PGS: fork lift trucks
Reggione Officine Meccaniche Italiane: passenger loading bridges
Silvani Anticendi: fire-fighting equipment
SITTI: communications equipment
SUTIC: chairs
Trastecnica: baggage and freight handling systems

Proavia

75 rue La Boetie, 75008 Paris, France

TELEPHONE: (1) 359 22 93

Member companies provide the following products:

Aerotechnique de Dreux-Vernouillet: power supplies
Aerazur: arresting barriers for aircraft
Alsthom Atlantique: power supplies
Société Aman: power supplies
Établissements Augier: ground lighting transformers
Barbier, Bernard et Turenne: airfield lighting
Biro SA: fire-fighting equipment
Les Cables de Lyon: power cables
Camiva: fire-fighting equipment
CEAT: power cables
CGF: airfield lighting cables
CIT-Alcatel: communications, meteorological equipment
Claude: obstruction lighting
Cie Électro-Mécanique: power supplies
CSEE: communications, power supplies, turnkey projects
DAPA Systems: communications
Elecma: communications, meteorological equipment, security system, power supplies
Enertec Schlumberger: meteorological equipment
Graniou: air traffic control equipment contractor
Holophane: airfield lighting
LCT: radar, meteorological equipment
LMT: navigational aid
Merlin Gerin: power supplies
Nardeux: communications equipment
Norelec-Sfere: control consoles, wind socks, electrical contractor
Omera-Segid: meteorological equipment
SAF: air traffic control consoles
Safare-Crouzet: communications equipment
SAGEM: communications equipment
Santerne: airfield lighting
SAT: navigation and test equipment
SEEE: communications equipment, power supplies
SELF: power supplies/airfield lighting contractor
Société Électro-Technique: lighting regulators
SIDES: fire-fighting equipment
Sofréavia/Sofréavia Service: consultants – planning, management project
Systelec: communications equipment
TEC: relay equipment for communications
Telerad: communications equipment
Thomson CSF: radar, navigation, communications, air traffic control equipment
UGEC: navigation/landing-aids contractor
Ulmer Aeronautique: airfield lighting

Summary of services

Company	Specialisation	Design, planning, systems engineering	Construction	Project management	Technical support	Airport management
Acculab	acoustics	x				
Aeroport de Paris	complete air-port projects	x	x	x	x	x
Airways Engineering Corporation		x				
Airwork Ltd		x		x	x	x
Amey Roadstone Construction Ltd			x			
Association of Consulting Engineers		x		x		
The Austin Company		x	x	x		
Aviation Planning Services Ltd		x		x		
R L Banks and Associates Inc		x				
Barton ATC Inc	atc facilities	x		x	x	
BP Trading Ltd	fuelling	x				
British Airports International		x		x	x	x
John F Brown Co Ltd		x				
Conder Buildings Overseas Ltd			x			
Costain International			x			
Dewberry and Davis		x				
Diemler and Diekemper Inc		x				
R M Douglas Construction Ltd		x	x			
John W Drake		x				
Drover Welch and Lindlan		x		x		
Edwards and Kelcey Inc		x		x		
ESG Elektronic System-GmbH	atc facilities/ information systems	x		x		
Federal Electric Corporation	atc facilities/ electronic systems	x		x	x	
Flaughafen Frankfurt/Main AG		x		x	x	x
Foundation Company of Canada Ltd		x	x	x		
Hanscomb Associates Inc	cost studies/ financial management	x				
Heliport Systems Inc	complete heli-port projects	x	x	x	x	x
CH2M Hill		x				
Hoyle, Tanner and Associates Inc		x				
Herbert H Howell		x				
HTB Inc		x		x		
IAL					x	x
John Laing Ltd			x			
Hugo Marom Aviation Consultants Ltd		x		x		
Maurice Co Ltd	airfield lighting	x		x		
The Mitre Corporation	atc systems	x				
Monenco Ltd		x	x	x		
Morrison-Maierle Inc		x				
Netherlands Airport Consultants BV (NACO)		x		x		
NORR Airport Planning Associates		x				

Company	Specialisation	Design, planning, systems engineering	Construction	Project management	Technical support	Airport management
Pan Am World Services Inc		x		x	x	x
Pavement Consultancy Services BV	pavements	x				
Pavement Specialists Inc	pavements		x			
Plessey Airports	complete airport projects	x	x	x	x	x
Prisma Universal Ltd	airfield marking		x			
Reynolds Smith and Hills Inc		x		x		
Royal Volker Stevin NV			x			
Sar-Assist Inc	safety/SAR equipment	x				
Schiphol Airport Authority						x
Scott Wilson Kirkpatrick and Partners		x		x		
Seri Renault Ingenerie	air freight terminals	x	x			
Shorgard Ltd	airfield lighting	x	x			
Skanska		x	x	x		
Skidmore, Owings and Merrill		x				
Smith Hinchman and Grylls Associates		x		x		
Societe Internationale de Telecommunications Aeronautique (SITA)		x			x	
Sofreavia/Sofreavia Service		x		x	x	
Software Sciences Ltd	atc systems	x		x		
Alan Stratford and Associates Ltd		x				
Tarmac Group			x			
Taylor Woodrow International Ltd			x	x		
Teleplan AB	atc facilities/ electronic systems	x		x		
Arnold W Thompson PC	terminal facilities	x				
Tippetts-Abett-McCarthy-Stratton (TAMS)		x		x		
TRA Airport Consultants		x				
Alistair Tucker Associates		x				
Wainwright Planning		x				
George Wimpey Ltd			x			

INTERNATIONAL ORGANISATIONS

Intergovernmental Organisations

Asecna
Agence pour la Sécurité de Navigation Aérienne en Afrique et a Madagascar, BP 3144, 32–38 Ave Jean Jaurès, Dakar, Senegal
TELEPHONE: 23 10 40 TELEX: 680 dakar

Icao
International Civil Aviation Organisation
PO Box 400, Place de l'Aviation Internationale, 1000 Sherbrooke Street West, Montreal, Quebec H3A 2R2, Canada
TELEPHONE: (514) 285 8219 TELEX: 05 24513

Regional offices
Middle East and Eastern African Office
16 Hassan Sabri, Zamalek, Cairo, Egypt
TELEPHONE: 650163, 65043 TELEX: 92459
European office
3 bis, villa Emile-Bergerat, 92200 Neuilly-sur-Seine, France
TELEPHONE: 747 95 73 TELEX: 610075
North American and Caribbean office
Apartado Postal 5-377, Av Thiers 251, Pios 5, Mexico 5, DF
TELEPHONE: 250 32 11 TELEX: 1777 598
South American office
Apartado 4127, Edif Corpac Zonal, Aeropuerto International Jorge Chavez, Lima 100, Peru
TELEPHONE: 51 5414, 51 5325 TELEX: 25689
African office
15 Boulevard de la République, BP 2356, Dakar, Senegal
TELEPHONE: 21 54 52 TELEX: 906676
Far East and Pacific office
Sala Santitham, Rajadamnoen Avenue, PO Box 614, Bangkok, Thailand
TELEPHONE: 581 5366, 28155 TELEX: 87969

INTELSAT
International Telecommunications Satellite Organization
490 L'Enfant Plaza, SW Washington DC 20024, USA
TELEPHONE: (202) 488 2300 TELEX: 85 2707

OMM/WMO
Organization Météorologique Mondiale
World Meteorological Organisation
Secretariat, Case Postale No 5, CH-1211 Geneva 20, Switzerland
TELEPHONE: (022) 34 64 00

Private Organisations

AACC
Airport Associations Co-Ordinating Council (Conseil de Co-ordination des Associations Aéroportuaires)
PO Box 125, CH-1215 Geneva 15, Switzerland
TELEPHONE: (022) 98 41 41 CABLE: aaccps

AACO
Arab Air Carriers Organisation
Arab Business Centre, Airport Boulevard, PO Box 11-7349, Beirut
TELEPHONE: 277710/2
CABLE: arabcarriers SITA beyxame

AFRAA
Association of African Airlines
PO Box 20116, Nairobi, Kenya
TELEPHONE: 338419, 338410 SITA nboxakq

AOCI
Airport Operators Council International
1700 K Street, NW Washington DC 20006, USA
TELEPHONE: (202) 296 3270 CABLE: aocihq

ATAF
Association Internationale de Transporteurs Aériens
12 rue de Castiglione, F-75001 Paris, France
TELEPHONE: (1) 260 32 35

CACAS
Civil Aviation Council of Arab States
50 El Khalifa el Mamoun Street, Manshiet el Bakry, Cairo
TELEPHONE: 896 418, 896 193 CABLE: cacas

EUROCONTROL
European Organisation for the Safety of Air Navigation
72 Rue de la Loi, 1040 Brussels, Belgium
TELEPHONE: 513 83 00 TELEX: 21173

FIATA
International Federation of Freight Forwarders Association
Box 177, 29 Brauerstrasse, CH-8026 Zurich, Switzerland
TELEPHONE: (012) 41 80 45 TELEX: 57278

IAAA
International Airforwarders and Agents Association
Suite 828, 910 17th Street NW, Washington DC 20006, USA
TELEPHONE: (202) 296 6660

IACA
International Air Carrier Association
PO Box 105 Airport, CH-1215 Geneva, Switzerland
TELEPHONE: 98 11 19 TELEX: 27724

IATA
International Air Transport Association
26 Chemin de Joinville, PO Box 160, 1216 Cointrin/Geneva, Switzerland
TELEPHONE: (022) 98 33 66 TELEX: 23391
Montreal office
PO Box 550, Place de l'Aviation Internationale, 1000 Sherbrooke Street West, Montreal, Quebec H3A 2R4, Canada
TELEPHONE: (514) 844 6311 TELEX: 05 267627

ICAA
International Civil Airports Association
Building 226, Orly Sud 103, 94396 Orly Aerogare, Cedex, France
TELEPHONE: 687 12 34 TELEX: 270803

ITA
Institute of Air Transport
4 rue de Solférino, 75007 Paris, France
TELEPHONE: 705 81 50 TELEX: 260717

WEAA
Western European Airports Association
Stuttgart Airport, Postfach 23 03 22, D-7000 Stuttgart 23, Federal Republic of Germany

National Organisations

AAAE
American Association of Airport Executives
2029 K Street NW, Washington DC 20006, USA
TELEPHONE: (202) 331 8994

ADV
Arbeitsgemeinschaft Deutscher Verkehrsflughaven eV (German Airports Authority)
Stuttgart Airport, D-7000, Stuttgart 23, Federal Republic of Germany
TELEPHONE: 0711 79011 TELEX: 724658

AOA
Aerodrome Owners Association
18 Orchard Street, Bristol, Avon BS1 5DX, England
TELEPHONE: 0272 292480

BAA
British Airports Authority
2 Buckingham Gate, London SW1E 6JL, England
TELEPHONE: 01-834 6621 TELEX: 919039

CAA
CAA House, 45/59 Kingsway, London WC28 6TE, England
TELEPHONE: 01-379 7311

CAB
Civil Aeronautics Board
Universal Building, 1825 Connecticut Avenue NW, Washington DC 20428, USA

FAA
Federal Aviation Administration
800 Independance Avenue SW, Washington DC 20591, USA
TELEPHONE: (202) 962 5166

ADDENDA

CANADA

SMI Series 7000 Self-propelled Snow-blower

SMI Series 7000 self-propelled snow-blower

The Series 7000 is a self-contained diesel snow-blower based on all-wheel drive chassis with a right-hand drive cab mounted over the main engine and an auxiliary diesel engine installed over the rear axle with the appropriate gearing powers. The snow-blower casts the snow a variable distance via a telescopic loading chute. Two models are available: the 7250 with a maximum capacity of 30 tons (30·4 tonnes) per minute and the 7200 capable of moving up to 35 tons (35·5 tonnes) a minute. Both feature automatic transmission and air brakes. The 7200 is available with either two- or four-wheel steering. The snow clearance machinery comprises two 1 foot 8 inch (0·51 metre) diameter horizontal conveyors in the 7250 or a single 3 foot 6 inch (1·07 metre) horizontal conveyor in the 7200; a six-bladed impeller and a spot casting chute which can be controlled in any direction over a 240 degree horizontal arc. Distance of cast can be controlled from 4 to 50 feet (1·22 to 1524 metres). Dual steering vanes are offered as optional equipment on the 7250 model.

Characteristics

MODEL	7250	7200
MAIN ENGINE	Diesel	Diesel
OUTPUT	202 hp (150·6 kW)	202 hp (150·6 kW)
AUXILIARY ENGINE	Diesel	Diesel
OUTPUT	335 hp (249·8 kW)	335 hp (249·8 kW)
CLEARING		
width	8 ft 6 in (2·59 m)	8 ft 6 in (2·59 m)
performance	1630 tonnes/h	1905 tonnes/h
MAX SPEED	45 mph (72 km/h)	45 mph (72 km/h)
IMPELLER DIAMETER	3 ft 6 in (1·07 m)	3 ft 6 in (1·07 m)
CONVEYOR DIAMETER	2 × 1 ft 8 in (0·51 m)	3 ft 6 in (1·07 m)
HEIGHT TO TOP OF		
CUTTING KNIVES	6 ft 10 in (2·08 m)	8 ft 2 in (2·49 m)
diameter	1 ft 10 in (0·56 m)	1 ft 10 in (0·56 m)
casting distance	4–50 ft (1·22–15·24 m)	4–50 ft (1·22–15·24 m)
CHUTE HEIGHT	10 ft 9 in (3·28 m)	10 ft 9 in (3·28 m)
TURNING RADIUS	23 ft 9 in (7·24 m)	
2-wheel steering		30 ft 8 in (9·35 m)
4-wheel steering		24 ft 6 in (7·47 m)
WHEELBASE	12 ft 1 in (3·68 m)	12 ft 1 in (3·68 m)
UNIT DIMENSIONS		
length	26 ft 10 in (8·18 m)	27 ft 7 in (8·4 m)
width	8 ft 2 in (2·49 m)	8 ft 2 in (2·49 m)
height	10 ft 7 in (3·23 m)	10 ft 7 in (3·23 m)
WEIGHT	23 840 lb (10 850 kg)	25 240 lb (11 500 kg)

Status

Production

Manufacturer

SMI Industries Ltd, 2055 Bennett, Montreal 404, Quebec, Canada

---

DENMARK

Bruel and Kjaer Type 2232 Sound Level Meter

A new sound level meter with a digital display has been released by Bruel and Kjaer.
The type 2232 measures A-weighted sound pressure levels using either slow- or fast-time constants over a dynamic range of 34 to 130 dB(A) in two 60 dB measuring ranges. The level is displayed on a large digital display and two modes of operation permit either the maximum rms level measured during the preceding second to be automatically updated and displayed, or a rms 'maximum hold' reading to be obtained by manual reset. An analogue dc output is provided for recording purposes.

Fitted with a high sensitivity prepolarised condenser microphone, type 4176, the sound level meter offers up to 30 hours' operation from one set of cells. In addition to displaying the measured level, the digital display indicates overload, underrange and low battery conditions and a visual check of all display segments when the instrument is first switched on.

Status

Production. The type 2232 was introduced in January 1982

Manufacturer

Bruel and Kjaer, 18 Naerum Hovedgade, 2850 Naerum, Denmark

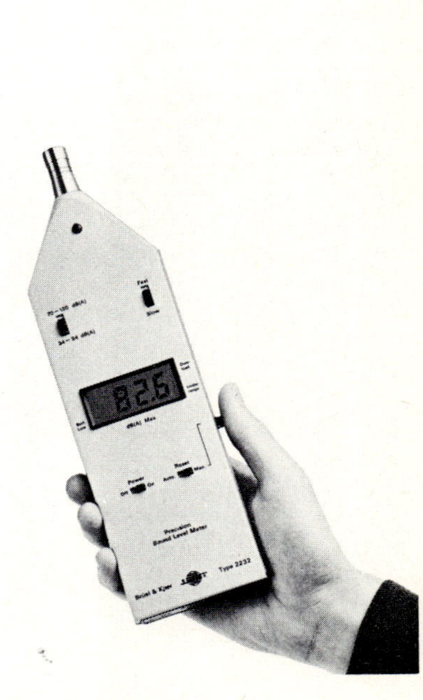

Bruel and Kjaer type 2232 sound level meter

FINLAND

Erikoiskori Jet Major Airport Crash Tenders

Two versions of the Jet Major are available; Jet Major 8500 mounted on a Mercedes Benz chassis and Jet Major 1000 on a Volvo chassis. Both are six-wheel drive vehicles.
The fire-fighting equipment consists of a roof-mounted monitor which is operated electrically from the driver's cab. The tender is equipped with a stainless-steel water tank and an acid-resistant foam tank and a Pi-Matic 300 HD automatic foam proportioning system.

Characteristics

MODEL	Jet Major 8500	Jet Major 1000
CHASSIS	Mercedes Benz 2632 AK	Volvo F12
DRIVE	6 × 6	6 × 6
TANK CAPACITY		
water	7000 litres	9000 litres
foam	900 litres	1100 litres
WATER PUMP	6000 litres at 10 bar	6000 litres at 10 bar
MONITOR		
output	4000 litres at 10 bar	4000 litres at 10 bar
range	70 m	70 m

Status

Erikoiskori has delivered 12 Jet Major crash tenders to the Finnish National Board of Aviation. Jet Major vehicles are also in service with the Swedish National Board of Aviation

Manufacturer

Erikoiskori Oy, Iisalmi, Finland

Jet Major 1000

FRANCE

Gallet Baggage Handling Equipment

Gallet specialises in the production of baggage handling equipment. For check-in facilities the company undertakes the design and development of the equipment offering: counters with check-in conveyors and automatic weighing; collecting conveyors; elevators and spiral conveyors, (both inclined plane and vertical arrangement); waiting stations on power-controlled baggage circuits with selective sorting and automatic extraction; baggage receivers on the aircraft side with either powered-roller tables or rotary rings for baggage storing and covered platforms for baggage trolleys. In the passenger baggage claim area, equipment available includes: baggage receiver table on the aircraft side; belt conveyor for baggage removal; elevators and spiral conveyors (both inclined plane and vertical arrangement); baggage distribution loops in passenger hall comprising carousels; rings with mobile platforms; control systems; protective guards to prevent baggage slip or loss, and fixed and hinged power-controlled baffles.
All systems can be automated with power and control or remote-control cubicles or operated by a computer.

Status

Gallet has installed baggage handling equipment at 11 airports in France and at airports in Angola, Cameroon, French Guiana, Guadeloupe, Indonesia, Ivory Coast, Jibuti, Martinique, Morocco, Spain and USA

Manufacturer

Gallet SA, 13 rue Georges Bouzerait, 92120 Montrouge, France

GERMANY, FEDERAL REPUBLIC

Expresso Type 898071 Baggage Trolley

Expresso has introduced a three-wheeled trolley incorporating a low loading area and an automatic braking system.
The brakes, which are adjustable, are claimed to hold the trolley in position with a load of 80 kg on slopes up to 13 degrees. The trolley is constructed in steel tubing and mounted on a single castor at the front and two non-swivelling wheels at the rear. It is designed to be stacked. Other features include a basket for hand luggage, advertising panel and a protective profile around the bottom frame. A hooped-mast model (-1078) is available on request.

Characteristics

TYPE	898071	898071-1078
CAPACITY	200 kg	200 kg
DIMENSIONS		
length	1·06 m	1·06 m
width	0·65 m	0·63 m
height	1·15 m	2 m
WEIGHT	28 kg	30 kg

Status

Production. Introduced in early 1982

Manufacturer

Expresso Material Handling Equipment, Ochshauser Strasse 11-25, D-3500 Kassel-B, Federal Republic of Germany

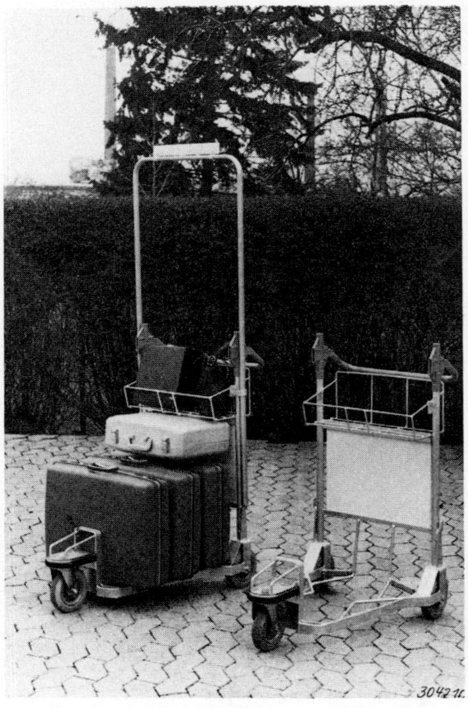

Expresso type 898071 baggage trolleys

Schorling 5000 Suction Runway Sweeper

Schorling has developed a suction runway sweeper for civil and military airports. Powered by a Mercedes Benz V8 diesel engine, the vehicle has a sweeping system comprising a central suction hood and two side hoods. The latter can be extended either side providing a maximum swept width of 5 metres with a total cleaning capacity of up to 250 000 square metres an hour. All hoods have an integrated 400 mm diameter roller brush. Two 890 mm diameter impellers provide 500 cubic metres a minute capacity and the system is capable of picking up 70 by 180 mm particles weighing up to 1 kg. Dust control is effected by water injection. Controls for the sweeping equipment are operated from the cab.

Characteristics

ENGINE: Mercedes Benz OM 422 V8 diesel
OUTPUT: 280 hp (206 kW) at 2300 rpm
MAX CLEANING SPEED: 50 km/h
CLEANING WIDTH
  central hood: 2·2 m
  with one side hood: 3·7 m
  two side hoods: 5 m
CAPACITIES
  cleaning: 250 000 m^2/h
  hopper: 8 m^3
  water tank: 1700 litre
UNIT DIMENSIONS
  length: 8·3 m
  width: 2·5 m
  height: 3·4 m
WEIGHT: 19 000 kg

Schorling 5000 suction runway sweeper

Status

Production. In service at Hanover Airport, Federal Republic of Germany

Manufacturer

Schorling International GmbH, PO Box 911060, Schorlingstrasse 3, D-3000 Hanover 91, Federal Republic of Germany

INDIA

HAL Surveillance and Precision Approach Radars

Two radars developed by Hindustan Aeronautics Ltd have been announced, an air route surveillance radar ARSR-1 and a precision approach radar PAR-1.

The air route surveillance radar operates in the 23 cm band with a maximum range of 160 nautical miles (300 km). The radar is equipped with inphase and quadrature mti which processes the entire information content of the signal, eliminating the losses inherent in single component mtis. The vertical pattern of the antenna is a modified cosec-squared with additional high angle coverage and antenna tilt which can be adjusted manually between 0·5 and 6 degrees. There is provision for displaying ssr video and adf information on the display screen.

The precision approach radar comprises a 3 cm band transmitter and a low noise receiver. It has a range of approximately 20 nautical miles (55 km) and provides azimuth and elevation information at a 1 second data rate. The manufacturer claims that the digital mti gives outstanding performance in a clutter environment. Depending on landing requirements, the controller can select any of the 14 glide-slope angles at either end of the runway. In addition the radar is designed for linear, elliptical or circular, continuously variable polarisation.

Status

HAL completed development of both radar systems in early 1982 at a cost of 203 million rupees for the ARSR-1 and 8 million rupees for the PAR-1. The systems are available for export

Characteristics

MODEL	ARSR-1	PAR-1
APPLICATION	Air route surveillance	Precision approach
FREQUENCY	1250–1350 MHz	9250–9450 MHz
PEAK POWER	2·2 Mw (5 Mw option)	225 kW
PRF	360 Hz	
ANTENNA		
rotational speed	5 or 10 rpm	
scan rate		0·746 cps
GLIDE-ANGLE COVERAGE		1–7·5°

HAL ARSR-1 air-route surveillance radar antenna

Manufacturer

Hindustan Aeronautics Ltd, Indian Express Building, Vidhana Veedhi, Private Bag 5150, Bangalore 560001, India

HAL PAR-1 precision approach radar antenna

NETHERLANDS

Hycom Central Hydraulic Test System

Hycom is offering a central hydraulic test system for use in hangars testing aircraft hydraulic circuits: pressure testing, internal leak testing, bleeding, filling and draining the aircraft system. The company claims numerous advantages for its test system: ease of use, lower noise (less than 65 dB) and high accuracy, to provide rapid and precise testing. The basic Hycom design comprises a hydraulic power unit, a test trolley, a set of hydraulic pipelines, an electrical relay unit and electric cables. Greater flexibility can be obtained by operating several systems simultaneously in the maintenance workshops. For example, an installation of four fixed hydraulic power units and four test trolleys allows simultaneous testing of the four hydraulic circuits of a Boeing 747; such an installation can also be used to work on several aircraft at the same time. The complete system is controlled from the test trolley. Various safety devices protect the personnel, the aircraft and the equipment itself.

The manufacturer recommends that, if possible, the hydraulic power units and the electrical equipment should be installed in the hangar basement which keeps the workshops clear and also provides a reduction in the noise level. The system allows pressure control between 35 and 210 bars approximately, with pressures up to 280 bars being available for tests at a reduced flow.

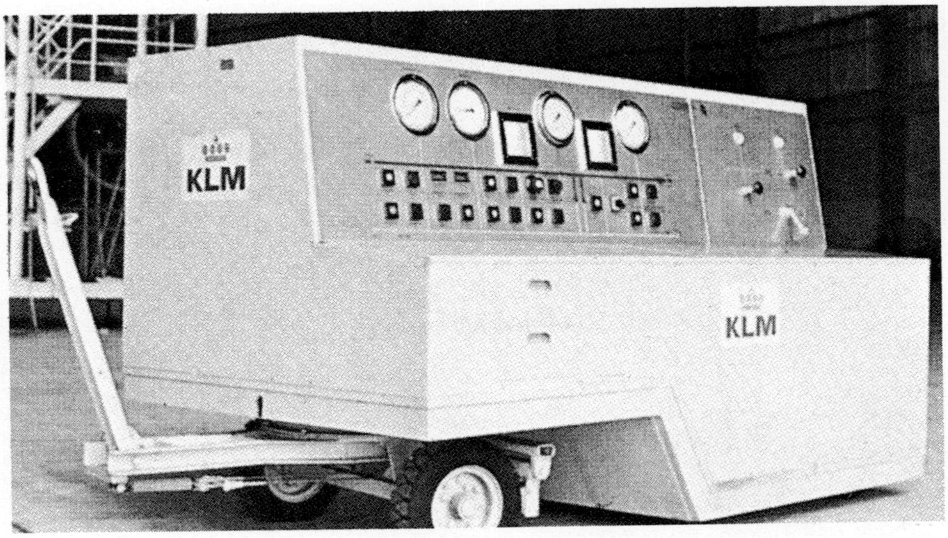

Hycom hydraulic test trolley

Status
Production. The system entered service with KLM in early 1982

Manufacturer
Hycom BV, Kanaal Zuid 114, PO Box 1079, NL 7301 BH Apeldoorn, Netherlands

UNITED KINGDOM

Ross Britec Electric Tractor

The Britec electric tractor manufactured by Ross Auto is suitable for towing baggage and freight trailers. Rated at 3000 lb (1360 kg) draw-bar pull, the tractor is capable of towing 25 000 lb (11 340 kg) on level ground at 8 mph (12 km/h) and a similar load up a 5 per cent gradient at 4 mph (6 km/h). Unladen the vehicle has a maximum speed of 18 mph (29 km/h). Pc wer is supplied by a standard series type traction motor which is fan-ventilated producing 6·2 hp (4·6 kW) with a solid-state scr controller permitting infinitely variable control. The battery comprises 36 lead acid cells (72 volts) providing 329 ampere hour capacity at the 5-hour rate of discharge. Other features include hydraulically-actuated brakes on all four wheels, all weather metal cab with driver's and optional passenger seat, and lighting for use on public roads. Towing hitches are fitted at the front and rear of the vehicle.

Characteristics

MOTOR: electric
OUTPUT: 6·2 hp (4·6 kW)
BATTERY: 72 V
CAPACITY: 329 Ah (5-h rate)
DRAW-BAR PULL: 3000 lb (1360 kg)
WHEELBASE: 3 ft 10 in (1·17 m)
TURNING RADIUS: 9 ft (2·74 m)
UNIT DIMENSIONS
length: 9 ft 2 in (2·719 m)
width: 4 ft 4½ in (1·32 m)
height: 5 ft 9 in (1·75 m)

Ross Britec electric tractor

Status
This tractor was used in trials by TWA at London Heathrow Airport, England

Manufacturer
Ross Auto Engineering Ltd, Banastre Road, Southport, Lancashire, England

UNITED STATES OF AMERICA

Alco Aircraft De-icing and Power Cleaning Unit

Alco Engineering International has added to its range of de-icing and power cleaning equipment, designated D-ICR and POWR-KLEENR, systems used to service all types of aircraft. In addition to the models 50, 100 and 250, details of which are contained in the Environmental Control section of the book, Alco has introduced the models 5, 25 and 350 to complete the line.
The model 5 is a 2 foot (0·61 metre) by 4 foot (1·22 metre) hand-cart mounted unit that will pass through a standard doorway. It has a 20 US gallon (76 litre) fibreglass tank with three heaters. The pumping system capable of 400 psi (27·5 bar) at 3 US gallons (11·4 litres/minute) is driven by a 3 hp (2·2 kW)Briggs and Stratton engine.
The model 25 offers the same features as the model 50, but is designed to be mounted on a steel deck 3 foot 11 inches (1·2 metres) by 6 feet (1·83 metres) that will fit into the box of a standard size pickup truck.
The model 350, 1½-ton truck mounted unit, is similar to the model 250 with the additional features of a 1000 US gallon (3785 litre) stainless steel tank. The pumping system is capable of producing 400 psi (27·5 bar) at 25 US gallons (95 litres/minute) and a gas heater is fitted producing 1 000 000 Btu (1055 MJ).

Status
Introduced in early 1982

Manufacturer
Alco Engineering International Inc, 6170 Olson Memorial Highway, Minneapolis, Minnesota 55422, USA

Alco model 5 D-ICR and POWR-KLEENR

Alco Fire Fighter Airport Fire-fighting Unit

Alco Engineering has announced the addition of a fire-fighting unit to its line of aviation products. The Fire Fighter was designed specifically for the small operators. It is a light-weight unit which can be towed by any vehicle equipped with a 2-inch (51 mm) towing point. The unit contains a 335 US gallon (1268 litre) fibreglass tank. The 7 hp (5·2 kW) Briggs and Stratton engine provides 400–600 psi (27–41 bars) at 10 US gallons/minute (38 litres/minute) with a down-stream chemical injector. Equipment on the unit includes a fire axe, shovel and broom, fire extinguisher and a 10 lb (4·5 kg) back-pack fire pump.
The unit, which has three 2·5 kW, 240 volt electric heaters, is equipped with flotation tyres and is available painted in either red or yellow.

Status
Introduced in early 1982

Manufacturer
Alco Engineering International Inc, 6170 Olson Memorial Highway, Minneapolis, Minnesota 55422, USA

Alco Fire Fighter

Caterpillar T30C-TC60C Lift Trucks

Caterpillar has developed the T30-TC60 series lift trucks with capacities ranging from 1500 to 3000 kg and offered with a choice of three power options.
The Caterpillar 1404 petrol or lpg engines are available with three transmission options: mechanical with oil clutch, hydrostatic with foot control and power shift with hand or foot control. The Perkins 4.2032 diesel engine has a power shift transmission.
The Caterpillar 1404 lpg and petrol engine standard features include electronic ignition, a full flow oil filter and a 9·9 litre reservoir for service intervals of 500 hours, and a high capacity cooling system. A full range governor designed specifically for lift truck operations provides optimum control of engine output.
The Perkins 4.2032 diesel engine has been developed and tested for lift truck use. Its squish-lip piston combustion chamber is claimed to provide a slower burning of fuel due to lower maximum cylinder pressures, with reduced noise levels and minimum vibration in the engine and lift truck frame.
The single-speed power shift transmission with Caterpillar controls permit mobile power reversals, loaded or empty at speeds up to 8 kilometres an hour, through modulated clutch engagement with an electric transmission control lever.

Caterpillar T-series lift truck, showing access to diesel engine

Status
Announced in March 1982

Manufacturer
Caterpillar Tractor Co, Engine Division, 100 NE Adams, Peoria, Illinois 61629, USA

D and D 400 Hz Ground Power Cable Reel

D and D Engineering has developed a new ground power cable reel, which is power-operated and will authomatically retrieve and store up to 60 feet (18·3 metres) of cable after an aircraft has been serviced. The cable reel and companion take-up unit which can be mounted under the outer end of any loading bridge, is designed to keep the ground power cable assembly out of the path of field traffic until power is needed. The cable reel and power assembly is offered with or without cable.

Status
Announced late 1981, this equipment is marketed by Burton Electrical Company of El Segundo, California, USA

Manufacturer
D and D Engineering Inc, 10849 Mid West Industrial Boulevard, St. Louis, Missouri, USA

Emergency One Airport Crash Tender

Emergency One airport crash tender

Emergency One has produced a tender for crash/fire/rescue operations at airports. The vehicle is mounted on a purpose-built chassis, powered by a turbo-charged Cummins engine with an Allison five-speed automatic transmission and all-wheel drive.

Foam is supplied by a round-the-pump proportioning system with a multi-metering valve, and an American single stage, gear-driven centrifugal pump.

The fire-fighting equipment comprises roof and bumper-mounted monitors, both controlled from the cab. An electric-rewind hose reel contains 100 foot (30·5 metres) of 1 inch (25·4 mm) diameter hose capable of providing 60 US gallons/minute (227 litres/minute) of foam or water. Vehicle protection is provided by two under-truck nozzles each having a capacity of 60 US gallons/minute (227 litres/minute).

Characteristics

ENGINE: Cummins VTA 903 turbocharged
POWER: 50 hp (37 kW) at 2800 rpm
DRIVE: 4 × 4
ACCELERATION
0 to 80 km/h: 25 s
MAX SPEED: 65 mph (105 km/h)
TANK CAPACITY
water: 1000 US gallons (3785 litres)
foam: 120 US gallons (454 litres)
WATER PUMP: American single stage; output 1250 US gallons/minute at 250 psi (4731 litres/minute at 17 bar)
MONITOR
roof-mounted: 500 US gallons/minute (1893 litres/minute)
bumper-mounted: 250 US gallons/minute (946 litres/minute)
HAND-LINE: 1 × 60 US gallons/minute (227 litres/minute)
VEHICLE PROTECTION
2 under truck nozzles total output: 120 US gallons/minute (454 litres/minute)

Manufacturer

Emergency One Inc, 1701 SW 37th Avenue, Ocala, Florida, USA

Directory

Abacus Municipal Limited
Oddicroft Lane, Sutton-in-Ashfield, Nottinghamshire NG17 5FT England
Tel: 0623 511111
Telex: 377000

Aberglen Orbits Structures
Aberglen Industrial Complex, Woodside Way, Eastfield, Glenrothes, Fife KY7 4PB, Scotland
Tel: 0592 773266
Telex: 727760

Abex Corporation
Jetway Division, 3100 South Pennsylvania Avenue, PO Box 9368, Ogden, Utah 84409, USA
Tel: (801) 731 1500
Telex: 388400

Access Equipment Limited
Maylands Avenue, Hemel Hempstead, Hertfordshire, England
Tel: 0442 60101
Telex: 826756 access g

Acculab
3201 Ridgewood Drive, Columbus, Ohio 43220 USA
Tel: (614) 876 5108

ACS Engineering Limited
Crew Lane, Southwell, Nottinghamshire, England
Tel: 0636 812723
Telex: 377362

Adtec Teoranta
Gibbstown, Navan, County Meath, Ireland
Tel: 0461 4183
Telex: 31842 til-el

Advanced Ground Systems Engineering Corporation
3270 East 70th Street, Long Beach, California 90805, USA
Tel: (213) 634 9010
Twx: 910 346 7745

AEG Telefunken
Industriestrasse 29, D-2000 Wedel (Holst), Federal Republic of Germany
Tel: (04103) 702-1
Telex: 2189520

AEG Telefunken
Radio and Radar Systems Operating Group, Elisabethenstrasse 3, Postfach 1730, 7900 Ulm, Federal Republic of Germany
Tel: (0731) 1921
Telex: 712723

Aerazur
58 boulevard Galieni, 92130 Issy Les Moulineaux, France
Tel: 554 92 80
Telex: 270887 F

Aeritalia
Soc. Aerospaziale Italiana pA, Caselle Torinese (TO)-CAP 10072, Italy
Tel: (011) 991 362
Telex: 210086 aitcea i

Aero-Docks Limited
Lower William Street, Northam, Southampton, SO9 2DN, England
Tel: 0703 34366
Telex: 47106

Aeroacoustic Corporation
4876 Victor Street, Jacksonville, Florida 32207, USA
Tel: (904) 731 3577

Aerodome Industries
2716 George Washington Boulevard, Wichita, Kansas 67210, USA
Tel: (316) 684 6266

Aeroenergy Systems Incorporated
20740 Marilla Street, PO Box 368, Chatsworth, California 91311, USA
Tel: (213) 341 3436
Telex: 691548

Aeronautica Macchi, SpA
Via Sanvito Silvestro 80, 21100 Varese, Italy
Tel: 83100
Telex: 380070 aviomacc

Aeroport de Paris
Orly Sud 103, 94396 Orly Aerogare Cedex, France
Tel: (331) 884 52 43
Telex: 200376

Aeroport de Paris
291 boulevard Raspail, 75675, Paris 14, France
Tel: 320 15 00
Telex: 270803

Aero Systems Engineering Incorporated
358 East Fillmore Avenue, St Paul, Minnesota 55107, USA
Tel: (612) 227 7515
Telex: 297043

Aerzen, Babb Company SA
1 Rue Lord Byron, 75008 Paris, France
Tel: 563 67 34
Telex: 641160F

AFA-Minerva (EMI) Limited
Extinguishing Division, Security House, Grosvenor Road, Twickenham, Middlesex TW1 4AB, England
Tel: 01-892 4422
Telex: 265518

AI Security
London Road, Pampisford, Cambridge CB2 4EF, England
Tel: 0223 834420
Telex: 817536

Air-A-Plane Corporation
3040 Virginia Beach Boulevard, PO Box 7245, Norfolk, Virginia 23509, USA
Tel: (804) 622 5761
Telex: 823417

Air Marrel
Division of Bennes Marrel SA, Paris-Orly,
Sales:
ZI du Delta, 13-15 rue du Pont des Halles, 94526 Rungis Cedex, France
Tel: (1) 687 31 81
Telex: Marelva 204148 F
Head Office:
ZI Saint-Étienne-Bouthéon, BP 56-42160 Andrézieux-Bouthéon, France
Tel: (77) 55 08 30
Telex: 330657

Air Marrel SA
Bat 5720, PO Box 10232, Aeroport Charles de Gaulle, Roissey 95703, France

Airport Equipment Company AB
s 16211 Vallingby, Sweden
Tel: 08 87 14 32
Telex: 14065-ICT

Airport Ground Equipment
54 rue Pasteur, 94120 Fontenay sous Bois, France
Tel: (1) 875 10 40
Telex: 230445 F

Airport Ground Equipment Italia
via campoleone tenuta 13, 04010 camponeone, Italy
Tel: (06) 9342151/2/3
Telex: 614568

Airways Engineering Corporation
1700 North Moore Street, Suite 810, Orlington, Virginia 22209, USA
Tel: (703) 522 4050
Telex: 440097

Airwork Limited
Bournemouth-Hurn Airport, Christchurch, Dorset, BH23 6EB, England
Tel: 02016 2271
Telex: 41282

Ajax Fire Protection
Cruquiusweg 118, 1019 AK Amsterdam, Netherlands
Tel: 020 654001
Telex: 14311

Alco Engineering International Incorporated
6182 Olson Memorial Highway, Golden Valley, Minnesota 55422, USA
Tel: (612) 542 9590

Alreso Elektronica BV
Zeemanstraat 62, 1506 CV Zaandam, Netherlands
Tel: 075 176871

Aluminium Alloy Fabrications Limited
Forsyth Road, Sheerwater, Woking, Surrey GU21 5SB, England
Tel: 048 62 5744
Telex: 859571 alifa g

AMA Private Limited
PO Box 556, Canda Building, D Naoroji Road, Bombay 1, India
Tel: 265271/2/3

Amalgamated Wireless (Australasia) Limited
422 Lane Cove Road, North Ryde, 2113, New South Wales, Australia
Tel: (02) 888 8111
Telex: AA20623

Amerace Limited
10 Esna Park Drive, Markham, Ontario, L3R 1E1, Canada
Tel: (416) 495 6000
Telex: 06986807

American Science and Engineering Incorporated
955 Massachusetts Avenue, Cambridge, Massachusetts 02139, USA
Tel: (617) 868 1600
Telex: 921458

American Sign and Indicator Corporation
2310 Fancher Way, Spokane, Washington 99206, USA
Tel: (509) 535 4101
Telex: 5107731839

Amey Roadstone Construction Limited
Sutton Courtenay, Abingdon, Oxford OX14 4PP, England
Tel: 023 582 811
Telex: 83496

Angloco Limited
Station Road, Batley, West Yorkshire, England
Tel: 0924 476421
Telex: 53220

Angus Fire Armour
Thame Park Road, Thame, Oxfordshire, England
Tel: 084 421 4545
Telex: 83425

Armco
Middletown, Ohio 45043, USA

Armfield Engineering Limited
Ringwood, Hampshire, England
Tel: 042 54 2405
Telex: 418295 arlico g

Arvico International Incorporated
5553 North West 36th Street, Miami, Florida 33166, USA
Tel: (305) 887 7694
Telex: 51 5066
Facility: 5329 Georgia Avenue, West Palm Beach, Florida 33405, USA

Assmann GmbH
5 Industriestrasse, PO Box 1147, D-6380 Bad Homburg VDH1, Federal Republic of Germany
Tel: (06172) 1061
Telex: 0415158

The Association of Consulting Engineers
Alliance House, 12 Caxton Street, London SW1H 0QL, England
Tel: 01-222 6557
Telex: 268312

Ateliers Reunis SA
29 rue de Lattre, 67300 Schiltigheim, France
Tel: (88) 33 20 60
Telex: 870987

Atlas Copco Airpower NV
Communications Department, Boomsesteenweg 957, B-2610 Wilrijk, Belgium
Tel: 031 87 68 70
Telex: 31118 copab b

Audix Limited
Station Road, Wenden, Saffron Walden, Essex CB11 4L9, England
Tel: 0799 40888
Telex: 817444

SA Augereau (Heuliez)
route de Chartres, 28160 Brou, France
Tel: (37) 98 01 91
Telex: 790170 F

Augier SA
2° Casier, Ilot J, Zone Industrielle Carros, 06510 Carros, France
Tel: (93) 87 71 71

Auriga Lighting Systems Limited
Sawyers, Horns Road, Hawkhurst, Kent TN18 4XH, England
Tel: 05805 3165

The Austin Company
3650 Mayfield Road, Cleveland, Ohio 44121, USA
Tel: (216) 382 6600
Cable: Austinclegen

Auto Diesels Braby Limited
Cowley Mill Road, Uxbridge, Middlesex UB8 2QG, England
Tel: 0895 38262
Telex: 263835

Autophon AG
CH-8036 Zurich/Switzerland, Steinstrasse 21, Switzerland
Tel: (01) 35 85 35
Telex: 53838

W and T Avery Limited
Export Division, Smethwick, Warley, West Midlands B66 2LP, England
Tel: 021-558 1112
Telex: 336490

Avialift Products Limited
Municipal Airport, Southend-on-Sea, Essex, England
Tel: 0702 48691
Telex: 99405

Aviation Planning Services Limited
800 Dorchester West, Montreal PQ, H3B 1X9, Canada
Tel: (514) 878 4388
Telex: 05560725

Aviogel Airport Ground Equipment Italia Srl
via Nettunense km 7200, 00040 Cecchina Rome, Italy
Tel: (06) 9341601
Telex: 614568
via Campoleone tenuta 13, 04010 Campoleone Aprilia (LT), Italy
Tel: (06) 924355

Avitech Incorporated
2 Higland Street, Port Chester, New York 10573, USA
Tel: (914) 939 4050
Telex: 996634

Avon Industrial Polymers (Melksham) Limited
Bumpers Way, Bristol Road, Chippenham, Wiltshire, SN14 6NF, England
Tel: 0249 56241

AWA New Zealand Limited
Communications Division, PO Box 50-248, Porirua, New Zealand
Tel: (64) 4 375069
Telex: NZ 31001

Aydin Monitor Systems
502 Office Center Drive, Fort Washington, Pennsylvania 19034, USA
Tel: (215) 646 8100
Telex: (510) 6611520

Balteau SA
rue de Magnee 54, B-4610 Beyne-Heusay, Belgium
Tel: (3241) (041) 58 56 60
Telex: Balto B 41 268

JC Bamford Excavators Limited
Rocester, Staffordshire ST14 5JP, England
Tel: 0889 590312
Telex: 36154

RL Banks and Associates Incorporated
900 17th Street North West, Washington DC 20006, USA
Tel: (202) 296 6700

Barbier, Benard and Turenne
82 rue Curial (19e), Paris, France
Tel: 03 94 90
Telex: BBT 670804 F

Barrel Lighting Company Limited
Unit 3, Parsonage Farm, Forest Hall Road, Stansted, Essex, England
Tel: 0279 814918
Telex: 817662

Barton ATC Incorporated
4986 Gateway Drive, Medina, Ohio 44256, USA
Tel: (216) 725 6776

Bauer/Electro Incorporated
Farmington Industrial Park, Farmington, Connecticut 06032, USA
Tel: (203) 677 9707
Telex: 7104231748

Becker Flugfunkwerk
Postfach 1980, D-7550 Rastatt, Federal Republic of Germany
Tel: (07222) 121
Telex: 0781271

Becker Lifts Limited
Ealing Road, Alperton, Wembley, Middlesex HA0 4PA, England
Tel: 01-903 8333
Telex: 923586

Belfort Instrument Company
1600 South Clinton Street, Baltimore, Maryland 21224, USA
Tel: (301) 342 2626

Bendix Corporation
Communications Division, East Joppa Road, Baltimore, Maryland 21204, USA
Tel: (301) 823 2200
Telex: 87824

Berkel International BV
Keileweg 5, PO Box 6018, Rotterdam 3022 AA, Netherlands
Tel: (010) 76 46 00
Telex: 23313

Bertin et Cie
BP 3, 78370 Plaisir, France
Tel: (3) 056 25 00
Telex: 696231 F

Beumer Maschinenfabrik KG
Oelder Strasse 40, D-4720 Beckum, Switzerland
Tel: (02521) 24 1
Telex: 089411

Binz GmbH
Zollstrasse 2, Postfach 1120, D-7073 Lorch/Württ, Federal Republic of Germany
Tel: (07172) 6011
Telex: 0724884

Biro SA
7 rue Steinlen, 75018 Paris, France
Tel: 255 52 50
Telex: 660588 F

Bizerba Weighing Machine Works
Postfach 1140, D-7460 Balingen 1, Federal Republic of Germany
Tel: (074 33) 121
Telex: 0763622

Blast Deflectors Incorporated
PO Box 1090, Los Gatos, California 95030, USA
Tel: (408) 354 3305

Blue Giant Equipment Corporation
2323 Kenmore Avenue, Buffalo, New York 14207, USA
Tel: (716) 873 0550
UK: Blue Giant, Scott Lane, Chew Stoke, Bristol BS18 8XF, England
Tel: 027589 2076

BMA Braunschweigische Maschinenbauanstalt
PO 3225 Am Alten Bahnhof 5, 3300 Braunschweig, Federal Republic of Germany
Tel: (0531) 8040
Telex: 952456

SA Andre Boet
26 rue Paul Doumer, Villeneuve d'Ascq 59650, France
Tel: (20) 91 92 29
Telex: Boetnor 160815

Bomford and Evershed Limited
Salford Priors, Evesham, Worcestershire WR11 5SW, England
Tel: 078988 3383

Boschung Mecatronic Limited
Engineering Works, Schmitten/FR CH 3185, Switzerland
Tel: (037) 36 24 36

Bowser Filtration Limited
Progress House, 21 Progress Way, Croydon, Surrey CR0 4XD, England
Tel: 01-681 0428

BP Trading Limited (Air BP)
Britannic House, Moor Lane, London EC2Y 9BU, England
Tel: 01-920 6286
Telex: 888811

Brackett Aircraft Company Incorporated
5015 Roadrunner Drive, Falcon Field, Mesa, Arizona 85205, USA
Tel: (602) 832 2770

British Airport Equipment Group
Stowe House, 5 Bishopgate Street, Leeds LS1 5DY, England
Tel: 0532 441911
Telex: 55293

British Airports International
8 Little College Street, London SW1P 3SH, England
Tel: 01-222 8592
Telex: 912266

British Association of Airport Equipment Manufacturers and Services
17 Castle Street, High Wycombe, Buckinghamshire HP13 6RU, England
Tel: 0494 30430

John F Brown Company Incorporated
3057 Erie Avenue, Cincinnati, Ohio 45208, USA
Tel: (513) 321 6080

Bruel and Kjaer
18 Naerum Hovedgade, 2850 Naerum Hovedgade, 2850 Naerum, Denmark
Tel: (452) 80 05 00
Telex: 37316 bruka dk

Buini and Grandi SpA
via Pier Crescenzi 13, 40131 Bologna, Italy
Tel: (051) 23 20 05
Telex: 511493 BG BO I

Buko Wire Goods Limited
Queensway Estate, Glenrothes, Fife KY7 5QJ, Scotland
Tel: 0592 752281
Telex: 72298

Bunce (Ashbury) Limited
Ashbury, Swindon, Wiltshire SN6 8LW, England
Tel: 079 371 212

Burndept Electronics Limited
St Fidelis Road, Erith, Kent DA8 1AU, England
Tel: 03224 41155
Telex: 896299

Cable and Wireless PLC
Head Office, Mercury House, Theobalds Road, London WC1R 4NA, England
Tel: 01-242 4433
Telex: 23181
Projects Division, Lion House, 72-75 Red Lion Street, London WC1R 4NA, England
Tel: 01-242 4433
Telex: 262545

Camiva
73230 Saint-Alban-Leysse, (Chambery), France
Tel: (79) 33 93 77
Telex: 320334 F

Caproni Vizzola
via Montecchio 2, 21010 Vizzola Ticino/va, Italy
Tel: (0331) 230 826
Telex: 332554 Caviz 1

Cardion Electronics
LI Expressway, Woodbury, New York 11797, USA
Telex: 5102212179

Carmichael Fire and Bulk Limited
Gregory's Mill Street, Worcester, Worcestershire WR3 8BE, England
Tel: 0905 21381

Caterpillar Tractor Company
Engine Division, 100 North East Adams, Peoria, Illinois 61629, USA

CDN Research and Development Limited
460 Wyecroft Road, Oakville, Ontario, Canada
Tel: (416) 842 2703
Telex: 06982308

Cella SpA
25040 Mandolossa/Brescia, Italy
Tel: (030) 2770161/2/3
Telex: 300087 cella I

Cesab Carrelli Elevatori
via Persicetana Vecchia 10, 40132 Bologna, Italy
Tel: (051) 727 350
Telex: 511064

Cescom Electronics Limited
Harrow Road, Leytonstone, London E11, England
Tel: 01-534 7613
Telex: 888941 Icci g attention cescom

Cheshire Fire Engineering Limited
Road One, Industrial Estate, Winsford, Cheshire CW7 3PJ, England

David Cheverton (Sales) Limited
The Parade, Cowes, Isle of Wight PO31 7QD, England
Tel: 0983 297111
Telex: 86466

Chinetti Sea
Piazza Republican 10, Varese 21100, Italy
Tel: (0332) 283350
Telex: 340499 chivar i

Ch2m Hill
1301 Dove Street, Suite 800, Newport Beach, California 92660, USA
Tel: (714) 975 1000 and (213) 628 5764

Christie Electric Corporation
20665 Manhattan Place, Torrance, California 90501, USA
Tel: (213) 320 0808

Chubb Fire Security Limited
Fire Vehicles Division, Pyrene House, Sunbury-on-Thames, Middlesex, England
Tel: 093 27 85588
Telex: 261402/929737 chbfyr-g

Cit-Alcatel
Division DTM Department, 10 rue Louis-Lormand-78-La Verriere, BP 19-78320 Le Mesnil Saint Denis, France
Tel: (062) 41 28

Cit-Alcatel
33 rue Emeriou, F 75725, Paris Cedex 15, France
Tel: (1) 577 10 10
Telex: 250927F

Societe Claude
Division Protection and Systems, 3 imp de la Blanchisserie, Boite Postale 56, 51052 Reims Cedex, France
Tel: (26) 06 07 70
Telex: Claudcc-Reims 830401F

Clyde Machines Incorporated
Box 194, Glenwood, Minnesota 56334, USA
Tel: (612) 634 4504
Telex: 9105782392

Cochran Airport Systems
PO Box 1336, Castroville, California 95012, USA
Tel: (408) 663 2500

Cochran Western Corporation
1215 Hansen Street, Salinas, California 93905, USA
Tel: (408) 758 4461

Columbus Jack Rop
1000 South Front Street, Columbus, Ohio 43206, USA
Tel: (614) 443 7492
Telex: 245472

Compagnie de Signaux et d'Entreprises Electriques
2 a 8, rue Caroline, 75850 Paris Cedex 17, France
Tel: 620 22 24
Telex: 204325

Siege Social, 17 place Etienne-Pernet, 75738 Paris, Cedex 15, France
Tel: 533 74 44
Telex: csee 203926 F

Computer Engineering Limited
Wallace Way, Hitchin, Hertfordshire, England
Tel: 0462 52731

Computer Terminal Systems
65 South Service Road, Planview, New York 11803, USA
Tel: (516) 752 1965

Concrete Utilities Limited
Lower Road, Great Amwell, Ware, Hertfordshire SG1 2HZ, England
Tel: 0920 2272
Telex: 81398 Cuware

Conder Buildings Overseas Limited
Montfort House, Botley Road, Romsey, Hampshire, England
Tel: 0734 517000
Telex: 477275/477283

Conrac Corporation
Systems-West Division, 1600 South Mountain Avenue, Duarte, California 91010, USA
Tel: (213) 359 9141
Telex: 675468 (conrac swd)
and
560 Sylvan Avenue, Englewood Cliffs, New Jersey 07632, USA
Tel: (201) 871 4990
Telex: 135798 conrac egwd

Conrac Elektron GmbH
Postfach 60, Industriestrasse 18, D 6992 Weikersheim, Federal Republic of Germany
Tel: 0 79 34/70 71
Telex: 074231 elecon

Henry Cooch Limited
PO Box 40, Sevenoaks, Kent TN15 8LN, England
Tel: 0732 884484

Cossor Electronics Limited
The Pinnacles, Elizabeth Way, Harlow, Essex, CM19 5BB, England
Tel: 0279 26862
Telex: 81228

Costain International Limited
111 Westminster Bridge Road, London SW1 7UE, England
Tel: 01-928 4977
Telex: 8811804 Cosdon G

Craig Systems
360 Merrimack Street, Lawrence, Massachusetts 01842, USA
Tel: (617) 688 6961
Telex: 7103420765

Crest Manufacturing Incorporated
9120 Barton Street, Overland Park, Kansas 66214, USA
Tel: 913 492 7800

Crompton Electricars Limited
Crown Avenue, Dukestown, Tredegar, Gwent NP2 4EF, Wales
Tel: 043 525 2921

Crouse-Hinds Company
Wolf and Seventh North Street, PO Box 4999, Syracuse, New York 13221, USA
Tel: 315/477 7000

Cryogenic Applications
1806 South San Gabriel Boulevard, San Gabriel, California 91776, USA
Tel: (213) 283 0231

John Curran Limited
PO Box 72, Curran Road, Cardiff CF1 1TE, Wales
Tel: 0222 20641
Telex: 497492

D and D Engineering Incorporated
10849 Mid West Industrial Boulevard, St Louis, Missouri, USA

Dahms and Company
PO Box 2049, Luneburg, 2120, Federal Republic of Germany
Tel: (04131) 32011
Telex: 02-182101 GWLD

Dale Electric of Great Britain Limited
Electricity Buildings, Filey, Yorkshire YO14 9PJ, England
Tel: 0723 514141
Telex: 52163

Danline Brushes Limited
Nebo Road, Llanrwst LL26 0SE, North Wales
Tel: 0492 640651
Telex: 61346

W S Darley and Company
2000 Anson Drive, Melrose Park, Illinois 60160, USA
Tel: (312) 345 8050

Datasaab AB
Interactive Data Systems, S-175 86 Jarfalla, Sweden
Tel: 46 8 36 28 00
Telex: 17892 dasaab s

Datom Directors Incorporated
3645 Inglewood Avenue, Redondo Beach, California 90278, USA
Tel: (213) 973 7358

Dayton Communications Corporation
2971 Bluefield Avenue, Dayton, Ohio 45414, USA
Tel: (513) 278 7937
Telex: 810-459 1703

De Vore Aviation Corporation
6104 B Kircher Boulevard, North East, Albuquerque, New Mexico 87109, USA
Tel: (505) 345 8713
Telex: 660436

Defiant Weighing
Vestry Estate, Otford Road, Sevenoaks, Kent TN14 5EL, England
Tel: 0732 50755
Telex: 957154

Desmi
Radiovagen 5, S-135 48 Tyreso, Sweden
Tel: (46) 08 712 00 55
Telex: 10902

Deutsche Geratebau GmbH
4796 Salzokotten, Postfach 1140, Federal Republic of Germany

Dewberry and Davis
8401 Arlington Boulevard, Fairfax, Virginia 22081, USA
Tel: 703 560 1100

Dexion-Comino International Limited
Hunton Park, Hunton Bridge, King's Langley, Hertfordshire WD4 8QD, England
Tel: 09277 66161
Telex: 923403

DI/AN Controls Incorporated
944 Dorchester Avenue, Boston, Massachusetts 02125, USA
Tel: 617 288 7700
Telex: 7103331732

Dictaphone Corporation
120 Old Post Road, Rye, New York 10580, USA
Tel: (914) 967 7300
Telex: 137384

Didsbury Engineering Limited
Clifton Works, Manor Road, Levenshulme, Manchester, England
Tel: 061-224 6224
Telex: 667811

Diemler and Diekemper Incorporated
715 Loda Drive, Cincinnati, Ohio 45245, USA
Tel: (513) 752 6200
and
1435 Powhatan Street, Alexandria, Virginia 22314, USA
Tel: (703) 549 5727

Digital Acoustics Incorporated
1415 East McFadden, Suite F, Santa Ana, California 92705, USA
Tel: (714) 835 4884

Digivision Limited
Wenlock Way, Leicester, Leicestershire, LE4 7HU, England
Tel: 0533 763221
Telex: 341382

Dima SpA
Pirelli Group, 20092 Cinisello Balsamo (MI), Viale Matteotti n. 53, Italy
Tel: (02) 6188314/6188301
Telex: 310135 piremi I/dima

Dole Division James Dole Corporation
1400 Industrial Way, Redwood City, California 94063, USA
Tel: (415) 364 7440
Telex: 348363

Dornier GmbH
Postfach 1420, 7990 Friedrichshafen 1, Federal Republic of Germany
Tel: (075 45) 8 38 93
Telex: 07342090

Dorman Smith Traffic Products Limited
Rufford Road, Crossens, Southport PR9 8LA, England
Tel: 0704 30911
Telex: 677167 dorman g

F L Douglas (Equipment) Limited
Village Road, Arle, Cheltenham, Gloucestershire GL51 0AB, England
Tel: 0242 27921
Telex: 43182

R M Douglas Construction Limited
306 George Road, Birmingham B23 7RZ, England
Tel: 021-356 4888
Telex: 338399

John W Drake
1815 Woodland Avenue, West Lafayette, Indiana 47906, USA
Tel: 317 463 5882

Drexel Industries Incorporated
PO Box 248, Maple Avenue, Horsham, Pennsylvania 19044, USA
Tel: (215) 672 2200
Telex: 902020

Driclad Limited
152 Staplehurst Road, Sittingbourne, Kent ME10 1XS, England
Tel: 0795 24422

Drover, Welch and Lindlan
Suite 1200, 3225 North Central Avenue, Phoenix, Arizona 85012, USA
Tel: (602) 264 9731

Duff Norton Company
Box 32605, Charlotte, North Carolina 28232, USA
Tel: (704) 588 0300
Telex: 575188

Dunlop Transportation Systems Limited
Crown Works, Clayton Road, Hayes, Middlesex, England
Tel: (01) 573 0457
Telex: 23903

Eagle Bob Tail Tractors Incorporated
PO Box 8, Grosse Ile, Michigan 48138, USA
Tel: (313) 671 8181
Telex: 234215 Bobtail Groi

Eastern Airmotive Industries (USA) Limited
115-117 London Road, St Albans, Hertfordshire AL1 1LR, England
Tel: 0727 321158/9
Telex: 298753

Economical Airport Architecture
EAA GmbH, Stiefelholz 3, 3050 Wunstorf 1, Federal Republic of Germany
Tel: (05031) 130 04
Telex: D 0924511 eaa

Edbro Limited
Airport Equipment Division, Lever Street, Bolton, Lancashire BL3 6DJ, England
Tel: 0204 28888
Telex: 63196

H W Edgehill Equipment Limited
Station Road, Hook, Hampshire RG27 9JB, England
Tel: 025 672 2121
Telex: 858835

Edmac Associates Incorporated
PO Box 391, 333 West Commercial Street, East Rochester, New York 14445, USA
Tel: 716 385 1440

Edwards and Kelcey Incorporated
70 South Orange Avenue, Livingston, New Jersey 07039, USA
Tel: (201) 994 4520

Edwards and Kelcey International Incorporated
One World Trade Centre, Suite 8721, New York, New York 10048, USA
Tel: (212) 466 1383

Eisenwerk Schroder GmbH
Postfach 500260, D-8000 Munich 50, Federal Republic of Germany
Tel: (089) 147115
Telex: 05213809

Elecma Division Electronique de la Snecma
22 Quai Calliéni, 92150 Suresnes, France
Tel: (1) 772 81 84
Telex: 620385

Electro Dynamic Construction Company Limited
St Mary Cray, Orpington, Kent BR5 2ND, England
Tel: 0689 27551
Telex: 896101

Electronic Corporation of Israel
88 Giborei Israel Street, Tel Aviv 67891, Israel
Tel: 255181 5
Telex: 033578

Elektro Spezial, Philips GmbH
Hans-Bredow-Strasse 20, Postfach 448740, D-2800 Bremen 44, Federal Republic of Germany
Tel: (0421) 42871
Telex: 0245268

Elektrosignal
Omnipol Foreign Trade Corporation, Nekaizanka 11, CSR-112, 21 Prague 1, Czechoslovakia
Tel: (02) 268 261/8
Telex: 121299

Elgin Leach Corporation
2 North La Salle Street, Chicago, Illinois 60602, USA
Tel: (312) 236 0728
Telex: 254146

Elta Electronics Industries Limited, Israel Aircraft Industries Limited
Ashdod, Israel
Tel: (055) 31155
Telex: 031807 Elta Il

Emergency One Incorporated
1701 South West 37th Avenue, Ocala, Florida, USA
Tel: (904) 237 1122
Telex: 567412

Enertec Schlumberger
Enertec Division, 1 rue Nieuport BP 54, 78140 Velizy-Villacoublay, France
Tel: (3) 946 96 50
Telex: 698201F

Equisa
Corretera de Barcelona 34400, Alcalá de Henares, Madrid, Spain
Tel: 889 27 62-6
Telex: 42828 Equi-E

Erikoiskori Oy
Yrittajanti 13, 74130 Iisalmi, Finland
Tel: (977) 22911
Telex: 4431

Esavian Limited
Esavian Works, Stevenage, Hertfordshire, SG1 2NX, England
Tel: 0438 3355
Telex: 82189

ESG Elektronik System-Gesellschaft mbH
Vogelweideplatz 9, D-8000 Munich 80, Federal Republic of Germany
Tel: (089) 92161
Telex: 522594

E-Systems Incorporated
1501 72 Street North, PO Box 12248, St Petersburg, Florida 33733, USA
Tel: (813) 381 2000
Telex: 0523455

J L Eve Construction
Trakway Division, Coxmoor Road, Sutton in Ashfield, Nottinghamshire, England
Tel: 0623 515333
Telex: 377731

Evers and Wall Limited
Lambourn Woodlands, Newbury, Berkshire RG16 7RK, England
Tel: 0488 71444
Telex: 848658

Expresso Material Handling Equipment
Ochshauser Strasse 11-25, D-3500 Kassel-B, Federal Republic of Germany
Tel: (0561) 54022

Fabrica Italiana Trainatori
Via Zumbini 36, 20143 Milan, Italy
Tel: (02) 470 530/470 598

Fair Manufacturing Company
401 South 5th, Menno, South Dakota 57045, USA
Tel: 387 2389

Faudi Feinbau GmbH
Postfach 1267, D-6370 Oberursel 1, Federal Republic of Germany
Tel: (06171) 56031
Telex: 410745

Federal Electric Corporation
621 Industrial Avenue, Paramus, New Jersey 07652, USA
Tel: (201) 967 0123
Telex: 134458

Feedback Instruments Limited
Park Road, Crowborough, Sussex TN6 2QR, England
Tel: 089 26 3322
Telex: 95255

Fenamec Limited
Sutton Road, Hull, Humberside, England
Tel: 0482 781211
Telex: 52437

Ferguson Door Company Incorporated
1937 West 169th Street, Cardena, California 90247, USA
Tel: (213) 321 2970
Telex: 181255

Fernau Avionics Limited
Holywell Hill, St Albans, Hertfordshire AL1 1HS, England
Tel: 0727 31215
Telex: 894570

Ferranti Computer Systems Limited
Bracknell Division, Western Road, Bracknell, Berkshire RG12 1RA, England
Tel: 0344 3232
Telex: 848117

Wythenshawe Division, Simonsway, Wythenshawe, Manchester M22 5LA, England
Tel: 061-499 3355
Telex: 668084

Ferranti Offshore Systems Limited
Scott House, 475/485 Union Street, Aberdeen AB1 2DB, Scotland
Tel: 0224 50282
Telex: 73369 foslab g

Ferranti Packard Electronics Limited
6030 Ambler Drive, Mississauga, Ontario L4W 2P1, Canada
Tel: (416) 624 3020
Telex: 06961437

FFV Maintenance Division
S 73200 Arboga, Sweden
Tel: 46 589 800 00
Telex: 73246 ffvu s

Findlay Irvine Limited
Industrial Division, Bog Road, Penicuik, Midlothian EH26 9BU, Scotland
Tel: 0968 72111
Telex: 727501

Fire Fighting Enterprises (UK) Limited
67A Victoria Street, Windsor, Berkshire SL4 1EM, England
Tel: 075 35 54534
Telex: 28604

Fleming Steel Company
PO Box 1487, New Castle, Pennsylvania 16101, USA
Tel: 412 658 1511

Flomaster Division
Portec Incorporated, PO Box 589, Royal Gorge Industrial Park, Canon City, Colorado 81212, USA
Tel: (303) 275 7471

Flughafen Frankfurt/Main AG
Airconsult, D-6000 Frankfurt am Main 75, Federal Republic of Germany
Tel: (49/611) 690 2038
Telex: 413384

Fluid Transfer Limited
Nailsworth Mills Estate, Avening Road, Nailsworth, Gloucestershire, GL6 0BT, England
Tel: 045383 3381
Telex: 43401 flutra g

FMC Corporation
Airline Equipment Division, 1115 Coleman Avenue, PO Box 145, San Jose, California 95103, USA
Tel: (408) 289 2342
Telex: 346441

FN (Industry) Comet SA NV
Chaussée de Perwez 301, B-5003 Namur 1, Belgium
Tel: (081/22) 82 01
Telex: 59110

Forest City Electric Company Limited
Cawdor Street, Manchester Road, Walkden, Manchester M28 5HF, England
Tel: 061-7908241
Telex: 669034 Forest G

Forest City Signs Limited
Park Road, Timperley, Altrincham, Cheshire WA14 5QX, England
Tel: 061 969 0441

The Foundation Company of Canada Limited
1 Yonge Street, Toronto, Ontario M5E 1E8, Canada
Tel: (416) 361 4611
Telex: 06524213

Fox Manufacturing Company
PO Box 34, 106-128 Woodpark Road, Smithfield, New South Wales 2164, Australia
Tel: (02) 632 5888
Telex: AA 25061 Fox

Fratelli Papa SRL
Via Gramsci, 57, 20032 Cormano, Milan, Italy
Tel: 61 95 141
Telex: 333655

Frink Canada
777 Laurel Street, Cambridge, Ontario N3H 4S3, Canada
Tel: (514) 326 4000
Telex: 05828508
and
Henri Bourassa boulevard East, Montreal Nord, Quebec H1G 5W9, Canada
Tel: (519) 653 6234
Telex: 06959347

Frink Sno-Plows
205 Webb Street, Clayton, New York 13624, USA
Tel: 315 686 5531

Fuji Heavy Industries Limited
Subaru Building, 2-7 Nishishinjuku 1-chome, Shinjuku-ku, Tokyo 160, Japan
Tel: (03) 347 2500/2529
Telex: 2322268

Gallet SA
13 rue Georges Bouzerait, 92120 Montrouge, France
Tel: (1) 253 54 54

Garrett GmbH
Frankfurter Strasse 51, Postfach 1150, D-6096 Raunheim am Main, Federal Republic of Germany
Tel: (06142) 405 1
Telex: 415734

Garrett Manufacturing Limited
255 Attwell Drive, Rexdale, Ontario M9W 5BB, Canada
Tel: (416) 675 1411
Telex: 06989142

Garsite Products Incorporated
10 Grand Boulevard, Deer Park, New York 11729, USA
Tel: 516 1010
Telex: 967749

GEC Electrical Products Limited
Aviation Division, Boughton Road, Rugby, Warwickshire, CV21 1BU, England
Tel: 0788 2144
Telex: 311335

GEC Mechanical Handling Limited
Spencer Works, Melksham, Wiltshire SN12 8AX, England
Tel: 0225 702732
Telex: 44392 gecmec

Gematronik GmbH
Raffeisenstrasse, 4040 Neuss 21, Rosellen, Federal Republic of Germany
Tel: (02107) 5065
Telex: 08517757

Genave Incorporated
4141 Kingman Drive, Indianapolis, Indiana 46226, USA
Tel: (317) 546 1111

German Airport Equipment
PO Box 2049, 2120 Lueneburg, Federal Republic of Germany
Tel: (04131) 32011
Telex: 02182101

Glasdon Limited
Olympic Modular Building Division, Industrial Estate, Poulton Le Fylde, Blackpool, Lancashire, FY6 8JW, England
Tel: 0253 891131
Telex: 677288

Gloster Saro Limited
Road Vehicle Division, Hucclecote, Gloucester, Gloucestershire, GL3 4AD, England
Tel: 0452 69321
Telex: 43134

Goodwood Data Systems
150 Rosamond Street, Carleton Place, Ontario K0A 1JQ, Canada
Tel: (613) 257 3610
Telex: 0533593

Gotz Rubner GmbH
Postfach 47, D-8744 Mellrichstadt 1, Bayem, Federal Republic of Germany
Tel: (09776) 441
Telex: 0672850

W W Grinder Incorporated
2957 North Market, PO Box 4059, Wichita, Kansas 67204, USA
Tel: (316) 838 4229

Gulf and Western Advanced Development and Engineering Center
101 Chester Road, Swarthmore, Pennsylvania 19081, USA
Tel: (215) 544 7600
Telex: 834728

EC Hallam Engineering (Leicester) Limited
Smeeton Westerby, Leicester, Leicestershire LE8 0LT, England
Tel: 053 753 2330
Telex: 342167

Hamech Limited
Central Trading Estate, Staines, Middlesex TW18 4UT, England
Tel: 0784 56321
Telex: 266359

Hanscomb Associates Incorporated
La Salle-Wacker Building, 221 North La Salle Street, Chicago, Illinois 60601, USA
Tel: (312) 984 1750
Telex: 253600

Hants and Sussex Aviation Limited
The City Road, Portsmouth, Hampshire PO3 5PJ, England
Tel: 0705 694981
Telex: 86403 Hansus G

Harford Engineering Company Limited
Dereham Road, Norwich, Norfolk NR5 0TU, England
Tel: 0603 743211
Telex: 975216

Harlan Corporation
PO Box 15159, Chrysler and Stanley Road, Kansas City, Kansas 66115, USA
Tel: (913/342) 5650
Telex: 42378 harlan ksck

Harper Trucks Incorporated
PO Box 12330, Wichita, Kansas 67277, USA
Tel: (316) 942 1381

Harris Corporation
RF Communications Division, 1680 University Avenue, Rochester, New York 14610, USA
Tel: (716) 244 5830
Telex: 978464

Hastings Deering Limited
Parramatta Road, Lidcombe, New South Wales 2141, Australia
Tel: 6480111
Telex: AA20810

Hatehof Tankers and Refuellers Builders Limited
Aviation Center, Israel Export Institute, PO Box 29732, Tel Aviv, Israel
Tel: 35613

Hawker de Havilland Australia Pty Limited
PO Box 78, Lidcombe, New South Wales 2141, Australia
Tel: 6490111
Telex: AA20214

Hazeltine Corporation
Commack, New York 11725, USA
Tel: (516) 462 5100

Heliport Systems Incorporated
Airport Park, Morristown, New Jersey 07960, USA
Tel: (201) 540 0011
Telex: 220883T aur

Henderson PC Limited
Romford Essex RM3 8UL, England
Tel: 040 23 45555
Telex: 262482 pch rom g

Hindustan Aeronautics Limited
Indian Express Building, Vidhan Veedhi, Private Bag 5150, Bangalore 560001, India
Tel: 75004/6
Telex: 043266

HML Engineering Limited
St John's Road, Isleworth, Middlesex TW7 6NE, England
Tel: 01-560 3011
Telex: 933887

Hobart Brothers Company
Power Systems Division, 1177 Trade Road East, Troy, Ohio 45373, USA
Tel: (513) 339 6011
Telex: 8104562907

Hollandia Kloos International
Industrieweg 2, 2920 AA Krimpen a/d Ijssel, PO Box 48, Netherlands
Tel: (01807) 10333
Telex: 21571 hocon nl

Hollandse Signaalapparaten BV
PO Box 42, 7550 GD Hengelo, Netherlands
Tel: 31 74 488111
Telex: 44310 sign nl

Holophane
156 boulevard Haussmann, 75008, Paris, France
Tel: 763 11 70
Telex: 280261 olofane

Honeywell Information Systems Incorporated
Honeywell Plaza, Minneapolis, Minnesota 55408, USA
Tel: (612) 870 2209

Houchin Limited
Chart Road, Ashford, Kent TN23 1DZ, England
Tel: 0233 23211
Telex: 96178

Hovey and Associates
2378 Holly Lane, Ottawa, Ontario K1V 7P1, Canada
Tel: (613) 731 1200
Telex: 0534922

Herbert H Howell
232 Northwood Avenue, St Louis, Missouri 63105, USA
Tel: (314) 725 8810

Hoyle, Tanner and Associates Incorporated
One Technology Park, Londonderry, New Hampshire 03052, USA
Tel: (603) 669 5420

HTB Incorporated
1411 Classen Boulevard, Box 1845, Oklahoma City, Oklahoma 73101, USA
Tel: (405) 525 7451

Hunter Douglas Europe BV
2 Piekstraat, PO Box 5072, 3008 AB Rotterdam, Netherlands
Tel: (010) 84 44 44
Telex: 2889928694

Hunting Hivolt Limited
Riverbank Works, Old Shoreham Road, Shoreham-by-Sea, Sussex BN4 5FL, England
Tel: 079 17 4511/8
Telex: 87466

Hycom BV
Kanaal Zuid 114, PO Box 1079, NL 7301 BH Apeldoorn, Netherlands

Hydraroll Limited
No 2 Factory, Industrial Estate, Gaerwen, Anglesey, Gwynedd LL60 6HR, Wales
Tel: 024 877 454
Telex: 61502

Hydro-Geratebau GmbH and Company
Industriegebier, D-7616 Biberach/Baden, Federal Republic of Germany
Tel: (07835) 639
Telex: 7524524

Hydro-Hygiene Services Limited
Air Division, St Peter's Road, Furze Platt, Maidenhead, Berkshire SL6 7QU, England
Tel: 0628 34134

Hyster Europe Limited
PO Box 54, Basing View, Basingstoke, Hampshire RG21 2HQ, England
Tel: 0256 61171
Telex: 858384

IAL
Aeradio House, Hayes Road, Southall, Middlesex UB2 5NJ, England
Tel: 01-843 2411
Telex: 24114

IBI Group (Monenco Limited)
156 Front Street West, Toronto M5J 2L6, Canada
Tel: (416) 596 1930

IBM United Kingdom Limited
International Airlines Support Centre, 42 The Centre, Feltham, Middlesex TW13 4AU, England
Tel: 01-751 8422
Telex: 934517

Idman Oy
PO Box 769, Helsinki 10, SF-00101, Finland
Tel: 523500
Telex: 124539

Impression Enrigistrement de Resultats
12 rue de Sebastopol, 92400 Courbevoie, France
Tel: 334 30 20
Telex: 620289

Impulsphysik GmbH
Suelldorfer Landstrasse 400, PO Box 56010, D-2000 Hamburg 56, Federal Republic of Germany
Tel: (040) 81 80 11
Telex: 02189514

Industrial Acoustics Company
Walton House, Central Trading Estate, Staines, Middlesex TW18 4XB, England
Tel: 0784 56251
Telex: 25518

Industrial Service Outlets Limited
Foundry Lane, Horsham, West Sussex RH13 5QE, England
Tel: 0403 60366
Telex: 877748

Industrie Face Standard SpA
Viale L Bodio 33, 20158 Milan, Italy
Tel: (02) 3772
Telex: 331157

Infinetics Incorporated
Friskem Division, 201 Vanderer Avenue, PO Box 2330, Wilmington, Delaware 19899, USA
Tel: (302) 658 2471
Telex: 5106662332

Inter Engineering BV
PO Box 483, 5600 AL Eindhoven, The Netherlands
Tel: 33 40 448845
Telex: 51121

International Airport Equipment Group Taymech Aviation Limited
Taywood Road, Northolt, Middlesex UB5 6RL, England
Tel: 01-578 2366
Telex: 24428

International Harvester Company
Payline Group, 600 Woodfield, Schaumburg, Illinois 60196, USA
Telex: 288430
Twx: 910 291 3828

Interscan Australia Pty Limited
PO Box 339, Ryde, New South Wales 2112, Australia
Tel: 807 4222
Telex: AA20961

Intex Limited
King Constantine Square No 1, Nea Smirni, Athens, Greece
Tel: 9348730
Telex: 219092, 221516

Intra-Alcatel
26 rue Malakoff, 92600 Asnieres, France
Tel: (1) 790 65 72
Telex: 610718

Ion Track Instruments Incorporated
109 Terrace Hall Avenue, Burlington, Massachusetts 01803, USA
Tel: (617) 272 7233
Telex: 7103321808

IPD/PEA
Divisao de Electronica, Sao Jose dos Campos, CP 6001, 12220, Brazil
Tel: (0123) 21 1311

The Iso-Speedic Company Limited
Charles Street, Warwickshire CV34 5LR, England
Tel: 0926 42534
Telex: 311761

Italian Airport Equipment
via Conca del Naviglio, 37 20123 Milan, Italy
Tel: 02 832 58 51
Telex: 320595

ITT Aerospace/Optical Division
3700 East Pontiac Street, Fort Wayne, Indiana 46803, USA
Tel: (219) 423 9636

ITT Avionics Division
500 Washington Avenue, Nutley, New Jersey 07110, USA
Tel: (201) 284 2171
Telex: 133361, 133432

ITT Business Systems
Hollingbury, Brighton, East Sussex BN1 8AN, England
Tel: 0273 507111

ITT Gilfillan
7821 Orion Avenue, PO Box 7713, Van Nuys, California 91409, USA
Tel: (213) 988 2600
Telex: 9104951787

Jasmin Electronics Limited
St Matthew's Way, Leicester, Leicestershire LE1 2AA, England
Tel: 0533 581 28
Telex: 341840 Jasmin G

Kalmar Motor AB
Franska vagen 18, PO Box 73, S-38101 Kalmar, Sweden
Tel: + 46 480 27678

Keene Corporation
Bohannon Avenue, Greeneville, Tennesee 37743, USA
Tel: (615) 638 8156
Telex: 557491

Kersey Manufacturing Company
Division of Figgie International Incorporated, PO Box 151, Bluefield, Virginia 24605, USA
Tel: (703) 326 1121

Walter Kidde and Company Incorporated
Belleville Division, 675 Main Street, Belleville, New Jersey 07109, USA
Tel: 201 759 5500

King and Taylor (Commercial) Limited
The Wharf, Godalming, Surrey GU7 1JE, England
Tel: 048 68 5503

Klauer Manufacturing Company
PO Box 59, Dubuque, Iowa 52001, USA
Tel: (319) 582 7201

Kone Instrument Division
PO Box 2, SF-02321, Espoo 32, Finland
Tel: 8019411
Telex: 122729

Kornylak Corporation
400 Heaton Street, Hamilton, Ohio 45011, USA
Tel: (513) 863 1277
Telex: 21 4327 Kornylak Hton

R L Kuss and Company Incorporated
1331 Broad Avenue, Findlay, Ohio 45840, USA
Tel: (416) 741 0300
Telex: 06965551

Kwikform Limited
Waterloo Road, Birmingham B25 8LE, England
Tel: 021-706 3399
Telex: 339442

Laboratoire Central de Telecommunications
18–20 rue Grange-Dame-Rose (LCT), BP40, 78140 Velizy-Villacoublay, France
Tel: (1) 946 96 15
Telex: 698892F

Lacre Limited
Hatfield Road, St Albans, Hertfordshire AL4 0JT, England
Tel: 0727 64422
Telex: 8811163 Lacre G

John Laing Limited
Page Street, Mill Hill, London NW7 2ER, England
Tel: 01-959 3636
Telex: 263271

Lansing Limited
Kingsclere Road, Basingstoke, Hampshire RG21 2XJ, England
Tel: 0256 3131
Telex: 858120

Laser Engineering (Development) Limited
7–11 Curtain Road, London EC2A 3LU, England
Tel: 01-247 9961
Telex: 28183

K J Law Engineers Incorporated
23660 Research Drive, Farmington Hills, Maryland 48024, USA
Tel: 478 3150
Telex: 231179

Leigh Instruments Limited
Navigation Systems, 7 Bovis Drive, Pointe Claire, Quebec H9R 4WR, Canada
Tel: (514) 695 8130
Telex: 05821529

Leyland Vehicles Limited
Passenger Vehicle Division, Bow Lane, Leyland, Preston PR5 1SN, England
Tel: 07744 24241
Telex: 67515

Lift A Loft Corporation
400 East Jackson Street, Muncie, Indiana 47305, USA
Tel: (317) 288 3691

Limess Maschinenfabrik AG
Flughafenstrasse 22, CH-8302 Kloten, Switzerland
Tel: 01/813/70 33
Telex: 52964

Linde AG
Werksgruppe Guldner, 8750 Aschaffenburg, Postfach 62, Federal Republic of Germany
Tel: 010 496021341
Telex: 0034188010

Listavia Limited
13 Woodman Court, Godalming, Surrey GU2 2BT, England
Tel: 048 68 28932
Telex: 896691

Litton Amecom
5115 Calvert Road, College Park, Maryland 20740, USA
Tel: (01) 864 5600
Telex: 710/8269650

Lockheed Electronics Company Incorporated
1501 US Highway 22, CS No 1, Plainfield, New Jersey 07061, USA
Tel: (201) 757 1600

Dipl Ing A Lodige GmbH Maschinenfabrik
Postfach 20.48, Rathernaustrabe 19, D-4750 Paderborn, Federal Republic of Germany
Tel: (05251) 703/1
Tel: 0936768/69

McCormick-Morgan
1415 Evans Avenue, San Francisco, California 94124, USA
Tel: (415) 826 9559

Magline Incorporated
503 South Mercer Street, Pinconning, Michigan 48650, USA
Tel: (517) 879 2411

Magnasync/Moviola Corporation
5539 Riverton Avenue, PO Box 707, North Hollywood, California 91603, USA
Tel: (213) 763 8441
Telex: 673199 magna/movio/las

Magnetic AB
PO Box 20036, 161 20 Bromma, Stockholm-Bromma, Sweden
Tel: 29 04 60
Telex: 19037

Manairco Incorporated
PO Box 111, Mansfield, Ohio 44901, USA
Tel: (419) 524 2121

Marconi Radar Systems Ltd
Writtle Road, Chelmsford, Essex CM1 3BN, England
Tel: 0245 67111
Telex: 99108

Hugo Marom Aviation Consultants Limited
50 Levi Eshkol Street, Tel-Aviv, Israel
Tel: (03) 411224
Telex: 35770 coin il Israel
and
Hugo Marom Aviation Consultants International Incorporated
28903 Hollowbrook Avenue, Agoura, California 91301, USA

Massey Ferguson (UK) Limited
Industrial Machinery Division, PO Box 62, Banner Lane, Coventry CF4 9GF, England
Tel: 0203 465211
Telex: 31655

Mather and Platt Limited
Fire Engineering Division, Park Works, Manchester M10 6BA, England
Tel: 061-205 2321
Telex: 667193

Mathews Conveyer Company
Division of Rexnord Canada Limited, Box 30, Port Hope, Ontario LIA 3WI, Canada
Tel: (416) 885 2491
Telex: 06981275

C Maurice Company Limited
Southmill Road, Bishop's Stortford, Hertfordshire CM23 3DH, England
Tel: 0279 52231
Telex: 817130

Mechola Engineering
Mobile Post, Bikat Hayarden, Israel
Tel: (065) 88086

M E L
Division of Philips Electronic and Associated Industries Limited, Defence and Avionic Systems, Manor Royal, Crawley, West Sussex RH10 2PZ, England
Tel: 0293 28787
Telex: 87267

Merlin Gerin
rue Henri-Tarze, 38000 Grenoble, France
Tel: (76) 57 60 60
Telex: 320842

Mero Raumstruktur GmbH
Steinachstrasse 5, Postfach 6169, D-8700 Wurzburg, Federal Republic of Germany
Tel: 0931/41 03 258
Telex: 068630 mero d

Messerschmitt-Bolkow-Blohm Gmbh
Advanced Transport Systems, Postfach 80 12 65, 8000 Munich 80, Federal Republic of Germany
Tel: (089) 60 00/1
Telex: 52870

Metalite Aviation Lighting
Winster Grove, Birmingham B44 9EJ, England
Tel: 021-360 2222
Telex: 338024

Meteorology Research Incorporated
464 West Woodbury Road, 91001 Altadena, California, USA
Tel: (213) 791 1901
Telex: 675421

Metric Systems Corporation
736 North Beal Street, Fort Walton Beach, Florida 32548, USA
Tel: (904) 242 2111
Telex: 702447

Meyer Products Incorporated
18513 Euclid Avenue, Cleveland, Ohio 44112, USA
Tel: (216) 486 1313

Microturbo
Chemin du Pont de Rupe, BP 2089, 31019 Toulouse Cedex, France
Tel: (61) 70 07 77
Telex: 531442

Middlehurst Limited
99–103 Boyn Valley Road, Maidenhead, Berkshire SL6 4EJ, England
Tel: 0628 28044

Midland Tankers Division Hartford Motors (Oxford) Limited
Seacourt Tower, West Way, Oxford OX2 0JP, England
Tel: 0865 49966
Telex: 83277

Mingay's (Chigwell) Limited
70 Bentfield Road, Stansted, Essex CM24 8HT, England
Tel: 0279 815673

The Mitre Corporation
Metrek Division, 1820 Dolley Madison Boulevard, McLean, Virginia 22102, USA
Tel: (703) 827 6000

M L Aviation Company Limited
White Waltham Aerodrome, Maidenhead, Berkshire SL6 3JG, England
Tel: 062 882 3361
Telex: 847568

Monenco Limited
PO Box 6088, Station A, 2045 Stanley Street, Montreal H3C 3Z8, Canada
Tel: (514) 286 3636
Telex: 05560735

Montgomerie Reid Limited
Bramley, Basingstoke, Hampshire RG26 5DW, England
Tel: 0256 881444
Telex: 858678

Montreal Engineering Company Limited
PO Box 6088, Station B, Montreal H3C 3Z8, Canada
Tel: (514) 286 3636
Cable: Monenco

Moody SI Limited
251 St Louis Street, Terrebonne, Quebec J6W 1H5, Canada
Tel: (514) 471 3711
Telex: 05267551

Morrison-Maierle, Incorporated
910 Helena Avenue, PO Box 6147, Helena, Montana 59604, USA
Tel: (406) 442 3050
Cable: Mmhelenamt

R W Munro Limited
Cline Road, Bounds Green, London N11 2LY, England
Tel: 01-368 4422
Telex: 24130

NACO
Netherlands Airport Consultants BV, 115 Jan Van Nassaustraat, 2596 BS The Hague, Netherlands
Tel: (070) 264131
Telex: 33511

Nardeux
La Vallee du Parc, ZI-37600 Loches, France
Tel: (47) 593232
Telex: 750808 F

Nautical Electronic Laboratories Limited
Hackett's Cove, RR No 1, Tantallon, Halifax County, Nova Scotia B0J 3JO, Canada
Tel: (902) 823 2233
Telex: 01922552

Nederlandse Kunststof Industrie bv (NKI)
PO Box 222, 5100 AE Dongen, Netherlands
Tel: 01623 17 200
Telex: 54578

Negretti and Zambra (Aviation) Limited
73—77 Lansdowne Road, Croydon, Surrey CR9 2HP, England
Tel: 01-688 3426
Telex: 267708

Neoplan Gottlob Auwarter GmbH and Company
Vaihinger Strasse, D-7000 Stuttgart 80, Federal Republic of Germany
Tel: (0711) 7835/1
Telex: 72 55 871

Nerion AS
Brekkeveien 22, PO Box 2 Kjelsaas, Oslo 4, Norway
Tel: (472) 237380
Telex: 72483 nerio n

New England Industries Pty Limited
PO Box 669, Tamworth, New South Wales 2340, Australia
Tel: (067) 66 6056 AH
Telex: Titam AA63315

Nippon Electric Company Limited
NEC Building, 33-1 Shiba-Gochome, Minato-ku, Tokyo 108, Japan
Tel: 454 1111
Telex: Necktokj 22686

Nolte EMI BV
5600 Ax Eindhoven, Netherlands
Tel: (040) 810935
Telex: 51168

Nolte Mastenfabriek BV
Dr A Philipsweg 2, 6026 RA Maarheeze, Eindhoven, Netherlands
Tel: 049 591735
Telex: 51168

Norr Airport Planning Associates
40 University Avenue, Toronto M5J 2G3, Canada
Tel: (416) 977 1266
Telex: 06523255

Norwich Coachworks Limited
Burton Road, Norwich NR6 6AX, England
Tel: 0603 49221
Telex: 975174

NVF Company
Container Division, PO Box 160, Kennett Square, Pennsylvania 19348, USA
Tel: (215) 444 2800

Oaf-Graf and Stift AG
Brunner Strasse 72, A-1211 Vienna, Austria
Tel: (0222) 39 35 35/422
Telex: 07/4172 affl

Officine Fresia
Via Trento e Trieste 30, 17017 Millesimo (Savona), Italy
Tel: 56065/56066

Officine Meccaniche Aeronautiche OMA SpA
Aeroport Foligno, 06034 Foligno, Italy
Tel: (0742) 674256/7
Telex: 660087

Officine Meccaniche Italiane SpA
Via Vasco Agosti, 27, Casella Postale n 31, 42100 Reggio Emilia, Italy
Tel: (0522) 41341
Telex: 530665

Officine Panerai SRL
Piazza G Ferraris 2, 50131 Florence, Italy
Tel: (055) 579304
Telex: 573144 panfiri

Officine Romanazzi SpA
Via Tiburtina 1072, 00156 Rome, Italy
Tel: (06) 43 621
Telex: 610594 roma ro

Officine Viberti SpA
Via C Matteotti 99, 10042 Nichelino, Italy
Tel: 62 44 44
Telex: 221516 vibert i nichelino

OMC Lincoln
920 North 21st Street, Lincoln, Nevada 68501, USA
Tel: 402/475 9581
Telex: 484359

Omera-Segid
49 rue Ferdinand-Berthoud, BP 68, 95101 Argenteuil Cedex, France
Tel: (3) 982 09 42
Telex: 696797 F

Oshkosh Truck Company
PO Box 2566, Oshkosh, Wisconsin 54903, USA
Tel: (414) 235 9150

OTE Divisione Della Si El SpR
Via E Barsanti, 8, 50127 Florence, Italy
Tel: (055) 41 09 21
Telex: 570276 ote-i

Outokumpu Oy
Engineering Division, POB 27, SF-02201 Espoo 20, Finland
Tel: 358 0 4211
Telex: 121053

Over-Lowe Company
2767 South Tejon Street, Englewood, Colorado 80110, USA
Tel: (303) 761 3760

Pan Am World Services Incorporated
666 Steamboat Road, Greenwich, Connecticut 06830, USA
Tel: (03) 661 1637
Telex: 643536

Park Air Electronics Limited
Northfields, Market Deeping, Peterborough PE6 8LG, England
Tel: 0778 345434
Telex: 32456

Parkin Partnership Architects Planners (Monenco Ltd)
55 University Avenue, Toronto M5J 2H7, Canada
Tel: (416) 863 9360
Cable: Parkintor

Parmeko Limited
Percy Road, Leicester, Leicestershire LE2 8FT, England
Tel: 0533 832287
Telex: 341689

Patek Philippe SA
39 rue Marziano 1211, Geneva 24, Switzerland
Tel: (022) 42 85 40
Telex: 27299

Pavement Consultancy Services BV
Kleine Tocht 23, 1507 CB Zaandam, Netherlands
Tel: (075) 31 1321
Telex: 19331

Pavement Specialists Incorporated
4405 Reese Drive, S-R 114 a Esters Road South, PO Box 61327, Dallas/Fort Worth Airport, Texas 75261, USA
Tel: (214) 258 0660

Paxman Diesels Limited
Paxman Works, PO Box 8, Colchester Essex CO1 2HW, England
Tel: 0206 5151
Telex: 98151 gecpdc g

Peat Marwick and Partners (Monenco Ltd)
PO Box 31, Commerce Court Postal Station, Toronto M5L 1B2, Canada
Tel: (416) 863 3515
Cable: Veritatem

Pelletier Industrie
22-24 Allée des Jachères, Sofilic 431, 94263 Fresnes, France
Tel: (1) 668 76 76
Telex: 201094F

Pentaport Sweden AB
Kungsportsavenyn 20, S-411 06 Gothenburg, Sweden
Tel: (031) 81 14 94/40 88 66

Perusyhtyma Oy Ara
Vahdontie 31, Turku, Finland
Tel: 35821 383 111
Telex: 62305 62252 ara sf

Perren Fire Protection
Murdock Road, Bicester, Oxfordshire OX6 7PP, England
Tel: 08692 44411
Telex: 837380

Petbow Limited
Sandwich, Kent, England
Tel: 0304 613311
Telex: 96329

Petitjean and Company Limited
Teesside Industrial Estate, Dukes Way, Thornaby, Cleveland TS17 9LT, England
Tel: 0642 766242
Telex: 58383 tripjan g

Pettibone Mercury Corporation
4700 West Division Street, Chicago, Illinois 60651, USA
Tel: (312) 772 9300

PGS, SpA
29010 Cadeo Piacenza, Italy
Tel: (0523) 50321
Telex: 530134 pgs 1

Philips Electronics Limited
601 Milner Avenue, Scarborough, Ontario M1B 1M8, Canada
Tel: (416) 292 5161
Telex: 06525100

Philips Elektronikindustrier AB
S-175 88 Jarfalla, Sweden
Tel: 46 758 10000
Telex: 10943

NV Philips' Gloeilampenfabrieken
Electro Acoustics Division, Eindhoven, Netherlands
Tel: 40 791111
Telex: 51121

Light Division, Eindhoven, Netherlands
Tel: 40 791111
Telex: 51121

Philips Telecommunicatie Industrie BV
PO Box 32, 1200 JD Hilversum, Netherlands
Tel: 35 899111
Telex: 43712

Photo Scan
Dolphin Estate, Windmill Road, Sunbury on Thames, Middlesex TW16 7HG, England
Tel: 093 27 89741

Max Pietsch
D-3014 Laatzen, 1 Augsburger Strasse 1, Postfach 1125, Federal Republic of Germany
Tel: 86 20 77
Telex: 922434

Pilcher-Greene Limited
Consort Way, Victoria Gardens, Burgess Hill, West Sussex RH15 9NA, England
Tel: 044 46 57079
Telex: 877102

Anton Piller (UK) Limited
Chesterton Lane, Cirencester, Gloucestershire GL7 1YE, England
Tel: 0285 67721
Telex: 43654

Pindair Limited
Unit 14/15 Quay Lane, Gosport, Hampshire, England
Tel: 07017 87830
Telex: 86210 skima

Plessey Airports Limited
Addlestone, Weybridge, Surrey KT15 2PW, England
Tel: 0932 47282
Telex: 8812217

Plessey Radar
Plessey Electronic Systems Limited
Addlestone, Weybridge, Surrey KT15 2PW, England
Tel: 0932 47282
Telex: 262329

Power Lifts Limited
Hadley Works, Caxton Way, Holywell Industrial Estate, Watford, Hertfordshire WD1 8TJ, England
Tel: 0923 27724

Prismo Universal Limited
Rowfant, Crawley, West Sussex RH10 4NF
Tel: 0342 714949
Telex: 95190

Proavia
75 rue La Bóetie, 75008 Paris, France
Tel: (1) 359 22 93

Pyrotechnics
8 Ridgedale Avenue, Cedar Knolls, New Jersey 07927, USA

Racal-Decca Navigator Limited
Ground Systems Division, Burlington House, Burlington Road, New Malden, Surrey KT3 4NW, England
Tel: 01-942 2488
Telex: 22891

Racal-MESL Radar Limited
Preston Road, Linlithgow, West Lothian EH49 6EU, Scotland
Tel: 050 684 5151
Telex: 727650

Racal Messenger Limited
5 Bennett Road, Reading, Berkshire, England
Tel: 0734 85181
Telex: 848011

Racal Recorders Limited
Hythe, Southampton, Hampshire SO4 6ZH, England
Tel: 0703 843265
Telex: 47600

Rapistan Lande BV
PO Box 18 5460, AA Veghel, Netherlands
Tel: 041 3085515
Telex: 50186

Raytheon Canada Limited
PO Box 1619, 400 Philip Street, Waterloo, Ontario, N23 4K6, Canada
Tel: (519) 885 0110
Telex: 06955431

Raytheon Company
Equipment Division, Boston Post Road, Wayland, Massachusetts 01778, USA
Tel: (617) 358 2721

Equipment Division Air Traffic Control Directorate, Boston Post Road, Sudbury, Massachusetts 01776, USA
Tel: (617) 443 9521
Telex: 948422

Raytheon Data Systems Company
1415 Boston-Providence Turnpike, Norwood, Massachusetts 02062, USA
Tel: (617) 762 6700
Telex: 924408

RCA Automated Systems
PO Box 62, Routes 3 and 588, Burlington, Massachusetts 01803, USA
Tel: (617) 272 4000

Rediffusion Radio Systems Limited
Broomhill Road, London SW18 7JQ, England
Tel: 01-874 7281
Telex: 264029

Rediffusion Industrial Services Limited
214 Red Lion Road, Surbiton, Surrey KT6 7RP, England
Tel: 01-397 5133
Telex: 929989

Rediffusion Reditronics Limited
La Pouquelaye, St Helier, Jersey, Channel Islands, UK
Tel: 0534 30321
Telex: 4192341

Redpath Dorman Long International Limited
PO Box 26, Dormbridge House, 16 St Cuthbert's Street, Bedfordshire MK40 3JU, England
Tel: 0234 63141
Telex: 825054

Reel SA
100 avenue Général Leclerc, 69300 Caluire, France
Tel: (7) 823 18 13
Telex: 310758F

Rees Instruments Limited
Westminster House, Old Woking, Surrey GU22 9LF, England
Tel: 048 62 62221
Telex: 859679

Reliance-Mercury Limited
Mile Cross Works, Gibbet Street, Halifax, West Yorkshire HX1 4JQ, England
Tel: 0422 58141
Telex: 517306

Remtec Incorporated
933 rue Simard, Chambly, Quebec J3L 4L9, Canada
Tel: (514) 658 6671
Telex: Egwcham 055 61653

Rens Manufacturing Company
83868 North Pacific Highway, PO Box 37, Creswell, Oregon 97426, USA
Tel: (503) 895 2172

Research Engineers Limited
Osman Road, Shoreditch, London N1 5RD, England
Tel: 01-739 7811
Telex: 261973 re ltd g

Reynolds, Smith and Hills
4019 boulevard Centre Drive, PO Box 4850, Jacksonville, Florida 32201, USA
Tel: (904) 396 2011

RFD Inflatables Limited
Catteshall Lane, Godalming, Surrey GU7 1LH, England
Tel: 04868 4122
Telex: 859233

Richards-Wilcox of Canada Limited
Box 5060, London, Ontario N6A 4M4, Canada
Tel: (519) 432 1811
Telex: 0645668

Ritron Incorporated
148 West Carmel Drive, PO Box 798, Carmel, Indiana 46032, USA
Tel: (317) 846 1201

Riwosa SA
Witikonerstrasse 80, CH-8032 Zurich, Switzerland
Tel: (01) 53 45 55/56
Telex: 55659 riwosa-ch

RLG and Company
12964 Moorpark Street, Studio City, California 91604, USA
Tel: (213) 872 1738

Martin Roberts Limited
Sittingbourne, Kent ME10 3JH, England
Tel: 0795 76161
Telex: 965639 doors g

Rockwell International
Collins Division, 400 Collins Road North East, Cedar Rapids, Iowa 52406, USA
Tel: (319) 395 5910

Röder Präzision GmbH
6073 Egelsbach/Kreis Offenback/M, Flugplatz Postfach 1120, Federal Republic of Germany
Tel: 0 61 03/4 31 35
Telex: 415023 roepr

Rohde and Schwarz GmbH and Company KG
Muhldorfstr 15, D-8000 Munich 80, Federal Republic of Germany
Tel: (089) 4129-1
Telex: 523703

Root Sprint Scraper Company
527 West North Street, Kalamazoo, Michigan 49007, USA
Tel: (616) 382 2025

Rosemount Engineering Company Limited
Durban Road, Bognor Regis, Sussex PO22 9OX, England
Tel: 0243 863121
Telex: 86218

Rosenbauer KG
A4021 Leoanding bei Linz, PO Box 176, Paschingerstrasse 90, Austria
Tel: (0 72 22) 52 4 76
Telex: 021271

Dr Justus Rosenhaven GmbH
2 Norderstedt 3, Kohfurth 19, Postfach 3540, Federal Republic of Germany
Tel: (040) 5 23 10 18
Tel: 2174 225 rhg d

Ross Auto Engineering Limited
Banastre Road, Southport, Lancashire, England
Tel: 0704 35757
Telex: 677574

Royal Volker Stevin NV
PO Box 2695, Costmaaslaan 71, 3000 CR Rotterdam, Netherlands
Tel: (010) 331533
Telex: 23486

John Rusling Limited
The Old Hall, Station Road, Newport, Salop TF10 7HU, England
Tel: 0952 811779
Telex: 35207

S and S Corporation
Route 3, PO Box 70, Cedar Bluff, Virginia 24609, USA

Engineering Centre, 4201 Lee Highway, Bristol, Virginia 24201, USA
Tel: (703) 466 2242

Saab Scania AB
Saab Car Division, S-61181 Nykoping, Sweden
Tel: (0155) 44 000
Telex: 64018

Saf
45 rue des Boulets, 75011 Paris, France
Tel: (1) 371 24 40

Safare-Crouzet
BP 171, 06005 Nice Cedex, France
Tel: (93) 84 72 79
Telex: 460813

Sangamo Weston Incorporated
Data Systems Division, Industrial Systems Group, PO Box 3041, Sarasota, Florida 33578, USA
Tel: (813) 371 0811
Telex: 052890

Santerne
23–25 rue due Depot - 62033 Arras Cedex, BP 948, France
Tel: (21) 21 00 04
Telex: Santara 820127 f

Sar-Assist, Incorporated
One Island Lane, Grenwich, Connecticut 06830, USA
Tel: (203) 869 1322

Satt Electronics
Industrial Division, PO Box 9034, S-200 39 Malmo, Sweden
Tel: 22 20 00
Telex: 32422

Saval Kronenburg BV
Parallelweg 3, Postbus 1, 5320AA Hedel, Netherlands
Tel: (076) 222081
Telex: 54369

Saxby, Department Tracma
40 rue de l'Orillon, 75526 Paris, Cedex 11, France
Tel: 357 65 30
Telex: 220554

Scanray Corporation
1526 West 240th Street, Harbor City, California 90710, USA
Tel: (213) 534 4370
Telex: 686233 astro hrbo

Fritz Scharfenstein
Buchenweg 8, 5239 Nister/Hochenburg, Federal Republic of Germany
Tel: (02662) 6817
Telex: 863536 steup d

Carl Schenck AG
Postfach 40 18, D-1600 Darmstadt 1, Federal Republic of Germany
Tel: (06151) 8821
Telex: 419611

Schiphol Airport Authority
PO Box 7501, 1118 ZG Schiphol Airport, Netherlands
Tel: 020 5172582
Telex: 14653

Schmid Telecommunication Zurich
Rieterstr 6, 8002 Zurich, Switzerland
Tel: (01) 2012353
Telex: 53240

Ing Alfred Schmidt GmbH
7822 St Blasien, Schwarzwald, Federal Republic of Germany
Tel: 07672/373
Telex: 07721213

Schorling International GmbH
PO Box 911060, Schorlingstrasse 3, D-3000 Hanover 91, Federal Republic of Germany
Tel: (0511) 44490
Telex: 922592

Schroder Eisenwerk
Postfach 500260, D-8000 Munich 50, Federal Republic of Germany
Tel: 089/1471 15
Telex: 05215809

Scott Wilson Kirkpatrick and Partners
Scott House, Basingstoke, Hampshire RG21 2JG, England
Tel: 0256 61161
Telex: 858805

Secre - Societe D'etudes et de Constructions Electroniques
214–216 rue du FBG Saint-Martin, 75483 Paris Cedex 10, France
Tel: (93) 86 50 00
Telex: Granice 470 696 F

Selenia Industrie Elettronich Associate SpA
Civil Radar and Systems Division, Via Tiburtina Km 124002 - 00131 Rome, Italy. P O Box 7083 - 00100 Roma
Tel: (0039 6) 43601
Telex: (43) 613690, 613649 selrom i

Seri Renault Ingenierie
2 Avenue du Vieil-Etang, 78340 Bois d'Arcy, France
Tel: (1) 043 81 34
Telex: 695426F

Servo Corporation of America
Hicksville, New York 11802, USA
Tel: (516) 938 9700
Telex: 5102211872

Shelter-Span (UK) Limited
PO Box 72, Maidenhead, Berkshire SL6 2EQ, England
Tel: 0628 29032
Telex: 847591

Shin Meiwa Industry Company Limited
5-25 Kosone-Cho, Nishinomiya-Shi, Hyogo-Ken, Japan
Tel: (0798) 47 0331

Tokyo Office, Shin-Ohtemachi Building, 2-1, Ohtemachi 2 chome, Chiyoda-ku, Tokyo
Tel: (03) 279 3531
Telex: 2222 431 smicair t j

Shorgard Limited
Airfield Lighting Division, 22A Manor Place, Edinburgh EH3 7DS. Scotland
Tel: 031-226 3901
Telex: 95452

Shorrock Security Systems Limited
Special Services Division, Shadsworth Road, Blackburn, Lancashire BB1 2PR, England
Tel: 0254 63644
Telex: 635151 shorok g

Siderpali SpA
40056 Crespellano (Bologna), Italy
Tel: (051) 964041
Telex: 510218 palii

Sides Société Industriele Pour le Developpment de la Sécurité
182 rue de Trignac, 44600 Saint-Nazaire, France
Tel: (40) 66 77 77
Telex: 700 955 sides

Siemag Rosenkaimer GmbH
PO Box 44, Hochstrasse 33, D-5653 Leichlingen 1, Federal Republic of Germany
Tel: (02175) 3051
Telex: 8515875 sr

Sietam
42/48 ave du President Kennedy, 91170 Viry-Chatillon, France
Tel: (6) 996 91 80
Telex: 690503F

Silver Instruments
2346 Stanwell Drive, Concord, California 94520, USA
Tel: (415) 676 5900
Telex: 9104815794 silver cncd

Simon Engineering Dudley
Dudley, West Midlands DY1 2HA, England
Tel: 0384 70180
Telex: 338568

Siraga
BP 14, 36500 Buzancals, France
Tel: (54) 84 09 30
Telex: Siraga 750120 f

Skanska
International Division, S-21102 Malmo, Sweden
Tel: (46) 401 440 00
Telex: 32247

Skidmore, Owings and Merrill
33 West Monroe Street, Chicago, Illinois 60603, USA
Tel: (312) 641 5959
Telex: 254337

and at the following
400 Park Avenue, New York 10022, USA
Tel: (212) 759 2121

One Maritime Plaza, San Francisco, California 94111, USA
Tel: (415) 981 1555

900 South West Fifth Avenue, Portland, Oregon 97204, USA
Tel: (503) 226 1431

1201 Connecticut, North West, Washington DC 20036, USA
Tel: (202) 828 0700

334 Boylston Street, Boston, Massachusetts 021116, USA
Tel: (617) 247 1070

70 Universal City Plaza, Universal City, California 91608, USA
Tel: (213) 760 8300

5251 Westheimer, Houston, Texas 77056, USA
Tel: (713) 965 9350

1675 Broadway, Denver, Colorado 80202, USA
Tel: (303) 825 3100

Smarte Carte Incorporated
1777 Stillwater Street, White Bear Lake, Minnesota 55110, USA
Tel: (612) 429 3614

SMI Industries Division, Caelter Enterprises Limited
2055 Bennett Avenue, Montreal H1V 2T2, Quebec, Canada
Tel: (514) 255 2883
Telex: 05828753

Smith, Hinchman and Grylls Associates Incorporated
455 West Port Street, Detroit, Michigan 48226, USA
Tel: (313) 964 300

Snowblast Division, Caelter Industries Incorporated
4695 Ironton Street, Denver, Colorado 80239, USA
Tel: (303) 371 7800
Telex: 45821

Societe Internationale de Telecommunications Aeronautiques, Sita
Registered Office: 16 Avenue Henri Matisse, 1140 Brussels, Belgium
Administrative Headquarters: 112 Avenue Charles de Gaulle, 92522 Neuilly-sur-Seine
Tel: 758 13 22
Telex: 620575

Socrat
2 and 4 rue Küss, 75640 Paris Cedex 13, France
Tel: 588 63 58

Sodern
1 avenue Descartes, 94450 Limeil-Brevannes, France
Tel: 569 96 00
Telex: 203475F

Sofreavia
75 rue de la Boetie, 75008 Paris, France
Tel: 359 22 93
Telex: 641360 soffavia

Sofreavia Service
75 rue de la Boetie, 75008 Paris, France
Tel: 359 00 56 57
Telex: 640288 sofaser

Software Sciences Limited
Special Systems Division, Abbey House, 282/292 Farnborough Road, Farnborough, Hampshire GU14 7NB, England
Tel: 0252 544321
Telex: 858228

Solari and c/udine SpA
Via Gino Pieri, 29, 33100 Udine, Italy
Tel: (0432) 43241/479361
Telex: 450155

Soudure et Mecanique Appliquees du Val Notre-Dam
50 rue de la Fosse aux Loups, 95100 Argenteuil, France
Tel: 982 65 96

Sovam
9 rue de Téhéran, 75008 Paris, France
Tel: 755 96 20
Telex: 660698 F

Sovex Marshall Limited
Cavendish Drive, Carlton, Nottingham NG4 3DY, England
Tel: 0602 249271
Telex: 37149

Speciali Impianti Telescriventi Telefonici Interni Sas
Via Desiderio 24/26, 20131 Milan, Italy
Tel: 29 68 94
Telex: 293371

Sperry Univac Division
PO Box 500, Blue Bell, Pennsylvania 19424, USA

Sperry Vickers
Asser House, Airport Service Road, Portsmouth Airport, Portsmouth PO3 5QD, England
Tel: 0705 697341
Telex: 86749

SRA Communications AB
Torshamngatan 21-23, PO Box 1, S-163 00 Spanga-Stockholm, Sweden
Tel: (08) 752 1000
Telex: 13545 sra s

SSIH Equipment SA
rue Stampfli 96, Ch-2500 Bienne 4, Switzerland
Tel: (032) 41 09 11
Telex: 34 289

Standard Elektrik Lorenz AG
Telecommunications Group, Franckstrasse 8, D-7141 Ludwigsburg, Federal Republic of Germany
Tel: (07141) 1241
Telex: 7211215

Oy Starck-Johann and Company AB
PO Box 107, SF-00811, Helsinki 81, Finland
Tel: (90) 7556 433
Telex: 122951 loatn st

Stearns, Airport Equipment Division, Stearns Manufacturing Company Incorporated
PO Box 646, Crowley, Texas 76036, USA
Tel: (817) 294 2020
Telex: 758541

Steelcons AB
Bryggaregatan 9, 231 00 Trelleborg, Sweden
Tel: (410) 15545
Telex: 32926 famab

Stevens Aircraft Ground Equipment Corporation
30 North Prospect Avenue, Lynbrook, New York 11563, USA
Tel: (212) 656 7982/(516) 887 1426
Telex: 645011

Stewart and Stevenson Services Incorporated
4516 Harrisburg Boulevard, PO Box 1637, Houston, Texas 77001, USA
Tel: (713) 923 2161
Telex: 762301

Still GmbH
International Division, PO Box 602340, D-2000 Hamburg 60, Federal Republic of Germany
Tel: (040) 731121
Telex: (0) 211141

Stinar Corporation
3255 Sibley Memorial Highway, St Paul, Minnesota 55121, USA
Tel: (612) 454 5112

Alan Stratford and Associates Limited
Fairoaks Airport, Chobham, Surrey
Tel: 09905 7822
Telex: 858348

STS Zweigniederlassung der Dorrier System GmbH
Postfach 1724, D-7990 Friedrichshafen 1, Federal Republic of Germany
Tel: (07541) 25058
Telex: 0734206

Sulzer Brothers
Zurcherstrasse 9, CH-8401, Winterthur, Switzerland
Tel: (052) 81 11 22
Telex: 22318

Sulzer Brothers (UK) Limited
Farnborough, Hampshire GU14 7LP, England
Tel: 0252 44311
Telex: 858771

Svenska Skumsläcknings Aktiebolaget
PO Box 32, 4421 21 Kungälu, Sweden
Tel: (0303) 15070
Telex: 21217 skumab s

Systelec
92 Avenue Jean-Jaures, 92120 Montrouge, France
Tel: 735 07 21

Tadiran Israel Electronics Industries Limited
3 Hashalom Road, PO Box 648, Tel Aviv 61000, Israel
Tel: (03) 267 272
Telex: 033537

Tarmac Group
PO Box 8, Ettingshall, Wolverhampton WV4 6JP, England
Tel: 0902 41101
Telex: 338544

Taylor Woodrow International Limited
Western House, Western Avenue, London W5 1EU, England
Tel: 01-997 6641
Telex: 23503
and
78 Mill Point Road, PO Box 91, South Perth, Western Australia
Tel: (367) 7177
Telex: 92673
and
PO Box 138, 10 Abebe Village Road, Iganmu, Lagos, Nigeria
Tel: 844066/7/8
Telex: 20117
and
1704/1706 Bangunan Fitzpatrick, 86 Jalan Raja Chulan, PO Box 2556, Kuala Lumpur 05-10, Malaysia
Tel: (603) 42844
Telex: 30975

Taylor Woodrow Towell Company (LCC)
PO Box 253, Muscat, Sultanate of Oman
Tel: Muttrah 70 4866
Telex: 3224

Taylor Woodrow Towell Company (LCC)
Teamwork Saudi Arabia, PO Box 10767, Riyadh, Saudi Arabia
Tel: 62388

Taymech Aviation Equipment Limited
345 Ruislip Road, Southall, Middlesex UB1 2QX, England
Tel: 01-578 2366
Telex: 24428

Tecnavia SA
Lugano Airport, 6982 Agno, Switzerland
Tel: (091) 59 34 02
Telex: 840009

Telecommunications Radioelectriques et Telephoniques TRT
88 rue Brillat Savarin, 75 640 Paris 13e, France
Tel: (33) 15 89 77
Telex: 250838F

Teledyne Inet
2750 West Lomita boulevard, Torrance, California 90509, USA
Tel: (213) 325 5040
Telex: 677228

Teledyne Sprague Engineering
19300 South Vermont Avenue, Gardena, California 90248, USA
Tel: (213) 327 1610
Telex: 910 3466943

Telefonbau und Normalzeit
Mainzer Landstrasse 128-146, Postfach 4432, D-600 Frankfurt-M 1, Federal Republic of Germany
Tel: 2661
Telex: 411141

AB Teleplan
PO Box 1310, S-171 25 Solna, Sweden
Tel: 98 10 00
Telex: 17587 telplan s

Telerad
rue Sainte-Marguerite, 64600 Anglet, France
Tel: (59) 24 10 28
Telex: 57008F

Tellus Maskin AB
S-186 00 Vallentuna, Sweden
Tel: (0762) 24125
Telex: 10227 tellus s

Terma Elektronik AS
Hovmarken 4, DK-8520 Lystrup, Denmark
Tel: (06) 22 20 00
Telex: 68109 terma dk

Terra Corporation
3520 Pan American Freeway North East, Albuquerque, New Mexico 87107, USA
Tel: (505) 884 2321

Tesla
Pardubice, Czechoslovakia
Exports handled by:
Omnipol Foreign Trade Corporation
Nekaizanka 11, CSR-112, 21 Prague 1, Czechoslovakia
Tel: (02) 268261/8
Telex: 121299

Texas Instruments
Equipment Group, PO Box 6015, Dallas, Texas 75222, USA
Tel: (214) 238 3401
Telex: 470900

Thermal Control Company Limited
Hydraulic Engineering Division, 136-140 Old Shoreham Road, Hove, East Sussex BN3 7BW, England
Tel: 0273 772302
Telex: 87303

Arnold W Thompson, PC
316 Elwood Avenue, PO Box 221, Hawthorne, New York 10532, USA
Tel: (914) 769 4456

Thompson Tankers
NEI Clarke Chapman Engineering Limited, Great Bridge Road, Bilston, West Midlands, England
Tel: 0902 43141
Telex: 337329

A/S Alfred Thomsen,
Askov, DK-6600 Vejen, Denmark
Tel: (455) 36 07 00
Telex: 50517 epoke

Thomson - CSF
Division DRS-TVT: 1 rue des Mathurins, BP10, 9222 Bagneux, France
Tel: (657) 13 65
Telex: 204780F
Export: 40 rue Grange-Dame-Rose BP 34, 92360 Meudon-La-Forêt, France
Tel: 630 23 80
Telex: 270375

Thorn EMI Lighting Limited
Airfield Lighting Division, Commercial House, Lawrence Road, Tottenham, London N15 4EG, England
Tel: 01-802 3151
Telex: 893024 tllt ot g

Thyssen Henschel
Henschelplatz 1, 3500 Kassel, West Germany
Tel: (0561) 801 6274
Telex: 99793

Tippets-Abbett-McCarthy-Stratton (TAMS)
1101 15th Street, North West Washington DC 20005, USA
Tel: (202) 296 4371
Telex: RCA 248677

Tissmetal Lionel-Dupont
Departement Manutention Teleflex, 32-34 rue Robert-Witchitz, BP 229, 94203 Ivry-sur-Seine Cedex, France
Tel: 681 21 21
Telex: 270324

TRA Airport Consultants
215 Columbia Street, Seattle, Washington 98104, USA
Tel: (206) 682 1133
Tel: 321298 tra sea

TRA Consultants, Incorporated
McCarran International Airport, PO Box 11-9, Airport Station, Las Vegas, Nevada 98111, USA
Tel: (702) 736 2947

TRA/FARR
313 PE" Street, Anchorage, Alaska 99501, USA
Tel: (97) 277 2641

Transact International
835 Hope Street, Stamford, Connecticut 06907, USA
Tel: (203) 348 8765
Telex: 965943

Transequip (London) Limited
Space Waye, North Feltham Trading Estate, Feltham, Middlesex TW14 0TH, England
Tel: 01-890 0788
Telex: 264993 teqldn g

Trastecnica Rapistan
20093 Cologno Monzese (Milan), v. Brunelleschi 7, Italy
Tel: (02) 2543816
Telex: 333654 traste

Trepel AG
Division Airport Equipment, Alte Schmeize, Trepel Haus, Postfach 130205, D-6200 Wiesbaden 13, Federal Republic of Germany
Tel: (0 61 21) 2091
Telex: 04186368

Trilectron Industries Incorporated
300 Ninth Avenue, Hawthorne, New Jersey 07506, USA
Tel: (201) 423 4005
Twx: 7109954414

Tronair Incorporated
S-1740 Eber Road, Holland, Ohio 43528, USA
Tel: (419) 866 6301
Telex: 8104402839

Trump Incorporated
PO Box 368, Plattsmouth, New Brunswick 68048, USA
Tel: (402) 296 2115

Alistair Tucker Associates
39 Charing Cross Road, London WC2H 0AW, England
Tel: 01-439 3631
Telex: 267678 aircon g

Turbomach Division
Solar Turbines International, 4400 Ruffin Road, PO Box 80966, San Diego, California 92138, USA
Tel: (714) 238 5500
Twx: 901 335 1502

Ulmer Aeronautique
12 rue Klock, 12 rue Curton, 92110 Clichy, France
Tel: 737 60 64
Telex: 210550

United Technologies
Hamilton Standard Division, Windsor Locks, Connecticut 06096, USA
Tel: (203) 623 1621

United Tractor Company
116 North 15th Street, Chesterton, Indiana 46304, USA
Tel: (219) 926 1186

Unitron Incorporated
2826 West Kingsley Road, PO Box 2159, Garland, Texas 75041, USA
Tel: (214) 271 4531
Telex: 730644

Universal Aviation Support Company Limited
232-234 Tolworth Rise South, Tolworth, Surrey KT5 9NE, England
Tel: 01-337 2028
Telex: 917054

Universal Conveyor Company Limited
Humberstone Lane, Leicester, Leicestershire LE4 7JT, England
Tel: 0533 693485
Telex: 341355

Vaisala Oy
PL 26, SF 00421, Helsinki 42, Finland
Tel: (358) 0 890933
Telex: 122832

Valk Manufacturing
PO Box 218, Carlisle, Pennsylvania, USA
Tel: (717) 766 0711

Vandeputte Fils and Cie
Département Électronique, 278 Chaussée Fernand Forest, 59203 Tourcoing, France
Tel: (20) 94 92 77
Telex: Univool 810958

Van Hool NV
B-2578, Lier-Koningshooikt, Belgium
Tel: Antwerp (31) 82 15 00
Telex: 3179

Vaughan Associates Group
Valair Division, Shrewsbury Works, Harlescott Lane, Shrewsbury, Shropshire SY1 3AG, England
Tel: 0743 3033
Telex: 35264

Vaughan Systems and Programming
The Maltings, Hoe Lane, Ware, Hertfordshire SG12 9LR, England
Tel: 0920 2282
Telex: 81516

Verolme Elektra BV
Elektraweg 5, PO Box 33, 3140 AA Maassluis, Netherlands
Tel: (31 1899) 13033
Telex: 21569 vems nl

Walter Vetter GmbH and Company
Ringstrasse 28, D-7012 Fellbach, Postfach 2080, Federal Republic of Germany
Tel: (0711) 58 90 41
Telex: 07254496

Vickers Fluid Power
Vickers PLC, PO Box 8, Swindon, Wiltshire SN3 4RA, England
Tel: 0793 823241
Telex: 44317

VJ Engineering
17 Morse Road, Whitnash, Leamington Spa CV31 2LH, England
Tel: 0926 38774 24163

Wainwright Planning
PO Box 11569, Montgomery, Alabama, USA
Tel: (205) 284 1797

Weathermeasure Division
Systron-Donner Corporation, PO Box 41257, Sacramento, California 95841, USA
Tel: (916) 481 7565
Telex: 377310

Weldwork Cargo Systems Limited
Pulborough Way, Green Lane, Hounslow, Middlesex TW4 6DE, England
Tel: 01-572 2896
Telex: 937735

Western Trailers
22 The Green, Stoke Gifford, Bristol BS12 6PD, England
Tel: 0272 694273
Telex: 444389

Westinghouse Electric Corporation
PO Box 1897, Baltimore, Maryland 21203, USA
Tel: (301) 765 3788
Telex: 87828

Konrad Wiedemann
3002 Wedemark 1, Allerhop 16, Postfach 1220, Federal Republic of Germany
Tel: 05130/3001
Telex: 0924673

Wilcox Electric Incorporated
Northrop Corporation, 1400 Chestnut, Kansas City, Missouri 64127, USA
Tel: (816) 231 0700
Telex: 042258

George Wimpey Limited
International and Engineering Division, Hammersmith Grove, London W6 7EN, England
Tel: 01-748 2000
Telex: 22436

Wollard Aircraft Equipment Incorporated
Heath Tecna Corporation, 6950 North West 77th Street, Miami, Florida 33166, USA
Tel: (305) 592 5450
Telex: 578833
Cable: waesair

Wulfsberg Electronics Incorporated
11300 West 89th Street, Overland Park, Kansas 66214, USA
Tel: (913) 492 3000
Tel: 424242 wulfsberg olpk

Wysepower Limited
Drove Road, Everton, Sandy, Bedfordshire SG1 2HX, England
Tel: 0767 50011
Telex: 826849

Zellinger Fahrzeugfabrik
Traun A 4050, Austria
Tel: (7229) 4491
Telex: 21903

Zem-Air Division
Walter E Zemitzsch Incorporated, 2705 Olive Street, St Louis, Missouri 63103, USA
Tel: (314) 371 3020

Zwicky Engineering Limited
Skyhi Division, Wokingham, Berkshire RG11 2RY, England
Tel: 0734 782500
Telex: 848478

INDEX